3rd Global Conference on Ecological Environment and Civil Engineering (GCEECE 2021)

Held online due to COVID-19

E3S Web of Conferences Volume 293 (2021)

Guangzhou, China
25-27 June 2021

Editor:

Antonio Caggiano

ISBN: 978-1-7138-3483-0

Printed from e-media with permission by:

Curran Associates, Inc.
57 Morehouse Lane
Red Hook, NY 12571

Some format issues inherent in the e-media version may also appear in this print version.

For additional information, please contact EDP Sciences – Web of Conferences at the address below.

EDP Sciences – Web of Conferences
17, Avenue du Hoggar
Parc d'Activité de Courtabœuf
BP 112
F-91944 Les Ulis Cedex A
France

Phone: +33 (0) 1 69 18 75 75
Fax: +33 (0) 1 69 28 84 91

contact-edps@webofconferences.org

Additional copies of this publication are available from:

Curran Associates, Inc.
57 Morehouse Lane
Red Hook, NY 12571 USA
Phone: 845-758-0400
Fax: 845-758-2633
Email: curran@proceedings.com
Web: www.proceedings.com

TABLE OF CONTENTS

ECOLOGICAL ENVIRONMENT PROTECTION AND POLLUTION CONTROL

ENVIRONMENTAL ENERGY AND CIVIL ENGINEERING AND WATER CONSERVANCY CONSTRUCTION

SUSTAINABLE RESOURCE DEVELOPMENT AND GREEN ENERGY SAVING

Preface

The 2021 3rd Global Conference on Ecological Environment and Civil Engineering (GCEECE 2021) was held on June 25-27, 2021 as a virtual conference due to the growing concerns over the coronavirus outbreak (COVID-19), and in order to protect the well-being of our attendees, partners, and staff as our number one priority. The primary goal of the conference is to promote research and developmental activities in Ecological Environment and Civil Engineering. Another goal is to promote scientific information interchange between researchers, developers, engineers, students, and practitioners. GCEECE 2021 welcomes researchers, engineers, scientists and industry professionals to an open forum where advances in the field of Ecological Environment and Civil Engineering can be shared and examined. The conference is an ideal platform for keeping up with advances and changes to a consistently morphing field.

This scientific event brings together more than 140 national and international researchers in Ecological Environment and Civil Engineering. On top of the local participants coming from different national universities, international participants are also registered from different countries, namely UK, Australia, Germany, Iran, Malaysia. During the conference, the conference model was divided into three sessions, including oral presentations, keynote speeches, and online Q&A discussion. In the first part, some scholars, whose submissions were selected as the excellent papers, were given about 5-10 minutes to perform their oral presentations one by one. Then in the second part, keynote speakers were each allocated 30-45 minutes to hold their speeches.

In the keynote presentations part, we invited four professors as our keynote speakers. The first keynote speakers, Prof. Antonio Caggiano, from Darmstadt University of Technology/CONICET and University of Buenos Aires Germany and Argentina was invited to present his talk *Micro-to-meso scale research of eNeRGy-FOAM PCM cement-based composites for green buildings.* Prof. Jonathan Adams, from Nanjing University, UK was our second keynote speakers. He presented a talk: *Changes in marine plastic biofilms under ocean acidification.* Prof. Baoyu Gao, from Shandong University, China. He delivered a wonderful speech: *Application of Forward Osmosis technology in the desalination of brackish water and the advanced treatment of municipal reclaimed water.* Prof. Guohua Chang, from Lanzhou City University, China as our finale keynote speakers. She presented a talk: *Phytoremediation of phenol by Hydrilla verticillata (L.f.) Royle and associated effects on physiological parameters.*

The proceedings are a compilation of the accepted papers and represent an interesting outcome of the conference. Topics include but are not limited to the following areas: Environmental Pollution & Management, Environmental Protection, Environmental Ergonomics, Civil Engineering, Water Supply and Drainage Engineering and more related topics. All the papers have been through rigorous review and process to meet the requirements of International publication standard.

We would like to acknowledge all of those who supported GCEECE 2021. The help and contribution of each individual and institution was instrumental in the success of the conference. In particular, we would like to thank the organizing committee for its valuable inputs in shaping the conference program and reviewing the submitted papers.

Committee of GCEECE 2021

Committee member

Conference Chairs
Prof. Tianpeng Gao, Xi'an University, China
Prof. Antonio Caggiano, Darmstadt University of Technology, Germany

Conference Co-Chairs
Prof. Baoyu Gao, Shandong University, China
Prof. Y.Q. He, Tongji University, China

Local Committees
Prof. Ying Fu, University of Jinan, China
Prof. Defu Liu, Ocean University of China, China
Prof. Qinghua Chen, China university of Petroleum, China
Prof. Zhanfeng Dong, Chinese Academy for Environmental Planning, China
Prof. Xiangbin Zeng, Huazhong University of Science and Technology, China

International Technical Committees
Prof. Zhao Liu, Chang'an University, China
Prof. Zhijun Zhang, University of South China, China
Prof. Biao Lei, Beijing University of Technology, China
Prof. Wenmei Ji, Anhui Jianzhu University, China
Prof. Zhang Wenjie, Guilin University of Technology, China
Prof. Carlos Caceres, The University of Queensland, Australia
Prof. Lam Kin Che, The Chinese University of Hongkong, HK
Prof. Jinglong Bu, Hebei United University, China
Prof. Yongchang Chen, Beijing University of Technology, China
Prof. Beidou Xi, Chinese Research Academy of Environmental Sciences, China
Prof. Guanghua Wang, Wuhan University of Science and Technology, China

edp sciences

Web of Conferences

Statement of Peer review

In submitting conference proceedings to *Web of Conferences*, the editors of the proceedings certify to the Publisher that

1. They adhere to its **Policy on Publishing Integrity** in order to safeguard good scientific practice in publishing.
2. All articles have been subjected to peer review administered by the proceedings editors.
3. Reviews have been conducted by expert referees, who have been requested to provide unbiased and constructive comments aimed, whenever possible, at improving the work.
4. Proceedings editors have taken all reasonable steps to ensure the quality of the materials they publish and their decision to accept or reject a paper for publication has been based only on the merits of the work and the relevance to the journal.

Title, date and place of the conference

2021 3rd Global Conference on Ecological Environment and Civil Engineering
June 25-27, 2021
Guangzhou, China

Proceedings editor(s):

Prof. Antonio Caggiano, Darmstadt University of Technology, Germany

Date and editor's signature

Antonio Caggiano

Impacts of long-term nitrogen addition on nitrous oxide in a temperate grassland

Si Chen[1,4], Tianpeng Gao[2,3], Tianxiang Hao[4], Kaihui Li[5], Xuejun Liu[2]*

[1] Lanzhou City University, Lanzhou 730070, China.

[2]School of Biological and Environmental Engineering, Xi'an University, Xi'an 710065, China

[3]Engineering Center for Pollution Control and Ecological Restoration in Mining of Gansu Provinc-e, Lanzhou City University, Lanzhou 730070, China

[4] Key Laboratory of Plant-Soil Interactions of MOE and Beijing Key Laboratory of Farmland Pollution Prevention and Remediation, College of Resources and Environmental Sciences, China Agricultural University, Beijing 100193, China.

[5] Key Laboratory of Biogeography and Bioresource in Arid Land, Xinjiang Institute of Ecology and Geography, Chinese Academy of Sciences, Urumqi 830011, China

Abstract: Atmospheric nitrogen (N) deposition has increased dramatically due to increased human activities since the industrial revolution. However, it is still unclear what the responses of soil nitrous oxide (N_2O) is to long-term elevated N deposition in a temperate grassland. Here, we conducted an *in situ* field experiment to investigate these responses to long-term high N addition on a temperate steppe in Inner Mongolia, China, from April 2017 to October 2018. Soil N_2O emissions significantly increased by long-term N addition, use of structural equation modeling (SEM) showed that topsoil (0-5 cm) NH_4^+-N content was the most important limiting factor for N_2O emission. Our results indicate that long-term high N addition showed a significantly increase in N_2O emission in this temperate grassland.

1 Introduction

Anthropogenic nitrogen (N) enrichment and global warming are primary components of global change. N deposition to terrestrial ecosystems has increased from 17.4 Tg N yr^{-1} in late 1860s to 187 Tg N yr-1 in the 2005, and is expected to increase to 200 Tg N·yr^{-1} in 2050[1]. This in natural ecosystems has many negative effects on ecosystem services: biodiversity loss, nutrient imbalance, and increased greenhouse gases (GHGs) emissions[2,3]

In Inner Mongolia grassland occupies a large area of northern China and 13% of the global grassland area[4]. However, models of N_2O emission with the N deposition shows N_2O emission increased as exponential and unimodal response[5,6], when the N reached biological threshold, N_2O emission transitions to the N saturation stage, at this stage, higher N could not increase N_2O emission, including in some temperate and subtropical grasslands[7].

N saturation model believes that in natural ecosystems, enhanced N deposition would produce more soil N_2O[8], but recent study on steppe showed that influence of significant climatic factors on N_2O depends on the ability to alleviate N_2O emission limited by soil "nitrogen deficiency"[9,10]. Therefore, the effects of long-term N addition on N_2O emission are still uncertain. In another way, N enrichment was always accompanied by the changes in soil properties like soil acidification, available N and C states, as well as vegetation of ecosystem. Therefore, it is also valuable to find the complex response of N_2O fluxes of soil to variations in multiple indirect factors caused by N deposition of grassland[11].

Therefore, we set up a long-term *in stiu* experiment in Inner Mongolia, China in 2005, which has caused severe soil acidification after 13 years of N fertilization, and significant changes in soil properties[12]. To assess how these changes have affected soil N_2O emission, we measured N_2O fluxes with high frequency from May 2017 to October 2018, with the objectives of: (i) quantifying the response of the N_2O emission to long-term increased N addition, and (ii) identifying the main controlling factors of N_2O caused by long-term increased N addition.

2 Experimental methods

2.1 Site description

This study was conducted in a semi-arid temperate grassland in Duolun County (116°17E, 42°02N, elevation 1324 m), northeastern of Xilingol of Inner Mongolia, China. The rainfall and air temperature average 316 mm and 3.3°C between 2006 to 2016, and 81% of total

* Corresponding author: Xuejun Liu, Emil: liu310@cau.edu.cn

precipitation in growing season (May 10th to August 30th; Fig.1(a)). The local N deposition has increased to 15.4 kg N ha^{-1} yr^{-1} in 2015, consist with 58% as $NH_4^+$and 42% as NO_3^-. 0-10 cm layer soil pH and bulk density are 7.12 and 1.31 g cm^{-1}. Soil total N and total P contents are 12.3 g kg^{-1} and 1.3 g kg^{-1}, organic carbon and C:N are 0.28 g kg^{-1} and 21, respectively.

2.2 Experimental treatments

The experiment was designed with three N addition treatments of four replicates that were arranged randomly. Plot area was 5×5 m with a 1 m wide buffer zone. Ammonium nitrate was applied as foliar fertilizer at simulated N deposition rates of N_0 (0 kg N ha^{-1} yr^{-1}), N_{30} (30 kg N ha^{-1} yr^{-1}) and N_{60} (60 kg N ha^{-1} yr^{-1}). The fertilizer was divided into 2 equal portions, and sprayed evenly in mid-June and mid-July every year (15th June and 17th July in 2017). N_2O was measured from May 2017 to October 2018.

2.3 Measurements method

Static closed chamber technique was used to measure N_2O fluxes. The chambers were made of removable top and previously-installed stainless steel (into soil for 10 cm). Sampling was conducted between 10:00 am and 12:00 pm. Gases fluxes were collected once or twice a week. Gas samples were injected into 50 ml, and were analyzed using a gas chromatograph (GC; Agilent 7890A, Agilent Technologies, Santa Clara, CA, USA) equipped with an electron capture detector for determination of N_2O[13,14].

Fresh soil samples were extracted in 0.01 mol L^{-1} $CaCl_2$ solution and shaken for 1 hour. The suspension was analyzed for nitrate and ammonium concentration using a continuous-flow auto-analyzer (Seal AA3, Germany). Dissolved organic carbon (DOC) was extracted from the fresh soil sub-samples by shaking with deionized water, and filtered at 0.45 μm and the filtrate analyzed with a TOC analyzer (multi N/C 3100, Jena, Germany). Soil pH was measured using a pH Meter (Seven Easy, Mettler-Toledo, Switzerland). Soil water content and temperature in the 0-5 cm layer were measured using a TDR probe.

3 Statistical analyses

Treatment enrichment effects on soil pH, NO_3^--N, NH_4^+-N content, DOC, total emission of N_2O were assessed using a least significant difference test (LSD, $P < 0.05$). In addition, the relationships between fluxes of N_2O with soil moisture, soil NO_3^--N and NH_4^+-N contents, pH were analyzed using structural equation models (SEM). One-way repeated measures analysis of variance was used to analysis. All statistical analyses were conducted using SPSS (version 20.0) with statistically significant differences set at $p < 0.05$. SEM analyses were carried out using AMOS 22.0. All figures were drawn using Sigmaplot (version 13.0).

4 Results

Long-term high N addition increased signifcnatly soil NO_3^--N, DOC content (except in N_{30} Treatment), and decreased pH significantly by long-term N addition, but has no effect on NH_4^+ content.

Small N_2O emissions were observed from N_0 plots, ranging from 0.59 to 16.32 μg N m^{-2} h^{-1} (average annual 4.45 ± 0.34 μg N m^{-2} h^{-1}) (Fig. 1b). Long-term increased N addition significantly increased the N_2O emissions ($P < 0.001$) (Fig 1b; Table 3), with average annual emissions from N_{30} and N_{60} plots of 9.49 ± 0.79 and 20.35 ± 3.23 μg N m^{-2} h^{-1}, respectively (Fig 1b). The SEM showed that soil moisture had a significantly positive correlation with soil N_2O emissions both directly and indirectly, and that soil NH_4^+-N content was the most important influencing factor (Fig. 2). Notably, neither soil pH or DOC had a significant impact on N_2O emissions (Fig. 2).

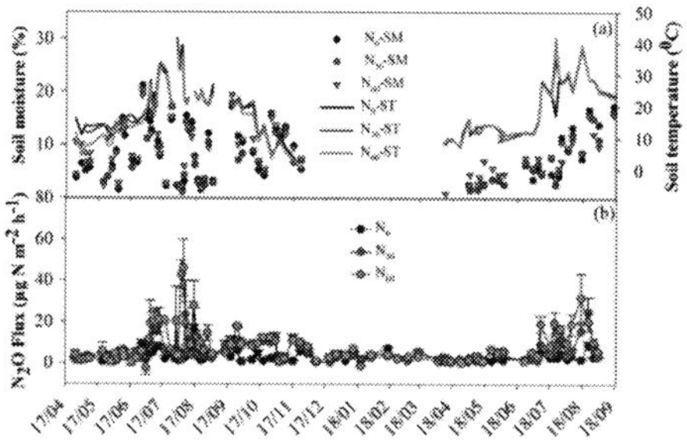

Figure 1. The Variation from April 2017 to October 2018 in soil moisture and soil temperature (°C) (a), and fluxes in N_2O (b) (mean, n=4) in an Inner Mongolian grassland.

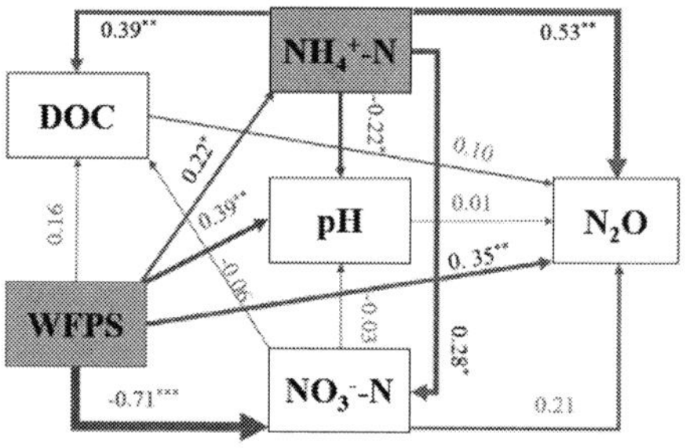

χ2=3.068, p=0.794, df=1, RMSEA = 0.00, AIC = 52.07

Figure 2. The effects on N_2O emission (n=28) of soil moisture, temperature, pH, soil NO_3^--N content and NH_4^+-N content as tested by structural equation modeling (SEM). Single-headed arrows showed that the impact of different significant controls on N_2O emission was analyzed. The blue arrows represent positive effects, and red arrows negative effects. The width of the arrows indicated the strength of the relationship. The numbers are standardized path coefficients, which show the significance of the variables in the model. Goodness-of-fit statistics for the model are shown below the model. *, ** and *** indicate significant effect at $P < 0.05$, $P < 0.01$, and $P < 0.001$, respectively.

Table1. The variation in NO_3^--N, NH_4^+-N, soil pH, dissolved organic carbon (DOC) of topsoil (0-5 cm) with the treatments in August 2017. Values in the same column with different letters indicate significant differences (P < 0.05)

Treatments	NH_4^+-N (mg kg^{-1})	NO_3^-N (mg kg^{-1})	pH	DOC (mg kg^{-1})
N_0	5.30±2.82a	11.65±5.28a	6.32±0.09b	68.46±4.01a
N_{30}	5.97±0.95a	17.29±2.05a	5.91±0.07a	75.83±8.89a
N_{60}	6.79±1.30a	19.82±3.90b	5.94±0.35a	82.59±8.06b

Table2. Repeated measures ANOVA analysis of variance with F and P values of the total effects of long-term nitrogen (N) addition on N_2O emissions in long-term N addition; n indicates sample size.

Source of variation	N_2O		
	df	F	P
Date	1.834	9.861	0.009**
N	1	101.127	<0.001**
Date*N	1.834	2.603	0.142

5 Discussion

N_2O emissions of the temperate grassland (Fig. 1b; Table 3), acted as a small source of N_2O and the cumulative emission of N_2O from N_0 plots is 0.37 kg N ha^{-1} (Fig. 1b), higher than that from other ungrazed steppes in Inner Mongolia (below 0.28 kg N ha^{-1})[16]. N_2O emissions was triggered by the day following fertilization, and the peaks lasted about 20 days and the emissions fell to the baseline (Fig. 1b), which was related to the dynamics of soil mineral N (Table 1). The SEM showed that NH_4^+-N was the most important controlling factor for N_2O emission rather than NO_3^--N or pH (Fig. 2), indicated that N_2O emission was mainly as byproduct of soil ammonia oxidation. Dry weather and the sandy texture (sand 62.8%) had been keeping soil WFPS below 30% (Fig.1a) over the observed period, in that conditions N_2O by denitrification produced much less (WFPS≤70%)[17]. Recent studies suggested that in a typical steppe, long-term nitrogen fertilization and soil pH could have contrasting effects on ammonia oxidizing microorganisms[18], our result showed that pH had less influence to N_2O, that was primarily due to N-reduced pH did not dramatically shifted N inducing function and diversity of ammonia oxidizers of grassland[19].

6 Conclusion

Atmospheric N deposition has dynamically increased in recent years, which will significantly affect the N_2O balance in grassland ecosystems. Long-term (13 year) increased N addition significantly increased N_2O emission. Soil NH_4^+-N content and soil moisture were the most important limiting factors for N_2O emissions, pH had less influence to N_2O. We conclude that N_2O emission would be enhanced by long-term increased N deposition, which will have an important effect for climate change with elevating atmospheric N deposition in future.

Acknowledgments

This work was supported by the Doctoral Research Fund of Lanzhou City University (LZCU-BS2019-43), Natural Science Foundation of China(NSFC 318601 76), Key Research and
Development Program of Gansu Province (20YF3FA037), Key Research and Development Program of Shanxi Pr ovince (2020ZDLSF0606) and XAWLKYTD012 the Natural Science Foundation of Gansu Province, China (21JR1RA319). Chinese Key State Research and Development Programme (2017YFC0210100), China National Funds for Distinguished Young Scientists (Grant No. 41425007), and the UK Newton Fund through the BBSRC project 'China Virtual Joint Centre for Improved N Agronomy (CINAG)' (BB/N013468/1)

References

1. Galloway, J. N., Townsend, A. R., Erisman, J. W., Bekunda, M., Cai, Z. C., Freney, J.R., Martinelli, L. A., Seitzinger, S. P., Sutton, M. A. (2008). Transformation of the nitrogen cycle: recent trends, questions, and potential solutions. Science, 320, 889-982.
2. Aber, J., McDowell, W., Nadelhoffer, K., Magill, A., Berntson, G., Kamakea, M., McNulty, S., Currie, W., Rustad, L., Fernandez, I. (1998). Nitrogen saturation in temperate forest ecosystems - Hypotheses revisited. Bioscience, 48, 912-93.
3. Lu, X. K., Mao, Q. G., Gilliam, F. S., Luo, Y. Q., Mo, J. M. (2015). Nitrogen deposition contributes to soil acidification in tropical ecosystems. Global Change Biology, 20, 3790-3801.
4. Wang, Y. S., Xue, M., Zheng, X. H., Ji, B. M., Du, R., Wang, Y. F. (2005). Effects of environmental factors on N2O emission from and CH4 uptake by the typical grasslands in the Inner Mongolia, Chemosphere, 58(2), 205-215.
5. Hao, T. X., Song, L., Goulding, K., Zhang, F. S. and Liu, X. J. (2018). Cumulative and partially recoverable impacts of nitrogen addition on a temperate steppe. Ecology Application, 28, 237-248.
6. Zhang, W., Liu, C. Y., Zheng, X. H., Fu, Y. F., Hu, X. X., Cao, G. M., ButterbachBahl, K. (2014). The increasing distribution area of zokor mounds weaken greenhouse gas uptakes by alpine meadows in the Qinghai–Tibetan Plateau. Soil Biology and Biochemistry, 71, 105-112.

7. Smith, K. A., Ball, T., Conen, F., Dobbie, K., Massheder, J. R. (2003). Exchange of greenhouse gases between soil and atmosphere: interactions of soil physical factors and biological processes. European Journal of Soil Science, 69(1), 10-20.
8. Zhang, W., Mo, J. M., Yu, G. R., Fang, Y. T., Li, D. J., Lu, X. K., Wang, H. (2008). Emissions of nitrous oxide from three tropical forests in Southern China in response to simulated nitrogen deposition. Plant and Soil, 306(1-2), 221-236.
9. Aronson, E. L., Helliker, B. R. (2010). Methane flux in non-wetland soils in response to nitrogen addition: a meta-analysis. Ecology, 91, 3242-3251,.
10. Yue, P. Li, K. H., Gong, Y. M., Hu, Y. K., Mohammat, A., Christie, P., Liu, X. J. (2016). A five-year study of the impact of nitrogen addition on methane uptake in alpine grassland. Scientific Reports, 6, 32064.
11. Tilman, D., Knops, J., Wedin, D., Reich, P., Ritchie, M., Siemann E. (1997). The influence of functional diversity and composition on ecosystem processes. Science, 277(5330), 1300-1302.
12. Zeng, J., Liu, X. J., Song, L., Lin, X. G., Zhang, H. Y., Shen, C. C., Chu, H. Y. (2016). Nitrogen fertilization directly affects soil bacterial diversity and indirectly affects bacterial community composition. Soil Biology and Biochemistry, 92, 41-49.
13. Xu, W., Zhao, Y.H., Liu, X J., Dore, A. J., Zhang, L, Liu, L., Cheng M. M. (2018). Atmospheric nitrogen deposition in the Yangtze River basin: Spatial pattern and source attribution. Environmental Pollution, 232, 546-555.
14. Liu, C. Y., Zheng, X. H., Zhou, Z. X., Han, S. H.,Wang, Y. H., Wang, K., Liang,W. G., Li, M., Chen, D. L., Yang, Z. P. (2010). Nitrous oxide and nitric oxide emissions from an irrigated cotton field in Northern China, Plant and Soil, 322, 123-134.
15. Jiang, X., Cao, L., Zhang, R. (2014). Changes of labile and recalcitrant carbon pools under nitrogen addition in a city lawn soil. Journal of Soils Sediments, 14, 515–524.
16. Shang Q. Y., Yang X. X., Gao C. M., Wu P. P., Liu, J. J., Xu, Y. C., Shen, Q. S., Zou, J. W., Guo, S. W. (2011). Net annual global warming potential and greenhouse gas intensity in Chinese double rice-cropping systems: a 3-year field measurement in longterm fertilizer experiments. Global Change Biology, 17(25), 2196-2210.
17. Zhu, X., Burger, M., Doane, T. A. and Horwath, W. R. (2013). Ammonia oxidation pathways and nitrifier denitrification are significant sources of N2O and NO under low oxygen availability. Proceedings of the National Academy of Sciences of United States of America, 110(16), 6328-6333.
18. Ying, J. Y, Li, X. X., Wang, N. N., Lan, Z. C., He, J. Z., Bai, Y. F. (2017). Contrasting effects of nitrogen forms and soil pH on ammonia oxidizing microorganisms and their responses to long-term nitrogen fertilization in a typical steppe ecosystem. Soil Biology and Biochemistry. 107, 10-18]
19. Chen, S., Hao, T. X., Goulding K., Misselbrook, T., Liu, X. J. (2019). Impact of 13-years of nitrogen addition on nitrous oxide and methane fluxes and ecosystem respiration in a temperate grassland. Environmental Pollution, 252, 657-681.doi.org/10.1016/j.envpol.2019.03.069

Changes on available phosphorus and total nitrogen contents from soil around roots of two desert plants

Guo-Hua Chang[1], Wang-jiang Wang[1], Zhuo-Xin Yin[1], Bin Yue[1], Jin-xiang Wang[1,*], and Tian-peng Gao[1,2,*]

[1]School of Geography and Environmental Engineering, Lanzhou City University, The Engineering Research Center of Mining Pollution Treatment and Ecological Restoration of Gansu Province, Gansu 730070, China

[2]College of Biological and Environmental Engineering, Xi'an University, Xi'an 710065, China

Abstract. In order to understand variation characteristics of the soil available phosphorus and total nitrogen (TN) content around the roots of typical desert plants *Peganum harmala* Linn. (PHL) and *Asterothamnus centrali-asiaticus* (AC), which are widely distributed in Jinchang mining area of Gansu Province, the soil samples around the roots of the two plants were collected, and the contents of available phosphorus and TN were analyzed. The results showed that the change trend of soil available phosphorus content was not the same in the range of 0-60 cm around the roots of PHL and AC. In most areas the change of available phosphorus content in the soil from around the root of PHL was U-shaped, and the content of phosphorus around the roots of AC increased firstly and then decreased, while the change trends of TN content around AC was decreased gradually.

1 Introduction

Phosphorus and nitrogen play an important role in the growth and metabolism of plants. To determine available phosphorus and nitrogen in soil is helpful for people to understand the supply of phosphorus and nitrogen in soil and to improve plant growth [1]. The absorption and utilization rate of nutrients, the demand for nutrients, and N and P nutrient levels provided by soil in the growing environment of plants are important factors which cause the differences in growth rate, reproductive output and competitiveness among plants [2, 3]. With the rapid growth of plant, it needs to require a large amount of rRNA for synthesizing protein, and rRNA contains a large amount of P, which would cause low C / P ratio and N / P ratio in the plants or tissues with high growth rate [3]. The contents of C, N and P and the ratio of elements in plant body could not only reflect the status of nutrient absorption and utilization, but also indicate the relative growth rate and nutrition strategy of plants [2-5]. The study on plant enrichment of heavy metals in nickel copper deposit area of Gansu Province showed that *Peganum harmala* Linn. (PHL) and *Asterothamnus centrali-asiaticus* (AC) were widely distributed here and could enrich copper and nickel [6]. The two plants are typical local desert plants. In Gansu province, Jinchang city has rich mineral resources and rich nickel ore reserves. The exploitation and utilization of the mine not only brings economic benefits to the local, but also brings many environmental problems [7]. It is needed to control soil pollution and restore vegetation in mining area for cherishing and rationally utilizing land resources, but also for environmental protection.

Peganum harmala is a perennial herb, which has the typical characteristics of xerophyte, such as developed root system, deeply split leaves and thin lobes, and strong drought resistance. In addition, the plant grows luxuriantly and plays an important role in wind resistance, sand fixation and soil erosion prevention, and improving desert environment in Northwest China [8-10]. In recent years, the research on the *Peganum harmala* mainly focused on the stress resistance, active components and pharmacological activities [11, 12].

Asterothamnus centrali-asiaticus is a semi shrubby desert plant, which appears in the desert and desert steppe zone, and is mainly found in loose sandy alluvial and proluvial soil. This plant has strong characteristics of drought resistance, cold resistance, barren resistance, sand burial resistance and wind erosion resistance, which plays an important role in maintaining the desert grassland ecosystem [13].

In order to better understand the growth characteristics of the two plants in the region, the content of available phosphorus and TN in the soil around the roots of the two plants was measured for providing a reliable basis for the recovery of vegetation in the arid nickel-copper mining area.

2 Materials and methods

Soil samples were collected from Jinchang City, Gansu Province [6]. 7 sampling areas were selected (A, 38° 28' 43.89"N, 102° 14' 00.50" E; B, 38° 28' 52.14"N, 102° 14'

* Corresponding author: zkgtp@163.com

25.44" E; C, 38 °27' 17.77"N, 102° 13' 21.69" E; D, 38 °30' 57.5"N, 102° 09' 4.2" E; E, 38° 39' 16.23"N, 102° 16' 46.32" E; F, 38° 29' 27.6"N, 102° 09' 10.3" E; G, 38° 33' 49.3"N, 102° 08' 55.8" E). Soil samples (0 ~ 20 cm) were collected at the horizontal direction of 0, 20, 40, and 60 cm from the root of the two plants. The collected soil samples were dried indoors, and stones and debris were removed from samples, and then crushed. 2.500 g of the sample after 1 mm sieving was weighed. Then 0.5 mol·L^{-1} $NaHCO_3$ (pH = 8.5) solution was added into the sample, followed by 30 min oscillation at constant temperature (25 ± 0.5 ℃), and then the sample was filtered immediately with a 0.22 μm phosphor-free membrane. The available phosphorus content in the filtrate was determined by Mo-Sb Anti-Spectrophotometer [14]. The available phosphorus content in the soil samples was expressed as mg·kg^{-1}, and the results were expressed as the mean ± standard deviation (SD). Soil samples were digested with concentrated sulfuric acid, and total nitrogen was determined by K9840 Automatic Kjeldahl nitrogen analyzer [15].

3 Results and discussion

3.1 Changes on available phosphorus in soil around plants

In the above areas, the content of phosphorus varied from 1.496 mg·kg^{-1} to 38.619 ± 2.293 mg·kg^{-1} (Fig 1 and 2). The content of available phosphorus in most samples is relatively low based on the classification standard of soil available phosphorus content in soil of China [16].

Changes on available phosphorus in soil around the roots of PHL from different areas were showed in Fig. 1. Within the 60 cm growth radius of PHL in A area, the phosphorus content in the soil from the horizontal direction around the plant roots decreased first and then increased, and the lowest content of phosphorus was 1.496 mg·kg^{-1} at 20 cm. In B area, the highest phosphorus content was 20.961 ± 0.677 mg·kg^{-1} within the growth radius of 60 cm around the root of PHL. In area C, the phosphorus content around PHL decreased first and then increased, and the lowest phosphorus content was 8.742 ± 0.730 mg·kg^{-1} at 40 cm. The content of phosphorus around PHL in area D decreased gradually with increasing distance, and the highest content of phosphorus was 8.411 ± 1.239 mg·kg^{-1} at 0 cm. The maximum value of phosphorus content around PHL was 24.218 ± 1.077 mg·kg^{-1} at 60 cm in area E. The content of phosphorus around PHL showed a gradually decreasing trend in area F. Within the growth radius of 60 cm of PHL in G area, the phosphorus content decreased first and then increased, and the lowest value was 8.251 ± 0.251 mg·kg^{-1} at 40 cm.

It can be seen that in most areas, the content of available phosphorus in the soil samples around the roots of PHL decreased first and then increased with the increase of the distance, and the content of phosphorus around the roots of the plant was relatively high. The phenomenon may be closely related to the microbial activities in the rhizosphere soil of the plant. Cheng et al. researched that the characteristics of bacterial communities inrhizosphere soil of *P. harmala* in a desert steppe in Baiyin, Gansu Province, and their results showed that there were significant differences in the composition and structure of rhizosphere microbial community from soil of *P. harmala* and its surrounding desert steppe soil. The number and diversity index of microorganisms in rhizosphere soil were significantly higher than those in non rhizosphere soil. The dominant bacteria in rhizosphere soil were consisted of 30.01% Actinobacteria, 23.98% Proteobacteria, 11.53% Bacteroidetes and 10.19% Acidobacteria, and the dominant soil bacterial community in the surrounding desert steppe were composed of 55.05% Actinobacteria, 21.11% Proteobacteria and 6.07% Acidobacteria [9]. The result indicated that desert plants could increase the number and activity of rhizosphere microorganisms, promote the accumulation and transformation of soil nutrients, and play an important role in maintaining the function of desert ecosystem.

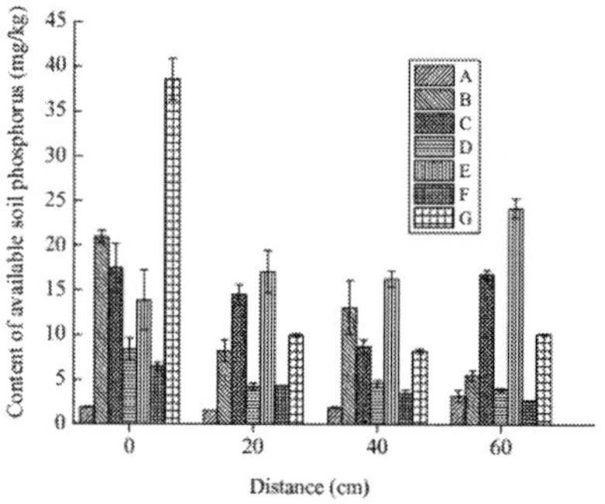

Figure 1. Changes on available phosphorus in soil around the roots of PHL from different areas.

Changes on available phosphorus in soil around the roots of AC were showed in Fig. 2. In the above areas, AC was not found within 100 meters of the collection point in area A. The average phosphorus content near AC in area B was 22.875 mg·kg^{-1}. The phosphorus content near the plant in area C was the lowest among all areas, with an average of 5.677 mg·kg^{-1}.

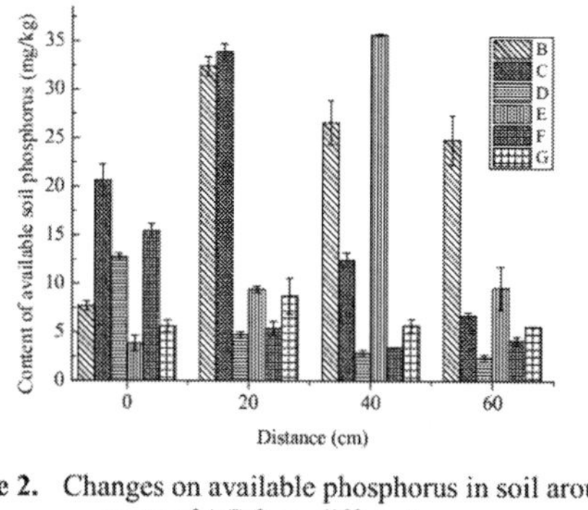

Figure 2. Changes on available phosphorus in soil around the roots of AC from different areas.

The results showed that the change trend of the mean value of phosphorus content from all areas around PHL decreased firstly and then increased with increasing distance, with the lowest phosphorus content of 8.051 ± 0.838 mg·kg^{-1} at 40 cm; and that of the mean value of phosphorus content from all areas around AC increased first and then decreased with increasing distance, with the highest value of phosphorus content 15.760 ± 0.844 mg·kg^{-1} at 20 cm (Fig 3). The content of available P at 60 cm around PHL was close to that of AC, and their values of available P were 9.511 ± 0.434 and 8.868 ± 0.979 mg·kg^{-1}, respectively (Fig. 3).

The average values of all the measured points from all the areas of these two plants were close to each other, with the values of 10.379 ± 0.891 mg·kg^{-1} for PHL and 12.519 ± 0.827 mg·kg^{-1} for AC. The results also indicated that the variation characteristics of available phosphorus content around the roots of different plants were different, which might be closely related to the characteristics of plant roots and the characteristics of microorganisms related to the roots of plants.

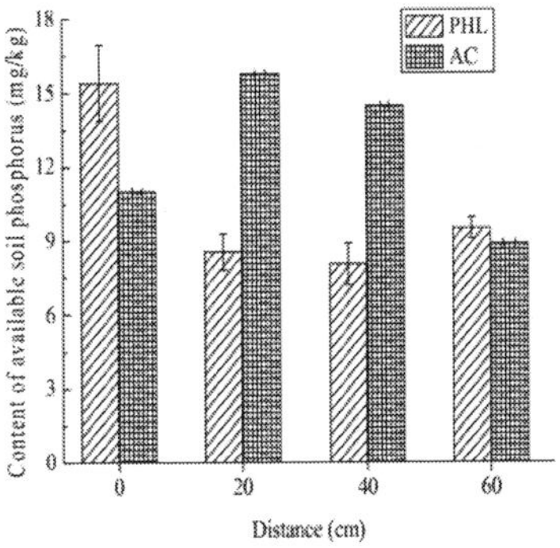

Figure 3. The average values of phosphorus content in soil around the two plants from all the areas.

3.2 Changes on TN in soil around plants

In the above areas, the content of TN varied from 0.11% to 0.020% (Fig 4 and 5). The content of TN in E area which is located in desert area is relatively lowest and there is the highest TN content in C area which is close to farmland.

Changes on TN in soil around the roots of PHL from different areas were showed in Fig. 4. Within the 60 cm growth radius of PHL in A area, the TN content in the soil from the horizontal direction around the plant roots decreased first and then increased, and the lowest content of phosphorus was 0.37% at 40 cm. In area C, the phosphorus content around PHL decreased first and then increased, and the lowest phosphorus content was 0.082% at 20 cm. The maximum value of phosphorus content around PHL was 0.092% at 0 cm around PHL in D area (at tailing dam). The content of phosphorus around PHL showed a gradually decreasing trend with increasing distance, in area F and G areas.

Changes on TN in soil around the roots of AC were showed in Fig. 5. The average TN contents in area C (near farmland) and D areas were higher than other areas. The TN contents near the plant in most areas decreased with increasing distance.

The results showed that the change trend of the mean TN content from all areas around PHL decreased firstly and then increased with increasing distance, with the lowest TN content of 0.043% at 40 cm; while the TN content around AC decreased gradually with increasing distance, and the highest content of TN was 0.063% at 0 cm (Fig 6). While, the average TN values of all the measured points from all the areas of these two plants were no significant difference between the two plants.

The variation characteristics of nutrients such as nitrogen and phosphorus around plant roots are closely related to the growth characteristics of plant roots and the interaction characteristics between roots and microorganisms. The variation characteristics of nitrogen and phosphorus around the roots of different plants may be different.

Li et al. found that the rhizosphere soil and the non-rhizosphere soil of seven typical plants from the Loess Plateau had significant differences in both soil microbial biomass and basal respiration, and the contents of organic carbon and nitrogen in rhizosphere soil were higher than those in non rhizosphere soil [17].

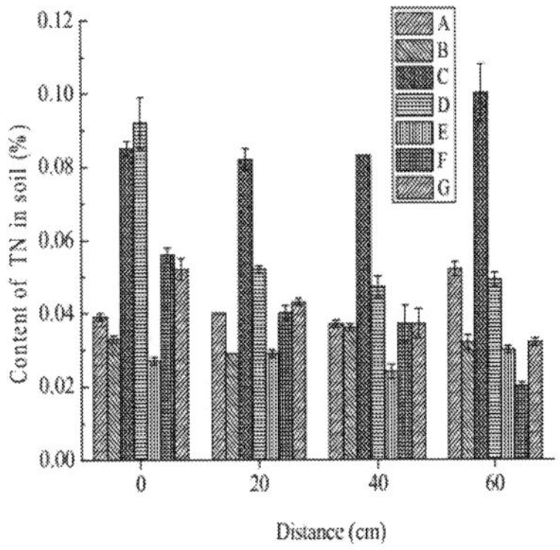

Figure 4. Changes on TN in soil around the roots of PHL from different areas.

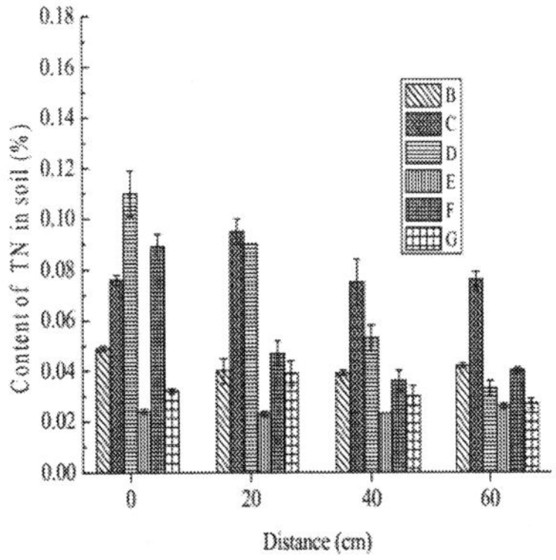

Figure 5. Changes on TN in soil around the roots of AC from different areas.

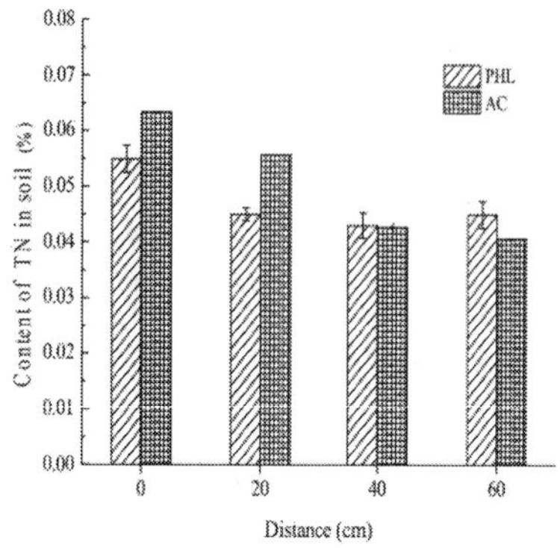

Figure 6. The average values of TN in soil around the two plants from all the areas.

4 Conclusions

On the whole, the content of soil available phosphorus in the study area is in the lower middle level. Within the range of 60 cm around the roots of the two plants, the change trend of average available phosphorus content in soil samples around AC is not the same as that of PHL. Around the two roots of AC and PHL, both of the highest values of TN content showed at 0 cm, and the former was higher than that of the latter. Generally, the change trends of TN content was similar with the available phosphorus content in soil around the roots of PHL, while the change trends on TN content around AC was not the same with the available phosphorus content in soil.

Acknowledgment

This research was funded by the National Natural Science Foundation of China (31860176), Science and technology project of Gansu Province (20YF3FA037), Shaanxi Province (2020ZDLSF06-06), Research Project of Universities from Gansu (2019B-169, 20JR5RA210 and 20JR10RA286), and the project of XAWLXYKYTD012.

References

1. Y. Luo, Y. Chen, Q.W. Peng, K.H. Li, A. Mohammat, W.X. Han. Nitrogen and phosphorus resorption of desert plants with various degree of propensity to salt in response to drought and saline stress, Ecological Indicators, **125** (2021), 107488.
2. C. Liu, Y. Wang, Wang N., G.X. Wang. Advances research in plant nitrogen, phosphors and their stoichiometry in terrestrial ecosystems: a review. Chinese Journal of Plant Ecology, **36** (2012), 1205–1216.
3. W. Chen, J.H. Wang, X.Y. Chen, N.H. Guo, J.L. Su, X.Y. Zhou, R. Huang. The carbon, nitrogen and phosphorus stoichiometric characteristics of invasive species *rhynchelytrum repens* and their nutrition strategy. Journal of Ecology and Rural Environment, **36** (2020), 1293–1300.
4. B.O.L. Demars, A.C. Edwards. Tissue nutrient concentrations in freshwater aquatic macrophytes: high inter-taxon differences and low phenotypic response to nutrient supply. Freshwater Biology, **52** (2010), 2073–2086.
5. S.J. Ren, G.R. Yu, B. Tao, S.Q. Wang. Leaf nitrogen and phosphorus stoichiometry across 654 terrestrial plant species in NSTEC. Chinese Journal of Environmental Science, **28** (2007), 2665–2673.
6. G.H. Chang, T.P. Gao, Q. Zhang, Y.Q. Chen, X.S. Tai, M.K. Chen. Study on the characteristics of heavy metals concentrated in native plants of Jinchuan mining area. Environmental Science and Engineering, **1** (2019), 842–849.
7. H.X. Ding, Z.R. Nan, X.W. Liu, Y. Li, S.L. Wang, S. Qin, Z.J. Zhao. Characteristics of selected heavy metal pollution in suburb cropland, Jinchang City, Gansu, China. Journal of Agro-Environment Science. **27** (2008), 2183–2188.
8. J. Ma, X.L. Wang. The species and distribution of genus *Peganum* L. in the desert area of China. J. Desert Res., **18** (1998), 131–136.
9. L. Cheng, J.X. Chen, Y.L. Li, Y.G. Wang, Y. T. Zhou, Y. Zhang. Diversity of bacterial communities in rhizosphere soil of *Peganum harmala* L. in desert steppe. Arid Zone Research, **35** (2018), 977-983.
10. M.Z. Xia, Y. Zhang, J.Y. Yu, F.Q. Zhang. Transcriptome analysis for leaves of *Peganum* mulstecutm. Guihaia, **39** (2019), 1–11.
11. M.N. Ansari. Assessment of antidiarrheal, antispasmodic and antimicrobial activities of methanolic seeds extract of *Peganum harmala* L. (Nitrariaceae). Journal of Pharmaceutical Research International, 2020, 74–82.
12. Z.N. Wu, N.H. Chen, Q. Tang, S. Chen, Z.C. Zhan, Y.B. Zhang, G.C. Wang, Y.L. Li, W.C. Ye. β-carboline alkaloids from the seeds of *Peganum harmala* and their anti-HSV-2 virus activities. Organic Letters. **22** (2020), 7310–7314.
13. Y.Q. Chai, Z.Z. Cao, B.R. Liu. Community characteristics of *Asterothamnus centrali-asiaticus* and *Gymnocarpos przewalskii* in the Subei Desert of Hexi Corridor, Gansu. Scientia Silvae Sinicae. **47** (2011), 166–171.
14. Ministry of agriculture of the people's Republic of China, the agricultural industry standard of the people's Republic of China NY / T1121.7-2014 Soil testing—Part 7: Method for determination of available phosphorus in soil. Beijing, 2014.
15. National environmental protection standard of the People's Republic of China. Soil quality—Determination of total nitrogen—Modified Kjeldahl method (HJ717-2014), 2014.
16. National Soil Census Office. Soil in China. Beijing: China Agricultural Publishing House, 1995.

17. G.H. Li, Q.F. Chen, Y.M. Huang, S.S. An, F.L. Zheng, L.D. Chen. Soil microbial biomass C, N, P and basal respiration in rhizosphere soil of typical plants on the Loess Plateau. Acta Ecologica Sinica, **30**, (2010), 976-983.

https://doi.org/10.1051/e3sconf/202129301003

Influence of curvature change on soil arching effect in underground space development

Binbin Xu[1,2,3,4*]

[1]Tianjin Port Engineering Institute Co. Ltd. of CCCC First Harbor Engineering Co. Ltd., Tianjin, 300222, China

[2]CCCC First Harbor Engineering Co. Ltd., Tianjin, 300461, China

[3]Key Lab. of Geotechnical Engineering of Tianjin, Tianjin, 300222, China

[4]Key Lab. of Geotechnical Engineering, Ministry of Communication, Tianjin, 300222, China

Abstract. This paper mainly studies the behavior of axisymmetric radial interaction between soil and structure under radial stress. Considering the possible factors in the interaction, such as the soil arching effect caused by the finite curvature, a simplified model is established. The influence of curvature change on the stress field, strain field, displacement field and stress-strain relationship is analyzed. Finally, the status of various factors in the axisymmetric radial interaction and the influence of their changes on the interaction are proposed.

1 Research background

With the rapid development of economy and the rapid growth of population, urban land is more tense and traffic is more crowded. This forces the housing construction and road traffic to develop to the high altitude and underground, and the high-rise building has almost become the symbol of modern city. For example, Sears building in Chicago is 443m in height, Kuala Lumpur twin tower building in Malaysia is 450m in height, Shanghai Jinmao Building in China is 42lm in height, and Shanghai global finance building is 460m in height. The appearance of these high-rise buildings undoubtedly puts forward more strict requirements for foundation treatment.

When human beings increase the height of ground buildings to increase living space, they also focus on the development of underground space. With the rapid development of modern underground engineering, there are many kinds of typical projects. All kinds of underground power stations are growing rapidly. Among them, the number of underground hydropower generation has exceeded 400 in the world, with a power generation capacity of more than 4.5 billion watts. The construction of underground power station is a huge underground project. Rogo hydropower station in the former Soviet Union has 5.1 million cubic meters of earthwork and 1.6 million cubic meters of concrete. 294 tunnels and chambers have been excavated, with a total length of 62 km. A large number of underground storage rooms have been built all over the world, and their construction technology has been constantly innovated[1-3]. At present, the development and utilization of urban underground space has become an important part of urban construction. Some developed countries have gradually integrated underground commercial streets, parking lots, underground railways and underground pipelines into a multi-functional underground complex.

The interaction between soil and structure is essentially the interaction between different media. It is an interdisciplinary research topic involving soil mechanics, structural mechanics, numerical simulation calculation and many other disciplines[4-6]. It involves many emerging research hotspots in the field of mechanics, such as boundary nonlinearity, interface problems and so on. At the same time, it is very important for energy, construction, water conservancy, municipal, military, electric power, etc As well as transportation and many other industrial sectors, it is also the key to project safety and quality. Based on the above importance of soil and structure in all aspects, it has been highly valued and widely studied by scholars and research institutions at home and abroad since ancient times.

2 Contact surface treatment method

Due to the great difference between the soil and the structure in the mechanical properties such as stiffness and strength, under certain force conditions, in addition to the transfer of force, there may be discontinuous deformation behaviors such as relative displacement, such as relative dislocation or cracking, between the soil and the structure, If the finite element method for continuum is directly used to solve the problem, there will be large errors, even serious distortion of the calculation results. For this kind of problem, at present, the soil and structure are usually regarded as a continuous whole, and different

* Corresponding author: 109930595@qq.com

https://doi.org/10.1051/e3sconf/202129301003

constitutive models are adopted for the soil or structure, while some treatments are carried out for the interface between the soil and structure.

a) Direct method: direct method is to transform contact problem into a constrained quadratic programming problem.

b) Contact mechanics method: the contact mechanics method is to treat the interface between soil and structure as a kind of boundary conditions, in which Lagrange multiplier method and penalty function method are most commonly used. The multiplier and penalty value in Lagrange multiplier method are equivalent to the contact stress or contact force on the interface, and this method can accurately meet the boundary conditions, However, the diagonal element of the stiffness matrix will be zero, resulting in no solution. Penalty function method overcomes the disadvantage of ill conditioned Lagrange multiplier method, but it can only approximately satisfy the contact boundary conditions. In order to overcome their shortcomings and make full use of their advantages, a modified Lagrange multiplier method was developed later. The common advantage of this kind of method is that it can directly obtain more real contact stress or contact force. The contact mechanics method is usually used to simulate the problems of rock joints and fissures. There are some limitations in the simulation of soil, so the application in this field is still relatively small.

c) Interface element method: the interface element method is introduced between soil and structure. It is a special finite element with very thin thickness or close to zero thickness, which plays a transitional role between soil and structure. This method is based on the preassumed constitutive relationship of interface, and adjusts the parameters in the element constitutive model by incremental and iterative means, The stress deformation relationship is approximately simulated. Because of its clear concept and easy operation, it is widely used in the field of geotechnical engineering.

Numerical simulation is a new way for human beings to understand the world. Since theoretical analysis and scientific experiments, numerical simulation has become the most important means for human beings to understand the world. It is mainly used to solve the following two kinds of problems: the problem that it is impossible to carry out experiments, and the problem that the cost of experiments is too high. At the same time, it integrates the characteristics of theoretical analysis and scientific experiment. Numerical simulation / simulation is no longer limited to scientific calculation, and is widely used in scientific research, engineering and production.

3 Computational model

3.1 Model simplification

Under the action of positive symmetrical load, the angle and horizontal displacement of the positive symmetric structure cannot occur. So only one quarter of the structure can be taken for analysis, as shown in Figure 1.

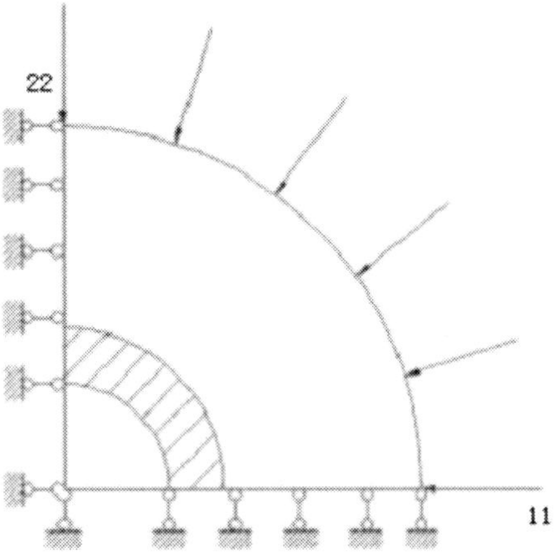

Figure 1. 1/4 model

3.2 Influence of different curvature

In the field of soil mechanics, soil arching is used to describe a phenomenon of stress transfer, which is realized by exerting the shear strength of soil. Terzaghi confirmed the existence of soil arching effect through movable door test [8].

Terzaghi's definition of soil arching effect is: "when a part of the supporting soil yields, the yield soil will move out of its original position, and the relative movement of the yield soil and the adjacent static soil will be hindered by the shear stress between the two parts of the soil, because the shear stress resistance tends to keep the yield soil in its original position, The phenomenon that the earth pressure transfers from the yield region to the adjacent static region is usually called soil arching effect. Sometimes, when the yield soil moves more than the adjacent soil, the soil arching effect will also occur According to this definition, the soil arching effect includes two parts: the decrease of the earth pressure at the yield position and the increase of the earth pressure near the yield position.

But the conclusion is discussed under the condition of zero curvature. In this problem, under the action of axisymmetric static earth pressure of the surrounding soil, the curvature of the structure studied is finite, and there will be different performances. Different structural supports require different stresses when yield, and the external of the yield structure may produce soil arch effect. On the one hand, the yielding soil bears the confining pressure of the part, but the pressure is gradually reduced, which makes the stress in the yield soil decrease, on the other hand, the soil near the outer side of the yielding soil will bear some pressure due to the soil arch effect, And with the obvious soil arching effect increases gradually, and the arching effect is very obvious, because the supporting structure itself is an arch structure.

Consider selecting different model sizes to represent different curvatures, and reduce the size in turn according to the proportional relationship to obtain three groups of data, as shown in Table 1.

Table1. Model size

	R1 (m)	R2 (m)	R3 (m)
NO.1	3	5	25
NO.2	2	3.2	25
NO.3	1	1.6	25

3.3 Influence of grid meshing

In FEM, different grid technology will have a certain impact on the results. Generally speaking, the finer the grid, the more accurate the calculation results will be, but the longer the calculation time will be. At the same time, we can also improve the mesh density of a certain part according to the accuracy of the results, in order to get more accurate solutions.

In the study of soil structure interaction, the structure and the soil near the interface will be the focus of consideration. The thickness of the structure is relatively small relative to the soil, and the grid density can not be the same as that of the soil. Otherwise, the changes in many details will not be so obvious. Only the structural grid is subdivided, Only in this way can we deeply consider the variation law of unknown variables in the contact surface and inside the structure; The soil near the contact surface and the surrounding soil can not use the same grid division, the general rule should be from the inside to the outside, and the grid density of the soil gradually decreases.

4 Result analysis

Under the confining pressure of 5MPa, the influence of different curvature is compared. At this time, each model has five different material compositions. If the results are obtained and all the analysis is carried out, at least five times should be carried out, but actually it is not necessary. The key problem is to make clear the purpose of the comparison in this step: only the influence of different curvature is considered, so the interference of other factors should be excluded. According to the above analysis, the comparison should choose between the two different model sizes with the least material influence.

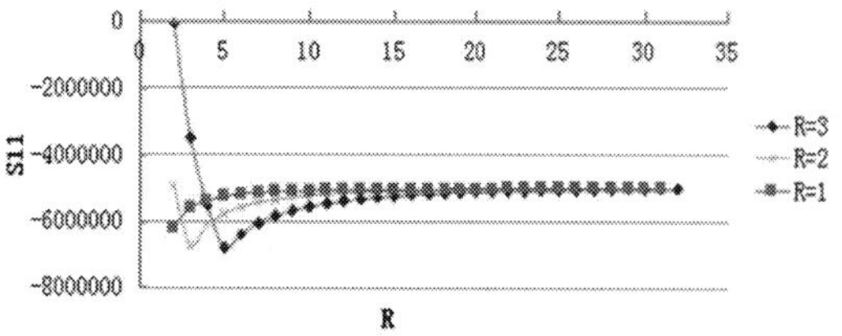

Figure 2. Comparison of S11 under different curvature

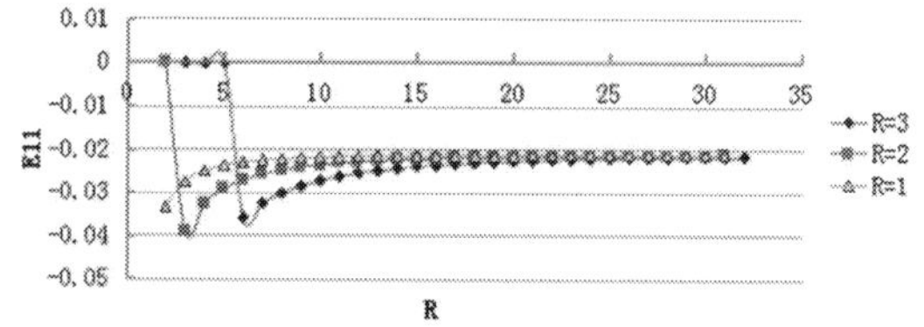

Figure 3. E11 ratio of elastic material with different curvature

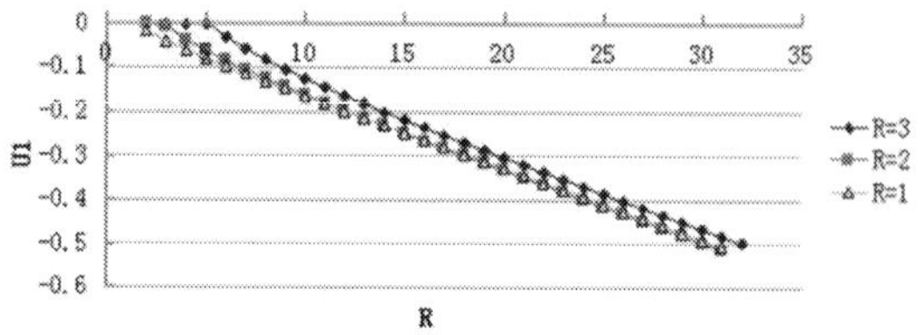

Figure 4. Comparison of U1 under different curvature

From R = 3m to R = 1m, it can be found that the initial amount of stress, strain and displacement increases gradually in the structure. When r = 3M, the minimum stress in the structure is zero, that is, it is not affected by the pressure from the soil, and the changes of strain and displacement also conform to this conclusion; When R = 1m, the initial stress in the structure can exceed 6Mpa, and the initial strain can reach 0.04, which is very large. The variation of R = 2m between R = 3M and R = 1m is between the two curvatures.

Taking a general view of the stress variation law along the radial direction in each model, the distribution and size of stress in the structure and its distribution and size in the soil are very different, so the stress change rate at the interface often has a sudden change, which is related to the mechanical properties of the interface, But at the same time, we should also see the important role of soil arching effect.

When the material can yield, it will yield first along the radial direction at the outside of the structure, and produce strain to the inside, which leads to the contact gap between the soil and the structure in contact with the structure becoming larger, so the soil will bear the pressure of another layer of soil close to it according to its own arch. It can be predicted that when the n+1 soil layer bears the pressure from the n+1 soil layer, the acting force and reaction force will not exist on the two contact surfaces, because they are not "close" contact in the traditional sense. The certain self-bearing capacity of the n+1 soil layer leads to the earth pressure transmitted to the n+1 soil layer being less than the force originally transmitted to the n+1 soil layer, In this way, the boundary which is divided into many concentric circles will be generated in the contour map, but it is not possible in practice. The premise of soil arching effect is that one of the contact parts will yield or have a large displacement, which will lead to the change of the original contact and the redistribution of stress.

In this model, the prediction of soil arch effect is the same as the actual location, which is located near the soil surface. Because of the huge difference between the elastic modulus of soil and structure, the large deformation of the soil near the soil will cause a certain range of soil mass to form a closed circle and bear the surrounding soil pressure. The degree of exertion of soil arch effect can be theoretically expressed by the degree of stress change in soil before and after the effect occurs

$$\frac{\Delta S_{11}}{\Delta R} \quad (1)$$

The following tables show the changes of the earth pressure before and after the soil arching effect under different curvature of various materials.

Table2. Stress changes of *R*=3m contact surface at 5MPa confining pressure level

r (m)	elastic	plastic	linear strain-hardening	logarithmic strain power hardening	rigid
1.95	-6.77e6	-6.75e6	-6.65e6	-6.76e6	-6.42e6
2.60	-6.37e6	-6.35e6	-6.35e6	-6.36e6	-6.08e6
3.25	-6.04e6	-6.02e6	-6.02e6	-6.03e6	-5.85e6

Table3. Stress changes of *R*=2m contact surface at 5MPa confining pressure level

r (m)	elastic	plastic	linear strain-hardening	logarithmic strain power hardening	rigid
0.68	-6.70E6	-6.39E6	-6.39E6	-6.70E6	-6.42E6
1.37	-6.08E6	-5.94E6	-5.94E6	-6.08E6	-6.08E6
2.05	-5.75E6	-5.68E6	-5.68E6	-5.75E6	-5.85E6

Table4. Stress changes of *R*=1m contact surface at 5MPa confining pressure level

r (m)	elastic	plastic	linear strain-hardening	logarithmic strain power hardening	rigid
0.00	-6.17E6	-6.17E6	-6.17E6	-6.17E6	-6.42E6
0.68	-5.60E6	-5.60E6	-5.60E6	-5.60E6	-6.08E6
1.37	-5.37E6	-5.37E6	-5.37E6	-5.37E6	-5.85E6

In each curvature radius, with the increase of radius, the change rate of stress relative to radius is gradually reduced, which shows that there is soil arching effect around the yielding soil at the interface, but in order to compare the degree of soil arching effect under different curvature conditions, it is necessary to conduct in-depth and detailed analysis on the change law of stress along the radial direction under each curvature.

5 Conclusions

A general model is established by FEM. In the analysis of the influence of curvature change factors on the results, we can see that:

1) For the models with different curvature conditions, the scaled model changes not only the curvature, but also the thickness of the structure, so the stiffness will also decrease. Some models with material mode will not converge due to too small stiffness. Therefore, the first problem that should be paid attention to after increasing the curvature is to reduce the external stress level of the model. In addition, under the condition of finite curvature, there will be obvious soil arching effect. On the one hand, at a higher stress level, the yield of soil near the contact surface is very obvious, and the adjacent outer soil layer can bear more stress, and the change rate of stress along the radial direction will be very obvious; On the other hand, the structure itself is an arched structure, and the constraint of boundary conditions makes it easier to form layered support arch between the soil, which can play the role of soil arch.

2) For the model with the same material and different curvature, the larger the stress level is, the smaller the stress attenuation is, the greater the stress transferred to the structure will be, and the more unfavorable the bearing capacity of the structure will be; On the contrary, if the stress level is small, the stress attenuation will be great, and even the stress transferred to the structure will be reduced to zero, which will be beneficial to the bearing capacity of the structure.

References

1. Y.R. Ren, Y.B. Liu, X.Y. Gu, Mechanics in engineering **26**, 6 (2004)
2. W.P. Li, Chinese journal of geotechnical engineering **22**, 4 (2000)
3. Y.R. Ren, Y.B. Liu, China offshore platform **21**, 3 (2006)
4. Y.R. Ren, Y.B. Liu, X.Y. Gu, China offshore platform **16** (2001)
5. R. F. Goodman, R. L. Taylor, Journal of Soil Mechanics and Foundation Engineering Division, **94** （1968）
6. J. Wang, G.L. Li, X.Q. Li, G. Hu, GongluJiaoTongKeJi, **13**, 2 (2017)
7. T. H. Lu, F. B. Bao, Journal of hydraulic engineering 2 (2000)
8. K. Terzaghi, Theoretical soil mechanics, Jhon Wiley and Sons, (1943)

Analysis of The Migration and Accumulation Process of Nitrate-nitrogen Pollutants in The Unsaturated Zone of Soil

Xiangcai Li[1], Ping Jing[1]*

[1]College of Geography and Environmental Sciences, Tianjin Normal University, Tianjin 300387, China

Abstract. In order to understand the migration process of nitrate-nitrogen pollutants in the shallow unsaturated zone of the soil, the Tianjin coastal dredger and typical fluvo-aquic soils in North China were used as the research objects. The Hydrus-1D software was used to establish a numerical model to simulate nitrate. The pollution process of nitrogen pollutants in the unsaturated zone of soil, the results show that: during the migration process of the two kinds of shallow vadose zones of soil, the nitrate-nitrogen pollutants migrate downward at the maximum concentration before reaching the maximum concentration. After the maximum concentration, the concentration of nitrate nitrogen in the soil no longer increases; due to the difference in the average particle size of the soil, the migration rate of nitrate nitrogen in fluvo-aquic soil is significantly greater than that in artificial dredger soil. Nitrate nitrogen is in the simulation period completely passing through the simulated soil layer, the fluvo-aquic soil is completely passed through by nitrate nitrogen at 2d, and the artificial dredge fill is completely passed through by nitrate nitrogen at 2.5d.

1 Introduction

The Scholars have conducted a large number of studies on the migration and accumulation of nitrogen pollutants in the soil. Weng Lingyun et al. [1] have conducted field experiments in Huangtuo Village, Qingyuan County, Hebei Province for a period of 6 years since 2010. In the continuous field experiment, the effects of different nitrogen application contents on the accumulation and leaching of soil nitrate nitrogen were studied; Zhang Xuexue et al. [2] studied the irrigation and silt soil under different irrigation water amounts and fertilization conditions based on the soil column simulation method, results show: the amount of nitrogen applied has the greatest impact on the cumulative effect of nitrate-nitrogen migration in the soil; Yang Ruxin [3] et al. used numerical simulation analysis to study the soil moisture-nitrate nitrogen dynamic transport law in different fertilization conditions and irrigation intensity scenarios. In recent years, numerical simulation analysis method has become one of the main methods to study the accumulation and change of nitrogen pollution in soil due to its advantages such as good economy, fast speed and good visualization effect [4].

In this study, Hydrus-1D was used to simulate and compare the one-dimensional vertical migration and accumulation process of nitrate-nitrogen pollutants in the typical artificial dredger fill soil in Tianjin coastal area and the typical fluvo-aquic soil in the unsaturated zone of the North China Plain, and analyze their accumulation characteristics and differences.

2 Model establishment

2.1 The basic equation

The basic equation of water movement does not consider the horizontal flow and lateral flow of water, but only considers the one-dimensional vertical migration process of water, the classic Richards equation is used to describe the basic migration process of water in Hydrus-1D, Richards equation is as follows:

$$\frac{\partial\theta}{\partial t} = \frac{\partial}{\partial z}\left[K\left(\frac{\partial h}{\partial z} + \cos\alpha\right)\right] - s \qquad (1)$$

$$k(h,z) = k_S(z)k_r(h,z) \qquad (2)$$

Where: θ is the soil volumetric water content cm^3/cm^3; t is the length of time for simulation; α is the angle between the direction of water flow and the vertical direction. In this paper, only the one-dimensional vertical movement of water is considered, so α=0; Z is the soil depth cm; h is the pressure head of the soil (cm); s is the source-sink term (cm^3/cm^3t^{-1}); k_r is the relative hydraulic conductivity; k_S is the saturated hydraulic conductivity

2.2 Soil moisture transport equation

The soil moisture transport model is used to describe the process of moisture transport in the soil. Hydrus-1D provides six soil water transport models including single-porous media model, dual-porosity/dual-permeability media model, etc. The classic van Genuchten-Mualem model in the single-porous media model is used for water transport in the soil. The simulation prediction of the

* Corresponding author: jingpping@sina.com

migration process does not consider the hysteresis effect of moisture.

The van Genuchten-Mualem equation is as follows:

$$\theta(h) = \begin{cases} \theta_\gamma + \frac{\theta_s - \theta_\gamma}{[1+|\alpha h|^n]^m} & h < 0 \\ \theta_s & h \geq 0 \end{cases} \quad (3)$$

$$k(h) = k_s s_e^l \left[1 - (1 - s_e^{1/m})^m\right]^2 \quad (4)$$

$$m = 1 - 1/n, \quad n > 1 \quad (5)$$

Where: θ_s is the saturated soil moisture content; θ_γ is the residual soil moisture content; a, m, and n are the empirical parameters of the soil moisture characteristic curve; s_e is the effective soil moisture content; k_s is the saturated hydraulic conductivity; l is the soil pore connectivity parameters, usually take the average value 0.5

2.3 Solute transport equation

This study uses the convection-dispersion equation to describe the solute transport process, and does not consider adsorption and various reactions

$$\frac{\partial \theta c}{\partial t} + p\frac{\partial s}{\partial t} = \frac{\partial}{\partial x}\left(\theta D \frac{\partial c}{\partial x}\right) - \frac{\partial qc}{\partial x} - s \quad (6)$$

In the formula: c is the liquid concentration of the solution, mg/cm^3; ρ is the soil bulk density; s is the solute solid phase concentration; D is the comprehensive dispersion coefficient; q is the volume flow flux density; S is the source-sink term.

2.4 Initial conditions and boundaries

This simulation process only considers the one-dimensional vertical transport of nitrate nitrogen in the soil, and does not consider its transformation process, so the upper boundary of the water flow model is set to a constant flux atmosphere Atmospheric Boundary Condition with Surface Layer, the lower boundary is set to Free Drainage; the upper boundary of the solute transport model is set to 10mg/L Constant Concentration Boundary Condition, and the lower boundary is set to Zero Concentration Gradient Boundary, the height of the simulated soil column model is 60cm, the initial pressure head at the top is set to 0cm, and the initial pressure head at the bottom is set to -60cm, uniformly distributed from the top to the bottom from 0 to 60cm.

2.5 Grid division and time step

During the simulation process, the 60cm simulated soil column was divided into 1cm equidistantly, and observation points were set at 20cm, 40cm, and 60cm soil depth to study nitrate nitrogen pollutants. Migration in the vertical direction; the total simulation time of this study is 4d, the initial time step is 0.01d, the minimum time step is 0.001d, and the maximum time step is 0.1d. This simulation process outputs a total of 6 time node data, respectively 0.5d, 1d, 1.5d, 2d, 2.5d, 4d to study the changes of nitrate nitrogen pollutants in the soil over time.

The iteration information of this simulation process uses the default values of the Hydrus-1D model.

2.6 Determination of soil hydraulic parameters

The main parameters involved in this simulation process are soil hydraulic characteristic parameters, because the conversion process of nitrate nitrogen is not considered. The basic physical and chemical data of the coastal artificial dredger fill used in this simulation was obtained by Ge Feiyuan et al. and use the geotechnical test method. The percentage of sand (0.05~0.5mm) is 2.7%, and the percentage of powder (0.002~0.05mm) is the percentage is 83.3%, and the percentage of clay particles (0.0001~0.002mm) is 14% [5]; the basic data of the typical fluvo-aquic soil in North China is obtained through experiments by Guo Zifan et al., and the percentage of sand particles (0.05~0.5mm) is 26%, the percentage of powder particles (0.002~0.05mm) is 67.5%, and the percentage of clay particles (0.0001~0.002mm) is 6.5%[6]. Using the Neural network prediction function in the Hydrus-1D water flow module to calculate the soil hydraulic characteristic curve, input the proportion of sand, silt, and clay to obtain the required soil hydraulic characteristic curve parameters for the simulation（see Table 1）.

Table1. soil simulation parameter

Soil	θ_γ	θ_s	α	n	K_s	l
Dredger fill	0.0692	0.4774	0.0069	1.6245	17.06	0.5
fluvo-aquic	0.078	0.43	0.036	1.56	24.96	0.5

3 Simulation results and analysis

3.1 Vertical change of soil nitrate nitrogen

The concentration of nitrate nitrogen increases rapidly within 0-2.5d in the two soils at different depths, and becomes stable after reaching the maximum concentration. After 2.5d, the concentration no longer increases(see Figure 1, 2). That is, when the reclaimed water is used for agricultural irrigation, the nitrate-nitrogen pollutants in the reclaimed water migrate and accumulate in the soil is increased to the maximum value at different depths in the unsaturated zone. After the cumulative concentration of nitrate reaches the pollutant concentration, the concentration of nitrate nitrogen pollutants in the soil layer will not increase in the continuous irrigation time.

At the three observation depths(20cm,40cm,60cm), the concentration change trends of fluvo-aquic soil and artificial dredger fill are basically the same. At a depth of 20cm, the concentration of nitrate nitrogen changes from 0 to 99% of the solute concentration. The fluvo-aquic soil lasts 0.6964 days, and the artificial dredger fill lasts 1.1175d. At the depth of 40cm and 60cm, the fluvo-aquic soil also reached the maximum solute concentration significantly earlier than the artificial dredger fill soil. It shows that the migration rate of nitrate-nitrogen pollutants in fluvo-aquic soil is significantly greater than that of artificial dredger fill. The reason for this phenomenon may be due to the fact that the proportion of silt and clay particles in the artificial dredger fill soil is smaller than that of the fluvo-aquic soil. The soil particles are smaller and the gaps between soil particles are smaller, which has a stronger hindering effect on water flow and solute transport. This makes the migration rate of nitrate nitrogen in artificial dredger fill lower.

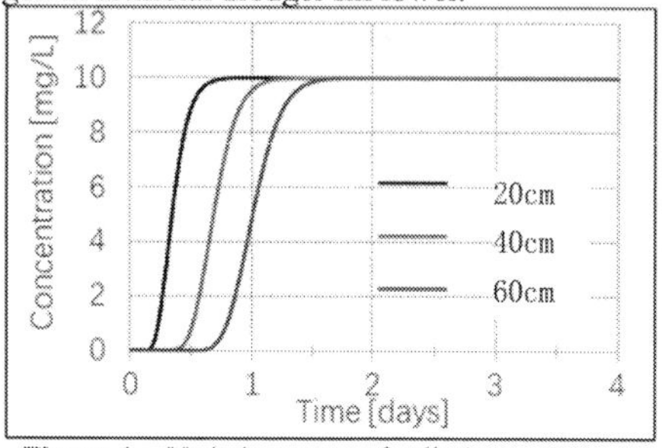

Figure 1. Variation curve of soil nitrate nitrogen concentration in fluvo-aquic soil

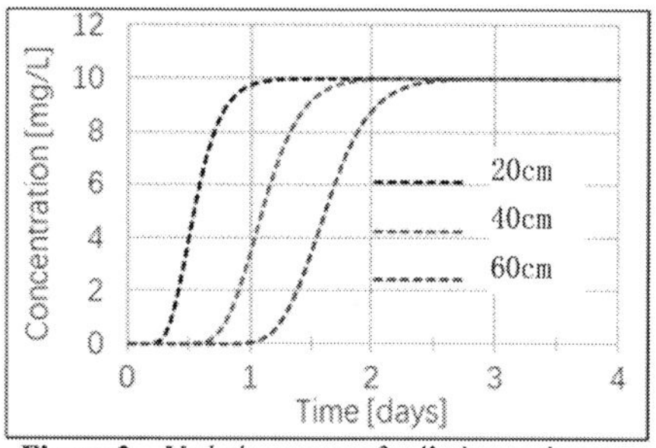

Figure 2. Variation curve of soil nitrate nitrogen concentration in artificial dredger fill soil

In order to further verify this point of view, the sandy loam soil is selected for simulation verification again,with larger average soil particles, and the sandy loam data is selected from the average sandy loam data that comes with the Water Flow Parameters module of the Hydrus-1D model(see Table 2).

Table2. soil simulation parameter

Soil	θ_r	θ_s	α	n	K_s	l
sandy loam	0.065	0.41	0.0075	1.89	106.1	0.5

At the three observation depths, the concentration change trends of sandy loam soil, dredger fill soil, and fluvo-aquic soil are basically the same. At a depth of 20 cm, the concentration of nitrate nitrogen changes from 0 to 99% of the solute concentration. The sandy loam soil lasts 0.1506 days, which is earlier than dredger fill soil and fluvo-aquic soil reach the maximum solute concentration. At the other two observation depths, sandy loam soil also reaches the maximum solute concentration significantly earlier than dredger fill soil and fluvo-aquic soil. The migration rate of nitrate-nitrogen pollutants in the unsaturated zone of soil is negatively correlated with the average soil particle size of the soil itself. This conclusion is more consistent with the research results of Wang Xiaodan et al. on the migration of the "three nitrogens" in the unsaturated zone of the Guanzhong Basin, Shaanxi[7]. .

If the appearance of nitrate nitrogen concentration in each soil layer is taken as the starting time, and the maximum concentration of nitrate nitrogen is reached as the ending time, then the time change of nitrate nitrogen pollutants from appearing and reaching the maximum concentration in each soil layer depth is obtained. The specific results are shown in the table3 and 4.

Table3. The time for the artificial dredger fill soil

Depth (cm)	20	40	60
Start time(d)	0	0.1519	0.353
maximum value time (d)	1.735	2.5627	3.2556
Last time(d)	1.735	2.4153	2.9026

Table4. The time for the fluvo-aquic soil

Depth (cm)	20	40	60
Start time(d)	0.2196	0.124	0.254
maximum value time (d)	1.0827	1.593	2.0207
Last time (d)	1.0608	1.469	1.7667

From Tables 3 and 4, it can be seen that the time step required for nitrate concentration to reach the maximum value in the soil layer will increase with depth, that is, the concentration migration rate slows down with the increase of depth, the reason is mainly due to the change of soil moisture at different depths.

3.2 Time series changes of soil nitrate-nitrogen concentration in the section of unsaturated zone

3.2.1 Analysis of concentration changes in artificial dredger fill soil

During the simulation process, the soil in the vadose zone of artificial dredging soil was completely penetrated by the nitrate nitrogen solution during the simulation period. In T2 time, the nitrate nitrogen had migrated to the bottom boundary, and the bottom boundary reached nitrate nitrogen pollution at T5 with the maximum concentration 10mg/L. Before T5, the nitrate nitrogen concentration in the soil gradually decreases from the top to the bottom of the vadose zone profile of the artificial dredging soil(see Figure 3). After T5, the nitrate nitrogen concentration in the entire vadose zone reaches the maximum value at 10mg/L.

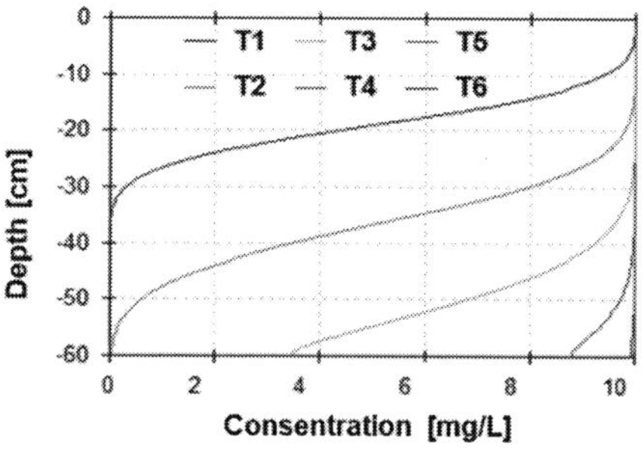

Figure 3. Variation curve of soil nitrate nitrogen concentration at different depths

(Note: T1, T2, T3, T4, T5, and T6 refer to 0.5d, 1.0d, 2.5d, 2.0d, 2.5d, 4d.)

3.2.2 Analysis of concentration changes of typical fluvo-aquic soil

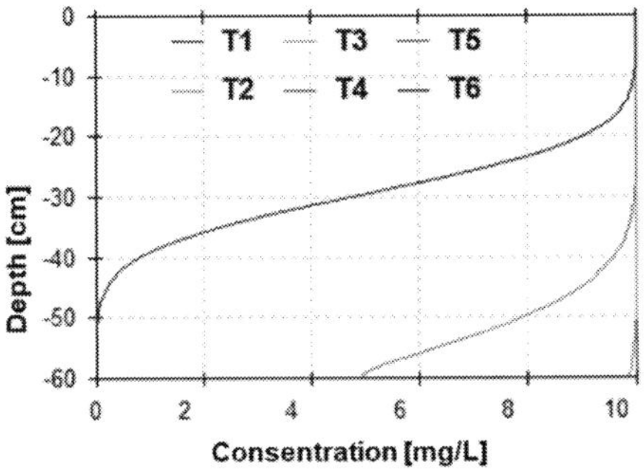

Figure 4. Variation of soil nitrate nitrogen concentration with depth

(Note: T1, T2, T3, T4, T5, T6 refer to 0.5d, 1.0d, 1.5d, 2.0d, 2.5d , 4d)

On the profile of the unsaturated zone in typical fluvo-aquic soil at different times, it can be seen from Figure 3 that during the simulation process, the soil in the typical fluvo-aquic zone of North China was completely penetrated by the nitrate nitrogen solution during the simulation period, and the bottom boundary was close to the maximum concentration value at T3, and reached the maximum concentration of nitrogen pollutants at T4 is 10 mg/L. Before T4, the nitrate nitrogen concentration in the typical fluvo-aquic soil zone profile from top to bottom in North China gradually decreases(see Figure 4). After T4, the nitrate nitrogen content in the entire unsaturated zone reached the maximum and stabilized at 10mg/L.

During T1 time, the migration depth of nitrate nitrogen in fluvo-aquic soil was significantly greater than that in artificial dredger fill; at time T2 and T3, nitrate nitrogen in both soils had migrated and diffused to the bottom boundary, but The cumulative concentration at the bottom boundary of the fluvo-aquic soil is significantly greater than the cumulative concentration at the bottom boundary of the dredger fill soil, indicating that the vertical migration rate of nitrate-nitrogen pollutants in typical fluvo-aquic soil is greater than that in artificial dredger fill soil.

4 Conclusion

During the migration process of nitrate-nitrogen pollutants in the vadose zone of artificial dredger fill soil, the migration and accumulation process in the soil at various depths will migrate and accumulate at the maximum concentration before the pollutant concentration is reached. After reaching the pollution source concentration, the nitrate nitrogen concentration in the soil does not continue to increase. That is, in the case of a longer scale, if the concentration of the nitrate nitrogen pollution source remains unchanged, the maximum concentration of the nitrate nitrogen pollutant concentration in the soil layer at each depth can eventually reach the pollution source concentration. Therefore, when the reclaimed water is used in the soil irrigation process, the nitrate nitrogen content must be strictly controlled, because the nitrate nitrogen concentration in the water will directly affect the nitrate nitrogen in the soil; the particle diameter of the soil itself will affect the nitrate nitrogen pollutants in the soil. The smaller the average soil particle diameter, the smaller the porosity between soil particles, the worse the transport of water and solute, and the slower the migration rate of nitrate-nitrogen pollutants in the soil; The migration rate in the soil is negatively related to the depth of the soil layer. The greater the depth of the soil layer, the slower the migration rate of nitrate nitrogen.

Acknowledgments

Authors wishing to acknowledge the financial support from Key Project of Tianjin Natural Science Foundation (18JCZDJC10090)

References

1. WENG L Y， YANG X Q， LYU M J， etc. Characteristics of soil nitrate accumulation and leaching under different long-term nitrogenapplication rates in winter wheat and summer maize rotation system. Chinese Journal of Applied Ecology, 2018,29(08):2551-2558.
2. ZHANG X K， BAI J Y， YAN H X. Effects of Irrigation and Nitrogen Application on Nitrate Transport in Different Soils[J]. Water Saving Irrigation，2020,(02):83-87
3. YANG R X， ZHANG H， XU W J,etc. Nitrate migration in vegetable cultivation soil under different fertilization and irrigation scenarios[J]. Environmental Engineering, 2018,36(07):133-137.
4. DU Q Q, YIN Z H, ZUO R,etc. Migration process simulation of ammonia nitrogen in contaminated site. China Environmental Science, 2017,37(12):4585-4595.
5. GE F Y， LIU J L， LI L W. Numerical simulation of pollutant transport in aeration zone of coastal reclamation land area based on Hydrus-1D[J]. Resource Information and Engineering，2020,35(04).
6. Guo Z F, Xu L, Guo H L, Zhou J Y, etc. Nitrogen accumulation and leaching charateristics of different aggregates in typical farmland soils in North China [J]. Chin J Appl Environ Biol, 2020, 26 (6): 1350-1358.
7. WANG X D, FENG W, WANG W K etc. Migrating and Transforming Rule of Nitrogen in Unsaturated Zone in Guanzhong Basin Based on HYDRUS -1D Model [J]. Geographical survey and research, 2015,38(04):291-298+304.

https://doi.org/10.1051/e3sconf/202129301005

Evaluation Index System of Marine Ecological Environment

Congcong Xing[1,2,3], Jing Li[1,2], Xiuren Li[4], and Bei Zhao[1,2*]

[1] North China Sea Environmental Monitoring Center, State Oceanic Administration, Ministry of Natural Resources, Qingdao 266033, China

[2] Shandong Provincial Key Laboratory of Marine Ecology and Environment & Disaster Prevention and Mitigation, Qingdao 266061, China

[3] Hainan Key Laboratory of Marine Geological Resources and Environment, Hainan Marine Geological Survey Academy, 570206, Haikou, China

[4] College of Oceanic and Atmospheric, Ocean University of China, Qingdao 266100, China

Abstract. Based on the model of ecological environment evolution in the process of the economic development, this paper constructed the evaluation index system of marine ecological environment .The results were shown as follows: the evaluation index system of marine ecological environment consists of 1 object layer, 3 element layers (state of marine ecological environment pressure of marine ecological environment and response of marine ecological environment), and 15 factor layers. Index weight was calculated through combination weighting approach of subjective and objective evaluation method of analytic hierarchy process(AHP) and entropy method .The evaluation indicator system and evaluation method have general applicability and strong operability.

1 Introduction

With the rapid development of marine economy, the marine ecological environment is also confronted with great pressures and challenge [1-8] . Grossman et al. [9], Kenneth et al. [10] discussed the relationship between environment and economic growth then put forward the theory of Environmental Kuznets Curve (EKC); Martinez et al. [11], Kildow et al. [12] put forward the ecological environment restoration measurements of different coastal areas to promote the marine ecosystem assessment and conduct comprehensive, coordinated and sustainable development. Quicoy et al. [13] discussed ecological environment carrying capacity of the coastal tourist industry. Song et al. [14] explained the stability of marine ecological environment under the optimal control of switching forward system; Wang et al. [15] analysed the coordination in marine economic development and marine ecological environment of Guangdong province by using coupling coordination degree mode; Di et al. [16] verified the stress relationship of marine ecological environment, economy and society in the coastal areas of China; Chen et al. [17] discussed the interaction and the evolution tendency between marine economic growth and marine environmental stress.

Based on the previous research work and the model of ecological environment evolution in the process of the economic development, this paper constructed the evaluation system of ecological environment to promote the scientific estimation of the marine ecological environment in China. It can provide reference for the protection of marine ecological environment and the optimal allocation of marine resources in China.

2 Coordinated development model

Marine economic development and ecosystem conservation are opposite as well as united. Based on the previous research work, the relationship between marine economy and ecological environment was shown as follows (Fig.1).

As the development degree metric, $y = \frac{1}{2} - x^3$,$y = \frac{3}{4} - x^3$ and $y = 1 - x^3$ denote the maximum economy development level under pressure of support for ecological environment protection and economic development. As the coordination degree metric, $y = x$, $y = x^3$, $y = x^{\frac{1}{3}}$ divides the coordination degree into four parts: strong pressure of economic development, moderate pressure of economic development, moderate support for ecological and environmental protection, strong support for ecological and environmental protection. Arrow 1 is the trend of economic development degree; arrow 2 is the trend of coordination degree.

* Corresponding author: 737711303@qq.com

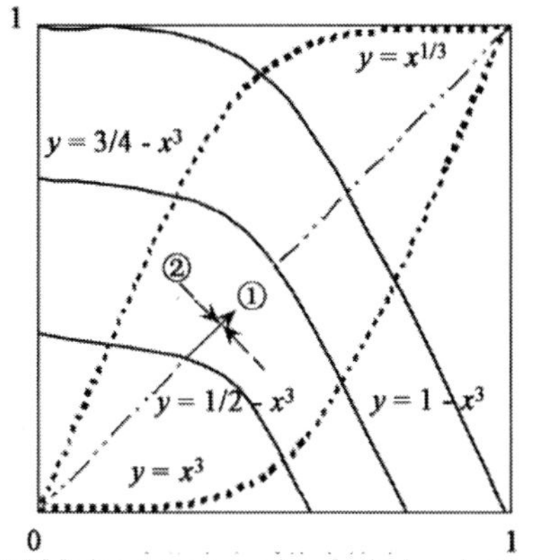

Fig.1 Marine economy and ecological environment coordinated development model

3 Evaluation index system

Based on the previous research work, this paper constructed the evaluation index system of ecological environment through combination weighting approach of subjective and objective evaluation method of analytic hierarchy process (AHP) and entropy method following the principles of systematicness, comprehensiveness and maneuverability. The evaluation index system of marine ecological environment consists of one object layer, three element layers(state of marine ecological environment, pressure of marine ecological environment, response of marine ecological environment) and fifteen factor layers (Table 1).

Table 1 Evaluation Index System of Marine Eecological Environment

Object layer	Element layers	Factor layers
Marine Eecological Environment developmental level	state of marine ecological environment (F_1)	marine biological diversity(I_1)
		per capita saltpan area(I_2)
		per capita coastal wetland area(I_3)
		marine natural reserves area ratio(I_4)
		per capita marine sand exploitation quantity(I_5)
	pressure of marine ecological environment (F_2)	polluted oceanic area ratio(I_6)
		erosion coast ratio(I_7)
		industrial solid wastes emissions per unit of land area(I_8)
		industrial wastewater emissions per unit of land area(I_9)
		direct economic losses from marine disasters per unit coastline(I_{10})
	response of marine ecological environment (F_3)	per capita gross ocean product(I_{11})
		marine employed person ratio(I_{12})
		marine scientific research personnel ratio(I_{13})
		per capita marine environmental protection input(I_{14})
		per capita marine infrastructure possession(I_{15})

State of marine ecological environment, which expressed the quantity and quality of marine ecological resources, consists of five factors: marine biological diversity (I_1), per capita saltpan area (I_2), per capita coastal wetland area (I_3), marine natural reserves area ratio (I_4) per capita marine sand exploitation quantity (I_5). The pressure of marine ecological environment, which expressed the pressure of human activities on marine resources and environment consists of five factors: polluted oceanic area ratio (I_6), erosion coast ratio (I_7), industrial solid wastes emissions per unit of land area (I_8), industrial wastewater emissions per unit of land area (I_9), direct economic losses from marine disasters per unit coastline (I_{10}). The response of marine ecological environment, which expressed the support of human activities to marine ecological environment protection, consists of five factors: per capita gross ocean product (I_{11}), marine employed person ratio (I_{12}), marine scientific research personnel ratio (I_{13}), per capita marine environmental protection input (I_{14}), per capita marine infrastructure possession (I_{15}).

4 Evaluation index system

The evaluation methodologies in this paper consist of three steps: standard processing of evaluation index, weight definition of evaluation index and marine ecological environment development index calculation.

4.1 Standard processing

Fifteen evaluation indexes of the evaluation system have different dimensions. And numerical values of positive indexes and negative indexes have different meanings. So this paper standardized the evaluation indexes using range method. The calculation formula is as follows:

$$X'_{ij} = \begin{cases} \frac{X_{ij}-X_{ij}^{min}}{X_{ij}^{max}-X_{ij}^{min}}, & \text{positive indexes} \\ \\ \frac{X_{ij}^{max}-X_{ij}}{X_{ij}^{max}-X_{ij}^{min}}, & \text{negative indexes} \end{cases} \quad (1)$$

X'_{ij} is the evaluated value of indexes, X_{ij} is the actual value of indexes; X_{ij}^{max} is the maximum value of indexes and X_{ij}^{min} is the minimum value of indexes.

4.2 Weight definition

The common evaluation methodologies include objective weighting method and subjective weighting method. In this paper, index weight was calculated through combination weighting approach of subjective and objective evaluation method of analytic hierarchy process (AHP) and entropy method. The comprehensive weight can be used to evaluate the development level of marine ecological environment and provide a basis for evaluation of the marine ecological environment development level.

4.2.1 Analytic hierarchy process (AHP)

Analytic hierarchy process (AHP) consists of four steps: hierarchy establishment, expert evaluation, consistency test and weight definition. The calculation formula is as follows:

$$A_{\omega} = \gamma_{max}\omega \quad (2)$$

$$CI = \frac{\gamma_{max}-n}{n-1} \quad (3)$$

$$CR = \frac{CI}{RI} \quad (4)$$

$$\omega_i = \sum_{j=1}^{m} a_j b_{ij} \quad (5)$$

A is judgment matrix; γ_{max} is the maximum eigenvalue; ω is eigenvector; CI is consistency index; RI is random consistency index; CR is consistency ratio; n is matrix dimension; ω_i is combination weight; m is the number of criterion layer index; a_j is the weight of criterion layer j to object layer; b_{ij} is the weight of index i to criterion j.

4.2.2 Entropy method

Entropy method consists of four steps: determining the index weight, calculating the index entropy value, calculating the coefficient of variation, calculating the weight. The calculation formula is as follows:

$$\rho_{ij} = \frac{a_{ij}}{\sum_{j=1}^{n} a_{ij}} \quad (6)$$

$$e_j = -k\sum_{i=1}^{m} \rho_{ij} \ln \rho_{ij} \quad (7)$$

$$g_j = 1 - e_j \quad (8)$$

$$\varphi_j = \frac{g_j}{\sum_{j=1}^{n} g_j} \quad (9)$$

ρ_{ij} is the weight of index i; e_j is the entropy, $k = \ln m$, $i = 1,2,3,\dots,m$; $j = 1,2,3,\dots,n$, suppose when $\rho_{ij} = 0$, $\rho_{ij} \ln \rho_{ij} = 0$; g_j is the coefficient of variation; φ_j is the weight.

4.2.3 Comprehensive weight definition

We can get comprehensive weight through combination weighting approach of subjective and objective evaluation method of analytic hierarchy process (AHP) and entropy method. The calculation formula is as follows:

$$W_i = \eta\omega_i + (1 - \eta\varphi_i) \quad (10)$$

ω_i is the weight of index calculated by analytic hierarchy process (AHP), φ_j is the weight of index calculated by entropy method, W_i is the comprehensive weight of index, η is adjustment parameter, $\eta = 0.5$ in this paper.

4.3 Index calculation

The evaluation system of ecological environment model was constructed as follows:

$$T = \sum_{j=1}^{n} P(F_{ij})\, I_j \quad (11)$$

T is the index of the marine ecological environment development level; $P(F_{ij})$ is the combination weight value of the element layer; I_j is the evaluation value of indexes.

5 Conclusions

The evaluation indicator system and evaluation method constructed in this paper have general applicability and strong operability, which can be widely applied to marine ecological environment development level evaluation of specific ecosystem (island, coastal zone and so on). It can also be widely applied to marine ecological environment development level evaluation of one coastal region or different coastal regions. The results we got from the evaluation indicator system and evaluation method constructed in this paper can provide a basis for evaluation of the marine ecological environment development level, so that decisions can be made to improve the efficiency of marine resource utilizing and protect the marine ecological environment.

Acknowledgements:

This study was sponsored by the North Sea Bureau, Ministry of Natural Resources (Grant No. 202102), the Shandong Provincial Key Laboratory of Marine Ecology and Environment & Disaster Prevention and Mitigation (Grant No. 201801), and the Hainan Key Laboratory of Marine Geological Resources and Environment (Grant No. HNHYDZZYHJKF008) .

References

1. Zhu Chaoqi. 2019. Introduction to the Themed Issue on New Advances in Marine Environmental Engineering in China. Journal of Marine Environmental Engineering, 10(3): 163-164.
2. Duarte C M. 2014. Global change and the future ocean: a grand challenge for marine sciences. Frontiers in Marine Science, 1: 63.
3. Jones P J S. 2013. A governance analysis of the Galápagos Marine Reserve. Marine Policy, 41: 65-71.
4. Yuanbin Fu, Yue Qi, Jing Kang, Ji Yu, Gongbo Ma, Chuanjun Wang and Yongguang Sunn. 2019. Coastal Wetland Vulnerability under the Influence of Human Activities: A Case Study of the Liaohe Estuary. Journal of Marine Environmental Engineering, 10(3): 181-193.
5. Ji-Yu C, Shen-Liang C. 2002. Estuarine and coastal challenges in China. Chinese Journal of Oceanology and Limnology, 20(2): 174-181.
6. Chuanjun Wang, Yingchao Wu, Ke Cao, Guangquan Chen, Wenquan Liu, Yuguang Wang and Yueyin Cai. (2019). Seawater Intrusion in the Coastal Area of Liaodong Bay. Journal of Marine Environmental Engineering, 10(3): 211-223.
7. Nguyen Thi The Nguyen and Grettel Gretchen. 2018. Regional Risk Zonation of Environmental Pollution on Marine and Coastal Zone]. Journal of Marine Environmental Engineering, 10(2): 97-107
8. Du Xing, Sun Yongfu, Song Yupeng, Zhu Chaoqi. 2021. In-Situ Observation of Wave-Induced Pore Water Pressure in Seabed Silt in the Yellow River Estuary of China. Journal of Marine Environmental Engineering, 10(4): 243-255.
9. Grossman G M, Krueger A B. 1994. Economic Growth and the Environment. Quarterly Journal of Economics, 110(2): 353-377.
10. Kenneth Arrow, Bert Bolin, Robert Costanza, Partha Dasgupta, Carl Folke, C.s. Holling, Bengt-owe Jansson, Simon Levin, Karl-gÖran MÄler, Charles Perrings And David Pimentel. 1996. Economic growth, carrying capacity, and the environment. Environment and Development Economics, 1(1), 104-110.
11. Martínez M L, Intralawan A, Vázquez G, Pérez-Maqueo O, Sutton P, and Landgrave R. 2007. The coasts of our world: Ecological, economic and social importance. Ecological economics, 63(2-3), 254-272.
12. Kildow J T, Mcilgorm A. 2010. The importance of estimating the contribution of the oceans to national economies. Marine Policy, 34(3): 367-374.
13. Quicoy A R, Briones N D. 2009. Beach carrying capacity assessment of coastal ecotourism in Calatagan, Batangas, Phlippines. Journal of Environmental Science & Management, 12(2): 11-26.
14. Song Liang, Lian Dongyan. The stability of marine ecological environment under the optimal control of switching forward system. Arabian Journal of Geosciences, 14(7): 1-6.
15. Wang Xianli, Jiang Guoqiang, Wang Junli. 2015. Evaluation of coordinated development of marine envirnment- economy system based on multivariate statistical methods. Marine Environmental Science, 34(5): 777-791.
16. Di Qianbin. Guo Yali. 2016. The stres rdlationship verifiation systeminteraction and coordination of the sustaina ble development of marineeconomy measure. Marine environmental science, 35(3): 453-459.
17. Chen Qi, Li Jingmei. 2015. Analysis on the decoupling relationship between marine environmental stress and marine economic growth in China. Marine environmental science, 34(6): 827-833.

https://doi.org/10.1051/e3sconf/202129301006

Advances in Studies on Nitrogen and Phosphorus Removal by Microalgal-Bacterial Consortia

Ruimin Mu[1], Qingyang Guo[1], Feng Qi[1,*], Yantian Jia[1], Qianya Meng[1], and Gejiang Yu[1]

[1]School of Municipal and Environmental Engineering, Shandong Jianzhu University, Jinan 250101, China

Abstract. With the development of life and industry, the nutrients in sewage increased gradually. The emerging symbiotic system of algal and bacteria has remarkable effect in removing nutrients such as nitrogen and phosphorus. In this paper, the influence of nitrogen and phosphorus on bacteria-algal consortia and the absorption mechanism of nitrogen and phosphorus by the interaction of bacteria-algal consortia were analyzed, and a variety of methods for studying bacteria-algal consortia were summarized, mainly using isotope tracer technology to study the research results of bacteria and algae absorbing nitrogen and phosphorus in water. This method is of great significance for analyzing the mechanism of the treatment of nitrogen and phosphorus by the bacterial-algal symbiosis system from the microscopic point of view.

1 Introduction

With the development of industrial level and the improvement of people's living quality, water pollution is becoming increasingly serious. The excess nutrients in water, such as nitrogen and phosphorus, can lead to eutrophication of water, affecting aquatic ecological balance and human health[1]. Therefore, it is of great significance to improve water quality and reduce treatment costs. At present, the commonly used denitrification process is activated sludge process, which uses nitrification and denitrification in activated sludge bacteria to remove nitrogen under aerobic and anoxic reaction conditions respectively. The phosphorus removal process relies on the release and absorption of phosphorus by phosphorous accumulating microorganisms under anaerobic and aerobic conditions respectively[2]. The conventional treatment techniques above can remove most suspended solids and organic pollutants, but the removal effect of nitrogen and phosphorus is not very ideal[3].

Studies have shown that the microalgal-bacterial consortia is an emerging wastewater treatment technology with lower cost and better treatment effect[4], because the interaction between bacteria and algae can improve the removal rate of nitrogen and phosphorus by the symbiosis. The growth of bacteria and algae depends on nutrients such as nitrogen and phosphorus, and their combination can produce synergistic effects, improving the nitrogen fixation potential and the ability to assimilate phosphorus[5,6]. The interaction between bacteria and algae is very complex, including nutrient exchange, signal transduction and so on[7]. For example, the signal molecule AHL mediated quorum sensing can change the related performance of microorganisms and regulate the community structure[8], and the signal molecule indole acetic acid (IAA) is also positively correlated with the removal rate of ammonia nitrogen in wastewater[9].

At present, microalgal-bacterial consortia have been used in a wide range of fields, including livestock and poultry sewage, saline marine sewage and domestic sewage, which can serve for the removal of nutrients and the commercial production of microbial metabolites. The study on the absorption of nitrogen and phosphorus by microalgal-bacterial consortia is of great significance for guiding the degradation of nutrients and the purification of wastewater[10]. There are many techniques for studying microalgal-bacterial consortia. Among them, stable isotope analysis of 15N is an established technology to determine the source of nitrogen and the contribution of different sources to microorganisms, as well as to track the whereabouts of nitrogen in microorganisms[11]. More and more scholars are interested in using this technology to study the dynamic behavior of nutrients in organisms and water[12]. However, few reviews have focused on the use of isotope tracer techniques to study the removal of nitrogen and phosphorus by microalgal-bacterial consortia. In this paper, the absorption of nitrogen and phosphorus by the interaction of microalgal-bacterial consortia and the application prospect of isotope tracer technique in consortia are introduced.

2 Removal effects of nitrogen and phosphorus on microalgal-bacterial consortia

Nitrogen (N) and phosphorus (P) are converted by various valence states and combined states under biological action in nature. Bacteria and algae are two types of organisms closely related to each other in the water ecosystem. Microalgae are photosynthetic autotrophic

* Corresponding author: qifeng@sdjzu.edu.cn

microorganisms, which can convert nitrogen, phosphorus and other inorganic substances in sewage into organic matter through photosynthesis, and at the same time release O_2 for the growth of bacteria, Figure 1. Under the action of microorganisms, nitrogen in the water is converted to ammonia, and in the presence of oxygen, ammonia nitrogen in the water is converted to nitrate nitrogen by nitrifying bacteria[13]. Aerobic bacteria in activated sludge oxidize and decompose organic matter and produce nitrate which can be used by microalgae to promote their photosynthesis and growth and reproduction[14].

Activated sludge can remove phosphorus by releasing phosphorus under anaerobic conditions and absorbing phosphorus under oxygen conditions respectively. However, phosphorus can be used to synthesize ATP by absorption and assimilation of microalgae[15], and phosphate can promote the growth of the community structure of microalgae and phosphorus accumulating bacteria[16].

The concentration of nitrogen and phosphorus had significant effects on the growth and community structure of the microalgal-bacterial consortia. Different combinations of algae species and strains have different nutrient removal efficiency in wastewater. For different sewage, the dominant communities of microorganisms change, and the communities interact with each other to form the appropriate microalgal-bacterial consortia[17]. Due to the "luxury uptake" of nutrients by microalgae and bacteria, the uptake of N and P by microalgal-bacterial consortia increased as the initial N and P concentration increased, far exceeding the stoichiometric N and P required for the corresponding increase in biomass[18]. However, the experiment proved that too high N/P ratio would slow down the growth rate of algal cell biomass and decrease the bacterial density in different degrees in the later period of the experiment[19].

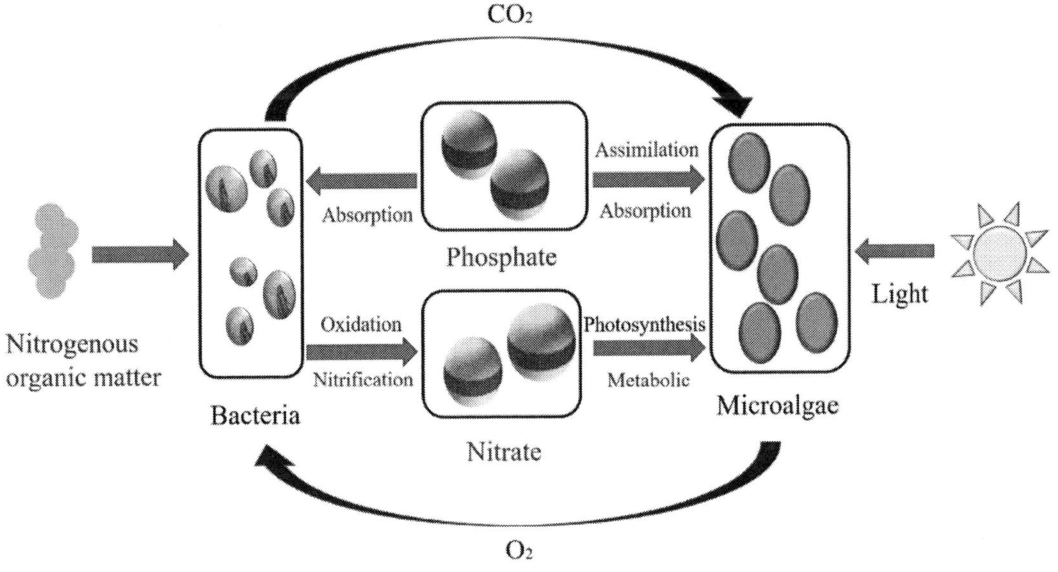

Figure 1. Nitrogen and phosphorus in wastewater are removed by microalgal-bacterial consortia.

3 N and P absorption by interaction of microalgal-bacterial consortia

Some scholars believe that there are a lot of interactions between microalgae and heterotrophic bacteria mediated by nitrogen and phosphorus in the state of poor nutrition[20]. For example, nitrogen-fixers provide inorganic nitrogen to scenedesmus, Escherichia coli provides inorganic phosphate to host microalgae, and Pseudomonas maintains phosphorus concentrations suitable for Microcystis aeruginosa through phosphorus transfer[21]. The treatment effect of algae or activated sludge on nutrients and COD in wastewater is far less than that of microalgal-bacterial consortia. Choi found that the removal efficiency of total N and total P in the bacterial-algal co-culture reactor was 3.4 and 6.5 times higher than that in the pure nitrification reactor[22]. In the treatment of nutrients in sewage by algae alone, due to the photoinhibition, algae autotrophic growth is slow, and the loss of activated sludge that provides carbon source, CO2 only comes from the air, photosynthesis of algae is restricted, resulting in low efficiency of algae removal of nutrients. Both algae and nitrifying bacteria in activated sludge can use ammonia and phosphate in sewage as common substrates[22]. In addition, when the ratio of microalgae and activated sludge was 5:1, the removal efficiency of total N and total P was the highest, which were 95.8% and 93.5%[23]. The N and P removal rates of the bacterial-algal symbiosis system are summarized in Table 1.

Algae and bacteria not only exist symbiosis, but also inhibit and compete. Bacteria can inhibit the growth of algae by competing with nitrogen and phosphorus nutrient elements, which is one of the effective ways to control fresh water blooms and maintain aquatic ecological balance.

Table1. Summarization of N and P removal rates in wastewater by microalgal-bacterial consortia.

Bacteria	Microalgae	Type of wastewater	TN	TP	NH_4^+-N	Reference
Brevundimona bullata	Chlorella	Light polluted water	64%	95.4%		[9]
Pseudomonas panipatensis			68%	65.7%		
Photosynthetic bacteria	Chlorella albumen	Artificial sewage	95%			[31]
Activated sludge	Chlorella	Wastewater from livestock and poultry breeding		95.1%	83.2%	[2]
Associated bacteria	Freshwater algae	Sanitary sewage		92.3%	95.6%	[4]
Activated sludge	Hybrid algae	Artificial sewage		83.3%	94.2%	[24]

4 Advances in techniques for the study of microalgal-bacterial consortia

With the increasing research on microalgal-bacterial consortia, the selection of appropriate techniques and specific tools plays an important role in the research. The selection of different research methods is helpful to the stable establishment of the bacteria-algal co-culture system, so that it can better serve the sewage treatment and the protection of aquatic ecological balance. In the macroscopic analysis, the research is mostly through physical and chemical detection methods. By measuring the contents of nitrogen and phosphorus before and after treatment, the removal efficiency of microalgal-bacterial consortia on nitrogen and phosphorus in wastewater was analyzed.

In the microscopic aspect, the research methods include microscopy, flow cytometry, cell sorting, omics and secondary ion mass spectrometry.

4.1 Microscopy

Microscopic observation of consortia can be used to obtain details of their cell morphology and composition. Microbial composition of mixed samples of bacterial-algal systems and water samples can be observed by microscope, and the increase of the number of dominant algae species and bacteria can be observed, and it can also be proved that nitrogen and phosphorus in sewage have been removed[24].

4.2 Flow cytometry and cell sorting

Flow cytometry and cell sorting are also microscopic techniques that have been used to measure specific cellular components of microalgae and bacteria in bacterial-algal systems. For example, analysis of polyphosphate-accumulating bacteria can be done by staining polyphosphate particles with fluorescent dyes DAPI and tetracycline hydrochloride by flow cytometry[5].

4.3 Omics

Metagenomics methods have also been applied to study the community structure of bacteria and algae, and to study the changes of dominant communities by sequencing genes, so as to improve the efficiency of consortia in wastewater treatment. A series of methods including genomics, transcriptomics, proteomics and metabolomics have been applied to the study of bacteria-algal symbiosis[5].

4.4 Secondary ion mass spectrometry

Secondary ion mass spectrometry (SIMS), such as isotope ratio mass spectrometry (IRMS), can be used to study the function and metabolism of individual microorganisms in a microbial community at the cellular level. Nanomaterials combined with fluorescence in situ hybridization were used to study and image the metabolic activity and absorption rate of microbial cells in a complex microbial population[25]. The multi-isotope imaging mass spectrometry (MIMS) showed the binding level of 15N in bacteria. This method can be used to study bacteria in the natural environment and explore the activity of nitrogen nutrient symbionts in algae and bacterial cell tissue. It will be a powerful tool for studying the effect of nitrogen fixation in the future[26].

Isotope tracer technology is to use the mass spectrometer or element analyzer, radioactive isotope as a tracer, used to study the distribution, migration, enrichment rules of a substance[25]. The combination of isotope labelling experiments with other methods such as nanotechnology will open up new directions for the analysis of microbial communities and their metabolic functions[27].

DE - Bashan to the isotope labeling algal-bacterial system experiment, will mark the party and on the other side of the untagged culture together, and then use the secondary ion mass spectrometry analysis of each element transfer between microbial cells and bacterial cell adhesion over the surface of the algal cells and release metabolite, proved that the nitrogen molecules by switching between two kinds of microbes, and a symbiotic interaction that promotes the growth of the microalgal-bacterial consortia[28]. Stable isotope analysis of nitrogen (14N15N) can determine the source of nitrogen in complex environments and analyze the contribution ratio of different sources[11]. For example, Cao used stable isotopes to track the source and contribution ratio of organic matter in sediments and analyze the causes of eutrophication of water bodies, which has indicative significance for the protection of ecological environment[29]. Han et al. added stable ^{15}N isotopes to the experiment, which showed that microalgae would preferently absorb and assimilate reduced ammonia nitrogen in water, followed by oxidized nitrate nitrogen, and different algae species had different rates of nitrogen absorption[30]. Wang et al. used stable isotopes to investigate the absorption rates of ammonia nitrogen, nitrate nitrogen and urea nitrogen by bacteria-algal combination under different temperature, light and salinity conditions[31]. Based on the stable isotope fractionation effect, Meng et al. established a model for the value of δ15N in activated sludge and the denitrification efficiency of the whole system, which improved the efficiency of sewage treatment and was of great significance for the denitrification process of water plants[32].

Different forms of phosphorus can be tracked by isotope labeling, but only 31P is stable isotope, and the P-O bond is relatively stable chemical bond[33], so the use of phosphate oxygen isotope can be used to study the direction of phosphorus in aquatic ecosystems[12]. Elsbury et al. used phosphate oxygen isotope tracer techniques to distinguish between different sources of phosphate and their relative contributions[34]. Based on the phosphate oxygen isotope tracer technique, McLaughlin et al. used $\delta^{18}O_p$ as a tracer to study the phosphorus cycling in the water body of San Francisco Bay[35].

5 Conclusion

The ability of microalgae to remove nutrients such as nitrogen and phosphorus in wastewater was combined with the ability of bacteria to degrade pollutants, and construct microalgal-bacterial consortia to achieve water purification and aquatic ecosystem balance is considered to be a very promising treatment technology, which is cheaper than traditional wastewater treatment technology, and has been widely studied. The interaction between algae and bacteria and their metabolic functions can be further studied by means of microscopy and omics. The use of advanced analytical tools also provides more comprehensive and detailed information for the construction of microalgal-bacterial consortia that can efficiently remove nitrogen and phosphorus. The application of isotope tracer technique to study the specific trends of nitrogen and phosphorus in microalgal-bacterial consortia can further explore the interaction and removal mechanism between the two elements from the microscopic molecular dimension, which is more meaningful than the study of removal rate alone. This method is helpful to improve the efficiency of wastewater treatment and to construct suitable microalgal-bacterial consortia for wastewater with different N and P concentrations. However, more research is needed.

Acknowledgement

This review was funded by "Youth Innovation Technology Project of Higher School in Shandong Province, grant number 2019KJD003; Shandong Province Natural Science Foundation, grant number ZR2019MEE104; Shandong Province Water Conservancy Research and Technology Extension Project, grant number SDSLKY201812.

References

1. H.ZENG, Q.LIU, C.XIAO, Optimization of microalgae-bacteria symbiosis system for the advanced treatment of livestock wastewater using response surface methodology,*Journal of Agro-Environment Science*,**39(6)**,1368-1379 (2020)
2. H.LIANG,Y.ZHANG,Study on the removal of nitrogen and phosphorus from sewage by Co-culture of immobilized sludge and immobilized Chlorella, *Guangxi Sciences*,**26(2)**,215-221 (2019)
3. B.GAO,J.CHENG,Y.CUI, Study on the purification of domestic sewage by immobilized alga with fibrous materials and its influencing factors, *Jiangsu Agricultural Sciences*, **46 (11)** ,276-279(2018)
4. M.GAO,Study on Removal of Nitrogen and Phosphorus and Heavy Metals from Wastewater by Alga-bacteria System.(2016)
5. I.A.Perera,S.Abinandan,S.R.Subashchandrabose,K. Venkateswarlu,R.Naidu,M.Megharaj,Advances in the technologies for studying consortia of bacteria and cyanobacteria/microalgae in wastewaters, *Crit Rev Biotechnol*,**39 (5)**, 709-731 (2019)
6. N.Renuka,A.Guldhe,R.Prasanna,P.Singh,F.Bux,Micr oalgae as multi-functional options in modern agriculture: current trends, prospects and challenges, *Biotechnol Adv*,**36 (4)**, 1255-1273(2018)
7. A.Kouzuma, K.Watanabe, COB, Exploring the potential of algae/bacteria interactions, *Curr Opin Biotechnol*,**33**, 125-9(2015)
8. Y.PANG, Z.WU, T.AN, Z.SUN, M.GOU, Quorum Sensing Mediated by Signal Molecule AHL in Wastewater Biological Treatment, *Technology of Water Treatment*,**46 (11)**(2020)
9. T.CHEN,The Study on Application and Mechanism of Bacteria-Algae Symbiosis and Bioaugmentation Technology in Wastewater Treatment .(2017)

10. S.R.Subashchandrabose,B.Ramakrishnan,M.Megharaj, K.Venkateswarlu, R.Naidu, Consortia of cyanobacteria / microalgae and bacteria: biotechnological potential, *Biotechnol Adv*,**29(6)**, 896-907(2011)
11. M.J.Ochoa-Izaguirre,M.F.Soto-Jimenez, Variability in nitrogen stable isotope ratios of macroalgae: consequences for the identification of nitrogen sources, *J Phycol*,**51 (1)**, 46-65(2015)
12. Q.CHU,J.FAN,G.FENG, New Advance of The P Isotopic Tracing In The Ecology And Resource-Environmental Sciences: A Review,**26 (05)**, 828-837(2012)
13. P.ZHAO, The ecological respond of microorganisms to environment in different nitrogen polluted water, (2016)
14. M.DIAO,J.LIU,W.WANG,W.HUA,S.WU,S.SUN，Y.CHENG, Impact of algae-bacteria symbiotic system on wastewater treatment and biomass accumulation of microalgae, *Environmental Engineering*, **36 (03)** , 8-12(2018)
15. Y.Collos,A.Vaquer,P.Souchu, Acclimation of Nitrate Uptake by Phytoplankton To High Substrate Levels, *Journal of Phycology*, **41 (3)**, 466-478(2005)
16. R.Mu,Y.Jia,G.Ma,L.Liu,K.Hao,F.Qi,Y.Shao,Advances in the use of microalgal-bacterial consortia for wastewater treatment: Community structures, interactions, economic resource reclamation, and study techniques , *Water Environ Res* ,(2020)
17. H.WANG , W. WANG , Y. SHENG,Influences of sewage treatments technologies on microalgae and bacteria community structures, *China Environmental Science*, **38(10)**, 3761-3766(2018)
18. V.Ashok,A.Shriwastav,P.Bose, Nutrient removal using algal-bacterial mixed culture, *Appl Biochem Biotechnol* ,**174 (8)**, 2827-38(2014)
19. Q.LIU,Establishment of stable culture and optimization of symbiltic system of bacteria and algae,(2018)
20. B.Zhang,W.Li,Y.Guo,Z.Zhang,W.Shi,F.Cui,P.N.L.Lens,J.H.Tay, Microalgal-bacterial consortia: From interspecies interactions to biotechnological applications, *Renewable and Sustainable Energy Reviews* ,**118**(2020)
21. L.Jiang, L.Yang, L.Xiao, X.Shi, G.Gao, B.Qin, *Hydrobiologia*,Quantitative studies on phosphorus transference occuring between Microcystis aeruginosa and its attached bacterium (Pseudomonas sp.),**581(1)**,161-165 (2007)
22. K.J.Choi, T.H.Han, G.Yoo, M.H.Cho, S.J.Hwang, Co-culture Consortium of Scenedesmus dimorphus and Nitrifiers Enhances the Removal of Nitrogen and Phosphorus from Artificial Wastewater, *KSCE Journal of Civil Engineering*, **22 (9)**,3215-3221(2017)
23. Y.Su, A.Mennerich, B.Urban, Synergistic cooperation between wastewater-born algae and activated sludge for wastewater treatment: influence of algae and sludge inoculation ratios, *Bioresour Technol*,**105**, 67-73(2012)
24. K.HAO,W.CHEN,R.MU,F.QI,H.WANG,Study on the removal of nitrogen and phosphorus by bacteria and algae system,*Journal of Shandong Jianzhu University*, **34 (05)**, 50-54(2019)
25. D.Gao,X.Huang,Y.Tao, A critical review of NanoSIMS in analysis of microbial metabolic activities at single-cell level, *Crit Rev Biotechnol*,**36(5)**, 884-90(2016)
26. C.Lechene,F.Hillion,G.M.Mahon,D.Benson,A.M.Kleinfeld,J.P.Kampf,D.Distel,Y.Luyten,J.Bonventre,D.Hentschel,K.M.Park,S.Ito,M.Schwartz,G.Benichou, G.Slodzian, High-resolution quantitative imaging of mammalian and bacterial cells using stable isotope mass spectrometry, *Journal of Biology*,**5**(2006)
27. N.Musat,B.Adam,M.M.M.Kuypers,Nano-Secondary Ions Mass Spectrometry (nanoSIMS) coupled with in situ hybridization for ecological research.(2017)
28. L.E.de-Bashan,X.Mayali,B.M.Bebout,P.K.Weber,A.M.Detweiler,J.P.Hernandez,L.Prufert-Bebout,Y.Bashan, Establishment of stable synthetic mutualism without co-evolution between microalgae and bacteria demonstrated by mutual transfer of metabolites (NanoSIMS isotopic imaging) and persistent physical association (Fluorescent in situ hybridization), *Algal Research*,**15**,179-186(2016)
29. H.CAO , Z.ZHANG, W. HE, X.HUANG, X.PU,Study on the Change process of the ecological environment in the Zhanjiang Bay through the organic carbon isotope technique,*Journal of Ocean Technology*, **39 (01)** 57-63(2020)
30. F.HAN,Z.ZHAO,D.LI,L.ZHANG,The use of ^{15}N to measure nitrogen uptake of phytoplankton, *Oceanologia ET Limnologia Sinica*, **50 (04)**,811-821(2019)
31. P.WANG, Effects of temperature, salinity and light intensity on nitrogen absorption of bacteria-algae system, (2020)
32. N.MENG, P.JIN，H.JIN，X.LI, Evaluation of the nitrogen removal performanceinactivated sludgeprocess based on nitrogen stable isotope fractionation principle, *Acta Scientiae Ccircumstantiae*,**38(5)**,1828-1834 (2018)
33. C.ZHANG, H. JI, Z.CHE,The application of phosphate oxygen isotope in organophosphorus degradation, *Journal of Shandong Agricultural University*,**49 (06)** ,1008-1014(2018)
34. K.E.Elsbury,A.Paytan,N.E.Ostrom,C.Kendall,M.B.Young,K.Mclaughlin,M.E.Rollog,S.Watson, Using Oxygen Isotopes of Phosphate to trace phosphorus sources and cycling in lake Erie,*Environmental Science&Technology*, **43(9)**,3108-3114(2009)
35. K.McLaughlin,C.Kendall,S.R.Silva,M.Young,A.Paytan, Phosphate oxygen isotope ratios as a tracer for sources and cycling of phosphate in North San Francisco Bay, California, *Journal of Geophysical Research*,**111**(2006)

https://doi.org/10.1051/e3sconf/202129301007

Reservoir characteristics and control factors of Tongbomiao formation in Hongqi sag

Wu Haibo[1], Zhou Yue[1*], and Chen Hao[1]

[1]Exploration and Development Research Institute of Daqing Oil Field Co. Ltd, Daqing City, Heilongjiang Province, 163712, China

Abstract. Hongqi sag is located in the north of Beiehur depression in Hailar Basin, which is a secondary structural unit in the peripheral depression of Hailar Basin, with great exploration potential. Based on SEM, conventional thin sheet, cast thin sheet, mercury pressure and physical property data, this paper discusses the reservoir characteristics, types and control factors of high quality reservoir in Tongbomiao formation, Hongqi sag, so as to provide a strong geological basis for the exploration of oil and gas in this area. The reservoir space of Hongqi sag is mainly composed of four types: primary intergranular pore, micro fracture, feldspar dissolution pore and debris dissolution pore. It is of the type of ultra low porosity and low permeability. Although Tongbomiao formation in Hongqi sag is of very low porosity and low permeability type, high quality reservoir is still developed, and its reason is mainly controlled by sedimentary environment, diagenesis and structural fracture.

1 Introduction

The characteristics and formation mechanism of clastic rocks in continental fault depression basin have been a hot issue, and many scholars have done more work in these aspects. Hailar Basin is a complex multi-stage superimposed fault basin. In recent years, there are many discussions on the physical control factors of the reservoir in Hailar Basin[1][2][3]. Hongqi sag has experienced multi-stage structural transformation and superimposed fault depression, and the influencing factors of reservoir physical properties are complex and diverse, and the distribution law is not clear. With the development of exploration, the physical properties of clastic rock reservoir are the important factors to control oil and gas. Therefore, it is of great significance to find high quality reservoirs. Many scholars have made important progress in the formation environment, diagenesis, pore type and physical characteristics of reservoirs. However, there are still some controversies about the main factors of high quality reservoirs in cupbimiao formation in Hongqi sag. Therefore, based on SEM, conventional thin sheet, cast thin sheet, mercury pressure and physical data, this study advances the reservoir characteristics and main control factors of cupbimiao formation in Hongqi sag The detailed analysis provides a strong geological basis for the exploration of oil and gas in the area.

2 Regional overview

Hongqi sag is located in the north of beierhu depression in Hailaer basin. It is distributed in northeast direction, adjacent to bayanshan uplift in the East and cuogang uplift in the West. It can be divided into 9 structural units: concave hill, cuogang uplift, Hongqi ranch sag, shutenor sag, central uplift fault zone, Hongxi inversion structural zone, shaoerbaoge slope zone, bayanshan uplift and wuxingdui uplift. It is characterized by "East-West zonation and North-South zonation", with an area of about 1000km^2 and a maximum burial depth of 3400m.

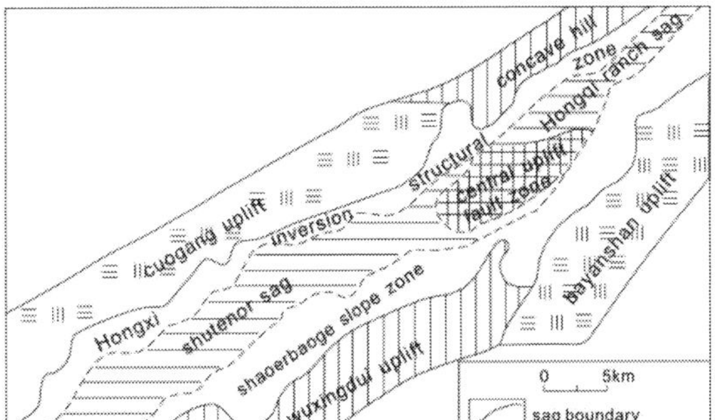

Figure 1. Structural unit division and stratigraphic histogram of Hongqi sag in Hailaer Basin.

The strata can be divided into Tamulangou formation, Tongbomiao formation, Nantun Formation, Damoguaihe Formation and Yimin formation from bottom to top. The Tamulangou formation is mainly composed of pyroclastic rocks and dark mudstone; the Tongbomiao and Nantun formations are mainly composed of fine sandstone, glutenite and dark mudstone; the Damoguaihe Formation is mainly composed of grey mudstone; the Yimin formation is composed of grey black shale interbedded with siltstone and fine sandstone; the Qingyuangang

* Corresponding author: 47310372@qq.com

formation is mainly composed of brownish yellow glutenite. Among them, Tongbomiao formation shows good oil and gas, which is the research focus of the sag (Fig. 1).

3 Petrological characteristics of reservoir

Through the analysis of thin section identification data, the reservoir rock types of Tongbomiao formation in Hongqi sag are various and the heterogeneity is strong. The reservoir rock types are mainly feldspar lithic sandstone and lithic sandstone (Fig. 2a). The content of quartz in clastic components is relatively low, and the content of feldspar and cuttings is relatively high, with the average volume fraction of feldspar accounting for 23.8% of the total clasts; the average volume fraction of quartz is 16.2%; the content of cuttings is relatively high, with the average volume fraction of 49.8%. The composition of cuttings is complex, mainly including magmatic rock and sedimentary rock cuttings. Among them, the content of magmatic rock cuttings is relatively high, accounting for 82% of the total volume content of relative cuttings. The magmatic debris are mainly trachyandesite, tuff, trachyte, andesite and basalt. The content of andesite and basalt is relatively high (Fig. 2b).The interstitial materials are chlorite, kaolinite, carbonate and siliceous. The average volume fraction of the interstitial materials is 10.5%. The cement is mainly carbonate cementation, a small amount of siliceous cementation, kaolin and authigenic chlorite formed after feldspar dissolution are relatively more; the miscellaneous base is mainly argillaceous and pozzolanic, which are mixed and accumulated between particles, so it is difficult to define their relative content.

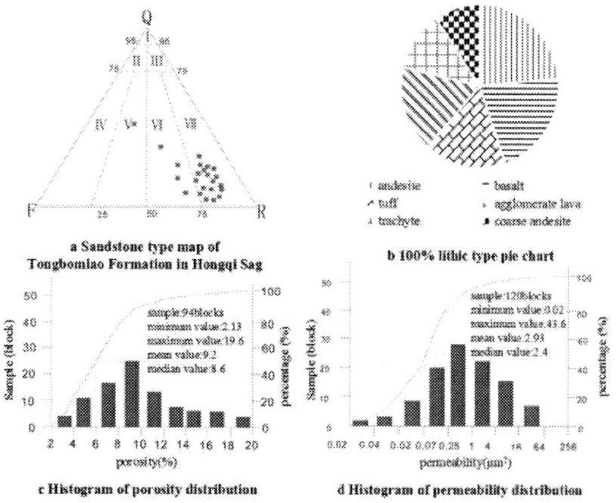

Figure 2. Sandstone types and porosity permeability istogram of Tongbomiao formation in Hongqi sag.

The reservoir rocks are poorly sorted, with sub angular and sub circular roundness, various cementation modes, mainly pore cementation, contact cementation and basement cementation, and low compositional maturity and structural maturity; the support type of rocks is particle support; the point line contact is the main contact between particles. Through thin section and scanning electron microscope, it is found that there are various pore types in the reservoir, including primary intergranular pore, feldspar solution pore and cuttings solution pore, and a small amount of clay mineral intergranular pore and fracture pore (Fig. 3). Among them, primary intergranular pores account for 18% of the total pore body, feldspar dissolved pores account for 36% of the total pore body, cuttings dissolved pores account for 37% of the total pore body, microfractures account for 7% of the total pore body, and clay mineral intergranular pores account for 2% of the total pore body. According to the national oil and gas industry standard, the porosity of Tongbomiao formation reservoir in Hongqi sag mainly ranges from 6 to 19.6%, with an average of 9.2% (Fig. 2c), and the permeability mainly ranges from 0.02 to 43.6x10^{-3}μ m^2, with an average of 2.93x10^{-3}μm^2 (Fig. 2d), which is an ultra-low porosity and low permeability reservoir.

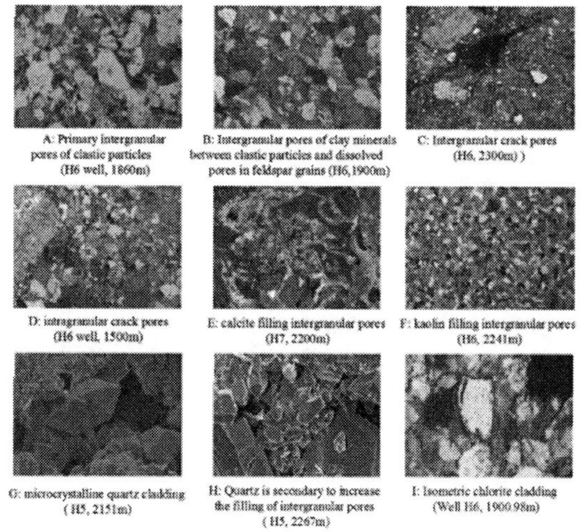

Figure 3. Pore types of Tongbomiao formation in Hongqi sag.

4 Analysis on controlling factors of high quality reservoir

Although Tongbomiao formation in Hongqi sag is an ultra-low porosity and low permeability reservoir, high quality reservoir is still developed, which is mainly controlled by sedimentary environment, diagenesis and structural fractures.

4.1 Depositional environment

Sedimentation controls the physical properties of the original reservoir. Under the action of different sedimentary environments, the particle size, rounding, sorting and impurity content of the sediments are obviously different under different sedimentary dynamic conditions. Therefore, the sedimentary environment will have a direct impact on the physical properties of rocks. The results show that porosity is negatively correlated with shale content. Low argillaceous content, strong hydrodynamic conditions, good sand sorting and grinding, strong anti compaction ability, which is conducive to the preservation of primary pores, more conducive to the generation of secondary pores in the later stage, and easy to form high-quality reservoirs. The reservoir of Tongbomiao formation in Hongqi sag mainly develops fan delta system. The physical properties of underwater distributary channel microfacies in fan delta front and above water distributary channel microfacies in fan delta plain are better than those of sheet sand and sublacustrine fan in fan delta front.

4.2 Diagenesis

Diagenesis includes the influence of compaction, cementation, chlorite film, pyroclastic materials on dissolution, which controls the formation, evolution and distribution of reservoirs[4][5]. The diagenesis of the reservoir in the study area mainly includes compaction, cementation and dissolution. According to the vertical variation of clay minerals, authigenic minerals and pore types, the Tongbomiao formation in Hongqi sag is in the middle diagenetic stage A-B (Fig. 4).

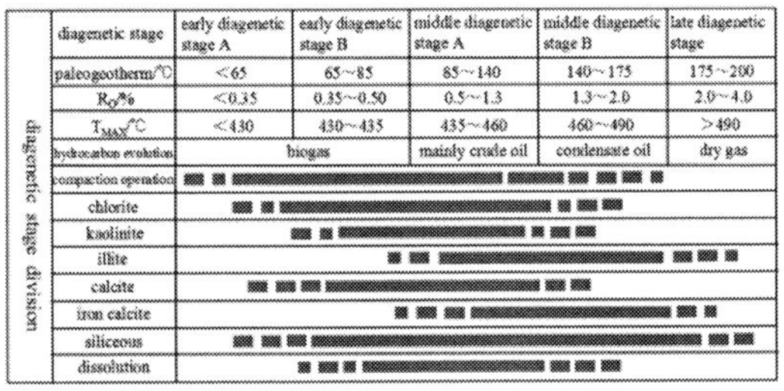

Figure 4. Pore types of Tongbomiao formation in Hongqi sag.

Compaction is the main diagenesis to reduce the reservoir physical properties, and it is a destructive diagenesis to the reservoir physical properties. The buried depth of Tongbomiao formation in Hongqi sag is between 1800 and 3400m, with an average of 2600m. The maturity of rock composition and structure are low, and the rock compression capacity is weak. The rock particles are mainly point line contact, some of them are concave convex contact, and the intergranular pores are damaged. At the surface contact of rock particles, the carbonate produced by early cementation can effectively reduce the damage of compaction to the original reservoir, support and reduce the damage of compaction to the original pores, so that the original reservoir can be effectively preserved.

The reservoir of Tongbomiao formation in Hongqi sag is mainly composed of carbonate, kaolin and chlorite. According to the experimental analysis, there are two forms of carbonate cementation: the filling of intragranular pores; under the microscope, the pyroclastic particles are replaced by calcite, and the pyroclastic particles are dissolved to form intragranular dissolution pores; the calcite carbonate fills intragranular dissolution pores in the form of clumps and grains; Intergranular pore filling, carbonate recrystallization is massive, filling the primary intergranular pore, which has a destructive effect on the primary intergranular pore (Fig. 3E, F). Kaolin is cemented in intergranular solution pores, and alkaline feldspar dissolves to form kaolin. Kaolin recrystallizes and distributes in the pores in the form of film, strip and block. In intergranular pores, kaolin and volcanic ash are mixed together, and the boundary is not clear, so it is difficult to distinguish, which has a destructive effect on the primary intergranular pores. Chlorite cement is mainly attached to the surface of particles in the form of tufts and leaves. When the chlorite content is less than 16%, the porosity increases with the increase of chlorite content; When the content is more than 16%, the porosity increases and decreases. As there is no chlorite film at the contact of particles, the chlorite film is formed after strong compaction, which has a protective effect on the primary pores of the reservoir. Microscopic observation shows that there are a large number of intergranular pores in authigenic chlorite with good connectivity, which can be used as an effective reservoir space (Fig. 3G).

The reservoir of Tongbomiao formation in Hongqi sag is affected by compaction and diagenesis, and the loss of primary pores is large. The dissolution pores can effectively improve the physical properties of the reservoir, which is constructive diagenesis. The pyroclastic materials are easy to be dissolved, and the secondary pores are mainly formed by the dissolution of secondary glass and crystal fragments. Tuffaceous is easy to form dissolution pores and effectively improve reservoir physical properties (Fig. 3C, D). In the process of compaction and diagenesis, with the increase of burial depth, pyroclastic materials have relatively strong anti compaction ability, so that the primary pores can be effectively preserved. With the gradual increase of organic matter maturity, a large number of organic acids and acid gases are released, and the acidity of formation fluid is gradually enhanced. Pyroclastic materials are easy to form secondary pores under the condition of acidic water with good fluidity (Fig. 3E).

4.3 Structural fracture

Microfractures produced by tectonism are favorable for improving reservoir physical properties. Hongqi sag has experienced multi-stage tectonic movement. A large number of fault systems developed in the sag. The fault system is not only the channel and reservoir space for oil and gas migration, but also the favorable place and infiltration channel for dissolution. The dissolved matter can be removed with the fluid, which is conducive to the development of secondary pores. The microfracture reservoir in Hongqi sag is mainly caused by tectonism. The overall extension direction of microfracture in the study area is the same as that of the fault system. A few of microfractures are controlled by the strike of regulating faults, mainly in NE direction, with an opening of 20-50 μ m, and mainly in grain margin fractures. The reservoir in the study area has the characteristics of dual media, strong reservoir heterogeneity, and the physical properties of the reservoir with the development of micro fractures have been significantly improved, which has an important contribution to the reservoir. Micro fractures not only improve the physical properties of reservoir effectively, but also can be used as effective reservoir space.

5 Conclusion

The types of Tongbomiao formation in Hongqi sag are feldspathic and lithic sandstone. The pore types include primary intergranular pores, feldspar dissolution pores and debris dissolution pores, and a small amount of intergranular and fractured pores of clay minerals. The porosity is mainly distributed in the range of 6-19.6%, the average is 9.2%, the permeability is mainly distributed in 0.02-43.6x10^{-3} μ m^2, with an average of 2.93x10^{-3} μ m^2. It is a reservoir of very low porosity and low permeability type.

Although Tongbomiao formation in Hongqi sag is of very low porosity and low permeability type, high quality reservoir is still developed, and its reason is mainly controlled by sedimentary environment, diagenesis and structural fracture. The reservoir of Tongbomiao formation in Hongqi sag mainly develops fan delta system. The physical properties of underwater distributary channel microfacies in fan delta front and above water distributary channel microfacies in fan delta plain are better than those of sheet sand and sublacustrine fan in fan delta front. According to the vertical changes of clay minerals, authigenic minerals and pore types, Tongbomiao formation in Hongqi sag is in the middle diagenetic evolution stage a-b. Hongqi sag has experienced multi-stage tectonic movement. A large number of fault systems developed in the sag. The fault system is not only the channel and reservoir space for oil and gas migration, but also the favorable place and infiltration channel for dissolution. The dissolved matter can be removed with the fluid, which is conducive to the development of secondary pores.

References

1. Zhang Fangdong, Wang Xuejun, Rao Lei, etc. Main Controlling Factors of Effective Reservoir in the Second Section of Bayingebi of Chagan Sag (Chinese:Journal of Chengdu University of Technology: Science & Technology Edition) **36** （3）: 295-300 (2014)
2. Li Wentao. Sequence stratigraphic characteristics and favorable zone prediction of Nantun Formation in Bayanhushu sag, Hailaer Basin (Chinese:Journal of Yangtze University, Natural Science Edition) 15（3）: 1-9 (2018)
3. Zhang Haiqiao. Evaluation of the hydrocarbon source rock and prediction of the favorable zone in Hongqi Sag of Hailar Basin (Chinese:Petroleum Geology & Oilfield Development in Daqing) （2）: 21-27 (2020)
4. Sun Fajie, Yang Lan. 2016 Geochemistry and tectonic setting of basalts in Tamulangou formation in northern Daxinganling (Chinese:Heilongjiang Science and Technology Information)（2）: 13 (2016)
5. Liu Xuewen. 2017 Greater Khingan Range Chaoyuan-Kun NECCHI area middle Jurassic Tamulangou group Geochemical characteristics and genesis (Chinese:World Nonferrous Metals)（7）: 97-98 (2017)

Flora of Seed Plants in Jinyin Lake National Wetland Park

Xinyang Zhang[1], Jin Cai[1], Huan Dai[1]

[1]Wuhan Institute of Design and Sciences, 430205 Wuhan, China

Abstract. Using line method and typical sample method, the plant resources of the Jinyin Lake National Wetland Park were investigated in detail. There were 62 families, 113 genera and 135 species of seed plants. 4 families and more families were the dominant families of seed plants in this area. 1 genera were the main reasons for the diversity of seed plants in this area. The 62 families can be divided into 9 areal types and 5 forms, of which tropical families accounted for 61.36% of the total number of non-cosmopolitan families. The 113 genera can be divided into 13 distribution types and 2 variants, of which the temperate distribution is 61.39% of the non-cosmopolitan genus. The 135 species can be divided into 7 distribution types, mainly in the subtropical monsoon climate area, accounting for 45.19% of the total species. The research results provide basic information and scientific basis for the protection and management of plant diversity resources in the Jinyin Lake National Wetland Park.

1 Preface

Wetland is the most important ecosystem on earth. The primary task of wetland biodiversity is plant diversity. Flora is the result of the development and evolution of plants under certain geographical and historical conditions. It is of great significance for regional plant composition and plant species diversity. Jinyin Lake National Wetland Park is the largest urban wetland park in Wuhan. It provides a variety of habitats for plants, animals and microorganisms in the park. It plays an important role in biodiversity protection, regional environment regulation and urban ecology improvement [1].

In recent years, there were few studies on Jinyin Lake National Wetland Park, and there was a lack of research on the flora diversity. The research team conducted a detailed investigation and floristic diversity research on the plant resources, aiming to provide basic information and scientific basis for the protection and management of plant diversity resources in Jinyin Lake National Wetland Park.

2 Research area and methods

2.1 Overview of research area

Jinyin Lake National Wetland Park built in 2001 year. It is the first national urban wetland park in Hubei Province. It is located in Jinshan Avenue, Dongxihu District, Wuhan City, Hubei Province. It covers an area of 77 hectare, including 17 hectare of peninsula land and 60 hectare of lake. The wetland area accounts for 91% of the wetland park. This is a natural ecological country wetland park dominated by aquatic plants. It is also the experimental base for the introduction and planting of aquatic plants in Wuhan Garden Research Institute.

At present, there were 135 species of plants in the park. The wetland park has a subtropical monsoon (humid) climate with an annual average temperature of 17.6°C, annual precipitation of 1300 mm. The annual frost free period is 240 days and the total sunshine hours is 2000 hours. The soil can be divided into 8 soil types, 17 subclasses, 56 soil genera and 303 soil species, of which paddy soil accounts for 45.23%, yellow brown soil 24.77%, fluvo aquic soil 16.97%, red soil 11.22%, etc [2-3].

2.2 Survey methods

The plant resources of Jinyin Lake National Wetland Park were investigated in detail by using route method and typical sample method. Multiple investigation lines cover the whole park, recording all plant species on the line, and selecting the sections with rich vegetation to conduct detailed investigation with typical sample method [4-6].

In the process of investigation, specimens of uncertain and important plants were collected, morphological and habitat characteristics were recorded, and photos were taken for later identification. Through the investigation data processing and specimen identification. Based on the related research data of Jinyin Lake National Wetland Park, the final plant list was determined and the floristic diversity of seed plants were analyzed [7-9].

* Corresponding author: xinyangyy@qq.com

https://doi.org/10.1051/e3sconf/202129301008

3 Research results and analysis

3.1 Basic composition of flora

There were 135 species of seed plants in Jinyin Lake National Wetland Park, belonging to 62 families and 113 genera. The seed plants in this area accounted for 31% of the total families, 8.34% of the total genera and 2.38% of the total species. It accounted for 17.37% of the total families, 3.51% of the total genera and 0.47% of the total species of seed plants in China. As the terrestrial area of the reserve is only 17 hectare, the plant species were relatively rich.

3.2 Composition of floristic families

There were 62 families of seed plants in Jinyin Lake National Wetland Park. Among them, only 1 family contained more than 10 species, including 14 genera and 18 species, accounting for 13.33% of the total families. There were 5 families with 4-9 species, accounting for 7.94% of the total families, including 26 genera and 33 species. There were 22 families with 2-3 species, accounting for 34.92% of the total families, including 50 species of 39 genera. There were 34 families with only one species, accounting for 55.56% of the total families, including 34 genera and 34 species.

It can be seen that although there were only 6 families with 4 or more species, accounting for 9.52% of the total number of families, the number of species accounts for 37.78% of the total number of species, more than one third of the total number of species, which were the dominant families in the flora. They were Rosaceae, Gramineae, Leguminosae, Oleaceae, Magnoliaceae and Liliaceae.

3.3 Composition of floristic genera

There were 113 genera of seed plants in Jinyin Lake National Wetland Park. They were *Magnolia*, *Acer* and *Ligustrum*, accounting for 2.52% of the total genera. They contained 9 species, accounting for 6.67% of the total. There were 106 genera with only one species, accounting for 89.08% of the total genera and 78.52% of the total species. Such as *Elaeocarpus*, *Liriodendron*, *Phyllostachys*, *Cynodon*, *Broussonetia*, *Pittosporum*, *Saxifraga*, *Zanthoxylum*.

3.4 Distribution types of floristic families

The 62 families of seed plants in Jinyin Lake National Wetland Park can be divided into 9 areal types and 5 varieties (Table1). There were 18 cosmopolitan families, accounting for 29.03% of the total families. The families with more species were Rosaceae, Oleaceae, Gramineae, Leguminosae, Rubiaceae, Saxifragaceae, Lythraceae, etc. Some families are mainly temperate or most of them are distributed in temperate zone, and have some temperate component properties.

There were 27 families in tropical distribution (type 2-7, Table2), accounting for 43.55% of the total families and 61.36% of non world distribution families. Among them, 20 families (74.07%) were dominant, such as Euphorbiaceae, Apocynaceae, Anacardiaceae, Guttiferaceae and Malvaceae.

There were 15 families in temperate distribution (8-14 types), accounting for 24.19% of the total families and 34.09% of the non world distribution families. Among them, 13 families (86.67%) were dominant, such as Caprifoliaceae, Pinaceae, Liliaceae, Fagaceae, Aceraceae and Salicaceae.

One family was endemic to China, accounting for 1.61% of the total families, and 2.27% of the non world families. It is Ginkgoaceae.

In addition, there was one family (16 types) which is not distributed in China. *Araucaria cunninghamii,* an exotic species, was widely used in some parks in Wuhan.

Table1. Distribution types of seed plants families in Jinyin Lake National Wetland Park

Code	Distribution types	Number of families	Percentage (%)
1	Cosmopolitan	18	29.03
2	Pantropic	17	27.2
2-2	Tropical Asia, Africa & South America disjuncted	1	1.61
2S	Tropical Asia, Africa & South America disjuncted	2	3.23
3	Tropical Asia & Tropical America disjuncted	3	4.84
4	Old World Tropics	2	3.23
5	Tropical Asia to Tropical Australia	1	1.61
6d	South Africa (mainly the Cape of Good Hope) disjuncted	1	1.61
8	North Temperate	4	6.45
8-4	North Temperate & South Temperate disjuncted	8	12.9
8-5	Eurasia & Temperate South America disjuncted	1	1.61
9	East Asia & North America disjuncted	2	3.23
15	Endemic to China	1	1.61
16	Australia, New Zealand, New Caledonia, North Kodak New Guinea to the Philippines and Temperate South America disjuncted	1	1.61
Total		62	100

Table2. Tropical distribution of seed plants families in Jinyin Lake National Wetland Park

Family	Distribution types code	Number of genera	Number of species
Cannaceae	2	1	1
Flacourtiaceae	2	1	1
Euphorbiaceae	2	2	2
Apocynaceae	2	2	2
Anacardiaceae	2	2	2
Theaceae	2	1	1
Guttiferaceae	2	1	2
Celastraceae	2	1	1

 https://doi.org/10.1051/e3sconf/202129301008

Sapindaceae	2	1	1
Pontederiaceae	2	1	1
Meliaceae	2	1	1
Rutaceae	2	1	1
Lauraceae	2	1	1
Marantaceae	2	1	1
Bignoniaceae	2	1	1
Palmae	2	1	1
Malvaceae	2	1	2
Iridaceae	2-2	1	2
Podocarpaceae	2s	1	1
Amaryllidaceae	2s	2	2
Elaeocarpaceae	3	1	1
Araliaceae	3	1	1
Aquifoliaceae	3	1	3
Pittosporaceae	4	1	1
Musaceae	4	1	1
Cycadaceae	5	1	2
Ericaceae	6d	1	2

Table3. Temperate distribution of seed plants families in Jinyin Lake National Wetland Park

Family	Distribution types code	Number of genera	Number of species
Caprifoliaceae	8	1	1
Pinaceae	8	1	1
Liliaceae	8	4	4
Platanaceae	8	1	1
Cupressaceae	8-4	2	2
Buxaceae	8-4	2	2
Hamamelidaceae	8-4	3	3
Fagaceae	8-4	1	1
Geraniaceae	8-4	1	1
Aceraceae	8-4	1	3
Cornaceae	8-4	1	1
Salicaceae	8-4	1	2
Berberidaceae	8-5	3	3
Taxodiaceae	9	2	2
Magnoliaceae	9	2	5

3.5 Distribution types of floristic genera

The 113 genera of seed plants in Jinyin Lake National Wetland Park can be divided into 13 areal types and 2 varieties (Table 4). There were 12 cosmopolitan genera, accounting for 10.62% of the total genera, such as *Oxalis*, *Lythrum*, *Pycnostachys*, *Euonymus*, *Geranium*, *Sabina*.

There were 33 genera in tropical distribution (type 2-7), accounting for 29.20% of the total genera and 32.67% of non world distribution genera. The dominant genera were pantropic, accounting for 33.33% of the total number of tropical genera, such as *Sapium*, *Hibiscus*, *Phragmites*, *Zanthoxylum*, *Xylosm*, *Albizia*, *Taxodium*.

There were 62 genera in temperate distribution (8-14 types), accounting for 54.87% of the total genera and 61.39% of non world distribution genera. The dominant genera were wide distribution and varieties in the north temperate zone, included 20 genera, accounting for 32.26% of the temperate genera, such as *Cedrus*, *Platanus*, *Salix*, *Acer*, *Prunus*, *Rhododendron*, *Iris*.

Three genera were endemic to China, accounting for 2.65% of the total genera, and 2.97% of the non world genera. They were *Koelreuteria*, *Metasequoia* and *Ginkgo*. *Ginkgo biloba* and *Metasequoia glyptostroboides* are known as living fossils. Among them, *Metasequoia glyptostroboides* was selected as city tree by Wuhan City, while *Ginkgo biloba* was selected as city tree by Chengdu City. They have an important position in this area.

Table4. Distribution types of seed plants genera in Jinyin Lake National Wetland Park

Code	Distribution types	Number of families	Percent age (%)
1	Cosmopolitan	12	10.62
2	Pantropic	11	9.73
3	Tropical Asia & Tropical America disjuncted	4	3.54
4	Old World Tropics	7	6.19
5	Tropical Asia to Tropical Australia	5	4.42
7	Tropical Asia(Indo-Malesia)	6	5.31
8	North Temperate	19	16.81
8-4	North Temperate & South Temperate disjuncted	1	0.88
9	East Asia & North America disjuncted	14	12.39
10	Old World Temperate	14	12.39
11	Temperate Asia	1	0.88
12	Mediterranea, West Asia to Central Asia	2	1.77
14	East Asia	10	8.85
14SJ	Sino-Japan	4	3.54
15	Endemic to China	3	2.65
Total		113	100

3.6 Distribution types of floristic species

There were 135 species of seed plants in Jinyin Lake National Wetland Park, which can be divided into 7 distribution types. The subtropical monsoon climate species accounted for 45.19% of the total species. In addition, due to the introduction of plants in the park, there were 3 species of coniferous forest in frigid zone and 1 species of plateau mountain. Although they do not belong to the origin of central China, they had been fully adapted to the local climate due to long-term cultivation and geographical evolution.

There were 27 national distribution species, accounting for 20%, such as *Koelreuteria paniculata*, *Prunus blireana*, *Amygdalus persica* var. *Persica*, *Melia azedarach*, *Photinia serrulata*, *Sabina chinensis*, *Buxus sinica*, *Nerium indicum*, *Punica granatum*. There were 19 tropical monsoon climate species, accounting for 14.07%, such as *Trachycarpus fortunei*, *Liquidambar formosana*, *Sophora japonica*, *Albizia julibrissin*, *Cycas revoluta*, *Jasminum mesnyi*. There were 61 subtropical monsoon climate species, accounting for 45.19%, such as *Cinnamomum camphora*, *Chaenomeles sinensis*, *Pyrus calleryana*, *Malus halliana*, *Platanus acerifolia*, *Salix babylonica*. There were 3 temperate monsoon climate species, accounting for 2.22%. They were *Amygdalus persica*, *Prunus cerasifera* and *Cerasus serrulata*. There were 21 temperate continental climate species, accounting for 15.56%, such as *Sophora japonica*, *Ligustrum lucidum*, *Ilex cornuta*, *Acer palmatum* var. *Palmatum*, *Ligustrum japonicum*, *Acer palmatum*. *Cedrus deodara*, *Araucaria cunninghamii* and *Podocarpus macrophyllus* were coniferous forest in frigid zone species, accounting for 2.22%. *Populus euphratica* was plateau mountain species, accounting for 0.74%.

4 Conclusion and discussion

There were 135 species of seed plants belonging to 113 genera and 62 families in Jinyin Lake National Wetland Park. Herbaceous plants and arbors accounted for a large proportion, and shrub species were relatively scarce. Families with four or more species were the dominant families in the flora, and genera with one species were the main reasons for the genus diversity of seed plants.

The 62 families of seed plants can be divided into 9 areal types and 5 varieties. The tropical families accounted for 61.36% of the non world families, and were dominant in the flora, with the pantropic distribution and its forms being the main ones. Temperate families accounted for 34.09% of the total number of non world families, and were mainly distributed in the north temperate zone. It was indicated that the floristic family was mainly composed of tropical elements and part of temperate elements.

The 113 genera of seed plants can be divided into 13 areal types and 2 varieties. The tropical genera accounted for 32.67% of the non world genera, and the pantropic type was the main one. The temperate genera accounted for 61.39% of the non world genera, and the north temperate genera were widely distributed and their forms were dominant. It showed that the flora was mainly temperate.

In general, the floristic composition of Jinyin Lake National Wetland Park was diverse, with obvious differentiation of families and genera. The Floristic and geographical characteristics of seed plants families were consistent with the local subtropical monsoon (humid) climate. The floristic characteristics of the genus were quite different from the local climate. This was due to the fact that most of the plant species in the park were artificially introduced and cultivated, resulting in the flora geographical type of the genus contrary to the local climate. It was suggested that native plants should be the main plant species in the future. Do not introduce a large number of alien species, as well as species that do not conform to the local climate characteristics and floristic geographical characteristics. To protect the original plant germplasm resources and realize the plant species diversity and ecosystem stability.

Acknowledgments

This research was financially supported by the Teaching Research Project of Hubei Provincial Universities (2017505 and 2018496) and the Scientific Research Project of Wuhan Institute of Design and Sciences (K201905 and K201915).

References

1. X. Y. Zhang, S. J. He, T Rong, H Dai, *E3S web of conferences* (EDP Sciences, **143**, 02040, 2020)
2. S. J. He, Z. K. Yu, HAS, 5, 104-109 (2019)
3. J. Liu, *Proceedings of 2013 Beijing Society of Landscape Architecture* (Science and Technology Literature Press, 2013)
4. X. Y. Zhang, R. X. Liu, L. Y. Zhu, PSJ, **31**, 5, 477-484 (2013)
5. X. Y. Zhang, S. Y. Tan, H. D. Pang, JAU(NSE), 2, 103-108 (2020)
6. X. Y. Zhang, J. H. Shang, S. J. He, STE, 21, 92-98, (2019)
7. Z. Y. Wu, Z. K. Zhou, D. Z. Li, ABY, **25**, 3, 245-257 (2003)
8. Z. Y. Wu, H. Sun, Z. K. Zhou, *Floristic geography of seed plants in China* (Science Press, 2011)
9. L. Z. Chen, *Flora and Vegetation Geography in China* (Science Press, 2017)

Study on the dilemma and solution of rural ecological environment governance

Hong Wang[1,*], Yu Ding[1]

[1*]Wuhan University of Science and Technology, Wuhan, China

[1]Wuhan University of Science and Technology, Wuhan, China

Abstract. Since the 19th National Congress of the Communist Party of China put forward the strategy of rural revitalization, the benign governance of ecological environment is the key to rural revitalization, and also the realistic carrier of Rural Revitalization. The rural ecological environment is faced with a complex reality, characterized by scattered ecological environment management area, complex types, congenital deficiency of management conditions and lack of power in the late management and promotion. These realistic representations lead to a variety of realistic governance dilemmas. Aiming at the specific realistic ecological governance dilemmas, this paper explores the corresponding solutions to solve the problems of rural ecological environment governance, highlight the beautiful rural ecological environment state, consolidate the achievements of Rural Revitalization ecological governance, and realize the sustainable development of rural ecological environment.

1 Introduction

In China's rural work, the Rural Revitalization Strategy makes clear the important position of ecological environment construction in rural areas. Ecological environment protection and development has always been an important part of Rural Revitalization. With the proposal of rural revitalization, rural governance has been carried out step by step. However, the rural ecological environment governance is facing some difficulties, such as weak basic ecology in rural areas, and even historical over opening up in some areas, which leads to frequent rural ecological environment problems, Because the good ecological environment supply has the attribute of public goods, and in the perspective of traditional development mode, there is a dilemma between environment and economy. Therefore, in this dilemma, the effect of rural ecological environment governance is not ideal. How to fundamentally solve the problem of rural ecological environment governance and realize the good and livable rural environment is the original intention of this study. To explore effective governance solutions is conducive to improve the efficiency of rural ecological environment governance and realize the construction of rural ecological civilization.

2 Realistic representation of rural ecological environment governance

2.1 The area of rural ecological environment management is scattered

First of all, the agricultural environmental governance area of cultivated land in China is widely distributed, mainly concentrated in rural areas. In the agricultural governance of cultivated land area of rural ecological environment, due to the basic characteristics of wide area, it presents the decentralized representation of governance area, which directly leads to the dispersion of governance effect, It makes the governance of cultivated land agriculture in different areas have different problem orientation, and we must formulate appropriate solutions according to different problems, which increases the refinement of the work in terms of optimizing the ecological environment of rural cultivated land. Secondly, the ecological treatment of water pollution in rural areas also shows the characteristics of scattered area. Moreover, China's per capita water resources is only 1 / 4 of the world's per capita water resources, which is a typical water shortage country. Water resource is an important ecological factor in agricultural production. In the process of industrialization and urbanization, more and more high-quality water resources are allocated to non-agricultural industries and urban and rural areas. Such decentralized governance areas make it difficult for some

* Corresponding author: wanghonglsx@163.com

https://doi.org/10.1051/e3sconf/202129301009

regions to effectively guarantee agricultural irrigation water.

2.2 The types of rural ecological environment governance are complex

In China's rural ecological environment governance, the types of governance need to be more complex. In the agricultural governance of cultivated land, according to the China land and resources bulletin, by the end of 2016, the cultivated land area in China was 134956600 hectares. In 2015, due to construction occupation, disaster damage, ecological conversion of farmland, agricultural structure adjustment and other reasons, 336500 hectares of cultivated land were reduced, and 293000 hectares of cultivated land were increased through land consolidation and agricultural structure adjustment, with a net decrease of 43500 hectares in 2015. From 2009 to 2016, the area of cultivated land decreased from 135.3846 million hectares to 134.95 million hectares, a decrease of 4346 million hectares or 0.32%. In such a realistic situation, industrialization, urbanization and urbanization tend to be obvious, and the governance of cultivated land and agricultural environment in rural ecological environment is more complex.

In the treatment of water pollution, China is a country with shortage of water resources, which shows the following characteristics: uneven distribution of water resources in time and space, mismatching of water and soil resources in space, coexistence of three different water shortage forms of resources, engineering and water quality, and coexistence of four phenomena of dirty water, more water, less water and muddy water. Therefore, there is a complicated situation in the water pollution treatment.

2.3 The governance of rural ecological environment is inherently inadequate

The congenital deficiency of rural ecological environment governance is caused by our country's traditional intensive cultivation mode and the basic congenital state of shortage of cultivated land resources. According to the China land and resources bulletin, as of the end of 2016, China's cultivated land area was 134956600 hectares. In 2015, 336500 hectares of cultivated land was reduced due to construction occupation, disaster damage, ecological conversion of farmland, agricultural structure adjustment and other reasons, and 293000 hectares of cultivated land was increased through land consolidation and agricultural structure adjustment, with a net decrease of 43500 hectares in 2015. From 2009 to 2016, the cultivated land area decreased from 135.3846 million hectares to 134.95 million hectares, a decrease of 4346 million hectares or 0.32%.

2.4 The Lack of strength in the late stage of rural ecological environment management

The rural ecological environment governance is based on the existing governance, but the existing assessment results show that some governance incentive measures adopted by the Ministry of environmental protection of China have not achieved ideal results, basically showing a weak state of governance in the later stage. In 2012, the Ministry of environmental protection officially began to evaluate the ecological performance of counties enjoying transfer payment, and the evaluation results are shown in Table 1. In Table 1, from 2012 to 2013, the implementation of the incentive measures of ecological compensation funds did not achieve good results. The continuous increase of investment did not bring about significant changes in the county ecological environment quality. On the contrary, the number of areas with improved ecological environment quality decreased from 58 to 26, and the proportion of basically stable areas increased rapidly from 84.071% to 90.987%, showing a state of constant and retrogression. However, there were some achievements in the quality of County Ecological Governance in 2015, and there was a great retrogression in 2017, which basically made it difficult to achieve a virtuous cycle of ecological governance.

File Name	Number of counties assessed	Number of counties with changes in ecological environment quality (proportion)		
		Getting better	Variation	Basically unchanged
Bulletin on the state of China's environment in 2012	452	58 (12.832%)	14 (3.097%)	380 (84.071%)
Bulletin on the state of China's environment in 2013	466	26 (5.579%)	16 (3.433%)	424 (90.987%)
Bulletin on the state of China's environment in 2014	492	69 (14.024%)	68 (13.831%)	355 (72.154%)
Bulletin on the state of China's environment in 2015	512	103 (20.117%)	65 (12.695%)	344 (67.188%)
Bulletin on the state of China's environment in 2017	723	57 (7.884%)	81 (11.204%)	585 (80.913%)
Bulletin on the state of China's environment in 2018	818	78 (9.535%)	93 (11.369%)	647 (79.095%)

Figure 1. Assessment form for ecological quality of counties enjoying transfer payment of national key ecological function areas in China from 2012 to 2017

3 Problems to be solved in rural ecological environment management

3.1 Investment gap of rural ecological environment management

There is a big gap in the investment of rural ecological environment governance, which is mainly reflected in the capital investment. In terms of capital investment, due to the dual structure of urban and rural areas in China, the thinking pattern of "attaching importance to the city and neglecting the countryside" is widespread. Instead of considering the countryside and the city in an equal framework, the countryside is placed under the city. Under this basic concept, there is an obvious gap in the investment in rural ecological environment governance. In 2017, the investment in village construction in China was 916.764 billion yuan, with an average investment intensity of 1.72 million yuan per administrative village. Among them, the municipal public facilities investment is

252.946 billion yuan, accounting for 27.59% of the village construction investment. In the municipal public facilities, the investment in drainage and environmental sanitation was 59.967 billion yuan, accounting for 6.54% of the investment in village construction. It can be seen from the above data that there is still a big gap in the investment of rural ecological environment, especially in many specific rural basic ecological environment governance.

3.2 Extensive governance mechanism of rural ecological environment

The extensive governance mechanism of rural ecological environment refers to the imperfection of governance mechanism, mainly refers to the imperfection of maintenance mechanism and guarantee mechanism of ecological environment governance. The maintenance mechanism of ecological environment governance will play an effective role in the early, middle and later stages of ecological environment governance, especially in the middle and later stages.

3.3 The rural ecological environment management technology is insufficient

The sources of rural environmental pollution are diversified and dispersed, and it is difficult to achieve rapid rural environmental governance with existing single technology. From the perspective of pollution sources, rural pollution includes domestic pollution, agricultural production pollution and industrial production. In the past, most policies and researches focused on industrial pollution. Because of enterprises, it is relatively simple to lock the main body of pollution and the type of pollution, and the pollution is harmless

Chemical treatment is relatively easy. However, domestic pollution mainly comes from rural daily life and environment. It is also relatively easy to supervise and manage poultry raising. The hardest part is soil pollution. And the accompanying water pollution, etc,The specific performance in the rural pollution monitoring technology and treatment technology. In order to achieve agricultural pollution control, pollution control technology is urgently needed, and the role of technical control plays an important role in practice.

3.4 Lack of supervision on rural ecological environment governance

All kinds of work of rural environmental supervision are not strict, and even there is environmental supervision corruption. The specific performance is that there is a lack of reasonable resource investment planning in the rural ecological environment management investment; When the government pays more attention to the environment, the supervision of environmental governance is relatively strict. Once the government's environmental supervision is over, the supervision of environmental governance is relatively loose. In addition, in the rural environmental supervision, due to the characteristics of diversification and dispersion of rural pollution sources, environmental monitoring needs to fully understand the characteristics of rural soil, water and biological resources, which requires high investment and capacity of monitoring, and is difficult, which is not conducive to the follow-up ecological environmental governance.

4 Suggestions on sustainable development methods of strengthening rural ecological environment governance

4.1 Strengthen the propaganda work of the concept of beautiful rural ecological environment

In view of the rural ecological environment governance work, the most fundamental idea is to pay attention to the rural ecological environment governance. Both the basic citizens and the relevant government agencies should strengthen the concept propaganda of the beautiful ecological environment, and strengthen the understanding of the role of the benign ecological environment. On the basis of strengthening the understanding, we can fundamentally change the wrong cognition and go deep into the rural ecological environment governance work.

4.2 Strengthen the investment of rural ecological environment management

The realization of benign governance mainly depends on the system and funds. When strengthening the rural ecological environment governance, in order to solve the actual funding gap in all aspects of ecological governance, it is necessary to increase the investment in rural ecological environment governance, so as to provide the financial basis and guarantee for the actual effective governance.

4.3 Provide technical support for rural ecological environment management

Providing rural ecological technology support is an important measure to realize the benign governance of rural ecological environment. The government should increase the investment in rural environmental governance, improve the rural environmental monitoring facilities and improve the technical level of rural environmental monitoring by inclining the capital investment to the introduction and promotion of modern technology. In addition, increase the research and development of rural ecological environment treatment technology, and strengthen technical communication.

4.4 Establishing a long-term governance mechanism of rural ecological environment

The establishment of long-term governance mechanism of rural ecological environment mainly includes operation management mechanism, participation mechanism and supervision mechanism. First of all, in the operation and management mechanism section, explore and establish different mechanisms according to the important areas of rural ecological environment governance in different regions. Through the purchase of services or other ways, establish an effective operation management mechanism to ensure the normal operation of infrastructure and achieve its sustainability. Secondly, in terms of participation mechanism, the objects of governance involve farmers, enterprises, governments, scientific research institutions, etc. One of the keys to achieve the effect of rural ecological environment governance is to clearly define the boundary of the role of different stakeholders, as well as the ways of participation and cooperation, and then establish an effective participation mechanism. Finally, on the basis of the establishment of the supervision mechanism, it should be equipped with the corresponding evaluation mechanism, which is based on the premise of evaluation and guided by the purpose of supervision, and the two should cooperate with each other to finally realize the effective implementation of the supervision mechanism.

References

1. Wen Feng an. Modernization of Rural Ecological Governance: importance, governance experience and development path in the new era [J]. Academic journal, 2020 (3).
2. Ren Yong. Issues needing attention in the modernization of ecological environment governance system and governance capacity [n]. China environment daily, November 19, 2019.
3. Yang Xue feng, Li Shuang, Wang Miaofeng. Community construction in rural environmental governance [J]. Open guide, 2019 (6): 81-84.
4. Wu er. Rural Ecological Environment Governance Dilemma and solution path under the background of Rural Revitalization -- from the perspective of social change （ J ）. Anhui Agricultural Science, 2020.48 (5): 359-262 + 275.
5. Liu Jianwei. Study on the concept, necessity and Countermeasures of national ecological environment governance modernization [J]. Journal of Fujian provincial Party School of CPC, 2014 (9).
6. Zhou Xin. Public participation in the construction of modern environmental governance system [J]. Journal of Harbin Institute of Technology (SOCIAL SCIENCE EDITION), 2020 (2).

Alleviation of Cd-induced growth improvement by exogenous melatonin pre-treatment in *Abutilon theophrasti* Medicus

Tianpeng GAO[1,5*], Jingwen FU[2], Mingbo ZUO[2], Haoming Wang[3], Xiao-Xiao Liu[1], Hai-Juan Li[1], Yuan-Yuan Shen[1], Yingli YANG[2], Dang-hui XU[4], and Yueqing He[3]

[1]School of Biological and Environmental Engineering, Xi'an University, Xi'an710065, China

[2]College of Geography and Environmental science, Northwest Normal University, Lanzhou 730070, China

[3]School of Chemical and Biological Engineering, Lanzhou Jiaotong University, Lanzhou 730070, China

[4]State Key Laboratory of Grassland Agro Ecosystems/School of Life Science, Lanzhou University, Lanzhou 730000, Gansu, China

[5]College of Geography and Environmental Engineering, Engineering Center for Pollution Control and Ecological Restoration in Mining of Gansu Province, Lanzhou City University, Lanzhou 730070, China

Abstract. The present work was aimed to study the effect of melatonin pre-treatments on oxidative stress modulation, cadmium (Cd) accumulation, and tolerance in *Abutilon theophrasti* Medicus under the hydroponic system. Application of substances that can modulate the harmful effects of Cd on plant yield and reduce its accumulation in the edible parts is of particular importance. Therefore, the Abutilon pre-treated with 0.1mM melatonin were exposed to 200μM, 400μM, 800μM Cd. Our results showed that 0.1mM melatonin led to positive effects on 200μM and 400μM Cd tolerance, including a significant increase in root activity andperoxidase (POD) activity and a significant decrease in malondialdehyde (MDA) content. Withthe help of melatonin, the enrichment factor (BCF) significantly decreased, reducing the content in the ground and roots. At least in our experiment conditions, this information appears to be useful for healthy food production.

1 Introduction

Among heavy metals, cadmium was a highly carcinogenic metal that could cause toxic reactions even at low concentrations[1-2]. In plants, yellowing and growth shunt caused by Cd toxicity were easily to find symptoms. Higher toxicity would inhibit plant growth and lead to plant necrosis[3]. The toxicity of Cd affected plants by inhibiting carbon fixation and reducing chlorophyll content and photosynthetic activity[4]. Cd exposure in the soil induced osmotic stress in plants by minimizing the relative water content, stomatal conductance and transpiration of leaves, which led to plant physiological damages[5]. The toxicity of cadmium could lead to the excessive production of reactive oxygen species (ROS), and lead to damage to plant membranes and destruction of cell biomolecules and organelles[6]. Cadmium also reduced the absorption of iron and zinc by plants, leading to yellowing of leaves[7]. Generally, Cd interfered with the transportation and absorption of Ca, P, Mg, K, and Mn.

Melatonin was a pleiotropic molecule, which had multiple functions in plants. It was mainly considered primarily an antioxidant with important actions in the control of reactive oxygen and nitrogen species, among other free radicals, and harmful oxidative molecules present in plant cells[8]. In addition, melatonin also participated in a variety of physiological functions, such as growth, rooting, seed germination, photosynthesis, and protection of plants from abiotic and biotic stresses. Due to the diversity of its effects, melatonin was also considered to be the master regulator of plants. Exogenous application of melatonin can improve the resistance of plants to abiotic stress, reduce cell damage, and increase the activity of antioxidant enzymes SOD, POD, CAT and APX[9-10]. Therefore, the protective effect of melatonin in abiotic stress might be an ideal target to improve the ability of plants to resist biotic stress. Researchers[11] used melatonin to treat Cd stress in tomato (*Solanum lycopersicum*), and obtained melatonin-induced antioxidant activity, phytochelatin biosynthesis and Cd partitioning played a key role in enhancing the tolerance to Cd stress effect.

Here, we took Abutilon seedlings as the research object, and explored the effects of melatonin on its physiology and enrichment function under cadmium stress by adding 0.1mmol/L melatonin to pre-treatment, in order to provided a scientific basis for phytoremediation of soil cadmium pollution, increased certain economic benefits while remediating cadmium contaminated soil, and ensured safe production.

* Corresponding author: zkgtp@163.com; ffjjww0224@163.com

2 Methods and materials

2.1 Design of experiment

The seedling growth test was carried out in the greenhouse of Xi'an University. We put the sterilized Abutilon seeds in a hydroponic device filled with vermiculite, incubate for 7 days in a 25°C incubator. Then selected seedlings with the same height, transferring to the artificial simulated climate room with 1/2 Hoagland nutrition solution. Maintain aeration during the cultivation process, the photoperiod was 12h/12h (day/night), the temperature was 22°C, the relative humidity was 80%, and the light intensity was 3000lx, the nutrient solution was changed every 3 days. In order to eliminate the influence of cadmium ions in the water, ultrapure water was used to configure the nutrient solution throughout the hydroponic process, and the pH of the culture solution was adjusted to 5.8~6.0 with 0.1mol/L HCl or 1mol/L KOH.

After 15 days of seedling growth, add 0.1 mM melatonin to the nutrient solution for treatment. After 24 hours of treatment, different concentrations of $CdCl_2$ were added to the 5L nutrient at 8:00the next morning for stress. The experiment set up 8 treatments: control group T0 (0 MEL, 0 Cd), Cd stress group CK200, CK400, CK800 (0 MEL, 200Cd; 0 MEL, 400Cd; 0 MEL, 800Cd), control plus melatonin group T0 (0.1 MEL, 0 Cd) and melatonin plus Cd^{2+} stress groups T200, T400, T800 (0.1 MEL, 200Cd; 0.1 MEL, 400Cd; 0.1 MEL, 800Cd). Plant samples were collected for analysis and testing at 5 days of stress.

2.2 Indicators and test methods

The content of malondialdehyde (MDA), soluble protein (SP), chlorophyll (Chl) and peroxidase (POD) were determined according to the TTC method for measuring root activity[12] and Chen method[13].

Referring to the method of Wei[14], the root system and the upper part of the root of the abutilon seedling were separated and dried at 105° to constant weight. The dried tissue was ground and weighed. The 0.25 g sample was digested inadigester (Labtech Digiblock ED54, China) with 10 mL HNO_3. The obtained digestive liquid was refrigerated, transferred to a 50mL volumetric flask, and diluted to the corresponding volume by adding 1% HNO_3. After dilution of the liquid by membrane filtration (0.45μm) to clean the pipe.The concentration of Cd was determined by ICP-MS (Agilent 7500X, USA). In acid digestion and Cd content determination, the standard product (GBW07603 GSV-2) was used as quality control product to analyze the content of Cd in different samples.The content of Cd in the standard was 0.38 $mg\cdot kg^{-1}$, and the analysis result was 0.376±0.011 $mg\cdot kg^{-1}$.

Transfer Factor (TF) = Heavy metal content in the above-ground part (mg/kg) / heavy metal content in the underground part (mg/kg)

Enrichment Factor (BCF) = Heavy metal concentration in plants (mg/kg) / heavy metal concentration in Nutrient solution (mg/kg)

2.3 Data processing and analysis

Statistical analyses and visualization were performed in SPSS 23.0, Origin 2018 for Windows. The data were mean ± standard error of three repeats. LSD test was selected ($P<0.05$) for significance analysis of differences.

3 Results and discussions

3.1 Effect of melatonin on activityof root under Cadmium Stress

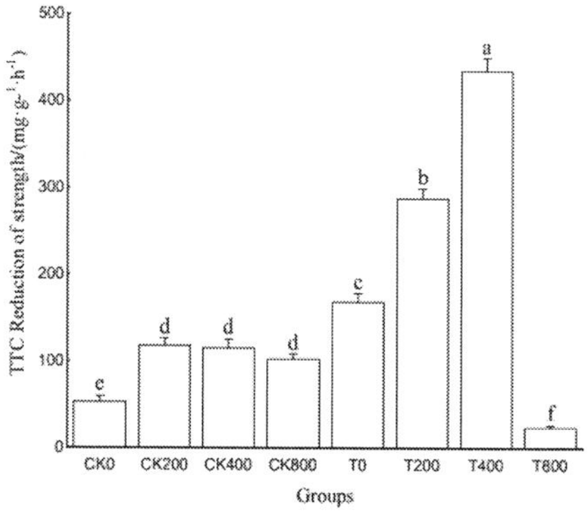

Figure 1. Effects of melatonin on the root activity of *Abutilon theophrasti* Medicus under Cd^{2+} stress

Note: Different lowercase letters indicate that there were significant differences between different stress concentrations ($P<0.05$), the same as below.

As shown in Figure 1, after 5 days of Cd stress treatment, root activity first increased and then decreased with increasing Cd stress treatment. Compared with CK400, the root activity of T400 increased by 276.58%, and reached the maximum and the root activity of T800 decreased by 76.79%. The results show that 0.1 mM melatonin pre-treatment significantly ($P<0.05$) increased the root activity of seedlings under low concentration stress, which meant melatonin had an activating effect on the growth and physiological of abutilon roots.

3.2 Effect of melatonin on content of SP under cadmium stress

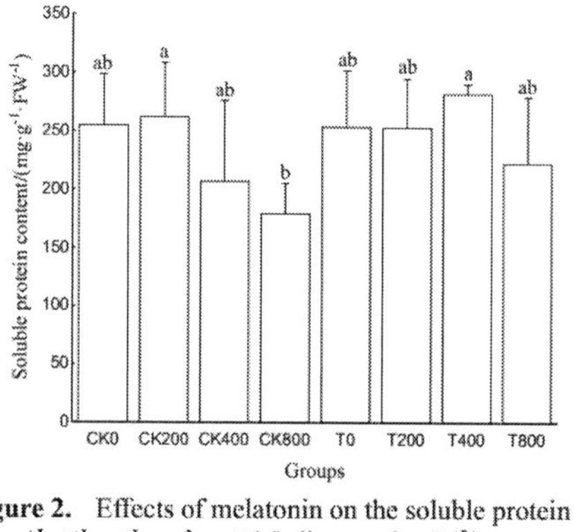

Figure 2. Effects of melatonin on the soluble protein of *Abutilon theophrasti* Medicus under Cd^{2+} stress

Protein, as a biological macromolecule in cells, was one of the most important basic materials for the structure and function of organisms. As shown in Figure 2, after 5 d Cd stress without melatonin, the SP content first increased and then decreased, and reached its maximum at CK200. After 5 d Cd stress withmelatonin, the trend was the same as above, and reached its maximum at T400. It showed that melatonin pre-treatment increased SP content, and worked best on 400μM Cd stress, by 36.41%.

Soluble protein was also an important osmotic adjustment substance in plants. Its content could show the strength of plant metabolism. Plants under adversity actively accumulate to improve plant osmotic balance. The study found that the SP content of abutilon seedlings increased first and then decreased under Cd stress, which might be due to an emergency response under adversity conditions. Melatonin treatment had no significant effect on the increase of soluble protein content.

3.3 Effect of melatonin on content of MDA under cadmium stress

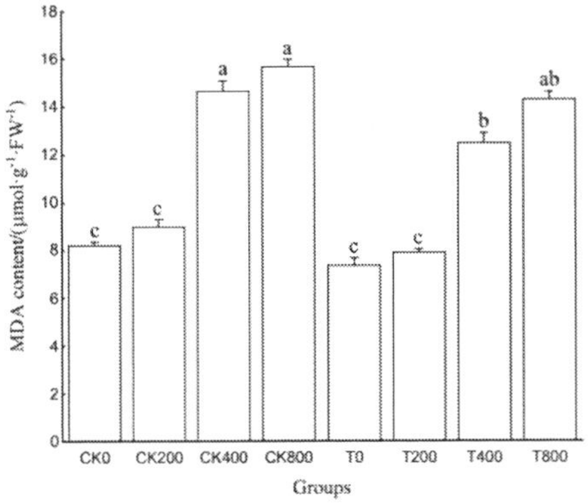

Figure 3. Effects of melatonin on the MDA content of *Abutilon theophrasti* Medicus under Cd^{2+} stress

MDA was a hallmark of cell membrane destruction. As shown in Figure 3, in the plant without melatonin pre-treatments, 400 μM and 800 μM Cd stress respectively increased MDA as much as 1.79-folds and 1.91-folds compared to the control. Compared to the 400μM Cd group, 0.1mMmelatonin pre-treatments resulted in MDA from 14.64 mmol·g^{-1}FW^{-1} to 12.47 mmol·g^{-1}FW^{-1} in the plants, significantly decreased 14.82%.

Once subjected to stress, the balance of reactive oxygenmetabolism in plants will be disrupted, leading to an increase inROS content that causes membrane peroxidation and damageto the biofilm system. But melatonin was a substance with strong antioxidant capacity. It reduced oxidative damage by directly cleared ·OH and H_2O_2 molecules in cells and kept ROS at a relatively low level[15]. In this study, 0.1mM exogenous melatonin were applied to the MDA content under the different stresses of Cd, indicating that Cd stress destroys the biofilm system, 0.1mM exogenous melatonin on plants reduced MDA content under Cd stress, thereby reducing cell damage and enhancing the ability of plants to resist abiotic stress. Melatonin had a certain stabilizing effect on alleviating the increase of MDA content caused by oxidative damage.

3.4 Effect of melatonin on activity of POD under cadmium stress

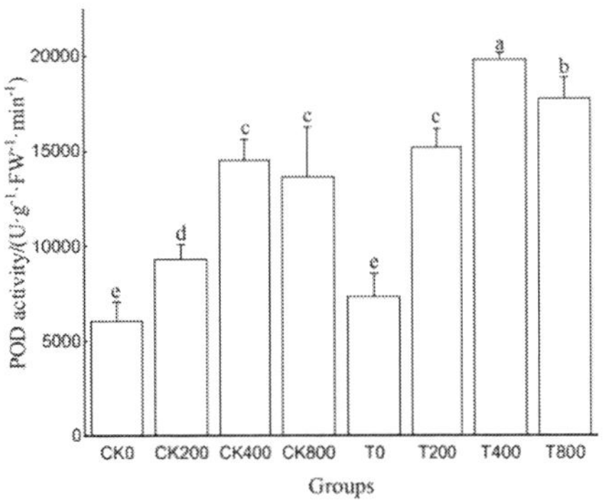

Figure 4. Effects of melatonin on POD content of *Abutilon theophrasti* Medicus under Cd^{2+} stress

Melatonin pre-treatment stimulated the activity of POD in leaves ofAbutilon seedlings under Cd stress. POD activity was the highest T400, and significantly higher than CK400 36.32%. T200 and T800 respectively higher than same concentration control group 63.46% and 29.94%. According to the results, 0.1 mM melatonin had the best effect on 200μM Cd stress.

POD played an important role in scavenging reactive oxygen free radicals in plants. Due to its relatively high activity and sensitive response, POD had been regarded as anconsequence substance to maintain the balance of active oxygen system in the plants.In this experiment, POD activity increased significantly under Cdstress, indicating that the addition of 0.1mMexogenous melatonin showed similar changes in POD activity under different concentrations of Cd stress. It might be that POD corresponding gene transcription was not inhibited. Generated H_2O_2 further converted into H_2O and O_2, so that H_2O_2 did not damage to the plants, guaranteeing the regular growth. T200 had the most positive effect on pre-treating melatonin. This was consistent with the conclusion of Zhang[16] that exogenous melatonin could improve POD enzyme activity of rice seedlings under AS stress.The above results further indicated that 0.1mM exogenous melatonin had a certain inhibitory effect on membrane lipid peroxidation of abutilon seedlings under Cd stress, further maintaining the stability of active oxygen system of the plant to a certain extent.

3.5 Effect of melatonin on accumulation under cadmium stress

Table1. Variation of cadmium accumulation in each part

Groups	Cadmium content in the ground (mg/kg)	Cadmium content in roots (mg/kg)

CK200	6.83±0.08Bf	87.27±0.58Ae
CK400	40.71±0.60Ba	189.65±0.71Ac
CK800	40.38±1.51Bb	315.27±0.64Aa
T200	15.33±0.36Be	58.28±5.31Af
T400	22.64±0.44Bd	108.73±1.10Ad
T800	28.31±0.33Bc	204.78±1.16Ab

Note: Different lowercase letters indicate significant differences among treatment groups in the same column. Different majuscule letters indicate significant differences among the ground and roots, the same as below ($P<0.05$).

Inaccordance with Table 1, Cd significantly increased ($P<0.05$) the Cadmium content in the ground and in the roots. The Cadmium content in the roots significantly higher than in the ground. This suggested that Cd mainly accumulated in the roots, only few shifted upward. It might be due to the roots directly contact with soil and a large number of microorganisms to heavy metal ions had good absorption and fixation, thus could enhance root of Cd enrichment capability. Gu[17] found that melatonin reduced the accumulation of Cd^{2+} by up-regulating the transcription of PDR8 and HMA4 and down-regulating the transcription of Nramp6.It was speculated that melatonin might have a similar mechanism in the response to Cd stress, which needed to be further proved.Under the stress of 400μM Cd, the plants with 0.1mM melatonin pretreatment showed a highest decline ($P<0.05$) in the Cd concentration as much as 44.39% and 42.67% in the ground and roots in comparison with the control (the same Cd concentration without melatonin), respectively.

3.6 Effect of melatonin on enrichment and transfer coefficients under cadmium stress

Table2. Changes of cadmium enrichment and transfer coefficients

Groups	Enrichment Factor	Transfer Factor
CK200	6.83±0.05b	0.08±0.00d
CK400	8.36±0.05a	0.21±0.00b
CK800	6.45±0.04c	0.13±0.00c
T200	5.34±0.37d	0.26±0.03a
T400	4.77±0.02e	0.21±0.01b
T800	4.23±0.03f	0.14±0.00c

In accordance with Table 1, Cd significantly decreased ($P<0.05$) the enrichment factor in the ground and in the roots. 200μM Cd significantly increased the transfer factor. With pre-treatment of 0.1mM melatonin, the minimum BCF and TF were determined in 800μM Cd stress. But without melatonin, the minimum TF was determined in 200μM. T400 showed the most obvious BCF decline, was 42.94%. Planting abutilon on medium contaminated soil effectively helped make more use of Cd contaminated soil and improve economic benefits. It seems that under Cd stress, melatonin could reduce the accumulation of cadmium in the plant, and increase the content of SP and the activity of antioxidant enzymes to resist the damage of Cd.

4 Conclusions

(1) Pretreatment with 0.1mM melatonin significantly increased plant root activity, POD activity, chlorophyll and soluble protein content, andreduced MDA content. Therefore, melatonin could improve the tolerance of abutilon to cadmium.

(2) The 0.1mM melatonin pretreatment reduced the accumulation capacity of abutilon. Although it increased TF slightly, TF was significantly lower than BCF. Only a small amount of cadmium was transferred to the ground of the plant, and most of it accumulated in the root. It showed that melatonin had a certain effect on the healthy and economic production of abutilon.

Acknowledgments

The research was funded by the National Natural Science Foundation of China (31860176), the Key Research and Development Program of Shanxi Province (No. 20YF3FA037) , the Key Research and Development Program of Shanxi Province (No. 2020ZDLSF06-06) and XAWLKYTD012. We were grateful to all anonymous reviewers whose comments improved the quality of the manuscript.

References

1. M. Benavides, S. Gallego, M.Brazilian Journal of Plant Physiology,**17** (2005)
2. A. Khan, S. Khan, M. Khan. Environmental Science and Pollution Research, **22**(2015)
3. C. Hermans,J. Chen,F. Coppens.New Phytologist, **192** (2011)
4. S. Gallego, L. Pena, R. Barcia. Environmental Experimental Botany, **83** (2012)
5. M. Rizwan, S. Ali, T. Abbas. Ecotoxicology Environmental Safety, **130**(2016)
6. T. Abbas, M. Rizwan,S. Ali. Environmental Science Pollution Research,**25** (2017)
7. Z. Xu, Q. Li, P.Yang. Environmental Pollution,**224** (2017).
8. C. Pape, K. Lüning.Trends Plant Science, **41**(2006)
9. H. Kaur, S. Bhatla. Nitric Oxide, **59**(2016)
10. G.Cui, X.Zhao,S. Liu.Plant Physiology and Biochemistry, **118**(2017).
11. M.Nawaz, Y.Jiao, C.Chen. Journal of Plant Physiology, **220** (2018)
12. F.Hou. Experimental Course in Plant Physiology, (2015)
13. J. Chen, X. Wang.Experimental Guide to Plant Physiology, (2006)

14. T.Wei, X. mLv, H.Jia. Journal of Environmental Management, **214**(2018)
15. U.Fasih, L.Cai, A.Jeffrey.Ecotoxicology and Environmental Safety, **211** (2021)
16. S. Zhang, Y. Huang, S. Bao. Chinese Journal of Ecology, **39**(2020)
17. Q.Gu, Z. Chen, X. Yu. Plant Science, **261** (2017)

Ecological planning strategies for urban parks based on biodiversity conservation and promotion: a case study of Qingxiushan Forest Park in Nanning City

Shuang Ming [1,*], Qin Du [1]

[1]School of Tourism and Landscape Architecture, Guilin University of Technology, 541006 Guilin, Guangxi, China

Abstract. Urban biodiversity plays an important role in maintaining urban ecosystem, beautifying urban environment, improving economic benefits and human well-being, and urban park, as an important habitat of urban biology, is an important part of urban biodiversity. Taking Qingxiu Mountain Park in Nanning City as an example, this paper analyzes the main problems and challenges of biodiversity conservation and promotion, and puts forward a scientific reference basis for the promotion and protection of biodiversity in Qingxiu Mountain Park in Nanning City.

1 Introduction

1.1 The concept of urban biodiversity and its importance

Urban biodiversity refers to the collection of wild animals, plants, microorganisms and other biological species in the physical environment of a city. The promotion of urban biodiversity plays an important role in maintaining the urban ecosystem, beautifying the urban ecological environment, improving economic benefits and human well-being[1]. In 2002, world leaders pledged to significantly reduce the rate of biodiversity loss by 2010 through the Convention on Biological Diversity, Despite extensive response measures, the rate of biodiversity loss does not seem to slow down[2]. Among them, many papers pointed out that biodiversity and species composition present obvious urban-rural gradient change, This suggests that urbanization is one of the most important drivers of species decline, homogenization and extinction compared to other human activities.

1.2 Research on Biodiversity of Urban Parks

Urban park is fragmented habitat patch preserved in the rapid progress of urbanization, with good landscape resources, diverse landforms, rich wildlife species and other characteristics. Park is often an important habitat of urban wildlife,is an important part of the urban ecosystem, so the protection and promotion of park biodiversity is an important link to improve the quality of urban environment,to create a good living environment. Chen Bo et al. pointed out that urban parks and suburban parks often have highly diverse habitats,By comparing the methods of park biodiversity assessment in China and Belgium, they proposed a set of index framework suitable for park biodiversity assessment in China[3]. Wang Min et al. proposed a framework for biodiversity design of urban parks based on the classification system of habitat units by analyzing the cases of park biodiversity improvement at home and abroad[4]. Hao Sheng et al. evaluated the biodiversity of Minghu National Wetland Park in Liupanshui by taking Minghu National Wetland Park as an example, confirmed that the level of biodiversity was improved after the park was designed and completed, and put forward optimization strategies and suggestions for the promotion and protection of biodiversity in the park[5].Zhao Caijun et al. took Perth, Australia as an example, introduced the design, construction and management of King's Park and Bird's Park, and summarized the strategies and methods to improve the biodiversity of urban parks[6]. In addition, the biodiversity design of more domestic urban parks is mainly to achieve the purpose of habitat diversity and biological species diversity by utilizing plant characteristics, matching plant community configuration and building biological habitats within the scope of urban public green space or wetland.

2 Regional Overview

2.1 Geographic location and Qingxiu Mountain Forest Park

Nanning City is located at 22°13' ~ 23°32'N, 107°45' ~ 108° 51 'E. Its topography is typical of mountains, hills and basins, with an average elevation of 76.5 meters. Nanning has a mild climate, abundant rainfall, and an average annual temperature of 21.7°C, belonging to the subtropical monsoon climate, which provides good living conditions for organisms, and has the reputation of China's Green City. As of February 2021, the city has

* Corresponding author: 1017045714@qq.com

jurisdiction over 7 districts, 4 counties and 1 city, with a total area of 22,112 square kilometers, including 372 square kilometers of built-up areas. The city has a permanent population of 7.3448 million and an urban population of 4.6788 million, accounting for an urbanization rate of 63.7 percent.

Qingxiu Mountain Park is located on the banks of Yongjiang River in Nanning City, Guangxi Zhuang Autonomous Region. Scenic spots include three mountain ranges, namely Phoenix Ridge, Fengyi Ridge and Qingxiu Mountain, with an altitude of 82-289 meters. The planned protected area of Qingxiu Mountain Scenic Spot is 13.54 square kilometers, the core protected area is about 5.86 square kilometers, and the forest plant park is about 7.68 square kilometers. The unique landscape resources and abundant animal and plant resources make Qingxiu Mountain known as the green lung of Nanning City, which plays an important role in the urban green space system planning of Nanning City.

2.2 Status of biological resources

The favorable climatic conditions and advantageous geographical location provide a good living environment for the flora and fauna resources in Qingxiu Mountain. According to the investigation of researchers, there are 174 families, 752 genera and 1431 species of vascular plants in Qingxiu Mountain Scenic Area, including 21 families of ferns. There are 36 genera and 48 species, 8 families, 14 genera and 42 species of gymnosperms, 145 families, 702 genera and 1,341 species of angiosperms. There are 997 species of cultivated plants, belonging to 156 families and 535 genera[7]. Among them, along the Qingxiu Mountain Ancient Road with less human intervention is a very rare area of native plant distribution in the city center, where you can see the naturally distributed plant communities of Nanning. Qingxiu mountain wildlife resources are relatively abundant, in terms of birds, currently in mountain has four kinds of wild animals under state protection (category ii) was observed Crested Goshawk（Accipiter trivirgatus）, Asian Barred Owlet（Glaucidium cuculoides）, Collared Scops Owl （Otus bakkamoena）Greater coucal (Centropus sinensi)in the breeding. In addition, Mount Qingxiu is also an important breeding, habitat and foraging ground for many other birds,Examples such as Hill Blue-flycatcher(Cyornis banyumas), Ashy Minivet (Pericrocotus divaricatus), Lesser Shortwing(Brachypteryx leucophris), White-bellied Yuhina(Erpornis zantholeuca), etc. Japanese Robin (Erithacus akahige) have been observed wintering here for years. In terms of amphibians, reptiles, insects, Tri-rope beauty snake(Coelognathus radiatus), Dorsal-striped Opposite-fingered Treefrog(Chiromantis doriae), and Original-tail gecko(Hemidactylus bowringii) are common wild animals in Qingxiu Mountain. These data all show that Qingxiu Mountain Park has abundant wild animal and plant resources, and protecting and improving the biodiversity of Qingxiu Mountain is an indispensable part of the planning and design.

Table1. List of Common Birds in Qingxiu Mountain.

Order	Family	pecies	scientific name	Protection level
Accipitriformes	Accipitridae	Crested Goshawk	Accipiter trivirgatus	National two levels of protections animals
Strigiformes	Strigidae	Asian Barred Owlet	Glaucidium cuculoides	National two levels of protections animals
Strigiformes	Strigidae	Collared Scops Owl	Otus lettia	National two levels of protections animals
Cuculiformes	Cuculidae	Greater coucal	Centropus sinensis	National two levels of protections animals
Passeriformes	Old World babbler	Hill Blue-flycatcher	Cyornis banyumas	
Passeriformes	Campephagidae	Ashy Minivet	Pericrocotus divaricatus	Key protected animals
Passeriformes	Vireonidae	White-bellied Yuhina	Erpornis zantholeuca	
Passeriformes	Turdidae	Japanese Robin	Erithacus akahige	
Passeriformes	Muscicapidae	Lesser Shortwing	Brachypteryx leucophrys	

3 Challenges and problems

3.1 invasive alien species

Biological invasion has become one of the most important factors that threaten biodiversity in the world,Due to the lack of natural enemies and strong reproductive capacity,invasive alien species often bring threats to native species and destroy the balance of ecosystem and species diversity,Its threat to biological species has attracted enough attention in the global scope. In recent years, the invasion of alien species has become increasingly harmful to the biodiversity of China, which has threatened the ecological security of China.

According to Chen Ting's statistics,there are 787 local plants (plants distributed in Guangxi) in Qingxiu Mountain, 72 plants introduced from other provinces of China, and 257 plants introduced and cultivated from abroad[8]. The results showed that the exotic plants in Qingxiu Mountain accounted for a large part of the total plants. Among them, there are a large proportion of exotic plants in the botanical gardens such as cycad gardens and palm gardens, and these exotic plants are likely to bring other exotic wild animals and plants with them during the introduction process, which will bring serious threats to the native species.

3.2 Habitat fragmentation

Due to the acceleration of urbanization in Nanning, the change of land use around Qingxiu Mountain Park and the influence of various garden building activities in Qingxiu Mountain Park, the wild animals and plants in the park are facing the problem of habitat loss and fragmentation. Hard squares and roads, as well as the application of artificial materials, have resulted in the continuous shrinking of biological habitat, loss of living and reproduction space, and a continuous decline in the quantity and quality of biological diversity.

3.3 Artificial interference

Qingxiu Mountain Park, a 5A scenic spot, receives more than 2.6 million visitors annually, including 300,000 during the Spring Flower Festival. All these data show that the tourists in Qingxiu District are so crowded that artificial interference is more serious, which is not conducive to the reproduction and communication of some sensitive wild animals. In addition, forest fires and environmental pollution caused by human factors are also one of the main factors causing the loss of biodiversity.

4 case study

The High Line in New York City was built in 1930 as an elevated route to transport meat supplies. It was designed to improve traffic congestion and safety in southwest Manhattan so that meat and milk could be transported into the city. Therefore, it was also called the transportation lifeline. After 1980 the lifeline fell into disuse and threatened to be demolished by the development of the transportation industry. The Hudson River and native birds brought seeds and landscape to the High Line. In order to preserve the natural High Line landscape, local travel journalists and entrepreneurs created a nonprofit organization to protect the High Line -- Friends of the High Line. Through the joint efforts of government, citizens, designers, entrepreneurs and philanthropists, the High Line park was saved and opened to the public in 2009. The High Line is 2.33 kilometers long, 9 meters above the ground and 9 meters wide on average. It runs through 22 blocks with 11 entrances and exits. Elevated park a total of more than 500 kinds of plants, shrubs, perennial trees and flowers and the plants attract bees, butterflies, birds and other insects, local Banks with a bee agencies, bees in the park, by raising bees, park and nearby need pollination on the growth of the plant are better, they produce honey also through the charity sale activity, used to run in the park. It is estimated that between 2007 and 2027, the High Line will generate $1.4 billion in tax revenue for New York City and indirectly generate $5 billion in development in the surrounding area. In addition, the construction of the elevated park achieved great social benefits, friends of the contour with the development of volunteer the nature of the many activities, such as biodiversity, share park plant experience, art exhibitions, family entertainment, education and so on a series of public activities, good interactive activities for a long time to also make many new yorkers to volunteer to give back to the park. More broadly, the Friends of the High Line formed the High Line Parks Network to share the success of the High Line with other cities. So far, 19 US cities have joined the network and set up their own public space projects, such as the renovation of space under the bridge in Miami.

5 Ecological planning strategies

5.1 attach importance to the water space in the park

Drainage space such as lakes and rivers have high biological richness, is one of the important habitats, aquatic animals and plants and amphibians. In the ecological planning and design of parks, park managers should pay attention to the blue space such as lakes, ponds and wetlands, and actively utilize the water system of parks to create habitat space suitable for the survival, reproduction and communication of aquatic animals and amphibians. At present, the water environment in Qingxiu Mountain is dominated by lakes, with large spatial distribution and poor connectivity, which is not conducive to water exchange, resulting in the decline of water ecosystem quality and reduction of aquatic organisms. In the ecological planning of Qingxiu Mountain, all waterscape resources in the park should be integrated, communicate and connect the existing water system, enhance the connectivity of the water system in the park, form the blue network of Qingxiu Mountain Park, and enrich the habitat space of aquatic animals and plants. Secondly, the combination of blue and green is conducive to the integration of natural resources in the park, ensuring the environment and quality of the habitat, and thus improving the regional species richness. The landscape area of qingxiu district should make full use of the biological advantages of the biggest river in Nanning, yongjiang river and its tributaries, make landscape according to the water, and distribute green transition zone on the skeleton of the blue space in the park, so as to form the biological living space in the park with blue and green fusion.

5.2 level protection for focus species

Qingxiu Mountain is an important habitat for Guangxi District Key Protected Animals, National I Class

Protected Animals, Class II Key Protected Animals and Plants, and Endangered Animals and Plants,It is very necessary to focus on the protection of these precious animals and plants. In the main action plan of ecological protection of Qingxiu Mountain, the status quo and spatial distribution of the rare plant and animal resources in the park should be sorted out and summarized, and the methods of listing and explaining the rare plants, drawing red lines and hierarchical protection should be adopted. For the rare wild animals in the park, the protection red line should be drawn to reduce human interference, and the shelter effect of plants should be used to create the wildlife habitat in the park, so as to ensure the safety of the endangered wildlife habitat. In terms of food sources, attention should be paid to the water and food sources of wild animals in the park, water features should be built near the habitats, and some tree species that wild animals like to eat should be selected for planting, so as to attract more wild animals to enter the habitats for foraging and communication.

5.3 Choose native species to enrich plant communities in the park

In order to build a thousand-year-old cycad garden, peach garden, palm garden and other special plant gardens in Qingxiu Mountain Park, many plants are introduced and cultivated from abroad. This practice not only poses an invisible threat to the native species in the garden, but also Introduced species do not adapt to the local climate and do not grow well, which wastes a lot of human and financial resources. Local species, such as Mangifera indica L, Artocarpus heterophyllus Lam,and Dracontomelon duperreanum Pierre should be adopted as far as possible in plant community construction, so as to enhance the seasonal landscape changes of plants and improve the stability of regional ecosystem.

5.4 Applied edge effect

Edge effect refers to the phenomenon that the number and species of organisms in the ecotone of the ecosystem are more than those in the adjacent ecosystem due to the influence of foraging, communication and reproduction of organisms. In the ecological planning of Qingxiu Mountain, the principle of edge effect should be actively used to transform the junction of the two ecosystems and improve the biodiversity of the edge of the ecosystem by means of growth, broadening and zigzagzing. By optimizing the edge of Qingxiu Mountain Yanyu Lake, Tianchi Lake and other lakes, we can widen the lakeshore zone and transform the curvature of the lake shoreline, so as to increase the heterogeneity of the lake edge and improve the level of biodiversity.

5.5 Improving methods for monitoring diversity of species

At present, biodiversity monitoring is mainly applied to large protected areas and national parks, while biodiversity investigation on urban public green space is less. Therefore, Qingxiu Mountain should improve the monitoring methods of biological species and adopt a variety of methods to protect and monitor the wildlife in the park. Carry out civic science education activities, use civic science to record classification and monitor wildlife data in the park, lay a solid foundation for various scientific researches in the park, and combine multi-professional experts to understand the resources of wild animals and plants in Qingxiu Mountain,Conduct long-term, large-scale, and professional investigations.

6 Conclusion

Biological diversity of urban parks is an important part of urban biological diversity, facing the severe challenge of biodiversity protection and promotion of urban parks, this paper combined with the actual situation of Qingxiu Mountain Park, put forward attach importance to the water space in the park,level protection for focus species,Choose native species to enrich plant communities in the park,Applied edge effect Improving methods for monitoring diversity of species. There are five strategies to improve the local biodiversity. It is expected to provide useful reference for the ecological planning of urban parks.

References

1. FULLER.R.A,IRVINE.K.N,DEVINEWRIGHT.P.Ps ychological benefits of greenspace increase with biodiversity, BIOLOGY LETTERS, 3(4): 390-394. 2007.
2. BUTCHART S H M, WALPOLE M, COLLEN B, Global Biodiversity: Indicators of Recent Declines,SCIENCE,328(5982):1164-1168,2010.
3. B.Chen,Z.Y.Bao,Indicators for biodiversity assessment in urban and suburban parks, Biodiversity, 1:169-176,2003.
4. M.Wang, Yan.Wang. Biodiversity design for urban parks,Landscape architecture, 47-52,2014.
5. S.HAO, C.L.Wang, H.W.Lin. Biodiversity design and assessment of urban wetland parks: A case study of Minghu National Wetland Park in Liupanshui,Acta Ecologica Sinica, 395967-5977, 2019.
6. C.J.Zha,Planning and Design of Urban Garden and Green Space Oriented by Protecting Urban Biodiversity: A Case Study of Perth, Australia, 2011 Conference on Urban Development and Planning, 2011.
7. J.D.Li, J.Pang, T.P.He, Flora of Qingxiu Mountain Scenic Area in Nanning City. Anhui Agricultural Sciences, 38: 10422-10425,2010.
8. T.Chen, Study on plant diversity in qingxiu mountain scenic spot of nanning city Urban Ecology,Guangxi University, 2005.

Toxicity of Cypermethrin-chlorpyrifos and Butachlor to Tadpole of *Quasipaa boulengeri*

Zhang Xiaoqin[1*], Zhao Changju[1], and Wu Haoyang[1]

[1]College of Resources and Environment, A'Ba Teachers University, Wenchuan, Sichuan Province 623002, China.

Abstract: The tadpole of *Quasipaa boulangeri* was taken as the research object in this paper. The toxic effects of cypermethrin-chlorpyrifos and butachlor on tadpole of *Q.boulangeri* were studied by acute toxicity experiment and subchronic toxicity experiment in order to master the toxicity of different types of pesticides on tadpole of *Q.boulangeri* and provide scientific basis for amphibian protection. The results of acute toxicity experiment were analyzed by the method of probability unit regression analysis. The results of the acute toxicity experiment were analyzed by means of the probability-unit regression analysis method, and the LC_{50} of cypermethrin-chlorpyrifos and butachlor to the 2-month-old frog tadpoles at 24h, 48h, 72h and 96h were 0.419, 0.357, 0.346L, 0.332 and 1.183, 1.108, 0.925 and 0.876(mg/L), respectively. In the subchronic toxicity experiment, we observed the accumulation effect of different concentrations of pesticides on tadpoles through the statistics of the growth and development of tadpoles exposed to the drugs for a long time. The results suggested that the body length and body weight of tadpoles increased slowly with the increase of time and concentration of pesticide in the solution during the experiment.

1 Introduction

With the continuous development of agriculture, the use of a large number of pesticides has not only caused serious pollution to the natural environment, especially water resources, but also had a certain impact on the growth and development of aquatic organisms and amphibians. Some studies have proved that the decline of amphibian population is closely related to pesticide pollution[1]. Liu Ying's experimental results showed that six pesticides, including triazophos, chlorpyrifos, acetochlor and butachlor, had high toxicity to *Microhyla pulchra* and tadpole of *Bufo Melanostictus Schneider*[2]. He Zhigang's study on the use of pesticides in paddy fields in the living environment of *Pelophylax nigromaculatus* showed that pesticides had some effects on the growth and development, tissue structure and genetic material of tadpoles[3]. Geng studied the effects of dichlorvos and butachlor on four species of batrachians through acute toxicity experiments. It was found that there were great differences in the sensitivity of different species of tadpoles to pesticides[4]. Howe exposed *Lithobates pipiens* to herbicides and found that herbicides cause deformities of tadpoles' tails, tissues and organs[5]. *Quasipaa boulengeri*, which a rare amphibian endemic to China, is commonly known as stone frog, Bangbang fish and so on and named *Paa tibetanus*[6]. This paper mainly studies the acute toxic effects of common pesticide cypermethrin-chlorpyrifos and herbicide butachlor on *Q.boulangeri* tadpoles and the subchronic toxicity effects of herbicide butachlor on *Q.boulangeri* tadpoles. It not only provides a scientific basis for the protection of wild *Q.boulangeri* tadpoles, but also has a positive impact on environmental protection.

2 Material and Methods

2.1 Reagent

The pesticides used in the experiment were chlorpyrifos chlorpyrifos with 50 percent total active component of emulsion, including chlorpyrifos 45 percentand cypermethrin 5 percent, which were produced by Jiangsu Lanfeng Bio-Chemical Co.,Ltd. The other, butachlor, is made by Jiangsu Lvilai Co., Ltd., which has an active ingredient of 60 percent dairy oil.

2.2 Animals

The tadpoles selected in this experiment were purchased from the professional cooperative of *Q.boulangeri* breeding in Linxidi, Ya'an City. The tadpole is the completion stage of 25 Gill cover fold[7].

* Corresponding author: qingerzx1982@163.com

https://doi.org/10.1051/e3sconf/202129301012

3 Acute toxicity of cypermethrin-chlorpyrifos and butachlor to *Q.boulangeri* tadpoles

3.1 Preliminary test

During the experiment, the water temperature was maintained at 22-25°C and the tadpoles were fed in a circular basin and stopped feeding 24 hours before the start of the experiment. Four concentration gradient experimental groups and one blank group were set up, and 10 tadpoles of *Q.boulangeri* were placed in each concentration group. The experiment lasted for 96 hours, and the 24-hour total death concentration (LC_{100}) and 96-hour non-death concentration (LC_0) of the two reagents were selected respectively.

3.2 Acute toxicity test

This experiment adopts the method of semi-purified water biological test[8]. The test solution was changed every 24 hours during the experiment. Five concentration gradient groups and one blank group were set up. The cypermethrin-chlorpyrifos group was treated with 0.15, 0.30, 0.45, 0.60, 0.75mg/L respectively, and the butachlor test concentration was 0.198, 0.948, 1.700, 2.300, 3.084mg/L. Ten tadpoles of uniform size, healthy and active were put into each group and were not fed during the experiment. The death number of tadpoles of *Q.boulangeri* were recorded at 24h, 48h, 72h and 96h after the beginning of the experiment. The death standard was that there was no blood supply to the heart of tadpoles. The poisoning symptoms of tadpoles were recorded and the dead individuals were cleaned up in time.

3.3 Results and analysis of acute toxicity test

3.3.1 Poisoning symptoms

No dead individuals were found in the tadpoles of *Q.boulangeri* in the blank group and the activity performance was normal during the experiment. The tadpoles of the *Q.boulangeri* do irregular strenuous exercise after a few seconds of contact with venom. After a period of time, it was observed that the excrement of the high concentration group was more than that of the low concentration group, and the tadpoles showed a certain adaptability mostly resting at the bottom of the water and some turn over or ventral side up at the bottom of the basin. The death of *Q.boulangeri* tadpoles in the two test solutions were accompanied by liver enlargement and some of the tadpoles in the butachlor group showed tail bending deformity. The tail of the dead tadpoles in the cypermethrin-chlorpyrifos group showed varying degrees of internal bleeding in the figure 1, while in the butachlor group, there was more slight bleeding in the liver as shown in figure 2.

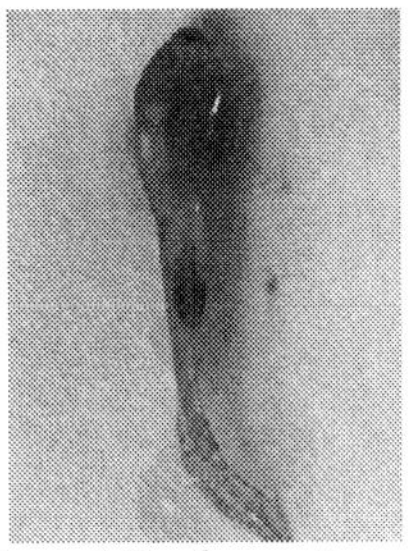

Figure 1 Death symptoms of *Q.boulangeri* tadpoles in cypermethrin-chlorpyrifos groups.

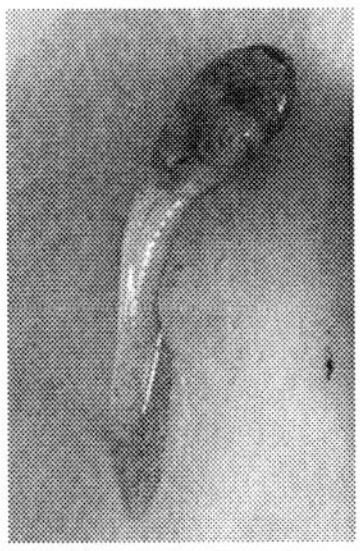

Figure 2 Death symptoms of *Q.boulangeri* tadpoles in butachlor groups.

3.3.2 Result analysis

The death of tadpoles of cypermethrin-chlorpyrifos in each treatment group was shown in Table 1 after 96 hours of exposure. Probit probability unit regression analysis[9] was used to analyze the drug concentration of cypermethrin-chlorpyrifos, butachlor and the cumulative death number of tadpoles. The results obtained by probability unit method were used to find the semi-lethal concentrations of cypermethrin-chlorpyrifos, butachlor to *Q.boulangeri* tadpoles for 24h, 48h, 72h, 96h and the safe concentrations were calculated. The results are shown in Table 2. The safe concentration (SC) = $0.1\times LC_{50}$ (96h).

Table 1 and Table 2 showed that the safe concentrations of *Q.boulangeri* tadpoles treated with cypermethrin-chlorpyrifos and butachlor were 0.0332 mg/L and 0.0876mg/L, respectively. The LC_{50} value decreased with the extension of the experimental time and the toxic ability gradually increased with the increase of exposure time. The toxicity grade is divided into four grades: $LC_{50} \leq 0.1$ as virulent toxicity grade, $0.1 < LC_{50} \leq 1.0$ as high toxicity grade, $1.0 < LC_{50} \leq 10$ as medium toxicity grade, $LC_{50} > 10$ as low toxicity grade according to the 96h-LC_{50}(mg/L) standard of pesticides to fish in the environmental safety assessment test criteria of chemical pesticides in China[10], the cypermethrin-chlorpyrifos and butachlor are highly toxic pesticides to *Q.boulangeri* tadpoles.

Table 1 Acute toxicity results of cypermethrin-chlorpyrifos and butachlor *Q.boulangeri* tadpoles.

Pesticide	Concent ration	The total number of tadpoles	Accumulated number of dead tadpoles			
	mg/L	number	24 h	48 h	72 h	96 h
blank control group	0	20	0	0	0	0
cypermethrin -chlorpyrifos	0.15	20	0	0	0	0
	0.30	20	5	6	6	8
	0.45	20	14	18	18	18
	0.60	20	17	19	20	20
	0.75	20	19	20	20	20
butachlor	0.198	20	0	0	0	0
	0.948	20	3	11	14	14
	1.700	20	13	17	18	19
	2.300	20	14	19	20	20
	3.084	20	17	20	20	20

Table 2 Regression analysis of acute toxicity of cypermethrin-chlorpyrifos and butachlor to *Q.boulangeri* tadpoles.

Pesticide	Time (h)	Regression equation	Significance (*p* value)	Medium lethal concentration(mg/L)
cypermethrin-chlorpyrifos	24	p=-2.633+6.289x	0.479	0.419
	48	p=-3.414+9.562x	0.320	0.357
	72	p=-.449+12.875x	0.995	0.346
	96	p=-.971+11.966x	0.958	0.332
butachlor	24	p=-2.050+1.131x	0.197	1.183
	48	p=-2.013+1.816x	0.248	1.108
	72	p=-2.168+2.345x	0.174	0.925
	96	p=-2.407+2.747x	0.340	0.876

Note: P. Unit of probability of death; X. Logarithm of concentration of experimental agents. Pearson goodness of fit significance $P > 0.05$, indicating that the model can fit the experimental data well.

4 Subchronic toxicity of butachlor to *Q.boulengeri* tadpoles

4.1 Subchronic toxicity test

The concentrations of each group were 0.1, 0.2, 0.3 and 0.4mg/L, respectively. Each group was divided into 3 parallel groups and corresponding control groups. 15 tadpoles of *Q.boulangeri* with uniform size, health and strong mobility were placed in each group.

4.2 Experimental results and analysis of subchronic toxicity

The results of two-way without repetitive analysis of variance of the average body length and body weight of the tadpoles exposed to butachlor for 0d, 5d, 10d, 15d and 20d and the control group were shown in Table 3 and the results of one-way ANOVA were shown in Figure 3 and Figure 4. In the subchronic toxicity experiment, the average±standard deviation method was used to represent the body length and body weight of the tadpoles. One-way analysis of variance (One-way ANOVA) was used to compare the changes of the data of each experimental group in each time period, and the two-way analysis of variance was used to analyze all the data and significant level $\alpha = 0.05$. The data were processed by SPSS 25.0 software. It can be seen from Table 3 that there is a significant difference in time on the body length and weight of *Q.boulangeri* tadpoles at 0.05 level, but there is no significant difference in concentration on the body length of *Q.boulangeri* tadpoles. It can be seen from Figure 3 and Figure 4 that the body weight of *Q.boulangeri* tadpoles increases slowly with the increase of pesticide concentration in the test solution.

In summary, it can be concluded that during the experiment, with the extension of exposure time and the increase of pesticide concentration in the test solution, the inhibition on the growth of body length and body weight of *Q.boulangeri* tadpoles was more obvious.

Table 3 The test results of main effect of butachlor concentrationand time on body length and weight of *Q.boulangeri* tadpole

Variable	Source	Intersubjective effect test: Degree of freedom	Mean square	*F*	Significance (*p* value)
length	time	4	0.541	6.204	0.000**
	concentration	4	0.113	1.300	0.269
weight	time	4	0.031	3.659	0.006*
	concentration	4	0.030	3.597	0.007*

Note: $P < 0.05$ indicates significant difference.

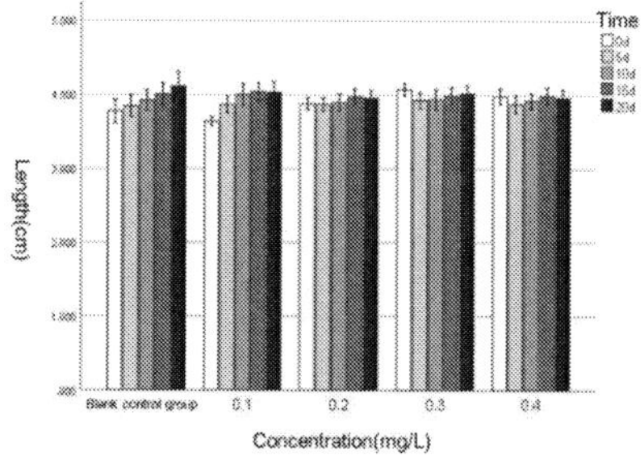

Figure 3 Variation of body length of *Q.boulangeri* tadpoles with different concentrations of butachlor in 20 days.

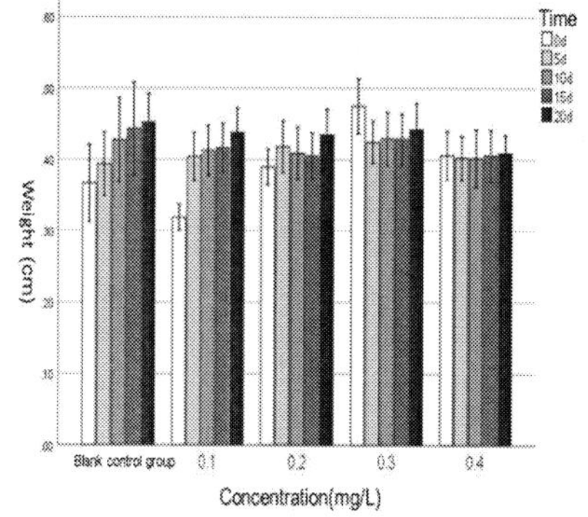

Figure 4 Changes of body weightof *Q. boulangeri* tadpoles with different concentrations of butachlor in 20 days.

5 Conclusion

The results of acute toxicity test showed that cypermethrin-chlorpyrifos and butachlor were highly toxic pesticides to *Q.boulangeri* tadpoles. The results of subchronic toxicity test showed that low concentration of butachlor test solution could promote the growth of body length and body weight of *Q.boulangeri* tadpoles in a short time, while high concentration of butachlor test solution could inhibit the growth of body length and body weight of *Q.boulangeri* tadpoles. The higher the concentration of pesticide, the more obvious the inhibitory effect.

https://doi.org/10.1051/e3sconf/202129301012

In summary, it can be seen that the effects of pesticides on amphibian tadpoles are not only shown in the short-term acute toxic effects, but also have potential effects on the growth and development of tadpoles with long-term low concentrations of pesticides. Therefore, the application of pesticides in agriculture should be strictly controlled to avoid the toxic effects of pesticides on amphibians and other aquatic organisms. It also urges people to scientifically and standardize the use of pesticides in daily production practice to minimize the impact on the environment and aquatic organisms.

Acknowledgements

This work was supported by the grants from project of A'ba Normal University (No. ASA21-06) to Xiaoqin Zhang and from project of A'ba Normal University (No.ASC20-04) to Xuemei Xian.

References

1. Cheek AO, Bollinger JE and Pider CV (1999) Alteration of Leopard Frog (*Rana pipiens*) Metamorphosis by the Herbicide Acetochlor *J. Archive of Envionmental Contamination Toxicology*.37 pp 70-7.
2. Ying L, Yan H and Lei J (2015) Acute Toxicity of Six Pesticides Used in Rice Field on Tadpole of Microhyla pulchra and Bufo melanostictus *J. Agrochemicals*.54(11) pp 825-7.
3. Zhigang H, Yuanan W, Dongwu W, Yongfu X, Jinglong L and Ming Z (2018) Effects of pesticides on the toxicity of Pelophylax nigromaculatus in Paddy Field *J. Grain Oil And Feed Technology*.(02) pp 27-31.
4. Geng BR, Yao D, Xue QQ (2005) Genotoxicity of the pesticide dichlorvos and herbicide butachlor in Rhacophorus megacephaslus tasdpoles *J. Acta Zoologica Sinca*.51(3) pp 447-454.
5. Howe C, Berrili M and Pauli D (2008) The Acute and Chronic Toxicity of Glyphosate-Based Pesticides in Northern Leopard Frogs *J. Methods*.(2008) 1-7.
6. Shibo X, Jingming X, Fan Y, Hong C, Zhen W and Chuan X (2020) Research Progress on the Biological Characteristics and Resource Conservation of Quasipaa boulengeri *J. Chinese Journal of Zoology*.55(01) pp 105-9.
7. Ermi Z (1990) Introduction of a frog embryo and Stages of tadpole development *J. Bulletin of Biology*.(01) pp 13-5.
8. Yongxin Z and Zongshe Z (1989) Test method for toxicity of aquatic organisms *M. China Agriculture Press*. pp 5-143.
9. Cuiping L, Minyao W and Hongyuan W (2012) LC_{50} Caculated by Kochi, Probit Analysis and Linear Regression Methods *J. Progress in Veterinary Medicine*.33(09) pp 89-92.
10. (2004) Test criteria for environmental safety assessment of chemical pesticides *R. The ministry of agriculture of the People's Republic of China*.GB##-2004.

Comparison of Soil Moisture Dynamics in Fixed Dune and Sandy Grassland on Korqin Sandy Land

Shuxia Yao[1], Chuancheng Zhao[1*]

[1]Lanzhou City University, Lanzhou, Gansu, 730070, China

Abstract. Based on the fixed-point observation data of soil moisture in different soil layers (0~150 cm) of fixed dunes and sandy grassland in the semi-arid region of Horqin Sandy Land, in the growing season (May to September) from 2005 to 2014, we conducted a comparative study on the seasonal, inter-annual, and spatial dynamics of soil moisture in the two types of sandy land. The results revealed that: (1) in the growing season, the soil moisture in sandy grassland was significantly higher than that in fixed dunes, and the multi-year average soil moisture in both types of sandy land was higher in July than in other months; (2) from 2005 to 2014, the average soil moisture in both types of sandy land was relatively high in 2010, and the average soil moisture in sandy grassland increased gradually with each year; specifically, there was a significant increase in soil moisture in June, while the average soil moisture in fixed dunes did not show an obvious increase; (3) the soil moisture in fixed dunes and sandy grassland generally showed a trend of "first increase, then decrease, and finally increase again", with the increase in soil depth, and this change was more obvious for sandy grassland.

1 Introduction

As an important component of the hydrological cycle of the Earth's surface system, soil moisture regulates the most basic spatial pattern and process of the land surface system, and connects a series of hydrological, ecological, climatic, and geological processes in the basin, as well as the storage, migration, and transformation of material and energy [1-3]. As an important water source of ecosystems in arid and semi-arid regions, soil moisture directly affects and restricts the growth, distribution, and stability of plants [4-7]. Due to the influence of climate, soil, and other factors, soil moisture is often in a state of dynamic change [8] and varies greatly with different ways of land use.

The Horqin Sandy Land is one of the most severe desertification regions in China. In recent decades, due to the impact of climate change and agricultural development, the water resource constraints in the region have gradually become increasingly prominent. Frequent droughts and declining groundwater level have become important issues affecting the desertification control process [9-10]. Horqin Sandy Land has a fragile ecological environment, however, there have been few comparative studies on soil moisture dynamics in different types of sandy land, and there is also a lack of continuous and systematic observation of soil moisture [11]. This paper uses the fixed-point observation data at different soil depths of fixed dunes and sandy grassland in the semi-arid region of Horqin Sandy Land for 10 consecutive growing seasons, and systematically compares and analyzes the seasonal, inter-annual, and spatial dynamics of soil moisture in these two types of sandy land, with the aim to provide a scientific basis for the effective use and management of soil moisture in this region.

2 Materials and Methods

2.1 Study Area Description

The study area is located in the southern part of the Horqin Sand Land in eastern Inner Mongolia, China (42°55′ N, 120°42′ E). This area has a temperate continental semi-arid monsoonal climate. The average annual temperature is 6.5 °C, and annual precipitation is 366 mm. The average annual wind speed is 3.4 m s^{-1}. Soils are sandy, light yellow, have loose structure, and are particularly susceptible to wind rosion [12].

2.2 Study Methods

In the study area, one special fixed dune and one sandy grassland were selected as the experimental plot, the main vegetation type of sandy grassland was *Artemisia scoparia* and *Setaria viridisthe*, and the main vegetation type of fixed dune was *Artemisia halodendron* and *Caragana microphylla*. In each plot, three 2m-deep neutron probes (CNC503DR) were buried at equal distances to obtain the soil volumetric water content (%) of 8 soil layers, the lower limit of the soil layer is 10 cm, 30 cm, 50 cm, 70 cm, 90 cm, 110 cm, 130 cm and 150 cm.

* Corresponding author: zhao_chch1978@163.com

From 2005 to 2014, each growth season (May to September), the soil volumetric water content (%) was conducted once every 7 days and added after rainfall. The soil moisture of each soil layer was measured for 164 times.

Precipitation data for the same period was obtained from the meteorological station of the nearby Naiman Desertification Research Station.

3 Results

3.1 Comparison of monthly changes in SM

By comparing the changes in multi-year monthly average soil moisture and precipitation at different soil depths of fixed dunes and sandy grassland in the growing seasons from 2005 to 2014 (Fig. 1), it was found that the multi-year monthly average soil moisture of the two types of sandy land did not change much from month to month, however, it was higher in July and lower in May and September for both land types, which was mainly due to the seasonal distribution of precipitation. The soil moisture in the study zone mainly came from precipitation [13], and the monthly average precipitation during the test period was higher in July and relatively lower in May and September (Fig. 1). Overall, the soil moisture in sandy grassland was significantly higher than that in fixed dunes in each month, which is consistent with existing research results [14].

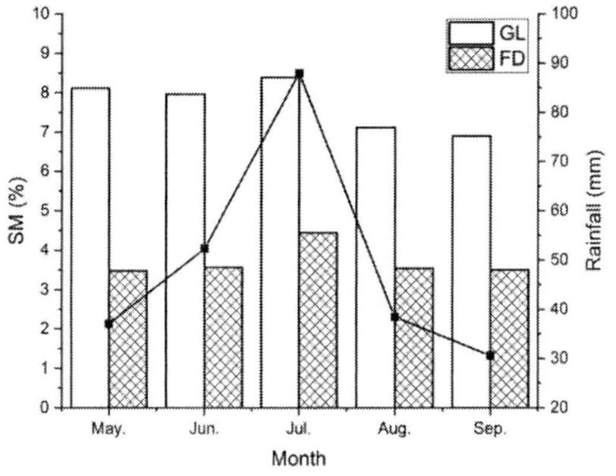

Figure 1. Average SM (%) of GL and FD and Rainfall (mm) change over month

3.2 Comparison of inter-annual changes in SM

The comparison of inter-annual changes in average soil moisture in different soil layers of fixed dunes and sandy grassland in Fig. 2 (F) shows that, from 2005 to 2014, the average soil moisture in sandy grassland increased gradually with each year, while the average soil moisture in fixed dunes did not show an obvious increase. The average soil moisture in both types of sandy land was relatively high in 2010, but the annual average precipitation in 2010 was not the highest among all these years. This shows that the change in soil moisture was not only affected by rainfall, but rather resulted from the comprehensive effects of rainfall, soil surface evaporation, plant transpiration, and other factors.

As can be seen from the inter-annual variation in average soil moisture in different soil layers of the two types of sandy land, from May to September (Fig. 2 (A~E)), the soil moisture in sandy grassland was still significantly higher than that in fixed dunes in each month of the growing season, and the soil moisture in fixed dunes in each month showed small inter-annual variations. However, the soil moisture in sandy grassland in each month showed varying degrees of increase with each year, specifically, the soil moisture in June had a significant increase. From the comparison of soil moisture in different months, the soil moisture in September (Fig. 2 (E)) for the two types of sandy land was lower than that in other months, which was mainly due to high evaporate-transpiration and low precipitation in the growing season. In addition, the soil moisture in fixed dunes in September had the minimum range of inter-annual variation, and showed little difference in other years except for a relatively low level in 2008.

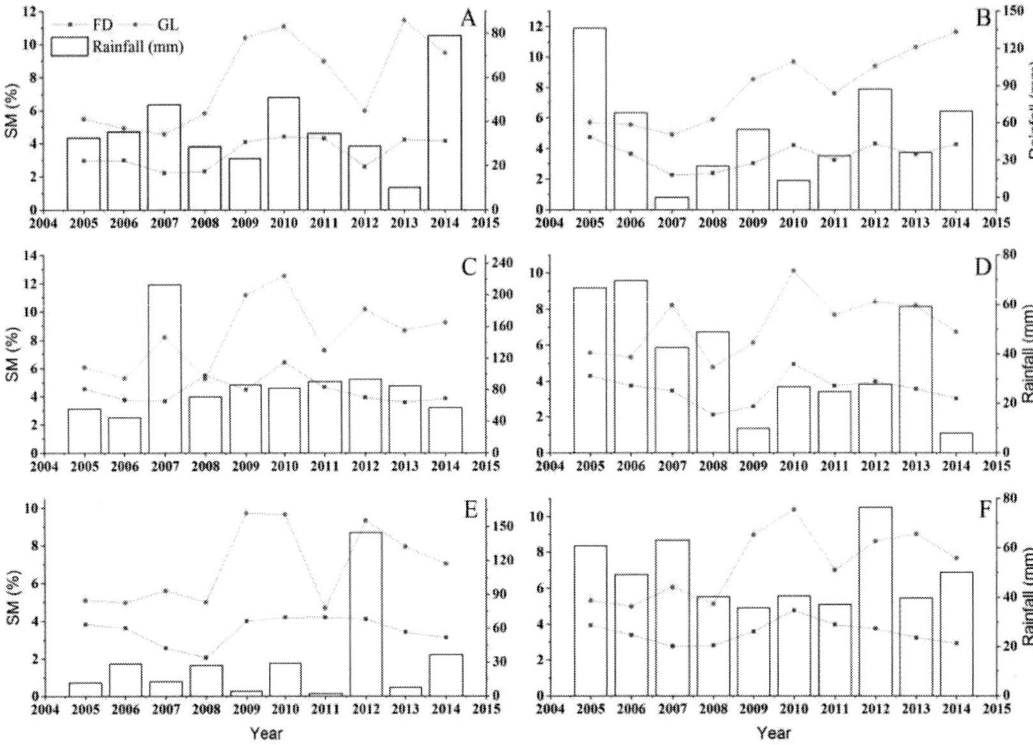

Figure 2. Average SM (%) of GL and FD and Rainfall (mm) change over year

A-May., B-Jun., C-Jul., D-Aug., E-Sep., F-Average

3.3 Comparison of spatial changes in SM

The analysis of the change in average soil moisture with the soil profile of fixed dunes and sandy grassland in the growing season from 2005 to 2014 shows that (Fig. 3), the soil moisture in both, sandy grassland and fixed dunes, showed an inversed S-shaped change with the increase in soil depth, and this change was more obvious for sandy grassland. For the two types of sandy land, the soil profile with a depth range of 0–150 cm can be divided into four layers: 0–30 cm, 30–70 cm, 70–130 cm, and 130–150 cm. The soil moisture in the 0–30 cm and 70–130 cm layers was relatively low, while that in the 30–70 cm and 130–150 cm layers was relatively high. The soil moisture in the two types of sandy land generally showed a trend of "first increase, then decrease, and finally increase again" with an increase in soil depth, which is consistent with existing research results [14-15].

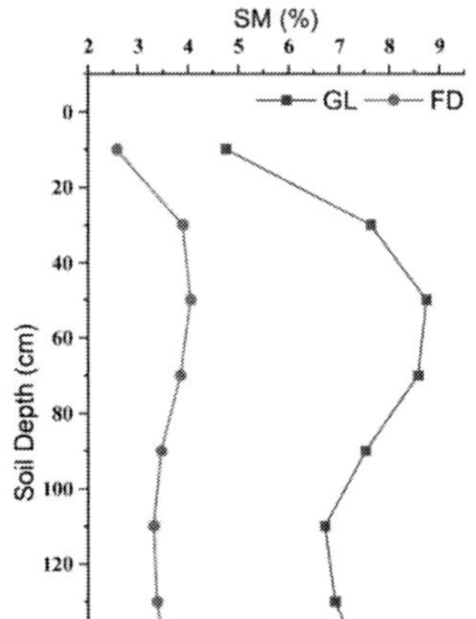

Figure 3. Average SM (%) of GL and FD change over soil depth

4 Conclusions

(1) In the growing season of study area, the soil moisture in sandy grassland was significantly higher than that in fixed dunes and the multi-year average soil moisture was higher in July than in other months.

(2) From 2005 to 2014, the average soil moisture was relatively high in 2010, and the average soil moisture in sandy grassland increased gradually with each year.

(3) With the increase in soil depth, the soil moisture in two sandy lands were all generally showed a trend of "first increase, then decrease, and finally increase again", and this change was more obvious for sandy grassland.

Acknowledgments

This study was supported by the National Natural Science Foundation of China (No. 31960273; No. 41771087). The authors also would like to acknowledge the helpful comments and suggestions from anonymous referees.

References

1. Modeling Soil Processes: Review, Key Challenges, and New Perspectives. Vadose Zone Journal, 2016, **15**(5), doi:10.2136/vzj2015.09.0131
2. Q. Zhu, K.H. Liao, X.M. Lai, et al. A review of soil water monitoring and modelling across spatial scales in the watershed, Progress in Geography, **38**(8):1150-1158 (2019)
3. J.J. Dai, X.P. Zhang, D.Q. Lv, et al. Dynamic of Soil Water in Cinnamomum camphora Forest in the Red Soil Hilly Region of South China. Research of soil and water conservation, **26**(4):123-131 (2019)
4. G.L. Wu, G.H. Ren, D. Wang, et al. Above- and below-ground response to soil water change in an alpine wetland ecosystem on the Qinghai-Tibetan Plateau, China. Biogeosciences Discussions, **8**(4):120-127 (2011)
5. Z. Huang. Research on soil water balance of typical artificial grassland and natural grassland in semi-arid area of the Loess Plateau (2017)
6. Y. Su, L. Chen, Y.F. Li, et al. Temporal and spatial changes of different types of soil moisture in desert steppe. Arid Zone Research, (6) (2018)
7. X.Y. Zhou, T.T. Yang, A.Z. Wang, et al. Distribution characteristics of soil moisture before and after rainfall in dune-dune lowland in Horqin Sandy Land. Chinese Journal of Ecology, (01):157-163 (2017)
8. F. Zhang, F.P Zeng, H. Du, et al. Response of soil moisture to precipitation characteristics in different land use patterns in Karst peak-cluster depression area. Ecological Science, **38**(5): 38-43 (2019)
9. X.Y. Zhao, C.M. Zhang, X.A. Zuo, et al. Challenges faced by desertification land restoration in Horqin Sandy Land. Journal of Applied Ecology, (07):1559-1564 (2009)
10. Y.L. Li, X.Y. Zhao, X.P. Liu, et al. Research on desertified land and its management to promote ecological restoration and sustainable development of agriculture and animal husbandry in the northern agricultural and pastoral areas. Bulletin of the Chinese Academy of Sciences, **034**(007):832-840 (2019)
11. W. Feng,W. Li , W.B. Yang, et al. Dynamic Response of Soil Moisture to Rainfall in the Pinus sylvestris var. mongolica Sand Fixation Forest of Mu Us Sandy Land. Research of soil and water conservation, **26**(3):101-107 (2019)
12. S.K. Wang, X.Y. Zhao, T.H. Zhang, et al. Effects of several shrubs on rhizosphere microorganisms in Horqin Sandy Land. Arid Area Resources and Environment, , **026**(005):140-144 (2012)
13. J.F. Reynolds, P.R. Kemp, K. Ogle, et al. Modifying the'pulsereserve' paradigm for deserts of North America: precipitation pulses, soil water and plant responses. Oecologia, 141, 194-210 (2004)
14. S.X. Yao, T.T. Zhang, C.C. Zhao, et al. Spatiotemporal heterogeneity of soil moisture in different types of sandy land in Horqin area. Journal of Soil and Water Conservation, **26**(001):251-254 (2012)
15. C.Y. Zhao, Y.C. Wang. Research on the Temporal and Spatial Dynamics of Soil Moisture in the Marginal Zone of Desert and Oasis. Journal of Soil and Water Conservation, **19**(1): 124-127 (2005)

 https://doi.org/10.1051/e3sconf/202129301014

Analysis on Spatial and Temporal Evolution Characteristics of National Land Structure of the Lanzhou-Xining Urban Agglomeration in the Upper Reaches of the Yellow River

Xiaoying Nie[1*], Kaiming Li[1], Wanzhuang Huang[1] and Xin Li[1]

[1]College of Geography and Environmental engineering, Lanzhou City University, Lanzhou 730070, China

Abstract. Based on the land utilization data of the Lanzhou-Xining urban agglomerations in 2000, 2005, 2010 and 2015, using the methods of transfer matrix, the paper makes a quantitative analysis of the spatial and temporal evolution process, characteristics and laws of various types of national land space from 2000 to 2015. The results are as follows. 1) Urban construction space, industrial and mining construction space and rural living space continued to increase, while agricultural production space showed a decreasing trend, and green ecological space and other ecological space showed an overall increasing trend from 2000 to 2015. 2) The characteristics of national land space transfer are as follows: the transition is relatively active from 2000 to 2005 and from 2010 to 2015, mainly from green ecological space and agricultural production space to urban construction space, rural living space and industrial and mining construction space.

1 Introduction

Urbanization is the process of population concentration in cities and towns and thus promotes the development of cities and towns [1], and it is also the inevitable outcome of the development of industrialization to a certain stage. As the main form of urbanization and an important carrier of modernization construction, urban agglomeration gathers important resources, industries and economy in specific regions, and becomes a key regional spatial form to realize the intensive utilization of regional spatial resources and promote the sustainable development of urbanization in the new era. Land resources are the material basis and carrier space for urbanization development, and will inevitably be largely occupied with the rapid advancement of urbanization [2]. Due to the spread of urban land, the land use efficiency is low and the ecological environment gradually deteriorates, thus affecting the quality of China's urbanization development and the healthy operation of the economy [3]. Unreasonable land use structure and methods, deterioration of environmental quality and deterioration of the ecological environment in the process of urbanization, will inevitably have a significant impact on the structure and landscape pattern of territorial spatial development, which restricts the sustainable development of human existence and the harmony and stability of social economy. However, urban agglomeration construction is an effective urbanization mode to solve the disorderly development of land space under the background of rapid urban development.

Connecting the spatial structure of land and land use/land cover change is a new perspective to study the spatial evolution of land. The spatial pattern of land use and its dynamic evolution have always been the focus and hot issue of geography, land use science and ecology. With the strengthening of global consciousness and system concept, many scholars have carried out research on land use in different regions by using different methods, models and software from different angles. Topics and models specially used for the study of land use and spatial pattern have been formed abroad [4-5]. With the rapid development of multi-temporal and multi-resolution aerial and aerospace remote sensing sensors, it provides a solid data basis for multi-scale spatial land use change process monitoring [6]. The comprehensive development of remote sensing, geographic information system and global positioning system has become the technical guarantee for the study of land use/land cover change [7-8]. Domestic scholars in time tracking the international trend of land use/land cover research, from the land cover change mechanism, the mechanism of land use change and regional and national land use survey in three key areas, mainly includes the dynamic process of land use change, driving factors and the analysis of the driving mechanism and ecological environment effect [9-14]. In general, the evaluation methods and contents of land use change are diverse at home and abroad, but there are few studies on the characteristics of spatial-temporal pattern evolution based on the carrier of territorial space.

The Lanzhou-Xining urban agglomeration is the most densely populated area and with relatively good economic foundation in Hehuang area, which taking the Yellow River and huangshui as the link, Lanzhou and Xining is the center. Based on geographical proximity and accessibility of traffic, the Lanzhou-Xining urban agglomeration forms the only geographical unit with watershed characteristics and relatively complete in the

*Corresponding author: nxy1005@163.com

https://doi.org/10.1051/e3sconf/202129301014

western region. Because of its special geographical conditions and economic base, the sustainable development and regional coordination of the Lanzhou-Xining urban agglomeration have always been the focus of the whole society. From the perspective of realizing the goal of land space development and promoting the formation of main functional zones, this study divides the land space of Lanzhou-Xining Urban Agglomeration into three types: urban space, agricultural space and ecological space. With the help of ArcGIS10.2 software platform, the spatiotemporal evolution process, characteristics and laws of different kinds of land space in the study area from 2000 to 2015 were quantitatively analyzed by using the models and methods such as dynamic attitude of land space and transfer matrix. Therefore, the temporal and spatial evolution characteristics of land spatial structure in Lanzhou-Xining City Agglomeration region under the background of rapid urbanization are revealed, which provides a foundation for further research on urban sustainable development.

2 Study area

The Lanzhou-Xining Urban Agglomeration is located in the upper reaches of the Yellow River, between the ecological barrier of the Qinghai-Tibet Plateau and the sand control belt in northern China, its main body includes the relatively dense urban areas in Gansu Province with Lanzhou as the core and the relatively dense urban areas in Qinghai Province with Xining as the core. The population of this region is relatively dense, and the water and soil resources are relatively good, which plays an irreplaceable role in maintaining the overall situation of national security and ecological security in China. Since Hehuang area is the basic bearing area of Lanzhou-Xining City Agglomeration, this study combined with the natural geographical environment of Hehuang area, From the perspective of geographical proximity and integrity, on the basis of referring to the spatial scope of preliminary research of Gansu Province and Qinghai Province on Lanzhou-Xining City Agglomeration Planning and the administrative division scope of Lanzhou-Xining City Agglomeration Development Plan approved by the State Council in February 2018, defined the range of Lanzhou-Xining urban agglomeration as large Hehuang area, which covers 53 counties (districts) of Gansu and Qinghai, such as Lanzhou(including Lanzhou New District), Baiyin, Dingxi, Linxia Prefecture, Tianzhu County, Hezuo, Xiahe County, Lintan County, Zhuoni County, Xining, Haidong, Menyuan County, Haiyan County, Qilian County, Gonghe County, Guide County, Tongren County, and Jianzha County. The total area of the study area is 155,700 km2, including 85,200 km2 in Gansu, accounting for 18.78% of the land area of Gansu and 72,300 km2 in Qinghai, accounting for 10.03% of the land area of Qinghai.

3 Materials and methods

3.1 Data

The data of land use in the study area came from the Data Center for Resources and Environmental Sciences, Chinese Academy of Sciences(http://www.resdc.cn). Including 1 km raster remote sensing monitoring data generated from 1:100,000 scale vector data grid in 2000, 2005, 2010 and 2015, and there are 6 primary land use types and 25 secondary land use types. Because the different land space has the different natural background condition, the different main body function and the different development way. According to the National Plan for Functional Zones -- Building an Efficient, Coordinated and Sustainable Development Pattern of Territorial Space, the Plan for Functional Zones of Gansu Province and the Plan for Functional Zones of Qinghai Province, space of national land can be divided into different types from the Angle of development mode, development content, main function and product supply, and the reflection of each kind of space in the development of space of national land constitutes the structure of space of national land. From the perspective of promoting the formation of main functional areas, combining with the land use remote sensing image interpretation data of the study area, the land space of Lanzhou-Xining Urban Agglomeration was divided into urban space, agricultural space and ecological space. Among them, urban space includes urban construction space and industrial and mining construction space, agricultural space includes agricultural production space and rural living space, ecological space includes green ecological space and other ecological space.

3.2 Methods

3.2.1 Analysis method of national land spatial structure change

(1) Range of change in the national land spatial structure

$$U_{b\text{-}a} = U_b - U_a \tag{1}$$

$$U = (U_b - U_a)/T \tag{2}$$

Where: $U_{b\text{-}a}$ is the overall range of change; U is the annual mean range of change; U_a is the initial number of a certain type of land space; U_b is the final number; T is the time span.

(2) Speed of change in the national land spatial structure

①Single national land space dynamic degree

$$K = \frac{U_b - U_a}{U_a} \times \frac{1}{T} \times 100\% \tag{3}$$

Where: K is the dynamic degree of the spatial types of a single national land in the study period.

②Comprehensive dynamic degree of national land space

$$LC = \frac{\sum_{i=1}^{n} \Delta LU_{i-j}}{2\sum_{i=1}^{n} LU_i} \times \frac{1}{T} \times 100\% \quad (4)$$

Where: LC is the Comprehensive dynamic degree of national land space; LU_i is the national spatial area of class i land at the starting time; ΔLU_{i-j} is the absolute value of area transformed from category i national land space to other types in the research period.

3.2.2 The transition matrix of national land space

The mathematical expression was as follows:

$$S_{ij} = \begin{bmatrix} S_{11} & S_{12} & \cdots & S_{1n} \\ S_{21} & S_{22} & \cdots & S_{2n} \\ \cdots & \cdots & \cdots & \cdots \\ S_{n1} & S_{n2} & \cdots & S_{nn} \end{bmatrix} \quad (5)$$

Where: S_{ij} is the state of national land space at the beginning and the end of the research; n is the number of national land space types.

4 Results and discussion

4.1 Characteristics of dynamic change of national land spatial structure

Combined with the interpretation data of remote sensing images of land use in 2000, 2005, 2010 and 2015, the land spatial use status maps of the four periods in the study area were obtained by using the reclassification method in ArcGIS10.2 (Fig.1), and the area changes of different national land spatial types in four periods were respectively counted (Table 1). Based on this, the paper analyzes the change process of spatial distribution of different types of national land in different periods in the study area. In terms of spatial distribution, the proportion of green ecological space in the study area is more than 72% in the whole study period, that is, the main space of the study area is green ecological space. The agricultural production space occupies the second place in the study area, with a proportion of about 18% in each period. Next is other ecological space, area proportion is above 8% in each period. The areas of urban construction space, industrial and mining construction space and rural living space are roughly the same, and the total proportion of the three kinds of national space is less than 2%, and the overall proportion of this relationship changes little in the four years.

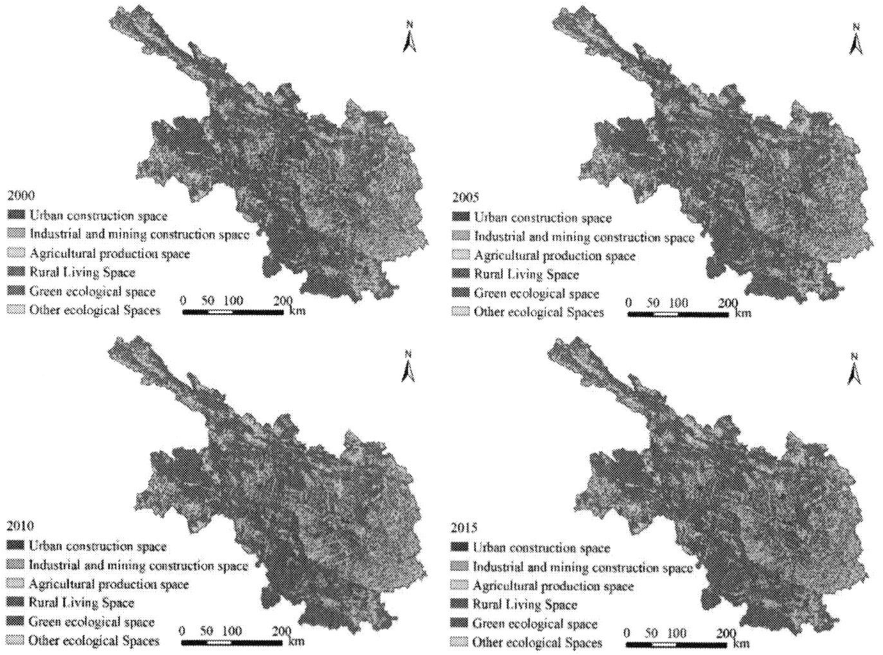

Figure 1. Current situation map of national land space utilization in different periods in Lanzhou-Xining urban agglomeration area.

Table1. The quantity variation of national land spatial structure in different periods in Lanzhou-Xining urban agglomeration area.

	Year	Urban construction space	Industrial and mining construction space	Agricultural production space	Rural Living Space	Green ecological space	Other ecological Spaces
Area Percentage (%)	2000	0.17	0.06	18.39	0.94	72.34	8.10
	2005	0.20	0.06	18.13	0.97	72.52	8.12
	2010	0.21	0.06	18.10	0.97	72.53	8.13
	2015	0.32	0.15	17.91	1.05	72.45	8.13
Range of change (km^2)	$U_{00\text{-}05}$	55	3	-395	37	271	29
	$U_{05\text{-}10}$	8	5	-49	5	20	11
	$U_{10\text{-}15}$	168	129	-293	131	-129	-6
	$U_{00\text{-}15}$	231	137	-737	173	162	34
	U	15.40	9.14	-49.14	11.54	10.80	2.67
Single type dynamic degree (%)	$K_{00\text{-}05}$	4.23	0.67	-0.28	0.51	0.05	0.05
	$K_{05\text{-}10}$	0.51	1.08	-0.04	0.07	0.00	0.02
	$K_{10\text{-}15}$	10.4	26.33	-0.21	1.76	-0.02	-0.01
	K	5.92	10.15	-0.17	0.80	0.01	0.02

In terms of the range of change, the spatial changes of various types of national land were relatively stable from 2000 to 2015. The space for urban construction, industrial and mining construction and rural living space continues to increase, while the space for agricultural production shows a decreasing trend, and the green ecological space and other ecological space show an overall increasing trend. Urban construction space and rural living space increased considerably, the average annual added value is 15.40 km2 and 11.54 km2; The second is green ecological space and industrial and mining construction space, with an average annual added value of 10.80 km2 and 9.14km2. In terms of the total amount change, the area and range of the largest change was agricultural production space, which decreased by 737 km2; The second is urban construction space, with an increased area of 231km2. The reason is mainly due to the influence of rapid urbanization, the rapid urban expansion, the obvious expansion of built-up areas, occupy a large number of arable land and basic farmland.

According to the dynamic degree model of the single national land space, the K values reflecting the changes of different national land space in different periods are calculated. From the average level, except for the negative annual dynamic degree of agricultural production space, all other types of national land space are positive. The average annual dynamic degree of agricultural production space was -0.17%, which showed a decreasing trend. Urban construction space and industrial and mining construction space increased significantly, with an average annual dynamic degree of 5.92% and 10.15%, this indicates that the increase of other construction land is obvious and the development intensity of national land space is great. The calculation results of comprehensive dynamic degree in territorial space show that it is 0.044% from 2000 to 2005, 0.012% from 2005 to 2010, and 0.048% from 2010 to 2015. This indicates that the change of territorial spatial structure has been relatively slow since 2000, but the overall change of territorial spatial structure has been improved since 2010.

Table2. National land transfer matrix of Lanzhou-Xining urban agglomeration(km^2).

Period	Pattern	UCS	ICS	APS	RLS	GES	OES	Roll-out
2000-2005	UCS	260	0	0	0	0	0	0
	ICS	7	83	0	0	0	0	7
	APS	28	7	27836	47	388	29	499
	RLS	11	0	2	1433	4	0	17
	GES	9	3	102	6	111328	14	134
	OES	0	0	0	1	13	12471	14
	summation	55	10	104	54	405	43	671
2005-2010	UCS	315	0	0	0	0	0	0
	ICS	0	93	0	0	0	0	0

https://doi.org/10.1051/e3sconf/202129301014

	APS	4	4	27844	10	78	0	96
	RLS	3	0	0	1481	3	0	6
	GES	1	1	44	1	111665	21	68
	OES	0	0	3	0	7	12504	10
	summation	8	5	47	11	88	21	180
2010-2015	UCS	322	0	0	0	1	0	1
	ICS	1	97	0	0	0	0	1
	APS	120	40	27535	119	44	33	356
	RLS	4	1	3	1483	1	0	9
	GES	44	81	52	21	111485	70	268
	OES	0	8	8	0	93	12416	109
	summation	169	130	63	140	139	103	744

Annotation: UCS stands for urban construction space; ICS stands for industrial and mining construction space; APS stands for agricultural production space; RLS stands for Rural Living Space; GES stands for green ecological space; OES stands for other ecological Spaces.

4.2 Spatial transformation of national land spatial structure

According to the Markov transfer matrix model, the raster calculator in ArcGIS10.2 software was used to analyze the remote sensing data of the study area, and the spatial change transfer matrix of the study area was obtained (Table 2). It can be seen that, the conversion range of national land space types in the study area was small from 2000 to 2015. Green ecological space and agricultural production space had the largest transfer area, and the main transfer direction was urban construction space, rural living space and industrial and mining construction space, while the transfer area of other types of national land space was small. According to the transformation of different types of national land space, the spatial differentiation characteristics of national land space change in different periods in the study area were analyzed. The change of spatial structure of national land was relatively active from 2000 to 2005, and the conversion area of agricultural production space to green ecological space was relatively large, which was mainly distributed in Anding District, Weiyuan County, Zhang County, Min County and Tongren County. The spatial transformation of national land is not obvious during 2005-2010, and it is difficult to effectively identify the degree of spatial agglomeration in this period. From 2010 to 2015, the area of agricultural production space converted into urban construction space and rural living space was relatively large. The urban construction space is mainly distributed in Yongdeng County and Gaolan County, namely the new district of Lanzhou, while the rural living space is mainly distributed in Huangzhong County and Chengbei District.

5 Conclusions

Economic development is inevitably accompanied by changes in spatial structure. As an important material basis to reflect economic structure, the evolution process and pattern characteristics of national land spatial structure have important influence on the transformation of development mode, optimization of resource allocation and realization of sustainable development. This study systematically analyzed the evolution characteristics of land spatial structure in Lanzhou-Xining Urban Agglomeration, and the results showed that: (1) In terms of quantity change, the main space of the study area is green ecological space, and the changes of all kinds of national land space are relatively stable during the whole study period. The space for urban construction, industrial and mining construction and rural living space continues to increase, while the space for agricultural production shows a decreasing trend, and the green ecological space and other ecological space show an overall increasing trend. From the perspective of time, the change rate was relatively slow from 2000 to 2010, and the overall change of national land spatial structure improved after 2010. (2) In terms of the transfer direction, the transfer was relatively active during 2000-2005 and 2010-2015, and almost stagnated during 2005-2010. The large transfer area was mainly green ecological space and agricultural production space, and the main transfer direction was urban construction space, rural living space and industrial and mining construction space.

As in Lanzhou-Xining urban agglomeration development plan for approval and implementation, and the promotion of "ecological protection and high-quality development of the Yellow River Basin" as a major national strategy, the strategic position of Lanzhou-Xining urban agglomeration is further improved, and its national spatial structure pattern of stability and security become maintain the important foundation of the development and prosperity of the northwest region. Due to national spatial land involves the natural and social economic factors such as the land resource, water resources, ecological environment, population, transportation and industry and so on, it can reflect the interaction and interrelation between land type and function in a certain region. Focusing on the development of land resources into the main functional area, this paper analyzes the spatial-temporal variation characteristics of the national land spatial structure in the study area, and provides a new perspective for the research on the evolution of the land spatial pattern of other urban agglomerations. However, due to the limitations of cognitive level and basic data, we did not analyze the evolution characteristics and internal mechanism of territorial spatial structure in combination with other factors. This will be the next research plan.

Acknowledgments

This research was supported by the Doctoral Research Fund of Lanzhou City University (LZCU-BS2019-16), the Natural Science Foundation of Gansu Province(20JR10RA296) and the National Natural Science Foundation of China (41661014).

References:

1. D D Lu, M X Chen, Acta Geographica Sinica, **70**, 179-185 (2015)
2. X W Liu, D X Zhang, B M Chen, Acta Geographica Sinica, **63**, 301-310 (2008)
3. W F Pu, Science and Technology & Innovation, 154-155 (2016)
4. E F Lambin, B L Turner, H J Geist, et al., Global Environmental Change, **11**, 261-269 (2001)
5. S Pauleit, R Ennos, Y Golding, Landscape & Urban Planning, **71**, 295-310 (2005)
6. G Gutman, A C Janetos, C O Justice et al., Journal of Environmental Management, 5-21 (2004)
7. P S Roy, A Roy, R Murtugudde, et al., Journal of the Indian Institute of Science, **90**, 489-502 (2010)
8. S F Fonji, International Microbiology the Official Journal of the Spanish Society for Microbiology, **16**, 45-52 (2013)
9. Z L Bai, A M Bao, J Zhao, et al., Bulletin of Soil and Water Conservation, **32**, 172-177 (2012)
10. X F Sun, T X Yue, Z M Fan, Acta ecologica sinica, **32**, 6440-6451 (2012)
11. J Y Wang, Y S Liu, Journal of Natural Resources, **24**, 1458-1466 (2009)
12. W X Zhou, P J Shi, Y N Wang, et al., Arid Zone Research, **34**, 232-241 (2017)
13. P J Shi, Z J Wang, C F Liu, Acta ecologica sinica, **34**, 4361-4371 (2014)
14. C Sun, L Gong, Y G Ma, et al., Journal of Northwest A&F University (Natural Science Edition), 1-10 (2020)

Analysis of metal element contents in soil sample of different altitude from lead-zinc tailing mine mount in Longnan City, Gansu Province

Zhuoxin Yin[1], Yao Zheng[1], Zhengzhou Wang[1], Bin Yue[1] and Guohua Chang[1,*]

[1] College of Geography and Environmental Engineering, Lanzhou City University, Jiefang Road 11 Anning District Lanzhou City, China

Abstract. Contents of metal elements in soil covered lead-zinc tailing mine should be determined after long-term exposured with tailing ores, results should be compared with the requirements of metal contents in National Standards in order to measure the the application level of land. The investigation of this paper chose the representative soil samples of different altitude from multiple soil samples collected from the soil covered lead-zinc tailings mine mount in Cheng County, Longnan City, Gansu Province in China. XRF analysis was used to determine the kind and content of metal elements of the soil sample. The data were compared to make sure the changes of metal elements contents of soil samples with the altitude. The result found that the maximum content of Fe in each sample, and the content of Fe is quite different of the content of other metal elements in samples, while the samples had relatively low contents of Pb and Zn. Content of Cu in soil samples is lower than the Risk Screening Value of Soil Pollution in Construction Land, while contents of Pb and Cr were higher than that in Risk Screening Value of Soil Pollution in Construction Land.

1 Introduction

Longnan City was located in southeastern part of Gansu Province in China, and its geographical latitude and longitude coordinates were between N 32°38' ~ 34°31' and E 104°1' ~ 106°35'. Longnan City was the exclusive Changjiang River basin region in Gansu Province, the only subtropical humid climate zone [1]. Local area of Longnan City had the East Asian monsoon climate with significant vertical zoning. The precipitation was closely related to local topography, the terrain change and hydrothermal distribution had great influence on production and living. It was rich in water, biological and mineral resources [2]. The altitude of Chengxian County in Longnan City was between 750 m and 2377 m [3-5]. As the miscellaneous metal elements in lead-zinc tailings mine could migrate to natural surrounding environment in long-term open storage of tailings, contents of metal elements in soil covered lead-zinc tailings mine should be tested. Soil samples investigated in this paper were collected from the soil covered lead-zinc tailings mine mount of altitude from the top to bottom of 1074 – 1248 m in Anfanggou Ditch of Chengxian County in Longnan City, Gansu Province, China. Contents of heavy metal in soil were not only affected by the eluviation and efflorescence during soil-forming process, but also influenced by plant absorption, enrichment, naturalization and other factors. However, various human activities impacted metal contents in soil of different functional areas as well.

Infiltration and diffusion of metal elements in ores as the result of the stacking of lead-zinc tailing mine. And also the metal elements in tailing ore could be dissolved by the acid in acid mine drainage water, and entered the water environment as the mine drainage and rainfall, or directly entered the soil, which could cause heavy-metal pollution of soil directly or indirectly. Soil heavy metal pollution may cause the phenomena of soil quality degradation of existing and potential soil source, deterioration of ecology and environment.

Scientific analysis of metal elements contents measured by XRF in collected soil sample to get clear about the spatial distribution pattern and pollution intensity of metal contents in soil samples from soil covered lead-zinc tailing mine, it was important to discover the main reason of heavy-metal contamination, to determine the location of pollution sources and pollution prevention and control [6].

X-ray fluorescence analysis method was used in the research process to determine the kind and content of metal elements of the soil sample with particle size of 2 mm after treatment and screening, the data obtained wer analyzed and discussed. The determined original data were analyzed, and the contents of metal elements in different soil samples were classified by element type, and the average content of each metal element in all collected soil sample were calculated.

* Corresponding author: cgh@lzcu.edu.cn.

2 Methods and Materials

2.1 Method for collecting soil samples

The lead-zinc tailings mine area in Chengxian County, Longnan City, where the soil samples were collected was built along the mountside, the spatial construction of tailings mine was laminated like terrace, with strip-stepped cross section. We collected soil samples of different altitude from the foot to the top of the mount in different heights of the lead-zinc tailing mine in the strip terrace plane obliquely cut from the side of the tailing mine mount. When the soil samples were collected, the weeds and gravel and other debris on the soil surface were pushed aside, the undersoil was digged with shovels and the mixed stones were picked out, the depth of digging was between 1 cm to 10 cm. Acquired soil samples were put into the sealed plastic sample bags, marking the acquisition time and location, and recording the geographical position of sampling collecting sites with hand-held GPS. Then the according sampling collecting time was compared with GPS records to determine the altitude and coordinates of sampling collecting sites.

2.2 Soil sample treatment and determination method of metal elements contents

Table1. Coordinates of soil sampling sites.

Sample No.	Altitude(m)	Latitude(N)	Longitude(E)
1	1075	33°50'22.59"	105°43'53.82"
2	1153	33°50'26.70"	105°44'00.66"
3	1199	33°50'29.90"	105°44'04.33"
4			
5	1246	33°50'35.92"	105°44'04.06"
6			
7	1248	33°50'34.55"	105°44'05.12"
8			

Collected soil samples were unified screening after natural drying, and the hand-held X-ray fluorescence (XRF) heavy metal analyzer(S1-TITAN, BRUKE) was used to determine the contents of metal elements in soil samples, the data used in this paper were measured results of soil samples through 2 mm sieve. Eight soil samples selected for the paper were collected from different altitudes, and were numbered in accordance with low to high altitude. The coordinate positions of sites for soil samples are listed in Table 1 by number. Some soil samples were collected in the different sites of the same coordinate.

Table2. Contents of metal elements in soil samples.

Sample No.	Contents of Metal Elements (mg·kg^{-1})			
	Cr	Mn	Fe	Ni
1	27.33	426.33	20666.67	39.67
2	——	419.67	25000.00	13.67
3	83.67	704.33	37666.67	6.33
4	59.33	131.00	10679.33	41.33
5	79.33	353.67	25000.00	13.00
6	19.67	232.33	8666.67	33.33
7	74.00	189.33	6450.67	——
8	93.00	246.00	10217.67	0.00
Average	62.33	337.83	18043.46	21.05
Sample No.	Contents of Metal Elements (mg·kg^{-1})			
	Cu	Zn	W	Pb
1	48.67	295.67	7	177.33
2	67.33	586	9.67	292
3	48	129.67	——	40.33
4	62.67	2645	15	1349
5	38	54.67	4.33	——
6	39.33	55.67	——	——
7	106	828	17.33	644
8	28	53.33	10.67	16.67
Average	54.75	581.00	10.67	419.89

3 Results and Discussion

3.1 Comparison of Determination Results of types and contents of Metal Elements in Soil Samples

Comparison of samples showed the colour of soil samples changed by brown gradually deepened with the

https://doi.org/10.1051/e3sconf/202129301015

decreasing altitude, from light-gray of soil sample in mount top to black-brown of soil sample in mount bottom. According to comparation of the XRF determination results of the types and contents of metal element in soil samples, 11 kinds of metal elements were analyzed in 8 soil samples, most of them were heavy-metal elements. In addition to Pb in the IVA main group and the sixth period of the periodic table, the others were all the transition metal, as the Cr, Mn, Fe, Ni, Cu and Zn in the forth period, the Ag, Cd in the fifth period, and W, Au, and Hg in the sixth period. While contents of some metal in some soil samples were not determined by XRF detector, only Ag was measured in Sample 7, Hg was measured in Sample 3, Cd and Au were measured in Sample 8. However, the other 8 kinds of metal elements were nearly measured in all 8 soil samples. The contents of various metal elements are listed according to the sample number, and the results are shown in Table 2. According to comparation of the XRF determination results of the types and contents of metal element in soil samples, eleven kinds of metal elements were analyzed in eight soil samples. In addition to Pb in the IVA main group and the sixth period of the periodic table, the others were all the transition metal, as the Cr, Mn, Fe, Ni, Cu and Zn in the forth period, the Ag, Cd in the fifth period, and W, Au, and Hg in the sixth period. While contents of some metal in some soil samples were not determined by XRF detector, only Ag (8.33 mg·kg^{-1}) was measured in Sample 7, Hg (9.67 mg·kg^{-1}) was measured in Sample 3, Cd (3.33 mg·kg^{-1}) and Au (45.67 mg·kg^{-1}) were measured in Sample 8. However, the other 8 kinds of metal elements were nearly measured in all 8 soil samples. The contents of various metal elements are listed according to the sample number, and the results are shown in Table 2.

To compare the contents of metal elements in soil samples covering lead-zinc tailings mine, according to the data in the Table 2, in all 8 soil samples, the content of Fe (37666.67 mg·kg^{-1}) was the highest, next was zinc(2645.00 mg·kg^{-1}), lead (1349 mg·kg^{-1}), maganese(704.33 mg·kg^{-1}), copper (106.00 mg·kg^{-1}), chromium (83.68 mg·kg^{-1}), nickel(41.33 mg·kg^{-1}) and tungsten (17.33 mg·kg^{-1}). Although the soil samples were collected from lead-zinc tailing mines, the content of Zn and Pb was lower than that of Fe. While the soil samples with the highest Zn and Pb contents were not Samples 1 and 8 collected from the top and foot of the tailings mount, but the Sample 4 from the sampling site with altitude (1199 m) at the middle of mountainside, and soil sample with the highest content of Fe, Mn and Cr is Sample 3 in the middle elevation of the tailiing mount. The results showed that contents of heavy metal measured in soil covered lead-zinc tailing mine had large amplitude of variation and heterogeneous distribution.

Table3. National Standard Requirment for contents of metal elements.

National Standard	**Contents of Metal Elements(mg·kg^{-1})**				
		Cr	Ni	Cu	Zn
Risk Control Standard for Soil Contamination of Agricultural Land	Risk Screening Values	150	60	50	200
	Risk Intervention Values	800	—	—	—
Risk Screening Values For Soil Contamination of Development Land	Land Category I	3.0	150	2000	—
	Land Category II	5.7	900	18000	—
Risk Intervention Values for Soil Contamination of Development Land	Land Category I	30	600	8000	—
	Land Category II	78	2000	36000	—

National Standard	**Contents of Metal Elements(mg·kg^{-1})**				
		Cd	Hg	Pb	
Risk Control Standard for Soil Contamination of Agricultural Land	Risk Screening Values	0.3	0.5	70	
	Risk Intervention Values	1.5	2.0	400	
Risk Screening Values For Soil Contamination of Development Land	Land Category I	20	8	400	
	Land Category II	65	38	800	
Risk Intervention Values for Soil Contamination of Development Land	Land Category I	47	33	800	

The reason for the highest contents of metal elements were found in Samples 3 and 4 collected from tailings hillside probably was leaching effect made the migration of transition metal Elements from the high altitude soil to low altitude soil. And because of the change of pH, the migration rates of metal ions with different atomic numbers and atomic weights in soil are different, thus lead to differences in the content of metal elements in soil samples. However, the contents of Pb and Zn in lead-zinc tailings mine are lower than that in raw ore, so although

Samples were collected from soil covered lead-zinc tailing mine, the contents of Pb and Zn is lower than that of Fe. Since Fe has a high content in soil samples as an associated element of a variety of transition metal ores [4, 5], maybe because iron has different oxidation state and was easy to form various compounds.

3.2 Comparation of metal elements contents in soil samples with that in National Standard

Land could be divided into agricultural land and development land by the usage category. Different types of land have different standards requirements for metal content. As the contents of metal elements in National Standard had some differences under different conditions, the mimimus value of risk screening values and risk intervention values for soil contamination of agricultural land in Risk Control Standard for Soil Contamination of Agricultural Land were selected in Table 3. Comparing the measured contents of metal elements in soil samples with the screening value and control value of relevant heavy metal contents standards in National Standards [6, 7] in Table 3, the comparison results were as follows. The city development land could be classified into Land Category I and II, while Land Category I was mainly Residential land in urban development land, Land Category II was mainly industrial land in urban development land.

In all soil samples, contents of Cr were all lower than that in risk screening values and risk intervention values for soil contamination of agricultural land, but higher than that in risk screening values for soil contamination of development land; while the maximum Cr content was a little higher than the risk intervention values for soil contamination of development land (Land Category II). There were no requirements for contents of Mn and Fe in soil National Standards. Content of Ni in soil samples were all under that of land in National Standard.

Contents of Cu in some soil samples were a little higer than that in Risk Screening Values for Soil Contamination of Agricultural Land, but all lower than the Risk control standard for soil contamination of development land [7, 8]. Contents of Zn in some soil samples were higher than the Risk Control Standard for Soil Contamination of Agricultural Land, while Risk Control Standard for Soil Contamination of Development Land did not require content of Zn. Contents of Pb in some soil samples were higher than that in Risk Control Standard For Soil Contamination Of Development Land, risk screening values and risk intervention values for soil contamination of development land (Land Category I), but they were all lower than that in Risk Intervention Values For Soil Contamination Of Development Land (Land Category II).

4 Conclusion

In Summary, results of analysis of metal element Contents in Soil Sample of different altitude from lead-zinc tailings mine mount in Anfanggou Dintch of Bijiashan Mountion, Longnan City, Gansu Province were listed as follows.

1) The results of the comparison of types and contents of metal elements in soil samples from the soil covering the lead-zinc tailing mine showed that in eight metal elements determined in soil samples, the contents of Fe were highest, while contents of W were lowest, and contents of Pb and Zn were moderate. The maximum contents of metal elements were in soil samples in the middle altitude of tailing mine mount. Eight kinds of metal elements were determined in most soil samples, but Ag, Cd, Au and Hg were only determined in three soil samples.

2) Comparing the Metal Contents in Soil Samples with that in National Standards, the results showed that the contents of Cr, Ni and Pb were both lower than that in Risk Control Standard for Soil Contamination of Agricultural Land, while the content of Cr was a little higher than that in risk screening values for soil contamination of development land. Contents of Cu and copper were a little higher than that in Risk Control Standard for Soil Contamination of Agricultural Land. Contents of Pb were higher than that in risk intervention values for soil contamination of development land (Land Category I), but lower than that in Land Category II. For the soil samples determined other metal elemts, contents of Cd and Hg were higer than that in Risk Control Standard for Soil Contamination of Agricultural Land, while content of Hg was lower than that in Risk Control Standard for Soil Contamination of Development Land.

Acknowledgement

This research was funded by the Scientific Research Projects in Colleges and Universities from Gansu (2019B-169) and the Doctoral Research Fund of Lanzhou City University (LZCU-BS2019-26).

References

1. K. Wei. Lanzhou University. (2014)
2. Y. Lu. Gansu Metallurgy. **31**, 01 (2009)
3. H. Liu. Gangsu Science and Technology. **31**, 17 (2015)
4. D. Peng. Contributions to Geology and Mineral Resources Research. **04** (1998)
5. C. Wang. Gold Science and Technology. **05** (2001)
6. C. A. Cambardella, T. B. Moorman, J. M. Novak, T. B. Parkin, D. L. Karlen, R. F. Turco, A. E. Konopka. Soil Sci. Soc. Am. J. **58** (1994)
7. Ministry of Ecology and Environment, State Administration of Market Supervision and Administration. GB 15618——2018. (2018)
8. Ministry of Ecology and Environment, State Administration of Market Supervision and Administration. GB 36600——2018. (2018)

Analysis of the Current Wastewater Recycling Standards

Zhang Xiaoxin[1], Huang Jin[1*], Lin Ling[1], Wang Xiuteng[1], Liu Jing[1], Fang Fei[1]and Zhang Sinan[1]

[1]China National Institute of Standardization, 100191 Beijing, China

Abstract. This study summarized the current development of wastewater recycling standards in China by analyzing published policies, existing national, industrial and local standards, as well as related activities in typical provinces. Meanwhile, further prospects for standardization in the area were also suggested in order to promote this industry.

1 Introduction

The theme of World Water Day 2021 is Valuing Water. According to the statistics of World Bank, the world suffers an economic loss of up to USD250 billion each year due to insufficient water supply or sanitation services, and 675,000 people die prematurely due to poor water quality, or poor environmental health or personal hygiene, with some countries suffering GDP losses as high as 7%. The water crisis led to change of people's traditional ideas about water management, and constant exploration and appreciation of value of different forms of water with innovative technologies and management models. Among the endeavors, wastewater recycling is most significant.

In January 2021, ten ministries and commissions including the National Development and Reform Commission jointly issued Guiding Opinions on Promoting Wastewater Recycling, pointing out that the wastewater, which meets the quality standards after harmless treatment, will replace conventional water used for industrial, municipal, domestic, ecological, agricultural and groundwater recharging purposes as a recycled water, and extraction of other resources and energy from wastewater is of great significance for optimizing the structure of water supply, diversifying water resources, alleviating the contradiction between water supply and demand, reducing water pollution, and ensuring water ecological safety. The guideline called for expedited improvement of policies and standards, and incorporation of recycled water into the urban water supply system. It proposed development and revision of local standards on water pollutants discharge, and differentiated requirements and measures for pollutants discharge and control. It also urged development of technical specifications and requirements on use of recycled water as ecological replenishment water, timely revision of standards on classification of quality of wastewater used for other purposes, as well as development and revision of standards on relevant equipment, engineering and operations.

2 Current situation of standardization

2.1 Existing standards on wastewater

Existing national, industrial and local standards on wastewater recycling are shown in Table.1.

Table1. Existing Standards on Wastewater Recycling

Serial No.	Standard No.	Standard name	Type
1	GB/T 18919-2002	Urban Wastewater Recycling-Classification	National standard
2	GB/T 18920-2002	Urban Wastewater Recycling-Quality of Urban Miscellaneous Water	National standard
3	GB/T 18921-2002	Urban Wastewater Recycling-Quality of Water for Scenic Environment Use	National standard
4	GB/T 19772-2005	Urban Wastewater Recycling-Quality of Water for Groundwater Recharging	National standard
5	GB/T 19923-2005	Urban Wastewater Recycling-Quality of Waster for Industrial Use	National standard
6	GB 20922-2007	Urban Wastewater Recycling - Quality of Water for Farmland Irrigation	National standard

* Corresponding author: huangjin@cnis.ac.cn

7	GB 20922-2007	Urban Wastewater Recycling - Quality of Water for Farmland Irrigation	National standard
8	GB/T 25499-2010	Urban Wastewater Recycling - Quality of Water for Landscape Irrigation	National standard
9	GB/T 22597-2014	Determination of Chemical Oxygen Demand in Recycled Water - Potassium Dichromate Method	National standard
10	GB/T 37905-2019	Quality of Recycled Water - Determination of Chromium - Polarography and Voltammetry	National standard
11	GB/T 37906-2019	Quality of Recycled Water - Determination of Mercury - Mercury Vapourmeter Method	National standard
12	GB/T 37907-2019	Quality of Recycled Water - Determination of Sulfide and Cyanide - Ion Chromatography	National standard
13	GB/T 27681-2011	Specification on Zero Discharge and Recycling of Cooling Water During Melting and Casting of Copper and Copper Alloy	National standard
14	GB/T 29773-2013	Specification on Recycling of Wastewater from Copper Concentrators	National standard
15	GB/T 30888-2014	Technical Specifications on Treatment and Recycling of Textile Wastewater by Membrane Method	National standard
16	GB/T 31189-2014	Technical Specification on Recovery of Phosphorus from Metal Phosphating Wastewater	National standard
17	GB/T 38224.1-2019	Evaluation of Heavy Metal Wastewater Treatment and Recycling Technology Part 1: Procedure and Approach	National standard
18	GB/T 38224.2-2019	Evaluation of Heavy Metal Wastewater Treatment and Recycling Technology Part 2: Index System	National standard
19	FZ/T 01107-2011	Quality of Recycled Water Used for Textile Dyeing and Finishing Industry	Industry Standard
20	TB/T 3007-2000	Quality Standard for Recycled Water Used for Railway Industry	Industry Standard
21	CJJ 252-2016	Technical Specifications on the Operation, Maintenance and Safety of Urban Water Recycling Plants	Industry Standard
22	DL/T 5483-2013	Code for Design of Advanced Treatment of Recycled Water from Thermal Power Plants	Industry Standard
23	HG/T 3923-2007	Quality Standard for Recycled Water Used for Circulation Cooling	Industry Standard
24	HG/T 4325-2012	Determination of Calcium and Magnesium Contents in Recycled Water - Atomic Absorption Spectrometry	Industry Standard
25	HG/T 4326-2012	Determination of Nickel, Copper, Zinc, Cadmium and Lead Contents in Recycled Water - Atomic Absorption Spectrometry	Industry Standard
26	HG/T 4327-2012	Determination of Total Iron Content in Recycled Water-Spectrophotometric Method	Industry Standard
27	SL 368-2006	Quality Standard for Recycled Water	Industry Standard

28	SL 760-2018	Guideline for Compilation of Specifications on Urban Wastewater Recycling	Industry Standard
29	JT/T 645.1-2016	Wastewater Recycling in Highway Service Areas Part 1: Water Quality	Industry Standard
30	JT/T 645.2-2016	Wastewater Recycling in Highway Service Areas Part 2: Technical Requirements for Treatment Systems	Industry Standard
31	JT/T 645.3-2016	Wastewater Recycling in Highway Service Areas Part 3: Operation and Management Requirements for Treatment Systems	Industry Standard
32	HG/T 5309-2018	Technical Specifications on Treatment and Recovery of Electroplating Copper-containing Wastewater	Industry Standard
33	HG/T 5361-2018	Technical Specifications on Treatment and Recovery of Coking Desulfurization and Decyanation Wastewater	Industry Standard
34	HJ 2019-2012	Technical Specifications on Treatment and Recycling of Wastewater from Iron and Steel Industry	Industry Standard
35	YS/T 1169-2017	Technical Specifications on Treatment and Recycling of Wastewater from Secondary Lead Production	Industry Standard
36	YS/T 1404-2020	Technical Specifications on Recycling of Lead Smelting Wastewater	Industry Standard
37	YS/T 1405-2020	Specifications on Treatment and Recycling of Lead-Zinc Beneficiation Wastewater	Industry Standard
38	YS/T 740-2010	Determination of Caustic Alkalinity in Wastewater from Alumina Industry	Industry Standard
39	T/CSES 01—2019	Guideline for Evaluation of Benefits of Recycled Water	Group Standard
40	T/GIEHA 007—2018	Qualification Levels of Service Agencies Engaged in Industrial and Commercial Circulating Water Treatment	Group Standard
41	T/JX 003—2018	Specifications on Engineering Design of Wastewater Treatment and Recycling of Reclaimed Water from Water Jet Looms	Group Standard
42	T/JX 002—2018	Specifications on Construction and Operation of Reclaimed Water Treatment Station in Water Jet Loom Industry	Group Standard
43	T/JX 001—2018	Quality Requirements for Reclaimed Water in Water Jet Loom Industry	Group Standard

2.2 Current situation of standardization in selective provinces of China

2.2.1 Hebei Province

The government of Hebei Province has released three local standards - Standards on Water Pollutants Discharged to Daqing River Basin, Standards on Water Pollutants Discharged to Ziya River Basin, and Standards on Water Pollutants Discharged to Heilonggang and Yundong Basins. In 2020, the Standards on Discharge by Sewage Treatment Plants in Hebei Province, which have stricter requirements than national standards, was officially implemented, with the treatment costs of up to 2 yuan per ton. These discharge standards are however not in harmony with the requirements for surface water quality across the province (assessed based on the cross section of water functional zones). The large difference in water quality is a hindrance to wastewater recycling. More standards on the treatment and recycling of wastewater from livestock and poultry breeding, and on the technology, equipment, and safety evaluation of recycled water for agricultural irrigation need to be developed. At the same time, it is necessary to develop an institutional framework guiding the market access of water-saving appliances. Continuous efforts are needed to build national and provincial green parks and green plants, and to guide and lead industrial enterprises in energy and water conservation, clean production, pollution prevention and control, and comprehensive utilization of resources.

2.2.2 Yunnan Province

The government of Yunnan Province has issued regulations on the protection of key lakes, and set water quotas and indexes for building sponge cities. As a result, the water quality of all sewage treatment plants in Dianchi Lake Basin conforms to the requirements for Class III-IV water bodies in GB 3838-2002 - Environmental Quality Standards for Surface Water, except for the total nitrogen content. Sewage treatment plants in other regions are currently in the process of upgrading the water quality from level B to level A. In the Water Quota in Yunnan Province, the government made clear the standards on water consumption intensity in production and life by specifying 1,135 water quotas for 143 industry sectors, including industrial production, farming and tourism. The nonconformity between the discharge standards and the requirements for surface water quality across the province, assessed based on the cross section of water functional zones, still exists. The large difference in water quality is a hindrance to wastewater recycling. It is necessary to develop standards on assessment of safety risk of recycled water for river replenishment (to Dianchi Lake), impact of recycled water for agricultural irrigation, and impact of long-term irrigation on shallow groundwater. Lijiang City, as a tourist city, needs more standards on treatment of wastewater from the catering industry and on building of a sponge city. Future efforts should be focused on the development of standards on equipment, products, technology, process, monitoring and testing in relation to disposal of river sediment (Dianchi), and centralized and decentralized wastewater treatment, to facilitate building of a sponge city.

2.2.3 Zhejiang Province

All sewage treatment plants in Zhejiang Province had completed upgrading by the year end of 2017, and starting from January 1, 2018, all urban sewage treatment plants in the province have been complying with level-A effluent standards. The Discharge Standards of Major Water Pollutants for Urban Sewage Treatment Plants issued by Zhejiang government (Zhejiang Standards, Surface Water Class 4) has been followed since January 1, 2019. In 2016, the local standard, Agricultural Water Quota of Zhejiang Province (DB33/T769-2016) was promulgated, specifying quotas for a total of 73 products in 13 industries. Meanwhile, 77 sets of water quotas have been revised for the textile and paper industries over the past two years under a joint dynamic quota revision mechanism. There is the same problem that the discharge standards are not in harmony with the requirements for surface water quality, assessed based on the cross section of water functional zones. Some sewage plants further purify the effluent by connecting the tail water to the ecological wetland, but no relevant standards on the technical processes and water quality are available. With the water recycling in industrial parks and wastewater collection in villages and towns in full swing, it is necessary to develop standards on water recycling performance, water quality, wastewater treatment appraisal and supervision by a third party.

2.2.4 Gansu Province

The Water Resources Department of Gansu Province issued the Implementation Measures of Gansu Province to Control the Use of Water Resources (G. Sh. Z.Y. F. [2019] No. 41). The Regulations of Gansu Province on Water Conservation, which was reported to the provincial People's Congress and the provincial government in 2017 is still under legislative survey today. It is urgent to harmonize the relevant indicators in the recycled water utilization standards and the natural water quality standards to reduce pollution caused by drainage of sewage treatment plants to natural water bodies and the environment. Due to the lower standards and lower quality indexes for ecological replenishment water, the discharged water and recycled water are still pollutants to natural water bodies (among the five classes of inferior water). Therefore, to enhance the wastewater discharge standards and water quality indexes is vital to protecting the ecological environment of water bodies.

3 Prospects for standardization of wastewater recycling

The wastewater recycling in China is still in its infancy, meaning that it has not been fully developed and the utilization level of recycled water is low, and thus is unable to meet the need of building a Beautiful China. In order to promote the development of the industry, and address such issues as water shortage, pollution of water environment and damage to water ecology, future efforts shall be focused on the following aspects.

1. Establishing a standard system for wastewater recycling and improving top-level design. A stand system that is well-balanced, perfectly structured, coordinated, open and inclusive, and where the government plays a guiding role and the market players play a dominant role, shall be established, to guide the efforts in establishing standards on technical process, equipment, products, management and services in wastewater treatment and recycling in urban and rural areas.

2. Accelerate development of key standards on wastewater recycling. The wastewater discharge standards should be harmonized with the recycled water quality requirements by taking into account the current standards and improvements needed for wastewater recycling in the regions. Technical standards on key equipment, materials, and water quality monitoring and testing methods in relation to wastewater cycling should be developed; standards on municipal or domestic water safety risk assessment should also be put in place to make recycled water more acceptable to the public; in terms of recycled water from breeding, standards on safety assessment of

https://doi.org/10.1051/e3sconf/202129301016

farmland irrigation technology, health and safety, and impact of long-term irrigation on groundwater should be developed.

3. Conducting wastewater recycling pilot demonstration projects. Key enterprises, parks and regions may be selected to pilot wastewater recycling according to a standardization management system developed based on national, industrial, local and group standards, and their model of success may be replicated in other regions to regulate and promote the development of the wastewater recycling industry in China.

4. Contributing to development of international standards on wastewater recycling. Contribution to development of intentional standards may be made by proposing marketable and competitive wastewater recycling technology or management methods or by leading or participating in the process; more Chinese technical standards should be translated into foreign languages, and advanced foreign standards should be adaptively applied, to form a standard development model of absorption, digestion and innovation. Coordination with the International Organization for Standardization, the international certification authorities and international peers should be strengthened, to enable cross-regional coordination and mutual recognition of standards with other countries.

References

1. National Development and Reform Commission. Guiding Opinions on Promoting Wastewater Recycling (2021)
2. S Gen-xing, D Qian, Y Fan, G Jinxin, Z Bolang, L Jianping. J. Shanxi Uni. Sci. Tech, **3**(2014).
3. Z Xiaoxin, H Jin, L Lin, L Shuming. W Yueping, Z Xinheng. 2019 IOP Conf. Ser.: Earth Environ. Sci. **267** (2019).
4. Z Chunyuan, Z Yong. J. China Water Resources, **1**(2020)
5. M Tao, L Jiufu, P Anbang, et al. J. Advances in Water Science. **31(6)**(2020)

Research on Evaluation of Ecological Carrying Capacity of Qinghai Province Based on PSR Model

Xiaoyong Li[1], Haifa Jia[1,2*], and Falan Zhao[1,2]

[1] College of Civil Engineering, Qinghai University, Xining, Qinghai, 810016, China

[2] Qinghai Provincial Key Laboratory of Energy-saving Building Materials and Engineering Safety, Xining, Qinghai, 810016, China

Abstract. Quantitative research and evaluation of ecological carrying capacity in Qinghai Province is in line with the requirements of the 13th five-year plan, and is of practical significance to the construction of ecological civilization in Qinghai Province. Based on the PSR model, this paper constructs the evaluation index system of ecological carrying capacity from three aspects of pressure, state ,and response, uses the entropy method to determine the index weight coefficient, and comprehensively evaluates the changing trend of ecological carrying capacity since the 12th Five Year Plan of Qinghai Province from the perspectives of economy, society, resources, and environment. The results show that: from 2011 to 2019, the comprehensive evaluation of ecological carrying capacity in Qinghai Province shows a trend of first decreasing and then increasing, the response index evaluation increases greatly, the state index evaluation shows a fluctuating growth, and the pressure index is relatively stable. Based on this, the PSR model and entropy method are typical and quantifiable, which can provide a reference for Qinghai provincial government to control the ecological environment, study ecological sustainability and delimit the ecological red line.

1 Introduction

The early exploration of carrying capacity was mainly in the field of ecology. In 1921, Parker and Burgess put forward the concept of carrying capacity in the Journal of ecology. Carrying capacity refers to the highest limit of the possible number of individual organisms in a specific environment[1]. Later, the term carrying capacity was extended to many fields, and some related concepts appeared, such as population carrying capacity, resource carrying capacity, environmental carrying capacity, ecological carrying capacity, and so on[2]. The establishment of the theoretical model lays a foundation for the quantitative study of ecological carrying capacity. At present, the quantitative research method of ecological carrying capacity is gradually becoming mature[3].

Ecological carrying capacity is an important indicator of sustainable ecological development. At present, quantitative research methods for evaluating ecological carrying capacity have been widely used at home and abroad[4]. A large number of scholars and experts have explored the ecological carrying capacity of regions at various levels. At the national level, Zhang Wenbin[5] et al., based on the perspective of self-purification capacity and pollution level of the atmosphere, water, and soil, constructed the evaluation framework of ecological carrying capacity by using the decoupling model of ecological support and pressure, and measured and evaluated the ecological sustainability of China. At the provincial level, Wang yening[6] et al., based on the statistical data of resources and social economy, used the three-dimensional ecological footprint model to conduct dynamic research on 34 provincial administrative regions in China, refined the balance factor and yield factor of different types of land, depicted the spatial and temporal characteristics of per capita ecological footprint breadth and ecological footprint depth from 2009 to 2016, and discussed and analyzed the capital utilization pattern and its socio-economic influencing factors among provinces. At the city level, Zhu Yulin[7] et al., based on the PSR model, used data to construct a safety early warning evaluation index system for the ecological carrying capacity of the Chang-Zhu-Tan urban, and calculated the Chang-Zhu-Tan urban agglomeration from 2006 to 2015 using the grey-weighted correlation theory ecological carrying capacity safety index, and the safety alertness of ecological carrying capacity in each year was judged.

2 Overview of the study area

Qinghai Province is located in the northwestern inland of China. It borders Gansu to the northeast, Xinjiang to the west, Tibet to the southwest, and Sichuan to the southeast. It has abundant water resources and is known as the "Water Tower of China". Qinghai Province has many ecological function zones, the most famous ones are the Sanjiangyuan, Qilian Mountains, and Qinghai Lake ecological function zones, which provide important ecological protection for China and neighboring countries. Therefore, the ecological function areas of Qinghai

*Corresponding author's e-mail: jiahf2014@qhu.edu.cn

Province are of great significance in China and even Southeast Asia.

3 Research methods and data

3.1 PSR model

The PSR model was first applied to Environmental Science in 1991 when OCEO used the PSR model to build an environmental assessment index system to achieve the purpose of linking the various indicators[8]. Later generations will continue to perfect this model, and the PSR model has been widely recognized and applied internationally[9]. This paper uses the PSR model to evaluate the ecological carrying capacity of Qinghai Province to link the environment, economy and society. It has the characteristics of more comprehensiveness, objectivity, validity and authenticity, and can provide a more scientific reference for the ecological construction of Qinghai Province.

3.2 Construction of ecological carrying capacity evaluation index system and index selection

Based on the PSR model, the index system is divided into three levels: target level, comprehensive level, and index level, which can more directly reflect the influencing factors of ecological carrying capacity in Qinghai Province. The comprehensive layer is composed of pressure, state and response of the PSR model, and the index layer is composed of 24 selected indexes. The relationship among pressure, state, and response are of great significance to the study of ecological carrying capacity in Qinghai Province. Table 1 shows the evaluation index system of ecological carrying capacity in Qinghai Province.

3.3 Determination of indicator weights

The entropy method is a more objective method to determine the weight of the comprehensive evaluation index. Jia Haifa[10] et al. used the entropy method to determine the index weight when measuring and evaluating the development level of ecological civilization in Qinghai Province from 2008 to 2018. This article will not repeat the calculation process of the entropy method. If you need a detailed calculation process, please contact the author to obtain it. The index weights of this article are given in Table 1.

3.4 Data sources

The data in this article are mainly derived from the 2011-2016 "Qinghai Statistical Yearbook", "China Statistical Yearbook", "Qinghai Province Environmental Status Bulletin" and other materials and documents, as well as the Qinghai Provincial People's Government. Individual missing data are calculated using the more scientific grey forecasting method.

Table1. Evaluation index system of ecological carrying capacity in Qinghai Province

Comprehensive layer	Index layer	Index attribute	Weight
Pressure	Sulfur dioxide emissions (10000 tons)	Negative	0.0533
	Total energy production (10000 tons of standard coal)	Negative	0.0390
	Total energy consumption (10000 tons of standard coal)	Negative	0.0456
	Number of geological disasters (time)	Negative	0.0200
	Urban registered unemployment rate (%)	Negative	0.0510
	Urban sewage discharge (10000 m^3)	Negative	0.0461
	Proportion of primary industry in GDP (%)	Negative	0.0386
State	Contribution rate of secondary industry (%)	Positive	0.0426
	Built area (km^2)	Positive	0.0379
	Forest coverage (%)	Positive	0.0379
	Total cultivated land resources (ha)	Positive	0.0299
	Total water resources (billion cubic meters)	Positive	0.0412
	Total annual precipitation (mm)	Positive	0.0449
	Proportion of tertiary industry in GDP (%)	Positive	0.0411
	Wetland area (10000 ha)	Positive	0.0268

Response	Number of students in higher education (10000)	Positive	0.0447
	Library collection (10000 copies)	Positive	0.0466
	Proportion of social security in fiscal expenditure (%)	Positive	0.0404
	Proportion of education expenditure in financial expenditure (%)	Positive	0.0435
	Proportion of energy conservation and environmental protection expenditure in financial expenditure (%)	Positive	0.0478
	Harmless treatment rate of municipal solid waste (%)	Positive	0.0307
	Urban sewage treatment rate (%)	Positive	0.0561
	Accumulated soil erosion control area (1000 ha)	Positive	0.0556
	Afforestation area of that year (10000 ha)	Positive	0.0388

4 Results and analysis

4.1 Comprehensive evaluation of ecological carrying capacity in Qinghai Province

Based on the PSR model and using the entropy method to calculate the weights of various indicators, it can be concluded that the comprehensive score of Qinghai Province's ecological carrying capacity from 2011 to 2019 shows a trend of first decreasing and then increasing, with an average annual growth rate of 8.8%. Among them, the annual growth rate of the pressure index evaluation score is 7.1%, the annual growth rate of the state index evaluation score is 10.5%, and the annual growth rate of the response index evaluation score is 14.5%. The comprehensive evaluation index of the annual ecological carrying capacity of Qinghai Province and its comprehensive evaluation index is shown in Figure 1.

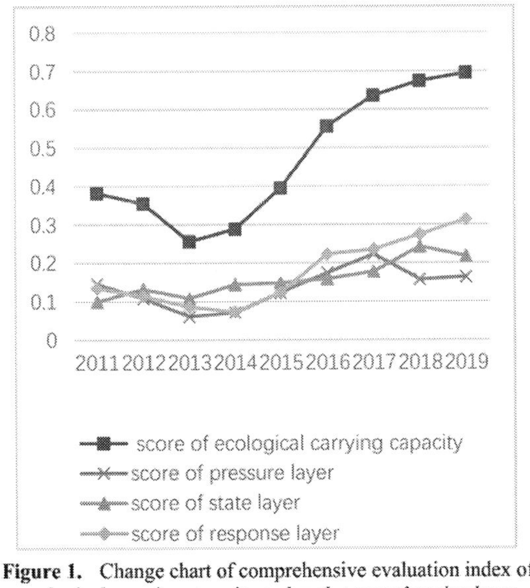

Figure 1. Change chart of comprehensive evaluation index of ecological carrying capacity and each comprehensive layer from 2011 to 2019

4.1.1 Evaluation of pressure indicators

From 2011 to 2019, the ecological pressure of Qinghai Province will be relieved, and the comprehensive score of ecological carrying capacity will show an overall growth trend. Among them, sulfur dioxide emissions, urban registered unemployment rate, and urban sewage emissions accounted for the largest proportion, accounting for 18.2%, 17.4%, and 15.7% respectively. From 2011 to 2019, sulfur dioxide emissions showed a linear reduction, with a cumulative reduction of 67700 tons, with a reduced rate of 43.1% compared with 2011. In 2016, the reduction rate reached 25% compared with the previous year, with a reduction of 37100 tons. This benefited from Qinghai province's efforts to promote the elimination of small coal-fired boilers and regulate the coal business market, to achieve the constraint and control of emission sources. With the acceleration of urbanization, urban population pressure, and urban sewage discharge is growing. After 2015, the discharge of urban sewage has been effectively controlled. Compared with 2015, the discharge of sewage in 2019 will be reduced by 1.48 million cubic meters.

4.1.2 State index evaluation

The comprehensive score of ecological bearing capacity state index in Qinghai province increased in a fluctuating way from 2011 to 2019, among which the annual total precipitation, the contribution rate of secondary industry, and the total water resources accounted for the largest proportion of the state indicators, 14.9%, 14.1%, and 13.6% respectively. Since 2011, the total GDP of Qinghai Province has been increasing, the proportion of the secondary industry in the total GDP is still very high, accounting for more than 66% every year, and the proportion of the tertiary industry is only 0.51% every year. The good phenomenon is that the contribution rate of the secondary industry continues to decline, with an average annual decline of 3.4%. It shows that the government is gradually upgrading and optimizing the industrial structure, which is beneficial to the carrying capacity of

Qinghai Province and the development of social harmony and stability.

4.1.3 Evaluation of response indicators

From 2011 to 2019, a comprehensive evaluation of the ecological bearing capacity response index of Qinghai Province shows a trend of decreasing first and then increasing, with a large growth rate of 132% in 2019 compared with 2011. The proportion of the major ones is the ratio of urban sewage treatment, cumulative soil erosion control area, and energy conservation and environmental protection expenditure, accounting for 13.9%, 13.6%, and 11.8% respectively. The obvious growth trend of response indicators shows that Qinghai provincial government attaches great importance to improving the ecological environment, actively responds to the pressure of ecological bearing capacity, puts forward various policies to protect ecological bearing capacity and implements the policies to the actual situation, and effectively improves the ecological civilization construction of Qinghai Province from the aspects of environmental economy and society.

5 Conclusion and discussion

Based on the PSR model, this paper selects 24 indexes from three angles of economic, environmental, and social, and uses the entropy method to determine the index weight to comprehensively evaluate the ecological carrying capacity of Qinghai Province. The results show that: from 2011 to 2019, the comprehensive evaluation of the ecological carrying capacity of Qinghai Province shows a trend of first decreasing and then increasing, and the later increasing range is large, which indicates that the ecological carrying capacity of Qinghai Province has been significantly improved. Among them, the response index evaluation performance is the most outstanding. Qinghai Province has made a good response to the current situation of ecological carrying capacity and the existing pressure. The state index evaluation shows a fluctuating rise, but the rise is small and slow, which indicates that the current situation of ecological carrying capacity in Qinghai Province is improving, but there are some obstacles; the pressure index evaluation shows a fluctuating change, specifically in the energy production and consumption and urban sewage discharge with the urbanization process is also rising, leading to the increase of ecological pressure. However, the pressure index is relatively stable, which shows that Qinghai Province has to take some measures to affect the ecological carrying capacity.

Acknowledgements

The authors gratefully acknowledge financial support from the Youth Research Fund Project of Qinghai University (Grant Nos. 2020-QSY-12).

References

1. Guo Xiurui, Mao Xianqiang, and Ran Shenghong, Research Progress of Domestic Environmental Carrying Capacity. China Population, Resources and Environment, p. 29-31(2000)
2. F Zhiming and LI Peng, The origin and development of the concept of carrying capacity: a discussion based on the perspective of resources and environment. Journal of Natural Resources, 33(9): p. 1475-1489(2018)
3. Liu Renzhi and Wang Chengwen, Research on Environmental Carrying Capacity Evaluation Technology. Journal of Applied Basic Science and Engineering, 17(S1): p. 92-101(2009)
4. Huang, W., et al. Study on the development and layout mode of oasis small town based on farmers' willingness survey: A case study of Zongzhai Town in China. in IOP Conference Series: Earth and Environmental Science.(2019)
5. Zhang Wenbin, Hu Jian, and Ma Yiming, Evaluation of China's ecological carrying capacity from the perspective of decoupling support and pressure. Economic Geography, 40(02): p. 181-188(2020)
6. Wang Yining, Zhou Qiang, and Wang Haowei, A dynamic study on the three-dimensional ecological footprint of 34 provincial administrative regions in China. Acta Ecologica Sinica, 40(18): p. 6434-6444(2020)
7. Zhu Yulin, Li Mingjie, and Gu Ronghua, Research on the safety early warning of the ecological carrying capacity of the Changzhutan urban agglomeration based on the pressure-state-response model. Resources and Environment in the Yangtze River Basin, 26(12): p. 2057-2064(2017)
8. Huang Jingnan, Ao Ningqian, and Xie Yuhang, Overview and Outlook of International Common Development Index Framework. International Urban Planning, 34(05): p. 94-101(2019)
9. Wang Guoping, et al., Construction of comprehensive disaster risk assessment index system for national parks based on PSR model. Acta Ecologica Sinica, 39(22): p. 8232-8244(2019)
10. Jia Haifa, Shao Lei, and Luo Shan, Comprehensive evaluation of ecological civilization in Qinghai Province based on entropy method and coupling coordination degree model Ecological Economy, 36(11): p. 215-220(2020)

Analysis and Diagnosis of Water Environment Pollution Sources of Tonggu Lake in Fuling District, Chongqing City

Xiaolei Zheng [1,*],Tao Yang[2], Gaoxiang Liu[3],

[1]Chongqing Jianzhu college, Associate professor, Chongqing, China, 400039

[2]Tianjin University, PhD; Chongqing Fuling District Xincheng District Management Committee, Senior engineer, Chongqing,China, 408000

[3]College of Environment and Ecology, Chongqing University, PhD, Chongqing, China, 400045

Abstract. As an important urban lake in Fuling District, Chongqing City, there is water blooms in Tonggu Lake, which affect the ecological environment. This study analyzed and studied the external and internal status of water pollution through site survey, water quality testing and analysis, and laboratory data analysis, and diagnosed the water environment of Tonggu Lake, which is helpful for providing strong data support for the follow-up solution to the problem of water bloom in Tonggu Lake.

1 Research Background

Urban lakes is extremely important for maintaining the safety of water, ecology and environment, promoting ecological balance, and improving the living environment of the people[1]. Tonggu Lake, located in Fuling District, Chongqing City, is an important part of the green barrier construction in Fuling District. However, the water quality of Tonggu Lake is generally poor, the water supply for the lake is scant, and the water replacement cycle is long due to the acceleration of urbanization in recent years, lacking of effective protection of the lake in the development process. At the same time, due to the input of mustard waste water and non-point source pollution, the water body of Tonggu Lake has turned black, and the growth potential of algae is large, and water blooms are common in summer. In order to improve the water environment management of Tonggu Lake, actively rectify the lake eutrophication and river black and odor problems, promote the improvement of the ecological environment, the research is specially conducted to provide reference for other urban lake pollution management.

2 Analysis of current situation of water environment in Tonggu Lake

Tonggu Lake, as a typical urban lake, there is a problem of poor fluidity due to closed water body in general.

Figure 1. Aerial view of Tonggu Lake

2.1 Current status of external source pollution

The main sources of replenishment for the Tonggu Lake are the Zhujiagou Reservoir in upstream area and natural precipitation, including runoff from farmland. The Zhujiagou effluent is at Level A Standard (TP 0.15 mg/L; TN 1.13 mg/L). When incoming water pipeline from Zhujiaku is damaged, for the external sewage and rainwater or initial rainwater on roads that have not been intercepted, as well as the farmland runoff, the nutrient elements such as N, P, etc. carried by them are destroyed by the ecosystem, and the hydro-fluctuation belt that loses the function of plant interception enters Tonggu Lake, which form the external source pollution to Tonggu Lake.

In order to investigate the impact of external source pollution on the water environment quality of Tonggu Lake, the surface water quality of different waters in Tonggu Lake was sampled and tested at different points mainly for the content of TP, COD, TN, NH4+-N, NO3-N and Chla[2]. According to the water quality detection and analysis of different sampling points, and according to the spatial distribution of COD, TP, TN and other

* Corresponding author: xiaoleitaotao@126.com

 https://doi.org/10.1051/e3sconf/202129301018

indicators under the influence of external water replenishment, it is found that the COD gradually decreases along the water inlet direction. For external water supply that flows from Zhujiagou Reservoir through green land and farmland into Tonggu Lake, Non-point source pollution and runoff entangled by its agricultural activities are the main reasons for the higher TP content in Tonggu Lake. And it leads to the phenomenon that the TN, COD, NH4+-N, and Chla values diffuse into the lake along with the inlet, and the pollutants are high at the inlet end and low in the middle of the lake. The Chla value in the surface water is much higher than the overlying water, which shows that external source pollution is the main source of eutrophication of Tonggu Lake.

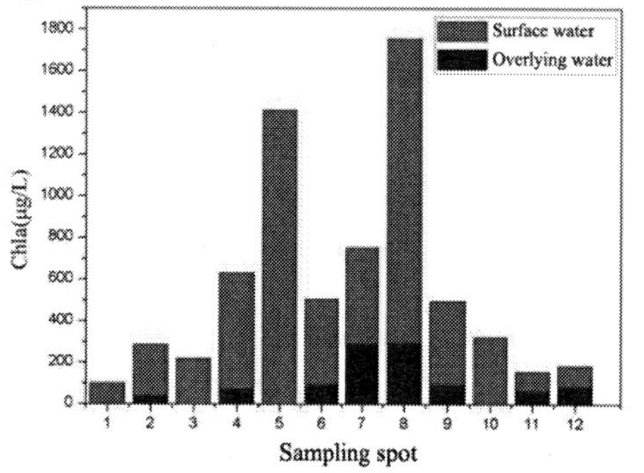

Figure 2. Chla content in the surface water and the overlying water

The external source pollution of Tonggu Lake is located around the lake and in the exposed and fluctuating zones around the lake that are prone to seasonal and time fluctuations in the water level of the lake. Through on-site surveys, most of the vegetation is missing, and there are large-scale hydro-fluctuation belt around the waters of the Tonggu Lake wetland, which greatly reduces the ability to intercept external non-point source pollution.

In general, the spatial distribution of water quality in Tonggu Lake is uneven, which may be related to the influx and diffusion of external water. The surface water quality at the mouth of the lake has deteriorated at the most, and the water quality in the center of the lake is less affected by external pollution. At the same time, higher nutrient levels indicate that the Tonggu Lake is seriously polluted, and farmland runoff and external water supply have a greater impact on the water quality of the lake area. There is a large area of hydro-fluctuation belt around the water area, which greatly reduces the interception capacity of non-point source pollution. The Chla content of surface water is much higher than that of overlying water, and both exceed the standard for eutrophication. Therefore, corresponding measures need to be taken to reduce the impact of external pollution on the water quality of Tonggu Lake.

2.2 Current status of internal source pollution

Internal pollution is the main pollution source of rivers and lakes. Long-term external input and the deposition of biological residues have led to the deposition of a large amount of organic matter and nitrogen-containing substances at the bottom of the river. At the same time, the release of nutrients from the bottom mud plays a very important role in the reproduction and growth of phytoplankton[3][4]. The exchange of nutrients among the sediment, interstitial water and overlying water has an important impact on the nutrient load and the formation of eutrophication in lake water.

Tonggu Lake has a low water flow speed, slow water renewal, and high chlorophyll a content. Therefore, in order to understand the temporal and spatial distribution of nutrients among surface water, overlying water, and interstitial water, the sediments at different monitoring points are sampled and monitored and the situation of internal source pollution is analyzed. Through testing of samples in laboratory, it was found that the nutrient salt in Tonggu Lake showed a trend of interstitial water> overlying water> surface water. The release potential of nutrients in sediment is large, and the TN and ammonia nitrogen in interstitial water are significantly higher than the surface water. Among them, the relatively higher content of NH4+-N will be toxic to the fish in the lake. The measurement found that the content of NH4+-N in interstitial water in some waters is much greater than that of surface water and overlying water. The proportion of NO3-N content in this water has dropped significantly at this point, indicating that the content of organic matter in the sediments of the waters is relatively high with a trend of spreading to interstitial water continuously.

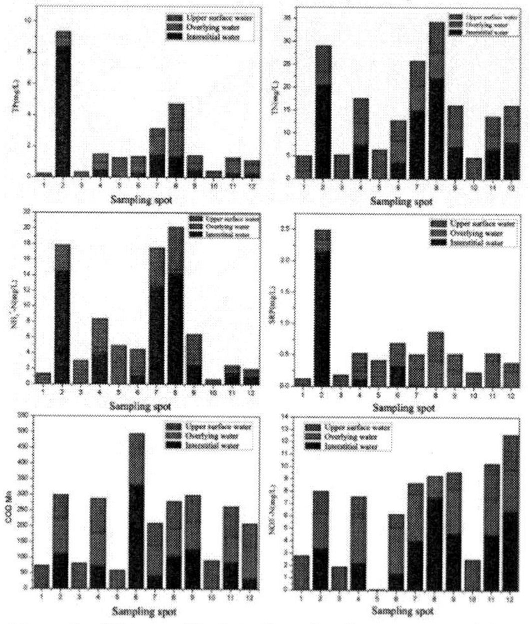

Figure 3. Water quality detection of surface water, overlying water and interstitial water

Through data analysis, the TP concentration of the interstitial water along the depth is greater than 4.5 mg/L, but there is no increasing trend along the depth, which is higher than the Class V standard of surface water (TP 0.2 mg/L). TN has an obvious increasing trend along the depth of the water body, and the maximum TN value can reach 45 mg/L within 35 cm of the surface layer, which is higher than the Class V standard of surface water (TN 2.0 mg/L). In addition, sediments also affect the content of heavy metals in lake water bodies, and the copper, chromium, and zinc are detected. The content of copper and lead increases along the depth of the water body, the content of chromium fluctuates along the depth of the water body, and the content of zinc decreases along the depth of the water body.

In summary: combined with the analysis of the sediments in Tonggu Lake, we can know from the water quality data that the contents of TP, TN, NH4+-N, NO3--N, COD and SRP are as follows: interstitial water> overlying water> surface water, of which TN and NH4+-N content in interstitial water is significantly higher than that of the surface layer, and the data in front of the dam is significantly higher than other sites. The release potential of nutrients in sediment is huge. In terms of the design of water environment treatment and water quality improvement project for Tonggu Lake, treatment measures should be proposed for internal pollution.

3 Diagnosis of water environment problems

3.1 Diagnosis of eutrophication

Through on-site sampling of Tonggu Lake, the water quality monitoring indicators of Tonggu Lake include COD, TP, TN, NH4+-N, NO3-N and Chla. The water quality results of the surface water at each sampling point are shown in Figure 4.

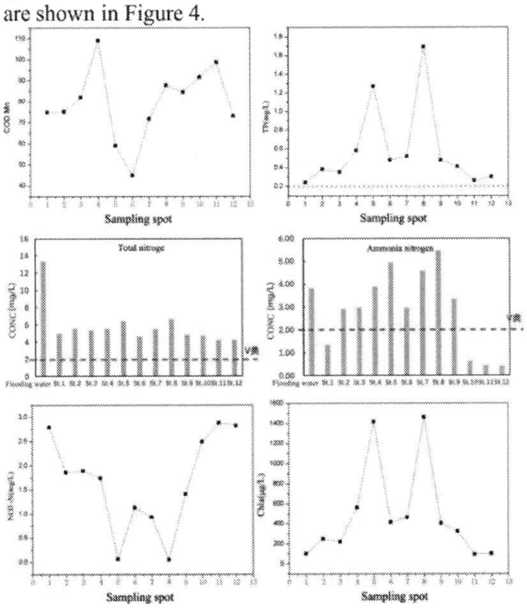

Figure 4. Surface water quality results of the water quality monitoring sampling points of Tonggu Lake

In the water quality indicators of Tonggu Lake, the concentration of TP is in the range form 0.22 mg/L to 1.70 mg/L, which is 1-9 times higher than the standard of surface water (0.2 mg/L); the concentration of TN is in the range of 4.0-13.5 mg/L, which is 2-7 times higher than the standard of surface water (2.0 mg/L); the concentration of NH4+-N ranges from 0.3 to 5.4 mg/L, and the concentration fluctuates significantly with spatial changes. The degree of conversion of nitrogen-containing organic matter is higher; the concentration range of chlorophyll content (Chla) in water is 80-150 μg/L, which far exceeds the eutrophication standard (Chla >10 μg/L).

The main phytoplankton in the water body are: colony Microcystis, Chlorella, Crescentia, and Pegasus, and there is the potential threat of algal blooms.

According to GB3838-2002 《Environmental Quality Standards for Surface Water 》[5] and the water environmental assessment of Tonggu Lake, the water quality of it is generally poor, and the water quality indicators of each sampling point are lower than the Class V standard. Among them, the surface water TP (total phosphorus) of the whole lake exceeds the water quality of surface water Class V standard by 1-9 times, and TN (total nitrogen) exceeds 2-7 times. TP is the priority pollutant of Tonggu Lake, and the Chla content far exceeds the eutrophication standard. There is a higher possibility of algal blooms. Through the monitoring and analysis of the water quality of Tonggu Lake, the lowest Chla content in the sample points was 92μg/L and the highest was 1463μg/L. It can be seen that the water quality of Tonggu Lake far exceeds the degree of eutrophication. Moreover, the dominant algae in Tonggu Lake are the Microcystis and Chlorella, which pose a huge potential threat of severe algal blooms.

3.2 Analysis of pollution structure

The pollution load of Tonggu Lake mainly comes from external pollution (external water replenishment and agricultural runoff) and internal pollution (the release of nitrogen and phosphorus from the sediment). Through the establishment of the total phosphorus dynamic model in Tonggu Lake and model calculations, a preliminary assessment of the external and internal pollution of the Tonggu Lake wetland was carried out. The results show that the total phosphorus emission from point sources, farmland non-point sources, etc. is 61.0 mg phosphorus/ square meter·day, and the total phosphorus emissions from sediment release and other internal sources are 13.5 mg phosphorus/ square meter·day. The impact of external pollution on the water quality of Tonggu Lake is greater than that of internal sediment release. Therefore, measures should be taken to control and reduce external pollution. We can further calculate through that: if the total phosphorus treatment emission from external sources reaches Class A standard, the water quality of Tonggu Lake can basically eliminate the inferior Class V (TP: 0.19 mg/L); if both internal and external pollution are controlled, the water quality of Tonggu Lake can be maintained at Class IV-V (TP: 0.15 mg/L).

According to the sampling and monitoring survey of external and internal pollution, the pollution caused by the sediment in the internal pollution is higher at the estuary of the Tonggu Lake lake than the estuary, and the the pollution caused by the sediment in the internal pollution is higher at the estuary of the Tonggu Lake lake than the estuary. This requires targeted treatment in the environmental improvement design of Tonggu Lake in the later stages of the environment aiming at the difference in pollution caused by external pollution and internal pollution in the geographical space of Tonggu Lake.

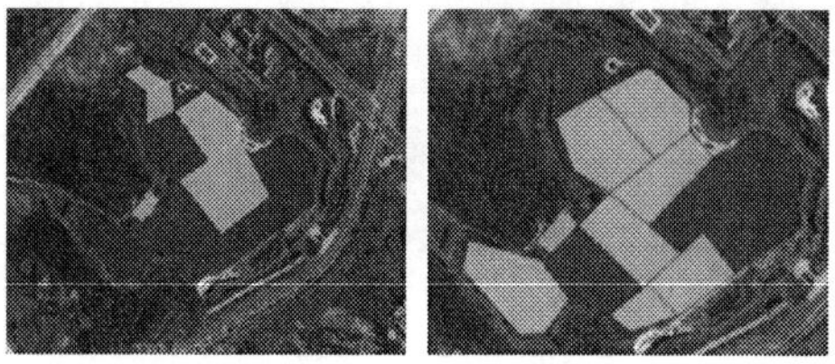

Figure 5. Zoning of external and internal pollution of Tonggu Lake

4 Conclusion

The water bloom phenomenon in Tonggu Lake has both external and internal causes. This study made a specific diagnosis. It is necessary to combine pollution characteristics of Tonggu Lake, and adopt external pollution control, internal pollution control, water quality improvement, water ecosystem construction and emergency measures to improve the water environment of Tonggu Lake in general.

Acknowledgments:

1. Fund Project: "Research on the Application of Prefabricated integrated pipe gallery in engineering" Science and Technology Research project of Chongqing Education Commission in 2018（KJQN201804307）

References

1. Mei Cai, Yuanyuan Wang, Lili Gong, Haiyang Xu. Thoughts and suggestisons on a new round of comprehensive management of water environment in Taihu Lake Basin［J］. Water conservancy planning and design,2021.2.
2. Yang Li. Analysis of water environment in Panjing River Basin and Research on Countermeasures [J]. Jouranl of Green Sicence and Technology, Volume 23 issue 2, 2021
3. Xinyue Pang. Ideas and measures of water environment control in Bidao construction of Guangzhou City[J]. Guangdong Water resource and hydropower,Phase 2, 2021
4. Jing Zhang, Hua Zheng, Mengnan He, Jinfeng Ma. Simulation of water environmental pollution and identification of key source areas in River Basin-A case study of Yiwu River Basin［J］. Journal of environmental engineering,Feb 2021.
5. General Administration of quality supervision, inspection and Quarantine of the people's Republic of China GB3838-2002 Environmental Quality Standards for Surface Water [S]. China Environmental Science Press, 4,2002

Synthesis of free-metal composites g-CN/GQDs with enhanced photocatalytic degradation of tetracycline

Kai Zheng[1], Jianan Chen[1], Xue Gao[1], Xiaoqing Cao[1], Shan Wu[1], and Jixin Su[1*]

[1] School of Environmental Science and Engineering, Shandong University, China

Abstract. Developing inexpensive and environmentally friendly no-metal photocatalyst is a promising strategy to addressing environmental issues. In this work, the metal-free catalyst g-CN/GQDs was synthesized through one-pot co-sintering method. The as-prepared photocatalysts showed enhanced photocatalytic activity for tetracycline (TC) degradation. Especially, the g-CN/GQDs-5 exhibited a nearly 30% improvement in TC degradation within 60 min of visible light irradiation compared to g-CN, and the kinetic constant was 1.8 times that of g-CN. The enhanced performance was attributed to the introduction of GQDs enhance the visible light absorption ability and suppress the recombination of photo-generated charges of g-CN. The capture experiment of active species showed that superoxide radicals played a major role in the degradation of TC in this system.

1 Introduction

Antibiotics, as an antibacterial medicine, have been widely applied to control infection diseases in humans and animals[1]. Particularly, tetracycline (TC), a class of broad-spectrum antibiotics with a parent nucleus of pentane, has attracted great attention in the last few decades owing to its wide range of application, hardly mineralization, and potential threats to water ecosystem and terrestrial[2-5]. Unfortunately, TC cannot be effectively removed by traditional treatment technology due to its chemical stability, good solubility, and non-biodegradability. Thus, it is meaningful to explore a proper way to sweep them. Semiconductor photocatalysis has been deemed as a possible and promising method to solve universal environmental predicaments, owing to the mild reaction conditions, low cost, and the stability of physicochemical properties[6-8].

Graphitic carbon nitride (g-CN), a metal-free visible responsive polymer photomaterial, has been widely studied in the removal of various contaminants, hydrogen production, and chemical synthesis, due to it has reliable chemical inertness, excellent thermal stability, proper band-gap and special visible absorption capacity[9-11]. However, the photocatalytic performance of g-CN are restricted by the fast recombination of photo-generated carriers and the poor light absorption capacity[12-14]. Hence, many methods including morphology control, the preparation of g-CN based heterojunction, deposition of noble metal and modified with carbon materials have been applied to enhance the photocatalytic performance of g-CN[15-17]. Carbon materials, as an earth-abundant resource, have been extensively explored to modified with g-CN. Among the many carbon materials, graphene quantum dots (GQDs) are a novel 0D fluorescent carbon nanomaterials that have <10nm lateral size and few-layer structure[18-20]. In the process of photocatalysis, GQDs can serve as an emulsion and an electron accepter to expand the range of visible absorption and reduce the recombination of photo-generated charges[21-23].

In this work, GQDs were prepared from 1,3,6-trinitropyrene by hydrothermal treatment, and the samples g-CN/GQDs-X was prepared through a one-pot co-sintering the mixture of GQDs and melamine. After modified with GQDs, the g-CN show more intensive visible light absorption and better transfer ability of photo-generated electrons. The prepared materials were characterized, and antibiotics TC was served as a model contaminant to test the photocatalytic ability. Meanwhile, the active species of TC degradation process were evaluated and a possible photocatalytic degradation mechanism were proposed.

2 Experimental

2.1 Synthesis of the catalysts

The GQDs were prepared from 1,3,6-trinitropyrene by hydrothermal treatment in accordance with previous reports[24]. The g-CN/GQDs samples was obtained by calcined a mixture of melamine and GQDs. Typically, 3g melamine was added to 50mL of DI water. Then, a certain amount of GQDs solution (10mg/mL) was dropped into the melamine suspension, and vigorously stirred 2h, then vacuum dried at 80°C to dewater. After that, the mixture was transferred to a covered ceramic crucible, calcined at 520°C in a muffle furnace for 2h with a heat-up speed of 5°C min^{-1}. The resultant samples were collected and

* Corresponding author: jxsu@sdu.edu.cn

milled into a fine powder. The final catalysts were named as g-CN/GQDs-X, where X represents the volume of GQDs solution added. The pristine g-CN was prepared by the same process without adding GQDs.

2.2 Characterization

Powder X-ray diffraction (XRD) patterns of as-prepared powders were analysed by Ultima IV X-Ray. The functional groups of prepared materials were collected through Fourier transform infrared spectrophotometer (FT-IR, Perkin-Elmer Spectrum BX00 spectrometer). The spectrophotometer (Shimadzu UV-2550) equipped with an integrating sphere accessory was used to obtain the diffuse reflectance spectra of the samples using $BaSO_4$ as a reference. Photoluminescence (PL) spectra of the samples were collected by using Edinburgh FL/FS900 spectrophotometer. Photocurrent response experiments were recorded at CHI 660E electrochemical workstation. Pt plate and Ag/AgCl electrode were used as counter electrode and reference electrode, respectively. And the synthesized samples dispersed on 1.5cm×1.5cm Indium tin oxide (ITO) glass were used as working electrode.

2.3 Photodegradation experiment

The photodegradation activity of the synthetic materials were study by monitoring the degradation of a TC solution upon irradiation with visible light (>400nm). The light source is a 300W Xenon lamp (CEL-HXF300) equipped with a 400nm cut off filter. Briefly, 50mg samples were added into the reaction breaker including 100ml TC solution (10mg/L). Before the photocatalytic, the samples were stirred in the dark for 30min to ensure the adsorption-desorption equilibrium. The reaction TC suspension (3mL) was taken out every 10min and filtered using a 0.45μm membrane filter, then analysed by a Shimadzu UV-2550 spectrophotometer to monitor the concentration of TC. After use, the samples were collected and washed several times with ultrapure water and ethanol for next experiment, and Four successive cycles of tetracycline degradation experiments were carried out.

3 Results and discussion

X-ray diffraction (XRD) patterns of the prepared materials were exhibited in Fig.1. GQDs show a clearly (002) signal at 26.4° corresponding to the typical XRD pattern of the pristine graphite[24]. In addition, it can be found that there are two peaks at $2\theta = 13.4°$ and 27.4° displayed in g-CN and g-CN/GQDs-X. The peak at 13.4° could be assigned to (100) diffraction planes of g-CN, attributed to the interplanar structural packing[25]. The peak at 27.4° corresponds to the (002) diffraction plane of g-CN, due to the stacking of the conjugated aromatic system[26]. However, no characteristic peak of GQDs was found in the g-CN/GQDs-X composite materials, which may due to the low GQDs content[27].

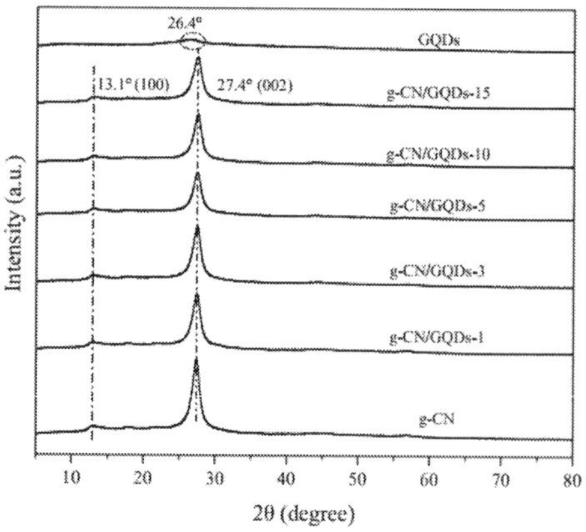

Figure 1. XRD of the prepared samples.

Fig.2 presents the FT-IR spectra of the synthetic materials. The GQDs contains abundant surface groups, among which the peak at 3400cm^{-1} corresponds to the stretching vibration of -OH group, and there is also vibration peak of C=C, C-H and C-OH. All the g-CN/GQDs-X samples showed some characteristic peaks at around 807 cm^{-1}, 1200~1600 cm^{-1} and 3210 cm^{-1}, respectively. The peak at 807 cm^{-1} reflected the breathing mode of the tri-s-triazine ring system, the peaks at 1200~1600 cm^{-1} was assigned to the characteristic stretching vibration modes of CN and C-N heterocycles, and the peak at 3210 cm^{-1} corresponds to the terminal N-H_2 or N-H stretching vibration[28-30]. Whereas, g-CN and g-CN/GQDs-X showed same FT-IR spectra, indicating that the addition of the GQDs has no influence on the functional groups of g-CN[27, 28].

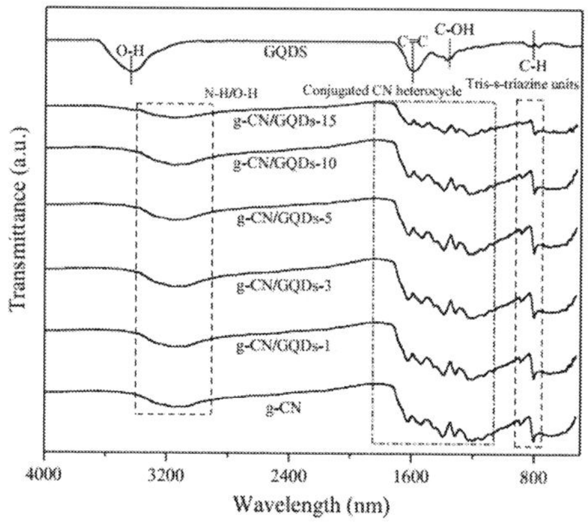

Figure 2. FT-IR of the prepared samples.

The morphologies of as-synthesized photocatalysts were characterized via TEM analyses. As show in Fig.3(a, b), the GQDs are dispersed uniformly and with a lateral size about 2~5nm. It can be discovered that the obvious crystalline lattices of GQDs in the high-resolution TEM picture, suggesting the GQDs has high crystallinity[24]. Fig.3(c) indicated that the g-CN was composed of stacked nanosized sheets. Compared with g-CN, the g-CN/GQDs-5 shows more broken and smaller lamellas (Fig.3(b)). Interestingly, no specific GQDs appearance was found in the electron microscope image of g-CN/GQDs-5, which may due to the low addition of GQDs and uniform distribution in the g-CN/GQDs-5[17, 25, 31].

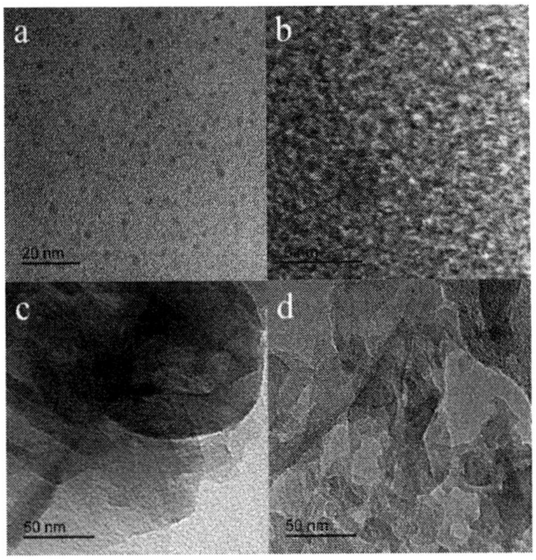

Figure 3. TEM images of (a, b) GQDs, (c) g-CN and (d) g-CN/GQDs-5.

The optical absorption performance of prepared samples was collected through UV–Vis diffuse reflectance spectroscopy. It can be seen from Fig.4, both g-CN and g-CN/GQDs-X samples present excellent absorption at the wavelength range of 200~450 nm. It is worth mentioning that the absorption property changes of g-CN/GQDs-X shows a slightly red shift. Besides, all the g-CN/GQDs-X samples have higher optical absorption in the full wavelength compared with g-CN, suggesting that the prepared g-CN/GQDs-X exhibit a better visible light utilization performance than that of g-CN, which may more conducive to photocatalysis process.

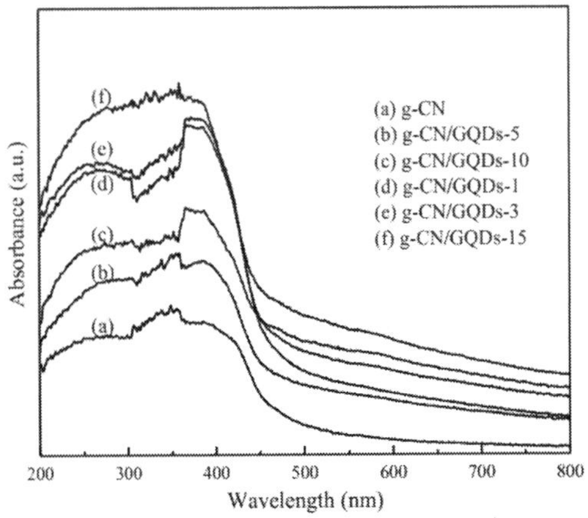

Figure 4. UV-Vis absorption spectra of the prepared samples.

Fig.5(a) shows the PL spectra of g-CN and g-CN/GQDs-5 samples at 340 nm excitation wavelength. The main peak of g-CN and g-CN/GQDs-5 is at ~466nm and the g-CN/GQDs-5 sample show much low PL intensity than g-CN, indicating that the latter possess the lower recombination rate of the charges. As show in Fig.5(b), the photocurrent density of g-CN/GQDs-5 was evidently higher than that of g-CN indicating the better photo-generated carrier separation ability.

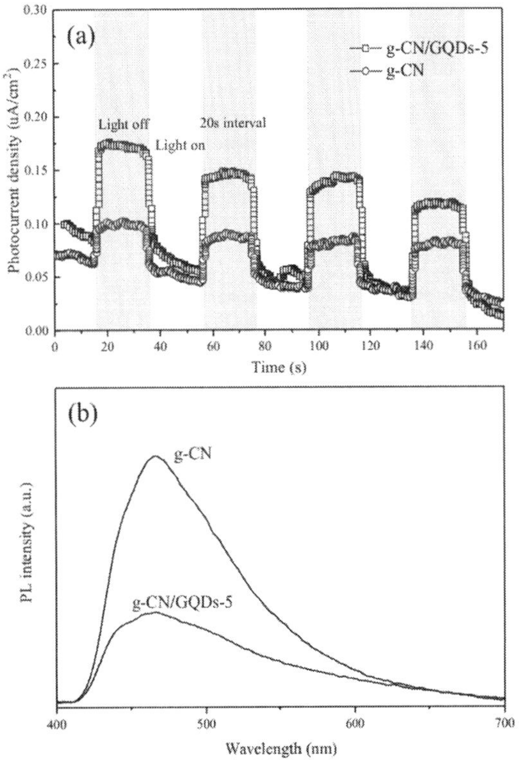

Figure 5. (a) The photocurrent response and (b) PL spectra of g-CN and g-CN/GQDs-5.

To investigate the photocatalysis performance of the prepared samples, the tetracycline (TC), a typical antibiotic, was choose as the targeted contaminant. It can be observed that all the samples were absorbed very little antibiotic under dark condition. In addition, there was almost no decomposition of TC under visible light irradiation without adding photocatalysts, suggesting the refractory nature of TC under visible light irradiation. After adding the as-prepared samples, the degradation rate of TC was significantly enhanced. As show in Fig.6(a), all the catalysts with GQDs modified show much better degradation rate of TC than g-CN (51%), which may be attributed to the introduction of GQDs into g-CN enhanced the visible light utilization performance and improved the separation of h^+ and e^-. As expected, the TC degradation rate of the optimal catalyst g-CN/GQDs-5 reached 78%, which had a nearly 30% improvement than that of g-CN. The pseudo-first-order models was used to study the kinetics of the photocatalytic TC degradation by prepared samples, the formula as follow:

$$\ln\frac{C_0}{C} = kt \quad (1)$$

where t represents the reaction time, C_0 and C represent the TC concentrations at 0 min and t min, k represents the reaction rate constant of pseudo-first-order kinetics. As

https://doi.org/10.1051/e3sconf/202129301019

show in Fig.6(b), the photocatalytic degradation curves of both g-CN and g-CN/GQDs-X follow pseudo-first-order kinetics (R^2>0.98), and the highest kinetic constants of g-CN/GQDs-5 were estimated to be 0.02427 min^{-1}, which were 1.8 times higher than that of g-CN (0.01644 min^{-1}).

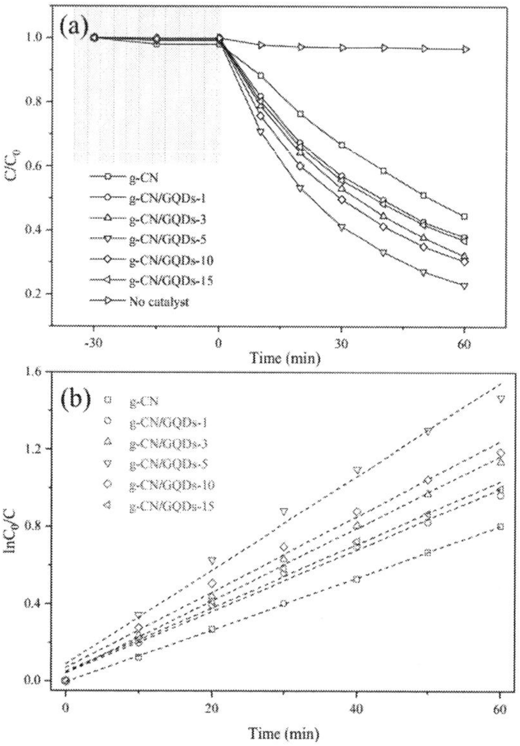

Figure 6. (a) Photocatalytic degradation performance and (b) pseudo-first-order kinetic fitting curves.

In the practical applications of photocatalyst, reusable performance is also crucially important. The g-CN/GQDs-5 composites of the TC degradation were reused four times under the same reaction conditions. As illustrated in Fig.7(a), the photocatalytic activity of g-CN/GQDs-5 does not show any significant decrease after four cycles, indicating that it has excellent stability and recyclability. To better understanding the impact of active species in the photo-degradation of TC, the scavengers EDTA, benzoquinone (BQ) or isopropanol (IPA) was respectively injected to the g-CN/GQDs-5 reaction system to trap the h^+, $\cdot O^{2-}$ and ·OH. From Fig.7(b), the degrading efficiency of g-CN/GQDs-5 was drastically reduced after adding BQ, suggesting the major impact of $\cdot O^{2-}$ in the photo-degradation of TC in g-CN/GQDs-5 system. It is worthy to point out that, IPA and EDTA shows little effect on the TC degradation, suggesting that ·OH and h^+ radicals also contribute to the degradation of TC.

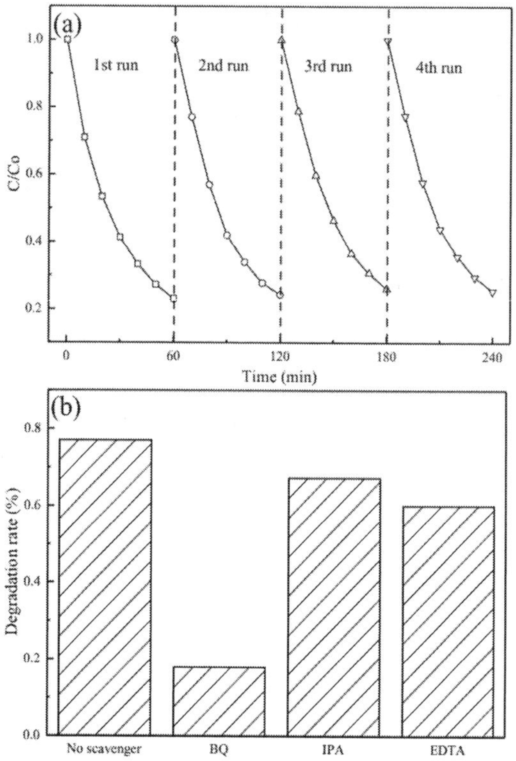

Figure 7. (a) Recycling performance and (b) activity capture experiment of g-CN/GQDs-5 sample.

Based on the experimental results, a possible mechanism of g-CN/GQDs-5 system is possessed. When the g-CN were exposed to visible light, electrons get excited and move to the conduction band, but a relatively large part of photo-generated electrons have been consumed on account of the rapid recombination of photo-generated e^- and h^+ pair. However, GQDs can act as an emulsion and an electron accepter to enhance light absorption capacity and reduce the photo-generated electron-hole pair recombination in the g-CN/GQDs-5 system, as evidenced by UV–Vis absorption spectra and PL spectra. In the photocatalytic process of g-CN/GQDs-5, the electrons of g-CN/GQDs-5 composite was excited from valence band to conduction band after exposed to visible light, and subsequently transfer to the tightly coupled GQDs rapidly. The electrons would react with the dissolved O_2 to generate $\cdot O^{2-}$ radicals, which played a major impact in the degradation of TC. Meanwhile, the $\cdot O^{2-}$ could reacted with H_2O to form H_2O_2 and further conversion to hydroxyl radicals to participated in the TC degradation process. Besides, the h^+ produced of g-CN/GQDs-5 also made contribution to decomposing TC into small molecules.

4 Conclusion

In this work, GQDs were prepared hydrothermal reaction of 1,3,6-trinitropyrene and g-CN/GQDs-X composites were prepared by calcined a mixture of melamine and GQDs. The prepared g-CN/GQDs-X show enhanced photo-degradation tetracycline activity as compared with the g-CN. The enhanced photocatalytic performance of g-CN/GQDs is mainly related to the rapid transfer and separation of photo-generated carriers, and the enhance of optical absorption capacity. This work provides a new way for the preparing of high efficiency carbon nitride-based photocatalyst by modified with quantum dots and for its application in environmental treatment.

References

1. Q. Q. Zhang, G. G. Ying, C. G. Pan, Y. S. Liu and J. L. Zhao, Environ Sci Technol **49** (11), 6772-6782 (2015).
2. S. Li and J. Y. Hu, J Hazard Mater **318**, 134-144 (2016).
3. S. Li, W. Z. Shi, W. Liu, H. M. Li, W. Zhang, J. R. Hu, Y. C. Ke, W. L. Sun and J. R. Ni, Sci Total Environ **615**, 906-917 (2018).
4. C. Reyes, J. Fernandez, J. Freer, M. A. Mondaca, C. Zaror, S. Malato and H. D. Mansilla, J Photoch Photobio A **184** (1-2), 141-146 (2006).
5. M. Ahmadi, H. R. Motlagh, N. Jaafarzadeh, A. Mostoufi, R. Saeedi, G. Barzegar and S. Jorfi, J Environ Manage **186**, 55-63 (2017).
6. S. Zhang, P. C. Gu, R. Ma, C. T. Luo, T. Wen, G. X. Zhao, W. C. Cheng and X. K. Wang, Catal Today **335**, 65-77 (2019).
7. Y. Zhang, M. J. Zhang, L. Tang, J. J. Wang, Y. Zhu, C. Y. Feng, Y. C. Deng, W. Z. He and Y. Hu, Chem Eng J **391** (2020).
8. P. P. Tun, J. T. Wang, T. T. Khaing, X. Y. Wu and G. K. Zhang, J Alloy Compd **818** (2020).
9. T. B. Nguyen, C. P. Huang, R. A. Doong, C. W. Chen and C. D. Dong, Chem Eng J **384** (2020).
10. M. Bellardita, E. I. Garcia-Lopez, G. Marci, I. Krivtsov, J. R. Garcia and L. Palmisano, Appl Catal B-Environ **220**, 222-233 (2018).
11. Z. H. Li, D. L. Huang, C. Y. Zhou, W. J. Xue, L. Lei, R. Deng, Y. Yang, S. Chen, W. J. Wang and Z. W. Wang, Chem Eng J **382** (2020).
12. D. L. Huang, Z. H. Li, G. M. Zeng, C. Y. Zhou, W. J. Xue, X. M. Gong, X. L. Yan, S. Chen, W. J. Wang and M. Cheng, Appl Catal B-Environ **240**, 153-173 (2019).
13. J. F. Zhang, J. W. Fu, Z. L. Wang, B. Cheng, K. Dai and W. K. Ho, J Alloy Compd **766**, 841-850 (2018).
14. D. H. He, C. Zhang, G. M. Zeng, Y. Yang, D. L. Huang, L. L. Wang and H. Wang, Appl Catal B-Environ **258** (2019).
15. P. Raizada, A. Sudhaik, P. Singh, P. Shandilya, V. K. Gupta, A. Hosseini-Bandegharaei and S. Agrawal, J Photoch Photobio A **374**, 22-35 (2019).
16. F. T. Yi, H. H. Gan, H. F. Jin, W. Y. Zhao, K. F. Zhang, H. X. Jin, H. N. Zhang, Y. X. Qian and J. Q. Ma, Sep Purif Technol **233** (2020).
17. B. T. Huang, J. B. He, S. Y. Bian, C. J. Zhou, Z. Y. Li, F. N. Xi, J. Y. Liu and X. P. Dong, Chinese Chem Lett **29** (11), 1698-1701 (2018).
18. X. T. Zheng, A. Ananthanarayanan, K. Q. Luo and P. Chen, Small **11** (14), 1620-1636 (2015).
19. S. Fang, Y. Xia, K. L. Lv, Q. Li, J. Sun and M. Li, Appl Catal B-Environ **185**, 225-232 (2016).
20. A. Ananthanarayanan, X. W. Wang, P. Routh, B. Sana, S. Lim, D. H. Kim, K. H. Lim, J. Li and P. Chen, Adv Funct Mater **24** (20), 3021-3026 (2014).
21. S. Kim, Y. J. Song and M. J. Heller, Adv Mater **29** (31) (2017).
22. L. Ai, R. Shi, J. Yang, K. Zhang, T. R. Zhang and S. Y. Lu, Small **2007523** (2021).
23. B. Li, W. Peng, J. Zhang, J. C. Lian, T. Huang, N. Cheng, Z. Y. Luo, W. Q. Huang, W. Y. Hu, A. L. Pan, L. Jiang and G. F. Huang, Adv Funct Mater **2100816** (2021).
24. L. Wang, Y. L. Wang, T. Xu, H. B. Liao, C. J. Yao, Y. Liu, Z. Li, Z. W. Chen, D. Y. Pan, L. T. Sun and M. H. Wu, Nat Commun **5** (2014).
25. A. L. Yuan, H. Lei, F. N. Xi, J. Y. Liu, L. S. Qin, Z. Chen and X. P. Dong, J Colloid Interf Sci **548**, 56-65 (2019).
26. C. Y. Zhou, P. Xu, C. Lai, C. Zhang, G. M. Zeng, D. L. Huang, M. Cheng, L. Hu, W. P. Xiong, X. F. Wen, L. Qin, J. L. Yuan and W. J. Wang, Chem Eng J **359**, 186-196 (2019).
27. W. Liu, Y. Y. Li, F. Y. Liu, W. Jiang, D. D. Zhang and J. L. Liang, Water Res **151**, 8-19 (2019).
28. C. Zhao, Z. Z. Liao, W. Liu, F. Y. Liu, J. Y. Ye, J. L. Liang and Y. Y. Li, J Hazard Mater **381** (2020).
29. F. Wei, J. D. Li, C. C. Dong, Y. J. Bi and X. J. Han, Chemosphere **242** (2020).
30. S. Z. Liu, H. Q. Sun, K. O'Donnell, H. M. Ang, M. O. Tade and S. B. Wang, J Colloid Interf Sci **464**, 10-17 (2016).
31. H. Zhang, L. X. Zhao, F. L. Geng, L. H. Guo, B. Wan and Y. Yang, Appl Catal B-Environ **180**, 656-662 (2016).

Evaluating pollution degree of oil spills on the sea surface

ZHOU Xuan[1*], KOU Jie[2], CHAO Hao[3] and WANG Taiyang[4]

[1]BaiLie School of Petroleum Engineering, Lanzhou City University, Lanzhou, China

[2]Shandong Key Laboratory of Oil & Gas Storage and Transportation Safety, China University of Petroleum (East China), Qingdao, China

[3]China Resources (Zhengzhou) Municipal Engineering Design & Research Insititute Co.Ltd, Zhengzhou, China

[4]Engineering Dept., East China Engineering Science and Technology Co.Ltd, Hefei, China

Abstract. This article analyses influencing factors of pollution degree of oil spills on the sea surface and establishes corresponding evaluation system, thus combines fuzzy comprehensive evaluation method and analytic hierarchy process to evaluate pollution degree of oil spill accidents in the Bohai Sea. Based on the evaluation system, oil spill accidents in Penglai 19-3 oilfield and Suizhong 36-1 oilfield central platform are classified as serious pollution and light pollution respectively, which is consistent with relevant institutions and scholars, proving the rationality of the evaluation system and parameter selected.

1 Introduction

In recent years, there have been several oil spill accidents in the Bohai Sea [1-5]. Oil spills into seawater, then drifts and diffuses under the influence of environmental dynamic fields, which causes pollution to the marine [6-10]. There are some differences among pollution degree of oil spill accidents to some extent [11-13]. Evaluation on pollution degree affects emergency response, hazard assessment and loss compensation [14]. Therefore, classification of pollution degree of oil spills on the sea surface is of significance to determine the corresponding emergency and respond to oil spill hazards efficiently and scientifically [15].

This paper applies a multi-level fuzzy comprehensive evaluation method to establishing pollution evaluation system of oil spills on the sea surface in section Evaluation System of Oil Spill Pollution. Section Membership of Evaluation Indexes elaborates membership degree of each evaluation index. Analytic hierarchy process is used to determine weight of each index in Section Weights of Evaluation Indexes. Pollution degree of oil spills on the sea surface is evaluated in Section Multi-level Fuzzy Comprehensive Evaluation and pollution degree of oil spill accidents in Penglai 19-3 oilfield and in Suizhong 36-1 oilfield is analysed in Section Case Analysis of Oil Spills on the Sea Surface.

2 Evaluation System of Oil Spill Pollution

At present, various methods [16-18] are widely used to evaluate pollution degree of oil spills on the sea surface, comprehensive evaluation method is adopted in the paper since it takes multiple effects of indicators into account synthetically [19].

Pollution degree of oil spills on the sea surface is related to factors as oil properties, spill location, meteorological and hydrological conditions as well as emergency measures [20]. Due to uncertainties and interaction in these factors, there is some ambiguity in evaluating pollution level [21], thus Fuzzy Comprehensive Evaluation (FCE) [22] is applied.

According to data source in weight calculation, methods to determine weights of indexes can be divided into subjective weighting method and objective method [23]. Subjective weighting methods include expert survey method (Delphi method) [24] and analytic hierarchy process (AHP) [25-26]. Objective weighting methods include principal component analysis [27], entropy method [28] and multi-objective programming method [29].

AHP this paper selected suits decision-making problems where evaluation indicators are complicatedly layered and the target values are difficult to describe quantitatively. The advantage of AHP lies in clear and systematic procedures as well as less data during analysis.

The main influence factors are determined and evaluation system of pollution degree of oil spills on the sea surface is established [30].

According to the evaluation system, there are 6 first-level evaluation factor sets and 23 second-level evaluation factor sets, first-level evaluation factor set represent as U={oil spill amount U_1, oil properties U_2, oil spill position U_3, meteorological factors U_4, hydrological factors U_5, emergency measures U_6}, second-level evaluation factor sets are U_2={toxicity U_{21}, persistence U_{22}, flammability U_{23}, viscosity U_{24}}, U_3={marine sensitive resources U_{31}, offshore distance U_{32}, shoreline type U_{33}}, U_4={wind

* Corresponding author: zhouxupc@163.com

speed U_{41}, wind direction U_{42}, visibility U_{43}}, U_5={current velocity U_{51}, current direction U_{52}, water temperature U_{53}, wave height U_{54}}, U_6={risk identification and control capabilities U_{61}, emergency resource support capabilities U_{62}, emergency processing capabilities U_{63}, emergency plan preparation capabilities U_{64}, emergency recovery capabilities U_{65}}.

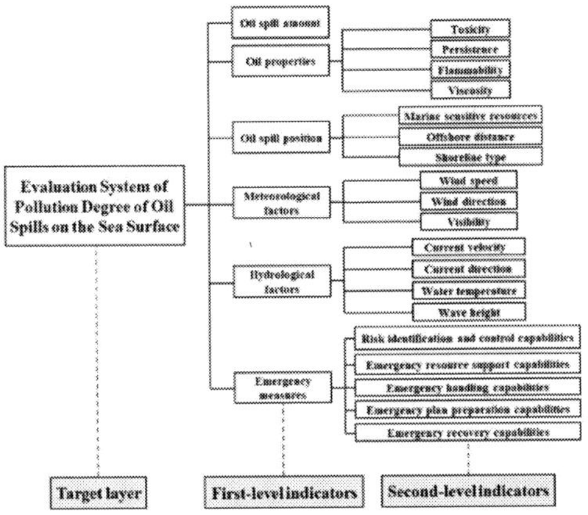

Figure 1. Evaluation system of pollution degree of oil spills on the sea surface

Pollution degree is divided into five levels and comment set is expressed as V= {V_1, V_2, V_3, V_4, V_5} in which V_j (j = 1, 2, ..., 5) represents five pollution levels, namely extremely light, light, medium, heavy and serious pollution respectively.

3 Membership of Evaluation Indexes

Membership of evaluation indexes [31] is quantitative description of comment set on factor set and normally determined by membership function or subset table of membership [32]. Among them, membership of sub-indicators such as oil spills, marine sensitivity and offshore distance is determined by establishing membership function curve and membership of other indexes is determined by constructing membership subset table.

3.1 Membership function of oil spill amount

Combining decree No.15 of the Ministry of Transport of the people's Republic of China on statistical measures for water traffic accidents (2014) with British method of dividing points for oil spill evaluation, a triangular and trapezoidal method is applied to determining membership function of oil spill amount.

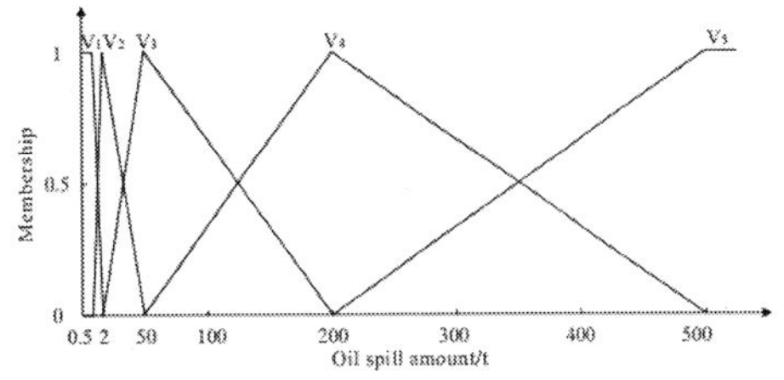

Figure 2. Membership function of oil spill amount

For example, membership function of oil spill amount t for extremely light pollution V_1 is

$$V_{1(t)} = \begin{cases} 1 & t \le 0.5 \\ 1-(t-0.5)/1.5 & 0.5 < t \le 2 \\ 0 & t > 2 \end{cases} \quad (1)$$

3.2 Subset table of sub-indexes of oil properties

Membership of oil properties as toxicity, persistence, flammability and viscosity are invested [33]. Toxicity of crude oil depends on its composition, corresponding proportion, physical and chemical properties of each composition [34]. According to hydrocarbon compounds, crude oil can be divided into paraffins, paraffins and naphthenes, naphthenes, naphthenes and aromatic hydrocarbons as well as aromatic hydrocarbons.

Persistence refers to the ability to remain in the water without being degraded [35]. In general, the greater density of oil spills, the stronger persistence and the longer retention time, the more serious damage to the marine. According to API°, crude oil can be divided into light (>34), medium (20~34), heavy (10~20) and extra thick crude oil (<10).

Flammability of oil spills brings potential danger to emergency operation and it depends on flash point of oil spills [36]. The lower the flash point, the more dangerous of oil spills. Accordingly, oil products can be divided into lubricating oil and grease (>125 °C), diesel and heavy oil (45~125 °C), kerosene (28~45 °C), solvent oil and gasoline (<28 °C).

Viscosity represents fluidity of oil spills. The greater the viscosity, the higher the membership value assigned to a higher pollution level [37]. On account of viscosity, crude oil can be divided into conventional oil (<100 mPa·s), heavy oil (100~10000 mPa·s), extra heavy oil (10000~50000 mPa·s) and super heavy oil (> 50000 mPa·s).

Table1. Membership subset of sub-indexes of oil properties

Sub-indexes of Oil properties	Weight	Level	Comment Set				
			V_1	V_2	V_3	V_4	V_5
Toxicity	0.4782	1	0.8	0.2	0	0	0
		2	0.2	0.6	0.2	0	0
		3	0	0.2	0.6	0.2	0
		4	0	0	0.1	0.6	0.3
		5	0	0	0	0.2	0.8

Persistence	0.1817	1	0.8	0.2	0	0	0
		2	0.2	0.6	0.2	0	0
		3	0	0.2	0.6	0.2	0
		4	0	0	0	0.2	0.8
Flammability	0.2321	1	0.8	0.2	0	0	0
		2	0.2	0.6	0.2	0	0
		3	0	0.2	0.6	0.2	0
		4	0	0	0	0.2	0.8
Viscosity	0.1080	1	0.8	0.2	0	0	0
		2	0.2	0.6	0.2	0	0
		3	0	0.2	0.6	0.2	0
		4	0	0	0	0.2	0.8

3.3 Subset table of indexes of oil spill location

Oil spill areas can be divided into sensitive and non-sensitive areas [38]. The richer the sensitive resources near the spill location, the more serious the pollution will be [39]. Once sensitive area is polluted, oil will cause heavy losses and huge ecological disasters [40].

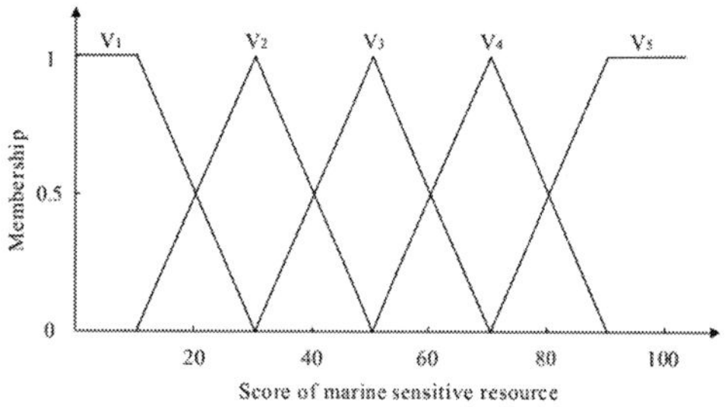

Figure 3. Membership function of marine sensitive resources

Take membership function of score x of sensitive resources for extremely light pollution V_1 as an example

$$V_1(x)=\begin{cases}1 & x\le 10\\ 1.5-x/20 & 10<x\le 30\\ 0 & x>30\end{cases} \tag{2}$$

The closer the spill location to shore, the higher the pollution degree [41]. Function curve of membership of offshore distance is illustrated.

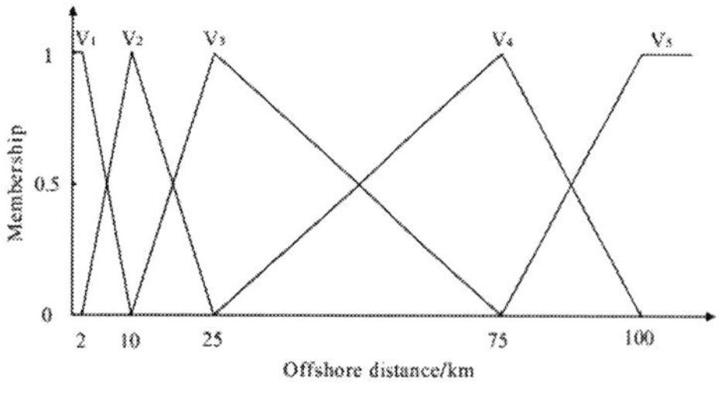

Figure 4. Membership function of offshore distance

For example, membership function of offshore distance d for extreme light pollution V_1 is

$$V_1(d)=\begin{cases}0 & d\le 75\\ (d-75)/25 & 75<d\le 100\\ 1 & d>100\end{cases} \tag{3}$$

With reference to Environmental Sensitivity Index (ESI) [42-43], coastlines can be divided into bedrock coast, gravel coast, plain coast, mangrove coast and coral reef coast in China and the membership subset is shown.

Table2. Membership subset of coastline types

Shoreline	V_1	V_2	V_3	V_4	V_5
1	0.8	0.2	0	0	0
2	0.2	0.6	0.2	0	0
3	0	0.2	0.6	0.2	0
4	0	0	0.2	0.6	0.2
5	0	0	0	0.1	0.9

3.4 Subset table of indexes of meteorological factors

Meteorological factors such as wind speed, wind direction and visibility influence pollution degree of oil spills on the sea surface [44].

Wind speed [45] determines wave size to some extent, promotes dispersion of oil spills, expands pollution area and adversely affects the recovery and treatment of oil spills, resulting in increased pollution level. According to wind speed, wind levels can be divided into level 1 (<1.5 m/s), level 2 (1.5~5.4 m/s), level 3 (5.4~10.7 m/s), level 4 (10.7~20.7 m/s) and level 5 (>20.7 m/s).

Wind direction [46] affects drift trajectory and diffusion direction of oil spills, and may cause serious pollution to coastal sensitive areas. According to the angle between wind direction and coastal sensitive area, wind direction is divided into three levels: reverse coastal sensitive area, a certain angle with coastal sensitive area and forward sensitive coastal area.

Visibility reflects atmosphere's transparency [47], which impacts satellite monitoring and emergency treatment, thus increasing pollution degree of oil spills. With reference to Horizontal Visibility Level, visibility at the sea surface is divided into level 1 (>4000 m), level 2 (1000~4000 m), level 3 (200~1000 m), level 4 (100~200 m) and level 5 (<100 m).

Table3. Membership subset of sub-indexes of meteorological factors

Sub-indexes of Meteorological Factors	Weight	Level	Comment Set				
			V_1	V_2	V_3	V_4	V_5
Wind speed	0.5278	1	0.8	0.2	0	0	0
		2	0.2	0.6	0.2	0	0
		3	0	0.2	0.6	0.2	0
		4	0	0	0.1	0.6	0.3
		5	0	0	0	0.1	0.9
		1	0.8	0.2	0	0	0
		2	0	0.1	0.6	0.3	0

		3	0	0	0	0.1	0.9
Visibility	0.1396	1	0.8	0.2	0	0	0
		2	0.2	0.6	0.2	0	0
		3	0	0.2	0.6	0.2	0
		4	0	0	0.2	0.7	0.1
		5	0	0	0	0.1	0.9

3.5 Subset table of hydrological factors

Hydrological factors as current velocity, current direction, water temperature and wave height affect pollution degree of oil spills on the sea surface [48].

Current velocity [49] increases oil diffusion and is not conducive to oil spill recovery, which aggravates oil spill pollution, and it can be divided into level 1 (<0.5 m/s), level 2 (0.5~1 m/s), level 3 (1~1.5 m/s), level 4 (1.5~2 m/s) and level 5 (>2 m/s).

Current direction and wind direction jointly determine drift direction of oil spills. Drifting to coastal sensitive resources may cause more serious harm to coastal ecological environment. According to the angle between current direction and coastal sensitive area, current direction is divided into 3 levels: reverse, at an angle and forward.

Water temperature [50] in the sea surface directly affects evaporation of oil spills. Volatilization of light components alters oil properties, which has a certain impact on other weathering processes. In addition, evaporation leads to a reduction in oil spill volume, thereby reducing its pollution degree. Water temperature can be divided into level 1 (>25 °C), level 2 (20~25 °C), level 3 (15~20 °C), level 4 (10~15 °C) and level 5 (-5~10 °C).

Wave disturbance [51] promotes stirring of oil spills and water, accelerates formation of emulsions and some oil break into small liquid clumps or droplets, increasing the difficulty of oil spill recovery. Wave disturbance is measured by wave height which can be divided into level 1 (<0.5 m), level 2 (0.5~1.5 m), level 3 (1.5~3 m), level 4 (3~9 m) and level 5 (>9 m).

Table4. Membership subset of sub-index of hydrological factors

Sub-indexes of Hydrological	Weight	Level	Comment Set				
			V_1	V_2	V_3	V_4	V_5
Current	0.420303	1	0.8	0.2	0	0	0
		2	0.2	0.6	0.2	0	0
		3	0	0.2	0.6	0.2	0
		4	0	0	0.2	0.6	0.2
		5	0	0	0	0.1	0.9
Current direction	0.1899	1	0.8	0.2	0	0	0
		2	0	0.2	0.6	0.2	0
		3	0	0	0	0.1	0.9
		1	0.8	0.2	0	0	0
		2	0.2	0.6	0.2	0	0
		3	0	0.2	0.6	0.2	0
		4	0	0	0.2	0.6	0.2
		5	0	0	0	0.2	0.8
Wave height	0.2685	1	0.8	0.2	0	0	0
		2	0.2	0.6	0.2	0	0
		3	0	0.2	0.6	0.2	0
		4	0	0	0.2	0.6	0.2
		5	0	0	0	0.2	0.8

3.6 Subset table of emergency measures

Emergency measures play an important role in controlling oil spills and reducing pollution [52]. Sub-indexes of emergency measures mainly include qualitative indicators such as risk identification and control capabilities, emergency resource support capabilities, emergency handling capabilities, emergency plan preparation capabilities and emergency recovery capabilities. Evaluation degree of emergency measures can be classified into excellent, good, medium, passing and poor. With reference to regulations as IMO's Manual on oil spill risk evaluation and assessment of response preparedness, Marine Environment Protection Law of the People's Republic of China and National Environmental Emergency Plan announced by General Office of the State Council as well as experienced scholars [53], membership subset for sub-indicators of emergency measures is shown.

Table5. Membership subset of sub-indexes of emergency measures

Sub-indexes of Emergency	Weight	Level	Comment Set				
			V_1	V_2	V_3	V_4	V_5
Risk identification and control capabilities	0.2157	1	0.8	0.2	0	0	0
		2	0.2	0.6	0.2	0	0
		3	0	0.2	0.6	0.2	0
		4	0	0	0.1	0.6	0.3
		5	0	0	0	0.1	0.9
Emergency resource support capabilities	0.2325	1	0.8	0.2	0	0	0
		2	0.2	0.6	0.2	0	0
		3	0	0.2	0.6	0.2	0
		4	0	0	0.1	0.6	0.3
		5	0	0	0	0.1	0.9
Emergency handling capabilities	0.3601	1	0.9	0.1	0	0	0
		2	0.1	0.6	0.3	0	0
		3	0	0.1	0.6	0.3	0
		4	0	0	0.1	0.6	0.3
		5	0	0	0	0.1	0.9
Emergency plan preparation capabilities	0.1918	1	0.8	0.2	0	0	0
		2	0.2	0.6	0.2	0	0
		3	0	0.2	0.6	0.2	0
		4	0	0	0.2	0.6	0.2
		5	0	0	0	0.2	0.8
Emergency recovery capabilities	0.2809	1	0.8	0.2	0	0	0
		2	0.2	0.6	0.2	0	0
		3	0	0.2	0.6	0.2	0
		4	0	0	0.2	0.6	0.2
		5	0	0	0	0.2	0.8

 https://doi.org/10.1051/e3sconf/202129301020

4 Membership of Evaluation Indexes

AHP is used to determine the weight of each index [54] and weights of oil properties and oil spill position are present respectively.

Table6. Weight of sub-indexes of oil properties

Oil Properties U_2	Toxicity U_{21}	Persistence U_{22}	Flammability U_{23}	Viscosity U_{24}	Weight
Toxicity U_{21}	1	2	3	4	0.4782
Persistence U_{22}	1/2	1	1/2	2	0.1817
Flammability U_{23}	1/3	2	1	2	0.2321
Viscosity U_{24}	1/4	1/2	1/2	1	0.1080

Table7. Weight of sub-indexes of oil spill location

Oil Spill Location U_3	Marine sensitive Resources U_{31}	Offshore distance U_{32}	Shoreline type U_{33}	Weight
Marine sensitive resources U_{31}	1	2	3	0.5499
Offshore distance U_{32}	1/2	1	1	0.2402
Shoreline type U33	1/2	1	1	0.2098

Table8. Weight of evaluation indexes of oil spill pollution degree

Indexes	U_1	U_2	U_3	U_4	U_5	U_6	Weight
U_1	1	2	3/2	6/5	6/5	3/4	0.2013
U_2	1/2	1	5/6	3/4	3/4	2/3	0.1098
U_3	2/3	6/5	1	3/4	3/4	5/6	0.1308
U_4	6/5	4/3	4/3	1	6/5	5/6	0.1722
U_5	5/6	4/3	4/3	5/6	1	5/6	0.1601
U_6	4/3	3/2	2	6/5	6/5	1	0.2259

Since the consistency ratio CR [55-56] is less than 0.1 for the weight of each index, according judgment matrix has good consistency, that is, the weight distribution is reasonable.

5 Multi-level Fuzzy Comprehensive Evaluation

In the fuzzy comprehensive evaluation, $\circ$ represents a fuzzy matrix synthesis operator in $B = A \circ R$ [57]. This paper selects a weighted average operator [58-59], which indicates all factors are balanced according to the weight and it is applicable to the effect of various factors.

Evaluation system for pollution degree of oil spills on the sea surface is a second-level fuzzy evaluation. Taking oil properties as an example to illustrate the process of the first-level fuzzy comprehensive evaluation, the membership matrix R_2 can be obtained from the subset table of toxicity, persistence, flammability and viscosity.

$$R_2 = \begin{bmatrix} r_{11}^2 & r_{12}^2 & r_{13}^2 & r_{14}^2 & r_{15}^2 \\ r_{21}^2 & r_{22}^2 & r_{23}^2 & r_{24}^2 & r_{25}^2 \\ r_{31}^2 & r_{32}^2 & r_{33}^2 & r_{34}^2 & r_{35}^2 \\ r_{41}^2 & r_{42}^2 & r_{43}^2 & r_{44}^2 & r_{45}^2 \end{bmatrix} \quad (4)$$

A_2=(A_{21}, A_{22}, A_{23}, A_{24}) represents the weight set of sub-indexes and the result matrix of fuzzy comprehensive evaluation B_2 for oil property U_2 is

$$B_2 = A_2 \circ R_2 = (b_{21}, b_{22}, b_{23}, b_{24}, b_{25}) = (A_{21}, A_{22}, A_{23}, A_{24}) \circ \begin{bmatrix} r_{11}^2 & r_{12}^2 & r_{13}^2 & r_{14}^2 & r_{15}^2 \\ r_{21}^2 & r_{22}^2 & r_{23}^2 & r_{24}^2 & r_{25}^2 \\ r_{31}^2 & r_{32}^2 & r_{33}^2 & r_{34}^2 & r_{35}^2 \\ r_{41}^2 & r_{42}^2 & r_{43}^2 & r_{44}^2 & r_{45}^2 \end{bmatrix} \quad (5)$$

Therefore, the total evaluation matrix R of the fuzzy comprehensive evaluation system can be obtained as

$$R = \begin{bmatrix} B_1 \\ B_2 \\ B_3 \\ B_4 \\ B_5 \\ B_6 \end{bmatrix} = \begin{bmatrix} A_1 \circ R_1 \\ A_2 \circ R_2 \\ A_3 \circ R_3 \\ A_4 \circ R_4 \\ A_5 \circ R_5 \\ A_6 \circ R_6 \end{bmatrix} \quad (6)$$

The second-level fuzzy comprehensive evaluation result is $B = A \circ R$, which is the overall evaluation result of pollution degree of oil spills on the sea surface, where A is the weight of the first-level indexes.

6 Case Analysis of Oil Spills on the Sea Surface

6.1 Pollution evaluation of oil spill in Penglai 19-3 oilfield

In Penglai 19-3 oilfield, there were two oil spill accidents on June 4 and 17 in 2011 when was off-season in the Bohai Sea and oil spill amount was at least 940 t [60-61]. The location is about 70 km from the nearest coast Longkou coast and the shoreline is sandy type. Coastal sensitive resources include nature reserves, marine special protection areas, natural landscapes, historical and cultural heritage areas, fishery waters, coastal tourist areas, port shipping areas and special utilization areas in adjacent marine. State Oceanography Bureau implemented three-dimensional all-weather surveillance and monitoring on the oil spill accident the first time. However, due to the deliberate delay of the responsible party for the accident, ConocoPhillips did not fully control the spill situation, the progress of oil spill investigation and plugging was slow, oil spills in platform B and C was basically controlled until June 19 and June 21 respectively.

At the time of oil spill accident, wind on the sea surface was about 5.7 m/s and there was a certain angle between wind direction and coastal sensitive resources. Current velocity was approximately 0.5~1 m/s and it was directly opposite coastal sensitive resources, wave height was less than 1m. The spilled oil is medium or heavy, mainly isoparaffin or cycloalkane but less n-paraffin, with low alkane content (32.8%~44.6%) and higher colloid content (18.9%~29.8%). API of ground crude oil is about 17.4, closed flash point is about 40 °C (open flash point is 75 °C), and its viscosity is about 60.42~3957 Pa·s (50 °C).

Secondary evaluation results B=(0.0353 0.1818 0.2358 0.1136 0.4375), according to the principle of maximum membership, Penglai 19-3 oil spill accident belongs to serious pollution, which is consistent with the result of Emergency Oil Exploration and Development Oil Spill Emergency Plan issued by State Oceanography Bureau on April 3rd, 2015.

6.2 Pollution evaluation of oil spill in Suizhong 36-1 oilfield

On November 26, 2002, an oil spill accident occurred on the central platform of Suizhong 36-1 oilfield in the Bohai Sea [62]. Referring to the method of determining the membership degree of emergency measures taken by Zhang Qian [63], this paper considers the emergency measures on the oil spill accident at a good level. Adopting the relevant data by Liu Jie [64], the second-level fuzzy comprehensive evaluation method is applied to evaluating the pollution degree of the oil spill and the total evaluation matrix is obtained as B= (0.2043 0.6071 0.2007 0.0476 0.0039). According to the principle of maximum membership, oil spill accident on the central platform of Suizhong 36-1 oilfield is classified as light pollution, which is consistent with the conclusion drawn by Liu Jie [64].

7 Conclusion

In this paper, multi-level fuzzy comprehensive evaluation method is applied to establishing pollution evaluation system for oil spills on the sea surface and analytic hierarchy process combining qualitative and quantitative analysis is applied to determining weights of each index.

(1) Evaluation system of oil spill pollution on the sea surface is established by multi-level fuzzy comprehensive evaluation method. There are 6 first-level evaluation factor sets and 23 second-level evaluation factor sets. Pollution degree of oil spills is divided into 5 levels, which are respectively extremely light, light, medium, heavy and serious pollution.

(2) Membership degree of sub-indexes such as oil spill amount, marine sensitive resources and offshore distance is determined by establishing membership function, while membership degree of other indexes is determined by constructing membership subset table.

(3) Analytic hierarchy process is applied to determining weight of each sub-index and the fuzzy operator in the multi-level fuzzy comprehensive evaluation is determined by a weighted average model, thus the sub-indexes can be balanced and considered in accordance with the corresponding weight.

(4) Based on the principle of maximum membership, oil spill accidents in Penglai 19-3 oilfield and in the central platform of Suizhong 36-1 oilfield are evaluated as serious pollution and light pollution respectively, which signifies feasibility and rationality of the evaluation system of pollution degree of oil spills on the sea surface.

References

1. Sun P Y , Gao Z H , Zhou Q , et al. Evaluation of the Oil Spill Accident in Bohai Sea, China[J]. Environ. Forensics,2009,10(4):308-316.
2. Van-Cuong Do, Hong-Xiang Ren. Simulation of Bohai sea tidal during Penglai 19-3 oil spill accident based on Mike21 model[C]. 2019 2nd Asia Conference on Energy and Environment Engineering (ACEEE),2019.
3. Jie G , Xin L , Qiang X .Characteristics of the Bohai Sea oil spill and its impact on the Bohai Sea ecosystem[J]. Chinese Science Bulletin, 2013, 58(019):2276-2281.
4. Li H , Li Y , Li C , et al. Bohai and Yellow Sea Oil Spill Prediction System and Its Application to Huangdao '11.22' Oil Spill Incident[J]. Iop Conference, 2017, 81:012054.
5. Wang Y , Lee K , Liu D , et al. Environmental impact and recovery of the Bohai Sea following the 2011 oil spill[J]. Environmental Pollution, 2020, 263(Pt B): 114-343.
6. Carmo D , Antunes J , Pinho , et al. Oil Spills in Coastal Zones: Predicting Slick Transport and Weathering Processes[J]. Open Ocean Eng. J., 2010, 3(3):129-142.
7. J. Antunes do Carmo, J.L. Pinho, J.P. Vieira. Oil Spills in Coastal Zones: Predicting Slick Transport and Weathering Processes[J]. Open Ocean Eng. J., 2010, 3:129-142.
8. Kameleh Aghajanloo, Moharam Dolatshahi Pirooz. The Simulation of the Oil Weathering Processes in Marine Environment[J]. 2011 International Conference on Environmental and Computer Science, 2011,19.
9. Xue J , Yu Y , Bai Y , et al. Marine Oil-Degrading Microorganisms and Biodegradation Process of Petroleum Hydrocarbon in Marine Environments: A Review[J]. Current Microbiology, 2015,71(2):220-228.
10. Stephanie E. Chang, Jeremy Stone, Kyle Demes and Marina Piscitelli. Consequences of oil spills: a review and framework for informing planning[J]. Ecology and Society, 2014,19(2):26.
11. Mostafawi N . How severely was the Persian Gulf affected by oil spills following the 1991 Gulf War?[J]. Environmental Geology, 2001, 40(10):1185-1191.
12. Yafeng, Guiqin, Zhiguo, et al. Research on the Assessment Method of Oil Spill Environmental Risks of Offshore Oil Facilities[J]. Aquatic Procedia, 2015.
13. Han Y , Nambi I M , Clement T P . Environmental impacts of the Chennai oil spill accident-A case study[J]. Sci. Total Environ., 2018, 626:795-806.

14. Okandan E , F Gümrah, Demiral B . Importance of assessment of oil pollution along Black Sea coast and Bosphorous Strait-Turkey[C]// International Seminar on Nuclear War & Planetary Emergencies Session. 2001.
15. Liu X , Wirtz K W . Decision making of oil spill contingency options with fuzzy comprehensive evaluation[J]. Water Resour. Manag.,2007,21(4): 663-676.
16. Lang Y H , Cheng F F , Wang N N . Fuzzy Comprehensive Evaluation of Oil Spills Pollution Level for Offshore Platform[J]. Appl. Mech. Mater., 2011, 71-78:3012-3015.
17. Zhou J , Xu L . The Damage Assessment of Marine Oil Spill Pollution Using Back-Propagation Neural Network[J]. Energy Procedia, 2011, 11:3116-3121.
18. Han D , Wang G Y , Zang T Y . Application of Fuzzy Decision Theory in the Oil-Gas Field Production and Management[J]. Appl. Mech. Mater., 2014, 488-489:1297-1300.
19. Mil'Shtein L M . Comprehensive Evaluation and Selection of Separators for the Oil and Gas Industry[J]. Chemical & Petroleum Engineering, 2014, 49(11-12):727-735.
20. Fingas M . Weather Effects on Oil Spill Countermeasures[J]. Oil Spill Science and Technology, 2011:339-426.
21. Dong Y . Factors influencing marine oil spills and cleanup operations[J]. Journal of Dalian Maritime University, 2009(S1):40-42.
22. Hongjun, You, Hongjie, et al. Forecasting the yield of oil field gas with the fuzzy comprehensive evaluation method and the genetic algorithm[J]. Chemical Technology, 2014, 9(1):4-7.
23. Zhao J , Jin J , Zhu J , et al. Water Resources Risk Assessment Model based on the Subjective and Objective Combination Weighting Methods[J]. Water Resour. Manag., 2016, 30(9):3027-3042.
24. Chan H K , Wang X . Hierarchical Model in Decision Making[J]. Fuzzy Hierarchical Model for Risk Assessment, 2013, 25-43.
25. Solnes J . Environmental quality indexing of large industrial development alternatives using AHP[J]. Environ. Impact Assess. Rev., 2003, 23(3):283-303.
26. Sahoo S , Dhar A , Kar A . Environmental vulnerability assessment using Grey Analytic Hierarchy Process based model[J]. Environ. Impact Assess. Rev., 2016, 56(JAN.):145-154.
27. Zhang J Y , Wang L C . Assessment of water resource security in Chongqing City of China: What has been done and what remains to be done? [J]. Nat. Hazards, 2015, 75(3):2751-2772.
28. Kang C , Feng J , Zheng Y . Study on Oil and Gas Pipeline Risk Assessment Based on Entropy Weight Fuzzy Comprehensive Evaluation Method[J]. Journal of Liaoning Shihua University, 2015, 35(5):27-30, in Chinese.
29. Wang X L , Xiao J Z . Multi-objective dynamic programming for the optimal operation of natural gas production and sales[J]. Computer & Communication Technologies in Agriculture Engineering International Confe, 2010, 3:240-243.
30. Lang Y H, Cheng F F, Wang N N. Fuzzy Comprehensive Evaluation of Oil Spills Pollution Level for Offshore Platform[J]. Appl. Mech. Mater., 2011,71-78:3012-3015.
31. Hao X , Xiao F . Fuzzy Synthetic Evaluation of Gas Station Safety[C]// Fuzzy Information and Engineering 2010 - Volume I, Proceedings of the 5th Annual Conference on Fuzzy Information and Engineering, ACFIE 2010, Sepember 23-27, 2010, Huludao, China. Springer Berlin Heidelberg, 2010.
32. Liang D , Liu D , Pedrycz W , et al. Triangular fuzzy decision-theoretic rough sets[J]. Int. J. Approx. Reasoning, 2013, 54(8):1087-1106.
33. Wang J , Liu C , Feng J , et al. Literature review of the effects of petroleum hydrocarbons on marine phytoplankton[J]. Environmental Pollution & Control, 2011, 33(4):81-86.
34. Paul J H , Hollander D , Coble P , et al. Toxicity and mutagenicity of gulf of Mexico waters during and after the deepwater horizon oil spill.[J]. Environmental Science & Technology, 2014, 48(6):3591-2.
35. Bejarano A C , Michel, et al. Oil spills and their impacts on sand beach invertebrate communities: A literature review[J]. Environmental Pollution, 2016,218(Nov.):709-722.
36. Zhou S D , Zheng F Z , Li T . Analysis on the Factors Influencing the Kinematic Viscosity of Petroleum Products[J]. Advanced Materials Research, 2012, 459:386-389.
37. Lefebvre-Chalain, Hélèn. Fifteen Years of Particularly Sensitive Sea Areas: A Concept In Development[J]. Ocean and Coastal Law Journal, 2016, 13(1):3-3.
38. Isah M N . The role of environmental impact assessment in Nigeria's oil and gas industry[J]. Waste Manage., 2012, 32(10):1989-1998.
39. 39. Mark A. Zacharias, Edward J. Gregr.Sensitivity and Vulnerability in Marine Environments: an Approach to Identifying Vulnerable Marine Areas[J]. Conserv. Biol.,2005,19(1):86-97.
40. Kawarada H , Baba E , Suito H . Effect of spilled oil on coastal ecosystems[C]// Proceedings of Fifth China-Japan Seminar on Numerical Mathematics. Department of Urban Environment Systems Chiba University, 1-33 Yayoicho, Inage-ku, Chiba 263-85SS, Japan, 2002
41. Deborah F M , Beegle-Krause C J , Rowe J , et al. Oil Spill Risk Assessment–Relative Impact Indices by Oil Type and Location[J]. Asascience Com, 2009.
42. Kankara R S , Arockiaraj S , Prabhu K . Environmental sensitivity mapping and risk

assessment for oil spill along the Chennai Coast in India[J]. Mar. Pollut. Bull., 2016, 106(1-2):95-103.

43. Li F , Zhang H , Chen L , et al. Influencing Factors of Oil Slick Movement on Tidal Rivers[C]// International Conference on Bioinformatics & Biomedical Engineering. IEEE, 2008.
44. Beegle-Krause C J , Lehr W J . Oceanographic and Meteorological Effects on Spilled Oil[M]. Handbook of Oil Spill Science and Technology, 2015:301-310.
45. Stringari C E , Marques W C , Mello L , et al. Modeling the wind influence in an oil spill along the southern Brazilian shelf[J].Engenharia Térmica (Thermal Engineering), 2012,11(1-2):100-109.
46. Brovchenko I , Kuschan A , Maderich V , et al. The modelling system for simulation of the oil spills in the Black Sea[J]. Oceanography, 2003, 69:586-591.
47. Schwartzberg H G . The movement of oil spills[J]. International Oil Spill Conference Proceedings, 1971(1):489-494.
48. Murray S P . The Effects of Weather Systems, Currents, and Coastal Processes on Major Oil Spills at Sea[J]. Pollut. Transfer Transp. Sea, 2018(2):169-227.
49. Hellmann H . Influence of water temperature on the behavior of floating crude oil[J]. Erdoel & Kohle, Erdgas, Petrochemie,1973,26(9):513-517.
50. Reisbig R L . Oil Spill Drift Caused by the Coupled Effects of Wind and Waves[R]. Missouri Univ-Rolla Div of Engineering Research, 1973.
51. Bechtel R D , Wickley-Olsen E , Boufadel M C , et al. The movement of oil at sea due to irregular waves[J]. International Oil Spill Conference Proceedings, 2008(1):943-947.
52. Liu H , Xiao Y , Gao X . Study on Environmental Supervision of Marine Oil Spill Pollution[J]. Ecological Economy, 2018,34(7):185-189, in Chinese.
53. Weiwei J, Wei A, Yupeng Z, et al. Research on Evaluation of Emergency Response Capacity of Oil Spill Emergency Vessels[J]. Aquatic Procedia, 2015, 3:66-73, in Chinese.
54. Zhao Y W, Yang Z F. Integrative fuzzy hierarchical model for river health assessment: A case study of Yong River in Ningbo City, China[J]. Commun. Nonlinear Sci. Numer. Simul., 2009, 14(4):1729-1736.
55. Alonso J A , Ma T L . A Statistical Criterion of Consistency in the Analytic Hierarchy Process[J]. International Conference on Modeling Decisions for Artificial Intelligence, 2005.
56. Alonso J A , Lamata M T . Consistency in the analytic hierarchy process: a new approach[J]. Int. J. Uncertainty Fuzziness Knowl.-Based Syst., 2006, 14(4):445-459.
57. Lowen R . Categorical methods and techniques in fuzzy mathematics[J]. Fuzzy Sets Syst., 1999, 104(3): 357.
58. Jiang X , Bao Z , Wang L . The coupled method Fuzzy-AHP applys to solve multi-criteria decision making problems. World Scientific and Engineering Academy and Society (WSEAS), 2009,8(11).
59. Wang J Q , Li X E , Chen X H . Hesitant Fuzzy Soft Sets with Application in Multicriteria Group Decision Making Problems[J]. Sci. World J.,2015:806-983.
60. Oil spill in Penglai 19-3 Oilfield[EB/OL]. https://www.conocophillips.com.cn/zh/ke-chi-xu-fa-zhan/zhi-li-yu-bo-hai-de-zhang-qi-fa-zhan/peng-lai19-3you-tian-yi-you-shi-jian/
61. State Oceanic Administration announces investigation of oil spill accident in Penglai 19-3 Oilfield[EB/OL].http://www.gov.cn/gzdt/2011-07/05/content_1899673.htm
62. The platform fault of Suizhong 36-1 oilfield of CNOOC has been eliminated and the oil film is being cleaned[EB/OL].https://www.chinanews.com/ny/2011/07-13/3177919.shtml
63. Zhang Q , Liu L , Pan B Y , et al. Marine oil-spill pollution assessment model based on fuzzy comprehensive evaluation[J].Marine Science,2016,40(5):97-102, in Chinese.
64. Liu Jie. Research on the method of the level assessment of oil spill pollution on Bohai offshore platform[D]. Ocean University of China, 2010, in Chinese.

 https://doi.org/10.1051/e3sconf/202129301021

The thinking and practice of Marine Ecological Conservation Red Line Delineation in China

Bei Zhao[1], Yanrong Zhou[1], Nana Liu[1], Kun Yang[1], and Jing Li[1*]

[1]North China Sea Environmental Monitoring Center, State Oceanic Administration, China

Abstract: China has established and implemented marine ecological conservation red line (MECRL) system to strengthen the bottom line constraints and ensure ecological security. This paper introduces the concept of MECRL and the process of overall consideration of the national MECRL delineation, which is defined "spatial-division + differential control measures". The MECRL in Bohai Sea pilot and delineation of MECRL nationwide are introduced. At present, MECRL accounts for about 30 percent of the total coastal management area, providing a good institutional guarantee for protecting marine ecology and guiding the good formation of a scientific marine spatial layout in china.

1 Introduction

China is rich in marine resource and marine biodiversity, with a total area of 3 million square kilometers under its jurisdiction, including Bohai Sea, Yellow Sea, East China Sea and South China Sea. There are mangroves, coral reefs, coastal wetlands, seagrass beds, islands, bays, estuaries and other types of marine ecosystems across the three climatic zones of warm temperate zone, subtropical zone and tropical zone. Marine resources and marine biodiversity are very rich, providing an important guarantee for the sustainable development of coastal economy and society. However, in recent years, due to multiple factors such as coastal development, population aggregation and resource and environment pressure, a large number of Marine space resources in coastal areas have been occupied, and the pressure on Marine ecosystem has been under increasing pressure. According to the continuous monitoring results of the Bulletin on the State of China's Marine Ecology and Environment, the water quality of some marine areas in China is still not optimistic, and the function of regional marine ecosystem has been degraded. It is urgent to establish marine environmental protection policies based on regional management and by means of Ecological Conservation Red Line, guided by the concept of marine ecological civilization, aiming at harmony between human and sea, so as to safeguard the safety and health of China's Marine ecology.

2 The concept of marine ecological conservation red line

The concept of red line originated in urban planning, which refers to a prohibited area. The red line is legally, mandatory, and no building or structure shall exceed the building red line [1]. There is no concept of Ecological Conservation Redline abroad [2], which is an institutional innovation in the field of ecological protection in China.

The red line was first introduced into the ecological field in 2000 when Gao Jixi put forward the red line control scheme [3] in ecological planning in Anji, Zhejiang Province. In 2012, The State Oceanic Administration launched the delineation of MECRL in bohai sea, aiming to implement a stricter marine ecological protection system.

MECRL refers to the marine ecological space with special important ecological functions, which must be strictly protected, including extremely important ecological functions areas, sensitive and fragile ecological areas, such as biodiversity conservation, coastal ecological stability. MECRL is an important part of China's pattern of marine space development and protection, which is also an important basis for the unified implementation of marine space use control and ecological protection and restoration. China attached great importance to the establishment of ecological conservation red line system, and has put forward clear requirements for the work of ecological conservation redline in important ecological protection strategies.

In 2014, Article 29 of the Environmental Protection Law of the People's Republic of China (Revised and adopted at the eighth meeting of the Standing Committee of the 12th National People's Congress on April 24, 2014) clearly stipulates that "China set the ecological conservation red line in key ecological functional areas, sensitive areas and vulnerable areas of the ecological environment, and implement strict protection."

In 2016, the revised Marine Environmental Protection Law of the People's Republic of China incorporated MECRL system into the basic law of marine

* Corresponding author: bhjczx_lijing@ncs.mnr.gov.cn

environmental protection, and MECRL was officially established at the legal level[4].

3 The role of marine ecological conservation red line

MECRL with the strictest marine ecological protection system protects the national marine ecological health and ecological safety bottom line, and together with marine function zoning and marine main function zoning plans to improve the marine spatial planning system in China. Once delineated, MECRL will become an insurmountable space protection line, which will implement the strictest environmental access and control measures, and will serve as the basis for China to build a framework for marine ecological security.

Firstly, The MECRL strengthens marine ecological protection and strictly protect important areas of Marine ecological function, such as important estuaries, important coastal wetlands, mangroves, coral reefs, and seagrass beds, which effectively restore the damaged and degraded ecosystem, strengthen the protection of marine biodiversity and ensure that the ecological functions are not degraded.

Secondly, reclamation and other activities affecting the protected objects within the redline are prohibited. Through spatial constraints on the use of the sea, strict prevention and control shall be carried out at the source, so as to reduce the interference of human activities on the objects under the protection of the red line, ensure the ability to provide Marine ecosystem services in a sustainable manner, and ensure that the area of the red line is not reduced

Thirdly, from the perspective of ecosystem function, we should severely crack down on human activities that illegally damage marine ecosystem functions, strengthen marine ecological protection, renovation and restoration, maintain the marine ecological environment in the marine ecological protection red line area, restore damaged and degraded ecosystems, and manage and protect the ecological red line. Specific measures include strengthening surveillance and monitoring, strict supervision and inspection, strict responsibility assessment, carrying out early warning and prompt, and establishing law enforcement inspection, supervision and accountability mechanisms.

Fourthly, the evaluation results will be included in the assessment of marine ecological civilization construction objectives. The assessment results shall be incorporated into the natural resources balance sheet. The environmental quality and the total amount of marine natural resources within the marine ecological protection red line reflected in the balance sheet shall not show a trend of gradual decline.

4 Basic train of thought of MECRL delineation

4.1 Overall consideration

The national MECRL follows the following principles:

MECRL takes into account both conditions of natural resources, the ecological environment and geographical location, and the need for sustainable economic and social development of the country and region. On the basis of maintaining the marine ecological function and the capacity of resources and the environment, the marine ecological protection and marine environmental quality bottom line should be defined by zoning, and all kinds of activities damaging the protected objects in MECRL areas should be strictly prohibited. At the same time, the demand for sustainable development should be taken into account, so as to leave room for future marine industry and social and economic development.

According to the characteristics and protection requirements of the marine ecosystem, MECRL should be delineated, and differentiated control measures should be formulated to effectively protect the national important ecological function areas, marine ecological sensitive areas and marine ecological fragile areas.

Equal attention should be given to ecological protection and restoration. Ecologically sensitive and fragile areas and important ecosystems should be included into the category of MECRL so as to limit the expansion of industries that damage dominant ecological functions and follow the path of ecological and economic development. Ecosystems that are already damaged and need to be remedied and restored should also be included in MECRL to curb further ecological damage.

The delineation of MECRL should be effectively linked with the issued national and provincial marine function zoning, national marine major function zoning plans, national strategic planning and other sea-related zoning and planning, with the focus on marine ecological protection. The management of MECRL area should be stricter than other zoning and planning. The delineation of MECRL in cross-provincial and cross-municipal coastal waters should be coordinated and cohesive.

4.2 Delineation path

Domestic scholars have studied the delineation of MECRL from aspects of the capacity of marine resources and environment, marine spatial planning technology, etc.[5-7]. MECRL system currently implemented in China takes into account the differences in the spatial distribution of marine ecological functions, regional marine development and utilization, etc. By delineating MECRL, the behaviors that destroy the marine ecological function will be effectively restrained.

Firstly, MECRL should be identified. Based on data collection and combined with the distribution of important ecological function areas, ecologically sensitive areas and ecologically fragile areas throughout the country, importance assessment of marine ecological function should be carried out. Important estuaries, important coastal wetlands, islands under special protection, marine conservation areas, natural landscapes and historical and cultural relics, concentrated distribution areas of rare and endangered species, important fisheries areas, important coastal tourism areas, important sandy shorelines, protected sea areas of sand sources, mangroves, coral reefs and seaweeds and other extremely important areas should be identified.

Secondly, the special scale of MECRL should be determined. From the perspective of ecological protection, the greater the area, the better the protection effect will be. However, the larger the area, the greater the pressure will be on local economic development and financial transfer payment. At present, the delineation plans of MECRL have been released by coastal provinces and cities, mainly relying on the relevant technical specification documents, fully considering the management requirements and the current situation in different marine environmental resources, from the perspective of operability and practicability for MECRL. For some specific areas, it is necessary to analyze and delineate them according to the specific circumstances. At the same time, the coordination analysis with national and provincial marine function zoning, development strategic planning of coastal areas, marine economy development planning and other relevant national strategic planning should be done well, so as to identify the conflict areas that do not affect the development and utilization of ecological functions, and finally determine the scope.

Thirdly, the zoning management control measures should be developed.

Due to the particularity of ocean space, MECRL is different from other red line and is under differentiated control measures. Through spatial management constraints on the use of the sea, The utilization of marine resources that has an impact on the MECRL is prohibited, and gradually improve the ecological capacity of the marine resources and environment. According to different types of MECRL, differentiated control measures should be implemented from two aspects of general management requirements and detailed utilization control requirements, and the MECRL should be managed by its delineation and classification, and protected by the importance of the level. In principle, human activities are prohibited in the core areas and buffer zones of marine protection areas within MECRL. In other areas, productive construction activities and other activities that have an impact on protected objects are strictly prohibited. At the same time, special management measures can also be formulated for MECRL that has special requirements for certain periods of time.

5 Practice of MECRL Delineation

In July 2012, China launched the work of delineating Bohai Sea MECRL, which is the first time that China has applied ecological conservation red line to the ocean. At the end of December 2013, Shandong provincial government issued the Delineation Plan for the Bohai marine ecological conservation redline Area of Shandong Province, which is valid from 2013 to 2020. The MECRL area is delineated 6534.42 square kilometers from the total 16313.9 square kilometers of sea area, making it the first province to implement the MECRL system in Bohai Sea in China. By July 2014, the delineation of the MECRL in Bohai Sea had been completed, including Hebei province, Tianjin city, the Bohai management areas of Shandong province and Liaoning province, and preliminary results showed good results. A total of 156 areas have been delineated under the Bohai Sea MECRL, effectively protecting more than 800 kilometers of natural shoreline, marine conservation areas, important estuaries and wetlands, and important fisheries areas, covering an area of more than 14,000 square kilometers. It will provide important technical support for the implementation of the strictest environmental protection policy, the establishment of MECRL system, and the promotion of marine ecological civilization in Bohai Sea.

The practice and application of MECRL delineation in Bohai Sea have been conducted on a trial basis and achieved good results, laying a good foundation for the comprehensive promotion of the MECRL delineation in China and the establishment of MECRL system. Since June 2014, China has been working on the national MECRL delineation. By the end of 2016, China had basically completed the delineation of MECRL in 11 coastal provinces and cities. By the end of 2017, MECRL had been issued and implemented. MECRL has been accounted for about 30 percent of the total coastal areas under management.

After the MECRL was delineated, its work shifted to strict compliance. China has implemented strict control measures in MECRL. Firstly, regular ecological monitoring and evaluation have been carried out. Through monitoring and evaluation, the ecological conditions of the protected targets within the MECRL can be found out, so as to give early warning in case of continuous ecological decline; Secondly, the key habitats and damaged ecosystems within the MECRL should be protected and restored, so as to improve their ecological services and restore the health and stability of the ecosystem. Thirdly, MECRL has been applied to Marine administrative licensing and administrative law enforcement, implement the strictest space access system, strictly control the access to the sea for projects, strictly protect and manage the use of coastline, and strictly punish human activities within the MECRL that violate the control measures. Fourthly, enhance red line quality performance appraisal, and carry out quality accountability for those who violate the control requirements, break the red line and cause ecological damage to the red line. Through the above measures, it will ensure that the MECRL is well guarded and managed.

6 subtotal

MECRL delineation method is generally operable and practical, but the difficulty lies in how to determine the boundaries of different classification scientifically and reasonably. There is no denying that due to the regional differences and the limitations of human cognitive level, in MECRL on the delineation of boundary, this method has certain limitations, such as important estuaries, important fisheries areas, sand protection areas can only be determined qualitatively with a delineation principle, which should be determined after specific analysis based on the practical situation. MECRL area delineated is not static, but a process of dynamic adjustment. With the improvement of cognitive level, the state of ecological restoration, resource and environment capacity in MECEL area, the delineation method will be improved and perfected on the basis of continuous practice.

MECRL, as an important component of the pattern of space development and protection, is the core element and mandatory content of space planning, and also an important basis for the unified implementation of space use control and marine ecological protection and restoration. At present, China is further optimizing the ecological protection red line, including MECRL. China's ecological red line will further optimize the spatial distribution pattern of MECRL based on the principle of "delimiting as much as is necessary and ensuring as much as is necessary" on the basis of identifying gaps and coordinating conflicts between marine protection and utilization.

Acknowledgement

In the process of writing this article, we got the guidance and help from my colleagues, and we would like to express our deep thanks to them.

Thanks for the support of China National Offshore Oil Corporation (CNOOC) Marine Environment and Ecological Protection Public Welfare Foundation.

Fund name: Study on the coordination of spatial layout of Marine energy (wind power, oil and gas) development and utilization.

Fund number: CF-MEEC/TR/2020-05.

References

1. Li S J, Yang X, Wang J K. Study on policy frame of marine ecological protection red line. Marine Environmental Science, 2016, **35**(2): 306-310.
2. Lin Y, Fan J F, Wen Q, Liu S X, Li B Y. Primary exploration of ecological theories and technologies for delineation of ecological redline zones. Acta Ecologica Sinica, 2016, **36**(5): 1244-1252.
3. Huang W, Zeng J N, Chen Q Z, Du P, Tang Y B, Yang H. Preliminary research on the zoning method of the marine ecological redline: a case study of Hainan Province. Acta Ecologica Sinica, 2016, **36**(1): 268-276.
4. Hu B, Chen Y. The development of the marine eco-redline to the China's marine eco-security legal system. Chinese Journal of Maritime Law, 2018, **29**(4): 94-101.
5. Liu S X, Wang Y, Wang R R, Yan J S, Fan J F, Wen Q. Study on the delimitation method of the overloading area of marine resources and environment in the red line area. Marine Science Bulletin, 2017, **36**(5): 497-503.
6. Chen G L, Hu W J, Chen B, Ma Z Y, Liao J J. Marine spatial planning using in marine ecological red line zoning— —a case study of dongshan county. Journal of Applied Oceanography, 2017, **36**(1): 6-15.
7. XU Y, LIANG B, BAO C G, LAN D D, YU C Y, MA M H. Research on the index system and the technical methods of ecological red line division for the Bohai Sea. Marine Science Bulletin, 2013, **32**(4): 361-367.

 https://doi.org/10.1051/e3sconf/202129301022

An eco-friendly and effective method to fabricate flame retardant cotton fabrics

Xiaotao Zhang [1*], Chunrong Zhou[2], Haifeng Pan [3]

[1] School of Architectural Economics and Engineering Management, Hubei Business College, Wuhan, 430079, China;

[2] School of intelligent manufacturing and automobile, Chongqing Vocational College of Transportation, Chongqing 402247, China

[3] Faculty of Engineering, China University of Geosciences, Wuhan 430074, People's Republic of China

Abstract: To reduce the flammability of cotton fabrics, an environmentally benign, simple and effective method was proposed. Coatings composed of phosphoguanidine /ATMP /alginate were deposited on the surface of the cotton fabrics through layer-by-layer (LbL) assembly. FTIR spectra indicate that the layer-by-layer deposition process is successfully carried out on the surface of cotton fabrics. TGA suggested that LbL coating can significantly increase the char residue. The results of MCC and vertical flame tests showed that the LbL coatings can remarkably enhance the fire safety properties of the cotton fabrics and a higher concentration of ATMP (2 wt%) can bring better effect of flame retardancy.

1 Introduction

Cotton fabrics is one of the most important natural textile fibers used to produce apparel, home furnishings, and industrial products, however, the low limiting oxygen index and combustion temperature (360-425 °C) makes the material highly flammable. Once ignited, cotton fabrics will burn rapidly, and the flame spreads quickly, potentially causing fatal burns within 15 s of ignition [1]. According to National Fire Protection Association statistics, there are approximately 17,300 mattress and upholstered furniture fires per year[2], which result annually in approximately 871 civilian deaths, 2230 civilian injuries, and $824 million in property losses.

Therefore, various flame retardant additives have been used to modify combustion characteristics of textiles, including cotton fabrics. Halogenated and boron-containing additives are widely used to improve the fire resistance of cotton fabrics, such as tris (2-chloroethyl) phosphate (TCEP), tris (1,3-dichloroisopropyl) phosphate (TDCPP) and tris (1-chloro-2-propyl) phosphate (TCPP)[3]. Despite their effectiveness, the potential toxicity and environmental problems emerge from cotton fabrics storage, transportation and combustion. Recently, organophosphorus compounds are considered as suitable alternatives to halogenated flame retardants because of the advantages of environmentally friendly and high efficiency. Many studies have been conducted to investigate flame retardant mechanism of phosphorus nitrogen synergistic [4-7]. However, the preparation process reported is relatively complex, which could not meet the large-scale industrial production. Therefore, developing environmentally friendly, high efficiency and convenient operation techniques will be the best solution to flame retardant cotton fabrics for soft furnishings.

Layer-by-layer (LbL) assembly is a method to fabricate multifunctional, multilayer thin films on the surface of a substrate[8]. By submersing a substrate into positively and negatively charged polyelectrolyte and/or nanoparticle solutions, a positive–negative pairing layer is developed and comprised electrostatic, van der Waals, H-bonding, and covalent bonds[9]. Recently, the LbL assembly method was used to fabricate flame retardant coatings on various polymer matrices, including textiles, plastic plaques, and flexible polyurethane foams [10–12].

In this study, three environmentally benign polyelectrolytes, phosphoguanidine, amino trimethylene phosphonic acid (ATMP) and alginate were deposited on cotton fabrics through layer-by-layer (LbL) assembly. The thermal stability and fire suppression of the coated and uncoated cotton fabrics were investigated. The mechanism of the enhanced fire safety of the coated cotton fabrics was proposed. It is anticipated that this multilayered coatings will provide a simple and green solution to reduce the fire hazards of cotton fabrics.

2 Experimental

2.1 Materials

Standard (untreated) cotton fabrics used in this study was received from Hengtai Cotton Fabric Factory of Jinzhou City. ATMP (50 wt %) were supplied by Shanghai Dibai Chemical Technology Co., Ltd. Phosphoguanidine and alginate were supplied by Sinopharm Chemical Reagent Co., Ltd.

* Corresponding author: smiletao@163.com

2.2 Preparation of Solution

Firstly, Phosphoguanidine solutions were prepared by adding 1.0 wt% Phosphoguanidine to deionized water. Then ATMP solutions were prepared by adding 1.5 wt% and 2.0 wt% ATMP. Finally, the alginate solutions were prepared by adding 0.4 wt% alginate to deionized water. All the solutions were vigorously stirred for 1 h until the polyelectrolyte was completely dissolved.

2.3 Preparation of LbL assembly coating

The standard size cotton fabrics were alternately sprayed by the phosphoguanidine solutions, ATMP solutions and alginate solutions. Every spray should ensure that the solution is evenly distributed on the cotton fabrics surface. Before spraying the next solution, the treated cotton fabrics should be placed in the vacuum blast drying oven 20 min to ensure it is completely dry. The preparation process is shown in Fig. 1 and the treated samples were marked as shown in Table 1.

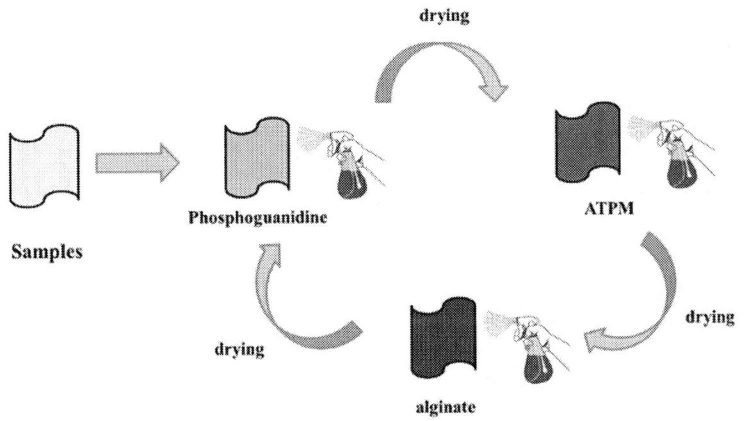

Figure 1. Scheme of layer-by-layer assembled deposition on the cotton fabrics.

Table1. Content of Phosphoguanidine /ATMP/ Alginate on the Surface of cotton fabrics

sample	Phosphoguanidine(wt%)	ATMP(wt%)	Alginate(wt%)	number of layers
F1	1.0	1.5	0.4	1
F2	1.0	1.5	0.4	2
F3	1.0	2.0	0.4	3
F4（untreated）	0	0	0	0

3 Characterization

Fourier-transform infrared (FTIR) spectra of cotton fabrics were obtained on a Nicolet 6700 FTIR instrument. Samples were mixed with KBr powder and pressed into tablets, which was then placed in a ventilated oven. The transition mode was used and the wavelength range was set from 4000 to 500 $cm^{-1.}$

The thermogravimetric analyzer (TGA) of samples under nitrogen atmosphere was examined on a TGA-Q5000 apparatus (TA Company, USA) from 50 to 700 °C at a heating rate of 20 °C/min.

Heat release rate (HRR) were measured in a microscale combustion calorimeter (GOVMARK MCC-2). Samples of approximately 5 mg were heated at a rate of 1 K/s in a nitrogen stream flowing at 80 cm 3 /min. The degradation products of the sample in nitrogen atmosphere were mixed with a 20cm^3 /min stream of pure oxygen prior to entering a 900 °C combustion furnace.

Vertical flame tests were performed according to ASTMD 6413-08 using a vertical burning tester (CZF-3, Nanjing Jiangning Analytical Instrument Factory, China). The samples (300 mm 9 76 mm), held 19 mm over the Bunsen burner, were first exposed to the flame for a period of 12 s and then rapidly removed.

4 Results and discussion

4.1 Characterization of cotton fabrics

In order to provide a clear comparison, FTIR spectra of F3 and F4 cotton fabrics are depicted in Fig. 2. As shown in the Fig.2, compared to the F4 cotton fabrics, the bands at 1320 cm^{-1} ~ 1070 cm^{-1} of the F3 sample become smoother, which is probably attributed to the superposition of stretching vibration of P=O and the C-N. The peak located at 870 cm^{-1} is assigned to stretching vibration of N-H. Moreover, the new characteristic peak at 1628 cm^{-1} and 1398 cm^{-1} appears that is corresponding to the antisymmetric and symmetric stretching vibration peaks of COO-. It is indicated that the layer-by-layer deposition process is carried out on the surface of F3 cotton fabrics.

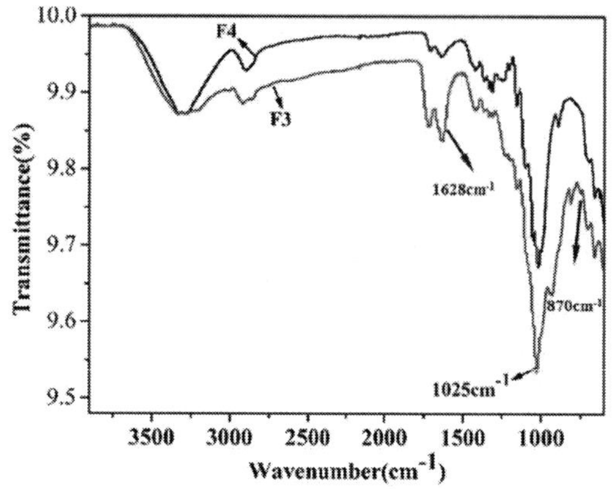

Figure 2. ATR-FTIR spectra of F3 and F4 cotton fabrics

4.2 Thermal properties

TGA provides direct information about the thermal stability and degradation mechanism by measuring the weight loss of the sample as a function of temperature. Fig. 3 present the TGA curves of the uncoated and coated cotton fabrics. It is can be seen from Fig.3, the pure cotton fabrics decomposition occurred in two main steps.

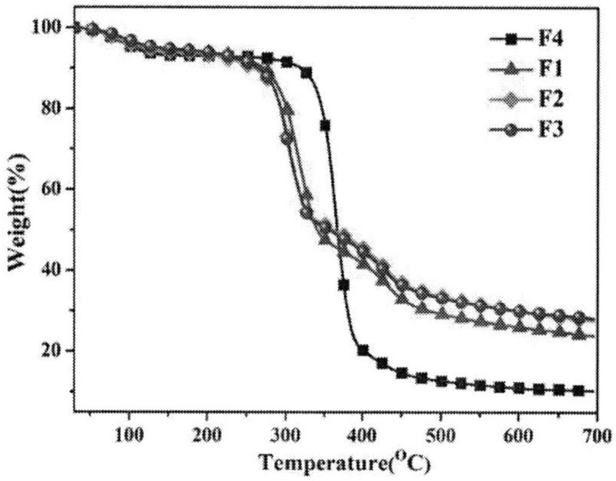

Figure 3. TGA curves of untreated and treated cotton fabrics

The first region around 360 °C is the main pyrolysis region, which shows significant degradation weight loss. And it is attributed to the dehydration and decarboxylation reactions .The second region around 400 °C is attributed to the further decomposition of residual char formed in the first region [13]. Meanwhile, for the treated cotton fabrics, the initial decomposition temperatures obviously decrease, which due to that the facile catalytic dehydration and cross-linking of cotton fabrics by phosphoric acid generated during pyrolysis of phosphorus-containing compounds. Another obvious phenomenon is that The coated cotton fabrics show higher solid char in the high temperature range 400 °C-650°C. It is indicated that the formation of stable intumescent chars in the first stage, which can efficiently retard the diffusion of the oxygen and heat into underlying matrix.

4.3 Flammability properties

Microscale combustion calorimetry is an effective approach to obtain the information regarding combustibility and fire hazard of materials. Fig. 4 shows heat release rate (HRR) and curves of all samples that measured by MCC. It is can be seen from Fig.4, all coated cotton fabrics (F1-F3) have lower peak HRR and compared with untreated one (F4), indicating the improvement of flame retardancy for coated cotton fabrics. Another obvious result is that two distinct peaks can be found in the treated cotton fabrics HRR curve. The first HRR peak (285-293 °C) was associated with thermal decomposition of guanidine phosphate and sodium alginate in flame retardant components. The second HRR peak (433-440 °C) was probably attributed to that the intumescent flame retardant system composed of sodium alginate and phosphate promotes charring and retards combustion reaction.

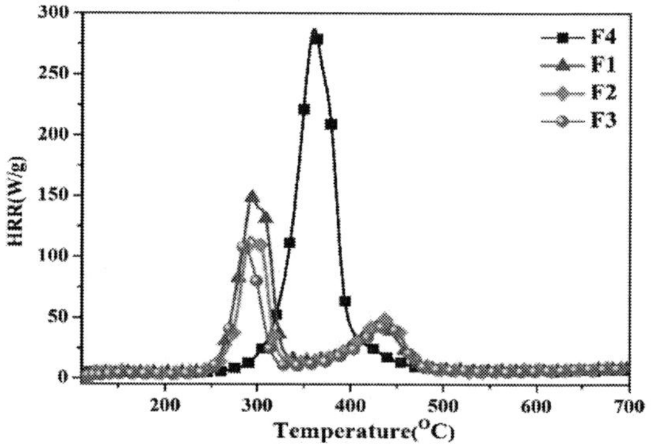

Figure 4. HRR curves of all cotton fabrics sample

The vertical flame test can be employed to evaluate the flammability of samples in the presence of flame spread. Fig 5 shows images of untreated and treated coated cotton fabrics after vertical flame test. As shown in Fig. 5, for untreated cotton fabrics (F4), there is no char left after the vertical flame test, indicating that the untreated cotton sample was completely consumed. For the F2 cottons, it was found that the char residues of these samples maintained their original shape. As for the F3 sample, the collected char residue is integral nearly with little slits on the bottom side and the char length is only 14.0 cm, which is less than the F3 sample. It indicated that a higher concentration of ATMP (2 wt%) can result in more content of ATMP on cotton fabrics and have highly effective flame retardancy.

https://doi.org/10.1051/e3sconf/202129301022

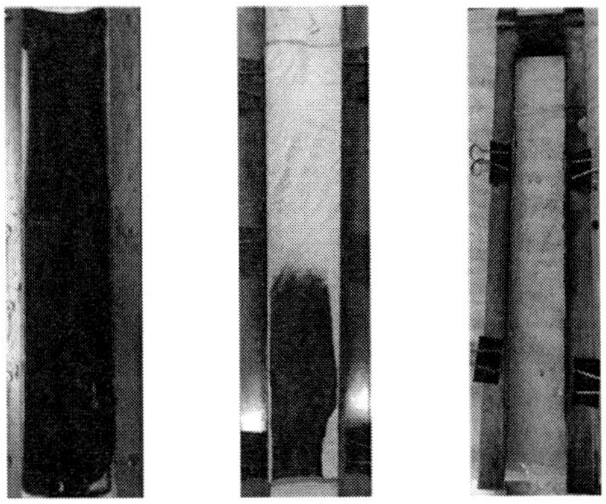

Figure 5. Vertical flame tests for untreated and treated cotton fabrics

5 Conclusion

Coatings composed of phosphoguanidine /ATMP /alginate were successfully deposited on the surface of cotton fabrics using a simple way that is layer by layer approach to improve the fire safety of the cotton fabrics. Coating analysis by TGA suggested that phosphoguanidine/ATMP/alginate LbL coating can significantly increase the char residue. Microscale combustion calorimeter results showed that the LbL coatings can remarkably enhance the fire safety properties of the cotton fabrics. Vertical flame test result show that a higher concentration of ATMP (2 wt%) can result in more content of ATMP on cotton fabrics and it will obtain highly effective flame retardancy.

Acknowledgements

This research was supported by Nature Science Foundation of Hubei Province under Grant No.2019CFB366.

References

1. Oulton, D. P. Fire-Retardant Textiles. In Chemistry of the Textiles Industry; Carr, C. M., Ed.; Springer-Verlag: Berlin, 1995; Chapter 3.
2. Ahrens M. Home fires that began with upholstered furniture. National Fire Protection Association, Quincy. 2020
3. H. M. Stapleton, S. Klosterhaus, A. Keller, P. L. Ferguson, S. VanBergen, E. Cooper, T. F. Webster and A. Blum, Environ. Sci. Technol., 2011, 45, 5323–5331.
4. Zheng T , Wang W , Liu Y . A novel phosphorus-nitrogen flame retardant for improving the flame retardancy of polyamide 6: Preparation, properties, and flame retardancy mechanism. Polymers for Advanced Technologies, 2021(2).
5. J Liu, Dong C , Zhang Z , et al. Durable flame retardant cotton fabrics modified with a novel silicon–phosphorus–nitrogen synergistic flame retardant[J]. Cellulose, 2020, 27(15):9027-9043.
6. Chi Z , Guo Z , Xu Z , et al. A DOPO-based phosphorus-nitrogen flame retardant bio-based epoxy resin from diphenolic acid: Synthesis, flame-retardant behavior and mechanism. Polymer Degradation and Stability, 2020, 176:109151.
7. He L , Xie B , Lai C, et al. Synthesis of Phosphorus-Nitrogen-Halogen Ternary Systems Flame Retardant and Its Application in PP. Engineering Plastics Application, 2016.
8. Kotov N A . Multilayer Thin Films: Sequential Assembly of Nanocomposite Materials. 2003
9. Kim Y S , Harris R , Davis R . Innovative Approach to Rapid Growth of Highly Clay-Filled Coatings on Porous Polyurethane Foam. Acs Macro Letters, 2012, 1(7):820-824.
10. Xue C H , Wu Y , Guo X J , et al. Superhydrophobic, flame-retardant and conductive cotton fabrics via layer-by-layer assembly of carbon nanotubes for flexible sensing electronics[J]. Cellulose, 2020, 27(6).
11. Majka T M , Witek M , Radzik P , et al. Layer-by-Layer Deposition of Copper and Phosphorus Compounds to Develop Flame-Retardant Polyamide 6/Montmorillonite Hybrid Composites. Applied Sciences, 2020, 10(14):5007.
12. Zhang Y , Yang W . Synthesis and characterization of PEDMCD as a flame retardant and its application in epoxy resins. RSC Advances, 2021, 11.
13. Formation of self-extinguishing flame retardant biobased coating on cotton fabrics via Layer-by-Layer assembly of chitin derivatives. Carbohydr Polym, 2015, 115:516-524.

Research on ecological compensation of urban residual space to promote social integration

Yu Chen[1], Longwei Zhang[1]*, Chao Wang[1], Xue Ding[1]

[1]School of Architecture and Urban Design, Shenyang Jianzhu University, 110168 Shenyang, China

Abstract. The continuous urbanization leads to a large number of land lost farmers pouring into the city, and the social integration of these people has become an important research topic of urban renewal. By means of field exploration and in-depth visit, this paper investigates and collates the data of some typical urban residual spaces where new industrial workers live, and finds that these spaces have common spatial characteristics, such as single spatial level, poor ecological environment and lack of landscape facilities. Then, by using statistical analysis and structural model, the comparative analysis of different types of space, and finds that the use of ecological compensation regulation can form a reasonable spatial layout, high-quality ecological effects and rich service facilities, so as to improve the spatial quality of urban residual space and provide high-quality space environment for new industrial workers, Promote social integration, social integration of workers in new industries.

1 Background

In recent years, with the rapid development of China's urbanization, the central area of the city continues to expand to the periphery of the city. Some farmers have lost the land they depend on to survive, and become "three noes" with "no land for farming, no employment, and no share of subsistence allowances" [1]. They become new industrial workers and stay in the city, hoping to integrate into the city and share resources. However, the lag of urban development does not consider the living space of this kind of people. They are trapped in the dilemma of weak information, weak development and weak resources. In order to remove the social restrictions, they stay in the public space of the city through the way of group gathering to obtain urban resources. This way of life breaks the rules of space and transforms the living and entertainment space originally provided for urban residents into the commercial gathering space mainly in the form of primitive economy - "urban residue space". This form and the life style of urban residents contradict with each other, and the activity space of urban residents is greatly limited. Common urban residual space, including passive idle corner green space, dark roadside green belt, messy space under the bridge and so on. The urban residual space studied in this paper is a dilapidated space which is gradually occupied by new industrial workers due to the lack of management and maintenance [2]. However, this does not mean that the "space monopoly" of new industrial workers should be denied, and the "space monopoly" behavior should not be solved by administrative means of simply driving them away. We should provide them with a warm city home through urban renewal. Therefore, how to intervene in this social problem and play a positive role of landscape designers is the main research content of this paper.

2 Regional environmental characteristics of "urban residue space"

2.1 Environmental data collection of "urban residue space"

A large number of new industrial workers seek new vitality in the "urban residue space", and spontaneously form a certain scale of new industrial workers cluster. From August 1 to August 31, 2020, the author conducted field survey, data sampling and in-depth interview on more than 60 "urban residue spaces" in Shenyang in three time periods (7:00-8:00, 11:00-12:00, 16:00-17:00) every day. Through the regional population structure, ecological characteristics and facilities characteristics of the information to conduct in-depth research. By means of in-depth interviews, this paper summarizes the behavior characteristics and basic needs of workers and people in new industries, and lists six representative spaces.

2.2 Analysis of environment and crowd characteristics of "urban residue space"

"Urban residue space" is generally located in the old urban areas, where supporting infrastructure is scarce and landscape service facilities are backward. Compared with the urban space in the new urban area, the cost of living is lower, and it is easier to generate natural economic and commercial forms (Table 1).

* Corresponding author: z_lw@sjzu.edu.cn

According to the survey results, we find that these "urban residue spaces" have the following characteristics:

1) From the perspective of spatial location, new industrial workers tend to choose the areas with high density of urban residential areas, avoid the main roads, commercial and office clusters, and choose the areas surrounded by urban secondary roads and urban branch roads. These areas have a large population base, convenient transportation, wide vision, and avoid large-scale transient crowd.

2) In terms of the ecological effect of space, the greening rate of most areas is low and the spatial level is single. After the new industrial workers settle in, they spontaneously gather on the street for a long time, resulting in a single population in the space. Only a few areas with high greening rate and complete space facilities have diversified population structure.

3) In terms of space facilities, most of the areas have no space facilities, the parking of electric vehicles in the area is disorderly, and the crowd in the area has no space to block out the wind and the sun; workers in new industries are standing on the street without corresponding commercial support facilities. The interior of the venue is open, with few people and lack of entertainment and leisure facilities.

3 Analysis on environmental suitability of "urban residual space"

3.1 Research assumptions

R: As the main evaluation index, public space attraction includes (R1) user structure, (R2) population size, (R3) work and rest time. Q: Public space location suitability index includes (Q1) spatial location, (Q2) traffic condition, (Q3) spatial area. H: Ecological environment suitability index includes (H1) greening rate, (H2) green space form, (H3) arbor and shrub coverage. S: The convenience index of public space facilities includes (S1) the number of facilities and (S2) the distribution of facilities.

3.2 Data fitting analysis

The attraction degree of public space is an important standard to determine the quality of space. Therefore, this element is used as a reference condition and is fitted with other relevant factors to find the following characteristics (Figure 1).

Table1. Spatial and crowd characteristics of "urban residue space"

Site name	Site area (HA)	Residential density(%)	Number of participants	Travel time	Greening rate (%)	Space facilities
Butterfly Garden	0.3	40	New industry workers 25-30	7	20%	Chair Pavilion
			city dweller 15-20	1		
Nameless Plaza	0.76	35	New industry worker35-50	6	10%	Nothing
Model Worker Park	1.3	80	New industry worker50-90	8	70%	Chair Pavilion
			Second hand peddler of urban residents 10-12	3		
			city dweller 25-28	7		
Tieyou Staff Square	4.6	40	New industry worker20-30	7	45%	Nothing
			Second hand peddler of urban residents 3-5	2		
			city dweller 14-18	3		
Minyue Plaza	1.6	85	New industry worker50-80	8	60%	Chair Sports facilities
			city dweller 24-33	4		
Shuangyong Forest	8	75	New industry worker30-60	8	65%	Nothing
			city dweller 20-30	5		

1) The location suitability of public space is the main factor affecting the attraction of public space. When the density of residential areas around the urban public space is more intensive, the number of new industrial workers is larger, and the demand of urban residents for urban public space is stronger.

2) Ecological environment suitability also significantly affects the attraction of public space. Through data analysis, it is found that the higher the greening rate, the larger the number of new industrial workers and urban residents, and the longer the activity time; the lower the greening rate, the more single the crowd structure and the shorter the activity time.

3) The convenience of public space facilities is an important factor affecting the attractiveness of public space. Diversified space facilities can attract a large number of people. In the area with poor space facilities, the structure of the crowd is single, the scale is small, and the residence time is relatively short.

3.3 Model verification and calculation results

Using Amos software tools to verify and modify the initial model, the relationship between the correlation degree of each element is displayed, and the correlation degree is "Q > H > S". For ordinary citizens, the degree of associ-ation is

"Q > H > S". In the selection of public space, the needs of the two groups are basically the same. Therefore, high quality regional location, reasonable green layout and optimized public space facilities build a bridge between new industrial workers and urban residents.

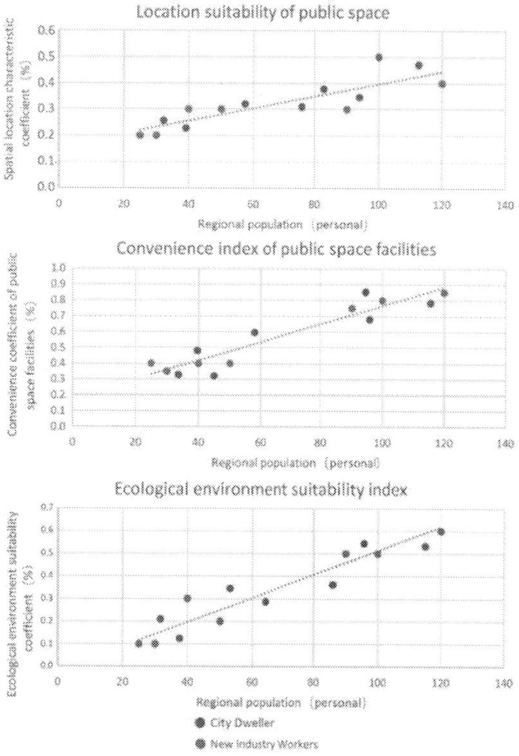

Figure 1. Fitting simulation of urban spatial attractiveness between new industrial workers and urban residents

4 Technical strategies for ecological renewal of "urban residue space"

4.1 Reorganization of ecological resources and environmental network architecture

The main reason for the over concentration of "urban residue space" population is that the ability of new industrial workers to obtain employment information is limited. They can only form the regional identity through the agglomeration effect to form the "information source". Therefore, we should open up the exclusive space belonging to this group in the city, and distribute it in the important space nodes of the city to form a urban recruitment market space network by region, scale and level [3].

Through the way of "urban acupuncture", the "urban leftover space" and the "urban corner space" form a whole urban space network, and a large number of new industrial workers are dispersed into a number of small-scale groups, which are distributed in the urban space network. This can not only reduce the adverse impact of new industrial workers' group aggregation on urban space, but also expand the service radius of new industrial workers and provide business services for more old urban areas. Then, the urban spatial network used by new industrial workers is nested with the geographic positioning function of the Internet, so that urban residents can establish a sustainable and feedback recruitment market cluster of new industrial workers through the positioning system of travel tools, which is convenient for urban residents to locate the recruitment market of new industrial workers through travel software (Figure 2).

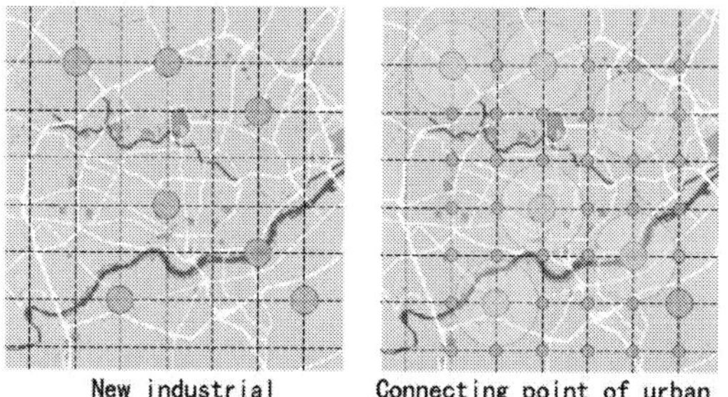

Figure 2. Connection point setting of new industrial workers gathering

4.2 Construction of ecological scene and cross of various activities

The construction of ecological scene can not only reasonably plan the layout of urban residual space, but also give rich vitality to the original rigid site; improve the regional ecological effect and create environmental microclimate; create a rich scene space, enrich the forms of venues and activities, and form cross activity forms, so as to provide a strong guarantee for the social integration of new industrial workers.

1) Reasonable planning of spatial layout. The main problem of urban residual space lies in the high homogeneity of space, the lack of independent activity space for new industrial workers and urban residents. Through the intervention of green plants, the two groups of new industrial workers on the street and leisure entertainers inside the venue are created with their own independent functional areas. The long-term stable operation mode provides stable commercial value and close geographical relationship for new industrial workers. This independent consumption scene shapes a higher quality spatial level through the renewal principle of "micro intervention and microcirculation", attracts more homogeneous types of people, and enable the urban space to produce sustainable vitality [4].

2) Improve the regional ecological effect. The green space with rich layers can absorb and reflect the solar radiation and reflection function, which can better block the rapid air circulation and mitigate the adverse effects of daily temperature changes; a large amount of water absorbed by plants from the soil wetted the surrounding environment through transpiration of leaves, forming a high-quality microclimate environment. The dust caused by urban traffic can be reduced by effective canopy

coverage; through the effective absorption and transfer of pollution particles by plant leaves, it plays the role of dust retention and dust collection; through the effective thickness of forest wind field structure, particles are prevented from entering the urban residual space. The above measures can improve the regional ecological environment and promote the sustainable development of urban residual space (Figure 3).

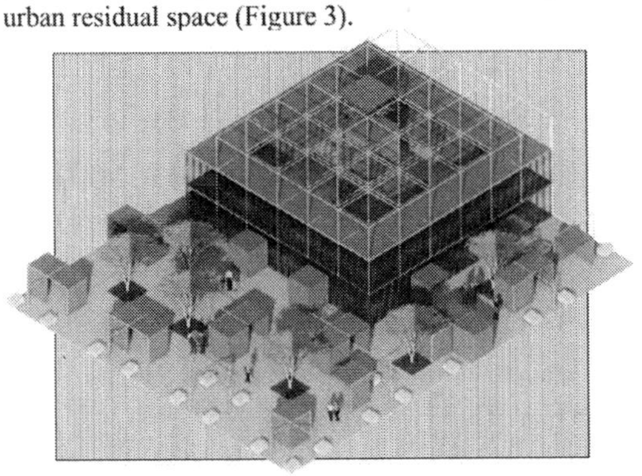

Figure 3. Landscape intention generation of embedded scene

4.3 Building ecological module and implanting digital technology

The main reason for the phenomenon of spatial separation is that new industrial workers have no independent spatial boundaries in urban space[5]. Therefore, it is very necessary to establish a special use space for this group. Exclusive space can not only show their professional characteristics, but also form a regional identification system, reflecting the diversified characteristics of urban landscape At the same time, it can better limit the commercial space of new industry workers and eliminate the interference to other space of the site[6]. However, the spatial forms of "urban residual space" are different, and it is impossible to complete the renovation of all spaces through a unified structure. Therefore, this paper considers introducing modular landscape structures to meet the needs of spaces with different characteristic. The "adaptive" unit framework and various activity modules are nested to form the overall "landscape module". The framework of free allocation and diversified combination form diversified landscape facilities module through unified standardized construction mode. It provides more selectivity and full adjustability for landscape facilities module, so as to adapt to different design conditions and functional requirements.

5 Summary

The continuous urbanization has caused a large number of new industrial workers to rush into cities. How to provide a comfortable space environment and make it better interact with urban residents has become an urgent problem in urban renewal. In this paper, a large number of "urban residual space" space problems are effectively combed, summarized and summarized. It is found that improving the ecological quality of space is an effective means to promote the interaction between new industrial workers and urban residents. This paper takes the new industrial workers as the target, ecological compensation as the means, and urban space as the medium, and puts forward the following corresponding renewal strategies: firstly, planning a reasonable urban ecological network, providing a high-quality space system for the new industrial workers, and dispersing the new industrial workers crowded in the urban core space into the "urban corner space". At the same time, we should establish a regional identification system to form an independent urban commercial space cluster belonging to new industrial workers. Secondly, through the way of ecological compensation, ecological restoration is carried out in the new industrial workers' gathering area to provide more ecological environment for "urban residue space". Through the embedding of green form, more reasonable functional zoning will be created to form the new industrial workers' commercial space, urban residents' use space and the integration and transition space of a variety of people. Finally, through the embedded multi-functional modular landscape structure, the new industrial workers' commercial landscape module, urban planting landscape module and urban residents' sports and leisure landscape module are formed, which provides various service functions for all kinds of people in the urban residual space, thus creating a strong guarantee for the social integration of new industrial workers.

Acknowledgments

The authors gratefully acknowledge the support of Scientific Research Plan of the Educational Department of Liaoning Province(Grant lnjc202003).

References

1. Huang Kaiping, Qin Zhi, Wang Li. Countermeasures for the resettlement of landless farmers in the construction of new urban areas [J]. China market, 2020 (36): 13-15
2. Lu Luguang. From differentiation and isolation to harmonious communication: Research on urban social communication [J]. Academia, 2005 (03): 106-114
3. Zhang Wenwu, Ou Xi. Agglomeration effect, congestion cost and urban spatial structure evolution [J]. Journal of Hebei University of economics and trade, 2020,41 (01): 28-38
4. Huang Lu. Thinking on urban public landscape design strategy of caring for vagrants [J]. Anhui architecture, 2017,24 (06): 8-10 + 82
5. Xin shanchao, Wang Zhiqiang. Speculation on the construction of modular connection -- Research on Architectural "design build" based on modular system [J]. Journal of western human settlements, 2016,31 (06): 23-28
6. Qiu Baoxing. Symbiosis and eco city [J]. Urban planning, 2013,37 (09): 9-16 + 50

The ability of UHPC prepared with unpurified sea sand and seawater to protect embedded steel against corrosion

Zuqi Wang[1,2,3,*], Jing Wang[1,2,3], Yongxiang Zhou[1,2,3], Putao Song[1,2,3], Qingfeng Guan[1,2,3], Zhiren Zhou[1,2,3]

[1]Institute of Building Materials, China Academy of Building Research, Beijing 100013, China

[2]State Key Laboratory of Building Safety and Built Environment, Beijing 100013, China

[3]National Engineering Research Center of Building Technology, Beijing 100013, China

Abstract. Preparing concrete with unpurified sea sand and seawater is of great significance for the construction in coastal areas short of river sand and pure water. However, unpurified sea sand and seawater contain a lot of chloride ions which may cause steel corrosion problems. In this paper, unpurified sea sand and seawater are used to prepare ultra-high performance concrete (UHPC), and the potential differences between the steel bars with protective layer thickness of 5mm and 15mm and the $Cu/CuSO_4$ reference electrode under two types of dry-wet cycle and the chloride ion concentrations at different depths from UHPC specimen surface after 40 dry-wet cycles are tested, the UHPC specimens are broken to observe corrosion of embedded steel bars. The results show that UHPC prepared with unpurified sea sand and seawater has good ability to protect embedded steel reinforcement against corrosion under dry-wet cycle conditions in lab.

1 Introduction

China has a vast sea area with coastlines of 18,000 kilometres, and abundant sea sand reserves [1,2]. If unpurified sea sand and seawater can be used as construction sand in coastal areas, it can save freshwater resources for desalinating sea sand and reap huge economic and ecological benefits. However, unpurified sea sand and seawater have a high concentration of chloride ions. Chloride ions can accelerate the corrosion of steel bars and reduce the bond between the steel bars and concrete, which will cause great safety risks to construction engineering [3-6]. In coastal areas in China such as Ningbo and Zhoushan in Zhejiang Province, there have been many engineering damage cases caused by unpurified sea sand [7]. Therefore, how to safely and economically use unpurified sea sand has become a problem for construction engineers to think about.

It is generally necessary to purify the sea sand before concrete preparation. Chloride ion in the unpurified sea sand will migrate to the surface of steel bars in concrete, which may cause the damage of the passive film on the surface of steel bars. If there is appropriate amount water and oxygen on the surface of the steel bars after the failure of the passive film, the steel bar will be corroded. However, high density and low porosity are the most significant characteristics of UHPC. It may be difficult for chloride ion to migrate in UHPC and the content of water and oxygen in UHPC may be very low, thus providing technical feasibility for the preparation of UHPC with unpurified sea sand and seawater. Whether UHPC prepared with unpurified sea sand and seawater has enough ability to protect steel reinforcement against corrosion must be studied before application in practical engineering. Research on the ability of UHPC prepared with unpurified sea sand and seawater to protect steel bars against corrosion has not been reported yet.

The half-cell potential method is commonly used to estimate the risk of corrosion of steel bars in concrete [8]. It is non-destructive, simple and cheap. Herein, we therefore designed an improved half-cell potential method to compare the half-cell potential for different dry-wet cycles, different numbers of cycles, different layer thicknesses and different UHPC proportions. Then we broke the specimens after a certain number of dry-wet cycles to observe the corrosion of steel bars and measure the water-soluble chloride ion concentrations at different depths from surface of UHPC specimens

2 Experimental Procedures

2.1Materials and mix proportions

2.1.1 Materials

A typical commercial ordinary Portland cement (P·O 42.5, produced by Jidong Cement Plant, China) was used in this study with mechanical properties detailed in Table 1.

* Corresponding author: zuqiwang@126.com

Table 1. Mechanical properties of P·O 42.5 cement

Compressive strength/MPa		Flexural strength/MPa	
7d	28d	7d	28d
34.4	50.06	7.95	8.96

The unpurified sea sand was obtained from Fujian Province, China, which is shown in Fig. 1. Sieve analysis and physical properties were carried out on the sea sand specimens according to Chinese code JGJ 206-2010 as shown in Fig. 2 and table 2.

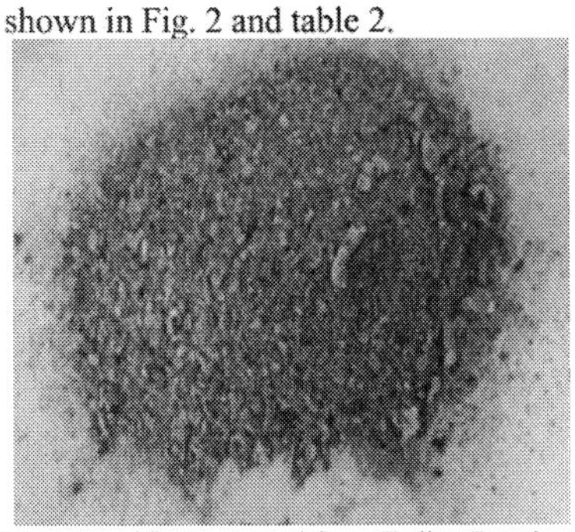

Fig. 1. The unpurified sea sand from Fujian Province, China

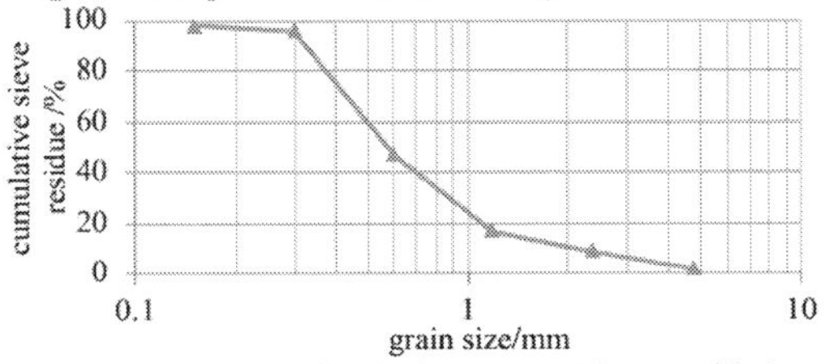

Fig. 2. The grain size distribution curve of the unpurified sea sand

A commercial quartz sand was used to compare with the sea sand, which has the same grain size distribution as that of the sea sand and SiO_2 content not less than 99.0%.

Two kinds of water were used. One is fresh water, and the other is 3.5% (mass fraction) NaCl solution to simulate sea water.

Other materials involved in the experiments include silica fume(properties as shown in table 3),viscosity reducer CABR-J1 produced by CABR Building Materials co., Ltd(properties as shown in table 4), polycarboxylic acid superplasticizer with solid content of 39% and water reduction rate of 30% and defoamer with solid content of 30% which were both produced by Tianjin Yueming Technology Development Co., Ltd, and steel fiber with length of 13 mm, diameter of 0.2mm and tensile strength of 2850 MPa produced by Beijing Huicheng Foundation Co., Ltd.

2.1.2 Mix proportions

Four mixes were cast and cured in steam at 70 °C for 48 hours. Mix proportions and compressive strength are as shown in Table 5.

Table 2. Physical properties of the unpurified sea sand

Apparent density (kg/m³)	Bulk density (kg/m³)		Mud content (%)	Crush value (%)	Moisture content (%)	Fineness modulus	Chloride ion content (%)	Shell content (%)
	Loose	Close						
2520	1370	1490	0.64	3.53	0.34	1.09	0.073	10.02

Table 3. Properties of silica fume

SiO_2 content /%	Specific surface area /(cm²/g)	Water demand ratio /%	Water content /%	Ignition loss /%	28d activity index /%
94.6	2.09×10^4	111	0.4	1.17	120

Table 4. Properties of viscosity reducer CABR-J1

Specific surface area/(m²/kg)	Water demand ratio/%	Activity index/%		Emptying time ratio of inverted slump cylinder/%	Chloride ion content/%
		7d	28d		
1159	88	92	123	42	0.057

Table 5. Mix proportions and compressive strength (kg/m³)

Sample number	P·O 42.5 cement	Silica fume	Viscosity reducer CABR-J1	Sand	Water*	Superplasticizer	Defoamer	Steel fiber	Compressive strength (MPa)
SWZ	756	162	162	1080 (quartz sand)	173(Fresh water)	43.2	0.54	0	119.5
HYZ	756	162	162	1080(sea sand)	173 (NaCl solution)	43.2	0.54	0	112.1
HYSZ	744	159	159	1063(sea sand)	170 (NaCl solution)	42.5	0.53	117	141.8

* Containing water in superplasticizer and defoamer.

https://doi.org/10.1051/e3sconf/202129301024

2.2 Experiments

2.2.1 Specimen preparations

A typical hot-rolled plain steel bar (HPB300) with length of 100mm were prepared and cleaned in accordance with CECS 13-2009 *Fiber Concrete Test Method Standard*. A mold of 100mm×100mm×100mm were used. Two thin plastic plates with two circular holes as shown in Fig. 2 were placed inside the mold close to both ends. Two steel bars were inserted into the circular holes of the plates to ensure that the steel bars are parallel with protective layer thickness of 5mm and 15mm. High performance concrete was mixed with a twin-shaft mixer, poured into the mold. After concrete was hardened, one exposed ends of the steel bars were connected to the wires and the other ends were sealed with epoxy. The specimens were cured in steam at 70 °C for 48 hours and dried at 80±5°C for 24h.

Simultaneously, specimens without steel bars with the same mix proportions were formed. Some were placed in the standard curing room, and the others were dry-wet cycled synchronously with the steel-containing specimens.

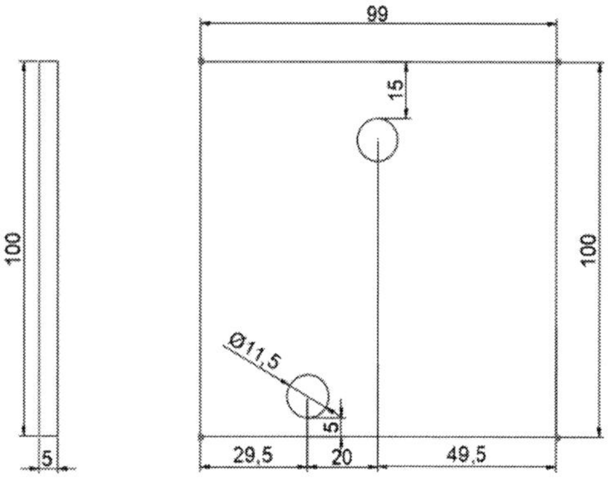

Fig. 3. Dimensions of thin plastic plates (mm)

2.2.2 Experiments

Two types of dry-wet cycles (recorded as "S4D3"and "S2D1") were applied to the specimens. Each dry-wet cycle of "S4D3" was defined as 4-day soaking in 3.5% (mass fraction) NaCl solution and 3-day drying in 60°C±5°C oven. Each dry-wet cycle of "S2D1" was defined as 2-day soaking in 3.5% (mass fraction) NaCl solution and 1-day drying in 60°C±5°C oven. Potential differences between the steel bars and Cu/CuSO4 reference electrod (CSE) were mearused using a multimeter every 4 cycles (for S4D3) or every 5 cycles (for S2D1). After 40 cycles for S2D1, specimens were broken, and corrosion area of each steel bar were test. The concentrations of water-soluble chloride ion at different depths in the specimens without steel bars were tested according to JGJ/T 322 *Technical specification for test of chloride ion content in concrete*.

3 Results and discussion

3.1 Potential differences between the steel bars and CSE

3.1.1 S4D3 cycle (4-day soaking and 3-day drying for each cycle)

Potential differences between the steel bars and CSE for different number of cycles are shown in Fig. 4~Fig. 6.

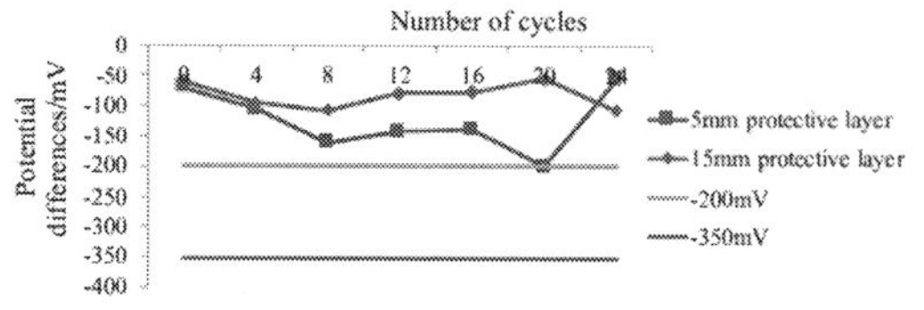

Fig. 4. Potential differences of steel bars with different protective layer thicknesses and CSE of SWZ (for S4D3 cycle)

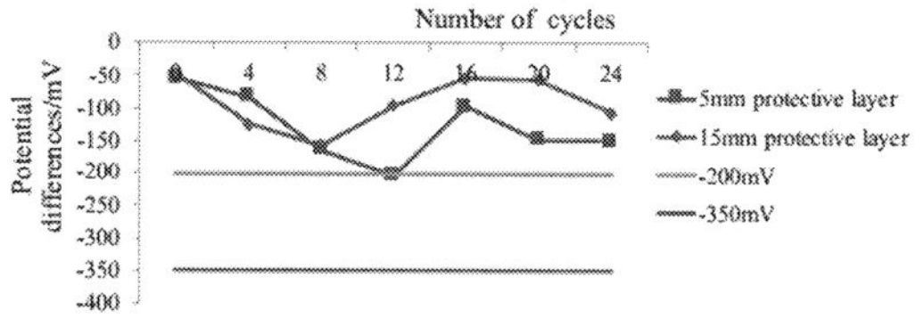

Fig. 5. Potential differences of steel bars with different protective layer thicknesses and CSE of HYZ (for S4D3 cycle)

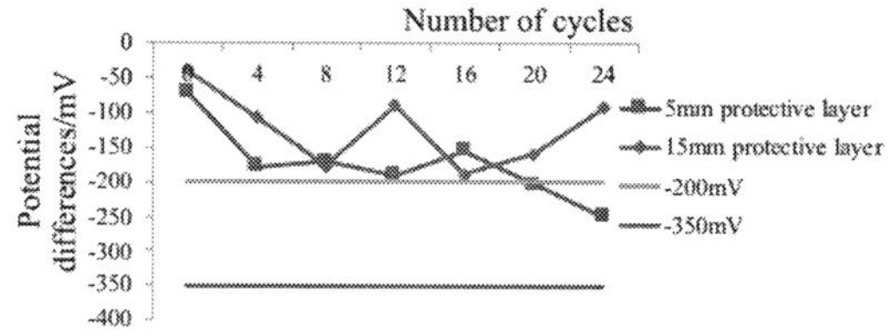

Fig. 6. Potential differences of steel bars with different protective layer thicknesses and CSE of HYSZ (for S4D3 cycle)

As the Fig.s above show, there is no obvious difference in the change trend of the potential differences of different proportions of UHPC, potential differences fluctuate with the number of cycles, mostly potentials of steel bars with 5mm protective layer are positive than that of steel bars with 15mm protective layer, most of potentials are positive than −200mV CSE and all the potentials are positive than −350mV CSE. According to ASTMC876-2015[9], if potentials are more positive than−200mV CSE, there is a greater than 90 % probability that no reinforcing steel corrosion is occurring, if potentials are in the range of −200mV to −350mV CSE, corrosion activity of the reinforcing steel in that area is uncertain, if potentials are more negative than−350mV CSE, there is a greater than 90 % probability that reinforcing steel corrosion is occurring. Therefore, there is a greater than 90 % probability that no steel bar corrosion is occurring in SWZ, HYZ and HYSZ for 40 cycles.

3.1.2 S2D1 cycle (2-day soaking and 1-day drying for each cycle)

Potential differences between the steel bars and CSE for different number of cycles are shown in Fig. 7~Fig. 10.

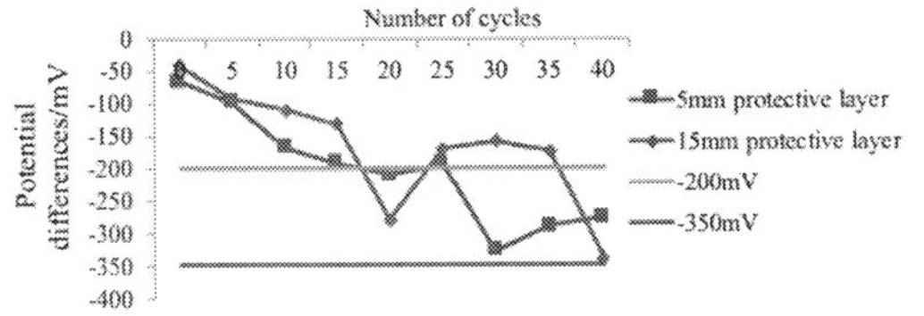

Fig. 7. Potential differences of steel bars with different protective layer thicknesses and CSE of SWZ (for S2D1 cycle)

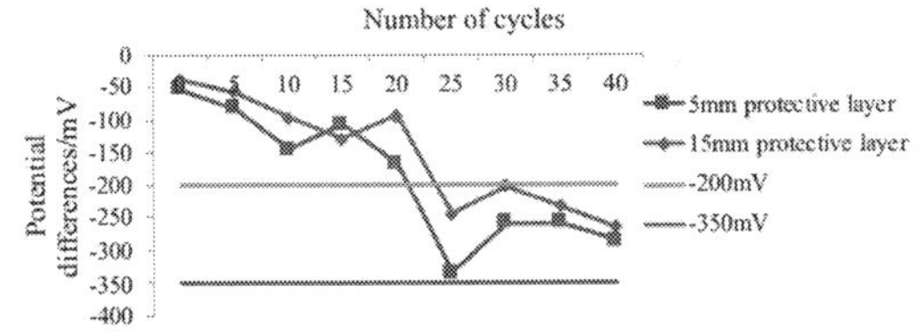

Fig. 8. Potential differences of steel bars with different protective layer thicknesses and CSE of HYZ (for S2D1 cycle)

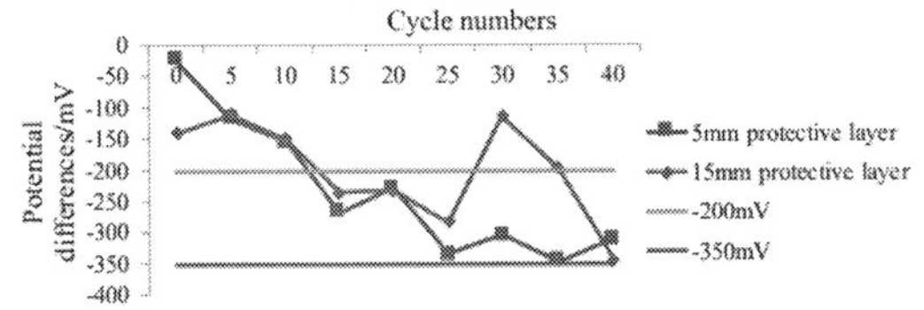

Fig. 9. Potential differences of steel bars with different protective layer thicknesses and CSE of HYSZ (for S2D1 cycle)

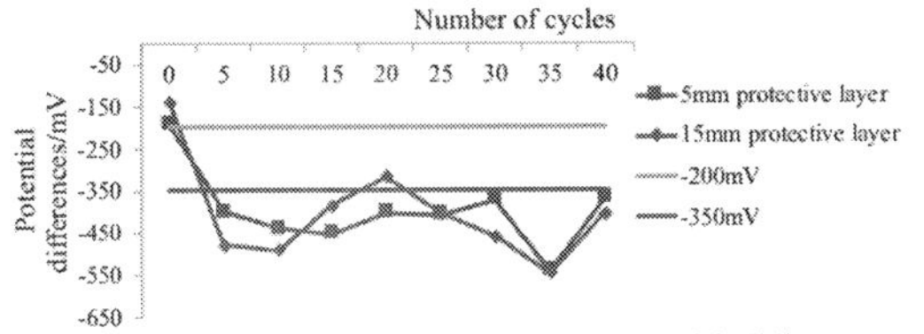

Fig. 10. Potential differences of steel bars with different protective layer thicknesses and CSE of artificially fractured HYSZ (for S2D1 cycle)

It can be seen from Fig. 7 that after 5 S2D1 cycles, the potential differences between steel bar with 5mm and 15mm protective layer and CSE in artificially fractured HYSZ are all negative than -350mV, which means that there is a greater than 90 % probability that steel bars are corroding. It's because the artificial cracks caused chloride ions to quickly enter the specimen and reach the surface of the steel bars, causing corrosion of the steel bars. Compared to S4D3 cycle, the potential differences between steel bars and CSE for S2D1 cycle have a larger fluctuation range. Potential differences in SWZ, HYZ and HYSZ generally show a downward trend with the number of cycles. In addition, the number of cycles corresponding to the first time when the potential difference in SWZ, HYZ and HYSZ is negative than -200 mV is 20 ,25 and 15 respectively. After then, the potential differences are always positive than -350 mV. Therefore, at the beginning of the S2D1 cycle, the probability of steel corrosion is very low. As the cycle continues, the probability of steel corrosion becomes uncertain.

After 40 cycles, all the specimens were broken. The steel bars in artificially fractured HYSZ were severely corroded and the corrosion areas of the two steel bars both exceeded 90%, as shown in Fig. 8 and 9. And in SWZ, HYZ and HYSZ, no steel corrosion occurred. The steel bars in HYSZ were shown in Fig. 11 and 14.

Fig. 11. The steel bar with 5mm protective layer in artificially fractured HYSZ after 40 S2D1 cycles

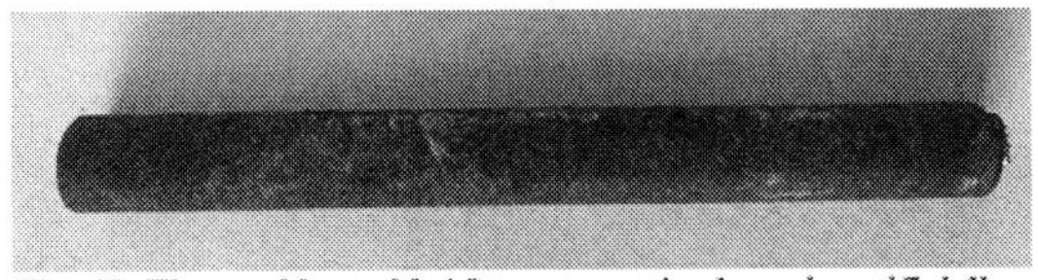

Fig. 12. The steel bar with 15mm protective layer in artificially fractured HYSZ after 40 S2D1 cycles

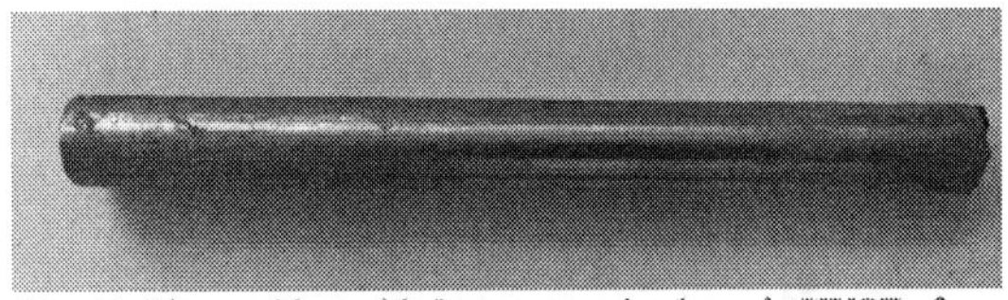

Fig. 13. The steel bar with 5mm protective layer in HYSZ after 40 S2D1 cycles

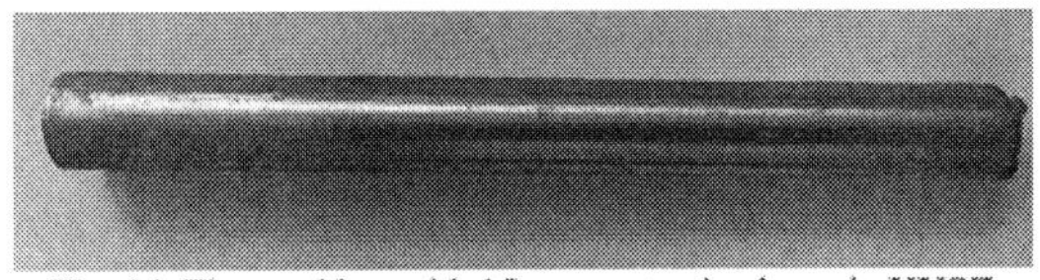

Fig. 14. The steel bar with 15mm protective layer in HYSZ after 40 S2D1 cycles

3.2 Water-soluble chloride ion concentration at different depths

The water-soluble chloride ion concentrations at different depths in UHPC are shown in Fig. 15.

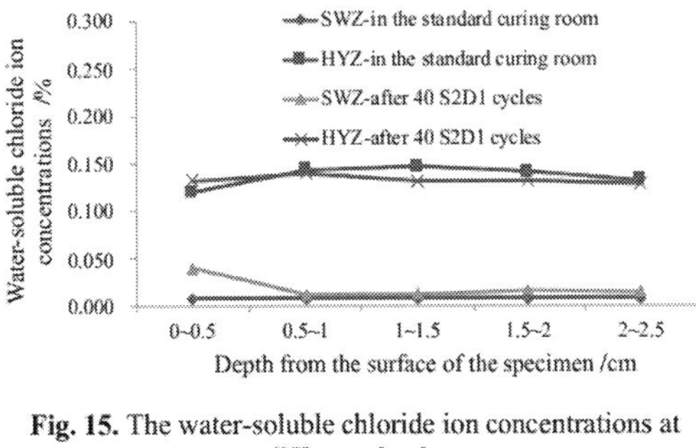

Fig. 15. The water-soluble chloride ion concentrations at different depths

It can be seen from Fig. 15 that the water-soluble chloride ion concentrations in the HYZ specimens are significantly higher than that in the SWZ specimens. Because HYZ specimens are prepared with NaCl solution, and some of the chloride ions from NaCl solution are not converted into bound chloride ions, but still exist as water-soluble chloride ions.

In addition, after 40 cycles of S2D1, the water-soluble chloride ion concentrations at different depths of the HYZ specimens are basically the same and in the range of 0.12% to 0.14%. After 40 cycles of S2D1, the water-soluble chloride ion concentration on the surface of the SWZ specimen is higher, and the water-soluble chloride ion concentrations at the other depths of the SWZ specimen are basically the same, and equivalent to the chloride ion concentrations in the SWZ specimen in the standard curing room, approximately around 0.01%. Because the UHPC specimens after steam curing are dense, even after dry-wet cycles, chloride ions can not penetrate into the specimens.

4 Conclusions

(1) Unpurified sea sand can be used to prepare ultra-high performance concrete (UHPC). The compressive strength of the UHPC which has a proportion of cement, viscosity reducer, silica fume, unpurified sea sand and seawater is 0.7:0.15:0.15:1.0:0.16 by weight, appropriate amount of superplasticizer and defoamer and 1.5% volume of steel fiber can reach 141.8MPa after being cured by 70°C steam for 48h.

(2) The steam-cured UHPC prepared with artificial seawater and sea sand is dense enough to prevent external chloride ions from penetrating into the concrete, and there is no enough water and oxygen on the steel surface even if there are a certain amount of chloride ions from seawater and unpurified sea,the steel will not corrode. So UHPC prepared with seawater and unpurified sea sand has a good ability to protect steel reinforcement against corrosion under dry-wet cycle conditions in lab.

References

1. X. Gu and M. Tian, China Water Transport **11**, 245 (2011).
2. X. Cao, Z. Ling, and J. Guo, Journal of Mianyang Normal University **27**, 121 (2008).
3. Y. Shi, D. Wang, and Z. Wu, Engineering Mechanics **27**, 212 (2010).
4. W. P. S. Dias, G. A. P. S. N. Seneviratne, and S. M. A. Nanayakkara, Construction and Building Materials **22**, 1377 (2008).
5. L. Bian, S. Song, and F. Li, China Concrete and Cement Products 11 (2012).
6. C. Fang, K. Lundgren, L. Chen, and C. Zhu, Cement and Concrete Research **34**, 2159 (2004).
7. H. Li, The Study on Emergency Management of “Sea Sand House,” Dissertation, Huaqiao University, 2012.
8. V. Leelalerkiet, J.-W. Kyung, M. Ohtsu, and M. Yokota, Construction and Building Materials **18**, 155 (2004).
9. G01 Committee, Test Method for Corrosion Potentials of Uncoated Reinforcing Steel in Concrete, ASTM International.

https://doi.org/10.1051/e3sconf/202129301025

Summary of research on pretreatment methods of recycled aggregate

Song Yufeng[1,*], Zhang Enqiang[1]

[1]China Railway 12th Bureau Group Co., LTD, 030024, China, Shanxi

Abstract: The recycled aggregate obtained by simple crushing and screening of waste concrete has a rough surface and many protruding edges and corners, and most of the aggregate surface is attached with old mortar, when directly used as aggregate in concrete, the water demand is large, the concrete strength and elastic modulus are low, and the durability is poor. Therefore, proper treatment of recycled aggregate is required.This article summarizes the research status of the pretreatment methods of mechanical strengthening, heating grinding, solution impregnation, cement slurry impregnation, carbonization treatment, nano-modification, microbial modification and water cleaning etc., and simply analyzes the advantages and disadvantages of various treatment methods, which provide a reference for the research of recycled aggregate processing methods.

1 Introduction

With the continuous development of the construction industry and the frequent occurrence of natural disasters such as earthquakes, more and more construction waste has accumulated, occupying a large amount of land resources, and also causing certain pollution to the environment. At the same time, China is a country with a large consumption of sand and gravel. In the past few years, the amount of sand and gravel has exceeded 15 billion tons per year. In 2020, the amount of sand and gravel has reached 17.826 billion tons. Natural sand and gravel that can be mined are getting less and less. The sand and gravel resources in some areas are very scarce, and the price of sand and gravel is constantly rising, which directly affects the progress and cost of project construction. Therefore, recovering sand and gravel aggregates from construction waste is the current focus of the construction industry.

A large amount of waste concrete will be produced after building disintegration, the recycling aggregate (RA) refers to the concrete aggregate obtained after the waste concrete is broken, graded, cleaned and adjusted according to the requirements of coarse aggregate particle grading according to the national standard[1].

The recycled aggregate particles prepared by simple crushing and screening processes have rough surface and prominent edges and corners[2], meanwhile, the surface of the aggregate is wrapped with much old hardened cement mortar, which is the main difference between the recycled aggregate and natural aggregate[3]. The quantity and quality of the attached hardened old cement mortar are the main factors affecting the physical properties of recycled aggregates[4], the old cement mortar has a porous micro-structure and many micro-cracks, which can adversely affect the strength of the interfacial transition zone when used to produce concrete[5]. At the same time, in the crushing process of waste concrete, a large amount of mechanical damage will be accumulated. Compared with natural aggregate, recycled aggregate has high water absorption rate, large porosity, low apparent density and high crushing value[6,7]. When directly used to produce concrete, more water is required, the hardened concrete has low strength and elastic modulus, meanwhile, the durability indexes such as permeability resistance, frost resistance and chloride ion permeability resistance are lower than ordinary concrete[8]. Therefore, the recycled aggregate needs to be pretreated before it is used.

2 Pretreatment method of recycled aggregate

Through appropriate pretreatment methods, the performance of RA can be improved and enhanced. The common strengthening and modification treatment methods include physical strengthening 、 chemical strengthening、carbonization treatment、nano-material modification、microbial deposition and water cleaning methods.

2.1 Physical strengthening method

Physical reinforcement refers to the further treatment of the RA obtained by physical means. The improvement of RA performance is mainly reflected in two aspects: one is to improve the aggregate grain shape through impact and friction, it can reduce the crushing value of aggregate, reduce the stress concentration inside

* *Song* Yufeng：1246699749@qq.com

concrete, and reduce the probability of crack. The second is to further reduce the old cement mortar wrapped on the aggregate surface, reduce porosity and water absorption of RA, and improve the roughness of aggregate surface. There are many specific methods of physical strengthening, there are roughly two ways, one is the mechanical strengthening method through the collision and friction between particles and machinery or between particles. The other is the heating grinding method which decomposes cement mortar and grinds it by heating. There are many specific methods of physical strengthening, which can be roughly divided into two methods. One is the mechanical strengthening method through collision and friction between particles and machinery or between particles, and the other is to decompose cement mortar and grind it by heating.

2.1.1 Mechanical strengthening

The mechanical strengthening method generally uses ball mills, rod mills and other mechanical equipment to process RA. It uses the collision and friction which between particles or between mechanical equipment and particles to remove the old mortar attached from the surface and grind off the protruding edges and corners, thereby improving the quality of RA.Among them,the vertical eccentric wheel high speed grinding method and horizontal rotary grinding method in Japan and the particle shaping and strengthening method proposed by Professor Li Qiuyi in China are the obvious methods to improve the effect.

①Vertical eccentric high speed grinding method and horizontal rotary grinding method

Japan is relatively advanced in the process of processing RA. It mainly uses vertical eccentric and horizontal rotary devices. The working principle of the vertical eccentric wheel high-speed grinding method is shown in Figure 1. The preliminarily broken concrete block is poured into the bucket, and the eccentric wheel rotates at high speed to squeeze and grind the material to reduce the old cement paste adhering to the surface of the aggregate, and to grind off the more prominent edges and corners on the aggregate, thereby improving the performance of RA[9].

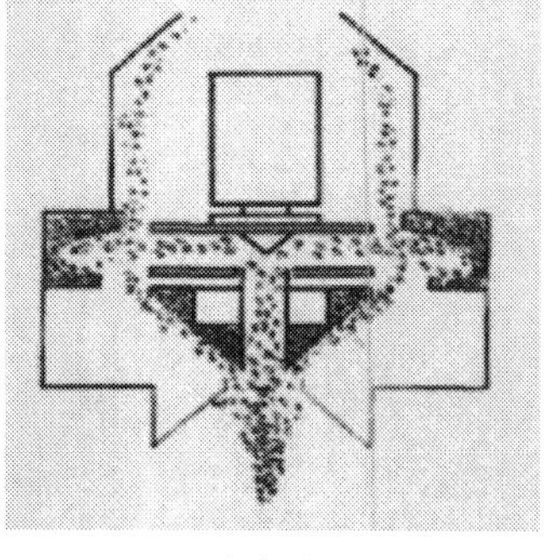

Fig. 1 Vertical eccentric high speed grinding device

The horizontal rotary grinding device is shown in Figure 2. It is similar to the inclined spiral conveyor, except that the spiral blade is transformed into a spiral belt with grinding blocks, and a large number of wear-resistant liners are also arranged on the inner wall of the casing. And at the top of the spiral belt is equipped with a cone that turns opposite to the spiral belt to increase the grinding effect on the material[10].

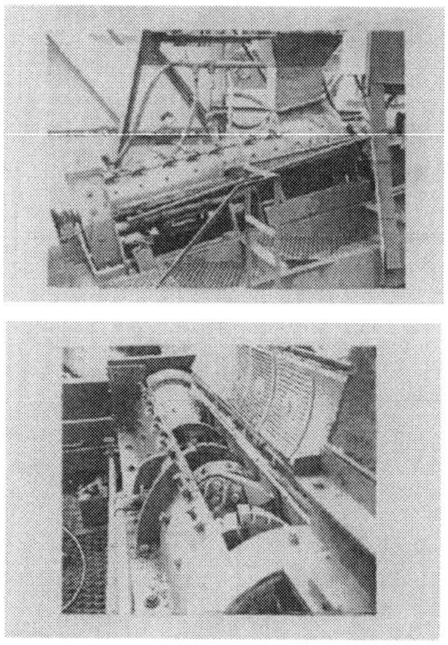

Figure 2 Horizontal rotary grinding equipment

②The particle shaping and strengthening method

The particle shaping and strengthening method[11] is proposed by Li Qiuyi. It uses the collision and friction between high-speed moving particles to effectively improve the particle size and remove the old mortar attached to the surface of the RA. The device is shown in Figure 3. When working, recycled aggregate is added to the machine from the upper port, part of the material enters the interior of the impeller through the top, and is ejected at high speed by the centrifugal action of the impeller when it rotates. The remaining materials fall along the periphery of the impeller and collide with the ejected materials to achieve the purpose of crushing. The crushed material repeatedly hits, cuts and rubs in the cavity many times, so as to achieve the purpose of shaping and strengthening.

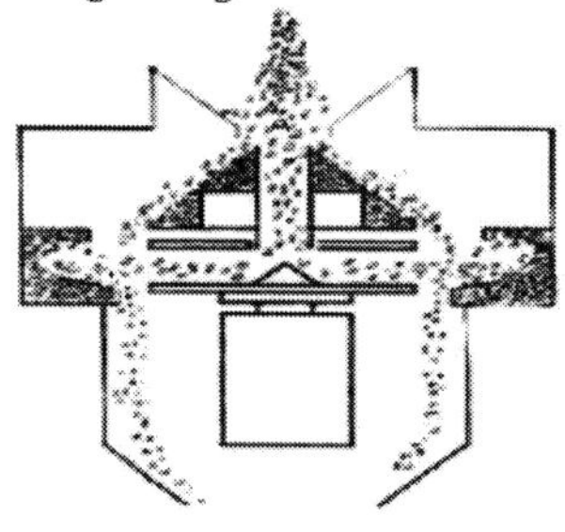

Figure 3 Structure of particle shaping equipment

2.1.2 Heating grinding method

Under the action of high temperature, the calcium hydroxide in the cement mortar will be decomposed into

https://doi.org/10.1051/e3sconf/202129301025

calcium oxide by heating, and the CSH gel will be dehydrated and deteriorated, so the strength of the mortar will decrease. At the same time, the thermal expansion coefficient of coarse aggregate and mortar is different,the thermal expansion coefficient of cement mortar is larger. When the temperature rises, the volume expansion of cement mortar is larger than that of coarse aggregate.When the temperature rises, the volumetric expansion of cement mortar is larger than that of coarse aggregate, compressive stress is formed in the transition zone between the aggregate and mortar. When the temperature is lowered, tensile stress is formed. The existence of tensile stress and compressive stress will make the interface between aggregate and mortar easier to be destroyed[12]. At high temperatures, the decrease in the strength of cement mortar and the presence of the interface stress between the aggregate and the mortar are all conducive to the separation of the old mortar from the recycled aggregate. Heating tools mainly include microwave ovens and muffle stoves, after heating, the separation of mortar and aggregate is not complete, and mechanical equipment is generally required for grinding.

Sui Yuwu's research[13] found that when heating and grinding recycled aggregates, the heating temperature is lower than 300°C, which cannot achieve the effect of separating the old mortar and aggregate. While when heating temperature higher than 500°C, it will cause high temperature damage to the aggregate and significantly increase energy consumption.

Akbarnezhad[14] used high-power microwave (10kw) to heat the RA. The experiment found that thermal stress will be generated between the interface of attached mortar and aggregate during heating,which will delaminate the mortar and aggregate,However, it is found that too high power will cause high temperature damage and cause a certain degree of damage to the original aggregate.

Xiao Jianzhuang[15] used low-power microwaves to cyclically heat the recycled aggregates. The aggregates were heated at 800W for 5 minutes, then taken out and placed in cold water for cooling. After cooling, they continued to be placed in a microwave oven and circulated back and forth. Experiments have found that microwave heating cycle is better than mechanical grinding, and the water absorption and crushing value decrease more obviously. The 28d strength of the recycled aggregate concrete after microwave strengthening is similar to that of ordinary concrete.

The general process of the heating grinding method is as fpllow: after the waste concrete is initially crushed, the crushed materials are put into a heating device for heating treatment to dehydrate and embrittle the cement stone attached on the aggregate, and then the aggregate and cement stone are separated by grinding of the mechanical device. The heating grinding method can significantly improve the performance of RA. Some Japanese factories use the heating grinding method, heating for 60 minutes and then grinding for 120 minutes, which can significantly reduce the amount of mortar attached on the RA surface, and meanwhile can significantly reduce the porosity and water absorption of the RA.

On the basis of heating grinding, the selective heating grinding method is also proposed. Bru[16] adopted selective heating and grinding method, microwave heating treatment was carried out selectively at the junction of aggregate and old mortar, combined with mechanical grinding, it obtained good quality recycled aggregate.

Under the action of mechanical force, the old cement paste is removed, and finally high-purity crushed stone is obtained. However, the technology consumes more energy and resources, requires complex equipment, which leads to higher cost, and the recycled aggregate is often worn seriously, which is not conducive to the long-term performance of the concrete.Nevertheless, mechanical processing of recycled aggregates is still the main method of pretreatment of RA.

2.2 Chemical strengthening method

Chemical strengthening method is to soak RA in a specific chemical solution, through the action of the physical properties of the solution or the reaction between the solution and the cement stone composition, to achieve the purpose of strengthening the RA.The mechanism of solution impregnation can be divided into three categories, one is to weaken the old mortar, make the old mortar deteriorate and easy to separate from aggregate, the second is to strengthen the old mortar, repair crackles and fill cracks; the third is to improve the surface properties of the aggregate, such as forming a film on the surface of the aggregate, which can well reduce the water absorption. According to the difference of chemical solution, it can be divided into inorganic solution impregnation, organic solution impregnation and cement slurry impregnation.

2.2.1 inorganic solution impregnation

①acid solution impregnation

The hydration products of the cement in the recycled aggregate can react and dissolve in the acid solution, which is beneficial to the separation of the old mortar and the primary aggregate. Acid solution impregnation is a simple modification method, which can adjust the modification effect of recycled aggregate by adjusting the type, concentration and impregnation time of acid. The commonly used acids are hydrochloric acid, sulfuric acid, phosphoric acid and glacial acetic acid.

Tam V W[17] comprehensively studied the dissolution of the attached mortar after being impregnated with three different acid solutions of hydrochloric acid, sulfuric acid, and phosphoric acid for 24 hours.Experiments have found that various of acids can react with the cement hydration products attached to the RA, which will dissolve the old mortar, so it can reduce the water absorption rate and improve the compressive and flexural strength of the recycled aggregate concrete. Among them, the hydrochloric acid works best.

Hydrochloric acid has a relatively obvious modification effect on RA[18], the concentration of impregnation is generally 0.1-0.5mol/L, and the impregnation time is less than 3 days. When using acid solution to modify RA, if over-treatment (too high concentration or too long immersion time), it will weaken the surface of the aggregate and increase the number of pores, which may damage the original aggregate. Güneyisi E[19] used SEM to analysis and found that the 0.1 mol/L HCl can better remove the adhered mortar and some loose substances on the surface of the RA, and at the same time, it will not cause obvious damage to the aggregate surface.

Impregnation with acid solution also has disadvantages. It will generate much salt substances such as sulfate ions and chloride ions, which will affect the durability and permeability of recycled aggregate concrete, while glacial acetic acid (acetic acid) is used, the acetate ion generated will not affect the durability of concrete.Compared with other strong acids, acetic acid has obvious advantages, including lower cost, safer operation, and cleaner process.Acetic acid can react with the hydration products in the attached mortar to produce calcium acetate, and calcium acetate can improve the early performance of concrete. Pan Z[20] found that when the acetic acid concentration is 0.2mol/L and the soaking time is 24h, the performance of the recycled aggregate was improved the most. Using the soaked solution as the mixing water could slightly extend the setting time of the cement slurry and increase the initial fluidity and fluidity loss, from a practical perspective, the impact is within an acceptable range.

②Sodium silicate solution (water glass) impregnation

When soaking RA with sodium silicate solution (water glass), on the one hand, sodium silicate can react with Ca(OH)2 in the mortar to form hydraulic calcium silicate colloid, on the other hand, silica acid gel will precipitate after the sodium silicate solution hardens.These two substances can fill the capillary pores, repair the micro cracks on the surface, and form a smooth and hard coating on the surface of the aggregate,which can reduce the water absorption and crushing value of the aggregate.

Gesoğlu's[19] research found that prepreg treatment of the water glass has a very obvious effect on reducing the water absorption. It will form a smooth, strong and dense external coating on the surface of the aggregate, and the water absorption of RA will reduce to about 2%.

Yang L X found[21] that the water absorption rate of RA would decrease with the increase of sodium silicate concentration, but the excessive concentration of sodium silicate would thicken the concrete mixture, thus increasing the water consumption of recycled aggregate concrete. At the same time, he studied the influence of the drying method and soaking time of sodium silicate on the water absorption of RA,and found that the water absorption first increases and then decreases as time goes by,soaking for 1 hour has the largest decrease in water absorption, and drying at 25°C has a more significant effect on reducing the water absorption of RA than drying at 60°C.

Pre-treatment of reclaimed aggregate with sodium silicate solution can significantly reduce water absorption, but the introduction of sodium ions will increase the risk of alkali aggregate reaction, which can be inhibited by adding silica powder

2.2.2 Impregnation with organic solution

①Organic polymer emulsion

Organic polymer emulsion has good bonding performance and can be cured in a short time. After RA is immersed in the emulsion, because the organic polymer has good cohesive force and reactive group activity, it can fill or inhibit the pore structure and micro-cracks. Some organic polymers (such as polyvinyl alcohol emulsion and silicon-based polymers) are hydrophobic and can form a hydrophobic film on the surface of the aggregate, thereby it can reduce the water absorption of RA.Commonly used organic polymer emulsions are: polyvinyl alcohol (PVA), vinyl acetate (EVA), styrene butadiene rubber (SBR), and silane coupling agent solutions and so on.

Kou[22] used PVA solution to treat RA, and the physical and mechanical properties of RA had been significantly improved. The results showed that as the PVA concentration increases, the water absorption rate of RA decreases, the recommended concentration of PVA impregnation is 10%. PVA emulsion has obvious precipitation effect on recycled aggregates, but due to its large molecules, it is not easy to enter the inside of the aggregates, so it needs to be pressed in with a vacuum pump under the action of external pressure when using[23]. The recycled aggregate can also be treated by mixing polymer emulsion, such as mixing acid solution and EVA emulsion, they do not interfere with each other, which can not only improve the surface of the aggregate, but also reduce the water absorption rate.

In addition, using polymer emulsion to modify the recycled fine aggregate can improve the flexural strength of the mortar, and the flexural and compressive strength ratio is also improved, which is close to the strength level of the mortar prepared by ordinary fine aggregate[24].

②Silicone waterproofing agent

Under the action of water and carbon dioxide in the air, the silicone waterproofing agent self-polymerizes to form a layer of silicone waterproof membrane, which can block part of the pores of the recycled aggregate. In addition, it is hydrophobic and can penetrate into the surface of the recycled aggregate to a certain depth, so ,it increases the contact angle of the surface of the recycled aggregate with water and reduces the water absorption rate of the recycled aggregate[25].

2.2.3 Cement slurry impregnation

①Pure cement slurry impregnation

The recycled aggregate is immersed in a plain cement slurry prepared with a higher-strength cement, and then dried.The plain cement slurry penetrates into the aggregate, through adsorption and hydration reaction, it can fill pores, repair cracks and strengthen the old mortar, thereby it can improve the internal structure of

RA. However, some scholars[15] have found that the performance improvement of recycled aggregate concrete is not very obvious with pure cement slurry immersion through experiments .

②Cement mixed slurry impregnation

In addition to pure cement slurry, ultrafine minerals (fly ash, silica fume etc.), waterproofing agent (FeCl3 waterproofing agent, Kim powder waterproofing agent etc.) or calcium sulfoaluminate expansion agent can also be added to the cement,which Formulated into different cement mixed slurry to soak RA. The particle size of the added material is generally one or two orders of magnitude smaller than that of the cement particles, which can fill the space between the cement particles and improve the compactness of the recycled concrete.Du Ting[26] used four different slurries, plain cement slurry, cement slurry mixed with silica fume , cement slurry mixed with Kim powder, and cement slurry mixed with fly ash , to soak RA. It was found that these kinds of grout can improve the apparent density and aggregate strength of RA to a certain extent, and the treatment of cement mixed with Kim powder has the best effect on strengthening the compressive strength of recycled concrete.

Pozzolanic materials such as fly ash and silica fume have active ingredients, which can cause secondary hydration reactions, to further fill and repair the interior of recycled aggregates, and improve the durability of recycled concrete. Tam's[28] research found that silica fume treatment can improve the micro-structure of the transition zone between the old-new mortar and the aggregate-old mortar in recycled concrete.In general, the efficiency of pozzolanic slurry treatment depends on the particle size, the calcium hydroxide content in the adhesive mortar, the alkalinity of the pore solution and the activity of the pozzolanic material[29].

Impregnating RA with cement slurry will form a layer of slurry on the surface of the aggregate, increasing the thickness of the mortar. While increasing the apparent density and improving the internal structure, it will increase the water absorption rate of RA. However, the mechanical properties of the material are not improved much. In addition, when using recycled aggregates to configure concrete, pozzolanic materials can also be mixed in the form of mineral admixtures

2.3 Carbonization treatment

Carbonization treatment refers to the strengthening of RCA with CO_2. It is based on the carbonization of concrete and CO_2 mineralization and storage technology. CO_2 can react with $Ca(OH)_2$ and C-S-H in the mortar attached to the recycled aggregate, and the reaction products are mainly calcium carbonate and silica gel[30]. The main reaction equation is as follows:

$$Ca(OH)2(s)+CO2\rightarrow CaCO3(s)+H2O(l) \quad (1)$$

$$C\text{-}S\text{-}H(s)+CO2(g)\rightarrow CaCO2(s)+SiO2\cdot nH2O(s) \quad (2)$$

The reaction products fill between the pores, increasing the apparent density of RA, reducing the porosity, improving the compactness of the attached slurry, making the RA structure denser and reducing its water absorption. In addition, CO_2 can also react with the mineral components (C_2S, C_3S) in the cement particles that are not involved in the hydration reaction in the attached mortar, which increase the volume of the solid phase[31].

Under the condition that the CO_2 concentration exceeds 99% and the pressure is 0.1bar, kou had studied the influence of different carbonization curing time (6, 12, 24, 48, 72h) on RA, and found that the longer the carbonization curing time, the lower the water absorption of RA, the better the aggregate quality.But when 72h compared with 48h, the properties of RA changed little[32].In the process of curing concrete by CO_2, there is a maximum value for the improvement of concrete performance[33].

CO_2 reinforced recycled concrete aggregate is affected by humidity, CO_2 concentration, pressure, time and temperature as well as the properties of recycled aggregate (water content, particle size, aggregate strength). After carbonization of aggregate, the mechanical properties and durability of recycled aggregate concrete can be significantly improved.

When CO_2 strengthens recycled concrete aggregates, it is affected by humidity, CO_2 concentration, pressure, time and temperature, and the properties of recycled aggregates (water content, particle size, aggregate strength).After carbonizing of RA, it can obviously improve the mechanical properties and durability of recycled aggregate concrete[34].

Carbonization treatment is a new type of aggregate strengthening treatment technology, which can significantly improve the surface properties of RA and improve the performance of recycled concrete.However, the technology has not yet been put into actual production, further research is needed on the internal structure of the concrete and the corrosion of the steel bar with carbonization treatment. .

2.4 Nano material modification treatment

Nano-materials refers to the ultra-fine materials with a size of 1-100nm, it has the characteristics of small size, strong permeability, large specific surface area, and many active atoms. Take active nano SiO_2 as an example,when mixed into concrete, it can fill the pore of aggregate, compact structure; and it is active and can react with calcium hydroxide to form C-S-H gel; while it has a crystal nucleation effect, which connects the sparsely distributed hydration products into a denser network structure. Thereby, it can optimize ITZ and improve the strength and durability of concrete[35]. When the recycled aggregate is soaked in the slurry which mixed with nano material ,it can fill the voids, optimize the surface of the aggregate, finally improve the performance of the aggregate.

Some scholars use nano-SiO_2 solution to soak the recycled aggregates. Experiments have found that the durability of the modified recycled aggregate concrete is improved, and the mechanical properties are close to

https://doi.org/10.1051/e3sconf/202129301025

natural concrete.The flexural and shear strength of the beam members which made of the recycled aggregate modified by nanomaterials are improved to a certain extent[36,37].

Nanomaterials can also be mixed into recycled concrete in the form of admixtures. Yang Qing[38] mixed high-activity nano-SiO_2 and nano-modified mineral admixtures into recycled concrete. The results show that the combination of the two materials can better fill the microscopic pores in the transition zone between the new and the old interface, increase the strength of the ITZ, and enhance the internal compactness of the concrete.For that, the strength and impermeability of recycled concrete are improved

Nanomaterials can reduce the amount of cement, improve the strength and durability of recycled concrete, especially the early performance, these are all based on the optimization of microstructure and the promotion of hydration reaction by nanomaterials. While, nano-materials are easy to agglomerate and are difficult to stir evenly, which has a great impact on the workability of the fresh concrete. Moreover, nanomaterials are generally more expensive. so the use of nanomaterials to pretreat recycled aggregates cannot be promoted and applied, and the content of nanomaterials is generally less than 7% of the total amount of cementitious materials[39,40].

2.5 Microbial mineralization and bio-deposition technology

Microorganisms can produce calcite through mineralization and deposition during metabolism (carbonate mineral, CaCO3) during metabolism. Boquet[41] discovered this phenomenon for the first time in 1973, in his experiment, it was found that soil bacteria could induce the deposition of calcium carbonate minerals. In 1999, Stocks[42] applied the phenomenon of Microbial mineralized deposits to the repair of concrete cracks. Microbial mineralization and bio-deposition technology is a new technology for treating RA. It uses the minerals produced by microbial bio-mineralization to fill or bond the pores of RA, so it can improve the pore structure and repair the micro-cracks on the surface to strengthen the material[43,44].Generally, after the appropriate strains are grown with the culture medium, they are formulated into a bacterial solution to soak the RA with a treatment seal.

The basic advantage of it is that the bacteria is small in size and can enter the micro-pores that cement or fly ash cannot enter. Grabiec's[45] research found that microbial treatment technology can improve the surface structure of RA and reduce the water absorption of recycled concrete aggregates. Zhu[46,47] used DSM8715 strain culture solution to soak the recycled coarse and fine aggregates. Through XRD and SEM experiments, it was found that the strains would form calcite-type calcium carbonate precipitates on the aggregates, which reduced the water absorption rate and crushing index of the aggregates. Wu[48] treated RA with pseudobacillus (DSM8715) and found that the smaller the RA's particle size, the more the water absorption of the aggregate decreases, and the less the aggregate crush value decreases. It may be related to the proportional formation of calcium carbonate on the surface of the aggregate

While microbial mineralization and bio-deposition technology improves the quality of recycled aggregates, it does not pollute the environment and consumes less energy. At present, this microbial deposition process has been applied to a certain extent in the repair and durability improvement of building materials[49]. However, the process of microbial mineralization and deposition to generate calcium carbonate is a slow growth and metabolism process, when it is used to modify the recycled aggregate, it is difficult to penetrate into the aggregate which takes a long time period, generally about 20 days. So in order to popularize the application, we need to study the method of shortening the time

2.6 Water cleaning treatment

The water cleaning treatment method is mainly to wash the RA with water, which can wash away the impurities, loose particles and old mortar with weak adhesion in the aggregate. It is generally treated by rinsing or soaking in circulating water.

A method of water cleaning is mentioned in the literature[50]. The recycled aggregate is placed in the rotating mill drum, the volume of which is about one third of the volume of the rolling mill drum. The drum rotates at about 60r/min for 10 to 15 minutes, and then soaks aggregate with water. Compared with the general water cleaning method, this method can improve the performance of aggregate better. Katz[51] used the repeated ultrasonic washing method to wash the recycled aggregate in a water bath, and found this method can remove the weakly adhered mortar, the compressive strength of the recycled aggregate concrete after the ultrasonic modification treatment increased by 7%. Meanwhile, it was found that compared with high-grade concrete, low-grade concrete would produce more fine particles when it is broken, which requires more cleaning cycles to clean the recycled aggregates.ELhakam[52] tried to soak RA in water for up to 30 days to improve the quality of RCA. The results showed that after immersion treatment, the 7d and 56d compressive strength of concrete increased by 4.8% and 15.7% respectively.

Compared with other modified and strengthened methods, water cleaning is easier to achieve, and the operation is simple, but it can only wash away loose loose particles and mortar with weak adhesion, and the optimization performance is limited. At the same time, if there is no thorough recycling process, it will consume much water resources and cause certain pollution to the environment. Water cleaning treatment methods can generally be used in combination with other treatment methods.

3 conclusion

At present, there are many modification methods of RA, among which physical method is the most commonly used, but it has high energy consumption and great wear on the recycled aggregate. The chemical method has a simple process, but the influence of some possible reactions on the durability is not clear, so it has not been applied in engineering practice on a large scale. I believes that the incorporation of pozzolan materials can effectively fill the pores of the mortar on the recycled aggregate and the voids in the interface transition zone, causing secondary hydration reactions, improving the interface transition zone of recycled aggregate concrete, and the cost is low. It is very suitable for improving the performance of recycled aggregate and recycled concrete. I believe that adding pozzolanic material can effectively fill the pores of the adhesive mortar on the recycled aggregate and the gap of the ITZ, which can lead to the secondary hydration reaction ,it can improve the ITZ of the recycled aggregate concrete, and the cost is low, so it is a very suitable method for improving the performance of the recycled aggregate and recycled concrete.

Nano-material modification is relatively expensive, and it is impossible to get large-scale application. Carbonization treatment is a very potential method, which can not only improve the quality of recycled aggregate to improve the mechanical properties and durability of recycled concrete, but also reduce carbon emissions. It has high economic and environmental benefits. However, there are still problems such as the collection of industrial CO2, and it is still in the experimental research stage. The process of microbial modification has no pollution and low energy consumption, although it takes a long time, it is still a method worthy of further study.

At present, most of the experiments and studies on the modification of recycled aggregate focus on the study of mechanical properties and various short-term indicators. We should consider the long-term use performance of recycled concrete and conduct further research on its strength stability and durability.

References

1. Su Wenyang, Hong Li, Li he Shan . Engineering and Construction . 2020, 34(06): 1197–1200.
2. Yao Yao, Liu Hongmei, Wang Jianjun, Mile, Li Huanhuan..Journal of Materials Science and Engineering. 2019(02)Page:339-344
3. Ye Ting. The Performance of Modified Pervious Asphalt Concrete with Recycled Coarse Aggregate.[D] Nanchang University ,2020.
4. Gutiérrez A. International RILEM Conference on the Use of Recycled Materials in Building and Structures. RILEM Publications SARL, 2004: 536-544.
5. Xiao Jianzhuang, Liu Qiong, Li Wengui. Journal of Qingdao Technological University. 2009(04) Page:24-30
6. Yi Chao. The Study on the Resource Recovery of Construction Waste for Recycled Aggregate Concrete[D].Jinan University,2014.
7. Peng GaiFei, Huang Yanzhu, Zhang Jiufeng. Journal of Building Materials. 2012(01) Page:80-84
8. Li Qiuyi Concrete recycled aggregate [M]. China Architecture & Building Press.2011.2
9. Zhan Xiangyu, Gao Ruize. Construction Wall Innovation & Building Energy-Saving. 2019(12) Page:67-69
10. Li Qiuyi, Li Yunxia, Zhu Chongji, Tian Li.Concrete. 2006(01) Page:74-77
11. Li Qiuyi, Li Yunxia, Zhu Chongji. Materials Science and Technology. 2005(06) Page:579-581+585
12. Wu Xubiao, Zhou Guangli. Journal of Guizhou University of Finance and Economics. 2020(04) Page:147-150
13. Sui Yuwu, Tian Qingbo, Yue Xuetao, Zhang Fengqing. Journal of Shandong Jianzhu University. 2015(06) Page:544-549
14. A. Akbarnezhad, K.C.G.Ong, M.H.Zhang, et al. Construction and Building Materials, 2011, 24(8):3469-3479.
15. Xiao Jianzhuang, Wu Lei, Fan Yuhui. Concrete. 2012(07) Page:55-57
16. Bru K, Touzé S, Bourgeois F, et al. International Journal of Mineral Processing, 2014, 126: 90-98.
17. Tam V W Y, Tam C M, Le K N. Resources, Conservation and Recycling, 2007, 50(1): 82-101.
18. Kim Y, Hanif A, Kazmi S M S, et al. Journal of cleaner production, 2018, 191: 339-349.
19. Güneyisi E, Gesoğlu M, Algın Z, et al. Construction and Building Materials, 2014, 64: 172-183.
20. Pan Z, Wang S, Liu Y, et al. Journal of Cleaner Production, 2019, 235: 866-874.
21. Yang L X, Yao-Li Q, Fang Q Q, et al. 3rd International Conference on Material Engineering and Application (ICMEA 2016). Atlantis Press, 2016: 181-186.
22. Kou S C, Poon C S. Cement and Concrete Composites, 2010, 32(8): 649-654.
23. Wei Yipeng, Ye Qiaoling, Guo Zirui. China Southern Agricultural Machinery. 2019(04) Page:198+231
24. Wang Ziming, Pei Xuedong, Wang Zhiyuan. CONCRETE. 1999(02) Page:44-47
25. Chen Deyu, Yuan Wei, Liu Huan. New Building Materials. 2009(02) Page:20-23
26. Du Ting, Li Huiqiang, Wu Xianguo.New Building Materials,2002(03):6-8.
27. S.C. Kou; C.S. Poon (2012).Enhancing the durability properties of concrete prepared with coarse recycled aggregate. , 35(none), 69–76.
28. Vivian W.Y. Tam X.F. Gao; C.M. Tam (2005). Microstructural analysis of recycled aggregate concrete produced from two-stage mixing

approach. , 35(6), 1195–1203.

29. Singh, L.P.; Karade, S.R.;Bhattacharyya, S.K.; Yousuf, M.M.; Ahalawat, S. (2013).Beneficial role of nanosilica in cement based materials – A review. Construction and Building Materials, 47, 1069–1077.
30. Li Linkun, Liu Qi, Ma Zhong, Peng Bo. Thermal Power Generation. 2021(01)Page:94-103
31. Shi C, Li Y, Zhang J, et al. Journal of Cleaner Production, 2016, 112: 466-472.
32. Kou S C, Zhan B, Poon C S. Cement and Concrete Composites, 2014, 45: 22-28.
33. Zou Qingyan. Carbon dioxide curing concrete technology research [D].Central South University,2008.
34. Pu Yunhui, Shi Xiaoshuang, Wang Qingyuan, Li Wenyuan, Fu Ling.Concrete,2020(06):65-69.
35. Yan Hongsheng. Experimental Study on Mechanical Properties of Recycled Coarse Aggregate Concrete Strengthened with Nano SiO2[D].Qingdao Technological University,2018.
36. Zhang H, Zhao Y, Meng T, et al. Journal of Materials in Civil Engineering, 2016, 28(2): 04015094.
37. Zhu Yongnian, Zhang Hongru, Meng Tao, Zhao Yuxi. Concrete 2014(07):138-144.
38. Yang Qing, Qian Xiaoqian, Qian Kuangliang, Wang Zhangfu, Zhou Tanggui. Journal of Materials Science and Engineering. 2011,29(01):66-69+130.
39. Wang Yujie, Liu Binghua. Shandong Communications Technology. 2014(02):39-41.
40. Meng Tao, Qian Kuangliang, Qian Xiaoqian, Zhan Shulin. Rare Metal Materials and Engineering. 2008, 37(S2):667-669.
41. Boquet E, Boronat A, Ramos-Cormenzana. Nature, 1973, 246(5434): 527-529.
42. Stocks-Fischer S, Galinat J K, Bang S S. Soil Biology and Biochemistry, 1999, 31(11): 1563-1571.
43. Zhang Xiaotong. [D].Qingdao Technological University,2016.
44. Wu Chunran. [D].Qingdao Technological University,2018.
45. Grabiec A M, Klama J, Zawal D, et al. Construction and Building Materials, 2012, 34: 145-150.
46. Zhu Yaguang, Wu Yankai, Wu Chunran. China Concrete and Cement Products,2017(12):93-96.
47. Zhu Yaguang, Wu Chunran, Wu Yankai, Xu Peizhi. Concrete,2018(07):88-92.
48. Wu C R, Zhu Y G, Zhang X T, et al. Cement and Concrete Composites, 2018, 94: 248-254.
49. N.K. Dhami, A. Mukherjee, M.S. Reddy, Biofilm and Microbial applications in Biomineralized concrete, in: Jong Seto (Ed.), Adv. Top. Biominer, InTech, 2012,pp. 137–164.
50. Mistri Abhijit; Bhattacharyya, Sriman Kumar; Dhami, Navdeep; Mukherjee, Abhijit; Barai, Sudhirkumar V. (2020). Construction and Building Materials, 233(), 117894
51. Katz, Amnon (2004).Treatments for the Improvement of Recycled Aggregate. Journal of Materials in Civil Engineering, 16(6), 597–603.
52. Ali Abd Elhakam; Abd Elmoaty Mohamed; Eslam Awad (2012).Influence of self-healing, mixing method and adding silica fume on mechanical properties of recycled aggregates concrete.,35(none)

 https://doi.org/10.1051/e3sconf/202129301026

Discussion and analysis of environmentally friendly refrigerant R404A as a substitute for R22 with different circuit number

Xue Wang[1,*], Yu Zhang[1], and Ruoyu Wang[1]

[1]Shandong Huayu University of Technology, College of Energy and Environmental Engineering, 253034, Dezhou, Shandong, China

Abstract. The refrigerant R22 has great damage to the ozone layer. R404A is currently the mainstream environmentally friendly refrigerant approved and recommended by most countries in the world, and has zero damage to the ozone layer. Based on this, the performance of heat exchangers of R22 and R404A with different inlet qualities with the circuit number were simulated and compared by EVAP-COND software. The results show that the capacity of the evaporator with R404A was greater than that with R22. Moreover, the difference in the capacity of the evaporator with R410A and R22 drops first, and then it rises. The flow path with the number of circuits 2 had the largest heat exchange rate. The heat exchange rates of R22 and R404A are 4.40kW and 4.50kW, respectively. The heat exchange amount of R404A was larger than that of R22, and the difference in heat exchange amount decreases first and then increases with the increase of branches, and the maximum difference was 4.41%. R404A can be used as a substitute for R22 and had a higher heat exchange rate. It is more environmentally friendly and saves energy.

1 Introduction

Heat exchanger is a kind of heat exchange equipment widely used in industrial fields such as refrigeration, air conditioning, and chemical industry.The finned tube heat exchanger was a heat exchange element that increases the heat transfer area on the outside of the tube to improve the heat transfer capability of the air side. It has the advantages of low manufacturing cost, simple structure and strong applicability. With the development of industrial production, higher requirements are put forward for the performance of heat exchangers. In order to further strengthen the heat transfer capacity of the finned tube, it was necessary to conduct an in-depth study on the fin structure and arrangement. The number of loops had a particularly significant influence on the heat exchanger.

Many researchers have done various researches. Huang Dong et al. simulated and compared the performance of the heat exchanger with the number of circuits when R22 and R410A refrigerants were used in the indoor heat exchanger.[1] Wang Yunli et al. studied the flow of the four-row fin-tube indoor heat exchanger for evaporation and condensation when the air inlet state and air volume, the structural parameters of the heat exchanger, and the refrigerant mass flow ware constant. The influence law of heat transfer performance.[2] Liu Jinping et al. studied the effect of branch number and tube row number on heat transfer oefficient and flow characteristic under the condition of the same geometry size of condenser, air inlet temperature and flow rate was analyzed. The results show that pressure drop decreases as branch numberincreases.[3] Yang Ting et al. used computer simulation technology under the conditions of the same geometrical structure and the same air inlet state and flow rate of the air-conditioning finned tube condenser. The influence of the number of loops and the number of tube rows on the heat transfer and flow characteristics of the finned tube condenser was studied.[4]

Previous research mainly focused on R22 working fluid. R404A is one of the main alternative working fluids of R22. It is necessary to study the difference between the number of branches on the R22 and R404A condensation and evaporation dual-use phase change heat exchangers, laying a foundation for R404A to replace R22. Theoretical basis. The refrigerant R404A complies with the standards of the US Environmental Protection Organization EPA, SNAP and UL, and complies with the A1 safety level category of the American Society of Heating, Refrigeration and Air-Conditioning Engineers (this is the highest level and harmless to humans). The fin spacing is very small, and it is more difficult to measure the gas-liquid two-phase flow. It is difficult and the workload to conduct synchronous experimental research on the local characteristics of the heat transfer of the fin-tube heat exchanger from both sides of the air and refrigerant. Simulation can better obtain the local characteristic parameters of the heat exchanger. As a result, this article intends to comprehensively improve the efficiency of heat pump air conditioning cooling and heating, and to study the feasibility of R404A as a substitute for R22. In this paper, for R22 and R404A refrigerants, simulation software was used to simulate the fin-tube heat exchangers with different numbers of circuits, and the heat transfer of the heat exchangers under different inlet

* Corresponding author: tlwx1994@163.com

qualities of the refrigerants was analyzed to explore the number of circuits exchange.

2 Simulation procedure and Restriction

2.1 Simulation procedure

This article adopts the software EVAP-COND designed by Domanski of NIST in the United States. It had undergone more than ten years of development and improvement. It had a strong representative in the numerical simulation of finned tube heat exchangers and had a wide range of international influences. EVAP-COND adopted a tube-by-tube calculation scheme, that is, each heat exchange tube was used as an independent calculation unit. The state and flow of refrigerant and air need to be input separately. For multi-branch arrangement, it was necessary to calculate each branch one by one.[5] In a branch, the calculation was carried out according to the sequence of piping layout and refrigerant flow. The exit parameter of the refrigerant in the former tube was the population parameter of the latter tube. For the bifurcation in the branch, first calculate one of the branches to the exit, and then back to the bifurcation to calculate the other branches.

2.2 Restriction

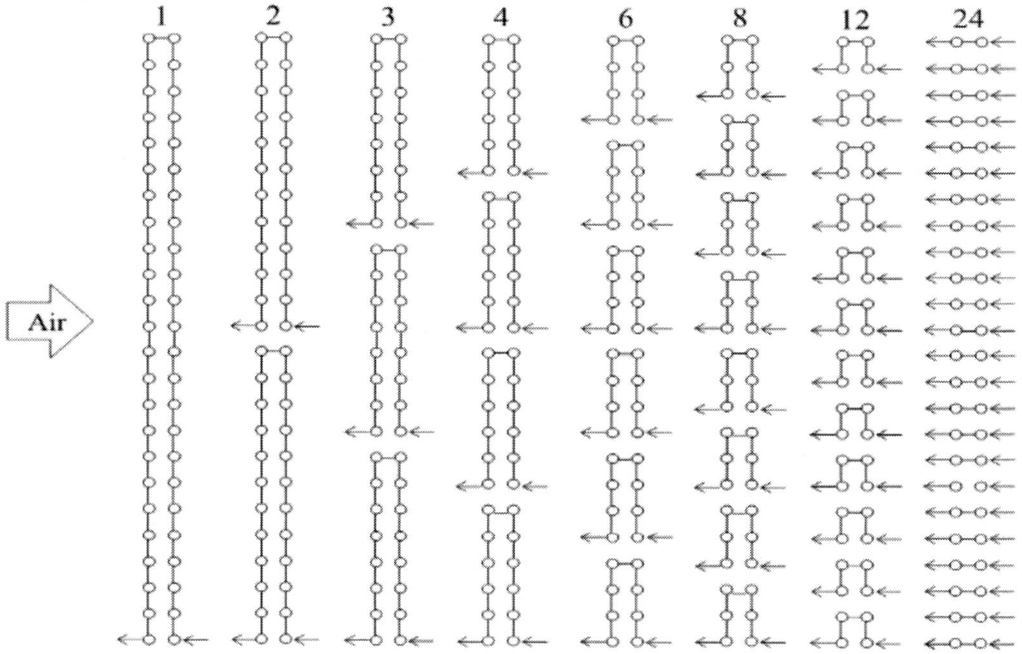

Figure 1. Refrigerant Circuitry.

This paper was studied the difference between the number of circuits on the performance of R22 and R404A refrigerants in the heat pump evaporator, and laid a theoretical foundation for R404A to replace R22. The circuit layout of the refrigerant and the air flow direction were shown in Figure 1. The number of circuits was 1, 2, 3, 4, 6, 8, 12, and 24 respectively. Explore the influence of the number of circuits on the performance of the heat exchanger.

The outlet pressure and temperature of the heat exchanger will not only affect the flow rate of the compressor, but also reflect the evaporation temperature and the size of the superheated section in the heat exchanger. Therefore, the refrigerant outlet state of the heat exchanger was taken as the simulated operating conditions, as shown in Table 1. The inlet quality reflects the state that the refrigerant enters the evaporator after being throttled by the throttling part. The evaporator mainly transforms heat through the evaporation phase of the refrigerant. The greater the dryness, the more flashed gas, and the evaporator can be used. The less effective refrigerant amount, the worse the heat transfer capacity, the lower the dryness, the better the heat transfer capacity.

Table1. Simulation conditions.

Refrigerant		
Inlet quality	Outlet saturation temperature/ °C	Inlet saturation temperature/ °C
0.1-0.3	5	5
Air inlet		
Dry bulb temperature/°C	Relative humidity/%	
27	50	

The dry and wet bulb temperatures of the inlet air adopt the data of the standard refrigeration conditions in the national standard. In order to make the research results reference and comparative, the structure parameters of the heat exchanger were the same as those in the literature[6], as shown in Table 2.

Table2. Parameters of evaporator.

Parameters	Number
Tube length/mm	400
Tube size/mm	9.52×0.35
Pipe spacing/mm	25.4
Row spacing/mm	22
Number of tubes	3×24
Fin thickness/mm	0.11
Fin pitch/mm	1.5
Air flow/m²/min	25

3 Simulation results and analysis

3.1 Simulation results of R22 and R404A

In this simulation, the outlet saturation temperature was 5°C, and the dryness of the refrigerant was variable. Figures 2 and 3 showed the changes in the heat exchange capacity of different circuit numbers when the inlet quality of different refrigerants R22 and R404A changed.

It can be seen from Figure 2 that the heat exchange amount of R22 first rised and then decreased as the number of circuits increased. When the number of circuits was 2, the heat exchange was the largest, and when the number of circuits was 24, the heat exchange was the smallest. And the curve with loop number 1 intersected with the curve with loop number 6. The inlet quality of the refrigerant had a greater impact on the 24 branches and less on the heat exchangers of other circuits. When the number of circuits were 2, 3, 4 and 6, the heat exchange amount increased as the inlet quality of the refrigerant increased. There was only 1 circuit and the heat exchange amount decreased with the increase in the inlet quality. The number of circuits 8, 12, and 24, the change in heat exchange was not monotonous.

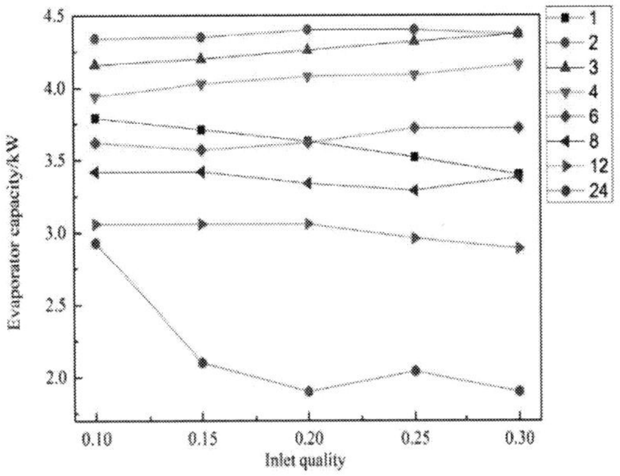

Figure 2. R22 heat exchange.

It can be seen from Fig. 3 that the heat exchange rate of R404A varies with the number of circuits in the same trend as R22, it rised first and then falled. Inlet quality had a small effect on heat transfer, with an overall upward trend, but the flow path with the number of loops shown a downward trend, and the changes in the 12 and 24 branches were irregular.

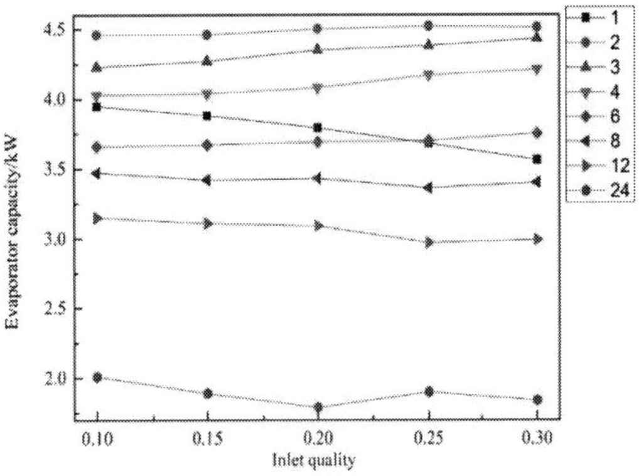

Figure 3. R404A heat exchange.

The Evaporator capacity of all methods increased and then decreased as the dryness increased. The main reason was that the total heat transfer coefficient first increased and then decreased with the change of inlet quality. In the first stage, the disturbance between the bubble and the fluid and between the bubble and the bubble generated by the boiling will intensify the heat transfer effect. Immediately, the total heat transfer coefficient decreased slightly. This was due to the fact that as the refrigerant continues to evaporate during the heat exchange process, the number of bubbles increased, and many small bubbles gradually aggregate to form a block or even a film, which inhibits the refrigerant and the tube wall. Heat transfer. Finally, in the heat exchange process, the liquid refrigerant in the pipe enters the lack of liquid, and the phenomenon of partial evaporation occurs. At the same time, the moisture desorption coefficient on the air side was reduced, and the heat exchange effect on the inner and outer sides was affected, resulting in a sharp decrease in the coefficient.[7]

3.2 Comparison results of R22 and R404A

When used as an evaporator, the difference in heat exchange between R410A and R22 was affected by the heat transfer temperature difference and the heat transfer coefficient, and the heat transfer temperature difference is determined by the inlet temperature of the refrigerant and the outlet temperature of the air. The outlet state of the evaporator is certain.[8]

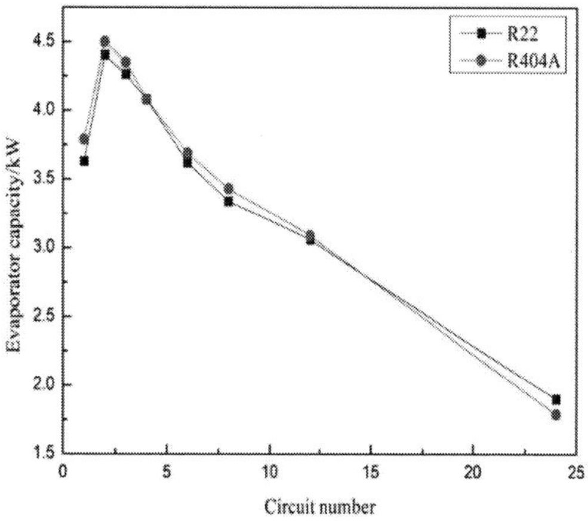

Figure 4. Comparison of R22 and R404A.

In this simulation, the inlet quality value was 0.2, which compares the heat exchange rate of R22 and R404A. It can be seen from the figure that the changes of the two working fluids first increased and then decreased with the increase of the number of circuits, and the difference in heat exchange first decreases and then increases with the increase of the number of circuits. And the heat transfer capacity of R404A was basically higher than that of R22. When the number of circuits was 1, the heat exchange amount of R404A was 4.41% higher than that of R22. When the number of circuits was 2, 3, 6, and 8, the heat exchange amount of R404A was about 2% higher than that of R22. When the number of circuits was 4, they The heat exchange was equal. When the number of circuits was 24, the heat transfer capacity of R22 was 4.33% higher than that of R404A.

4 Conclusion

This paper compares the performance of the evaporator with the number of loops when the refrigerant is R22 and R404A under the same conditions of the air inlet state and flow, the geometrical structure of the heat exchanger, and the pipe arrangement. The conclusions are as follows:

(1) The heat exchange amount first increases and then decreases with the increase in the number of circuits. The flow path with the number of circuits 2 has the largest heat exchange amount. The heat exchange amounts of R22 and R404A are 4.40kW and 4.50kW, respectively.

(2) Inlet quality has less influence on heat exchange. When the inlet quality value is 0.2, the heat exchange rate of R404A is about 2% higher than that of R22 in most flow paths. The difference of their heat exchange first decreases and then increases with the increase of the number of circuits.

(3) The heat exchange amount of R404A is larger than that of R22, with a maximum difference of 4.41%. It can be seen that R404A can not only be used as a substitute for R22, but also has better performance than R22.

Acknowledgement

This work was supported by Shandong Huayu University of Technology and Dezhou's Key Laboratory of High-efficiency Heat Pump Air Conditioning Equipment and System Energy Saving Technology (No.26).

References

1. Huang Dong, Li Quanxu. Performance Comparison of Evaporator with Circuit Number Change in Heat Pump Using R22 and R410A. Journal of Xi'an jiaotong university. J. **43** (2009)
2. Wang Yunli, Yang Bing, Li Ming. Influence of circuit number on performance of R410A indoor heat exchanger. Proceedings of the 6th China Refrigeration and Air Conditioning Industry Information Conference. C. (2012)
3. Liu Jinping, Zheng Qingling. Numerical Simulation of Performance of Fin-and-tube Condenser with Different Structural Arrangement in Air-conditioning Systemand. Refrigeration. J. **28** (2009)
4. Yang Ting. Experiment study and theoretical analysis on the performance of fin packed condensers. Shanghai Jiao Tong University. D. (2014)
5. Fu Yongjun. The Application of EVAP-COND in Rotor Compressor Matching. Journal of Shunde Polytechnic. J. (2016)
6. Wang Qiang, Liu Yanlong, Liu Zuyi, et al. An Optimized Design and Experimental Research on Finned-tube Evaporator with Nonuniform Air Distribution. Journal of Refrigeration. J. **37** (2016)
7. Li Xia, Weng Wenbing, Mo Pengcheng, et al. Study on heat transfer performance of fin-tube heat exchanger with different section dryness. China Water Transport. J. (2019)
8. Gao Yuping, Shao Shuangquan, Si Chunqiang, et al. Heat Transfer and Pressure Drop Characteristics of R717 Flow Boiling inside a Horizontal Small Diameter Plain Tube. Journal of Refrigeration. J. (2018)

Metabolic activity of an anammox population affected by major environmental factors

Yifu Rao[1], Jia Meng[1*], Yajie Tian[1] and Jianzheng Li[1]

1State Key Laboratory of Urban Water Resource and Environment, Harbin Institute of Technology, 73 Huanghe Road, Harbin 150090, China

Abstract: Anammox is known as a cost-effective approach to nitrogen removal from ammonium-rich wastewater. However, the eco-physiological characteristics of anammox population have not well been understood so far, which has obstructed the practical application of anammox process. In the present research work, an enrichment culture of anammox population was collected from an expanded granular sludge bed (EGSB) reactor, and the anammox activity was evaluated by various pH, temperature and reactant density, progressively. The results showed that the optimal ecological amplitude of pH and temperature for the anammox population was ranged from 7.0-7.5 and 30°C-35°C, respectively. To make anammox process more efficient, the density of NH_4^+-N and NO_2^--N should be around 10.0 and 4.0 mmol·L^{-1}, respectively. It was found that anammox population was more sensitive to NO_2^--N density than NH_4^+-N density, and then NO_2^--N density was proposed as a key indicator to practicing anammox process in engineering. With the findings, the present research will be significant in practicing annamox process.

1 Introduction

Water eutrophication has been widely considered as a major environmental issue all around the world. For the restoration of natural water environment, discharge standards of ammonia (NH_4^+-N) and total nitrogen (TN) have become stricter than ever before in China. Though combined anoxic-aerobic (A/O) processes have been extensively practiced in engineering for nitrogen removal from various wastewater, the managers are still suffering from the high engineering investment and operating costs. The A/O processes are constructed based on the traditional theory of biological denitrification. In other words, NH_4^+-N has to be oxidized to nitrite (NO_2^--N) or nitrate (NO_3^--N) in the aerobic reactor, which would be returned to the anaerobic reactor for heterotrophic denitrification [1,2]. Therefore, the traditional nitrification-denitrification processes not only requires higher energy consumption but also exhibits a limitation in TN removal when the organics in wastewater could not meet the requirement of heterotrophic denitrification [3]. Though the next developed shortcut nitrification-denitrification process has obviously reduced the energy consumption in wastewater treatment process, the nitrogen removal efficiency is still unsatisfactory [4].

The novel anammox is considered as the most cost-effective biological nitrogen removal process in which no organic carbon source and aeration are required [5]. In the anammox process, anammox bacteria use NH_4^+ and NO_2^- as the electron donor and electronic acceptor, respectively, to form NO_3^- and gaseous nitrogen (N_2) under anaerobic conditions, as expressed as the following equation [6].

$$NH_4^+ + 1.32NO_2^- + 0.066HCO_3^- + 0.12H^+ \rightarrow$$
$$1.0N_2 + 0.26\ NO_3^- + 0.066CH_2O_{0.5}N_{0.15} + 2.03H_2O$$

Though aeration is still required to oxidize partial NH_4^+ to NO_2^- in ammonium-rich wastewater treatment, practicing anammox has attracted an increasing interesting all around the world because of the advantages as follows [7-11]. 1) The nitrogen removal efficiency is no longer restricted by organic carbon source in wastewater; 2) energy consumption is remarkably reduced because of the less oxygen supplement; 3) less excess sludge will be produced in the treatment process.

However, anammox bacteria are very difficult to be enriched in organic wastewater treatment processes because of their slow growth [12], though widely distributed in natural environment. Therefore, it is significant for practical application to understand the environmental factors affecting the metabolic activity of anammox bacteria [13-18]. Unfortunately, any pure culture of anammox bacteria has not been obtained so far, which is a serious impediment to understanding the eco-physiological characteristics of anammox bacteria [19].

* Corresponding author: jiameng@hit.edu.cn

In the present research work, an enrichment culture of anammox bacteria was obtained from an expanded granular sludge bed (EGSB) reactor. And then the effects of pH, temperature and density of reactants (NH_4^+-N and NO_2^--N) as the key environmental factors were investigated on the activity of annamox population by batch tests. With the findings of the optimal conditions for annamox population, the present research will be helpful to practice annamox in engineering.

2 Materials and methods

2.1 Basic medium for anammox bacteria

The basic medium for the enrichment of anammox bacteria was an inorganic salt culture medium composed of (per liter): $(NH_4)_2SO_4$ 0.33 g, $NaNO_2$ 0.35 g, $KHCO_3$ 1.50 g, KH_2PO_4 0.03 g, $CaCl_2 \cdot 2H_2O$ 0.14 g, $MgSO_4 \cdot 7H_2O$ 0.20 g, and microelement solution 1.0 mL. The pH of the medium was above 7.0 naturally. The microelement solution including per liter: EDTA 5.00 g, $FeSO_4$ 5.00 g, $ZnSO_4 \cdot 7H_2O$ 0.43 g, $CoCl_2 \cdot 6H_2O$ 0.24 g, $MnCl_2 \cdot 4H_2O$ 0.99 g, $CuSO_4 \cdot 5H_2O$ 0.25 g, $NiCl_2 \cdot 6H_2O$ 0.19 g, H_3BO_3 0.014 g, $Na_2MoO_4 \cdot 2H_2O$ 0.22 g, $NaSeO_4 \cdot 10H_2O$ 0.21 g.

2.2 Enrichment culture of anammox population for batch tests

The enrichment culture of anammox bacteria for batch tests was collected from an expanded granular sludge bed (EGSB) reactor. Fed with the basic medium containing NH_4^+-N and NO_2^--N of the same 20.0 $mmol \cdot L^{-1}$, the EGSB had been continuously operated for over 270 days with a constant hydraulic retention time of 8 h at 30°C. Removal of NH_4^+-N and NO_2^--N in the last steady phase was kept at about 91.5% and 98.3%, respectively. The enrichment culture was brick-red granules with a specific NH_4^+-N and NO_2^--N removal of about 2.90 and 3.48 $mmol \cdot g^{-1} \cdot d^{-1}$ in terms of mixed liquid volatile suspended solids (MLVSS).

The enrichment culture collected from the EGSB was washed for 3 times with $NaHCO_3$ buffer (pH 7.1) to remove the residual matrix and then used as inoculum for the batch tests. Biomass in the inoculum was about 5.85 $g \cdot L^{-1}$.

2.3 Experimental process

Effect of pH, temperature and the density of NH_4^+-N and NO_2^--N on the metabolic activity of anammox population in the enrichment culture were evaluated by batch tests, progressively. All of the batch tests were conducted in 50 mL anaerobic bottles. Each of the bottle was load with 28 mL basic medium and then inoculated with 2 mL of the pretreated enrichment culture.

With the same 5.0 $mmol \cdot L^{-1}$ of NH_4^+-N and NO_2^--N in the basic medium, nine pH gradients within 5.0-9.0 were tested firstly. At the obtained optimal pH, seven temperature gradients ranged from 4°C-50°C were further evaluated. With the obtained optimal pH and temperature, effect of reactants densities, i.e. the concentration of NH_4^+-N and NO_2^--N, on the metabolic activity of anammox population were finally investigated. Seven NH_4^+-N gradients within 2.0-16.0 $mmol \cdot L^{-1}$ and eight NO_2^--N gradients within 0.5-6.0 $mmol \cdot L^{-1}$ were tested with a constant NO_2^--N or NH_4^+-N of 5.0 $mmol \cdot L^{-1}$, respectively.

Each of the tests was conducted in triplicate. All of the constructed anaerobic bottles were incubated for seven days in a constant temperature air bath shaker (HZQ-C, Harbin Donglian Electronic Technology Development Co. Ltd., China) at 120 rpm. Except the temperature tests, the other anaerobic bottles were incubated at 30°C.

2.4 Analytical methods

Supernatant in each of the anaerobic bottles was sampled once per six hours for the determination of pH, NH_4^+-N, NO_2^--N and NO_3^--N. In addition to MLVSS, all of the analysis items were also conducted following the standard methods [20]. Metabolic activity of the enrichment culture in each of the anaerobic bottles was evaluated by specific anaerobic ammonium removal rate in terms of MLVSS ($mmol \cdot g^{-1} \cdot d^{-1}$), which was calculated based on the data collected from the logarithmic phase, normally in the first day.

3 Results and discussion

3.1 Effect of pH on the activity of anammox population

It is well known that pH is one of the most important environmental factors on the growth and metabolism of microorganisms. The enrichment culture of anammox population was collected from the EGSB and the influence of pH on the anammox activity was firstly investigated by batch tests. The results (Fig. 1) showed that a weakly alkaline environment was more favourable to anammox activity in the enrichment culture. As shown as Fig. 1, the anaerobic ammonium removal rate was gradually enhanced with the pH increased from 5.0 to 7.5. When the pH was over 8.0, the anaerobic ammonium removal rate was decreased remarkably. Though anammox activity could be kept by the enrichment culture at the pH of as low as 5.0 or as high as 9.0, the optimal ecological amplitude was identified as 7.0-8.0 with the anaerobic ammonium removal rate ranged from 0.14-0.17 $mmol \cdot g^{-1} \cdot d^{-1}$. Similarly, Daverey et al. reported that the maximum specific anammox activity was observed at pH of 7.5 [21]. Moreover, a wider optimum pH range of 6.5-9.3 was also found for the growth and activity of anammox bacteria [22]. In summary, anammox bacteria are sensitive to pH changes, and pH control is important during process operation.

 https://doi.org/10.1051/e3sconf/202129301027

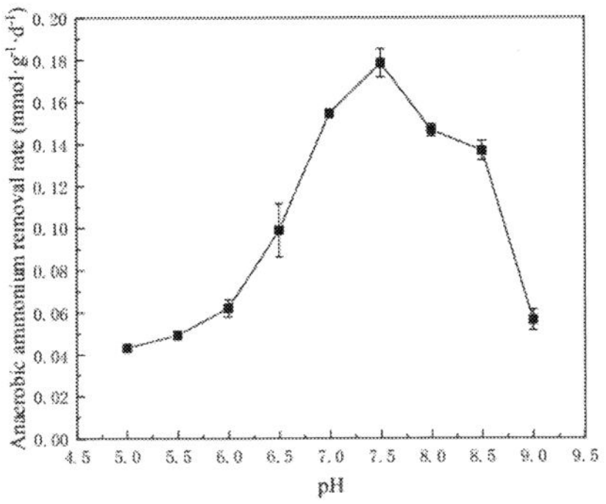

Fig. 1. Anaerobic ammonium removal rate of the enrichment culture with the increase of pH

3.2 Effect of temperature on the activity of anammox population

As another important environmental factor on the growth and metabolism of microorganisms, the effect of temperature on anammox activity in the enrichment culture was further evaluated at pH 7.0. The results (Fig. 2) showed that pH also had an obvious effect on the anammox population. As illustrated in Fig. 2, the anaerobic ammonium removal rate of the enrichment culture was improved rapidly along with the temperature increased from 4°C to 30°C. When the temperature was increased from 30°C to 35°C, the anaerobic ammonium removal rate was slightly decreased from 0.19 to 0.18 mmol·g^{-1}·d^{-1}. However, a sharp decrease in anaerobic ammonium removal rate was observed when the temperature was further increased to over 40°C. The inhibition of anammox activity at the higher temperature was likely due to biomass lysis. Also operated at 30°C, the anaerobic ammonium removal rate in the EGSB was as high as 2.90 mmol·g^{-1}·d^{-1} (Section 2.2). The results suggested that the enzymes involved in anammox were most active at the temperature around 30°C [23]. This is consistent with the previous studies reporting that the optimal temperature for most anammox species used in wastewater treatment are between 30°C and 40°C [24]. As known, the optimum temperatures for the two typical anammox species i.e. *Candidatus Brocadia anammoxidans* and *Candidatus Kuenenia stuttgartiensis* were reported to be 40°C and 37°C, respectively [24]. It can be concluded that the dominant anammox bacteria in the enriched anammox population could be *Candidatus Kuenenia stuttgartiensis*.

In addition, temperature along with pH also affect the concentration of free ammonium (NH_3) and free nitrous acid (HNO_2) which can strongly inhibit the anammox process. However, the mechanisms of the anammox activity affected by NH_3 and HNO_2 are still unclear. Furthermore, the threshold values of NH_3 and HNO_2 for suppressing anammox bacteria are various reported in previous studies [22]. Therefore, in next step study, the inhibition mechanisms and threshold values of the NH_3 and HNO_2 need to be further investigated.

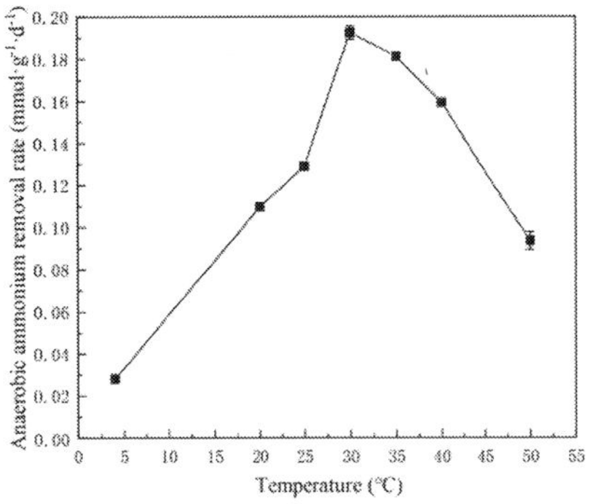

Fig. 2. Anaerobic ammonium removal rate of the enrichment culture with the increase of temperature

3.3 Effect of reactant density on the activity of anammox population

NH_4^+ and NO_2^- are the dominant reactants served as electron donor and electronic acceptor, respectively, in anammox reaction [6]. And it is known that the specific substrates have an outstanding effect on the growth and metabolism of microorganisms. Therefore, the effects of NH_4^+-N or NO_2^--N density on the metabolic activity of anammox population were further evaluated at the optimal pH 7.0 and temperature 30°C.

Fig. 3 illustrated the effect of NH_4^+-N density on the anaerobic ammonium removal rate of the enrichment culture with a constant NO_2^--N density of 5.0 mmol·L^{-1}. The results indicated that the anaerobic ammonium removal rate of the enrichment culture was remarkably enhanced from 0.29 to 0.64 mmol·g^{-1}·d^{-1} with the NH_4^+-N density increased from 0.5 to 10.0 mmol·L^{-1}. When the NH_4^+-N density was further increased to 12.0 and 16.0 mmol·L^{-1} step by step, the anaerobic ammonium removal rate was severely dropped to 0.59 and 0.46 mmol·g^{-1}·d^{-1}, respectively. The results suggested that the NH_4^+-N density of 10.0 mmol·L^{-1} was more favorable for conducting anammox process.

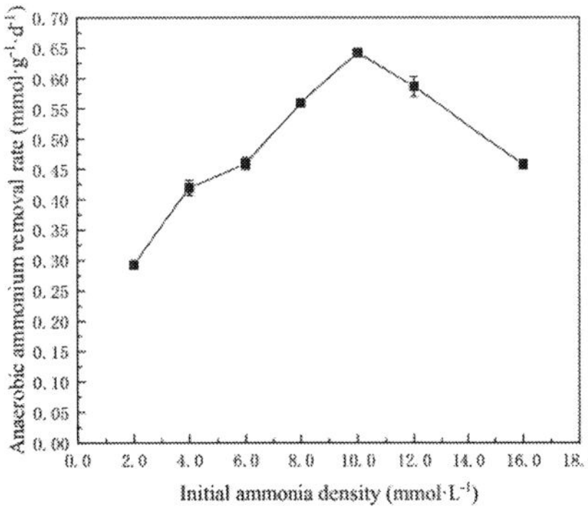

Fig. 3. Anaerobic ammonium removal rate of the enrichment culture with the increase of NH_4^+-N density

Contrary to the NH_4^+-N density, the NO_2^--N density less than 2.0 mmol·L^{-1} had no obvious effect on the anammox activity of the enrichment culture (Fig. 4). However, the anaerobic ammonium removal rate was sharply improved from 0.09 to 0.24 mmol·g^{-1}·d^{-1} when the NO_2^--N density was enhanced from 2.0 to 2.5 mmol·L^{-1}. When the NO_2^--N density was further increased, no obvious improvement of anaerobic ammonium removal rate was observed anymore. Even an obvious decrease of anaerobic ammonium removal rate was found when the NO_2^--N density was further enhanced to 6.0 mmol·L^{-1}. The highest anaerobic ammonium removal rate of 0.27 mmol·L^{-1} came up to the NO_2^--N density of 4.0 mmol·L^{-1}.

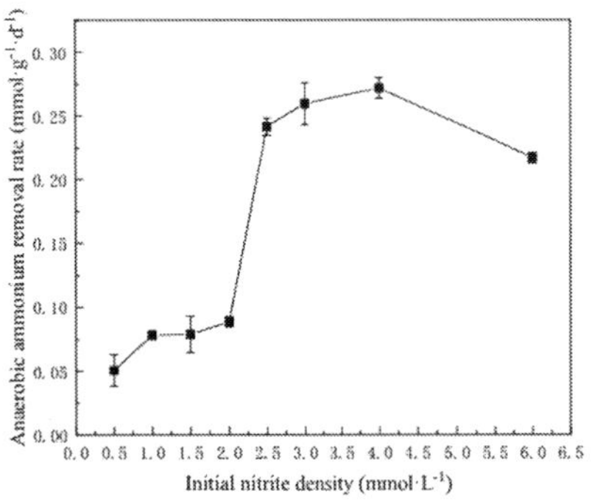

Fig. 4. Anaerobic ammonium removal rate of the enrichment culture with the increase of NO_2^--N density

The above results showed that anammox population were much more sensitive to NO_2^--N density (Fig. 4) than that to NH_4^+-N density (Fig. 3). It has been reported that NO_2^--N could function as a strong biological inhibitor of anammox population [25]. Thus, NO_2^--N density was proposed as a key indicator to practicing anammox process in engineering.

4 Conclusion

In this study, the effects of pH, temperature and reactant density on the activity of annamox population were investigated by batch tests. Based on the experiment results, following conclusions can be drawn:

(1) The optimal ecological amplitude of pH and temperature for the anammox population was ranged from 7.0-7.5 and 30°C-35°C, respectively.

(2) Anammox process would be more efficient with the NH_4^+-N and NO_2^--N density of 10.0 and 4.0 mmol·L^{-1}, respectively.

(3) NO_2^--N density was proposed as a key indicator to practicing anammox process in engineering because anammox population was more sensitive to NO_2^--N density than NH_4^+-N density.

Acknowledgment

The authors acknowledge the financial support from the National Natural Scientific Foundation of China (51778171).

References

1. Delre A, Monster J, Scheutz C. Sci Total Environ, 605, 258-268 (2017)
2. Volcke E I P, Van Loosdrecht M C M, Van Voorthuizen E M, et al. Water Sci Technol, 67, 2350-2355 (2013).
3. Carrera J, Vicent T, Lafuente J. Process Biochem, 9,2035-2041 (2004)
4. Gao Da-Wen, Peng Yong-Zhen, Zheng Qing-Zhu. China Water & Wastewater, 18, 13-18 (2002)
5. Strous M, Pelletier E, Mangenot S, et al. Nature, 440, 790-794 (2006)
6. Kartal B, Maalcke W J, De Almeida N M, et al. Nature, 479, 127-130 (2011)
7. Kartal B, Rattray J, van Niftrik L A, et al. Syst Appl Microbiol, 30, 39-49 (2007)
8. Strous M, Fuerst J A, Kramer E H M, et al, Nature, 400, 446-449 (1999)
9. Kartal B, van Niftrik L, Slickers O. et al. Rev Environ Sci Bio, 3, 255-264 (2004)
10. Schmid M, Twachtmann U, Klein M. et al. Syst Appl Microbiol, 23, 93-106 (2000)
11. Schmid M, Walsh K, Webb R, et al. Syst Appl Microbiol, 26, 529-538 (2003)
12. Schmid M C , Risgaard-Petersen N , Vossenberg J , et al. Environ Microbiol, 9, 1476-1484 (2010)
13. van Niftrik L, Geerts W J C, Van Donselaar E G, et al. J Struct Biol, 161, 401-410 (2008)
14. Sinninghe Damsté J S, Strous M, Rijpstra Wic, et al. Nature, 419, 708-712 (2002)
15. van Niftrik L, Geerts W J C, van Donselaar E G. et at. J Bacteriol, 190, 708—717 (2008)
16. van Niftrik L A, Fuerst J A. Fems Microbiol Lett, 233, 7-13 (2004)
17. Sinninghe Damsté J S, Rijpstra W I C. Schouten S. et al. Org Geochem, 35, 561-566 (2004)
18. Ranray J E, van de Vossenberg J, Hopmans E C. et al. Arch Microbiol, 190, 51-66 (2008)
19. Strous M, Kuenen J G, Fuerst J A, et al. Anton Leeuw, 81, 693-702 (2002)
20. Washington, DC, APHA (2005)
21. Daverey A, Pang C C, Dutta K, et al. Int Biodeter Biodegr, 102 (2015)
22. Tomaszewski M, Cema G, A Ziembińska-Buczyńska. Chemosphere, 203-214 (2017)
23. Dosta J, Femandez I, Vazquez-p J R,et al. J Hazard Mater, 63, 2446-2448 (2007)
24. Strous M M, Kuenen J, Jetten M. Appl.environ.microbiol, 65,3248-50 (1999)
25. Dapena-mora A, Fernandez I, Campos J L, et al. Enzyme Microb Tech, 40, 859-865 (2007)

Isolation and identification of a high-efficiency phenol-degrading bacteria and optimization of its degradation conditions

Haili Sun [1]*, Tianpeng Gao[1,2]*, Guohua Chang[1], Xisheng Tai[1], Ruiqi Yang[1], Qing Zhang[1], Bin Yue[1]

[1]College of Geography and Environmental Engineering, Lanzhou City University, Lanzhou Gansu 730070, China

[2]School of Biological and Environmental Engineering, Xi'an University, Xi'an 710065, China

Abstract: Phenol is widely used in China, it not only pollutes the environment, but also accumulates toxic substances in the human body through the food chain, further harming humans. In this experiment, a strain of high-efficiency low-temperature degradation phenol bacteria B5 was selected from the soil contaminated by organic matter of Lanzhou. Through research methods such as Gram staining observation, DNA extraction, PCR amplification, sequencing and comparison, it was found that this strain was Pantoea agglomerans. Through the subsequent optimization of degradation conditions, it was found that the B5 strain can degrade 500mg/L of phenol to 24.8mg/L in 36h. The ability to degrade phenol is stronger between pH5.5-pH6.0, and the ability to degrade phenol is higher in a medium containing 4-8g/L sodium chloride. This research can provide certain theoretical guidance for phenol degradation.

1 Introduction

Phenol is a simple phenolic organic weak acid, which is a common organic pollutant in industrial wastewater such as petroleum, fertilizer, paper making, rubber[1]. Phenol has strong toxic effects on many organisms. It has carcinogenic, teratogenic and mutagenic effects, so it has been included in the list of key pollutants in many countries [2]. Phenol can be directly toxic to the human body through contact with the skin and mucous membranes. It can chemically react with the proteins in the cell protoplasm to cause cell inactivation. Low-concentration phenol solution can only denature protein, while high-concentration phenol solution can coagulate protein and eventually cause tissue necrosis [3].

In addition to industrial wastewater, feces and nitrogen-containing organic matter will also produce phenolic compounds during the decomposition process. Therefore, a large amount of fecal sewage discharged from cities is also an important source of phenol pollution in water bodies[4]. After the water quality is polluted by phenol-containing wastewater, it will have many serious consequences, such as the destruction of the oxygen balance of the water body, etc. The arbitrary discharge of phenol-containing wastewater leads to deterioration of water quality, which seriously affects people's physical and mental health, and also restricts the sustainable development of cities. Therefore, it is urgent to seek effective environmental protection measures to treat phenol-containing wastewater.

Biological treatment of phenol-containing wastewater mainly uses the metabolism of microorganisms to convert harmful phenolic pollutants into non-toxic substances [5,6], without secondary pollution, and has a large amount of treatment. Therefore, it has become the first choice for the harmless treatment of phenol-containing wastewater in China. The phenol-degrading bacteria that have been obtained include algae (Alga Ochromonas danica) [7], rhizobia (Rhizobia) [8], yeast (Yeast Trichosporon cutaneum) [9], and Pseudonomonas sp. [10], etc. However, for the treatment of high-concentration phenol-containing wastewater, it is necessary to dilute the phenol concentration before entering the water to reduce the toxic effect on the organism, resulting in a substantial increase in the treatment cost. If the tolerance of microorganisms to phenol can be improved and the dilution factor can be reduced, it will be of great significance to the treatment of high-concentration phenol-containing wastewater. In this experiment, a microbial strain B5 that can efficiently degrade high-concentration phenol at low temperature was screened and isolated from the organic polluted soil of chemical plants, which has certain guiding significance for the treatment of high-concentration phenol-containing wastewater at low temperature in the future.

2 Materials and methods

2.1 Main instruments

Microscope, spectrophotometer, constant temperature incubator, shaker, PCR machine, etc.

* Corresponding author: Haili Sun and Tianpeng Gao, Emil: lzusun@163.com(H. S.) , zkgtp@163.com(T. G.)

2.2 Source of strain isolation

The source of the strain used in this experiment came from a chemical plant in Lanzhou, which was polluted by organic matter throughout the year. Soil microorganisms are mainly distributed on the surface of the soil and are sensitive to changes in the surface soil environment, so this experiment takes a 0-10cm soil layer, removes the surface litter layer when sampling, and then uses the quarter method to sample. Put the collected soil sample into a sterile sealed bag and put it in the refrigerator at -20°C for later use.

2.3 Medium formula

(1) Select the medium: K_2HPO_4 0.5725 g•L^{-1}, KH_2PO_4 2.9575 g•L^{-1}, $MgSO_4$ 0.25 g•L^{-1}, $(NH_4)_2SO_4$ 2.5 g•L^{-1} , $CaCl_2$ 0.0825 g•L^{-1}, NaCL 0.25 g•L^{-1}, KCl 0.25 g•L^{-1}, Agar 10-20 g•L^{-1}, pH 6.8, sterilize at 121°C for 20 minutes, and then add phenol 100ml•L^{-1}. (Note: All media should be protected from light during operation, because phenol is easily decomposed into pink or red benzoquinone when exposed to light.)

(2) Re-screening medium: K_2HPO_4 0.5725 g•L^{-1}, KH_2PO_4 2.9575 g•L^{-1} , $MgSO_4$ 0.25 g•L^{-1} , $(NH_4)_2SO_4$ 2.5 g•L^{-1}, $CaCl_2$ 0.0825 g•L^{-1}, NaCl0.25 g•L^{-1}, KCl 0.25 g•L^{-1}, Agar 10-20 g•L^{-1}, pH 6.8, sterilize at 121°C for 20 minutes, and then add phenol 500ml•L^{-1}.

(3) Acclimatization medium: peptone 10g•L^{-1}, yeast extract 5g•L^{-1}, NaCl 5 g•L^{-1}, pH 6.8, sterilized at 121°C for 20 minutes, and then 500ml•L^{-1} phenol is added.

3 Experimental methods

3.1 Screening and isolation of strains

Take 10 mL of contaminated soil sample in a 50 ml sterile centrifuge tube, add 30 ml of sterile water and mix thoroughly. After standing for a few minutes, on a sterile workbench, use a pipette to draw 200 μL of supernatant, spread it evenly on the selection medium with a spreading rod, and then place it at a constant temperature of 25°C and 200r/min Cultivate for 72 h in an incubator. Select a single colony according to the shape, size, color, gloss, softness, transparency, edge shape, swelling degree, and surface smoothness of the colony grown in the culture medium, and inoculate it with the plate streak method on the newly sterilized selection medium, culture for 72 h in a constant temperature incubator at 25°C. Further purify the strain, repeat the above step, and separate and purify the single colony again 3 times by the plate streaking method. Through multiple screening and separation, two pure cultures of phenol-degrading bacteria with the best growth and the largest colony area were finally selected.

3.2 Preservation of strains

There are many methods for preservation of bacteria, including: liquid paraffin method, low temperature regular transplantation method, sand preservation method, liquid nitrogen ultra-low temperature preservation method. In this experiment, the liquid paraffin method was used to preserve the selected phenol-degrading bacteria.

Firstly, on a sterile workbench, inoculate the most efficient phenol-degrading strains screened out on the re-screening medium, and incubate them in a shaker at 25°C and 200 r/min for 36 hours. Microscopic observation shows that the concentration of bacterial turbidity is about $2*10^8$CFU/ml, it can be made into bacteria liquid. Then, draw 1.0 mL of bacterial solution and 0.5 mL of 30% glycerol in a 1.5 mL cryostat tube for strains. Save three copies of each strain, label them and store them in a refrigerator at -80°C for subsequent experiments.

3.3 Molecular identification of bacterial species

(1) PCR amplification

In this experiment, two universal primers (27F, 1492R) were used for PCR amplification, and the reaction system used was 25μL each. The PCR reaction system is shown in Table 1:

Table1. PCR reaction system

Name	Dosage(μL)
Phanta Max Master Mix	12.5
Primer 1 (27F)	1
Primer 2 (1492R)	1
Bacteria	1
ddH_2O	9.5

(2) 16S rDNA sequencing comparative analysis

The PCR amplification product of the strain was sent to Xi'an Kinke Jersey Biotechnology Co., Ltd. for analysis and sequencing, and the 16S rDNA sequence fragment of the strain was obtained. The measured sequences were compared and analyzed in NCBI's BLAST program, and the strains with the highest similarity were selected.

3.4 Determination of phenol degradation ability

(1) Correlation test between cell concentration and phenol degradation ability

The strain was cultured in the acclimation medium to the logarithmic growth phase (OD_{600}=0.5-1), and the bacteria were collected by centrifugation at 7000rpm for 5min. Resuspend the bacteria in a re-screened liquid medium containing 500 mg/L of phenol according to the wet reuse of the bacteria, and the final bacteria content reached 1g/L, 2g/L, 4g/L, 8g/L, 16g/L. Then they were cultured in a constant temperature incubator at 25°C and 200 r/min, and samples were taken every 12 hours to monitor the phenol content.

(2) Correlation test of phenol degradation ability of strains under different pH conditions

The strain was cultured in the acclimation medium to the logarithmic growth phase (OD_{600}=0.5-1), and the bacteria were collected by centrifugation at 7000rpm for 5min. Resuspend the bacteria in the re-screened liquid medium with different pH (pH 4.5, 5.0, 5.5, 6.0, 6.5, 7.0) to make the content of the bacteria in the medium 8g/L. Then they were cultured in a constant temperature incubator at 25°C and 200 r/min, and samples were taken every 12 hours to monitor the phenol content.

(3) Correlation test of strain phenol degradation ability under different salt concentration conditions

The strain was cultured in the acclimation medium to the logarithmic growth phase (OD_{600}=0.5-1), and the bacteria were collected by centrifugation at 7000rpm for 5min. Resuspend the bacteria with acclimation media containing different sodium chloride concentrations (sodium chloride concentrations are 0.5g/L, 1g/L, 2g/L, 4g/L, 8g/L, 16g, 32g/L), The cell content in the culture medium was 8 g/L. Then they were cultured in a constant temperature incubator at 25°C and 200 r/min, and samples were taken every 12 hours to monitor the phenol content.

3.5 Determination of phenol content

Through ultraviolet-visible spectrophotometric analysis and HPLC C-18 column, with methanol, formic acid, and double-distilled water in a ratio of 33:8:1 (volume ratio), preparatory experiments to determine the internal standard.

4 Results and analysis

4.1 Strain isolation and morphological observation

After separation and purification, 15 strains of phenol-tolerant bacteria were obtained at 18°C. After further screening, a strain that grows best at low temperature and has good degradation performance on phenol was finally determined, and it was named BF5.

4.2 Strain identification

(1) Detection and analysis of PCR products of 16S rDNA

The amplified PCR products were measured by an agarose gel electrophoresis instrument and observed with an ultraviolet gel imager. The gel electrophoresis pattern of PCR products is shown in Figure 3.5. The electrophoresis bands are clear and there is no dragging phenomenon, indicating that these PCR amplified products can be used in subsequent experiments.

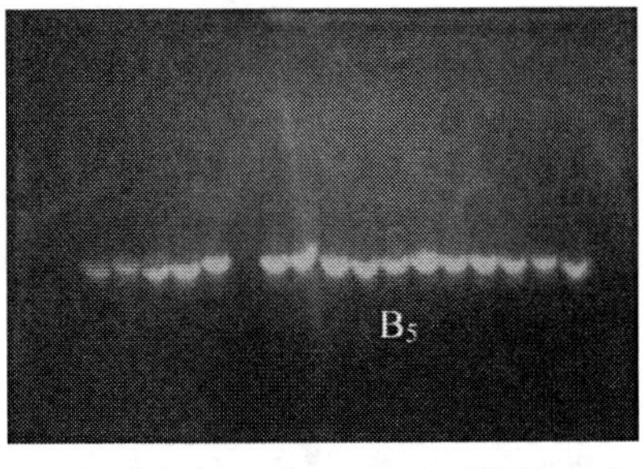

Figure 1. Gel electrophoresis map of PCR products

(2) Sequencing and analysis of 16S rDNA amplification products

After the PCR amplification is completed, the amplified products are sent to Xi'an Kinke Jersey Biotechnology Co., Ltd. for analysis and sequencing, and the 16S rDNA sequence fragment of the strain is obtained. The measured sequence was compared and analyzed with 16S rDNA gene sequence on the BLAST program on the NCBI website, and it was determined that the degrading bacterium B5 was *Pantoea agglomerans*.

(3) Phylogenetic tree

Use analysis software such as MEGA and Clustal X1.38 to analyze the sequence homology to construct a phylogenetic tree, as shown in the figure 2.

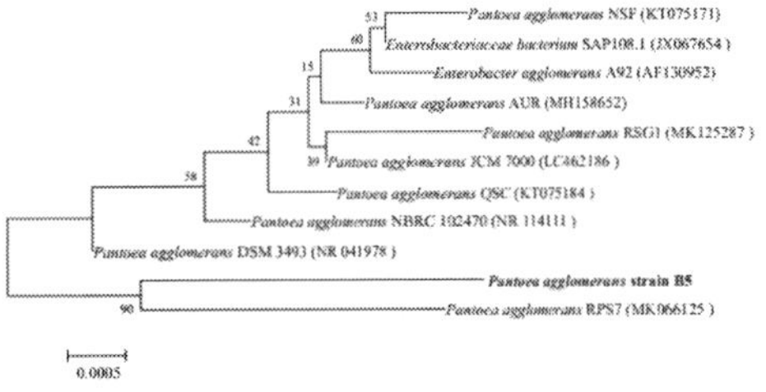

Figure 2. Phylogenetic tree of *Pantoea agglomerans* strain B5

4.3 Optimization of phenol degrading bacteria conditions

(1) Correlation test between cell concentration and phenol degradation ability

The strains were cultured in the acclimation medium to the logarithmic growth phase (OD600=0.5-1), and the bacteria were collected by centrifugation at 7000rpm for 5min. Resuspend the bacteria in a re-screened liquid medium containing 500 mg/L of phenol according to the wet reuse of the bacteria, and the final bacteria content reached 1g/L, 2g/L, 4g/L, 8g/L, 16g/L. Then it was cultured in a constant temperature incubator at 25°C and 200 r/min, and samples were taken every 12 hours to monitor the phenol content. Figure 3 shows that the B5 strain with an initial concentration of 8g/L can degrade 500mg/L of phenol to 24.8mg/L in 36h.

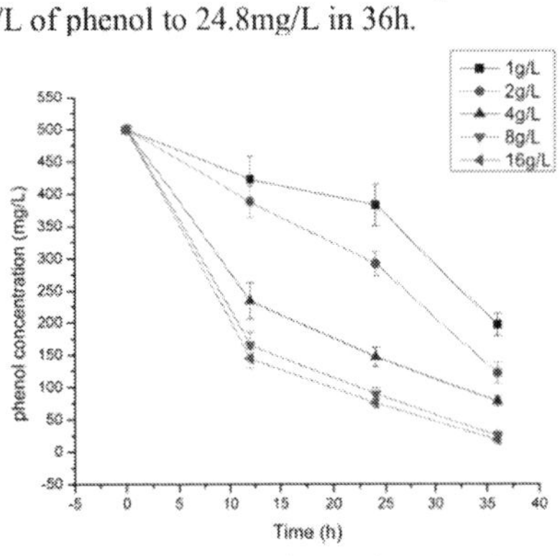

Figure 3. The degradation ability to the phenol

(2) Correlation test of phenol degradation ability of strains under different pH conditions

The strains were cultured in the acclimation medium to the logarithmic growth phase (OD600=0.5-1), and the bacteria were collected by centrifugation at 7000rpm for 5min. Resuspend the bacteria in re-screened liquid media with different pH (pH 4.5, 5.0, 5.5, 6.0, 6.5, 7.0) to make the content of the bacteria in the medium 8g/L. Then it was placed in a constant temperature incubator at 25°C and 200 r/min, and samples were taken for 24 hours to monitor the phenol content. Figure 4 shows that the B5 strain has a strong ability to degrade phenol between pH 5.5 and pH 6.0.

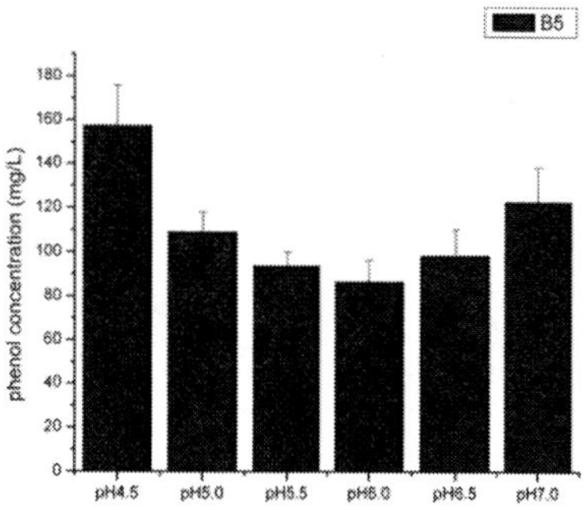

Figure 4. The degradation ability to the phenol in different pH

(3) Correlation test of strain phenol degradation ability under different salt concentration conditions

The strains were cultured in the acclimation medium to the logarithmic growth phase (OD600=0.5-1), and the bacteria were collected by centrifugation at 7000rpm for 5min. Resuspend the bacteria with acclimation media containing different sodium chloride concentrations (sodium chloride concentrations are 0.5g/L, 1g/L, 2g/L, 4g/L, 8g/L, 16g, 32g/L), The cell content in the culture medium was 8 g/L. Then it was placed in a constant temperature incubator at 25°C and 200 r/min, and samples were taken for 24 hours to monitor the phenol content. Figure 5 shows that the B5 strain has a higher ability to degrade phenol in a medium containing 4-8g/L sodium chloride, but has a relatively low ability to degrade phenol in a sodium chloride medium greater than 8g/L.

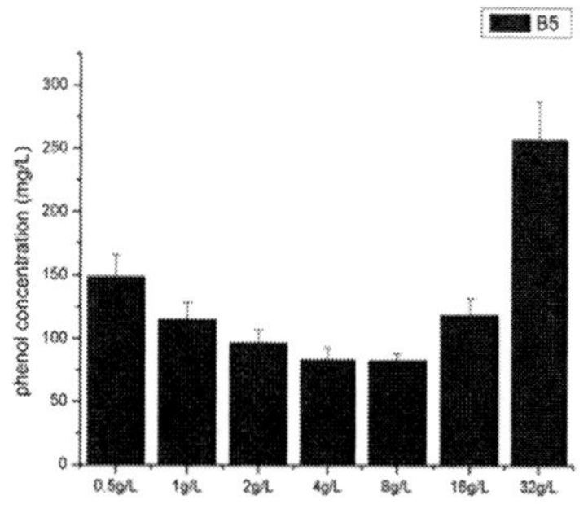

Figure 5. The degradation ability to the phenol with different salt

5 Conclusion

In this experiment, we isolated and screened a microbial strain B5 that can use phenol as the sole carbon source and has the ability to efficiently degrade phenol from the soil contaminated by organic matter all year round. Gram staining confirms that it is a Gram-negative cocci, and 16S rDNA sequence identification. It is determined that the bacteria is *Pantoea agglomerans*. Through the subsequent optimization of degradation conditions, it was found that B5 can degrade 500mg/L of phenol to 24.8mg/L in 36h, and the ability to degrade phenol between pH5.5-pH6.0 is strong, and the strain contains 4-8g/L. Sodium chloride has a higher ability to degrade phenol in the medium. This research can provide certain theoretical guidance for phenol degradation.

Acknowledgments

The research was funded by the Natural Science Foundation of Gansu Province (20JR5RA207), National Natural Science Foundation of China (NSFC 31860176), Key Research and Development Program of Gansu Province (20YF3FA037), Research projects of Gansu University (2018B-062), Key Research and Development Program of Shanxi Province (2020ZDLSF06-06) and XAWLKYTD012. We are grateful to all anonymous reviewers whose comments improved the quality of the manuscript.

References

1. MAO Z, YU C, XIN L. Enhancement of phenol biodegradation by Pseudochrobactrum sp. through ultraviolet-induced mutation[J]. Molecular Sciences, 2015, 16: 7320-7333.
2. Wu Zhengyong, Zhao Gaofeng, Zhou Huaidong, etc. The distribution characteristics and potential risks of phenols in the surface water of the Three Gorges Reservoir during the wet season [J] . Environmental Science, 2012, 33(8): 2580-2585.
3. Zhang Zijie. Drainage Engineering (Volume 2). Beijing: China Construction Industry Press, 1996: 115−128.
4. Zhang Fangxi, Jin Chengji. Treatment and Utilization of Phenolic Wastewater. Beijing: Chemical Industry Press, 1983: 48−55.
5. Xiao Zailiang, Lan Shi, Tian Xi, Screening of strong phenol degrading bacteria and the effect of temperature on degradation[J]. Journal of Safety and Environment, 2010, 10(4): 20-22.
6. ZHI Y L, XIAO J G, HUI L, et al. High-throughput screening for a moderately halophilic phenol degrading strain and its salt tolerance response [J]. Molecular Sciences, 2015, 16: 11834-11848.
7. Jian Fangling, Huang Kaina. Preliminary study on the pretreatment of synthetic phenolic resin wastewater. Zhong Yan. Journal of Agricultural Technology College, 2000, 13(1): 40−44.
8. Rodriguez I, Liompart MP, Cela R. Solidphase extraction of phenols. Journal of Chromatography A, 2000, 885(1/2):291−304.

9. Andrzej W, Trochimczzuk, Michael Streat. Highly polar polymericsorbents characterization and sorptive properties towards phenol andits derivatives. Reactive and Functional Polymers, 2001, 46(3): 259−271.
10. Fewson CA. The identity of the gramnegative bacterium NCIB8250. J Gen Microbiol, 1967(48): 107−110.

Determination of effective medicinal components in Lonicera macranthoides flower buds

Jing Cai[1], Xinyang Zhang[1], Huan Dai[1]

[1]Wuhan Institute of Design and Sciences, Wuhan 430205

Abstract. This study is to determination of active medicinal components in Lonicera macranthoides. The results of High-performance liquid chromatography showed that the contents of chlorogenic acid were as follows: Baiyun: 6.297% , jincuilei: 5.293% , yincuilei: 5.288% , common variety: 7.455% , common variety (after flowering) : 4.140% . The contents of Luteolin in different Lonicera macranthoides varieties were as follows: Baiyun: 0.0264% ; Jincuilei: 0.0573% ; Yincuilei: 0.0347% ; common variety: 0.0394% ; common variety (after flowering) : 0.0197% .

1 Introduction

In the 2005 edition of the Chinese Pharmacopoeia, one of the bases for distinguishing flos Lonicerae from Flos Lonicerae japonicae was "analysis of chemical constituents, the active constituents of Flos Lonicerae are mainly Luteolin glycosides, but the other varieties contain little chlorogenic acid, there have also been a number of reports in recent years of differences in the contents of Lonicera Japonica and honeysuckle, in order to further determine the market value of Lonicera macranthoides, in this experiment, the contents of Luteolin and chlorogenic acid in the flower buds of Lonicera macranthoides were determined[1-2]. Chlorogenic acid, alias: Chlorogenic acid, CAFFEIC tannic acid, 3-caffeoylquinic acid, formula:C16H18O9.

Luteoloside (English name: Luteoloside, alias name: luteolin-7-O-glucoside); Cynaroside, molecular formula: $C_{21}H_{20}O_{11}$[3].

2 Materials and methods

2.1 Experimental materials and apparatus

2.1.1 Experimental materials

This experimental material is Lonicera macranthoides dry flower bud, from Hunan Longhui honeysuckle science and technology development center. The four common varieties are Jincuilei, Yincuilei, Baiyun and Longhui (before and after flowering) . The control samples, chlorogenic acid (batch number: 110753-200413) and Luteolin (batch number: 111720-201106; 99.3%) were all from China Institute of Pharmaceutical and biological products.

2.1.2 Equipment and chromatographic conditions

P230IIhigh-pressure constant-current pump; UV230IIultraviolet-visible detector; Zwii Column Incubator; EC2006 Data Chromatographic Workstation; Diamonsil C18 column. Octadecyl Silane Bonded Silica Gel was used as Filler, acetonitrile (a)-0.1% phosphoric acid solution (b) was used as mobile phase gradient elution:

Table1. Gradient elution program

Time(min)	Mobile phase A (%)	Mobile phase B(%)
0→5	14	86
5→15	14→28	86→72
15→27	28→25	72→75
27→30	25→14	75→86

The flow rate of the mobile phase was 1.0 Ml/min, the column temperature was 30 °c, the detection wavelength was 327 nm (0-12 Min) , 350 nm (12-30 Min) and the injection volume was 10μl.

2.2 Assay

2.2.1 Preparation of Reference Solution

This experiment take 19.10mg of chlorogenic acid reference substance, put it in a brown measuring flask, add 50% methanol to make 0.382mg/ml, 0.0382mg/ml solution, that is (stored below 10°C). Take 10.58mg of luteolin reference substance, add 50% methanol to make 0.01mg/ml, 0.001mg/ml solution, that is (stored below 4°C).

* Corresponding author: Jing Cai email: 84016963@qq. com

 https://doi.org/10.1051/e3sconf/202129301029

2.2.2 Preparation of Test Solution

This experiment take about 0.5g of this product powder (passed through the No. 4 sieve), accurately weigh it, place it in a stoppered conical flask, add 50% methanol 50ml accurately, weigh it, and ultrasonically treat it (power 250W, frequency 30kHz) for 30 minutes. Let it cool and weigh it again. Use 50% methanol to make up the lost weight. Shake well and filter. Accurately measure 5ml of the filtrate and place it in a 25ml brown measuring flask. Add 50% methanol to the mark and shake well to get it.

2.2.3 Investigation of linear relationship

In this experiment, precisely aspirate 0.0382 mg/ml chlorogenic acid reference substance 1μl, 3μl, 0.382mg/ml chlorogenic acid reference substance 2μl, 5μl, 8μl, 13μl, 18μl, 26μl, 28μl, respectively according to the chromatographic conditions, and use the reference substance to inject samples The amount (μg) is the abscissa and the corresponding peak area is the ordinate to obtain the standard curve of chlorogenic acid.

In the meantime, Precisely draw 0.001050594 mg/ml luteolin reference substance 1μl, 8μl, 12μl, 16μl, 0.01050594 mg/ml luteolin reference substance 4μl, 5μl, 8μl, 10μl, 12μl, 16μl, 20μl, 25μl, 30μl, respectively according to the chromatographic conditions For sample injection, take the injection amount of the reference substance (μg) as the abscissa and the corresponding peak area as the ordinate to obtain the standard curve of luteolin.

2.2.4 Precision test

In this test, Precisely draw 10μl of the same test solution and repeat the injection 6 times to obtain the RSD of the peak area of chlorogenic acid and luteolin.

2.2.5 Stability Test

In this test, at 0 h, 3 h, 6 h, 9 h, 12 h, 15 h and 1 d, 2 d, 3 d, 4 d, and 5 d after the preparation of the test product, 10μl of the test product solution was precisely drawn into the sample. Calculate the RSD of the peak area of chlorogenic acid and luteolin.

2.2.6 RepeatabilityTest

In this test, take the same sample, accurately weigh 6 parts, prepare the sample solution according to the method, inject the sample according to the chromatographic conditions, determine the content of chlorogenic acid and luteolin, and obtain the RSD of the peak area of chlorogenic acid and luteolin.

2.2.7 Sample recovery test

In this test, take 0.5g of the same sample powder of Lonicera japonica, 6 parts, accurately weighed, and accurately add 1ml of reference substance chlorogenic acid (1.0mg/ml) and 2ml of luteolin (0.1mg/ml) respectively, and prepare according to the method. The test solution was tested for content, and the average recovery rate of chlorogenic acid and luteolin was calculated.

2.2.8 Determination of sample content

The prepared test solution was subjected to gradient elution according to the gradient elution procedure in Table 1, and then the content of the sample was calculated according to the linear regression equation of the peak area of the sample and the corresponding reference substance.

3 Results and analysis

3.1 Investigation results of linear relationship

Chlorogenic acid-linear relationship investigation: According to the system adaptability test method, enter the chlorogenic acid reference substance of different concentrations, and then calculate the linear correlation number of chlorogenic acid. The regression equation is: y = 2482.2219 x + 164.7491; r = 0.9999, and the linear range is 0.0382~10.696μg.

Table2. Analysis of linear relationship on chlorogenic acid and luteolin glycosides

Chlorogenic-acid injection volume (ul)	Injection volume	Peak area	Luteolin injection volume (ul)	Injection volume	Peak area
0.1	0.0382	104.11	0.1	0.001050594	2.67
0.3	0.1146	307.69	0.8	0.008404752	20.56
2	0.764	2041.21	1.2	0.012607128	30.62
5	1.91	4977.65	1.6	0.016809504	41.71
8	3.056	7945.76	4.0	0.042023760	106.28
13	4.966	12609.52	5.0	0.052529700	132.38
18	6.876	17424.46	8.0	0.084047520	212.21
26	9.932	24581.39	10	0.105059400	262.70
28	10.696	26691.11	12	0.126071280	312.44
			16	0.168095040	433.58
			20	0.210118800	535.19
			25	0.262648500	675.17
			30	0.315178200	806.19

 https://doi.org/10.1051/e3sconf/202129301029

Investigation of luteolin-linear relationship: According to the system adaptability test method, different concentrations of luteolin reference substance were added, and then the linear correlation number of luteolin was calculated. The regression equation is: y=2565.4213x-2.4465; r=0.9999, and the linear range is 0.001050594~0.3151782μg.

The linear relationship investigation results show that the linear relationship between chlorogenic acid and luteolin is good.

3.2 Precision test results

Precisely draw 10uL of the same test solution and repeat the injection 6 times. The results are shown in Table 3. The RSD of the peak areas of chlorogenic acid and luteolin were 0.3% and 2.1%, respectively. It shows that the precision of the detection method in this experiment is good.

Table3. Precision results

	Chlorogenic acid peak area	Luteolin peak area
	3242.12	15.08
	3249.38	14.36
	3265.84	14.25
	3261.25	14.52
	3262.95	14.85
	3265.68	14.55
RSD	0.00301339	0.021260992

3.3 Stability test results

At 0, 3, 6, 9, 12, 15 h and 1, 2, 3, 4, and 5 days after the preparation of the test product, 10 uL of the test product solution was precisely drawn and injected. The results are shown in Table 4. The intraday RSDs of the peak areas of chlorogenic acid and luteolin were 0.3% and 2.0%, respectively; the RSDs of the peak areas of chlorogenic acid and luteolin during the day were 0.4% and 2.1%, respectively. It shows that the stability of the test product during the test is good.

Table4. Stability Test Results

Time	Chlorogenic acid peak area	Luteolin peak area	Time	Chlorogenic acid peak area	Luteolin peak area
0 h	3265.68	15.09	1 d	3245.68	14.95
3 h	3249.38	14.76	2 d	3248.26	14.56
6 h	3265.84	14.56	3 d	3255.84	14.25
9 h	3261.25	14.52	4 d	3251.25	14.31
12 h	3262.95	14.5	5 d	3220.95	14.25
15 h	3240.06	14.22	RSD	0.00420428	0.020761881
RSD	0.003220575	0.020020526			

3.4 Repeatability test results

Take the same sample, accurately weigh 6 parts, and prepare the sample solution. The results are shown in Table 4-5: The RSD of chlorogenic acid and luteolin were determined to be 1.4% and 3.8%, respectively, indicating that the experimental method has good repeatability.

Table5. Repeatability Test Results

Sample weight	Luteolin peak area	Luteolin peak area
0.5067	3487.07	15.58
0.4917	3384.82	14.36
0.4964	3416.94	14.25
0.5049	3475.02	14.52

https://doi.org/10.1051/e3sconf/202129301029

0.4760	3276.91	14.85
0.4696	3232.17	14.55
0.4835	3327.76	15.08

3.5 Sample recovery rate

Take 0.5 g of the same gray felt honeysuckle sample powder, 6 parts, accurately weighed, and accurately add 5 mL of the reference substance chlorogenic acid (0.382mg/mL) respectively, and prepare the test solution according to the method under the item, and carry out the content determination and calculation The average recovery rate of chlorogenic acid was 98.3%, and the RSD was 0.37%.

Table6. Sample recovery Test Results

Sample weight	Luteolin peak area	Recoveries (%)
0.5014	5330.34	100.12
0.4987	5301.30	99.78
0.4998	5312.88	98.98
0.5087	5407.60	100.09
0.4887	5195.01	101.12
0.5167	5492.64	99.99

3.6 Results of sample content determination

Prepare the test solution of each sample according to the method below, inject the sample according to the chromatographic conditions below, measure each sample 3 times, record the chromatogram and calculate the peak area according to the external standard method. The content of chlorogenic acid and luteolin in 5 different origins of Lonicera japonica is shown in Tables 7 and 8. The HPLC diagrams of the reference substance and the sample solution are shown in the figure below. The results of high performance liquid chromatography showed that the chlorogenic acid content of each Lonicera japonica varieties were as follows: Baiyun: 6.297%, RSD0.0137; Jin Cui Lei: 5.293%, RSD 0.0060; Yin Cui Lei: 5.288%, RSD 0.0029; common species : 7.455%, RSD0.0065; common species (after flowering): 4.140%, RSD0.0151. The content of luteolin in each Lonicera japonica variety is as follows, Baiyun: 0.0264%, RSD 0.0079; Jin Cui Lei: 0.0573%, RSD 0.0053; Yin Cui Lei: 0.0347%, RSD 0.0010; Common species: 0.0394%, RSD 0. 0073; common varieties (after flowering): 0.0197%, RSD0.0031.

Table7. Content of chlorogenic acid in different species

Species	Chlorogenic acid (%)	RSD
Baiyun	6.297	0.0137
Jin Cuilei	5.293	0.0060
Silver Cuilei	5.288	0.0029
Common species	7.455	0.0065
Common species (after flowering)	4.410	0.0151

Table8. Content of Luteoloside in different species

Species	Luteoloside (%)	RSD
Baiyun	0.0264	0.0079
Jin Cuilei	0.0573	0.0053
Silver Cuilei	0.0347	0.0010
Common species	0.0394	0.0073
Common species (after flowering)	0.0197	0.0031

4 Conclusion and discussion

The results of this experiment show that the chlorogenic acid content of the flower buds of Lonicera japonica is far higher than that required for genuine honeysuckle. However, the content of luteolin is not as good as that of genuine honeysuckle. However, chlorogenic acid and luteolin have their own different pharmaceutical values[4-5].

The dry processing technology of honeysuckle flower buds directly affects the content of chlorogenic acid and luteolin in the flower buds. At present, the drying processing techniques used for the flower buds of the gray felt hair honeysuckle include sun drying, drying, shade drying and fumigation[6]. Comparing the effects of sun drying, shade drying, steaming and sulfur fumigation on the content of chlorogenic acid in Lonicera japonica, the results show that the content of chlorogenic acid in the shade drying is the highest. Compared with the other three methods, there are extremely significant differences. The sulfur fumigation method Second, but it is also significantly better than steaming and direct drying. The content of chlorogenic acid in honeysuckle dried by sun drying is the lowest. The sun-drying is divided into steaming, fumigating, frying, and raw drying. Studies have shown that steaming and fumigation are better, while the content of chlorogenic acid and luteolin is lower in raw sun-dried products. Studies have shown that honeysuckle produced by sulfur fumigation and drying methods not only has good appearance quality, but also has high chlorogenic acid content.

The samples in this experiment are divided into two drying methods, one is drying and the other is drying. The results showed that the raw sun exposure method greatly reduced the content of the effective ingredients in the flower buds of Lonicera japonica. Moreover, the dried flower buds are obviously darker, which also affects the appearance quality. The dried flower buds are green in appearance and good in quality.

Author

* Corresponding author: Jing Cai (1986-) , female, from Macheng, Hubei Province, lecturer, email: 84016963@qq.com, Research Direction: Utilization of GERMPLASM resources and its research.

Acknowledgments

This research was financially supported by the Provincial Teaching Research Project (2018496) , (2017505) ; Scientific Research Project of Wuhan Institute of Design and Engineering (K201905) ; Wuhan Institute of Design and Engineering, school-level quality courses (2017 YK1116).

References

1. Yan Chen. Determination of chlorogenic acid in processed products of Lonicera macranthoides by HPLC. Identification of Flos Lonicerae, Flos LONICERAE and Flos LONICERAE. 2011, (31)23-24
2. Xugang Sun . Electronic Journal of Clinical Medical Literature. Determination of chlorogenic acid in zipu tongfeng mixture by 2018(a 4)
3. Sha Yang.Determination of Luteolin, chlorogenic acid and their adulterants in honeysuckle granules by HPLC.Hwaseo Journal of Pharmacy. 2021,36(01)
4. Zhongjun Liu .Determination of chlorogenic acid in Zipu tongfeng mixture by HPLC.Hwaseo Journal of Pharmacy. 2016(06)
5. Feiyan Wu, Jianguo Zeng. Identification and identification of Lonicera Japonica and Lonicera Japonica [j] .Chinese herbal medicine. 2014(08)
6. Jinhua Zhu. Simultaneous determination of nine active components in leaves of Lonicera Japonica thunb by HPLC.Chemical Research. 2017(01)

Detection and analysis of surface currents in western Guangdong by compact surface wave radar

zili Li[1], and Junjie Li[1,*]

[1] College of Electronic Engineering, Guangxi Normal University, Guilin 541004, China

Abstract. In order to study the detection performance of compact surface wave radar on ocean surface current in the west of Guangdong Province, firstly, the bistatic radar is set up to collect ocean echo data on the west coast of Guangdong Province. Then, by using the Multiple Signal Classification (MUSIC) algorithm, the comparative analysis results of ocean surface current information and the actual measurement results of DVS Current Meter are obtained. The root mean square error of flow velocity less than 10cm/s and the root mean square error of flow direction less than 25 degrees were obtained. It shows that compact surface wave radar can achieve good results and has good detection capability in the west of Guangdong, china.

1 Introduction

West Guangdong Sea, located in the southwest of the north of South China Sea, includes the sea from the west of Pearl River Estuary to the north of Hainan Island [1]. The coastal area of the west of Guangdong is a sea area closely related to the contemporary economic development, and its current state has an important influence on mariculture, material transportation and offshore engineering in this area. In order to effectively sup-port the development of Marine economy, it is necessary to carry out real-time, continuous and comprehensive monitoring of ocean surface current, and then provide accurate multi-time and multi-scale Marine environmental forecast [2].

At present, high frequency surface wave radar is the main tool capable of measuring ocean surface current state in a large range and with high precision, nearly 500 high frequency radar systems have been set up around the world[3]. These radar systems can be divided into two types in antenna form: array HF surface-wave radar based on beam-forming technology and compact HF surface-wave radar based on direction finding technology . Among them, the phased array antenna is used as the receiving antenna for the array high frequency surface wave radar represented by the WERA system (Germany). The receiving antennas of portable high frequency surface wave radars, represented by Seasonde system (USA) and OSMAR-S system (China), adopt monopole cross loops. However, the antenna array of array HF surface wave radar is relatively large, which brings the disadvantages such as large area, difficulty in erection and high power consumption[3]. However, compact portable radar does not need to specially open up a site for its antenna. It is relatively easy to locate, build and maintain the station, and has strong adaptability to installation. It is the main direction of the development of high frequency surface wave radar. In the international high frequency surface wave radar market, portable high frequency surface wave radar occupies more than 90% of the market share .

The OSMAR-S series compact high frequency surface wave radar developed by China has been tested in many sea areas to detect ocean surface currents. In this paper, the performance of the Omar-S portable radar system is tested by the current measurement data of the current meter, and the trial situation of the Omar-S portable radar system in the sea area of the west of Guangdong is studied, so as to provide a useful reference for the application of the compact surface-wave radar in the sea area.

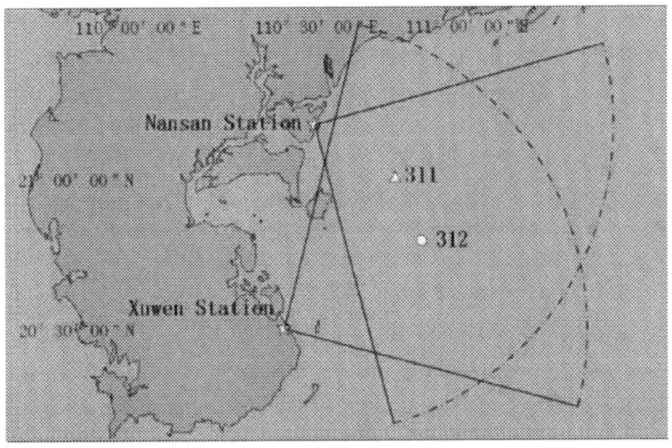

Figure 1. Location map of radar and current meter.

2 Overview of Compact Surface Wave Radar System

OSMAR-S is a coherent scattering radar system, which mainly consists of transmit-ting antenna, receiving antenna, receiver, transmitter and display terminal [9]. The trans-mitting antenna of the system is a whip antenna and the receiving antenna is a monopole crossed ring antenna. After the receiving signal is preprocessed, the

* Corresponding author: 61419008@qq.com

radial velocity of the ocean surface is calculated by combining with the system parameters. The radial flows in different directions of the same measuring point are obtained by the bistatic radar, and the vector flow field of the measured point is obtained by combining the vector synthesis method [10]. The operating frequency of the radar is 13±0.5MHz; the maximum detection range is 100 km; the radial range resolution is 2.5km; the velocity resolution is 5 cm/s; and the angular resolution is 2.5°.

In this experiment, the bistatic radar system was used to install the high frequency surface wave radar in Nansan Town(110°35′16″E 、 21°8′13″N) and Xuwen County (110°29′49″E 、 20°32′46″N) of Zhanjiang City, Guangdong Province. As shown in Figure 1, the triangle and the circle indicate the specific positions of two DVS current meters 311 and 312, which are located at 110°49′43″E, 20°59′25″N and 110°53′48″E, 20°47′40″N. The five-pointed star is the location of two radar stations, respectively named as Nansan Station (located in Nansan Town, Zhanjiang City) and Xuwen Station (located in Xuwen County, Zhanjiang City). The two fan regions are the theoretical coverage range of electromagnetic waves of the two radar stations, with a radius of 100 km. The radar flow measurement area is the common coverage range of the two radar stations.

3 Extraction of surface flow

3.1 Extraction of radial flow velocity

In the experiment, the scattered echo signals from the ocean surface are received from the radar station according to the principle of interrupted frequency modulated continuous wave. The received echo signals are separated into different range units, and the sum and average of the echo power received by each channel in the coherent accumulation time are obtained to obtain the original echo data detected by the radar. The echo spectrum of each range element is obtained by two fast Fourier transform of the original data of a single channel, and then the spectrum of all range elements is combined to get the range Doppler spectrum of the whole detection area [5] .

The velocity information of ocean current is determined by the frequency shift of first-order internal frequency points relative to Bragg frequency, so it is necessary to determine which effective frequency points can be used to retrieve ocean current. The echo in a distance element is not only the first order echo of ocean waves but also the high order echo and other noises. In order to determine the first-order peak frequency points, it is usual to define the maximum frequency offset of the first-order peak by setting a maxi-mum probed velocity, which is obtained from the historical data in the observed area. Then the noise power intensity is calculated according to the noise region. Finally, the preset threshold value of SNR (dB) is used to select the first-order peak frequency points that meet the requirements of SNR. The process of determining the effective frequency points of the first order peak in the Doppler spectrum is shown in Fig.2, where the SNR threshold is set at 10dB, and the two gray areas above the SNR threshold in the figure are the final determined first-order peak frequency points. Of course, there are other methods to select the first-order peak, such as the difference spectrum method and the first-order peak separation method based on image processing technology[13], but the method used here to select the first-order peak frequency points by setting the historical maximum flow rate and the SNR threshold is still the most convenient and the most widely used.

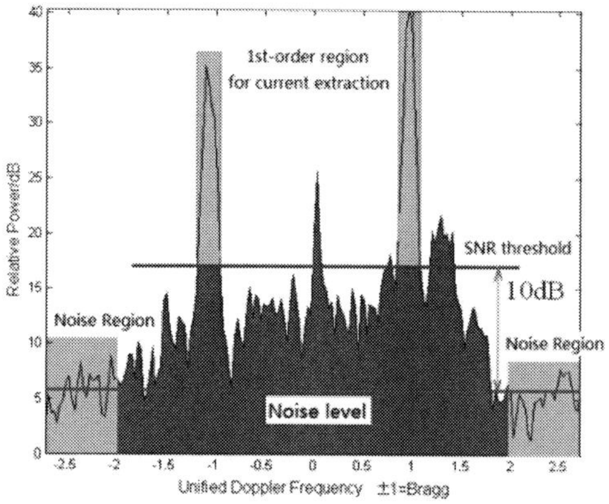

Figure 2. Schematic diagram of first order peak extraction.

After the first-order spectrum containing the ocean surface current information is obtained, the magnitude of the radial velocity can be calculated. The corresponding relationship between the frequency offset of the corresponding spectral point and the ocean surface radial velocity is as follows:

$$\Delta f = 2v / \lambda \quad (1)$$

Where is the operating wavelength of the radar. The radial velocity can be calculated by Equation (1).

3.2 Extraction of radial flow direction

After the size of the radial flow is obtained, the next step is to determine the radial flow orientation. Spatial spectrum estimation in array signal processing is a very popular field of study, when the signal into the antenna array, as a result of the existence of array element spacing, the antenna receives signals exist fixed phase difference, there is a phase difference is frequency, azimuth estimation became the frequency spectrum estimation, so you can use the classic spectrum estimation and modern spectrum estimation algorithm to estimate azimuth[6].

MUSIC (Multiple Signal Classification) directional algorithm has a common applicability, can deal with arbitrary shape, arbitrary direction characteristics of antenna array, can give a good result in practical application, the sea echo theory shows that the first-order sea echo Signal is Gaussian Signal, and on the time and location are quick to related, so the MUSIC algorithm is applied to the radial flow calculation.

When the number of signal sources N is less than the number of antenna array elements M, the mathematical model of the narrow-band far-field signal is:

$$X(t) = A(\theta)S(t) + N(t) \qquad (2)$$

X is the antenna receiving data, A is the antenna array type, and N is the noise. The covariance of the received data is:

$$R = E\left[XX^H\right] = AE\left[SS^H\right]A^H + \sigma^2 I = AR_S A^H + \sigma^2 I \qquad (3)$$

Where R_s is the covariance of the signal source, $AR_S A^H$ is the signal part, and $\sigma^2 I$ is the noise part.

The eigendecomposition of R can be obtained as follows:

$$R = U_S \Sigma_S U_S^H + U_N \Sigma_N U_N^H \qquad (3)$$

U_S is the signal subspace, U_N is the noise subspace, Σ_S and Σ_N are the corresponding eigenvalues of the signal and noise respectively.

Since U_S and U_N are orthogonal after eigendecomposition, and the span of U_S and the guide vector $a(\theta)$ of the signal is the same space, so:

$$a^H(\theta)U_N = 0 \qquad (4)$$

We are dealing with data of finite length, so we can only get the maximum likelihood estimate of R:

$$\hat{R} = \frac{1}{L}\sum_{i=1}^{L} X_i X_i^H \qquad (5)$$

Estimation $\hat{U}_S$ and $\hat{U}_N$ of signal subspace and noise subspace can be obtained by eigendecomposition of $\hat{R}$. Due to the presence of noise, $a(\theta)$ and $\hat{U}_N$ cannot be completely orthogonal, but there is a minimum value, which MUSIC algorithm determines the azimuth by searching:

$$P(\theta) = \frac{1}{a^H(\theta)\hat{U}_N \hat{U}_N^H a(\theta)} \qquad (6)$$

The Angle θ corresponding to the peak value of the spatial spectral function is the direction of the radial flow to be estimated. After the size and direction of the radial flow are obtained, the radial flow field diagram of the whole detected sea surface can be obtained. Fig.3 shows the radial flow diagram at 7:00 on September 25, 2017 of the south-central three stations of the bistatic radar.

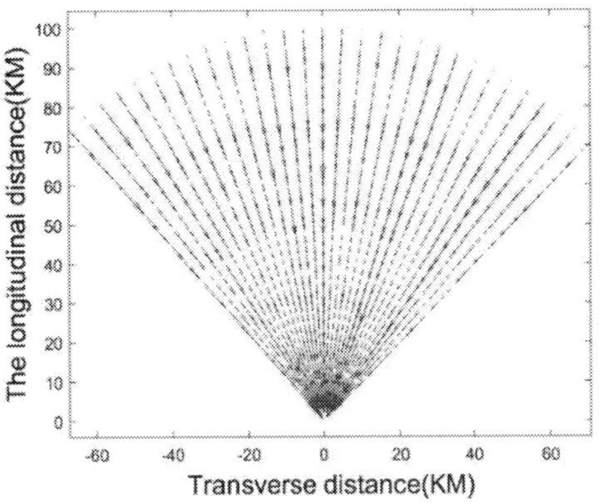

Figure 3. vector synthesis schematic.

3.3 Synthesis of vector flow

In double station radar to detect the overlapping area according to the longitude and latitude to construct the public grid, respectively of radial flow with the same session each stance radar site contains a single point of the radial velocity information is distributed to the corresponding mesh point, so that each public grid point contains two messages from different position of the radial velocity[7]. According to the two radial flow information, a vector flow information is synthesized to obtain the vector velocity information at the point of the common grid.

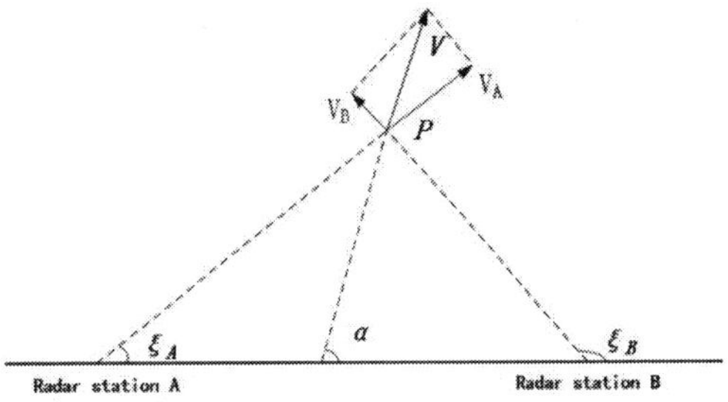

Figure 4. vector synthesis schematic.

The relationship between the direction and velocity of the vector flow and the direction and magnitude of the radial flow is shown in Fig.4. V , α are the velocity and di-rection of the measured vector point P; V_A , ξ_A are the radial velocity and direction at P measured by station A; V_B , ξ_B are the radial velocity and direction at P measured by station B. After the radial flow of a single station is obtained, the final velocity V and flow direction α can be obtained by the following formula:

$$V_A = V\cos(a - \xi_A) \qquad (4)$$

$$V_B = V\cos(a - \xi_B) \qquad (5)$$

$$a = \arctan\frac{V_B\cos\xi_A - V_A\cos\xi_B}{V_A\sin\xi_B - V_B\sin\xi_A} \qquad (6)$$

 https://doi.org/10.1051/e3sconf/202129301030

$$V = \frac{V_B}{\cos(a - \xi_B)} \tag{7}$$

Combined with the above vector synthesis method, the radial flow diagram of the two stations can be synthesized to obtain the vector flow diagram. Fig.5 shows the vector flow diagram combined by two stations at 10:00 on September 25, 2017.

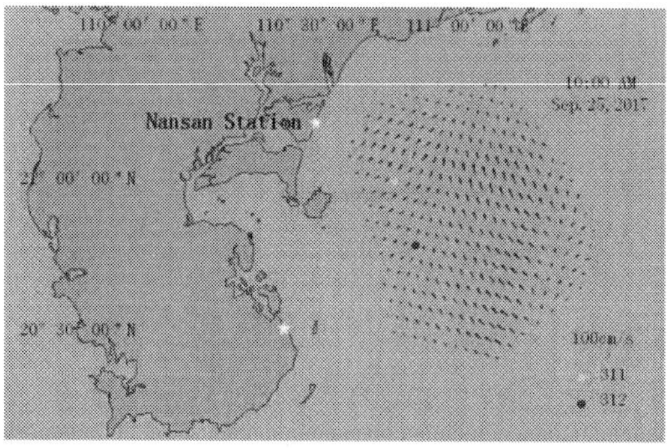

Figure 5. Composite vector flow diagram.

4 Result Analysis

4.1 Comparison scheme

In this experiment, the real-time data collected by TRDI company's DVS single-point current meter and the surface current data generated by remote sensing data were com-pared and analyzed. In general, the change of ocean surface current is slow. the bistatic compact surface-wave radar system synthesizes a vector current map every 20 minutes, and combines the four data of two single-point current meters into a group for comparison with the data of remote sensing detection.

4.2 Result analysis

The measured DVS and radar values from 0:00 to 12:00 on September 25, 2017 were selected for the analysis of the experimental results. Figure 6 and Figure 7 are the comparison diagram of the velocity and direction detected by the fixed-point current meter and radar at two places. It can be seen from the figure that the velocity and direction detected by radar have a high degree of agreement with the measured value. The correlation coefficients between the velocity measured by surface wave radar and the measured values of 311 and 312 were 0.8586 and 0.8475, and the root mean square errors were 6.6cm/s and 6.4cm/s, according to the results collected by current meter. The correlation coefficients between the flow direction measured by surface wave radar and the measured values of current meter 311 and 312 are 0.8844 and 0.9642, and the root mean square errors are 24.05°and 22.75°. The velocity and direction values detected by the compact high frequency surface wave radar system are in good agreement with those measured by the fixed-point current meter, and the overall changing trend is consistent.

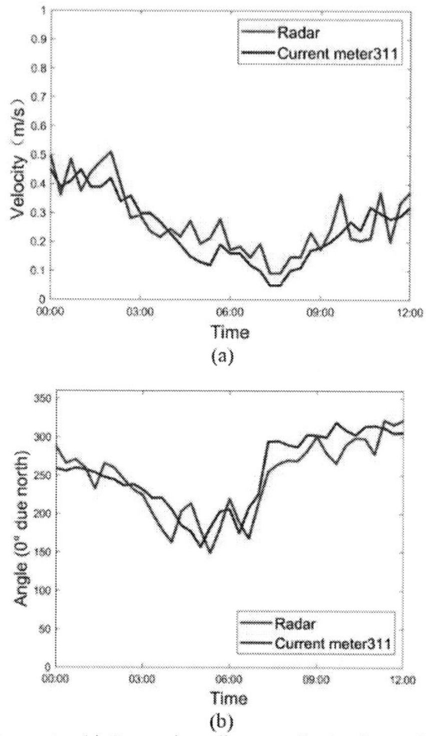

Figure 6. (a) Comparison diagram of radar data and 311 current meter rate values; (b) Comparison of radar data and 311 current meter Angle values.

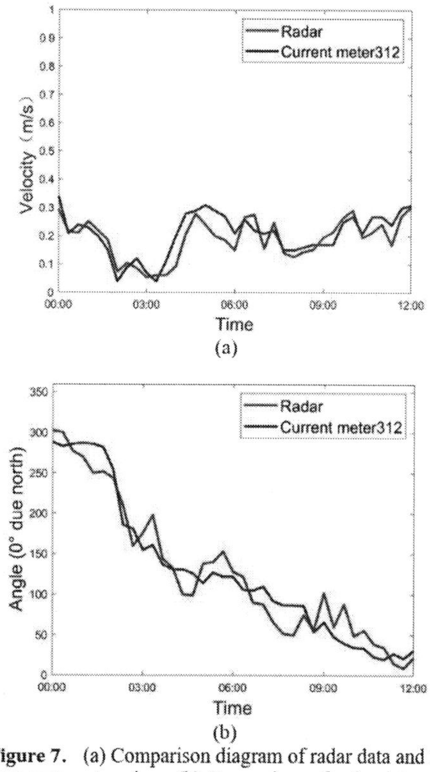

Figure 7. (a) Comparison diagram of radar data and 312 current meter rate values; (b) Comparison of radar data and 312 current meter Angle values.

5 Result Analysis

The contrast experiment of surface current detection in the sea area west of Guang-dong by the compact high frequency surface wave radar adopts the ocean surface current detection equipment and the bistatic radar to make synchronous observation and com-parison in the same sea area, which fills in the blank of the previous experimental data. With the continuous development of radar detection technology, real-time monitoring of large area Marine environment has been realized, which provides a large amount of sur-face current information for intensive study of the ocean. The detection results of Omar-S in this paper are in good agreement with the observation results of other standard equip-ment, satisfy the requirement of large area real-time detection of ocean surface current, and verifying the practical performance of the compact surface-wave radar system.

References

1. X. Lingling, C. Ruixiue, S. Qingtong. Progress of Study on Coastal Circulation near the Shore of Western Guangdong. J. Journal of Guangdong Ocean University. 2012, 32, 94-98.
2. X. Xiaobai, W. Yuanqing, L. Hong.Data verification of high frequency ground wave radar and its application in typhoon monitoring. J. RADOR & ECM. 2020, 40, 43-48.
3. Y. Hua, J. Huifeng, G. Qingqing. Summertime surface current characteristics in the southern region of Subei radial sand ridges detected by HF surface wave radar. J. Acta Oceanologica Sinica. 2019, 41, 1-11.
4. W. Biyang, L. Zili, Z. Hao. Sea Surface Currents Detection at The Eastern China Sea by HF Ground Wave Radar OSMAR-S. J. Acta Electronica Sinica. 2009, 37, 2778-2782.
5. C. Jiang, W. Xiongbin, W. Biyang. Ocean Surface Current Detection by HF Ground Wave Radar with Portable Antenna System. J. Wuhan Univ. (Nat. Sci. Ed.). 2005, 51, 105-108.
6. J. Lijie, W. Biyang, Z. Hao. HFSWR with two-element cross-loop monopole antenna array. J. Huazhong Univ. of Sci. & Tech. (Natural Science Edition). 2017, 45, 6-10.
7. S. Fang, W. Yuanqing, L. Hong. Review of Developments and Applications Status of High-Frequency Radar. J. Hydrographic Surveying and Charting. 2018, 38, 22-30.

 https://doi.org/10.1051/e3sconf/202129301031

Influence on indoor comfort due to the application of Covid-19 natural ventilation protocols for schools at subtropical climate during winter season

Sergio Gómez Melgar[1*], Antonio Sánchez Cordero[1], Marta Videras Rodríguez[1], José M Andújar Márquez[1]

[1]TEP192 Control y Robótica, Escuela Técnica Superior de Ingeniería, Universidad de Huelva, CP. 21007 Huelva, Spain.

Abstract. Thermal comfort and energy performance of our buildings has been seriously improved during the last decades with the introduction of better envelope airtightness and mechanical ventilation systems. The arrival of Covid-19 has forced us to accept new natural ventilation protocols to improve the air quality inside buildings over other characteristics of the buildings. As a consequence, it brings some negative effects, like a reduction in the control of indoor dry bulb temperature (T), and an increase in energy demand for heating and air conditioning. To evaluate these influences during a winter season, a standard classroom from a school located at subtropical climate has been selected to be monitored with air quality and comfort devices. The experiment considers two different scenarios in 2018, and 2021. At them, data have been compared with several comfort ranges from international standards and the recommendations from the Spanish Government, obtaining opposite results. The results from 2018 showed that the comfort T ranges were adequate, but CO_2 concentrations rise over 2000 ppm which was unhealthy. On the contrary, in 2021 when the Covid-19 natural ventilation protocol is activated, the CO_2 concentration becomes less than 700 ppm. Finally, the suggested natural ventilation protocol was proved as an efficient measure to improve air quality, but it reduces the performance of the building in terms of comfort and energy demand. Therefore, several alternative ventilation methods should be considered to keep both energy air quality building's performance.

1 Introduction

Children spend inside schools an important part of their lives, and many often, the comfort conditions of these buildings seem to be inconvenient. Many assessment methods have been recently released to provide comprehensive evaluations about comfort and social sustainability inside buildings [1]. Even in many modern countries, most of the schools are old, poorly insulated, and with phase out facilities [2]–[4]. Therefore, parameters such as energy performance, air quality, T and relative humidity stay far from those suggested in best practices guides in most of the cases. Several studies have reviewed the indoor conditions of schools and kindergarten with unsatisfactory conclusions [2], [6].

In some cases, indoor air quality may be improved with natural ventilation, but some problems may arise in those schools located in polluted environments. There, ventilation has to be carefully considered in combination with filtering and mechanical systems [7]–[9].

Since the recent spread of the disease Covid-19, the improvement of air quality has appeared with more relevance than before. It is now a matter of fact that Covid-19 is mainly spread by airborne transmission [10], [11]. Unfortunately, there are no efficient methodologies to measure the concentration of Covid-19 in the air now. However, some alternative methodologies can be applied to approximate the idea of infection risk looking at the concentration of CO2 in the air. Additionally, it has been proved as an easy and cheap method to measure the indoor air quality [12]. Because of the proved relationship between Covid-19 and the CO2, decision makers has set regulations to increase natural ventilation as the best option to reduce Covid-19 indoor airborne transmission [13]. In particular, the Spanish government set some ventilation protocols that should be considered in all buildings, but specifically in those for education [14]. These protocols stablish the need to maintain natural ventilation during all the time children are inside the classroom. They set 12.5 l/sec per person as the minimum exterior air renovation [15], but recognize that some problems may arise with extreme T in winter and summer. The effectiveness of this ventilation protocol can be evaluated using CO_2 measuring devices.

Unfortunately, the generalized use of exterior air for natural ventilation can lead to comfort problems, such as noise, pollution, exposure to wind gust, inadequate relative humidity, but most important uncomfortable T [16]. Additionally, this ventilation protocol is producing extra energy demand for the heating, ventilation, and air conditioning (HVAC) systems. Obviously, these effects vary depending on the climate of the site, and it may be less important in warm subtropical climates than others [17].

Correspondence author: sergomel@uhu.es

https://doi.org/10.1051/e3sconf/202129301031

This paper aims to compare the performance of the air quality and comfort parameters between 2 different scenarios: the first one with free natural ventilation with openings controlled by teachers according to their comfort experience, whilst the second provides natural ventilation during all time the time scholars are inside the room. Finally, some improvements to the current ventilation protocol can be suggested to increased comfort, that would probably reduce HVAC primary energy demand.

2 Material and Method

2.1 Case study

The classroom monitored is in a public school in the city center of Sevilla (Spain) at 37.39 north latitude, and 5.98 east longitude. Figure 1, shows the aerial view with the north-south orientation, with the classroom position marked in red. The original building was built in 1946, and since then it was enlarged and modified several times according to the evolution of the educational trends. The envelope of the building is poorly insulated. The HVAC system is based: a central controlled system with high temperature hot water radiators for winter conditions, and a decentralized air condition with air-to-air split units for summer conditions. There is no mechanical ventilation system for air renovation.

Figure 1. Aerial view, Google [18]

Figure 2. shows the studied classroom within the building from the north east view

Figure 2. South east bird view

.Figure 3 describes the geometry and the specific position of openings, HVAC systems, and monitoring devices. The size of the room is 8.0x6.2x3.7 m^3. Windows are located at south and north facades, providing full cross ventilation. The facade includes a set of white louvres to reduce solar radiation in the cooling season. The radiators (in red color) are located at the north façade, under the windows. The AC system (in blue color) is in the south east corner, see Figure 3.

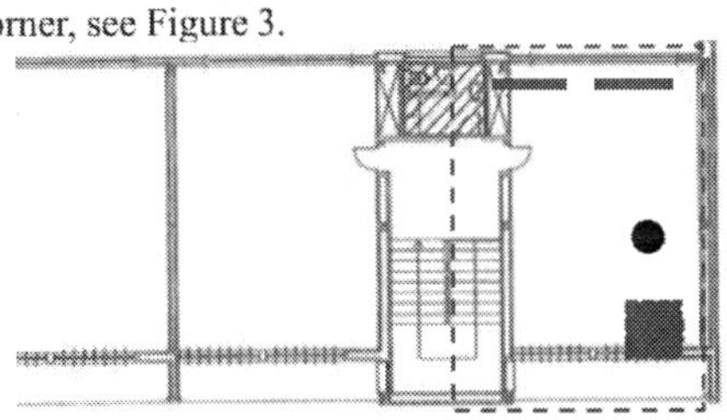

● Air quality monitoring system
● Heating ● AC system

Figure 3. Classroom layout with key elements in colour.

2.2 Materials and Method

According to Figure 3 the air quality monitoring systems (in black color) were located at the east facade at 0.9 m above de floor. Figure 4 shows the position of the devices. All children in the classroom look at the opposite direction to reduce the unwanted effect of close breathing. The position of the devices is also intentionally separated from windows to reduce the local effect of the openings.

Figure 4. Classroom view.

Figure 5 shows the devices located at the back of the classroom. Described as “1”, the device installed in 2017, set by the University of Burgos [4]. It measures and record T in ºC, RH in % and CO_2 in ppm every 10 minutes. However, at the end of 2020, another device described as “2” in Figure 5, the EGVOC-165 Air quality Monitor with a NDIR CO_2 sensor was installed to provide remote access. It measured several parameters, including T, RH and CO_2 concentration, every 1.5 sec. The “2” device was connected to a mobile WIFI system: 4G LTE M7200 from tp-link, described as “3” in Figure 5.

Figure 5. Monitoring devices "1, 2, 3"

Climate data have been obtained from a close weather station registered in Weather Underground [19]and collected manually on the 26th of January 2021.

With all data collected from both scenarios: 2018, and 2021, results graphs have been obtained and compared with those comfort ranges suggested by the literature review. Air quality and CO_2 concentration in ppm, have been compared with 800 ppm as the highest recommendable value to reduce Covid-19 airborne infection as the Spanish government has set [14]. Temperature comfort ranges have been defined according to methodology defined in ASHRAE 55_2017 (see equations (1), (2) and (3)) for all buildings and based in the concepts of adaptivity and 80% of acceptability [20]. The mean outdoor temperature used for calculations has been set by the average between 2018 and 2021.

$$Tc = (17.6 + 0.31To), \quad (1)$$

Considering the 80% of acceptability

$$Tc1 = (Tc - 2.5) \quad (2)$$

and $$Tc2 = (Tc + 2.5) \quad (3)$$

3 Results

At any regular day of both scenarios, the heating system worked from 08:00 to 14:00, whilst the children stayed at school from 09:10 to 13:45, but with a short recess from 11:30 to 12:00, see Figure 6.

On the contrary, natural ventilation was different in 2021 from 2018. According to Covid-19 ventilation protocol, the natural ventilation is working from 09:10, when the children get in the classroom, until 14:00, when they leave (see Figure 6). There is no strict evidence about the ventilation proceeding in 2018, but according to teacher's survey, most of the time the openings were close at that time and they freely operate depending on their comfort perception.

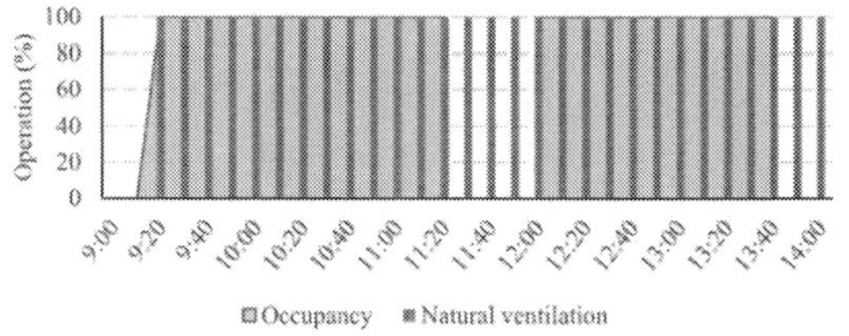

Figure 6. Operation schedule in a 2021 typical winter day.

3.1 Air quality

Air quality data is shown in Figures 6 and 7 The data have been obtained from one typical winter day in January. Figure 6 contains 2018 data, whilst Figure 3. contains 2021 data. Graphs in Figure 7 and 3.3 contain CO_2 concentration in ppm compared with those security limits set by other authors: 800 ppm for Covid-19 security protocols, and 1000 ppm according to ASHRAE 62.1 recommendations [21].

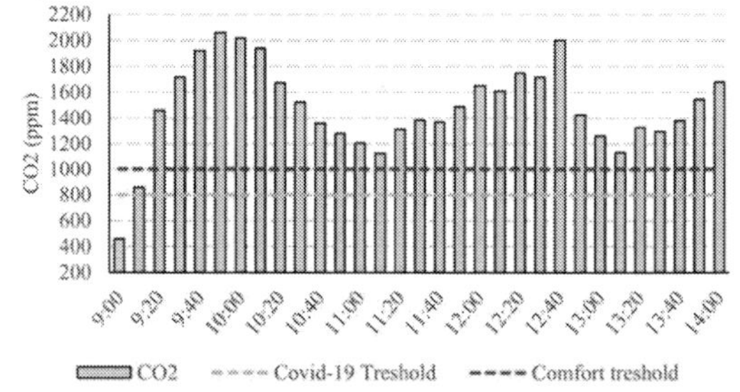

Figure 7. CO_2 ppm in a typical winter day, 26th January2018.

According to Figure 7, data from 2018 were as follows: at 09:00 the CO_2 was 460 ppm, but as soon as the children arrived to the room it started to rise rapidly until 2061 ppm at 09:50. Since 10:00, the CO_2 was reduced until 1125 ppm at 11:10. When the children came back from the break time at 12:00, the CO_2 concentration was 1649 ppm, and it kept rising until 2002 ppm at 12:40. Since then, the CO_2 started to decrease to 1133 ppm at 13:10. Finally, the CO_2 increased again until 1681 ppm at 14:00. According to these data, the CO_2 levels stayed over security levels for all the considered time, except from at 09:10 when children arrived at the classroom.

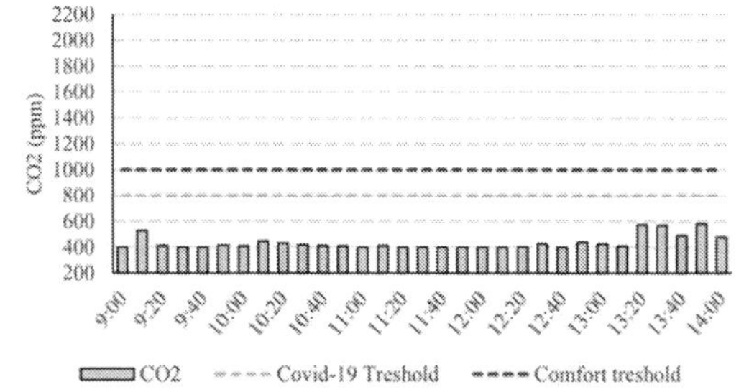

Figure 8. CO_2 ppm in a typical winter day, 26th January 2021

Figure 8 shows data from 2021 as follows: at 09:00 the CO_2 was 400 ppm or less and it remained under 500 ppm until 13:20, and always under 600 ppm. Therefore, the concentration of CO_2 remained always under security

https://doi.org/10.1051/e3sconf/202129301031

limits when the openings in the classroom were fully open during all the morning.

3.2 Temperature & comfort

T and adaptive T comfort ranges are shown in Figures 9 and 10 for the selected typical winter days in January 2018 and 2021. The T comfort ranges have been set into 18.4°C and 23.1 °C for both scenarios.

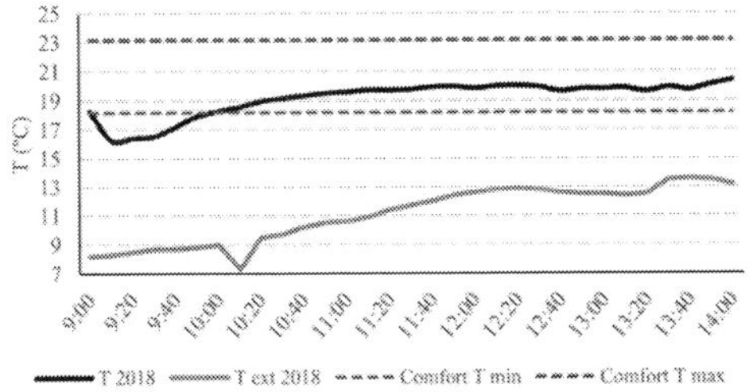

Figure 9. T°C in a typical winter day, 26th January 2018, compared with adaptative ASHRAE comfort range

As seen in Figure 9, in 2018, at 09:00 T was 18°C, but it decreased fast until 16°C at 09:20 when doors got open. Since then, the T increased steadily until 20°C at 11:50 and it remain with small variations until the end of the studied period at 14:00. Therefore, the T stayed between comfort conditions from 10:00, whilst the exterior T (T ext) was always under comfort, from 8 °C, to 14°C.

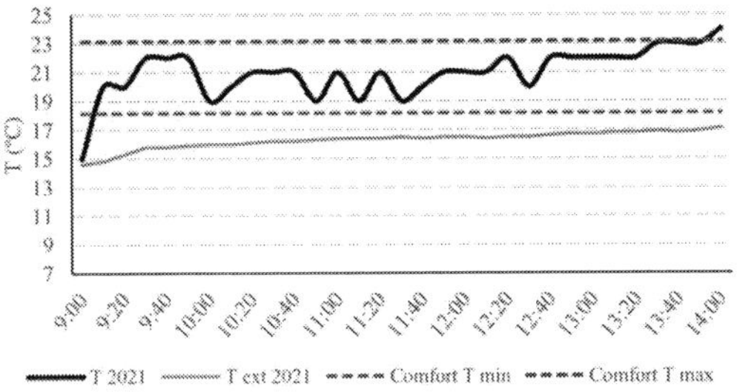

Figure 10. T°C in a typical winter day, 26th January 2021, compared with adaptative ASHRAE comfort range

In 2021, see in Figure 10, at 09:00 the T was 15°C, but it rose fast until 22°C at 09:30. From 09:30 to 13:50 it was maintained within the comfort ranges, between 19 and 22 °C. On the contrary T ext was steadily raising from 15 °C at 09:00 to 17.00 °C at 14:00, which was close to the lower comfort range.

In Table 1 a summary of T from both scenarios can be seen with hourly and average data. Additionally, the variation from T and T ext is presented for both scenarios. In 2018, the variation between T and T ext was 8.32 °C, whilst in 2021, it was 4.12 °C.

Finally, when CO_2, and T, are considered together, the 2018 scenario was out of comfort for all the period from 09:00 to 14:00, whilst the 2021 scenario was under comfort from 09:10 to 13:50.

Table1. T deviation between scenarios 2018 and 2021.

	09.00	10.00	11.00	12.00	13.00	14.00	Aver.
2018							
T	18.24	18.24	19.56	19.81	19.78	20.37	19.37
T ext.	8.20	9.00	10.60	12.60	12.50	13.20	11.02
T-Text	10.04	9.26	8.96	7.21	7.28	7.17	8.32
2019							
T	15.00	15.00	19.00	21.00	21.00	22.00	24.00
T ext.	14.60	14.60	16.00	16.40	16.50	16.70	17.10
T-Text	0.40	0.40	3.00	4.60	4.50	5.30	6.90

4 Discussion

Materials and Methods proposed in Section 2 have been adequate for the purpose of this research. However, the EGVOC-165 Air quality Monitor should be replaced for another system to allow more precision in T data necessary for comfort calculations.

The results shown in Section 3, describe the improvement on air quality from scenario 2018 to 2021, but with some considerations. According to Figure 7 and Figure 8, the concentration of CO_2 has been drastically reduced from a range between 1200 ppm to 2000 ppm, to less than 600 ppm. This is less than the 800 ppm CO_2 concentration suggested by the authorities [14]. Therefore, the result of the suggested ventilation protocol has demonstrated its adequacy to reduce the Covid-19 airborne infection risk. On the contrary, the excess of uncontrolled natural ventilation is producing fast variations in T, and it may have also provided an increase in heating energy demand. According to Figure 9 and Figure 10 both scenarios provide indoor T between comfort ranges during most of the time, except at 09:10, when children get into the classroom and the space needs some time to achieve a steady state situation.

If T data are compared only from 10:00, some details must be described: In 2018, the T in Figure 9 shows slight variations between 18.24 °C to 20.37 °C, whilst in 2021 the T in Figure 10 shows higher variation in both directions between 19.00 to 24.00. Unfortunately, T data cannot be solely discussed because T ext is quite different in the scenarios. As described in Table 3.1, the average T is quite similar in both cases with only 0.96 °C difference, but T ext is 5.22°C higher in 2021. Additionally, the variation between T and T ext is 4.20°C lower in 2021. Therefore, to provide a more precise comparison between ventilation scenarios, it would be necessary to repeat the experiment in a winter day with similar conditions to those in 2018. There would probably be more difficulties to maintain T between the ASHRAE T comfort ranges [20].

Heating energy demand has not been considered in this research to be focused on comfort, but according to the comments from the management staff of the school, the heating system is obviously working more time at a higher rate in 2021 than it did in 2018. Therefore, the energy demand should be higher even with less indoor T.

To increase the performance of the building, some improvements could be done to the ventilation protocol. Openings can be reduced until the CO_2 concentration rise to 800 ppm. Mechanical ventilation with heat recovery and high efficiency filters can replace natural ventilation [5]. Air exhaust should be considered upwards over the ceiling to avoid indoor contamination [10]. Additionally, results can be affected by the influence of windows, the close presence of children and their movement, the existence of mask protection protocols and the effect of air gusts into the classroom, among others. Because of that, Additional monitoring devices could be installed in different positions and heights to have a complete description about the air displacement [22].

5 Conclusion

This research has demonstrated the adequacy of natural ventilation as an effective method to reduce CO_2 concentration, which is associate in the Covid-19 airborne infection risk.

However, this seems to provide a negative effect on thermal comfort and heating energy demand. Therefore, alternative ventilation methods should be considered to maintain the air quality and ventilation effectiveness, but also comfort and energy performance of the buildings. These alternative methods could be based on two different strategies:

- An improvement version of the natural ventilation protocol which can provide real time evaluation of both, weather conditions and indoor air quality, to adjust the natural ventilation protocol.
- An additional introduction of mechanical ventilation systems with high efficiency filters to maintain the proper air quality levels and heat recovery units to keep T within comfort ranges, but with less heating and cooling energy demand.

Acknowledgments

The authors wish to acknowledge the collaboration of the school anonymous staff at the data gathering.

References

1. A. S. Cordero, S. G. Melgar, and J. M. A. Márquez, "Green building rating systems and the new framework level(s): A critical review of sustainability certification within Europe," *Energies*, vol. 13, no. 1, pp. 1–25, 2019.
2. M. M. Andamon, P. Rajagopalan, J. Woo, and R. Huang, "An investigation of indoor air quality in school classrooms in Victoria, Australia," *Proc. Int. Conf. Archit. Sci. Assoc.*, vol. 2019-Novem, pp. 497–506, 2019.
3. G. de Gennaro *et al.*, "Indoor air quality in schools," *Environ. Chem. Lett.*, vol. 12, no. 4, pp. 467–482, 2014.
4. Plataforma de Edificación Passivhaus and Universidad de Burgos, "Proyecto de monitorización de colegios. Estudio completo e informe de conclusiones." pp. 1–261, 2020.
5. S. G. Melgar, M. Á. Martínez Bohórquez, and J. M. Andújar Márquez, "Uhumebr: Energy refurbishment of existing buildings in subtropical climates to become minimum energy buildings," *Energies*, vol. 13, no. 5, 2020.
6. T. D. Tran *et al.*, "Assessment of air quality in school environments in Hanoi, Vietnam: A focus on mass-size distribution and elemental composition of indoor-outdoor ultrafine/fine/coarse particles," *Atmosphere (Basel).*, vol. 11, no. 5, 2020.
7. L. Stabile, M. Dell'Isola, A. Russi, A. Massimo, and G. Buonanno, "The effect of natural ventilation strategy on indoor air quality in schools," *Sci. Total Environ.*, vol. 595, pp. 894–902, 2017.
8. L. Zhang *et al.*, "Characteristics of air pollutants inside and outside a primary school classroom in Beijing and respiratory health impact on children," *Environ. Pollut.*, vol. 255, 2019.
9. T. Salthammer *et al.*, "Children's well-being at schools: Impact of climatic conditions and air pollution," *Environ. Int.*, vol. 94, pp. 196–210, 2016.
10. L. Morawska and D. K. Milton, "It Is Time to Address Airborne Transmission of Coronavirus Disease 2019 (COVID-19)," *Clin. Infect. Dis.*, vol. 71, no. 9, pp. 2311–2313, 2020.
11. A. L. Zhang, Y. Wang, and M. J. Molina, "Erratum: Identifying airborne transmission as the dominant route for the spread of COVID-19 (Proceedings of the National Academy of Sciences of the United States of America (2020) 117 (14857-14863) DOI: 10.1073/pnas.2009637117)," *Proc. Natl. Acad. Sci. U. S. A.*, vol. 117, no. 41, pp. 25942–25943, 2020.
12. Z. Peng and J. L. Jimenez, "Exhaled CO2 as COVID-19 infection risk proxy for different indoor environments and activities," *medRxiv*, pp. 1–9, 2020.
13. V. and A. C. A. Heating, Federation of Eropean Heating, "REHVA COVID-19 guidance document: How to operate HVAC andother building servicesystems to prevent thespread of the coronavirus(SARS-CoV-2) disease(COVID-19) in workplaces," 2020.
14. Ministerio de Sanidad Gobierno de España, "Medidas de prevención, higiene y promoción de l salud frente a COVID-19 para centros educativos en el curso 2020-2021," vol. 119, no. 3, pp. 670–673, 2020.
15. Ministerio de Sanidad, "Sistema de climatización de edificios y locales para la prevención de la propagación del SARS-CoV-2," p. 15, 2020.
16. F. Jomehzadeh *et al.*, "A review on windcatcher for passive cooling and natural ventilation in buildings, Part 1: Indoor air quality and thermal comfort assessment," *Renew. Sustain. Energy Rev.*, vol. 70, no. December 2016, pp. 736–756, 2017.
17. S. G. Melgar, M. Á. M. Bohórquez, and J. M. A. Márquez, "UhuMEB: Design, construction, and management methodology of minimum energy buildings in subtropical climates," *Energies*, vol. 11, no. 10, 2018.

18. G. [online] Maps, "Colegio Sor Angela de la Cruz," 2021. [Online]. Available: https://www.google.es/maps/@37.3989395,-5.9836181,253m/data=!3m1!1e3. [Accessed: 25-Feb-2021].
19. W. underground S. Network, "La Barqueta weather station." [Online]. Available: https://www.wunderground.com/dashboard/pws/ISEVIL38. [Accessed: 26-Feb-2021].
20. The American Society of Heating, Refrigerating and Air-Conditioning Engineers, "ANSI/ ASHRAE Standard 55-2013. Thermal Environmental Conditions for Human Occupancy." New York, 2010.
21. The American Society of Heating, Refrigerating and Air-Conditioning Engineers, "ANSI/ASHRAE Standard 62.1-2019. Ventilation for Acceptable Indoor Air Quality." Atlanta, 2019.
22. R. K. Bhagat, M. S. Davies Wykes, S. B. Dalziel, and P. F. Linden, "Effects of ventilation on the indoor spread of COVID-19," *J. Fluid Mech.*, vol. 903, 2020.

Study on collapsibility of typical structural loess by multi-parameter comparative analysis

Liu Hongcheng[1], Lu Changwei[1*], Wang Yinxia[2], Song Yi[1], Guan Xiangfeng[1], Fu Rui[1], Cao Jingfeng[1], Yang Zhao[3]

[1]Jilin Province Water Conservancy and Hydropower Survey and Design Institute, China

[2]Jilin Jianyuan Design Group Co. Ltd, China

[3]Henan Daming Architectural Design Co. Ltd, China

Abstract: To study the collapsibility of typical loess and its influencing factors in different areas, the samples in Jingyang, Lanzhou and Yili were studied. The correlation between initial moisture content, dry density, composition of particle size, structural parameters and the coefficient of subsidence is analyzed. The results show that: the coefficient of collapsibility is negatively correlated with the initial moisture content and dry density. In the experiments of multiple groups, the collapsibility coefficient has a certain correlation with the clay content, but not a uniform correlation with the particle content. The collapse coefficient is approximately positively correlated with the composite structure potential.

1 Introduction

Loess is a special soil formed in the Quaternary Period and widely distributed in the world. The total area of loess in China is about 640,000 km^2, of which about 60% is collapsible loess [1]. The special material composition and structure of loess make it have special engineering properties, such as collapsibility. These special properties are easy to cause a variety of geological disasters and engineering problems, such as slope instability, foundation settlement and tunnel collapse.

Due to the influence of hydrogeological conditions and climate, the engineering properties of loess in different areas are obviously different, which increases the complexity of engineering problems in loess distribution areas. Therefore, the study of loess collapsibility and its controlling factors in different areas is of great significance to the prevention and control of geological disasters and the safety of engineering construction in loess distribution areas. Many engineering problems in loess area are directly caused by loess collapsibility deformation and its mutation. In view of these problems, some scholars have explored the calculation model of unsaturated loess collapse deformation, and studied the influence of discontinuous distribution of collapsible units on the law of collapse deformation [5][6]. However, the engineering properties of loess are the result of many factors. It is the key to study the collapsibility of loess to clarify the influencing factors and the effects of each factor. In recent years, many achievements have been made in this field. Some scholars have explored the influence of physical and mechanical properties such as water content, dry density, plastic index and pressure on loess collapsibility, and the results show that the collapsibility coefficient has a good correlation with the above parameters [7]. By comparing the collapsibility characteristics of undisturbed and compacted loess, some scholars explored the influence of structure on collapsibility from a macro perspective [13]. Some other scholars have qualitatively and quantitatively analyzed the types of loess microstructure and the influence of various experimental parameters on collapsibility from the microscopic perspective [14][15].

Most of the existing studies take loess in a certain area as the research object and explore the influence of one or several factors on collapsibility. However, the collapsibility of loess is very regional, so the loess in different areas cannot be generalized. Therefore, this study takes the loess in Jingyang, Lanzhou and Yili as the research object to analyze the effects of particle size, density, humidity and structure on loess collapsibility, in order to provide a certain theoretical basis for loess collapsibility research, engineering construction and geological disaster prevention and control in different loess areas. Therefore, this paper takes loess samples from Jingyang, Lanzhou and Yili as the research objects to analyze the effects of particle size, density, humidity and structure on loess collapsibility, in order to provide certain theoretical basis for loess collapsibility research, engineering construction, geological disaster prevention and control in different loess areas.

* Corresponding author: Lu Changwei, lcw_jsy@163.com

2 Materials and methods

2.1 Materials

Soil samples were collected from Jingyang County, Shaanxi Province (E108°47'51.37", N34°50'32.53"), Lanzhou City, Gansu Province (E103°52'37.16", N35°53'49.71") and Yili City, Xinjiang Province (E81°43'27.16", N43°49'38.31"). Samples were taken in exploratory Wells ranging from 1 m to 19 m at the sampling sites. The basic physical properties of each soil sample were tested, and the results were shown in Table 1. As can be seen from the data in the table, with the increase of sampling depth, the moisture content and dry density of soil samples at the three sampling locations showed an increasing trend. In general, the dry density of the loess in Jingyang is similar to that in Lanzhou, while the dry density of the loess in Yili is slightly lower, but its plasticity index is higher, which may be caused by the higher clay content.

Table1. Basic physical properties of soil samples

Location of samples	Number of soil samples	Depth	Density	Natural moisture content	Dry density	Void ratio	Index of plasticity
		m	Gs	ω/%	ρ_d/g·cm^{-3}	e	I_p
Jingyang	JY01	1	2.71	11.8	1.32	1.053	12
	JY02	3		13.2	1.37	0.978	13
	JY03	5	2.71	14.4	1.35	0.993	13
	JY04	7		14.1	1.39	0.950	11
	JY05	9	2.71	15.8	1.38	0.964	14
	JY06	11		16.3	1.42	0.908	12
	JY07	13	2.71	17.5	1.43	0.895	15
	JY08	15		18.1	1.45	0.869	13
	JY09	17		18.2	1.43	0.895	11
	JY10	19	2.71	17.8	1.47	0.844	12
Lanzhou	LZ01	1	2.70	7.8	1.32	1.045	8
	LZ02	3		8.9	1.34	1.015	7
	LZ03	5	2.70	9.6	1.35	1.000	8
	LZ04	7		10.9	1.33	1.030	9
	LZ05	9	2.71	12.4	1.37	0.978	8
	LZ06	11		14.7	1.41	0.915	10
	LZ07	13	2.70	15.3	1.43	0.888	7
	LZ08	15		16.0	1.46	0.849	9
	LZ09	17		16.2	1.45	0.862	8
	LZ10	19	2.71	15.9	1.48	0.831	8
Yili	YL001	1	2.71	8.8	1.26	1.151	15
	YL02	3	2.71	10.6	1.25	1.168	14
	YL03	5		12.9	1.28	1.117	12
	YL04	7	2.71	14.9	1.28	1.117	12
	YL05	9		14.0	1.30	1.085	16
	YL06	11	2.70	16.2	1.31	1.061	14
	YL07	13		17.7	1.35	1.007	13
	YL08	15	2.71	18.4	1.38	0.964	15
	YL09	17		18.9	1.41	0.922	17
	YL10	19	2.71	18.5	1.45	0.869	15

2.2 Methods

Grain size, density, humidity and structure are the important physical properties of soil, which reflect the essential characteristics of soil and are the important factors affecting the collapsibility of loess. Therefore, four parameters, including particle size composition, dry density, initial water content and comprehensive structural potential, were selected to characterize the physical properties of loess, and the correlation between these parameters and the coefficient of collapsibility was explored. The test method of each part and the sample preparation process are as follows:

2.2.1 Influence of initial moisture content on the coefficient of collapsibility

The moisture content experiment was carried out in three groups, which were named as "JY Group", "LZ Group" and "YL Group" respectively. The moisture content of each sample was changed, and the collapsibility coefficient was measured by the double-line method. In order to ensure the same dry density, JY03, LZ03 and YL07 samples were selected for the test. Each group contained 9 samples, and the water content was controlled at 5%, 7%, 9%, 11%, 15%, 20%, 25% and 30%. For samples with lower moisture content than natural moisture content, air drying method was adopted to reduce moisture content to the target value. On the contrary, water film transfer method was used. The type of triple consolidation instrument used in the collapse test is GDG-4S, the sample area is 30 cm^2, and the height is 2 cm. The loading grades are 12.5 kPa, 25 kPa, 50 kPa, 100 kPa, 200 kPa, 300 kPa, 400 kPa.

2.2.2 Experimental study on the effect of dry density on collapsibility coefficient

As the dry density of samples at each sampling point changes with the increase of depth, the moisture content of each sample is controlled to be 7%. The collapsibility coefficient was measured by double - line method.

2.2.3 Experimental study on the effect of particle size composition on collapsibility coefficient

In order to determine the particle size composition and collapsibility coefficient of soil samples in each area and analyze their correlation, it is necessary to control the same moisture content and dry density of the samples. Therefore, the collapsibility test results of several groups of samples with similar dry density in 2.2.2 were selected for the analysis in this part, and the particle size composition of each sample was measured by laser particle size analyzer. The sample numbers of each group are as follows: (a) JY01, LZ01, YL06 (b): JY02, LZ05, YL08 (c): JY03, LZ03, YL07 (d): JY06, LZ06, YL09 (e): JY08, LZ09, YL10.

2.2.4 Experimental study of structural influence on the coefficient of collapsibility

The comprehensive structural potential is used to characterize the structural strength of loess [16], and its definition is as follows:

$$m_p = \frac{m_1}{m_2} = \frac{S_s/S_o}{S_o/S_r} = \frac{S_r \cdot S_s}{S_o^2} \quad (1)$$

In the formula, S_o, S_s and S_r are respectively the deformation of undisturbed sample, undisturbed saturated sample and remolded sample under a certain pressure.

Therefore, lateral confining compression tests were carried out on the original samples, remolded samples with different water content (water content and dry density were the same as the original samples) and the original saturated samples, and the water content Settings, experimental instruments and loading levels were the same as those in test 2.2.1. In this part, only the confined compression experiments of the remolded samples and the original saturated samples were carried out, and the data of 2.2.1 were used for the confined compression experiments and the collapse coefficient of the original samples. Since the comprehensive structural potential of the sample changes gradually with the increase of pressure, the maximum value of the comprehensive structural potential of the sample is selected for analysis in this paper.

3 Results and Discussion

3.1 Correlation between collapsibility coefficient and initial moisture content

The initial moisture content of the samples at each sampling point was changed to measure their collapsibility coefficient, and the results were shown in Fig. 1. It can be seen from the figure that under the natural moisture content, the collapsibility coefficient of Jingyang soil sample (JY03) is 0.023, indicating that it is a slightly collapsible loess. The collapsibility coefficients of Lanzhou and Yili soil samples (LZ03 and YL07) are 0.051 and 0.053, respectively, indicating that both of them are moderately collapsible loess.

As shown in Figure 1, the collapsibility coefficient of soil samples from Jingyang, Lanzhou and Yili is negatively correlated with the initial water content. When the water content is low, the decrease rate of the collapsibility coefficient is faster. This is because the increase of the initial moisture content of samples reduces the matric suction between the soil particles, the cementing effect of the carbonate cements is weakened, and the compressibility of samples is enhanced. However, for the same kind of soil, the change of water content does not affect the total deformation (including compression deformation and collapse deformation) because of its constant porosity ratio. Therefore, when the initial water content increases, the collapsibility of the sample decreases and the collapsibility decreases. When the initial water content is high, the influence of the change of water content on matric suction and cementation is reduced, so the change of the coefficient of collapsibility is small when the water content is high. In addition, the overall collapsibility coefficient of Jingyang soil sample is slightly lower than that of Lanzhou soil sample, and the collapsibility coefficient of Yili soil sample is the strongest. However, when the initial water content is 30%, the collapsibility coefficient of the three soil samples is basically the same. In addition, on the whole, the collapsibility coefficient of Jingyang soil sample is slightly lower than that of Lanzhou soil sample, and the collapsibility of Yili soil sample is the strongest. However, when the initial moisture content is 30%, the collapsibility coefficients of the three are basically the same.

https://doi.org/10.1051/e3sconf/202129301032

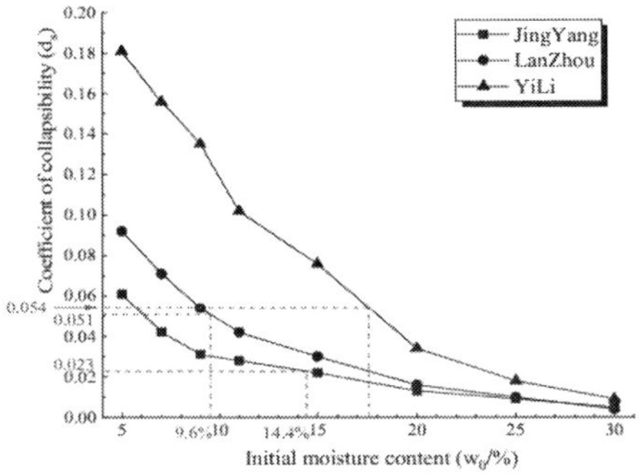

Figure 1. Variation curve of the sample's collapsibility with the initial moisture content at each sampling point

3.2 Correlation between collapsibility coefficient and dry density

The dry density of each sample changes gradually with the increase of the buried depth. Therefore, the correlation between the collapsibility coefficient and the dry density is analyzed by measuring the collapsibility coefficient of samples with different buried depths at each sampling point. The results are shown in Figure 2. As can be seen from Fig. 2, the collapsibility coefficient of each sample is in approximately linear negative correlation with the dry density. The analysis shows that the collapsible deformation of loess depends on the size of the collapsible space. When the dry density of soil is small, its structure is looser. In this case, a large number of pores will be filled with soil particles after the structure is destroyed, resulting in large collapsible deformation and strong collapsibility.

By comparing the test results of soil samples from the three regions, it can be seen that the overall collapsibility coefficient of soil samples from Yili is the highest, followed by that from Lanzhou, and the collapsibility of soil samples from Jingyang is slightly lower than that from Lanzhou. The study on the collapsibility of the loess in Yili area shows that the loess in this area was formed in a late age, was less affected by the humid and hot climate, and its natural structure was less damaged, and its structure was dominated by point contact and overhead arrangement, with less cementation. This structural feature of the Yili loess leads to its strong collapsibility [17].

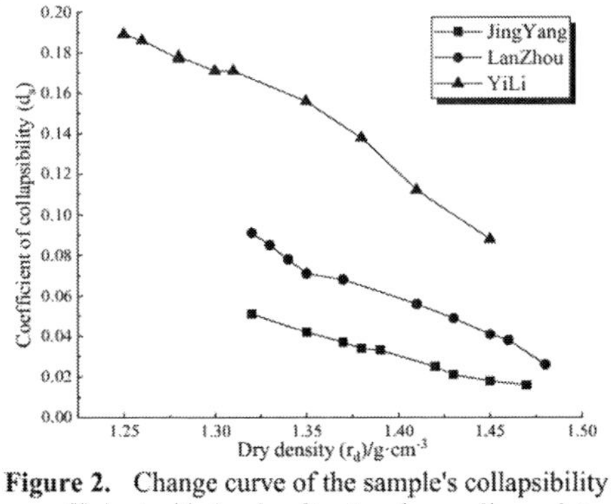

Figure 2. Change curve of the sample's collapsibility coefficient with dry density at each sampling point

3.3 Correlation between collapsibility coefficient and particle size composition

There is no obvious uniform correlation between the collapsibility coefficient and the particle content of each sample. According to the analysis of the reasons, the grain size composition of the same loess does not change significantly before and after the collapsibility, but the collapsibility of the original soil samples is greatly different from that of the post-collapsibility soil samples. It can be seen that the particle size component does not have an effect on the collapsibility alone.

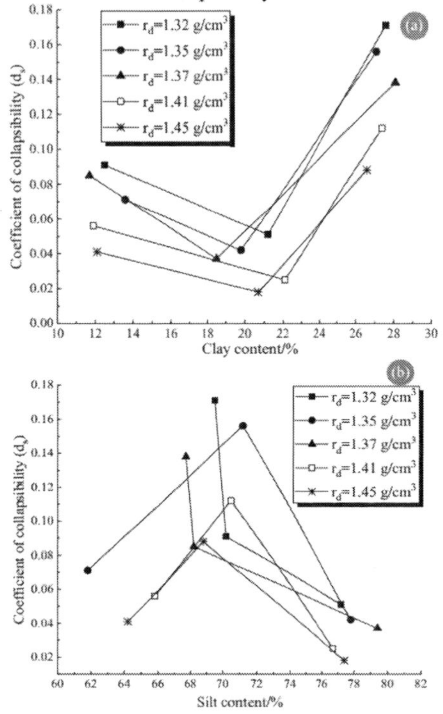

Figure 3. Relation curve between sample collapsibility coefficient and particle size composition at each sampling point:(a) Collapsibility coefficient - clay content;(b) Collapsibility coefficient - powder content

Fig. 4 shows the microstructure photos of sample JY01 before and after collapsibility. It can be seen from the figure that there are many overhead pores in the original sample, and the skeleton particles are mainly silt particles, while the clay particles are mostly aggregates or attached to the surface of larger particles. After the occurrence of collapsibility, there are still many small clay particles attached to the surface of larger particles. The clay particles existing in the form of adhesion have poor mobility and will not become the main material to fill the pores. The pores are mostly filled by larger clay particles, fine silts or unbroken aggregates. Therefore, when the content of large clay particles (which exist alone or are generated by the fragmentation of aggregates) in the soil is high, and the content of aggregates and fine fine particles with strong cementing effect is high, the filling effect of overhead pores or macropores is strong in the process of collapsibility, and the collapsibility deformation is small.

From the above analysis, it can be found that the effect of particle size composition on collapsibility is the result of the interaction between particle size composition, clay form and pore characteristics. Therefore, there is no good regularity between the content of clay and silt and the collapsibility.

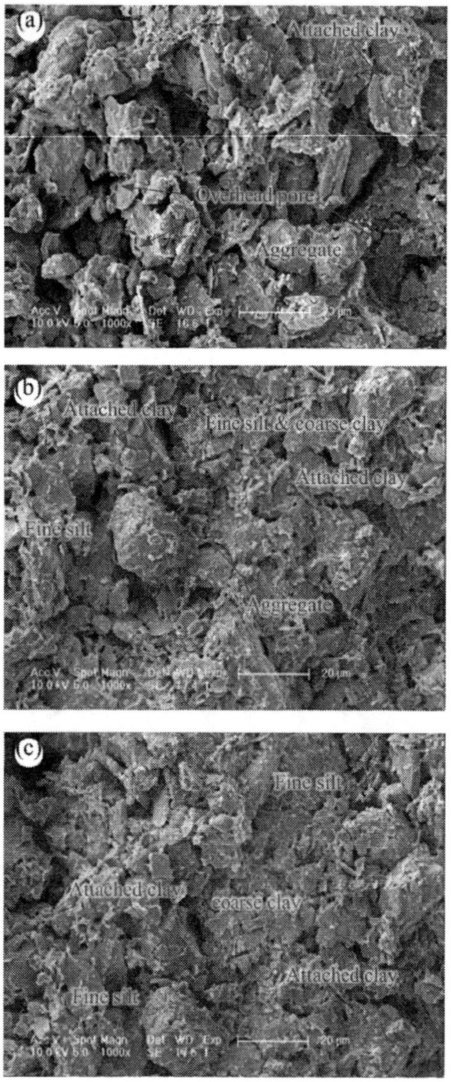

Figure 4. Microstructure photos of JY1 sample before and after wetting:(a) The original sample;(b) Sample after wetting;(c) Sample after wetting

3.4 Correlation between collapsibility coefficient and composite structural potential

In different water content states, the relationship curve between the collapsibility coefficient of samples at each sampling point and their comprehensive structural potential is shown in Fig. 5. It can be seen from the figure that the coefficient of collapsibility is in approximately linear positive correlation with the comprehensive structural potential. The samples at each sampling point showed similar rules. When the comprehensive structural potential is small, the collapsibility coefficient of the samples at the three sampling points is basically the same. When the comprehensive structural potential reaches 23.8, the collapsibility coefficient of the Yili soil sample is obviously higher than that of the Lanzhou soil sample, and with the continuous increase of the comprehensive structural potential, it is expected that the gap will continue to increase. Studies have shown that the comprehensive structural potential is the comprehensive embodiment of soil connection (stability) and arrangement (variability). When the variability is strong, the soil particle bond is destroyed, and the soil structure is in a larger unstable state. Strong structural soil must have strong stability and variability [16]. According to the analysis, when the dry density of soil is small and the content of overhead porosity and macroporosity is high, the soil has strong variability, and the collapsibility deformation is also large. Therefore, the collapsibility coefficient is positively correlated with the comprehensive structural potential.

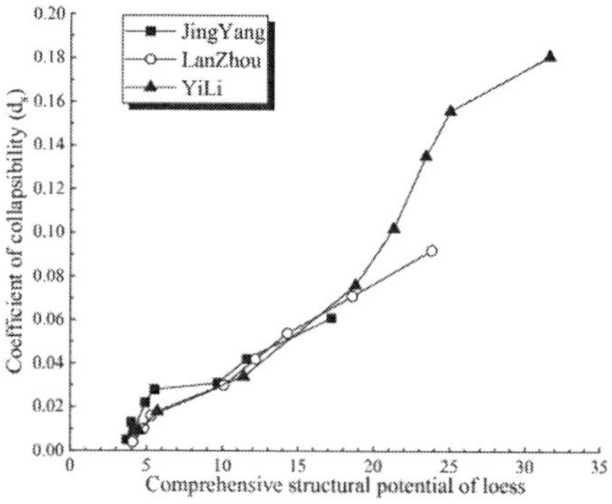

Figure 5. Relation curve of sample collapsibility coefficient and composite structural potential at each sampling point

4 Conclusions

Through the collapsibility tests of loess in Jingyang, Lanzhou and Yili under different conditions, the influences of initial moisture content, dry density, particle size composition and structural parameters on loess collapsibility were explored and the reasons were analyzed. The main conclusions are as follows:

1. The collapsibility coefficient of loess in each area is negatively correlated with the initial water content, and the decrease rate of the collapsibility coefficient is greater when the initial water content is lower. The reason is that when the initial water content is high, the compressibility of soil is enhanced, the collapsibility deformation is reduced, and the collapsibility coefficient is also reduced.
2. There is a negative linear correlation between the collapsibility coefficient and dry density. On the whole, the collapsibility of Yili loess is the strongest, while that of Jingyang loess is the weakest. When the dry density of soil is large, the collapsible space decreases and the collapsibility weakens.
3. The collapsibility has a certain correlation with the clay content. When the clay content is about 20%,

the collapsibility is the weakest. There is no obvious uniform correlation between collapsibility and silt content. It is concluded that the effect of particle size composition on collapsibility is based on the characteristics of pore structure, and is also related to the form of clay particles.

4. The analysis shows that the strong structural soil has high content of overhead pores and macropores, so it has strong variability and strong collapsibility. Therefore, the collapsibility coefficient of loess in each area is approximately linear positive correlation with the comprehensive structural potential.

References

1. Liu Zudian. Loess mechanics and engineering [M]. Xi'an: Shaanxi Science and Technology Press, 1996.
2. Fang Xiangwei, Shen Chunni, Li Chunhai, et al. Collapsible deformation properties of Q2 loess in Pucheng of Shaanxi province [J]. Rock and Soil Mechanics, 2013, 34(S2):115-120.
3. Wang Jian. Study on the deformation law of undisturbed loess under true triaxial compression and humidification[D]. Xi'an: Xi'an University of Technology, 2020.
4. Fang Jinjin, Feng Yixin, Yu Yongqiang, et al. Humidification and deformation characteristics of undisturbed loess under true triaxial conditions [J]. Rock and Soil Mechanics, 2020, 41(4): 1235-1246.
5. Hou Xiaokun. Research on Deformation Characteristics of Loess Unsaturated Collapsibility[D]. Xi'an: Chang'an University, 2015.
6. Zheng Jianguo, Deng Guohua, Liu Zhenghong, et al. Influence of discontinuous distribution of loess collapsibility on collapse deformation[J]. Chinese Journal of Geotechnical Engineering, 2015, 37(1):165-170.
7. Mi Wenjing, Zhang Aijun, Guo Minxia. Research on the Deformation Characteristics of Yili Loess Collapsible[J]. Jiangxi Journal of Agriculture, 2019, 31(12):49-53.
8. Niu Lisi, Zhang Aijun, Wang Yuguo, et al. Compression and collapsibility characteristics of Yili loess under changes of humidity and density[J]. Journal of Hydroelectric Engineering, 2020.
9. Li Ping, Li Tonglu. The relationship between physical properties of loess and collapsibility and its engineering significance[J]. Journal of Engineering Geology, 2007, 15(4):506-511.
10. Ma Yan, Wang Jiading, Peng Shujun, et al. The relationship between collapsibility of loess and soil properties and its prediction model[J]. Bulletin of Soil and Water Conservation, 2016, 36(1):120-128.
11. Wu Xiaopeng, Zhao Yonghu, Xu Anhua, et al. The relationship between collapsibility of loess and its physical and mechanical indexes and evaluation methods[J]. Journal of Yangtze River Scientific Research Institute, 2018, 35(6):75-80.
12. Li Yongwei, Zhang Peiheng, Li Zhenxiao. Collapsibility of Lishi Loess in Shanxi[J]. Journal of Engineering Geology, 2014, 22(sl):395-399.
13. Mu Qingyi, Dang Yingjie, Dong Qi, et al. Comparative test study on water holding characteristics and collapsibility of undisturbed and compacted loess[J]. Chinese Journal of Geotechnical Engineering, 2019, 41(8):1496-1504.
14. Cui Jingyu, Zhang Wuyu, Xie Banglong, et al. Collapsibility and microscopic mechanism of undisturbed loess at different depths in Xining area[J]. Chinese Journal of Geotechnical Engineering, 2019, 41(s2):249-252.
15. Fang Xiangwei, Ou Yixi, Shen Chunni, et al. Research on the influencing factors of Q2 loess collapsibility[J]. Journal of Water Resources and Architectural Engineering, 2016, 14(1):49-54.
16. Xie Dingyi, Qi Jilin. A new way to study soil structure and its quantitative parameters[J]. Chinese Journal of Geotechnical Engineering, 1999, 21(6):651-656.
17. Zhang Aijun, Xing Yichuan, Hu Xinli, et al. Analysis of influencing factors of strong self-weight collapsibility of Ili loess[J]. Chinese Journal of Geotechnical Engineering, 2016, 38(0z2):117-122.

Research on uplift model test of single pile in sand

Liuqing Huang[1], Huibiao Ma[2], Qiang Li[2], Ling Hu[1], Xinyi Li[1]

[1]School of Shipping and Maritime, Zhejiang Ocean University, Zhoushan City, Zhejiang Province, China

[2]Ocean Engineering and Equipment College, Zhejiang Ocean University, Zhoushan City, Zhejiang Province, China

Abstract: This paper mainly uses the method of model test to study the ultimate uplift bearing characteristics of a single pile with different length-to-diameter ratios, mainly from the distribution laws of the uplift bearing capacity of the pile, the axial force of the pile and the side friction of the pile perform analysis. The model pile is made of PVC pipe, and resistance strain gauges are attached to both sides of the pipe. The strain value of the PVC pipe under different load conditions is measured to obtain the working behavior of the pile during the process of pulling out the pile. The result shows that the model test data is in good agreement with the numerical simulation.

1 Introduction

The uplift pile foundation is an important form of foundation for buildings and structures, and is widely used in high-voltage power lines, communication towers and highways.

Regarding the research on uplift piles, a large number of foreign scholars have made relevant experimental and theoretical results. Rao et al[1] carried out a model experiment of uplift piles in sand. The study showed that as the pile length, the friction coefficient of the pile-soil interface and the sand density increase, the bearing capacity of the pile will also increase accordingly. Dash et al[3] applied a compressive load to 36 steel pipe model piles, and conducted experiments on the influence of the ultimate pull-out bearing capacity of the model piles. The study showed that the greater the pressure applied, the lower the ultimate pull-out bearing capacity. Krabbenhoftfenxile et al[4] analyzed the static load experiments of bored cast-in-place uplift piles with two diameters, and discussed the factors affecting the average value of the pile side friction. Luo Yaowu et al[5] studied the influence of the pile-soil interface on the uplift pile in the sand foundation through the method of model experiment. Research shows that the rougher the pile-soil interface, the greater the lateral pressure coefficient around the pile. Zhou Junpen et al[6] obtained the pile shaft axial force and the pile side friction resistance and the load applied on the pile top through the on-site uplift static load experiment of the mini-single pile. Huang feng et al[7] used the finite element analysis method to discuss and analyze the pile side friction resistance under the uplift loading mode of the uplift pile, and obtained the theoretical solution of the relationship between the uplift pile load and displacement.

This paper mainly adopts indoor model test method, chooses 1.75m and 1.5m model piles, analyzing and studying the displacement, pile axial force and pile side friction of a single pile under the uplift load. Obtaining the change law of the axial force of the pile body and the side friction resistance of the pile during the uplifting process. Then, based on the ABAQUS finite element software, the load-displacement curve of the pile and the axial force of the pile body in the model test were verified, and it has good regularity.

2 Test overview

2.1 Experimental setup

The model box used in the experiment is welded with a 10mm steel plate, and the length, width, and height are 2m, 1m, and 3m respectively. The experimental device is shown in Figure 1.

Figure 1. Model box

2.2 Preparation of model sand

The sand used in the experiment was collected from the sea sand around Zhoushan . The filling of the model box

Corresponding author: huangliuqing199506@163.com

 https://doi.org/10.1051/e3sconf/202129301033

is roughly divided into three parts. The bottom is laid with 0.3m thick gravel composed of different particle sizes as the filter layer, and geotextile is laid on the filter layer. Fill the top with fine sand in layers, and compact the soil layer every 25cm in thickness.

2.3 Model pile

The test pile is made of PVC pipe with a diameter of 32mm and an outer wall thickness of 2mm. A set of 120 ohm resistance strain gauges are symmetrically pasted on one side of the PVC pipe at regular intervals. The outside of the strain gauges were coated with waterproof glue. The wires were drilled into the PVC pipe and sealed with 706 silicone rubber. The exterior is with insulating waterproof tape to prevent seepage. Finally, the bottom of the pipe is sealed with cement mortar grouting.

2.4 Loading method

Figure 2. Schematic diagram of anti-pull device

In this test, a vertical uplift load was applied to the model piles through wire ropes, pulley blocks, loading trays and weights, as shown in the figure2. The whole test process adopts the slow sustaining load method to load. The displacement of the pile after pulling up is recorded by the observation displacement sensor, and the pulling displacement of the pile is recorded by the observation dial indicator for 5, 10, 30, and 60 minutes after the load is applied. After the application, when the pull-up displacement reaches a relatively stable value, the next level of loading is performed.

3 Model test results and analysis

3.1 Analysis of the law of load settlement

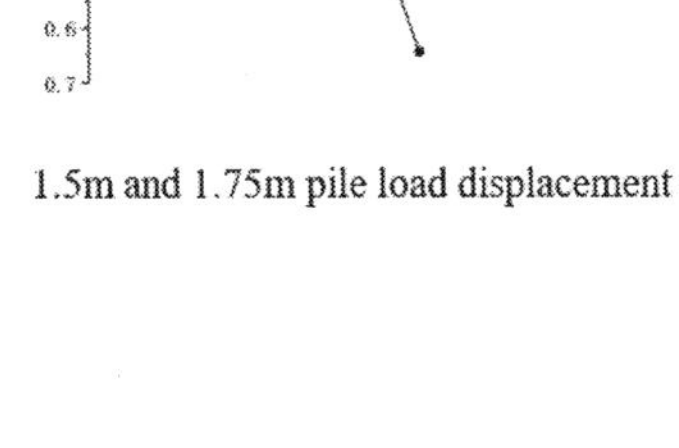

Figure 3. 1.5m and 1.75m pile load displacement curve

Figure 3 shows the load-displacement curves of different pile lengths in this experiment. It can be seen from the figure that when the load is small, the load-displacement curves of the two piles are relatively smooth. As the load increases, the slope of the curve begins to increase. From the gradual change to the steep change, the lower soil mass changes from elasticity to plasticity. Both the pile and the surrounding soil have a certain amount of displacement, until the pile reaches the ultimate uplift bearing capacity, the displacement of the pile increases rapidly.

3.2 Analysis of pile shaft axial force

The 1.75m pile is selected as a representative for analysis. According to the pile strain gauges, the pile body's performance during the uplift process is measured. For the strain value, calculate the axial force of the pile body according to the following formula, and obtain the distribution law of the axial force value at each pile section with the pile depth as shown in the figure 4.

$$\delta_i = \frac{N_i}{A} = E\varepsilon_i \quad (1)$$

$$N_i = EA\varepsilon_i \quad (2)$$

Among them:N_i is the axial force of the i section of the pile; E is the elastic modulus of the pile; ε_i is the average strain of the pile at the section;A is the cross-sectional area of the pile.

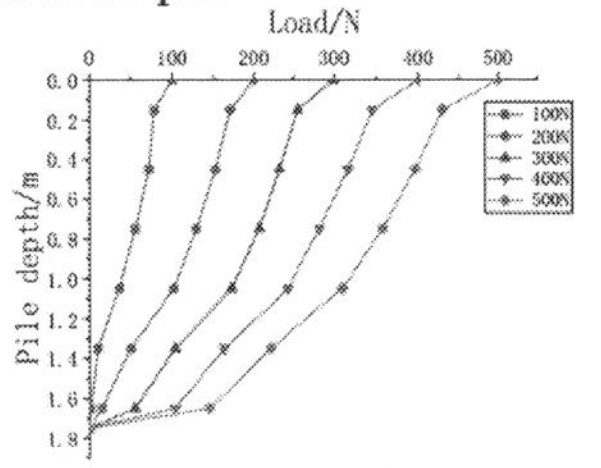

Figure 4. Axial force curve of 1.75m pile length

During the whole uplifting process, the axial force of the pile body gradually decreases with the increase of depth. The closer to the top of the pile, the greater the axial force, and the axial force at the end of the pile is zero. The axial force distribution of the uplift pile is not evenly distributed along the length of the pile, but exhibits different deceleration rates. It is mainly manifested in the small change of the axial force on the upper part of the pile and the larger change of the axial force on the lower part. It shows that the axial force of the uplift pile will gradually transfer to the lower part of the pile with the increase of the uplift force, and it will be borne by the soil under the pile.

3.3 Analysis of pile body friction

The formula for calculating the side friction resistance of the pile is

$$q_{si} = \frac{N_{i+1}-N_i}{A_{[illegible]}} \quad (3)$$

Where:$q_{si}$is the pile side friction resistance (KN);N_i is the axial force of the pile body at point i; P_{i+1} is the axial

 https://doi.org/10.1051/e3sconf/202129301033

force of the pile body at point i+1; A_{mi} is the surface area of the pile body from point i+1 to point i.

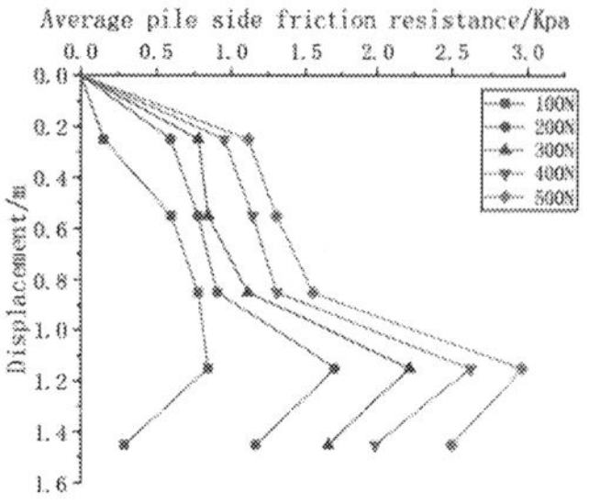

Figure 5. The side friction curve of 1.75m pile length

It can be seen from Figure 5 that in different loading stages, the axial force curve of the pile shows different deceleration rates, which reflects the change in the side friction resistance of the pile. Generally speaking, as the uplift load increases, the pile side friction resistance increases first and then decreases. As the load increases, the side friction resistance of the pile changes more and more, until the ultimate pull-out bearing capacity of the pile is reached, and the pile-soil begins to debond. At this time, the pile produces a large displacement until the pile is pulled out.

4 Numerical simulation analysis of model test

In this chapter, the ABAQUS finite element analysis software is used to simulate and analyze the pile-soil model of the uplift pile. The ABAQUS finite element software performs very well in solving nonlinear problems, and is especially suitable for research in geotechnical engineering[8].

4.1 Establishment of calculation model

This model is mainly composed of two parts: soil and pile. The soil model is selected according to the size of the model test 2m*1.5m*3m, the diameter of the pile is 0.032m, and the length is 1.5m. The soil model adopts the Mohr Coulomb model, and the pile adopts the linear elastic model.

4.2 Analysis of calculation results

After the model is established, the uplift load is applied to the top surface of the pile according to the step-by-step loading method, and the numerical simulation calculation results are compared with the model test results. As shown in Figure 6, it can be seen from the figure that there are some deviations between the load-displacement curve of the model test and the curve of the numerical simulation, but the two are basically consistent in the elastic stage and has good regularity.

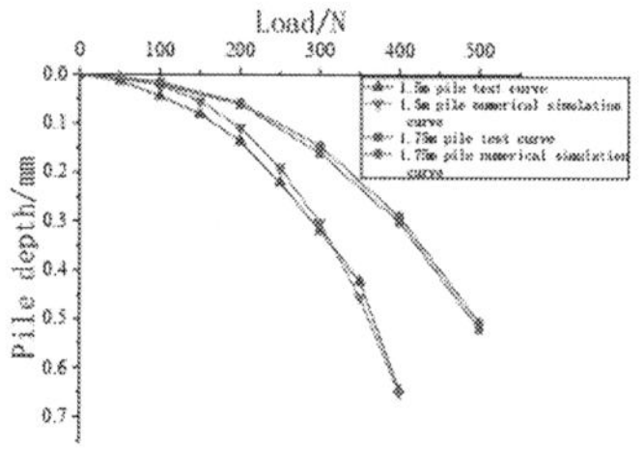

Figure 6. Comparison curve of experiment and numerical simulation

5 Conclusion

In this paper, the load-displacement curve of single pile in sand and the lateral friction resistance of the pile shaft are studied by means of model experiments and numerical simulations.

(1) The axial force distribution curve of a single pile mainly decreases with the increase in depth, until it reaches the bottom of the pile, the axial force of the pile is 0.

(2) The side friction resistance of the pile changes smoothly when the load is small. As the load increases, the slope gradually becomes larger, and the side friction resistance of the pile changes more and more big until it reaches the ultimate uplift bearing capacity of the pile, and the pile soil begins to debond.

Renference

1. KANAKAPURA S.SUBBA RAO,K.H. VENKATESH. UPLIFT BEHAVIOUR OF SHORT PILES IN UNIFORM SAND[J]. The Japanese Geotechnical Society,1985,25(4).
2. B. K. Dash, P. J. Pise. Effect of Compressive Load on Uplift Capacity of Model Piles[J]. Journal of Geotechnical and Geoenvironmental Engineering,2003,129(11).
3. Sven KrabbenhoftS. Krabbenhoft,Allan AndersenA. Andersen,Lars DamkildeL. Damkilde. The tensile capacity of bored piles in frictional soils[J]. Sven KrabbenhoftS. Krabbenhoft; Allan AndersenA. Andersen;Lars DamkildeL. Damkilde,2008,45(12).
4. Luo Yaowu, Hu Qi, Ling Daosheng, Chen Zheng, Chen Yunmin. Model test study on the influence of pile-soil interface characteristics on the bearing characteristics of uplift piles in sandy soil[J].Rock and Soil Mechanics,2011, 32(03) :722-726+732.
5. Zhou Junpeng, Huang Xuefeng, Liu Zilong, Zhu Zhonghua. Research on the mechanical characteristics of the miniature uplift pile foundation[J].Engineering Investigation, 2017, 45(02): 19-24.
6. Huang Feng, Li Guangxin, Zheng Jiqin. Finite element calculation study of pile side friction resistance of single pile under compressive and uplift loads[J]. Engineering Mechanics, 1999(06): 97-101+69.

7. Wu Jiangbin, Wang Weidong, Huang Shaoming. Numerical analysis of the uplift bearing characteristics of equal-section piles and expanded-bottom piles[J].Rock and Soil Mechanics, 2008(09): 2583-2588.
8. Fei Kang, Zhang Jianwei. The application of ABAQUS in rock mass engineering[M]. Beijing: China Water Power Press, 2010.

Research on environmental impact assessment system and legalization in China

Chen Mingtsung [1]* Feng Zhi'an [2] Qian Wei [2] Zhang Yujing [2]

[1]Guangzhou College of Commerce, Guangzhou Guangdong, 511363, China
[2]Nanfang College Guangzhou, Guangzhou, Guangdong, 510970, China

Abstract. In order to solve the conflict between economic growth and environmental protection, the environmental impact assessment system used by western countries has been gradually introduced into China. It is a scientific and professional assessment system, which is regarded as the best tool to solve the conflict between economic growth and environmental protection. The implementation of the environmental impact assessment system can undoubtedly prevent some construction projects from having serious adverse impact on the environment. Through the comparison and selection of the feasibility schemes, the environmental impact of some construction projects can be minimized. Therefore, the environmental impact assessment system is regarded as an important pillar and effective legal system for the implementation of predictive environmental policy, which has attracted more and more attention in the world. In order to implement the strategy of sustainable development, prevent the adverse impact on the environment caused by the implementation of planning and construction projects, and promote the coordinated development of economy, society and environment, China has formulated the environmental impact assessment law. The law came into force on September 1, 2003 and was amended for the second time on December 29, 2018.

1 Preface

With the rapid growth of economy and the democratization of politics, people pay more and more attention to the environment. The living environment space that people need is not only based on economic production, but also environmental protection and cultural identity. In the past, China's development experience, with economic growth as its main goal, has created a world-famous economic miracle. However, many waves of environmental protest have not disappeared with the progress of science and technology, and the recent protests are impressive. In order to solve this problem, China continues to maintain economic development and introduces the environmental impact assessment system of western developed countries as a weapon to solve the contradiction between economic growth and environmental protection. China is one of the first developing countries to implement the environmental impact assessment system of construction projects. In 1979, the eleventh session of the fifth NPC Standing Committee passed the environmental protection law of the people's Republic of China (for Trial Implementation), which established the environmental impact assessment of construction projects as a legal system for the first time. The environmental protection laws formulated in succession contain the principles and regulations of environmental impact assessment of construction projects.

2 Connotation of environmental impact assessment system

Environmental impact assessment (EIA) refers to the method and system of analyzing, predicting and evaluating the environmental impact that may be caused by the implementation of planning and construction projects, proposing countermeasures and measures to prevent or mitigate adverse environmental impact, and tracking and monitoring. Environmental impact assessment is the most important step in environmental risk management, the most important part of which is risk assessment, which is a comprehensive assessment of the various threats faced by the subject matter. The following will use risk assessment formula, risk assessment array and risk assessment process to clarify the connotation of risk assessment and quantify it as much as possible, so as to determine the priority and select the right measures should be taken. Generally speaking, it is to analyze the possible impact on the environment after the project is completed and put into operation, and put forward countermeasures and measures for pollution prevention and control. Importance and significance of environmental impact assessment:

(1) It provides scientific basis for decision-making of development and construction activities;

(2) To provide scientific basis for the rational distribution of economic construction;

*Corresponding author's e-mail: 306470714@qq.com

(3) In order to determine the direction and scale of economic development in a certain region, and to provide scientific basis for formulating regional economic development planning and corresponding environmental protection planning;

(4) It provides the basis for formulating environmental protection countermeasures and scientific environmental management;

(5) Promote the development of related environmental science and technology.

2.1 Risk evaluation formula

Risk = Threat × Probability × Vulnerability ×Criticality ($R = T \times P \times V \times C$)

In this formula, item "T" is whether the subject matter of the risk assessment will face the threat. Its value is 0 or 1. Item "P" is the objective possibility of the threat. Most of the data can be obtained through public data. Some data collection and analysis work should be done in advance. Item "V" is the vulnerability degree of the subject matter to the threat. Some vulnerability detection services can provide objective observation values. However, unless it is extremely high-tech, it is usually the most known one, so the rule of thumb is also very good. Item "C" refers to the impact on the overall operation of the subject matter if it suffers from the threat actually.

2.2 Risk assessment array

The risk assessment array is to help us easily judge the priority of risk management. Of course, the higher the risk, the more risk management is needed. This study believes that such a phased and sequential thinking process can help enterprise risk managers to do systematic evaluation and reference for future management priorities.

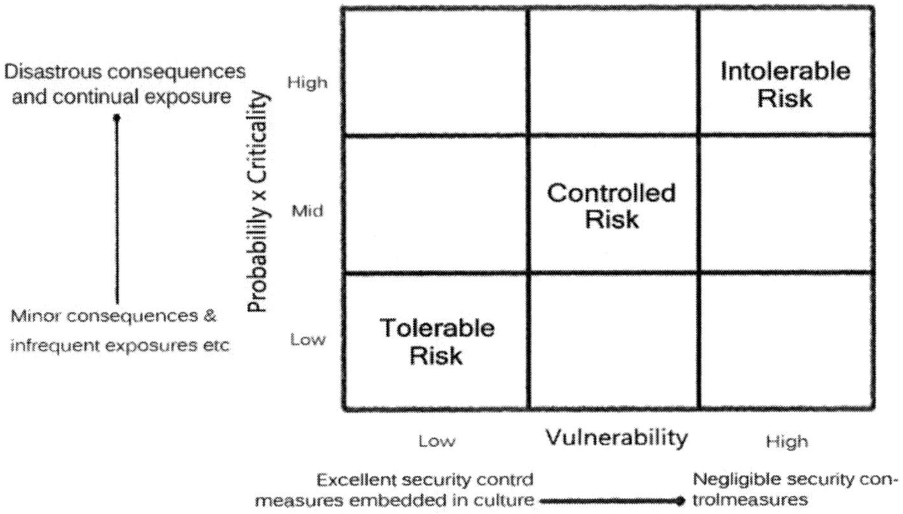

Figure 1. Risk assessment matrix

2.3 Risk assessment process

Risk assessment is an important process of risk management. The process of risk assessment defined by management internationalization standard is: risk assessment is the whole process of risk identification, risk analysis and risk assessment. In the iso31000 risk management standard, five basic problems need to be solved in the risk assessment process are clarified:

- What is the status quo? What can happen? Why did it happen?
- What are the consequences? What is the impact on the goal?
- How likely are these consequences to occur?
- Are there any factors that can reduce the risk consequences and the possibility of risk?
- Is the risk level tolerable or acceptable? Is there a need for further response?

In all kinds of complex system engineering, advanced risk assessment system is used to increase the safety and reliability of the project, especially in the field of life, environment or machine operation. Risk assessment has a long history in nuclear industry, aerospace, petroleum, railway, military and other industries. In addition, the application of risk control and risk assessment in medical, hospital and food industry is continuously promoted. Because of the difference of financial decision-making, environment, ecology or public health risk, there are different risk assessment methods among different industries.

3 The legalization of environmental impact assessment system in China

Environmental Impact Assessment Law of China has been passed and implemented for many years in 2003. Major development cases must be reviewed by environmental impact assessment. The development of environmental impact assessment system mainly goes through the following three stages.

The first stage is the establishment stage. The concept of environmental impact assessment was first put forward in 1973, and the environmental protection law (Trial) issued in 1979 institutionalized and legalized the environmental impact assessment. The basic contents and procedures of environmental impact assessment are specified in the administrative measures for environmental protection of capital construction projects issued in 1981. After modification, the administrative measures for environmental protection of construction projects was promulgated in 1986 to further clarify the scope, content, management authority and responsibility of environmental impact assessment.

The second stage is the development stage. In 1989, the official environmental protection law was promulgated. Article 13 of the law states: "the construction of projects that pollute the environment must comply with the provisions of the state on environmental protection management of construction projects. The environmental impact statement of a construction project must evaluate the pollution caused by the construction project and its impact on the environment, stipulate the prevention and control measures, be pre examined by the competent department of the project and submitted to the competent department of Environmental Protection Administration for approval in accordance with the prescribed procedures. After the environmental impact report is approved, the planning department can approve the design assignment of the construction project. " In 1998, the State Council promulgated the regulations on the administration of environmental protection of construction projects, which further improved the legislative specifications of the environmental impact assessment system. At the same time, the scope of application, the opportunity of assessment, the procedure of examination and approval, and the legal liability of environmental impact assessment have been greatly modified. In March 1999, the State Environmental Protection Administration issued the management measures for the qualification certificate of environmental impact assessment of construction projects, which set China's environmental impact assessment on a professional road.

The third stage is the improvement stage. In view of the deficiencies of the regulations on the administration of environmental protection of construction projects and the needs of the development of the new situation, the environmental impact assessment law, which was implemented on September 1, 2003, can be said to be a new milestone in the development history of China's environmental impact assessment system and a sign of the improvement of China's environmental impact assessment.

The fourth stage is the perfection stage. On December 29, 2018, the Environmental Impact Assessment Law of the people's Republic of China was amended and shall come into force as of the date of promulgation. The following are the specific amendments to the Environmental Impact Assessment Law: (1) the competent administrative department of environmental protection is revised to the competent department of ecological environment; (2) the qualification examination and approval supervision of environmental assessment companies (technical units) is cancelled, and the construction units with ability can also prepare the environmental impact assessment report (form) on their own (3) The construction unit shall be responsible for the contents and conclusions of the EIA report (table), and the technical unit shall bear the corresponding responsibility in case of errors. For new credit supervision, the EIA approval department shall record the relevant illegal information of the establishment unit, the establishment host and the main establishment personnel into the social credit archives, and put it into the national credit information sharing platform and the national enterprise credit information publicity system for public announcement (4) The responsibility of false contents and incorrect conclusions in the EIA report (table) is the most important part of the revision of the EIA law (5) If the compilation unit has the above-mentioned illegal acts, the compilation host and main compilation personnel are prohibited to engage in the compilation of environmental impact report and environmental impact report form within five years; If a crime is constituted, criminal responsibility shall be investigated in accordance with the law, and the compilation of environmental impact report and environmental impact report form shall be prohibited for life.

4 Conclusion

Environmental impact assessment is a process, which focuses on the prevention function of environmental impact assessment before decision-making and development and construction activities. After the decision-making or the start of development and construction activities, through the implementation of the environmental monitoring plan and continuous research, the environmental impact assessment is still continuing, continuously verifying its assessment conclusions, and feeding back to the decision makers and developers, further modifying and improving its decision-making and development and construction activities. In order to embody the role of EIA, we must adhere to the sustainable development strategy and the concept of circular economy in the organization and implementation of EIA, strictly abide by the relevant laws, regulations and policies of the state, and achieve scientific, fair and practical. However, the controversy has not disappeared, and many environmental protest movements are still emerging. When the decision-maker's consideration is to develop the single oriented economy, the risk assessment is only a reference to standardize the decision-making behavior, and the underestimation of risk will become normally. Unfortunately, in the era of increasing globalization, risk and science are moving towards a more undemocratic way to construct its relative system, and the more unlikely the “prevention first” environmental protection policy will appear, which is an important topic that we must continue to pay attention to. The environmental impact assessment law was implemented in 2003. Its legislative purpose is to implement the sustainable development strategy, prevent the adverse impact on the environment caused by the implementation of planning and construction projects, and promote the coordinated development of economy, society and environment. During the 15 years of promulgation and implementation, this law has played an important role in preventing environmental pollution and ecological damage The purpose of amending EIA law is to achieve the goal of reducing environmental pollution from the source. Of course, in order to make EIA system play a better role, it is necessary to strengthen planning EIA, promote its participation in comprehensive decision-making and play a substantial role; Public participation should be strengthened, and social supervision should be used to prevent arbitrary intervention in EIA; Strengthen the rule of law to form the restriction and supervision of administrative power, and further promote the reform of administrative system.

REFERENCES

1. Barker, A., C. Wood. (1999): An evaluation of EIA system performance in eight EU countries, *Environmental Impact Assessment Review*, 19, pp.387-404.
2. Chambers, J.C. (2002): Community participation in brownfields redevelopment, In: T. S. Davis, ed., *Brownfields: A Comprehensive Guide to Redeveloping Contaminated Property*, 2 ed, Chicago: American Bar Association, pp.243-253.
3. Fischer, F. (1996): Risk assessment and environmental crisis: Toward an integration of

   Science and participation. In: S. Campbell and S.S. Fainstein, eds., *Reading in Planning Theory*, Cambridge: Blackwell Publishers Inc, pp. 485-506.
4. Gismondi, M. (1997): Sociology and Environmental Impact Assessment, *Canadian Journal of Sociology*22(4), pp.457-479.
5. Plough, A., S. Kirmsky. (1987) :The emergence of risk communication studies: Social and political context, *Science, Technology, & Human Values*, 12(3&4), pp. 4-10.
6. Tzoumis, K., L. Finegold. (2000):Looking at the quality of draft environmental impact statements over time in the United States: Have ratings improved?, *Environmental Impact Assessment Review*,20,pp.557-578.
7. Winner, L. (1988): Do artifacts have politics? In: M. E. Kraft and N. J. Vig, eds., *Technology and Politics*, Durham: Duke University Press, pp.33-53.
8. Wynne, B.(2005):Risk as globalizing democratic discourse? Framing subjects and citizens. In: M. Leach, I. Scoones and B. Wynne, eds., Science and Citizens: Globalization & the Challenge of Engagement, London: Zed Books, pp. 66-82.

https://doi.org/10.1051/e3sconf/202129301035

Research on Site Selection and Capacity Determination of Mobile Energy Storage in Extreme Scenarios Considering the Behaviour of Decision

Shiyu Deng[1,*], Chang Liu[1], Wenxia Liu[1] and Zongqi Liu[1]

[1] North China Electric Power University, School of Electrical and Electronic Engineering, 102206 Beijing, China

Abstract. In order to ensure that the power supply can be restored quickly and efficiently under extreme conditions, an evaluation and decision-making method for mobile energy storage site selection and capacity planning considering the behaviour of decision makers is proposed. The prospect value is calculated based on the prospect theory to describe the bounded rationality and loss aversion of decision makers, and the attribute weight is calculated based on the relative closeness, and then the comprehensive value is obtained. Based on this, we can choose a mobile energy storage installation plan that is more in line with the psychology of decision makers.

1 Introduction

Vigorously developing renewable energy has become an important strategic consensus for all countries in the world to deal with environmental pollution and resource depletion. Energy storage technology is recognized as one of the most effective measures to solve the problem of renewable energy consumption and improve the safe and economic operation of the power system.

Due to the strong flexibility of the energy storage device, mobile energy storage can be dispatched to provide stable power support for lifeline loads in extreme disaster scenarios, and to minimize the power outage loss of important loads.

The literature [1] takes into account the uncertainty of load forecasting, configures energy storage with the goal of minimizing energy storage investment costs and operation and maintenance costs, and comprehensively considers the power purchase costs, operation and maintenance costs, switching costs, losses of the distribution network and load penalty fees to coordinate and optimize measures such as energy storage operation, demand side management, and network reconfiguration. The literature [2] proposed a two-stage scheduling framework based on pre-positioning and real-time allocation for mobile energy storage, which realized the improvement of resilience. The literature [3] comprehensively considers the line reinforcement cost, energy storage investment cost and load loss cost, and coordinates and optimizes the line reinforcement and energy storage configuration with the goal of minimizing the overall cost, and the two work together to improve the emergency response capability of the distribution network. Literature [4] proposed a mixed integer programming model to solve the problems of mobile energy storage dispatching and distribution system operation at different time scales, road and grid coupling, etc., to improve the resilience of the distribution network by optimizing the dynamic dispatch of mobile power sources. However, measures to improve the resilience of the grid are usually costly. In terms of the benefits of the distribution network, the existing methods only consider the effect of improving the resilience, without considering the investment margin and the economy under normal operation, which may lead to excessive investment. Therefore, it is necessary to propose a reasonable energy storage configuration method to achieve a balanced improvement of resilience and economy.

In summary, in order to improve the accuracy of the results, this paper introduces the prospect theory into the decision-making process of mobile energy storage selection and designation capacity planning. First, this paper establishes a comprehensive index system through the analysis of the power supply recovery capacity and economics of mobile energy storage; secondly, we calculate the importance of the index based on the relative discount, and normalize the calculation of the index weight; finally, the schemes are sorted and selected based on the prospect theory multi-index decision-making method.

2 Index system construction

The planning scheme of energy storage includes the capacity configuration and installation location of energy storage. The investment cost of energy storage can be obtained through the planning scheme, including the initial investment cost and the operation and maintenance cost. At the same time, the capacity configuration of energy storage can be embodied as a constraint in the

* Corresponding author: dengsy@ncepu.edu.cn

operation of energy storage. The evaluation indicators of the plan include not only the investment cost, but also the operating results under the plan, that is, the economic benefits obtained from normal operation, and the ability for self-healing recovery.

2.1 Economic indicators

1) The investment cost of energy storage

The investment cost of mobile energy storage is mainly divided into two parts, The investment cost of mobile energy storage is mainly divided into two parts, namely the initial investment cost and the operation and maintenance cost. The initial investment cost refers to the total amount of capital invested at the beginning of the construction of the distributed energy storage system project. Operation and maintenance cost refers to the cost investment of energy storage system due to the equipment loss, aging or the need for maintenance or replacement after the operation parameters reach the scrap standard in the total service life.

$$M_f = C_{inv} + C_{op} \tag{1}$$

$$C_{inv} = \sum_{j \in N} C_{inv,j} = \sum_{j \in N} \eta_{n_j} \left(C^P_{inv,j} P_j + C^E_{inv,j} E_j + C^F_{inv,j} \right) \tag{2}$$

$$C_{op} = \sum_{j \in N} C_{op,j} = \sum_{j \in N} \eta_{n_j} \sum_{i=1}^{n_j} C_{opj,i} P_{j,i} \tag{3}$$

In the formula, C_{inv} is the initial investment cost, and C_{op} is the operation and maintenance cost. N represents the total number of installed energy storage; n_j refers to the operating life of energy storage battery j; represents the capital recovery coefficient; P and E represent the power and capacity of energy storage respectively; $C^P_{inv,j}$, $C^E_{inv,j}$, $C^F_{inv,j}$ represent the unit power cost and unit capacity cost of energy storage j, respectively and fixed installation costs.

2) Benefits from the participation of energy storage in normal operation

The benefits of energy storage participating in normal operation are mainly obtained through peak shaving and valley filling.

$$W = \sum_{1}^{365} \sum_{i=1}^{24} (P_{i,dis} - P_{i,ch}) p_i \tag{4}$$

2.2 Resilience index

The evaluation of the quality of the self-healing recovery plan is not only related to the degree of recovery of important loads, but also related to the network status and fault characteristics after the fault recovery. Therefore, the strength of recovery requires comprehensive consideration of multiple indicators. The resilience indicators constructed in this paper are:

1). Proportion of isolated nodes

$$K_{iso} = \frac{N_{iso}}{N} \tag{5}$$

In the formula, N_{iso} represents the number of isolated nodes, and N represents the total number of nodes.

2). Fault recovery time T

3). Switch operation times K

4). Loss of power load L

The power loss load is calculated according to the priority of the load, and the important load is restored first.

$$L = \sum_{i \in N_L} \sum_{j \in N_{Li}} c_{ij} x_{ij} L_{ij} \tag{6}$$

Among them, N_L represents the number of power-loss nodes, N_{Li} represents the number of loads on node i, NK represents the number of switches, c_{ij} is the importance of the jth load in node i, and x_{ij} represents the power supply status of the jth load in node i, in which x_{ij}=1 means the load has regained power, and x_{ij}=0 means the load has failed to regain power.

3 Decision and evaluation model of energy storage scheme based on Prospect Theory

Prospect theory believes that decision-makers have limited rationality in actual decision-making [5-], and with changes in external factors, the preferences of decision-makers will also change. When gains appear, decision makers tend to avoid risks. At this time, the value function of decision makers is a concave function; when there is loss, decision makers often show a risk preference, and then the value function is convex. Among them, the gains and losses are relative values relative to the standard set by the decision maker.

3.1 Prospect value function

Based on different reference points, the prospect value of different solutions can be calculated, namely:

$$V_i = \sum_{j=1}^{m} \pi(p_j) v(\Delta x_j) \tag{7}$$

Among them, V_i is the prospect value of the planning scheme Ni; $v(\Delta x_j)$ represents the prospect value function of the scheme Ni at the jth attribute value and the reference point difference Δx_j; π represents the probability weight function, and its size represents the influence of the attribute weight p_i on the entire foreground.

Where, the value function is as follow:

$$v(\Delta x_j) = \begin{cases} (\Delta x_j)^\alpha, \Delta x_i > 0 \\ -\lambda(-\Delta x_j)^\beta, \Delta x_i < 0 \end{cases} \tag{8}$$

Among them, α and β respectively represent the degree of unevenness of the value function relative to the gain and loss of the image, which measures the degree of preference of decision-makers for risk-taking, and $0<\alpha$, $\beta<1$; the larger the value of λ ($\lambda>1$), the more averse the decision-maker is to loss, the more sensitive the decision-maker is relative to the equivalent loss. A large number of experimental results show that the values of $\alpha=\beta=0.88$ and $\lambda=2.25$ are consistent with empirical data.

The probability weight functions when facing gains and losses are

$$\pi(p)^{+}=\frac{p^{\chi}}{\left[p^{\chi}+(1-p)^{\chi}\right]^{1/\chi}} \tag{9}$$

$$\pi(p)^{-}=\frac{p^{\delta}}{\left[p^{\delta}+(1-p)^{\delta}\right]^{1/\delta}} \tag{10}$$

In the formula, $\pi^{+}(p)$ represents the non-linear probability weight formula of the return; $\pi^{-}(p)$ represents the non-linear probability weight formula at the time of loss, χ is the size of the risk when representing the return, and δ represents the risk attitude coefficient at the time of loss. The two coefficients indicate the degree of profit preference and risk aversion of the decision maker. A large number of experimental results show that $\chi=0.61$, $\delta=0.69$ is more consistent with the actual situation.

3.2 Construction of Prospect Value Matrix

There is an energy storage planning scheme set $N=\{N_1, N_2, ..., N_n\}$. In this paper, each indicator is used as the attribute of the scheme, and the attribute set of each scheme is A = {A1, A2, ..., Am}, and then the effect vector of the planning scheme N_i under the attribute set A is

$$x_i=\left(x_{i1},x_{i2},\cdots,x_{im}\right) \tag{11}$$

The sample matrix of the effect of scheme set N on attribute set A is

$$X=\left(x_1,x_2,\cdots,x_n\right)^T=\begin{bmatrix} x_{11} & x_{12} & \cdots & x_{1m}\\ x_{21} & x_{22} & \cdots & x_{2m}\\ \cdots & \cdots & \cdots & \cdots\\ x_{n1} & x_{n2} & \cdots & x_{nm}\end{bmatrix} \tag{12}$$

Normalize it as

$$r_{ij}=\frac{x_{ij}-\max\limits_{1\le i\le n}\{x_{ij}\}}{\max\limits_{1\le i\le n}\{x_{ij}\}-\min\limits_{1\le i\le n}\{x_{ij}\}} \tag{13}$$

The positive and negative ideal points are

$$R^{+}=\left\{r_1^{+},r_2^{+},\cdots,r_m^{+}\right\}$$
$$R^{-}=\left\{r_1^{-},r_2^{-},\cdots,r_m^{-}\right\} \tag{14}$$

Where,

$$r_j^{+}=\max_{1\le i\le n}(r_{ij})$$
$$r_j^{-}=\min_{1\le i\le n}(r_{ij}) \tag{15}$$

Calculate the value function of each attribute according to the formula (8). Then calculate attribute weights based on relative closeness [6] and the smaller the relative closeness, the greater its contribution. Attribute weight can be calculated as follows.

$$\omega_i=\varphi_i/\sum_{i=1}^{n}\varphi_i \tag{16}$$

According to the above, the prospect value of each program can be calculated. And according to the value of V_i to sort the optimal scheme.

4 Case analysis

In order to comprehensively evaluate the resilience and economy of the mobile energy storage configuration plan, the overall idea of this article is shown in figure 1.

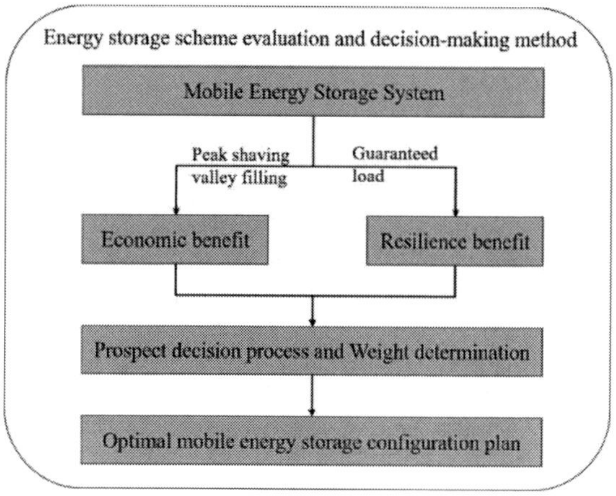

Figure 1. The overall assessment framework.

Taking a certain area as an example, in order to more effectively respond to extreme events, it is planned to select an optimal mobile energy storage configuration plan from 3 alternative points. The decision indexes A_1-A_6 are the investment cost, the benefit from the participation of energy storage in normal operation, proportion of isolated nodes, fault recovery time, the switch operation times and the loss of power load. Based on historical experience and local conditions, decision makers make an evaluation of each attribute, as shown in Table 1.

Table1. Energy storage installation scheme evaluation matrix.

	A_1	A_2	A_3	A_4	A_5	A_6
N_1	79.97	31.61	0.31	3.15	3	54.79

N_2	150.32	59.52	0.42	6.04	2	62.03
N_3	171.73	67.99	0.47	5.73	2	64.17

According the above formulas, calculate the schemes prospect value matrix and the result is shown in Table 2.

Table2. The prospect value matrix.

	A_1	A_2	A_3	A_4	A_5	A_6
N_1	-1.12	0.5	-1.13	-1.12	0.5	0.5
N_2	0.08	-0.75	-0.48	-0.04	-1.01	-0.76
N_3	0.51	-1.1	0.52	0.294	-1.12	-1.13

Calculate the decision weights $\omega 1$ = 0.2467, $\omega 2$ = 0.1126, $\omega 3$ = 0.120, $\omega 4$ = 0.0122, $\omega 5$ = 0.1558, $\omega 6$ = 0.3373. Then calculate the comprehensive prospect valueV_1=-0.4689, V_2=-0.3512, V_3=-0.1586. Therefore, the optimal sorting scheme can be obtained as N_3.

5 Conclusion

Based on the prospect theory, this paper considers the psychological behaviour of decision-makers, and proposes a decision-making method for energy storage location and capacity. The different parameters in the value function of the prospect theory and the decision weight function indicate the different attitudes of decision makers to gain and risk. In decision-making based on prospect theory, decision makers adopt different parameters and different reference points in the value function and decision weight function according to their own preferences, and the prospect values of the obtained schemes may be different, and the ranking results may also be different.

On the one hand, evaluate the plan by establishing an indicator system that comprehensively considers the recovery capacity and economics. On the other hand, using the value function in the prospect theory and the attribute importance index based on the relative closeness to calculate the comprehensive value of the scheme, considering the risk preference of the decision maker in the face of gains and losses, the evaluation results are more in line with the psychological behaviour of the decision makers. Through the analysis of calculation examples, the rationality and feasibility of the method applied to the problem of energy storage site selection and capacity determination are verified. The method proposed in this paper has certain reference value for the research on the selection and evaluation of energy storage planning schemes.

References

1. Yang Huoming. Optimization method of energy storage system location and capacity considering the resilience of distribution network[D]. Shanghai Jiaotong University, 2018.
2. Gao Peng. Research on fault recovery method of active distribution network under extreme weather[D]. Yanshan University, 2019.
3. LEI Shunbo, WANG Jianhui, CHEN Chenl. Mobile emergency generator pre-positioning and real-time allocation for resilient response to natural disasters[J]. IEEE Transactions on Smart Grid. 2016,9(3):2030-2041.
4. LEI Shunbo, CHEN Chen, HUI Zhou. Routing and scheduling of mobile power sources for distribution system resilience enhancement[J]. IEEE Transactions on Smart Grid,2019,10(5):5650-5662.
5. Tversky K A. Prospect Theory: An Analysis of Decision under Risk[J]. Econometrica, 1979, 47(2):263-291.
6. Yongwei Yang. Distance-based multi-attribute decision-making method on improved situational fuzzy sets[J]. Fuzzy Systems and Mathematics, 2021, 35(02):107-114.
7. Hu Jun-hua and Zhou Yi-wen. Prospect theory based multi-criteria decision making method[C]. 2009 Chinese Control and Decision Conference, 2009,2930-2934.
8. Jiefang Liu, Sifeng Liu and Zhigeng Fang, Storage reliability evaluation model based on grey target theory combined with prospect theory[C]. 2015 IEEE International Conference on Grey Systems and Intelligent Services (GSIS), 2015, 374-377.
9. J. Yang, B. Yuan, F. Zhang, F. Luan, G. Lu and Z. Xu, Research on the Application Prospect of Energy Storage Technology in Energy Internet[C]. 2019 IEEE 3rd International Electrical and Energy Conference (CIEEC), 2019,1608-1612.
10. Liuxin Chen, Nanfang Luo. Pythagorean fuzzy multi-criteria decision-making based on prospect theory[J]. Systems Engineering-Theory and Practice. 2020,40(03):726-735.

https://doi.org/10.1051/e3sconf/202129302001

Research on Different Applications of SMA in Seismic Resistance of Building Structures

Zhang Zhe[1,2,a]

[1] College of Civil Engineering, Huaqiao University, Xiamen, China

[2] 501, Unit 2, Building 7, Wenhuayuan, Economic Development Zone, Zhangjiakou City, Hebei 075000, China

Abstract. SMA(Shape memory alloys) are unique materials with many excellent properties, such as excellent wear resistance, corrosion resistance, high power/weight ratio and biocompatibility, high damping, shape memory effect, super elasticity, etc. In civil engineering, scientists try to use the advantages of SMA by making it into different components and they have already made some types of components that are SMA bracing, SMA constrains RC columns, SMA ring springs and base isolation bearing. All of these components can make contributions to seismic resistance, therefore, this article mainly focuses on the the contribution of SMA bracings and SMA constrains RC columns to the seismic resistance of the structure and the relationship between the improvement of their own performance and SMA.

1 INTRODUCTION

Earthquake is a kind of sudden, destructive and frequent natural disaster, whose most of the energy is consumed through the plastic deformation of the building structure, leading to the destruction of the building structure. China is located between the Pacific Rim Seismic Belt and the Eurasian Seismic Belt, the plate movement is frequent, is a typical earthquake-prone country.

Furthermore, studies have shown that the world has entered a period of high seismic activity. The traditional seismic technology thought takes resistance as the central idea, when the building appears crack, the building does not collapse to ensure the function and safety. Construction industry seismic target of China generally covers three target that were buildings in small earthquake is not damaged, buildings in middle earthquake can be repaired, buildings in big earthquake does not fall [1]. The traditional seismic method increases the seismic performance of the building structure by increasing the stiffness and strength of the building structure, but often leads to the increase of the overall cost of the building structure, the difficulty of construction, and the effect is not ideal. Therefore, structural designers no longer blindly pursue to increase the overall stiffness of the building structure, but through the use of some other means to increase the seismic performance of the building structure. For example, there have been many researches on the contribution of shape memory alloys to earthquake resistance, which will be discussed in detail below.

2 OVERVIEW OF SMA

2.1 DEFINITION OF SMA

SMA(Shape memory alloys) are unique materials that can deform and return to a predetermined shape when unloaded or heated, and have been discovered for nearly a hundred years[2]. In 1932, Swedish scientist Ölander first observed the phenomenon of martensite increasing and decreasing with temperature In Au-Cd alloy. Since then, researchers have successively found this phenomenon In Cu-Zn alloy, Cu-Sn alloy and In-Tl alloy[3].

Nowadays, we have found that shape memory alloys have many excellent properties, such as excellent wear resistance, corrosion resistance, high power/weight ratio and biocompatibility, high damping, shape memory effect, super elasticity, etc.[4] In the field of civil engineering, we mainly use its high damping and super elasticity to weaken the influence of earthquake and improve the seismic capacity of building structures.

2.2 APPLICATIONS OF SMA

Scientists have long been studying the application of shape memory alloys in high-end and sophisticated fields. Now, SMA has been widely used in many aspects of our life. It is acknowledged that SMA is found in many important instruments in aerospace, automotive, and biomedical applications in Tab.1.

[a]E-mail:1812102042@stu.hqu.edu.cn

Table1. Applications of SMA in different fields.

Field	Application
Aerospace	Fuselage, wings, landing gear, engine, etc.[5]
Automotive	Radiator, fan clutch, engine, fuel system, piston ring, supercharger, transmission control, brake, bumper, rearview mirror, airbag, hood, etc.[6]
Biomedical Application	Stent, orthodontic appliance, medical catheter, endoscope, laparoscope, file, surgical thread, artificial bone, artificial joint, artificial organ, fixed splint heart repair device, thrombosis filter, artificial blood vessel, etc.[7]–[10]

In particular, in the field of civil engineering, the two most famous applications of SMA in the field of architecture are the bell tower of the church of San Giorgio[11] in Italy and the roof of the church of San Francesco[12]. The bell tower of the San Giorgio church, which was severely damaged by an earthquake in October 1996, was repaired by connecting four pre-stretched steel rods with four hyperelastic SMA rods, which were inserted from the top to the bottom of the tower Fig.1. In June of the following year, another earthquake occurred in the area, and the repaired clock tower remained intact, providing the strongest evidence of the seismic performance of the hyperelastic SMAdamper.

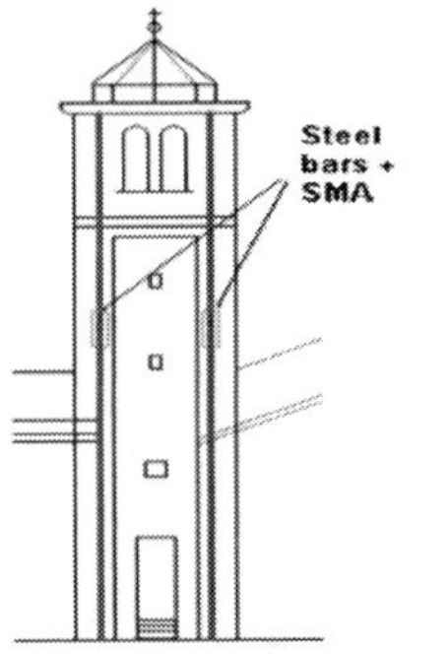

[11](a)Damping mounting position

(b)The restored clock tower

Figure 1. (c)Damper after installation

3 SMA SEISMIC COMPONENT FORMS

3.1 SMA BRACINGS

3.1.1 DEFFERENT FORMS OF SMA BRACINGS

Adding SMA bracings to the structure can improve the stiffness and damping ratio of the structure systems and reduce the seismic response. However, after adding SMA energy dissipation device, the reduction of structural response is not only related to the performance of the energy dissipation device itself, but also closely related to the arrangement of the energy dissipation device and its coordination with the structure. At present, the main forms of diagonal brace and chevron brace[13]. The deformation of this additional damper is less than or equal to the inter-layer deformation of the structure. In the case of large structural deformation, the damper still cannot fully play its role. In order to increase the deformation of the damper and make it play a better role, some researchers put forward new energy dissipation bracings, such as scissors bracings and reverse bracings[14] Fig.2.

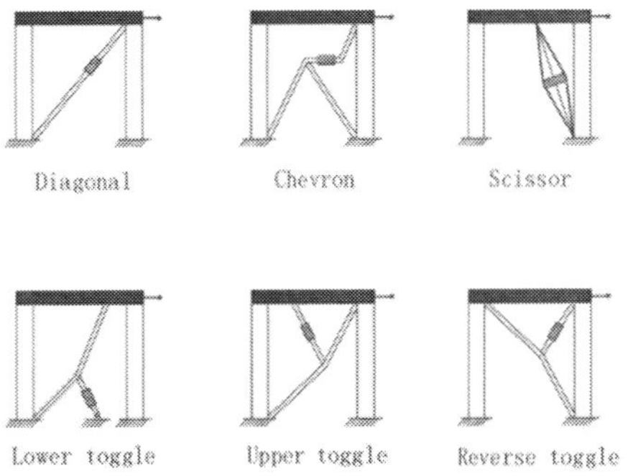

Figure 2. Different forms of bracings

3.1.2 SINGLE DIAGONAL BRACING IS USED IN THE FRAME STRUCTURE

Shi et al.[15] adopted the single diagonal bracing mode in their experiment and established the mode with OpenSees software, and achieved good experimental results. Therefore, we also adopted the experimental model in SHI

 https://doi.org/10.1051/e3sconf/202129302001

and applied it to the experiment of Hu et al[16] for analysis Fig.3.

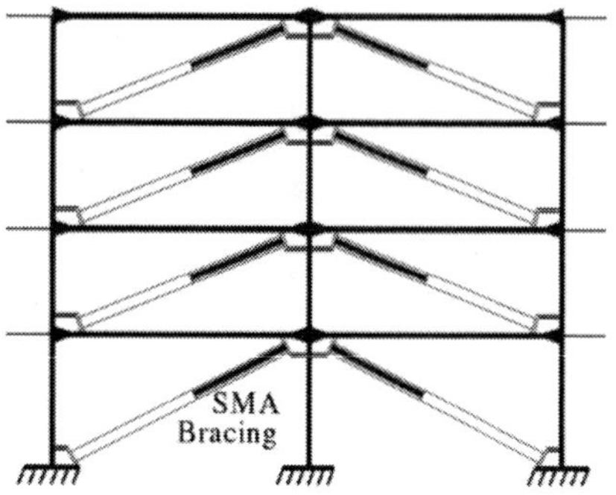

Figure 3. Single diagonal bracing in frame construction

3.1.3 A TYPE OF STRUCTURE OF SMA BRACING

Gu et al.[17] used a new self-reset SMA bracing which is consisted of the high-strength bolt(1), the subpanel(2), the sliding screw(3), SMA(4), the nut(5), the slip sheet(6), the left panel(7), the setting-up piece(8), the butyl rubber piece(9) and the right panel(10) Fig.4. A long round hole is respectively opened at the same position on the left and right sides of the middle of the front subpanel, the back subpanel and the right panel. The circular holes of bolts are respectively arranged on the same vertical axis of the front subpanel, the bank subpanel and the right panel. A butyl rubber gasket is arranged on both sides of each panel at the two long round holes, and then a slip gasket is arranged on the side far away from each panel of the butyl rubber gasket, and the front subpanel, the back subpanel and the right panel are fixed connected through the slip screw, and the slip screw is arranged in the long round hole near the end of the other side of the long round hole. The fixed gaskets are respectively arranged on both sides of each panel at the round hole of the bolt, and are fixed and connected by high strength bolts. After the bolts and screws are fixed, the SMA wire material is wound on the two sliding screws respectively to form the new self-reset SMA bracing.

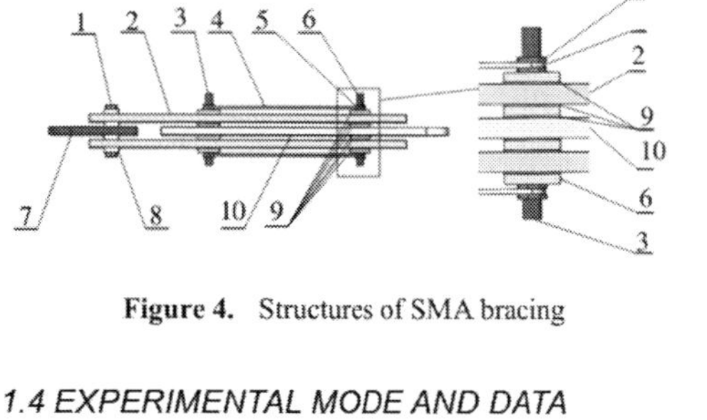

Figure 4. Structures of SMA bracing

3.1.4 EXPERIMENTAL MODE AND DATA

I extracted five groups of experiments from several groups of data in Hu's[16] experiment for analysis. The first group of tests was a bracing model without SMA wire as a control test. The second test was carried out with 16 laps of SMA wire (8 laps on each side) with a cross sectional area of 25.12 mm^2 and a pretension of 1 KN. The third test was carried out with 28 laps of SMA wire (14 laps on each side) with a cross sectional area of 43.96 mm^2 and a pretension of 1 KN. The fourth test was carried out with 28 laps of SMA wire (14 laps on each side) with a cross sectional area of 43.96 mm^2 and a pretension of 3 KN. The fifth test was carried out with 28 laps of SMA wire (14 laps on each side) with a cross sectional area of 25.12 mm^2 and a pretension of 5 KN. The influence of SMA cross-sectional area on the bracing performance can be obtained through the experiments of group 2 and 3. The influence of SMA pretension on the bracing performance can be obtained through the experiments of group 3, 4 and 5 in Tab.2.

Table2. The experimental group

Specimen number	1	2	3	4	5
specimen name	SCB-0-0	SCB-16-1	SCB-28-1	SCB-28-3	SCB-28-5
SMA laps	0	16	28	28	28
SMA proportion/mm^2	0	25.12	43.96	43.96	43.96
SMA pretension force/KN	0	1	1	3	5

3.1.5 THE EFFECT OF SMA PROPORTION ON PERFORMANCE

The comparison between test 2 and test 3 shows that increasing the SMA area can effectively improve the bearing capacity of the component and slightly reduce the residual deformation of the component. In addition, increasing the SMA area has no obvious effect on the anti-slip bearing capacity Fig.5.

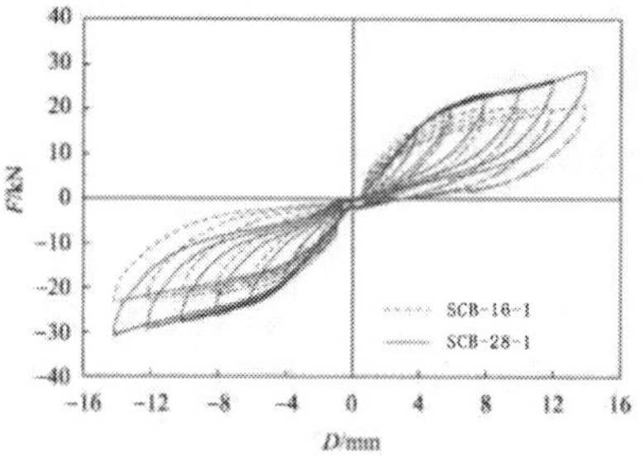

Figure 5. [18] effect of SMA proportion on performance

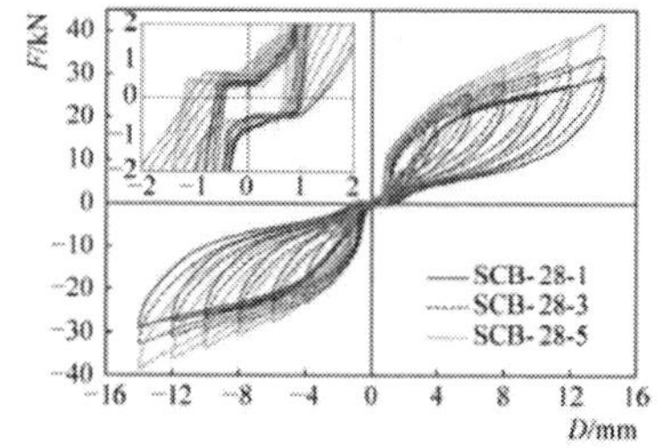

Figure 6. [16] effect of SMA pretension on performance

3.1.6 EFFECT OF SMA PRETENSION ON PERFORMANCE

Compared with the curves of 1 KN, 3 KN and 5 KN, it can be seen that increasing the SMA pretension can improve the bearing capacity of the component, but the hysteretic curve obtained is more kneaded and the hysteretic area is smaller, which is mainly because increasing the pretension can make the SMA have higher stress under the same displacement. However, it has no obvious influence on the resistance sliding bearing capacity and residual deformation (the residual deformation deviates when the prestress is 5 KN, but the positive and negative residual deformation are basically the same, which may be caused by construction errors), and it has a good restoration ability Fig.6.

3.1.7 ENERGY-DISSIPATING CAPACITY

The corresponding energy dissipation value gradually increases with the increase of the displacement of each specimen, that is, the specimen can provide stable energy dissipation capacity. In addition, by comparing the two groups of experiments, we can find that increasing the amount of SMA can effectively improve the energy consumption capacity. By comparing the tests of 3, 4 and 5 groups, it can be found that the increase of SMA pretension also contributes to the energy consumption Fig.7.

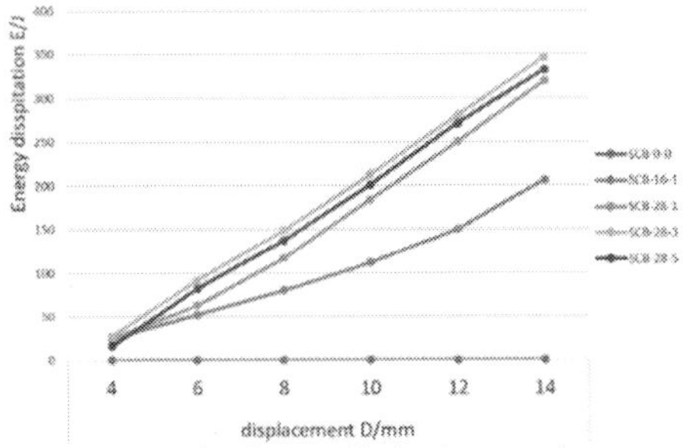

Figure 7. Energy-dissipating capacity

3.2 SMA CONSTRAINS RC COLUMNS

3.2.1 EXPERIMENTAL DESIGN

Pan[19] and Qian[20] have conducted relevant experiments in this regard. Both of them constrained RC columns with SMA wires, and used similar instruments to load the columns, observe and record the experimental situation. But there were some differences in their experiments such as the size of the columns and the reinforcement were not identical, though not identical. The restraint mode of SMA wire is different. Pan uses SMA wire to restrict the height of part of the column, while Qian is completely restricted by SMA wire from the bottom to the top of the column. Some detailed experiments are shown in Tab.3.

Table3. The experimental group

	Pan	Qian	
Control group	SMA wire not reinforced(0)		
Quantity of SMA	SMA winding is 300mm, the interval is 20mm	In prestress is 2 KN	54 windings(c)
			72 windings(d)
			108 windings(e)
Methods of reinforcement	Active reinforcement (a)	/	
	Passive reinforcement (b)		
Prestress/KN	/	In 72 windings	0(f)
			1(g)
			2(d)
			4(h)

Figure 8 is a schematic diagram of the experimental instruments of Pan and Qian. The difference between the two has the column height is different, the SMA restraint range is different, the steel bar selection and the reinforcement is different. I will not study the experiment in detail, but specifically analyze the SMA wire constraints on the column energy consumption.

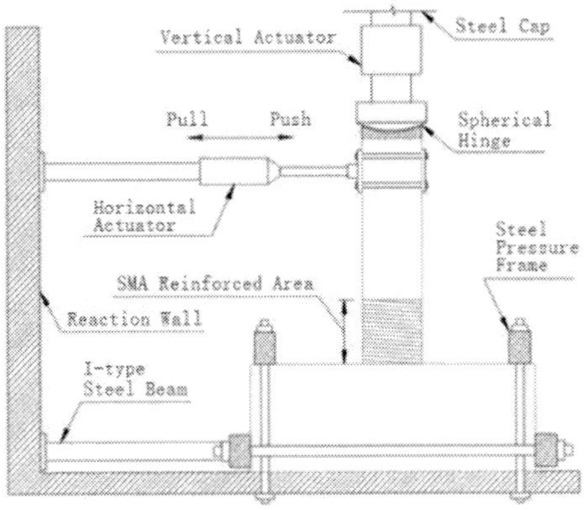

Figure 8. Experimental mode

3.2.2 ENERGY-DISSIPATING CAPACITY

I have summarized the experiments of Pan and Qian together and divided them into groups 0~ H, of which group 0 is the control experiment of A and B. As shown in Figure 9, each group respectively in the displacement of 5 mm, 10 mm, 15 mm, 20 mm, 25 mm, 30 mm, 35 mm and 40 mm, reflected their energy consumption by intuition.

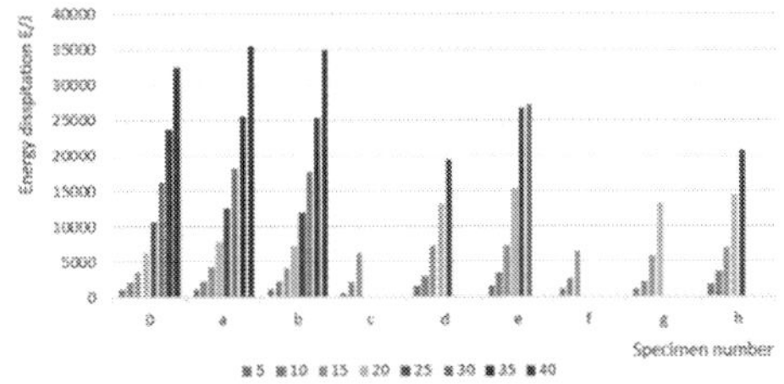

Figure 9. Energy-dissipating capacity

From the general point of view of the experiment of group 0~H, although the experimental conditions and methods are not exactly the same, under the same displacement conditions, the amount of energy consumption is at the same order of magnitude and theoretically correct. Therefore, it is feasible and meaningful to compare them together. With the increase of SMA placement rate, the energy dissipation capacity of all the specimens increased, and with the increase of displacement, the energy dissipation capacity increased more obviously.

As can be seen from groups 0, a and b in the figure, when the horizontal displacement reaches 10mm, the accumulated energy dissipation of the three specimens is almost the same, indicating that the energy dissipation of all specimens is basically the same before the occurrence of major oblique cracks. After 120mm, the accumulative energy consumption of the reinforcement pieces is higher than that of the original specimen. Moreover, when the horizontal displacement is 40 mm, the accumulative energy consumption of the reinforcement pieces A and B is 3554J and 34891J, respectively, which is increased by 9% and 7% compared with that of the original specimen 32609J. It can be found that under the same displacement amplitude, SMA wire has a certain increase in the energy dissipation of pier column, and it shows that the energy dissipation capacity of the active reinforcement test piece is better than that of the passive reinforcement test piece, but the degree of improvement is limited. This is because the current SMA wire winding distance is relatively large, which leads to the small confining pressure provided to pier concrete. Therefore, the influence of different SMA wire interval can be considered in subsequent studies.

From the F, G, D and H groups, the energy dissipation capacity of RC pier with non-prestressed SMA wire is weaker than that of the prestressed specimen, because the SMA wire in the prestressed specimen forms a lateral constraint on the RC pier earlier and participates in the energy dissipation of the specimen earlier. The energy dissipation capacity of SMA wire with pretension of 2% and 4% is stronger than that with pretension of 1%, but the energy dissipation capacity of SMA wire with pretension of 4% is weaker than that of sample with pretension of 2%. This is because the wire begins to harden when the SMA strain rate reaches 6%.

3.2.3 OTHER CAPACITIES

In addition to the energy dissipation capacity, the SMA wire reinforcement has a more significant effect on the strength attenuation control of the pier under cyclic load, which improves the bearing capacity, ductility, stiffness degradation and residual deformation of the pier. I will not go into details here.

3.3 OVERVIEW OF OTHER COMPONENT FORMS

In addition to the above two SMA application methods, there are many other SMA components proposed by scholars, which can also improve the seismic performance of building structures.

SMA ring spring is a kind of efficient shock absorption functional element[21], which is composed of an interconnected SMA outer ring and an inner ring of fine steel. Deformation mainly concentrates on the outer ring of SMA, while the inner ring of fine steel remains elastic. SMA ring spring can change the axial (or bending) force mechanism into the circum-axial force mechanism with overload protection characteristics, which solves the problems of anchorage, fracture prevention and other aspects, and breaks through the problem of low utilization rate of SMA materials.

In addition, SMA components can also be installed in the beam-and-column joints to make the overall structure self-reset, ensuring that other important components (such as beams and columns) basically remain elastic under the action of the earthquake, so that the repair work and costs after the earthquake are minimized. At present, the research of SMA screw joint and SMA end-plate joint has attracted much attention. Sepulveda et al.[22] conducted experiments on the self-reset beam-and-column joints with copper-based SMA screws, and the results showed that the joints had certain hyperelastic behavior and energy dissipation capacity, and no performance degradation occurred within the displacement Angle of 3%.

The base isolation bearing can effectively reduce the seismic response of multi-story buildings and Bridges, and the self-reset and energy dissipation capacity of the isolation bearing can be further improved by using SMA components. The isolation bearing based on SMA technology mainly has two types SMA- rubber bearing and SMA- friction bearing. For SMA-rubber bearing, the SMA wire (or stranded wire) is interspersed between the upper and lower end plates, and the stretching of the SMA wire is driven by the relative displacement of the upper and lower end plates. Dezfuli et al.[23] and Shinozuka et al.[24] concluded through finite element analysis that such bearings had superior self-reset performance and energy dissipation capacity. Gur et al.[25] studied the isolation effect of SMA-rubber bearing under near-fault earthquake action, and believed that SMA-rubber bearing has advantages over ordinary lead-rubber bearing in three indicators of acceleration, maximum structural displacement and residual displacement. Ozbulut et al.[26] considered the influence of outdoor temperature difference on SMA-friction bearing, and found that if 20 °C was taken as the reference temperature, the influence of fluctuation of 20 °C on the maximum displacement of bearing was not more than 13%, and the influence on acceleration was negligible at about 2%.

4 CONCLUSIONS

There is no doubt that SMA has many unique advantages, and there is a lot of room to play in the seismic aspect of building structures. However, there are still many problems to be solved in the research and application of SMA. Firstly, the manufacturing cost needs to be reduced. At present, the SMA processing and preparation technology in the market are not mature, which not only causes its high cost, but also brings the material reliability problem, so it is still unable to realize the large-scale application of products[4]. In addition to this, steel-concrete composite structure is the main form of modern steel structure system, but there is little research on the performance of SMA composite joints and systems, and further consideration should be given to the deformation coordination of the structure and the damage of concrete floor. Also, the lack of research on the temperature-sensitive characteristics of SMA members and their effects on the seismic response of the whole structure is crucial to the development of the applicable conditions of SMA members including the range of indoor and outdoor working temperatures. From the perspective of energy consumption, whether the SMA component can be used as an energy consumption component alone is questionable, which needs to be analyzed from the aspects of functionality and economy. Besides, due to the degradation effect of SMA materials under large cyclic deformation, the post-disaster strengthening strategies and the ability to resist secondary disasters (such as aftershocks, etc.) of SMA systems have not been fully studied[27].

REFERENCES

1. Code for seismic design of buildings, vol. GB 50011-2. 2010.
2. L. I. N. Cheng-xin and F. Wang, "Review of shape memory alloy application status NIU," J. TIANJIN Univ. Technol., vol. 36, no. 4, pp. 1–6, 2020.
3. J. Mohd Jani, M. Leary, A. Subic, and M. A. Gibson, "A review of shape memory alloy research, applications and opportunities," Mater. Des., vol. 56, pp. 1078–1113, 2014, doi: 10.1016/j.matdes.2013.11.084.
4. Y. Jiannan, "A Review of Shape Memory Alloys: Mechanical Behavior and Application," Chinese J. Solid Mech., 2021.
5. D. J. Hartl and D. C. Lagoudas, "Aerospace applications of shape memory alloys," Proc. Inst. Mech. Eng. Part G J. Aerosp. Eng., vol. 221, no. 4, pp. 535–552, 2007, doi: 10.1243/09544100JAERO211.
6. D. Stoeckel, "Shape Memory Actuators for Automotive Applications," Mater. Des., vol. 11, no. 6, pp. 302–307, 1991.
7. D. Mantovani, "Shape memory alloys: properties and biomedical applications," Jom, vol. 52, no. 10, pp. 36–44, 2000, doi: 10.1007/s11837-000-0082-4.
8. S. R. Oh et al., "A comparison of nickel-titanium rotary instruments manufactured using different methods and cross-sectional areas: ability to resist cyclic fatigue," Oral Surgery, Oral Med. Oral Pathol. Oral Radiol. Endodontology, vol. 109, no. 4, pp. 622–628, 2010, doi: 10.1016/j.tripleo.2009.12.025.
9. R. Pfeifer, C. W. Müller, C. Hurschler, S. Kaierle, V. Wesling, and H. Haferkamp, "Adaptable orthopedic shape memory implants," Procedia CIRP, vol. 5, pp. 253–258, 2013, doi: 10.1016/j.procir.2013.01.050.
10. [10]G. Lim, K. Park, M. Sugihara, K. Minami, and M. Esashi, "Future of active catheters," Sensors Actuators, A Phys., vol. 56, no. 1–2, pp. 113–121, 1996, doi: 10.1016/0924-4247(96)01279-4.
11. M. Indirli, M. G. Castellano, P. Clemente, and A. Martelli, "Demo-application of Shape Memory Alloy devices:the rehabilitation of the S.Giorgio Church Bell-Tower," vol. 4330, pp. 262–272, 2001.
12. F. M. Mazzolani and A. Mandara, "Modern trends in the use of special metals for the improvement of historical and monumental structures," Eng. Struct., vol. 24, no. 7, pp. 843–856, 2002, doi: 10.1016/S0141-0296(02)00023-8.
13. Z. Deng, "New SMA Energy Dissipation Brace Systems," J. Beijing Univ. Technol., vol. 34, no. 5, pp. 511–515, 2008.
14. B. M. C. Constantinou, P. Tsopelas, A. Member, W. Hammel, A. N. Sigaher, and S. Member, "TOGGLE-BRACE-DAMPER SEISMIC ENERGY DISSIPATION SYSTEMS By," no. February, pp. 105–112, 2001.
15. F. Shi, "Study on the Seismic Performance of Steel Frame Structure Equipped with Novel SMA Bracing System." .
16. S. J. Hu, Q. Gu, G. Q. Jiang, and J. G. Xiong, "Experimentral study on seimeic performance for an innovative self-centering sma brace," Gongcheng Lixue/Engineering Mech., vol. 38, no. 1, 2021, doi: 10.6052/j.issn.1000-4750.2020.02.0087.
17. G. U. Qi, S. Guquan, H. U. Shujun, G. U. O. Qi, and J. Jun, "Mechanical properties and design method of an innovative re- centering brace with low friction and SMA," 2019, doi: 10.13764/j.cnki.ncdg.2019.03.011.
18. S. Hu, Q. Gu, J. Jiang, G. Song, and J. Xiong, "Hysteresis performance and simplified mechanical model of an innovative self-centering SMA brace," Jianzhu Jiegou Xuebao/Journal Build. Struct., vol. 41, pp. 66–75, 2020, doi: 10.14006/j.jzjgxb.2020.S1.008.
19. S. Pan, R. Yue, H. Hui, and Y. Zhu, "Experimental Study on Seismic Performance of Active Confined RC Piers with NiTiNb-SMA Wires," Hunan Daxue Xuebao/Journal Hunan Univ. Nat. Sci., vol. 47, no. 7, pp. 93–101, 2020, doi: 10.16339/j.cnki.hdxbzkb.2020.07.011.
20. H. Qian, Z. Ren, J. Xiong, and L. Kang, "Experimental investigation on seismic performance of RC pier columns confined with shape memory alloy wires," Tumu Gongcheng Xuebao/China Civ. Eng. J., vol. 53, pp. 265–271, 2020, doi: 10.15951/j.tmgcxb.2020.s2.040.
21. C. Fang, M. C. H. Yam, A. C. C. Lam, and Y. Zhang, "Feasibility study of shape memory alloy ring spring systems for self-centring seismic resisting devices," Smart Mater. Struct., vol. 24, no. 7, p. 75024, 2015, doi: 10.1088/0964-1726/24/7/075024.
22. J. Sepúlveda, R. Boroschek, R. Herrera, O. Moroni, and M. Sarrazin, "Steel beam-column connection using copper-based shape memory alloy dampers," J. Constr. Steel Res., vol. 64, no. 4, pp. 429–435, 2008, doi: 10.1016/j.jcsr.2007.09.002.
23. F. Hedayati Dezfuli and M. S. Alam, "Performance-based assessment and design of FRP-based high damping rubber bearing incorporated with shape memory alloy wires," Eng. Struct., vol. 61, pp. 166–183, 2014, doi: 10.1016/j.engstruct.2014.01.008.
24. M. Shinozuka, S. R. Chaudhuri, and S. K. Mishra, "Shape-Memory-Alloy supplemented Lead Rubber Bearing (SMA-LRB) for seismic isolation," Probabilistic Eng. Mech., vol. 41, pp. 34–45, 2015, doi: 10.1016/j.probengmech.2015.04.004.
25. and S. C. Sourav Gur, Sudib K. Mishra, "Performance assessment of buildings isolated by shape-memory- alloy rubber bearing: Comparison with elastomeric bearing under near-fault earthquakes," Struct. Control Heal. Monit., vol. 21, pp. 449–465, 2014, doi: 10.1002/stc.
26. O. E. Ozbulut and S. Hurlebaus, "Evaluation of the performance of a sliding-type base isolation system with a NiTi shape memory alloy device considering

temperature effects," Eng. Struct., vol. 32, no. 1, pp. 238–249, 2010, doi: 10.1016/j.engstruct.2009.09.010.

27. C. Fang, W. Wang, and Y. Chen, "State-of-the-art for application of superelastic shape memory alloy in seismic resistant steel structures," Jianzhu Jiegou Xuebao/Journal Build. Struct., vol. 40, no. 7, pp. 1–12, 2019, doi: 10.14006/j.jzjgxb.2018.C168.

Test of AC - 20 asphalt mixture ratio on the runway of Da Lian Airport

Shuaituan Tian[1,2], Lei Guo[1,2,*], Jixiang Su[1,2], Kunyang Zhao [1,2]

[1] China Airport Construction Group Co., Ltd, Beijing, 100101, China

[2] Beijing Super-Creative Technology Co., Ltd, Beijing, 100621, China

Abstract. After years of use, there's oil loss on the asphalt pavement . the bond strength of aggregate decreases . Loose threshing is serious. In order to improve the pavement performance, Extend the service life of pavement , Improve the safety of aircraft operation .we test the AC-20 modified asphalt mixture on the runway of Da Lian Airport from many factors , including raw materials, admixtures, gradation design, optimum ratio of oil to stone, then we find the best ways to resolve it .

1 Introduction

After years of use, there's oil loss on the asphalt pavement . the bond strength of aggregate decreases . Loose threshing is serious. In order to improve the pavement performance, Extend the service life of pavement , Improve the safety of aircraft operation .we test the AC-20 modified asphalt mixture on the runway of Da Lian Airport from many factors , including raw materials, admixtures, gradation design, optimum ratio of oil to stone, then we find the best ways to resolve it.

2 Raw materials

2.1 Aggregate

There are limestone aggregates, we test the aggregate indexes in accordance with the relevant regulations, the test results are shown in the table below

Table1. The Basalt Coarse aggregate technical indicators[1].

Test items		Test results	Technical requirements	Test method
Bulk specific gravity	15~25mm	2.715	≥2.50	T 0304
	10~20mm	2.676		
	5~10mm	2.643		
	3~5mm	2.693		
Apparent specific gravity	15~25mm	2.771	≥2.50	T 0304
	10~20mm	2.753		
	5~10mm	2.775		
	3~5mm	2.693		

From the chart, we can see all the technical indicators of the limestone coarse aggregate meet the specification requirements, and they can be used in the design and engineering.

Table2. The Basalt Coarse aggregate Particle gradation[2].

mesh size /mm	P (%)							
	15～25mm		10～20mm		5～10mm		3～5mm	
	Technical requirements	Test results	Technical requirements	Test results	Technical requirements	Test results	Technical requirements	Test results
31.5	100		—		—		—	
26.5	95～100	100	100	100	—		—	
19	—	52.9	95～100	100	—		—	
16	—	11.9	—	98.5	—	100	—	
13.2	0～15	0.8	—	80.2	100	100	—	
9.5	—	0.4	0～15	11.9	95～100	98.8	100	100

*Email Address: 420069269@qq.com;

4.75	0~5	0.4	0~5	0.3	0~10	17.1	85~100	97.8
2.36	—	0.4	—	0.3	0~5	1.0	0~25	29.3
1.18	—	0.4	—	0.3	—	1.0		11.1
0.6	—	0.4	—	0.3	—	1.0	0~5	5.3
0.3	—	0.4	—	0.3	—	1.0		2.9
0.15	—	0.4	—	0.3	—	1.0		2.1
0.075	—	0.4	—	0.3	—	1.0		0.6

From the chart, we can see some passing rate of limestone coarse aggregate can not meet the requirements of specification. We should strictly control the quality of the aggregate in the construction in order to make the passing rate and meet the specification requirements.

Table3. The Basalt FINE aggregate technical indicators[3].

Test items	Technical requirements	Test results
		0~3mm
Apparent specific gravity (g/cm^3)	≥2.50	2.671
Sand equivalent (%)	≥60	80
Sturdiness (%)	≤12	4.5

Table4. The Basalt FINE aggregate Particle gradation

mesh size (mm)	P (%)	
	0~3mm	
	Technical requirements	Test results
9.5		100
4.75	100	100
2.36	85~100	92.3
1.18		56.1
0.6	20~50	37.8
0.3		19.2
0.15		12.1
0.075	0~15	2.9

From the chart, we can see some passing rate of limestone fine aggregate can not meet the requirements of specification. We should strictly control the quality of the aggregate in the construction in order to make the passing rate and meet the specification requirements.

2.2 Filler

The filler is milled limestone powder, they all meet the technical requirements, and the test results are shown in the table below.

Table5. The Mineral filler technical indicators

Test items		Technical requirements	Test results
Apparent specific gravity (g/cm^3)		≥2.50	2.669
Water content (%)		≤1	0.6
Particle gradation	<0.6mm	100	100
	<0.15mm	90~100	99.5
	<0.075mm	75~100	85.2
Hydrophilic coefficient		≤1	0.7

From the chart, we can see filler technical indexes meets the specification requirements, it can be used in the design and engineering.

2.3 Asphalt

There are Cnooc modified asphalt, We test the asphalt indexes in accordance with the relevant regulations, they all meet the technical requirements.

Table6. Cnooc modified asphalt performance test results[4].

Test items		Technical requirements	Test results
Penetration (25℃, 100g, 5s) (0.1mm)		≤50	36.3
Softening point (℃)		>80	88.3
Ductility (5cm/min, 10℃) (cm)		>40	48.3
Equivalent softening point T_{800} (℃)		>50	61.0
Equivalent brittle point $T_{1.2}$ (℃)		<-13	-18.4
Flash point (COC) (℃)		>250	277
Density (25℃) (g/cm^3)		Actual measurement	1.018
Elastic recovery (15℃)		>80	92.3
Filmy heating operational test163℃/5h	Mass loss (%)	<1	-0.04
	Penetration ratio (%)	>70	75.6
	Ductility (10℃) (cm)	>30	40.8

Test results show that the modified asphalt technology indexes meet the requirements.

2.4 Anti-rutting agent

Add 0.5% of the asphalt mixture quality anti-rutting agent can gain higher dynamic stability of mixture. Anti rutting agent basic indexes are tested in table below.

Table7. Anti rutting agent basic indexes test results

Test items	Technical requirements	Test results
Density (g/cm^3)	0.92~1.1	0.96
Melt flow rate (190℃, 2.16kg) (g/10min)	≥3	7~9
Water content (%)	<2	0.5
Softening point (℃)	140~170℃	145

From the chart, we can see anti-rutting agent of all the indicators meet the technical requirements and it can be used in the design and engineering.

2.5 Fiber

Fiber can get oil Absorption and reinforcement. Add 0.4% of the asphalt mixture quality fiber can gain higher dynamic stability and water damage resistance of mixture. Fiber agent basic indexes are tested in table below.

Table8. Fiber basic indexes test results

Test items	Technical requirements	Test results
Diameter (μm)	10~25	12.85
Length (mm)	6±1.5	6
Tensile strength (MPa)	≥500	546
Elongation at break (%)	≥15	20
Heat resistance (℃)	≥210℃, no change in volume	no change in volume

From the chart, we can see fiber of all the indicators meet the technical requirements and it can be used in the design and engineering.

3 Mix design of AC-20 asphalt mixture

3.1 Aggregate gradation design

The mix design is adopted for the mineral aggregate gradation of skeleton dense structure. The matching is in table.

Table9. Ac-20 ratio of mineral aggregate gradation[5].

specifications	15~25 mm	10~20 mm	5~10 mm	3~5 mm	0~3 mm	Filler
The percentage (%)	15	30	20	4	30	6

Synthetic mineral aggregate gradation is shown in table and figure.

Table10. AC-20 mineral synthesis aggregate gradation

Mesh size (mm)	19	16	13.2	9.5	4.75	2.36	1.18	0.6	0.3	0.15	0.075
upper	100	90	80	72	58	46	34	27	20	14	8
lower	95	75	62	52	38	28	20	15	10	6	4
middle	97.5	82.5	71	62	48	37	27	21	15	10	6
synthetic	95.3	90.7	84.1	63.4	43.5	35.2	23.6	17.4	10.5	7.7	4.3

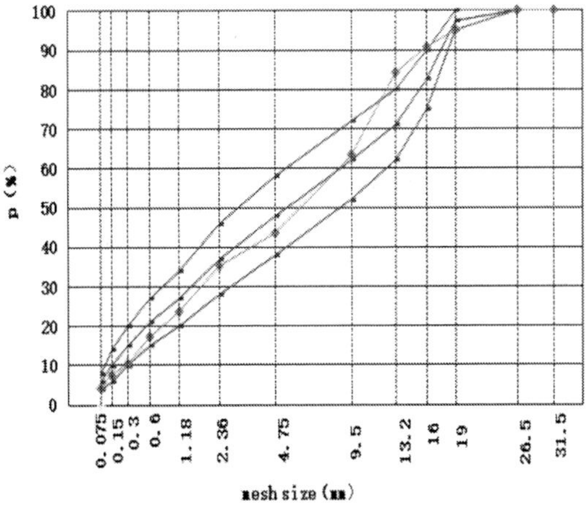

Figure 1. AC-20 grading curve

3.2 The determination of the optimum proportion

We select 5 asphalt aggregate ratio of Marshall test and calculate their physical indicators in order to determine the optimum proportion , the test results are shown in the table and figure below.

Table11. The Marshall test results of different asphalt aggregate ratio

Asphalt-aggregate ratio (%)	Theoretical density (g/cm³)	Bulk density (g/cm³)	VV (%)	VMA (%)	VFA (%)	MS (KN)	FL (0.1mm)
3.2	2.580	2.414	6.5	13.8	53.2	17.9	29.8
3.7	2.562	2.438	4.8	13.3	63.7	19.5	38.4
4.2	2.543	2.444	3.9	13.5	71.3	17.0	32.1
4.7	2.525	2.467	2.3	13.1	82.5	15.7	45.0
5.2	2.508	2.470	1.5	13.4	88.7	15.0	46.6
Technical requirements	—	—	3~5	≥13	65~75	>9	20~40

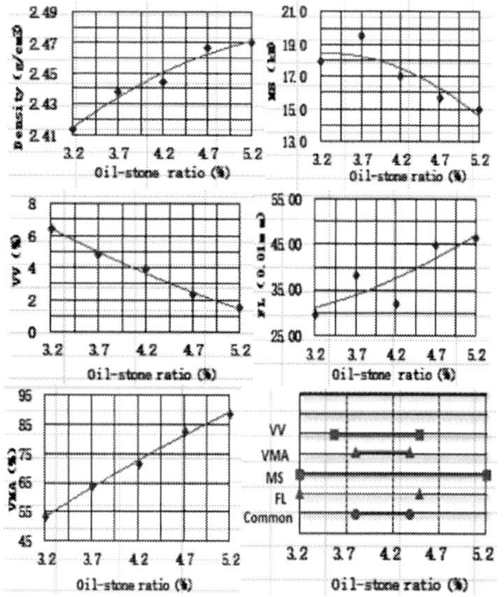

Figure 2. The determination of AC-20 optimum proportion

According to the requirements of the relevant specification, calculated the optimum proportion of 4.1%.

3.3 Road performance verification

We make road performance of verification AC-20 asphalt mixture in the optimum asphalt aggregate ratio of 4.1%, the test results are shown in the table below.

Table12. The Road performance verification test results

Test items	Test results	Technical requirements
MS (KN)	17.5	>8
FL (0.1mm)	33.4	20~40
MSo (%)	91.1	≥80
TSR (%)	92.8	≥80
DS (time/mm)	17077	≥10000
Cw (mL/min)	No seepage	—

MS is strength index, MSo and TSR are water damage resistance index, DS is high temperature performance index, Cw is water permeability index.

From the above test results, all the indexes can meet the requirements of related technologies in the optimum proportion 4.1%.

Adding anti-rutting agent quality of 0.5% asphalt mixture, dynamic stability of asphalt mixture arrives at 17077 times/mm, MS arrives at 17.5 (KN) , FL arrives at 33.4 (0.1mm) , MSo arrives at 91.1%, TSR arrives at 92.8%, Cw arrives at no seepage. They all meet the design requirements. We can see that the mixture has formed the skeleton dense structure from the profile of the specimen, it meets with the skeleton dense type AC asphalt mixture design intent.

4 Conclusion

For the analysis of AC - 20 mix proportion design of asphalt mixture, in order to improve the high temperature performance of asphalt mixture, we can adopt the following measures.

1) We should strictly control each index of the aggregate, all indicators are within the specification limits, including aggregate, filler and asphalt. The important indicators of aggregate are bulk specific gravity, apparent specific gravity, sand equivalent, sturdiness and p%; The important indicators of filler are apparent specific gravity, water content, Particle gradation and hydrophilic coefficient; The important indicators of asphalt are penetration, softening point, ductility, equivalent softening point, equivalent brittle pointT1.2, flash point, density, elastic recovery and filmy heating operational test163°C/5h.

2) We should be reasonable to add additives, such as anti-rutting agent and fiber. The important indicators of anti-rutting agent are density, melt flow rate (190°C, 2.16kg) , water content and softening point; The important indicators of fiber are diameter, length, tensile strength, elongation at break and heat resistance.

3)it is better to choose skeleton dense type of grading, such as 15~25mm: 10~20mm: 5~10mm: 3~5mm: 0~3mm:fiber=15:30:20:4:30:6.

References

1. The civil aviation administration of China. JTG F40-2004. Specification for asphalt concrete pavement construction of civil airports [S].
2. The ministry of communications highway engineering science institute. JTG F40-2004. Technical specification for construction of highway asphalt pavement [S].
3. The ministry of communications highway engineering science institute. JTJ E20-2011.Standard test methods of bitumen and bituminous mixtures for highway engineering [S].
4. The ministry of communications highway engineering science institute. JTG E42-2005. Test methods of aggregate for Highway Engineering [S].
5. Xiaoming,Huang. Asphalt and asphalt mixture [M].

https://doi.org/10.1051/e3sconf/202129302003

Kaczmarz method for saddle point systems

Jinmei Wang[1], Lizi Yin[1], and Ke Wang[2,*]

[1]School of Mathematical Sciences, University of Jinan, Jinan 250022, P.R. China

[2]Department of Mathematics, Shanghai University, Shanghai 200444, P.R. China

Abstract. The Kaczmarz method is presented for solving saddle point systems. The convergence is analyzed. Numerical examples, compared with classical Krylov subspace methods, SOR-like method (2001) and recent modified SOR-like method (2014), show that the Kaczmarz algorithm is efficient in convergence rate and CPU time.

1 Introduction

A block 2 × 2 linear system of the form

$$\begin{pmatrix} A & B \\ B^{\mathrm{T}} & 0 \end{pmatrix}\begin{pmatrix} x \\ y \end{pmatrix} = \begin{pmatrix} b \\ q \end{pmatrix} \tag{1}$$

where $A \in \mathrm{R}^{m\times m}$ and $B \in \mathrm{R}^{m\times n}$, called a saddle point system, arises in a wide variety of technical and scientific applications such as constrained optimization, the finite element method to Stokes equations, fluid dynamics and weighted linear least squares problem [2]. In 2001, Golub, Wu and Yuan [6] proposed the SOR-like method for solving symmetric augmented linear system (1). Then, many improved numerical methods are suggested [1, 10, 13]. Benzi, Golub and Liesen [2] also gave a review on numerical solution of saddle point problems in 2005. Pan, Ng and Bai [9] developed a deteriorated positive-definite and skew-Hermitian splitting (DPSS) preconditioner for nonsymmetric saddle point system (1), based on which, a series of DPSS-type methods [3, 4, 12] are investigated.

The Kaczmarz algorithm is a popular iterative projection method [11] as it is simple to implement and the convergence is superior to classical splitting iterative methods such as SOR-like method. Many authors studied Kaczmarz methods for solving linear systems [7, 8, 11]. In this paper, the Kaczmarz algorithm is presented for symmetric saddle point system (1) with numerical comparisons to classical Krylov subspace methods [5] and splitting iterative methods SOR-like [6] (2001) and modified SOR-like [12] (MSOR-like, 2014).

The rest of this paper is organized as follows. In Section 2, the Kaczmarz method is presented and the convergence is analyzed. In Section 3, numerical examples are provided to show the effectiveness and efficiency of the algorithm. In Section 4, concluding remarks are drawn. Throughout the paper, I denotes the identity matrix.

2 Kaczmarz method

For saddle point problem (1), where $A \in \mathrm{R}^{m\times m}$ is symmetric positive definite, and $B \in \mathrm{R}^{m\times n}$ is of full column rank, Kaczmarz method can be proposed as follows,

---

KACZMARZ METHOD. *Given initial vectors* $x_0 \in R^m$ *and* $y_0 \in R^n$, *for* $k = 0, 1, 2, \cdots$, *the following iterative scheme is taken,*

$$\begin{cases} x_{k+1} = x_k + \dfrac{q^{(i_k)} - (B^{\mathrm{T}})^{(i_k)} x_k}{\left\|(B^{\mathrm{T}})^{(i_k)}\right\|_2^2}\left((B^{\mathrm{T}})^{(i_k)}\right)^{\mathrm{T}}, \\ y_{k+1} = y_k + \dfrac{(b - Ax_{k+1})^{(j_k)} - B^{(j_k)} y_k}{\left\|B^{(j_k)}\right\|_2^2}\left(B^{(j_k)}\right)^{\mathrm{T}}, \end{cases} \tag{2}$$

where $i_k=(k \bmod n)+1$, $j_k=(k \bmod m)+1$, $(\cdot)^{(i_k)}$ *and* $(\cdot)^{(j_k)}$ *denote the* i_k*th row and* j_k*th row of a matrix.*

The convergence result for method (2) is as the following theorem.

Theorem 2.1. *Suppose that saddle system (1) is consistent. Then the iteration sequence* $\left\{\begin{pmatrix} x_k \\ y_k \end{pmatrix}\right\}_{k=0}^{\infty}$, *generated by the Kaczmarz method (2) starting from an initial guess* $\begin{pmatrix} x_0 \\ y_0 \end{pmatrix}$ *with* x_0 *in in the column space of B and* y_0 *in the column space of* B^{T}, *converges to the unique least-norm solution* $\begin{pmatrix} x^* \\ y^* \end{pmatrix} = \begin{pmatrix} A & B \\ B^{\mathrm{T}} & 0 \end{pmatrix}^{\dagger}\begin{pmatrix} b \\ q \end{pmatrix}$ *of (1), where* $(\cdot)^{\dagger}$ *indicates the Moore-Penrose inverse of a matrix. Moreover, the solution error for the iteration sequence is*

* Corresponding author: kwang@shu.edu.cn

$$\|x_{k+1}-x^*\|_2^2 \le \left(1-\frac{\lambda_{\min}(BB^{\mathrm{T}})}{\|B\|_F^2}\right)\|x_k-x^*\|_2^2,$$

$$\|y_{k+1}-y^*\|_2 = \sqrt{1-\frac{\lambda_{\min}(B^{\mathrm{T}}B)}{\|B\|_F^2}}(\|y_k-y^*\|_2 + \sqrt{\lambda_{\max}(A^{\mathrm{T}}A)}\|x_k-x^*\|_2),$$

where $\lambda\min(\cdot)$ and $\lambda\max(\cdot)$ are the the smallest and largest nonzero eigenvalues of a matrix.

3 Numerical experiments

Two numerical examples are discussed using Kaczmarz algorithm (2) compared with the classical SOR-like method [6], recent MSOR-like method [12] and Krylov subspace methods with preconditioner

$$P=\begin{pmatrix}\hat{A} & 0\\ 0 & \hat{S}\end{pmatrix} \quad (3)$$

where $\hat{A}$ = diag(A) and $\hat{S} = B^{\mathrm{T}}\hat{A}^{-1}B$ [5] for solving symmetric saddle system (1). All runs are performed in MATLAB 7.12 on an Intel Core CPU 2.80 GHz (8.00 GB RAM) Windows 7 system.

The first example is for solving weighted least squares problem [10].

Example 3.1. The weighted linear least squares problem

Find: x^* so that: $\|Bx^*-b\|_{A^{-1}}^2 = \min_{x\in R^n}\|Bx-b\|_{A^{-1}}^2$, induces the following saddle point system

$$\begin{pmatrix}A & B\\ B^{\mathrm{T}} & 0\end{pmatrix}\begin{pmatrix}z\\ x\end{pmatrix}=\begin{pmatrix}b\\ 0\end{pmatrix}, \quad (4)$$

where $z=A^{-1}(b-Bx)$.

With $A=\mathrm{tridiag}(1,2,1)\in R^{m\times m}$ and $B=I_m$, the Kaczmarz method is applied to problem (4). Take $Q=B^{\mathrm{T}}B$ as in [6] and the optimal parameter ω_b for SOR-like and MSOR-like methods. The numerical results are listed in Table 1, where the norm of absolute residual vectors is defined as

$$r=\sqrt{\|b-Bx_k-Az_k\|_2^2+\|B^{\mathrm{T}}z_k\|_2^2},$$

the initial vector is taken as

$$\begin{pmatrix}z_0\\ x_0\end{pmatrix}=0,$$

and all runs terminate if $r\le 10^{-7}$.

Table 1 shows that Kaczmarz method converges faster than MSOR-like and SOR-like methods with the optimal parameter ω_b, that is, Kaczmarz method needs much fewer iterations and much less CPU time and and has much higher precision than MSOR-like and SOR-like methods which have similar convergence.

Next example in [1] is for solving the Stokes problem.

Example 3.2. Consider Stokes problem: find **u** and **p** such that

$$\begin{cases}-\mu\Delta\mathbf{u}+\nabla\mathbf{p}=f, & \text{in }\Omega,\\ \nabla\mathbf{u}=g, & \text{in }\Omega,\\ \mathbf{u}=0, & \text{on }\partial\Omega,\\ \int_\Omega \mathbf{p}(x)\,dx=0,\end{cases} \quad (5)$$

where $\Omega=(0,1)\times(0,1)\subset R^2$, $\partial\Omega$ is the boundary of Ω, Δ is the componentwise Laplace operator, **u** is a vector-valued function representing the velocity, and **p** is a scalar function representing the pressure. By discretizing (5) with the upwind scheme, the following saddle point system of linear equations is obtained:

$$\begin{pmatrix}A & B\\ B^{\mathrm{T}} & 0\end{pmatrix}\begin{pmatrix}\mathbf{u}\\ \mathbf{p}\end{pmatrix}=\begin{pmatrix}f\\ g\end{pmatrix}. \quad (6)$$

In the experiment, the right hand side of f and g is chosen such that the exact solution of (6)

$$\begin{pmatrix}\mathbf{u}^*\\ \mathbf{p}^*\end{pmatrix}=\begin{pmatrix}1\\ 1\\ \vdots\\ 1\end{pmatrix}\in R^{m+n},$$

with

$$A=\begin{pmatrix}I\otimes T+T\otimes I & \mathbf{0}\\ \mathbf{0} & I\otimes T+T\otimes I\end{pmatrix}\in R^{2q^2\times 2q^2},$$
$$B=I\in R^{2q^2\times 2q^2},$$

and

$$T=\frac{1}{h^2}\cdot\mathrm{tridiag}(-1,2,-1)\in R^{q\times q},$$

Table 1: Iterations (IT), CPU time (CPU) and absolute residual (r) for Example 3.1

$m(n)$	Kaczmarz			MSOR-like			SOR-like		
	IT	CPU	r	IT	CPU	r	IT	CPU	r
20	20	0.0004	7.7×10^{-15}	152	0.0027	9.6×10^{-8}	165	0.0031	9.7×10^{-8}
200	200	0.0177	5.7×10^{-12}	1746	0.2118	9.9×10^{-8}	1733	0.2077	9.9×10^{-8}
2000	2000	29.2106	3.8×10^{-9}	18404	627.1635	1.0×10^{-7}	18386	626.4071	1.0×10^{-7}

where $\otimes$ denotes the Kronecker product, $h=\frac{1}{q+1}$ is the discretization mesh size, and the relative error is defined as

$$\mathrm{ERR}=\frac{\sqrt{\|\mathbf{u}_k-\mathbf{u}^*\|_2^2+\|\mathbf{p}_k-\mathbf{p}^*\|_2^2}}{\sqrt{\|\mathbf{u}^*\|_2^2+\|\mathbf{p}^*\|_2^2}}.$$

With $m = n = 2q^2$, the initial vector

$$\begin{pmatrix} \mathbf{u}_0 \\ \mathbf{p}_0 \end{pmatrix} = 0$$

and all runs terminated if $\mathrm{ERR} \leq 10^{-7}$, the numerical results are listed in Table 2, where $Q = B^{\mathrm{T}}B$ and $\omega = \omega_b$ (the optimal) for SOR-like and MSOR-like.

From the numerical results in Tables 2 and 3, it can be seen that the Kaczmarz method is superior to MSOR-like and SOR-like methods with the optimal parameter ω_{opt} and the classical Krylov subspace methods, which is also shown from Table 1 in Example 3.1. Obviously, the Kaczmarz method requires fewer iterations and less CPU time and has higher precision than MSOR-like and SOR-like methods.

Table 2: Iterations (IT), CPU time (CPU) and absolute residual (r) for Example 3.2

$m(n)$	Kaczmarz			MSOR-like			SOR-like		
	IT	CPU	r	IT	CPU	r	IT	CPU	r
242	483	0.0327	0	1281	0.1222	1.0×10^{-7}	4120	0.1383	1.0×10^{-7}
648	1295	0.1522	0	2987	6.6767	1.0×10^{-7}	9680	3.9641	1.0×10^{-7}
1250	2499	2.1855	0	5419	48.9726	1.0×10^{-7}	19786	32.5917	1.0×10^{-7}

Table 3: Krylov subspace methods for Example 3.2

$m(n)$	242			648			1250		
method	IT	CPU	r	IT	CPU	r	IT	CPU	r
bicg	55	0.0789	3.7×10^{-8}	186	1.3151	3.2×10^{-8}	401	9.0979	5.9×10^{-8}
bicgstab	139	0.1394	4.6×10^{-8}	772	4.4161	9.0×10^{-8}	-	-	-
bicgstabl	60	0.1362	4.4×10^{-8}	276	3.3722	9.2×10^{-8}	623	24.9574	9.8×10^{-8}
cgs	-	-	-	-	-	-	-	-	-
gmres	-	-	-	-	-	-	-	-	-
lsqr	9	0.0539	8.0×10^{-11}	10	0.4050	3.5×10^{-10}	-	-	-
minres	1602	0.5708	9.9×10^{-8}	3115	8.4035	9.9×10^{-8}	5114	48.5307	1.0×10^{-7}
pcg	-	-	-	-	-	-	-	-	-
qmr	57	0.0903	1.1×10^{-8}	189	1.3779	4.0×10^{-8}	414	9.5015	9.9×10^{-8}
symmlq	4084	1.8005	9.6×10^{-8}	5062	13.3150	9.6×10^{-8}	4089	39.1793	9.3×10^{-8}
tfqmr	-	-	-	-	-	-	-	-	-

All tests use Matlab built-in iterative solvers and the preconditioner is taken as (3).

4 Conclusion

In this paper, the Kaczmarz method is suggested for solving symmetric saddle point system, which is a popular iterative projection method and does not need multiplications of matrix and vector (different to classical splitting iterative method), thus easy to implement and more efficient. Numerical examples show that the Kaczmarz algorithm is much faster than the classical Krylov subspace methods and splitting iterative methods SOR-like and recent MSOR-like and has higher precision for a special case where $B = I$. As for general cases, additional techniques are expected, which will be presented in further work.

Acknowledgement

This work was supported by 2019 Shandong Postgraduate Education Quality Curriculum Project, Jinan University Graduate Teaching Reform Research Project (jdyy903), Jinan University Teaching Reform Research Project (jzc1903), and Key Scientific Research Project of Colleges and Universities in Henan Province (20B110012), China.

References

1. Z.-Z. Bai, B.N. Parlett, Z.-Q. Wang, On generalized successive overrelaxation methods for augmented linear systems, Numerische Mathematik 102 (2005) 1-38.
2. M. Benzi, G.H. Golub, J. Liesen, Numerical solution of saddle point problems, Acta Numerica 14 (2005) 1-137.
3. Y. Cao, Regularized DPSS preconditioners for non-Hermitian saddle point problems, Applied Mathematics Letters 84 (2018) 96-102.
4. Y. Cao, J.-L. Dong, Y.-M. Wang, A relaxed deteriorated PSS preconditioner for nonsym-metricsaddle point problems from the steady Navier-Stokes equation, Journal of Computational and Applied Mathematics 273 (2015) 41-60.
5. J. Chiu, L. Davidson, A. Dutta, J. Gou, K.C. Loy, M. Thom, D. Trenev, Efficient and robust solution strategies for saddle-point systems, University of Minnesota, Institute for Mathematics and Its Applications (2014). Retrieved from the University of Minnesota Digital Conservancy. http://hdl.handle.net/11299/181496.

6. G.H. Golub, X. Wu, J.-Y. Yuan, SOR-like methods for augmented systems, BIT 41 (2001) 71-85.
7. Y. Liu, C.-Q. Gu, Variant of greedy randomized Kaczmarz for ridge regression, Applied Numerical Mathematics 143 (2019) 223-246.
8. Y.-Q. Niu, B. Zheng, A greedy block Kaczmarz algorithm for solving large-scale linear systems, Applied Mathematics Letters 104 (2020) 106294.
9. J.-Y. Pan, M.K. Ng, Z.-Z. Bai, New preconditioners for saddle point problems, Applied Mathematics and Computation 172 (2006) 762-771.
10. Y. Wei, X. Yu, R. Zhang, Preconditioned conjugate gradient method and generalized successive over relaxation method for the weighted least squares problems, International Journal of Computer Mathematics 81 (2004) 203-214.
11. J.-J. Zhang, A new greedy Kaczmarz algorithm for the solution of very large linear systems. Applied Mathematics Letters 91 (2019) 207-212.
12. J. Zhang, C. Gu, A variant of the deteriorated PSS preconditioner for nonsymmetric saddle point problems, BIT 56 (2016) 587-604.
13. B. Zheng, K. Wang, Y. Wu, SSOR-like methods for saddle point problems, International Journal of Computer Mathematics 86 (2009) 1405-1423.

https://doi.org/10.1051/e3sconf/202129302004

Subsection construction technology of rectangular concrete independent column with super high cross section

Guangjun Li[1,2], Tianfang Mo[1,2], Ningbo Xu[1,2], Weixiong Zhang[1,2], Hanwen Lu[3*], Deyuan Deng[4], Mingbo Yi[1,2]

[1] China Construction Fourth Engineering Division Corp. Ltd, Guangzhou, Guangdong, 510430, China

[2] China Construction 4th Engineering Bureau 6th Corp.Limited, Hefei, Anhui, 230011, China

[3] Foshan Univerisity, Foshan, GuangDong, 518054, China

[4] China Construction Steel Engineering Co. Ltd, Shenzhou, 518052, GuangDong, China

Abstract. With the progress of engineering technology, the ability of design and construction has been significantly improved, which the number of long-span construction and long-span Bridges and long-span steel structures are increasing more and more. As the main supporting members of Bridges and steel structures, the quality assurance of rectangular reinforced concrete independent columns is particularly important. In the implementation process of No.18 Maintenance Hangar Project of GAMECO Aircraft Maintenance Facility Phase III in Guangzhou Baiyun International Airport of China Southern Airlines, through research and practice, our company applied the subsection construction technology of super high section rectangular concrete independent columns, which can not only guarantee the quality and forming effect of the column body, but also reduce the input of formwork materials. Improve the utilization rate of formwork and scaffold effectively, and then reduce the input of turnover materials, and achieve remarkable results.

1 Introduction

Guangzhou Baiyun International Airport China Southern Airlines GAMECO Aircraft Maintenance Facility Phase III Maintenance Hangar Project No. 18 consists of the Hall of No. 18 Maintenance Hangar and the 4-floor Production Auxiliary Building. The structural form of hangar hall is reinforced concrete column and lattice roof. Six wide-body aircraft and five narrow-body aircraft can be placed in the hangar hall for overhaul and airline maintenance at the same time. The span is 95m+144m+95m, the depth is 107.65m, the building height is 44.2m, and the lower string elevation of the roof truss is 30m. The 26 large section independent columns in the hangar hall are the supporting points of the steel grid.

2 Construction content

In the deepening process of composite steel formwork, the combination and separation of two integral plates can be applied to independent columns of different section sizes, so that the composite steel formwork of one independent column size can be applied to independent columns of various types and sizes, and the input of steel formwork material can be reduced. The section construction of the super-high independent column ensures that the lower part is reinforced with 500mm high steel formwork, and the upper part is hoisted with the combination of 5500mm high steel formwork and 500mm high steel formwork, which greatly facilitates the installation and adjustment process of the steel formwork.

The overall steel formwork is quick, convenient to install and has many times of turnover use. At the same time, it is equipped with four sets of 500mm high reinforced steel formwork, which can implement a set of steel formwork at the same time with three independent columns, greatly improving the use efficiency of steel formwork and greatly reducing the input of manpower and material resources[1]. The use of composite steel formwork has better bearing capacity, stiffness and stability, the concrete pouring process has little effect on the formwork verticality, concrete forming effect and good sense, can achieve the effect of plaster-free.

3 Process principle

3.1 Selection of template

(1) Formed wood template weight is small, convenient for workers to disassemble, do not rely on mechanical equipment, and wood template price is low, less cost input. The main disadvantage is that the wood formwork itself is relatively low stiffness, which is not conducive to the control of the verticality of the independent column, and the appearance of the molding is poor. In addition, the loss after repeated use is large, and it is difficult to control the

* Corresponding author: luhanwen@fosu.edu.cn

number of turnover. And the installation efficiency is low, the progress is slow.

(2) The whole steel template is used for the molding steel die, which has large stiffness and good forming effect. And the steel die is bolted, quick installation and disassembly. Disk type steel tube scaffold erection, dismantling speed is fast, good stability. Its main drawback is that the weight of the steel die is large, and the mechanical equipment used for hoisting is large. The formalized steel die is only suitable for one size, the column section of different size needs to increase the number of steel die, and the steel die itself to design and manufacture, the cost is high. Next, dish buckle type steel tube scaffold also wants price to compare scaffold of steel tube of fastener type expensive.

(3) The combined steel die can not only guarantee the quality of the column body and the forming effect, but also has less input of formwork material compared with the standard steel die[2]. The installation and disassembly speed is fast, and the turnover can be carried out quickly[3]. With the socket type plate buckle steel pipe scaffold, the utilization rate of template and scaffold can be effectively improved,and adaptability of structure distribution is strong[4], and then the investment of turnover materials can be reduced.

Comprehensive analysis of the construction selection of combined steel mold construction.

3.2 Selection of composite steel formwork

Taking the No.18 Maintenance Hangar Project of GAMECO Aircraft Maintenance Facility Phase III of Guangzhou Baiyun International Airport of China Southern Airlines as an example, the total number of independent columns in the hangar hall is 26. Column Z1~Z4 is made of C30 concrete, longitudinal reinforcement is made of HRB500 steel bar, stirrup is made of HRB400 steel bar.

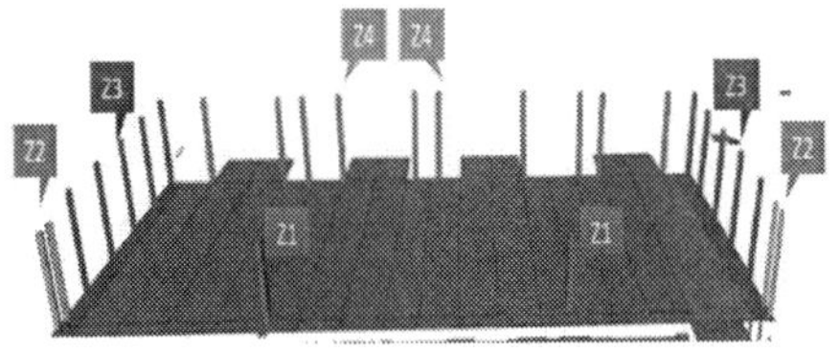

Figure 1. Schematic diagram of independent concrete column with bars

According to the calculation to determine the thickness of the steel plate, the main rib, back bar spacing and model. A piece of integral plate is formed through processing, and the four integral plates are combined to assemble the steel template of rectangular column with cross-section. The panel is made of 6mm steel plate, the flange is made of 12x100mm flat iron, the main rib is made of 10# channel steel, the ribs are configured according to the spacing of 300x300, and the connecting hole is a long hole with a diameter of 22x28. Back carrying with 14# double channel steel, using M32 fine gadolinium rebar reinforcement, back bar spacing design is 700.

There are 4 types of independent columns, and the largest number of them are Z3 and Z4. In the combination design of templates, a large number of templates are given priority, and two sets of templates for Z3 and Z4 are determined by comprehensively considering the impact of construction period. Then use the template of Z3 and Z4 for size design and assemble to become Z1 and Z2. The horizontal combination design is as follows:

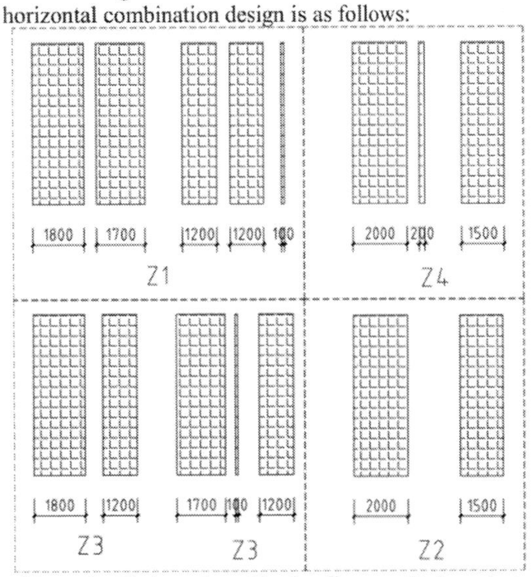

Figure 2. Horizontal combination diagram of independent column template

Through the combination of horizontal formwork, only 1 set of Z1 and 2 sets of Z4 formwork are finally needed to meet the construction schedule requirements of independent columns.

3.3 Subsection construction of super-high section independent column

The maximum height of the independent column is 30.13m, and the minimum height is 22m. Steel mould is used for construction. In order to ensure the forming effect and casting quality, the height of each section column should be controlled at 5-8m. In order to ensure the construction safety of independent columns, the operating frame shall be controlled below 8m. Finally, it is confirmed that the independent columns shall be divided into sections of every 6m for construction. The specific sections are shown in the figure below:

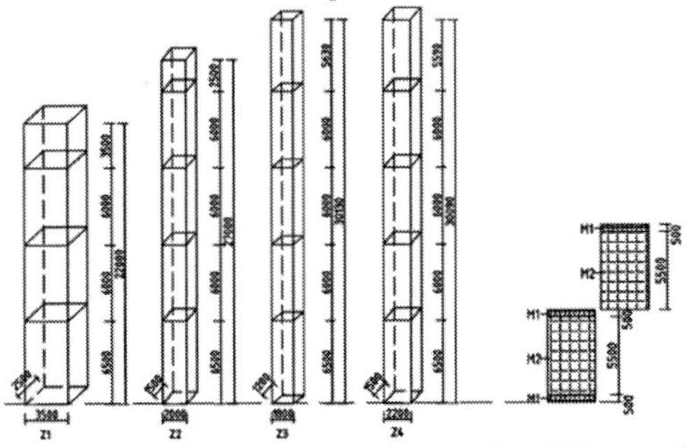

Figure 3. Schematic Diagram of Independent Column and Formwork in Vertical Segment

In order to ensure the stability of the formwork in the concrete pouring process, the segmental construction method is adopted in the construction process. The segmental construction steel formwork is configured as 500mm high steel formwork +5500mm high steel formwork +500mm high steel formwork. In section construction, it is ensured that the lower part is reinforced with 500mm high steel formwork, and the upper part is hoisting the overall uncast formwork combined with 5500mm high steel formwork and 500mm high steel formwork. The design, configuration, construction and assembly of prefabricated components are planned with the overall concept[5].

3.4 Verticality control of ultra-high cross section independent column

In order to ensure the verticality of the independent column, the concrete pouring of the first section of the independent column is particularly important. Considering the difficulty of adjusting the verticality and fixing in the construction process, the first section is poured 500mm high first. The vertical degree of the steel formwork is adjusted by the lifting cone in the section construction process[6], and then the four angles of the independent column are measured by the total station, and the vertical degree of the steel formwork is adjusted by the total station meter.

4 Key points of construction

4.1 Deepen design of steel formwork

(1) Considering that a set of templates and water 3 pillar of large cross section of reinforced concrete independent construction, determine the combination of steel template component model must start with the biggest cross-section of reinforced concrete column is calculated independently, ensure the steel template has enough bearing capacity, stiffness and stability, can reliably under the lateral pressure of concrete and construction load.

(2) Considering that a set of steel formwork is suitable for independent columns with different section sizes, the joint of the plate must be tight in the deepening process to prevent the concrete from forming a larger joint line, which affects the appearance of the forming.

(3) At the same time, when three independent columns are constructed with water flow, the casting height of each section is 6000mm. Considering that the 500mm high reinforced steel formwork cannot be removed in the process of formwork transport, four sets of 500mm high reinforced steel formwork must be equipped at the same time in the construction process.

4.2 Flow construction

Before the construction of the independent columns, a special schedule for the independent columns is arranged, and three independent columns are constructed in parallel according to a set of steel formwork, taking into account factors such as labor force, mechanical equipment and materials. After the completion of the layout of the plan, the site management and operation personnel to disclose, strictly according to the plan implementation.

Due to the characteristics of large independent column construction, a professional construction team is selected. The construction staff of this professional team are all skilled in the construction technology of reinforcing bar binding, formwork installation and concrete pouring. Therefore, each set of formwork can be assigned two groups of construction personnel, one group is specialized in reinforcing bar binding, one group is specialized in installing formwork and concrete pouring.

4.3 Separate column section construction

(1) In the construction of the first 500mm high independent column, jack is used to straighten it. After the adjustment is completed, leveling mortar should be applied outside the root of the column. The mortar should be leveled by the ruler to avoid the leakage of slurry at the root of the column in the process of concrete pouring.

(2) The use of day pump pouring concrete, the use of hoses to extend the length of the tube into the independent column pouring, pouring height is not more than 2m.

(3) Correct the steel formwork. After the wire is pulled, do not remove it. In the process of pouring concrete, a special person is set to observe whether the formwork is deformed or not and adjust the formwork in time.

Figure 4. Adjusting perpendicularity of jack

(4) The spacing between the operating frame and the independent column structure is guaranteed to be between 600mm and 800mm, which is convenient for the overall lifting and transportation of the steel formwork, but the next step of the construction surface must take protective measures.

(5) In the process of independent column casting, layered casting shall be carried out according to the thickness of each 300mm ~ 400mm.

(6) before casting a section, the junction must be gouged.

(7) In the process of sectional casting, 500mm high steel formwork must be dismantled late to serve as the supporting resting point of the formwork to be cast in the previous section, so as to facilitate the installation and adjustment of the steel formwork.

4.4 Verticality adjustment of independent column construction

Hoisting the formwork to vertical hoisting steady up and down[7], the initial assembly template is adjusted by lifting cone, avoid loose shaking[8]. After the initial installation is completed, the coordinate positions of the four corner points of the independent column are measured by the total station, and the steel template is fine-tuned according to the total station data.

The position of the measuring point of the total station is unchangeable, and the measurement must be carried out in the same point until the completion of the pouring of the independent column.

The monitoring personnel must be present throughout the casting process and periodically review the position of the steel formwork.

4.5 Steel formwork removal

The steel formwork shall be lifted and dismantled by truck crane as a whole. In the dismantling process, the bolts shall be loosened after the steel formwork is locked by the lifting ring, and individual bolts shall not be loosened before the lifting and dismantling. In the process of lifting and dismantling, no secondary collision can be produced, so as to avoid the situation that the independent column does not lack the edge and drop the Angle.

4.6 Mass concrete curing and temperature monitoring

After the removal of the steel formwork, a special person shall be arranged for film mulching and watering maintenance, and the maintenance time shall not be less than 14 days.

Temperature monitoring does not need to be separately embedded temperature measuring tube, can be used to draw the screw indwilled hole temperature measurement, the heating phase send a person responsible for measuring every 2 hours, every 3 hours in the cooling phase measurement, 8 days later when the temperature difference between inside and outside measured is less than 25°, can stop the temperature measurement. If the temperature difference is still greater than 25°, continue to monitor.

5 Quality control

5.1 Quality standard

The quality standard meets the current national and industrial standards

5.2 Quality requirements

(1) The verticality of the template installation is allowed to deviate 10mm, and the field control is within 5mm.

(2) The position of the template mounting axis is allowed to deviate 5mm.

(3) the surface flatness of the template is 5mm, and the field control is within 2mm.

(4) Permissible deviation of vertical degree of independent column H/30000+20mm, and field control is within 10mm.

(5) The allowable deviation of section size is +10mm, -5mm, and the field is controlled within 2mm.

6 Conclusion

Segmental construction of ultrahigh large section rectangular concrete independent column in Guangzhou Baiyun International Airport Southern Airlines GAMECO Aircraft Maintenance Facility Phase III Maintenance Hangar Project No. 18 This has greatly improved the use efficiency of steel formwork and achieved remarkable results. During the construction process, the steel formwork is hoisted as a whole, and the material utilization rate is improved, without generating additional construction waste and dust. It has good comprehensive benefits. Compared with the traditional standard composite steel formwork, the plate split assembly of the formwork can be applied to the construction of separate columns with different section sizes. This project has four independent columns with different section sizes. During the construction process, the total number of steel formwork is configured according to different section sizes, among which the steel formwork can be divided into multiple sets of steel formwork with section sizes. The traditional standard composite steel formwork needs about 53.6 tons of steel formwork materials according to the same effect. The formwork input of this project is reduced by about 20.6 tons, and the economic benefit is obviously improved. The integral steel formwork is quick, convenient to install and has many times of turnover, which can meet the requirements of 3 independent columns at the same time, greatly reducing the manpower and material input. The use of steel formwork concrete forming effect and good sense, can achieve the effect of plastering, plastering engineering quantity, unit price package labor package material greatly reduced.

References

1. WANG Jiangang. Integrated Construction Technology of Reinforced Formwork for Cast-in-place Concrete Independent Column[J]. Construction Technology, 2020,49(21):104-105+109.
2. GB/T 50214-2013, Technical code for composite steel-form[S].
3. JIANG Lu , XIE Xiaomeng. Development and Application of Prefabricated Partially Encased

 https://doi.org/10.1051/e3sconf/202129302004

Composite Shear Wall Structure[J]. Construction Technology, 2020,49(15):22-24+37.

4. XUE Qing, LIU Dong, ZHANG Yuanzhi, ZHAO Ji, CHEN Yuhang. Technology Application of Modularized Prefabricated Super High-rise Formwork System[J]. Construction Technology, 2019,48(08):22-26.
5. Liu lingling. Application of Aluminum Alloy Formwork in Assembly Building Construction[J]. Chinese & Overseas Architecture, 2019(08):213-215.
6. GB 50026-2007, Code for engineering surveying[S].
7. Yao Jialiang, Xie Qichao, Yin Bo, Liu Nan. Key Control Points Analysis of Steel Formwork from Selection to the Whole Process of Hoisting[J]. Construction Technology, 2016,45(S2):633-636.
8. XIE Zhong-xin, YAO Xing-shu, CHEN Xing-you. Application of Aluminum Alloy Formwork in Assembly Construction[J]. Intelligent Building & Smart City, 2020(09):131-132+135.

Risk assessment of water and mud inrush in a tunnel in Chongqing based on analytic hierarchy process

Yong Wang[1,*]

[1]The First Municipal Design Institute, Chongqing Architectural Design Institute Co.,LTD, 40015 Chongqing, China

Abstract: Water and mud inrush is one of the typical geological hazards during tunnel construction in karst area. In weight of water or mud inrush risks of karst tunnels are studied and the index system for water and mud inrush risk evaluation is established based on analytic hierarchy process. The analysis shows that geological structure, formation lithology and groundwater level are the main control ingredients of water or mud inrush risk； geographic and geomorphic conditions, attitude of rock formation are the minor significant factors： surrounding rock mass classification , rainfall are the least important elements. Aiming at a tunnel in Chongqing, combining with the actual geological situation of the tunnel site, the evaluation model is used to evaluate the risk of water and mud inrush in each section, which provides a certain theoretical basis for the prevention and control of water and mud inrush disaster during the tunnel construction.

1 Introduction

With the promotion of the strategy of "Western Development" and the construction of the Belt and Road Initiative, the Tunnel construction in China is gradually transferring to the western mountainous and karst areas with extremely complex terrain and geological conditions. It also presents the development trend of "large buried depth, long tunnel line, high stress, strong karst, high water pressure, complex structure, frequent disasters", and water and mud inrush disasters are becoming more and more serious. The water and mud inrush of tunnels in karst strata has always been considered as a worldwide problem. Once the disaster occurs, the tunnel will be submerged, the construction will be interrupted, and the heavy casualties and economic losses will be caused. And even some projects will be forced to stop or change lines[1].

Accurate evaluation and prediction of the risk of water and mud inrush in tunnel is an effective method to prevent water and mud inrush disaster and reduce its impact on tunnel construction. How to accurately evaluate the risk of water inrush in tunnel design and construction stage has gradually become one of the key technical problems of tunnel construction in karst area[2-6]. At home and abroad, many scholars have done a lot of research on the risk analysis and application of water and mud inrush in karst tunnels in recent years. Mao Bangyan[7] and others put forward the risk staging system of water and mud inrush in tunnels; Han Xingrui[8,9] puts forward an expert evaluation system for karst water gushing in tunnels; Li Shucai[2-6] and others have continuously improved the analytic hierarchy process (AHP) and applied it to the risk assessment of water inrush in different projects, and achieved good results. Zhang Qingsong[10] and others have carried out the research on the theory and method of water inrush risk assessment of high-risk karst tunnels, and established the quantitative assessment method of water inrush risk and the four-color disaster warning mechanism; Kuang Xing[11] and others established a geological disaster system for karst water inrush during tunnel construction in karst areas, and used a fuzzy comprehensive evaluation model to evaluate the risk of water inrush and mud inrush disasters in karst tunnels. In addition to using empirical models for evaluation, some progress has been made in using machine learning models, such as BP neural network model[12] to evaluate the risk of water and mud inrush in karst tunnels.

Analytic hierarchy process (AHP) is a simple and practical decision-making method, which can make the decision-making process hierarchical and quantitative according to certain rules. It is a decision-making method for multiple schemes or objectives. At present, the AHP is widely used in the risk assessment of water and mud inrush in karst tunnels, and has achieved good results[3-6,13-17]. A Tunnel in Chongqing is a deep and extra long mountain tunnel. The complex engineering geological and hydrogeological environmental conditions of the tunnel site are extremely easy to cause geological disasters such as water and mud inrush, which profoundly affects the safety of the construction and use of the tunnel. Therefore, the prevention and control of water and mud inrush is one of the important problems to be solved in the process of tunnel construction. It is the key to accurately evaluate the risk of water and mud inrush in each section of tunnel. Based on the analytic hierarchy process (AHP), this paper will choose appropriate evaluation indexes to evaluate the risk of water and mud inrush in the tunnel, so as to better prevent and control the water and mud inrush disaster in the tunnel and ensure the safety of the tunnel.

* Corresponding author: 851537298@qq.com

https://doi.org/10.1051/e3sconf/202129302005

2 Establishment of evaluation index system for water and mud inrush risk of tunnel in karst area

2.1 Selection of risk influencing factors

According to the basic geological characteristics of the tunnel site and the existing disaster-causing factors, combined with the research results of experts and scholars, we selected the formation lithology, geological structure, geographic and geomorphic conditions, groundwater level, attitude of rock formation, the level of surrounding rock, rainfall for seven evaluation factors, and divided the risk levels (levels are divided into I - IV, including I for the minimum risk, IV for the highest risk), and established an index system for risk evaluation of water and mud inrush of the tunnel in karst area.

2.1.1 Geological structure

Geological structural conditions not only control the migration and enrichment process of groundwater, but also determine the channel type, smoothness and water source strength of water gushing in tunnel area. When the fault structure connects the surface water, water rich cavity, underground river or high pressure aquifer with the tunnel, large-scale water gushing or water inrush disasters are usually formed[18]. Geological structure includes fault structure, fold structure and monoclinic structure. Due to the formation reasons of fault fracture zone, there are many kinds of strata lithology, geological structure development, serious rock weathering, broken rock mass, abundant groundwater supply, and complex geological conditions. After tunnel excavation, the underground hydraulic path is easy to change, and water and mud inrush are prone to occur under the joint action of in-situ stress and groundwater. In the fold structure, the rock mass in the syncline core is subject to horizontal compression, the fractures are relatively developed, the integrity and strength of the rock mass are poor, and the rock strata on both sides incline to the core, which easily makes the groundwater gather in the core, resulting in the relatively developed karst in the syncline core area. The rock mass in the core of the anticline is affected by horizontal rip, the fracture develops more and the integrity of rock mass is poor, which is convenient for the infiltration of surface water and underground water, so it is beneficial to the development of karst to some extent. In the water bearing rock group of monoclinic stratum, the difference of karst development degree is significant due to the influence of rock thickness and rock inclination. The steeper the rock inclination is, the smaller the catchment area exposed on the surface is, and the smaller the probability and scale of water inrush during tunnel crossing is. According to whether there are any unfavorable geological structures such as hidden rivers, water-filled mud-filled cavities, water-conducting faults, karst pipeline nearby the tunnel, Xu Zhenhao and others divided it into four grades according to the disaster causing strength and weakness:

Table1. Grade division of unfavorable geology

The grade of disaster causing	**definition**	**The level of risk**
Strong disaster causing	There are spme large water-conducting structures above the floor near the tunnel, or some large pressurized water-conducting structures below the floor near the tunnel	I
Medium disaster causing	There are some medium water-conducting structures above the floor near the tunnel, or some medium pressurized water-conducting structures below the floor near the tunnel	II
Weak disaster causing	There are some small water-conducting structures above the floor near the tunnel, or some l large pressurized water-conducting structures below the floor near the tunnel	III
No disaster causing	There is no adverse geological phenomenon that can cause water inrush near the tunnel	IV

2.1.2 Formation lithology

The soluble rock stratum is the place where the groundwater is stored and transported. The chemical and mineral composition of the rock is closely related to the solubility of the rock. The differentiation caused by the difference of chemical composition and structure of soluble rock is the material basis of karst development, resulting in different enrichment space of groundwater, which determines the intensity and scale of water filling in the tunnel[18]. According to the solubility of stratigraphic lithology, the documents [3-4] divides it into four levels: strong, medium, weak and non-soluble rock. See Table 2 for the specific division.

Table2. Grade division of formation lithology for water inrush

The grade of soluble rock	**Rock types**	**The level of risk**
strong soluble rock	thick - medium thick stratified pure limestone, ancient siliceous cemented dolomite, Carbonaceous and asphaltic limestone	I

medium soluble rock	thick - medium thick marble, thick - medium thick argillaceous limestone, thick - medium thick dolomite	II
weak soluble rock	thin marble, thin argillaceous limestone and thin dolomite	III
non-soluble rock	Non-soluble rocks are generally considered as reliable relative impermeable and non-karst layers in the area of carbonate rocks	IV

2.1.3 Groundwater level

Groundwater is the material basis and power source of karst water inrush[3-4]. The karst hydrodynamic vertical zoning can be divided into: vadose zone or vertical circulation zone, seasonal variation zone, saturated zone or horizontal circulation zone and siphon circulation zone. The vadose zone is located above the phreatic surface, and the surface water moves vertically downward through fissures. The karst forms are mainly funnel, sinkhole and shaft; In the seasonal variation zone, the groundwater movement alternates periodically with seasonal changes. The very active groundwater activities lead to the most developed karst in this zone, often forming large complex karst caves, underground rivers, underground lakes and so on. Similar to the seasonal variation zone, the groundwater activity in the shallow saturated zone is more intense, and karst development is more intense. Groundwater activities in deep saturated zone and subzones are gentle, but most of them are pressurized, and also have the risk of water and mud inrush disaster. Li Liping[3-4] took the height difference h between the groundwater level and the tunnel floor as the factor determining the risk degree of water inrush, as shown in Table 3:

Table3. Grade division of groundwater table

The range of h	definition	The level of risk
$h \geq 60$ m	If karst water is exposed, the instantaneous water inrush is large and fast, which often causes heavy casualties and property losses	I
$30\ m \leq h < 60$ m	Medium disaster causing	II
$0\ m \leq h < 30$ m	The seepage instability or overall extrusion process of karst fillings takes a long time and the disaster causing ability is weak	III
$h < 0$ m	No risk of water and mud inrush	IV

2.1.4 Geographic and geomorphic conditions

Trough valley, karst depression, precipitation funnel, sinkhole and other surface karst forms are the main connecting channels of underground karst system, surface karst system and atmospheric precipitation, and the combined confluence capacity of different surface karst forms is obviously different[5]. The negative landform such as trough, sinkhole and precipitation funnel almost penetrates the surface water into the ground and has strong water collecting capacity, while the open low valley or slope land has weak water collecting capacity[18]. Yang Yanna[18] proposed that three methods can be used for quantitative evaluation of topography and geomorphology, that is, according to the surface precipitation infiltration conditions, according to the influence of surface karst morphology combination on the confluence intensity, and according to the slope gradient of surface precipitation infiltration conditions. According to the catchment capacity and karst development degree of the terrain, Xu Zhenhao[3-4] and others divided the negative terrain into four levels, as shown in Table 4:

Table4. Grade division of groundwater table

Classification	Characteristics of Geographic and geomorphic	The level of risk
Large negative terrain	It has strong water collecting capacity. In the underground karst system, the groundwater often has the characteristics of large amount of water and high pressure, and the positive feedback of karst development is strong	I
Medium negative terrain	The catchment capacity is the second, and karst development degree is the second	II
Small negative terrain	The capacity of water collection is poor, and the development degree of underground karst is generally poor	III
No negative terrain	The intensity of groundwater activity is low and generally speaking, the ability of water and mud inrush to cause disaster is very weak	IV

2.1.5 The attitude of rock formation

The development of rock dip is closely related to groundwater movement and karst development. The permeability of underground strata is anisotropic, and the permeability along the strata is much greater than that perpendicular to the strata. The flow pattern of groundwater, the degree of karst development and the distribution of karst water disasters are closely affected by the occurrence of rock strata. According to the analysis of statistical data, Li Liping[3-4] thought that the most favorable attitude of rock stratum for karst development is the wing of syncline or anticline with dip angle of 25 ° to 65 ° and divides the dip angle of rock stratum into four levels, but the dip angle is not continuous after the classification. Zhou Zongqing [21] and others corrected the dip angle of rock stratum, as shown in Table 5:

Table5. Relationships between strata inclination and risk grade of water inrush

dip angles of rock stratum	Correction value of rock dip angle	The level of risk
25°～ 45°, 45°～ 65°	25°～ 65°	I
10°～25°, 25°～ 80°	10°～ 25°	II
80°～ 90°	0°～ 10°	III
0°～ 10°	0°～ 10°	IV

2.1.6 The level of surrounding rock

The intact strength and mechanical properties of rock mass are closely related to the risk of water and mud inrush in tunnel. If the surrounding rock has good integrity, high strength and less harmful structural planes, it has strong deformation resistance and is not easy to cause water and mud inrush. The surrounding rock with poor integrity, low strength and more harmful structural planes has weak anti deformation ability, especially the weak zones (bodies) such as fault fracture zone and filling cavity often produce seepage instability or overall extrusion failure under the action of groundwater. If the construction is slightly improper, it can cause water and mud inrush and other disasters[3]. Hou Dongsai[23] studied the relationship between rock mass integrity coefficient (k_V), Poisson's ratio (P) and risk level of water inrush, as shown in Table 6:

Table6. Relationships among integrity coefficient of rockmass, Poisson's ratio and grade of water inrush

Kv	P	The level of risk
<0.37	<0.15	I
0.37-0.51	0.1-0.25	II
0.50-0.74	0.2-0.3	III
>0.74	0.3-0.45	IV

2.1.7 Rainfall

Surface water is an important recharge source of groundwater. In rainy season or when there is heavy rainfall, the surface water is replenished, the water quantity rises rapidly, and the water quantity increases. At the grounding end of pipeline flow, the risk of water inrush in tunnel will increase. At the same time, the surface water will be collected into the karst cave or underground river through the seepage of cracks, which will recharge the groundwater and increase the probability of water inrush. According to the average annual rainfall, the precipitation intensity is divided into four levels, as shown in Table 7:

Table7. Relationships between average annual rainfall and risk grade of water inrush

the average annual rainfall/mm	The level of risk
<400	I
400-800	II
800-1200	III
>1200	IV

2.2 Comprehensive evaluation on the risk of water and mud inrush in tunnel

According to the actual situation of the tunnel site and the summary of the influencing factors proposed in Section 2.1, the comprehensive evaluation table of the risk of water and mud inrush in the tunnel is obtained. The risk of the evaluation model is divided into four grades, and S is the corresponding expert rating. The criteria for determining different grades are shown in Table 8:

Table8. The factors and grading standard

	Influencing factors	The level of risk			
		I (80≤S≤100)	II (50≤S <80)	III (20≤S <50)	IV (0≤S<20)
Risk assessment of water and mud inrush in Karst Tunnel A	Geological structure B_1	strong disaster causing	medium disaster causing	weak disaster causing	no disaster causing
	Formation lithology B_2	strong soluble rock	medium soluble rock	weak soluble rock	No-soluble rock
	Groundwater level B_3	h ≥ 60 m	30 m ≤ h < 60 m	0 m ≤ h < 30 m	h < 0 m
	Geographic and geomorphic conditions B_4	large negative terrain	medium negative terrain	small negative terrain	no negative terrain

	The attitude of rock formation B_5	25° ~ 65°	10° ~ 25°	0° ~ 10°	0° ~ 10°
	The level of surrounding rock B_6	Kv<0.37 or P<0.15	0.37≤ Kv ≤ 0.51 or 0.1≤ P ≤0.25	0.50≤ Kv ≤0.74 or 0.2≤ P ≤0.3	Kv >0.74 or 0.3≤ P ≤0.45
	the average annual rainfall B_7	>1200mm	800mm - 1200mm	400mm - 800mm	<400 mm

2.3 The weight analysis of influencing factors

The weight analysis and risk assessment classification of risk factors for water inrush mud in karst tunnel are based on the analytic hierarchy process. Firstly, the judgment matrix P is constructed. Then, the factor weight vector ω, the maximum eigenvalue λ_{max} and the random consistency ratio CR are calculated respectively by formula (1) - (4).

$$\overline{\omega} = \left(\prod_{j=1}^{n} P_{ij}\right)^{1/n} \quad (i = 1,2,\dots, n) \tag{1}$$

$$\omega_i = \frac{\omega_i}{\sum_{j=1}^{n} w_j} \quad (i = 1,2,\dots, n) \tag{2}$$

$$\lambda_{max} = \left(\sum_{j=1}^{n}((A\omega)_i/\omega_i)\right)/n \tag{3}$$

$$CR = \{(\lambda_{max} - n)/(n-1)\}/RI \tag{4}$$

Where A is the judgment matrix, n is its order, RI is the average random consistency index value of the judgment matrix, and for the 1-9 order judgment matrix, RI value is: 0, 0, 0.58, 0.90, 1.12, 1.24, 1.32, 1.41, 1.45.

2.3.1 Construct judgment matrix

The judgment matrix represents the comparison of the relative importance between a certain factor at the upper level and its related factors at this level. Assuming that the factors of level A are related to B1, B2……Bn in the next level, the judgment matrix is generally in the following form:

A	B_1	B_2	……	B_n
B_1	b_{11}	b_{12}	……	b_{1n}
B_2	b_{21}	b_{22}	……	b_{2n}
……	……	……	……	……
B_n	b_{31}	b_{n2}	……	b_{nn}

b_{11}, … … ,b_{nn} is the relative importance of each evaluation factor, which can be determined by the 1-9 scale method (Table 9) [20] ,and through the comparison between the two factors, the value is determined.

Table9. The significance of 1-9 scale

Intensity of Importance	sense
1	It means that two factors are equally important
3	One factor is moderately more important than the other
5	One factor is strongly more important than the other
7	One factor is very strongly more important than the other
9	One factor is extremely strongly more important than the other
2,4,6,8	The median value of the above adjacent judgment
Reciprocal of above	If b_{ij} is obtained by comparing factor i with j, then $b_{ij} = 1 / b_{ij}$ by comparing factor j with i

2.3.2 Weight analysis of each influencing factor

Based on the statistics of engineering examples and theoretical analysis, and referring to the existing research results, the risk evaluation matrix of water and mud inrush in karst tunnels is constructed:

$$P_{A\text{-}B} = \begin{bmatrix} 1 & 2 & 2 & 4 & 6 & 7 & 7 \\ 1/2 & 1 & 1 & 2 & 3 & 4 & 4 \\ 1/2 & 1 & 1 & 2 & 3 & 4 & 4 \\ 1/4 & 1/2 & 1/2 & 1 & 2 & 3 & 3 \\ 1/6 & 1/3 & 1/3 & 1/2 & 1 & 2 & 2 \\ 1/7 & 1/4 & 1/4 & 1/3 & 1/2 & 1 & 1 \\ 1/7 & 1/4 & 1/4 & 1/3 & 1/2 & 1 & 1 \end{bmatrix}$$

According to formula (1) - (4), the matrix weight vector $\omega_{A\text{-}B}$ = [0.364, 0.189, 0.189, 0.111, 0.062, 0.043, 0.043], the maximum eigenvalue λ_{max} = 7.061, the consistency index CI = 0.010, C_r = 0.008 < 0.1, which meets the consistency condition. From the calculation results, it can be seen that geological structure, formation lithology and groundwater level have the greatest influence on water inrush, and the sum of weights accounts for 73.7% of all factors, followed by geographic and geomorphic conditions and the attitude of rock formation, with the sum of weights accounting for 17.7% of all factors, and finally the level of surrounding rock and rainfall, with the sum of weights accounting for 8.6% of all factors.

3 Risk assessment of water or mud inrush of a tunnel in Chongqing

3.1 Risk assessment

A tunnel in Chongqing is a separate and independent two-hole tunnel. The left tunnel is 7635.00m long and the right tunnel is 7656.00m long. The maximum buried depth is 1296.50m. It is a deep buried extra long mountain tunnel. The tunnel site belongs to the high and middle mountain landform of structural dissolution erosion, and there are three faults, four anticlines and three synclines in the area, with strong tectonic compression and broken rock mass. The strata through the tunnel are composed of the lower Triassic Jialingjiang Formation, Daye Formation, Permian, Silurian and Ordovician strata. The lithology is mainly composed of limestone, shale, siltstone and so on. The rocks are soft rock ~ hard rock, and The level of surrounding rocks are V ~ III. The groundwater in the tunnel site is mainly carbonate rock fissured karst water and fissured karst water. Due to the uneven development of karst, the distribution and enrichment of groundwater are also uneven. Therefore, concentrated water gushing may occur in the local karst pipeline development area during construction.

Therefore, the stratum lithology and geological structure of the tunnel are complex, and the tunnel body mainly passes through the soluble rock section. There are rich water and ponding in the fault, the contact zone between soluble rock and non soluble rock, and the syncline core where the tunnel site is located. During the tunnel construction, the probability of encountering large karst caves, underground rivers, dissolution funnels and other hidden karst forms is high, which is very easy to induce water and mud inrush in the tunnel. Therefore, it is necessary to strengthen the prediction of water and mud inrush disaster and take measures in advance to ensure the construction safety.

The tunnel is divided into four sections according to the mileage pile number, and the influencing factors of water and mud inrush risk in each section are analyzed and evaluated respectively, as shown in Table 10.

Table10. Evaluation index analysis of each section

	Geological structure B1	Formation lithology B2	Groundwater level B3	Geographic and geomorphic conditions B4	The attitude of rock formation B5	The level of surrounding rock B6	the average annual rainfall B7
K46+495～K50+300	strong disaster causing	strong soluble rock	$h \geq 60$ m	large negative terrain	35°～60°	Kv=0.30 P=0.25	1261.4mm
K50+300～K53+050	strong disaster causing	strong soluble rock	$h \geq 60$ m	large negative terrain	25°～77°	Kv=0.30 P=0.25	1261.4mm
K53+050～K53+725	weak disaster causing	weak soluble rock	$30\ m \leq h < 60\ m$	medium negative terrain	35°～45°	Kv=0.64 P=0.3	1261.4mm
K53+725～K54+150	strong disaster causing	medium soluble rock	$30\ m \leq h < 60\ m$	large negative terrain	35°～68°	Kv=0.5 P=0.21	1261.4mm

The risk classification of water and mud inrush in tunnel can be obtained by summation of the product of expert score and factor weight (formula 6). The form of score vector is shown in formula (5):

$$F = [f_1\ f_2\ \cdots f_i \cdots f_n] \tag{5}$$

$$S = F \cdot w \tag{6}$$

Fig. 1. Side gushing water

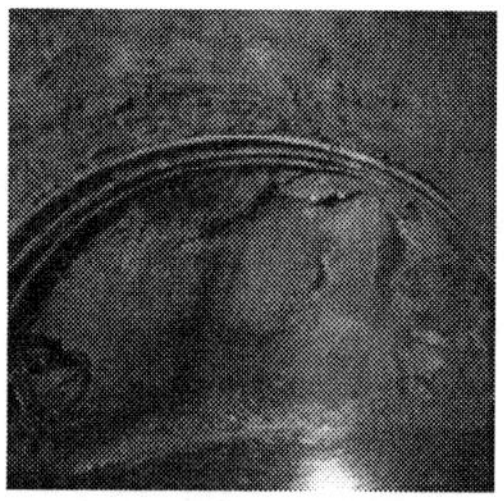

Fig. 2. Muddy water gushes from top

Fig. 3. A lot of muddy water

Fig. 4. Water gushes out of the excavation face

The expert score vector of K46 + 495-K50 + 300 section is f = [87 95 90 93 78 85], which can be substituted into equation (6) to get *S*= 89.33, and the risk possibility of water and mud inrush is grade I. The section passes through two faults, and the stratum lithology is mainly limestone and dolomite. The proposed tunnel is located in the seasonal variation zone and vertical circulation zone. The karst development is strong on both sides of the fault zone and planation surface, mainly manifested as high level horizontal karst cave, karst trough, depression and sinkhole, and deep karst development. Large water and mud gushing may occur in the groundwater enrichment section under the condition of strong karstification and rainstorm.

The expert score vector of K50 + 300 ~ K53 + 050 section is f = [95 92 94 95 91 78 85], which can be substituted into equation (6) to get *S*= 92.83. The risk possibility of water and mud inrush is grade I. The core of the Syncline in this section is developed with tension fissures, which cross faults and have the condition of developing underground river. The lithology is mainly limestone and dolomite. The proposed tunnel is located in the seasonal variation zone and vertical circulation zone. The karst development has a long history, and the karst forms are diverse. The depressions and water holes on the high-level planation plane are developed with a large scale. In case of strong karst development and rainstorm, large water and mud gushing may occur in the section of groundwater enrichment.

The expert score vector of K53 + 050 ~ K53 + 725 section is f = [35 33 65 70 93 32 85], which can be substituted into equation (6) to get S= 50.02. The risk possibility of water and mud inrush is grade III. This section is a one-way slope with developed surface longitudinal gullies. The lithology of the stratum is mainly shale and siltstone, which is a regional relative aquifuge and a weak water rich rock formation. The possibility of large water and mud inrush is small.

The expert score vector of K53 + 725 ~ K54 + 150 section is f = [85 70 65 87 90 45 85], which can be substituted into equation (6) to get S= 77.28. The risk possibility of water and mud inrush is grade II. The fold core and other geological structures in this section are developed, the rock mass is broken, and the lithology is mainly limestone. The water content of the water bearing rock group is extremely uneven, and the water yield is quite different. Water inrush and mud inrush may occur in rainy season.

3.2 The validation

There are many water and mud inrush accidents of different scales occurred (Fig.1~Fig.4) during the excavation and construction of the tunnel. For example, when the tunnel was repaired to K48+340-K48+352.8, mud inrush and water inrush occurred at the arch waist on the right side of the tunnel entrance, with a square amount of about 400m^3. After that, mud inrush occurred again, resulting in extrusion deformation of I-beam and water inrush with pressure. When the construction length of the hand face is K48+617, there is water seepage in the deepened gun hole of the left vault, and the water pressure is small. When the construction reaches YK48+622, there is water gushing in the left vault, the aperture is about 0.5-0.7 meters, and the water inflow is 3000m^3/h. When the face of the hand is excavated to K51+081, water gushing appears in the left hole of the exit. The jet distance is as long as 30m, and the water inflow reaches 410m^3/h. Many water and mud inrush accidents occurred during inclined shaft construction. The evaluation results of water and mud inrush in tunnel are in good agreement with the field excavation results, which verifies the rationality and feasibility of the risk evaluation method of water and mud inrush in tunnel based on hierarchical method.

4 Conclusion

(1) Based on statistics and analytic hierarchy process (AHP), the influencing factors and factor weights of water and mud inrush in karst tunnel are analyzed. The results show that the main controlling factors of water and mud inrush are geological structure, formation lithology and groundwater level; geographic and geomorphic conditions, attitude of rock formation are the minor significant factors; surrounding rock mass classification, rainfall are the least important elements.

(2) The risk of water inrush and mud inrush can be quantified by the product of expert score and factor weight, which can be used to grade the risk of water inrush in karst tunnels under different hydrogeological conditions or in different sections of karst tunnels.

(3) The risk assessment method proposed in this paper is used to evaluate and classify the risk of water and mud inrush in each section of a tunnel in Chongqing, and the evaluation results are consistent with the actual excavation situation. Therefore, this method can provide a theoretical basis for the project design and construction organization

https://doi.org/10.1051/e3sconf/202129302005

design of the tunnel to reduce the risk of water and mud inrush.

References

1. Li Shu-cai, Xu Zhen-hao, Huang Xin, et al.CHIN J ROCK MECH ENG, **37 (5)**, 1041-1069(2018)
2. Hou Dongsai, Zhang Xiao, Wang Lei. TUNNEL CONSTR, **37(6)**, 691-699(2017)
3. Xu Zhenhao, Li Shucai, Li Liping,et al. ROCK SOIL MECH, **32(6)**, 1757–1765(2011)
4. Li Liping, Li Shucai, Chen Jun, et al. CHIN J ROCK MECH ENG, **30(7)**, 1345–1354(2011)
5. Zhou Zong-qing, Li Shu-cai, Li Li-ping, et al. ROCK SOIL MECH, **34(3)**, 818-826(2013)
6. Li Shucai, Zhou Zongqing, Li Liping, et al.CHIN J ROCK MECH ENG, **32(9)**, 1858-1867(2013)
7. Mao Bang-yan, Xu Mo, Jiang Liang-wen. CARSOL SIN, **29(2)**, 183—189(2010)
8. Han Xing-rui. CARSOL SIN, **23(3)**,213-218(2004)
9. Du Yu-chao, Han Xing-rui, Li Zhao-lin.CARSOL SIN, **28(3)**, 281—287(2009)
10. Zhang Qing-song, Li Shu-cai, Han Hong-wei, et al.J SHANDONG UNIV (ENG SCI), **39(3)**, 106—110(2009)
11. Kuang Xing, Bai Ming-zhou, Wang Cheng-liang, et al. J HIGHW TRANSP RES DEV, **27(11)**, 100—103(2010)
12. Yang Zhuo, Ma Chao. TUNNEL CONSTR, **36 (11)**, 1337-1342(2016)
13. Zeng Zhifa, Song Bo, Li Tianbin, et al. J CHINA FOREIGN HIGHW, **31(4)**, 213-216(2011)
14. Jiang Guoyun.CHIN J UNDERGROUND SPACE ENG, **8 (2)**, 274-279(2012)
15. Shu Sen.RAILW STANDARD DESIGN, **59(4)**, 72-78(2015)
16. Mao Zhengjun, Yang Shaozhan, Zhu Yanyan, et al.J RAILW SCI ENG, **14 (6)**, 1332-1339(2017)
17. Liu Dongbo.RAILW CONSTR TECHNOL, **08**, 36-39+69(2018)
18. Yang Yanna, Cao Huaping, Xu Mo. GEOSCI, **29(2)**, 414-420(2015)
19. Wang Deming. SHANDONG UNIV(2017)
20. SAATY T L. EUR I OPER RES, **48(1)**, 9–26(1990)

Research on the Structural Design of Academic Report Complex Building in a University

Haibo Ren[1]

[1]China Architecture Design & Research Group, Beijing 100044, China

Abstract. This project is the construction project of the academic report comprehensive building of the Central Youth League School. The main function of the project is the academic report room, including an 800-person lecture hall, a 300-person multi-function lecture hall, a 500-person multi-function lecture hall, an exhibition hall, etc., and three underground floors. There are three floors above the ground, and one floor is partially high. There are 12.8, 13.8, 15.7, 18.2, 18.6, 20.8, 27.8 meters in the sub-regions of the roof structure elevation. This article introduces the structural system, the judgment of structural out-of-limit situations, prestress design, steel truss design, suspension steel structure design, cantilevered large-span concrete beams, etc., through a variety of structural calculation and analysis methods, and find out structural weaknesses, and adopt Corresponding seismic strengthening measures, the calculation results show that the structure has good seismic performance and can meet the requirements of the code.

1 Project Overview

This project is the construction project of the academic report comprehensive building of the Central Youth League School. The total construction area is about 31,500 square meters, the above ground construction area is about 10,500 square meters, and the underground construction area is about 21,000 square meters. The main function of the project is the academic report room, including 800 people. The lecture hall, the multi-functional lecture hall for 300 people, the multi-functional lecture hall for 500 people, the exhibition hall, etc., have three floors underground, three floors above ground, and part of the first floor is full-height. Basic wind pressure: w0=0.45kN/m2 (recurrence period of 50 years); Fortification intensity: The seismic fortification intensity of this project is 8 degrees, the design basic earthquake acceleration value is 0.20g, and the design earthquake is grouped into the second group. The soil category of the construction site is Class II, and the characteristic period value is 0.40s. Building seismic fortification category: key fortification category (category B); design service life: 50 years, safety level of building structure: second class. The effect diagram and structure model are shown in Figure 1, Figure 2.

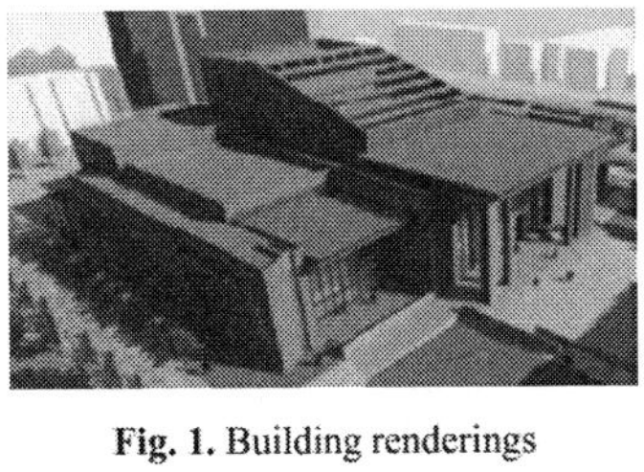

Fig. 1. Building renderings

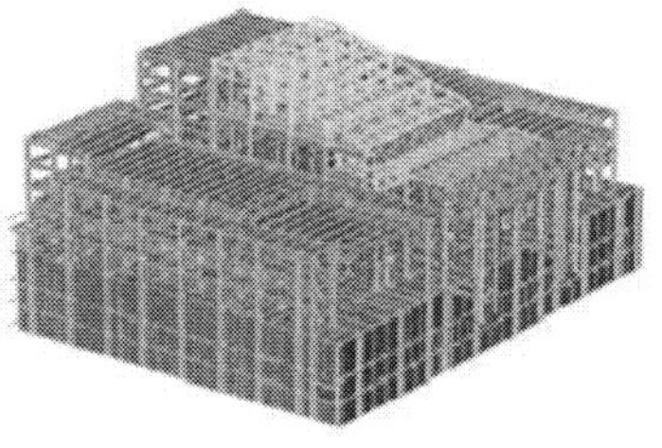

Fig. 2. Structural model

2 Structural system

2.1 Structure selection

In the structural design, the architectural function and plan layout should follow the simple, regular and symmetrical structural layout principles as far as possible in the structural plan and vertical layout. The main structure adopts the form of reinforced concrete frame structure with few walls. The 800-person lecture hall adopts steel truss structure. The truss is connected to the main structure with hinged supports, and the large-span space adopts a prestressed concrete beam structure. The construction length of this project is super long. In order to ensure the use function and the effect of the building facade, expansion joints are not set on the ground and underground, and measures are taken to ensure that the structural cracks under temperature stress meet the specification requirements.

* Corresponding author: renhb@cadg.cn

2.2 Seismic level of components

The seismic fortification intensity of this project is 8 degrees, the building seismic fortification category is the key fortification category, and the structure type is frame structure. The seismic fortification level is increased by one degree. The results are as follows[1]:

Table 1. Earthquake resistance rating.

Structural part	Seismic rating
-3-story frame	Level 3
-3-story frame	Level 2
-3-story frame	Level 1
Above-ground frame	Level 1

2.3 Calculation problems of single-story through-height columns

The software can automatically determine the one-way spanning column, and calculate the reinforcement according to the appropriate calculated length coefficient and column length. But for the overhanging beam on the column, the software judges it to be the constraint of the frame column and judges it as the floor. This is unreasonable. You should manually modify the calculated length coefficient to ensure the accuracy of the calculation.

The design analysis of the story-through high-column is based on three considerations, and the envelope is used for reinforcement. (1) For calculations under frequent earthquakes in the overall model, the seismic resistance level of the seismic structural measures should be increased by one level, according to the second level; (2) Take the single model, input the load effect, optimize the calculation length coefficient, and use the PK software to check; (3) In the overall model, use moderate earthquake elasticity to check column reinforcement.

2.4 Judgment of structure

According to the 2015 edition of the judging standards for the technical points of the special review of seismic fortification of over-limit high-rise buildings, the height of this project is not over-limit, but there are the following irregularities[2]:

(1) Reverse irregularities. Considering accidental eccentricity, the torsional displacement ratio is 1.37, which is greater than 1.2;

(2) Eccentric layout. The eccentricity is greater than 0.15;

(3) The floor is not continuous. The opening area is greater than 30%, and the staggered level is greater than the beam height.

(4) Intermittent components. There are many conversion columns, but the conversion column is converted on the embedded layer of the main structure (first floor ground), so this irregularity can not be counted.

The above two irregularities (1) and (2) belong to the same category and will not be counted repeatedly. Based on the above judgment, this project belongs to two irregular structures and is not an over-limit project.

2.5 The overall calculation results of the structure

The overall calculation results are shown in Table 2:

Table 2. Main calculation results

		YJK	Remarks
Vibration shape (s)	First cycle	0.8225 (Y)	< 0.85
	Second cycle	0.7687 (X)	
	Third cycle	0.6697 (T)	
	Cycle ratio	0.814	
Layer displacement angle	One-way earthquake in X direction	1/626 (n=5)	< 1/550
	One-way earthquake in Y direction	1/595 (n=5)	
Ratio of maximum displacement to average displacement (considering 5% accidental eccentricity)	X direction earthquake	1.21	Prescribed horizontal force
	Y direction earthquake	1.31	
Base shear force (kN) (shear weight ratio)	One-way earthquake in X direction	37204.74(5.602%)	X≥3.20% Y≥3.20%
	One-way earthquake in Y direction	38333.65(5.772%)	
Stiffness to weight ratio	X direction earthquake	52.879	>10
	Y direction earthquake	46.394	>10

3 Important and difficult points of structural design

3.1 Slow bonding technology and its application in large-span prestressed beams

The slow-bonding prestressed tendons are composed of prestressed tendons, slow gel adhesive, and outer protective surface layer. In the early stage, the slow-adhesive adhesive is equivalent to the anti-corrosion grease of the unbonded prestress, which has certain fluidity and good adhesion to the prestressed tendons. With the passage of time, after the cementitious material gradually solidifies, it is closely occluded with the concrete, and the prestressed tendons cannot slide freely in the concrete. The slow bonding prestress produces the bonding effect of the bonded prestress, good ductility and seismic performance, and at the same time has the advantages of convenient construction of the unbonded prestress, and overcomes the disadvantages of the complicated construction process of the bonded prestress.

Due to the needs of building functions, there are a large number of large-span beams (greater than 15 meters) in this project, and the effect of using prestress is obvious. After setting the prestress, the deflection and cracks meet the requirements of the specification. The typical span of the prestressed beam is: 19.5 m, 16.8m, etc., the section size of the 19.5m large-span prestressed beam is 450mmx1150mm, the deflection control limit is L/300, and the crack limit is controlled according to 0.2mm. The main layout positions are as shown in Figure 3:

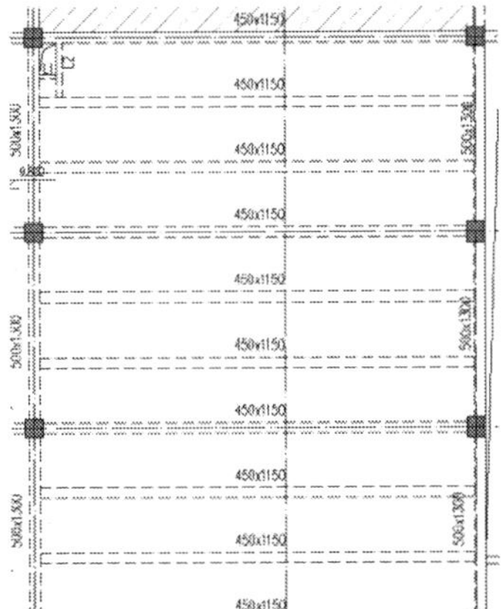

Fig. 3. Layout of large-span prestressed beams

3.2 Steel truss and hanging steel structure

The main structure is a concrete frame structure. The roof of the 800-seat lecture hall adopts a steel truss structure with a span of about 30m. It adopts a deck-type flat steel truss, and the support is a fixed hinged steel support. The truss is equipped with horizontal supports, tie rods and vertical supports. The roof slab is a steel truss floor deck, and the lower chord layer has tie rods, horse road hangers, etc. The splicing of steel components adopts the method of rigid connection[3]. The main truss is considered based on 1/500 of the arch, and the layout of the steel frame is shown in Figure 4 above.

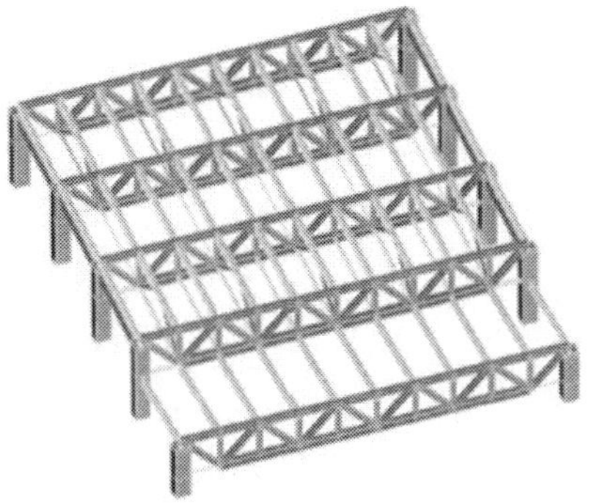

Fig. 4. Layout of steel truss roof structure

The truss members are all made of hot-rolled H-section steel, and the stress ratio of the members is controlled below 0.9. The calculation results of the steel truss are shown in Figure 5. The steel truss roof has a hanging steel structure on the lower chord of the C and D axis steel truss. The steel structure shall consider the construction sequence and the lifting sequence during calculation and analysis.

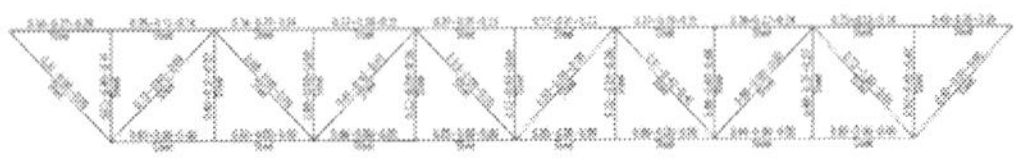

Fig. 5. Calculation results of steel truss

3.3 Stand cantilever beam

The entrance hall requires column-free space, which leads to the oblique cantilever beam of the grandstand above the auditorium. The length of the horizontal cantilever is 8.4m, and the beam section is 600x1300/800, which extends inward for one span. It adopts slow bonding prestress and strict control. The deflection and cracks are calculated to meet the requirements of the specification, and the cantilever beam layout is shown in Figure 6.

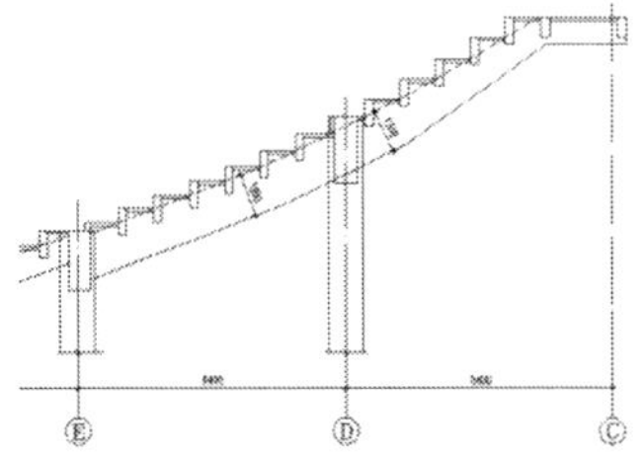

Fig. 6. The cantilever beam of the grandstand

3.4 The frame beams of the auditorium are haunched

The size of the frame column surrounding the large space is 900x1100, and the frame beam width is 400. Due to the wall layout requirements of the building, the frame beam is attached to the outer skin of the column, and the eccentricity of the beam column does not meet the requirements of Article 6.1.7 of the "Technical Regulations for High-rise Building Concrete Structures", see Figure 7. Shown. The frame beam needs to be haunched. After calculation, the horizontal haunched layout is shown in Figure 8.

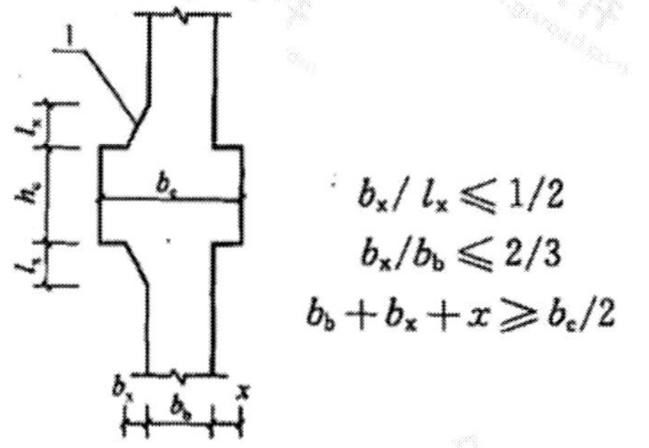

Fig. 7. beam level haunching requirements

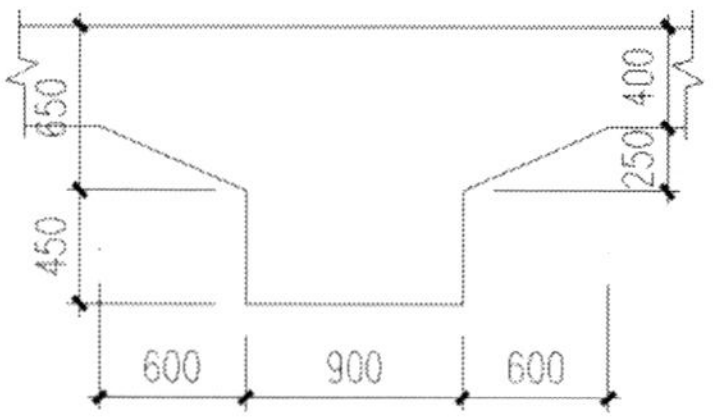

Fig. 8. Beam horizontal haunching situation

4 Conclusion

This project is a public building with complex functions and spatial relationships. In the conceptual design stage, the overall structural system and layout are carefully considered and appropriately optimized to make it have good structural performance. Through the overall calculation, the weak parts of the structure are found, and corresponding seismic strengthening measures are taken. The analysis shows that the structure has good seismic performance, and many indicators perform relatively well, which can meet the requirements of the specification. In engineering, slow bonding technology is used for large-span beams and cantilever beams to effectively control deflection and cracks. The steel truss members are carefully analyzed to obtain the optimal member cross section. The hanging structure should be set in the calculation Good construction sequence, and implement it during actual construction.

References

1. GB 50011-2010 Code for Seismic Design of Buildings [S]. Beijing: China Building Industry Press, 2010
2. JGJ 3-2010 Technical Specification for High-rise Building Concrete Structure [S]. Beijing: China Building Industry Press, 2010
3. GB 50017-2017 Steel Structure Design Standard [S]. Beijing: China Building Industry Press, 2018
4. Lin Chaowei, Mao Chaojiang The simulation analysis of the construction of super high-rise suspended steel structure in Shenwan Huiyun Center Journal of Jiangsu Institute of Architecture and Technology[J]. 2018.18(4): 13-18
5. Wang Fengzhe Multi-layer suspension steel structure construction technology Building construction [J]. 37 (12): 1406-1408

Coupling and Coordination Analysis of Comprehensive Benefits of Water Conservancy Scenic: Taking Qingpu District of Shanghai as an Example

Liu Xianyu [1], Yang Xuejun [2,*], and Pan Jun [3]

[1]Tongji University, Institute of New Rural Development, 201804, Shanghai, China

[2]Tongji University, Institute of New Rural Development, 201804, Shanghai, China

[3]Tongji University, Institute of New Rural Development, 201804, Shanghai, China

Abstract. Based on the domestic comprehensive evaluation methods for water conservancy scenic spots, water environment, and water resources, this paper constructs a comprehensive benefits evaluation index system for water conservancy scenic from both ecological and social benefits. Then, it takes the Qingpu District of Shanghai as the research object, using the entropy method to calculate the comprehensive benefits index of the two major subsystems of ecology and society from 2009 to 2019. Based on that, it concludes that the overall coupling and coordination level of the ecological and social benefits of water conservancy scenic slowly rises at this stage but remains the low-level coupling coordination stage. Finally, from the perspective of improving the comprehensive benefits of water conservancy scenic, it puts forwards the development of Qingpu District towards the beautiful prospect of the harmonious co-existence between human and nature.

1 Introduction

According to the "Administrative Measures for Water Conservancy Scenic Areas" issued in 2004, water conservancy landscape resources refer to waters (water bodies) and associated shores, islands, forests and grasses, buildings, and other natural and cultural landscapes that are attractive to people. Through scientific and reasonable manual intervention to achieve the optimization of comprehensive benefits, areas with characteristic water conservancy and scenic resources background conditions, can better play the potential value in the construction of ecological civilization, and transform towards the goal of "the harmony between man and water".

2 Overview of the research area

Qingpu District is located in the southwest of Shanghai, as well as the lower reaches of Lake Tai and the upper reaches of River Huangpu. The territory is flat with dense rivers and a superior ecological environment. The 21 natural lakes of Shanghai, including Lake Dianshan, are located in this district(Fig.1). The water source protection area of the upper reaches of the River Huangpu and the Jinze Reservoir are important sources of drinking water in Shanghai. In addition, there are countryside ecological corridors such as the Qingxi Country Park, the Lake Lanlu Forest, and the River Taipu Forest. They are important ecological spaces and ecology barriers in Shanghai. Among them, the Lake Dianshan area is rich in water conservancy scenery resources and has a unique location advantage. As early as 2006, it was selected as one of the sixth batches of national *Water Parks*.

Qingpu is rich in surface water resources, but due to its low-lying terrain, especially the water net polders located in the central and southern low-lying hinterlands of Qingpu, the water level is often above the water level during the flood season. At the same time, the contradiction between the imbalance of water resource allocation and the demand for water resources is prominent, and the contradiction between water quality-type water shortage and sustainable development is prominent[1]. Therefore, how to make use of water resources after rational use of water resources has a certain research value.

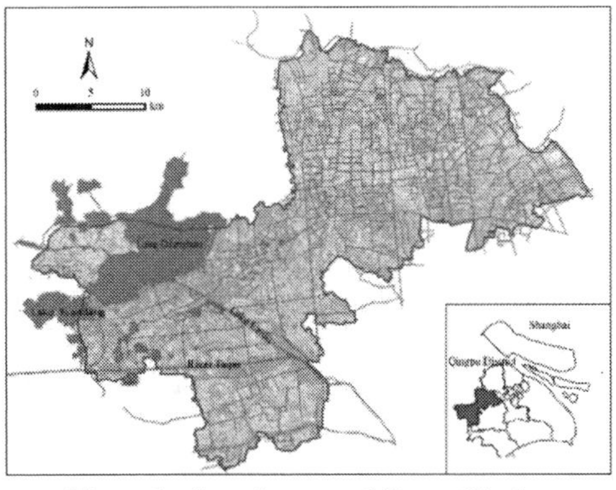

Figure 1. Location map of Qingpu District.

* Corresponding author: espi@tongji.edu.cn

https://doi.org/10.1051/e3sconf/202129302007

3 Data processing and calculation

3.1 Data source

To ensure the completeness and accuracy of statistical data, we choose years from 2009 to 2019 as the research period. The data mainly comes from *Shanghai Statistical Yearbook 2010-2020*, *Qingpu District Statistical Yearbook*, *Qingpu District Ecological Environment Status Bulletin*, relevant information publicly released from the Water Affairs Bureau, and Ecological Environment Bureau of Qingpu District. The missing data in individual years is supplemented by Lagrange Interpolation Polynomial.

3.2 Establishment of evaluation index system

3.2.1 Reference of related evaluation methods

When most domestic scholars evaluate water conservancy scenic, they mainly refer to the"Water Conservancy Scenic Area Evaluation Standards"(SL300-2013). Taking a specific water conservancy area as the research object, they have a comprehensive evaluation from the aspects of landscape resources, development and utilization conditions, environmental protection evaluation, and management [2]. In the selection of evaluation methods, most scholars use a combination of qualitative and quantitative methods of on-site investigation and questionnaire interviews based on collecting and collating literature data [3]. For example, Wang Li[4] uses the Analytic Hierarchy Process to select evaluation indicators from the five aspects of viewing, ecology, environment, culture, and science according to the characteristics of water conservancy landscape resources in Guangxi, to construct water conservancy landscape resource value evaluation model. Meanwhile, she takes River Chengbi Water Conservancy Scenic Spot as an example for evaluation.

Coupling originally refers to the association and coordination of energy transmission between two or more circuits in physics, and then is gradually used to describe the degree of mutual influence and interaction between two or more systems. In recent years, many scholars have applied coupling analysis to the empirical study of water environment and social development such as urbanization [5-6]. In addition to the state of the water environment, some indicators that reflect the state of the ecological environment are also water conservancy to some extent. The embodiment of natural ecological benefits of landscape resources also has reference value. The social benefits of water conservancy scenery include not only the intuitive changes after the construction of water conservancy projects and the improvement of water transportation and transportation conditions, but also the invisible promotion of the development of agriculture, cultural tourism, and other industries. Especially in Qingpu District, water conservancy is the lifeblood of agriculture, and water culture also constitutes the shape of Qingpu's tourism resources, laying the foundation for the urban development of Qingpu District as "Shangshan City".

3.2.2 Construction of Comprehensive Benefit Evaluation Index System

Follow the principles of systematicity, representativeness, and quantification, this paper selects 19 indicators to construct an evaluation index system for the comprehensive benefits of water conservancy and landscape resources from both ecological and social benefits. Ecological benefits are composed of three aspects: comprehensive environmental status, water environment quality, and greening environment. Social benefits are composed of four aspects: cultural tourism, water conservancy project construction, water transportation, and agricultural development.

Table 1. Comprehensive benefits evaluation index system.

Serial number	Target layer	System layer	Index layer			
			First level indicator	Secondary indicator	Indicator attributes	Entropy weight
1	Overall benefits of water conservancy scenic	Ecological benefits	Comprehensive environmental status	Biggest water level change	minus	0.0837
2				Flood season accounts for annual precipitation	minus	0.0784
3				Air quality index good rate	plus	0.0711
4	Overall benefits of water conservancy scenic	Ecological benefits	Water environmental quality	Water environmental quality	minus	0.0801
5				Comprehensive pollution index of main rivers	minus	0.1145
6				Comprehensive trophic state index of Dianshan Lake	minus	0.0752

Serial number	**Target layer**	**System layer**	**Index layer**			
			First level indicator	**Secondary indicator**	**Indicator attributes**	**Entropy weight**
7				Average industrial wastewater discharge	minus	0.1171
8				Urban sewage treatment rate	plus	0.1991
9			Green environment	Green area rate	plus	0.0919
10				Water conservation forest area	plus	0.0889
11		Social benefits	Cultural tourism	Receive tourist visits throughout the year	plus	0.0800
12				Number of visitors to cultural science exhibitions	plus	0.2674
13			Water conservancy project construction	The population of water conservancy dykes	plus	0.1191
14				Green coverage rate of embankment	plus	0.0706
15				The proportion of the area with guaranteed harvest during droughts and floods in the effective irrigation area	plus	0.0786
16			Water transportation	Water transport volume	plus	0.0816
17				Water transport turnover	plus	0.0792
18			Agricultural development	Per capita gross agricultural product	plus	0.1210
19				Production of aquatic products per aquaculture area	plus	0.1025

3.3 Entropy method data processing

3.3.1 Standardized processing

Since the unit, magnitude, and attribute of the original value of each indicator in the evaluation system are different, in order to eliminate the influence of the above factors on the evaluation result, the original value of each indicator is first standardized according to the normalization method. In other words, the absolute value of the original data of each indicator is converted into a relative value through standardization, so that different indicators are more comparable. Positive indicators indicate that larger indicators are more conducive to the coupling between ecological benefits and social benefits of water conservancy scenic, while reverse indicators indicate that larger indicators are less conducive to the coupling between comprehensive benefits. Namely:

Positive index（"+"）:

$$\mathbf{Y}_{ij} = \frac{\mathbf{X}_{ij} - \min \mathbf{X}_{ij}}{\max \mathbf{X}_{ij} - \min \mathbf{X}_{ij}} \tag{1}$$

Negative index（"-"）:

$$\mathbf{Y}_{ij} = \frac{\max \mathbf{X}_{ij} - \mathbf{X}_{ij}}{\max \mathbf{X}_{ij} - \min \mathbf{X}_{ij}} \tag{2}$$

$\mathbf{Y}_{ij}$ is the standardized value of the index item in the year; $\mathbf{X}_{ij}$ is the original value of the index item in the year, $\max \mathbf{X}_{ij}$ and $\min \mathbf{X}_{ij}$ are the maximum and minimum values in the original value of the index item in the year (i is the year; j is the quantity of each system index.)

3.3.2 Setting of indicator weight

Since the natural logarithm is needed to calculate the information entropy, the index value cannot be 0, so the translation processing of 0.01 is required, that is, let $\mathbf{H} = \mathbf{Y} + d$, where d is a positive number that makes it slightly greater than 0, and 0.01 is here.

$$\mathbf{P}_{ij} = \frac{\mathbf{H}_{ij}}{\sum_{i=1}^{n、m} \mathbf{H}_{ij}} \quad (0 \le \mathbf{P}_{ij} \le 1) \tag{3}$$

3.3.3 Entropy calculation

Calculate the entropy value of the jth index.

$$\mathbf{E}_j = -\frac{1}{\ln n} \sum_{i=1}^{n、m} \mathbf{P}_{ij} \ln(\mathbf{P}_{ij}) \, (0 \le \mathbf{E}_j \le 1) \tag{4}$$

3.3.4 Calculation of difference coefficient

Directly subtract the entropy value of this indicator from 1 to get the coefficient of difference.

$$\mathbf{G}_j = 1 - \mathbf{E}_j \tag{5}$$

3.3.5 Calculate the weight of the jth indicator in each system

$$\mathbf{W}_j = \frac{\mathrm{G}j}{\sum_{i=1}^{m} \mathrm{G}j} \tag{6}$$

3.3.6 Comprehensive evaluation index calculation

Multiply the obtained index weight and the ratio of the data of the jth year on the evaluation index to obtain the comprehensive score of each evaluation index(Tab.1).

$$\mathbf{U}_{ij} = \sum_{i=1}^{n} \mathbf{W}_j \mathbf{P}_{ij} \tag{7}$$

3.4 Calculation of coupling degree and coupling coordination degree

In order to measure and evaluate the degree of coordinated development of ecological and social systems in the comprehensive benefits of water conservancy scenic, the coupling degree and coupling coordination degree are introduced. The calculation formula is:

$$\mathbf{C} = \left[\frac{\mathbf{U}_1 \times \mathbf{U}_2}{\left(\frac{\mathbf{U}_1 + \mathbf{U}_2}{2}\right)^2} \right]^{1/2} \tag{8}$$

$\mathbf{C}$ is the coupling degree, $\mathbf{C}_1$ and $\mathbf{C}_2$ are the comprehensive coordination indexes of ecological and social subsystems respectively. $\mathbf{C} \in [0,1]$, the larger the value of $\mathbf{C}$, the stronger the coupling effect between systems is. Although the degree of coupling can quantitatively analyze the degree of mutual influence between the various subsystems, it cannot reflect the comprehensive development level of the research system, and it is difficult to measure the overall effectiveness of the research system [7]. Therefore, a coupling coordination degree model must be constructed. The calculation formula is:

$$\mathbf{D} = \sqrt{\mathbf{C} \times \mathbf{T}}, \quad \mathbf{T} = \alpha \mathbf{U}_1 + \beta \mathbf{U}_2 \tag{9}$$

$\mathbf{D}$ is the coupling coordination degree and the value range is [0,1]. α and β are undetermined parameters, and the sum is 1. Since the ecological and social aspects of the comprehensive benefits of water conservancy and landscape have the same status, both α and β are taken as 1/2 in this study.

4 Result analysis

4.1 Evolution characteristics of the comprehensive development index of water conservancy scenic

4.1.1 Ecological benefits

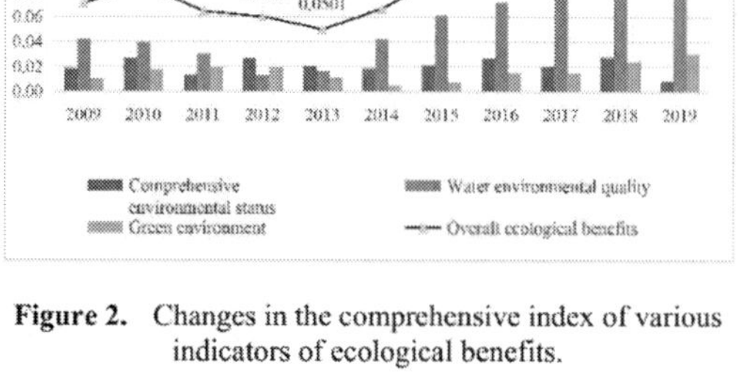

Figure 2. Changes in the comprehensive index of various indicators of ecological benefits.

It can be seen from Fig.2 that since 2014, the comprehensive index of water environment quality in Qingpu District has increased year by year, driving the increase in the overall index of the ecological benefit system. The comprehensive index of the greening environment has experienced two stages——decrease in the early stage and then increase, and it shows an increasing change year by year after 2014. There is no obvious change rule in the comprehensive environmental quality comprehensive index, indicating that the state of the natural environment is greatly affected by nature.

In order to fully realize the water environment governance goal of "clear water, green shores, smoothness, beautiful scenery, and ecology", in recent years, Qingpu District has taken advantage of the construction of the national water ecological civilization city to vigorously rectify and protect the water environment and promote the River and Lake System in an all-round way. Meanwhile, it pays more attention to water conservation, safety protection, and greening construction. On the whole, the environmental quality is showing an improvement trend, and the effect of water control and district development is gradually showing, but the overall level is still not high.

4.1.2 Social benefits

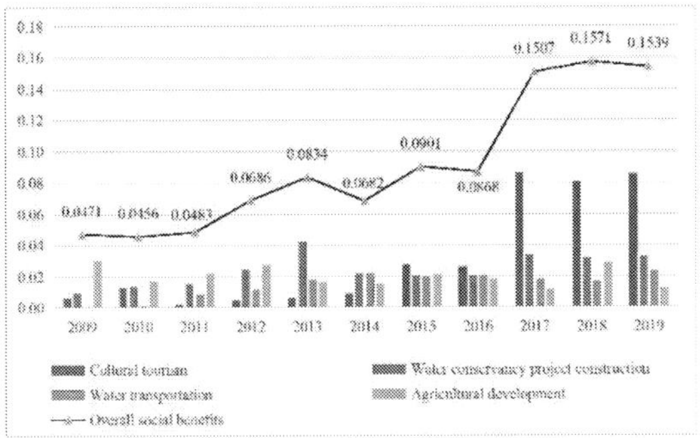

Figure 3. Changes in the comprehensive index of various indicators of social benefits.

It can be seen from Fig.3 that the social benefits of the water conservancy scenery in Qingpu District are showing an upward trend. From 2009 to 2019, cultural tourism, water conservancy project construction, and water transportation have developed rapidly. Especially from 2016 to 2017, the annual increase in the number of tourists and the number of visitors to cultural science exhibitions nearly doubled, driving social benefits to achieve explosive growth. In contrast, water conservancy scenery has a weak supporting role in agricultural development, and the per capita gross agricultural production value and the output of aquatic products per unit of aquaculture area fluctuate repeatedly in a small range. Of course, this is also related to the impact of agricultural production statistics on the fluctuations in the prices of agricultural products that year.

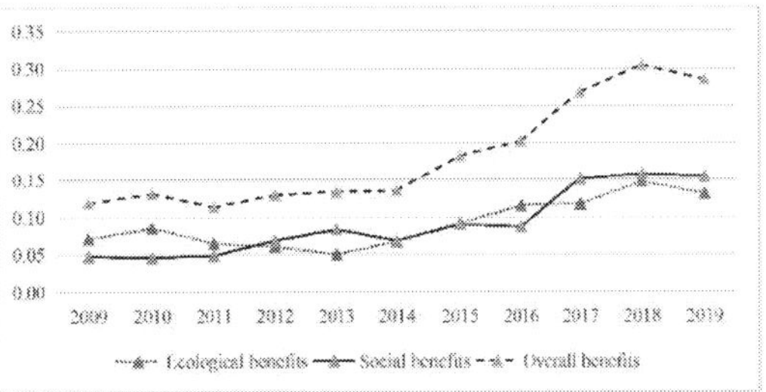

Figure 4. Change trend of the comprehensive index of comprehensive benefits system.

(Fig.4)For Qingpu, a hub town that connects Shanghai with Jiangsu and Zhejiang, water transportation has great utility value, which is reflected in the rapid growth of water transportation volume and water transportation turnover. In addition, in recent years, the Qingpu District Water Affairs Bureau has focused on promoting the reform and development of water conservancy, the water conservancy disaster prevention and reduction system has been continuously improved, the construction of rural water conservancy infrastructure has continued to be consolidated, and the construction of water ecological civilization has accelerated, providing a strong force for the sustainable economic and social development of water conservancy support and protection of the entire district.

4.2 Coupling analysis of comprehensive benefits of water conservancy scenic

Regarding the classification of ecological civilization coordination types in the relevant research literature[8], according to the degree of coupling coordination, the degree of coupling coordination of the various subsystems of comprehensive benefits of water conservancy and landscape resources is divided into 4 stages, namely $0 < \mathbf{D} \leq 0.3$, $0.3 < \mathbf{D} \leq 0.5$, $0.5 < \mathbf{D} \leq 0.7$, $0.8 < \mathbf{D} \leq 1$, respectively corresponding to the low-level coupling coordination stage, the antagonist stage, the primary coordination stage, and the high-level coupling stage.

From the calculation results, it can be seen that the coupling degree is above 0.9, indicating that the social and ecological benefits of the water conservancy scenery in Qingpu District are close to each other, which is a very strong coupling. The level of coupling coordination is between 0.23 and 0.39, and the overall shift is in the direction of optimization. During the period from 2009 to 2014, the coordination index fluctuated slightly but the amplitude was small. Starting from 2015, the coordination state has moved from a moderately unbalanced stage to a mildly unbalanced stage, and the degree of coupling coordination has shown a slow upward trend. It can be seen that the construction of ecological civilization related to water conservancy and scenery in Qingpu District has achieved certain results in the past ten years, but it is still in the low-level coupling and coordination stage, and there is still a lot of room for development in the future.

Table 2. Evaluation results and coupling coordination degree of comprehensive benefits of water conservancy scenic in Qingpu District.

Years	Ecological benefits index	Social benefits Index	Comprehensive benefits Index	Coupling degree	Coupling coordination degree	Coordination status
2009	0.0713	0.0471	0.1184	0.9790	0.2408	Moderate disorder
2010	0.0855	0.0456	0.1311	0.9525	0.2499	
2011	0.0647	0.0483	0.1130	0.9895	0.2365	
2012	0.0602	0.0686	0.1289	0.9979	0.2536	
2013	0.0501	0.0834	0.1335	0.9684	0.2542	
2014	0.0669	0.0682	0.1351	1.0000	0.2599	
2015	0.0918	0.0901	0.1820	1.0000	0.3016	Mild disorder
2016	0.1152	0.0868	0.2020	0.9901	0.3162	
2017	0.1170	0.1507	0.2677	0.9920	0.3644	
2018	0.1466	0.1571	0.3037	0.9994	0.3896	
2019	0.1307	0.1539	0.2847	0.9967	0.3767	

5 Conclusion and Outlook

5.1 Research conclusion

Under the development goal of building an ecological region of rivers and lakes with characteristics of history, culture, and environment in southern China, the research on the water conservancy scenic of Qingpu District, especially the measurement and evaluation of comprehensive benefits, has high practical significance and application value. However, the current related research lacks investigation about quantitative coupling and coordination. Therefore, this study selects 19 index factors from two aspects of ecological benefit and social benefit, uses the entropy method to comprehensively evaluate the water conservancy scenic benefits of Qingpu District from 2009 to 2019, and uses the coupling coordination degree model to evaluate the comprehensive benefits subsystem. The following conclusions were drawn from the comprehensive measurement and analysis: from 2009 to 2019, the overall benefits of water conservancy scenic in Qingpu District showed an upward trend, especially after 2014. The social benefit slowly rises amidst fluctuations, and the ecological benefit goes through two phases of change, which first drops and then rises. During the period from 2009 to 2014, the coordination index fluctuated slightly. Starting from 2015, the coordination state has changed from the stage of moderate imbalance to the stage of mild imbalance.

Objective analysis, this study uses the entropy method to determine the weight of each evaluation index, which can reduce the error caused by human subjective factors. Since there is not much quantitative statistical data at present, it is relatively difficult to obtain other index data. In addition, the connotation of comprehensive benefits of water conservancy scenic is relatively rich, and the index system and research perspectives still need to be further improved. In the later stage, further in-depth research can be conducted from the perspective of the spatial differences of internal areas within Qingpu District.

5.2 Outlook

With the implementation of the River and Lake System in various places in recent years, the concept of harmonious coexistence between man and nature has gradually sunk deep into people's minds. As the model for the construction of beautiful rivers and lakes, water conservancy scenic has gradually increased its role in regional tourism. Qingpu District is included in the Yangtze River Delta Integration Demonstration Zone in 2019 because it happens to be at the junction of the three provinces and cities of Shanghai, Jiangsu, and Zhejiang. Among them, Zhujiajiao Town and Jinze Town are the first start-up areas. In recent years, under the goal of promoting the ecological quality construction of the demonstration area, Qingpu District has not only created new highlands of ecological value such as Lake Dianshan and Lake Yuandang, but the urban water system landscape of the water system park around the city has also gradually taken shape. Qingpu Water Gate and Shangshan Square have become "Instagram Worthy Locations". Meanwhile, the "blue pearl chain" leisure landscape belt planning has been taken into the agenda. With expectations of further improving the comprehensive benefits of water conservancy and scenery in Qingpu District, and strengthen the penetration and integration between ecological and social systems, this article puts forward the following suggestions.

First, it is important to make full use of Qingpu's existing ecological, cultural, and location advantages to strengthen the building of cultural tourism brands. At present, the characteristics of scenic spots in the area around Lake Dianshan are not outstanding enough, especially the brand influence advantage of the three ancient towns in Qingxi has not yet been established. The water culture core of "Shangshan City" should be deeply explored to form a tourism brand with international influence. Second, strengthen the comprehensive utilization of the waterfront area. The current ratio of construction land to ecological space in Qingpu District is

3:7, which means the man-land relationship is suitable. The rural landscape resources of the water village are rich, and there is a large amount of waterfront space to be developed. Only by implanting characteristic waterfront functions in the construction of ecological landscape and highlighting the characteristics of water conservancy, can it become an important window for displaying water conservancy and publicizing water conservancy to the public.

References

1. Y. Zhao, F. Ye. Rural Economy and Science. **28(14)**: 1+3(2017)
2. L. L. An. Haihe Water Resources. **227(1)**:31-33(2021)
3. C. Feng, Q. Xue, F. Xu, et al. China Water Resources. **798(12)**: 50, 57-60(2016)
4. L. Wang, H. J. Ren, H. Huang, et al. Technology Innovation and Application. **(01)**: 142-145(2016)
5. A. L. Chen, T. Lin, Y. P. Zhong, et al. Energy and Environment. **(01)**:8-11 (2021)
6. R. H. Yuan, Y. Q. Zang. Journal of Economics of Water Resources. **39(02)**:1-8+95. (2021)
7. M. Q. Wang, J. D. Wang, J. S. Liu, et al. Chinese Journal of Applied Ecology. **20(01)**:170-176. (2009)
8. H. F. Jia, X. Qing. Jiangsu Agriculture Sciences. **49(02)**:217-222. (2021)

https://doi.org/10.1051/e3sconf/202129302008

Simulation Analysis of Vertical Movement of Nitrate Nitrogen in Vadose Zone of Fluvo-aquic Soil Based on HYDRUS-1D

Ping Jing[1], Hui Zhang[1]*

[1]College of Geography and Environmental Sciences, Tianjin Normal University, Tianjin 300387, China

Abstract. Wastewater reuse has significant social, economic and ecological benefits, and has become an important measure to alleviate the water crisis. In this paper, in the simulation analysis of soil nitrate nitrogen, the method of combining soil column leaching and model simulation is used to test the accuracy of the model. The Pearson correlation coefficient is as high as 0.99 and the Nash coefficient is also over 0.9. It shows that the simulation results of the simulation model are credible. The Hydrus-1D model is used to visually simulate the change process of Tianjin fluvo-aquic soil over time. In the simulation, the process of solute migration in the 6-18 cm of the ground surface is basically the same, and they all reach a stable state about 4 hours, and show a linear increase trend four hours ago, and then the concentration remains stable and unchanged.

1 Introduction

The migration law of pollutants in the unsaturated zone has become an important content of groundwater pollution risk assessment[1]. Research on the migration mechanism and cumulative effects in the soil vadose zone can provide early warning of potential groundwater environmental pollution. In this paper, through indoor soil column experiments, the physical and chemical parameters of the fluvo-aquic soil and the vertical leaching process data are obtained, the simulation parameters are verified through the measured data, and then the Hydrus-1D model is used to simulate the vertical migration process in the vadose zone. The simulation results are input into the simulation software as the initial conditions for the next simulation to realize the quantitative simulation of the cumulative effect of multiple sewage irrigation and predict the vertical distribution law in the vadose zone.

2 Materials and methods

Unsaturated zone is the soil medium through which crop irrigation water can flow, and is the core area where surface water, soil water and atmospheric water transform into each other [2]. The landform type of Tianjin sewage irrigation area is a subregion of marine low plain, which belongs to a warm zone with semi-humid continental monsoon climate [3]. Fluvo-aquic soil is a basic soil type widely distributed in alluvial plains in Tianjin. The process of water and solute transport in the unsaturated zone of soil is often difficult to carry out field trials under natural conditions, because many uncertain factors have a great influence on the accuracy of the data. The simulation of the vertical migration process by the indoor soil column leaching device is simple and easy to implement. It can provide verification data for model simulation, invert the simulation analysis parameters, and determine the parameter values of the fluvo-aquic zone.

2.1 Test design

The indoor test soil column is a transparent organic glass column with an inner diameter of 3cm, a height of 50cm, and a total filling soil of 30cm. A sampling location is set at the bottom of the soil column. The diameter of the soil column used in the study is very small, it will take a short time during the leaching experiment and can fully infiltrate the soil with a short time. in order to prevent wall flow, and avoid the leaching solution staying in the soil for a short time and not fully infiltrating the soil, petroleum jelly is used on the pipe wall to prevent edge dominant flow.

Filling the bottom of the tube with clean glass beads with different diameters of less than 5cm to prevent water flow from disturbing the bottom soil, and then fill the soil sample according to the actual soil bulk density of 1.5g/cm3. After every 10cm of filling, the surface of the soil is compacted and then the filling is continued to ensure uniform soil bulk density and no clearly stratified. Fill the soil up to 30cm and then fill with 5cm clean glass beads of different diameters to prevent the surface soil from disturbing by adding water. The upper device is a Markov bottle with graduations, which is used to control the water head at the top of the soil column to ensure a constant water head during the test. The lower end is connected to a bottom funnel, and a rubber tube is connected to the receiving bottle to receive the leachate.

* Corresponding author: zhang_hui69@163.com

2.2 Sampling and testing

The test soil was taken from the typical sandy fluvo-aquic soil in Jinnan District, Tianjin. The soil samples were taken from the 0-40cm soil layer section and brought back to the laboratory for processing. After the plant roots and debris are removed, the soil sample is naturally dried and crushed. The soil sample is sieved with a 2mm aperture sieve, and the soil is discarded by the quarter method. Finally, about 4 kg of soil sample is left for the filling soil column and used to determine the soil background value.

Samples were taken at 1, 2, 3, 6, and 18 h after the start of the single sewage irrigation experiment, and every 1h after the start of the cumulative experiment, and the content in the sample solution was measured with an ultraviolet spectrophotometer.

In order to understand the characteristics of leaching loss in different time periods in a single leaching process, a single process sampling method was used to detect the soil column leaching solution. Sampling is taken every 1 hour after the start of leaching, and sampling can be extended for several hours after three samplings, because there is little difference in concentration after long-term leaching.

3 Model construction

Hydrus-1D is a professional and free water environment simulation software developed by the American Salt Soil Laboratory. The software contains a one-dimensional (1D) finite element model to simulate the movement of water and solutes in porous media Process, support interactive graphical interface and discretization of soil profile. The software can perform data pre-processing, structured and unstructured finite element mesh generation, and graphical display of the results. The software covers the hysteresis effects of crop root water absorption and soil water holding capacity, and can set different constant or non-constant boundary conditions, which has good applicability [4].

3.1 The basic equation of water movement

The simulation analysis of water in the vadose zone of fluvo-aquic soil mainly focous on the change law of the one-dimensional vertical migration process. It is assumed that the gas phase plays a very small role in the liquid flow. In this case, the one-dimensional uniform motion process of water in the vadose zone medium is described by the modified Richards equation [5].

3.2 Soil water transport equation

Soil hydraulic model is an important method to describe the soil water transport characteristic curve. There are six soil hydraulic models commonly used to analyze the soil water characteristic curve: (1) van Genuchten-Mualem model; (2) air Enter the van GenuchtenMualem model with a value of -2 cm; (3) modified van Genuchten-type equations; (4) Brooks and Corey equations; (5) Kosugi's log-normal distribution model; (6) dual porosity model. Hydrus-1D provides the first five types of soil hydraulic models. Since the van Genuchten-Mualem model has a better fit for coarse soil and clay soil, it is currently the most widely used, so the van Genuchten-Mualem model is used for soil moisture transport [6].

3.3 the solute transport equation

The properties of the solute transport equation in the vadose zone are relatively stable, and it is easy to move with the irrigation water from the surface to the groundwater. The improved convection dispersion equation is used to simulate the solute transport process [7].

$$\frac{\partial \theta c}{\partial t}+\rho\frac{\partial s}{\partial t}=\frac{\partial}{\partial z}(\theta D\frac{\partial c}{\partial z})-\frac{\partial qc}{\partial z}-\Phi \quad (1)$$

In the formula, θ is the volumetric water content ($cm^3 \cdot cm^{-3}$); c is the liquid concentration of the soil and filler solute ($g \cdot cm^{-3}$); t is the simulation time (h); ρ is the soil bulk density ($g \cdot cm^{-3}$); s is the solid phase concentration of solute in soil and filler ($g \cdot g^{-1}$); D is the comprehensive dispersion coefficient, representing molecular diffusion and hydrodynamic dispersion, reflecting the diffusion and dispersion mechanism of solute molecules in soil and filler; q is volume Flow flux density ($cm \cdot h^{-1}$); Φ is the source-sink term, which represents the various first, zero and other solutes occurring between the three phases of solid, liquid, and gas during the vertical migration process response ($g \cdot cm^{-3} \cdot s^{-1}$).

3.4 Model parameter settings

Changes in soil hydraulic parameters have important effects on the output variables of HYDRUS-1D, such as soil water potential and cumulative infiltration. The measured soil media data is used to determine the parameters, and the artificial neural network that comes with the software is used to analyze the physical properties of the soil to obtain the simulation parameters used in the model analysis. The model parameters can be quickly calibrated and the accuracy of model simulation can be improved.

The pedotransfer function (PTF) based on neural network is used to predict soil water retention parameters and saturated hydraulic conductivity (K_s) based on structural information. According to the micrographs of soil particles, the composition ratio of sand, powder and clay is counted, and the soil composition ratio data is input into the initial estimation module of soil hydraulic parameters (see Figure 2), and the optimized parameters of the soil simulation can be directly obtained.

3.5 Simulation boundary conditions

In the boundary condition setting of the simulation model, for the upper boundary of solute transport, the solute flux boundary is selected, and the lower boundary is selected as the zero gradient boundary condition to calculate the soil leaching loss. The soil layer is divided into 101 units in the Z-axis direction, the time step is hours, the

https://doi.org/10.1051/e3sconf/202129302008

minimum time step is set to 0.001 h, and the maximum time step is 0.1 h. In order to better analyze the migration law in the vadose zone, this simulation sets the background value in the vadose zone as the simulated background value, and the upper boundary water concentration is set to 40mg/L.

3.6 Model verification

In order to evaluate the simulation accuracy of the model, through the correlation analysis of the experimental data and the simulation data, by which the validity of the model prediction results can be judged. Nash-Suttcliffe simulation effect coefficient (NSC), root mean square error RMSE and Pearson correlation coefficient r were used to analyze and evaluate the accuracy of the simulation results. The value of NSC ranges from negative infinity to 1, and the value of NSC is close to 1, indicating that the model simulation effect is good, and the reliability of model analysis is high; when the value of NSC is close to 0, it indicates that the simulation result is close to the average level of the observed value, indicating the model prediction result is credible in general, but the simulation error of the process is large; when the NSC value is far less than 0, the model prediction result is not credible.

RMSE is used to measure the accuracy of simulation results, and NSC is used to measure the calculation deviation. The specific calculation formula is as follows:

$$RMSE=\sqrt{\frac{1}{N}\sum_{i=1}^{N}\left(x_i'-x_i\right)^2} \quad (2)$$

$$NSC=1-\frac{\sum_{i=1}^{N}\left(x_i-x_i'\right)^2}{\sum_{i=1}^{N}\left(x_i-\bar{x}_i\right)^2} \quad (3)$$

$$r=\frac{\sum_{i=1}^{N}(x_i-\bar{x}_i)(x_i'-\bar{x}_i')}{\sqrt{\sum_{i=1}^{N}(x_i-\bar{x}_i)^2\sum_{i=1}^{N}(x_i'-\bar{x}_i')^2}} \quad (4)$$

The root mean square error RMSE can be calculated based on the measured concentration value at 30 cm of the soil column and the simulated value, in order to measure the accuracy of the simulated value. The smaller the RMSE, the higher the simulation accuracy, and the closer the model predicted to the actual situation. The NSC value is usually between 0-1. The larger the NSC, the smaller the error between the simulated value and the experimental data, and the smaller the data error. When the NSC value is greater than 0.75, the simulation result is acceptable. The Pearson correlation coefficient r is used to analyze the degree of linear correlation between two variables. The value of r ranges from -1 to +1. If r>0, the two variables have a positive correlation; if r<0, It shows that the two variables have a negative correlation. The larger the absolute value of r, the stronger the correlation between the two undesirables. If r=0, it means that there is a non-linear correlation between the two variables. Close to 1 or -1 is called a strong correlation [7].

4 Analysis of results

The results of this study are divided into two parts. The first part is to verify the agreement between the measured values of the soil column leaching process and the simulated values of Hydrus-1D to test the feasibility of the simulation parameters and determine whether the simulation results can be used for simulating. The process analysis of sewage irrigation in the soil field; the other part is the Hydrus-1D model to study the vertical migration law in Tianjin fluvo-aquic soil, quantitatively analyze the vertical distribution law of solute, and visually simulate the change process in the soil over time.

In the study, five observation points were set up for the vertical division of the vadose zone, with buried depths of 6, 12, 18, 24, and 30 cm respectively. The distribution of solutes in the vertical direction was described in a vertical equidistant manner. In the process of model simulation, all the data of different buried depths are output to reveal the law of solute movement over time on the section of the vadose zone.

4.1 Accuracy analysis of simulation results

According to the previous model verification method, the Nash-Suttcliffe simulation effect coefficient (NSC), root mean square error RMSE and Pearson correlation coefficient r are calculated respectively, and the accuracy of the simulation results is analyzed and evaluated.

According to the measured value of the soil column leaching experiment and the model simulation value(see Figure 1), the relevant error analysis was calculated respectively. The results are shown in Table 1. The Pearson correlation coefficient and the Nash coefficient are as high as 0.99 or more, indicating that the simulation accuracy is very high, and the parameter selection of the model conforms to the soil Ratio of particle size

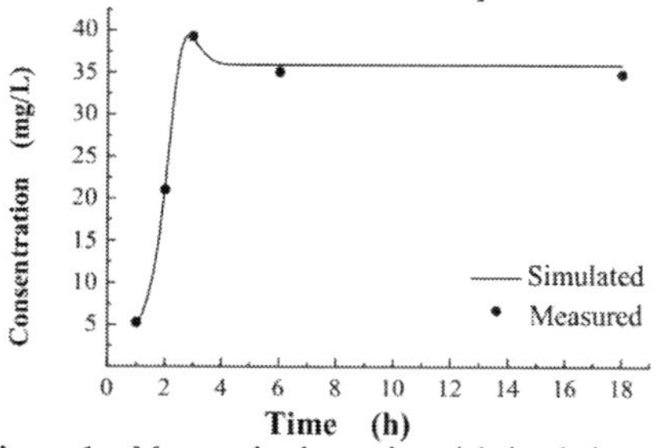

Figure 1. Measured value and model simulation value

Table1. Error analysis of measured value and simulated value

parameter	result
Pearson	0.9978
nsc	0.995806
RMSE	0.810939

4.2 Simulation process analysis

The concentration increases rapidly from the beginning of irrigation, and the adsorption of pollutants by the soil gradually decreases. As the adsorption decreases, different depths reach the maximum adsorption value after 1-3 hours after leaching, and the desorption causes the pollutant concentration to reach the peak. After reaching the peak value, it decreased slightly, and the adsorption reached a stable state after 4 hours, and the solute maintained a certain stable value at different depths (see Figure 2). It shows that after the solute is saturated with water in the vadose zone, the adsorption and desorption reach an equilibrium state, and the concentration of soil water remains stable.

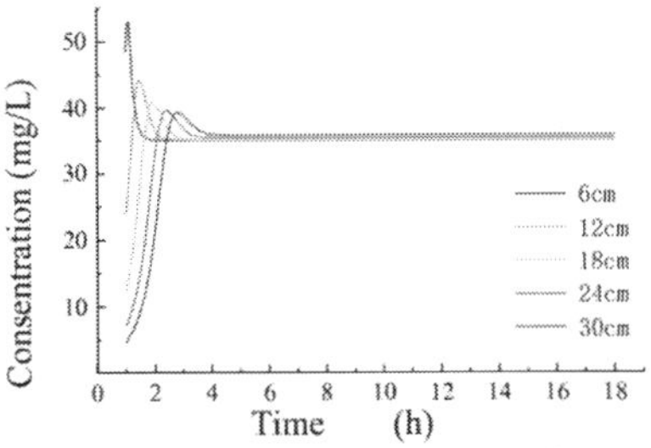

Figure 2. Model simulation process diagram

4.3 Vertical concentration change analysis

In the sequential process analysis using Hydrus-1D software, it can be clear that the leaching concentration value after the leaching process is stable and almost unchanged, and the most complicated process is mainly concentrated in the first 4 hours. Therefore, in order to determine vertical concentration change process, a vertical change distribution map can be displayed during the simulation process based on the simulation analysis data (see Figure 3).

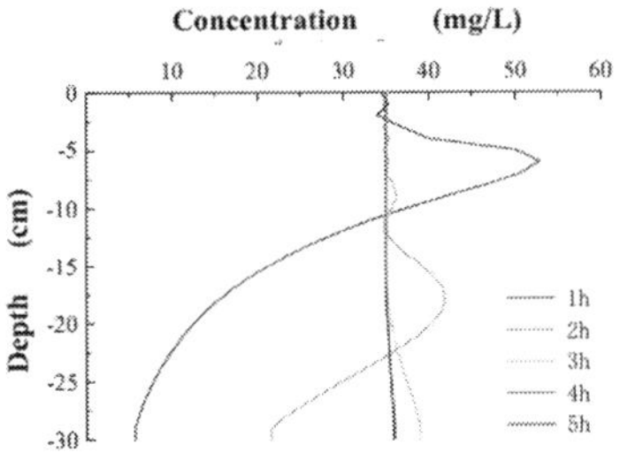

Figure 3. The vertical simulation process

During the first 1 to 2 hours of leaching, there is a process where the concentration does not change with time in the vertical direction of the soil, and then there is an increasing peak value. This process shows that there is a time course for the soil to enter the leaching solution, and then the value of the upper layer of the soil is increased, and then the concentration decreases due to the leaching loss. Up to 3 hours, the concentration has basically no longer changed in the vertical section of the soil.

5 Conclusion

In the soil simulation analysis, the method of combining soil column leaching and model simulation is used to test the accuracy of the model. Among them, the Pearson correlation coefficient value is greater than 0.8, especially the value of the first two times is as high as 0.99. The simulation accuracy is very high, and the Nash coefficient is also above 0.9, indicating that the simulation results of the simulation model are credible.

Hydrus-1D is used to simulate the vertical migration process of soil aeration zone, and the two modules of Hydrus-1D water flow and solute migration are used to predict. The law of solute migration at 6-18cm on the surface is basically the same, both in about 4 hours to reach a steady state, the concentration increases rapidly from the beginning of irrigation, and the soil's adsorption of pollutants gradually decreases. As the adsorption decreases, different depths reach the maximum adsorption value after 1-3 hours after leaching, and the desorption will increase and the concentration of pollutant reach the peak.

Acknowledgments

Authors wishing to acknowledge the financial support from Key Project of Tianjin Natural Science Foundation (18JCZDJC10090)

References

1. WANG S Q, ZHENG W B, KONG etc. Spatial distribution characteristics of nitrate in shallow groundwater of the agricultural area of the North China Plain[J]. Chinese Journal of Eco-Agriculture, 2018, 26(10).
2. LAI X M, LIAO K H, ZHU Q, etc. Feture analysis of soil water leakage and leaching of nitrogen and phosphorus in typical farmland of the Taihu Lake Basin based on the Hydrus-1D model[J]. Resources and Environment in the Yangtze River Basin, 2015, 24(09).
3. GE F Y, LIU J L, LI L W. Numerical simulation of pollutant transport in aeration zone of coastal reclamation land area based on Hydrus-1D[J]. Resource Information and Engineering, 2020,35(04)
4. Van Genuchten, M. Th. A closed-form equation for predicting the hydraulic conductivity of unsaturated soils, Soil Sci. Soc. Am. J., 1980, 44:892-898.
5. van Genuchten, M. Th. Convective-dispersive transport of solutes involved in sequential first-order decay reactions, Computers & Geosci., 1985, 11(2):129-147.
6. WANG X D, FENG W, WANG W K etc. Migrating and Transforming Rule of Nitrogen in Unsaturated Zone in Guanzhong Basin Based on HYDRUS -1D Model[J]. Geographical survey and research, 2015,38(04):291-298+304.
7. ZHANG Y H, ZHOU B B, CHEN X P etc. Effect of nano carbon on soil moisture movement characteristics in the loess slopes Based on Hydrus simulation[J]. Research of Soil and Water Conservation, 2020,27(01):132-138.

https://doi.org/10.1051/e3sconf/202129302009

Research on the influence of boulder powder content on concrete performance

Guangcheng Meng[1,*]

[1] China Road and Bridge Co., Beijing 100011, China

Abstract. To solve the problem of environmental pollution caused by the accumulation of granite powder and the shortage of traditional mineral admixtures, the influence of the amount of granite powder on the mechanical properties of concrete was studied by replacing cement with different amount of granite powder Different amount of granite powder can be used to prepare concrete with satisfactory performance. When the amount of granite powder is small (not more than 5%), granite powder will not reduce the compressive strength of concrete, or even slightly improve the compressive strength of pure cement concrete. When the amount of granite powder is more than 5%, the compressive strength of concrete will gradually decrease; when the amount of granite powder is more than 5%, the compressive strength of concrete will gradually decrease. The elastic modulus of concrete decreased, and the electric flux increased with the increase of the amount of admixture.

1 Introduction

With the increase in the scale of national infrastructure construction and the advancement of the "One Belt One Road" initiative, concrete raw materials, especially mineral admixtures, are in short supply in many regions. Fraud and adulteration of mineral admixtures (granulated blast furnace slag powder, fly ash, silicon ash, etc.) often occur, which greatly affects the quality of engineering construction. At present, many scholars are devoted to the research of new supplementary cementitious materials, such as limestone powder [1–3], metakaolin [4,5] and steel slag powder [6,7], to cope with current shortage of traditional mineral admixtures.

Phnom Penh-Sihanoukville Expressway (PPSE) is an important golden channel connecting Phnom Penh, the capital of Cambodia, to Sihanouk, the largest seaport in China and an external port. The construction of PPSE is of great significance for promoting the local economic development of Cambodia and implementing the "One Belt One Road" initiative. The starting point of PPSE is at the intersection of National Highway No. 4 and Phnom Penh City Circle Line in Cambodia, and the end point is on the No. 4 road on the edge of Sihanoukville. It is about 8.6km away from Sihanoukville Port to the west and about 5.5km away from Westport Economy, and it is about 8.5km from Sihanoukville Airport. The total length is 187.05Km. The whole line adopts Chinese highway construction standards and specifications. The design speed is 100Km/h, and the two-way four-lane.

The construction of PPSE also faces the problem of the lack of high-quality mineral admixtures and river sand resources. The project adopts the common boulder production machine-made sand along the project, and a large amount of rock powder of boulder is produced during the production process. If the boulder powder (BP) is recycled as mineral admixtures in engineering concrete production through research, not only can the pollution of the environment by the stacking of stone powder be solved, but also ordinary mineral admixtures can be replaced, and construction costs can be reduced. Hence, this paper studies the influence of different content of BP on concrete performance and provides basic research data support for the application of BP in engineering.

2 Materials and testing methods

2.1 Testing materials

In this paper, the test concrete is mainly prepared with the following raw materials:

(1) Cement: ordinary Portland cement whose initial setting is 161min, final setting is 242min, 3-day and 28-day compressive strength are 22.3MPa and 49.2MPa respectively. The chemical composition is shown in Table 1.

Table 1 Chemical composition of cement and boulders powder (BP) (wt.%)

Material type	CaO	Al_2O_3	SiO_2	SO_3	Fe_2O_3	K_2O	Na_2O	MgO	LOI
cement	72.27	3.98	18.77	3.73	0.21	0.43	0.06	0.21	0.82
Boulder powder (BP)	8.99	15.32	49.19	-	9.18	0.21	0.43	9.31	1.28

* Corresponding author: menggc@crbc.com

(2) Admixture: polycarboxylate superplasticizer (PS) with solid content of 16.2% and water reducing rate of 18.2% is selected.

(3) Boulder powder: the fluidity ratio of boulders powder (BP) from the preparation process of manufactured sand is 81%, and the 28d activity index is 60%.

(4) Fine aggregate: manufactured sand made by boulders is used, which has a fineness modulus of 2.9, a stone powder content of 9.6%, a methylene blue value of 1.1.

(5) Coarse aggregate: 5-20mm continuous graded crushed stone is used, and the apparent density of it is 2642kg/m^3, the bulk density is 1626kg/m^3, the water absorption is 1.3% and the crushing value is 6.3%.

2.2 Testing method

The slump and expansion of concrete were tested according to the Chinese standard *Standard for test method of performance on ordinary fresh concrete: GB/T 50080-2016*. According to the standard for the Chinese standard *Standard for test method of concrete physical and mechanical properties: GB/T 50081-2019*, the compressive strength of concrete is measured. The size of the test piece is 150mm×150mm×150mm. The raw materials of concrete were mixed in the mixer for 90s, and the test block was formed by vibration and placed in the room with temperature of 20°C and humidity of 90% for 24h.Then the mold was disassembled, and the concrete sample was transferred into the standard curing room until the testing age. According to the Chinese standard *Standard for test methods of long-term performance and durability of ordinary concrete: GB/T 50082-2009*, the electric flux of concrete was measured.

2.3 Mix proportion

The mix proportion of concrete contained BP was listed in Table 2. The content of BP accounted for 0%, 5%, 10%, 15% and 20% of cementitious materials, and the concrete slump is controlled within 180mm to 210mm by adjusting the dosage of PS.

Table 2. Mix proportion of concrete (kg/m^3)

Sample ID	Water-to-binder ratio	Water	Cement	Boulder powder (BP)	Content of BP(%)	Fine aggregate	Coarse aggregate
J-0	0.45	175.5	390.0	—	0	825	1015
J-5	0.45	175.5	370.5	19.5	5	825	1015
J-10	0.45	175.5	351.0	39.0	10	825	1015
J-15	0.45	175.5	331.5	58.5	15	825	1015
J-20	0.45	175.5	312.0	78.0	20	825	1015
J-25	0.45	175.5	292.5	97.5	25	825	1015

3 Result and discussion

3.1 Workability

The influence of BP content on the workability of concrete was shown in Fig. 1, and amount of PS with different content of BP was listed in Fig. 2. It could be found that the slumps of J-0 to J-25 were between 180mm to 210 mm, and the largest one was the J-10 sample with a BP content of 10% and the smallest one was the J-25 with a slump of 180mm. The slump and slump-flow of concrete first increased and then decreased with the increase of the BP content.

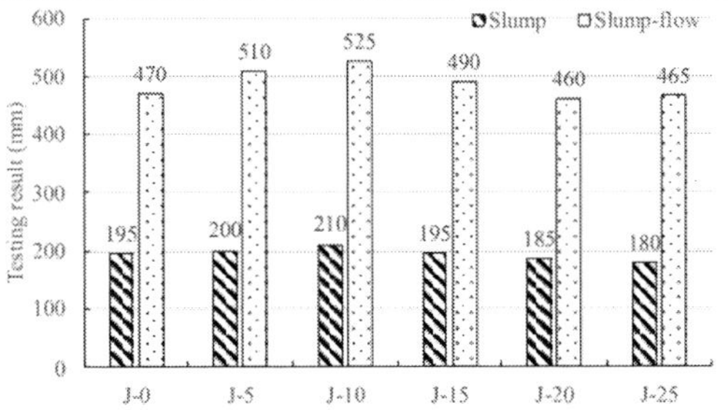

Fig. 1. Workability of concrete with different content of boulders powder.

Fig. 2 showed that the amount of PS is between 0.90% to 1.20% (the mass ratio of cementitious materials). The smallest dosage of PS was J-5 as well as J-10, and the largest one belonged to J-25, meaning that the usage of PS first decreased and then gradually increased with the increase of the BP content. The main reason was that the BP mixed into concrete contained some particle with a small particle size, which could fill the pores between cement particles and play a certain water-reducing effect. Secondly, the surface of the BP particle was relatively rough and sometimes contained a certain amount of mud powder. Under normal circumstances, the water demand was very large, and the PS would be adsorbed at the same time, which has a certain negative effect on the workability of the concrete. When the amount of BP was low, its water-reducing effect was more obvious, and the requirement of PS was reduced, while the flowability of fresh concrete was better. If the amount of BP was large, it water demand and the negative effect of adsorbing PS were prominent, and therefore the usage of PS increased and the slump as well as slump-flow of concrete became worse.

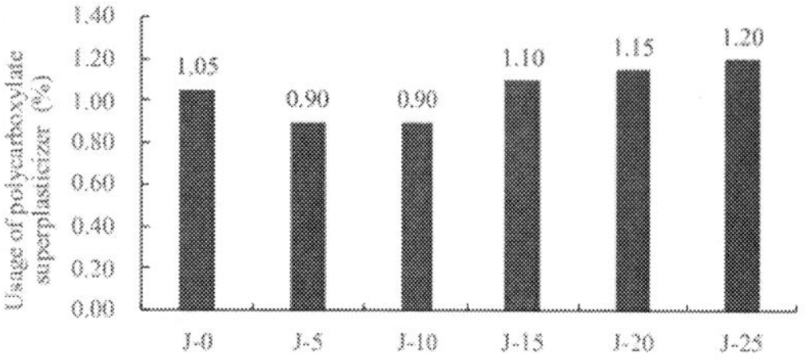

Fig. 2. Amount of polycarboxylate superplasticizer of concrete

3.2 Compressive strength

The influence of the amount of BP on the compressive strength of concrete was shown in Fig. 3. Taking J-0 as the controlled group, when the BP content was 5%, the strength of each curing age is slightly higher than that of the controlled group. As the BP content further increased, the compressive strength continued to decrease. In order to quantify the impact of BP on the strength of concrete, this section defined the ratio of the compressive strength of the testing group mixed with BP to the strength of the controlled group as the compressive strength ρ, as shown in equation (1). Based on the testing result in Fig. 3, ρ of different testing groups were calculated and the results were shown in Fig. 4.

$$\rho_i = (1 - f_{t,i}/f_{c,i}) \times 100 \qquad (1)$$

Where ρ_i was the compressive strength ratio of concrete in the curing age i (i= 3 days, 7days, 28 days or 56 days), $f_{t,i}$ was the compressive strength of the testing groups in the curing age i, $f_{c,i}$ was the compressive strength of the controlled strength in the curing age i.

The ρ of J-5 at different ages were all less than 0, indicating that the strength of J-5 was higher than that of J-0. As the amount of BP increased, the ρ value of concrete was positive and gradually increased, which meat that the compressive strength of concrete decreased more. When the amount of BP was 25%, the ρ of concrete at various ages were sequentially 29.8%, 18.8%, 18.2% and 12.3%. On the other hand, for concrete with the same content of BP, the addition of BP had a greater impact on the compressive strength of concrete at an early age (for example, 3 days). As the curing age increased, the adverse effect of adding BP on the strength of concrete gradually weakened.

Fig. 1 and Fig. 2 showed that when the content of BP is low (for example, 5%), the BP particles could fill in the voids of the concrete paste, so that the particle size distribution of concrete was more reasonable, and thus the fluidity was better. The compressive strength of J-5 was also enhanced and finally slightly higher than that of J-0. When the content of BP was further increased, the workability of fresh concrete became worse, and the air bubbles were difficult to smoothly discharged during the vibration forming process of the concrete, resulting in more initial micro-defects after the concrete was hardened. In the low hydration stage, the compressive strength of concrete was weakened to a higher degree. Secondly, BP is an inert material. After replacing part of the cement, the cement clinker content would decrease, and the amount of hydration products will be reduced at the same age. The size of the pores in the cement matrix was larger, the compactness was poor, and there were many voids and cracks distributed in the interface transition zone, leading to the destruction of concrete sample under lower load.

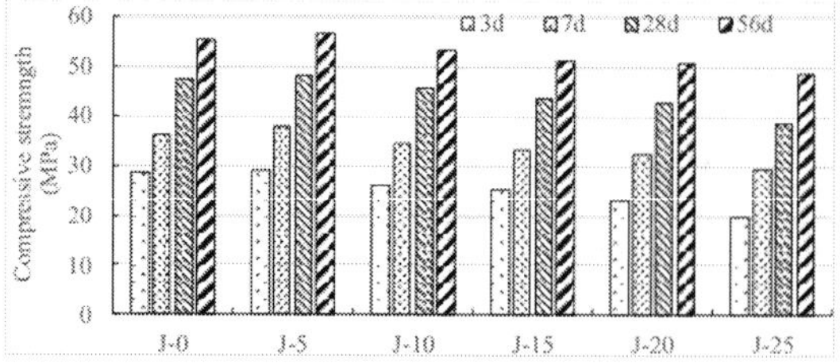

Fig. 3. Influence of boulders powder content on compressive strength of concrete

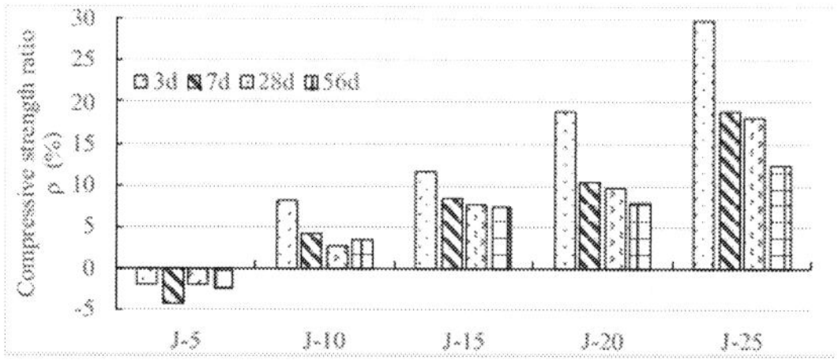

Fig. 4. Influence of boulders powder content on compressive strength ratio ρ of concrete

3.3 Elastic modulus

Fig. 5 showed the testing results of elastic modulus of concrete mixed with different content of BP at different curing age. The elastic modulus of concrete is the ratio of compressive stress and strain, which could measure the degree of deformation and is an important parameter for the design of concrete structures. When concrete has a high elastic modulus, the structures is more resistant to deformation. Similar to the compressive strength, when the BP content was low (5%), the elastic modulus of concrete at different ages was high than that of the controlled group, but as the BP content was increased, the concrete elastic modulus continued to decrease and the curing age prolonged could improve the elastic modulus of concrete to a lower extent.

The testing operation of elastic modulus of concrete is complicated and consume a long time, which brings inconvenience to engineering practice. Many researchers [8–11] had carried out series experiments to establish mathematical models of the relationship between the elastic modulus and compressive strength of concrete, including power function and polynomial function models, and try to use compressive strength to predict the elastic modulus of concrete. In this paper, a power function equation was used to establish a compressive strength-elastic modulus model, as shown in equation (2).

$$E_c = 12.715x^{0.2854} \qquad (2)$$

Where E_c was the calculated value of the elastic modulus of concrete containing BP, and f_c was the compressive strength of the concrete at the corresponding age.

Fig. 6 indicated that there was a strong correlation between the compressive strength and elastic modulus of concrete mixed with BP. After calculation, it was found that the difference between the predicted elastic modulus and the testing value in Fig. 5 was not higher than 5%, meaning that the mathematical model could accurately predict the elastic modulus of BP concrete.

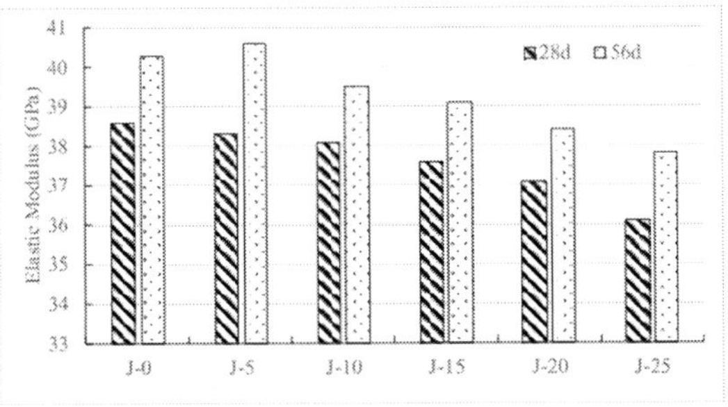

Fig. 5 Influence of boulders powder content on elastic modulus of concrete

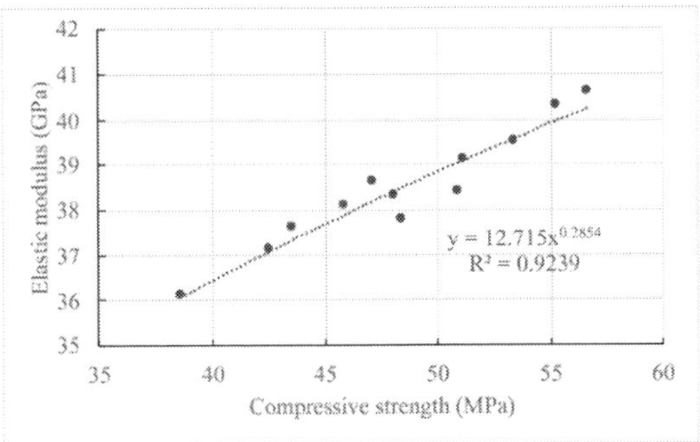

Fig. 6 The correlation of compressive strength and elastic modulus of concrete

3.4 Electric flux

The experimental results of the influence of BP on the electric flux of concrete were shown in Fig. 7. When the BP content did not exceed 10%, increasing the BP content was beneficial to reduce the electric flux of concrete, since the BP particle could improve the particle size distribution of the concrete sample through the filling effect and could also act as nucleation sites to promote the hydration process and improve the compactness of the concrete. If the content of BP was further increased, its inert effect would reduce the amount of hydration products, and the pores of the concrete matrix could not be effectively refined during the hydration process, so that the average size of the pores was too large and the total porosity was higher. Thus, the permeability of concrete increased, and the final results of electric flux increased accordingly.

On the other hand, prolonging the curing age could effectively reduce the electric flux of concrete and compensate for the negative effect caused by the excessively high content of BP. Fig. 7 showed that when the curing age was 28 days, the electric flux of all concrete was higher than 2000C, which was a medium permeability level[12]. When the curing age was extended to 56 days, the electric flux of all groups were between 1000C and 2000C, which was decreased to the low permeability level. Among them, the electric flux of J-10 was only 1106C, which was close to ultra-low permeability. In order to ensure that the performance of the concrete was not significantly reduced, combined with the testing results of workability, compressive strength and elastic modulus as well as electric flux of the concrete, the mixing content of BP should not exceed 10%.

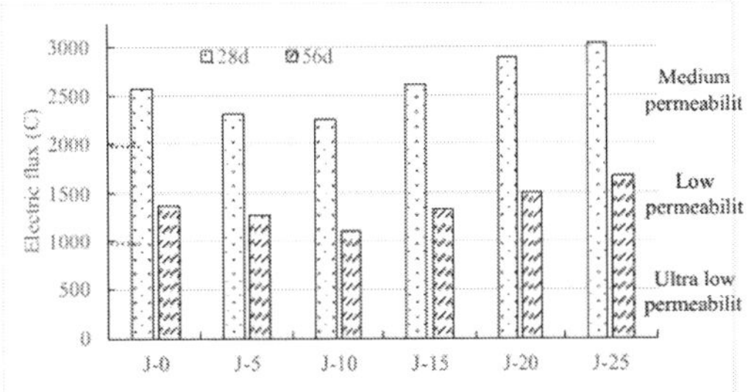

Fig. 7 Influence of boulders powder content on electric flux of concrete

4 Conclusions

To solve the problem of environmental pollution caused by the accumulation of boulder powder and the shortage of traditional mineral admixtures, the influence of boulder powder content on the mechanical properties of concrete was studied by replacing cement with different content of boulder powder, and the following conclusions were obtained:

(1) Different content of boulder powder could be used to prepare concrete with satisfactory workability. The usage of polycarboxylate superplasticizer could be reduced when the content of boulder powder was small (less than 10%), and increased as the content of boulder powder increased.

(2) When the content of boulder powder was low (not more than 5%), boulder powder would not reduce the compressive strength of concrete, and even slightly enhanced the strength at different curing age. If the content of boulder powder exceeded 5%, the compressive strength gradually decreased with the increase of content of boulder powder.

(3) With the increase of the content of boulder powder the elastic modulus of concrete showed a downward trend, and the electric flux first decreased and then gradually increased. Prolonging the curing age could effectively compensate for the negative effects caused by the excessively high content of boulder powder. In order to ensure that the performance of concrete was not significantly weakened, the content of boulder powder should be controlled within 10%.

https://doi.org/10.1051/e3sconf/202129302009

References

1. L. Hu and Z. He, Constr. Build. Mater. **262**, 119847 (2020).
2. J. Yu, H.-L. Wu, D. K. Mishra, G. Li, and C. K. Leung, J. Clean. Prod. **278**, 123616 (2021).
3. H. Du and S. D. Pang, Constr. Build. Mater. **264**, 120152 (2020).
4. K. Scrivener, F. Martirena, S. Bishnoi, and S. Maity, Cement. Concrete. Res. **114**, 49 (2018).
5. F. Avet and K. Scrivener, Cement. Concrete. Res. **136**, 106124 (2020).
6. D. Wang, Q. Wang, and Z. Huang, Composites Part B: Engineering **198**, 108207 (2020).
7. S. Zhuang and Q. Wang, Cement. Concrete. Res. **140**, 106283 (2021).
8. A. F. Hashmi, M. Shariq, and A. Baqi, Constr. Build. Mater. **283**, 122772 (2021).
9. C. Zhou, Z. Zhu, A. Zhu, L. Zhou, Y. Fan, and L. Lang, Constr. Build. Mater. **228**, 116809 (2019).
10. Y. F. Wu, S. M. S. Kazmi, M. J. Munir, Y. Zhou, and F. Xing, J. Clean. Prod. **264**, 121746 (2020).
11. J. Álvarez-Pérez, J. H. Chávez-Gómez, B. T. Terán-Torres, M. Mesa-Lavista, and R. Balandrano-Vázquez, Constr. Build. Mater. **241**, 118002 (2020).
12. R. Yang, R. Yu, Z. Shui, X. Gao, X. Xiao, D. Fan, Z. Chen, J. Cai, X. Li, and Y. He, J. Clean. Prod. **258**, 120673 (2020).

Design of Flying-Bird Type Cable Stayed Arch Bridge with Three-Leaf Rose Shaped CFST Arch Rib

Weisheng Wang[1*] and Jilin Wang[1]

[1] China Design Group Co., Ltd, Nanjing, China,210007

Abstract: According to the geometric figure of three-leaf rose curve, this paper puts forward a flying-bird type cable-stayed arch bridge with three-leaf rose curve shaped CFST arch rib, which has beautiful appearance and one bridge leads to three banks.The space cables are set between the three-leaf rose linear space arch ribs to resist negative bending moment, and the tie cables are set between the arch feet to resist positive bending moment. On the inclined pylon at the tail of the flying-bird type cable-stayed arch bridge, the stay-cables are arranged to balance the arch thrust, and the suspension-cables are arranged on the three leaf rose linear spatial arch ribs to hang the three fork bridge deck.The flying-bird type cable-stayed arch bridge with three-leaf rose linear CFST arch rib adopts spatial cable structure system, which has good structural stability.Combined with the actual project, the engineering parameters are designed, the Midas finite element analysis model is established, the internal force analysis and modal analysis are carried out, and the rationality of flying-bird type cable-stayed arch bridge with three-leaf rose linear CFST arch rib is verified.

1 Introduction

The Midland fork bridge in Michigan, USA, was built in 1981. It spans the intersection of the Chippewa river and the Tittabawassee river. The convenient transportation and unique shape make the bridge a famous local landscape.

Figure 1. Midland fork bridge

According to the landform of three-rivers-confluence, the structural shape of clover and the mathematical equation of three-leaf rose line, the design scheme of concrete filled steel tube flying swallow type cable-stayed arch bridge structure is adopted, and the stiffening beam of three fork highway bridge deck is suspended by suspension cable.

This paper studies the configuration of the three-leaf rose shaped CFST flying swallow type cable-stayed arch bridge, designs the engineering parameters, establishes the Midas finite element model, carries out the dynamic modal research, and demonstrates the structural rationality of the three fork shaped concrete-filled steel tube flying swallow type cable-stayed arch bridge in many aspects.

2 Configuration study

The arch rib of the three leaf rose line flying swallow type cable-stayed arch bridge is a three-dimensional spatial arch rib, and its plane forward projection is a three leaf rose line figure, and the side elevation projection of the arch rib is cosine function. According to the flying swallow type cable-stayed arch bridge, the three-dimensional space arch rib is bent and warped at the arch foot to form the cable-stayed bridge tower.

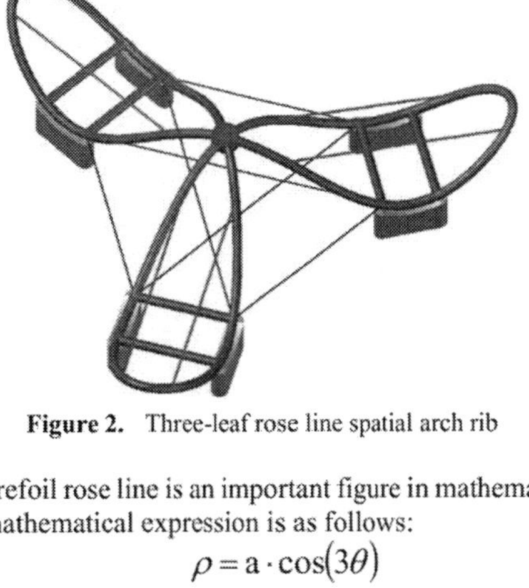

Figure 2. Three-leaf rose line spatial arch rib

Trefoil rose line is an important figure in mathematics, the mathematical expression is as follows:

$$\rho = a \cdot \cos(3\theta) \quad (1)$$

[*] Wang Weisheng: 289200225@qq.com

Where: parameter a is the envelope radius, which controls the length of the rose line leaves.

The three leaf rose line flying swallow type cable-stayed arch rib is installed with tie bar cable, and the cable between the arch ribs is installed to form a whole structure. After floating and dragging across the river, the sling is installed. The three leaf rose line flying swallow type cable-stayed arch rib suspends the three fork bridge deck, forming a three fork shaped steel tube concrete flying swallow type cable-stayed arch bridge structure.

The three branch concrete-filled steel tube flying swallow type cable-stayed arch bridge skillfully uses the trefoil rose line to construct the cable-stayed arch bridge, which has unique shape and beautiful shape. With smooth and convenient traffic and coordinated with the environment of three-rivers-confluence, it will become a beautiful scenic line in three-rivers-confluence. The self balanced structure system of flying-swallow type cable-stayed arch bridge is formed by setting the tie bar of flying-swallow arch bridge and the cable between arch ribs to balance the support thrust of arch bridge and greatly improve the structural stiffness. The scheme of dragging across the river simplifies the construction and is economical.

Figure 3. Three fork flying swallow cable stayed arch bridge

3 Parameter design

Three-fork bridge is a three-leaf rose shaped concrete-filled steel tube flying swallow type cable-stayed arch bridge. Its plane orthographic projection is a three-leaf rose line figure, the side elevation projection of the arch rib is a cosine function, and the envelope diameter of the three leaf rose line spatial arch rib is 200 meters.

Three leaf rose line spatial arch rib adopts reinforced concrete structure, the side elevation projection of arch rib is cosine function, the distance from the center point to the support is 60 meters, the 40 meter interval section after the support point is warped to form an arch cable-stayed bridge tower. The diameter of steel pipe is 2.5 meters, the wall thickness of steel pipe is 20 mm, and C40 concrete is poured inside, and the rise span ratio of arch rib is 1 / 4.

Ring section lane is set in the center of three-fork bridge. The inner diameter of the ring is 40m, the ring lane is 3 lanes, the bridge deck width is 12M, the straight section is 4 lanes, and the bridge deck width of straight section is 16 M. the circular section transition is adopted between the straight section and the circular section. The deck of three-fork bridge is steel box girder with a beam height of 1.5m.

The cable is set in the middle of the three fork concrete-filled steel tube flying swallow type cable-stayed arch bridge, and the cable-stayed cable is set at the tail to suspend the three branch stiffening beam bridge surface. The sling and cable-stayed are made of 1670mpa galvanized high-strength steel wire. The diameter of the sling and cable-stayed cable are 0.3m, the sling spacing is 20m, and the cable-stayed spacing is 16m.

There are six cables between the central arch rib and the cable-stayed bridge tower of the triple fork concrete-filled steel tube flying swallow type cable-stayed arch, one on the left and one on the left. The arch rib cable is made of 1670mpa galvanized high-strength steel wire, and the diameter of the cable is 0.5m. The cable is set to form the cable-stayed arch bridge structure system, reduce the unbalanced internal force of the support and improve the rigidity of the flying swallow type cable-stayed arch structure.

A total of 6 tied cables are set between the arch foot supports of the three branch concrete-filled steel tube flying swallow type cable-stayed arch. The cable at the arch rib foot adopts 1670mpa galvanized high-strength steel wire, and the diameter of the arch foot cable is 0.5m. The tie bar cable is set to form a flying swallow type arch bridge structure, which can reduce the unbalanced internal force of the support and improve the bearing capacity of the three fork shaped concrete-filled steel tube arch bridge.

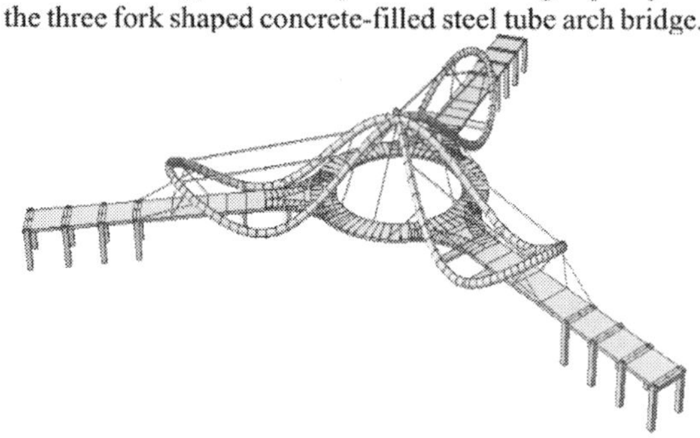

Figure 4. Midas model

4 Numerical analysis under vertical load

The bridge deck of main span is loaded with full load, the standard value of uniform load 5kn / m2 is adopted for additional dead load of bridge deck, and standard value of highway first-class load is adopted for bridge deck live load. The calculation results are as follows:

The calculation results are as follows:

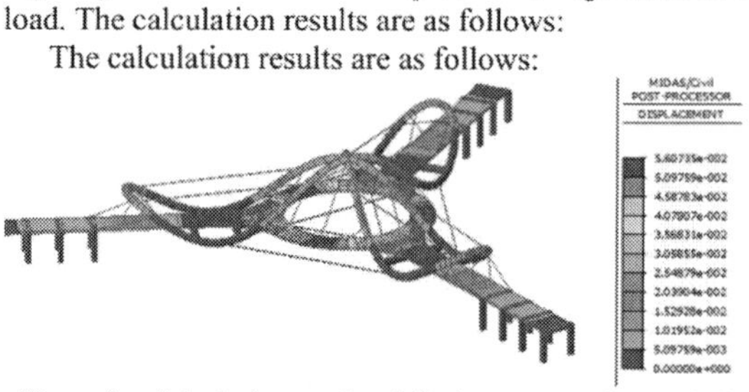

Figure 5. Calculation results of displacement under vertical load（dead load+live load）

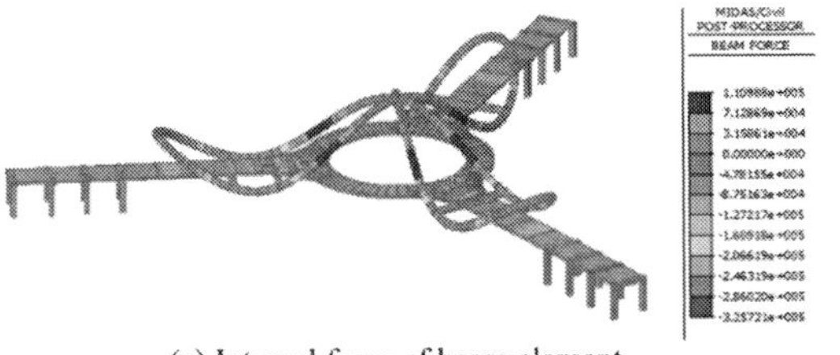

(a) Internal force of beam element

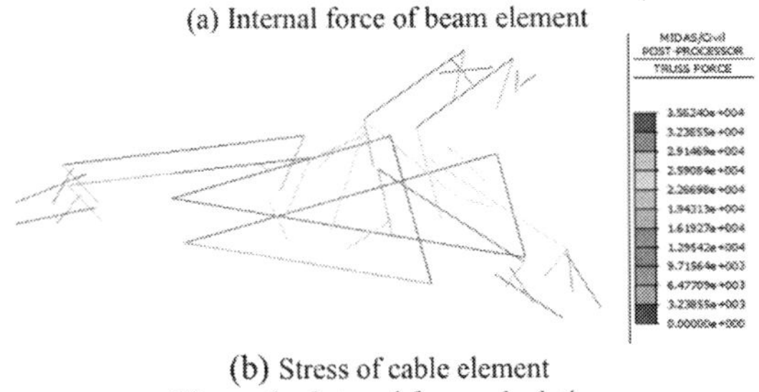

(b) Stress of cable element

Figure 6. Internal force calculation result(dead load+live load)

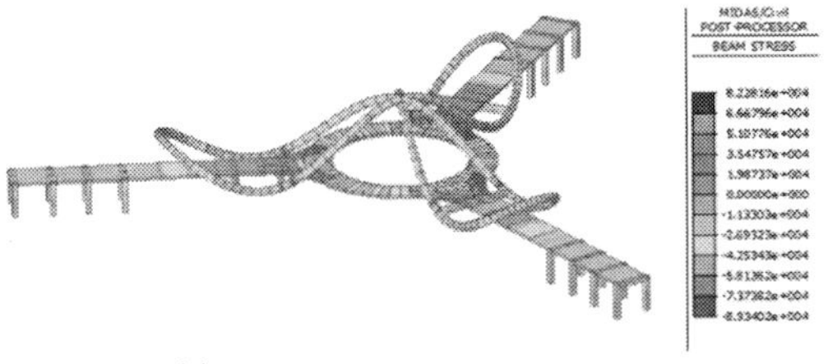

(a) Internal force of beam element

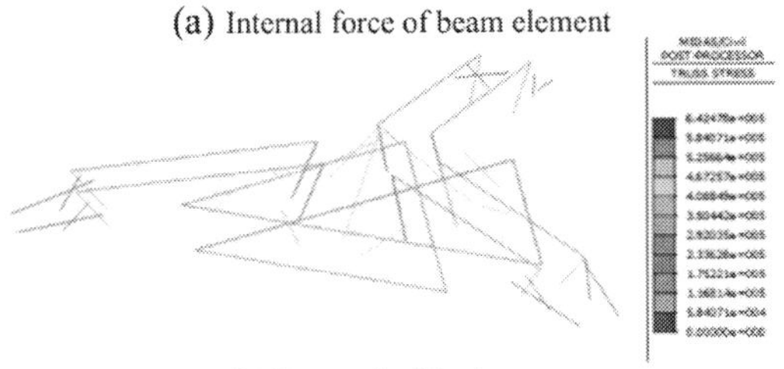

(b) Stress of cable element

Figure 7. Stress calculation results (dead load+live load)

The maximum vertical displacement occurs in the middle of the span, and the maximum displacement is 0.036 m, which meets the requirement of 1 / 500 limit specified in the code.

The maximum internal force of the cable is 35624.0kn, and the maximum stress of the cable is 642.5mpa, which meets the strength requirements. The maximum internal force of bridge arch is 325721kn, and the maximum stress of bridge arch is 89.3mpa, which meets the strength requirements. The concrete-filled steel tube, C40 concrete and q355 steel can meet the strength requirements.

5 Dynamic modal analysis

Based on Midas's nonlinear static analysis and modal analysis function, the natural vibration characteristics are analyzed. In order to avoid omitting any vibration mode, the sub block method is used to solve the characteristic equation.

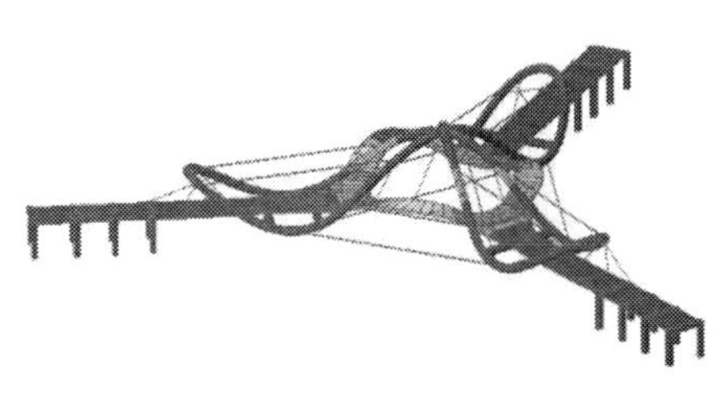

(a) 1st order mode（2.037HZ）

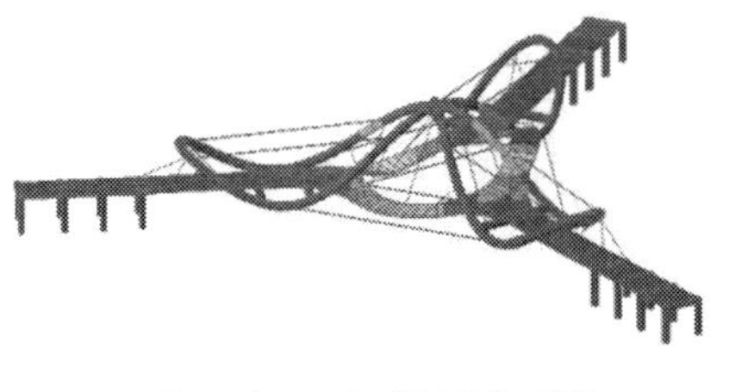

(b) 2st order mode（2.062Hz HZ）

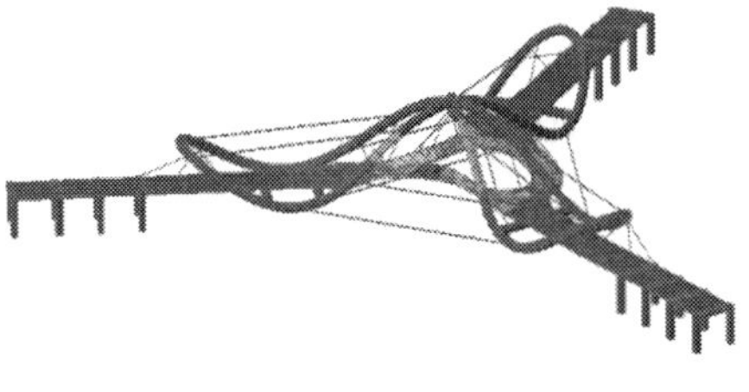

(c) 3st order mode（2.093HZ）

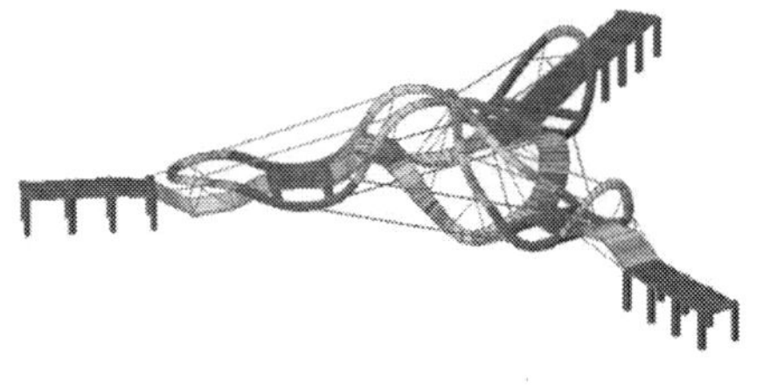

(d) 4st order mode（2.693HZ）

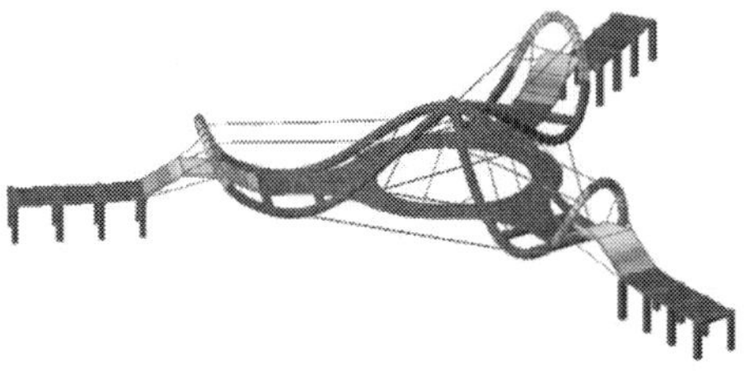

(e) 6st order mode（2.87HZ）

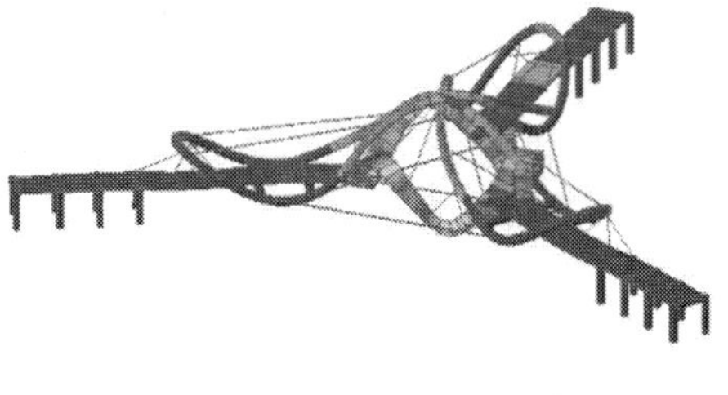

(f) 9st order mode（3.314HZ）

Figure 8. Dynamic mode

It can be seen from the calculation results in the above figure that the first mode is positive symmetrical side bending with frequency of 2.037hz; the ninth mode is torsional mode with frequency of 3.314hz.

On the whole, the vibration modes are dense, and obvious mode grouping phenomenon occurs. The first eight modes are mainly lateral bending and vertical bending vibration, and the positive symmetric torsional vibration mode does not appear until the ninth order. The torsion to bending frequency ratio of 1.63 is higher, and the wind resistance stability of the structure is better.

The triple fork shaped CFST flying swallow type cable-stayed arch bridge, with three legs, greatly improves the structural stability of the arch bridge and has good wind resistance stability.

6 Conclusion

In this paper, the configuration of three fork flying swallow type cable-stayed arch bridge is studied, the Midas finite element model is established, the calculation and analysis under vertical load are carried out, and the dynamic modal analysis is carried out

(1) A new type of three leaf rose shaped concrete-filled steel tubular flying swallow type cable-stayed arch bridge is proposed. Its plane orthographic projection is three-fork rose line figure, and the side elevation projection of arch rib is cosine function, which has unique shape and convenient transportation.

(2) Space cables are set between the three-leaf rose linear spatial arch ribs, and tie bar cables are set between the arch feet to facilitate the floating and dragging across the river. The suspension cables are suspended on the three bifurcated bridge deck, and the cable-stayed cables are set at the tail to form a space cable system. The thrust force of the middle arch and the reverse thrust of the tail warped cable-stayed arch bridge tower are balanced each other, forming a self balancing structural system, eliminating the unbalanced internal force of the support.

(3) The maximum vertical displacement occurs in the middle of the span, and the maximum displacement is 0.036 m, which meets the requirement of 1 / 500 limit specified in the code. The maximum internal force of the cable is 35624.0kn, and the maximum stress of the cable is 642.5mpa, which meets the strength requirements.

(4) The first eight vibration modes are mainly lateral bending and vertical bending vibration, and positive symmetry torsional vibration mode does not appear until the ninth order. The torsion to bending frequency ratio of 1.63 is higher, and the wind resistance stability of the structure is better.

References

1. Yiming, J.,Sanyuan, Z., Generalized rose line and its application [J]. Computer application research, 2004 (03): 170-171 (in Chinese)
2. Zhongfu, Y., Research on the horizontal thrust balance effect of tie bar on arch foot of flying swallow type concrete filled steel tube arch bridge [D]. Shaanxi: Chang'an University, 2009 (in Chinese)
3. Zong Z H, Jaistli B, Ge J P-Dynamic analysis ofa half-through Concrete-filled steel tubular arch bridge[J].Engineering structures.2005.
4. Yong, W., Mechanical analysis and experimental verification of special-shaped concrete filled steel tube arch bridge [D]. Sichuan: Southwest Jiaotong University, 2014(in Chinese)
5. Morcous, G., Hanna, K., Deng, Y., & Tadros, M. K. . Concrete-Filled Steel Tubular Tied Arch Bridge System: Application to Columbus Viaduct[J]. JOURNAL OF BRIDGE ENGINEERING, 1, 107. 2012.
6. Lei Z.,Xinxin H., Experimental study on natural vibration characteristics and dynamic response of cable-stayed flying swallow arch bridge [J]. Building engineering technology and design, 2015, (9): 1198-1198(in Chinese)
7. Min P., optimization of design parameters of cable stayed arch bridge [D]. Hunan: Hunan University of science and technology, 2008(in Chinese)
8. .Nakamura, S., Tanaka, H., & Kato, K. . Static analysis of cable-stayed bridge with CFT arch ribs [J]. Journal of Constructional Steel Research, 65(4), 776–783.2009.

The influence of soft clay saturation characteristics on unconfined compressive strength in Guangdong-Hong Kong-Macao Greater Bay Area

DaShu Guan[1,2,a], JiaXi Zheng[1], WenHao Huang[1*], JunZhi Zhong[1], WenWen Du[1], ZhiDe Luo[1] , and YinShi Zhou[1]

[1] Guangdong Polytechnic of Water Resources and Electric Engineering, Guangzhou, Guangdong 510925, China

[2] State Key Laboratory for Geomechanics & Deep Underground Engineering, China University of Minding and Technology, Xuzhou, Jiangsu221116, China

Abstract. In order to study the influence of saturated characteristics of soft clay on unconfined compressive strength, the soft clay of Guangdong-Hong Kong-Macao Bay Area is taken as the research object, comparing and contrasting on unconfined compression test of saturated clay and unsaturated clay in laboratory, studying the variation law of unconfined compressive strength and sensitivity of unsaturated and saturated soft clay. The test results show that: 1. During the failure of unsaturated soft clay samples, oblique fractures appear, showing brittle shear failure, while the saturated clay samples appear constant bulging in the middle, and finally a "cross" is broken in the central bulging part, showing plastic shear failure.2. The unconfined compressive strength of unsaturated clay is about 10kPa higher than that of saturated soft clay, and its corresponding sensitivity is also about 0.4 higher. It can be shown that the soil saturation has a certain influence on the soil strength. The difference between saturated soil and unsaturated soil is the existence of gas phase. To be more precise, the existence of gas phase in unsaturated soil, i.e. the existence of suction, makes the soil stronger and presents the brittle shear failure form, while the saturated soil basically presents the plastic shear failure form.

1 Introduction

Saturated soil has only two phases, i.e. solid phase and liquid phase (i.e. soil particles and water); unsaturated soil is generally recognized as four phases, namely solid phase (soil particles), liquid phase (water), gas phase (air) and shrinkage membrane (water air interface). The saturation of saturated soil is 1, and that of unsaturated soil is less than 1. Therefore, the two can convert each other, exhausting all the gas in saturated soil, so that the saturation becomes 1, that is to say, saturated soil. However, the water in the saturated soil is drained away, and the air enters into the soil, which makes the unsaturated soil become. Therefore, there are differences and connections between the two. Due to the existence of gas phase in unsaturated soil, the properties of soil are more complex than that of saturated soil. For the convenience of engineering research, most researchers will consider using saturated soil to do the experiment. Because the influence of gas and water air interface can be excluded in saturated soil, it can be regarded as a kind of ideal soil, so the research theory of saturated soil is also better Comprehensive improvement. However, the research on unsaturated soil is less. As a common soil in nature, the study of unsaturated soil is also considered from the aspects of deformation, strength and seepage. However, due to the existence of suction, the soil condition is more complex. And suction is part of the gas phase.

Suction is an important parameter of unsaturated soil properties and an important stress variable to characterize unsaturated soil properties. It is the premise and foundation of unsaturated soil properties research. The total suction of soil consists of matrix suction and osmotic suction. The matric suction is mainly affected by the change of soil water content or saturation, and the osmotic suction is mainly affected by the change of salt content. It is also an important factor affecting the strength characteristics of unsaturated soil. Domestic and foreign scholars have studied the influence of suction on the engineering properties of unsaturated soil [1~3]. Toll [4] proposed to separate the total stress and matrix suction to consider its shear effect on unsaturated soil, which also proved that the matrix suction affects the shear strength of soil through some transformation. Domestic scholars have also proved this idea through various indoor experiments [5~6] . Through unconfined compressive strength test, Jinfeng Li [7] proved that water and salt have influence on the permeability and suction of unsaturated soil.

Unconfined compressive strength test is an important index of soil mechanical properties [8].

In this paper, the soft clay in Guangdong, Hong Kong and Macao Bay area is taken as the research object.

[a]E-mail: 156230234@qq.com
* Corresponding author: WenHao Huang: E-mail: 2489202747@qq.com

 https://doi.org/10.1051/e3sconf/202129302011

Through the indoor unconfined compressive strength test, the stress-strain curve, unconfined compressive strength and sensitivity of saturated clay and unsaturated clay are obtained respectively to explore the difference of unconfined compressive strength between saturated clay and unsaturated clay. The results are expected to promote the further understanding of gas phase in soil, and promote the further development of saturated soil mechanics.

2 Sample preparation

2.1 Test materials

The test soil samples were collected from the soft clay of Shanwei and Foshan in Guangdong, Hong Kong and Macao Bay area respectively. The basic physical parameters of the soil were determined according to the standard for soil test methods. The results of the basic parameters are shown in Table 1.

Table1. Basic physical parameters of soil samples.

test number	Density /g·cm^2	natural water content /%	proportion	liquid limit /%	plastic limit /%	cohesive force /kPa	internal friction angle /°
14#	1.689	53.38	2.641	56.98	33.61	19.76	16.22
21#	1.625	53.37	2.639	59.18	34.43	19.24	16.54
B区	1.694	53.40	2.637	48.09	28.61	12.37	12.30
25-1	1.685	46.78	2.646	53.48	31.21	14.75	18.10
25-2	1.682	46.92	2.643	53.54	31.17	14.70	18.06
2K-24	1.673	30.55	2.647	47.69	27.72	9.40	23.50

2.2 Test scheme

Six triaxial soil samples were made by remolding cylinder, and then three of them were extracted and saturated to make saturated soil samples. Then, the unconfined compressive strength and its corresponding sensitivity are calculated according to the experimental data, and the stress-strain curve is drawn.

2.3 Test procedure

In this unconfined compressive strength test, the strain controlled unconfined pressure gauge was used to make the original soil into a triaxial soil sample with a diameter of 39.1mm and a height of 80mm.The axial strain rate should be 1%~3% per minute.When the axial strain is less than 3%, read once every 0.5% strain (or 0.4mm), read once every 1% strain (or 0.8mm) when the axial strain is equal to or greater than 3%.The test should be completed within 8~10min.

Measurement sensitivity, the damage will be immediately after the specimen to remove the surface of the coated with vaseline, add a few more than soil, pack in the plastic film with the hand rub knead, its structure, reshape into cylindrical, into the cylinder is reshaping, with metal plate sample to squeeze into equal sample and undisturbed sample size, density, and test according to the above steps.

2.4 Data processing and calculation

The axial strain of the specimen is as follows:

$$\varepsilon_1 = \frac{\Delta h}{h_0}$$

Average sample area:

$$A_a = \frac{A_0}{1-0.01e_1}$$

Axial stress of specimen:

$$\sigma = \frac{CR}{A_a} \times 100$$

Sensitivity:

$$\varepsilon_t = \frac{q_u}{q_u'}$$

3 Analysis of test results

3.1 Data sorting

The data measured from the test were substituted into the above formula to calculate the relevant data of unearthed samples respectively. The following are 6 classic samples, as shown in Table 2.

Table2. Unconfined compression test data of soil samples.

Soil sample number	category	Undisturbed soil unconfined compressive strength q_u /kPa	The remoulded unconfined compressive strength $q_u^{'}$ /kPa	sensitivitys
14#	Unsaturated soil sample	28.25	9.51	2.97
21#	Unsaturated soil sample	25.67	9.98	2.57
B 区	Unsaturated soil sample	21.36	7.54	2.83
25-1	Saturated soil sample	9.10	4.06	2.24
25-2	Saturated soil sample	7.66	1.85	2.15
2K-24	Saturated soil sample	11.86	7.49	1.58

Then the corresponding stress-strain curve is obtained by fitting software according to the recorded data of the test.

Figure 1 to Figure 6 are the stress-strain curves of the soil sample, in which ① is the unaltered soil sample and ② is the remolded soil sample:

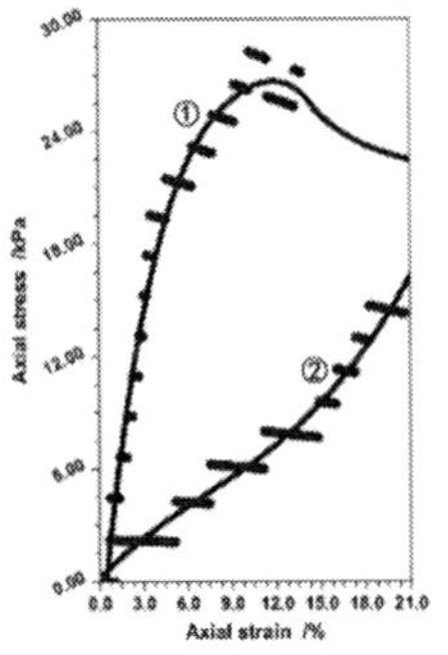

Fig.1. Sample no.14 unsaturated soil.

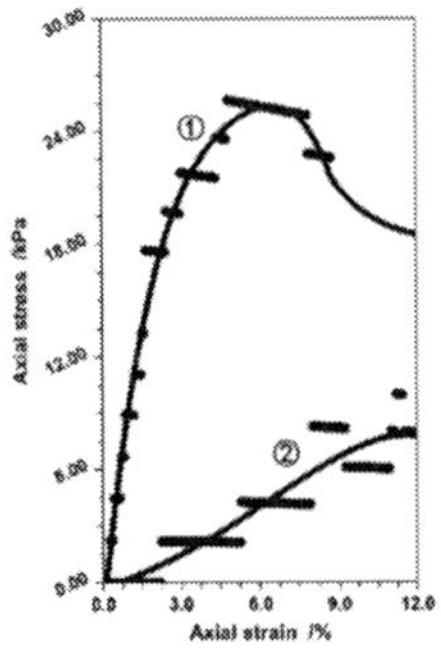

Fig.2. Sample no.21 unsaturated soil.

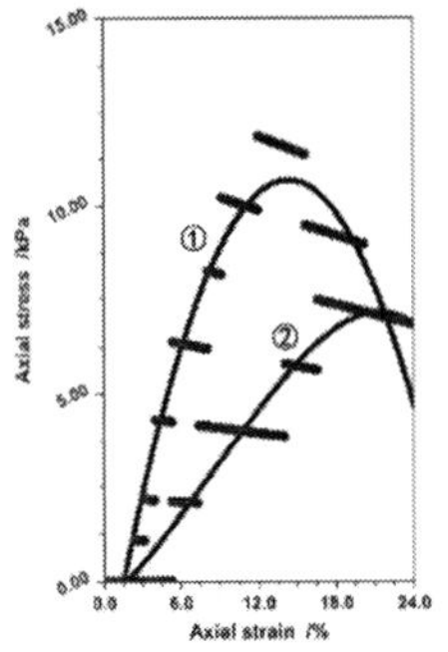

Fig. 3. Unsaturated soil saturated sample in area B.

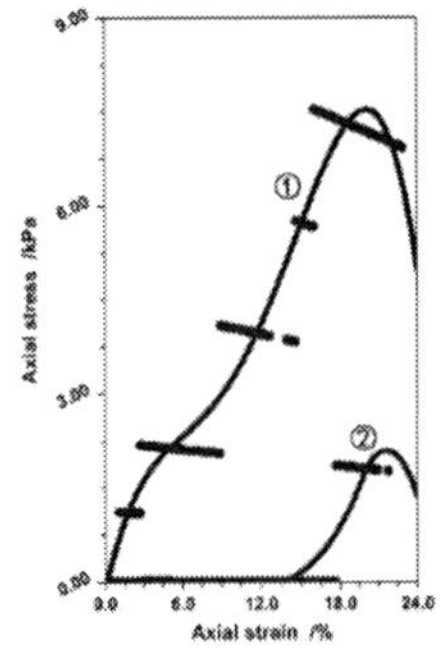

Fig. 4. Sample no No. 2-2 saturated soil.

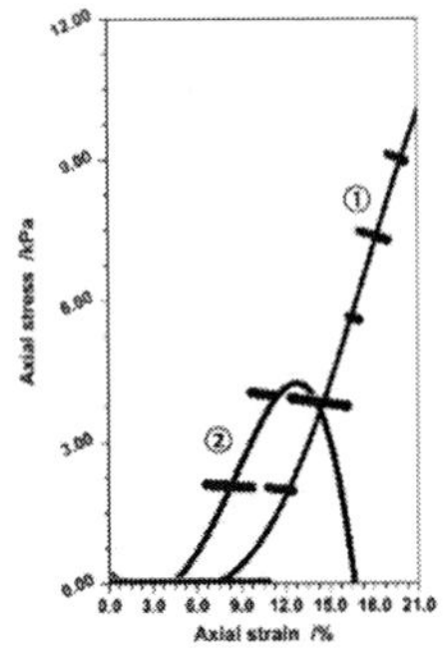

Fig.5. Sample of saturated soil.

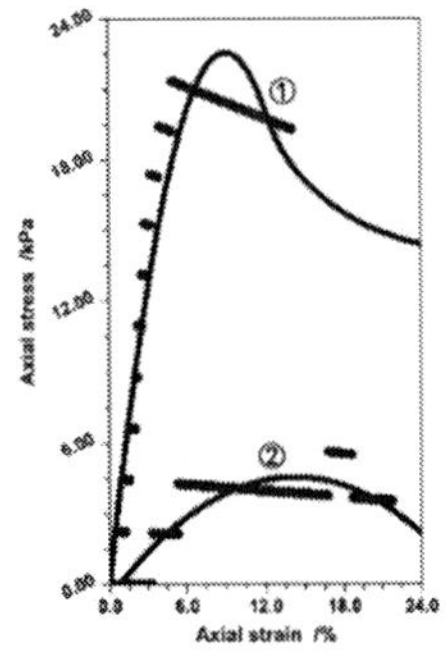

Fig.6. 2K-24 soil sample.

3.2 Analysis of Results

From 6 set of test data record and finishing in table 2, can see intuitive unsaturated undisturbed soil and remolded soil unconfined compressive strength than saturated undisturbed soil and remolded soil higher than that of the unconfined compressive strength of around 10 kPa, and its corresponding sensitivity is high, about 0.4 or so, which can explain the soil saturation and a certain influence on the strength of the soil moisture content. Both of the stress-strain curves fitted by the fitting software showed strain softening type with only peak strength, and both of them had a steep rise stage at the early stage of axial strain change, then experienced a peak, and finally experienced a steep fall stage after the peak. The difference is that the peak stage of the soil sample of unsaturated soil appears earlier than that of saturated soil, and the axial stress of unsaturated soil sample is generally about 17kPa higher than that of saturated soil.

3.3 Failure mode

The failure results of each soil sample after unconfined test are shown in the figure below.

Fig.7. 14# Failure of undisturbed soil.

Fig.8. 14# Reshaped soil destruction.

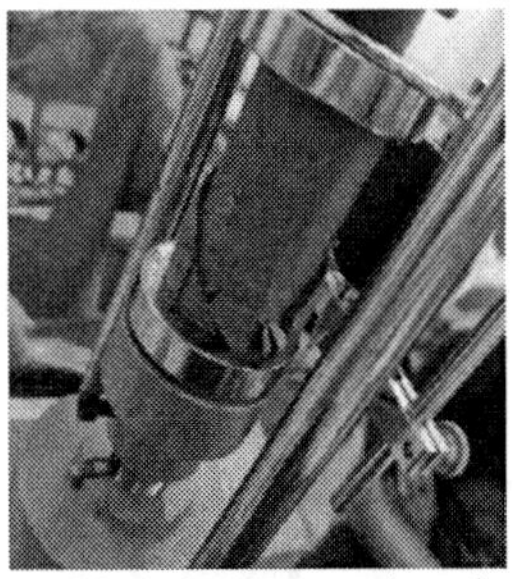

Fig.9. 21# Destruction of undisturbed soil.

Fig.10. 21# Reshaped soil destruction.

https://doi.org/10.1051/e3sconf/202129302011

Fig.11. 25-1 Failure undisturbed soil.

Fig.12. 25-1Remodeled soil destruction.

Fig.13. 25-2 Failure of undisturbed soil.

Fig.14. 25-2 Remolded soil destruction.

Fig.15. Failure of undisturbed soil in Area B.

Fig.16. Remolded soil failure in Area B.

Fig.17. Damage of 2K-24 undisturbed soil.

Fig.18. 2K-24 remolded soidamage.

4 Conclusion

Unsaturated soil by extraction after becoming saturated soil in the saturated, unconfined strength of the soil cut relatively than the original by about 10 kpa, sensitivity and relatively low around 0.4, existing in the research results show that in the unsaturated soil sample was particle pore in the extraction of gas saturation after most of the displacement into water, the soil strength is relatively weak. More specifically, the presence of suction in the soil is relatively higher than the strength of the soil sample filled with water in the pores, that is, the presence of gas phase in the unsaturated soil, that is, the presence of suction, has a certain influence on the strength of the soil. The presence of suction makes the soil stronger.

About the author

* First author: *DaShu* Guan (1983 -), male, PhD, senior engineer, mainly engaged in research and design of engineering geology, composite foundation and geological disaster prevention. Mobile :15913142403 ;
* Corresponding author: *WenHao* Huang (2001 -), male, mainly engaged in research of engineering geology and foundation treatment. Mobile: 13421528577;

Funded by the following funds:

1. Guangdong Polytechnic of Water Resources and Electric Power (cy0604zk06);
2. Scientific Research innovation Platform of Guangdong Polytechnic of Water Resources and Electric Power (cy0603yz02);
3. Special Fund for Science and Technology Innovation Strategy of Guangdong Province (Special Fund for "Climbing Plan", pdjh2021b0812).

References

1. Hongzhou Lin, Guangxin Li, Yuzhen Yu, et al. The influence of matrix suction on shear resistance of unsaturated soil [j]. Geotechnical mechanics, 2007, 28 (09): 1931-1936.
2. Kun Huang, Junwei Wan, Gang Chen, et al. Experimental study on the relationship between shear strength and water content of unsaturated soil [J]. Geotechnical mechanics, 2012, 33 (09): 2600-2604.
3. Qianling Meng, Hailin Yao, Lunfeng Qiu. Contribution of suction to shear strength of unsaturated soil [J]. Geotechnical mechanics, 2001, 22 (04): 423-426 + 431.
4. Toll D G. A framework for unsaturated soil behaviour[J]. Géotechnique, 1990, 40(1): 31-44.
5. Haotian Guo. Study on frost heaving failure mechanism of deep foundation pit in seasonal frozen area based on triaxial test of unsaturated soil [D]. Jilin Jianzhu University, 2018.
6. Tao Li, Bo Liu, Weihong Yang, et al. Experimental study on the influence of matrix suction on shear strength of remolded red clay [J]. Journal of China University of mining and technology, 2013,42 (03): 375-381. [77].
7. Jinfeng Li. Experimental study on influence of water and salt on suction and unconfined compressive strength of unsaturated soil [D]. Jilin University, 2020.

https://doi.org/10.1051/e3sconf/202129302011

8. Nanjing Institute of water resources research. Standard of soil test method: GB/T50123-1999 [S]. Beijing: China water resources and Hydropower Press, 1999:14-16.
9. Yuhan Li. Experimental study on influence of water content and clay content on strength of unsaturated soil [D]. East China Jiaotong University, 2014.
10. Ning Lu, William, J. likos, unsaturated soil mechanics [M], June 2015, higher education pres.

Application of J_2 Integral Method in Calculating Complicated Stress Intensity Factors on Three-dimensional Components

Long. Li[1], Yousheng. Deng[1]*, Liqing Meng[1], Lingtao Li[1], and Yunfang Zheng[1]

[1]College of Civil and Architectural Engineering, Xi'an University of Science and Technology, Xi'an 710054, China

Abstract. In this paper, a technique to determine complicated stress intensity factors on three-dimensional components, which based on the conservation law and the elementary mechanics is proposed, it only needs the geometric relationship between multiple singular stress fields from the crack section, and obtaind the relationship between the stress at different crack tips. In the expression of the stress intensity factor K, K is proportional to the stress σ at the crack tip, and we can get the supplementary equation of between different stress fields K according to the ratio of the stress at the crack tip, then use the *J*-integral method to calculate the stress intensity factors of different stress fields. In order to verify the feasibility of this method, a cracked R-fluted shells model was constructed. Under the action of the bending moment, the corner crack propagation is simulated through the reserved corner crack, and two crack tips with different stress fields are generated during the simulation. The experimental result indicates that the proposed method is effective for cracked R-fluted shells. It is also shown that the method has universal applicability for solving complex stress intensity factors on three-dimensional components.

1 Introduction

R-fluted shells are widely used in engineering structures, and cracks and similar cracks are common defects. The cracked R-fluted shells is a typical three-dimensional structure. Regarding infinite two-dimensional (2d) and three-dimensional (3d) elastomer crack problems, the closed solutions of SIFs can be conveniently given. However, for cracked slender R-fluted shells,which belongs to the three-dimensional (3d) problem with finite boundaries, In addition to numerical methods, it is difficult to get the analytical solutions. For cracked beams-like structures, much effort had been done to find the simple methods solving the SIFs. In 1986, a simple method for close approximation of stress intensity factor in cracked beams was proposed by Kienzler and Herrmann [1-2]. Kienzler and Herrmann assumed, without specific justification, that the strain energy release rate G for crack extension is equal to that for crack widening, thereby using Irwin's G-K relation to determine the SIFs. In 1990, Bazant [3] indicated that this assumption is approximately valid within a correction factor, and claimed this can only be determined through optimum fitting of the exact solution. On the basis of previous researchs, Gao and Herrmann [4] showed that this correction factor can be obtained through simple asymptotic matching with standard limiting crack solutions in 1992. They also made some corrections and clarifications for estimation of the stress intensity factor of axial-symmetrically cracked beams. Dunn et al.[5] in 1997, based on the elementary strength theory for cracked beams forwarded by Gao and Herrmann, presented a simple, closed-form expressions for stress intensity factors for cracked I-beams subjected to a bending moment. In 2006, by estimating the SIFs of cracked T-beams and bars, Ricci and Viola [6] presented an extension of a simple and convenient method formerly proposed by Kienzler and Herrmann. In 1998, Based on the three-dimensional (3d) conservation law and the virtual work principle, Xie et al. [7-8], proposed the J_2-integral theory, which greatly simplified the analysis of the stress intensity factor for cracked slender structures. The Xie's approach had been extended by Dotti et al. [9-10] as an important contribution to determine the mode-I stress intensity factor of the cracked thin-walled beam. Nobile [11-12], Kolitsch et al. [13], Livieri [14] and Alijani et al. [15], also made some efforts to utilize the simple engineering method to calculate the SIFs.

Since the J_2 integral has obvious simplicity and accuracy in solving the stress intensity factor of a three-dimensional cracked shell structure, it is widely used to solve the SIF of cracked hollow shaped pipes, pipes and other engineering components[16-18]. However, the J_2-integral can only provide one equation, which may be used to solve one unknown stress intensity factor, including some cases of multiple similar cracks with the same SIFs in a cross-section [19-20]. Obviously, it is not enough for engineering applications. Nearly all the cracked beams and slender structures concerning multiple and different singular stress fields in a cross-section had been sidestepped [21]. In the present work, based on the assumption that flat cross-sections remain plane in the elementary mechanics, some supplementary equations,

* Corresponding author: dengys2009@126.com

together with the J_2-integral, can be effectively used to solve the SIFs corresponding to multiple and different singular stress fields, such as the SIFs of slender R-fluted shells with periodic cracks, such as the circumferential corner cracks and inner corner cracks. It is worth noting that all the analysis and calculations of present method are within the scope of elementary strength of materials and the error is acceptable in practice.

2 Configuration of slender R-fluted shell

Fig. 1 gives a case of four periodic corner cracks in slender R-fluted shell under bending. There are eight crack tips with different singular stress fields, each of which has typical features of two-dimensional singular stress field in plane strain state. The x_2 - x_3 plane in Fig. 1 is a symmetry plane for cracked shell, and all loads act in this plane; hence the bending deflections will take place in this plane also. Obviously, when $l_b >> B >> t$ the deformation of the shell possesses the characteristics of the three-dimensional shell and the slender beam. Therefore, the stress intensity factors can be obtained by the conservation law and the bending theory in the material mechanics.

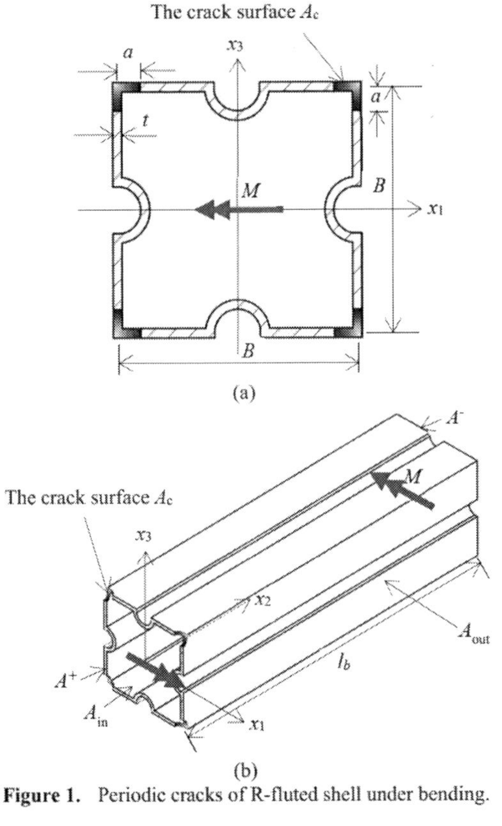

Figure 1. Periodic cracks of R-fluted shell under bending.

3 J_2-integral and stress intensity factors

Considering a three-dimensional strain field, for which the displacement vector ui depends on x_1, x_2, x_3 as shown in Fig. 2, from the conservation law, J_2-integral can be defined as [7].

$$J_j = \oint\!\!\oint_{\Omega} \left(w\boldsymbol{n}_j - T_i u_{i,j} \right) d\Omega = 0, \quad j = 1,2,3 \tag{1}$$

where Ω in Eq. (1) denotes an arbitrary closed surface in the solids without defects; w is the strain energy density; T_i is the stress vector acting on the outer side of Ω; $\boldsymbol{n}$ is the unit outward normal to Ω.

For a two-dimensional deformation field. along the path S_{ab} and S_{bc} , Eq. (1) yields[7]:

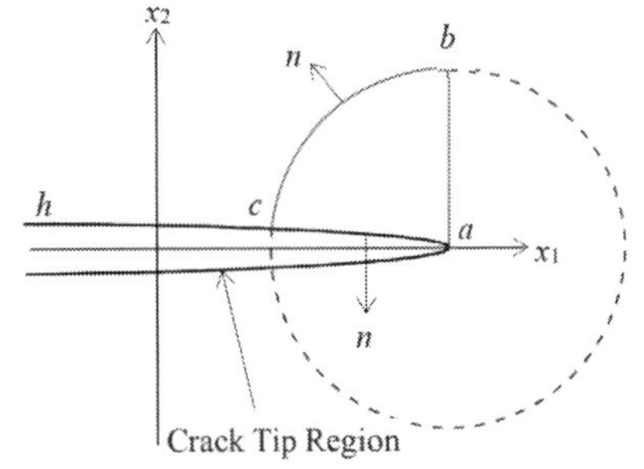

Figure 2. Integration path around crack tip region.

4 SIFs for a R-fluted shell with circumferential periodic cracks under bending

4.1 Configuration of circumferential periodic cracks in four corners of R-fluted shell

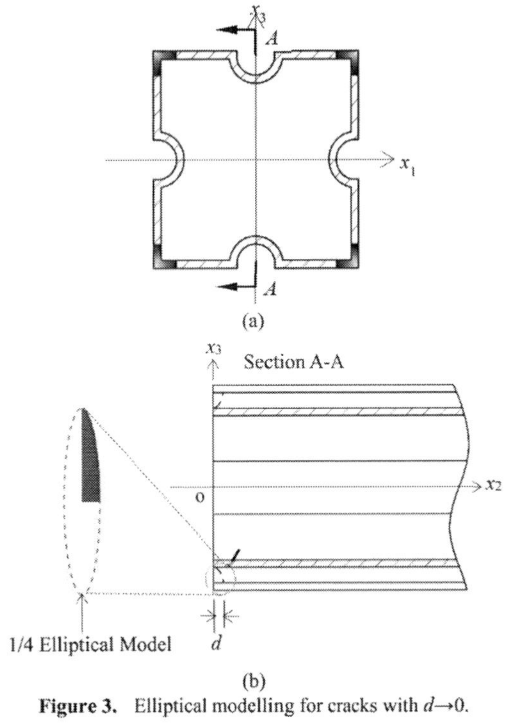

Figure 3. Elliptical modelling for cracks with $d \to 0$.

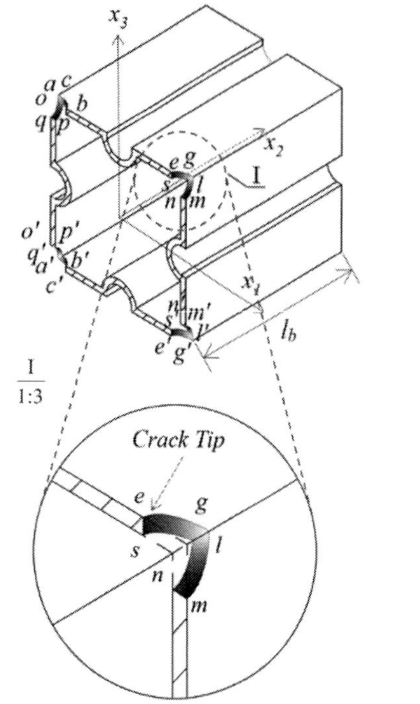

Figure 4. Local integral surfaces within the K-dominant regions.

A cracked R-fluted shell is shown in Fig. 1. The x_1, x_2 plane in Fig. 1 is a symmetrical plane of the shell, and all the loads act in this plane. t is the thickness of the shell, a is the length of the crack, where $l_b >> t$, l_b is the cross-section length of shell, B is the cross-section width of shell, M is the axial moment.

Fig. 1 shows a three-dimensional closed surface Aclosed

$$A_{closed} = A^{+} + A^{-} + A_{in} + A_{out} + A_c \tag{2}$$

where the symbol '+' denotes the cracked cross section;'-' denotes the remote uncracked cross section; Ain is the inner surface of shell; Aout is the outer surface of shell; A_c is the sum of the four crack surfaces. Since A_{in} and A_{out} is free surface, $T_i = 0$ and $n_2 = 0$. following results can be found

$$\iint_{A_{in}}\left(wn_2 - T_i u_{i,2}\right)dA = \iint_{A_{out}}\left(wn_2 - T_i u_{i,2}\right)dA = 0 \tag{3}$$

According to beam theory, for the surfaces A^- and A^+, it is not difficult to get

$$\iint_{A^-}\left(wn_2 - T_i u_{i,2}\right)dA = \overline{w}^- - M\varphi'^- = -\frac{M}{2}\varphi'^- \tag{4}$$

$$\iint_{A^+}\left(wn_2 - T_i u_{i,2}\right)dA = -\left(\overline{w}^+ - M\varphi'^+\right) = -\frac{M}{2}\varphi'^+ \tag{5}$$

where is the axial curvature of the R-fluted shell, is the strain energy density per unit length， n is unit outward normal to surface A in Eqs. (4) and (5). In the remote transverse cross-section, the curvature of the axis is

$$\varphi'^- = \frac{M}{EI} \tag{6}$$

where the moment of inertia I = B⁴/12-πD⁴/64-2(πD⁴/128-8πR⁴/9π²+πR²/2(B/2- 4R/3π)²) and D =2R. If the crack is formed from an 1/4 ellipse with $d \to 0$ as shown in Fig. 3, the R-fluted shell can be regarded as a variable cross-section beam, and the mean curvature at the cross section becomes

$$\phi'^{+} = \lim_{d\to 0}\frac{1}{d}\int_0^d \frac{M}{EI}\frac{dx_2}{1-\left[\frac{8t^3\left(\frac{a}{B}\right)\left(1-\left(x_2/d\right)^2\right)^{\frac{1}{2}}+48ta\left(1-\left(x_2/d\right)^2\right)^{\frac{1}{2}}}{B^3}+\left(\frac{a}{B}\right)64t\left(1-\left(x_2/d\right)^2\right)^{\frac{3}{2}}\right]\Big/ I} = \frac{M}{EI}\gamma_1\left(a/B\right) \tag{7}$$

Where by introducing the integral variable $\xi = x_2/d$,

$$\gamma_1\left(a/B\right) = \int_0^1 \frac{d\xi}{1-\left(\frac{a}{B}\right)\frac{\left(96t^3+576tB+768tB^3\right)\left(1-\xi\right)^{\frac{1}{2}}}{B^7-6\pi B^3R^4-3\pi B^5R^2+16R^3B^4}} \tag{8}$$

Additionally, in cracked cross sectional area, there are eight asymptotic singular stress fields next to the crack tip zone, i.e. , b, b', e, e', q, q', m, m' as shown in Fig. 4. for which the SIFs can be expressed by $K_{Ib} = K_{Ie} = -K_{Ib'} = -K_{Ie'}$, $K_{Iq} = K_{Im} = -K_{Iq'} = -K_{Im'}$, The crack surface A_c can be divided into K-dominants (A_{c1}) and free surface (A_{c2}), i.e., $A_c = A_{c1} + A_{c2}$., it is not difficult to get

$$\begin{aligned} A_{c1} &= A_{bc} + A_{eg} + A_{lm} + A_{pq} + A_{b'c'} + A_{e'g'} + A_{l'm'} + A_{p'q'} \\ A_{c2} &= A_{ab} + A_{es} + A_{ln} + A_{oq} + A_{a'b'} + A_{e's'} + A_{l'n'} + A_{o'q'} \end{aligned} \tag{9}$$

Taking the three-dimensional J_2-integral over A_{ac} as shown in Fig. 4 for example, by using the conservationlaw and Eqs. (1),then,

$$J_2\left(A_{ml}\right) = \iint_{A_{ml}}\left(wn_2 - T_i u_{i,2}\right)dA = t\int_{S_{ml}}\left(wn_2 - T_i u_{i,2}\right)ds = -t\frac{\left(1-\mu^2\right)K_{Im}^2}{2\pi E} \tag{10}$$

$$J_2\left(A_{eg}\right) = \iint_{A_{eg}}\left(wn_2 - T_i u_{i,2}\right)dA = t\int_{S_{eg}}\left(wn_2 - T_i u_{i,2}\right)ds = -t\frac{\left(1-\mu^2\right)K_{Ie}^2}{2\pi E} \tag{11}$$

Then, from Eqs. (9)-(11), the J_2-integral over A_c becomes

$$\begin{aligned} J_2\left(A_c\right) &= 8J_2\left(A_{ml}\right) + 8J_2\left(A_{eg}\right) + J_2\left(A_{c2}\right) \\ &= -t\frac{4\left(1-\mu^2\right)K_{Im}^2}{\pi E} - t\frac{4\left(1-\mu^2\right)K_{Ie}^2}{\pi E} - J_2\left(A_{c2}\right) \\ &= -t\frac{4\left(1-\mu^2\right)\left(K_{Im}^2 + K_{Ie}^2\right)}{\pi E} - \iint_{A_{c2}} w\,dA \end{aligned} \tag{12}$$

Substituting the Eqs. (3)-(5) and Eq. (12) into Eq. (1), the J_2-integral over the closed surface Aclosed can be expressed as

$$t\frac{4(1-\mu^2)(K_{Im}^2+K_{Ie}^2)}{\pi E}+\iint_{A_{\varepsilon 2}} w dA=\frac{M}{2}(\phi'^{+}-\phi'^{-}) \quad (13)$$

4.2 Supplementary equation of SIFs

Since J_2-integral provide only one equation that contains two different SIFs, as indicated by Eq.(13), more equation is needed to solve these SIFs. For a R-fluted shells under pure bending, the distribution of the normal stress is schematically shown in Fig. 5. based on the elementary mechanics and σ=My/I, at the crack tips *e* and m as shown in Fig. 4, σ_e / σ_m can be found as

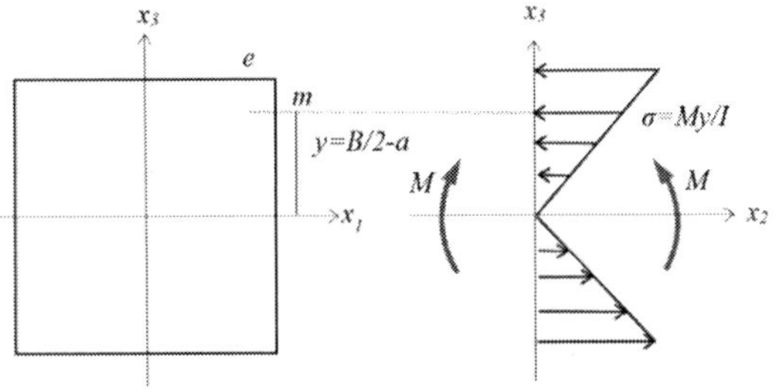

Figure 5. Distribution of the normal stress for a slender R-fluted shell under pure bending.

$$\frac{\sigma_e}{\sigma_m}=\frac{B}{B-2a} \quad (14)$$

Referring to above equations and $K_I=\sigma_0\sqrt{\pi B}$, K_{Ie} and K_{Im} can be expressed as

$$\frac{K_{Ie}}{K_{Im}}=\frac{B}{B-2a} \quad (15)$$

4.3 Normalized SIFs

Solving the Eqs. (13) and (15) and by using Eqs. (6) and (7), the J_2-integral over the closed surface can be expressed as

$$\frac{4t(1-\mu^2)K_{Ie}^2}{\pi E}+\iint_{A_{\varepsilon 2}} w dA=\frac{M^2}{2EI}\left[\gamma_1(a/B)-1\right] \quad (16)$$

where $\iint_{A_{\varepsilon 2}} w dA$ is a small quantity and can be neglected. Hence, $f_e(a/B)$ and $f_m(a/B)$ can be found as

$$f_e(a/B)=\frac{K_{Ie}}{\sigma_0\sqrt{\pi B}}=\left[\frac{\pi}{4t(1-\mu^2)}(\gamma_1(a/B)-1)\right]^{1/2} \quad (17)$$

$$f_m(a/B)=\frac{K_{Im}}{\sigma_0\sqrt{\pi B}}=\frac{B}{B-2a}\left[\frac{\pi}{4t(1-\mu^2)}(\gamma_1(a/B)-1)\right]^{1/2} \quad (18)$$

where $\sigma_0 = MB/2I$, $f_e(a/B)$ and $f_m(a/B)$ is the normalized SIFs of K_{Ie} and K_{Im} respectively.

The mode I SIFs K_{Ie} and K_{Im} were calculated also by using XFEM in the commercial software Abaqus, in which the C3D20 type element is employed. The slender R-fluted shell possesses both shell and beam characteristics, so the geometrical dimension of the slender R-fluted shell considered were l_b = 100 mm, B = 25 mm, t = 1 mm, R = 5 mm, E = 200 GPa and μ = 0.3. The comparisons of the value of FEM and present method are as shown in Table. 1 and Table. 2 , the comparisons on the results of Eq. (17) and Eq. (18) with the finite element analysis are as shown in Fig. 6.

Table1. Comparisons of the value of FEM and Present method(f_e)

a/B	0.05	0.1	0.15	0.2	0.25	0.3	0.35	0.4	0.45	0.5
FEM	0.025	0.056	0.162	0.270	0.370	0.520	1.000	2.300	6.012	11.040
Present method	0.026	0.054	0.130	0.260	0.340	0.500	0.910	1.800	5.100	10.110

Table2. Comparisons of the value of FEM and Present method(f_m)

a/B	0.05	0.1	0.15	0.2	0.25	0.3	0.35	0.4	0.45	0.5
FEM	3.21	5.19	7.01	8.09	8.74	8.78	8.19	7.05	4.15	-
Present method	3.71	5.76	7.50	8.19	8.94	8.58	8.00	6.85	3.95	0

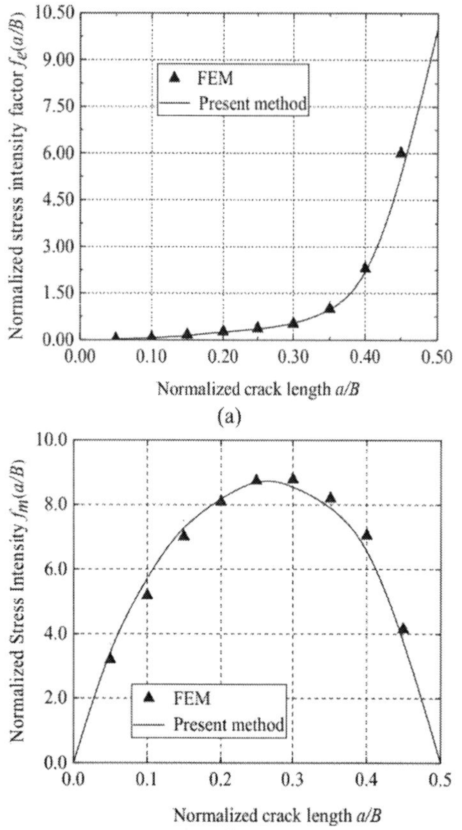

(a)

(b)

Figure 6. The normalized SIFs.

5 Conclusions

This paper draws the following conclusions:

(1)It is proved that J_2-integral and the assumption that plane cross-sections remain plane in the elementary mechanics is feasible for estimate the different SIFs for R-fluted shells with multiple cracks under bending.

(2)The simple relationship between different stress intensity factors can be found based on the elementary mechanics and σ=My/I, which provides an idea for solving the problem of stress intensity factors corresponding to complex stress fields.

(3) This method simplifies the solution of SIFs of cracked engineering structures with complex cracks and all analyses and calculations of the method are within the scope of elementary mechanics. This method is flexible and the error is acceptable in practice.

Acknowledgements

This research was financially sponsored by National Natural Science Foundation (51878554, 41672308) and Natural Science Foundation of Shaanxi (2018JZ5012).

References

1. R. Kienzler, G. Herrmann. Acta Mech, **62**, 37-46. (1986)
2. G. Herrmann, H. Sosa. Eng Fract Mech, **24**,889-894. (1986)
3. ZP. Bazant. Eng Fract Mech, **36**, 523-525. (1990)
4. H. Gao, G. Herrmann. Eng Fract Mech, **41**, 695-706.(1992)
5. M. L. Dunn, W. Suwito, B. Hunter. Eng Fract Mech. **57**, 609-615 (1997)
6. P. Ricci, E. Viola. Eng Fract Mech. **73**, 91-111(2006)
7. Y. J. Xie, H. Xu, PN. Li. Theor Appl Fract Mech.;**29**, 195-203 (1998)
8. Y. J. Xie. Int J Pressure Vessels Piping. **75**, 865-869. (1998)
9. FE. Dotti, VH. Cortínez, F. Reguera. Theor Appl Fract Mech. **67**, 38-45. (2013)
10. VH. Cortínez, FE. Dotti. Eng Fract Mech. **110**, 249-257. (2013)
11. L. Nobile. Theor Appl Fract Mech. **33**, 107-116. (2000)
12. L. Nobile. Theor Appl Fract Mech. **41**, 137-145. (2004)
13. S. Kolitsch, HP. Gänser, R. Pippan. Theor Appl Fract Mech. **92**, 167-177. (2017)
14. P. Livieri. Theor Appl Fract Mech. **84**, 119-128. (2016)
15. A. Alijani, MM. Abadi, A. Darvizeh, MK. Abadi. Arch Appl Mech. **88**, 875-895. (2018)
16. Y. J. Xie, X. H. Wang, Y. C. Lin. Fatigue Fract Eng M. **71**, 1501-1513. (2004)
17. H. Yuan, W.J. Liu, Y.J.Xie. Eng Fract Mech. **207**, 131-148. (2019)
18. W.J. Wang, Y.J. Xie. Acta Mech. **232**, 619-637. (2021)
19. Y. J. Xie, X. H. Wang, Y. Y. Wang. Int J Solids Struct. **44**, 4830-4844. (2007)
20. Y. J. Xie. Int J Solids Struct. **37**, 5189-5201. (2000)
21. Y. Z Chen. Int J Pressure Vessels Piping. **76**, 49-53. (1999)

 https://doi.org/10.1051/e3sconf/202129302013

New iterative methods for dense linear systems

Jinmei Wang[1], Lizi Yin[1], and Ke Wang[2,*]

[1]School of Mathematical Sciences, University of Jinan, Jinan 250022, P.R. China

[2]Department of Mathematics, Shanghai University, Shanghai 200444, P.R. China

Abstract. Solving dense linear systems of equations is quite time consuming and requires an efficient parallel implementation on powerful supercomputers. Du, Zheng and Wang presented some new iterative methods for linear systems [Journal of Applied Analysis and Computation, 2011, 1(3): 351-360]. This paper shows that their methods are suitable for solving dense linear system of equations, compared with the classical Jacobi and Gauss-Seidel iterative methods.

1 Introduction

Linear system plays an important role of applications in engineering and scientific computing such as boundary element methods, quantum mechanical problems and large least squares problems [1, 2, 7, 12, 16, 17]. Large sparse linear systems can be solved efficiently by iterative methods, especially those based on a Krylov subspace. However, for large dense linear systems, it is hard to develop good numerical methods. The linear systems usually have the following form of linear equations

$$Ax = b, \tag{1}$$

where A is nonsingular, x is unknown and b is known and nonzero. The case where A is dense can be solved numerically by direct methods, iterative methods and parallel methods [11].

Early in the study for solving (1), the main methods are direct methods [9, 10]. Then the iterative methods become more popular [5, 13]. Recent years, the parallel methods are more and more presented [4, 6, 8]. Many recent studies focus on exploiting the rank structures in the systems. The hierarchically semiseparable (HSS) representations are shown to be very useful for some dense problems such as Toeplitz matrices and certain discretized matrices (e.g., discretized integral equations and Schur complements in the factorizations of discretized PDEs). HSS matrices are closely related to other rank-structured representations such as sequentially semiseparable matrices and quasi-separable matrices [8]. For PCs (personal computers), the iterative methods are preferred.

Du, Zheng and Wang [3] suggested some new iterative methods for solving linear systems, and they showed that these methods, compared with the classical Jacobi and Gauss-Seidel methods, can be applied to more systems and have faster convergence. The new proposed methods are easy to construct and the convergence conditions are easy to check. They are convergent as long as the coefficient matrix is diagonally dominant, while the classical methods require that the matrix be either strictly diagonally dominant or irreducibly diagonally dominant. It was showed that the infinity norm of the iterative matrix of the new methods are less than or equal to that of the iterative matrix of the Jacobi method.

In this paper, the new methods presented by Du, Zheng and Wang [3] are discussed for solving dense linear systems. The theoretical results indicate that a dense system can be solved by the new methods when the coefficient matrix is diagonally dominant, while the classical methods require that the matrix is either strictly diagonally dominant or irreducibly diagonally dominant. And numerical examples are provided to further show that they are suitable and efficient for dense cases.

The rest of the paper is organized as follows. In Section 2, the new methods presented in [3] are briefly introduced. In Section 3, the new methods are discussed for dense linear systems. Section 4 provides numerical examples to show the effectiveness and efficiency of the new methods. The conclusions are in Section 5.

2 New iterative methods

For linear system (1), an iterative scheme can be made as follows [3],

$$x^{(k+1)} = Tx^{(k)} + D^{-1}b, \quad k = 0,1,2,\cdots, \tag{2}$$

where $T = D^{-1}E$ with splitting

$$A = D - E,$$

i.e., $E = -(A - D)$, and

$$D = \begin{bmatrix} a_{11} & & & & & \\ & a_{22} & & & & \\ & & \ddots & & & \\ a_{l1} & a_{l2} & \cdots & a_{ll} & \cdots & a_{ln} \\ & & & & \ddots & \\ & & & & & a_{nn} \end{bmatrix},$$

---
* Corresponding author: kwang@shu.edu.cn

here $1 \le l \le n$, and all other entries are zero.

The following theorem shows that the infinity norm of the iteration matrix of method (2) is less than or equal to that of the Jacobi method.

Theorem 2.1 [3]. *Assume that* $A = (a_{ij})_{n\times n}$, $\sum_{j=1,j\neq i}^{n}|a_{ij}| \le |a_{ii}|$, $|a_{ii}| > 0$, $i = 1,2,\cdots,n$, $n \ge 2$. *Let* T_J *and* T *be the iteration matrices of the Jacobi method and the method* (2). *Then* $\|T\|_\infty \le \|T_J\|_\infty$.

The convergence theorem for (2) is as below.

Theorem 2.2 [3]. *Let* $A = (a_{ij})_{n\times n}$, $a_{ij} > 0$, $\sum_{j=1,j\neq i}^{n}|a_{ij}| \le |a_{ii}|$, $n \ge 3$. *Then the iteration matrix* T *of the method* (2) *satisfies* $\rho(T) < 1$, *i.e., the iterative method* (2) *converges.*

For Theorem 2.2, there is the following remark addressed.

Remark 2.3 [3].

(1) The size n of the matrix A has to satisfy $n > 3$.

(2) The condition $a_{ij} > 0$ is necessary.

(3) Notice that the iterative method (2) converges even if the matrix A is just diagonally dominant.

In contrast, there is the following convergence theorem for Jacobi and Gauss-Seidel methods, which can be found in [14, 15].

Theorem 2.4. *If* A *is strictly diagonally dominant, or irreducibly diagonally dominant, then both Jacobi and Gauss-Seidel methods converge regardless of the choice of the initial guess* $x^{(0)}$.

Du, Zheng and Wang [3] suggested another method as follows,

$$x^{(k+1)} = \tilde{T}x^{(k)} + \tilde{D}^{-1}b, \quad k = 0,1,2,\cdots, \tag{3}$$

where $\tilde{T} = \tilde{D}^{-1}\tilde{E}$ with splitting

$$A = \tilde{D} - \tilde{E},$$

i.e., $\tilde{E} = -(A - \tilde{D})$, and

$$\tilde{D} = \begin{bmatrix} a_{11} & & & & & \\ & \ddots & & & & \\ & & a_{ii} & \cdots & a_{ij} & & \\ & & & \ddots & \vdots & & \\ & & & & a_{jj} & & \\ & & & & & \ddots & \\ & & & & & & a_{nn} \end{bmatrix},$$

here $1 \le i, j \le n$, and all other entries are zero.

The following is the convergence result for method (3).

Theorem 2.5 [3]. *Let* $A = (a_{ij})_{n\times n}$, $a_{ij} > 0$, $\sum_{j=1,j\neq i}^{n}|a_{ij}| \le |a_{ii}|$, $n \ge 3$. *Then the iteration matrix* $\tilde{T}$ *of the method* (3) *satisfies* $\rho(\tilde{T}) < 1$, *i.e., the iterative method* (3) *converges.*

Theorem 2.6 [3]. *Let* $A = (a_{ij})_{n\times n}$, $n \ge 3$, *satisfy*

(a) $\sum_{j=1,j\neq i}^{n}|a_{ij}| \le |a_{ii}|$, $a_{ii} \neq 0$, $i = 1,2,\cdots,n$;

(b) *There exist three different indexes* p, q, r *such that* $a_{ip} \neq 0$, $i = 1,2,\cdots,n$, *and* $a_{pq}a_{pr}a_{qq}a_{qr} > 0$.

If $\tilde{D}$ *and* $\tilde{E}$ *are chosen such that*

$$\tilde{D} = \begin{bmatrix} a_{11} & & & & & \\ & \ddots & & & & \\ & & a_{pp} & \cdots & a_{pq} & & \\ & & & \ddots & \vdots & & \\ & & & & a_{qq} & & \\ & & & & & \ddots & \\ & & & & & & a_{nn} \end{bmatrix}$$

and the other entries are zero, then $\rho(\tilde{D}^{-1}\tilde{E}) < 1$.

Remark 2.7 [3]. The following matrix A shows what condition (b) means,

$$A = \begin{bmatrix} a_{11} & \cdots & a_{1p} & & & & & & \\ & \ddots & \vdots & & & & & & \\ & & a_{pp} & \cdots & a_{pq} & \cdots & a_{pr} & & \\ & & & \ddots & \vdots & & \vdots & & \\ & & \vdots & & a_{qq} & \cdots & a_{qr} & & \\ & & & & & \ddots & \vdots & & \\ & & \vdots & & & & a_{rr} & & \\ & & & & & & & \ddots & \\ & & a_{np} & & \cdots & & \cdots & & a_{nn} \end{bmatrix},$$

where the shown entries are nonzero entries.

3 Discussion for dense linear system

Remark 2.3 suggests that the new iterative method (2) only converges for matrices with positive entries. Thus, the following theorem holds.

Theorem 3.1. *For an* $n \times n$ $(n \ge 3)$ *dense linear system* (1), *if the coefficient matrix* A *is diagonally dominant and all entries are positive, then it can be numerically solved by the method* (2).

Although Theorem 2.6 implies that the condition $a_{ij} > 0$ in Theorem 2.5 is not required, it is still true that the iterative method (3) is also suitable for dense linear system.

Theorem 3.2. *For an* $n \times n$ $(n \ge 3)$ *dense linear system* (1), *if the coefficient matrix* A *is diagonally dominant and all entries are positive, then it can be numerically solved by the method* (3).

In the next section, some numerical examples are provided to further illustrate that the methods (2) and (3) are suitable for solving dense linear system of equations (1) with diagonally dominant coefficient matrix. And the numerical results show that the new methods outperform the classical Jacobi and Gauss-Seidel iterative methods even though the coefficient matrix is strictly diagonally dominant.

4 Numerical experiments

In this section, three examples are given to illustrate the new methods (2) and (3) for solving dense linear system of equations (1). All performances run in MATLAB 7.12 with 2.93GHz CPU, 4.00GB RAM and 32-bit Windows 10 Professional. The initial guess is 0, and the stopping criterion is

$$\frac{\|r^{(k)}\|_2}{\|r^{(0)}\|_2} < 10^{-6},$$

where $r^{(k)}$ is the residual vector after k iterations. The numerical results are listed in Tables 1-3, where Jacobi, GS, I and II stand for Jacobi method, Gauss-Seidel method, method (2) and method (3), respectively.

Example 4.1. Consider $n \times n$ dense linear system with

$$A = (a_{ij})_{n\times n} = \begin{cases} a_{ij} = i + j, & i = j, \\ a_{ij} = 1, & i = 1, j = n, \\ a_{ij} = 2n - 1, & i = n, j = 1, \\ a_{ij} = \frac{1}{n}, & \text{others}, \end{cases} \quad 1 \le i, j \le n,$$

and $b = (1,0,\cdots,0)^T$.

Table 1: Iterations (IT), CPU time (t) and relative error (ERR) for Example 4.1

n	Jacobi			GS			I			II		
	IT	t	ERR	IT	t	ERR	IT	t	ERR	IT	t	ERR
1000	40	0.2	9.5e-7	20	0.4	9.4e-7	20	0.1	9.4e-7	20	0.1	9.4e-7
2000	40	0.9	9.5e-7	20	1.9	9.5e-7	20	0.4	9.5e-7	20	0.4	9.5e-7
3000	40	1.8	9.5e-7	20	3.5	9.5e-7	20	0.9	9.5e-7	20	0.9	9.5e-7
4000	40	3.5	9.5e-7	20	5.8	9.5e-7	20	1.8	9.5e-7	20	1.8	9.5e-7
5000	40	5.0	9.5e-7	20	8.6	9.5e-7	20	2.6	9.5e-7	20	2.6	9.5e-7
6000	40	7.3	9.5e-7	20	11.3	9.5e-7	20	3.7	9.5e-7	20	3.7	9.5e-7

Table 2: Iterations (IT), CPU time (t) and relative error (ERR) for Example 4.2

n	Jacobi			GS			I			II		
	IT	t	ERR	IT	t	ERR	IT	t	ERR	IT	t	ERR
1000	78	0.4	8.7e-7	47	0.9	9.7e-7	59	0.3	8.4e-7	78	0.4	8.7e-7
2000	90	1.8	8.4e-7	54	3.9	5.2e-7	68	1.3	8.1e-7	90	1.8	8.4e-7
3000	97	4.3	8.8e-7	62	10.0	7.3e-7	73	3.1	9.1e-7	97	4.1	8.8e-7
4000	102	8.2	9.3e-7	65	17.9	6.8e-7	77	6.2	8.9e-7	102	8.0	9.3e-7
5000	106	12.7	9.5e-7	67	27.0	7.9e-7	80	9.1	9.2e-7	106	12.5	9.5e-7
6000	110	19.0	8.7e-7	69	38.8	5.3e-7	83	13.1	8.3e-7	110	18.8	8.7e-7

Table 1 shows that the new iterative methods (2) and (3) have less iterations and CPU time than Jacobi method, and have less CPU time than Gauss-Seidel method.

Example 4.2. Consider $n \times n$ dense linear system (1) with

$$A = (a_{ij})_{n\times n} = \begin{cases} a_{ij} = \dfrac{1}{2} + \dfrac{1}{n-1} + \dfrac{i}{2n}, & i < j, \\ a_{ij} = \dfrac{1}{n-1}, & i > j, \\ a_{ij} = \displaystyle\sum_{k\ne i} a_{ik} + 2 + \frac{i}{n}, & i = j, \end{cases} \quad 1 \le i, j \le n,$$

and $b = (1,1,\cdots,1)^T$.

In this example, the new iterative method (2) has less iterations and CPU time than Jacobi method, and has less CPU time than Gauss-Seidel method does; the method (3) is as good as Jacobi method and better than Gauss-Seidel method, as seen from Table 2.

Example 4.3. Consider $n \times n$ dense linear system (1) with

$$A = (a_{ij})_{n\times n} = \begin{cases} a_{ij} = n, & i = j, \\ a_{ij} = a_{ji} = n - 1, & i = 1;\ j = n, \\ a_{ij} = \frac{1}{n-1}, & i = 1, n; j = 2, \cdots, n-1, \\ a_{ij} = 1, & \text{others}, \end{cases}$$

$$1 \le i, j \le n,$$

and $b = (1,2,\cdots,n)^T$.

Table 3: Iterations (IT), CPU time (t) and relative error (ERR) for Example 4.3

n	Jacobi			GS			I			II		
	IT	t	ERR	IT	t	ERR	IT	t	ERR	IT	t	ERR
100	4905	0.6	1.0e-6	600	0.2	1.0e-6	600	0.1	9.9e-7	600	0.1	9.9e-7
200	5000	1.1	7.8e-4	1169	0.9	1.0e-6	1169	0.3	9.9e-7	1169	0.3	9.9e-7
300	5000	1.8	0.0076	1724	2.7	1.0e-6	1724	0.6	1.0e-6	1724	0.6	1.0e-6
400	5000	2.3	0.0236	2271	5.9	1.0e-6	2271	1.1	1.0e-6	2271	1.1	1.0e-6
500	5000	3.7	0.0464	2812	11.1	1.0e-6	2812	1.8	1.0e-6	2812	1.8	1.0e-6
600	5000	4.3	0.0728	3347	20.5	1.0e-6	3348	2.8	1.0e-6	3348	2.8	1.0e-6

In Table 3, when it reaches the maximum iteration number 5000, Jacobi method does still not reach to the given precision and needs more CPU time. And the two new methods have same iterations to Gauss-Seidel method and less CPU time than Gauss-Seidel method.

5 Conclusion

In this paper, two new iterative methods are discussed for solving dense linear system, which is easy to establish and meet the convergence conditions. The theoretical results indicate that a dense system can be solved when the coefficient matrix is diagonally dominant, while the classical methods require either strictly diagonally dominant or irreducibly diagonally dominant. Numerical experiments show that the new methods are better than Jacobi and Gauss-Seidel methods.

Acknowledgement

This work was supported by 2019 Shandong Postgraduate Education Quality Curriculum Project, Jinan University Graduate Teaching Reform Research Project (jdyy903), Jinan University Teaching Reform Research Project (jzc1903), and Key Scientific Research Project of Colleges and Universities in Henan Province (20B110012), China.

References

1. V. Ayala and A. Da Silva, Central periodic points of linear systems, Journal of Differential Equations 272 (2021) 310-329.
2. C.A. Buzzi, J.C. Medrado and J. Torregrosa, Limit cycles in 4-star-symmetric planar piecewise linear systems, Journal of Differential Equations 268 (2020) 2414-2434.
3. J. Du, B. Zheng and L. Wang, New iterative methods for solving linear systems, Journal of Applied Analysis and Computation 11 (2011) 351-360.
4. S. Duminil, A parallel implementation of the CMRH method for dense linear systems, Numerical Algorithms 63 (2013) 127-142.
5. S. Duminil, M. Heyouni, P. Marion and H. Sadok, Algorithms for the CMRH method for dense linear systems, Numerical Algorithms 71 (2016) 383-394.
6. Y.-L. Jiang, Y.-W. Liu, K.K. Mei and R.M.M. Chen, A new iterative technique for large and dense linear systems from the MEI method in electromagnetics, Applied Mathematics and Computation 139 (2003) 157-163.
7. S. Li and J. Llibre, Phase portraits of piecewise linear continuous differential systems with two zones separated by a straight line, Journal of Differential Equations 266 (2019) 8094-8109.
8. X. Liu, J. Xia and M.V. de Hoop, Parallel randomized and matrix-free direct solvers for large structured dense linear systems, SIAM Journal on Scientific Computing 38 (2016) S508-S538.
9. S.M. Rump, Accurate solution of dense linear systems, part I: Algorithms in rounding to nearest, Journal of Computational and Applied Mathematics 242 (2013) 157-184.
10. S.M. Rump, Accurate solution of dense linear systems, part II: Algorithms using directed rounding, Journal of Computational and Applied Mathematics 242 (2013) 185-212.
11. F. Song and J. Dongarra, A scalable approach to solving dense linear algebra problems on hybrid CPU-GPU systems, Concurrency and Computation: Practice and Experience 27 (2015) 3702-3723.
12. J. Sun, S.-H.Wang and K.Wang, A new choice of the preconditioner of PHSS method for saddle point problems, Chinese Quarterly Journal of Mathematics 30 (2015) 555-561.
13. P.S. Theocaris and N. Karayanopoulos, An algorithm for the numerical solution of dense large general linear systems, Computers & Structures 14 (1981) 377-383.
14. R.S. Varga, Matrix Iterative Analysis, Prentice-Hall, Englewood Cliffs, New Jersey, 1962.
15. D.M. Young, Iterative Solution of Large Linear Systems, Academic Press, New York, 1971.
16. C. Zhang, B. Zheng and R. Su, Realizability of the normal forms for the non-semisimple 1:1 resonant Hopf bifurcation in a vector field, Communications in Nonlinear Science and Numerical Simulation 91 (2020) 105407.
17. C. Zhang, B. Zheng and P. Yu, Second-order normal forms for n-dimensional systems with a nilpotent point, Journal of Applied Analysis and Computation 10 (2020) 2233-2262.

Study on the adhesion performance of medium weathered igneous rock aggregate and asphalt

Xuxue Jiao[1], Yuanhong Tao[1], Longfang Ren[2,*]

[1]China Road and Bridge Co., Beijing 100011, China

[2]China Academy of Building Research, Beijing 100013, China

Abstract: In view of the current situation of the shortage of alkaline aggregate and the abundance of weakly acidic weathered igneous rocks around Nairobi, Kenya, in order to fully improve the adhesion performance of asphalt and acidic igneous rocks, this paper studies the effects of different anti-stripping agents separately and combined on the adhesion performance of asphalt aggregates. The results show that the anti-stripping agent can improve the anti-stripping property of the medium weathered igneous rock aggregate and asphalt, and the effect is better when the anti-stripping agent is combined.

1 Introduction

The state promotes the construction of "One Belt And One Road". Nairobi Expressway in Kenya is one of the important projects of infrastructure construction in East Africa. The rocks in Nairobi and its surrounding areas are mainly weakly acidic moderately weathered igneous rocks, other rock yards are far away. The medium weathered igneous rocks along the highway will be used in the project construction.

At present, asphalt pavement is widely used in road construction. With the extension of service life, the expressway faces different forms of damage, which will greatly reduce the pavement performance. Water damage is one of the main early diseases of asphalt expressway, which is easy to cause loose, exposed and poth of pavement. The adhesion between asphalt and aggregate is one of the important indexes to evaluate the resistance of asphalt pavement to water damage.

The asphalt material is slightly acidic, it has good adhesion to alkaline stone, but poor adhesion with acid stone. Under the action of water, the asphalt is easily stripping, which leads to water damage of asphalt pavement. Due to the scarcity of available alkaline stone around the project site, long-distance transportation will greatly increase the cost of the project, so this method is not advisable. Considering the rich reserves of weak acid igneous rocks around the project site, the anti-stripping treatment can be adopted to improve the adhesion between medium weathered igneous rocks and asphalt, and a large amount of igneous rocks around the site can be utilized to ensure the quality of the project and reduce the project cost.

Adding anti-stripping agent is one of the common methods to improve the adhesion of asphalt aggregate. Traditional inorganic anti-stripping agent include cement and slaked lime which are less costly but its applying process is complex and it is difficult to mix evenly. Chemical anti-stripping agents include amines and non-amines, mostly amines,This kind of anti-stripping agent mainly achieves the anti-stripping effect by combining its polar group with acid aggregate and its oil-friendly group with asphalt [1]. It has the advantages of easy to use and easy to mix evenly, but it also has the problem of high cost.

A great deal of research has been done on the application of anti-stripping agents at home and abroad. Tan Jizong [2] verified that granite asphalt mixture can reach the road performance index of limestone asphalt mixture after adding anti-stripping agent. The study of Zheng Xiaoguang [3] shows that cement instead of mineral powder can greatly improve the high temperature stability and water stability of asphalt mixture. The study of Wei Zepeng et al. [4] shows that the water stability of asphalt mixture can be improved when cement replaces 20%~40% mineral powder.

The study of anti-stripping agents at home and abroad mainly focuses on the improvement of pavement performance, water stability and temperature stability.In this paper, the stripping rate of asphalt on the surface of medium weathered igneous aggregate is studied by using quantitative analysis method and single or multiple anti-stripping agent for asphalt aggregate. It provides theoretical basis for water resistance measures of asphalt pavement.

2 Materials and Methods

2.1 Raw Materials

(1) Aggregate: the aggregate used in this experiment is medium weathered igneous rocks in Nairobi and its surrounding areas. The rock is processed into coarse and

* Corresponding author: 13051185882@163.com

https://doi.org/10.1051/e3sconf/202129302014

fine aggregate by laboratory small jaw crusher. The chemical composition of stone powder with particle size of 0-0.075mm was analyzed by chemical titration. According to GB/T 14685-2011, test the compressive strength, apparent density, soundness and crushing value of aggregates with particle size of 5-20mm. The technical indicators are shown in table 1.

(2) Asphalt: the asphalt used in this test is road petroleum asphalt 70 and SBS modified asphalt. The technical indicators are shown in table 2.

(3) Cement: The Cement used in this experiment is CEMI 42.5 which produced in Kenya, its setting time is initial 165 minutes, final solidified 267 minutes, the 3d compressive strength is 27.1 MPa, and the 28d compressive strength is 52.3 MPa.

(4) Slaked lime：The slaked lime used in this experiment is produced in Kenya, the effective contents of CaO and MgO were 68.2%.

(5) Amine anti-stripping agent: Its appearance is milky thick liquid, pH value is between 10 ~ 12, freezing point is less than 0°C, and its dosage is 0.4% of asphalt mass.

Table 1. Technical index of aggregate

chemical composition/%					compressive strength/MPa	apparent density, kg/m³	crushing value/%	soundness/%
SiO_2	Al_2O_3	Fe_2O_3	CaO	MgO				
65.39	13.18	12.27	2.66	0.82	95.2	2672	10.15	3.2

Table 2. Technical index of asphalt

type / indicato	25°C penetration /0.1mm	softening point/°C	Ductility (15°C，5cm/min)	Ductility (5°C，5cm/min)
petroleum asphalt 70[#]	64.2	49.2	146.1	——
SBS modified asphalt	52.6	86.9	——	42

2.2 Testing plan

First, without adding any anti-stripping agent, the igneous rock aggregate is directly combined with the two kinds of asphalt respectively; secondly, the igneous rock aggregate is pretreated by soaking in the saturated solution of cement or Slaked lime, and then compounded with two kinds of asphalt respectively; the third is to add 0.4% amine anti-stripping agent to the asphalt and mix well with igneous aggregate; the fourth is that igneous aggregate is pretreated by saturated solution of cement or Slaked lime, and then compounded with asphalt treated by amine anti-stripping agent. The effect of single or multiple anti-stripping agent on the adhesion of asphalt aggregate was studied.

2.3 Test methods

According to the method of T0616-1993 in JTU E20-2011, the boiling test was carried out. In order to make the test results more objective and accurate, the test improves the detection method on the basis of the standard test method, and obtains the asphalt stripping rate through formula calculation. The test method is as follows.

(1) 5 aggregates with particle size of about 16mm were taken from each group, and put them into the oven to dry, then cooled to room temperature. The mass of each particle was weighed and recorded as m_{0n} (n=1~5).

(2) The suspension device is used to hang each aggregate particle separately.

(3) Heat the asphalt in an oven at about 150 °C until it is molten, immerse the suspended aggregate particles in the asphalt, then take them out and hang them to dry to room temperature, The weighed mass is recorded as m_{jn} (n=1~5);

(4) Wrap the dried bituminous aggregate with gauze and put it in an ultrasonic wave filled with boiling water to heat it for 0~10min, then take it out and dry it;

(5) Take off the gauze, and then cut off the asphalt bonded on the gauze and the asphalt separated from the surface of the aggregate particles with a wallpaper knife, dry the remaining asphalt mixture to constant weight, and record the weight as m_{zn} (n = 1~5);

(6) The stripping rate of asphalt was calculated to judge the adhesion performance of asphalt and aggregate particles. the calculation method was shown in formula (1):

$$D_{in}=(m_{jn}-m_{zn})/(m_{jn}-m_{0n})\times 100\% \qquad (1)$$

In the formula:

D_{in} is the nth asphalt stripping rate for testing;

m_{0n} is the mass of the nth aggregate particle;

m_{jn} is the total mass of the nth asphalt mixture after adhesion;

m_{zn} is the mass of the nth adhesive asphalt particle after being immersed in water and shedding part of asphalt mixture.

3 Results and Analysis

3.1 Study on the adhesion property of igneous rock aggregate and petroleum asphalt 70[#]

According to the four test schemes in section 1.2, igneous rock aggregate and petroleum asphalt 70 are compounded to study the stripping rate of petroleum asphalt 70[#] on the surface of moderately weathered igneous rock aggregate under different boiling time and different anti stripping measures. The test results are shown in table 3:

Table 3. Anti-stripping test results of igneous rock aggregate and petroleum asphalt 70# under different anti-stripping measures

boiling time/min	stripping rate of asphalt film/%					
	Blank(A-1)	cement(A-2)	slaked lime(A-3)	amine anti-stripping agent(A-4)	cement+amine anti-stripping agent(A-5)	slaked lime+amine anti-stripping agent(A-6)
1	5	3	3	0	0	0
2	28	7	6	0	0	0
3	42	10	9	2	0	0
4	63	15	13	4	2	2
5	80	24	21	6	4	3
6	94	29	26	22	15	13
7	100	35	32	38	22	22
8	100	41	37	57	28	26
9	100	46	42	76	33	31
10	100	49	44	83	36	35

3.1.1 Effect of single anti-stripping agent on adhesion property of petroleum asphalt 70# aggregate

Under a single anti-stripping measure, the effect of different boiling time and different anti-stripping agents on the adhesion properties of igneous rock aggregate and petroleum asphalt 70# aggregate is shown in Fig. 1. It can be seen from Figure 1, with the extension of boiling time, the stripping rate of petroleum asphalt 70# on the surface of igneous rock aggregate increases gradually, and the stripping rate of asphalt aggregate without anti stripping agent increases rapidly with the extension of boiling time, and the asphalt film falls off completely in about 7 minutes, the adhesion performance is poor. However, the stripping rate of asphalt on the surface of igneous rock aggregate which was pretreated by cement and Slaked lime is obviously reduced, and the stripping resistance of Slaked lime is slightly better than that of cement. The main reason why both cement and lime can improve the adhesion of asphalt aggregate is that lime can react with organic acids in asphalt, the product makes the aggregate surface activate to improve the adhesion between the asphalt and the aggregate [5]. Meanwhile, slaked lime is beneficial to the formation of microporous adsorption of asphalt on the surface of aggregate and enhances the adhesion of asphalt and aggregate [6], the cement will absorb the moisture in asphalt and hydration reaction will occur, and then chemical adsorption will occur, forming a layer of film to enhance the adhesion between asphalt and aggregate [7].

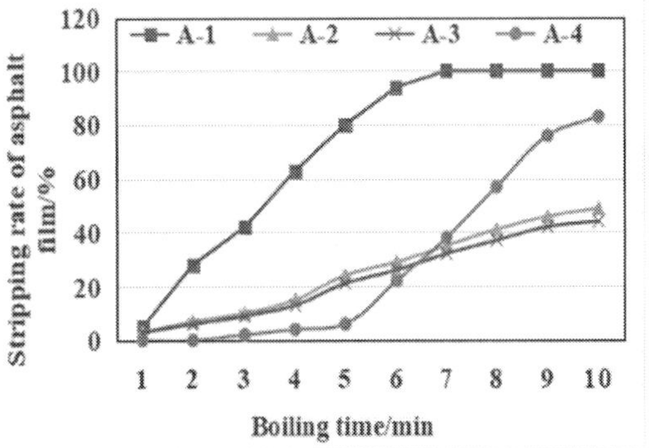

Fig. 1 Effect of different anti-stripping agents on the adhesion property of petroleum asphalt 70# aggregate

It can be seen from Fig. 1, the anti-stripping ability of amine anti-stripping agent is stronger than that of cement and Slaked lime before boiling for 6 minutes, with the extension of boiling time, the anti-stripping ability of amine anti-stripping agent decreases. Before boiling for 5 minutes, the stripping rate of asphalt film is low, and the anti-stripping ability is good and stable. After boiling for 5 minutes, the stripping speed of asphalt film is accelerated. The main reason is that the amine anti-stripping agent belongs to high molecular polymer material, which has poor heat resistance. With the extension of heating time, it is easy to decompose and lose part or all of its function.

3.1.2 Effect of multiple anti-stripping agent on adhesion performance of petroleum asphalt 70# aggregate

The influence of different boiling time and anti-stripping agent on the adhesion properties of igneous rock aggregate and petroleum asphalt 70# aggregate is shown in Fig. 2. It can be seen from Fig. 2, with the extension of boiling time, the stripping rate of petroleum asphalt 70# on the surface of igneous rock aggregate increases gradually, and the stripping resistance of asphalt aggregate with multiple anti-stripping agent is better than that of asphalt aggregate with single anti stripping agent.

Due to the poor heat resistance of amine anti-stripping agent, the stripping resistance of asphalt aggregate which compound mixed with anti-stripping agent decreases gradually with the extension of boiling temperature, After 5 minutes, the stripping rate increased and the stripping resistance decreased, however ,it is always better than adding anti-stripping agent alone.The adhesion performance of asphalt aggregate 70# mixed with (Slaked lime + amine anti-stripping agent) is better than that of asphalt aggregate mixed with (cement + amine anti-stripping agent).

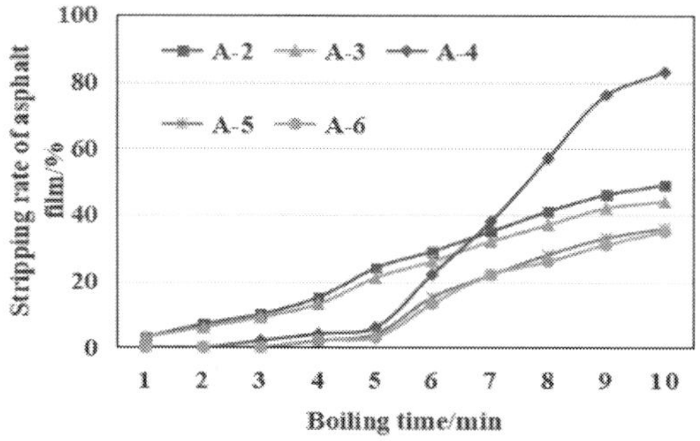

Fig. 2 Effect of multiple anti-stripping agent on adhesion performance of petroleum asphalt 70# aggregate

3.2 Study on adhesion properties of igneous aggregate and SBS modified asphalt

According to the four test schemes in section 1.2, igneous rock aggregate and SBS modified asphalt are compounded to study the stripping rate of SBS modified asphalt on the surface of moderately weathered igneous rock aggregate under different boiling time and different anti stripping measures. The test results are shown in Table 4.

3.2.1 Effect of single anti-stripping agent on adhesion of SBS modified asphalt aggregate

Under single anti-stripping measure, the effect of different boiling time and different anti-stripping agents on the adhesion properties of igneous rock aggregate and SBS modified asphalt aggregate is shown in Fig. 3. It can be seen from Fig. 3, with the extension of boiling time, the stripping rate of SBS modified asphalt aggregate on the surface of igneous rock aggregate increases gradually, and the stripping rate of asphalt aggregate without anti stripping agent increases rapidly with the extension of boiling time, and the asphalt film falls off completely in about 10 minutes, the adhesion performance is poor.

However, the stripping rate of asphalt on the surface of igneous rock aggregate which was pretreated by cement and Slaked lime is obviously reduced, and the stripping resistance of Slaked lime is slightly better than that of cement. Before boiling for 7.5 minutes, the anti-stripping ability of amine anti-stripping agent is better than that of cement and slaked lime. If the boiling time is extended, the anti-stripping ability of amine is lower than that of cement and slaked lime. For amine anti-stripping agent, boiling for 6 minutes is a turning point. Before, the stripping rate of asphalt film is low and the growth rate is slow, but after that, the stripping rate accelerates and the adhesion performance between SBS modified asphalt and igneous rock aggregate becomes worse.

Table 4 Anti-stripping test results of igneous rock aggregate and SBS modified asphalt under different anti stripping measures

boiling time/min	stripping rate of asphalt film/%					
	Blank(B-1)	cement(B-2)	slaked lime(B-3)	amine anti-stripping agent(B-4)	cement+amine anti-stripping agent(B-5)	slaked lime+amine anti-stripping agent(B-6)
1	3	0	0	0	0	0
2	16	0	0	0	0	0
3	28	3	3	0	0	0
4	40	8	7	2	0	0
5	52	13	11	6	2	2
6	63	19	17	11	5	4
7	77	26	25	23	13	12
8	86	33	31	35	20	18
9	96	39	36	49	25	23
10	100	45	42	62	29	27

https://doi.org/10.1051/e3sconf/202129302014

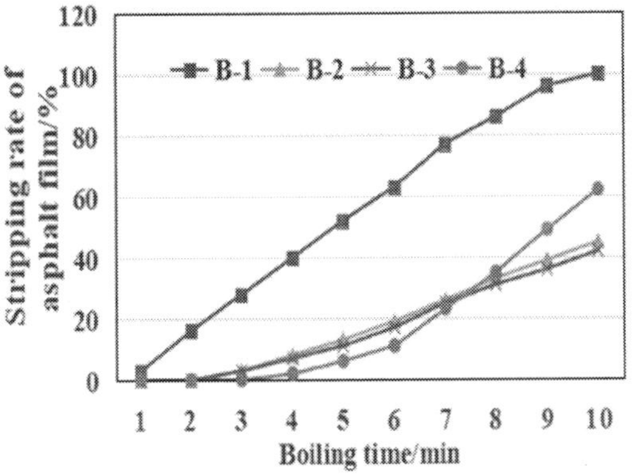

Fig. 3 Effect of different anti-stripping agents on adhesion of SBS modified asphalt aggregate

3.2.2 Effect of multiple anti-stripping agent on adhesion of SBS modified asphalt aggregate

The influence of different boiling time and anti-stripping agent on the adhesion properties of igneous rock aggregate and SBS modified asphalt aggregate is shown in Fig. 4. It can be seen from Fig. 4, with the extension of boiling time, the stripping rate of SBS modified asphalt on the surface of igneous rock aggregate increases gradually, and the stripping resistance of asphalt aggregate with multiple anti-stripping agent is better than that of asphalt aggregate with single anti stripping agent. For SBS modified asphalt aggregate mixed with multiple anti-stripping agent, the stripping rate of asphalt film has a turning point at 6 minutes. Before 6 minutes, the stripping rate increases slowly with boiling time, and after 6 minutes, the stripping rate increases rapidly. The adhesion performance of SBS modified asphalt mixed with (Slaked lime + amine anti-stripping agent) is better than that of asphalt aggregate mixed with (cement + amine anti-stripping agent).

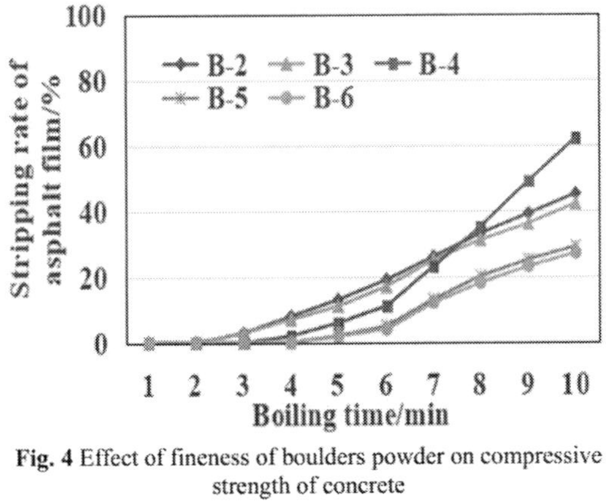

Fig. 4 Effect of fineness of boulders powder on compressive strength of concrete

4 Conclusion

(1) With the extension of boiling time, the stripping rate of asphalt on the surface of igneous rock aggregate increases gradually, and the adhesion between igneous rock aggregate and two kinds of asphalt is greatly improved by adding anti-stripping agent.

(2) Under the single anti-stripping measure, the anti-stripping ability of slaked lime is slightly better than that of cement for two kinds of asphalt aggregates, For the igneous petroleum asphalt $70^{\#}$ aggregate, the anti-stripping ability of amine anti-stripping agent is higher before boiling for 5 minutes, and gradually decreases with the extension of boiling time. For the igneous SBS modified asphalt aggregate, the anti-stripping ability of amine anti-stripping agent is higher before boiling for 6 minutes, and the anti-stripping ability decreases rapidly with the extension of boiling time.

(3) The results show that the igneous asphalt aggregate mixed with multiple anti-stripping agent has better adhesion than that mixed with single anti-stripping agent. The adhesion performance of asphalt aggregate mixed with (Slaked lime + amine anti-stripping agent) is better than that of asphalt aggregate mixed with (cement + amine anti-stripping agent).

References

1. Y. Wu, F. Wang, and Y. Zhang, J. China Foreign Highw. 36, 289 (2016).
2. J. Tan, West. China Commun. Sci. Technol. 21 (2018).
3. X. Zheng, Q. Yang, and W. Lv, J. Build. Mater. 480 (2005).
4. Z. Wei, Transp. Sci. Technol. 132 (2017).
5. S. Du and X. Suo, J. Xi'an Technol. Univ. **33**, 982 (2013).
6. D. Wu, The Research of Performance on Cement and Amine Anti-Stripping Material's Asphalt Mixture Pavement, Dissertation, Chongqing Jiaotong University, 2016.
7. Z. Wang, A. Sha, J. Xiao, and S. Du, J. Wuhan Univ. Technol. 31, 16 (2009).

Spatial simulation of population in Shandong Province based on night-time imagery and land cover data

Keyi Yang[1,*], Yunling Li[1], and Yang Liu[1]

[1] Shandong University of Science and Technology, Qingdao, Shandong, 266590, China

Abstract. Population spatial data can more truly express the actual distribution characteristics of the population, and provide data support for the regional environment and population development. Use Shandong Province as the research area, township-level census data, revised DMSP/OLS night-time data, and Globaland30 land cover data as data sources, partitions based on population agglomeration, and uses a stepwise regression method to build a population data spatial model. Use the model to simulate population density with a resolution of 100m. The experimental results show: Stepwise regression model good precision, the average relative error was 23.56%, and Root Mean Square Error, Mean Absolute Error are better than the other two public datasets. The simulation results are better than the two public datasets.

1 Introduction

The current population data is mainly obtained through the census or sampling survey, which is summarized level by level by administrative division units, which cannot reflect the spatial distribution of the population. The spatialized population data can provide key parameters for regional economic and environmental changes research [1].

As a hot issue, the research of population spatialization has developed rapidly, and many scholars have carried out research. Land use data can provide population distribution weight based on its own land type, but it cannot reflect the difference in population distribution between the same land type [2]. The night-time imagery has been proved to have a very high correlation with population distribution [3][4]. The combined use of the two effectively alleviated the problem of indistinguishable differences in population distribution of the same land type, then was later recognized by many scholars [5][6][7] analyze and verify and use in population spatialization modeling. Taking Shandong Province as an example, this article attempts to simulate the spatial distribution of population in a simpler and more computationally efficient way to provide basis and services for regional multi-faceted planning.

2 Area and Data

2.1 Study Area

Shandong Province is located on the east coast of China, as of the end of 2010, the results of the sixth census showed that the total population of Shandong Province was 9,572,719 million [8], accounting for about 7% of the total population of the country.

2.2 Data and Pre-treatment

All data use Asia North Albers Equal Area Conic. The regional scope is the vector boundary of the administrative division of Shandong Province. Raster data are resampled to 100m resolution. The data source and data preprocessing are as follows:

- The land cover data comes from the 2010 version of the global land cover data GlobeLand30 [9], which is reclassified into six types of land types: Farmland, Woodland, Grassland, Waters, Built-up area, and Unused land.
- The DMSP/OLS night-time lights data were downloaded from the National Geophysical Data Center (NGDC), which is the Stable Lights Data of non-radiative calibration at night in 2010.This data had Digital Number (DN) values ranging from 0 to 63 and the spatial resolution is about 850m. According to the DMSP/OLS night light data correction method proposed by ZhuoLi et al. [13], the EVI data is used to correct the night-time data to obtain the Enhanced Vegetation Index Adjusted Night-time Light Index (EANTLI).
- The township census data comes from Tabulation on the 2010 population census of the People's Republic of China by Township [8]. According to official information, some township boundaries were adjusted and population data were matched. In the end, 1908 township units were obtained.
- The MODIS-EVI data is extracted from the Level 3 grid data product (MOD13Q1) of the NASA Data Center. The data time is July and August 2010, and the spatial resolution is about 250 m.

* Corresponding author: nuoeryky@163.com

- The township boundary data comes from the township cadastral boundary in the administrative division of TianDitu (http://www.sdmap.gov.cn/), and the overall current situation is 2015.

- The 2010 WorldPop dataset with 100m resolution [11] and China's population spatial distribution kilometer grid dataset [12].

3 Methodology

3.1 Mapping Population

In this paper, referring to the classification standard of population agglomeration in China [14] and research needs, the study area is divided into 5 subareas (Table 1) to achieve the purpose of reducing regional differences and improving simulation accuracy.

$$JDD_i = \frac{(P_i/P_n)\times 100\%}{(A_i/A_n)\times 100\%} = \frac{P_i/A_i}{P_n/A_n} \quad (1)$$

Where JJD_i is the population concentration of i township; P_i is the population of i township(person); A_i is the land area of i township (km^2); A_n is the land area of the province (km^2); P_n is the total population(people) of the province.

Table1. Subareas of Shandong Province.

Subareas		Agglomeration	Township (number)
Sparse area	A	<0.5	258
Mean area	B	0.5-2	1287
Dense area	C_1	2-4	126
	C_2	4-15	126
	C_3	>15	111

Using GIS software, the number of lighting pixels and lighting values of the six types of land use types are counted according to the township, expressed as the area of lighted area NL, the area of non-lighted area NU, and the total light value LE. With the population data of the township census as the dependent variable, the area of lighted area NL, the area of non-lighted area NU, and the total value of light LE of each land use type are used as independent variables, and stepwise regression modeling is carried out. Keep the independent variables that are positive and significant at the 0.05 confidence level. And set the constant of the regression model to zero to satisfy the objective fact that there is no land and no population. The final model is:

$$P_i = \sum_{j=1}^{M}\left(A_j \times NL_{ij} + B_j \times NU_{ij} + C_j \times LE_{ij}\right)_{ij} \quad (2)$$

$$P_{ijk} = \sum_{j=1}^{M}\left(A_j \times NL_{ijk} + B_j \times NU_{ijk} + C_j \times LE_{ijk}\right)_{ijk} \quad (3)$$

In the two formulas, P_i is the population; *NL* stands for number of lit pixels and *NU* for the number of unlit pixels; *LE* denotes light emission; i is the index of different townships, j of various land cover types and k of different pixels; A, B and C are coefficients for each land cover type; and M is the count of different land covers.

$$P'_{ijk} = P_{ijk} \times \frac{\overline{P_i}}{P_i} \quad (4)$$

Where: P'_{ijk} is the final population of the grid; P_i, $\overline{P_i}$ are the simulated population and statistical population of the i-th township.

In the process of model construction, the independent variables and corresponding model coefficients required for modeling can be obtained through equation (2); the model coefficients can be substituted into equation (3) to obtain the simulated population on the pixel scale; use the township boundary Statistical formula (4) simulates the population at the pixel scale and compares it with the census population data to adjust the population on each pixel so that the final simulated population distribution data is consistent with the census data.

3.2 Accuracy Assessment

The simulated population and public data sets are counted by towns. The Relative Error (RE) between the simulated data and the census data is calculated; the Root Mean Square Error (RMSE) and the Mean Absolute Error (MAE) of the simulated results and the two public data sets are calculated. The accuracy of the model is evaluated through these three errors, and the relevant formula is as follows:

$$RE = \frac{P_i' - P_i}{P_i} \times 100\% \quad (5)$$

$$RMSE = \sqrt{\frac{\sum_{i=1}^{N}(P_i' - P_i)^2}{N}} \quad (6)$$

$$MAE = \frac{1}{N}\sum_{i=1}^{N}|P_i' - P_i| \quad (7)$$

Where P', P is the simulated and census population; i is the index of different townships; N is the count of townships.

4 Results

4.1 Population Density

The model results (Table 2) and the correlation analysis between land cover types and population (Table 3) are shown in the table.

Take the Shandong Province township division map as the base map and combine the census population data to form a population density distribution map with townships as the unit (Figure 1a). According to the above model, the population distribution simulation is carried out, and the simulation distribution map of the township population in 2010 is obtained (Figure 1b). The figure shows that the population density map is restricted to administrative divisions and cannot reflect the spatiality of population distribution. The spatialized population distribution data effectively avoids the restrictions of administrative boundaries, and the population distribution is more realistic: The population density changes within each township are better reflected, and the population is mainly concentrated on the Built-up area; The population is mainly concentrated in urban cents, which is significantly higher than the population density in rural areas, and gradually decreases to the surrounding area; The northern coastal areas are mostly affected by low-lying terrain, and most of them are saline-alkali land. The central and southern areas of Shandong and the peninsula are mostly affected by the terrain, mainly mountains and hills, which have a greater impact on population distribution and lead to sparse population distribution. To sum up, the population space simulation has a good effect on the detail description, which is difficult to show in the population density map drawn by the administrative boundary.

Table2. Population model in each subarea.

Subarea	Model	R^2
A	$0.048 \times LE_b + 3.053 \times NL_f + 2.027 \times NU_f$	0.915
B	$23.290 \times NL_b + 3.036 \times NL_f + 4.842 \times NU_f + 0.017 \times LE_f$	0.941
C_1	$24.034 \times NL_b + 12.133 \times NL_f + 0.025 \times LE_f$	0.959
C_2	$46.414 \times NL_b + 0.022 \times LE_b + 13.774 \times NL_f$	0.924
C_3	$115.087 \times NL_b + 0.055 \times LE_b$	0.920

b stands for Built-up area; *f* stands for Farmland.

Table3. The correlation between land use and population.

subarea	Farmland	Woodland	Grassland	Waters	Built-up area	Unused land
A	0.720**	0.173**	0.157**	0.015	0.288**	0.080
B	0.598**	0.133**	0.154**	0.161**	0.727**	0.109**
C_1	0.798**	0.157	0.181*	0.288**	0.661**	0.114
C_2	0.628**	-0.026	0.216*	0.164	0.877*	-0.026
C_3	0.302**	0.212*	0.098	0.099	0.898**	0.055

**Significantly correlated at 0.01 level (two-sided);*Significantly correlated at the 0.05 level(two-sided).

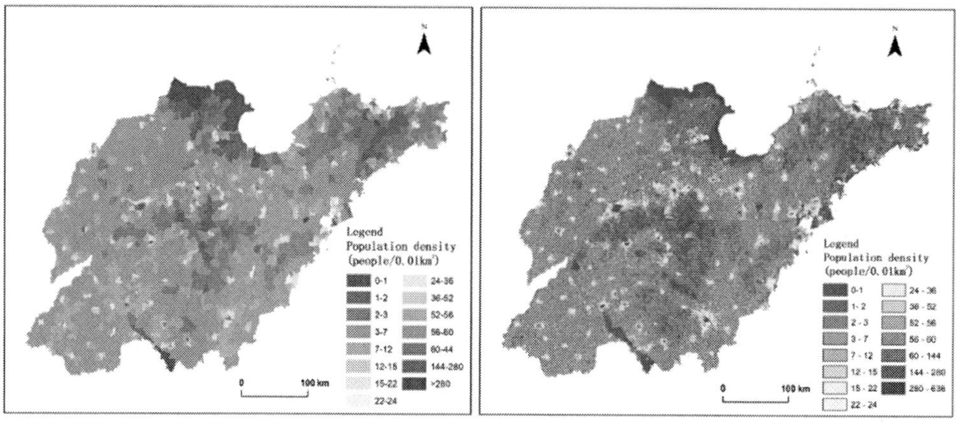

(a) Population density map (b) Population simulation map

Figure 1. Comparison of population density map and distribution simulation map

4.2 Accuracy Assessment

Calculate the Relative Error (RE) between the simulated data and the census data according to formula (5); calculate the Root Mean Square Error (RMSE) and Mean Absolute Error (MAE) between the simulated data and the public data set according to formula (6) (7).

The RE are shown in Table 4. The percentages of the five error intervals are 56.98%, 31.34%, 7.91%, 1.99%, 1.78%, and the average relative error is 23.56%. The RMSE and MAE are shown in Table 5. The RMSE and MAE of the simulated data in this paper are both smaller than the two public data sets. The comparison shows that the accuracy of the population spatial simulation data in this paper is better.

Table4. Relative error distribution interval statistics.

	Error Interval (Number of township)					Average relative error
	0~0.2	0.2~0.4	0.4~0.6	0.6~0.8	>0.8	
Simulation results	1087	598	151	38	34	23.56%

Table5. Error statistics.

	RMSE	MAE
Simulation data	14694	10031
WorldPop	23577	13329
Kilometer grid data	29790	18546

5 Conclusion

This paper draws the following conclusions during the process of spatial simulation of the population census at the township level in Shandong Province:

1) The correlation between various land types and population shows that land types such as cultivated land and man-made ground have strong indications for population distribution. Combining with night light data can better simulate population distribution. Spatial data can better reflect the differences in population distribution compared to administrative division statistics.

2) The R^2 of the stepwise regression models established in this paper is greater than 0.90. In the simulation data, there are 1685 towns and towns with a relative error of less than 0.4, accounting for 88.31%. The Root Mean Square Error and Mean Absolute Error of the simulated data are smaller than the two public datasets, which proves that the overall simulation effect is better.

3) The two kinds of population indicative data used in this study cover the whole world, the modeling method is simple and the calculation efficiency is high, it has good applicability and migration, and has certain reference value for generating large-scale population data sets in other regions.

References

1. D.R. LI, X. LI, An Overview on Data Mining of Nighttime Light Remote Sensing. Acta Geodaetica et Cartographica Sinica, 44(6): 591-601, (2015).
2. D.J. Briggs, J. Gulliver, D. Fecht, et al. Dasymetric modelling of small-area population distribution using land cover and light emissions data. Remote sensing of Environment, 108(4): 451-466, (2007).
3. C.D. Elvidge, K.E. Baugh, J.B. Dietz, et al. Radiance calibration of DMSP-OLS low-light imaging data of human settlements. Remote Sensing of Environment, 68(1): 77-88, (1999).
4. P. Sutton, D. Roberts, C. Elvidge, et al. Census from Heaven: An estimate of the global human population using night-time satellite imagery. International Journal of Remote Sensing, 22(16): 3061-3076, (2001).
5. C. Zeng, Y. Zhou, S. Wang, et al. Population spatialization in China based on night-time imagery and land use data. International Journal of Remote Sensing, 32(24): 9599-9620, (2011).
6. Q. GAO, K. ALIMUJIANG. Modeling the population spatial distribution of Tianshan north-slope urban agglomeration based on DMSP/OLS night lighting data. Northwest Population, 38(03):113-120, (2017).
7. M.M. WANG, J.L. WANG. Spatialization of township-level population based on nighttime light and land use data in Shandong province. Journal of Geo-information Science,21(5):699-709, (2019).
8. Population Census Office under the State Council. Tabulation on the 2010 population census of the People's Republic of China by Township. Beijing: China Statistics Press,(2012).
9. J. CHEN, J. CHEN, A.P. LIAO, et al. Concepts and key techniques for 30 m global land cover mapping. Acta Geodaetica et Cartographica Sinica,43(6):551-557, (2014).
10. National Oceanic and Atmospheric Administration (NOAA) National Centers for Environmental Information.Version 4 DMSP-OLS nighttime lights time series.https://www.ngdc.noaa.gov/eog/dmsp/downloadV4composites.html, 2021-02-11.
11. WorldPOP. China population 2010. https://www.worldpop.org/geodata/summary?id=24916.2021-02-11.
12. X.L. XU, China's population spatial distribution kilometer grid dataset. Data Registration and Publishing System of the Resource and Environmental Science Data Center of the Chinese Academy of Sciences. http://www.resdc.cn/DOI.2017.
13. L. ZHUO, X.F. ZHANG, J. ZHENG, et al. An EVI-based method to reduce saturation of DMSP/OLS nighttime light data. Acta Geographica Sinica,70(8): 1339-1350,(2015).
14. R.W. LIU, Z.M. FENG, Y.Z. YANG, et al. Research on the spatial pattern of population agglomeration and dispersion in China. Progress in Geography, **29(10)**:1171-1177,(2010).

 https://doi.org/10.1051/e3sconf/202129302016

Structural dynamic response of underground structure with vertical ground motion input

Tong Liu[1], Fuwang Xu[2], Qinghe Wang[3,*], Huaiyu Xu[3]

[1]School of Management, Shenyang Jianzhu University, Shenyang 110168, China.

[2]China Communications Tunnel Engineering Company Limited, Fujian 353100, China.

[3] School of Civil Engineering, Shenyang Jianzhu University, Shenyang 110168, China.

Abstract. Since many cases of structural damage in past earthquakes have been attributed to strong vertical ground motion, our understanding of vertical seismic load effects and their influence on seismic performance of subway station structure is limited. In this study, the Daikai subway station is taken as a typical example. A two-dimensional finite element model of both soil and structure was established using finite element software ABAQUS. Two input approaches of ground motion are considered, including the horizontal component alone and the vertical and horizontal motions simultaneously. Four groups of ground motion records are selected according to the site type of this station and scaled to the strong intensity which can make the station damage. Results show that the vertical seismic load increases the axial force of the column component apparently, while horizontal seismic load has little effect on axial compression ratio.

1 Introduction

Since the Kobe earthquake at Japan in 1995, enough attention has been paid to the seismic performance of underground structure. Researchers have carried out a series of studies on the seismic performance of various underground subway station structures [1-3]. For example, Huo et al.[1] conduct dynamic time-history analysis on the Daikai subway station using the finite element software ABAQUS, and systematically analyzed the force transfer mechanism and failure mechanism of both soil and structure. Zhuang et al.[2] conduct numerical simulation analysis on the nonlinear seismic response of the Dakai subway station using the finite element software ABAQUS. However, these research mainly focus on structural dynamic characteristic under horizontal or horizontal and vertical ground motion input, while the research on the comparison between these two cases is limited.

Besides, studies have shown that the damage of underground structure under seismic load is related to the large vertical ground motion, and the vertical ground motion component in some earthquake areas is greater than the horizontal ground motion component[4-6]. For example, the collapse of the middle layer of multi-storey building structure and the destruction of the central column of underground structure in the Kobe earthquake in 1995 are related to the vertical ground motion. Therefore, the influence of vertical ground motion on structural damage cannot be ignored.

In this paper, the Daikai subway station is taken as a typical example and the numerical model is developed using the finite element software ABAQUS. Four groups of ground motion records are selected according to the type of site, and the dynamic response characteristics of the structure under different ground motion input conditions are investigated through the ground motion amplitude modulation method of incremental dynamic analysis.

2 Numerical modeling

Based on the general profile of Daikai station[1], the cross section with 17m wide and 7.17m high, as shown in Figure 1, is taken into consideration. The cross section of central column is 0.4m by 1.0m, with space of 3.5m in longitudinal direction. The soil condition is shown in Table 1.

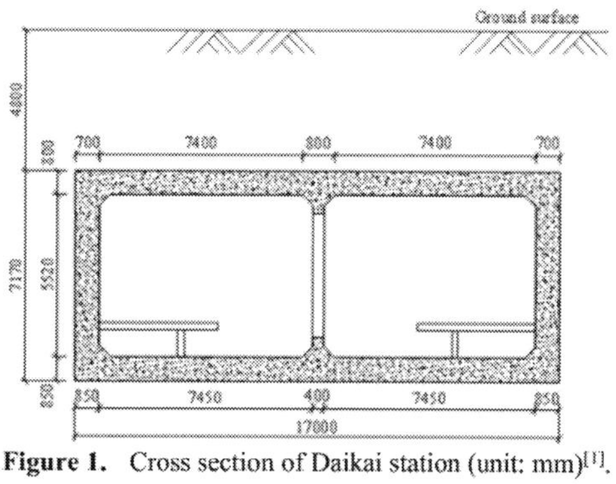

Figure 1. Cross section of Daikai station (unit: mm)[1].

* Corresponding author: wangqinghe@sjzu.edu.cn

A detailed two-dimensional model of Daikai station and the layered soil is established using finite element software package ABAQUS[7]. As illustrated in Figure 2, the numerical model is 1000m long and 58m high, among which the structural members are simulated through beam element(B21), and the Young's modulus and unit weight of concrete are 24GPa and 25kN/m^3, respectively.

The soil is simulated using four-node plain strain element (CPE4R) as well as the infinite element (CINPE4). For the detail statistics of numerical model and its verification, please refer to Liu et al. [8].

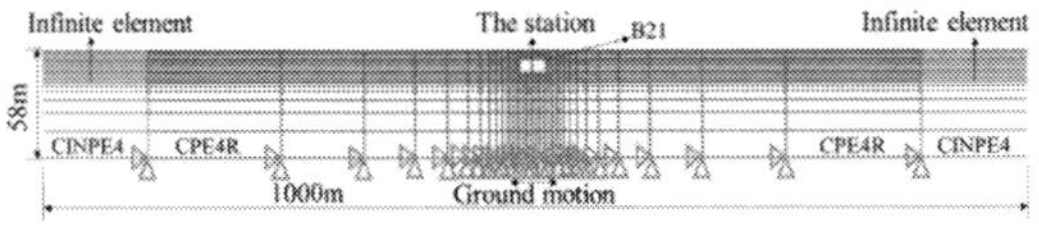

Figure 2. Numerical modeling of Daikai station.

3 Incremental dynamic analysis

To investigate seismic performance of subway station under vertical motion, four groups ground motion records are selected herein based on the soil condition, as illustrated in Table 2, during which Mi means horizontal ground motion and Vi means vertical ground motion. The simulated cases include horizontal motion input (H), horizontal and vertical motions input (H&V). The horizontal motions are scaled with peak velocity as 3, 5, 10, 20, 30, 40, 50, 60, 70, 80cm/s, and the vertical motion is 0.67 times of horizontal intensity. Besides, peak acceleration at the base of structure (PBA) is selected as intensity measure and maximum inter-story drift angle (θ_{max}) and axial compression ratio (α) are chosen as damage measure of incremental dynamic analysis (IDA). Then, with nonlinear dynamic time-history analysis, the seismic response under increasing seismic intensity can be obtained.

Table1. Soil parameter of Daikai station.

Sequence	Name	Depth (m)	Unit Weight (kN/m^3)	Cohesion (kPa)	Friction Angle (°)	Poisson's Ratio	Elastic Modulus (MPa)
1	Fill	0-1	19	20	15	0.33	101.308
2	Holocene clay	1-2	19	30	20	0.32	101.308
3	Holocene sand	2-4.8	19	1	40	0.32	147.840
4	Pleistocene sand	4.8-8	19	1	40	0.40	195.972
5	Pleistocene clay	8-17	19	30	20	0.30	290.342
6	Pleistocene gravel	17-	20	1	40	0.26	560.045

Table2. Ground motion records.

No.	Event	Station	Component	PGA(g)	PGV(cm/s)
M1	Chi-Chi, Taiwan, 1999	CHY101	CHY101-E	0.353	70.64
M2	Chi-Chi, Taiwan, 1999	CHY101	CHY101-N	0.439	115.0
V1	Chi-Chi, Taiwan, 1999	CHY101	CHY101-V	0.166	27.10
M3	Chi-Chi, Taiwan, 1999	TCU050	TCU050-E	0.147	36.91
M4	Chi-Chi, Taiwan, 1999	TCU050	TCU050-N	0.130	42.36
V2	Chi-Chi, Taiwan, 1999	TCU050	TCU050-V	0.088	42.16

M5	Chi-Chi, Taiwan, 1999	TCU065	TCU065-E	0.79	125.35
M6	Chi-Chi, Taiwan, 1999	TCU065	TCU065-N	0.575	92.13
V3	Chi-Chi, Taiwan, 1999	TCU065	TCU065-V	0.263	69.41
M7	Chi-Chi, Taiwan, 1999	CHY025	CHY025-E	0.162	50.92
M8	Chi-Chi, Taiwan, 1999	CHY025	CHY025-N	0.155	32.88
V4	Chi-Chi, Taiwan, 1999	CHY025	CHY025-V	0.173	38.03

4 Results

As shown in Figure 3, structural dynamic response under two ground motion input approaches are compared. At the initial stage of IDA curve development, the structure is in the elastic stage, and the curves almost coincide, which means that the two input approaches have little influence on structural dynamic response. However, when it comes to the elastic-plastic stage, the structural dynamic responses obtained under the two input methods are obviously different, during which the data of θ_{max} with H&V input is less than that of the H input.

As illustrated in Figure 4, the axial compression ratio of central column changes slightly with the increase of ground motion intensity under H input only, which varies between 0.3 and 0.5. However, when the structure under horizontal and vertical seismic load simultaneously, the axial compression ratio of central column increases obviously with the increase of the ground motion intensity. Moreover, the axial compression ratio of central column under H&V input is greater than that of H input, which varies from 8% to 93% range.

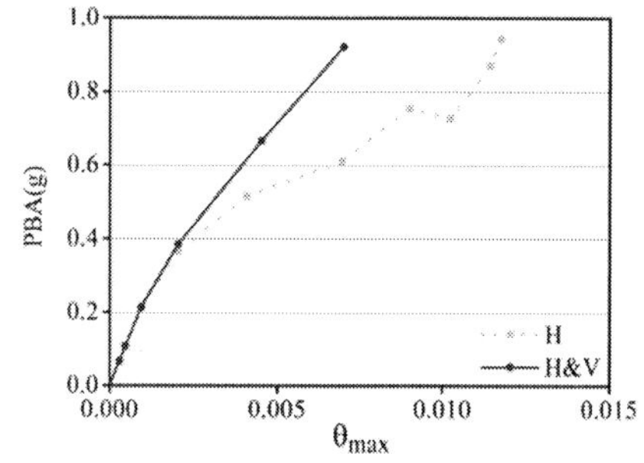

(a) CHY101

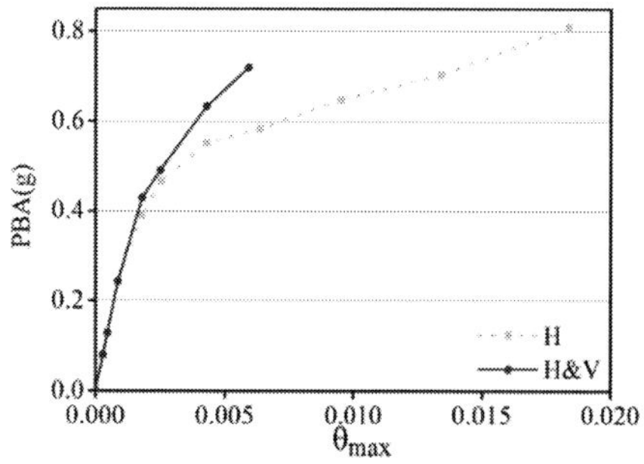

(b) TCU050

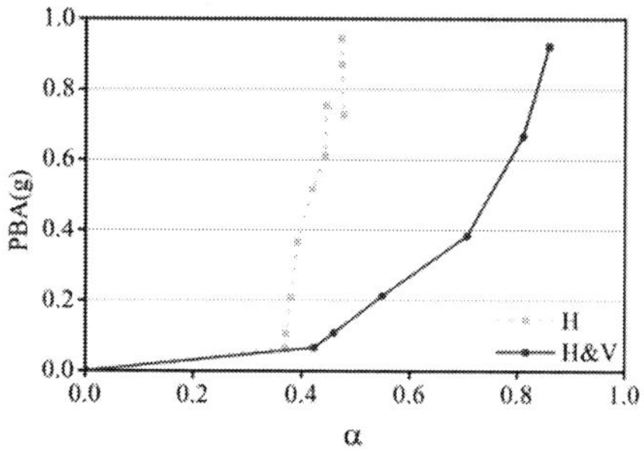

(a) CHY101

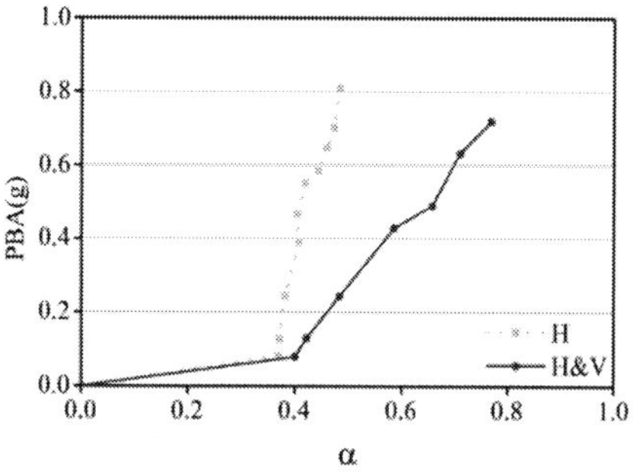

(b) TCU050

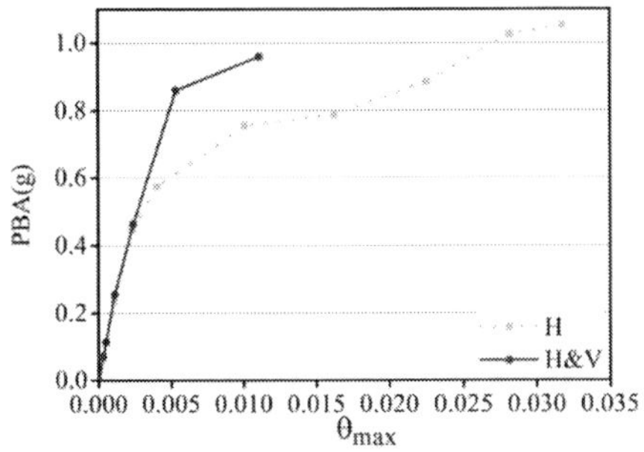

(c) TCU065

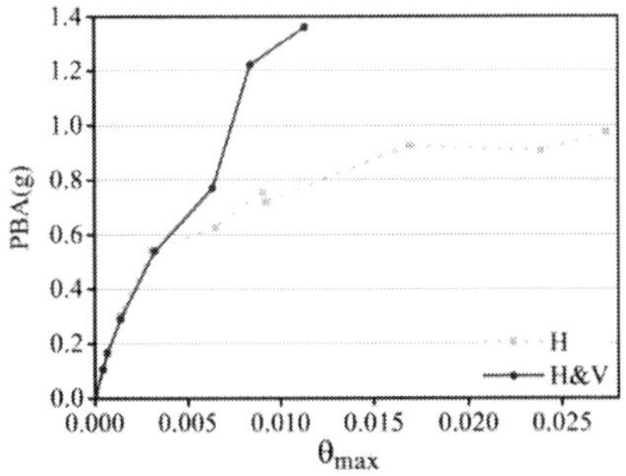

(d) CHY025

Figure 3. IDA curves with θ_{max} as DM.

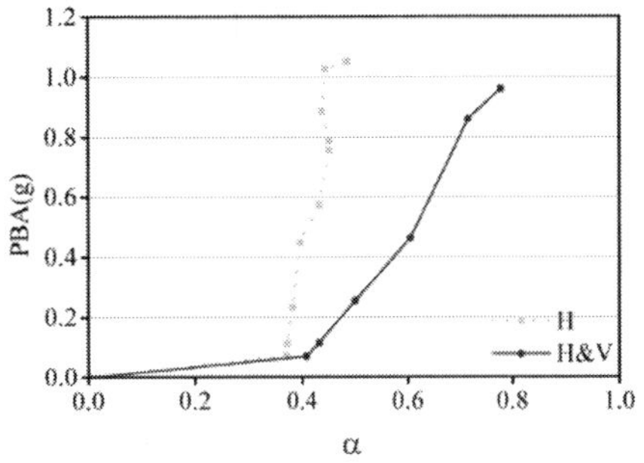

(c) TCU065

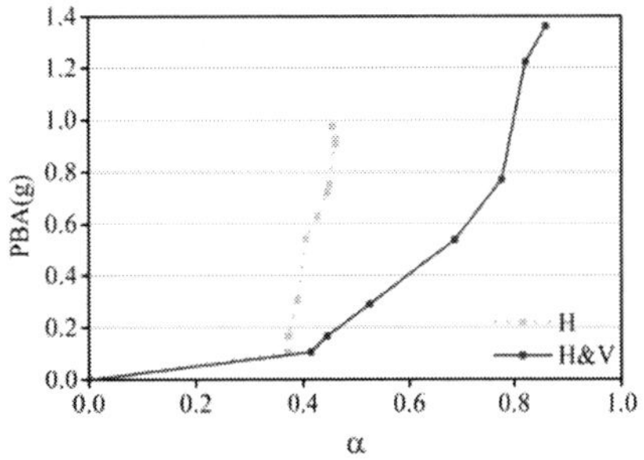

(d) CHY025

Figure 4. IDA curves with α as DM.

From the IDA results with θ_{max} and axial compression ratio as DM respectively, it can be summarized that the lateral deformation of structure is greatly influenced by the horizontal ground motion, and the axial compression ratio of the central column is mainly controlled by the vertical ground motion, and the horizontal ground motion has little effect on the axial compression ratio. Therefore, the axial compression ratio should be considered as an important index during the design of structural vertical member.

5 Conclusion

In this study, the Daikai subway station is taken as a typical example and its seismic performance under vertical ground motion is investigated through incremental dynamic analysis. The conclusions are as follows.

(1) Structural lateral displacement is mainly influenced by horizontal ground motion.

(2) The axial compression ratio of central column is controlled by vertical ground motion, but not the horizontal ground motion.

(3) It is suggested herein that the axial compression ratio should be taken as an important index during vertical member design.

Acknowledgements

The research work in this paper was supported by LiaoNing Revitalization Talents Program, China (XLYC1902027), by Doctoral Scientific Research Foundation, China (2019 - BS- 193), by Science and Technology Project of MHRUD, China (2019 - K - 054), by Shenyang Science and Technology Project, China (RC200143) and Tianjin Science and Technology Project, China (18ZXGDGX00030).

References

1. Huo H, Bobet A, Fernández G, et al. Load Transfer Mechanisms Between Underground Structure and Surrounding Ground: Evaluation of the Failure of the Daikai Station[J]. Journal of Geotechnical & Geoenvironmental Engineering, 2005, **131**(12): 1522-1533.
2. Zhuang H Y, Chen S G, Chen G X . Numerical Simulation and Analysis of Earthquake Damage of Daikai Metro Station Caused by Kobe Earthquke[J]. Rock and Soil Mechanics, 2008, **29**(1): 245-250 (in Chinese).
3. Du X L, Ma C, Lu D C, et al. Collapse Simulation and Failure Mechanism Analysis of the Dakai Subway Station Under Seismic Loads[J]. China Civil Engineering Journal, 2017, **50**(1): 53-62+69 (in Chinese).
4. Tao L J, Li J D, Wu B L, et al. Influence of the Vertical Seismic Wave on the Seismic Response of an Large-span High-section Y-shaped Column Subway Station[J]. China Earthquake Engineering Journal, 2015, **37**(3): 648-654 (in Chinese).
5. Guo S B, Zhao Y, Zhao Y T, et al. Dynamic Analysis of under ground Structures under the Vertical and

Horizontal Seismic Load[J]. Underground Space, 2002, **22**(4): 314-319 (in Chinese).

6. Chen C S, Qi C Z, Qian Q H, et al. Dynamic response of the Roof Slab of a Shallowly Buried Underground Structure Under Vertical Seismic Excitation[J]. World Earthquake Engineering, 2010, **26**(4): 86-92 (in Chinese).
7. ABAQUS, Users Manual V. 6.10-1. 2010. Dassault Systemes Simulia Corp., Providence, RI.
8. Liu T, Chen Z, Yuan Y, et al. Fragility Analysis of a Subway Station Structure by Incremental Dynamic Analysis[J]. Advances in Structural Engineering, 2017, **20**(7): 1111-1124.

Research and Analysis of concrete facilities' corrosion in coastal substaion of fujian

Yang Dishan [1], Zhang Jinbo[1], LIN Xi[1], Huang HuanXin [2]

[1]National Grid Fujian Electric Power Co., Ltd. Institute of Economics and Technology, Fuzhou 350012, China

[2]Power China Fujian Electric Power Engineering Co., Ltd., Fuzhou 350001, China

Abstract: Corrosion problems of outdoor concrete facilities was summarized through research on coastal substation of Fujian, such as serious corrosion of steel ring and weld in the joints of cement poles, exposure and corrosion of reinforcement, cracking and falling off of concrete cover. The information of concrete facilities' assessment reports was summarized, and it was found that concrete compressive strength of present age satisfied early design drawings. In addition, appraisal of reliability revealed that all facilities couldn't satisfy the requirements of current standards and seismic appraisal revealed that all facilities couldn't meet the seismic requirements. The analysis of concrete facilities' corrosion mechanism was made, and it was found that concrete facilities in coastal substation suffered triple effects of carbonation, chloride corrosion and stray current corrosion. The power failure schemes needed for present repairing and strengthening measures of concrete facilities lack of feasibility. Research on repairing and strengthening measures of concrete facilities that requiring no or small-scale power cut would be of great significance.

1 Introduction

Fujian outdoor substation frame has concrete frame and steel frame two schemes. With the improvement of economic level, the frame constructed after 2006 generally adopts steel frame, while the frame constructed before this is all concrete frame. The concrete truss column is generally made of precast annular cement pole, with longitudinal bars and spiral stirrups, and the truss beam is Angle triangular lattice beam. In recent years, with the increase of the renovation and expansion projects of old substations that have been in operation for more than 20 years, it has been found that the concrete frame columns of these old stations have serious corrosion problems, and the mechanical properties of the frame columns are not enough to support the power grid renovation plan, or even to guarantee the safety under the current operating conditions. In order to ensure the safe operation of the whole power grid, it is urgent to investigate and analyze the corrosion of the concrete frame of the substation.

2 Investigation of corrosion conditions

2.1 Category of Corrosion Problems

Time of Fujian province was put into operation for more than 20 years old substation site investigation found that the old station common concrete frame column in the steel and concrete pole joint weld serious rust, steel rust, concrete cover exposed burst off visible corrosion problems, these problems have seriously affect the structure reliability and safety.

The concrete truss column is assembled by butt welding of the steel ring of the rod head of the precast annular cement rod. The steel ring is the key force transfer joint of the concrete truss column. The steel ring and the weld seam are seriously corroded (Figure 1), which weakens the stressed area of the cross section at the joint and reduces the bearing capacity of the truss column. Prefabricated circular concrete pole configuration of longitudinal reinforcement and spiral stirrups to improve the bearing capacity of normal section and inclined section, section bar shaft cover off cause longitudinal reinforcement and stirrup the leakage and corrosion (Figure 2), weakened the reinforcement stress area, and rust particles can also cause the burst loss of concrete protective layer, forming " rebar corrosion steel corrosion protection layer burst loss -" the vicious circle. The on-site survey of the substation found that the concrete protective layer fell off the body of most cement poles (Figure. 3). This part of the bar is in the early stage of corrosion development, if left untreated, it will lead to the exposure of reinforcement and further corrosion, and the lack of protective layer will also cause adverse effects on the stability of longitudinal bars. The above-mentioned multiple problems often occur at the same time in a single cement pole. For example, the exposed steel bar means that the concrete protective layer has fallen off. Some rods are even found in the same place with serious corrosion of steel ring and weld seam and exposed steel bar (Figure. 4),

* Corresponding author: 550818454@qq.com

causing great hidden danger to the safe operation of the substation.

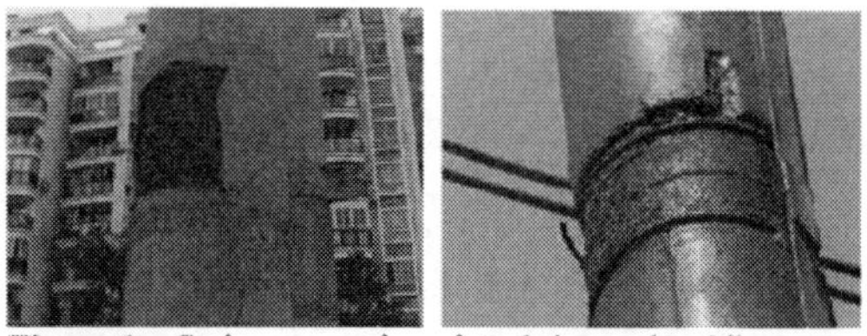

Figure 1. Serious corrosion of steel ring and welding seam

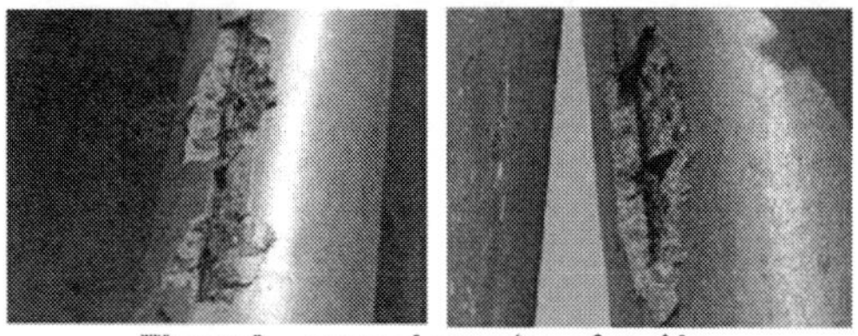

Figure 2. exposed corrosion of steel bar

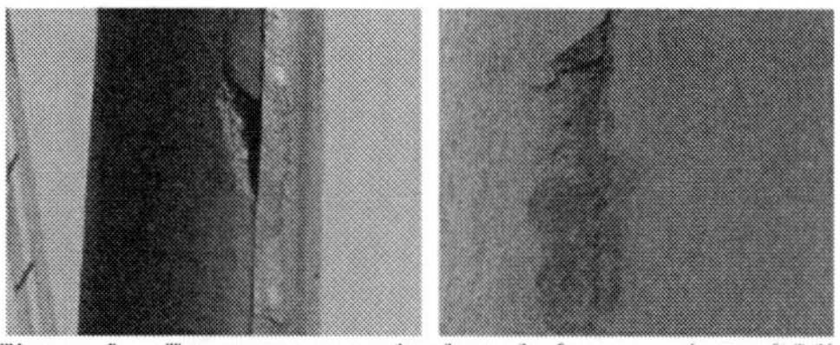

Figure 3. Concrete protective layer bulges, cracks and falls off

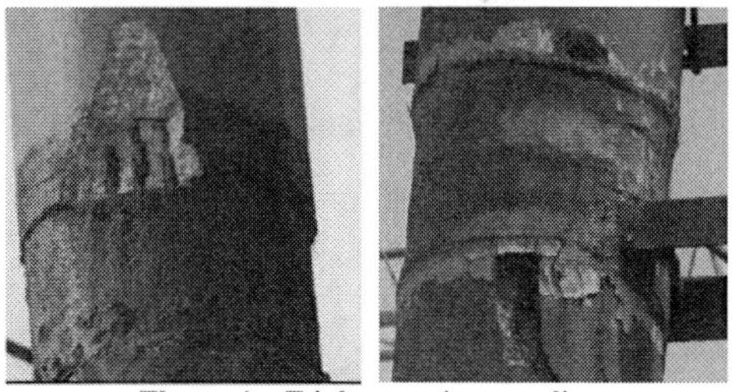

Figure 4. Triple corrosion coupling

2.2 Structural assessment of corroded concrete frames

The power company entrusted the professional testing center to carry out structural evaluation on the concrete frames of 220kV and 110kV substation that have been put into operation for more than 20 years. The evaluation contents include the use of rebound method to test the compressive strength of the concrete frame columns at present age, and the reliability evaluation and seismic performance evaluation of the frame based on the field survey results.

According to a total of 34 evaluation reports of the frame, the compressive strength of the concrete in the current age of the frame columns can all meet the requirements of the early design drawings, indicating that concrete carbonization does not directly affect the performance of concrete. The field survey results show that all the frames have the phenomenon of exposed corrosion of cement rod and steel ring. For example, in a 220kV frame in Quanzhou, the steel ring at the connection of cement pole of concrete frame column is generally corroded. The rod body generally has different degrees of exposed corrosion of steel bar and bulging and cracking of concrete protective layer. According to the "Industrial Building Reliability Evaluation Standard" [1], the testing center carries out reliability evaluation on 34 frames, among which 29 frames are rated as Level 4, that is, they do not meet the safety requirements of the current national standards, and have seriously affected the safe operation of the substation, and must take immediate measures; The reliability level of the five frameworks is rated as Level 3, that is, they do not meet the safety requirements of the current national standards and affect the safety, and measures should be taken; According to the Seismic Evaluation Standard for Structures [2], the seismic performance of 34 frames was evaluated, and none of the frames met the seismic requirements.

The evaluation report of the concrete frame reflects the risk of the transformer substation frame which has been in service for more than 20 years. All the frames do not meet the safety requirements of the current national standards. Measures should be taken to solve the corrosion problem of the concrete frame of the old station.

3 Analysis of corrosion mechanism of concrete cement pole

In general atmospheric environment, concrete carbonation is the main reason for the corrosion of concrete cement pole and steel bar. For the concrete frame column in the coastal substation, it is simultaneously subjected to the triple coupling effect of concrete carbonation, chloride ion corrosion and stray current corrosion.

3.1 Carbonization of concrete

Cement produces a large amount of calcium hydroxide in the hydration process, and the pore water of concrete is calcium hydroxide saturated solution with a pH value of about 12~13, showing strong alkalinity [3~5]. Due to the existence of high alkalinity concrete hole solution, the concrete concrete surface formed a stable and compact passivation film, has a good protective effect on the reinforcement. Carbon dioxide in the atmosphere diffuses through the pores of concrete to its interior, and reacts with calcium hydroxide to produce calcium carbonate, which is the carbonization of concrete. The reaction formula is as follows:

$$Ca(OH)_2 + CO_2 \rightarrow CaCO_3 + H_2O$$

The occurrence of concrete carbonization does not directly affect concrete, but generates $CaCO_3$ to reduce its pH value, which causes the destruction of the passivation film on the surface of the steel bar, so that O_2 in the air can directly contact the steel bar and oxygen absorption corrosion occurs. The chemical reaction formula is as follows:

$$Fe + O_2 + H_2O \rightarrow Fe(OH)_2$$

Fe(OH)2 will be further oxidized to Fe(OH)3, dehydration to form loose, porous red rust Fe2O3, under the condition of less oxygen, Fe(OH)2 oxidation is not complete to form black rust Fe3O4, the final corrosion products depend on whether oxygen is sufficient. The volume of rust is generally about 2.5 times that of iron. The pressure caused by the increase in volume makes the concrete protective layer bulge and crack, which further increases the contact between the steel bar and the outside world and intensifies the corrosion process of the steel bar. This vicious circle causes the damage of the corrosion of the steel bar to spiral out of control. Studies show that [6], when the pH of concrete pore fluid is 9.88, the reinforcement surface begins to form passivating film, and when the pH value reaches 11.5, the reinforcement surface forms a complete passivating film. The carbonation of concrete can reduce its pH value to below 9 and completely destroy the passivation film on the surface of the steel bar. It can be said that the carbonation of concrete is a prerequisite for the corrosion of the steel bar in the general atmospheric environment.

3.2 Chloride ion corrosion in salt spray environment

The concentration of chloride ion has an important relationship with the type of environment, and the concentration of chloride ion is higher in coastal areas. The existing concrete structure durability evaluation standard "[7] the chloride ion erosion environment can be divided into coastal atmospheric environment, atmosphere salt fog area and the change of water level area, splash zone, including offshore atmospheric environment is set to coast 0 ~ 1000 m, studies have shown that [8], salt fog environment chloride ion corrosion of concrete damage and even more serious than the seawater erosion.

The corrosion mechanism of chloride ion has two aspects: the destruction of passivation film and the electrochemical reaction of corrosion. Studies show that [9~11], when the chloride ion content on the surface of steel bars in concrete reaches a certain concentration, the passivation film on the surface will begin to destroy, and then lead to corrosion of steel bars. The destruction mechanism of chloride ions on the passivation film is consistent with that of carbonization. The chemical reaction between chloride ions and calcium hydroxide solution reduces the PH value of concrete hole solution, leading to the destruction of the passivation film and corrosion [12~17]. After the destruction of the passivation film, an electrochemical corrosion reaction occurs between chloride ions and the exposed cationic Fe2+ in the iron matrix, and the reaction formula is as follows:

$$Fe^{2+} + Cl^{-} + H_2O \rightarrow Fe(OH)_2 + 2HCl$$

From the corrosion mechanism analysis, chloride ions in the corrosion process as a catalyst, all into the concrete chloride ions, will play a destructive role over and over again. Studies have shown that [18], the corrosion rate of chloride ion on reinforcement is generally faster in the early stage, and decreases with the increase of corrosion products. When the internal structure of concrete is damaged, the corrosion rate will accelerate again. In addition to the destruction of the passivation film and catalytic electrochemical corrosion, chloride ion can also directly with the concrete in the $Ca(OH)_2$, $3CaO•Al_2O_3•6H_2O$ to form a much larger volume than the reactants of solid compounds, so that the concrete expansion crack.

3.3 Stray current corrosion

The current that is not transmitted according to the original set path is called stray current, and the corrosion caused by it is called stray current corrosion. In recent years, many scholars devoted to stray current corrosion effects of metro and light rail research think stray current corrosion on track than fatigue durability damage, chloride ion corrosion is more serious [20], ministry of organization compiled the "metro stray current corrosion protection technology standard" [21], specific measures to improve the subway stray current protection design and the technical level. Current transformer substation stray current corrosion in concrete frame has not yet been taken seriously, jian-yu Yang, etc. [22] study, stray current through concrete structure, under the conductive effect of reinforcement, current be brought into the drainage points, steel cathode area change current strength, current to achieve a certain strength, reinforced surface cathodic hydrogen evolution reaction, The hydrogen precipitated in the member inside the formation of pressure, so that the steel bar and concrete off, and the concrete protective layer expansion crack; In the anode part of reinforcement, the passing of stray current will accelerate the transport of ions, accelerate the occurrence of oxidation process, and promote the corrosion and expansion of reinforcement. The corrosion of rebar caused by stray current is essentially the same as the electrochemical corrosion caused by chloride ions, but the stray current is tens or even thousands of times larger than the electrochemical corrosion current caused by chloride ions numerically [23].

The concrete framework of substations built in coastal areas is corroded by the coupling of chloride ion and stray current at the same time. Studies show that when chloride ion coexists with stray current, chloride ion intensifies the stray current corrosion [24]. The limit concentration, existence form and migration rate of chloride ions that induce steel corrosion are also affected by stray current [25-26]. According to the study of Chen Xunjie et al. [27], the concentration of chloride ion in concrete increases with the increase of stray current density and erosion age in the chloride salt-stray current environment.

4 Analysis of reinforcement measures for concrete cement pole repair

4.1 Research progress on repair and reinforcement measures of cement poles

The existing repair and reinforcement measures for cement poles include those aimed at durability and those aimed at safety. The corrosion of concrete frame columns in substations is serious, and neither of the above two measures is necessary. Technical Specification for Durability Repair and Protection of Concrete Structures [28] puts forward guidance schemes for various durability problems. For various reinforcement corrosion, repair measures such as reinforcement rust resistance treatment, electrochemical re-alkalization, electrochemical desalination and cathodic protection can be adopted. For various cracks, repair methods can be divided into surface treatment method, filling sealing method, pressure grouting method; Concrete surface repair materials can be used interface treatment materials and repair mortar, surface protection materials can be used inorganic materials, organic polymer materials and composite materials. The Code for Design of Reinforcement of Concrete Structures [29] puts forward guidance schemes for various safety problems, which can be used to increase the bearing capacity of members, such as the reinforcement method of increasing section, the reinforcement method of outer section steel, the reinforcement method of bonded steel plate, and the reinforcement method of bonded fiber composite material. In recent years, studies on the repair and reinforcement of substation concrete frame columns [30-33] have mostly started from the above two aspects, and the repair and reinforcement measures have not been separated from the standard methods. Some scholars have done some studies from the perspective of new grouting materials and new reinforcement materials [34-35], and some scholars have proposed a scheme of overall replacement of the frame [36].

4.2 Difficulties in repair and reinforcement of cement pole

The existing research on the repair and reinforcement measures of concrete frame columns in substations has ignored a difficult point: the feasibility of power outage scheme. In general, the safety distance of the side of the 110kV frame column is 1.65m, and the safety distance of the side of the 220kV frame column is 2.55m[37]. During the construction period, no material or even the flying sparks or water spray is allowed to enter the electrified area, otherwise the air will be broken down by the high pressure and a major safety accident will occur. Existing framework column reinforcement plans need to build 10-20m high scaffolding, but also need to operate a variety of construction tools, it is impossible to avoid multi-interval power failure at the same time, the establishment of power failure plan is the existing substation concrete frame column reinforcement of the biggest difficulty. In the preparation of transformation plans for 220kV substations in Xiamen, Quanzhou and Putian, due to the difficulty in coordinating power failure plans with framework reinforcement and lack of constructibility, the total station reconstruction combined with the intermediate transition plan was finally chosen, resulting in a ten-fold increase in investment. The corrosion problem of the concrete frame of the old station has become increasingly prominent. The repair and reinforcement scheme without power failure or with small power failure scale will bring huge economic benefits and is of great significance.

5 conclusion

According to the investigation results of substations in coastal areas of Fujian Province that have been in operation for more than 20 years, there are widespread corrosion problems in concrete frame columns, such as severe corrosion of steel ring and weld at the connection of cement pole, exposed corrosion of steel bar, swelling, cracking and shedding of concrete protective layer. From the frame reliability evaluation report, these frames do not meet the safety requirements of the current national standards, do not meet the seismic requirements, the solution of concrete frame column corrosion problem has been imminent. Concrete carbonation, chloride ion corrosion in salt spray environment and stray current corrosion are the three major reasons for severe corrosion of concrete frame columns in coastal substations. Combined with the analysis of engineering experience, the existing repair and reinforcement measures of concrete frame columns require power failure schemes with low feasibility, and the study of repair and reinforcement schemes without power failure or with small power failure scale is of great significance.

References

1. Ministry of Housing and Urban-Rural Development of the People ' s Republic of China.Industrial Building Reliability Appraisal Standard : GB 50144-2019 [S].Beijing : *China Construction Industry Press*, (2019).
2. Ministry of Housing and Urban and Rural Construction of the People ' s Republic of China.Structure seismic appraisal standard : GB 50117-2014 [S].Beijing : *China Construction Industry Press*, (2014).
3. Xiao Jia,GOU Cheng-fu. Overview of the research for concrete carbonation[J]. *Concrete*, **01** (2010): 40-44, 52.
4. Xiao Weiwei, Zhang Jigang, Zhang Junbo, et al . Research review on concrete deterioration under the combined action of carbonation and chloride[J]. *china concrete and cement products*, **12** (2019): 24-28.
5. Wang Jianhua, Wang Zuhe, Mu Zhengao.Concrete steel corrosion and rust inhibitor use [J].*Shanxi architecture*, **36**,27 (2010): 54-55.
6. Hong naifeng. Concrete alkalinity and steel corrosion [J]. *Concrete and cement products*, **5** (1990): 16-18.
7. Ministry of Housing and Urban-Rural Development of the People 's Republic of China. Durability assessment criteria for existing concrete structures : GB / T 51335-2019 [S]. Beijing : *China Construction Industry Press*, (2019).
8. Wu Qingling, Yang Yihong, Fei Weiwei.Deterioration damage of concrete in salt spray corrosion and seawater erosion [J].*Concrete and cement products*, **11** (2014): 30-34.

9. Poursaee A. Temperature dependence of the formation of the passivation layer on carbon steel in high alkaline environment of concrete pore solution[J]. *Electrochemistry Communications*, **73** (2016): 24-28.
10. Scott A, Alexander M G. Effect of supplementary cementitious materials (binder type) on the pore solution chemistry and the corrosion of steel in alkaline environments[J]. *Cement & Concrete Research*, **89** (2016): 45-55.
11. Yao Senyuan, Yan Jiali, Liu Junzhe, etc.Research on the critical chloride ion concentration of steel bar in concrete [J]. Journal of NingBo university (NESS) ,**31**,03 (2018): 83-88.
12. WANG Y, NANUKUTTAN S, BAI Y, et al. Influence of combined carbonation and chloride ingress regimes on rate of ingress and redistribution of chlorides in concretes[J]. *Construction and Building Materials*, **140** (2017): 173-183.
13. G. Roventi, T. Bellezze, E. Barbaresi, et al. Effect of carbonation process on the passivating products of zinc in Ca(OH)2 saturated solution. **64**,11 (2013): 1007-1014.
14. Jon Williamson, O. Burkan Isgor. The effect of simulated concrete pore solution composition and chlorides on the electronic properties of passive films on carbon steel rebar, **106** (2016): 82-95.
15. WU Qun, LIU Yu,,DU Ronggui, LIN Changjian. Electrochemical study on the passivation effect of chloride ion on reinforcement in simulated concrete pore solution[J]. *Acta Metallurgica Sinica,* **03** (2008): 346-350.
16. H. Burak Gunay, Pouria Ghods, O. Burkan Isgor, et al. Characterization of atomic structure of oxide films on carbon steel in simulated concrete pore solutions using EELS. **274**(2013) : 195-202.
17. Li Xiaozhen, Liu Junzhe, Yan Jiali, etc. Effect of carbonation and chloride on steel bar passivation in concrete pore solution [J]. *Journal of Building Materials*, **23**, 01(2020): 224-229.
18. He Yong, Tang Xianxi, Sun Shuhu, etc. Experimental study on corrosion resistance of bridge reinforced concrete based on electrochemical corrosion [J]. *Concrete and cement products*, **9** (2018): 10-15.
19. F.C. Robles Hernández, G. Plascencia, Kevin Koch. Rail base corrosion problem for North American transit systems[J]. *Engineering Failure Analysis*, **16**, 1(2009):281-294.
20. Du Yingji, Ma Shaojun, Cai Yuebo, etc. Prediction of durability of concrete against stray current in subway engineering [J]. Journal of Water Conservancy and Construction Engineering, **01** (2003): 7-9.
21. Ministry of Housing and Urban-Rural Development of the People ' s Republic of China.Technical standard for stray current corrosion protection of subway : CJJT49-2020 [S].Beijing : *China Construction Industry Press*, (2020).
22. Yang Jianyu, Yi Weijian, Li Wensheng.Spurious current corrosion of reinforced concrete frame in substation [C]//Papers of the 2018 National Academic Annual Conference of Industrial Buildings (Volume) in Beijing, **48** (2018): 277-281.
23. Introduction to durability of concrete structures [M]. Shanghai : *Shanghai Science and Technology Press,*(2003).
24. Luca Bertolini, Maddalena Carsana, Pietro Pedeferri. Corrosion behaviour of steel in concrete in the presence of stray current.**49**, 3(2006): 1056-1068.
25. Wu Xiong. Study on the deterioration characteristics of concrete under the combined action of stray current and chloride ion [D]. Wuhan : *Wuhan University of Technology*, (2008).
26. Geng Jian. Study on the degradation mechanism of concrete in the environment of stray current and chloride ion [D]. Wuhan : *Wuhan University of Technology*, (2008).
27. Chen Xunjie, Ouyang Youngling, Qian Wenxun, etc. Influence of stray current on concrete corrosion in different environments [J]. *Journal of Water Conservancy and Water Transport Engineering*, **02**(2014): 33-37.
28. Ministry of Housing and Urban and Rural Construction of the People' s Republic of China.Durability Repair and Protection Technical Specification for Concrete Structures : JGJ / T 259-2012 [S].Beijing : *China Construction Industry Press*, (2012).
29. Ministry of Housing, Urban and Rural Construction of the People 's Republic of China. Code for Design of Concrete Structure Reinforcement:GB 50367-2013 [S]. Beijing : *China Construction Industry Press*, (2013).
30. Liang Jingming, Li Xinhai, Zeng Lingcheng, etc. Research on Reinforcement Technology of Concrete Frame for Substation Equipment [J]. *Electrical Technology*, **21**(2019): 115-117.
31. Xiang Mingrong. Study on basic mechanical properties and long-term mechanical properties of outdoor support for substation reinforced by fiber reinforced composite materials [J]. *Southern energy construction,***2**, S1(2015): 88-92.
32. Zou Chongxian. Study on Corrosion Mechanism and Anticorrosion Reinforcement Measures of Outdoor Brackets in Substations [J]. *Industrial Buildings*, **11**(2006): 111-114.
33. Dong Manling, Li Peishan, Chen Xinliang.Corrosion of concrete members in substation and its prevention [J].*Journal of Shandong Agricultural University (Natural Science Edition)*, **03**(2001): 323-326.
34. Luo yexiong. Reinforcement Analysis and Design of Concrete Structure Support in Substation [J].

Architectural Design Management, **31**, 11(2014): 85-87.

35. Li Wensheng, Chen Shizhi, Nie Ming, etc. Research on Reinforcement Design of Old Concrete Frame in Substation [J].*Guangdong Electric Power*, **29**, 9(2016).
36. Peng Lijuan. Structural support reinforcement of 220 kV substation [J]. *Rural electrification*, **03** (2015) : 12-14.
37. National Energy Bureau.Design specification of high voltage distribution equipment : DL / T 5352-2018 [S].Beijing : *China Plan Press*, (2018) .

https://doi.org/10.1051/e3sconf/202129302018

Research progress of rice husk ash in solidified soil

HAO Tong[1], LIU Qian[1], LENG Fa-Guang[2], QIAO Tian-Long[1]

[1] School of Civil Engineering zhengzhou University, Zhengzhou 450000, China

[2] China Academy of Building Research, Beijing 100013, China

Abstract. Rice husk ash is a kind of volcanic ash material with high silicon. This also provides a feasibility for rice husk ash as cement-based auxiliary cementing material. China is rich in rice husk resources, and the rice husk ash is stacked. Using rice husk ash (RHA) instead of traditional cementitious materials can effectively solve the environmental pollution caused by the accumulation of rice husk ash. Domestic and international scholar have added rice husk ash as mineral admixture to concrete, but the utilization rate is very low. In order to increase the added value of rice husk ash, the high silicon property of rice husk ash was used to solidify soil. The application range of solidified soil is wide, the requirements for materials are not high, and the functions are different, which can realize the extensive use of rice husk ash. This paper summarizes the literature on comprehensive utilization of rice husk ash at home and abroad, systematically expounds the physicochemical properties and production process of rice husk ash, introduces the application of rice husk ash in solidified soil, and points out the unsolved problems in this field, which provides a direction for the further development.

1 Introduction

China is a big rice country, according to rice production data released by the National Bureau of Statistics in 2020, rice production is 21. 86 million tons, rice contains about 20 % of rice husks[1]. Main Utilization Ways of Rice Husk[2-4]: One is to produce high value-added products based on the physical and chemical properties of rice husk. The second is to produce inorganic chemical products according to the silicon source in rice husk. The third is to use the high calorific value of rice husk as fuel for biomass power generation. In order to improve the added value of rice husk, the special physical and chemical properties of rice husk ash have entered people's vision.

Rice husk ash is commonly used to produce decontamination agent, adsorbent, desiccant, soil conditioner, cement of rice husk ash and raw materials of thermal insulation sheet[5-10]. In recent years, rice husk ash was added into soil as a curing agent to improve the performance of soil. Solidified soil is the use of some measures to consolidate the scattered soil into a solid with a certain strength[11].

Foreign research on rice husk ash started early. Ario muhammad[12]pointed out that rice husk ash is a kind of volcanic ash material, but because of its high silicon and lack of calcium, it is usually mixed with cement or lime, The rice husk ash, lime and clay are mixed, and the reinforced clay can be used as road base and bottom base materials. Muntohar[13]revealed that the mechanical properties and durability of rice husk ash reinforced soil were improved to varying degrees by adding some fiber materials. Domestic scholars initially added rice husk ash as admixture to cement concrete to study the impact on the performance of concrete. With the further understanding of the properties of rice husk ash, studies on the application of rice husk ash to soil reinforcement have gradually emerged. Yu Xiaoting[14] mixed rice husk ash into clay, and obtained that with the increase of rice husk ash content, the maximum dry density of the mixed soil decreased, and the optimum moisture content increased. Han[15] added rice husk ash to saline soil to study the engineering characteristics of improved soil.

2 Production technology of rice husk ash

2.1 Production equipment of rice husk ash

Compared with the traditional preparation of rice husk ash by muffle furnace combustion, the small-scale circulating fluidized bed is often used to produce rice husk ash, but the production efficiency is very low. Yu et al. [16]designed continuous rotary and intermittent box combustion furnaces to achieve semi-industrial production of rice husk ash. The rice husk ash obtained in this way has high volcanic activity. Liu Yuanchun, etc. [17]Liu Yuanchun et al. use large step inclined reciprocating grate semi-suspension combustion hot water boiler combustion rice husk, but rice husk is not easy to burn out, rice husk ash separation rate is low, high carbon

* Corresponding author: Liu qian 2904831806@qq.com

content. Ouyang Dong et al. [18] used his own research capacity of 1×10^3 L incinerator, furnace temperature and air content can be changed to produce high silicon content and high activity of rice husk ash. Chen Pei et al. [19]Chen Pei et al. used fluidized bed combustion pretreatment of rice husk, the rice husk ash high silicon high activity, loose particles, non-bonding condition. Jia Yu[20]adopts rotary floating combustion hot blast stove to overcome the characteristics of low combustion efficiency and high carbon content in grate and reduce pollution, which has good environmental benefits.

2.2 Combustion temperature of rice husk ash

The rice husk ash can be burned under controlled low temperature conditions to obtain high activity, containing a large number of amorphous SiO_2. When the temperature is too high, the crystal form of SiO_2 changes from amorphous to crystal, and the utilization value is lost. Gong Xiaoqiang[21]calcined rice husk ash at different temperatures of 400°C, 600°C and 800°C, which proved that with the increase of temperature, the surface unburned substances were less. At 400 °C, the outer surface of rice husk ash is basically complete. At 600 °C, the surface is full of spherical substances and relatively complete. Temperature to 800 °C, showing a molten shrinkage (see figure 1). Hongtao[22]Sun Qingwen[23]Jiang Xinhui[24]pointed out that when the combustion temperature is lower than 600 °C, amorphous SiO_2 is the main component in rice husk ash. Ouyangdong[3]compared the low-temperature combustion rice husk ash with silica fume, and obtained that the X-ray diffraction patterns of low-temperature rice husk ash and silica fume were similar. The two substances were composed of amorphous SiO_2, and they were mixed with Ca (OH) $_2$ according to the ratio of 3 : 1. XRD analysis showed that the ability of low-temperature rice husk ash to absorb Ca (OH) $_2$ could be equivalent to that of silica fume. Although scholars have a slight controversy on the optimal combustion temperature of rice husk ash, they have basically a relatively unified understanding of the calcination temperature that maintains the high activity of rice husk ash. It is generally believed that the ash obtained from the combustion of rice husk at about 600 °C contains more than 90 % amorphous SiO_2. When the temperature is above 650 °C, the content of crystalline SiO_2 begins to increase. When the temperature reaches 800 °C, amorphous SiO_2 content is only about 50 %. When the temperature is above 950 °C, amorphous SiO_2 almost disappears.

400°C

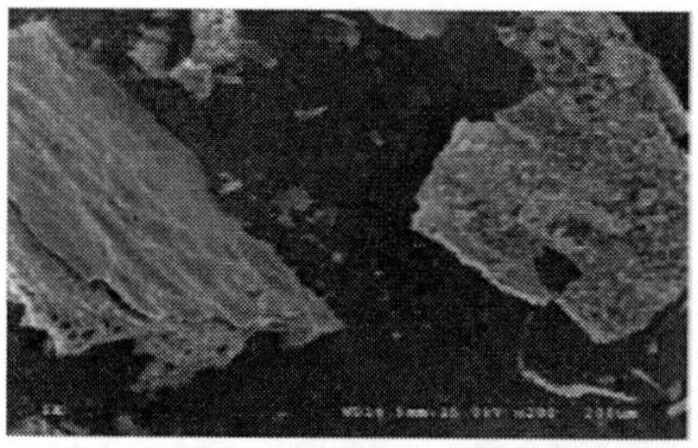

600°C

800°C

Figure 1. SEM images of rice husk ash samples (200 times)[21]

3 Properties of rice husk ash

3.1 Physical properties of rice husk ash

The physical properties of rice husk ash are affected by many factors. Relative density of rice husk ash ranged from 2. 05 to 2. 53 and specific surface area ranged from 500 to 1000 cm^2 / kg[25]. Kothandaraman[26] found that the gravity of rice husk ash could qualitatively indicate its crystalline state. The greater the gravity of rice husk ash was, the more the amorphous substance content was. Therefore, the gravity of rice husk ash should be controlled at 1. 1 – 1. 9 kN / m^3.

Most of the rice husk ash is gray black flake, irregular shape, which is due to the carbon content in the rice husk ash. Krishnarao [27]found that potassium in rice husk ash was an important cause of black particles. Potassium could melt the surface of SiO_2 and transform amorphous SiO_2 to crystalline form. When the combustion temperature is too high, potassium will eutectic with SiO_2, and melt on the surface of rice husk, which hinders the further reaction of carbon in rice husk. Gong Xiaoqiang [21]concluded that the color of rice husk ash would change from gray-black to gray-white with the increase of temperature. This is because with the increase of temperature, the combustion of rice husk ash is more sufficient, the combustible material decreases, and the carbon content decreases.

3.2 Chemical properties of rice husk ash

The chemical composition of rice husk ash varies with combustion equipment, combustion temperature, combustion time and geographical location. The main chemical composition of rice husk ash is SiO_2, and the content range is from 49. 7 % to 95. 41 %. The following

https://doi.org/10.1051/e3sconf/202129302018

is the chemical composition of rice husk ash obtained from some representative literatures. [26-36] （See table 1）.

Table1. Main chemical components of rice husk ash

Chemical composition	SiO_2 (%)	Al_2O_3 (%)	Fe_2O_3 (%)	CaO (%)	MgO (%)
Rungroj Piyaphanuwat	93. 18	0. 31	0. 26	0. 47	0. 23
N. F. Mohd Joharudin	87. 23	0. 35	0. 23	1. 74	1. 23
Indian rice mill	93. 26	4. 10	0. 08	0. 18	0. 00
Patel Kameshwar	82. 70	0. 58	0. 39	0. 75	0. 27
Tarun Kumar Naiya	84. 30	0. 30	0. 60	1. 40	0. 50
Hossein Moayedi	96. 2	-	0. 05	0. 13	0. 36
S. K. Antiohos	89. 05	0. 25	0. 41	0. 85	0. 81
J. G. Tapali	86. 50	0. 92	0. 56	1. 12	1. 30
Antiohos, S K	89. 47	0. 18	0. 25	1. 10	0. 44
Papadakis, V G	93. 15	0. 13	0. 18	0. 89	0. 40
A. Muthadhi • S.	89. 47	-	0. 62	2. 69	1. 16
	90. 20	0. 26	0. 03	0. 07	0. 03
R. V. Krishnarao	92. 20	0. 07	0. 01	0. 00	0. 01
	89. 80	0. 29	0. 04	0. 06	0. 03
	61. 90	1. 20	11. 4	5. 40	1. 40
Y. O. Abiodun	62. 60	3. 30	5. 10	6. 90	3. 40
	59. 50	2. 10	8. 80	7. 30	2. 20

3.3 Microstructure of rice husk ash

The microstructure of rice husk ash is divided into three layers. Chen Pei[37]used SEM to find that the outer surface of rice husk ash was dense, smooth, angular and orderly arranged. The interlayer was composed of vertically and horizontally crisscrossed plate-like structure with uniform spacing. The size of tens of microns was honeycomb pore structure. The inner surface was flat sheet structure (see Fig. 2). This is consistent with the findings of Zeng Xiaoping[5]，Kameshwar P[38]. Ou yangdong[39]used TEM to reveal that rice husk ash contains a large number of nano-scale pores (50nm) formed by the very close cohesion of SiO_2 gel particles in addition to micro-scale (10um) honeycomb pores, which is also the reason for the large specific surface area of rice husk ash(see Fig. 3).

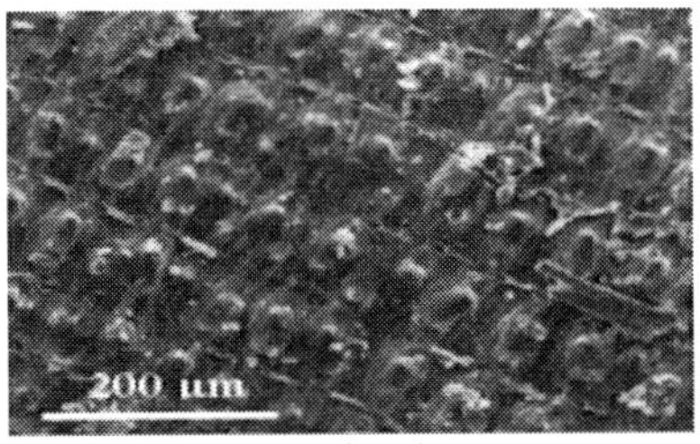

external surface

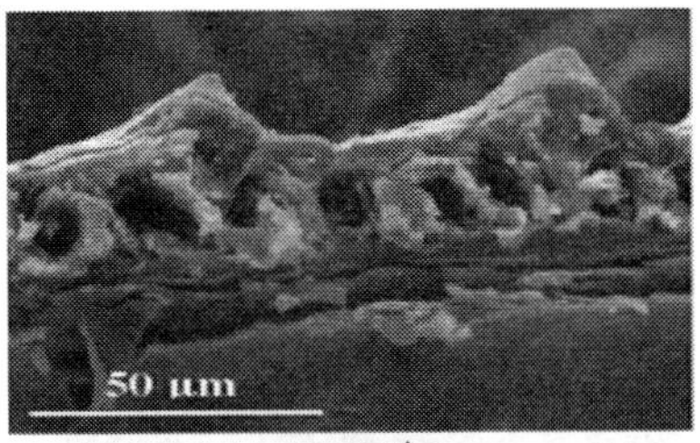

cross section

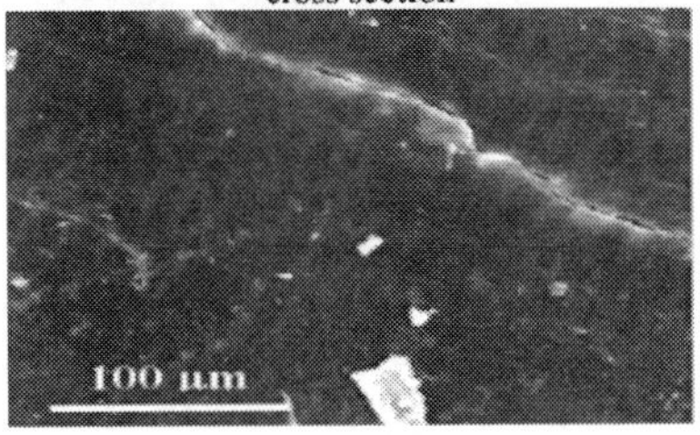

internal surface

Figure 2. Micromorphology of rice husk ash[37]

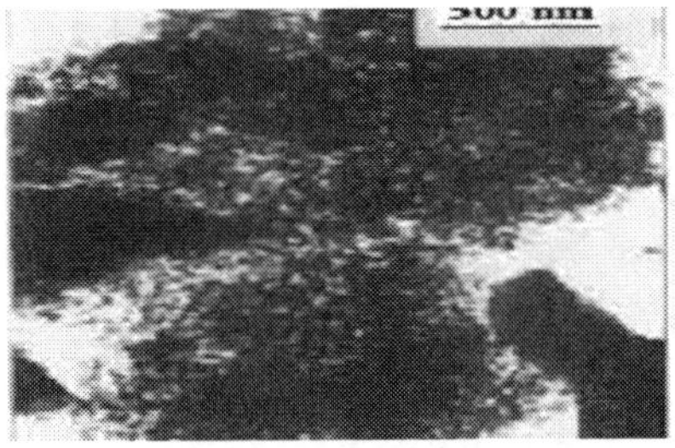
Rice - grained particles of rice husk ash particles

Granular particle magnification of rice

Figure 3. TEM images of rice husk ash

4 Application of Rice Husk Ash in Solidified Soil

4.1 Rice husk ash solidified soil as base

At present, there are abundant studies on the application of rice husk ash as admixture to concrete. In recent years, some scholars have begun to apply rice husk ash to soil solidification. The traditional pavement base material uses cement gravel. With the expansion of the scale of transportation infrastructure construction, the price of raw materials has risen rapidly, and excessive exploitation is not conducive to environmental protection. B. Suneel Kumar[40] studied the mixture of rice husk ash and lime to solidify cohesive soil, and concluded that the combination of 6 % lime and 10 % RHA (rice husk ash) improved CBR. Leonardo Behaket al. [41]used RHA and lime modified silty soil as pavement base. The soil base layer modified by RHA and lime showed good performance on low volume road, and also showed good performance after 20 months, which was better than the traditional pavement with fine sand structure. Muthu Lakshmi S. [42] used RHA and lime to enhance the strength of clay subgrade. Si Zehua[43] uses 20 % RHA + 10 % cement + 0. 3 % polypropylene fiber to solidify carbonated subgrade soil. SiO_2 in RHA inhibits the chemical reaction of cement hydration products and Cl- and SO_4^{2-} in saline soil, and reduces the formation of low-strength Friedel salt ettringite and ettringite. The compressive strength, CBR and resilient modulus of each grade of road base meet the requirements of the specification. Roy Aparna[44] used RHA and cement to solidify subgrade cohesive soil. The results showed that with the increase of RHA and cement ratio, the optimum moisture content and CBR increased, and the maximum dry density decreased. Basha et al[45]added RHA to cement stabilized soil, significantly improved the CBR value. Liu Yuyi[46]Guo Shuai[47] used RHA and carbide slag to solidify expansive soil at the mass ratio of 65:35, and carried out compaction, unconfined compression, CBR, shear strength, dry shrinkage and other experiments. It proved that the performance of rice husk ash- carbide slag composite cementitious material was excellent, and the performance of solidified expansive soil met the requirements of road base. Zhang Xinyu[48] uses 20 % steel slag + 5 % quicklime + 10 % RHA to improve expansive soil. The physical properties, mechanical properties and swelling and shrinkage of the improved soil are significantly improved. Subgrade paving can reduce materials and save costs. Anjani Kumar Yadav[49] used RHA, bagasse ash and cow dung ash to stabilize subgrade alluvial soil. After solidification, the CBR of soil was improved, the dry density was reduced and the maximum water content was increased. J. E. Sani[50]studied the influence of rice husk ash and sisal fiber on the laterite of road building materials.

4.2 Effect of solidified soil from rice husk ash on leaching of heavy metals

The solidification of heavy metal contaminated soil with rice husk ash not only brings economic benefits, but also usually improves the durability and sulfate resistance of solidified soil. In addition, the volcanic ash reaction will continue for a long time, resulting in a long-term stability effect. The fixation mechanisms of heavy metals are mainly adsorption, precipitation, complexation, encapsulation, etc[51]. Metals in sludge or soil can be combined into C-S-H to form C-S-M-H or to form complex silicates directly through M-O-Si bonds[52-62].

Rungroj Piyaphanuwatet al[63] used rice husk ash and carbide slag to stabilize and solidify zinc electroplating sludge. The weight percentage of carbide slag and rice husk ash was 60 : 40, and then the activators Na_2SiO_3 and Na_2CO_3 were added to improve the strength and durability of solidified sludge, which effectively reduced the leaching of metal substances. The treated electroplating sludge was used for soil landfill, which effectively reduced the pollution to the surrounding environment. Chun-Yang Yin et al[64]used OPC and RHA to solidify lead contaminated soil. The results showed that when the content of RHA increased from 0 to 30 %, the leaching rate of lead in the treated samples decreased accordingly, and the lead concentration decreased from 3. 70 mg / L to 1. 06 mg / L. From the XRD spectrum, it was concluded that lead nitrate was precipitated on the surface of C-S-H as a metal hydration phase or metal hydroxide, which reduced the solubility. Abraham C. F. Chiu[65]used rice husk ash instead of 20 % – 40 % cement to solidify artificially prepared heavy metal contaminated soil. The results showed that compared with pure cement solidified contaminated soil, the incorporation of rice husk ash significantly reduced the diffusion coefficient of Pb and Cu content samples. With the increase of rice husk ash content, the fixation effect of Pb was significant, but the effect of Cu was slightly poor.

5 Conclusions

The comprehensive utilization of rice husk ash has great prospects for development. It not only deals with industrial solid waste, but also greatly reduces the cost. Despite the high added value of rice husk ash, we still face several major challenges:

The supply of rice husk ash to meet the requirements is extremely low, and rice husk ash has higher requirements for combustion conditions. Underburning can make the carbon content of rice husk ash extremely high, and overburning can make the crystal morphology of rice husk ash from crystallization to amorphous.

In China, the utilization of rice husk ash started relatively late, and it is rarely applied to the ranks of solidified soil. The solidified soil of rice husk ash is not only lack of research on macro mechanical properties, but also lack of research on micro mechanism.

The performance of rice husk ash is easily affected by the types of rice husk, and there are few studies on the related properties of different types of rice husk ash.

The properties of rice husk ash are comparable to silica fume, but rice husk ash lacks relevant standards to guide our engineering application.

References

1. Pode, R. Renewable & sustainable energy reviews, **53**: page 1468-1485（2016）
2. ZhaoLi ming,Wang Shiqiang, Gu Chunmei, etc.Heilongjiang Agricultural Science,(03), page147-151（2017）
3. Lu Shiyang, Shenyang University of Chemical Technology. Page 65,(2019)
4. Liu Wenlong et al., University Chemistry, **33** (07) : page 90-95, (2018)
5. Zeng Xiaoping, Science and Technology, (21) : Page 120, (2018)

6. Xiao Lufei et al., Chemical World, **61** (07) : page. 457-464, (2020)
7. Yang Zhonghao, Heilongjiang Papermaking, **44** (03) : Page 6-8, (2016)
8. Mingyang et al., cement engineering, (03) : page 69-72, (2015)
9. Li and Cheng, Fujian Building Materials,(11) : pages 13 – 14, (2013)
10. Liu Yan, Jilin University. Page 155, (2013)
11. Zhou Yongxiang and Yan Peiyu, Journal of Railway Science and Engineering, **3** (4) : 35-40,（2006）
12. A, M. APreliminary Study. （2007）
13. Muntohar, A. S. , et al. Journal of materials in civil engineering, **25**(9): page 1260-1270,（2013）
14. Yu, 2019, Hubei University of Technology. Page 67.
15. Korea Aerospace, Lanzhou University of Technology. Page 101. (2020)
16. Yu et al., Journal of Wuhan University of Technology, (01) : page 15 – 18,(2003)
17. Liu Yuanchun et al., industrial boiler, (06) : page 26-29,(2009)
18. Ou yangdong, South China University of Technology, (1997)
19. Chen Pei, Wu O and Bie Rushan, Journal of Harbin Institute of Technology, **51** (03) : page. 46-54,(2019)
20. Jia Yu, Qin Feng and Xiao Yanmin, modern food, (02) : page 95-97,(2017)
21. Gong Xiaoqiang et al., Journal of Wuhan University of Light Industry, **37** (01) : page 51 - 55,(2018)
22. Li Hongtao et al., Journal of Agricultural Machinery, 44 (04) : page 131-136,(2013)
23. Sun Qingwen, Xu Kejing and Guo Yanqing, Chinese ceramics, **48** (07) :page 1-6(2012)
24. Jiang Xinhui, 2010, Harbin Institute of Technology.
25. Wang Zhiwen and Li Bixiong, Material Report, **34** (09) : Page 9003-9011.(2010)
26. Muthadhi, A. and S. Kothandaraman, Materials and structures, **43**(9): page 1303-1315.(2010)
27. Krishnarao, R. V. , J. Subrahmanyam and T. Jagadish Kumar,Journal of the European Ceramic Society, **21**(1): page 99-104,(2001)
28. Moayedi, H. A. H. M. , et al. Journal of Cleaner Production, **237**(2019).
29. Aliyu, M. K. , et al. ,Water and environment journal : WEJ, **34**(S1): page 229-238,(2020)
30. Piyaphanuwat, R. and S. Asavapisit, Environmental progress & sustainable energy,**37**(6):page 1965-1972,(2018)
31. Mohd Joharudin, N. F. , et al.**50**(3): page 283-288,(2019)
32. Kameshwar, P. , et al. Structural concrete : journal of the FIB, **22**(S1): p. E501-E513,(2021)
33. Naiya, T. K. , A. K. Bhattacharya and S. K. Das, Environmental progress & sustainable energy, **28**(4): page 535-546,(2009)
34. Antiohos, S. K. , et al. Construction & building materials, 49: page 455-463,(2013)
35. Antiohos, S. K. , V. G. Papadakis and S. Tsimas,Cement and Concrete Research, 61-62: page 20-27,(2014)
36. Abiodun, Y. O. and A. A. Jimoh, Nigerian journal of technology, **37**(1): page 71,(2018)
37. Chen, Harbin Institute of Technology. Page 142,(2019)
38. Kameshwar, P. , et al. journal of the FIB,(2020)
39. Ouyangdong and Chen Kai, Journal of Materials Science and Engineering, (05) , page 647-650,(2003)
40. B. Suneel Kumar, T. V. P. IJETT Journal,**11**, page 1(2014)
41. Behak, L. and M. Musso, Transportation Research Procedia, 18: page 93-99(2016)
42. Muthu Lakshmi, S.et al. Materials today : proceedings, (2021)
43. Siserhua, Jilin University. Page 91,(2020)
44. Roy, A. International Journal of Civil Engineering Research, **5**: page 49-54,(2014)
45. Basha, E. A. , et al. Construction & building materials, **19**(6), page 448-453(2005)
46. Liu Yuyi, China University of Mining and Technology. Page 221,(2019)
47. Guo Shuo, Highway Engineering, **45** (03), Page 210-215.(2020)
48. Zhang Xinyu.Road Engineering. Page 20-24, (2018)
49. Kumar Yadav, A. , et al. International Journal of Pavement Research and Technology,**10**(3):page 254-261,(2017)
50. Sani, J. E. , P. Yohanna and I. A. Chukwujama, Engineering sciences, **32**(1): page 11-18,(2020)
51. Tang Dengyong et al., China Rural Water Conservancy and Hydropower, (11) : page 68-72, (2017)
52. Wang, F. , et al. Journal of hazardous materials, 300, page 451-458,(2015)
53. Nowak, B. , P. Aschenbrenner and F. Winter, Fuel processing technology, 105: page 195-201,(2013)
54. Cubukcuoglu, B. and S. K. Ouki, Chemosphere, **86**(8): page 789-96, (2012)
55. Li, X. , et al. Waste Manag, **34**(12): page 2494-504,(2014)
56. Tan Jintao et al., China Environmental Science, **40** (07) ,page 3054-3060,(2020)
57. Lin Meishan, Chen Yu and Chen Ting, Journal of Wuhan University of Engineering, **42** (01), page 38-44 ,(2020)
58. Liu Ling and Guan Chang, science and technology innovation, (31) , page 42-43,(2019)
59. Liu Ling et al., Science and Technology Innovation, (30), page 55-56,(2019)

https://doi.org/10.1051/e3sconf/202129302018

60. Liu Qi et al., Chemical Research and Application, **31** (08) ,page 1482 - 1491,(2019)
61. Zuo Hai Qiang, industrial water and wastewater, **43** (04) : page 58 - 61,(2012)
62. Chiu, A. , et al. J Hazard Mater, 371, page 62-71,(2019)
63. Piyaphanuwat,R.and S. Asavapisit, Environmental progress & sustainable energy, **37**(6), page 1965-1972, (2018)
64. Yin, C. Y. , H. B. Mahmud and M. G. Shaaban, J Hazard Mater, **137**(3): page 1758-64,(2006)
65. Chiu, A. C. F. , et al. Journal of hazardous materials, 371, page 62-71,(2019)

A series of equivalent area circle yield criteria based on unified strength theory

Pei Zhou [1,2]

[1] Faculty of Engineering, China University of Geosciences, Wuhan 430074, China

[2] Engineering research center of rock-soil drilling & excavation and protection, Ministry of education, Wuhan 430074, China

Abstract: Unified strength theory considering the influence of the intermediate principal stress is widely used in geotechnical engineering, but the singularities bring inconvenience to the numerical calculation. A series of equivalent area circle yield criteria based on unified strength theory are derived. The parameters of the new yield criteria and Drucker-Prager criteria are discussed, and the flow vector coefficients of the new yield criteria are given. The new series of yield criteria are very convenient for numerical calculation and can be served as reference for the evaluation of the effects of strength theory.

1 Introduction

The strength theory of geotechnical engineering provides the criteria to analyze the yield and failure of geotechnical materials and structures, which also plays an essential role on stability research. At present, hundreds of strength criteria have been proposed, forming a rich theoretical system of strength theory. This theoretical system can be classified from different perspectives. According to the physical quantity of the expression, it can be divided into stress form, strain form and energy form. It can also be divided into linear strength theory and nonlinear strength theory, depending on the property of the yield function. According to the number of parameters, it can be divided into single parameter, double parameters, and multiple parameters. In addition, there are some other classification methods (e.g., according to the characteristics of the limit lines, the characteristics of the off-plane trace, the number of yield surfaces). The representative strength criteria used in geotechnical engineering mainly include: Tresca criterion, Mohr-Coulomb criterion, Mogi-Coulomb criterion, Huber-Mises criterion, Drucker-Prager criterion, Matsuoka-Nakai criterion, Zienkiewicz-Panda criterion, etc[1]. Among of which, Mohr-Coulomb criterion is the most widely used and most controversial strength theory in geotechnical engineering. The main problem is that it does not consider the intermediate principal stress. A large number of experimental studies have shown that the intermediate principal stress has a significant effect on the yield and failure of the material[2, 3]. Moreover, the use of different strength criteria has a greater impact on the calculation results[4]. Effect of strength theory in elastic-plastic analysis cannot be ignored, so finding a reasonable strength criterion of geotechnical materials and structures is a research hotspot in geotechnical engineering.

2 Analysis of unified strength theory

Yu Maohong has carried out a systematic study of strength theory and proposed the unified strength theory on the basis of the double-shear stress yield criterion, the double-shear strength theory and the generalized double-shear stress yield criterion[5]. The unified strength theory believes that the yield and failure of materials depend on the shear stress and normal stress on the section of the double-shear element, and when the two larger principal shear stresses and the corresponding normal stress on the double-shear element reach at the limit value, the material begins to yield or fail.

The mathematical expression is expressed as follows:

$$\begin{aligned} &F = \tau_{13} + b\tau_{12} + A(\sigma_{13} + b\sigma_{12}) = C \\ &for\ \tau_{12} + A\sigma_{12} \geq \tau_{23} + A\sigma_{23} \end{aligned} \tag{1}$$

$$\begin{aligned} &F' = \tau_{13} + b\tau_{23} + A(\sigma_{13} + b\sigma_{23}) = C \\ &for\ \tau_{12} + A\sigma_{12} \leq \tau_{23} + A\sigma_{23} \end{aligned} \tag{2}$$

Strength parameters of geotechnical materials are cohesion and internal friction angle, and the mathematical expressions of the unified strength theory expressed by principal stress and strength parameters are described as follows:

$$\begin{aligned} &F = \sigma_1 - \frac{1-\sin\varphi}{(1+b)(1+\sin\varphi)}(b\sigma_2 + \sigma_3) = \frac{2c\cos\varphi}{1+\sin\varphi} \\ &for\ \sigma_2 \leq \frac{1}{2}(\sigma_1 + \sigma_3) + \frac{\sin\varphi}{2}(\sigma_1 - \sigma_3) \end{aligned} \tag{3}$$

* Corresponding author: zhoupeicug@126.com

$$F' = \frac{1}{1+b}(\sigma_1 + b\sigma_2) - \frac{1-\sin\varphi}{1+\sin\varphi}\sigma_3 = \frac{2c\cos\varphi}{1+\sin\varphi}$$
$$for \ \sigma_2 \ge \frac{1}{2}(\sigma_1+\sigma_3) + \frac{\sin\varphi}{2}(\sigma_1-\sigma_3) \tag{4}$$

In above formula c is cohesion, φ is internal friction angle and b is the strength criterion parameter.

The mathematical expressions of the unified strength theory expressed by stress invariant are described as follows:

$$F = \frac{I_1}{3}(1-\alpha) + (2+\alpha)\frac{\sqrt{J_2}}{\sqrt{3}}\cos\theta + \frac{\alpha(1-b)}{1+b}\sqrt{J_2}\sin\theta = \sigma_t \tag{5}$$
$$for \ 0^\circ \le \theta \le \theta_b$$

$$F' = \frac{I_1}{3}(1-\alpha) + \left(\frac{2-b}{1+b}+\alpha\right)\frac{\sqrt{J_2}}{\sqrt{3}}\cos\theta + \left(\alpha + \frac{b}{1+b}\right)\sqrt{J_2}\sin\theta = \sigma_t \tag{6}$$
$$for \ \theta_b \le \theta \le 60^\circ$$

In which, $\alpha = \frac{1-\sin\varphi}{1+\sin\varphi}$ $\sigma_t = \frac{2c\cos\varphi}{1+\sin\varphi}$, $\sigma_c = \frac{2c\cos\varphi}{1-\sin\varphi}$, θ_b is the stress angle of the boundary.

Intermediate principal stress has taken unified strength theory into consideration by introducing the parameter b, which has been verified that respectively the yield criterion as $b = 0$ is an inner envelope and the yield criterion as $b = 1$ is an outer one in π plane for stable material[6].

Unified strength theory is a major breakthrough in the history of strength theory research. It is not a single strength criterion, but a collection of a series of strength criteria, covering the continuous space by changing the parameter b. Many strength criteria are special cases of unified strength theory. Therefore, unified strength theory has been widely used in geotechnical engineering (i.e., soil pressure, foundation bearing capacity, slope stability, underground engineering stability)[7].

There are singularities on the yield surface of unified strength theory and which makes numerical calculations inconvenient. Mathematical processing of singularities is troublesome[8]. In order to deal with the singularity of Mohr-Coulomb criterion, a series of Drucker-Prager criteria are proposed[9, 10]. At present, Drucker-Prager criteria are applied to numerical calculation software (e.g., ABAQUS, ANSYS, FLAC, MARC).

3 Equivalent area circle yield criteria

3.1 Derivation of the general equation

Unified strength theory expressed in the Haigh-Westergaard stress space (ξ, ρ, θ) is detailed as follows:

$$F = \frac{\sqrt{3}}{3}\cdot\xi\cdot(1-\alpha) + (2+\alpha)\cdot\frac{\rho}{\sqrt{6}}\cdot\cos(\theta) + \frac{\alpha\cdot(1-b)}{1+b}\cdot\frac{\rho}{\sqrt{2}}\cdot\sin(\theta) - \sigma_t = 0 \tag{7}$$
$$for \ 0^\circ \le \theta \le \theta_b$$

$$F' = \frac{\sqrt{3}}{3}\cdot\xi\cdot(1-\alpha) + \left(\frac{2-b}{1+b}+\alpha\right)\cdot\frac{\rho}{\sqrt{6}}\cdot\cos(\theta) + \left(\alpha+\frac{b}{1+b}\right)\cdot\frac{\rho}{\sqrt{2}}\cdot\sin(\theta) - \sigma_t = 0 \tag{8}$$
$$for \ \theta_b \le \theta \le 60^\circ$$

On the π plane, the vector radius with different stress angles can be easily obtained.

When $\theta = 0$, the conventional triaxial compression strength r_c is:

$$r_c = \frac{\sigma_t\cdot 6}{\sqrt{6}\cdot(\alpha+2)} \tag{9}$$

When $\theta = \pi/3$, the conventional triaxial extension strength r_t is:

$$r_t = \frac{\sigma_t\cdot 6}{\sqrt{6}\cdot(2\cdot\alpha+1)} \tag{10}$$

When $\theta = \theta_b = \tan^{-1}\frac{\sqrt{3}}{2\cdot\alpha+1}$, the vector radius r_b is:

$$r_b = \frac{\sigma_t\cdot 3\cdot(2\cdot\alpha+1)\cdot(b+1)\cdot\sqrt{\frac{3}{(2\cdot\alpha+1)^2}+1}}{\sqrt{6}\cdot\left(\left(\alpha^2+\alpha+1\right)\cdot b+\alpha^2+4\cdot\alpha+1\right)} \tag{11}$$

The equation of equal area:

$$\frac{1}{6}\cdot\pi\cdot r_e^2 = \frac{1}{2}\cdot r_c\cdot r_b\cdot\sin(\theta_b) + \frac{1}{2}\cdot r_b\cdot r_t\cdot\sin\left(\frac{\pi}{3}-\theta_b\right) \tag{12}$$

By solving the above equation, we can get:

$$r_e = \sqrt{\frac{6}{\pi}\cdot\frac{3\cdot\sigma_t^2\cdot(b+1)\cdot\sqrt{\frac{4\cdot\alpha^2+4\cdot\alpha+4}{(2\cdot\alpha+1)^2}}}{2\cdot(\alpha+2)\cdot\left(\left(\alpha^2+\alpha+1\right)\cdot b+\alpha^2+4\cdot\alpha+1\right)}} \cdot\sqrt{\left((\alpha+2)\cdot\sin\left(\frac{\pi-3\cdot\theta_b}{3}\right)+(2\cdot\alpha+1)\cdot\sin(\theta_b)\right)} \tag{13}$$

The equal area yield criterion can be written as:

$$F = \alpha\cdot I_1 + \sqrt{J_2} - k = 0 \tag{14}$$

In which, $k = \frac{r_e}{\sqrt{2}}$, $\alpha = k\cdot\frac{\sin(\varphi)}{3\cdot c\cdot\cos(\varphi)}$.

It is interesting that the form of the equal area yield criterion is the same as Drucker-Prager criteria, while the parameters are different.

Five Drucker-Prager criteria are proposed base on Mohr-Coulomb criterion (M-C) in geotechnical engineering which include M-C exterior angle circumcircle, M-C interior angle circumcircle, M-C inscribed circle, M-C equivalent area circle, M-C non-associated matching circle.

The analytical expressions of parameters α and k of yield criteria (DP1~5 and the new criteria) are given in Table 1.

Table1. Parameters' expressions of yield criteria

yield criteria	α	k
DP1	$\frac{(2\cdot\sin(\varphi))}{\sqrt{3}\cdot(3-\sin(\varphi))}$	$\frac{(6\cdot c\cdot\cos(\varphi))}{\sqrt{3}\cdot(3-\sin(\varphi))}$
DP2	$\frac{2\cdot\sin(\varphi)}{\sqrt{3}\cdot(3+\sin(\varphi))}$	$\frac{6\cdot c\cdot\cos(\varphi)}{\sqrt{3}\cdot(3+\sin(\varphi))}$
DP3	$\frac{\sin(\varphi)}{\sqrt{3}\cdot\sqrt{\sin(\varphi)^2+3}}$	$\frac{\sqrt{3}\cdot c\cdot\cos(\varphi)}{\sqrt{\left(4-\cos(\varphi)^2\right)}}$
DP4	$\frac{2\cdot\sqrt[4]{3}\cdot\sin(\varphi)}{\sqrt{2\cdot\pi\cdot\left(9-(\sin(\varphi))^2\right)}}$	$\frac{6\cdot\sqrt[4]{3}\cdot c\cdot\cos(\varphi)}{\sqrt{2\cdot\pi\cdot\left(9-(\sin(\varphi))^2\right)}}$
DP5	$\frac{\sin(\varphi)}{3}$	$c\cdot\cos(\varphi)$
New	$\frac{2\cdot\sin(\varphi)}{1+\sin(\varphi)}\cdot T$	$\frac{6\cdot c\cdot\cos(\varphi)}{1+\sin(\varphi)}\cdot T$

Note:
DP1: M-C exterior angle circumcircle
DP2: M-C interior angle circumcircle
DP3: M-C inscribed circle
DP4: M-C equivalent area circle
DP5: M-C non-associated matching circle
New: Equivalent area circle yield criteria

$$T=\sqrt{\frac{1}{\pi}\cdot\frac{(b+1)\cdot\left((\alpha+2)\cdot\sin\left(\frac{\pi}{3}-\theta_b\right)+(2\cdot\alpha+1)\cdot\sin(\theta_b)\right)\cdot\sqrt{\alpha^2+\alpha+1}}{(\alpha+2)\cdot\left(\left(\alpha^2+\alpha+1\right)\cdot b+\alpha^2+4\cdot\alpha+1\right)\cdot(2\cdot\alpha+1)}}$$

3.2 Calculation and analysis

The general equation is not a single criterion, and it contains a series of new criteria.

For c=0, the value of parameter α calculated by analytical expressions are given in Table 2.

Table2. Comparison of parameter α calculated by different yield criteria (c = 0)

yield	internal friction angle (degree)					
criteria	10	20	30	40	50	60
DP1	0.0709	0.149	0.231	0.315	0.396	0.469
DP2	0.0632	0.118	0.165	0.204	0.235	0.259
DP3	0.0576	0.112	0.160	0.201	0.234	0.258
DP4	0.0609	0.121	0.177	0.230	0.277	0.317
DP5	0.0579	0.114	0.167	0.214	0.255	0.289
New b=0	0.0609	0.121	0.177	0.230	0.277	0.317
New b=0.1	0.0623	0.123	0.181	0.234	0.281	0.319
New b=0.2	0.0635	0.125	0.184	0.237	0.283	0.321
New b=0.25	0.0641	0.126	0.185	0.239	0.285	0.322
New b=0.3	0.0646	0.127	0.187	0.240	0.286	0.323
New b=0.4	0.0657	0.129	0.189	0.242	0.288	0.324
New b=0.5	0.0666	0.131	0.191	0.245	0.290	0.326
New b=0.6	0.0674	0.132	0.193	0.247	0.292	0.327
New b=0.7	0.0681	0.134	0.195	0.248	0.293	0.328
New b=0.75	0.0685	0.134	0.195	0.249	0.294	0.329
New b=0.8	0.0688	0.135	0.196	0.250	0.295	0.329
New b=0.9	0.0695	0.136	0.198	0.252	0.296	0.330
New b=1	0.0701	0.137	0.199	0.253	0.297	0.331

As seen from table 2, the value of parameter α of DP4 is the same as of the new criterion for b = 0. It shows that the new criterion is reliable because the unified strength theory degenerates to Mohr-Coulomb criterion when b = 0. The surface of unified strength theory is convex when $0 < b < 1$, and it is non-convex when $b > 1$ or $b < 0$. While the new series of equivalent area circle yield criteria based on unified strength theory are always convex.

Some equations can be derived base on UST (unified strength theory) exterior angle circumcircle, UST interior angle circumcircle, UST middle angle circumcircle, UST inscribed circle, etc.

The new parameters α_e can be written as: $\alpha_e = t\cdot\alpha$, in which t is the area parameter. Therefore, the new general equation of a series of equivalent area circle yield criteria based on unified strength theory essentially includes these above equations by changing the value of the parameter t.

But it is unnecessary for which can lead to overestimate or underestimate of the strength of the material.

The flow vector of the strength criterion is:

$$\frac{\partial F}{\partial\sigma_{ij}}=C_1\cdot\frac{\partial I_1}{\partial\sigma_{ij}}+C_2\cdot\frac{\partial\sqrt{J_2}}{\partial\sigma_{ij}}+C_3\cdot\frac{\partial J_3}{\partial\sigma_{ij}} \quad (15)$$

In which: $C_1=\alpha_e$, $C_2=1$, $C_3=0$,

$\frac{\partial I_1}{\partial\sigma_{ij}}=[1,1,1,0,0,0]^T$,

$\frac{\partial\sqrt{J_2}}{\partial\sigma_{ij}}=\frac{1}{2\sqrt{J_2}}[S_x,S_y,S_z,2S_{xy},2S_{yz},2S_{zx}]^T$,

$$\frac{\partial J_3}{\partial \sigma_{ij}} = \begin{bmatrix} \left(S_y S_z - S_{yz}{}^2\right) + \frac{1}{3} J_2 \\ \left(S_x S_z - S_{xz}{}^2\right) + \frac{1}{3} J_2 \\ \left(S_x S_y - \tau_{xy}{}^2\right) + \frac{1}{3} J_2 \\ 2\left(S_{yz} S_{xz} - S_z S_{xy}\right) \\ 2\left(S_{xz} S_{xy} - S_x S_{yz}\right) \\ 2\left(S_{xy} S_{yz} - S_y S_{xy}\right) \end{bmatrix}.$$

The form of the new equation is similar to Drucker-Prager criterion. It is a cluster of strength criteria, rather than a single strength criterion, and Drucker-Prager criterion is a special case of the new strength criteria. The new yield criteria have no singularities; thus, it is suitable for numerical calculations. By changing the parameter b, a series of continuously changing strength criteria can be obtained.

4 Conclusion

Based on the in-depth analysis of existing strength criteria, a series of new strength criteria are proposed. And the parameters of new strength criteria are calculated and discussed in detail. The coefficients of the flow vector are given, which can be used to develop subroutines of finite element software for numerical calculation.

The new strength criteria are a series of equivalent area circle yield criteria based on unified strength theory through the continuous change of b. The new strength criteria have no singularities and which are smooth everywhere on the yield surface.

The equivalent area circle yield criteria are very convenient for numerical calculations, which can provide an important reference for evaluating the effect of strength theory.

References

1. YU M. Advances in strength theories for materials under complex stress state in the 20th Century[J]. Applied Mechanics Reviews, 2002,55(3): 169-218.
2. MOGI K. Effect of the intermediate principal stress on rock failure[J]. Journal of Geophysical Research, 1967,72(20): 5117-5131.
3. SHI W, ZHU J, LIU H. Influence of intermediate principal stress on deformation and strength of gravel[J]. Yantu Gongcheng Xuebao/Chinese Journal of Geotechnical Engineering, 2008,30(10): 1449-1453.
4. ZHENG Y, KONG L. Geotechnical plastic mechanics[M]. second. Beijing: China Building Industry Press, 2019.
5. YU M. Unified Strength Theory and Its Applications[M]. second. Berlin: Springer-Verlag Berlin and Heidelberg GmbH & Co. KG, 2017.
6. LU Z, GONG X. Verification of inner and outer envelopes for the yield curve of stable material in π plane[J]. Chinese Journal of Geotechnical engineering, 1997,19(5): 1-5.
7. YU M, LI J. Computational plasticity: With emphasis on the application of the unified strength theory[M]. Springer Science & Business Media, 2012.
8. ZIENKIEWICZ O C, TAYLOR R L, FOX D. The Finite Element Method for Solid and Structural Mechanics [M]. Seventh. Oxford: Butterworth-Heinemann, 2014.
9. DRUCKER D C, PRAGER W. soil mechanics and plastic analysis or limit design[J]. quarterly of applied mathematics, 1952,10(2): 157-165.
10. DENG C, HE G, ZHENG Y. Studies on Drucker-Prager yield criterions based on M-C yield criterion and application in geotechnical engineering[J]. Chinese Journal of Geotechnical engineering, 2006,28(6): 735-739.

Development Trend of Hotel Architectural Space in The Post-Epidemic Era

Zhang Yaoyao[1]

[1]Art & Design Department, Gingko College of Hospitality Management, Chengdu, Sichuan, 611743, China

Abstract. At the beginning of 2020, the epidemic of COVID-19 broke out rapidly in countries all over the world. The sudden epidemic poses a significant impact on the operation of China's industrial economy. With the changes in the social and economic environment, hotels and tourism seem to have been pressed the "pause button" in a second, and instantly entered the "extremely cold" moment. The hotel occupancy rate plummeted to the "freezing point". In view of the large-scale outbreak of the epidemic in various countries, many architects and urbanists are rethinking the development direction of the hotel construction model in the post-epidemic era.

1 Introduction

At the beginning of 2020, the COVID-19 epidemic spread widely in a very short time, affecting people's lives. Under this new type of safety problem, the use of part of the architectural space has changed. At the present time of the normalization of the pandemic, this new type of security problem will affect the design and development of architectural space. After the outbreak of the epidemic, people's consumption habits have changed, multi-functional, personalized, intelligent, interesting, boutique has become the only way for hotel transformation and upgrading. In addition, the impact of Internet economy on hotel supporting, product pattern, hotel multi-dimensional space building, digital contactless service, environmental protection and health will be the direction of hotel development in the future.

2 Architectural Space Design under the principle of Safety in the Post-epidemic era

2.1 Impact of epidemic on the use of architectural space

COVID-19 is transmitted mainly through droplets and contact, which requires people to maintain a safe social distance, leading to changes in social distance.**[1]** The change in social distance affects the way buildings are used. In architectural design, designers often use Corbusier's "Le's modulus ruler" as a measurement, using human body components such as stride and limb length to measure the modular scale of buildings, so as to make buildings meet the needs of people's life and communication.**[2]** However, the epidemic of COVID-19 has changed this common scale, increasing the distance of close contact and increasing online communication at the same time. This change of "human modulus" is bound to bring changes to the use of buildings.

2.2 Adaptive design

Under the influence of COVID-19, the way of production and life has changed. Under this new type of safety problem, the importance of adaptive design of buildings has been highlighted. Elastic design can be reflected in two aspects: spatial function and facility layout: in terms of spatial function, buildings can carry out formal design and spatial division. With regard to the layout of facilities, different requirements can be met from the aspects of building equipment, building materials, building technology, etc. **[3]** In short, the adaptive design under the new safety problem is to redesign the building in the present and future of the normalization of the epidemic, so that the space can meet the new needs caused by the epidemic.

Table1. Analysis of Adaptive Design process under New Safety problems [4]

Adaptive design process	
Space selection	Select a space with a space function to change the foundation
Function implantation	Realize the change of space function through the subdivision of space
Dynamic update	Re-divide the space according to the different needs of users, and continue to adjust according to changes in use and environmental changes

Lillian2211@163.com

2.3 Selection of adaptive space

Under the new security problems, not all spaces are suitable for elastic design, so how to choose a specific space for adaptive design is particularly important. The adaptive space under new security issues has the following characteristics:

(1) Sharing, which can be regarded as the degree of openness of the space under the new security problems. During the epidemic, the openness of part of the space changed, and the acceptance of the space to external users decreased, which will affect the users' use of space functions.

(2) Composition, which refers to the diversification of spatial functions. The compound of the hotel space can be reflected in whether the hotel faces the street or whether it is in the crowd gathering business district. The functional diversity of time can be reflected in whether the space has obvious time service attributes and whether it is only used in a specific time.

(3) Availability, which means that the transformation of spatial function can be carried out through the division of space, building equipment and building materials, for example, it is difficult to change the spatial function of some spaces with fixed facilities.

3 Selection of Hotel Architectural space Materials in Post-epidemic era

The inspiration of COVID-19 's epidemic situation to the future hotel space design is to pursue minimalist design in order to reduce the risk of indoor virus infection. In the face of infectious risks such as the coronavirus, designers should use healthy materials and indoor minimalist finished products in different design ways to make the hotel space safer and healthier.

According to the New England Journal of Medicine, Coronaviruses can survive for hours or even days on different surfaces. Scientists found that Coronaviruses survived in aerosol for a long time throughout the experiment, and survived longest on smooth surfaces. For example, Coronaviruses could survive 72 hours on the plastic surface of furniture, 72 hours on the surface of stainless steel appliances, and 24 hours on indoor paper, cardboard or clothes. The biggest surprise was the performance of copper: the virus disappeared on the surface of copper alloy utensils in only four hours.

Table2. Survival time of Coronaviruses on various materials

Material	Items	Duration (unit: hour)
Glass	Windows, mirrors, mobile phone screens, glass doors, etc.	96
Stainless Steel	Handrails, faucets, door handles, coat hooks, paper drawers, etc.	48-72
Plastic	Take-out food containers, light switches, mobile phone cases, elevator buttons, etc.	48-72
Paper	Newspapers, receipts, express boxes, toilet paper, etc.	24
Copper	Door handles, handrails, copper containers, etc.	4

3.1 Door and window rail material

When coronaviruses land on the hardest surface, they can survive for 2 to 5 days. But when they land on copper alloys such as copper and brass, they only last about four hours. Moreover, copper ion can be used as a disinfectant, anti-virus. This is one reason why old-fashioned doorknobs are made of brass. At the same time, hospitals also began to use a large number of copper products in the patient environment, so the use of copper products can be increased in hotel design.

3.2 Flooring material

Another inspiration from novel coronavirus to the hotel design is the use of ecologically healthy flooring materials. The first is to set the floor in the hotel room and refuse to lay the carpet, because the floor is easy to clean and easy to clean, avoiding dusty corners that are difficult to clean, and there is no hiding place for dust and insects. The main choices for ecologically healthy flooring materials are:

Vinyl flooring: this flooring material has always been a popular choice for environmentally friendly flooring. It has the advantages of warmth, dirt resistance and durability. It has solid flooring and its thin sheets and bricks.

Linoleum floor: made of natural materials such as cork oak, flaxseed oil, wood and resin. Because the floor is a pure green natural wood layer, it gives people a good feeling of living in nature, and when installed, maintained and used, it can enjoy the simplicity and convenience of reinforced composite wood flooring. It is a "real high-tech solid wood floor" that integrates the natural beauty of solid wood flooring and the firmness and durability of composite flooring.

Natural cork floor: the cork floor which is softened and dried by cork oak is completely natural and antibacterial. There are about 40 million air bags per cubic centimeter of cork, each of which is a small silent absorber, a small temperature insulator and a small flexible shock absorber. In the chemical composition of cork structure, cork does not contain sugar, starch and protein. Without protein, cork will not rot; without sugar and starch, there will be no food for bugs in cork, so using cork as hotel room floor material is naturally antibacterial and insect-proof.**[5]**

3.3 Concrete and tile materials

The requirements of COVID-19 epidemic on the hotel environment should not only be comfortable and clean, but also beneficial to the residential health care of antibacterial and detoxification, so the application of concrete and ceramic tile materials should also be committed to protecting the hotel building environment.

Photocatalytic concrete: the photocatalytic cement is prepared by adding human TiO2 into cement, which can produce photocatalytic reaction in ordinary sunlight. Photocatalytic concrete belongs to gas environmental material and health environmental material. It can catalyze sulfur dioxide and nitrogen oxides produced during combustion or emitted by motor vehicles, which are harmful to human health.

The gas is decomposed and removed, which plays a role in purifying the air; at the same time, it also has antibacterial functions such as sterilization and decontamination. This kind of cement not only has the characteristics of light weight, high temperature resistance and good flame retardancy, but also can produce high concentration of germicidal ions through the photocatalytic reaction of its own stored sunlight, so it has high antibacterial and antitoxic effect. it is a new material for hotel buildings in the post-epidemic era.

Antibacterial ceramic tile: it is on the basis of ordinary ceramic tile, through antibacterial technology, ceramic tile is endowed with antibacterial performance. Antibacterial ceramic tiles, not easy to breed bacteria, used in public places such as hotels, is very necessary to prevent bacterial infectious diseases.

4 Design Strategy of Hotel Architectural space in the Post-epidemic era

For the future development of the hotel after the COVID-19 epidemic, the society, investors, designers and consumers should think accordingly, especially for the current practitioners in the design industry, the responsibility is particularly important. Designers should re-examine the design ideas of architectural space and pay more attention to the layout of space, especially private space, so as to adapt to the design and consumption needs under the epidemic.

4.1 The principle of "people-oriented"

It is mainly reflected in the design and treatment of privacy space. Hotel architecture and courtyard landscape do not seek luxury but internal quality and cultural connotation, fully reflect the ritual sense of life, and increase a little more independent space. This is the "people-oriented" design consideration under the epidemic, at the same time, it can also improve the experience of guests and the quality of the hotel.

4.2 The mutual unity of function and form

The relationship between function and form is a pair of contradictions that integrate and depend on each other. In the hotel architectural design, we should comprehensively consider the relationship between the internal function and form, use various techniques to make the two combine with each other, and emphasize the formal beauty on the basis of satisfying the hotel function, so as to design the hotel building with complete functions, rich forms and beautiful appearance.**[6]**

4.3 The internal combination of technology and art

The development of modern science and technology has not only changed the conditions for the use of architecture, but also changed people's concept of architecture. Due to the application of new technology, modern architecture promotes people to carry out technological innovation and material iteration constantly. In modern architectural design, including hotel architectural design, the application and expression of technology has been regarded as an innovative expression of art, which is the internal combination of technology and art.

4.4 From contagion without boundaries to defense units

Virus transmission has no boundaries, and urbanization enhances the ability of virus transmission. In the treatment of hotel architectural space, it is suggested that reasonable consideration should be given to the integration and blocking of urban space, and to establish an effective space system to prevent and deal with emergencies through "defense unit". In the past, the disorderly spread of urban development has caused some problems, such as too high complexity, difficult to disassemble as a whole, too difficult to regulate space and so on. Therefore, it is suggested to design the hotel architectural space based on the defense unit, which will increase the feasibility of the hotel to deal with emergencies and the security of the city as a whole.

5 Conclusion

In recent years, hotel architectural design is getting rid of the single mode of traditional standardized star hotels in a stable and gradual manner. Specifically, more and more design tasks are focused on these guest experience-centered personalized boutique hotels, theme hotels, and luxury hotels. The design of hotel architecture is progressively divorced from the mainstream orientation of our architecture. From spatial sequence to artistic atmosphere, the interest pursued by hotel design is obviously different from that pursued by mainstream architecture: Hotel, as a leisure resort for people to enjoy the experience, has different emphasis and direction in the pursuit of space experience, which is different from the architecture which has inherited the ritual and commemorative architectural aesthetic genes since its birth.

In the long run, the COVID-19 pandemic will lead to changes in business models and consumption habits. [7]Accordingly, the construction of smart hotel, as a hotel with digital, intelligent and networked management through advanced information and communication technology, provides a goal and platform for the transformation and upgrading of the hotel industry.[8] Based on the smart hotel platform, the hotel industry may be able to build hotel formats with different themes and features, such as health-care hotels. It is generally believed that after this pandemic, consumers will inevitably favor it more. On the basis of the smart hotel platform, it is easy for convalescent hotels to realize the functions of remote booking, contactless facial scanning, room air disinfection, voice control room, guest sleep monitoring, toilet body fluid monitoring, online consultation, health care needs

https://doi.org/10.1051/e3sconf/202129302020

and so on through scientific and technological equipment with artificial intelligence as the core, which enables the hotel products no longer single, but more in line with the needs of consumers. Relying on the smart hotel technology platform, it is easy to achieve multi-format cross-border. If hotel products are linked to other industries, the hotel product line will be greatly enriched.

In the present and future of the normalization of the epidemic, how to make better use of different types of architectural space will become an issue of concern. Under the new security problems, designers should consider using Internet technology to release space under the premise of ensuring security, improve the flexibility of architectural space, make architectural space more anti-epidemic resources, and meet people's new use needs.

References

1. Wang Jiao'e, Du Delin, Wei Ye, Et Al. Research on The Spatial Diffusion Process and Model of the COVID-19 Epidemic [J]. Geographical Research, 2020, 39(07): 1450-1462.
2. Lu Yue. Application Of Le Corbusier's Modular System in Architecture [J]. Shanxi Architecture, 2015, 41(07): 18+39.
3. Zhou Qingwei. The Characteristic Expression of Flexible Design Concept in Architectural Design [J]. Ju She, 2019(23): 86.
4. Zhu Xiaofang, Chen Yawei, Dong Bei. Exploring The Layout of Community-Level Public Sports Facilities Under the Concept of Flexible Planning: Taking Zhenjiang Old Town as An Example [J]. Jiangsu Urban Planning, 2019(02): 16-22.
5. Wang Zhicheng, Werner Woolf, Jenny Woulfe.New Trends of residential building construction after epidemic situation[J]. Housing and real estate. 2020,(14)
6. Zhuang Zhaojian, Mei Qing.Rethinking the Design of Homestays in the Context of COVID-19:Taking Guangxi Practice as an Example[J]. Architecture and culture.2021(02)
7. Zheng Jianghuai, Fu Fu, Tao Jin. Impact of COVID-19 Epidemic on consumption economy and Countermeasures [J]. Consumption economy, 2020(03) : 1-10.
8. State Administration for Market Regulation, China National Standardization Administration. GB/T 37976-2019. General technical requirements for the interface of the Internet of Things smart hotel application platform [S]. Beijing: China Standard Press, 2019 (08)

Study on the technology of strengthening concrete by impregnation with modified sodium silicate compound strengthening fluid

Yang Dishan[1,*], Yan Tongyu[1], Huang Xiaoyu[1], Xiong Ying[1], Lin Xi[1], Zhou Hong[2] and Shi Jianguang[2]

[1]National Grid Fujian Electric Power Co., Ltd. Institute of Economics and Technology, Fuzhou 350012, China

[2]School of Architecture and Civil Engineering, Xiamen University, Fujian 361005, China

Abstract. It is the key to the concrete repair and strengthening construction that the base concrete has a certain quality and strength. Based on the strengthening mechanism of sodium silicate,the sodium silicate was modified to develop a kind of compound strengthening liquid. Through the contrast test results of concrete impregnation. The strength of C5 concrete before and after strengthening is increased by 42.5%, while the strength of C10 concrete is increased by 21.4%, and the strength of C15 concrete is increased by 10%.

1 Introduction

It is very common and serious that the durability damage of the outdoor concrete structure support in the coastal substation, which causes the aging and deterioration of concrete in different degrees[1,2]. There are four steps, base treatment, interface treatment, repair treatment and surface treatment in the repair process of durability damage[3]. The treatment of concrete base includes eliminating the defects such as floating slurry, weathering and spalling, and chiseling base to meet the requirements of certain roughness[4~7], so as to ensure that repair and reinforcement materials and base concrete bond strength enough to ensure that damage does not occur at the interface. But it is difficult the concrete base treatment construction as to cause dust and other environmental problems, especially for the construction of the outdoor concrete support of the substation. Therefore, it is effective technical measures to reduce or avoid the elimination, roughen and other concrete base if we can reinforce the concrete base directly after surface cleaning, improve the quality and strength of concrete base and meet the bonding conditions between repair or reinforcement materials. So, on the basis of summarizing the research results of sodium silicate reinforced concrete, the modified sodium silicate composite reinforced concrete was developed in this paper. Through the contrast test after concrete immersion, the reinforcement effect of impregnation was studied.

2 Research status of sodium silicate reinforced concrete technology

Sodium silicate solution state, generally colorless or slightly colored transparent or translucent viscous liquid. Molecular formula $Na_2O{\cdot}mSiO_2$, molar ratio of silica sand to alkali, i.e. the molar ratio of SiO_2 to Na_2O, determines the modulus M of sodium silicate. The modulus not only reflects the composition of sodium silicate, but also affects the physical and chemical properties of sodium silicate. Therefore, different modulus of sodium silicate has different uses[8]. Shi[9] found that sodium silicate sodium silicate can hydrolyze at room temperature, ionize the silica gel and dehydrate and condense to form a network structure with Si-O-Si as the framework. Fan[10] Using nano-silica soaking solution soak, PVA solution strengthening, soaking in the sodium silicate solution strengthening, sodium silicate solution immersion as well as PVA solution soak four kinds of strengthening methods for recycled aggregates for strengthening experiment, found that modified PVA solution and sodium silicate solution immersion of recycled coarse aggregate has good effect, the PVA solution concentration was 0.5%, concentration of sodium silicate solution was 5% when modified effect is best, recycled concrete 28-day strength increased by 7.2% and 23.7% respectively. In order to study the influence of inorganic surface treating agent on the surface performance and durability of airport pavement concrete, Li[11] used sodium silicate with lithium silicate to form composite sodium silicate inorganic surface treatment agent, the inorganic surface treatment agent is coated on the surface of concrete block. The test results showed that the surface hardness of concrete reached 9.0 after 7 days.

* Corresponding author: 550818454@qq.com

3 Research and development of modified sodium silicate composite reinforcement solution and experimental study on impregnating reinforced concrete

3.1 Research and development of modified sodium silicate composite reinforcement solution

The modification methods of sodium silicate mainly include enhancing the penetration ability of sodium silicate, increasing the curing strength of sodium silicate, and speeding up the curing time of sodium silicate. To enhance the penetration ability of sodium silicate is to reduce the viscosity of sodium silicate and make it easier to penetrate into the concrete. The penetration of liquid in concrete can be regarded as capillary water absorption, and the penetration depth of solution in capillary can be described by Equation (1).

$$x = \sqrt{\frac{\sigma r \cos\theta}{2\eta}}\sqrt{t} = k\sqrt{t} \quad (1)$$

Where, x is penetration depth; σ is surface tension; r is pore radius; θ is the contact angle; η is viscosity; T is time; K is the capillary coefficient or permeability coefficient.

It can be seen from Equation (1) that, under the same conditions, the penetration depth of the solution in concrete is determined by the contact angle and interfacial energy of the solution, and the relationship between the contact angle and interfacial energy can be expressed by Young's Equation (2).

$$\gamma_{sg} - \gamma_{sl} = \gamma \cos\theta_{lg} \quad (2)$$

Where: γ_{sg} is solid-gas interface energy; γ_{sl} is the solid-liquid interface energy; γ_{lg} is the liquid-gas interface energy; θ is the solid-liquid contact angle.

According to Young's equation (2), the contact angle is affected by the solid-liquid-gas interface energy, which is determined by the liquid's surface tension. In order for the liquid to be wetted at the solid interface, it is necessary to reduce the surface tension of the liquid. The addition of chemical surfactants can effectively reduce the surface tension of the liquid, Li[12] pointed out that the surfactant is a compound composed of two groups with different properties: hydrophilic group and hydrophobic group. Hydrophilic group is generally non-polar hydrocarbon chains, while hydrophobic groups are generally polar groups such as − OH, − COOH and − SO_3. When the surfactant is dissolved in water, its hydrophilic end will be attracted into the water molecules, and its hydrophobic end is repelled, so the molecules in the surfactant on the surface of water orientation, resulting in water solution surface tension greatly reduced, thereby greatly improving the wetting ability of the solution.

In addition, the penetration depth can be increased by reducing the viscosity of sodium silicate, and the viscosity can be adjusted by changing the concentration of sodium silicate. The influence of concentration and modulus on the viscosity of sodium silicate is plotted by Zhu[13]. The viscosity of sodium silicate increases linearly when it exceeds the critical value. Since the modified sodium silicate module used is 3.4 and the critical concentration is 37%, 2.7 liters of water should be added for dilution per liter of sodium silicate to minimize its viscosity while maintaining the concentration of sodium silicate as much as possible.

The curing strength of sodium silicate is generally related to the curing conditions and additives in sodium silicate. The addition of additives in sodium silicate can also improve the curing strength and bonding strength of sodium silicate. There are four curing ways of sodium silicate including heating hardening, microwave hardening, carbon dioxide hardening, alcohol and ester hardening[14]. The mechanism of hardening of sodium silicate by heating is mainly that the heating will dehydrate the sodium silicate, which will have different hardening processes at different temperatures. When the temperature reaches more than 100 degrees, the Si-OH bond in the sodium silicate will dehydrate and associate with each other and form Si-O-Si bond, so the sodium silicate will form a stable material with Si-O-Si chain structure. Microwave hardening is the use of microwave to accelerate the hardening of sodium silicate. The temperature will rise rapidly after the sodium silicate absorbs the energy of microwave. The silica molecules and water molecules in the sodium silicate will oscillate at a high speed at the same time. The colloidal particles will accelerate the movement and under the influence of heat, and quickly form a dense and small glass-like network structure of colloidal particles. CO_2 gas hardening refers to the rapid hardening of sodium silicate by blowing CO_2. Under the action of CO_2 gas, sodium silicate can react with CO_2 to form silica gel. At the same time, CO_2 has a good drying effect, which can quickly dry sodium silicate. The water of silica gel is evaporated and dehydrated to form solid SiO_2, which is finally condensed and hardened. The hardening of alcohols and esters refers to the addition of organic esters can react with sodium silicate, so that the modulus of sodium silicate increases, and further dehydration of sodium silicate occurs hardening. Studies have shown that adding polyols in sodium silicate can improve the bonding strength of sodium silicate after curing.

The mechanism is not exactly the same although the four methods can make the sodium silicate hardening. The compressive strength of sodium silicate obtained by microwave hardening method is higher, while that obtained by alcohol and esters hardening method has better bonding property. Li[15] found that adding a small amount of sodium tetraborate can improve the hardening strength of sodium silicate, and the bond strength increases with the increase of sodium tetraborate content. Sodium tetraborate or borax, molecular formula $Na_2B_4O_7 \cdot 10H_2O$ is a very important boron mineral and boron compound. It is usually a white powder containing colorless crystals and soluble in water. Sodium tetraborate presents a tetrahedral form because its structure formula is similar with sodium silicate, both of them are tetravalent boron atoms or silicon atoms connected with three oxygen atoms. Therefore, adding sodium tetraborate in sodium silicate can form a composite glass network

 https://doi.org/10.1051/e3sconf/202129302021

with sodium silicate and improve the integrity of the network, so as to improve the curing strength of sodium silicate.

It is necessary to control the dosage of curing agent because sodium silicate penetrating into concrete takes a certain time as well as adding curing agent can not make sodium silicate solidify too fast so as to lose liquidity. When sodium silicate is added to sodium silicate, it can react with the hydrolysate of sodium silicate. The hydrolysis reaction of sodium silicate is shown in equation (3).

$$Na_2SiO_3+3H_2O=2NaOH+Si(OH)_4 \quad (3)$$

The product $Si(OH)_4$ is also called orthosilicate, generally written as H_4SiO_4, but written as $Si(OH)_4$ is easier to understand. Adding sodium fluosilicate to sodium silicate can react with NaOH, the hydrolyzed product of sodium silicate. The reaction formula is shown in Equation (4).

$$Na_2SiF_6+6NaOH=Na_2SiO_3+6NaF+3H_2O \quad (4)$$

By combining the two equations above, the total equation can be obtained, as shown in Equation (5).

$$2Na_2SiO_3+Na_2SiF_6+6H_2O=6NaF\downarrow+3Si(OH)_4 \quad (5)$$

It can be concluded from the total reaction that after adding NA_2SiF_6, it reacts with NaOH, thus accelerating the hydrolysis of sodium silicate. Si in Na_2SiO_3 and Na_2SiF_6 all generates the orthosilicate monomer $Si(OH)_4$. $Si(OH)_4$ is an elastic colloid, and it is also the material that finally plays a bonding role. Orthosilicate can be dehydrated and condensed to form a network structure material with Si-O-Si chain. The NaF generated by the whole reaction is insoluble material, so if the sodium silicate is solidified in the concrete, the NaF generated can not only fill the gap after the solidification of sodium silicate, but also fill the pores in the concrete, which is of positive significance to enhance the curing strength of sodium silicate and concrete strength. Therefore, the addition of Na_2SiF_6 not only shortens the curing time of sodium silicate, but also improves the curing strength of sodium silicate.

Based on the control of sodium silicate concentration, the selection of surfactants and additives, and the control of curing agent dosage, the sodium silicate is modified to improve its permeability and wetting ability, and improve the curing strength, bond strength and concrete strength, so as to achieve the purpose of providing concrete durability and strength. The material preparation of modified sodium silicate composite reinforcing solution selected the sodium tetraborate product produced by Tianjin Zhiyuan Chemical Reagent Co., Ltd. The molecular formula is $Na_2B_4O_7\cdot 10H_2O$, the molecular weight is 381.38, and the purity is 99.5%. The molecular formula of sodium dodecyl sulfate produced by Beijing Jintong Letai Chemical Products Co., Ltd. is $C_{12}H_{25}SO_3Na$. Due to the negative charge of sulfonic acid group-SO3 on the left side, it belongs to anionic surfactant. The molecular formula of sodium fluorosilicate produced by Pingxiang Baishi Chemical Reagent Co., Ltd. is Na2SiF6 with purity of 99.0 %.

3.2 Manufacture of concrete test blocks and impregnating reinforced concrete

The main materials of concrete test block are silicate cement, water, coarse aggregate and fine aggregate. The cement is P·O32.5 ordinary Portland cement, which meets the national standard GB175-1999 <Portland cement, ordinary Portland cement>. The flexural strength of 3 d and 28 d is 4.0 MPa and 7.6 MPa, respectively. The compressive strength of 3 d and 28 d is 18.8 MPa and 37.4 MPa, respectively. The fine aggregate used is manufactured sand from Quanzhou, Fujian, and the apparent density is 2680 kg/m^3. According to the requirements of GB/T 14684-2011 <Sand for construction>, the percentage of sieve residue meets the requirements. The coarse aggregate is gravel from Fujian Xiamen, and the continuous gradation is 5~25 mm. According to the requirements of GB/T 14685-2011< Construction pebbles and gravels>, the cumulative screening percentage of gravel aggregate meets the standard requirements. Water is tap water. Three concrete blocks with different mix proportions of C5, C10 and C15 were designed. According to< Selection table of concrete sand ratio> in JGJ 55 - 2011 <Specification for mix design of ordinary concrete>, the sand ratio was determined to be 45%, and then the amount of coarse and fine aggregate was determined by the absolute volume method. Three kinds of concrete mix ratio see table 1.

Table1. Mix ratio of three kinds of concrete test blocks.

Concrete label	Material dosage/(kg/m^3)			
	cement	water	sand	the stone
C5	156	195	915	1, 118
C10	207	195	894	1, 092
C15	256	195	875	1, 069

In order to make the modified sodium silicate reinforcing liquid penetrate deep into the concrete as far as possible, the concrete test block was infiltrated by impregnation method. Three different strength concrete test blocks were put into the impervious PE plastic bag, and then the matched reinforcing liquid was added. The opening was closed to ensure that the reinforcing liquid was not evaporated, and the environment temperature was 20 °C and the relative humidity was 70% (figure 1).

Figure 1. Concrete test mould, test block and impregnation method.
(a) test mould; (b) test block; (c) Impregnation method.

4 Properties of concrete reinforced by modified sodium silicate composite reinforcing solution

4.1 Compressive strength of concrete reinforced with modified sodium silicate composite reinforcement solution

It can be seen from the comparison of the compressive strength of concrete reinforced by modified sodium silicate composite reinforcing liquid (Fig. 2, Table 2) that the strength of C5 grade concrete increases by 42.5%, that of C10 grade concrete increases by 21.4%, and that of C15 grade concrete increases by 10%. Sodium silicate is the liquid form of sodium silicate, which contains more silicate, and free Ca after entering the concrete2+. In addition, the three-dimensional mesh material with Si-O-Si chain is generated when the modified sodium silicate composite reinforcing liquid is solidified, so the reinforcing effect is better.

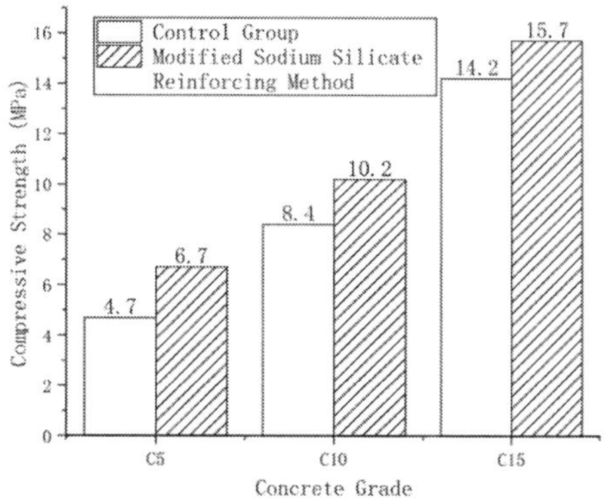

Figure 2. Comparison of compressive strength of concrete before and after reinforcement.

Table2. Compressive strength of concrete before and after reinforcement with modified sodium silicate composite reinforcement solution.

group	Concrete label	Test block number /MPa			Average/MPa
		1	2	3	
Modified sodium silicate composite reinforcing liquid before reinforcing	C5	4.8	4.8	4.6	4.7
	C10	8.6	8.1	8.6	8.4
	C15	14.7	13.8	14.0	14.2
Modified sodium silicate composite reinforcing liquid after reinforcing	C5	6.7	6.7	6.7	6.7
	C10	10.4	10.4	10.0	10.2
	C15	15.9	15.7	15.7	15.7

4.2 Microstructure of Modified Sodium silicate Composite Reinforcement

The microstructure shown in the concrete after reinforcement by modified sodium silicate composite reinforcement solution can be divided into two categories. The first category is the dense network structure formed by the condensation of integral sodium silicate itself, such as the left part of the scanning electron microscope after 10,000 times magnification (Fig. 3). The second is the C-S-H gel produced by the reaction of silicate in sodium silicate with $Ca(OH)_2$ in concrete. It can be seen that the reinforcement mechanism of sodium silicate for concrete substrate can be divided into two categories. The first is its own condensation curing, and the second is the reaction with calcium hydroxide that does not contribute to the strength in concrete to generate calcium silicate hydrate that plays a major role in the strength. From the analysis of the reinforcement mechanism and curing mechanism of modified sodium silicate composite reinforcement solution, the reinforcement can be divided into two aspects. The first aspect is the reaction of silicate ions in sodium silicate and calcium ions in concrete to generate calcium silicate. On the other hand, the self-curing of sodium silicate forms a polymer with Si-O-Si chain. Therefore, the modified sodium silicate composite reinforcement solution improves the internal microstructure of concrete and the reinforcement effect is good.

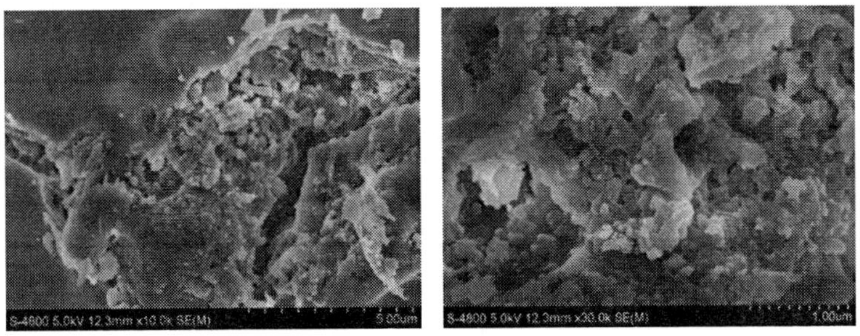

Figure 3. Microstructure of concrete reinforced by modified sodium silicate composite reinforcing solution. (a) 10000-fold; (b)300-fold

5 Conclusion

The sodium silicate was modified by controlling the concentration of sodium silicate, selecting surfactants and additives, and controlling the dosage of curing agent. The permeability and wetting ability were improved, and the curing strength, bond strength and concrete strength were improved. The modified sodium silicate composite reinforcement solution was prepared by adding sodium tetraborate, sodium dodecyl sulfate, sodium fluosilicate and other materials. Through the comparative test of concrete impregnation, it can be found that:

(1) Modified sodium silicate composite reinforcement solution has a very significant reinforcement effect on low-strength concrete. The strength of C5 concrete before and after reinforcement increased by 42.5%, C10 concrete by 21.4%, and C15 concrete by 10%.

(2) It can be seen from the microstructure of the reinforcement solution, it can be seen that the reinforcement effect comes from the reaction of silicate ions in sodium silicate and calcium ions in concrete to form calcium silicate, and the glass itself is solidified to form a polymer with Si-O-Si chain.

References

1. Yang Dishan, Lin Xi, Shi Jianguang et al. Research summary of durability damage and strengthening of outdoor concrete facilities in coastal substation [A]. Paper collection of 2020 Industrial Architecture Academic Exchange (Volume I) [C]. Zhongye Architectural Research Institute Co., Ltd : Industrial Architecture Journal, 2020 : 7.
2. Xia Xiaojian, Jin Yan, Qiao Hanwen, et al. Survey on Corrosion of Power Transmission and Transformation Equipment [J]. Corrosion science and protection technology, 2019,31 (02) : 121-127.
3. Hui Yunling, Yue Qingrui. The durability damage and repair of concrete structures [C], the sixth national conference on building identification and reinforcement. The National Technical Committee for Building Identification and Reinforcement Standards, 2002.
4. JGJ / T 238-2011, Technical specification for interface guniting on concrete base [S].
5. JGJ/T 259-2012, Technical specification for rehabilitation and protection of concrete structures durability [S].
6. GB 50367 - 2013, Code for design of strengthening concrete structure [S].
7. GB 50550-2010, Code for acceptance of construction quality of strengthening building structures [S].
8. Wang Kun, Kang Yong, Ai Jiang. Modification, Hardening Mechanism and Application Prospect of Sodium Silicate Materials [J]. Foshan ceramics, 2016,26 (05) :
9. Shi Caijun, Wang Yue, Pan Xiaoying, et al. Advances in Inorganic Surface Treatment of Concrete [J]. Materials reports, 2017,31 (13) : 113-119.
10. Fan Yuhui, Niu Haicheng, Zhang Xianggang. Experimental study on the modification of recycled aggregate concrete by nano-SiO_2 [J]. Concrete, 2017, 000 (007) : 92 – 95.
11. Li Xiaorong. Effect of inorganic surface treatment agent on performance of airport pavement concrete [J]. Hunan Communication Science and Technology, 2019,45 (03) : 71-74.
12. Li Yiming. Study on modification of sodium silicate and its application in protection of ancient ivory [D]. Chengdu University of Technology, 2007.
13. Zhu Chunxi, Lu Chen. Understanding Process of Sodium Silicate Hardening [J]. Inorganic Chemicals Industry, 2001 (01) : 22-25 + 2.
14. Hu Caiying, Zhou Yanming, Luo Hong, et al. The Analysis of the Solidification Mechanismes & Water Resistance Improvement Accesses of the Water-Glass [J]. Ceramics, 2011 (08) : 18-21.
15. Li Yiming, Ye Qiaoming. Study on the modification effect of sodium tetraborate on sodium silicate [J]. Guangdong Weiliang Yuansu Kexue, 2006 (10) : 66 – 68

Seismic vulnerability assessment of RC curved bridges piers subjected to bidirectional earthquake

Rui Li[1,*]

[1]CHANG'AN UNIVERSITY, highway school, 710064 Middle-section of Nan'er Huan Road Xi'an, China

Abstract. Curved beam bridge is a kind of irregular structure, which has the advantage of adapting to complex terrain, but it is more vulnerable to damage than regular bridge under earthquake. This paper investigates the vulnerability of curved continuous girder bridges under the action of bidirectional earthquake horizontal earthquake, and the difference of vulnerability between side pier and intermediate pier is analyzed. Fragility assessment is performed using an incremental dynamic analysis method subjected to a wide range of as-recorded sequences. A proper engineering demand parameter (EDP) which can result in the most probability of failure at bridges employed in this study is determined. Result indicates that only considering the unidirectional seismic input will underestimate the seismic response and potential damage of the structure, which is not accurate for the seismic performance evaluation of the bridge. Result also shows that the damage probability of the intermediate pier of the curved bridge is higher than that of the side pier, and the more serious the failure, the smaller the difference between the two piers.

1 Introduction

In the Wenchuan earthquake, the large relative displacement at the expansion joint of Baihua bridge caused the upper structure of the fifth continuous beam (with a radius of curvature of 66m) to fall and break, resulting in the overall collapse of the structure[1]. For the four circular ramp bridges of Huilan interchange, the piers close to the abutment are rigidly connected with the upper structure. Under various actions, bending shear failure occurs. The damaged piers present X-shaped fracture section, the longitudinal bars at the bottom of the piers buckle, and the confined concrete is crushed[2]. The seismic research of curved bridges began in 1971 when the San Fernando earthquake in Los Angeles caused the fall of two curved bridges on the interstate highway interchange. [3] and [4] respectively carried out detailed modeling analysis and shaking table test for the damaged double curve girder bridge, and put forward some suggestions for seismic design and calculation of curved bridge. The influence of curve radius and other factors on the seismic performance of curved bridges is studied by means of spectrum analysis or time history analysis[5] [6]. [7] studied the influence of curve radius on the damping effect of viscous damper of curved bridge.

In the seismic research of curved beam bridge, the actual input direction of seismic wave will have different angles with different Bearings, so there is no clear main direction to guide the seismic wave input to obtain the most unfavorable response of curved beam bridge. In the past, many researchers have discussed the problem of the most unfavorable direction. [8] used the rotation method to input one-way time history of a curved continuous beam bridge with 0-180 degrees every 10 degrees as an excitation direction, and obtained the maximum response angle of the middle pier in radial and tangential directions respectively. [9] gives the formula for calculating the most unfavorable input angle of curved girder bridge, and the difference between the calculation results and the rotation method is small. However, no matter what method is used, most of the conclusions are focused on the most unfavorable input angle of a certain bridge pier in a certain direction. Different focuses on the bridge often lead to different worst input angles. It is difficult to evaluate how to input seismic wave to make the bridge structure suffer the most adverse overall stress. With the development of performance-based seismic design method, the seismic performance of structural components in a specific site can be evaluated from the perspective of probability, and the seismic performance of the whole bridge structure can also be calculated. Therefore, taking the whole bridge vulnerability characteristics of curved girder bridge as the evaluation standard, a seismic input direction which makes the curved beam bridge most vulnerable to damage can be determined.

Curved beam bridge is a kind of irregular structure, which has the advantage of adapting to complex terrain, but it is more vulnerable to damage than regular bridge under earthquake. For the straight-line bridge, the most unfavorable response value of the straight-line bridge can be obtained when the seismic wave direction and the bridge axis direction are 0 degree and 90 degree respectively. However, for curved bridge, the direction of seismic wave is not the same as the angle of main beam in different positions. Therefore, the curved bridge does

* Corresponding author: 13572500825@163.com

https://doi.org/10.1051/e3sconf/202129302022

not have a certain longitudinal (transverse) direction, and it is necessary to consider bidirectional seismic input. This paper aims to evaluate the fragility curves of piers of curved bridges subjected to bidirectional earthquake.

2 Model

A continuous curved beam bridge has been considered in this study. With a total length of approximately 100 m, this four-span RC bridge comprises a continuous reinforced concrete multi-cell box girder deck supported on two abutments at both ends and three single column bents (piers) at Mid-spans. The axis of the bridge is a circular curve with a radius of curvature of 80m. The height of columns is approximately 10m with a circular cross-sectional geometry which is constant along the length of column with 1.6m-diameter. This pier is consisted of 30 longitudinal bars, 28mm-diameter and 12mm-diameter transverse reinforcement in the form of spiral. The cross section the bridge has been shown in Fig. 1.

Fig.1 Three dimensional model of curved bridge

Three-dimensional non-linear finite element models (FEM) are established in SAP2000. The RC deck has been modeled as an elastic straight beam column element; it is assumed that superstructure behaves elastically during the earthquake according to the assumption presented by[10]. The plate rubber bearing is simulated by plastic (Wen) connection unit. The bilinear model with yield point as the only turning point is chosen to simulate the plastic hinge in bridge piers, and the simplified skeleton curve is as shown in Fig. 2. Multiline plastic connection unit can be selected to simulate it. For curved beam bridge, the pier will be damaged in both transverse and longitudinal directions, so two bending degrees of freedom are set as non-linear, three translation and torsion are set as fixed.

3 Ground motion

The acknowledgements should be typed in 9-point Times, without title. Assuming that the soil type in this paper is a class II type, 10 seismic records are randomly selected after the near-field seismic waves are excluded from the Pacific earthquake engineering research center (PEER) database according to the shear wave velocity. Based on the statistics of the peak acceleration of the bidirectional component of a large number of seismic waves, [11] suggests that a seismic wave can be input into a bidirectional seismic wave at a peak acceleration ratio of 1:0.85. However, the as-record data show that the spectral characteristics of two horizontal components of the same seismic wave sometimes differ greatly. If the difference is ignored in the calculation, the calculation results are often quite different from the actual ones. Therefore, in this paper, when inputting bidirectional seismic waves, two actual seismic wave components are selected without adjustment, which is consistent with the as-record earthquake. The acceleration of the selected seismic wave is scaled in 0.1g increments to make its peak value from 0.1g to 1.0g, which is divided into 10 levels, with 100 seismic waves in total. Fig. 3 shows the initial response spectrum curves, respectively.

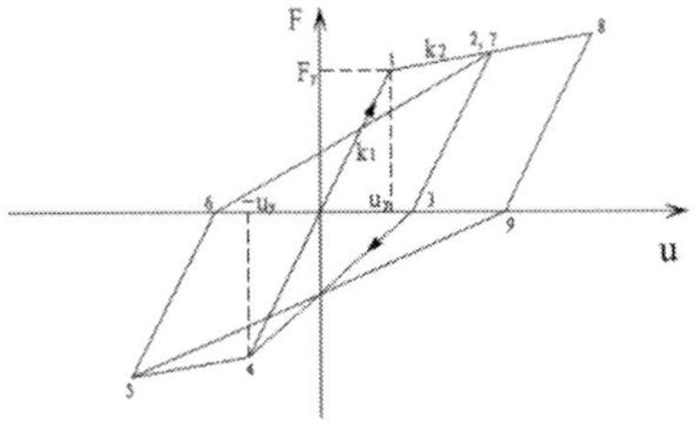

Fig.2 bilinear model curve

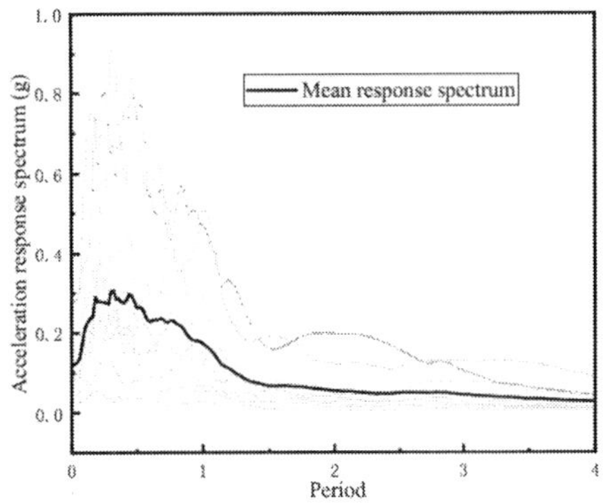

Fig.3 Initial response spectrum

4 Method

Fragilities can be calculated from the median and log-standard deviation values, with the assumption of lognormal distribution. The equation for this function is given by:

$$P_f = \phi[\frac{\ln(a(IM)^b) - \ln(LS)}{\sqrt{\beta_{EDP}^2 + \beta_{LS}^2}}] \quad (1)$$

where P is the probability of exceedance of the limit state based on ground motion intensity (IM), Φ is the probability density function of a normal distribution, IM is the ground motion intensity indexes, such as peak ground acceleration (PGA) or spectral acceleration, LS is limit value of damage index for each failure state of structure, EDP is engineering demand parameter, β_{EDP} and β_{LS} are the log-normal standard deviation of structural seismic response and limit state of components, respectively; a and b are fitted parameter values. When SA

https://doi.org/10.1051/e3sconf/202129302022

is used as vulnerability analysis index, a is 0.4, and when PGA is used as index, a is 0.5[12].

Curved beam bridge is also an irregular bridge, the influence of high-order vibration mode on the structural response cannot be ignored, so EDP defined by curvature ductility ratio for pier can more accurately reflect the deformation and failure of materials in the plastic hinge area of pier.

Considering the complexity of the structure and the small basic period, PGA, which is simple in concept, can be selected as the seismic intensity index to meet the calculation requirements.

5 Fragility curves of piers

This section compares the response difference of piers under unidirectional seismic excitation. Seismic excitation direction and bridge arrangement are shown in Fig. 4. According to the difference of seismic input direction, three working conditions are established.

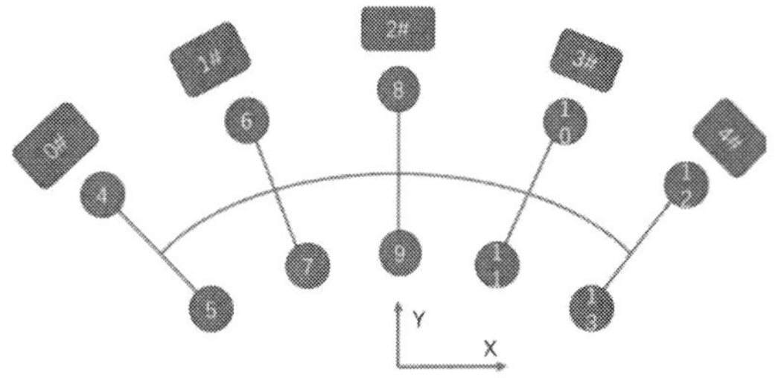

Fig.4 Schematic diagram of seismic wave input direction

Case 1 (longitudinal direction): the larger component of seismic wave will be input along X direction.

Case 2 (transverse direction): the smaller component of the seismic wave will be input along the Y direction.

Case 3 (bidirectional): the larger component of seismic wave along X direction and smaller component of seismic wave along Y direction, bidirectional input.

The comparison of the maximum curvature of 1# pier bottom and the fragility curves of 1# and 2# pier subjected to bidirectional earthquake are shown in Fig. 5 and Fig. 6 respectively.

It can be seen from Fig. 5 that with the increase of PGA, the comprehensive maximum curvature of pier bottom increases continuously under three cases. When the comprehensive maximum curvature reaches the limit value of slight damage, the increase speed of the comprehensive maximum curvature value increases significantly, because the plastic hinge area of the pier has been yielding and entered the plastic deformation stage. Before PGA reaches 0.7g, the maximum curvature of the pier under the condition of transverse input and longitudinal input is basically same. When PGA is greater than 0.7g, the maximum curvature under the condition of longitudinal input is gradually greater than that of the pier under the condition of transverse input. And with the increase of PGA, the curvature values of the three cases are more and more different. It can be seen that bidirectional seismic input will increase the maximum curvature of the pier bottom and increase the bending deformation of the pier.

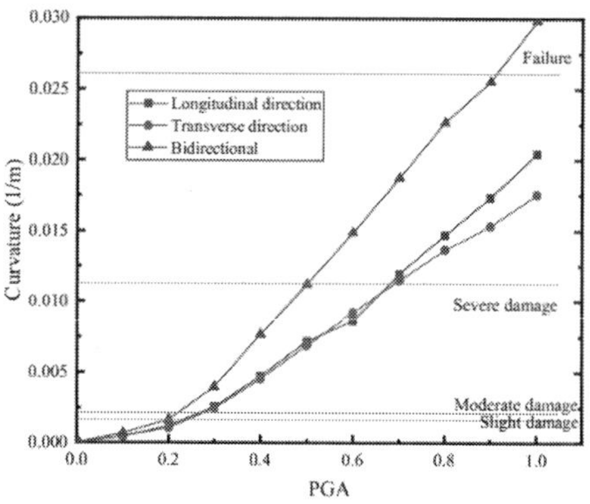

Fig.5 Comprehensive maximum curvature comparison of 1# piers

According to Fig. 6(a), the probability of bending failure of 1# pier increases with the increase of ground motion intensity. When PGA is less than 0.1g, the probability of failure is approximately 0, which indicates that 1# pier will not be destroyed under this condition. When PGA gradually increases from 0.1g to 0.4g, the probability of slight and moderate damage increases rapidly, and the two damages probability are very close, because the boundary value of dividing moderate failure and slight failure is very close, and the difference of curvature ductility coefficient is only 0.24. When PGA reaches 0.4g, the probability value of slight and moderate damage is close to 1, indicating that slight and moderate damage are inevitable, while the probability of severe damage and failure is still low, the probability of severe damage is 0.118 and failure is 0.002. With the further increase of PGA, the probability of slight and moderate damage tends to be stable to more than 0.99, and the probability of severe damage continues to increase, but the growth rate is significantly lower than that of slight and moderate damage. The growth rate of probability of failure is slower, indicating that the sensitivity of severe damage and failure to PGA growth is far lower than that of slight damage and moderate damage. The fragility curve of bending failure of 2#pier is similar to that of 1#pier. It is worth noting that the axial pressure at the bottom of the 2# pier is smaller than that of the 1# pier, so the curvature ductility coefficient of the 2# pier will be relatively large, resulting in the failure probability of the 2# pier is higher than that of the 1# pier, which shows that the failure of the 2# pier will be more serious under the action of bidirectional seismic wave.

The probability difference of 1# pier and 2# pier under each damage state is shown in the Fig. 7. The difference of four kinds of damage from slight damage to failure first reaches a peak state with the increase of PGA, and then decreases. The probability difference of four kinds of damage states reaches the maximum with the increasing of corresponding PGA in turn. This is because with the increase of PGA, the growth rate of failure probability of each level gradually increases to a peak value, then decreases and finally approaches to zero infinitely, and

The difference between the responses of 1# pier and 2# pier leads to the non-synchronization of the probability growth of the same damage state.

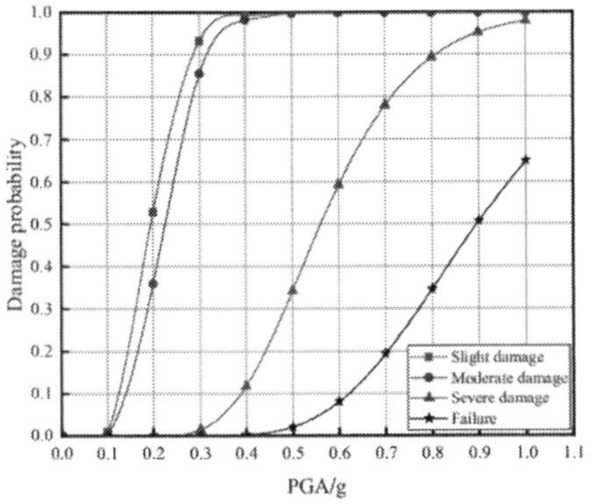

Fig. 6(a) Bending failure fragility curve of 1# pier

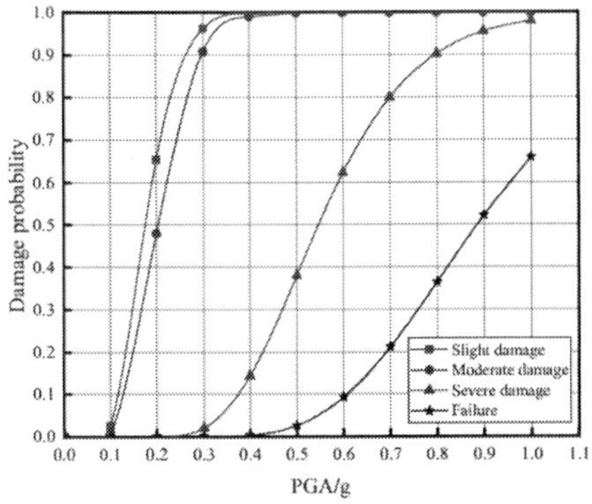

Fig. 6(b) Bending failure fragility curve of 2# pier

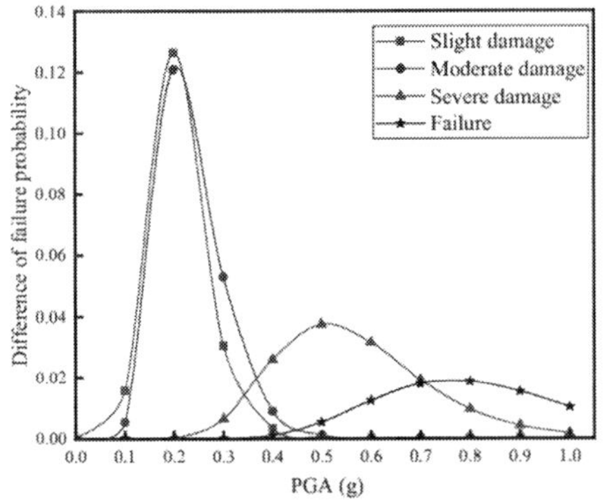

Fig.7 Difference of bending failure probability between 1# pier and 2# pier

According to Fig. 8, the probability of each damage state of 2# pier in bidirectional seismic input is significantly higher than that in unidirectional input, in which the probability values of longitudinal input and transverse input are very close, but the longitudinal input is slightly higher than that of transverse input. For the two cases of slight damage and moderate damage, the damage probability value of the bridge pier increases rapidly to 1 under unidirectional input and bidirectional input. It is worth noting that with the increase of ground motion intensity, the increase degree of bidirectional input to the probability of slight, moderate and severe damage increases first and then decreases compared with unidirectional input, while for failure, the greater the ground motion intensity is, the greater the aggravating effect of bidirectional earthquake is. For the VIII seismic fortification area, the probability of bridge pier at all damage states is less than 1% under the condition of E1 earthquake intensity. The probability of severe damage and failure of bridge pier under the condition of E2 earthquake intensity is less than 1%, and compared with the horizontal input, the probability of slight and moderate damage increases by 28% and 35% respectively. For the IX seismic fortification area, the possibility of severe damage and failure of piers is ignored under the condition of E1 earthquake intensity. Compared with the transverse input, the bidirectional input increases the probability of slight and moderate damage by 31% and 22% respectively. Under the condition of E2 earthquake intensity, the probability of slight, moderate, severe damage and failure of pier is increased by 0.3%, 1.3%, 31% and 3.3% respectively by bidirectional input compared with transverse input.

6 Conclusion

In this paper, a four-span curved continuous beam bridge is taken as an example to compare and analyze the structural response difference under different cases of unidirectional and bidirectional seismic input. The fragility curve of bridge pier under bidirectional seismic input is calculated by the method of probability seismic demand analysis based on IDA, and the development rule and difference of damage probability under bidirectional and unidirectional seismic input are compared and studied.

In the case of bidirectional seismic input, the damage probability of bridge pier is 0-40% higher than that of unidirectional input. With the increase of ground motion intensity, for the damage state other than failure, the aggravating effect increases first and then decreases. For failure state, the effect will become more and more severe with the increase of ground motion intensity. In high intensity area, the increasing effect of bidirectional seismic input on the probability of severe damage and failure state is more obvious. Therefore, only considering the unidirectional seismic input will underestimate the seismic response and potential damage

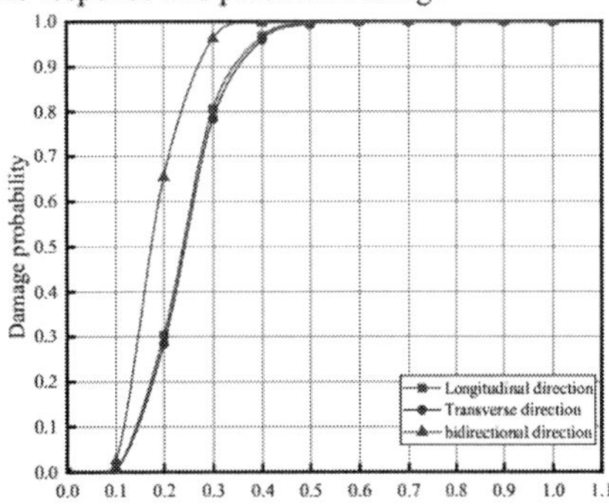

(a) Slight damage

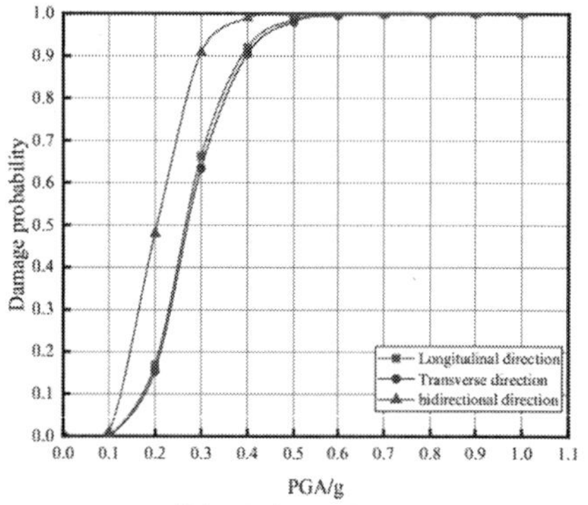

(b) Moderate damage

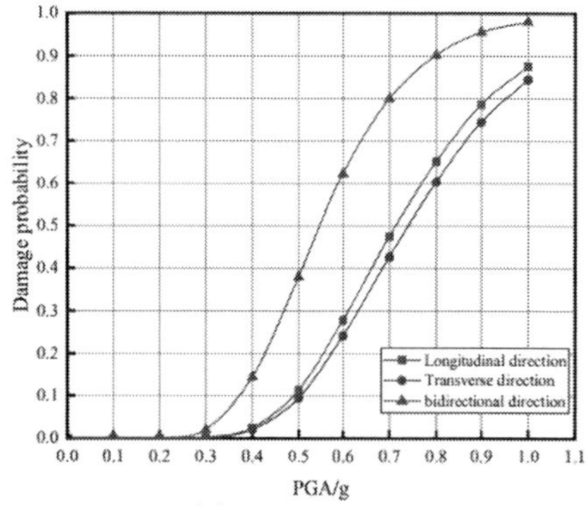

(c) Severe damage

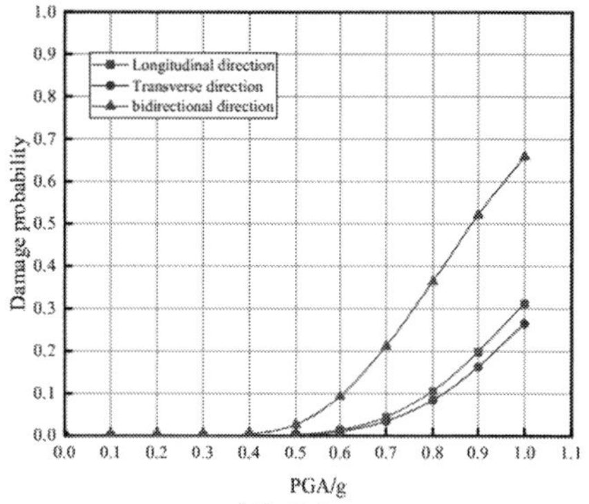

(d) Failure

Fig.8 Vulnerability comparison of unidirectional and bidirectional seismic input of 2# pier

of the structure, which is not accurate for the seismic performance evaluation of the bridge.

The damage probability of the intermediate pier of the curved bridge is higher than that of the side pier, and the more serious the failure, the smaller the difference between the two piers. When the ground motion intensity is less than 0.4g, the probability of slight and moderate damage of the intermediate pier is 13% higher than that of the side pier; when the seismic intensity is more than 0.4g, the probability of slight and moderate damage of the intermediate pier and the side pier is basically the same, but the probability of severe damage of the intermediate pier is less than 4%, and the probability of failure of the intermediate pier is less than 2%.

References

1. Ren, L., Characteristics of bridge damages in Wenchuan earthquake. Journal of Nanjing University of Technology, 2009.
2. Li, J., T. Peng, and Y. Xu, Damage investigation of girder bridges under the Wenchuan earthquake and corresponding seismic design recommendations. Earthquake Engineering and Engineering Vibration, 2008. **7**(4): p. 337-344.
3. Tseng, W.S. and J. Penzien, Seismic analysis of long multiple-span highway bridges. Earthquake Engineering & Structural Dynamics, 1975. **4**(1): p. 1-24.
4. Williams, D. and W. Godden, Seismic response of long curved bridge structures: Experimental model studies. Earthquake Engineering & Structural Dynamics, 1979. **7**(2): p. 107-128.
5. Zhu, D.S., S. Liu, and L.S. Yu, Research on Seismic Response of Curved Girder Bridges. China Journal of Highway and Transport, 2002. **15**(3): p. 42.
6. Yongliang, Z., Influence of curvature radius change on seismic response of urban curved bridge. World Earthquake Engineering, 2012.
7. Xingjun, Q.I., Influence of Variation of Radius of Curvature on Seismic Mitigation Effectiveness of Viscous Dampers for Curved Girder Bridge. Journal of Highway and Transportation Research and Development, 2012.
8. Junyan, W., Nonlinear Seismic Responses Research of Curved Girder Bridge. Dalian University of Technology, 2016.
9. Quan, W., Studies on Seismic Analysis of Large-span Bridges Subjected to Multi-component and Multiple-support Earthquake Excitations. Dalian University of Technology, 2008.
10. TG/TB02-01, Guidelines for seismic design of highway bridges, Beijing, China. 2008.
11. GB, "Code for seismic design of buildings." GB50011-2010, Beijing, China. 2010.
12. HAZUS99, F.E.M.A., Earthquake Loss Estimation Methodology: User's Manual. Federal Emergency Management Agency, Washington DC. 1999.

Influence of boulder machine-made sand powder on compressive strength of concrete

Pengtao Wang[1,*]

[1]China Road and Bridge Co., Beijing 100011, China

Abstract: In order to recycle the boulder powder produced in the process of manufactured sand production and reduce the cost of engineering concrete, this article studied the influence of boulders powder on the compressive strength of concrete. The results show that in the early stage of concrete test, the compressive strength of rock powder concrete is slightly lower than of fly ash and mineral powder concrete. With the development of curing age, the strength of boulders powder concrete developed slower. As the increase of boulders powder content, the compressive strength of different curing age gradually decreased, and it was suggested that the content of boulders powder should be controlled within 20% of cementitious materials mass. The positive effect of boulders powder fineness on the strength of concrete is limited, so it is suggested to use unprocessed collected boulders powder in the project, which is economical and environmentally friendly. With the adjustment of water-to-binder ratio, boulders powder can be prepared with different strength grades of concrete to meet the needs of engineering; the composite of boulders powder with traditional mineral admixtures, such as fly ash, and especially granulated blast furnace slag powder, can significantly improve the strength of concrete.

1 Introduction

Phnom Penh-Sihanoukville Expressway (PPSE) is an important golden channel connecting Phnom Penh, the capital of Cambodia, to Sihanouk, the largest seaport in China and an external port. The construction of PPSE is of great significance for promoting the local economic development of Cambodia and implementing the *"One Belt One Road"* initiative. The starting point of PPSE is at the intersection of National Highway No. 4 and Phnom Penh City Circle Line in Cambodia, and the end point is on the No. 4 road on the edge of Sihanoukville. It is about 8.6km away from Sihanoukville Port to the west and about 5.5km away from Westport Economy, and it is about 8.5km from Sihanoukville Airport. The total length is 187.05Km. The whole line adopts Chinese highway construction standards and specifications. The design speed is 100Km/h, and the two-way four-lane.

There are a large number of boulders on the hilltops and slopes around the planned route of the project, forming local boulder groups, and piles of boulders can be formed in the river to block the flowing water. The preliminary survey revealed that the irregularly distributed boulders generally have a diameter of 0.50 to 2.00m, with a maximum diameter of not less than 5m, and are hard and sub-circular in appearance, whose main minerals are quartz and feldspar. If these boulders are recycled into manufactured sand, it can reduce the cost of sand and gravel procurement and transportation, and reduce the occupation of arable land, which has significant economic and social benefits. Therefore, this article intended to study the influence of manufactured sand prepared by boulders and its gravel powder on the mechanical properties of concrete, aiming to promote the application of boulders manufactured sand and its gravel powder in engineering.

2 Materials and Methods

2.1 Raw Materials

In this paper, the test concrete is mainly prepared with the following raw materials:

(1) Cement: ordinary Portland cement whose initial setting is 161min, final setting is 242min, 3d and 28d compressive strength are 22.3MPa and 49.2MPa respectively. The chemical composition is shown in Table 1.

(2) Admixture: polycarboxylate superplasticizer with solid content of 16.2% and water reducing rate of 18.2% is selected. The specific surface area (Bertrand method) of fly ash (FA) and granulated blast furnace slag powder (GGBS) are 396 m^2/kg and 454 m^2/kg, and the density of them are 2.49 g/cm^3 and 2.84 g/cm^3. The chemical composition of FA and GGBS are shown in Table 1.

(3) Fine aggregate: manufactured sand made by boulders is used, which has a fineness modulus of 2.9, a stone powder content of 9.6%, a methylene blue value of 1.1; the fluidity ratio of boulders powder (BP) from the preparation process of manufactured sand is 81%, and the 28d activity index is 60%.

* Corresponding author: wangpt@crbc.com

(4) Coarse aggregate: 5-20mm continuous graded crushed stone is used, and the apparent density of it is 2642kg/m^3, the bulk density is 1626kg/m^3, the water absorption is 1.3% and the crushing value is 6.3%.

Table 1. Chemical composition of cement, FA and GGBS (wt.%)

Material type	CaO	Al_2O_3	SiO_2	SO_3	Fe_2O_3	K_2O	Na_2O	MgO	LOI
cement	72.27	3.98	18.77	3.73	0.21	0.43	0.06	0.21	0.82
Fly ash (FA)	1.03	28.05	48.91	-	-	2.15	2.65	0.80	3.28
Glast furnace slag powder (GGBS)	32.46	15.11	36.32	1.24	0.87	0.43	0.23	8.42	1.28

Table 2 Test mix proportion of contrast admixture type (kg/m^3)

No.	Water – to - binder ratio (W/B)	Water	Cement	Mineral admixture			Dosage of mineral admixture (%)	Specific surface area of BP(m^2/kg)	Fine aggregate	Coarse aggregate
				FA	GGBS	BP				
J1	0.45	175.5	292.5	97.5	—	—	25	-	825	1015
J2	0.45	175.5	292.5	—	97.5	—	25	-	825	1015
J3	0.45	175.5	292.5	—	—	97.5	25	205	825	1015
J4	0.45	175.5	370.5			19.5	5	205	825	1015
J5	0.45	175.5	331.5			58.5	15	205	825	1015
J3	0.45	175.5	292.5			97.5	25	205	825	1015
J6	0.45	175.5	292.5			97.5	25	451	825	1015
J7	0.40	160	300			97.5	205	205	825	1015
J8	0.35	150	322.5			100	30	205	828	1009
J9	0.47	178.6	266	38	—	76	30(10/0/20)	205	828	1009
J10	0.47	178.6	266	—	38	76	30(0/10/20)	205	828	1009
J11	0.47	178.6	266	22.8	22.8	68.4	30(6/6/18)	205	828	1009

2.2 Testing Methods

The compressive strength of concrete is tested according to the Chinese standard for test methods of physical and mechanical properties of concrete (GB/T 50081-2019). The size of the test piece is 150 mm×150 mm×150 mm. The raw materials of concrete were mixed in the mixer for 90s, and the test blocks were formed by vibration table. The test blocks were placed in the room with a temperature of 20 °C and a relative humidity of 60% for 24 h, and then the molds were removed, and the concrete samples were put into the standard curing room until the testing age.

In order to investigate the effect of BP on the mechanical properties of concrete, this article designed series mix proportion to prepared different concrete samples, just as shown in Table 2. The research variables in this experiment included the mineral admixture types, water-to-binder ratio, content and fineness of BP, and composite effect of mineral admixture.

3 Results and Analysis

3.1 Admixture types

Fly ash (FA), granulated blast furnace slag powder (GGBS) and boulder powder (BP) were respectively used as mineral admixtures. Under the same dosage of 25wt.% (the ratio of binder mass), the concrete performance is tested and compared, and the influence of BP on the compressive strength of concrete is studied. As shown in Table 2, the initial slump of concrete mixture is controlled at 180 ± 30mm.

The influence of different mineral admixtures on the compressive strength of concrete was shown in Fig. 1. According to the test results, the compressive strength of J3 is similar to that of J1 and J2 in the early test age, and the compressive strength of concrete contained BP is slightly lower than that of concrete with FA or GGBS. The 3-day compressive strength of J1, J2 and J3 were 19.1MPa, 22.1MPa and 20.2MPa, respectively. With the increase of curing age, the developing rate of compressive strength of J3 is significantly lower than that of J1 and J2. At 56 days, the compressive strength of J3 is significantly lower. The main reason is that BP is usually inert admixtures, which only plays a filling role in concrete and does not have pozzolanic activity with less contribution of the development of compressive strength. By contrast, the traditional mineral admixtures, FA and GGBS, can not only play a filling effect but also have certain pozzolanic activity, which can react with cement hydration products for secondary hydration, further compact concrete and improve the compressive strength of concrete [1–3]. In general, adding BP to prepare concrete sample will put an adverse effect on compressive strength of concrete, and it can be foreseen that when recycling BP to replace cement mass, the dosage of BP should be limited.

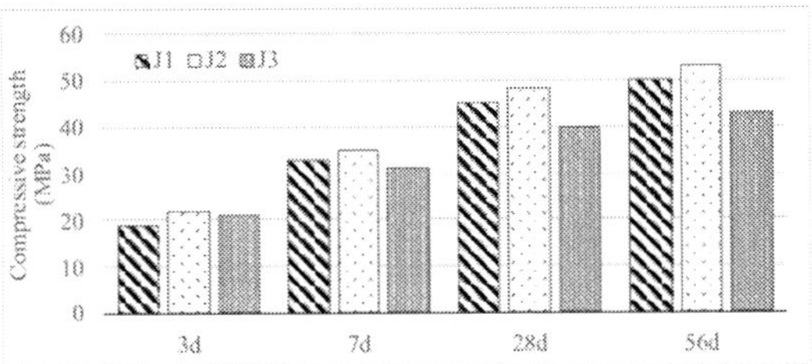

Fig. 1 Influence of admixture types on compressive strength of concrete

3.2 BP content

In order to further understand the effect of BP on compressive strength of concrete, the content of BP was set at 5%, 15% and 25% of binder mass, respectively, and then the compressive strength of concrete was measured. The mix preparations were shown in Table 2, and the initial slump of concrete mixture is controlled at 180 ± 30mm.

The influence of BP content on the compressive strength of concrete was shown in Fig. 2. From the test results, it can be seen that as the increase of BP content, the compressive strength at each curing age gradually decreased. The decrease range of compressive strength when the boulders powder content increased from 5% to 15% was smaller than that when the boulders powder content increased from 15% to 25%. The addition of BP has obvious weakening effect on the compressive strength of concrete. When its content was less than 15%, it exhibited little influence on the compressive strength before 28 days, especially when the curing age was 3 days, and it showed an obvious reduction effect on the compressive strength in later age. When BP content was 25%, the strength development trend of concrete was consistent with that of 15%, but the compressive strength was lower. The main reason was presumed to be that BP is inert powder and does not participate in the process of cement hydration reaction. Therefore, when the amount of BP is large, inert powder particles are more likely to become the initial micro defects in cement hydration products. The greater the amount of boulders powder, the more cement stone defects. It is suggested that the content of BP should be controlled within 20% of cementitious material mass.

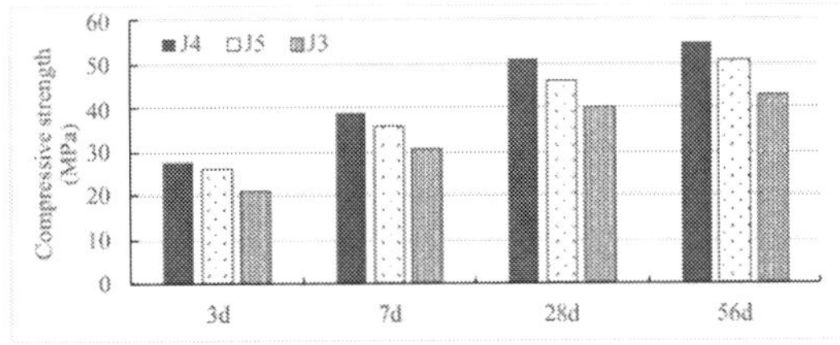

Fig. 2 Effect of boulders powder content on compressive strength of concrete

3.3 Fineness of BP

The BP particle collected in the process of manufactured sand production was generally coarse, in order to understand the influence of fineness of BP on the compressive strength of concrete and release the technical value of BP, this article designed two types of concrete sample with different fineness of BP which was obtained through physical grinding with specific surface area of 205m2/kg and 451m2/kg respectively. The mix proportion was shown in Table 2, and the slump of fresh sample was controlled at 180 ± 30mm.

The influence of fineness of BP on the compressive strength of concrete was shown in Fig. 3. The test results showed that the compressive strength of concrete increased with the increase of fineness of boulders powder. Compared with the concrete with specific surface area of 205m^2/kg, the compressive strength of J6 increased by 9.5%, 9.6%, 2.5% and 6.9% at the age of 3 days, 7 days, 28 days and 56 days, respectively. The finer the powder fineness is, the larger the specific surface area is, the more pores can be filled in the concrete, which makes the concrete structure more compact and improves the compressive strength of concrete. It meant that the finer the fineness of BP, the higher the compressive strength of concrete at each age. However, the improvement of compressive strength was not obvious. In order to achieve a certain fineness, the BP particle need to be ground by corresponding grinding equipment, which not only consumed working time, but also increased the cost. Considering comprehensively, it is not economical to adopt the technical measures of increasing the fineness of BP to improve the compressive strength of concrete [4–7]. In the actual production and application of concrete, if the stone powder produced by manufactured sand is used, it is recommended to use it directly instead of grinding.

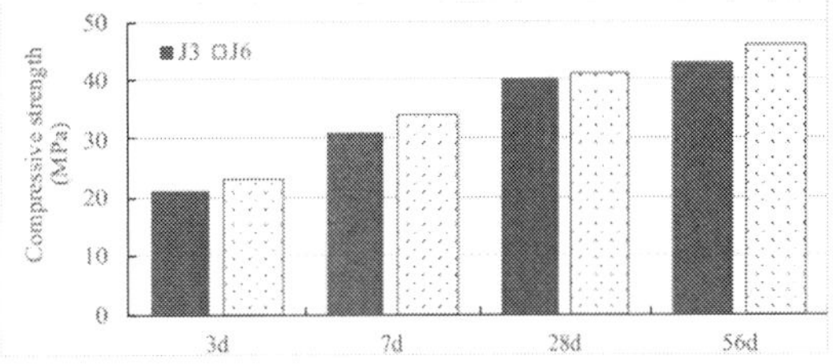

Fig. 3 Effect of fineness of boulders powder on compressive strength of concrete

3.4 Water-to-binder ratio

In the process of engineering construction, there are many strength grades of designed concrete. In order to meet the needs of different strength grades of concrete, the water-to-binder ratios of 0.45, 0.40 and 0.35 were used to study the influence of different water-to-binder ratio on the compressive strength of concrete contained BP. Different strength grades of concrete are prepared to broaden the application scope of boulder powder. The mix proportion was listed in Table 2, and the slump of fresh sample was controlled at 180 ± 30mm.

The influence of different water-to-binder ratio on the compressive strength of concrete with BP was shown in Fig. 4. The test results showed that the compressive strength of concrete increased with the decrease of water-to-binder ratio. When the water-to-binder ratio was 0.45, the 56-day compressive strength of concrete was 43.1MPa, reaching the strength requirements of C30 concrete ($f_{cu,0} \geq f_{cu,k}+1.645\sigma$, where the $f_{cu,k}$ is the designed strength grade, the $f_{cu,k}$ of C30 concrete is 30, and the σ of C30 concrete is 5.0, and the prepared strength $f_{cu,0}$ should reach 38.23MPa), When the water-to-binder ratio was 0.35, the 56d compressive strength of concrete is 63.0MPa, which meets the strength requirements of C50 concrete ($f_{cu,0} \geq f_{cu,k}+1.645\sigma$, where the σ of C50 concrete is 6.0, the prepared strength $f_{cu,0}$ should reach 59.87MPa). It could be seen that by adjusting the water-to-binder ratio, different strength grades of concrete required by general engineering can be configured. Therefore, in the construction process, BP can be considered as mineral admixture from the perspective of strength grade.

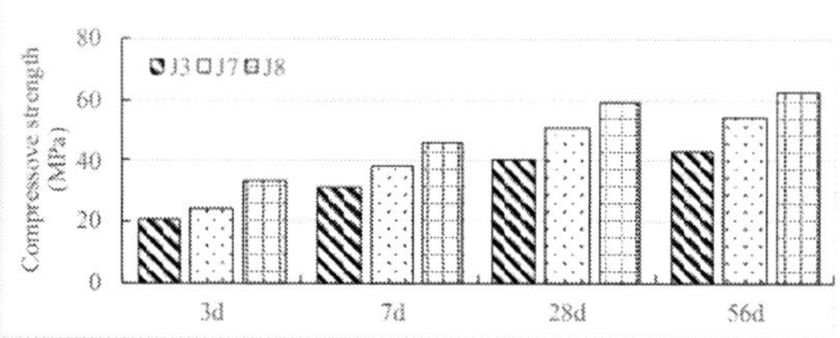

Fig. 4 Effect of water binder ratio on compressive strength of concrete

3.5 Compound mineral admixture

In the actual engineering construction process, in addition to the requirements for compressive strength, most projects have certain requirements for concrete durability, so it is necessary to add traditional mineral admixtures, such as FA and GGBS, in the concrete sample during the prepared process. Therefore, the composite addition of BP, FA and GGBS was studied in this section, and the proportion of composite addition and mix proportion were shown in Table 2. The initial slump of concrete mixture was controlled at 180 ± 30mm.

The influence of composite addition of BP and traditional mineral admixtures on the compressive strength of concrete was shown in Fig. 5. From the testing results, it could be seen that compounding BP and GGBS was more beneficial to enhance the compressive strength than combining BP and FA. At the age of 3d, 7d, 28d and 56d, the compressive strength of J10 was 12.0%, 11.1%, 8.9% and 12.0% higher than that of J9, respectively. Furthermore, the effect of 6:2:2 ternary composite of BP, mineral powder and fly ash on the compressive strength of concrete was the best, mainly since the composite of traditional mineral admixtures has a certain super superposition effect on particle size gradation and activity. It can be seen from Fig. 5 that the strength growth of the three groups of specimens was consistent throughout the age, the compressive strength growth rate from 3d to 7d was the largest, followed by 7d to 28d, and the growth rate of 28 to 56d was slowest. Fig. 5 also showed that the compressive strength of different mix proportions of concrete at different curing ages was in the order of J9 < J10 < J11. It could be found that the contribution of GGBS on compressive strength of concrete with BP was higher than that of FA, and the ternary composite of BP, GGBS and FA could improve the strength of concrete better than the combination of BP and a single mineral admixture.

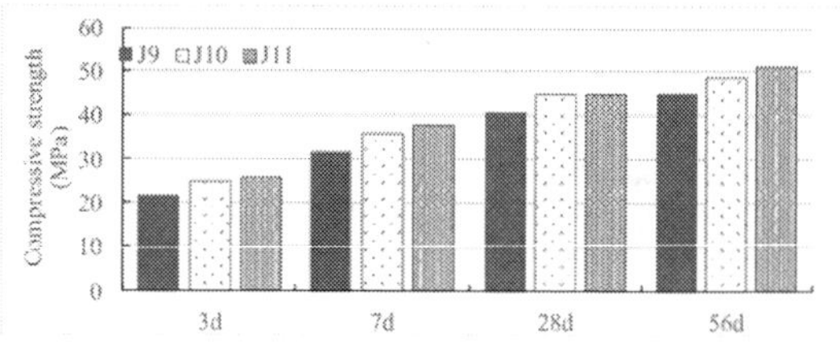

Fig. 5 Effect of different proportion of composite mineral admixture on compressive strength of concrete

4 Conclusion

Through the comparison of boulder powder (BP) with fly ash (FA) and granulated blast furnace slag powder (GGBS), the influence of different dosage and fineness, water-to-binder ratio and compounding traditional mineral admixtures on concrete strength was studied, and the conclusions could be drawn as follows:

(1) In the early curing age, the compressive strength of concrete with BP was similar to that with FA or GGBS. The strength of boulder powder concrete ws slightly lower than that of traditional mineral admixture concrete. With the development of curing age, the strength of boulders powder concrete develops slowly.

(2) With the increase of BP content, the compressive strength of each age gradually decreased. It was suggested that the content of BP should be controlled within 20% of cementitious materials mass; the fineness of BP showed limited effect on the improvement of compressive strength, meaning that raw boulders powder could be used in the project which is economical and environmentally friendly.

(3) Through the adjustment of different water-to-binder ratio, BP could be used to prepare concrete with different strength grades to meet the needs of engineering; and the composite of BP and traditional mineral admixtures FA or GGBS could significantly improve the strength of concrete, while the positive effect of the ternary composite of BP, GGBS and FA was better.

References

1. F. Dif, T. H. Douara, R. Zaitri, and M. Mouli, J. Build. Eng. **32**, 101835 (2020).
2. S. E. Mohammadyan-Yasouj and A. Ghaderi, Constr. Build. Mater. **252**, 119115 (2020).
3. R. Yang, R. Yu, Z. Shui, X. Gao, X. Xiao, D. Fan, Z. Chen, J. Cai, X. Li, and Y. He, J. Clean. Prod. **258**, 120673 (2020).
4. R. Choudhary, R. Gupta, R. Nagar, and A. Jain, Constr. Build. Mater. **269**, 121282 (2021).

https://doi.org/10.1051/e3sconf/202129302023

5. Q. D. Nguyen, A. Castel, T. Kim, and M. S. H. Khan, Cement. Concrete. Res. **139**, 106265 (2021).
6. S. Mundra, V. Agrawal, and R. Nagar, J. Build. Eng. **32**, 101534 (2020).
7. A. Alnuaim, Y. M. Abbas, and M. Iqbal Khan, Constr. Build. Mater. **274**, 121245 (2021).

 https://doi.org/10.1051/e3sconf/202129302024

The Study of Wind Turbine Blade Fatigue Test

Jiaqi Zheng[1,*], Qiming Yu[1], Botao Zhu[1], Changqing Wu[1], Yiling Huang[1], Haoliang Sa[1], Mengwei Wu[1], Shanqing He[2]

[1]Civil engineering and architectural institute, Wuhan Polytechnic University, Wuhan 430023, Hubei

[2]Wuhan Zhongjiao Qingchuan road and Bridge Technology Co, Ltd, Wuhan 430050, Hubei

Abstract. The fatigue test of wind turbine blade is an important means to verify the fatigue life of wind turbine blade. This paper analyses the problems existing in the fatigue test of wind turbine blade, focuses on the excitation mode, the relationship between excitation amplitude and vibration frequency and the vibration principle, and puts forward feasible solutions in practical operation.

1 Introduction

The design life of wind turbine blade is generally 20 years. How to verify its fatigue life in the blade design stage has been the focus and difficulty of blade design. It is well known that fatigue damages are one of the main causes of blade damage, which seriously affects the service life and safe operation of wind turbine blades. Since the theoretical method of blade fatigue life analysis is not yet mature, the results of theoretical calculation alone are often quite different from the actual results. Therefore, the fatigue test of wind turbine blade is a key link to test its life and quality. At the same time, the measured data are compared with the design data, which can effectively to verify the reliability of the theoretical analysis[1].

This paper emphasis on the theoretical analysis of the vibration principle of wind turbine blade fatigue tests, providing theoretical basis and practical guidance for testers engaged in blade fatigue tests[2].

2 Mode of excitation

2.1 Single point excitation

Single point excitation refers to the excitation mode with only one loading point. The excitation system is installed on a certain section of the blade. The centrifugal force generated by the rotation of the eccentric wheel of the excitation system imposes alternating constant amplitude loads onto the blade. By adjusting the rotational speed of the eccentric wheel and the mass of the eccentric wheel, the rotational frequency is close to the natural frequency of the blade, and then the resonance is generated to achieve the excitation effect. The single point excitation is shown in Fig.1. Single-point excitation method has low requirements on test equipment and test cost, which is the main method for fatigue test for wind turbine blades in China[3].

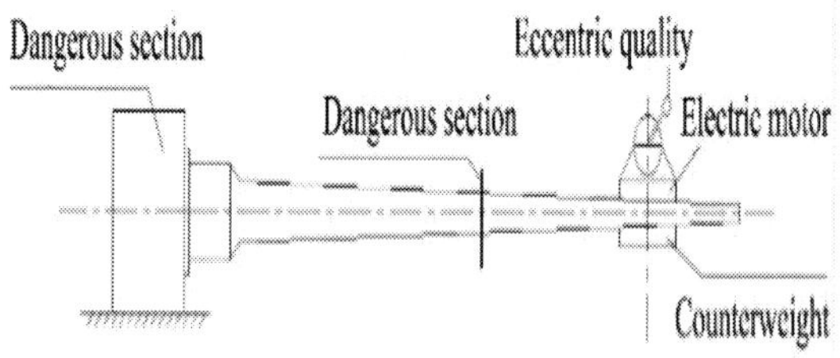

Figure 1. Diagram of single point excitation

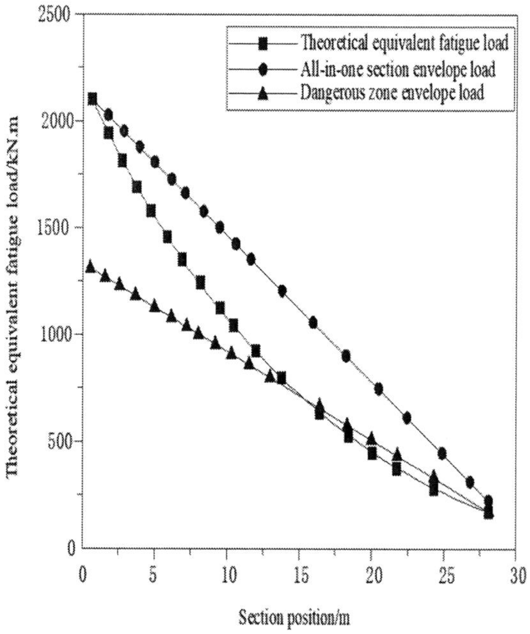

Figure 2. Load curves of each section under single point excitation

*Corresponding author: 1291689273@e-mail.org

 https://doi.org/10.1051/e3sconf/202129302024

Since there is only one loading point, the test load of each section is approximately linear with the section position, and the equivalent fatigue load of each section of the blade is concave curve with the section position. It can be seen from Fig. 2. that the difference between the test load of each section and the equivalent fatigue load is large. If the test load is enveloped from the root of the blade, the test load of the other sections will significantly exceed the equivalent fatigue load, and there is a risk that the service life of the blade cannot be effectively verified. Therefore, single-point excitation is generally used to conduct targeted tests on dangerous areas with small fatigue safety factors, or to verify the fatigue life of the blade in sections.

2.2 Multi-point excitation

Multi-point excitation refers to the test method of loading and exciting vibration through multiple loading points, mainly through hydraulic lifting equipment, and simultaneous loading and unloading of multiple sections of the blade at the same time to achieve the excitation effect. The schematic diagram is shown in Figure. 3. The multi-point excitation method requires high test equipment and test cost. At present, only a few foreign laboratories can carry out multi-point loading.

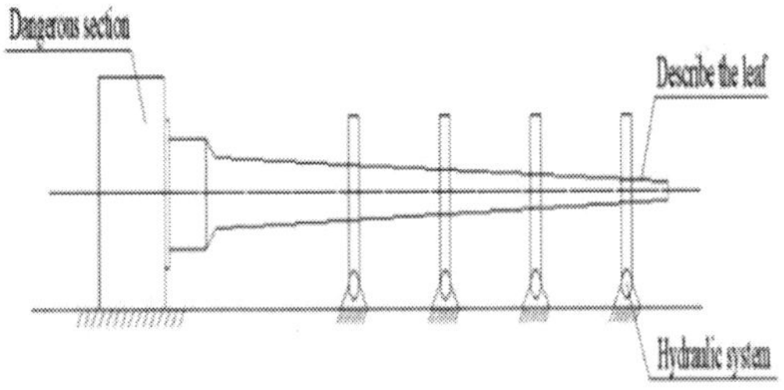

Figure 3. Schematic diagram of multi-point excitation

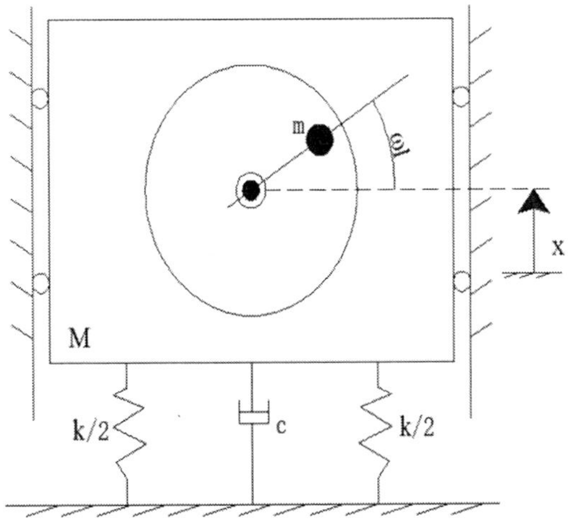

Figure 4. Mechanical model diagram of blade vibration

Because there are multiple loading points working together, and the position, load and vibration frequency of each loading point can be accurately adjusted, the test load of each section of the blade can be approximated to the equivalent fatigue load. Therefore, the multi-point excitation method is generally used for the equivalent fatigue load envelope test of the full section of the blade.

Multi-point excitation requires high accuracy on the test site and equipment, and it takes a long time to debug before the test. According to statistics, multi-point excitation tests requires at least one year for foreign experienced testing institutions, while single-point excitation only needs 1 year. The test can be completed within half a year from the preparation time of 2 months. This article analyses and studies the single-point excitation method commonly used in China.

3 Relationship between amplitude and number of vibrations

According to the principle of equal damage, using the S / N curves formula in GL2010:

$$N = \left[\frac{R_{k,t} + |R_{k,c}| - |2\gamma_{Ma}S_{k,M} - R_{k,t} + |R_{k,c}||}{2(\gamma_{Mb}/C_{1b})S_{k,A}}\right]^{10}$$

$$D = \sum \frac{n}{N}$$

After the transformation:

$$S_{KA} = \left[\frac{R_{k,t}+|R_{k,c}| - 2\gamma_{Ma}S_{k,M} - R_{k,t} + |R_{k,c}||}{[2(\gamma_{Mb})]N^{\frac{1}{10}}}\right]$$

Where "n" is the number of test vibrations; "N" is the number of vibrations allowed; "D" is fatigue damage; $S_{KA}$is the amplitude stress of dangerous section;$S_{k,M}$is the mean stress of dangerous section;$R_{k,t}$is the material partial coefficients.SoS_{KA}、$S_{k,A}$、$S_{k,M}R_{k,t}$、$R_{k,c}$are constants.

The relation between amplitude and vibration times is obtained by conversion as follows.

$$S_{KA^{10}} = \frac{c}{n} \quad (1)$$

It can be seen from the above equation that if the test amplitude is small, the vibration number increases exponentially. If the test amplitude cannot meet the requirements, the test time will be increased, which seriously affects the test progress and greatly increases the test cost. Therefore, how to effectively control the amplitude to reach the design value is the premise of whether the fatigue test can proceed smoothly.

4 Principle of amplitude regulation

According to the amplitude control principle of vibration mechanics analysis, the mechanical model of blade vibration is established by assuming that the fixed mass of the blade and the surge system is "M", the mass of the eccentric wheel is "m", the eccentricity is "e", the rotational angular velocity of the rotor is "ω", and the blade is regarded as two parallel springs with stiffness

"$k/2$". The damping in the vibration process is simplified as the effect of the damper with damping coefficient "c" in parallel with the spring on the system. As shown in Fig.5.

According to the literature.

Frequency-ratio is.

$$\lambda = \frac{\omega}{\omega_n} \tag{2}$$

The natural frequency of the system without damping are.

$$\omega_n = \sqrt{\frac{k}{M}} \tag{3}$$

The relative damping coefficient is.

$$\xi = \frac{c}{2\omega_n M} \tag{4}$$

Phase difference is.

$$\phi = tan^{-1}\frac{2\xi\lambda}{1-\lambda^2} \tag{5}$$

The amplitude of steady-state vibration is. $B = \frac{me}{M} \cdot \frac{\lambda^2}{\sqrt{(1-\lambda^2)^2+(2\xi\lambda)^2}}$ (6)

The curve families of phase difference "ϕ" and frequency "$\frac{MB}{me}$" ratio under different damping coefficients "ξ" drawn by Formulas are called phase frequency response curve and amplitude frequency response curve.

According to the amplitude-frequency response curve, when $\lambda = 1$ reaches resonance, the maximum amplitude are.

$$B = \frac{me}{2\xi M} = \frac{me\sqrt{k}}{c\sqrt{M}} \tag{7}$$

According to the phase-frequency response curve, the phase difference at resonance is 90° . Relationship between Blade Position and Rotary Arm Position is shown in Fig. 5.

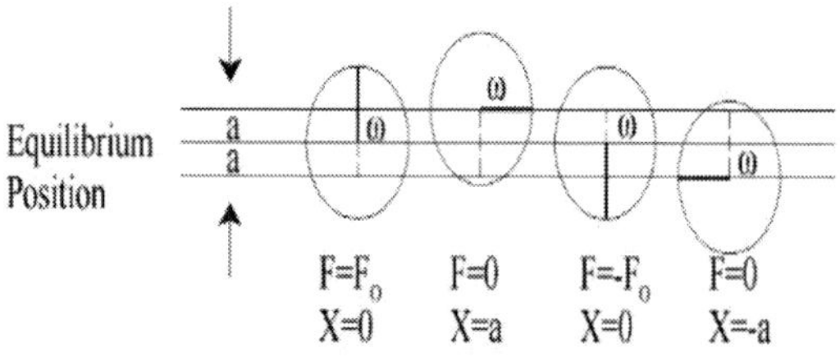

Figure 5. Relationship between Blade Position and Rotary Arm Position

Based on the above analysis, conclusions can be drawn. When the excitation point position is constant, namely, "k" is constant, the maximum amplitude of the blade is proportional to the mass "m" of the eccentric block, proportional to the cantilever length of the eccentric wheel "e", inversely proportional to the mass square $\sqrt{M}$ of blade and excitation system, and inversely proportional to the damping coefficient "c".

When the frequency of the motor is close to the natural frequency of the blade and the excitation system, the amplitude changes sharply.

When the blade reaches the maximum amplitude, the arm position is horizontal, and the blade reaches resonance.

5 Amplitude controls method

According to the above analysis and combined with the engineering practice, the following measures are taken. Adjust to increase the amplitude:

Adopt high-power motor, increase the mass m of eccentric block.

Under the premise of constant power, light motor is used to reduce the fixed mass M.

Saw off the tip part, reduce the fixed mass Y, and reduce the damping C.

High precision frequency converter is used to get closer to the resonant frequency.

The practice shows that the above methods can effectively adjust the blade amplitude and meet the requirement of theoretical amplitude.

6 Conclusion

Multi-point excitation is mainly used for the test of the full section envelope theory load, while single-point excitation is mainly used for the targeted test of the dangerous area, and has low requirements on the test equipment and cost. Therefore, it is the main method for the fatigue test of wind turbine blades in China[4].

The number of vibration is inversely proportional to the amplitude of 10 times. The premise of the fatigue test is to effectively adjust the amplitude adjustment.

According to the analysis results of vibration mechanics, the maximum vibration amplitude of the single-point excitation system on the blade is $B = \frac{me}{2\xi M} = \frac{me\sqrt{k}}{c\sqrt{M}}$. The fixed mass M and damping C can be adjusted by changing the motor power, counterweight mass, sawing off the blade tip, etc. At the same time, the high precision frequency converter is used to effectively control the amplitude, reaching the ideal level of fatigue tests[5].

Acknowledgement

This work was supported by the research of Wuhan Polytechnic University student scientific research project 2021(Funds, xsky2021138) and the research project of Wuhan Polytechnic University (NO 2021Y44)

References

1. Shi Kezhong, Zhao Xiaolu, Xu Jianzhong. Fatigue test of large wind turbine blades [J]. Journal of Solar Energy, Vol.32, No.8, PP.1264-1267, 2001
2. Lin Yuandong , Dynamic study on vibration of rotating system caused by eccentric mass[J]. Mechanical design and manufacture, No.5, PP.29-30, 2008
3. Wind turbine generator systems Part 23: Full-scale structural testing of rotor blades. EC61400-23[M]
4. Verein Deutsche Ingenieure. VDI2230 Part 1[M]
5. Wu Jianzhong, Ma Yi, Multi-point excitation fatigue loading test analysis of fan blade.2017, (12), 158-160+186
6. LLOYD G . Guideline for the certification of wind turbines edition 2010 [S]. Hamburg: Germanischer Lloyd, 2010.
7. VDI2230: Systematic calculation of high duty bolted joints Joints with one cylindrical bolt-Part 1 , 2003 [Z]. Berlin: Beuth Verlag GmbH, 2003.
8. Baihuichao, Su Shaochang, Zhang Wenwei, etc. Typi cal parameters for wind turbine blade embedded leaves effect of root connection bolts on bearing cap acity [J]. FRP / composite material, 2019 (9) : 38-43.
9. Pu Lianggui, Ji Gang. Mechanical Design (8th Editi on) [M]. Beijing : Higher Education Press, 2006.
10. Pretightening force and fatigue strength of mechanic al bolt connection [J]. Forging Machinery,1996,96 (4) : 44-46.
11. Zhao Shaobian, Wang Zhongbao.Anti-fatigue design methods and data [M].Beijing : Mechanical Engine ering Industry Press, 1997

Study on Crack Resistance of Hydraulic Roller Compacted Concrete Dam with Fly Ash Admixture

Xiaogang Li[1]

[1]China Northwest Water Conservancy & Hydropower Engineering Consulting Co., Ltd., Xi 'an, 710100, China

Abstract. With the continuous improvement of the level of construction industry in modern industry, at the same time, the requirements of the construction industry for the crack resistance of concrete dams are also higher and higher. In the preparation stage of the experiment, the corresponding experimental materials and equipment are provided. In order to ensure that the concrete can meet the requirements of workability and strength grade, the proportion of materials is designed; During the experiment, the activity characteristics of fly ash admixture are extracted, the particle size distribution of admixture is tested, and the fracture data of hydraulic RCC dam is monitored; The experimental results show that when the ratio of cement and fly ash is 0.7:0.1, the number, width and area of concrete cracks are the minimum, which proves that the fly ash admixture under this ratio is more suitable for construction engineering.

1 Introduction

With the development of concrete science, it is no longer difficult to prepare concrete with high strength and good working performance[1]. People gradually shift the focus of research to the durability of concrete. Whether concrete can be used as one of the most important building structural materials for a long time depends on whether it can become a high durability and environment-friendly green material [2]. The early cracking of concrete structure is a common phenomenon in engineering. The practice shows that the early cracking phenomenon of concrete has become a research hotspot in the field of concrete technology at home and abroad. The existing researches mainly focus on the volume stability and cracking of hardened concrete, and have made many achievements. The crack resistance of concrete is one of the most important performance indicators, which directly affects its application. Improving the crack resistance of concrete has always been the goal pursued by people [3]. How to improve the crack resistance of concrete is a very important project, which involves many aspects such as design scheme, construction technology, material selection, etc. the concrete mix proportion design is the basis of durability research, and the design objective of concrete mix proportion has changed from single strength index design to multi-objective design of workability, strength, crack resistance and durability.

2 Experimental preparation

2.1 Preparing experimental materials and equipment

Concrete is a general term for composite materials that aggregate is cemented into a whole by cementitious materials. According to the different cementitious materials can be divided into cement concrete, asphalt concrete, polymer concrete, etc. Among them, cement concrete is the most widely used and used building material in construction engineering[4]. The application of cement concrete can be traced back to the invention of Portland cement in the early 19th century. In the early stage, due to the need of mixing and pouring performance, the cement concrete with high water cement ratio and low strength was mostly used in engineering. However, the weakness of low tensile strength and poor crack resistance of concrete has not been improved[5]. In addition, compared with the low strength concrete, the hydration of high strength concrete consumes more water, and the water diffusion coefficient is smaller, so the internal humidity is more uniform, and the degree of uneven shrinkage is smaller[6] . C30, C50 and C80 are selected as the research objects, which represent ordinary concrete, medium high strength concrete and high strength concrete respectively. The experimental equipment is shown in Table 1.

* E-mail: Lxg0312@126.com

Table1. Experimental equipment

number	name	Manufacturer	model number
1	Microcomputer controlled electronic universal testing machine	China Building Materials Equipment Corporation	RGM-100A
2	Automatic mercury porosimeter	Shanghai future experimental equipment Co., Ltd	AutoPoreIV9500
3	Automatic elastic modulus tester	Hebei Dahong Test Instrument Co., Ltd	DTW18
4	scanning electron microscope	Hitachi Manufacturing Co., Ltd	S-4800
5	Concrete single side freeze thaw testing machine	Yingbei'er (Tianjin) measurement and control equipment Co., Ltd	IMYD-10
6	Forced single horizontal shaft concrete mixer	Shanghai Luda Testing Instrument Co., Ltd	SJD
7	Cement concrete standard constant temperature and humidity curing box	Hebei Dahong Test Instrument Co., Ltd	YH-40A
8	Electric blast drying oven	Tianjin taist Instrument Co., Ltd	101-3AB
9	Medical cryopreservation box	Qingdao Haier Special Electric Appliance Co., Ltd	DW-25W388
10	Vertical omnidirectional planetary ball mill	Nanjing NANDA Instrument Factory	QM-QX

The cement used in the experiment is P · O42.5R ordinary portland cement produced by Dalian cement plant, and the cement is placed in a dry and ventilated place to ensure that the cement will not be affected by moisture and caking[7-8]. In the process of concrete configuration, adding admixtures can improve the performance of concrete and reduce the cost.

2.2 Proportion of design materials

In this test, self mixing pumping concrete is required, and the slump is up to 200mm ±20mm. According to the mix proportion design method of ordinary concrete, the initial mix proportion is calculated, adjusted and trial mixed for many times[9-10], and finally three kinds of strength grade concrete mix proportion with strength value and slump value meeting the requirements are obtained, as shown in Table 2:

Table2. Concrete ratio

Dosage per cubic meter (kg/m^3), proportion	C30	C50	C80
Cement + fly ash	354+86 0.7:0.3	352+44 0.7:0.1	387+94 0.8:0.2
Sand	721 0.35	716 0.3	706 0.4
gravel	1082 0.6	1082 0.6	1058 0.5
water	220 0.4	220 0.4	198 0.4
Water reducing agent	4.6 1%	5.71 1.3%	7.728 1.5%
Slump	212	200	210
7d strength	20	29	30
28d strength	33	44	56

According to the standard test method, the average 28 day cube compressive strength of C30, C50 and C80 concrete is 33.48MPa, 48.36MPa and 87.33MPa respectively, meeting the strength requirements.

3 Experimental process

3.1 Extraction of activity characteristics of fly ash admixture

The core of fly ash has the following basic micro morphology: 1. Spherical particles, including floating beads, composite beads, iron rich micro beads and other basic components. The particle has a more regular external performance, the actual size has a significant difference, the comprehensive surface maintains a relatively smooth shape, belongs to one of the core components of the special dry discharged fly ash. 2. Irregular porous glass particles, the core of which is the related vitreous material, showing a special basic form of sponge and honeycomb .3. The obtuse angle core belongs to the special residue of related quartz particles, and there is no basic hydration activity. 4. Fine particles, which belong to relatively small existence, the core belongs to the basic debris of special particles and special aggregates of various particles. 5. Carbonaceous particles, which belong to related regular porous particles, are easy to be broken and then evolve into related porous debris.

3.2 Testing the particle size distribution of admixtures in hydraulic rolling

The type and size distribution range of admixtures can affect the related performance of concrete. Therefore, the reasonable selection of admixtures is one of the important preconditions to improve the performance of concrete. Use the sampling spoon to take the right amount of admixture in the beaker with the right amount of distilled water, and use the glass rod to stir, so that the diatomite can be evenly dispersed in the water. The formula for calculating the linear expansion coefficient of admixtures is as follows:

$$Q = D \cdot f\left(\frac{1}{k}\right) + \Delta\varepsilon \qquad (1)$$

In formula (1), D is the elastic modulus, f is the ultimate tensile value, k is the dry shrinkage value, and ε is the autogenous volume deformation value. But at the same time, the increase of concrete strength and elastic modulus will weaken the creep ability. When the ambient temperature changes, the formula of particle size change is as follows:

$$R = \frac{v\left(\varphi_i + m_j\right)}{t} \qquad (2)$$

In formula (2), v is the tensile strength of the admixture, φ is the tensile stress of the admixture, m is the tensile strength, t is the stress time, and i, j is the maximum and minimum deflection coefficient respectively. Generally, the higher the ambient temperature of concrete, the greater the creep.

3.3 Monitoring fracture data of hydraulic RCC dam

In the constrained ring test, three acoustic sensors are spaced 120 apart ° .It is mounted on the outer or upper surface of the annular specimen. When the shrinkage of concrete is restrained by the steel ring, the tensile strain is generated in the concrete, and then the elastic wave is released. With the continuous shrinkage, the elastic wave increases gradually. When the microcracks are formed and aggregated, the elastic wave increases rapidly. When the micro fracture evolves into a macro fracture, the elastic wave increases suddenly, which shows a sudden rise of the elastic wave curve. In order to quantitatively analyze the influence of specimen size and material parameters and calculate the crack displacement of concrete, the mathematical expression is as follows:

$$\beta = \frac{W - W\left|Q_s\right.}{p} \times 100\% \qquad (3)$$

In formula (3), W is the shrinkage time of concrete, Q is the displacement ratio of concrete, s is the restraint degree, and p is the shrinkage strain value. In the newly poured concrete, the water is filled in the gap between the cementitious material and the aggregate. In the process of concrete curing, the moisture is consumed by the hydration reaction of cement. At the same time, the moisture diffuses to the dry surface and loses to the environment through the dry surface until the environmental humidity keeps balance with the internal humidity of concrete. According to the water diffusion theory of concrete dam, the moisture presents obvious non-linear distribution along the direction of water diffusion, and the area near the dry surface has fast water loss and large drying shrinkage. In contrast, the area far away from the dry surface of the dam has slow water loss and small drying shrinkage. The non-uniform shrinkage along the direction of water diffusion caused by this will make the area near the dry surface appear larger tensile stress, while the area far away from the dry surface appear smaller tensile stress or even compressive stress. In the newly poured concrete dam, water is filled in the gap between the cementitious material and the aggregate. In the process of concrete dam maintenance, the water is consumed by the hydration reaction of cement. At the same time, the water diffuses to the dry surface and loses to the environment through the dry surface until the environmental humidity keeps balance with the internal humidity of concrete. Among the fracture properties of concrete, fracture energy is a key parameter. The formula of fracture energy is as follows:

$$F = \frac{U_0 + \frac{1}{2} l \cdot \theta_{max}}{y} \qquad (4)$$

In formula (4), U_0 is the integral area under the displacement curve of the loading point, l is the weight of the concrete, θ is the maximum displacement of the dam loading point measured, and y is the ligament area of the three-point bending beam specimen. According to the theory of concrete moisture diffusion, the moisture presents obvious non-linear distribution along the direction of moisture diffusion, and the area near the drying surface has fast moisture loss and large drying shrinkage. In contrast, the area far away from the dry surface has slow water loss and small drying shrinkage.

4 Analysis of experimental results

The length of the crack is measured with a steel ruler, and the straight-line distance between the two ends of the crack is taken as the crack length. When there are two cracks on a knife edge, the length of the two cracks is added and converted into one crack. The crack width is measured by a comprehensive crack measuring instrument with a magnification of 50 times, and the maximum width of each crack is measured. The average cracking area of concrete specimen is calculated according to the crack data measured 24 hours after concrete pouring. The formula of cracking area of concrete crack is as follows:

$$\eta = \frac{1}{2N}\sum_{i=1}^{N}\left(E_i \times M_i\right) \qquad (5)$$

In formula (5), N is the total number of cracks, E is the flat area, M is the maximum width of cracks, and i is the i -th crack. The experiment was carried out at (22 ± 1) ℃.The relative humidity was (65 ℃) ± 3) In a constant temperature and humidity chamber; After the concrete is poured into the mold, the concrete is leveled and the surface is slightly higher than the mold frame. Put the mold with concrete on the vibration table to control the vibration time to prevent over vibration and under vibration. After vibration, the concrete surface shall be leveled with trowel, and the aggregate shall not be exposed, and the surface shall be smooth. After the specimen is molded for 30min, adjust the position of the fan and the wind speed immediately, so that the wind speed at 100 ~ above the center of the specimen surface is (4 ± 0.5) m/s. The wind direction should be parallel to the specimen surface and crack inducer. The experimental time was calculated from the beginning of concrete mixing and adding water ± 5) H. According to the calculation results, the experimental results are shown in Figure 1-3:

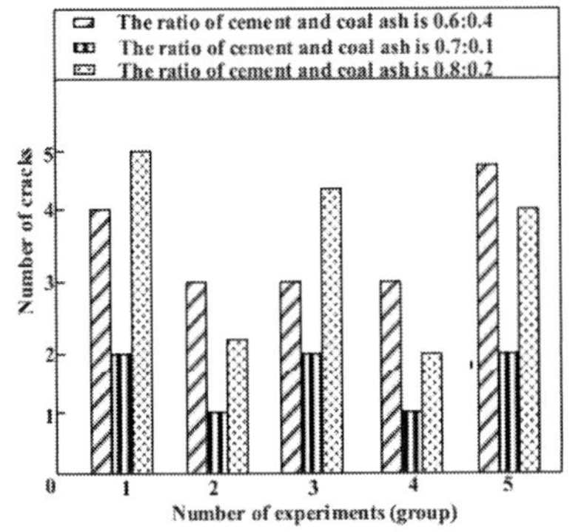

Figure 1. Number of concrete cracks

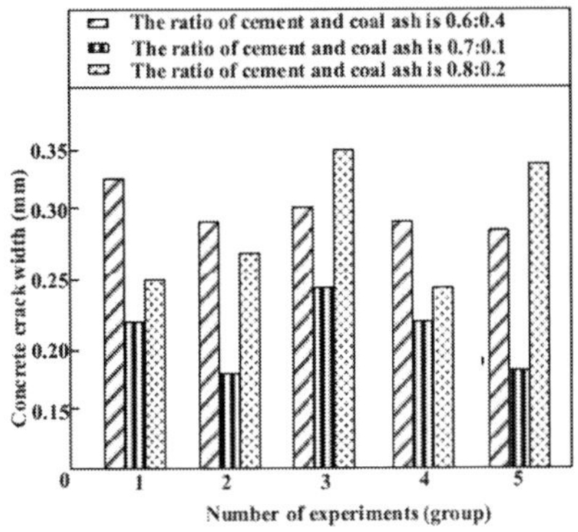

Figure 2. Width of concrete crack（mm）

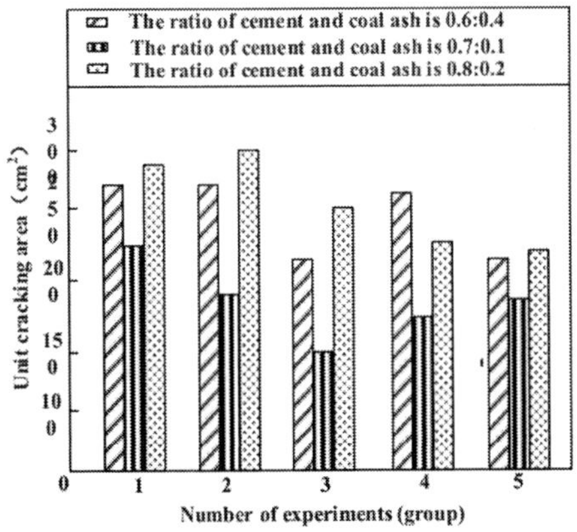

Figure 3. Cracking area of concrete（cm^2）

It can be seen from Figure 1 to figure 3 that when the ratio of cement and fly ash is 0.6:0.4, the average number of cracks is about 3, the average width of cracks is 0.334mm, and the cracking area is 234.16cm^2; When the ratio of cement and fly ash is 0.7:0.1, the average number of cracks is about 2, the average width of cracks is 0.206 mm, and the cracking area is 165.09 cm^2; When the ratio of cement and fly ash is 0.8:0.2, the average number of cracks is about 3, the average width of cracks is 0.328mm, and the cracking area is 229.31cm^2; The experimental results show that when the ratio of cement and fly ash is 0.7:0.1, the crack resistance of concrete is the best.

5 Conclusion

This paper studies the crack resistance of hydraulic roller compacted concrete dam with fly ash admixture, and obtains the corresponding experimental results, which enriches the academic literature on admixture and concrete. Due to the limited research conditions, the application research of other types of admixtures in concrete is not thorough enough, and will continue to improve in the future.

References

1. Thang N C， Tuan N V， Yang K H，et al. Effect of Zeolite on Shrinkage and Crack Resistance of High-Performance Cement-Based Concrete[J]. Materials, 2020, 13(17):3773.
2. Mizobuchi T， Ishizeki K， Sagawa T，et al. Study on the Influence of Minor Constituents in Blast Furnace Slag Rich Cement on the Thermal and Mechanical Properties of Concrete[J]. Journal of Advanced Concrete Technology, 2019, 17(1):46-61.
3. Weber M ， Thiele C . Correlation between compressive and tensile strength of old concretes: Applicability of the relationship described in EN 1992-1-1[J]. Structural concrete, 2019, 20(1):483-492.
4. Hassan A ， F Khairallah， Elsayed H ，et al. Behaviour of concrete beams reinforced using basalt and steel bars under fire exposure[J]. Engineering Structures, 2021, 238(3):112251.
5. Autrup F， Joergensen H B . Shear capacity of RC members without shear reinforcement: A modified crack sliding model[J]. Engineering Structures, 2021, 239(6):112147.
6. Heek P， Look K， Oettel V，et al. Bemessung von Stahlfaserbeton und stahlfaserbewehrtem Stahlbeton[J]. Beton- und Stahlbetonbau, 2021, 116(S1):2-12.
7. Li S， Zhao X . Automatic Crack Detection and Measurement of Concrete Structure using Convolutional Encoder-Decoder Network[J]. IEEE Access, 2020, PP(99):1-1.
8. Ding Y， Hao X， Men X . Effect of fiber on the crack width, tortuosity and permeability of

https://doi.org/10.1051/e3sconf/202129302025

concrete[J]. Fuhe Cailiao Xuebao/Acta Materiae Compositae Sinica, 2019, 36(2):491-497.

9. X Fan, Liu J , Hu S , et al. Influence of FRP Bonding Length on Fracture Parameters of Concrete Three-Point Bending Beam[J]. Jianzhu Cailiao Xuebao/Journal of Building Materials, 2019, 22(1):38-44.
10. Zhu Z , Tan G , Zhang W , et al. Preliminary Analysis of the Ductility and Crack-Control Ability of Engineered Cementitious Composite with Superfine Sand and Polypropylene Fiber (SSPP-ECC)[J]. Materials, 2020, 13(11).

Screening and Preliminary Identification of Soil Potassium and Phosphorus-hydrolyzing Bacteria WP1-3

Zhao Yongmei[1*], Wu Panpan[1]

[1]School of Biological and environment technology, Xi'an University, Xi'an 710065, China

Abstract. In order to screen the strains with Potassium and phosphorus-releasing bacteria from the soil, the potassium-dissolving medium was used for preliminary screening, and the transparent circle method was used for re-screening. The content of soluble potassium was determined by sodium tetraphenylborate method, and the content of phosphorus was determined by anticolorimetric method of molybdenum and antimony. The strains were identified by morphological, biochemical and molecular biology. The results showed that ten strains with potassium-solubilizing function were obtained by preliminary screening of potassium solubilizing medium. The re-screening by transparent circle method showed that WP1-3 strain had the strongest potassium-solubilizing ability, with D/d value of 1.33. WP1-3 strains of decomposition of soluble potassium content is 22.918 mg/L, decompose inorganic phosphorus and organic phosphorus ability is 4.443 mg/L and 22.408 mg/L, respectively. Identifying morphological and biochemical identification showed that WP1-3 strains was gram-negative bacteria, not produce spore, the nitrogen content of WP1-3 strain increased after detection and culture for 3d and 7d were 10.825 and 38.521 mg/L. The 16SrDNA size was 1351bp, the ribosomal database comparison showed that WP1-3 strain was 99.8% similar to Rhizobium. Therefore, the WP1-3 strain belongs to the rhizobia genus.

1 Introduction

Potassium and phosphorus are important nutrient elements in plants and are biological nitrogen fixation must element, also are the key to promote the metabolism of plant growth. Potassium and phosphorus play an important role in regulating cell osmosis, material transport, promoting enzyme activation, promoting photosynthesis, improving carbon dioxide assimilation rate, promoting the synthesis and transport of substances in crops, and enhancing the resistance and tolerance of plants. Most of the potassium and phosphorus in soil are found in minerals such as silicate aluminates and apatite. Therefore, 95% of the potassium and phosphorus in soil cannot be directly absorbed by plants [1].

Potassium-hydrolyzing bacteria and phosphorus-hydrolyzing bacteria are a kind of bacteria isolated from soil that can decompose aluminosilicate and apatite minerals. As microbial fertilizers, they can decompose insoluble silicate aluminates such as potassium feldspar and apatite. They can promote the transformation of insoluble nutrients such as potassium, phosphorus and magnesium into soluble nutrients, and increase the content of available nutrients in soil, promote crop growth and development, improve crop yield [2]. Therefore, it is of great significance to separate the bacterial strains with potassium-releasing and phosphorus-releasing effects from soil and prepare microbial fertilizer to improve the utilization rate of potassium-phosphorus in soil. This paper studied the process of extraction and identification of potassium-releasing strains from soil, which provided reference for the development of new microbial fertilizer.

2 Materials and Methods

2.1 Material

2.1.1 Sample materials

Take 100g of soil samples below 10cm of surface near Xi'an University, mix samples well, dry them and stored.

2.1.2 apparatus

Fluorescence microscope (OLYMPUS-CX31), Super clean workbench, Electronic balance, Vertical automatic electric pressure steam sterilizer (LDZX-40II-type), Water-jacket heating constant temperature incubator (PYX-OHS-40x-0), Spectrophotometer, Digital constant temperature water bath oscillator (SHA-C) and others were used.

2.2 Methods

2.2.1 Preparation of medium

* Corresponding author: xazym1118@163.com

Nutritional AGAR medium: peptone 10g, beef extract 3g, NaCl 5g, AGAR 15g, distilled water to 1000mL, pH7.2-7.4, sterilized and stored.

Potassium bacterial culture medium: glucose 10g, $CaCO_3$ 5g, Na_2HPO_4 0.2g, $MnSO_4 \cdot 7H_2O$ 0.2g, $CaSO_4 \cdot 2H_2O$ 0.2g, NaCl 0.2g, AGAR 18g, potassium feldspar 10g, distilled water to 1000mL, pH7.2, sterilized and stored.

Potassium bacterial liquid medium: sucrose 0.75g, potassium feldspar 10g, $(NH_4)_2SO_4$ 0.15g, Na_2HPO_4 0.30g, distilled water to 1000mL, pH7.2, sterilized and stored.

Nitrogen-free liquid medium: sucrose 50g, KH_2PO_4 2g, $MnSO_4 \cdot 7H_2O$ 0.5g, $FeCl_3$ 0.005g, $CaCO_3$ 0.1g, distilled water to 1000mL, pH7.0-7.5, sterilized and stored.

2.2.2 Primary screening and purification of strains

10g of soil samples were through a 100-mesh sieve and put into 90mL sterile water. After standing for 30min, they were shaken at 100rpm in a shaker at room temperature for 30min to get a 10^{-1} times diluted suspension. 0.1mL of 10^{-5} times diluted soil suspension was absorbed with sterile tubes and inoculated on potassium bacteria plate medium. After 5-7d of culture at 30℃, the single colonies with transparent circles were isolated and purified. After pure culture, the strains were transferred to nutrient AGAR medium, examined under microscope and numbered, then stored at 4℃.

2.2.3 Re-screening of potassium-hydrolyzing bacteria using transparent circle method

The Primary screened strains were inoculated on potassium bacteria solid plate medium and cultured at 28℃ for 5d. The potassium-lytic circle diameter (D/mm) and colony diameter (d/mm) of the strains producing transparent circles were measured with a ruler, and the ratio of colony diameter (D/d) was calculated to determine the potassium-lytic ability of the strains. The screened potassium-hydrolyzing bacteria with strong potassium-hydrolyzing ability were inoculated on nutrient AGAR medium and stored at 4℃.

2.2.4 Soluble potassium content of strain using sodium tetraphenylboron method

The strains were inoculated into the sterilized potassium bacterial liquid medium at the proportion of 3-6%, and the same volume of sterile water was used as the control. The seed liquid was prepared by shaking culture for 5-7d at 30℃ and 180rpm. The seed solution was centrifuged at 2000rpm for 20min in order to discard the bacteria and residue. Then, 1.0mL of formaldehyde-EDTA mask agent was added, shaken well, and 1.0mL of sodium tetraphyborate solution was added accurately and quickly with a pipetet. After standing for 15min, it was shaken again. According to the standard curve drawn by sodium tetraphyborate method, the zero point of the instrument was adjusted with blank seed reference solution, colorimetric determination was carried out, and the absorbance of OD at 420nm was recorded and the content was calculated.

2.2.5 Determination of soluble phosphorus in strains using Mo-Sb antimony anticolorimetric method

The strains were inoculated into 150mL liquid phosphorus fermentation medium with 3-6% inoculation amount, and cultured at 30℃ and 180rpm for 5-7d, with blank seed liquid as the control. Molybdenum-Sb storage solution, molybdenum-Sb antichromogener and 5mg/L phosphorus standard solution were prepared, and the phosphorus standard curve was drawn according to the molybdenum-Sb antichromogener method. The soluble phosphorus content of strain was determined, and the blank seed reference solution was used to adjust the zero point of the instrument to conduct colorimetric determination. The absorbance of OD at 700nm was recorded and the content was calculated.

2.2.6 Morphological identification and biochemical detection of WP1-3 strain

After the preserved strains were activated by the culture medium of potassium-digesting bacteria, the morphological characteristics of their colonies were observed, and then the colonies were selected for Gram staining to observe the morphology of the bacteria, and the biochemical identification of potassium-digesting bacteria was carried out according to the common manual of common bacteria.

2.2.7 16SrDNA identification of strain WP1-3

The identification process of the strain consisted of DNA extraction, PCR amplification, gel electrophoresis, purification and recovery, ligation, preparation of competent cells, transformation of ligation products, extraction and sequencing of plasmid, and comparison of 16SrDNA sequence on ribosome database after purification. The procedure was completed by Sangon Bioengineering Co., Ltd. (Shanghai). The reaction system of PCR was as follows: bacterial 16SrDNA universal primers 27 (AgTTTGATCMTGGCTCAG) and 1492 (GGTTACCTTGTTACGACTT) were selected to establish PCR amplification system for amplification. The PCR reaction system template (genomic DNA 20-50ng/nL) was 0.5μL, 10x buffer 2.5μL, dNTP (2.5mM) 1μL, enzyme 0.2μL, F(10μm) 1μL, R(10μm) 1μL and double steaming water to 25μL. The amplification procedures were as follows: 94℃ pre-denaturation for 4min, 94℃ denaturation for 45s, 55℃ annealing for 45s, 72℃ extension for 1min, 30 cycles in total, 72℃ extension for 10min, and the reaction products were terminated at 4℃.

2.2.8 Determination of nitrogen fixation capacity of WP1-3 strain

The strain WP1-3 was cultured in a nitrogen-free liquid medium for 3d and 7d, respectively. At the end, the nitrogen content in the medium was measured by microKjeldahl nitrogen determination method.

3 Results and Analysis

3.1 Preliminary screening and purification of strain

10 strains with strong potassium-hydrolyzing ability were screened out by culture on potassium releasing medium, which were WP1-1, WP1-2, WP1-3, WP2-1, WP2-2, WP3-3, WP3-2, WP4-1, WP4-2, etc.

3.2 Re-screening results

Transparent circle method is used to determine the releasing bacteria screening of soluble potassium circle diameter (D/mm) and colony diameter (d/mm), calculate the circle of soluble potassium and colony diameter ratio (D/d). According to the ratio, the (D/ d) of WP1-3 strain was 1.33, which was higher than that of other strains (Table 1). Therefore, the strain WP1-3 screened by the transparent circle method had the strongest potassium-hydrolyzing ability.

Table1. The potassium releasing ability of strains determined by transparent circle method

strain	Potassium-soluble circle diameter (D/mm)	Colony diameter (d/mm)	D/d
WP 1-1	67	57	1.33
WP 1-2	61	50	1.22
WP 1-3	69	52	1.33
WP 2-1	70	57	1.23
WP 2-2	63	55	1.15
WP 2-3'	68	57	1.19
WP 3-2	65	54	1.30
WP 3-3'	63	51	1.24
WP 4-1	66	55	1.20
WP 4-2	65	53	1.21

3.3 Potassium minerals decomposed

The standard potassium curve was y=0.0734x+0.002, R^2=0.9822. The determination results of potassium minerals decomposed by each strain are shown in Table 2. WP1-3 strain can decompose potassium minerals, and the decomposed potassium content is 22.591mg/L after comparing with the standard curve.

Table2. Results of potassium mineral decomposition

strain	Potassium content mg/L	strain	Potassium content mg/L
WP 1-1	12.632	WP 2-3'	18.325
WP 1-2	17.321	WP 3-2	20.262
WP 1-3	22.591	WP 3-3'	16.183
WP 2-1	18.115	WP 4-1	17.463
WP 2-2	16.921	WP 4-2	13.586

3.4 Phosphorus decomposition ability

The standard curves of inorganic phosphorus and organophosphorus were y=0.5731x, R^2=0.9992 and y=0.5417x, R^2=0.9958, respectively. Determination results of decomposition phosphorus content of each strain are shown in Table 3. Compared with the standard curve, the degradation of inorganic phosphorus and organic phosphorus by WP1-3 strain reached 4.443mg/L and 22.408 mg/L.

Table3. Results of phosphorus decomposition

strain	inorganic phosphorus content mg/L	organophosphorus content mg/L
WP 1-1	0.284	5.563
WP 1-2	0.676	0.389
WP 1-3	4.443	22.408
WP 2-1	0.115	0.546
WP 2-2	0.228	0.687
WP 2-3	0.109	10.356
WP 3-2	3.147	1.549
WP 3-3	1.086	3.896
WP 4-1	0.834	8.408
WP 4-2	2.456	0.789

3.5 Morphological observation and biochemical detection of WP 1-3 strain

According to its ability to dissolve potassium and phosphorus in soil, WP1-3 strain was screened out for morphological identification and biochemical detection. The results showed that WP1-3 colonies were round, rod-shaped bacteria were colorless and transparent, with neat edges, moist and viscous, and elastic. The results of Gram staining showed that WP 1-3 colonies were Gram-negative bacteria.

3.6 Identification of 16SrDNA strains of WP1-3

The 16SrDNA sequence analysis and comparison results of strain WP1-3 showed that the fragment size of the product amplified by PCR was 1351bp, and the 16SrDNA partial gene sequence obtained by sequencing (Seq ID

No.3) had 99.8% homology with the related fragment sequence of rhizobia published in GenBank.

Based on the results of biochemical identification and 16SrDNA sequence analysis, the strain WP1-3 of the present invention was identified as *Rhizobium sp*.

3.7 Nitrogen fixation ability of WP1-3 strain

The nitrogen content in the medium was measured by microKjeldahl nitrogen determination method. The results showed (Table 4) that the nitrogen content of WP1-3 strain increased after detection and culture for 3d and 7d.

Table4. Nitrogen content determination results

strain	3d nitrogen content mg/L	7d nitrogen content mg/L
WP 1-3	10.825	38.521

4 Conclusion and Discussions

Microbial fertilizers are also known as biofertilizers, in which the effective microorganisms play a key role in the fertilizer effect [3]. It has the function of improving soil structure and adjusting microecological balance. The effective microorganisms can convert organic or inorganic minerals that plants cannot or are difficult to use into substances that can be used. It can improve soil fertility by increasing the soil effective N, P, K content, can enhance plant resistance by producing various antagonists and beneficial metabolites, so it affects plant hormone balance and promotes plant growth. It can also replace chemical fertilizers, pesticides and herbicides and reduce the amount of water used. At the same time , it can reduce soil pollution, improve the quality of crops, increase the content of protein and other aspects of the role [4]. However, due to the lack of in-depth research on potassium-hydrolyzing bacteria and phosphorus-hydrolyzing bacteria, the kinds of microbial fertilizer are not rich and many problems cannot be solved [5].

The WP1-1 strain purified in this study is a kind of rhizobia which can be used to decompose potassium and insoluble phosphorus in soil. The strain originated from the soil and was ecologically safe. It was selected from potassium feldspar as potassium source, as the effective microorganisms, during the growth and reproduction process, it can convert insoluble K in silicate aluminates into soluble components and increase the source of P in soil by promoting the decomposition of insoluble phosphorus. Moreover WP1-1 strain also has the function of nitrogen fixation, provide the required nitrogen metabolism of plant growth.

To sum up, there is still a lot of work to be done to develop new microbial fertilizers with high efficiency, environmental protection and multi-function. In this study, the potassium-hydrolyzing microorganisms collected from the soil of Xi'an were isolated, and their phosphorus digesting effect was studied by the anticolorimetric method of molybdenum and antimony. According to the biochemical characteristics and 16S rRNA gene sequence analysis, the taxonomic status of the strains was determined. This study plays an important role in the development of microbial fertilizer made from phosphorus and potassium-hydrolyzing bacteria.

Acknowledgement

Sponsored by Xi'an Science and Technology Plan Project (No 2019KJWL03)

References

1. M.Andrews, S.Hodge, J.A. Raven, Ann App Biol 157 (2010)
2. Illmer P. Schinner F. Soil Biol Biochem (1992, 24)
3. Cordell D, Drangert JO, White S. Global Environmental Change 19 (2009)
4. Bin Wu,Zhen Yao Xia,Lin Lin Zhang,Shi Feng Guan,Ming Yi Li,Ke Yang,Wen Nian Xu. Advanced Materials Research 726 (2013)
5. U. Braun, W. Jehle. The Veterinary Journal,173 (2005)

Wind Resistant Analysis of 5000m Hyperbolic Parabolic Spatial Mixed Cable Suspension Bridge

ZhaoYang Chen[1*] and WenPing Xu[2]

[1]CSCEC AECOM CONSULTANTS CO., LTD, Lanzhou China,730000

[2]College of Civil Engineering, Southeast University, Nanjing China,210096

Abstract. Combined with the engineering background of a 5000-meter Strait Suspension Bridge, the configuration study of the hyperbolic parabolic space cable suspension bridge was carried out, and the ANSYS finite element analysis model was established to analyze and study the structural internal force and dynamic modal characteristics. The research shows that the hyperbolic parabolic space mixed cable suspension bridge has excellent spatial stiffness and wind stability performance, its torsional frequency and torsional frequency ratio are significantly improved, the critical wind speed of flutter is greatly improved. In order to further improve the wind stability of the hyperbolic parabolic space mixed cable suspension bridge, temporary wind resistance cable measures adopted during strong typhoons are proposed. This method can ensure that the 5000-meter-level hyperbolic parabolic space mixed cable suspension bridge has the ability to withstand 120 m / s rare extreme typhoons.

1 Introduction

As the span of the suspension bridge continues to increase, the structure tends to be softer, and its torsional frequency and torsion-bending ratio continue to decrease[1-2]. The key to the design is to check the stability against wind flutter under test wind speed [3-4].

The structural rigidity of the super-span suspension bridge mainly comes from the main cable[5-6]. Therefore, the key point to improve the overall rigidity of the suspension bridge structure should be placed on the main cable. The use of a three-dimensional spatial cable suspension bridge structure system can effectively improve the wind-resistant stability of super-span suspension bridges [7-8].

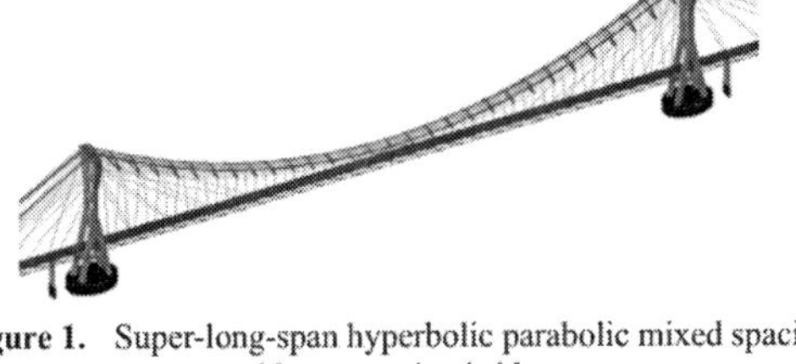

Figure 1. Super-long-span hyperbolic parabolic mixed spacial cable suspension bridge

In this paper, in conjunction with a 5,000-meter super-long-span suspension bridge, the configuration study of the super-long-span suspension bridge of the hyperbolic parabolic mixed space cable system was carried out. The ANSYS finite element analysis model was established to study the dynamic modal characteristics of the hyperbolic parabolic mixed space cable suspension bridge, and explore the advantages of the wind-resistant stability of the hyperbolic parabolic mixed space cable suspension bridge. This paper provides a theoretical basis for the construction of the super-long-span suspension bridge in the future.

2 Design Parameters

Figure 2. effect drawing of 5000 m hyperbolic parabolic spatial mixed cable suspension bridge

The super-span suspension bridge has a main span of 5000 meters, and adopts a single-span suspension bridge structure. The distance from the anchor to the tower is 1000 meters and the width of the deck is 60 meters. In order to meet the wind stability requirements of super-long-span suspension bridges, a new structural system of super-long-span suspension bridges with hyperbolic parabolic mixed space cable system was adopted.

The parallel steel wire rope adopts the prefabricated parallel wire strand method (PPWS method). The total length of the bridge is 5000m, and the vector span ratio is 1/11. The parallel steel wire rope sags 454.5 meters. The

*ZhaoYang Chen: 807105720@qq.com

parallel cable adopts the product of φ5.2mm high-strength steel wire cable of 2000MPa. A total of six steel cables are arranged on the whole bridge, and three cables are bundled together using a triangular method. Each main cable has a diameter of 1.6 meters.

The hyperbolic parabolic cable network adopts 4000MPa high-strength carbon fiber cables. The bridge has a total of 18 carbon fiber hyperbolic parabolic space cables with a diameter of 0.4 meters per main cable.

The curved steel beam with lower prestressed tension rods are set on the hyperbolic parabolic cable network and two parallel steel wire cables. The curved steel beam with a spacing of 120 meters and a cross-sectional size of 0.4 × 0.2 meters adopts the box-shaped arch structure. The wall thickness of the steel plate is 25mm, and the lower prestressed rods adopt two 750MPa high-strength prestressed steel rods with a diameter of 0.10 meters.

The distance between vertical wire slings is 40 meters. The steel wire sling adopts 1860MPa high-strength steel wire rope, and the diameter of the steel wire sling is 0.12m.

The curved steel beam is provided with diagonal carbon fiber suspension slings connected to the stiffening beam at a distance of 120 meters. The diagonal carbon fiber suspension slings adopt the 4000 MPa high-strength carbon fiber cable with a diameter of 0.12 meters.

The stiffening beam structure adopts a three-box streamlined steel box beam similar to the Messina Strait Bridge. The width of the stiffening beam is 60 meters, and the height of the stiffening beam is 15 meters.

The central area of the suspension bridge is provided with a concrete-filled steel tube arch central buckle to improve the integrity of the bridge and reduce the longitudinal displacement of the stiffening beam. The span of the arched central buckle is 150 meters and the height is 30 meters.

In this paper, ANSYS software was used to analyze the hyperbolic parabolic mixed space cable suspension bridge. A 5000m parallel cable suspension bridge model was established as a reference for comparative analysis.

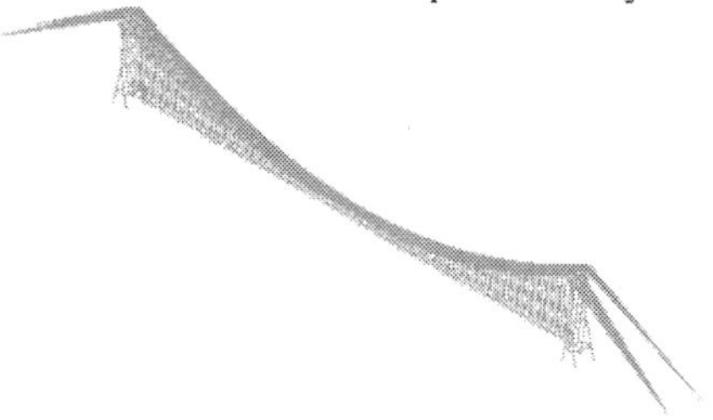

Figure 3. Hyperbolic paraboloid mixed space cable suspension bridge model

3 Dynamic modal analysis

The cable system of the space cable suspension bridge is complex, and it is difficult to find the shape automatically through the software. For the hyperbolic parabolic mixed space cable suspension bridge in this paper, MIDAS software was used to iteratively iterate to determine the bridge shape and cable initial stress. Then ANSYS was used for subsequent dynamic analysis. Fig.4 and Fig.5 show the representative vibration modes of the parallel suspension bridge and the hyperbolic parabolic mixed-space suspension bridge respectively. Table1 and Table2 show the vibration frequencies and modes of the two bridges.

Table1. Natural frequency and vibration mode of parallel cable suspension bridge

order	frequency (Hz)	Mode shape
1	0.023082	first-order positive symmetric lateral bending
2	0.036529	longitudinal drift
3	0.038249	first-order antisymmetric lateral bending
4	0.051164	first-order antisymmetric vertical bending
5	0.057903	lateral vibration of main cables
6	0.058462	positive symmetric lateral bending
7	0.065518	first-order positive symmetric vertical bending
8	0.072958	positive symmetric lateral bending
9	0.078812	lateral vibration of main cables
10	0.079882	first order positive symmetric torsion
11	0.080597	antisymmetric lateral bending
12	0.088853	positive symmetric vertical bending
13	0.097143	antisymmetric vertical bending
14	0.098884	first order antisymmetric torsion

Remarks: Parallel cable suspension bridge adopts traditional gate tower.
flexible central buckle for parallel cable suspension bridge.

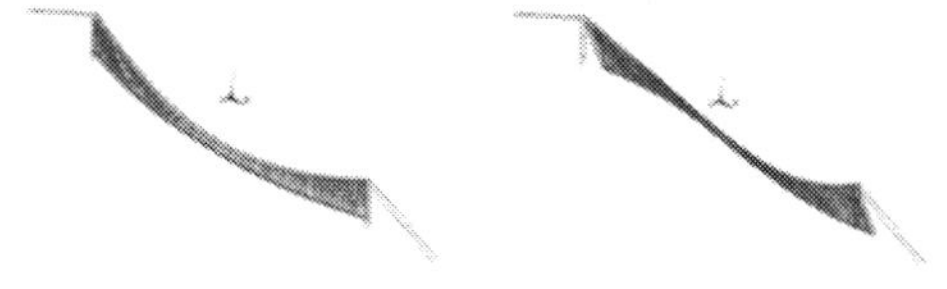

(1) 1st mode (0.02308Hz) (2) 2nd mode (0.03653Hz)
(Positive symmetrical lateral bending) (Antisymmetric lateral bending)

(3) 3rd mode (0.03825Hz) (4) 4th mode (0.05116Hz)
(Antisymmetric lateral bending) (Antisymmetric vertical bending)

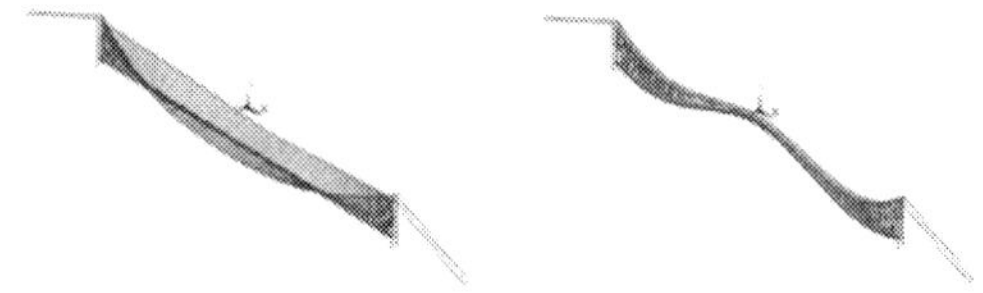

(5) 5th mode (0.05790Hz) (6) 7th mode (0.05846Hz)
(Lateral vibration) (Positive symmetric vertical bending)

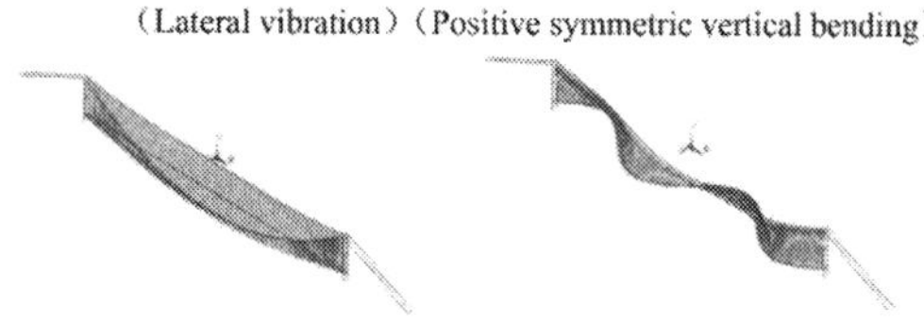

(7) 10th mode (0.07988Hz) (8) 14th mode (0.09888Hz)
(Positive symmetric torsion) (Antisymmetric torsion)

Figure 4. Representative vibration mode diagrams of parallel cable suspension bridge

Table2. Natural frequency and vibration mode of hyperbolic paraboloid mixed space suspension bridge

order	frequency (Hz)	mode shape
1	0.02786	first-order positive symmetric lateral bending
2	0.05487	first-order antisymmetric lateral bending
3	0.06769	first-order antisymmetric vertical bending
4	0.06912	first-order positive symmetric vertical bending
5	0.07170	positive symmetric lateral bending
6	0.07564	positive symmetric vertical bending
7	0.08017	antisymmetric lateral bending
8	0.08559	positive symmetric vertical bending
9	0.08560	antisymmetric lateral bending
10	0.08846	antisymmetric vertical bending
11	0.09374	antisymmetric vertical bending
12	0.09838	positive symmetric lateral bending
13	0.10340	positive symmetric lateral bending
14	0.10660	positive symmetric vertical bending
15	0.10828	antisymmetric vertical bending
16	0.12434	antisymmetric vertical bending
17	0.13448	antisymmetric lateral bending
18	0.14573	antisymmetric lateral bending
19	0.15824	antisymmetric vertical bending
22	0.18214	first-order positive symmetric torsion
29	0.25577	first-order antisymmetric torsion

Remarks: the space mixed cable suspension bridge adopts herringbone limb bridge tower.
Spatial mixed cable suspension bridge is provided with arched central buckle.

(1) 1st mode (0.02786Hz) (2) 2nd mode (0.05487Hz)
(Positive symmetrical lateral bending) (Antisymmetric lateral bending)

(3) 3rd mode (0.06769Hz) (4) 4th mode (0.06912Hz)
(antisymmetric vertical bending) (positive symmetric vertical bending)

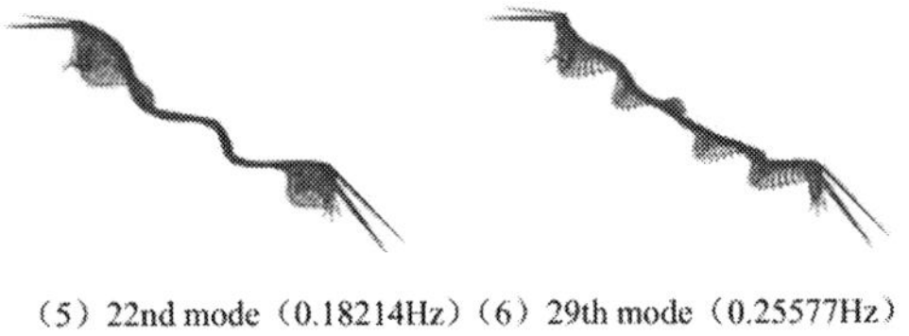

(5) 22nd mode (0.18214Hz) (6) 29th mode (0.25577Hz)
(Positive symmetric torsion) (Antisymmetric torsion)

Figure 5. Representative vibration mode diagrams of a hyperbolic parabolic mixed space suspension bridge

The above calculation results show that:

The first few vibration modes of the hyperbolic parabolic mixed space cable suspension bridge and the parallel cable suspension bridge are basically similar, and there are obvious vibration mode groupings. However, the space mixed cable suspension bridge adopts herringbone four-limbs bridge tower and is provided with an arch-shaped central buckle, which improves the longitudinal rigidity of the suspension bridge and avoids the longitudinal drift vibration mode.

The torsional modes of the parallel suspension bridge appear in the 10th and 14th orders, and the corresponding frequencies are 0.079882Hz (first-order positive symmetry) and 0.098884 Hz (first-order antisymmetric). The torsional vibration modes of the hyperbolic parabolic mixed space suspension bridge appear at the 22nd and 29th orders respectively, with corresponding frequencies of 0.18214Hz (first-order positive symmetry) and 0.25577 Hz (first-order antisymmetric).

The torsion-bending frequency ratios of the hyperbolic parabolic mixed space cable suspension bridge and the parallel suspension bridge are: hyperbolic parabolic space suspension bridge 2.69, parallel suspension bridge 1.56. Therefore, the hyperbolic parabolic mixed space cable suspension bridge has good torsion resistance.

4 Stability analysis of flutter

Flutter is a phenomenon of divergent instability of the structure under wind load. In this paper, the Selberg formula [11] was used to analyze the flutter stability of the suspension bridge.

$$V_{cr} = \eta_s \eta_\alpha \omega_t b_1 \sqrt{\frac{r}{b_1}\mu\left[1-\left(\frac{\omega_v}{\omega_t}\right)^2\right]} \qquad (1)$$

$$\mu = M/\pi\rho b_l^{\,2} \qquad (2)$$

In the formula:

η_s is the influence factor of the cross-sectional shape of the main beam.

η_α is the angle–of–attack effect coefficient . For a flat plate section at 0° wind angle of attack, η_s and η_α both take 1.0.

r is the inertia radius of the bridge section (including stiffening beam and main cable).

b_l is the half-bridge width of the stiffening beam section.

μ is the density ratio of bridge to air.

m is the mass density of stiffening beam and main cable.

ω_t and ω_v are the lowest order torsional circle frequency and the vertical circle frequency respectively.

Definition:

V_{cr}^m , V_{cr}^p—critical flutter wind speed of hyperbolic parabolic suspension bridge and parallel suspension bridge.

ω_t^m、ω_v^m，ω_t^p、ω_v^p—the lowest order torsional circle frequency and vertical bending circle frequency of the hyperbolic parabolic suspension bridge and parallel suspension bridge.

https://doi.org/10.1051/e3sconf/202129302027

From Equation 3, the calculation results of the ratio of the flutter critical wind speed of the two suspension bridges are as follows:

(1) In the parallel cable suspension bridge scheme, M=68440kg/m, $\omega_t^P = 0.5019\text{rad/s}$, ω_v^P =0.3215rad/s. After calculating $I = 2.34 \times 10^3 m^4$, $r = 18.55\text{m}$, $\mu = 19.76$, therefore, the critical flutter wind speed V_{cr}^P of the parallel cable suspension bridge is:

$$V_{cr}^P=0.8\times0.8\times0.5019\times30\times\sqrt{\frac{18.55}{30}\times19.79\times\left[1-\left(\frac{0.3215}{0.5019}\right)^2\right]}$$

=25.87m/s

(2) In the spatial hybrid cable suspension bridge scheme, M=69360kg/m, $\omega_t^m = 1.144rad/s$, $\omega_v^m = 0.4253rad/s$, $I = 2.34 \times 10^3 m^4$, $r = 18.55\text{m}$, $\mu = 20.03$, therefore, the spatial hybrid cable The critical flutter wind speed V_{cr}^m of the suspension bridge is:

$$V_{cr}^m=0.8\times0.8\times1.144\times30\times\sqrt{\frac{18.55}{30}\times20.03\times\left[1-\left(\frac{0.4253}{1.144}\right)^2\right]}$$

=71.78m/s

V_{cr}^m=71.78m/s, V_{cr}^P=25.87m/s. The critical flutter wind speed of the hyperbolic parabolic space hybrid cable system suspension bridge has been greatly improved, with an increase of 177%.

5 Temporary auxiliary anti-wind cable

At present, earthquakes cannot be predicted, but super strong typhoons can be accurately predicted. The three-level fortification principle of seismic design of structures are: no damage under small earthquakes, repairable under middle earthquakes, and no downfall in large earthquakes. It is recommended that the wind-resistant design of the super-long-span suspension bridge adopts the two-level fortification principle: no damage under the middle wind, and a temporary wind-resistant cable should be set under the strong wind.

The 5,000-meter hyperbolic parabolic spatial mixed cable-net suspension bridge in this paper can withstand the strong typhoon of level 18. However, the current global El Niño phenomenon is intensifying, and the intensity of typhoons is increasing. The design reference period of the ultra-long-span suspension bridge is 100 years. Therefore, it is necessary to make psychological preparations for encountering strong typhoons above 19.

This paper draws on the principle of ship anchor cable and puts forward the technical measures for emergency filing of temporary auxiliary wind-resistant cable. The use requirements of the 5,000-meter hyperbolic parabolic mixed space cable super-long-span suspension bridge are as follows:

1. In case of 8-12 level typhoon, the channel is not blocked and the bridge is used normally;
2. In case of 13-18 level typhoon, the channel is blocked and suspending bridge deck traffic;
3. In case of typhoon level 19 or above, activating temporary auxiliary wind resistant cables.

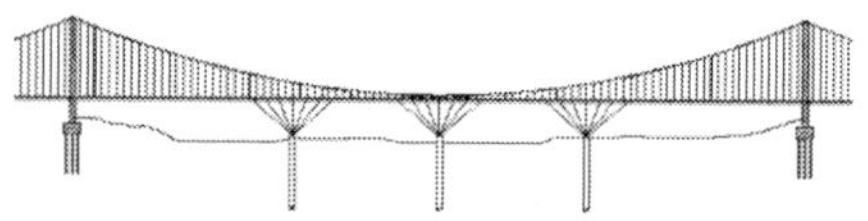

Figure 6. Temporary wind resistance cable

The construction method of temporary wind-resistant cable is as follows: ground anchor cable pile shall be constructed first. The top of the pile is connected with temporary carbon fiber wind-resistant cables. The carbon fiber wind-resistant cables are "deeply buried" in the seabed trench (the thin line is pulled to the bridge tower), which does not affect the channel operation.

When encountering strong typhoon weather, quickly pull the carbon fiber wind-resistant cable to the mid-span stiffener beam deck and fix it firmly on the stiffener beam deck to ensure that there is no slight structural damage to the super-large-span bridge. After the typhoon disaster, the wind-resistant cable will be pulled back to the original deep groove of the seabed to resume traffic.

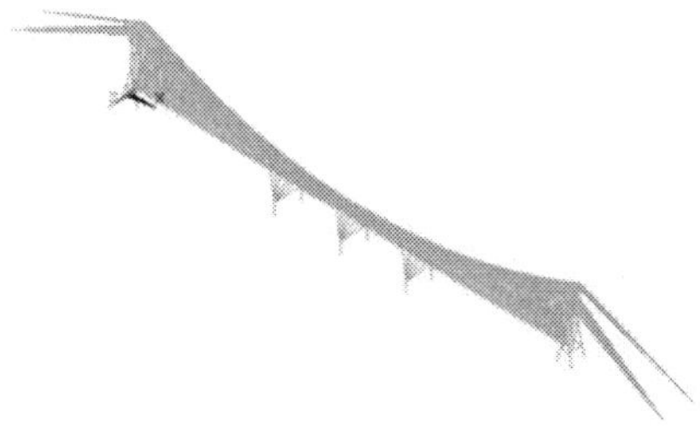

Figure 7. Model of hyperbolic parabolic mixed space cable suspension bridge with additional wind resistant cables

In this design, three ground anchor cable piles are set on each side of the 5000-meter-level strait suspension bridge. The diameter of the ground anchor cable pile is 4 meters, and the length of the ground anchor cable pile is 180 meters. The ground anchor cable pile has a longitudinal spacing of 600 meters and a horizontal spacing of 120 meters. The angle between the wind resistant cable and the vertical coordinate axis of the stiffening beam is about 30°. Each ground anchor cable pile is equipped with 7 carbon fiber wind resistant cables with a diameter of 0.5 meter which are connected with the stiffening beam to provide vertical and lateral restraint for the stiffening beam.

5.1 Dynamic modal analysis

In this paper, ANSYS software was used for finite element analysis, and the representative vibration modes of the hyperbolic parabolic mixed space cable suspension bridge with temporary wind-resistant cables can be obtained as shown in Figure 8.

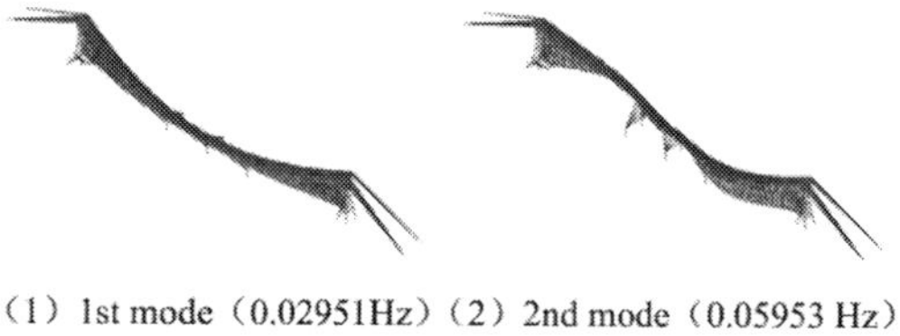

(1) 1st mode (0.02951Hz) (2) 2nd mode (0.05953 Hz)
(Positive symmetrical lateral bending) (Antisymmetric lateral bending)

(3) 3ed mode (0.07228Hz) (4) 4th mode (0.07347Hz)
(Antisymmetric vertical bending) (Positive symmetrical vertical bending)

(5) 27th mode (0.23829Hz) (6) 30th mode (0.29728Hz)

(Positive symmetric torsion) (Antisymmetric torsion)

Figure 8. Vibration mode of hyperbolic parabolic mixed space suspension bridge with temporary wind resistant cables

5.2 Analysis of critical wind speed of flutter

Through calculation, the following calculation parameters of the hyperbolic parabolic space hybrid cable suspension bridge can be obtained: η_s and η_α are both 0.8, and the calculation results of the flutter critical wind speed are as follows: After setting the temporary wind resistance cable, in the calculation, the temporary wind resistance cable should be considered. The vertical force component is equivalent to M=81540kg/m , I=2.34×10^{3}m^4, r=18.55m, μ=23.54, ω_t^L=1.4972rad/s, ω_v^L=0.4542rad/s.

Therefore, after the temporary windresistant cable is installed, the critical flutter wind speed V_{cr}^L of a 5000-meter spatial hybrid cable suspension bridge is:

$$V_{cr}^L=0.8\times0.8\times1.4972\times30\times\sqrt{\frac{18.55}{30}\times23.54\times\left[1-\left(\frac{0.4542}{1.4972}\right)^2\right]}$$

$$=104.5\text{m/s}$$

5.3 Comparative analysis

Table3. Comparison before and after installing temporary wind resistant cable for hyperbolic paraboloid suspension bridge

Comparison parameters	Hyperbolic parabolic suspension bridge	Hyperbolic parabolic suspension bridge with temporary wind-resistant cables
First-order lateral bending frequency (Hz)	0.02786	0.02951
First-order vertical bending frequency (Hz)	0.06912	0.07347
First-order positive symmetric torsion frequency (Hz)	0.18214 (22nd order)	0.23829 (27th order)
First-order antisymmetric torsion frequency (Hz)	0.25577 (29th order)	0.29728 (30th order)
Torsion ratio	2.69	3.30
Flutter critical wind speed (m/s)	71.78	104.5

The modal characteristics and critical wind speed of the hyperbolic parabolic mixed space cable suspension bridge before and after the installation of wind resistant cables are shown in Table 3.

The analysis shows that the provision of temporary auxiliary anti-wind cables can significantly increase the torsional frequency and torsional bending frequency ratio. The 27th order mode of the hyperbolic parabolic mixed space suspension bridge with temporary wind-resistant cables is a positively symmetric torsional mode with a frequency of 0.23829 Hz. The 30th order mode is an anti-symmetric torsional mode with a frequency of 0.29728HZ. The ratio of torsion frequency is as high as 3.30, which is 22.7% higher than that of the model without wind resistance cable. It can be seen that the provision of temporary auxiliary wind-resistant cables can significantly improve the wind-resistant stability of the suspension bridge.

The 5000-meter hyperbolic parabolic mixed space cable suspension bridge without a wind-resistant cable has a critical wind speed value of 71.78m/s, which can resist a strong typhoon of level 18. After the temporary auxiliary wind-resistant cable is activated, the critical wind speed of flutter can reach 104.5m/s, which can meet the requirements of resisting the 18th-level rare super strong typhoon. Therefore, the problem of wind resistance stability of the 5,000-meter super-long-span suspension bridge has been fundamentally solved, marking that mankind has the scientific and technological capability to build a 5,000-meter strait suspension bridge.

6 Conclusion

(1) The torsional modes of the parallel suspension bridge appear at the 10th and 14th orders, and the corresponding frequencies are 0.079882 Hz (first-order positive symmetry) and 0.098884 Hz (first-order antisymmetric). The torsional modes of the hyperbolic parabolic mixed-space suspension bridge appear in the 22nd and 29th modes respectively with corresponding frequencies of 0.18214Hz (first-order positive symmetry) and 0.25577Hz (first-order antisymmetric).The hyperbolic parabolic mixed space cable suspension bridge has good torsion resistance performance.

(2) In this paper, the dynamic modal comparison analysis of hyperbolic parabolic mixed space cable suspension bridge and parallel suspension bridge was carried out. The analysis shows that the torsion-bending frequency ratio of hyperbolic parabolic space suspension bridge is 2.69 and that of parallel suspension bridge is 1.56. The critical wind speed for flutter of hyperbolic parabolic space suspension bridge is 71.78m/s and that of parallel suspension bridge is 25.87m/s. The former has increased by 117% compared to the latter. Therefore, the hyperbolic parabolic mixed space cable suspension bridge space suspension bridge has good wind stability.

(3) In order to resist the rare super strong typhoon, technical measures to add temporary auxiliary anti-wind cables were proposed. After the temporary auxiliary wind-resistant cable is activated, the torsional frequency does not appear until the 27th mode (positive symmetric

torsional frequency is 0.23829 Hz), and the ratio of torsion-bending frequency is as high as 3.30. The flutter critical wind speed value reaches 104.5 m/s, which can resist the 18-level rare super strong typhoon and ensure the safety of the bridge.

(4) The technical solution of the hyperbolic parabolic mixed space cable suspension bridge structure can fundamentally solve the wind stability problem of the 5,000-meter super-long-strait suspension bridge, which indicates that mankind has the scientific and technological capability to build a 5,000-meter-level strait suspension bridge.

References

1. Xiang Haifan Ge Yaojun. On aerodynamic limits to suspension bridges[J]. China civil engineering journal,2005,38(01):60-70.(In Chinese)
2. Tang M-C. Super-long span bridges[J]. Structure & Infrastructure Engineering: Maintenance, Management, Life Cycle Design & Performance,2017,13(6):722-730.
3. Sato H, Hirahara N, Fumoto K, Hirano S. Full aeroelastic model test of a super long-span bridge with slotted box girder[J].Journal of Wind Engineering & Industrial Aerodynamics,2002,90(12):2023-2032.
4. Chen ze. Conceptual design and wind resistance research of the 5000m-span suspension bridge with CFRP cables[D]. Southwest jiaotong university,2016.(In Chinese)
5. Vu T-V, Kim Y-M, Lee H-E. Coupled flutter analysis of long-span bridges using full set of flutter derivatives[J]. KSCE Journal of Civil Engineering,2016, 20(4):1501-1513.
6. Zhang Xinjun. Study of the wind stability of long-span suspension bridges with spatial cable systems[J]. China civil engineering journal,2011,44(06):80-86.(In Chinese)
7. Tang, H., Li, Y., & Shum, K. M.. Flutter performance of long-span suspension bridges under non-uniform inflow[J]. Advances In Structural Engineering, 2018, 21(2): 201–213.
8. Abbas T, Kavrakov I, Morgenthal G. Methods for flutter stability analysis of long-span bridges: a review[J]. Bridge Engineering, 2017,170(4):271-310.

Design and Characteristic Analysis of a New Type of Hydraulic Pulsation Suppressor

Jun Yuan[1], Hang Jiang[1], Zheng-yan Fang[2], Tian-sheng Wang[1], Zhi Luo[1], Feng-xian Jiang[3] and Gui-hong Yin[1]

[1] Institute of Physics and Information Engineering, Zhaotong University, Zhaotong, Yunnan 657000

[2]School of Education and Science, Zhaotong University, Zhaotong, Yunnan 657000

[3]Business School, Yuxi Normal University Yuxi, Yunnan 653100

Abstract. In order to decrease the pressure pulsation of high-pressure hydraulic system and speed up the process of high-pressure hydraulic system, a new type of hydraulic pulsation suppressor is designed. The hydraulic pulsation suppressor is designed on the basis of the structural characteristics and attenuation performance of the expansion chamber attenuator and H-type muffler. The theoretical model of the modern hydraulic pulsation suppressor is established based on the fluid theory, and the attenuation performance of the new hydraulic pulsation suppressor is analyzed. The results demonstrate that the designed new hydraulic pulsation suppressor can attenuate the pressure pulsation with a pulsation frequency of 20 Hz~2000 Hz, and attenuate the frequency bandwidth; the attenuation effect is above 50dB, and the attenuation effect is ideal.

1 Introduction

With the development of hydraulic technology, the pressure of the hydraulic system is further increased. High pressure has made the pressure pulsation problem of the hydraulic system more prominent, which has been one of the important factors hindering the development of hydraulic technology. Therefore, reducing the pressure pulsation of the high-pressure hydraulic system is of great importance.

At present, hydraulic pulsation attenuators are mainly used in hydraulic systems to attenuate the pressure pulsations[1-5]. The hydraulic pulsation attenuator installed at the pump outlet mainly includes two types: resonance type and expansion chamber type. OU YANG Xiaoping [6] summarized the resonant hydraulic pulsation attenuator, explained the research progress and performance of various types of resonant hydraulic pulsation attenuators, and proposed that the resonance is best for compactness, high versatility, and strong frequency adaptability. The development trend of the research focusses of hydraulic pulsation attenuator. Resonant hydraulic pulsation attenuators are mainly used to eliminate low-frequency and intermediate-frequency narrow-band pressure pulsations. Wang Yan[7] and others focused on the types, research methods and progress of expansion chamber hydraulic pulsation attenuators, and on this basis, discussed the research trend of expansion chamber hydraulic pulsation attenuators, and proposed that expansion chamber hydraulic pulsation attenuators should be based on theoretical methods and structures. Improve and strengthen research. The attenuation characteristics of the expansion chamber hydraulic pulsation attenuator mainly depend on the expansion ratio and length. This type of hydraulic pulsation attenuator mainly attenuates the pressure pulsation in the high frequency band[8].

Built on the above analysis, a new type of hydraulic pulsation suppressor combining expansion chamber type and resonance type is designed. The original hydraulic pulsation suppressor combines the advantages of the expansion chamber hydraulic pulsation attenuator and the H-type hydraulic silencer. In this study, a theoretical analysis of the latest hydraulic pulsation suppressor was carried out by means of fluid dynamics; the insertion loss was used to evaluate the attenuation performance of the new hydraulic pulsation suppressor. The rationality of the modern hydraulic pulsation suppressor is verified.

2 The working principle of the new hydraulic pulsation suppressor

Figure 1 is a schematic diagram of the structure of the new hydraulic pulsation suppressor. The original hydraulic pulsation suppressor is composed of a rigid shell, two expansion chambers, twelve volume chambers, twelve slender mass chambers and two flow-limiting pipes. When the pulsation of the plunger pump enters the first expansion chamber from the inlet of the suppressor, the pulsation is attenuated for the first time; after the first decay, the pulsation enters the second expansion chamber with different parameters through the first restrictor pipe, and then attenuates again ; The second attenuated pulsation flows through the second restrictor pipe and then enters the first volume chamber through the four mass

* Corresponding author: 2451485983@qq.com

chambers; the hydraulic oil entering the first volume chamber is divided into two parts, one part is compressed, and the other part enters the first volume chamber. Two volume chambers; the same is valid for the hydraulic oil entering the second volume chamber; at this time, the hydraulic oil entering the third volume chamber is only compressed. The hydraulic oil in the twelve mass chambers moves with the pulsation of the piston, plays the role of liquid sense and fluid resistance, and consumes a part of the pulsation energy at the same time. The expansion chamber, mass chamber, and volume chamber has different parameters, which can be used for multiple frequency ranges. The pressure pulsation is attenuated many times. Through the reasonable design of the modern hydraulic pulsation suppressor, a new type of hydraulic pulsation suppressor integrating expansion chamber, restrictor tube, mass chamber and volume chamber is realized.

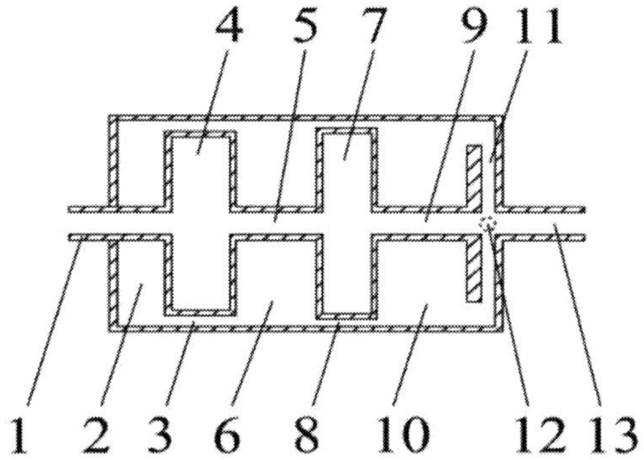

1. Suppressor inlet pipe 2. Volume chamber 3
3. Mass chamber 3 4. Expansion chamber 1 5. Flow restriction pipe 1 6. Volume chamber 2 7. Expansion chamber 2 8. Mass chamber 2 9. Restriction pipeline 2
10. Volume chamber 1 11. Mass chamber 1
12. Node 1 13. Suppressor outlet pipeline

Figure 1. Schematic diagram of the structure of the new hydraulic pulsation suppressor.

3 Mathematical model and attenuation characteristics analysis of the new hydraulic pulsation suppressor

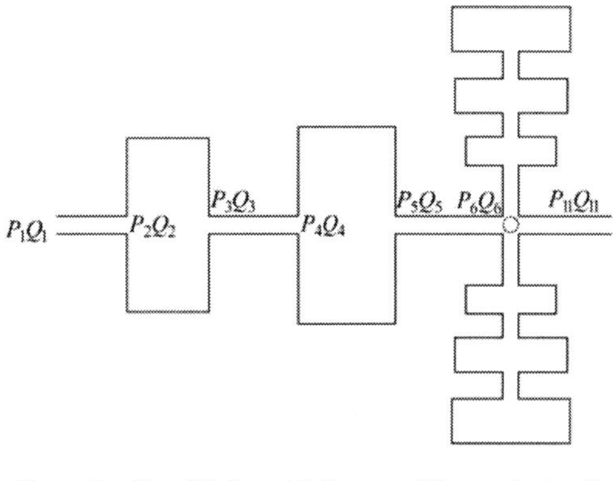

Figure 2. Simplified model diagram of the new hydraulic pulsation suppressor.

The simplified model of the new hydraulic pulsation suppressor is shown in Figure 2. In Figure 2, P_1, Q_1, and P_{12}, Q_{12} are the frequency domain dynamic pressure and flow at the inlet and outlet of the new hydraulic pulsation suppressor; P_6, Q_6, and P_{11}, Q_{11} are the frequency domain dynamic pressure and flow at the inlet and outlet of node 1; P_2 , Q_2, and P_3, Q_3 are the frequency domain dynamic pressure and flow at the inlet and outlet of the expansion chamber 1; P_4, Q_4, and P_5, Q_5 are the frequency domain dynamic pressure and flow at the inlet and outlet of the expansion chamber 2.

Figure 3 is an equivalent simplified model diagram of node 1. In Figure 3, Q_x and P_x $(x=7,8,9,10)$ are the frequency domain dynamic pressure and flow of node 1.

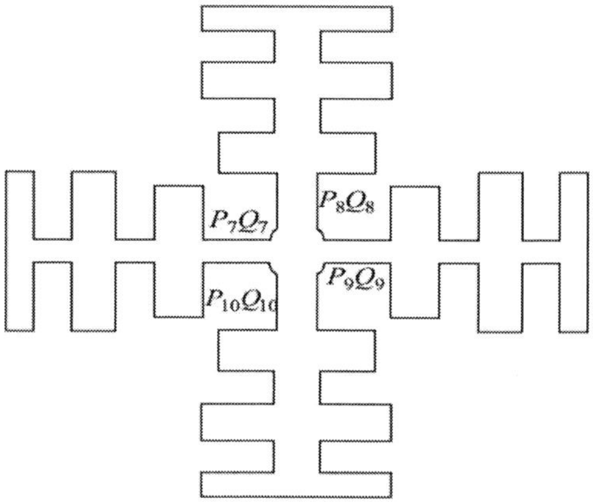

Figure 3. Simplified model diagram of node 1.

It can be written by node conditions:

$$Q_6=\sum_{x=7}^{10}Q_x+Q_{11} \tag{1}$$

$$P_6=P_7=P_8=P_9=P_{10}=P_{11} \tag{2}$$

Make

$$Z_x=P_x/Q_x \tag{3}$$

Where, $x=7,8,9,10$

Z_x is the branch point impedance of the three-stage H-type hydraulic silencer system. Figure 5 is a schematic diagram of the structure of a three-stage H-type hydraulic silencer. According to the literature [9], the expression of Z_x is:

$$Z_x=\frac{a_{11x}a_{23x}a_{32x}+a_{12x}a_{21x}a_{33x}-a_{11x}a_{22x}a_{33x}}{\rho csA_{1x}\left(a_{23x}a_{32x}-a_{22x}a_{33x}\right)} \tag{4}$$

Where, $a_{11x} = \rho A_{1x} l_{1x} s^2 + 8\pi\rho\upsilon l_{1x} s + K\dfrac{A_{1x}^2}{V_{1x}}$

$$a_{12x} = a_{21x} = -K\frac{A_{1x}A_{2x}}{V_{1x}}$$

$$a_{23x} = a_{32x} = -K\frac{A_{2x}A_{3x}}{V_{2x}}$$

$$a_{22x} = \rho A_{2x} l_{2x} s^2 + 8\pi\rho\upsilon l_{2x} s + KA_{2x}^2\left(\frac{1}{V_{1x}} + \frac{1}{V_{2x}}\right)$$

$$a_{33x} = \rho A_{3x} l_{3x} s^2 + 8\pi\rho\upsilon l_{3x} s + KA_{3x}^2\left(\frac{1}{V_{2x}} + \frac{1}{V_{3x}}\right)$$

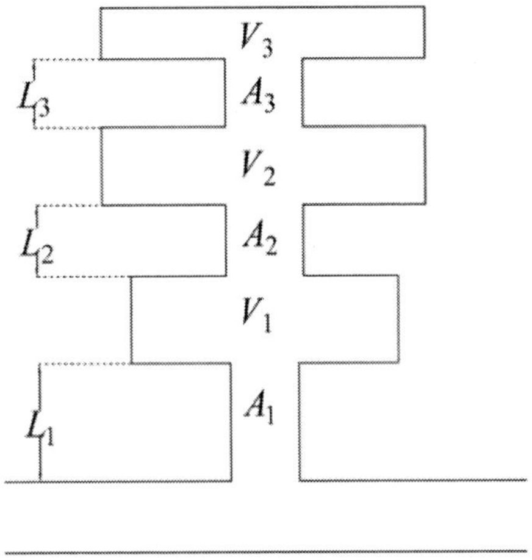

Figure 4. L is the length of the mass chamber; A is the cross-sectional area of the mass chamber; V is the volume of the volume chamber;

The principle diagram of the three-stage H-type hydraulic pulsation attenuator.

The transfer matrix from the inlet pipe of the hydraulic pulsation suppressor to node 1 is:

$$\begin{bmatrix} P_1 \\ Q_2 \end{bmatrix} = \begin{bmatrix} A_1 & B_1 \\ C_1 & D_1 \end{bmatrix} \begin{bmatrix} 1 & 0 \\ \frac{V_2}{Ks} & 1 \end{bmatrix} \begin{bmatrix} A_3 & B_3 \\ C_3 & D_3 \end{bmatrix} \begin{bmatrix} 1 & 0 \\ \frac{V_4}{Ks} & 1 \end{bmatrix} \begin{bmatrix} A_5 & B_5 \\ C_5 & D_5 \end{bmatrix} \begin{bmatrix} P_6 \\ Q_6 \end{bmatrix} \quad (5)$$

Where, V_2 is the volume of expansion chamber 1

V_4 is the volume of expansion chamber 2

For the pipeline, its dynamic characteristic equation is expressed as:

$$\begin{bmatrix} A_i & B_i \\ C_i & D_i \end{bmatrix} = \begin{bmatrix} ch[T(s)L_i] & Z_i(s)sh[T(s)L_i] \\ sh[T(s)L_i]/Z_i(s) & ch[T(s)L_i] \end{bmatrix} \quad (6)$$

Where, $Z_i(s) = \rho c^2 T(s)/(A_{xi}s)$ is the characteristic impedance of the pipeline

$T(s) = (s^2 + R_v s)^{0.5}/c$ is the pipe propagation constant

A_{xi} is the pipe cross-sectional area

L_i is the length of the pipe

$R_v = 32\mu/A_{xi}$

$i = 1,3,5$

From the formula (5):

$$\begin{bmatrix} P_1 \\ Q_1 \end{bmatrix} = \begin{bmatrix} A_{11} & B_{11} \\ C_{11} & D_{11} \end{bmatrix} \begin{bmatrix} P_6 \\ Q_6 \end{bmatrix} \quad (7)$$

Simplify (7) formula, there are

$$P_1 = A_{11}P_6 + B_{11}Q_6 = A_{11}P_{11} + B_{11}Q_6 \quad (8)$$

$$Q_1 = C_{11}P_6 + D_{11}Q_6 = C_{11}P_{11} + D_{11}Q_6 \quad (9)$$

From (1), (2) and (3), we get:

$$Q_6 = \sum_{x=7}^{10}\frac{P_x}{Z_x} + Q_{11} = P_{11}\sum_{x=7}^{10}\frac{1}{Z_x} + Q_{11} \quad (10)$$

From (9) and (10), we get:

$$Q_1 = C_{11}P_{11} + D_{11}\left(P_{11}\sum_{x=7}^{10}\frac{1}{Z_x} + Q_{11}\right) \quad (11)$$

Make $Y = Q_{11}/P_{11}$.

Substituting into (11) formula:

$$P_{11} = \frac{Q_1}{C_{11} + D_{11}\sum_{x=7}^{10}\frac{1}{Z_x} + D_{11}Y} \quad (12)$$

Incorporate the outlet pipe of the hydraulic pulsation suppressor into the dynamic impedance part. The dynamic characteristics of the pipeline and throttling load, the expression of load admittance Y can be obtained as [10]:

$$Y = \frac{2\Delta P_F C_{22} + Q_F D_{22}}{2\Delta P_F A_{22} + Q_F B_{22}} \quad (13)$$

Where, ΔP_F is the pressure loss of the throttle valve

Q_F is the flow rate of the throttle valve

A_{22}、B_{22}、C_{22}、D_{22} are the load-end pipelines to transfer matrix elements

The anti-vibration effect of the new hydraulic pulsation suppressor is evaluated by the insertion loss K_i. The insertion loss not only considers the characteristics of the pulsation attenuator device and the piping system, but also considers the influence of source impedance and load impedance. The insertion loss is defined as the pressure ratio of the pipe load end before and after adding the pulsation attenuator [11], namely

$$K_c = 20\lg\left|\frac{P'_{11}}{P_{11}}\right| \tag{14}$$

P'_{11} is the use of rigid hydraulic tubing of the same length as the pulsation attenuator to replace the frequency domain pressure at the load end of the new hydraulic pulsation suppressor.

P'_1, Q'_1 and P'_{11}, Q'_{11} are the dynamic frequency domain pressure and dynamic frequency domain flow rate at the inlet and outlet of the rigid straight hydraulic oil pipe after replacing the hydraulic pulsation attenuator with a rigid straight hydraulic oil pipe. The dynamic characteristics of the rigid hydraulic oil pipe are obtained according to the pipeline network calculation method. The equation is:

$$\begin{bmatrix} P'_1 \\ Q'_1 \end{bmatrix} = \begin{bmatrix} A_{33} & B_{33} \\ C_{33} & D_{33} \end{bmatrix} \begin{bmatrix} P'_{11} \\ Q'_{11} \end{bmatrix} \tag{15}$$

From the formula (15):

$$P'_{15} = \frac{Q'_1}{(C_{33} + D_{33}Y)} \tag{16}$$

For the plunger pump with flow disturbance source, it's working condition is stable, so

$$Q_1 = Q'_1 \tag{17}$$

Substituting (10), (16), and (17) into (14), the insertion loss of the new hydraulic pulsation suppressor can be expressed as:

$$K_c = 20\lg\left|\frac{C_{11} + D_{11}\sum_{x=7}^{10}\frac{1}{Z_x} + D_{11}Y}{C_{33} + D_{33}Y}\right| \tag{18}$$

4 Simulation analysis of the attenuation characteristics of the new hydraulic pulsation suppressor

Use MATLAB software to simulate the insertion loss of the original hydraulic pulsation suppressor. The basic parameters of the simulation are presented in table 1. The parameters of the throttle valve and hydraulic oil are given in table 2. The simulation frequency range is: 0 Hz~2000 Hz.

Table1. Basic parameter table of the new hydraulic pulsation suppressor.

Name	Physical parameter	Value
Inlet pipe	Length L_1/mm	100
	radius r_1/mm	12.5
Quality Room 7	Length L_7/mm	60
	radius r_7/mm	7
Quality Room 8	Length L_8/mm	60
	radius r_8/mm	6
Quality Room 9	Length L_9/mm	60
	radius r_9/mm	5
Quality Room 10	Length L_{10}/mm	60
	radius r_{10}/mm	4
Restrictor pipe 1	Length L_{x1}/mm	23
	radius r_{x1}/mm	12.5
Restrictor pipe 3	Length L_{x3}/mm	23
	radius r_{x3}/mm	12.5
Restrictor pipe 5	Length L_{x5}/mm	23
	radius r_{x5}/mm	12.5
Expansion room 1	Length L_{k1}/mm	120
	radius r_{k1}/mm	90
Expansion room 2	Length L_{k2}/mm	100
	radius r_{k2}/mm	90
Volume chamber 1	Length L_{v1}/mm	23
	Inner radius r_{n1}/mm	12.5
	Outer radius r_{w1}/mm	100
Volume chamber 2	Length L_{v2}/mm	23
	Inner radius r_{n2}/mm	12.5
	Outer radius r_{w2}/mm	100
Volume chamber 3	Length L_{v3}/mm	23
	Inner radius r_{n3}/mm	12.5
	Outer radius r_{w3}/mm	100
Quality Room 72	Length L_{72}/mm	100
	radius r_{72}/mm	7
Quality Room 82	Length L_{82}/mm	100
	radius r_{82}/mm	6
Quality Room 92	Length L_{92}/mm	100
	radius r_{92}/mm	5
Quality Room 102	Length L_{102}/mm	100
	radius r_{102}/mm	4
Quality Room 73	Length L_{73}/mm	120
	radius r_{73}/mm	7
Quality Room 83	Length L_{83}/mm	120
	radius r_{83}/mm	6
Quality Room 93	Length L_{93}/mm	120
	radius r_{93}/mm	5
Quality Room 103	Length L_{103}/mm	120
	radius r_{103}/mm	4
Load equivalent pipeline	Length L_f/mm	400
	radius r_f/mm	12.5

Table2. Basic parameter table of hydraulic oil and throttle valve .

Name	Physical parameter	Value
No. 32 hydraulic oil	Bulk modulus /MPa	1400
	Viscosity /(Pa • s)	0.00314
	density /(kg/m^3)	890
Throttle valve	Differential pressure /MPa	2.5
	Flow /(L/min)	160

According to the above formula and the basic parameters of the simulation, the insertion loss of the hydraulic pulsation suppressor is programmed in MATLAB software as showed in Figure 5.

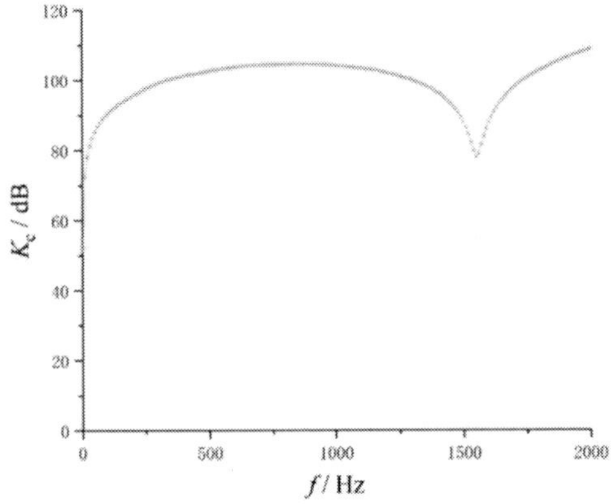

Figure 5. MATLAB simulation curve of insertion loss of the new hydraulic pulsation suppressor.

Figure 5 is the MATLAB simulation curve of the insertion loss of the original hydraulic pulsation suppressor. It can be obtained from the figure that the attenuation effect of the hydraulic pulsation suppressor is above 50 dB in the pulsation frequency range of 20 Hz to 2000 Hz, and has a sound attenuation effect. Compared with ordinary pulsation attenuators, the attenuation effect is better and the attenuation frequency band is wider.

5 Conclusion

(1) A new type of hydraulic pulsation suppressor with compact structure is designed, and the mathematical model of the hydraulic pulsation suppressor is reasonably deduced. According to theoretical analysis, the new hydraulic pulsation suppressor achieves the effect of triple attenuation of pulsation in a wide frequency band.

(2) Through the analysis of the simulation calculation results of MATLAB software, it is found that the new hydraulic pulsation suppressor has an ideal effect of attenuating pressure pulsation in the frequency range of 20 Hz to 2000 Hz, and achieves an efficient and broad-spectrum attenuation of pressure pulsation.

References

1. HARRISON K A, EDGE K A.Reduction of axial piston pump pressure ripple[J]. Proceedings of the Institution of Mechanical Engineers, Part I: Journal of Systems and Control Engineering, 2000, 214(1): 53-64.
2. XU Bing, CHEN Yuan, ZHANG Jun-hui. Current Researches and Progress on Vibration and Noise Reduction Technology of Axial Piston Pump[J]. Chinese Hydraulics & Pneumatics, 2014, (3): 1-12.
3. ZENG Xiangrong. Hydraulic noise control[M]. Harbin: Harbin Institute of Technology Press,1988.
4. KWONG AH M, EDGE K A.A method to reduce noise in hydraulic systems by optimizing pipe clamp locations[J]. Proceedings of the Institution of Mechanical Enginessrs, rat I: Journal of Systems and Control Engineering, 1998, 12(4): 67-280.
5. WANG Zhanlin. Aircraft high-pressure hydraulic energy system[M]. Beijing University of Aeronautics and Astronautics Press, 2004.
6. OUYANG Xiaoping, LI Lei, FANG Xu, YANG Huayong. Research Status and Prospects of Resonant-type Hydraulic Pulsation Attenuators[J]. Journal of Mechanical Engineering, 2015, 51(22): 168-175, 182.
7. WANG Yan, HAO Fengqian, GUO Shengrong, LU Yueliang, CHEN Jinhua. Research Status and Development Trend of the Expansion Chamber Pressure Pulsation Attenuator[J]. Machine Tool & Hydraulics 2015, 43(15): 180-186.
8. YANG Fan, DENG Bin, WANG Guozhi, ea lt. Fluid Acoustic Properties of Expansion Chamber Mufflers Based on Electro-hydraulic Analogy Method[J]. Machine Tool & Hydraulics, 2018, (14): 60-63.
9. GUAN Changbin, JIAO Zongxia. Modeling and Optimal Design of 3 Degrees of Freedom Helmholtz Resonator in Hydraulic System[J]. Chinese Journal of Aeronautics, 2012, 25(5): 776-783.
10. ZHANG Yin, YU Jun, LI Shen. Experimental research and CFD simulation of pressure pulsation attenuator[J].
Chinese Hydraulics & Pneumatics, 2011, (6): 47-50.
11. SHAN Changji. Structure Design and Attenuation Method of Hydraulic Mufflers[J]. Journal of Jiamusi University(Natural Science Edition), 2013, (5): 749-751.

 https://doi.org/10.1051/e3sconf/202129302029

High Temperature Performance of SBS and LM-S Modified Asphalt Mixtures by Triaxial Test

Tang-Baoli[1], Ren-yongqiang[1*], Chen-Xiangmei[1], Hou-Huifang[1], Liang-Jianping[1]

[1]Department of Civil Engineering, Ordos Institude of Technology, Ordos, Inner Mongolia, 017010, China.

Abstract. In order to study the high temperature performance of LM-S modified asphalt mixture and SBS modified asphalt mixture, repeated loading creep test was used to study the influence of temperature and deviatoric stress on the axial permanent deformation of the two kinds of asphalt mixture. At the same time, Permanent deformation, ε@5000, flow number FN and creep rate were select to evaluation of high temperature performance from different directions. The results show that the ε@5000 and creep rate are failed in the condition of high temperature and large deviatoric stress,so it has`t widely practicable. The flow number FN is also limited by the conditions, which leads to the distortion of the flow number at lower temperature and smaller deviatoric stress so it is not easy to direct used as the evaluation index. Axial permanent deformation can reflect the permanent deformation in different cycles which is an excellent index to evaluate the high temperature performance of the two kinds of asphalt mixture, it is recommended to use axial permanent deformation to compare the LM-S modified asphalt mixture and SBS modified asphalt mixture The experimental results show that the axial permanent deformation of the LM-S modified asphalt mixture is always less than that of SBS modified asphalt mixture,it indicating that the high temperature rutting resistance of the LM-S modified asphalt mixture is better than that of SBS modified asphalt mixture.

1 Introduction

As one excellent modified asphalt mixture, SBS modified asphalt mixture, has always been favored by road workers. DaquanSuns research shows that SBS modified asphalt has very good anti-aging property and Xu, Songs study also points out that SBS modified asphalt has great regeneration ability, and star type SBS is superior to linear type in high-temperature and anti-aging aspect[1]-[2]. But the SBS modified asphalt can only be used in middle and high class roads due to its high price, while a large number of middle and lower class roads still use base asphalt for the economical reason, which restricts SBS modified asphalt further development[3]-[4]. In order to solve the problems of SBS modified asphalt, such as easy segregation, high requirements in transportation and storage etc, the LM-S modified asphalt mixture which studied by road workers is the obtained corresponding asphalt mixture by dry-mixed. It could save investment cost greatly for no requirement of modified asphalt production equipment addition[5]. The modified asphalt mixture performance is stable, especially at high-temperature, which attracts more and more people' s attention[6]. Wang Chaohui and others of Chang'an University compared the impact of different mixing methods on road performance through indoor tests. The results showed that the high-temperature performance of asphalt mixture produced by dry mixing method is better than that of asphalt mixture produced by wet mixing method. In order to compare the high-temperature performance of LM-S modified asphalt mixture and SBS modified asphalt mixture, and to find suitable high-temperature performance evaluation indicators, the following research is carried out.

2 Test

2.1 Test materials

（1）Asphalt binder

The subject research uses two types of asphalt binders: SBS modified asphalt and base asphalt. The basic performance of two asphalt binders both meets the "Highway Engineering Asphalt and Asphalt Mixture Test Regulations" (JTG E20-2011). The specific test results are shown in Table 1 below.

Table1. Technical indicators of asphalt binder

Technical indicators	Unit	Test results	
		SBS modified asphalt	Asphalt
Penetration（25℃, 100g, 5s）	0.1mm	65.3	85
Ductility（5cm/min, 5℃）≥	cm	28	16
Softening Point）, ≥	℃	56	50
density（15℃）	g/cm³	1.031	1.028
Viscosity（135℃ kinetic viscosity）	Pa*s	2.6	1.3

First author: TANG Baoli(1980-), male, master degree, lecturer.
*Correspondent author: Ren-yongqiang, master degree, senior engineer.E-mail:ryongqiang@oit.edu.cn

(2) Mineral aggregate

The coarse aggregate used in the subject research is basalt, the fine aggregate is limestone machine-made sand, and the ore powder is finely ground from limestone. The basic performance index of the mineral also meets the requirements of the "Highway Engineering Aggregate Test Regulations" (JTG E42-2005). The test results are shown in Table 2, Table 3 and Table 4

Table2. Technical indicators of Coarse aggregate

Technical indicators	Requirements	Test results
Water absorption/%	≤2.0	0.65
Robustness/%	≤12.0	5.2
Apparent relative density/ (g/cm³)	≥2.60	2.849
Adhesion	Five grade	Five grade

Table3. Technical dicators of fine aggregate

Technical indicators	Requirements	Test results
Mud content/%	≤2.0	0.65
Robustness/%	≤12.0	5.2
Apparent relative density/ (g/cm³)	≥2.60	2.849

Table4. Technical dicators of Mineral powder

Technical indicators	Requirements		Test results
Apparent density/ (g/cm³)	2.5		2.702
Water content/%	1		0.8
Particle size range/%	<0.6	100	100
	<0.15	90～100	98
	<0.075	75～100	95
Appearance	Non-agglomerated		Non-agglomerated
Hydrophilic coefficient	<1		0.6
Plasticity index/%	<4		2.3
Heating stability	Recorded		No deterioration

2.2 Test method

(1) Test plan

Add modifier LM-S in the mixing process of base asphalt and aggregate to produce LM-S modified asphalt mixture (dry mix), SBS modified asphalt mixture still adopts conventional production methods. Adopt Marshall design method for mix design.

Table5. Program of test

Equipments	CONTROL UTM-100
Preload /KPa	20
Load mode and waveform	Loading 0.1s and unloading 0.9s, half sinusoid intermittent loading
Confining pressure/KPa	138
Deviatoric stress/Mpa	0.7、0.9、1.1
Test Temperature/°C	40、50、60
Test termination	N=10000 or 5% (NCHRP9-19)

(2) Triaxial repeated loading permanent deformation test method

The test specimens were kneaded and formed by the rotary compactor SGC rotary compactor, and the mixture specimens were formed into cylindrical specimens of Φ 150mm×175mm, and then cored and cut with a Φ100 water drill to obtain a cylindrical test piece of Φ100mm ×150mm. Finally, use sandpaper to polish to ensure that the surface unevenness of the test piece does not exceed 0.05mm. Place the test specimen at the corresponding test temperature for at least 5h but not more than 10h. Ensure that the temperature inside and outside of the specimen reaches the predetermined test temperature. Set the test parameters and start the test based on test plan requirements.

3 Results and analysis

3.1 The influence of temperature on permanent deformation

Since the permanent deformation under other deviator stresses is consistent with the change rule under the standard tire pressure of 0.7 MPa, for the convenience of comparison, only 0.7 MPa is used as an example for illustration. It can be seen from Figure 1 and Figure 2 that the permanent deformation at different temperatures: as the number of loading increases, the permanent deformation of SBS and LM-S modified asphalt mixtures will continue to increase, which means the more equivalent axes act on the road surface, the greater of permanent deformation of the pavement is, because a load must produce a certain amount of permanent deformation each time , the road surface actually shows the permanent deformation accumulated under a certain axis; the higher the temperature is, the greater the permanent deformation could be, and when the temperature exceeds 50 ℃, the permanent deformation will increase sharply and the specimen will be destroyed quickly in a short time. That is to say, temperature has a huge influence on permanent deformation. The higher the temperature, the greater the permanent deformation. This is due to the increase of temperature and the softening of cement, which makes the specimen easy to damage, especially when the temperature is close to the softening point of cement. Permanent deformation will increase rapidly until it breaks.

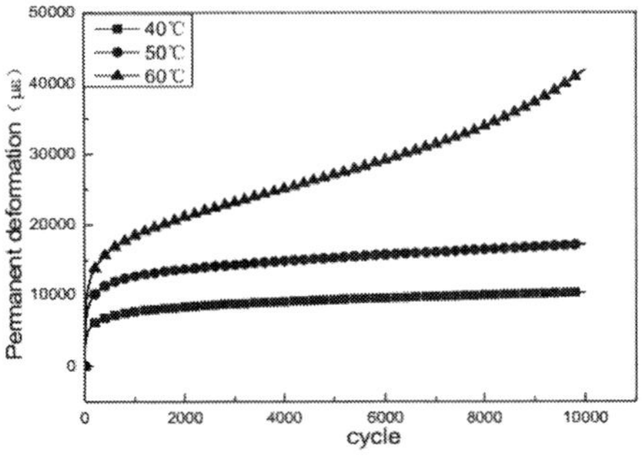

Figure 1. Permanent deformation of SBS@ 0.7MPa

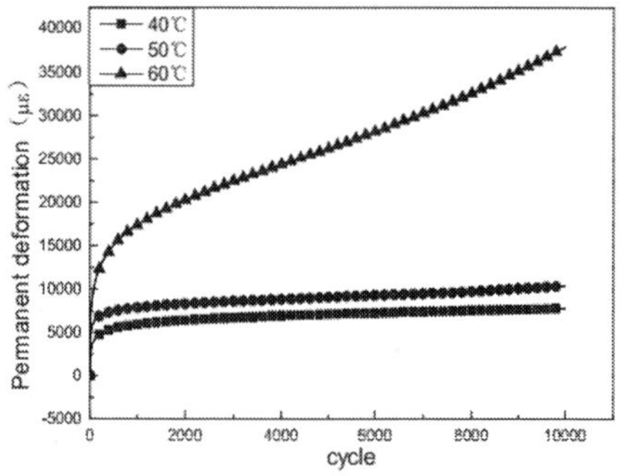

Figure 2. Permanent deformation of LM-S@ 0.7MPa

3.2 The influence of deviator stress on permanent deformation

Also because the permanent deformation at other temperatures is basically the same as the change at 40°C (only the third stage will appear in the permanent deformation curve of the mixture at higher temperatures) for the convenience of comparing the effect of deviator stress on the axial permanent deformation. Only take 40°C as an example. It can be seen from Figures 3 and 4 that the permanent deformation under different deviator stresses: as the number of loading increases, the permanent deformation of both SBS and LM-S modified asphalt mixtures will continue to increase, which is the same as the previous. The greater the deviator stress is, the greater the permanent deformation will be. When the deviator stress is 0.7 MPa, the permanent deformation curve is smooth. When the deviator stress reaches 1.1 MPa, the slope of the permanent deformation curve increases significantly compared with 0.7 MPa. Which means that the greater the axle load is, the greater the permanent deformation will be. This is because the asphalt mixture is a viscoelastic material, the axial strain will increase with the increase of stress, and the irreversible permanent deformation of the material with large strain will also be greater. In the end, a larger deviator stress shows a larger axial permanent deformation.

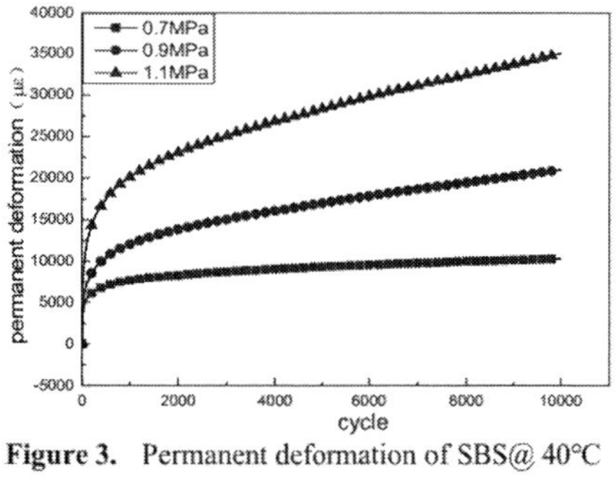

Figure 3. Permanent deformation of SBS@ 40°C

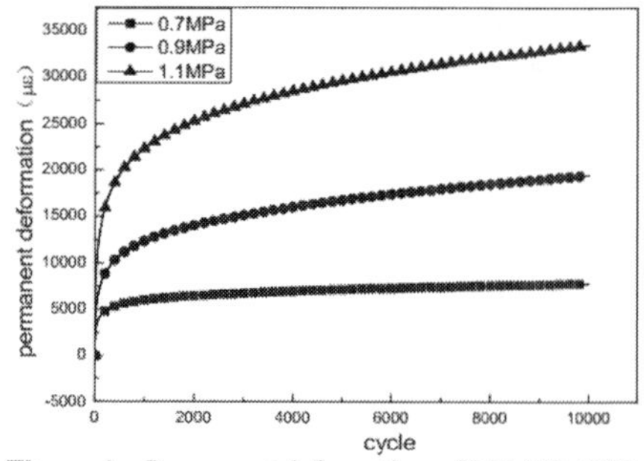

Figure 4. Permanent deformation of LM-S@ 40°C

3.3 Comparison of high-temperature performance of SBS and LM-S modified asphalt mixture

(1) Comparison of permanent deformation

Since the permanent deformation change rule under the combination of standard tire pressure 0.7MPa with 40°C and the combination of 1.1MPa with 50°C can represent the test results under all other combinations, for the convenience of comparison, only take 0.7MPa-40°C and 1.1MPa-50°C as examples to illustrate. It can be seen from the permanent deformation of the two asphalt mixtures in Figures 5 and 6: The permanent deformation of LM-S modified asphalt mixture is less than that of SBS modified asphalt mixture, and the greater of both temperature and deviator stress, the more significant effect can be shown. This indicates that the permanent deformation resistance of LM-S modified asphalt mixture is stronger than that of SBS. It is said that LM-S modified asphalt mixture is more advantageous than SBS in high-temperature areas and large axle load sections.

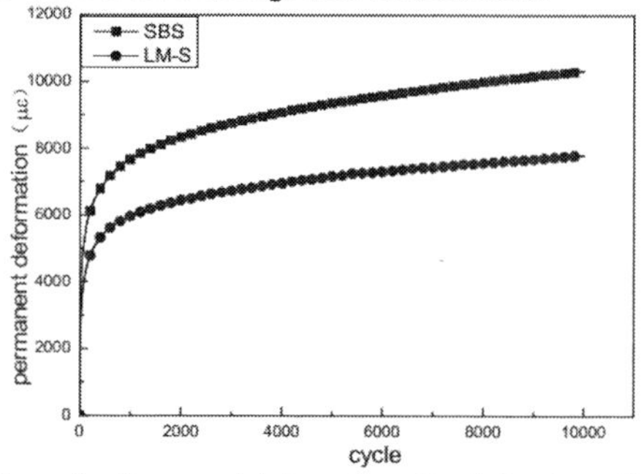

Figure 5. Permanent deformation of asphalt mixture@ 0.7MPa-40°C

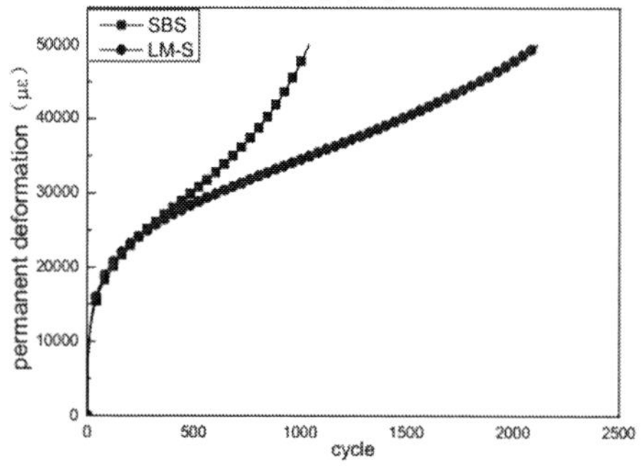

Figure 6. Permanent deformation of asphalt mixture@ 1.1MPa-60°C

(2) The comparison on ε @500

The research results show that the accumulated permanent deformation of the repeated loading test under 5000 cycles can well reflect the high-temperature anti-deformation ability of the asphalt mixture. It can be seen from the two asphalt mixtures ε@5000 in Fig. 7 that: as the deviator stress increases, the ε@5000 of two kinds of asphalt mixtures are constantly increasing, the greater the deviator stress is, the greater the permanent deformation of the asphalt mixture after 5000 cycles is, which also shows that the deviator stress increases while the the asphalt mixture performance at high-temperature will decrease; at the same time, it is found that ε@5000 will increase as the temperature increases, which means that the high-temperature deformation resistance of asphalt mixture will also decrease when temperature rise. Observation found that the two asphalt mixtures ε@5000 can be strictly distinguished only at 50°C. At 40°C, it is difficult to accurately judge the high-temperature performance of the two asphalt mixtures ε@5000. At 60°C, asphalt mixture will soon be destroyed under large deviator stress and ε@5000 even not exist. It is also difficult to determine the high-temperature performance of the asphalt mixture according to the size of ε@5000. This means that, only at 50°C, ε@5000 can be used as evaluation indicator of the high- temperature performance, which will have a more obvious effect, and it is difficult to achieve in higher or lower temperatures.

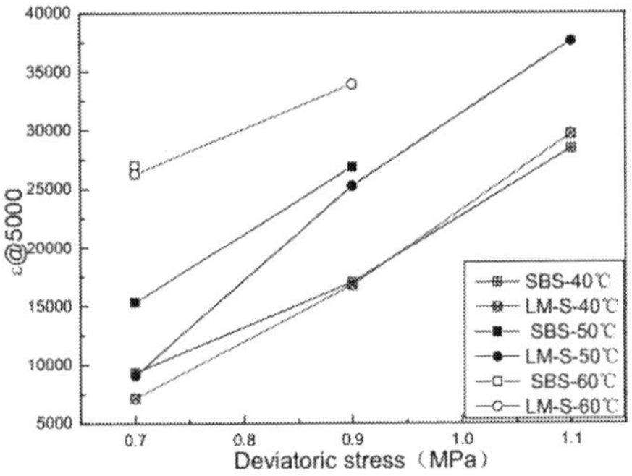

Figure 7. ε@5000 of asphalt mixture

(3) Comparison of flow number FN

The same research results show that for repeated loading tests with confining pressure, the flow number FN can be used as a basic index to evaluate the high-temperature performance of asphalt mixtures. It can be seen from the flow number FN of the two asphalt mixtures in Figure 8 that the flow number of the LM-S modified asphalt mixture is generally greater than that of the SBS under the same deviator stress and temperature combination, indicating that the high temperature of the LM-S modified asphalt mixture performance is better than SBS. With the increase of deviator stress, the flow number FN changes little at 40°C, and basically does not change with the deviator stress, in other word, the flow number is not sensitive to the change of deviator stress at 40°C. It becomes sensitive when the temperature reaches 50°C and the deviator stress goes to 0.9MPa or higher. When the temperature reaches 60°C, the flow number FN decreases rapidly with the increase of deviator stress, which means that the flow number is very sensitive to the change of deviator stress. Although we can draw the conclusion that the flow number of LM-S modified asphalt mixture is generally greater than that of SBS, since the flow number FN at 40°C is as high as 9500 or more, and the test termination limitation is only 10,000 cycles, the results are obtained within 10,000 cycles. The flow number is often not the real flow number. This is also the reason that the flow number is concentrated at about 9500 times at 40°C. At this time, the flow number cannot reflect the high-temperature performance of the actual road surface. While, at 60°C, the specimen will be destroyed quickly when the deviator stress is large, the flow number is very small at this time and the variability is large, which also cannot reflect the high-temperature performance of the asphalt mixture.

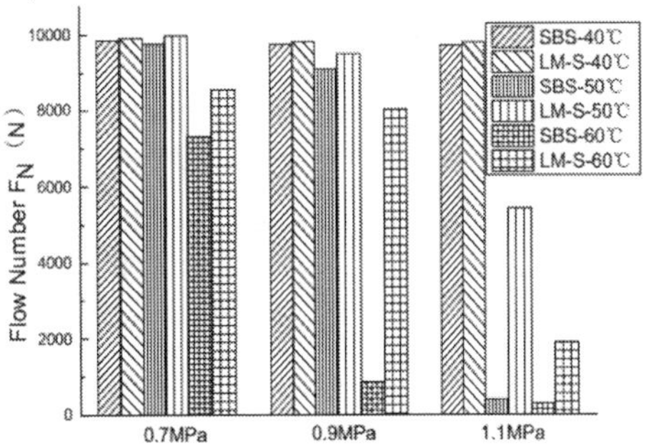

Figure 8. Flow number FN of asphalt mixture

(4) Comparison of creep rate

Creep rate has always been used as a basic indicator for evaluating high-temperature performance. From the creep rate of the two asphalt mixtures in Figure 9, it can be seen that with the increase of deviator stress, the creep rates of the two asphalt mixtures are increasing. When the temperature exceeds 50°C and the deviator stress reaches 0.9MPa, the creep rate increases rapidly, indicating that deviator stress increases, fluidity of the asphalt mixture increases, while the ability of deformation resistance at high temperature decreases. At the same time, the higher the temperature is, the greater the creep rate is, that is to say that while the ability of deformation resistance in high temperature decreases when temperature increases. However, when the temperature reaches 60°C and the deviator stress exceeds 0.7 MPa, the SBS modified asphalt mixture will be quickly destroyed. The second stage is very short and it is difficult to accurately determine the corresponding creep rate. Under the same temperature and the same deviator stress, the creep rate of LM-S modified asphalt mixture is slightly lower than that of SBS, indicating that the deformation resistance of LM-S modified asphalt mixture is slightly better than that of SBS modified asphalt mixture. This is because the binder viscosity of the LM-S modified asphalt mixture is slightly higher than that of the corresponding SBS, resulting in a stronger anti-deformation ability of the formed asphalt mixture, showing slightly better high-temperature performance than the SBS modified asphalt mixture.

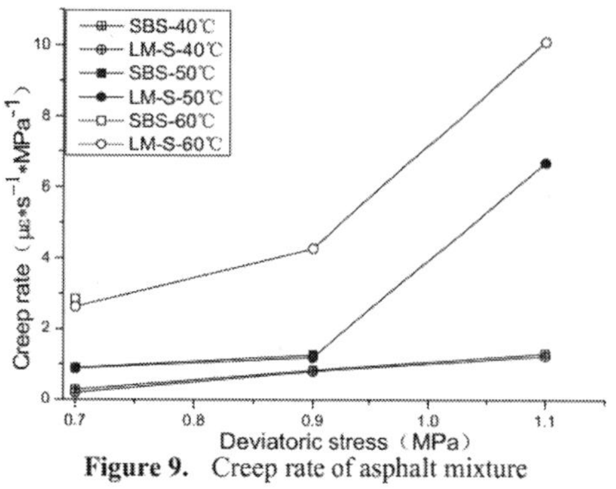

Figure 9. Creep rate of asphalt mixture

Through the evaluation of the high-temperature performance of the two asphalt mixtures by the different angle of four basic indicators: axial permanent deformation, ε@5000, flow number FN, and creep rate, the following can be concluded. The axial permanent deformation index reflects corresponding permanent deformation changing rules under different cycles. The changing law of deformation can significantly distinguish the permanent deformation of two kinds of asphalt mixtures in test condition, which can be considered as an excellent index to evaluate the high-temperature performance of asphalt mixtures. ε@5000 only takes accumulative permanent deformation corresponding to the first 5000 cycles, without considering the entire process of deformation; moreover, when the temperature and deviator stress are both large, the test specimen will be destoryed before reaching 5000 cycles, which means that the ε@5000 is not exist at all, so it is difficult to use this index to comprehensively measure the pros and cons of high-temperature performance of the two asphalt mixtures. Although the flow number FN exists in different conditions, but the largest number of loading cycles is 10,000. The flow number FN under the combination of lower temperature and small deviator stress is very large or even more than 10,000 originally, the test instrument can only take the number of loading times corresponding with the minimum value of the current permanent deformation rate as the flow number FN in calculating. In fact, this value is very different from the real flow number. Therefore, if the flow number FN be chosen as the evaluation index of high-temperature performance, there will be the distortion problem under low temperature and small deviator stress. The creep rate uses the slope of the creep curve in the stable phase as the evaluation index. The curve in the stable phase changes smoothly, which can indirectly reflect the anti-deformation ability of the asphalt mixture to a certain extent. It has a certain objective representativeness, but similarly to ε@5000, when the temperature and deviator stress are both large, the test specimen will be destroyed immediately before entering the stable phase, or the stable phase is too short to get the accurate value of the creep rate, so the creep rate is difficult to fully measure the high-temperature performance of two asphalt mixture.

4 Conclusion

(1)For LM-S and SBS modified asphalt mixtures, the corresponding permanent deformation under different cycles can be observed by the axial permanent deformation any time, which can be used as an excellent index for evaluating the high-temperature performance of the two asphalt mixtures.

(2) The ε@5000 and creep rate under higher temperature and larger deviator stress are limited by conditions, so the corresponding value does not exist or is difficult to find. Both are not suitable to be directly used as an evaluation index, but these two are recommended in lower temperature and smaller deviator stress, especially the creep rate.

(3) The flow number FN, due to the limitation of conditions, leads to the distortion of the flow number under lower temperature and smaller deviator stress, but it has better representativeness under larger deviator stress and temperature conditions.

(4) Considering the two asphalt mixtures used in the study comprehensively, this article finally recommends the axial permanent deformation as a basic index for evaluating high-temperature performance.

(5) Different indicators of test show that the high-temperature performance of LM-S modified asphalt mixture is better than that of SBS modified asphalt mixture. It is recommended to use LM-S modified asphalt mixture instead of SBS modified asphalt mixture to improve anti-rutting ability of pavement.

References

1. DaquanSun, WeiminLu. Investigation and Improvement of Storage Stability of SBS Modified Asphalt[J]. Liquid Fuels Technology, 2003, 21(5-6):901-910.
2. DaquanSun, WeiminLu. Investigation and Improvement of Storage Stability of SBS Modified Asphalt[J]. Liquid Fuels Technology, 2003, 21(5-6):901-910.
3. ZHENG Chuan-feng, WANG Lei, XU Ya zhi, ZHANG Li, FU Ji Experimental Study on Recycled SBS Asphalt Mixture for use in Surface Course[J]. Journal of Civil Architectural&Environmental Engineering, 2011,33(S1):143-147.
4. Yu Yin.SBS modified asphalt mixture performance[J]. Jiangxi building materials 2016(10):152.
5. DaquanSun, WeiminLu. Investigation and Improvement of Storage Stability of SBS Modified Asphalt[J]. Liquid Fuels Technology, 2003, 21(5-6):901-910.
6. Xu S, Yu J, Hu C, et al. Laboratory evaluation of rejuvenation effect of reactive rejuvenator on aged SBS modified bitumen[J]. Materials & Structures, 2017, 50(6):233.

7. Hu Fang. The Study on Anti-aging Propertiesof SBS Modified asphalt[J]. Shanxi Communications Technology. 2012(05):17-19.
8. Wang Chunjiang. Star SBS modified asphalt research and industrial production application[J]. ShandongChemical Industry. 2016,45(08):70-72.
9. Lei Min. Impact on Rubber modifier of the Properties of asphalt and mixture[D]. Wuhan University of Technology,Wuhan University ofTechnology, 2014.
10. Li Guanlong, Study on the performance of SBS/LM-S composite modified asphalt[D]. EastChina University of Science and Technology.2016
11. Wang Zhaohui, Fan Qiaojuan, Li Yanwei, et al. Effect of Mixing Process on Road Performance of Low-carbon Multi-functional Modified Asphalt Mixture [J]. Journal of China andForeign Highway;, 2016(6):234-238.
12. Fang H, Liu Q, Mo L, et al. Characterization of three-stage rutting development of asphalt mixtures[J]. Construction & Building Materials, 2017, 154:340-348.
13. Guo Yanmei,Ni Fujian. Analysis of NonlinearViscoelastic Response of Modified Asphalt Based on Repeated Creep[J].Journal of BuildingMaterials, ,2013,16(05):908-912
14. He Changxuan, Fan Yinghua Triaxial Permanent Deformation Test of Asphalt Mixture in Repeated Loading[J].Highway, 2010(06):181-184.
15. Zhang Yuqing, Huang Xiaoming.Triaxial Repeated Load Creep Test and Study on Deformation Characteristics of Asphalt Mixture at High Temperature and Heavy Load [J] Highway 2006(12):151-156.
16. Yao H, You Z. Performance of Micro- and Nano-Modified Asphalt Mixtures Through Flow Number and Moisture Susceptibility Evaluations[J]. Journal of Testing & Evaluation, 2017, 45(6):20150434.
17. Lu Hewei, Zhang Hongchao, Wang Jian, Guo Yinan. Research on Asphalt Mixture Stability at High Temperatures Applying MMLS3[J]. Journal of Building Materials, 2011,14(05):624-629.
18. Li Peilong, Zhang Zhengqi, Wang Binggang High Temperature Creep Deformation Behavior and its Mechanism of Asphalt Mixture[J].Journal of Building Materials,2012,15(03):422-426.
19. Li Zhaosheng, Tan Yiqiu, Wu Sigang, Yang Fuyu.The Effect of Freeze-thaw Cycle on Mechanical Properties of the Asphalt Mixture[J].Journal of Harbin Engineering University. ,2014,35(03):378-382.
20. Zhou Zhigang, Liu Xuguang, Luo Suping, ShaXiaopeng. Effect of Water intrusion on Performance of Asphalt Mixture[J].Journal of Central South University (Natural Science Edition), 2016,47(04):1359-1367.

https://doi.org/10.1051/e3sconf/202129302030

Seismic vulnerability assessment of multi-storey subway station structure

Tong Liu[1,*], Zhixin Li[2], Qinghe Wang[3], Huaiyu Xu[3]

[1]School of Management, Shenyang Jianzhu University, Shenyang 110168, China.

[2]Shenyang Urban Construction Tianhe Real Estate Company Limited, Shenyang 110168, China.

[3]School of Civil Engineering, Shenyang Jianzhu University, Shenyang 110168, China.

Abstract. The safety of underground structure under seismic load is an important basis of the normal operation of underground rail transit system. Structural seismic vulnerability assessment based on incremental dynamic analysis method can evaluate the probability of the structure exceeding a certain limit state under a specific seismic intensity in terms of probability. In this paper, the seismic vulnerability of a multi-storey subway station structure is evaluated using this method, and a two-dimensional finite element model of both soil and structure is established by finite element software ABAQUS. The vulnerability curve is obtained through incremental dynamic analysis and mathematical statistics. Based on this curve, the probabilities of the structure exceeding four seismic limit states are obtained under the design seismic intensity and the rarely occurred seismic intensity at 7 degree. Results show that this station may attain slight damage under design seismic intensity of 7 degree, and may attain life safety at the rarely occurred seismic intensity at 7 degree.

1 Introduction

In earlier years, the researchers generally agree that the seismic performance of underground structure is better than that of the surface structure. However, the Daikai subway station encountered M type failure mode during the Kobe earthquake at Japan in 1995[1,2]. Therefore, seismic performance of underground structure should obtain enough attention. With the continuous development of science and technology, underground space is fully exploited and the layer of underground subway station structure increases. Consequently , it is necessary to do research on the seismic performance of multi-storey underground subway station.

Among the approaches of evaluating seismic performance, seismic vulnerability assessment based on incremental dynamic analysis(IDA) can evaluate the probability of the structure exceeding a certain limit state under a specific seismic intensity in terms of probability[3,4]. At the same time, the method can fully take the randomness of seismic ground motion and structural dynamic response characteristics under strong earthquake into consideration. This method has been applied to the seismic vulnerability assessment of conventional underground structures[5,6].

Considered the advantages of seismic vulnerability assessment, in this study, the seismic vulnerability of multi-storey subway station is carried out. A two-dimensional numerical model of a four-storey three-span structure and the surrounding soil is developed

Table1. Soil parameters of the example structure.

Sequence	Name	Depth (m)	Unit Weight (kN/m^3)	Elastic Modulus (MPa)	Poisson's Ratio	Friction Angle (°)	Cohesion (kPa)
1	Man-made fill	0~0.5	1820	14	0.32	14.5	21
2	Man-made fill	0.5~1.3	820	14	0.32	14.5	21
3	Brownish yellow to grayish yellow silty clay	1.3~3.2	820	14	0.32	14.5	21
4	Grey silty silty clay	3.2~8.4	760	10	0.33	12.5	14
5	Grey silty clay	8.35~14.8	680	7	0.37	9.5	14

* Corresponding author: 262liut@tongji.edu.cn

https://doi.org/10.1051/e3sconf/202129302030

6	Gray clay	14.8~18.2	760	11	0.34	13.5	15
7	Gray silty clay	18.2~22.4	790	13	0.31	16.5	17
8	Gray silty clay interbedded with silt	22.4~29.2	800	17	0.29	18	16
9	Gray silty clay	29.15~32.3	840	16	0.31	16	24
10	Gray-green silty clay	32.3~33.4	980	26	0.26	19	34
11	Grass yellow to cyan sandy clay	33.4~39.1	900	33	0.26	29	5
12	Grass yellow to gray silt	39.05~66.6	870	43	0.24	31.5	3
13	Gray silty clay	66.6~71.6	820	17	0.23	14.5	26
14	Gray fine silty sand	71.55~85	890	45	0.25	34.5	0

using the finite element software ABAQUS. Through incremental dynamic analysis, the vulnerability curves at four limit states are obtained, based on which the exceeding probabilities corresponding to four limit states at specific seismic intensity are ensured.

2 Numerical modeling

2.1 Typical multi-storey subway station structure

As shown in Figure 1, a four-storey three-span subway station is considered as a typical example of multi-storey subway station. The first underground floor is the stereo garage, the second underground floor is the lobby floor, the third underground floor is the device floor, and the bottom floor is the island platform floor. The section width of the station is 23.6m, and the height is 29.11m. The size of central column at the first and second underground floors is 0.7m by 1.1m, while the size of central column at the third underground floor and the bottom floor is 0.7m by 1.4m, and the longitudinal space of the central column along the station is 8m. The buried depth of the station is 2.9m and the soil properties are shown in Table 1.

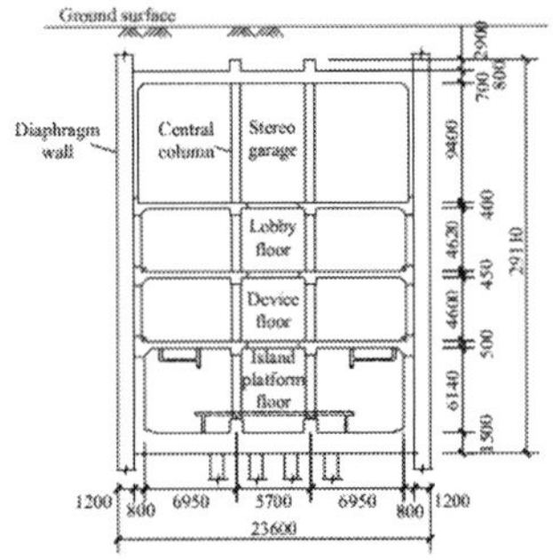

Figure 1. Cross section of the multi-storey subway station structure.

2.2 Numerical model

According to the general profile , a two-dimensional finite element model of soil and structure is established using finite element software ABAQUS[7], the dimension of which is 1000 m long and 85 m high, as shown in Figure 2. The structure members are simulated through beam element (B21) and the material properties of concrete and steel are illustrated in Table 2. The layered soil is simulated through 4-nodes plane strain element (CPE4R) and the quadrilateral plane strain infinite element (CINPE4). The interface between the soil and the structure is modeled as a frictional surface, among which coefficient of friction μ is equal to 0.4 and friction angle is 22°. Moreover, no cohesion exists between structure and soil and the ground motion is imposed at the bottom of the model.

Table2. Material Parameters of multi-storey subway station.

Component	Material					
	Concrete				Steel	
	Elastic Modulus (GPa)	Axial Tensile Strength (MPa)	Axial Compression Strength (MPa)	Possion's Ratio	Elastic Modulus (GPa)	Yield strength (MPa)
Central column	33.5	2.51	29.6	0.2	200	400
Others	31.5	2.20	23.4	0.2	200	335

 https://doi.org/10.1051/e3sconf/202129302030

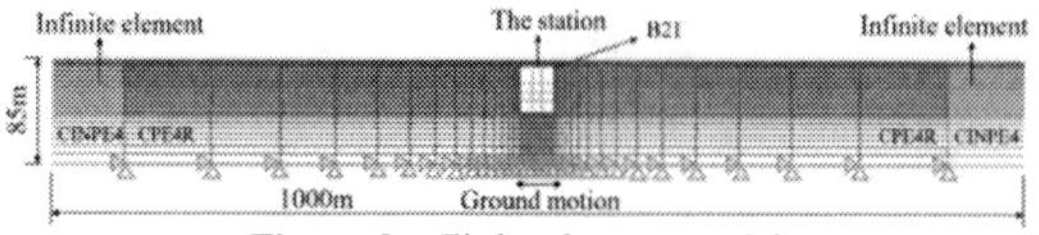

Figure 2. Finite element model

3 Seismic vulnerability assessment process

3.1 Incremental dynamic analysis

Based on the procedure of IDA[4], the ground motion records are selected according to the site type, as shown in Table 3. Peak acceleration at the bottom of structure (PBA) is selected as intensity measure(IM) and the maximum inter-story drift angle(θ_{max}) is chosen as the damage measure(DM). Besides, the ground motion records are scaled with Peak Velocity as 3, 5, 10, 20, 30,

40, 50, 60, 70, 80cm/s, and with nonlinear dynamic time-history analysis under each ground motion intensity,

the IDA curves can be obtained, as shown in Figure 3. The range of θ_{max} is 0 to 0.045. θ_{max} increases gradually with the increasing of ground motion intensity and apparent discrepancy exists among each IDA curve. To better statistic IDA data, the IDA curves are arranged into 16%, 50% and 84% fractile curves to analyze the IDA calculation results in the mean sense, as illustrated in Figure 4.

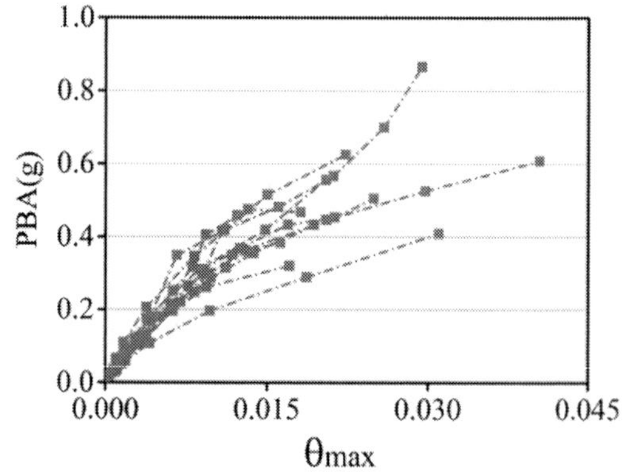

Figure 3. IDA curves

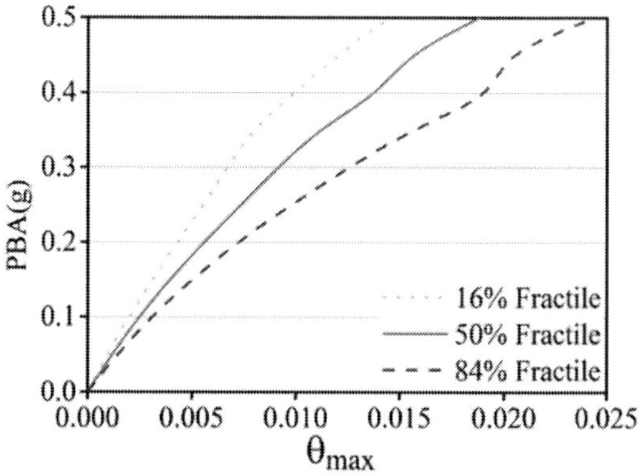

Figure 4. Fractile curves

Table3. Twelve ground motion records selected.

No.	Event	Station	Component	PGA(g)	PGV(cm/s)
M1	Imperial Valley-06, 1979	El Centro Array #3	E03140	0.267	47.97
M2	Imperial Valley-06, 1979	El Centro Array #3	E03230	0.223	43.29
M3	Loma Prieta, 1989	APEEL 2 - Redwood City	A02043	0.274	53.65
M4	Loma Prieta, 1989	APEEL 2 - Redwood City	A02133	0.220	34.12
M5	Loma Prieta, 1989	Foster City - Menhaden Court	MEN270	0.110	21.98
M6	Loma Prieta, 1989	Foster City - Menhaden Court	MEN360	0.119	20.93
M7	Loma Prieta, 1989	Treasure Island	TRI000	0.100	15.59
M8	Loma Prieta, 1989	Treasure Island	TRI090	0.160	33.20
M9	Superstition Hills-02, 1987	Imperial Valley Wildlife Liquefaction Array	IVW090	0.179	31.67
M10	Superstition Hills-02, 1987	Imperial Valley Wildlife Liquefaction Array	IVW360	0.208	36.21
M11	Northridge-01, 1994	Carson - Water St	WAT180	0.091	6.33
M12	Northridge-01, 1994	Carson - Water St	WAT270	0.088	8.32

3.2 Results of structural seismic vulnerability assessment

Seismic vulnerability assessment means calculate the probability of structural dynamic response exceeding a certain limit state LSi with different ground motion intensity levels, i.e $P(LSi\,|IM = im)$. Usually, we assume that DM for IM's condition distribution follows the lognormal cumulative distribution function. Then,

$$P(LS_i|IM = im) = P(DM > dm_i|IM = im) = 1 - \Phi\left(\frac{\ln dm_i - \mu_{\ln DM|IM=im}}{\sigma_{\ln DM|IM=im}}\right) \quad (1)$$

In formula (1), $\mu_{\ln DM|IM=im}$ is the mean of $\ln(DM)$ and $\sigma_{\ln DM|IM=im}$ is the standard deviation of $\ln(DM)$. Φ is the standard cumulative normal distribution function.

Based on the formula, we can obtain the seismic vulnerability curve with IM as horizontal axis, exceeding probability as vertical axis and the seismic vulnerability under specific ground motion intensity can be investigated. Based on the procedure mentioned above, the limit states of this station should be ensured firstly, which can refer to Liu et al.[5], as illustrated in Table 4.

Then, the vulnerability curve can be obtained using formula (1) and nonlinear regression analysis of MATLAB, as shown in Figure 5.

In this study, we mainly investigate structural seismic vulnerability under the design seismic intensity and the rarely occurred seismic intensity at 7 degree[8]. For design seismic intensity of 7 degree, PGA is equal to 0.1g with the exceeding probability in 50 years of 10%.

Table4. Thresholds at each limit state.

Limit state (LS)	Operational (OP)	Slight Damage (SD)	Life Safety (LY)	Collapse Prevention (CP)
θ_{max}	0.0011	0.0024	0.0065	0.0116

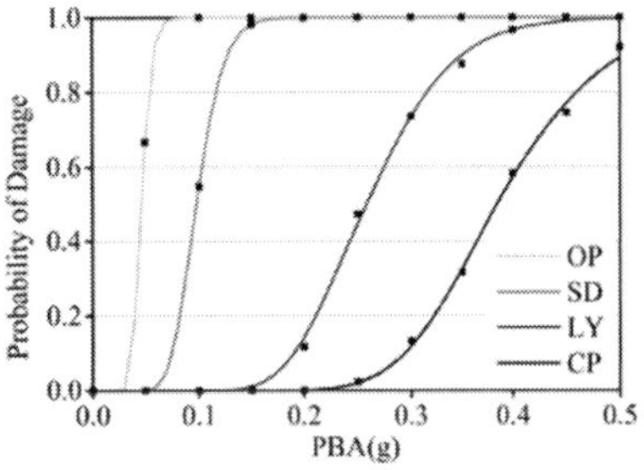

Figure 5. Seismic vulnerability curves

Table5. Exceeding probability at each limit state.

Seismic intensity	PGA(g)	PBA(g)	Limit state			
			Operational	Slight Damage	Life Safety	Collapse Prevention
Design seismic intensity at 7 degree	0.1	0.074	99.9%	11%	0%	0%
Rarely occurred seismic intensity at 7 degree	0.22	0.160	100%	99.2%	2.2%	0%

For rarely occurred seismic intensity of 7 degree, PGA is equal to 0.22g with the exceeding probability in 50 years of 2%. Then, we can calculate the corresponding mean value of PBA under specific seismic intensity and the exceeding probability of each limit state can be found through vulnerability curve in Figure 5, and the results are shown in Table 5.

It can be seen that the subway station structure may attain slight damage under design seismic intensity of 7 degree and the probability is 11%. When it comes to the rarely occurred seismic intensity at 7 degree, the subway station may attain life safety limit state with the probability of 2.2%. What's more, from the seismic vulnerability assessment result, the multi-storey subway station basically meets the design concept for underground structure that not damaged in medium earthquake and repairable in large earthquake.

4 Conclusion

In this study, seismic vulnerability assessment of multi-storey subway station is conducted based on incremental dynamic analysis. The results are as follows.

(1) There's apparent discrepancy of IDA curves with different ground motion input.

(2) The multi-storey subway station structure investigated may attain slight damage under design seismic intensity of 7 degree and the probability is 11%.

(3) When it comes to the rarely occurred seismic intensity at 7 degree, the subway station may attain life safety limit state with the probability of 2.2%.

Acknowledgements

The research work in this paper was supported by LiaoNing Revitalization Talents Program, China (XLYC1902027), by Doctoral Scientific Research Foundation, China (2019 - BS- 193), by Science and Technology Project of MHRUD, China (2019 - K - 054), by Shenyang Science and Technology Project, China (RC200143) and Tianjin Science and Technology Project, China (18ZXGDGX00030).

References

1. Iida H, Hiroto T, Yoshida N, et al. Damage to Daikai subway station[J]. Soils & Foundations. 1996, 283-300.
2. Huo H, Bobet A, Fernández G, et al. Load Transfer Mechanisms Between Underground Structure and Surrounding Ground: Evaluation of the Failure of the Daikai Station[J]. Journal of Geotechnical & Geoenvironmental Engineering, 2005, 131(12): 1522-1533.
3. Cornell C A. Progress and Challenges in Seismic Performance Assessment[J]. Peer Center News. 2000, 20(2): 130-139.
4. Vamvatsikos D, Cornell C A. Incremental dynamic analysis[J]. Earthquake Engineering & Structural Dynamics. 2002, 31(3): 491-514.
5. Liu T, Chen Z, Yuan Y, et al. vulnerability Analysis of a Subway Station Structure by Incremental Dynamic Analysis[J]. Advances in Structural Engineering, 2017, 20(7): 1111-1124.
6. Zhong Z L, Shen Y Y, Zhao M, et al. Seismic vulnerability assessment of the Daikai subway station in layered soil. Soil Dynamics and Earthquake Engineering, 2020, 132: 106044.
7. ABAQUS, Users Manual V. 6.10-1. 2010. Dassault Systemes Simulia Corp., Providence, RI.
8. GB50011 (2010). Code for Seismic Design of Buildings, China Architecture & Building Press, Beijing, China.

Construction and application of factory digital twin model based on BIM and point cloud

Guocheng Qin[1,*], Ling Wang[1], YiMei Hou[1], HaoRan Gui[1], and YingHao Jian[1]

[1]Chongqing Institute of Surveying and Monitoring for Planning and Natural Resources, Hi-Tech Service Trade Industrial Park No.64, Middle Section of Huangshan Avenue, Building N, YuBei District, Chongqing, China

Abstract. The digital twin model of the factory is the basis for the construction of a digital factory, and the professional system of the factory is complex. The traditional BIM model is not completely consistent with the actual position of the corresponding component, and it is difficult to directly replace the digital twin model. In response to this situation, relying on a certain factory project, the point cloud is used to eliminate the positional deviation between the BIM model and the factory during the construction phase, improve the efficiency and accuracy and reliability of model adjustment and optimization, and , realize the conversion from BIM model to digital twin model. A novel algorithm is developed to quickly detect and evaluate the construction quality of the local structure of the factory, so as to input the initial deformation data of the structure into the corresponding model and feed back to the construction party for improvement. The results show that the digital twin model, which is highly consistent with the actual location of the factory components, not only lays a solid foundation for the construction of a digital factory, but also further deepens the integration and application of BIM and point clouds.

1 Introduction

With the introduction of concepts such as Industry 4.0 and intelligent manufacturing, the digital factory has become an important strategic goal of many companies, and building a digital twin model is the first step in building a digital factory. Large factories are characterized by multiple specialties and complex systems. The BIM model of the factory created based on two-dimensional drawings in the design phase has a lot of professional interference. Although collision checking can temporarily avoid interference between professional systems, it is in the construction phase of the factory. Due to construction errors and random errors, the position deviation between the construction site and the BIM model is becoming more and more serious. The traditional "rework adjustment" has low efficiency and accuracy, and it is difficult to efficiently and systematically solve the existing problems of the model, and it is even more difficult to get a highly consistent digital twin model at the completion stage.

Li Lingjun et al. [1] introduced the application ideas of BIM technology through collaborative design and collision checking, and showed that BIM technology can help solve the coordination problems between various disciplines. Zeng Songlin et al. [2] relied on the technical transformation project of a cigarette factory to discuss the use of BIM technology in the pipeline synthesis and collision inspection of the factory, and pointed out that the application of BIM can effectively solve the problems in the two-dimensional design. Alberto Sánchez et al. [3] used accurate 3D models to integrate factory data into a collaborative work environment, enabling BIM and HBIM to play a role in the restoration of the old factory. Jinsol Kim et al. [4] introduced how to use BIM to reduce building energy consumption and realize a digital factory through two cases. Scholars at home and abroad did not integrate BIM and 3D laser scanning technology in their application research in digital factories, nor did they consider creating a model that is highly consistent with the actual positions of factory components, which plays a fundamental role in the construction of digital factories.

This article will focus on the technical integration of BIM and 3D laser scanning to improve the optimization of the factory's BIM model in the construction phase, adjust the efficiency and accuracy, and find the next stage of construction problems in advance. After that, three-dimensional scanning is used to detect and evaluate the small deformation of the factory floor to obtain a more useful and reliable digital digital twin model, which lays a solid foundation for the further construction of the digital factory and future operation and maintenance.[5].

2 Technical route

The factory project based on this article covers an area of about 10,000 square meters, and the professional systems are huge and complex. It mainly includes four majors: architecture, structure, electromechanical HVAC and process equipment, and each major is divided into several

* Corresponding author: guochengqin@foxmail.com

subsystems. The spatial positions of majors and sub-departments are arranged alternately.

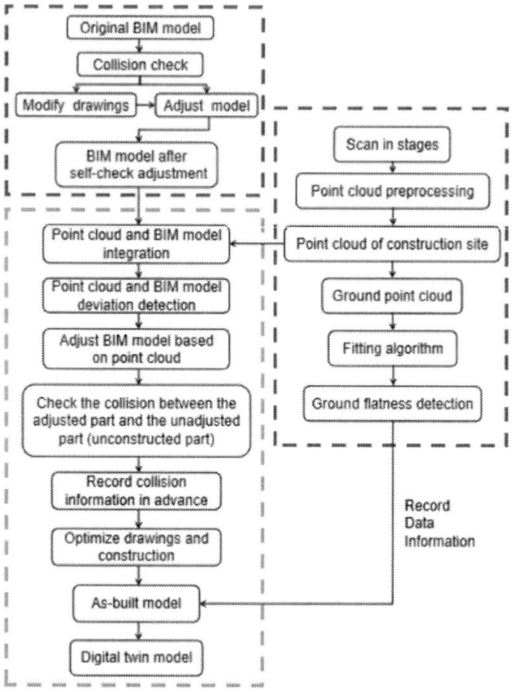

Figure 1. Technical flow chart

The main technical route of this article is shown in Figure 1. It mainly consists of three major components. The first is the BIM technology in the blue area, the second is the 3D laser scanning technology in the red area, and the third is the fusion application of the two in the yellow area. The three finally converge in the creation of the digital twin model.

3 Original model self-check

Before the start of construction, the 3D modeling software used to create the original BIM model of the factory was Bentley's AECOsim Building Designer (AECO) [6]. The software is more popular in the industrial field in Europe, and is often used in the design, analysis, construction, etc. of automotive manufacturing plants.

The original BIM model is based on two-dimensional drawings. Due to the limitations of two-dimensional drawings in three-dimensional visualization, it is difficult to avoid collisions and interferences between professional systems. Therefore, the original BIM model needs to be checked in turn by professional and sub-system. Regardless of whether it is a drawing design error or an error in the model creation process, adjustments are required. If it is a problem with the drawing, you need to rework the drawing and optimize the design first. The result of the partial collision before and after adjustment is shown in Figure 2.

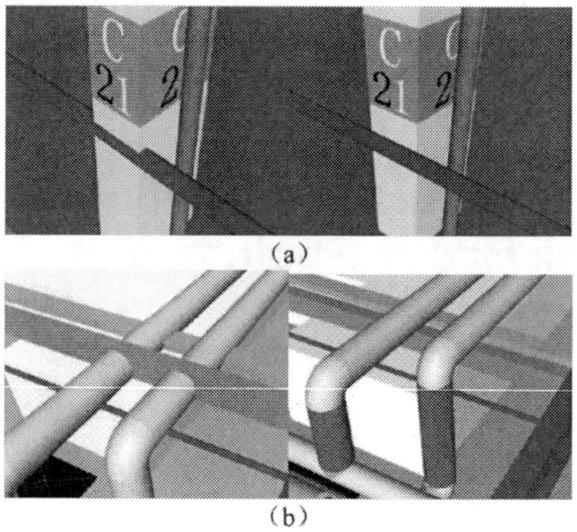

Figure 2. (a) Pipe and structural column collision (b) Bridge and pipe collision

4 Point cloud advanced collision check

4.1 Point cloud data collection

The 3D laser scanner used this time is FARO Focus3D X330, as shown in Fig. 3, it is a kind of precision measuring equipment, the scanned point cloud is as realistic as an image [7] .

Figure 3. Photos of the scene

In the construction phase, as shown in Figure 3, the original point cloud is obtained by three-dimensional scanning of the constructed parts, and the optimized factory point cloud is obtained after point cloud splicing, point cloud noise reduction, and point cloud sampling. Shown in Figure 4.

Figure 4. Factory point cloud

4.2 BIM model review and optimization

Integrate the collected factory point cloud data with the BIM model of the factory. The point cloud data and the BIM model are usually not in the same coordinate system. It is necessary to use the point cloud and multiple feature points of the corresponding model to match, as shown in Figure 5.

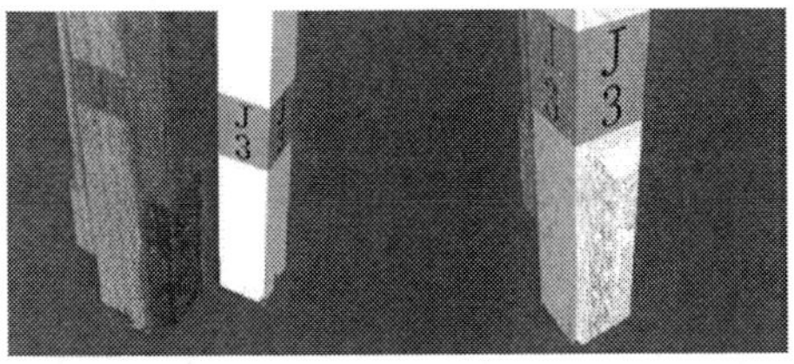

Figure 5. Structural column feature matching

After completing the integration of the point cloud and the BIM model, by checking the error-prone construction parts, the position deviation caused by the construction error can still be found. Under the condition that no other new interference collisions are added, the model is adjusted according to the point cloud, and some deviations are shown in Figure 6.

Figure 6. Geometric deviation between point cloud and model

After completing the model adjustment of the constructed part, the model is checked for collision again. The position of the model here has been consistent with the height of the point cloud, which is equivalent to the collision check of the point cloud and the model. This step is to check whether the parts that have been completed and adjusted for position collide with the parts that will be constructed next. If they happen, record them in advance, analyze the reasons, optimize drawings and construction plans, and realize advanced collision check of point clouds.

5 Deformation detection and digital twin model

5.1 Ground flatness detection

Due to the requirements of factory production and manufacturing, automated and intelligent self-moving machinery and equipment, or precision processing instruments, are used after production. This has high requirements for the flatness of the internal ground. Regardless of its efficiency or accuracy, traditional detection instruments have certain difficulties in detecting geometric micro-deformation in large-scale factories [8].

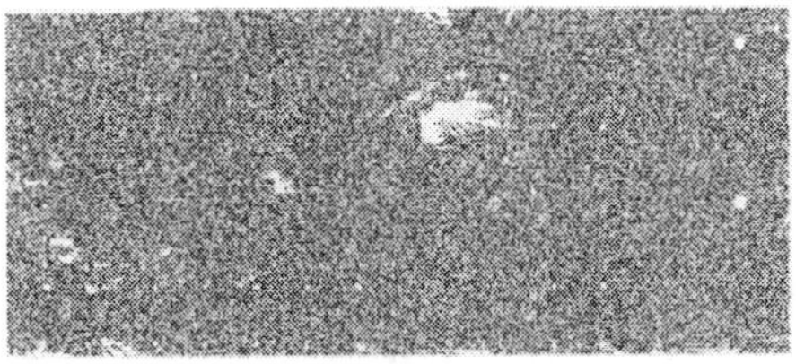

Figure 7. Ground point cloud

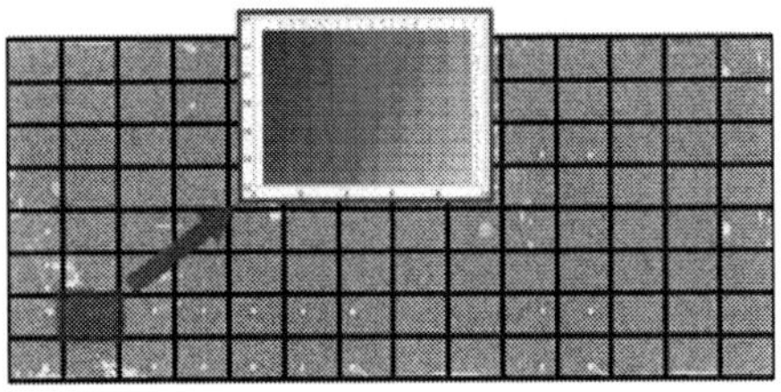

Figure 8. Plane diagram of best fitting of single grid

Segment the point cloud after preprocessing to obtain the ground point cloud, as shown in Figure 7. Aiming at the detection of the flatness of the super-large area of the ground point cloud, this paper studies and proposes a RANSAC-based flatness solving algorithm. The specific steps are shown in Figure 9 [9].

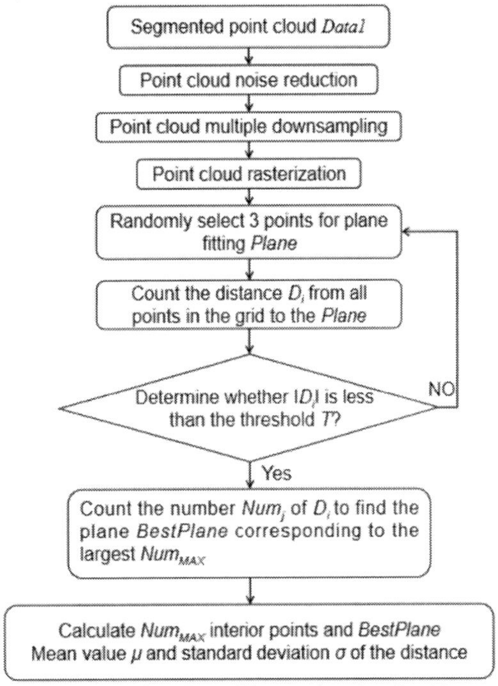

Figure 9. Flow chart of flatness detection algorithm

In addition to improving the operating efficiency of massive point cloud processing and calculation, the algorithm for rasterization, as shown in Figure 8, can also assist in positioning a large area of the ground, making it easier for construction personnel to find locations with poor flatness. Finally, the average value μi and standard deviation σi of all grids are statistically analyzed, focusing on the grids with larger average value and standard deviation, as shown in Figures 10 and 11.

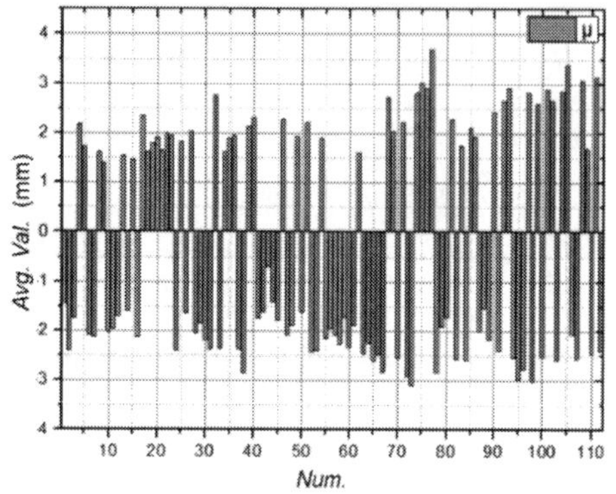

Figure 10. Average value of single grid

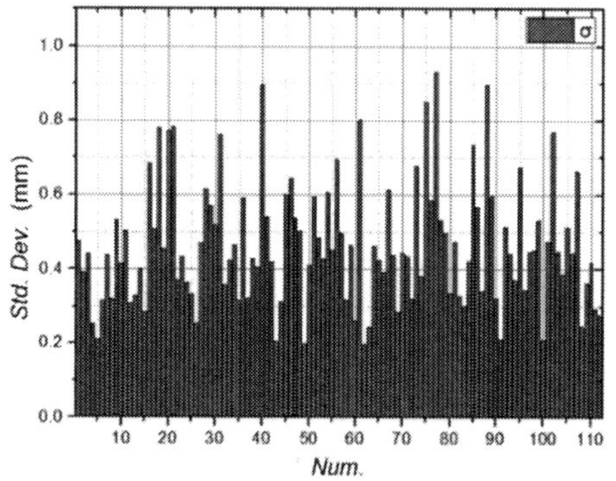

Figure 11. Standard deviation of single grid

5.2 Digital Twin Model

After completing the inspection of the ground flatness of the factory, the detected information is evaluated and then handed over to the construction party for reference and optimization of the construction plan, and the information is correspondingly entered into the digital twin model [10]. With the progress of the construction progress, the final digital model is obtained, as shown in Figure 12. The spatial position of each component of the model is consistent with that of the completed factory, and it carries the geometric deformation information of the factory floor, which is convenient for project management using BIM technology in the later stage [11].

Figure 12. Digital model of factory

6 Conclusion

In this paper, through the fusion application of Bim and 3D laser scanning technology, the original BIM model has been self checked and adjusted for many times. According to the scanning site point cloud, the BIM model has been adjusted in stages, and the collision between the adjusted BIM model and the non construction part model has been checked, so as to find out the problems ahead of time, and to facilitate the response measures of the construction scheme in advance.

By using the algorithm proposed in this paper, the scanned point cloud is quickly detected for the construction quality, the later construction is optimized, and the project management level is improved. The information obtained is entered as the initial deformation data of the structure into the corresponding component model, which is convenient for the later completion acceptance and operation and maintenance. Finally, the BIM model is transformed into a digital twin model, laying the foundation for the digital construction of the factory, and further strengthening the fusion application of BIM and 3D laser scanning technology.

Acknowledgments

Thanks to the owners and partners of the project platform.

References

1. L.J. Li, C. Lu, Wt. Trans. Eng. **06**, 147 (2019)
2. S.L. Zeng, Ent. Tech. Dev. (2013)
3. S. Alberto, G.G. Cristina, Z. Patricia, Appl. Sci. (2019)
4. Y.N. Zhang, J. Kim, G.T. University, J. Inf. Tech. Civil Eng. Arch.(2015)
5. M. Wei, Petro. Auto. **52**, 43 (2016)
6. S.N. Zhao, M. Intel. Prop. Publishing House. (2015)
7. Y. Wang, J.B. Qu, Y.C. You, Surv. Map. Bull. **S1**, 70 (2018)
8. X. Zhang, W. Xiang, T. Ding, T. Song, Yan. Riv. **51**, 129 (2020)
9. M.D. Li, S.P. Jiang, H.P. Wang, Surv. Map.Sci. **40**, 102 (2015)
10. Y. Chen, Bldg. Stru. **50**, 115 (2020)
11. C. Liu, Y.F. Lu, Q. Wen, Z.L. Fu, Sci. Tech. Eng. **20**, 6515 (2020)

Influence of Silica Fume on the Performance of Moderately Weathered Phonolite Concrete

Xiaowei Wu[1,*], Zhaoxi Tang[2]

[1]CCCC Highway Consultants CO.,Ltd. Beijing 100088, China

[2]CCCC Fourth Harbour Engineering Co.,Ltd., Guangzhou 510220, China

Abstract. The effect of silica fume on the performance, mechanical properties, durability and air tightness of airtight concrete was studied. The results showed that with the increase of silica fume content, the concrete performance will decrease, the compressive strength of airtight concrete will increase first, and reach the highest when the silica fume content was 8%, and then it will decrease gradually. With the increase of silica fume content, the corrosion resistance coefficient of concrete increased gradually, then decreased. The permeability coefficient of concrete decreases with the increase of silica fume content, and the air tightness of concrete was improved obviously; In the project, it was advisable to use silica fume to prepare airtight concrete, and the amount of silica fume should be controlled at 4% - 8%.

1 Introduction

The main line of the Kenya Nairobi Expressway Project has a total length of 27km and a design speed of 80 kilometres per hour. It is a national highway A-class highway with asphalt concrete pavement. The 15.710km front of the main line is mainly subgrade, and the rear 11.421km is mainly bridge. Except for the two-way six-lane roadbed standard width of 28.6m in the east ring to the south ring road, the remaining road sections adopt the two-way four-lane 21.6m wide A-class road standard. Rocks in Nairobi and the surrounding areas are dominated by igneous rocks, of which medium weathered sonic rock accounts for a relatively large proportion. Only sonic rock is available around the project. Other rock quarries are far away and costly, and the surrounding sonic rock has problems such as high water absorption and poor solidity [1,2].

The high water absorption rate and poor firmness of aggregate have certain negative effects on the performance of concrete, such as reducing concrete slump and rapid slump loss, resulting in relatively poor mechanical properties of concrete and poor frost resistance [3–7]. For this reason, many scholars have studied to improve the performance of this type of aggregate in the preparation of concrete. The main measures were aggregate pre-wetting treatment, aggregate surface shelling technology, the surface of the aggregate was impregnated with organic hydrophobic materials, and the surface of the aggregate was reinforced with carbon dioxide [8–12]. Referring to the current research results, this paper used higher activity silica fume as a mineral admixture to replace conventional admixtures to study the improvement of the performance of moderately weathered phonolite concrete and promote the application of moderately weathered phonolite in engineering.

2 Materials and methods

2.1 Raw materials

(1) Cement: P·O 42.5 ordinary Portland cement and its properties were shown in Table 1.

(2) Fly ash: II grade fly ash was used, and its fineness is 18.1%, water demand ratio was 101.3%, activity index in 7d was 75.9% and in 28d was 78.5%.

(3) Silica fume: the used silica fume has a silica content of 92.5%, a loss on ignition of 1.2%, a water demand ratio of 121.5%, a specific surface area of 16.8 m2/g, and a 7d activity index of 116%.

(4) Fine aggregate: the fine aggregate is the second zone medium sand with a fineness modulus of 2.6.

(5) Coarse aggregate: the coarse aggregate is processed from Kenyan weathered phonolite, with a continuous gradation of 5-20mm, and the specific technical indicators are shown in Table 2.

(6) Water-reducing agent: the polycarboxylic acid-based high-performance water-reducing agent was used, and its water-reducing rate is 27.5% and gas content is 3.1%.

2.2 Mix proportion

In order to experimentally study the influence of silica fume on the properties of moderately weathered phonolite concrete, the concrete mix proportion was designed. The amount of cementitious material was

* Corresponding author: 44103044@qq.com.

https://doi.org/10.1051/e3sconf/202129302032

398kg/m3, the water consumption was 163kg/m3, the water-binder ratio was 0.41, and the sand ratio was 45. %, silica fume content was designed to be 0%, 4%, 6%, 8%, 10% and 12%. The specific test mix numbers and corresponding concrete mix ratios were shown in Table 3.

Table 1 Physical properties of P·O 42.5 cement

Specific surface area (m^2/kg)	*Standard consistency* (%)	*Setting time* (min)		*Compressive strength* (Mpa)		*Flexural strength* (MPa)	
		Initial setting	Final setting	3d	28d	3d	28d
315	25.7	175	310	30.2	51.3	4.3	8.1

Table 2 Technical indicators of mid-weathered phonolite in Kenya

Technical indicators	Crush value (%)	Los Angeles wear rate (%)	Apparent density (kg/m^3)	Water absorption (%)	Robustness (%)
Testing results	21.6	24.1	2668	2.2	12.8

Table 3 Mix proportion of concrete sample

Sample	Cement (kg/m^3)	Fly ash (kg/m^3)	Silica fume (kg/m^3)	Silica fume content (%)	Fine aggregate (kg/m^3)	Coarse aggregate (kg/m^3)	Water (kg/m^3)
G0	318	80	0	0	824	1025	163
G4	318	64	16	4	824	1025	163
G6	318	56	24	6	824	1025	163
G8	318	48	32	8	824	1025	163
G10	318	40	40	10	824	1025	163
G12	318	32	48	12	824	1025	163

2.3 Testing methods

The airtightness of concrete was tested in accordance with *Standard for Acceptance of Concrete Works in Railway* TB10424-2018 and carried out under constant air pressure. The size of the test piece was a round table with an upper diameter of 175mm, a lower diameter of 185mm, and a height of 150mm. After removing the mold, the samples were cured in a standard curing room for 28 days and continued to air dry for 14-28 days. When the humidity of the specimen was balanced with the atmosphere, the air permeability test can be carried out by the lower air inlet method or the upper air inlet method. The air inlet pressure was 0.3MPa and 0.6MPa. After the voltage was stabilized, the air permeability was read and use to calculate the concrete air permeability coefficient.

According to *Standard for Test Methods of Long-Term Performance and Durability of Ordinary Concrete* GB/T50082-2009, the dry wet cycle sulfate erosion test was carried out on the samples. After forming, one part of the concrete was put into the dry wet cycle sulfate erosion testing machine for dry wet cycle test, and the other part was put into the standard curing room for standard curing. When the concrete dry wet cycle reached 150 times, the compressive strength of concrete under two environments was tested respectively, and the ratio of the strength of concrete under dry-wet cycles to the strength of the marked-cured concrete was taken as the anti-corrosion coefficient.

The test of compressive strength of concrete was conducted in consonance with *Standard for Test Method of Mechanical Properties on Ordinary Concrete* GB 50081-2019.

3 Results and discussion

3.1 Influence of silica fume on the workability of concrete

The work performance of concrete is very important to the construction quality of concrete, and it determines the construction quality of concrete to a certain extent. If the work performance of concrete is good, the construction is relatively easy, and the requirements for workers are low, and the construction intensity of workers is low, which is easy to improve the quality of concrete construction. If the work performance of concrete is poor, on the one hand, the construction is difficult, and workers are easy to slack off. On the other hand, to improve the difficulty of construction, illegal water addition is likely to occur, which will further cause a high risk of construction quality.

For this reason, the experiment researched the influence of silica fume on the working performance of moderate weathered phonolite concrete. The test results were shown in Table 4. It can be seen that with the increase of silica fume content, the performance of concrete showed a downward trend, except for the addition of silica fume. At 4%, the work performance of concrete changed little, and the content of other silica fume increased, and the decline of concrete work performance was more obvious. In terms of slump, the initial slump of the benchmark concrete was 205mm. When the content of silica fume was 4%, the initial slump was consistent with the benchmark concrete.

When the content of silica fume was increased to 6%, 8%, 10%, and 12%, the slump of the concrete decreases by 7.3%, 14.6%, 29.3%, and 36.6%, respectively, compared with the benchmark concrete. The change law of concrete expansion and air content with the increase of silica fume content was similar to that of concrete slump, and both showed a downward trend with the increase of silica fume content.

The fluidity loss of concrete was relatively obvious over time. The standard concrete slump loss 5mm after 1h, the expansion loss was 15mm, and the test group G4 with 4% silica fume content loss 15mm after 1h. The expansion loss was 20mm. It can be seen that after adding silica fume, the performance of the concrete will be more lost over time. The test groups G10 and G12 with higher silica fume content had a smaller initial slump of the concrete. After 1h, The concrete had shown a dry and hard state.

The main reason for the influence of silica fume on the performance of concrete was that on the one hand, the particle size of silica fume is small, and the same mass can replace fly ash to obtain more slurry, which has a beneficial effect on the performance of concrete; on the other hand, the particles silica fume with a smaller particle size can fill the pores between cement and fly ash particles, start a certain water reducing effect, and improve the performance of concrete. However, it is also because the particle size of silica fume is too small, the specific surface area is large, and usually has a large water demand. This effect reduced the working performance of the concrete. When silica fume was mixed into concrete at a low content, the negative impact of silica fume on the performance of concrete was still small, and the performance of concrete did not change much, and when the content of silica fume was high, it had a negative impact on the performance of concrete. The impact has become prominent, and the performance of concrete has decreased significantly. From the above research, it can be known that when using silica fume to prepare airtight concrete, the dosage should not be too high.

Table 4 The influence of silica fume on the performance of moderately weathered phonolite concrete

Sample	Initial moment			After 1h		
	Slump (mm)	Slump flow (mm)	Air content (%)	Slump (mm)	Slump flow (mm)	Air content (%)
G0	205	510	3.1	200	495	2.6
G4	205	490	3.2	190	470	2.8
G6	190	450	2.6	180	430	2.3
G8	175	400	2.2	165	—	1.7
G10	145	—	2.1	—	—	—
G12	130	—	1.7	—	—	—

3.2 Influence of silica fume on the compressive strength of concrete

Compressive strength is the most basic technical index of concrete, and it is also the most important technical index as a structural material. For this reason, the influence of different dosages of silica fume on the compressive strength of concrete was tested. The test results were shown in Fig. 1. It can be seen that with the increase of the amount of silica fume, the compressive strength of the moderately weathered phonolite concrete increased first and reached the highest when the amount of silica fume was 8%, and then gradually decreased.

At the 7d, the compressive strength of the benchmark concrete with a silica fume content of 0% was 29.7MPa, and the test groups G4, G6, G8, and G8 with a silica fume content of 4%, 6%, 8%, 10%, and 12% G10 and G12 were respectively 12.5%, 20.2%, 27.3%, 21.5% and 15.5% higher than the compressive strength of the benchmark concrete. At 28d, the compressive strength of reference concrete with 0% silica fume was 43.6 MPa, and the compressive strength of G4, G6, G8, G10 and G12 with 4%, 6%, 8%, 10% and 12% silica fume were 4.6%, 8.3%, 10.3%, 6.7% and 3.9% higher than that of reference concrete respectively.

The reason why the compressive strength of moderately weathered phonolite concrete first increased and then decreased with the increase of silica fume may be that the silica fume particle size is small, which can well fill the pores between relatively coarse particles such as cement and fly ash in concrete, so as to improve the compactness of concrete and the compressive strength of concrete. On the other hand, silica fume has certain pozzolanic activity in the alkaline environment of cement hydration, its secondary hydration reaction can occur to further improve the compressive strength of concrete. However, due to the large water demand of silica fume, when the silica fume content was too large, the higher water demand increased the viscosity and the forming difficulty of concrete, and then the compactness of concrete became poor during the forming process, which made the improvement trend of concrete compressive strength decrease.

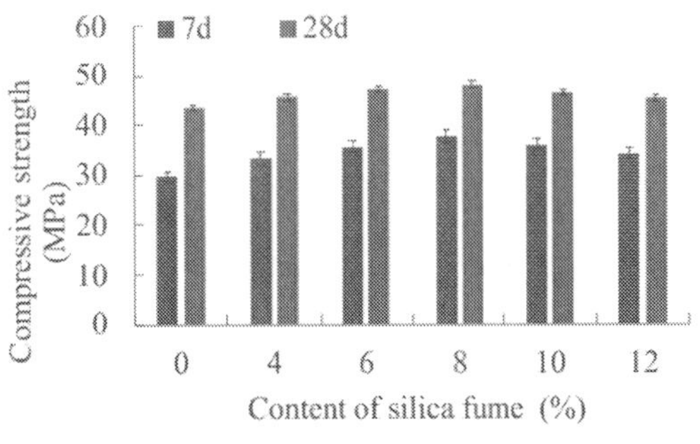

Fig. 1 Effect of silica fume on compressive strength of concrete

3.3 Influence of silica fume on the corrosion resistance coefficient of concrete

In addition to the working performance and mechanical properties, the durability of moderately weathered phonolite concrete is also very important, such as dry wet cycle, sulfate erosion, chloride ion erosion, freeze-thaw damage of concrete and other factors. Therefore, in addition to considering the working performance and mechanical properties to meet the design requirements, the durability of moderately weathered phonolite concrete needs to be studied.

Concrete durability refers to the ability of concrete to resist the action of environmental media and maintain its good performance for a long time, so as to maintain the safety and normal use of concrete structure. It usually includes salt crystal erosion, steel corrosion, concrete carbonation, freeze-thaw damage and so on. The influence of silica fume on the durability of concrete was studied by using dry wet cycle sulfate attack as a representative.

The influence of silica fume content on the corrosion resistance coefficient of concrete was shown in Fig. 2. From the test results in the figure, it can be seen that with the increase of silica fume content, the corrosion resistance coefficient of concrete first gradually increased, and then presented a downward trend. The corrosion resistance coefficient of concrete with 12% silica fume content was basically close to that of benchmark concrete. The corrosion resistance coefficients of G0, G4, G6, G8, G10 and G12 with 0%, 4%, 6%, 8%, 10% and 12% silica fume were 76.1%, 85.3%, 89.2%, 81.3%, 79.8% and 77.6% respectively. It can be seen that the corrosion resistance coefficient of concrete with 6% silica fume was the highest, and the resistance to wet and dry cyclic sulfate was the best. The corrosion resistance coefficients of concrete with 10% and 12% silica fume were close to those of reference concrete without silica fume. The effect of high content of silica fume on the resistance of concrete to dry wet cycle sulfate attack was small.

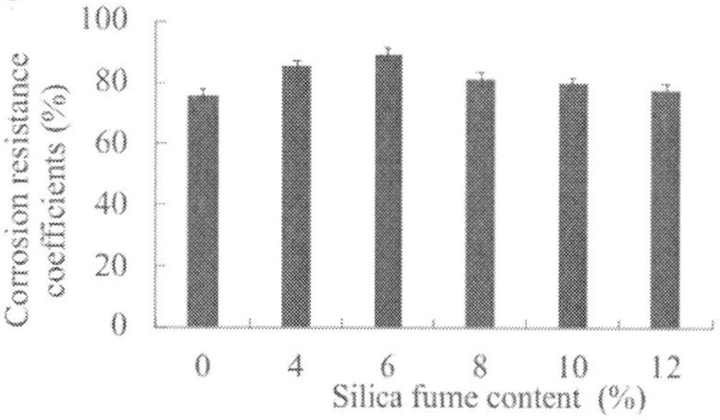

Fig 2 Influence of silica fume on the corrosion resistance coefficient of concrete

3.4 Influence of silica fume on the airtightness of concrete

The test results of the influence of silica fume content on the airtightness of concrete were shown in Fig. 3. It can be seen from the figure that with the increase of the amount of silica fume, the air permeability coefficient of concrete was significantly reduced, and the airtightness of concrete was significantly improved.

The air permeability coefficient of the benchmark concrete with a silica fume content of 0% was 1.825×10^{-12} cm/s, and the air permeability coefficients of G4, G6, G8, G10, and G12 samples were 1.12, 1.01, 0.355, 0.238, 0.105×10^{-12} cm/s, respectively.

When the content of silica fume exceeded 8%, the air permeability coefficient of concrete decreases exponentially. The reason why silica fume can well reduce the air permeability coefficient of concrete and improve the airtightness of concrete may be due to the extremely fine particle size and large specific surface area of silica fume. Although the content of silica fume was relatively small, the amount of particles mixed was very large, so it can fill the pores between cement and fly ash particles, reduce the pore size and improve the gas compactness. In addition, silica fume, fly ash or slag powder are all artificial pozzolanic materials, which all have a pozzolanic effect. The pozzolanic effect of silica fume is more significant than that of fly ash, with higher hydration activity, more hydration products, and greater contribution to the gas compactness of concrete.

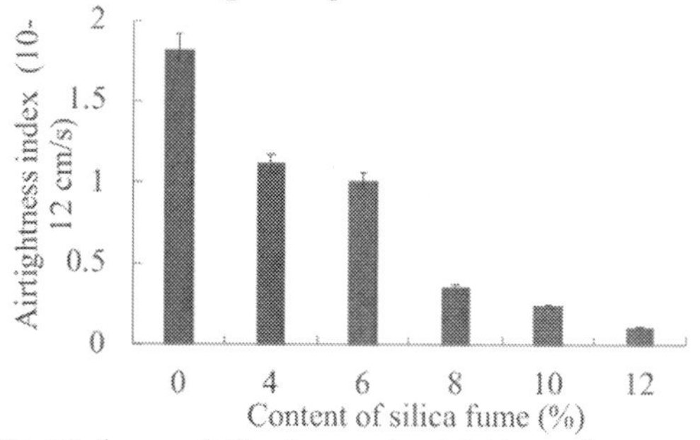

Fig. 3 Influence of silica fume on the airtightness of concrete

4 Conclusions

In view of the widespread existence of moderately weathered phonolite around the Nairobi expressway project in Kenya, which has certain adverse effects on the performance of concrete, this paper studied the effect of adding silica fume to improve the performance of concrete and promote the application of moderately weathered phonolite in concrete. The results were as follows:

(1) With the increase of silica fume content, the workability of concrete showed a downward trend. Except for 4% silica fume content, the change of concrete workability was small, and the decline trend of concrete workability was obvious with the increase of other silica fume content. The compressive strength of airtightness concrete increased at first and reached the highest when the silica fume content is 8%, and then decreased gradually.

(2) With the increase of silica fume content, the anti-corrosion coefficient of concrete increased first and then decreased. With the increase of silica fume content, the permeability coefficient of concrete decreased significantly, and the airtightness of concrete increased significantly.

(3) Considering the working performance, mechanical properties, durability and air tightness test results of moderately weathered phonolite concrete, silica fume can be used to prepare moderately weathered phonolite concrete in the project, and the content of silica fume was recommended to be controlled at 4%-8%.

References

1. C. Dong and Y. Wan, World Build. Mater. **42**, 25 (2021).
2. Z. Ma, X. Wen, Y. Xiao, X. Lei, and H. Guo, Water Resour. Hydropower Northeast China **37**, 51 (2019).
3. Y. Zhao, Y. Zhang, J. Xia, Y. Zhou, and K. Li, World Build. Mater. **39**, 9 (2018).
4. Z. Li, P. Guo, J. Xia, Y. Zhou, Y. He, and K. Li, Subgrade Eng. 121 (2018).
5. Z. Li, X. Meng, Y. Zhou, J. Xia, H. Chen, and K. Li, Concrete 95 (2018).
6. P. Bai, Y. Zhang, and C. Zhu, New Build. Mater. **45**, 5 (2018).
7. X. Wen, Y. Lei, J. Xia, Y. Zhou, and K. Li, New Build. Mater. **44**, 126 (2017).
8. C. Guo, C. Chen, Q. Zou, C. Zhu, G. Wang, and Z. Sun, China Concr. Cem. Prod. 27 (2017).
9. S. Ding, J. Liang, J. Xia, Y. Zhou, B. Gao, and K. Li, World Build. Mater. **38**, 16 (2017).
10. D. Qu, K. Li, J. Xia, Y. Zhou, and Y. Wang, New Build. Mater. **44**, 91 (2017).
11. K. Li, B. Zhang, D. Yang, L. Xiao, J. Xia, and Y. Zhou, Subgrade Eng. 114 (2016).
12. Z. Cui and J. Li, Bull. Chin. Ceram. Soc. **35**, 2396 (2016).

Effect of adding waste boulder powder on mechanical properties of green concrete

Xiaohui Li[1,*]

[1] CCCC Fourth Harbor Engineering Co., Ltd., Qingyuan 515000, Guangdong, China

Abstract. Reusing of stone powder formed in the production of manufactured sand is of great significance to environmental protection and resource utilization. In this paper, C30 and C40 environment-friendly manufactured sand concrete were prepared by adding 6%, 9%, 12% and 15% of waste boulder powder from manufactured sand production line into concrete and controlling the total weight of manufactured sand and rock powder to be constant. The influence of stone powder on concrete fluidity, compressive strength and elastic modulus were analysed. It was found that with the increase of stone powder content, the compressive strength of fresh paste and C30 concrete first increased and then decreased, while the elastic modulus of concrete and the compressive strength of C40 concrete continued to decrease. It is suggested that the content of rock powder should not exceed 9%.

1 Introduction

With the increase in the scale of national infrastructure construction and the advancement of the "One Belt One Road" initiative, concrete raw materials, especially mineral admixtures, are in short supply in many regions. Fraud and adulteration of mineral admixtures (granulated blast furnace slag powder, fly ash, silicon ash, etc.) often occur, which greatly affects the quality of engineering construction. At present, many scholars are devoted to the research of new supplementary cementitious materials, such as limestone powder [1–4], metakaolin [5–9] and steel slag powder [10–13], to cope with current shortage of traditional mineral admixtures.

Phnom Penh-Sihanoukville Expressway (PPSE) is an important golden channel connecting Phnom Penh, the capital of Cambodia, to Sihanouk, the largest seaport in China and an external port. The construction of PPSE is of great significance for promoting the local economic development of Cambodia and implementing the "One Belt One Road" initiative. The starting point of PPSE is at the intersection of National Highway No. 4 and Phnom Penh City Circle Line in Cambodia, and the end point is on the No. 4 road on the edge of Sihanoukville. It is about 8.6km away from Sihanoukville Port to the west and about 5.5km away from Westport Economy, and it is about 8.5km from Sihanoukville Airport. The total length is 187.05Km. The whole line adopts Chinese highway construction standards and specifications. The design speed is 100Km/h, and the two-way four-lane.

The construction of PPSE also faces the problem of the lack of high-quality mineral admixtures and river sand resources. The project adopts the common boulder production manufactured sand along the project, and a large amount of rock powder of boulder is produced during the production process. If the boulder powder is recycled as mineral admixtures in engineering concrete production through research, not only could the pollution of the environment by the stacking of stone powder be solved, but also ordinary mineral admixtures could be replaced, and construction costs could be reduced. Hence, this paper studies the influence of different content of boulder powder on mechanical performance of concrete with different water-to-binder ratio and provides basic research data support for the application of boulder powder in engineering.

2 Material and testing method

2.1 Raw material

The apparent density of manufactured sand is 2650 kg/ m^3, the loose bulk density is 1630 kg/m^3, the porosity of loose bulk is 41%, the content of stone powder is 10.2%, the fineness modulus is 2.94, the water demand ratio is 103%, the MB value is 0.75, and the maximum crushing index of single stage is 16.5%. The crushed stone used is composed of 5-10 mm, 10-20 mm and 31.5 mm graded crushed stone with mass ratio of 2:5:3. The content of needle flake particles is 0.6%, and the maximum crushing index of single stage is 8.7%.

The waste boulder powder was collected from the limestone manufactured sand production line and screened by 0.075mm screen. The test results showed that the fluidity ratio was 106%, the water demand ratio was 98.30%, the MB value was 3.8, the specific surface area was 900.21 kg/ m^3, the density was 2655 kg/ m^3, the water content was 0.41%, and the 7d activity index was 90%. Using P.O. 42.5 cement, standard consistency

* Corresponding author: 157906902@qq.com

27.1%, specific surface area 326m²/kg, initial setting time 161min, final setting time 246min, 3d strength 32.2MPa, 28d strength 57.5MPa. Type F class II fly ash is used, with fineness of 13.2% (45μm sieve residue) and 28d activity index of 74%. Using polycarboxylate water reducer, the solid content is 30.2%, and the water reducing rate is 28%.

2.2 Mix proportion

The influence of waste boulder powder on the performance of manufactured sand concrete was shown in Table 1. The water-to-binder ratio (W/B) of 0.44 and 0.38 were designed respectively to prepare C30 and C40 strength grade concrete and compare the influence of W/B. The total weight of manufactured sand and waste boulder powder in C30 was 798 kg/m³, and that in C40 was 814 kg/m³. The amount of rock powder in brackets after the weight of rock powder was 6%, 9%, 12% and 15% respectively.

Table 1 Mix proportion of influence of content of waste boulder powder on mechanical properties of concrete (kg/m³)

Sample ID	Cement	Fly ash	Manufactured sand	Waste boulder powder	Coarse aggregate	Water	W/B
C30-S-6	294	74	750	48 (6%)	1058	162	0.44
C30-S-9	294	74	726	72 (9%)	1058	162	
C30-S-12	294	74	702	96 (12%)	1058	162	
C30-S-15	294	74	678	120 (15%)	1058	162	
C40-S-6	341	85	765	49 (6%)	994	162	0.38
C40-S-9	341	85	741	73 (9%)	994	162	
C40-S-12	341	85	716	98 (12%)	994	162	
C40-S-15	341	85	692	122 (15%)	994	162	

2.3 Testing method

The slump, slump-flow and air content of concrete were tested in accordance with the Chinese standard *Standard for test method of performance on ordinary fresh concrete: GB/T 50080-2016*; The compressive strength and static compressive elastic modulus of hardened concrete (hereinafter referred to as elastic modulus) are tested in accordance with the Chinese standard *Standard for test method of concrete physical and mechanical properties: GB/T 50081-2019*. The size of compressive strength test specimen was 100 mm cube, and the test age was 3d, 7d, 28d and 56d. The test age of elastic modulus was 28d and 56d.

3 Result and discussion

3.1 Workability

Fig. 1 showed the effect of waste boulder powder on the workability of manufactured sand concrete under different W/B and stone powder content.

According to the change of slump, when the stone powder content increased from 6% to 15%, the slump of fresh paste first increases and then decreased. When the stone powder content in C30 concrete exceeded 12%, the slump decreased, and when the stone powder content in C40 concrete exceeded 9%, the slump began to decrease. The relationship between slump, air content and slump-flow of fresh paste were quantitatively described in Fig. 2. In order to characterize the correlation between fresh concrete performance and stone powder content, the quadratic function model was used to regress the relationship between concrete fluidity, air content and stone powder content in Fig. 3. It was found that there was a significant quadratic function relationship between concrete slump, slump-flow, air content and stone powder content under different W/B, which confirmed the rule that the fluidity in Fig. 1 first increases and then decreases.

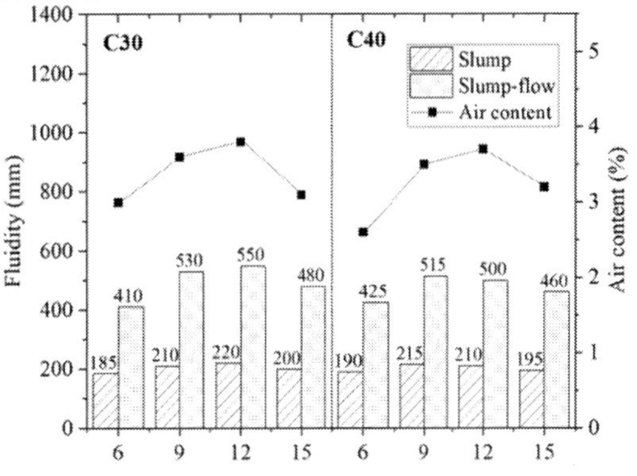

Fig. 1 Effect of waste boulder powder on workability of manufactured sand concrete

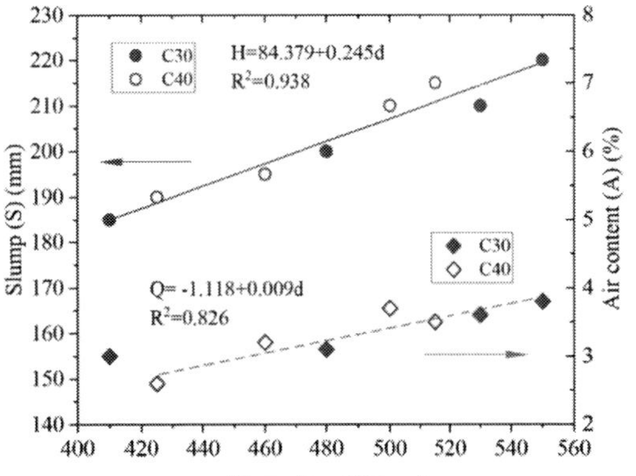

Fig. 2 Correlation of slump, air content and slump-flow of waste boulder powder concrete

As a kind of calcareous powder, the biggest difference between limestone powder and granite, basalt and other rock powder lay in its small adsorption performance. According to the research [14,15], the surface energy of calcium carbonate in limestone powder particles was about $230\times10^{-7}J/cm^2$. Low surface energy could reduce the water demand ratio of concrete with limestone rock powder, that is, reduce the water absorption of paste or have water reducing effect. Therefore, with the increase of calcareous rock powder content, the fluidity of concrete gradually increased, which was similar to the existing research results [16–18]. However, the MB value of rock powder used in this paper reached 3.8, which reflected that the rock powder contains more clay debris to a certain extent. When the content of stone powder exceeded a certain extent, the strong adsorption impurities absorb a large amount of free water, which weakened the fluidity of the paste. At the same time, the specific surface area of stone powder reaches 900 g/cm^3, and more mixing water is needed to wet the surface of stone powder particles to form a water film, resulting in the decline of concrete workability.

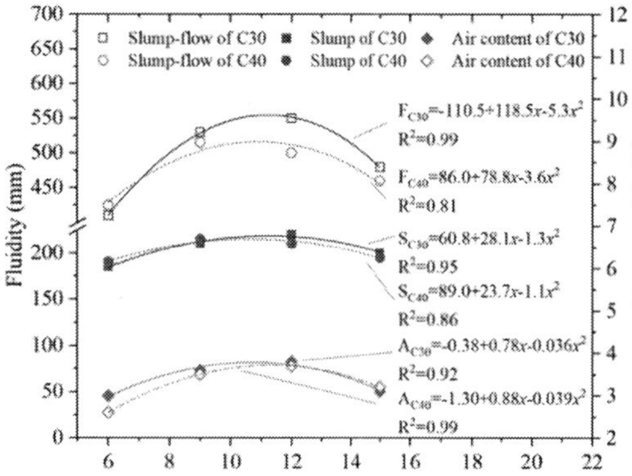

Fig. 3 Correlation between fluidity, air content and stone powder content of waste boulder powder concrete

3.2 Compressive strength

The experimental results of the influence of concrete prepared with waste boulder powder on its strength were shown in Figure 4. The solid line symbol represented C30 concrete, and the dotted line symbol represented C40 concrete.

With the increase of stone powder content, the strength of C30 concrete at different ages first increased and then decreased. The strength of C30-S-9 specimen was the highest at each age, and its 56d strength was 65.5MPa. Although the strength of C30-S-15 was low, its 56d strength could still reach 61.1MPa. It could be seen that the increase of stone powder content from 9% to 15% had little effect on the strength of C30 concrete. The strength of C40 concrete decreased with the increase of stone powder content. At 56d, the strength of C40-S-9 group was 70.8MPa, and that of C40-S-15 group was 65.6MPa. It could be seen that the weakening effect of stone powder content on the compressive strength of high strength concrete was more prominent. On the other hand, by comparing the strength data of concrete with different W/B at the same age, it could be seen that the change of W/B had little effect on the early strength of concrete at 9% and 12% dosage.

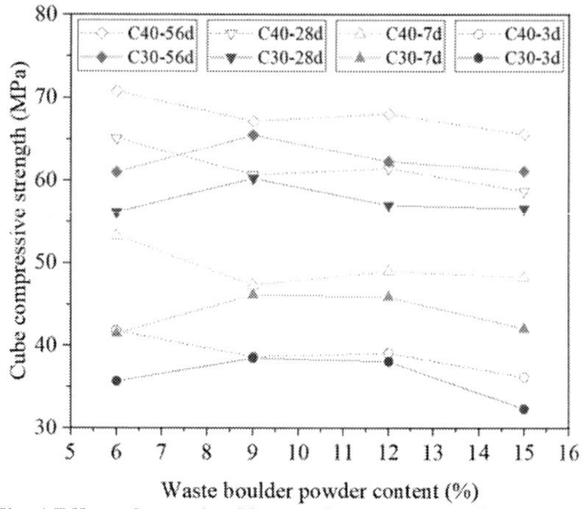

Fig. 4 Effect of waste boulder powder content on cube compressive strength of concrete

The influence mechanism of limestone powder on the mechanical properties of concrete has been widely studied, which could be divided into microcrystalline nucleation, micro aggregate and micro chemical action. Soroka et al. [19,20] found that calcium carbonate powder particles could absorb the ions of hydration products, make C-S-H grow continuously on their own particle surface, promote the migration of C3S particle surface ions in cement clinker to aqueous solution, thus accelerate the hydration of C3S, and significantly advance the nucleation time of $Ca(OH)_2$.

Secondly, the stone powder particles with the maximum particle size less than 75μm could fill the voids between the cementitious material particles, replacing the excess free water existing in the voids, optimizing the particle accumulation mode, reducing the water consumption and average pore size [21], and improving the compactness of the slurry; At the same time, its tiny particles could play a "Ball Effect", improve the fluidity of the fresh slurry, and reduce the risk of large pores due to insufficient vibration in the moulding process.

In addition, limestone manufactured sand powder has weak chemical activity. According to the research of K. D. Weerdt et al. [22], micro particles of calcium carbonate powder will dissolve carbonate ions into the slurry aqueous solution, which could react with aluminum phase in Portland cement to form hard single carbon calcium carboaluminate and semi-carbon calcium carboaluminate, which will consume Ca(OH)2, inhibiting the transformation of aft into AFm [23]. Finally, the average pore size of concrete paste is refined, and the strength of concrete is improved. These three kinds of actions make the early strength of manufactured sand concrete containing proper amount of stone powder increased rapidly. However, as a kind of inert material in the traditional sense, limestone rock powder itself could not provide outstanding cementitious and pozzolanic activity. Once the content of limestone rock powder was too high, its tiny particles will introduce too many low cohesive interface transition zone, which makes the slurry appear more thin and weak zone, resulting in the weakening of concrete hardening strength.

Therefore, for the concrete with different cementitious material dosage and strength requirements, there are differences in the optimal stone powder content of manufactured sand. As shown in Figure 4, the stone powder content of C30 concrete is about 9%, and that of C40 concrete is about 6%, which is close to the existing research [24].

3.3 Elastic modulus

The elastic modulus could reflect the constitutive relation of concrete after hardening, and it is an important mechanical property index of concrete. Table 2 showed the elastic modulus test results of waste boulder powder concrete. It could be seen that the elastic modulus of concrete will be slightly reduced when the content of manufactured sand powder increased from 9% to 15%. Increasing the content of stone powder was equivalent to replacing the dense sand particles with loose powder particles, that is, the volume of concrete slurry increased, the content of coarse particles which play the main role of skeleton decreased, and the elastic modulus decreased. On the other hand, when the W/B increased, hydrated products contained more calcium silicate hydrate gel, and the microstructure of the slurry was denser, which was conducive to enhancing the elastic modulus of the slurry. When the W/B increased, there were more pores in the microstructure, higher air content, lower compactness and smaller elastic modulus.

Elastic modulus and compressive strength were important basic parameters of concrete, which were closely related to W/B, material composition, curing age and microstructure. Considering the uncertainty of concrete test, the effective prediction probability model of concrete elastic modulus was summarized based on compressive strength. Referring to the Chinese Standard *Code for design of concrete structures (GB 50010–2010)*, formula (1) prediction model was adopted in many literatures.

$$E_c=10^5/(2.2+34.7/f_c) \quad (1)$$

Where f_c is the compressive strength of concrete, E_c is the elastic modulus.

There were also many researchers using the power function model of formula (2), and the relevant institutions in Norway (NS) and the United States (ACI) recommend using formula (3) ~ (5) respectively:

$$E_c=(a\times f_c^{0.5}+b)\times 10^4 \quad (2)$$

NS 3493: $E_c=9.5 f_c^{0.3}$ (3)

ACI 363: $E_c=3.32 f_c^{0.5}+6.9$ (4)

ACI 318: $E_c=4.73 f_c^{0.5}$ (5)

Where f_c and E_c have the same meanings as above. In equation (2), a and b are regression constants.

According to the test results and referring to equation (2), we used the power function model of equation (6) for regression analysis:

$$E_c=40.6+1.1\times10^{-41} f_c^{22.7} \quad (6)$$

It could be seen from Figure 5 that the regression curve of equation (6) was more consistent with the change rule of test data. In order to further analyze the calculation accuracy of formula (6) for the test results of each mix proportion specimen, the following prediction error was defined:

$$e_{E_c}=\frac{100}{\max\limits_{i=1,2,\cdots,N}(|E_{ci}|,|E_{ci}'|)}\sqrt{\frac{1}{N}\sum_{i=1}^{N}(E_{ci}-E_{ci}')^2} \quad (7)$$

Where, e_{Ec} was the prediction error of concrete elastic modulus, %; The subscript i referred to the i-th mix proportion in Table 1, and N was the total number of mix proportion, which was taken as 8; E_{ci} and E_{ci}' were the measured value and calculated value of the elastic modulus of the i-th mix proportion specimen. The final calculation result of e_{Ec} in equation (6) was 2.65%, in formula (1) ~ formula (5) was 14.04%, 9.09%, 21.71%, 21.10% and 11.79% respectively (a and b in formula (1) was 0.26 and 1.8 respectively). It could be seen that the power function model (6) proposed in this paper had higher prediction accuracy.

Table 2 Effect of manufactured sand powder content on elastic modulus of concrete (GPa)

Age/d	C30-S-9	C30-S-15	C40-S-9	C40-S-15
28	41.555	40.382	43.058	42.939
56	42.197	40.952	44.646	43.391

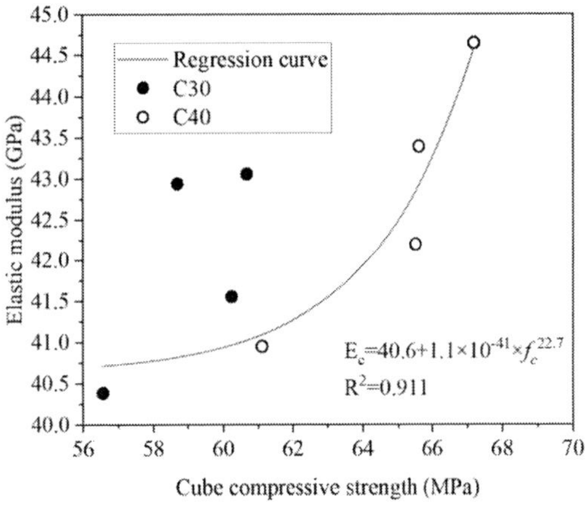

Fig. 5 Correlation between elastic modulus and cube compressive strength of concrete

4 Conclusion

In this paper, the waste boulder powder formed in the process of manufactured sand production was mixed into manufactured sand concrete by 6%, 9%, 12% and 15% respectively to prepare green environment-friendly concrete. The conclusions were as follows

(1) When the content of stone powder was more than 12%, the fluidity of fresh paste could be improved, and the fluidity will be weakened if the content of stone powder was further increased.

(2) With the increase of stone powder content, the elastic modulus of hardened paste decreased, the compressive strength of C30 concrete first increased and

then decreased, and the strength of C40 concrete continued to decrease.

(3) A mathematical model for predicting the elastic modulus of concrete by using the compressive strength of waste boulder powder was proposed, which had higher accuracy than the model in the existing specifications.

Acknowledgements

This research was funded by National Natural Science Foundation of China, grant number 51978408, and National Key R&D Program of China, grant number 2016YFC0701000.

References

1. Palm Sebastian, Proske Tilo, Rezvani Moien, Hainer Stefan, Müller Christoph, and Graubner Carl-Alexander, Constr. Build. Mater. 119, 308 (2016).
2. Proske Tilo, Rezvani Moien, Palm Sebastian, Müller Christoph, and Graubner Carl-Alexander, Cem. Concr. Compos. 89, 107 (2018).
3. Courard Luc and Michel Frédéric, Constr. Build. Mater. 51, 439 (2014).
4. Boubekeur Toufik, Boulekbache Bensaid, Aoudjane Kheireddine, Ezziane Karim, and Kadri El-Hadj, Constr. Build. Mater. 209, 215 (2019).
5. K. Scrivener, F. Martirena, S. Bishnoi, and S. Maity, Cem. Concr. Res. 114, 49 (2018).
6. Sánchez Berriel S., Favier A., Rosa Domínguez E., Sánchez Machado I. R., Heierli U., Scrivener K., Martirena Hernández F., and Habert G., J. Clean. Prod. 124, 361 (2016).
7. F. Avet and K. Scrivener, Cem. Concr. Res. 107, 124 (2018).
8. Dhandapani Yuvaraj, Sakthivel T., Santhanam Manu, Gettu Ravindra, and Pillai Radhakrishna G., Cem. Concr. Res. 107, 136 (2018).
9. Muzenda Tafadzwa Ronald, Hou Pengkun, Kawashima Shiho, Sui Tongbo, and Cheng Xin, Cem. Concr. Compos. 107, 103516 (2020).
10. S. Zhuang and Q. Wang, Cem. Concr. Res. 140, 106283 (2021).
11. D. Wang, Q. Wang, and Z. Huang, Compos. Part B Eng. 198, 108207 (2020).
12. Ji Yukun, Zhou Guoqing, Zhao Xiaodong, Wang Jianzhou, Wang Tao, Lai Zejin, and Mo Pinqiang, Cold Reg. Sci. Technol. 142, 25 (2017).
13. Wang Dengquan, Wang Qiang, and Xue Junfeng, Resour. Conserv. Recycl. 154, 104645 (2020).
14. W. D. Pratiwi, Triwulan, J. J. Ekaputri, and H. Fansuri, Constr. Build. Mater. 234, 117273 (2020).
15. J. Feng, F. Yang, and S. Qian, Constr. Build. Mater. 269, 121249 (2021).
16. L. Hu and Z. He, Constr. Build. Mater. 262, 119847 (2020).
17. J. Yu, H.-L. Wu, D. K. Mishra, G. Li, and C. K. Leung, J. Clean. Prod. 278, 123616 (2021).
18. H. Du and S. D. Pang, Constr. Build. Mater. 264, 120152 (2020).
19. I. Soroka and N. Stern, Cem. Concr. Res. 6, 367 (1976).
20. J. Péra, S. Husson, and B. Guilhot, Cem. Concr. Compos. 21, 99 (1999).
21. Y. Senhadji, G. Escadeillas, M. Mouli, H. Khelafi, and Benosman, Powder Technol. 254, 314 (2014).
22. K. De Weerdt, M. B. Haha, G. Le Saout, K. O. Kjellsen, H. Justnes, and B. Lothenbach, Cem. Concr. Res. 41, 279 (2011).
23. T. Matschei, B. Lothenbach, and F. P. Glasser, Cem. Concr. Res. 37, 551 (2007).
24. B. Li, G. Ke, and M. Zhou, Constr. Build. Mater. 25, 3849 (2011).

The project design of closure with cofferdam in a large reclamation project

JIN Hai-sheng[1], XU Ji[2] and GE Guo-chang [3*]

[1]Wenzhou Oufei development and Construction Investment Group Co. , Ltd. , Wenzhou 325025, China;

[2]Zhejiang Design Institute of water conservancy & Hydro-electric power, Hangzhou 310002, China;

[3]Zhejiang Guangchuan Engineering Consultation Co. , Ltd. , Hangzhou 310020, China

Abstract. The closure of the reclamation project generally requires that the sluice has been built and can be normally opened and closed for drainage. There is a large amount of follow-up work after closured of this project. In order to reach the safe flood control standard for the next typhoon season after the closure, reduce the flood control risk, and strive for sufficient time for the subsequent construction of the seawall, it is necessary to closure in advance under the condition that the sluice has not yet been opened. Therefore, this project demonstrated the feasibility of the closure with cofferdam based on the premise of think outside the box and theoretical calculation. The successful completion and acceptance of the reclamation project proved that the project is successful.

1 Preface

The reclamation project is an important part of the infrastructure to resist storm surge and flood control and moisture-proof engineering system in coastal areas such as Zhejiang, Seawalls can effectively reduce the hazards and losses caused by natural disasters such as typhoons, rainstorms and tides, and provide an important guarantee for the safety of people's lives and property [1]. The key link of seawall construction is "closure" , which determines the success or failure of the whole reclamation project. Therefore, the design and construction of closure gap is the top priority. The closure of the reclamation project generally requires that the sluice has been built and can be normally opened and closed for drainage to ensure waterlogging prevention in fenced areas[2].

This project is a large-scale reclamation project with long dike lines, deep coating surface, complex geological conditions and large wind and waves in the open sea. The outer protective surface of the seawall should be completed as far as possible before typhoon season in order to meet the safety flood control standards and reduce risk. In order to make the seawall settlement occur as early as possible and make the post-construction settlement as small as possible, and meanwhile to bought more time for the subsequent construction of the seawall, On the premise of mature conditions, proposed of " Closure With Cofferdam " innovatively through theoretical calculation, before the sluice has not yet been plugged in advance, ensured the smooth completion of the project, and also bought more time for the subsequent construction of the project .

2 Project summary

A reclamation project is a I grade project, the main building seawall and sluice are I grade buildings. The reclamation area is 66,400 mu, including 20. 33km seawall, divided into the north dike and the north section of the east dike; north 1# sluice , north 2# sluice and navigation hole, east 1# sluice , the total net width of the sluice is 168m; 1# partition dike, 2# partition dike, etc. The 100-year tide level is 5. 44m. The foundation of the seawall is mainly silt and silty soil from 30 m to shallow, especially the III_0 layer and the surface layer of 0. 5m~0. 8m is newly deposited drift mud, and the engineering geological conditions are extremely poor.

3 Calculation of water level in the enclosure area with cofferdam

3.1 Basic information of closure gap

The construction of this Cofferdam is the closure gap of 1# and 2# enclosure area, where the area of 1# enclosure area is 35,700 mu and the average bottom elevation is about -1. 50m; the area of 2# enclosure area is 30,700 mu and the average bottom elevation is about -1. 40m. The flood control standard of the closure gap is 4. 79m and its typical tide type during the 20-year high tide, and the standard of the closure is 4. 02m during the non-flood 10-

* Corresponding author: zjggc2007@163.com

year high tide and its typical tide type. According to the principle of water balance in and out of the inner harbor, the maximum flow velocity is 4. 38m/s at the width of 1000m for 1# closure gap and 4. 38m/s at the width of 900m for 2# closure gap.

3.2 Model calculation and analysis

3.2.1 Calculation method

The model calculation method adopts the non-constant flow of the river network, and the set of partial differential equations for the non-constant flow of the St. Venant nullah is transformed into differential equations by the implicit difference method, and then constitutes a large set of nonlinear equations with boundary conditions and initial conditions, which is solved by Newton iteration and Gaussian column principal elimination method to derive the water level and flow process at each calculation section. The basic equations Saint Venant set of partial differential equations are:

$$B\frac{\partial z}{\partial t}+\frac{\partial Q}{\partial s}=q \tag{1}$$

$$\frac{1}{g}\frac{\partial v}{\partial t}+\frac{\partial}{\partial s}\left(z+\frac{v^2}{2g}\right)+\frac{Q|Q|}{F^2K^2}=0 \tag{2}$$

Style (1), (2) in: Z, Q, F, V and K are expressed at a certain moment t and in a certain spatial location S section of the water level, flow, the corresponding overwater cross-sectional area, section average velocity and flow modulus; q for the unit river length side inlet flow.

3.2.2 Design Boundary Conditions

The design upper boundary of Rui'an City is taken to the sail swim of Wenruitang River, Longwan District is taken to the downstream of Tianhe East Reservoir, Tianzhuji Reservoir, Shuangao Reservoir, Yao Xi Reservoir, etc. , using the flow process. Design lower boundary taken to Ou River, Feiyun River and eastward discharge to the East China Sea exit gate, using the water level process.

Calculated initial value: Longwan District planning river normal storage level 2. 6m, planning ground elevation 4. 50m; Rui'an City planning river normal storage level 2. 6m, planning ground elevation 4. 00m ~ 4. 50m.

3.2.3 THE design of 2# enclosure area closure and calculation results

According to the analysis of local hydro tidal conditions and project progress, the time of closuring in 2# enclosure area was chosen on December 22, 2016, and the planned completion and opening of East 1# sluice was at the end of March 2017, so during the period of closuring in 2# enclosure area and completion and opening of East 1# sluice , the weir of East 1# gate was not demolished, and the 1# dike between 1# and 2# enclosure area was not opened, so as to analyze the impact of the discharge and non-discharge of flood water in Longwan II on the water level in the 2# enclosure area and the impact on the discharge of flood water in the upstream Longwan II and Wenrui Plain. 2# enclosure area during the closure of the closure gap is shown in Figure 1.

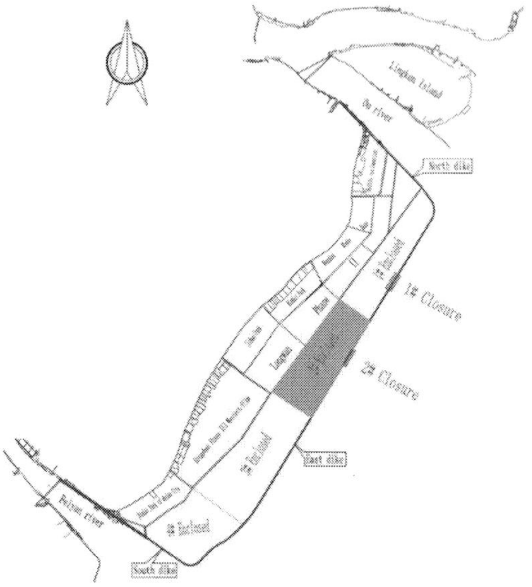

Figure 1. Schematic diagram of closure gap in 2# enclosure area during closure

(1) during the period of closuring in 2# enclosure area, Longwan II 3 # enclosure area and 4 # enclosure area flooding water does not enter oufei 2 # enclosure area, 2 # enclosure area closued on December 22, 2016, the starting water level within the enclosure area is tentatively set at 0. 50m, to the end of March 2017 East 1 # sluice water opening, 2 # enclosure area water level within the raised about 0. 1m to 0. 6m. If a rainfall of 5-year magnitude is encountered during the period, the water level in the 2# enclosed area will be raised by about 0. 44m to 0. 94m.

(2) during the period of closuring in 2# enclosure area, Longwan II 3 # enclosure area and 4 # enclosure area flooding water enter oufei 2 # enclosure area, 2 # enclosure area closued on December 22, 2016, the starting water level within the enclosure area is tentatively set at 0. 50m, to the end of March 2017 East 1 # sluice water opening, 2 # enclosure area water level within the raised about 0. 17m to 0. 67m. If a rainfall of 5-year magnitude is encountered during the period, the water level in the 2# enclosed area will be raised by about 0. 79m to 1. 29m.

(3) during the period of closuring in 2# enclosure area, considering Longwan District water system connected with Longwan II 3# enclosure area and 4# enclosure area, part of its waterlogged water into oufei 2# enclosure area, 2 # enclosure area closued on December 22, 2016, the starting water level within the enclosure area is tentatively set at 0. 50m, to the end of March 2017 East 1 # sluice

 https://doi.org/10.1051/e3sconf/202129302034

water opening, If a super standard flood is encountered during the period, the water level within the 2# enclosed area exceeds 2. 6m. To ensure the safety of flood control and drainage in Longwan District of Wenrui Plain and Rui'an City, open the 1# partition dike between 1# and 2# enclosure area, and bury 10 culverts of 1m diameter in the lower part of the coating surface, i. e. the area near Ou Fei East seawall, to communicate the waters between 1# enclosure area and 2# enclosure area, so that the flooded water in the enclosure area will be discharged outwardly through the d closure gap of 1# enclosed area.

3.2.4 THE design of 1# enclosure area closure and calculation results

According to the analysis of local hydro tidal conditions, engineering progress, etc. , the time of closuring in 2# enclosure area was chosen on March 7, 2017, until the 2017 typhoon season, the North 1# sluice and North 2# sluice cofferdam will not be demolished, 1# partition dike between 1# and 2# enclosure area will not be opened, considering that East 1# sluice of 2# enclosure area gate is completed and open to water, the flood water of Longwan II can be discharged into the East China Sea through 2# enclosure area via East 1# sluice , not into 1# enclosure area. analysis of the possible water level impact within the 1# enclosure area. 1# enclosure area closure gap during the closure of the schematic diagram is shown in Figure 2.

(1) during the period of closuring in 1# enclosure area, considering the completion and opening of East 1# sluice of 2# enclosure area, the flood water of Longwan II can be discharged into East China Sea through East 1# sluice of 2# enclosure area and not into 1# enclosure area, 1# enclosure area closured on March 7, 2017, the starting water level in the enclosure area is tentatively set at 0. 50m, and the water level in 1# enclosure area will be raised by about 0. 31m to 0. 81m before the 2017 flood season. If a 5-year rainfall is encountered during the flood season, the water level in the 1# enclosure area will be raised by 0. 66m to 1. 16m.

(2) If during the period encountered super standard flood, 1 # enclosed area water level exceeds 2. 6m, in order to ensure the safety of flood control and drainage in Longwan District of Wenrui Plain and Rui'an City, open the 1# partition dike between 1# and 2# enclosure area, and bury 10 culverts of 1m diameter in the lower part of the coating surface, i. e. the area near Ou Fei East seawall, to communicate the waters between 1# enclosure area and 2# enclosure area, so that the flooded water in the enclosure area will be discharged outwardly through the d closure gap of 1# enclosed area.

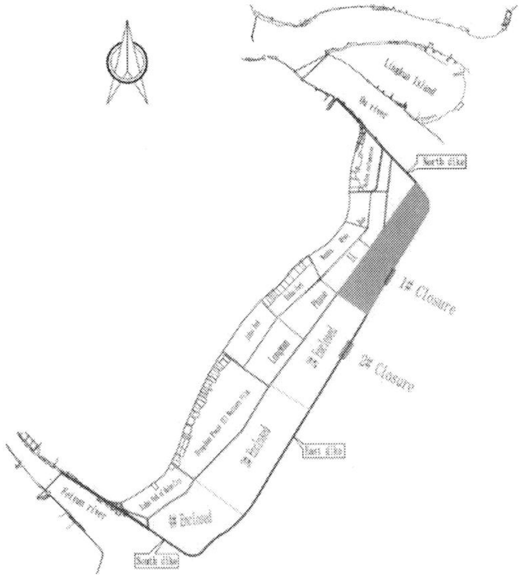

Figure 2. Schematic diagram of closure gap in 1# enclosure area during closure

3.2.5 Analysis of calculation results

Condition 1: Ou Fei 1# and 2# enclosed area is not closured, affected by the outer sea tide level, the 20-year flood water level within the enclosed area is 2. 53m ~ 3. 01m.

Condition 2: When the 2# enclosed area is closured, and the East 1# sluice is not open to water, on the one hand, it can not be affected by the tide level, on the other hand, Longwan II and Wenrui plain part of the flooded water can use the storage volume of the 2# enclosed area, so the water level of Tianhe, Sanjia, East Gate and Field Bridge is slightly reduced, and the flooded water level in the 1# and 2# enclosed area is slightly raised, and the raised range is 0. 04m ~ 0. 07m.

Condition 3: 1# enclosed area is closured, consider Longwan and upstream flood water does not enter 1# enclosed area, on the one hand through the East 1# gate discharge into the East China Sea, on the other hand can make full use of 2# enclosed area storage volume, the upstream Tianhe, Sanjia, East Gate, field bridge water level reduced 0. 01 ~ 0. 11m, 2# enclosed area through the East 1# sluice tide discharge, can reduce the highest water level of 2# enclosed area 0. 31m, Because 1# enclosed area does not bear the influence of the upper tourist water and tide, the water level in the enclosed area is reduced significantly.

Condition 4: On the basis of Condition 3, considering that 1# enclosed area bears part of the upstream flooding, the water level of the upstream Tianhe, Sanjia, Dongmen and Field Bridge can be further reduced by 0. 01m~0. 05m, the flooding water level of 2# enclosed area an be reduced by 0. 31m, and the water level of 1# enclosed area is raised by 1. 22m because it bears part of the upstream flooding.

In summary analysis, although sluice of the 1 # and 2 # enclosure area has not yet been opened to water (cofferdam still exists), the early stage only by 1 # closure gap, later by the East 1 # sluice drainage, theoretically can also meet the entire perimeter water level control requirements.

4 Emergency preparedness

If the period encounters super standard flood, the water level within the 1# enclosure area exceeds 2. 60m, in order to ensure the flood control and drainage safety of Longwan District of Wenrui Plain and Rui'an City, on the basis of 10 culverts of 1m diameter buried in the 1# partition dike, further excavate the 1# and 2# enclosure area between the 1# partition dike to -1. 00m elevation, and determine the width of excavation according to the strength of super standard flood, so that The flood water of the 1# and 2# enclosure area is discharged through the East 1# sluice to meet the drainage requirements of the whole area. Through the development of the above emergency plan, the subsequent construction of the project after the closure of the plug will provide more adequate protection.

5 Program implementation results

The 2# closure gap started to closured in October 2016 and was successfully closured in December 2016, and the East 1# sluice was opened to water in April 2017; the 1# closure gap started to closured in December 2016 and was successfully closured in March 2017, and the North 1# sluice and North 2# sluice were opened to water in July 2018. During the implementation of the program, the smooth progress of the project was ensured through water level monitoring in the perimeter area as well as contingency plans, and the main project was successfully completed and passed the acceptance in June 2019. The program with closure with cofferdam has advanced the construction period of this project by 12 months.

6 Conclusion

Usually, the reclamation project requires that the sluice has been completed and can be opened and closed normally for drainage before closuring. Due to the special nature of this large scale reclamation project, it is necessary to closured in advance. Therefore, this project broke through the conventional thinking and demonstrated the feasibility of the scheme of closure with cofferdam through theoretical calculations. During the implementation of the scheme, the water level monitoring in the polder area and the emergency plan ensured the smooth progress of the project. The successful completion of the polder project proved the success of the scheme and provided a favorable reference for the subsequent reclamation projects.

References

1. Huang Jinlin, Jiang Wei. Technical Supervision In Water Resoures,03，1-3，(2009).
2. Reclamation Bureau of Water Resources Department of Zhejiang Province. Hangzhou： Reclamation Bureau of Water Resources Department of Zhejiang Province，(2010).
3. Yuan Wenxi, Wu Lei, Xu Ji, et al. . Hangzhou： Zhejiang Design Institute Of Water Conservancy And Hydro-Electric Power ,(2012).
4. Yuan Wenxi, Wu Lei, Xu Ji, et al. S. Hangzhou： Zhejiang Design Institute Of Water Conservancy And Hydro-Electric Power ,(2014).
5. Chen Dechun, Wu Jiwei, Li Yu, et al. Study on closure technique for sea reclamation works[J]. Journal of Hohai University(Natural Sciences), 2002(5) : 67-70.

https://doi.org/10.1051/e3sconf/202129302035

Reliability Fuzzy Comprehensive Evaluation based on construction monitoring long-span arch truss

Kongliang Chen[1,*], Guoqing Yuan[1], Liankun Wang[1]

[1]School of Civil Engineering, Wuyi University, Jiang men, 529000, Guangdong, China

Abstract: The steel roof of the West Stand structure in Jiangmen Sports Center is a super-span space steel tube truss arch structure. In this article, the simulation analysis, and construction monitoring process, and the arch foot deformation monitoring is researched systematically, and the fuzzy reliability analysis is established based on the construction monitoring process. The monitoring results show that the roof structure is in healthy state, and the arch foot deformation meets the design requirements, which means that detailed simulation calculation provides theoretical basis for construction monitoring, and the measured data agree well with the theoretical data. Reliability Fuzzy Comprehensive Evaluation of structural monitoring adopts a layered weighting method by dividing the entire structural monitoring into several types of key monitoring indicators, corresponding analysis is carried out according to the calculation formula of the weight of various indicators to solve the reliability problem of structural monitoring. It is shown through analysis that this method can provide an effective basis for on-site safety construction, and the above conclusions can provide references for related projects in the future.

1 Introduction

The large-span complex space truss structure is widely used in various major buildings because of its beautiful appearance and reliable structure. However, Large-span steel structure has a large span and complex forces[1-4], and the potential for damage and destruction is particularly high during the construction phase. There are many hidden safety hazards in the construction of large-span steel structures. In the last two decades, some major safety accidents have occurred. A roof collapse accident occurred in the 2E terminal of Paris Charles de Gaulle International Airport in May 2004, causing many people were unfortunately killed, and a large factory building in Shanghai collapsed during construction in July 2009 with an area of about 20,000 square meters and the main steel structure of the platform collapsed, and a collapse accident occurred during the demolition of the Shenzhen Sports Center in July 2019. Therefore, monitoring the stress and deformation during the construction phase is of great importance for ensuring the structural safety[5-8]. And there are less research on construction monitoring[9-10] of large arch trusses at present. There are relatively few research results on long-span arch truss construction monitoring and corresponding reliability analysis.

In response to the above problems, the large-span arch truss has a novel structure and complex forces. It is still insufficient in on-site monitoring, measurement and early warning, which greatly reduces the role of monitoring and measurement on site construction guidance. Supported by on-site monitoring data and based on structural reliability design principles, a probabilistic early warning system for construction monitoring has been established, which provide references for related projects in the future.

2 Project Overview

The steel structure of the West Stand roof consists of arch trusses, truss beams, and canopies that cantilevered to the sides, and the material of the steel structure is Q345B, as shown in Fig.1. The cross section of the arch truss is an inverted triangle, and the arch feet at both ends are supported on the ground, of which the theoretical span between the arch feet is about 260m with 50m height from arch to ground, including the size of the concrete support. The arch truss is composed of round steel pipes tilted out of the field, which is supported on the roof truss beam by setting arch support rods, and one end of the roof truss beam is hinged to the lower concrete outer edge column, and the other end is hinged to the lower chord steel pipe of the arch frame. The cross section of the main truss arch is round steel pipe, and those of the cantilevered beam above the main truss arch is H-shaped steel beam, and the other rod sections are mainly round tubes. To ensure the overall rigidity of the roof, diagonal braces are set in the middle of the truss beams, and stable trusses are set at the mid-span and ends.

* Corresponding author: kongliangchen@126.com

Figure 1. The steel roof structure of West Stand

3 Analysis of construction technology

3.1 Construction process and simulation calculation

The entire arch truss is divided into 11 segments, and the arch truss of each segment is fixed to the temporary support by hoisting from the bottom to the middle of the stands on both sides, and then the radial truss is installed, which is assembled from both sides to the middle high altitude, as shown in Figures 2.

The structure of the west stand needs to be unloaded as a whole, and the unloading tonnage is about 1900t. For the purpose of ensuring the safety of the structure and the overall appearance, it is necessary to control the key points to descend simultaneously. However, the roof unloading organization is so complicated that it must have corresponding guarantee measures for the entire unloading process implemented according to the established plan, which requires calculating the actual value before the unloading phase. Therefore, the overall unloading must theoretically be calculated accurately for safety. Construction simulation calculations are performed using Midas finite element program. According to the parameters of the design and construction drawings, the main truss and other members are simulated by beam element, and supports are simplified as hinges, which is shown in Figure 3, and there are 5580 elements with a total of 30 sections.

Figure 2. Main truss installation

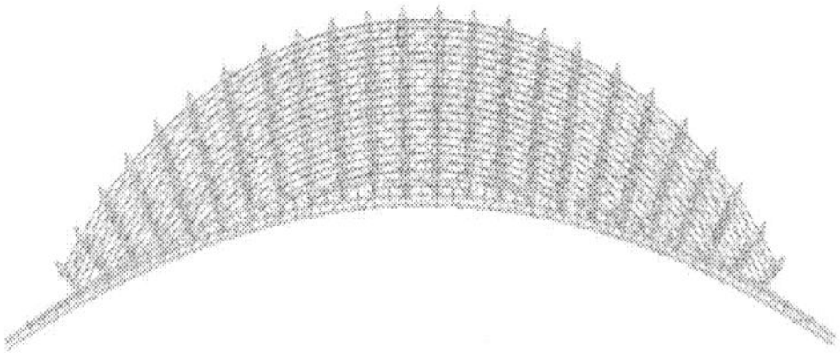

Figure 3. Calculation model

On the basis of the construction characteristics and construction monitoring requirements of the west stand roof structure, the finite element step-by-step formal installation calculation method is used. In theoretical analysis, the vertical displacement of unloading stage has a large change, while the horizontal displacement changes little. Therefore, Vertical displacement is the key stage of construction monitoring.

As shown in Figure 4, there are three stress monitoring sections for the main truss of the West Stand, which are section A, section B, and section C, and each rod has 4 sensors and each section has 12 sensors, and section A is at the vault, section B and section C are at the arch feet. The displacement monitoring points of the main arch are at the arch vault and arch feet, and one measuring point is on the vault, and 4 measuring points on each of the two arch feet, and S1,S2,J1,J2 are in left arch foot, and Q1,Q2,T1,T2 are in right arch foot. The radial truss displacement monitoring points are located at the junction of the radial truss and the column, and there are a total of 8 radial truss measuring points. The X-axis in the figure is basically coincident with the line connecting the centers of the two arch feet, and the Z-axis is vertical. There are 15 stages in the construction monitoring of the structure, and 18 stages in the deformation monitoring of the arch foot, and the 11th stage is the roof unloading stage, and the stress monitoring of the arch vault starts from the 7th stage.

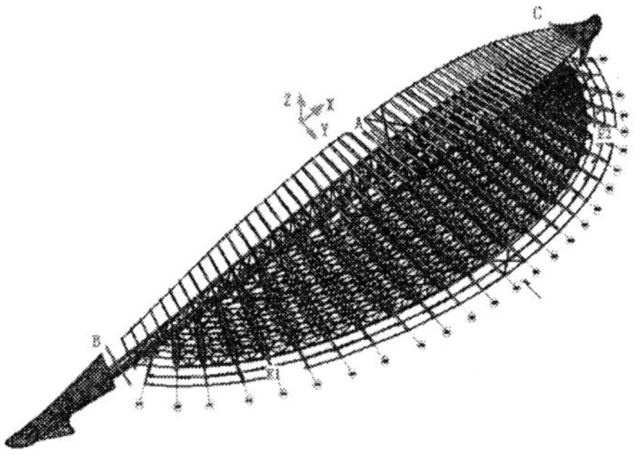

Figure 4. Schematic diagram of measuring points

Figure 5. Installation of the arch foot strain gauge

https://doi.org/10.1051/e3sconf/202129302035

Figure 6. Installation of arch vault strain gauge

The monitoring system of this project is mainly composed of sensor subsystem, data collection subsystem and so on, which can realize continuous collection of monitoring data, and continuous real-time monitoring. The sensor subsystem arranges sensors in key parts of the structure according to construction monitoring scheme, which is used to obtain the response information under load, and the stress installation of the arch top and arch foot is shown in Figures 5-6.

3.2 Stress and deformation monitoring results and analysis

The process from the closing of the arch truss to the removal of the support of the tire frame completes the transition from the temporary support tower load to the steel structure. This is a process of gradually transferring the permanent structure and redistributing internal forces. Due to more measured points and data, this article only selects part of the stress measured points for analysis, and finds out the corresponding laws.

It can be seen from figures 7-8 that the stress curve of the arch foot measurement point is basically uniform before unloading, the curve variation is in the range of about -10MPa ~ 10MPa, and the internal force change law of the arch foot measurement point is roughly similar. During the unloading of tire frame, the stress data suddenly increased sharply about 20 to 40MPa, but it is far less than the yield strength of the steel. After the unloading, the internal force of the rod is gradually transferred and the internal force is redistributed, but the overall change is not large. And the stress curve of the vault measurement point has obvious abrupt change during the unloading stage, and vault stress will gradually increase and finally stabilize, and this process is similar to the stress change of the arch foot.

As shown in Table 1, assuming tensile stress is positive, the stress data in the unloading stage have large variations, which is generally about 20 to 30 MPa. Moreover, the ratio of the measured stress data in most rods to the theoretical value is about 0.6 to 0.95, which is in a reasonable range, but there are also some points of data that are not in this range because of temperature or environmental impact. The comparison of measured stress and theoretical stress fully demonstrate the correctness of theoretical analysis.

Due to more measured points and data, this article only selects part of the displacement measured points for analysis, and finds out the corresponding laws.

It can be seen from the figures 9-10 that when the roof structure is unloaded, the horizontal deformation of the arch foot has a sudden change about 3 ~ 6mm, but the deformation of the arch foot slightly decreases and gradually stabilizes in the subsequent stage.

From the results in Table 2, it can be seen that the distance between the two measuring points of the arch foot is pushed outward after the tire frame is removed, and the distance between the two points has increased by about 4 ~ 9mm, but the actual monitoring result is smaller than the theoretical calculation, which shows that the arch foot stiffness is greater than the calculated value of the theoretical model.

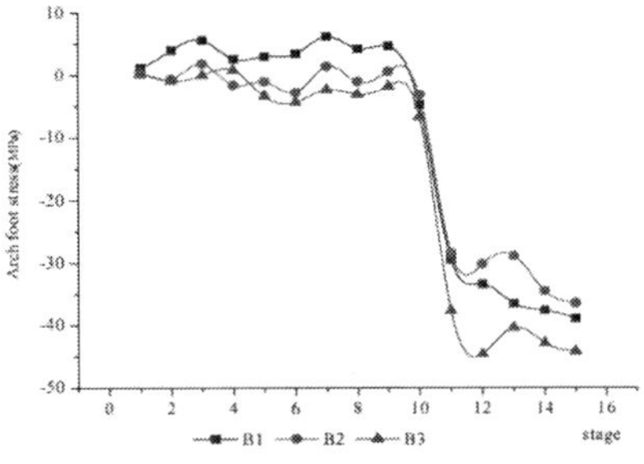

Figure 7. Stress change of left arch foot

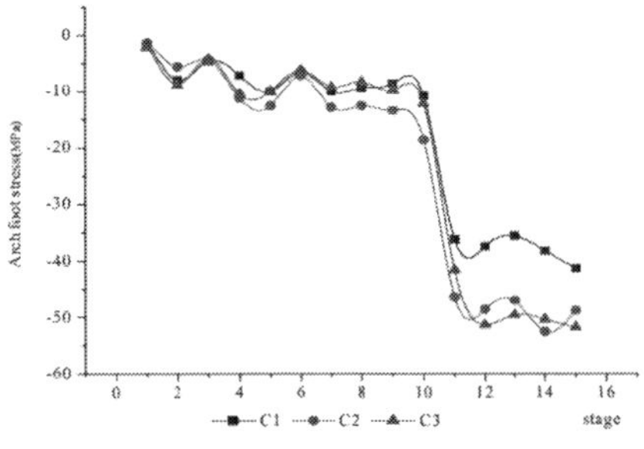

Figure 8. Stress change of the right arch foot

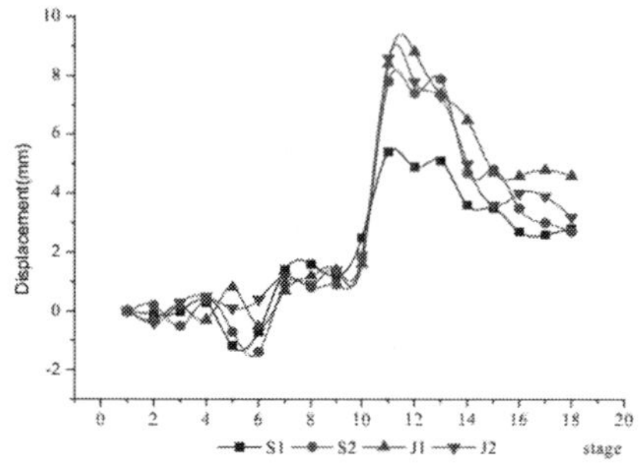

Figure 9. Deformation of the left arch foot

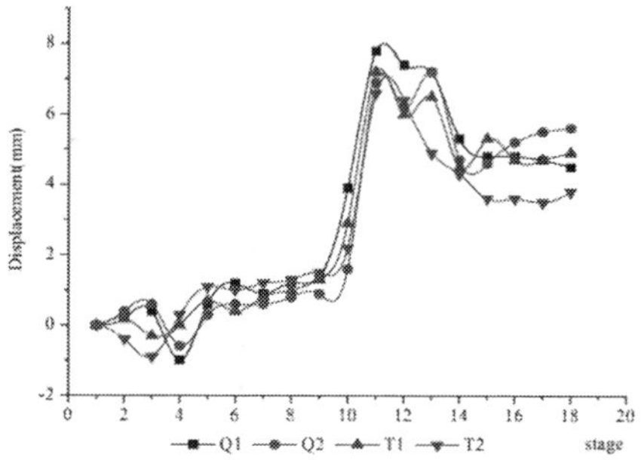

Figure 10. Deformation of the right arch foot

Table1. Comparison of stress values during unloading (Unit:MPa)

Serial number	position	Construction stress		Change in stress	Theoretical stress	Value comparison
		Stage 10(①)	Stage 11(②)	(②-①)	Stage 11(③)	②/③
Section A	A1	-10.4	-22.9	-12.5	-30.4	0.41
	A2	-14.7	-37.8	-23.1	-30.4	0.76
	A3	-11.2	-39.4	-28.2	-30.4	0.93
Section B	B1	-4.8	-29.5	-24.7	-40.6	0.61
	B2	-3.2	-28.4	-25.2	-40.6	0.62
	B3	-6.7	-37.6	-30.9	-40.6	0.76
Section C	C1	-10.8	-36.2	-25.4	-40.6	0.63
	C2	-18.6	-46.4	-27.8	-40.6	0.68
	C3	-12.2	-41.6	-29.4	-40.6	0.72

Table2. Deformation comparison before and after unloading (Unit: m)

Left	Right	Straight line	Distance between two measuring points		Measured value	Allowable value
			Before	After	ΔL	[ΔL]
S1	Q1	S1Q1	266.575	266.579	0.004	0.018
S2	Q2	S2Q2	264.526	264.535	0.009	0.018
J1	T1	J1T1	271.727	271.735	0.008	0.018
J2	T2	J2T2	273.505	273.511	0.006	0.018

4 Theoretical analysis of structural reliability

In the reliability fuzzy evaluation of structural monitoring, the whole can be divided into three types of indicators as shown in Figure 11. And fuzzy evaluation is the first category of indicators, and arch foot monitoring, main truss monitoring and radial truss monitoring are the second category of indicators, and the other are the third category of indicators. The first type of indicators can be obtained through expert survey methods to obtain the weights of corresponding factors, while the evaluation of other types of indicators can be obtained through the hierarchical weighting method based on the monitoring data information, and the key task is to determine the calculation method of the underlying monitoring indicators.

For the first type of indicators, the five technical conditions can refer to the specification[11], and the fuzzy evaluation level can be determined according to the calculated score. For the second index, a dimensionless model can be used to process the value according to the percentage system. For the third indicator, the monitoring data is the evaluation indicator of the data sequence, and it needs to be processed appropriately dimensionless to make the detection results of each measurement point comparable. The classification of the monitoring data is shown in Table 3.

Table3. Monitoring rating level

Rating	Level one	Level two	Level three	Level four	Level five
Points	80~100	60~79	40~59	20~39	0~19

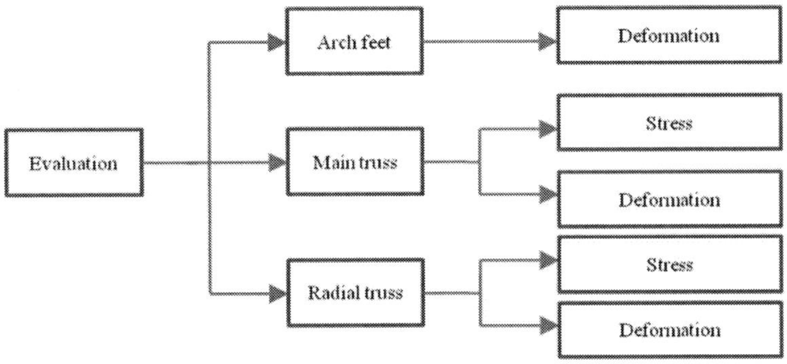

Figure 11. Three types of monitored indicators

The levels and deductions of the underlying indicators are shown in the table 4. Among them, "a" indicates that the monitoring data is less than the theoretical calculation value, and "b" indicates that it is greater than one time the theoretical calculation value and less than twice the theoretical calculation value, and "c" indicates that it is greater than twice the theoretical calculation value.

Table4. Deductible value of the indicator

Stress			
classification	a	b	c
Tolerance	<[σ]	<2.0[σ]	>2.0[σ]
Deduction value	0	15	40
Deformation			
classification	a	b	c
Tolerance	<Δ	<2.0Δ	>2.0Δ
Deduction value	0	15	40

The calculation result of the fuzzy evaluation is carried out by the analytic hierarchy process divided into three levels of indicators. The weight of the second-level indicators is determined according to the importance of the monitoring components. Among them, the monitoring of the arch foot and the main truss is the most critical determined by the expert measurement method. Therefore, the weights are based on the expert survey method, the values of these two parts are the largest, and the weights of the secondary indicators are 0.45, 0.45 and 0.1 respectively. The corresponding calculation formulas are as (1) ~ (5).

$$S=\sum_{i=1}^{3}A_i\cdot w_i \tag{1}$$

$$A_i=\sum_{j=1}^{2}B_j\cdot w_{ij}=\sum_{j=1}^{2}(100-D_j)\cdot w_{ij} \tag{2}$$

$$D_j=\sum P_k\cdot w_k \tag{3}$$

$$w_k=3.0\mu^3-5.0\mu^2+3.0\mu \tag{4}$$

$$\mu=\frac{P_k}{\sum P_k} \tag{5}$$

In the above formula, " S" is the comprehensive score, and "Ai" is the score of the second-level indicator, and "wi" is the second -level weight determined by the expert survey method, and "Bj" is the single score of the third-level indicator, and "wij" is the weight value of the third-level indicator which is the element contained in a single indicator of the second level, and "Dj" is the deduction value of the third-level indicator, and "Pk" is the deduction value of the third level indicator, and "wk" is the weight value which is the calculation formula for the weight value of the third level indicator, and μ is the proportion which is the ratio of the single deduction system of the third level indicator to the total deduction value.

And the following example is used to illustrate the calculation process of this method, and the data is shown in table 5, and combined with the previous limit deduction conditions, the fuzzy evaluation value of this example can be calculated as 82 points, and the rating is level one.

Table5. Example data

data	w11	w21	w22	w31	w32
1	b1<[Δ]	c1<[σ]	d1<[Δ]	e1<[σ]	f1<[Δ]
2	b2<[Δ]	c2<[σ]	d2<[Δ]	e2<[σ]	f2<[Δ]
3	b3>2.0[Δ]	c3<[σ]	d3<[Δ]	e3<[σ]	f3<[Δ]

The calculation process has been described through the above examples. The fuzzy evaluation value of the construction monitoring of the west stand structure can be calculated from the data in Table 1 and Table 2. In the process of solving, pay attention to the abnormal data collection or the situation that the data exceeds twice the theoretical value, and the data is substituted into the formula (1) -(5), the fuzzy evaluation value of the unloading stage can be calculated as 100, and the evaluation level is the first level.

5 Conclusion

Through the previous discussion in this article, the following conclusions can be drawn.

Through the real-time monitoring of structure construction stage, the results show that the stresses of arch vault and arch foot have a sudden change during the unloading phase, and the range is about 20MPa~40MPa, which is far less than the yield strength of the material, and the measured value during the unloading phase agrees well with the theoretical calculation value. The stress state of each measuring point is always in an elastic working state, and the bearing capacity of the entire structure is safe and meets the design requirements.

The horizontal deformation of the arch foot has a sudden change when the roof structure is unloaded, but the deformation of the arch foot slightly decreases and gradually stabilizes, and the deformation of the radial truss measurement points has not changed significantly, and the distance between the two measuring points of the arch foot is pushed outward after the tire frame is removed, but the actual monitoring result is smaller than the theoretical value, indicating that the structural stiffness meets the design requirements.

The analysis of the monitoring results shows that the monitoring method of the project is reasonable, the measured values are in good agreement with the theoretical calculation results, and the monitoring results accurately reflect the actual stress state, which is of great significance for ensuring the safety of the structure. The construction simulation analysis, construction monitoring process and monitoring results in this article can provide references for similar structures.

A fuzzy evaluation method is proposed by dividing the structure monitoring into several parts according to the expert survey method, and using the layered weighting method to score the entire monitoring structure, the scoring result can better reflect the structure construction Monitoring the status. This method is more practical and simple, which can better distinguish the total evaluation value of each construction monitoring stage, and provide reference for related research.

Acknowledgements

The paper supported by the Special Funds of Guangdong Science and Technology Department Project(2016A040403125) And
Research on Teaching Method Reform of Bridge Inspection Based on Practical Application" (Project Number: GDJX2017018)

References

1. X.H.Liu, M.Y.Wang, Y. Zhang. "Installation technique and stability checking calculation of top arch tube truss structure in Shenzhen kingkey financial center".Steel Construction. Vol.33. no.1. pp. 92-96,52. 2018.
2. H.F.Chen, X.W.Liui, X.J.Wu, W.D.Xia, S.P.Zhang. "Construction Technology of Partial Continuous Cumulative Unloading of Large-span Space Pipe Truss". Construction Technology, Vol. 46. No.15. pp.34-37,53. 2017.
3. X.D.Shao, G.He, X.J.Shen, P. Zhu,Y. B. Chen. "Conceptual design of 1000 m scale steel-UHPFRC composite truss arch bridge.Engineering Structures". Vol.226, No.1.PP.1-12.2021.
4. S. Afshan, M. Theofanous, J. Wang, M. Gkantou, L. Gardnere."Testing, numerical simulation and design of prestressed high strength steel arched trusses".Engineering Structures. Vol.183,No.15.pp510-522.2019.
5. W. Wang, M.F. Wang, S.H. XU, Z.Y. Wang, K.Zhang. "Research on second of large span arch pipe truss structure node during unloading process". Architecture Technology. Vol.47.No.4. pp.368-370. 2016
6. H.P.Xie, Z.C.Jin, Z.L.Luo. "Consturction of Large-span Inclined Arch Pipe Truss Installation. Construction Technology". Vol.46. No.7.pp.33-35.2017
7. Y. Mesda. "Analytical Study of the Cable-Truss Systems on the Glass Certain Walls with Vertical Uses". Engineering. Vol.5.No.10. pp.819-826.2013.
8. B.Wu, L.L. Zhang, Y. Yang, L.J. Liu, Z.J. Ni. Claudio Mazzotti. "Refined Time-Domain Buffeting Analysis of a Long-Span Suspension Bridge in Mountainous Urban Terrain.Advances in Civil Engineering". Advances in Civil Engineering. Vol. 2020.No.1.pp.1-22. 2020.
9. K.L. Chen, G.Q. Yuan, L.K. Wang, W.Z. Zhang, X.K. Wang. "Research on Construction Monitoring of Large-Span Steel Pipe Truss Structure". Open Journal of Civil Engineering. Vol. 9. No. 4. PP. 255-267.2019.
10. X.L.Wang, Z.H.Rong, B.X.Yang, C.D.Luo, Y.Zhou, X.Y.Wu. "Whole Process Simulation Analysis and health monitoring of spatial Pipe Truss Structure Gymnasinm". Journal of Architecture and Civil engineering. Vol. 34.No.2. pp. 18-25. 2017.
11. Ministry of Construction of the People's Republic of China.Technical code of maintenance for city bridge(CJJ99-2003). 2003

The design of super long span pedestrian landscape cable stayed bridge with sightseeing bridge tower

Tao Jiang[1] and BiQing Li[1*]

[1]Jiangsu Southeast Special Technical Engineering Co., Ltd, Nanjing, China,210008

Abstract. According to the design requirements of the super long-span pedestrian landscape cable-stayed bridge in Lake view Park, a kind of sightseeing tower-style pedestrian landscape three-tower cable-stayed bridge structure system was proposed. Sightseeing tower type bridge tower structure and duck egg arched bridge tower replace traditional cable-stayed bridge tower structure. Through the establishment of Midas finite element analysis model, dynamic modal analysis research was carried out. The analysis shows that: the sightseeing tower structure and the cable-stayed bridge tower structure are combined into one, and the two have good structural matching.

1 Introduction

With the development of the times, more and more people like to visit the Lake View Park. The sightseeing tower buildings and Landscape Bridges of the Lake View Park are also becoming increasingly important[1-2]. It is necessary to coordinate the two basic needs of the traffic bridge and the sightseeing tower in the scenic area [3-5]. It integrates the sightseeing tower building and landscape bridge in one, complementing each other in function, forming the landmark characteristic building of Lake View Park, and enriching the cultural connotation of the scenic spot [7-8].

The traditional portal pylon structure of cable-stayed bridge is reformed. The sightseeing tower type pylon structure and duck egg shaped arch pylon are used as pylon structure of cable-stayed bridge. The pedestrian landscape three pylon cable-stayed bridge of Lake View Park is built, killing two birds with one stone, forming a super long span pedestrian landscape three pylon cable-stayed bridge based on sightseeing tower type pylon structure.

In this paper, combined with the design of the super long span pedestrian landscape cable-stayed bridge of a lake view park, the structural configuration of the pedestrian landscape cable-stayed bridge with tower structure is studied, the engineering parameters are designed, the Midas finite element analysis model is established, and the internal force analysis and dynamic modal analysis under vertical load are carried out, so as to verify the sightseeing tower structure and cable-stayed bridge The matching rationality of the structure.

2 Research on geometric configuration

The ecological landscape pedestrian cable-stayed bridge in lake view park is a three tower cable-stayed bridge structure. The middle tower adopts the sightseeing tower tower tower structure of single leaf hyperboloid shape, and the two sides of the bridge tower structure adopt duck egg arch bridge tower.

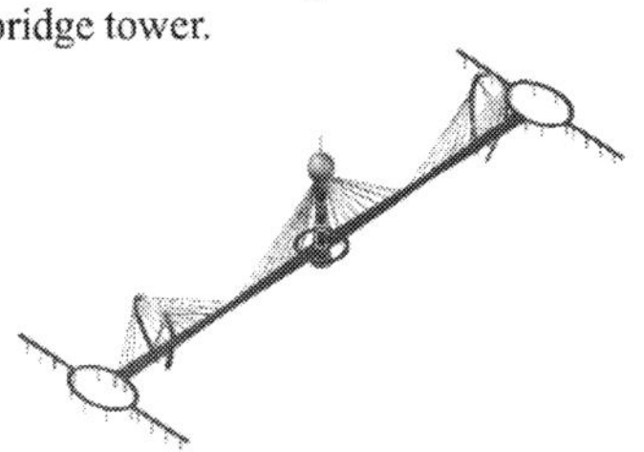

Figure 1. 3D drawing

The traditional three pylon cable-stayed bridge has no auxiliary transition piers on both sides of the middle pylon, which makes the cable-stayed bridge with flexible single column tower structure more flexible. The live load deflection of the main beam in the middle span of the three tower cable-stayed bridge is much larger than that of the conventional cable-stayed bridge. Therefore, the sightseeing bridge tower with higher structural stiffness is adopted in the middle tower of the design, which can greatly improve the vertical stiffness, reduce the live load deflection of the main beam in the middle span, meet the functional requirements of landscape and structural safety, and kill two birds with one stone.

Multiple stay cables are anchored to the top of the tower structure and duck egg arch structure in batches, and the two sides of variable cross-section elliptical arc single box and double chamber concrete box girder structure are suspended by multiple cables in batches.

* Li Biqing: umabank@163.com

https://doi.org/10.1051/e3sconf/202129302036

A giant circular concrete curved beam is set around the tower structure of the sightseeing tower, which is also used as a pedestrian sightseeing platform.

The pedestrian landscape cable-stayed bridge and the sightseeing tower share the same tower body and foundation, which greatly saves the investment in the reconstruction and construction of Lake View Park, and has remarkable economic benefits.

Figure 2. Effect Drawing

The sightseeing tower and the cable-stayed bridge tower are in one, with complementary functions, novel shapes and rich facades, which enhance the aesthetic effect of pedestrian landscape bridges. The curve structures of small waist-shaped sightseeing tower with single leaf hyperboloid shape, duck egg arch bridge tower and single leaf hyperboloid variable cross-section elliptical arc single box and double chamber concrete box girder structure complement each other and complement each other Beautiful and graceful.

3 parameter design

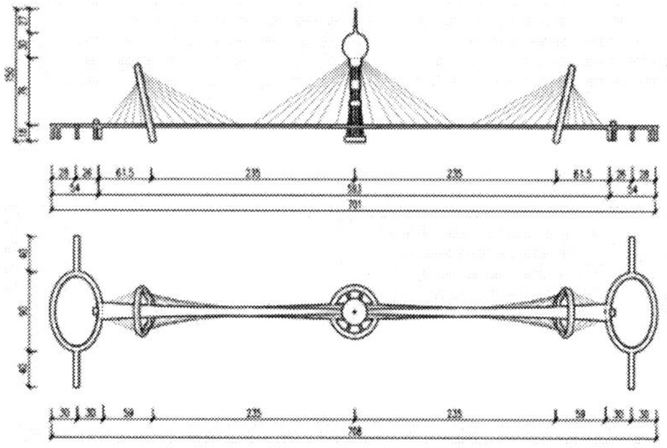

Figure 3. overall design

The main span of the three-tower cable-stayed bridge in a long-span pedestrian landscape in a lakeside park is 80 m +2×240 m +80 m. The middle bridge tower is a sightseeing tower structure with a single lobed hyperboloid shape, the two sides of the bridge tower are arched duck eggs, and the deck structure is a concrete box girder structure with variable section elliptic single box and double chamber.

The tower height of sightseeing cable-stayed bridge is 100.5m. The tower structure is composed of sightseeing tower base, sightseeing tower body and sightseeing tower top. The height of sightseeing tower base is 5.5m reception hall. The sightseeing tower body is a 75m high single leaf hyperboloid tube in tube structure. The top of sightseeing tower is an inverted conical shell steel tower top with a height of 20m.

The bottom of sightseeing tower type bridge tower adopts pile foundation, and the tower body adopts tube in tube structure of internal reinforced concrete core tube and external concrete-filled steel tube frame tube. The internal reinforced concrete core tube is equipped with spiral stairs and a circular elevator, and the inner cylinder diameter is 12 meters; the external steel tube concrete frame tube adopts single leaf hyperboloid shape, the outer tube diameter is 12-24 meters, and the frame tube is vertical The column is composed of eight hyperbolic concrete-filled steel tubular columns with a diameter of 1.4 meters and a wall thickness of 25 mm. The steel ring beam is connected with eight hyperbolic concrete-filled steel tubular columns. The steel ring beam is 900mm in diameter, 16mm in wall thickness and 6m in spacing between steel ring beams, forming a single leaf hyperboloid shape. The external concrete-filled steel tube frame tube is built in the inner and outer reinforced concrete core tube The concrete-filled steel tube frame tube is provided with multiple diaphragms with a spacing of 18 meters.

The Arch Tower of duck egg on both sides is composed of the top of parabola section and two columns with left and right elliptic arc sections. The highest part of the tower is 65 meters, the widest part of the tower is 40 meters, of which the width of the tower top of parabola section is 32 meters, the arch height is 7 meters, the height of the tower column in the elliptic arc section is 58 meters, the duck egg arch bridge tower adopts the square steel pipe with the side length of 5m, the steel pipe wall thickness is 18mm, C40 concrete is poured in the inner part, and the duck egg arch is used The pile foundation is adopted for the shaped bridge tower with an inclination angle of 24 °.

The variable cross-section elliptical arc single box double chamber concrete box girder structure is a half single leaf hyperboloid shape, one end is fixed in the tower body of the sightseeing tower, and the other end is placed on the duck egg shaped arch bridge tower body. The width of the bridge surface of the variable cross-section elliptical arc single box double chamber concrete box girder is 10-20 meters, the beam height is 2.5-4 meters, the section shape is elliptical arc box girder structure, and the top plate thickness of the concrete box girder is 250 The thickness of elliptical arc-shaped bottom plate of concrete box girder is 200-300 mm, the middle longitudinal web plate is corrugated steel web, and the diaphragms are reinforced concrete structure diaphragms, and the spacing between diaphragms is 60m.

The variable cross-section elliptical arc single box double chamber concrete box girder structure is a half single leaf hyperboloid shape, one end is fixed in the tower body of the sightseeing tower, and the other end is placed on the duck egg shaped arch bridge tower body. The width of the bridge surface of the variable cross-section elliptical arc single box double chamber concrete box girder is 10-20 meters, the beam height is 2.5-4 meters, the section shape is elliptical arc box girder structure, and the top plate thickness of the concrete box girder is 250 The thickness of elliptical arc-shaped bottom

https://doi.org/10.1051/e3sconf/202129302036

plate of concrete box girder is 200-300 mm, the middle longitudinal web plate is corrugated steel web, and the diaphragms are reinforced concrete structure diaphragms, and the spacing between diaphragms is 60m. The span arrangement of the three tower cable-stayed bridge in this design is 80 m + 2 × 240 m + 80 m unequal span cable-stayed bridge, and the side span is asymmetric cable-stayed bridge. The structural treatment methods are as follows: adopting variable cross-section and variable cross-section elliptical arc-shaped single box and double chamber concrete box girder structure to reduce the dead weight of mid span load; adopting inclined duck egg arch bridge tower, carrying out side span counterweight and adding side span ground anchor cable In order to adjust the unbalanced internal force of the side tower structure, the steel-making slag with a height of 0.8 m is filled in the stiffening beam of the 80 m side span concrete box girder.

There are 2 spans × 18 pairs of steel wire cables in the whole bridge. There are 7 pairs of cables on the left and right sides of the sightseeing bridge tower in the middle. The steel wire cables are 0.25m high-strength steel wire cables with a spacing of 15m. There are 7 pairs of cables inside the left and right duck egg shaped arch bridge towers, and the steel wire cables are 0.25m high-strength steel wire cables with a spacing of 15m. The outer sides of the two duck egg shaped arch bridge towers are four pairs of twisted back cables and steel cables The diameter of the cable is 0.3m, and the cable spacing is 20m. The prefabricated parallel strand method (PPWS method) is adopted for the parallel steel wire cable, and 1770mpa φ 5.2mm high-strength steel wire is used.

Midas finite element model of bridge structure is shown in the figure below.

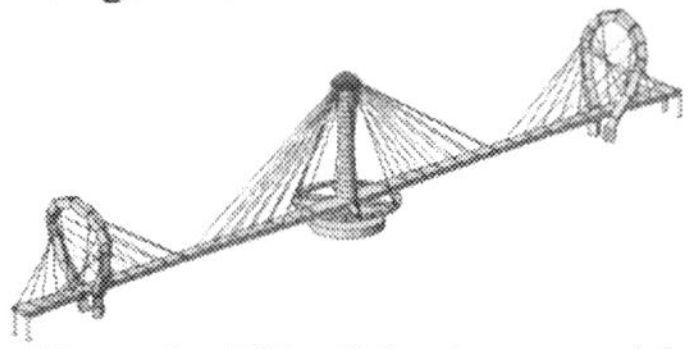

Figure 4. Midas finite element model

The center diameter of the giant circular concrete curved beam is 60 meters, the width of the giant circular concrete curved beam is 6 meters, and the beam height is 2.5 meters. The single box and single cell elliptical arc concrete box girder structure is adopted. The thickness of the concrete box girder top plate is 200 mm, and the thickness of the elliptical arc bottom plate of the concrete box girder is 160 mm.

The giant circular concrete curved beam is anchored on the top of the sightseeing tower by eight high-strength steel wire cables with a diameter of 0.2m. Six concrete tie beams are set between the giant circular concrete curved beam and the tower body to prevent the lateral deformation of the giant circular concrete curved beam.

4 Calculation results under vertical load

Considering that the stiffness of the middle span girder of the three tower cable-stayed bridge is the key problem in the design of the bridge, the structural calculation of the bridge is carried out under two load cases: the full span live load and the disadvantageous layout live load. The calculation results under vertical load are as follows:

(1) Full span live load case

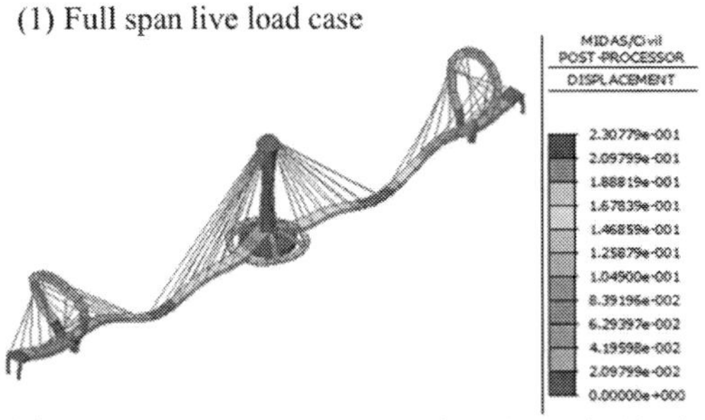

(a) Displacement calculation results under vertical load

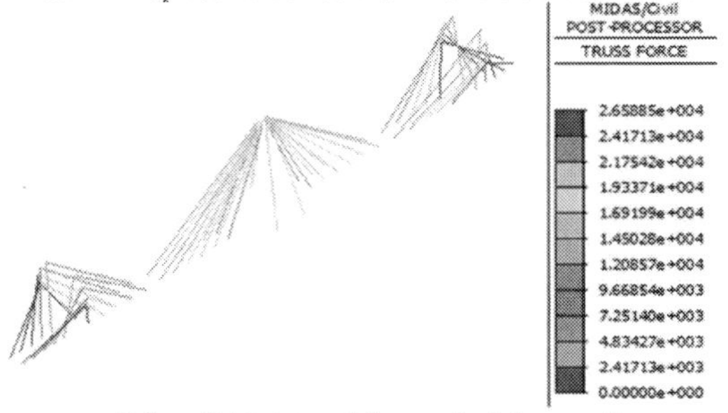

(b) Cable internal force calculation results

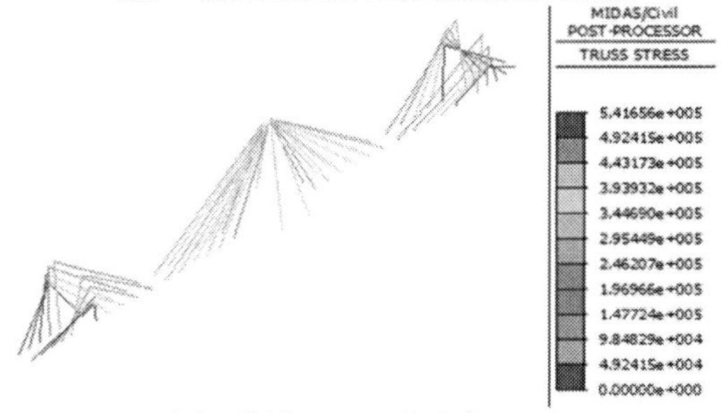

(c) Cable stress calculation results

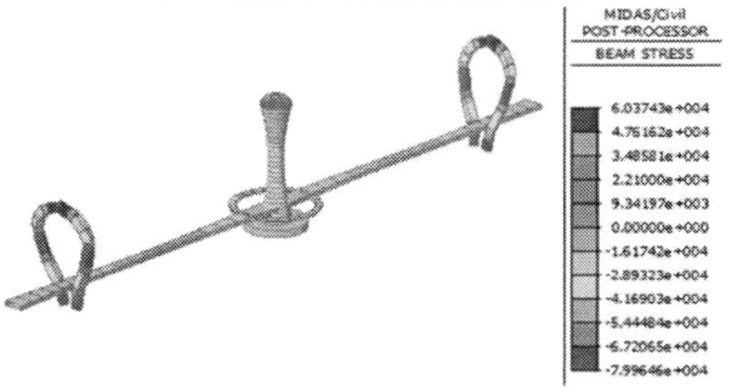

(d) Stress of tower beam element

Figure 5. calculation results of full span live load (dead+ Live)

It can be seen from the above calculation results that the maximum vertical displacement occurs in the middle of the span, and the vertical deflection under the vertical live load is 0.231m, which meets the requirement of 1 / 500 limit specified in the code.

The maximum internal force of the cable is 26588.5kn, and the maximum stress of the cable is 541.7mpa, which meets the strength requirements.

The maximum stress of tower beam element is 79.9mpa, which meets the strength requirements.

https://doi.org/10.1051/e3sconf/202129302036

(2) Considering the unfavorable layout condition of live load

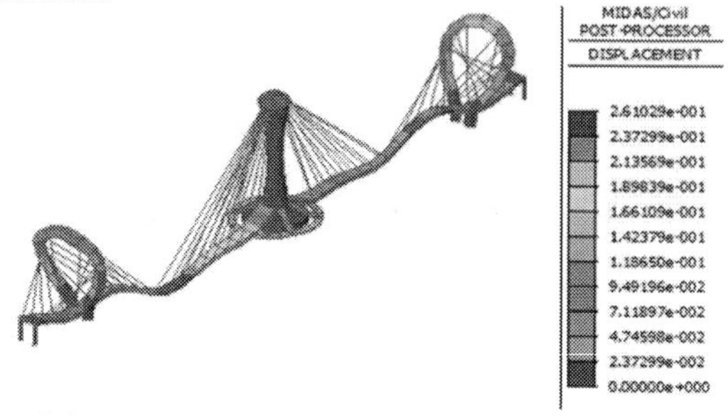

（a） Displacement calculation results under vertical load

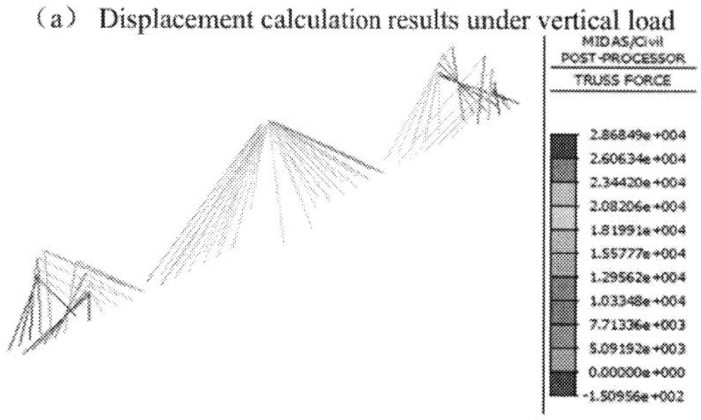

（b） Internal force of cable

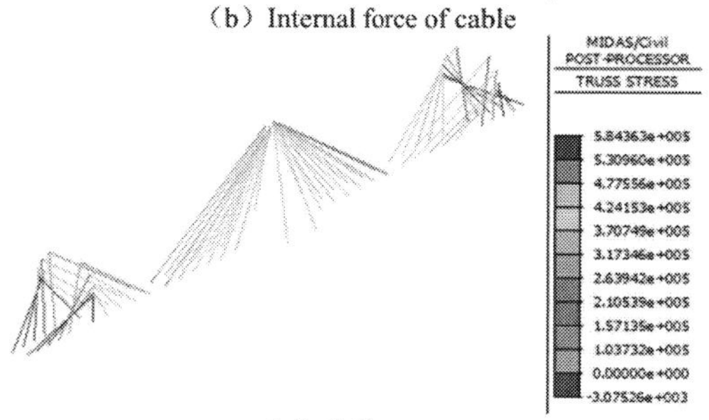

（c） Cable stress

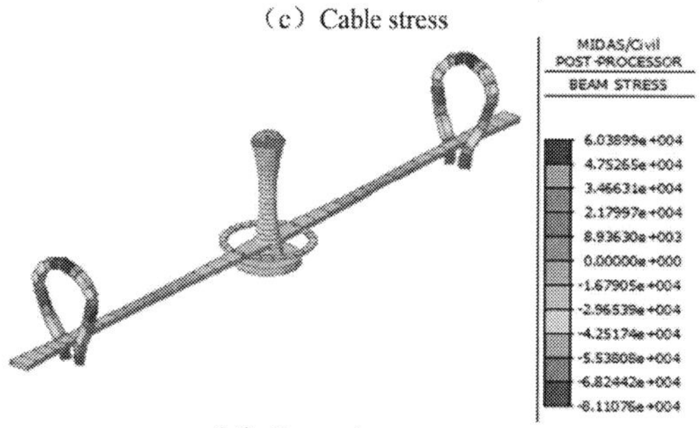

（d） Beam element stress

Figure 6. calculation results of unfavorable live load cases (dead+ Live)

From the above calculation results, it can be seen that the maximum vertical displacement of the bridge deck under the adverse layout live load (only the right one span is arranged with live load) occurs in the middle of the span, and the vertical deflection under the vertical live load is 0.261 m, which meets the 1 / 500 limit specified in the code.

The bridge deck without live load (the right one span) has a 0.105m upward inverted arch deformation in the middle of the span.

The lateral displacement of the middle tower is 0.018M, which meets the requirements of the code. The inclination ratio of the tower top after bending is 0.024%, which meets the requirements of the regulations. It shows that the tower stiffness of the sightseeing tower type bridge in the middle is large, which can avoid the weak tower structure problem of the single column middle tower of the three tower cable-stayed bridge.

The maximum internal force of the cable is 28684.9kn, and the maximum stress of the cable is 584.4mpa, which meets the strength requirements.

The maximum stress of the tower beam element is 81.1mpa, which meets the strength requirements.

There is no auxiliary transition pier on both sides of the middle pylon of the three tower cable-stayed bridge, which lacks the constraint help of the main beam and the cable tower stiffness, which makes the cable-stayed bridge with flexible single column tower structure more flexible. The live load deflection of the main beam in the middle span of the three tower cable-stayed bridge is much larger than that of the conventional cable-stayed bridge.

The key to the smooth implementation of the single column three tower cable-stayed bridge with weak tower structure is how to solve the stiffness problem of the main beam of the middle span. Therefore, the sightseeing bridge tower with larger structural stiffness is adopted in the design of the middle tower, which can greatly improve the vertical stiffness and effectively reduce the midspan deflection of the main beam of the middle span under the adverse live load. The combination of the sightseeing tower structure and the tower structure of the cable-stayed bridge Two for one, the two have good structural matching.

5 Modal calculation results

The main cable is the main load-bearing structure of cable-stayed bridge, and the internal force of main cable and sling is included in the form of initial tension in dynamic characteristic analysis and modeling. Based on Midas's nonlinear static analysis and modal analysis function, the natural vibration characteristics are analyzed. In order to avoid omitting any vibration mode, the sub block method is used to solve the characteristic equation.

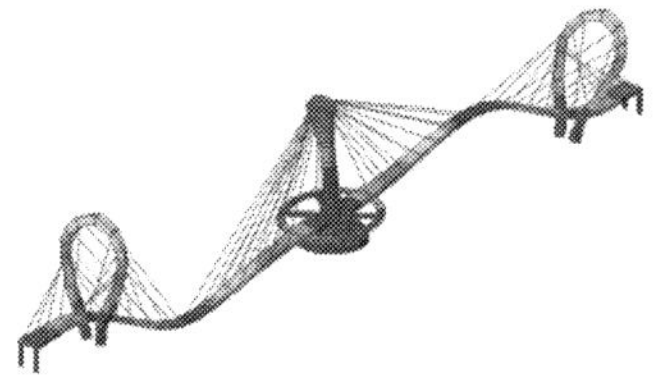

（a） 1st mode shape (antisymmetric vertical bending) (0.587hz)

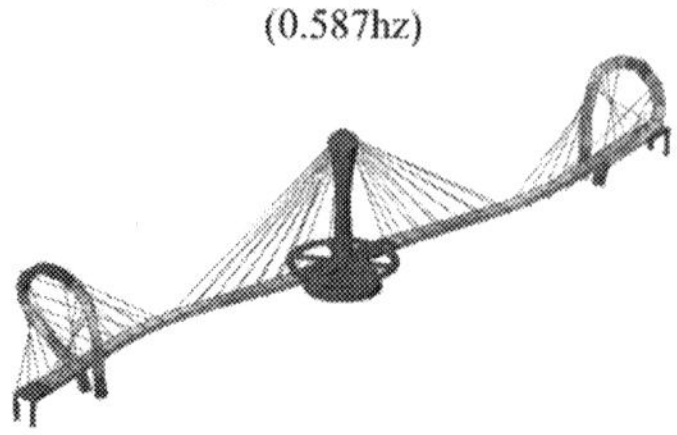

（b） 2nd mode shape (antisymmetric lateral bending of duck egg arch) （0.596Hz）

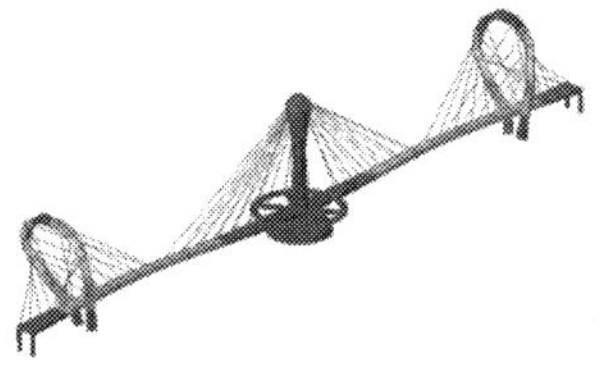

(c) 3rd order mode shape (duck egg arch positive symmetrical lateral bending) (0.603hz)

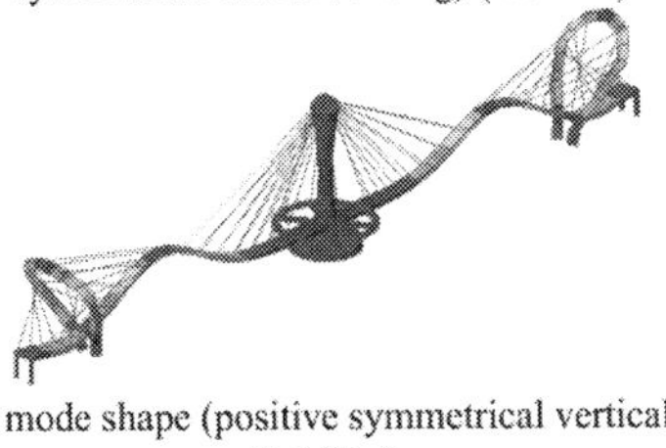

(d) 4th mode shape (positive symmetrical vertical bending) (0.640hz)

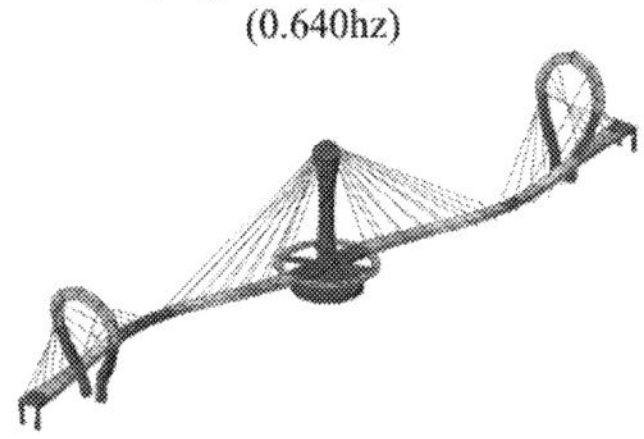

(e) 5th mode shape (antisymmetric lateral bending) (0.647hz)

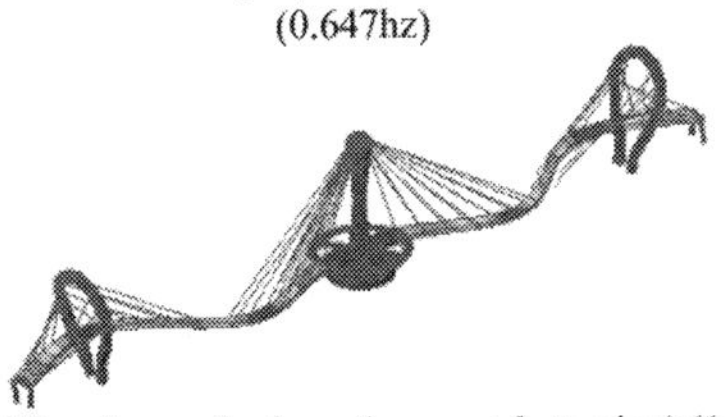

(f) 15th order mode shape (symmetric torsion) (1.389hz)

Figure 7. Typical dynamic modes

According to the modal analysis results, the first mode is antisymmetric vertical bending with frequency of 0.587hz; the second mode is antisymmetric lateral bending of duck egg arch bridge tower with frequency of 0.596hz; the fifth mode is antisymmetric lateral bending mode with frequency of 0.647 Hz and the first torsional frequency of 1.389 Hz (15 order). The analysis shows that the vibration modes of the bridge are dense, and obvious mode grouping phenomenon appears. The torsion frequency of the bridge is large, and the torsion to bending frequency ratio is 2.37, which indicates that the spatial stiffness of the bridge is good.

6 Conclusion

(1) The traditional portal pylon structure of cable-stayed bridge is reformed. The middle tower adopts the tower structure of sightseeing tower, and the tower on both sides adopts the duck egg shaped arch tower as the tower structure of the cable-stayed bridge. The three tower cable-stayed bridge with pedestrian landscape in Lake View Park is built with beautiful appearance, leisure buildings are set on the bridge deck, flowers and plants are planted on the bridge deck, and ecological protection is achieved.

(2) The large-scale giant sightseeing tower tower tower structure improves the middle tower stiffness, effectively improves the structural stiffness of the three tower cable-stayed bridge, solves the weak tower effect problem of the traditional single column middle tower structure, and reduces the deflection deformation under the adverse live load arrangement. The sightseeing tower structure and the cable-stayed bridge tower structure are combined into one, and the two have good matching.

(3) Under the action of vertical full distributed live load, the maximum displacement in the main span is 0.231 m, and the maximum stress of stay cable is 541.7 MPa, which meets the requirements of the code. Under the unfavorable live load, the maximum vertical displacement in the middle of the main span is 0.261m, the horizontal displacement of the middle tower is 0.018M (the horizontal displacement of the middle tower is very small), and the maximum stress of stay cable is 584.4 The analysis shows that the vertical displacement and deformation in the middle span of the bridge under unfavorable layout live load and full distributed live load are almost the same. The middle tower structure of the bridge adopts large-scale giant sightseeing tower tower tower structure, which can avoid the weak tower effect of the tower structure of the three tower cable-stayed bridge.

(4) According to the results of modal analysis, the first mode is antisymmetric vertical bending with frequency of 0.587Hz; the second mode is antisymmetric lateral bending of duck egg arch bridge tower with frequency of 0.596hz; the fifth mode is antisymmetric lateral bending mode with frequency of 0.647 Hz and the first torsional frequency of 1.389 Hz (15 order). The analysis shows that: the vibration modes are dense and obvious mode grouping phenomenon occurs. The torsion frequency of the bridge is large, and the torsion to bending frequency ratio reaches 2.37, which indicates that the spatial stiffness of the bridge is good.

References

1. M. Kuo, & M. Xiao-Lan, Towards a landscape-approach bridge design: Philosophy and practice[C]. 2011 International Conference on Electric Technology and Civil Engineering (ICETCE), 2011,7102–7106.
2. Lian Q , Yuan W C . The Concept of Harmony in Aesthetics of Urban Landscape Bridge[C]// International Conference on Measuring Technology & Mechatronics Automation. IEEE Computer Society, 2018:81-85.
3. L. Xiaotian. Design of pedestrian landscape bridge in Nanxun Jinxiang Lake Park [J]. World bridge, 2013,41 (3): 33-37
4. W. Rui, H. Dazhang, Z. Yanfeng. Mechanical analysis of steel and concrete mixed tower of Nanjing Qingao landscape bridge [J]. Shanghai highway, 2017(04):21-24+27+4.
5. Okamoto, Y., & Nakamura, S. Static and seismic studies on steel/concrete hybrid towers for multi-span

cable-stayed bridges[J]. Journal of Constructional Steel Research, 2011,67(2).

6. Y. Yong, L. Junqing. Design parameter analysis of three tower cable stayed bridge [J]. World bridge,2016,44(02):53-57.
7. Vladimír, Š., Michal, P., & Tomáš, P. A Dynamic Analysis of the Cable-Stayed Footbridge in Čelákovice Town[J]. Procedia Engineering, 2017,199, 2877–2882.
8. Pepi, C., Gioffre, , Massimiliano, Grigoriu, M. D. . Parameters identification of cable stayed footbridges using Bayesian inference[J]. Meccanica: An International Journal of Theoretical and Applied Mechanics AIMETA, 2019,54(9), 1403.

Influence of waste calcareous rock powder on concrete transmission performance

Jianping Cheng[1,*]

[1] The 4th Engineering Co., Ltd., of China Railway 12th Bureau Group, Xi'an 710021, Shangxi, China

Abstract. Reuse of stone powder formed in the production of manufactured sand is of great significance to environmental protection and resource utilization. In this paper, the waste calcareous rock powder formed in the manufactured sand production line was added into the concrete by 6%, 9%, 12% and 15%, and the total weight of manufactured sand and rock powder was controlled to be constant to prepare C30 and C40 environment-friendly manufactured sand concrete. The influence of rock powder on the compressive strength and transmission performance of concrete was analyzed. It was found that with the increase of stone powder content, the compressive strength of fresh paste and C30 concrete first increased and then decreased, while the concrete transmission performance and the compressive strength of C40 concrete continued to decline. It was suggested that the content of rock powder should not exceed 9%.

1 Introduction

In recent years, large-scale engineering construction makes the high-quality natural sand resources increasingly scarce. Natural sand mining has become a scarce resource after strict management and policy control. With the continuous development of manufactured sand production technology, the quality, supply and use technology of manufactured sand are steadily improved, which makes manufactured sand gradually become an irreplaceable key material for concrete production. In the process of manufactured sand production, a large amount of stone powder is inevitably produced during the crushing process of rock stone. According to the investigation, the preparation of one ton of manufactured sand product was accompanied by the formation of 150kg ~ 200kg stone powder [1–3]. If stacking or landfill treatment was adopted, it will not only occupy the site, cause dust flying and water pollution, but also greatly reduce the utilization rate of energy and manufactured sand resources. On the other hand, the application specification of manufactured sand allowed the use of manufactured sand with appropriate stone powder content to prepare concrete. However, the influence of stone powder content on the comprehensive performance of concrete need in-depth experimental analysis to form a comprehensive and reliable mix design method of manufactured sand concrete. In order to protect natural resources and environment and reduce the production cost of manufactured sand, it was of great significance to recycle manufactured sand powder to prepare green environmental protection concrete.

At present, Kishan et al. [4] prepared green concrete with granite waste stone powder and glass powder and observed that the water permeability and water absorption of hardened paste were significantly improved. Iman et al. [5] evaluated the influence of marble and granite waste dust on the corrosion resistance of steel bars in concrete by statistical analysis method and found that 10% granite powder and 10% marble powder can replace cement to improve the open circuit potential and electrochemical impedance of ordinary concrete. Ding [6] studied the influence of different water binder ratio on the content of limestone manufactured sand and stone powder and found that with the decrease of water binder ratio, the appropriate value of manufactured sand and stone powder content decreased. Yan et al [7] considered that the comprehensive performance of HPC is best when the manufactured sand contains 7.0 wt.% stone powder. Xie et al. [8] considered that if the content of manufactured sand and stone powder is more than 6%, the mechanical properties of concrete prepared will be lower than that of natural sand concrete and found that when the content of manufactured sand and stone powder increases, the mechanical properties of concrete first strengthen and then weaken. Sun [9] thinks that the best stone powder content of manufactured sand in C30 concrete is 10% ~ 15%, and the best stone powder content of manufactured sand in C50 concrete is 7% ~ 10%.

In order to improve the utilization rate of rock powder by-products formed in the production of manufactured sand, reduce the occupied area of the site, master the influence law of rock powder on the performance of concrete, and form the green design concept of sustainable development of manufactured sand concrete, this paper used the waste calcareous rock powder formed in the production of manufactured sand to be mixed into concrete. The influence of water binder ratio on mechanical strength and transmission performance of ordinary concrete was studied.

* Corresponding author: 1290451762@qq.com

2 Material and testing method

2.1 Raw material

The apparent density of manufactured sand is 2690 kg/ m^3, the loose bulk density is 1600 kg/m^3, the porosity of loose bulk is 41%, the content of stone powder is 10.2%, the fineness modulus is 2.94, the water demand ratio is 103%, the MB value is 0.75, and the maximum crushing index of single stage is 16.5%. The crushed stone used is composed of 5-10 mm, 10-20 mm and 31.5 mm graded crushed stone with mass ratio of 2:5:3. The content of needle flake particles is 0.6%, and the maximum crushing index of single stage is 8.7%.

The waste calcareous rock powder was collected from the limestone manufactured sand production line and screened by 0.075mm screen. The test results showed that the fluidity ratio was 106%, the water demand ratio was 98.30%, the MB value was 3.8, the specific surface area was 900.21 kg/ m^3, the density was 2655 kg/ m^3, the water content was 0.41%, and the 7d activity index was 90%. Using P.O. 42.5 cement, standard consistency 27.1%, specific surface area 326m^2/kg, initial setting time 161min, final setting time 246min, 3d strength 32.2MPa, 28d strength 57.5MPa. Type F class II fly ash is used, with fineness of 13.2% (45μm sieve residue) and 28d activity index of 74%. Using polycarboxylate water reducer, the solid content is 30.2%, and the water reducing rate is 28%.

2.2 Mix proportion

The influence of waste calcareous rock powder on the performance of manufactured sand concrete was shown in Table 1. The water-to-binder ratio (W/B) of 0.44 and 0.38 were designed respectively to prepare C30 and C40 strength grade concrete and compare the influence of W/B. The total weight of manufactured sand and waste calcareous rock powder in C30 was 798 kg/m^3, and that in C40 was 814 kg/m^3. The amount of rock powder in brackets after the weight of rock powder was 6%, 9%, 12% and 15% respectively.

Table 1 Mix proportion of influence of content of waste calcareous rock powder on mechanical properties of concrete (kg/m^3)

Sample ID	Cement	Fly ash	Manufactured sand	Waste calcareous rock powder	Coarse aggregate	Water	W/B
C30-S-6	294	74	750	48 (6%)	1058	162	0.44
C30-S-9	294	74	726	72 (9%)	1058	162	
C30-S-12	294	74	702	96 (12%)	1058	162	
C30-S-15	294	74	678	120 (15%)	1058	162	
C40-S-6	341	85	765	49 (6%)	994	162	0.38
C40-S-9	341	85	741	73 (9%)	994	162	
C40-S-12	341	85	716	98 (12%)	994	162	
C40-S-15	341	85	692	122 (15%)	994	162	

2.3 Testing method

The compressive strength of hardened concrete was tested in accordance with the *Standard for test method of mechanical properties on ordinary concrete (GB/T 50081-2019)*. The size of compressive strength test specimen was 100 mm cube, and the test age was 3d, 7 d, 28 d and 56 d. The transmission performance of concrete had a profound impact on the durability of concrete [10]. In order to analyse the impact of waste calcareous rock powder on the transmission performance of concrete, this paper evaluated the transmission performance of concrete based on the rapid chloride diffusion coefficient (D_{RCM}) and carbonation depth test, and the test operation followed the *Standard for Test Methods of Long-Term Performance and Durability of Ordinary Concrete (GB/T50082-2009)*, The test age of D_{RCM} was 28 days and 56 days, and the test age of carbonation depth was 3d, 7 d and 28 d.

3 Result and discussion

3.1 Compressive strength

The test results of the influence of concrete prepared with waste calcareous rock powder on its strength were shown in Table 2. The solid line symbol represents C30 concrete, and the dotted line symbol represented C40 concrete. With the increase of stone powder content, the strength of C30 concrete at different ages first increased and then decreased. The strength of C30-S-9 specimen was the highest at each age, and its 56d strength was 65.5MPa. Although the strength of C30-S-15 specimen was low, its 56d strength can still reach 61.1MPa. It can be seen that the increase of stone powder content from 9% to 15% had little effect on the strength of C30 concrete. The strength of C40 concrete decreased with the increase of stone powder content. At 56d, the strength of C40-S-9 group was 70.8MPa, and that of C40-S-15 group was 65.6MPa. It can be seen that the weakening effect of stone powder content on the compressive strength of high strength concrete was more prominent. On the other hand, by comparing the strength data of concrete with different water binder ratio at the same age, it can be seen that the change of water binder ratio has little effect on the early strength of concrete at 9% and 12% dosage.

Therefore, for the concrete with different cementitious material dosage and strength requirements, there were differences in the optimal stone powder content of manufactured sand. As shown in Table 2, the stone powder content of C30 concrete was about 9%, and that of C40 concrete was about 6%, which was basically close to the existing research [11,12].

Table 2 Effect of content of waste calcareous rock powder on cube compressive strength of concrete

Conetent of waste calcareous rock powder (%)	C30-3d	C30-7d	C30-28d	C30-56d	C40-3d	C40-7d	C40-28d	C40-56d
6	35.7	41.4	56.2	61.0	41.8	53.3	65.1	70.8
9	38.5	46.2	60.2	65.5	38.6	47.4	60.7	67.2
12	38.1	45.9	57.0	62.3	39.1	49.0	61.5	68.1
15	32.3	42.0	56.6	61.1	36.2	48.3	58.7	65.6

Table 3 Prediction model parameters of carbonation reaction of waste calcareous rock powder concrete

Age/d	C30-S-6	C30-S-9	C30-S-12	C30-S-15	C40-S-6	C40-S-9	C40-S-12	C40-S-15
a	0.021	0.889	0.479	0.634	0.389	-0.357	0.008	0.553
b	1.893	1.941	2.038	2.191	1.738	1.776	1.874	1.993
R^2	0.991	0.987	0.99	0.984	0.91	0.985	0.992	0.983

3.2 Rapid chloride diffusion coefficient

The amount of waste calcareous rock powder in each mix proportion was listed in detail in column 5 of Table 1. The test results of D_{RCM} of waste calcareous rock powder concrete cured to 28d and 56d were plotted in Fig. 1 with the weight of stone powder as the fan angle value. The larger the arc in the Fig. was, the higher the amount of stone powder was.

With the content of waste calcareous rock powder increased from 6% to 15%, the D_{RCM} of C30 and C40 concrete increased gradually. For example, the D_{RCM} of C30-S-6 to C30-S-15 samples at 56d were $5.6\times10^{-12}m^2/s$、$4.8\times10^{-12}m^2/s$ 、 $5.9\times10^{-12}m^2/s$ and $6.5\times10^{-12}m^2/s$. Compared with C30-S-6, the latter three groups increased by -8.74%, 5.88% and 6.67% respectively. The hydration activity of waste rock powder used in this paper was very low, and the density of concrete paste can only be improved in a limited range by filling effect and nucleation effect after mixing with concrete. However, due to the low bonding degree between the particles, hydration products and hindering the extension and intersection of whiskers between hydration products, the three-dimensional random distribution network skeleton structure of paste was weakened. It can be seen from Table 2 and Fig. 1 that C30-S-9 with 9% rock powder content had higher compressive strength and lower D_{RCM} value, which indicated that adding 9% rock powder in the scope of this study was more beneficial to the performance of concrete.

Fig. 1 also showed that prolonging the curing period of concrete can effectively reduce the chloride ion diffusion resistance of waste calcareous rock powder concrete, that was, reduce the transmission performance of concrete. With the age increasing from 28d to 56d, the D_{RCM} of C30 and C40 concrete with different amount of stone powder decreased significantly. For example, the 28d D_{RCM} of C40-S-15 group was $11\times10^{-12}m^2/s$, $6.1\times10^{-12}m^2/s$ at 56d. When the age was extended, cementitious particles will produce more calcareous silicate hydrates. Meanwhile, the fly ash contained in the slurry played a volcano ash effect, further producing C-S-H gel with high silica-calcareous ratio [13]. Increasing hydration products will fill in the slurry pores and then reduce the pore size and connectivity, making the paste structure denser, which was beneficial to improve the concrete's ability to resist chloride ion penetration.

It can be seen from Fig. 2 that the increase of concrete compressive strength was accompanied by the decrease of D_{RCM} value, which means that there was a good correlation between the compressive strength of waste calcareous rock powder concrete and the rapid chloride ion diffusion coefficient. Concrete strength was the macro reflection of internal micro compactness and pore structure. High compressive strength often required the paste to have a dense structure, while the transmission performance of dense paste was low, so the chloride ion diffusion rate was slow, and its diffusion coefficient was small [8,14–17].

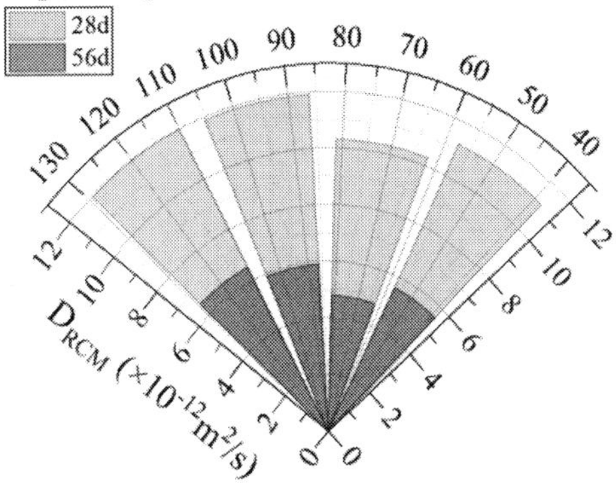

(a) C30 concrete

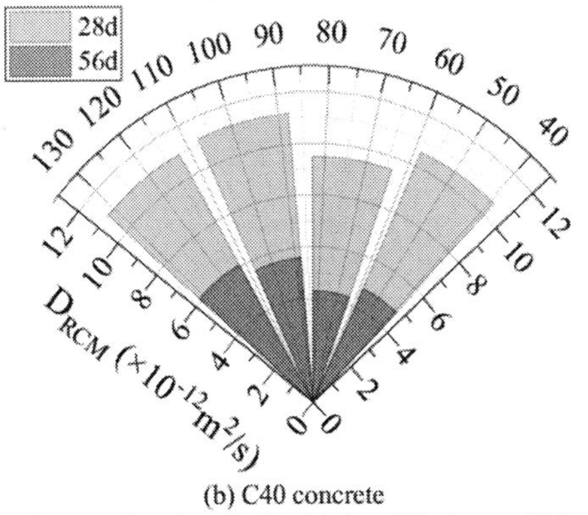

(b) C40 concrete

Fig. 1 Test results of rapid chloride ion diffusion coefficient of waste calcareous rock powder concrete

 https://doi.org/10.1051/e3sconf/202129302037

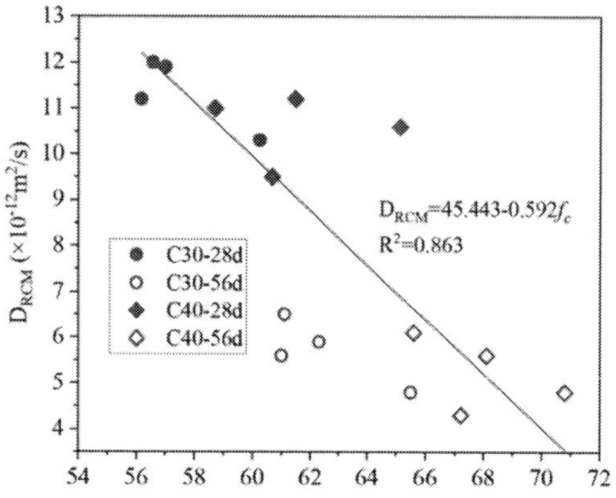

Fig. 2 Correlation between compressive strength of waste calcareous rock powder concrete and rapid chloride ion diffusion coefficient

3.3 Carbonation depth

In order to further analyze the influence of adding waste calcareous rock powder on the transmission performance of concrete, this paper used carbon dioxide gas as the invasion fluid medium for the test, and the relevant test results were shown in Fig. 3. With the extension of carbon dioxide invasion time, the carbonation depth of concrete with different strength grades increased gradually. On the other hand, the carbonation depth of concrete increased with the increase of stone powder content. When the age was 28 days, the carbonation depth of C30-S-6 to C30-S-15 groups were 6.2 mm, 6.5 mm, 6.8 mm and 7.3 mm respectively, and the increase rates of the latter three groups were 4.84%, 9.68% and 17.74% respectively compared with C30-S-6. Reducing the water binder ratio can counteract the negative effect of rock powder. For example, the carbonation depth of C40-S-15 group was 6.5mm on the 28th day, which was consistent with C30-S-9, and decreased by 10.96% compared with C30-S-15 group.

As a result of carbonation reaction, a series of reactions will take place in the carbonation zone of concrete surface, including the decrease of alkalinity of pore solution, the shrinkage of slurry and the formation of stress and strain gradient field, which will lead to the cracking risk of concrete surface. In order to form the anti-carbonation protection design method of waste calcareous rock powder concrete and slow down the adverse effect of carbonation reaction, the logarithmic model of equation (1) was used to make regression analysis on the test results in Fig. 3

$$D_{CO2}=b*ln(t+a) \quad (1)$$

Where t was the carbonation reaction time (d), a and b were the regression constants, and D_{CO2} was the prediction of carbonation depth. The parameters of carbonation depth prediction model of different mix proportion concrete in Table 1 were shown in Table 3, and the corresponding regression curve was shown in Fig. 3. It can be seen from Fig. 3 that the fitting curve of equation (1) was in good agreement with the variation law of the measured value of carbonation depth.

We used the quadratic function model to perform regression analysis on the correlation between the concrete carbonation depth and the cubic compressive strength, as shown in Fig. 4. It can be found that as the compressive strength of concrete decreased, the carbonation depth also decreased, that was, there was a good correlation between the two. Similar to the rapid chloride ion diffusion coefficient, the carbonation depth test reflected the difficulty of the transfer between the internal material of the hardened concrete specimen and the external environment from the side. The anti-carbonization performance of concrete was closely related to factors such as the type and amount of cement, the type of admixture, the water-binder ratio, and the admixtures. These factors can also have a significant impact on the compressive strength of concrete. For ordinary concrete, when the compressive strength was higher, it indicates that the degree of hydration of the paste was relatively high, and the sample has formed a denser pore structure, which can effectively delay the carbonization rate of the concrete.

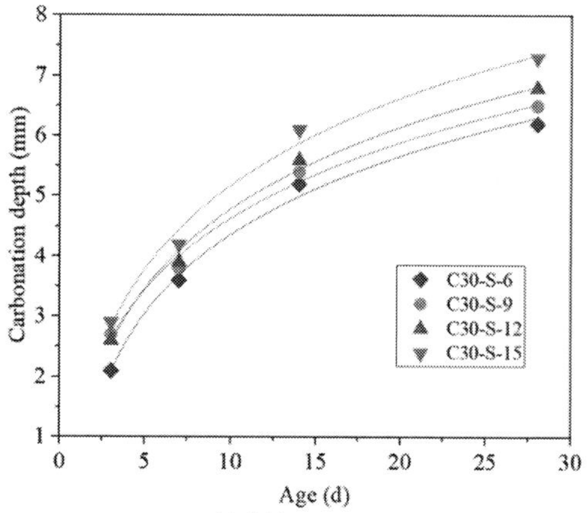

(a) C30 concrete

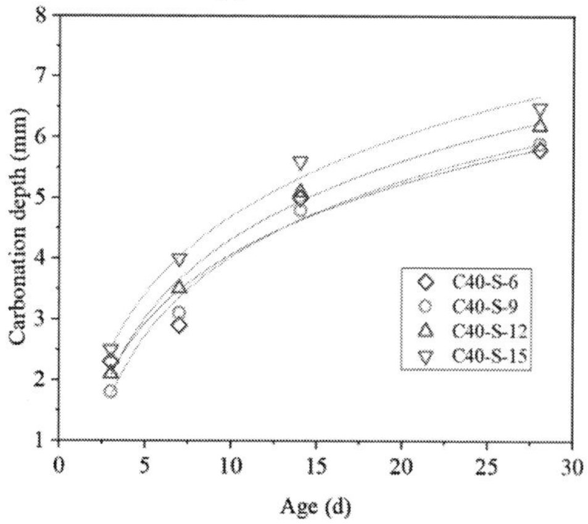

(b) C40 concrete

Fig. 3 Test results of carbonation depth of waste calcareous rock powder concrete

https://doi.org/10.1051/e3sconf/202129302037

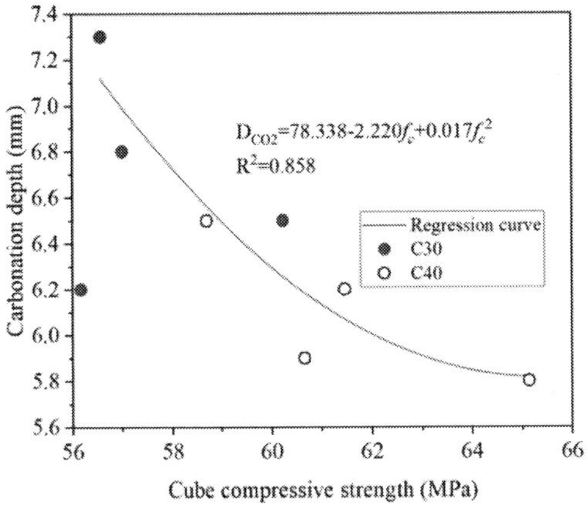

Fig. 4 Correlation between compressive strength and carbonation depth of waste calcareous rock powder concrete

4 Conclusion

In this paper, the waste calcareous rock powder formed in the production process of manufactured sand was mixed into the concrete of manufactured sand at 6%, 9%, 12% and 15% respectively to prepare green environment-friendly concrete. The influence of the rock powder on the concrete performance was analyzed by testing the cube compressive strength and transmission performance (including rapid chloride diffusion coefficient and carbonation depth). The conclusions were as follows:

(1) With the increase of stone powder content, the compressive strength of C30 concrete first increased and then decreased, while that of C40 concrete continued to decrease.

(2) The results show that the rapid chloride diffusion coefficient and carbonation depth increased with the increase of stone powder content, in which 9% stone powder content was the most favorable to improve the performance of concrete. Prolonging the age and reducing the water binder ratio can improve the transmission performance of concrete and make up for the adverse effect of too high stone powder content.

(3) Logarithmic function and polynomial function were used to analyze the correlation between rapid chloride diffusion coefficient, carbonation depth and compressive strength, respectively.

Acknowledgements

This research was funded by National Natural Science Foundation of China, grant number 51978408, and National Key R&D Program of China, grant number 2016YFC0701000.

References

1. X. Song, H. Dong, Z. Huang, J. Xia, Y. Zhou, and J. Wang, China Railw. 31 (2019).
2. R. Yang, R. Yu, Z. Shui, X. Gao, X. Xiao, D. Fan, Z. Chen, J. Cai, X. Li, and Y. He, J. Clean. Prod. **258**, 120673 (2020).
3. M. Davraz, H. Ceylan, İ. B. Topçu, and T. Uygunoğlu, Constr. Build. Mater. **165**, 494 (2018).
4. K. L. Jain, G. Sancheti, and L. K. Gupta, Constr. Build. Mater. **252**, 119075 (2020).
5. I. Taji, S. Ghorbani, J. de Brito, V. W. Y. Tam, S. Sharifi, A. Davoodi, and M. Tavakkolizadeh, J. Clean. Prod. **210**, 837 (2019).
6. H. Ding, J. Zhejiang Univ. Water Resour. Electr. Power **31**, 55 (2019).
7. J. Yan, X. Ye, Y. Li, S. Liu, B. Chang, and J. Yu, J. Mater. Sci. Eng. **37**, 828 (2019).
8. K. Xie, H. Wang, J. Xiao, and D. Yang, J. Archit. Civ. Eng. **36**, 31 (2019).
9. S. Sun, J. Water Resour. Archit. Eng. **17**, 160 (2019).
10. Z. Liu, Study on Methods of Accelerated Testing of Marine Concrete Durability Based on Simulating Environment and Service Life Prediction, PhD Thesis, Southeast University, 2006.
11. X. Chen, G. Yuguang, B. Li, M. Zhou, B. Li, Z. Liu, and J. Zhou, Constr. Build. Mater. **240**, 117953 (2020).
12. H. Li, F. Huang, G. Cheng, Y. Xie, Y. Tan, L. Li, and Z. Yi, Constr. Build. Mater. **109**, 41 (2016).
13. P. Chindaprasirt, C. Jaturapitakkul, and T. Sinsiri, Constr. Build. Mater. **21**, 1534 (2007).
14. A. Taha, W. Alnahhal, and N. Alnuaimi, Compos. Struct. **243**, 112277 (2020).
15. D. Niu, L. Su, Y. Luo, D. Huang, and D. Luo, Constr. Build. Mater. **237**, 117628 (2020).
16. Y. Guo, X. Hu, and J. Lv, Constr. Build. Mater. **223**, 142 (2019).
17. S. Kim, Y. kim, M. Usman, C. Park, and A. Hanif, J. Build. Eng. **33**, 101641 (2021).

https://doi.org/10.1051/e3sconf/202129302038

Test of SMA - 16 asphalt mixture ratio on the runway of Da Lian Airport

Shuaituan Tian[1,2] Kunyang Zhao [1,2,*], Lei Guo[1,2], Hanxun Wang[3]

[1] China Airport Construction Group Co., Ltd, Beijing, 100101, China

[2] Beijing Super-Creative Technology Co., Ltd, Beijing, 100621, China

[3] School of Engineering and Technology, China University of Geosciences (Beijing), Beijing, 100083, China

Abstract. After years of use, there's oil loss on the asphalt pavement . the bond strength of aggregate decreases . Loose threshing is serious. In order to improve the pavement performance, Extend the service life of pavement , Improve the safety of aircraft operation .we test the SMA-16 modified asphalt mixture on the runway of Da Lian Airport from many factors , including raw materials, admixtures, gradation design, optimum ratio of oil to stone, then we find the best ways to resolve it.

1 Introduction

After years of use, there's oil loss on the asphalt pavement , the bond strength of aggregate decreases . Loose threshing is serious. In order to improve the pavement performance, extend the service life of pavement , improve the safety of aircraft operation, we test the SMA-16 modified asphalt mixture on the runway of Da Lian Airport from many factors , including raw materials, admixtures, gradation design, optimum ratio of oil to stone, then we find the best ways to resolve it .

2 Raw materials

2.1 Asphalt

We use high modulus asphalt and test its performance indicators, we put the results on the below table .

Table1. High modulus asphalt performance test results

Test items		Technical requirements	Test results
Softening point(°C)		>80	88.3
Penetration(25°C, 100g, 5s) (0.1mm)		≦50	36.3
Ductility(5cm/min, 10°C)(cm)		>40	48.3
Filmy heating operational test163°C/5h	Mass loss (%)	<1	-0.04
	Penetration ratio (%)	>70	75.6
	Ductility(10°C) (cm)	>30	40.8
Equivalent softening point T_{800} (°C)		>50	61.0
Equivalent brittle point$T_{1.2}$(°C)		<-13	-18.4
Flash point(COC)(°C)		>250	277
Elastic recovery(15°C)		>80	92.3

From the results, it appears that all the performance indicators meet the technical requirements, we can use this asphalt.

2.2 Coarse aggregate and fine aggregate

(1) Coarse aggregate

There are basalt coarse aggregates (10~20mm 、5~10mm)and limestone(3~5mm) aggregates , We tests the aggregate indexes in accordance with the relevant regulations , the test results are shown on the table below.

Table2. The coarse aggregate technical indicators

Aggregate size	Test results	Test items
10~20mm	2.784	Bulk specific gravity
5~10mm	2.840	
3~5mm	2.747	
10~20mm	2.922	Apparent specific gravity
5~10mm	2.980	
3~5mm	2.907	

From the results, it appears the bulk specific gravity and the apparent specific gravity all meet the technical requirements, we can use them in the project.

*Email Address: zhao_kunyang@126.com

Table3. The coarse aggregate particle gradation

mesh /mm	P (%)					
	10~20mm		5~10mm		3~5mm	
	Technical requirements	Test results	Technical requirements	Test results	Technical requirements	Test results
19	95~100	100	—		—	
16	—	95.4				
13.2	—	46.9	100	100	—	
9.5	0~15	2.1	95~100	98.4	—	
4.75	0~5	0	0~10	0.9	85~100	97.8
2.36	—	0	0~5	0.1	0~25	29.3
1.18	—	0	—	0	—	11.1
0.6	—	0	—	0	0~5	5.3
0.3	—	0	—	0	—	2.9
0.15	—	0	—	0	—	2.1
0.075	—	0	—	0	—	0.6

From the results, it appears that the 4.75mm and 2.36mm passing rate of 3~5mm cannot meet the technical requirements. So we suggest that We should screen 3~5mm until the passing rate meet technical requirements.

(2) Fine aggregate

We use 0-3mm fine aggregate, we test their performance indicators, and put them on the below table.

Table4. The basalt fine aggregate particle gradation

mesh (mm)	P (%)	
	Technical requirements	Test results
4.75	100	100
2.36	85~100	92.3
1.18		56.1
0.6	20~50	37.8
0.3		19.2
0.15		12.1
0.075	0~15	2.9

From the results, it appears that 0~3mm can meet the technical requirements.

2.3 Filler

We use the limestone powder filler, we test its performance indicators, we put the results on the table 5.

Table5. The mineral filler technical indicators

Test items		Technical requirements	Test results
Hydrophilic coefficient		≤1	0.6
Particle gradation	<0.6mm	100	99.4
	<0.15mm	99.5	96.3
	<0.075mm	85.2	83.7
Apparent specific gravity (g/cm^3)		≥2.50	2.669
Water content (%)		≤1	0.7

From the results, it appears that all the performance indicators meet the technical requirements, we can use this filler.

2.4 Fiber

We use polyacrylonitrile fiber to test, its main technical indicators are shown in table 6.

Table6. Polyacrylonitrie fiber test results

Test items	Technical requirements	Test results
Diameter (μm)	10~25	12.9
Length (mm)	6±1.5	6.0
Tensile strength (MPa)	≥500	546
Elongation at break (%)	≥15	20

From the chart, we can see all the indicators of polyacrylonitrile fiber meet the specification requirements.

2.5 Anti-rutting agent

We use haichuan Anti-rutting agent, through results, Adding 0.5% anti-rutting agent of the asphalt mixture we can achieve the best test results. We test its performance indicators, and put the results on the below table.

Table7. Anti rutting agent basic indexes test results

Test items	Technical requirements	Test results
Density (g/cm^3)	0.9~1.1	0.96
Melt flow rate (190℃, 2.16kg) (g/10min)	≥3	8
Water content (%)	≤2	0.2
Softening point (℃)	140~170℃	141

From the results, it appears that all the performance indicators meet the technical requirement.

3 Mix design of SMA-16 asphalt mixture

3.1 The determination of aggregate gradation ratio

The aggregate gradation ratio of SMA-16 asphalt mixture is formed with five different raw materials .We choose three mix designs. All the mix designs and raw materials are put on the below table.

Table8. SMA-16 ratio of mineral aggregate gradation(%)

Specifications	10~20mm	5~10mm	3~5mm	0~3mm	filler
Coarse gradation	15	24	10	26	5
Middle gradation	15	24	10	26	5
Fine gradation	15	24	10	26	5

Table9. SMA-16 mineral synthesis aggregate gradation

Mesh size(mm)	coarse	middle	fine	upper	lower
26	100.0	100.0	100.0	100	100
19	100.0	100.0	100.0	100	100
16	97.6	97.7	97.7	95	90
13.2	73.5	75.0	75.6	70	60
9.5	50.6	53.6	54.6	50	40
4.75	22.2	26.1	28.1	26	20
2.36	18.7	21.8	23.0	22.5	18
1.18	15.4	17.6	18.3	18	14
0.6	12.9	14.6	15.0	15.5	12
0.3	9.2	10.3	10.5	13	10
0.15	7.6	8.5	8.6	11.5	9
0.075	5.9	6.4	6.5	10	8

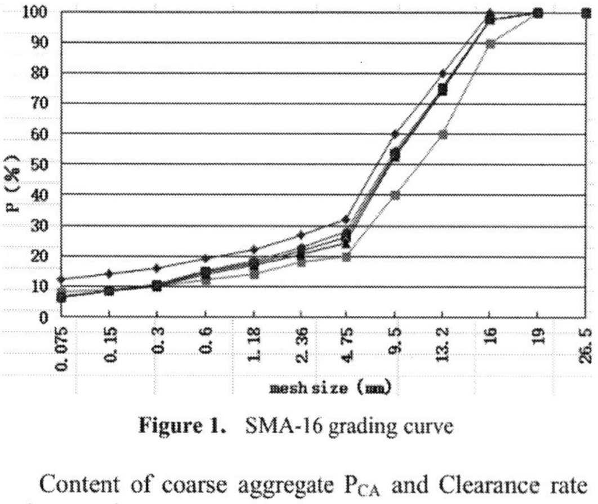

Figure 1. SMA-16 grading curve

Content of coarse aggregate P_{CA} and Clearance rate VCA_{DRC} of more than 4.75mm in three mixtures are tested in table 10.

Table10. The Marshall test results of different asphalt aggregate ratio

items	loose unit weight (g/cm³)	the passing rate of 4.75 mm (%)	bulk specific gravity of above4.75mm (g/cm³)	P_{CA}	VCA_{DRC}
coarse	1.720	22.2	2.799	0.734	38.55
middle	1.706	26.1	2.779	0.697	38.61
fine	1.700	28.1	2.773	0.678	38.69

Based on the experience of previous similar airport engineering, we use ratio of 6.0% as a first try oil-stone ratio in Marshall design method, and then mold specimens in accordance with the specification requirements, measure the physical indexes of the specimens. Specific data are shown in table 11.

Table11. Performance of the first grading

Test items	Coarse gradation	Middle gradation	Fine gradation
Bulk specific gravity(g/cm³)	2.465	2.487	2.503
Theoretical density(g/cm³)	2.569	2.563	2.567
VV(%)	4.0	3. 0	2.4
VMA(%)	17.04	16.06	15.47
VFA(%)	76.4	81.6	84.7
VCA_{mix}(%)	35.54	37.61	38.87
VCA_{DRC}(%)	38.55	38.61	38.69
MS(KN)	10.75	9.42	9.63
FL(0.1mm)	34.1	36.9	34.2

According to the relevant specifications, the coarse grading is the best grading.

3.2 The determination of the optimum proportion

We select 3 asphalt aggregate ratio of Marshall test and calculate their physical indicators in order to determine the optimum proportion, the test results are shown in the table12.

Table12. The Marshall test results of different asphalt aggregate ratio

Test items	oil-stone ratio(%)			Specification requirements
	5.7	6.0	6.3	
Bulk specific gravity(g/cm³)	2.458	2.465	2.466	—
Theoretical density(g/cm³)	2.580	2.569	2.558	—
VV(%)	4.7	4.0	3.6	3~4
VMA(%)	17.1	17.1	17.3	≥16.5

Test items	oil-stone ratio(%)			Specification requirements
	5.7	6.0	6.3	
VFA(%)	72.3	76.3	79.2	-
VCA_{mix}(%)	35.36	35.36	35.52	≤VCA_{DRC}
VCA_{DRC}(%)	38.55	38.55	38.55	—
MS(KN)	9.46	9.81	10.13	≥6
FL(0.1mm)	37.2	31.4	28.4	—

According to the requirements of the relevant specification, calculated the optimum proportion of 6.0%.

3.3 The optimum proportion of validation

In the optimum asphalt aggregate ratio of 6.0%, we make Marshall specimen and the dynamic stability of specimen, then test their Physical and mechanical performance. we put the results on the below table.

Table13. The Road performance verification test results

Test items	Test results
△M(%)	0.1
△S(%)	1.81
DS(time/mm)	12000
MSo(%)	89.3
TSR(%)	91.6
Cw(mL/min)	32.1
TD(mm)	1.1

MS is strength index, MSo and TSR are water damage resistance index, DS is high temperature performance index, Cw is water permeability index.

From the results, it appears that all the test results can meet the requirements of related technologies.

Adding anti-rutting agent quality of 0.5% asphalt mixture, dynamic stability of asphalt mixture arrives at 12000 times/mm, MSo arrives at 89.3%, TSR arrives at 91.6%, Cw arrives at 32.1 mL/min. They all meet the design requirements. We can see that the mixture has formed the skeleton dense structure rom the profile of the specimen, it meets with the skeleton dense type AC asphalt mixture design intent.

4 Conclusion

In combination with the runway of Da Lian Airport, in order to slove the serious problems of SMA - 16 asphalt mixture, we achieve these important ways.

First; Starting from the raw material, we should choose the qualified raw materials. including aggregate, filler and asphalt. The important indicators of aggregate are bulk specific gravity, apparent specific gravity, sand equivalent, sturdiness and p%; The important indicators of filler are apparent specific gravity, water content, Particle gradation and hydrophilic coefficient; The important indicators of asphalt are penetration, softening point, ductility, equivalent softening point, equivalent brittle pointT1.2, flash point, density, elastic recovery and filmy heating operational test163℃/5h.

Second: we should choose good aggregate gradation.such as 10～20mm: 5～10mm: 3～5mm: 0～3mm:fiber=15:24:10:26:5.

Third: In order to obviously improve the dynamic stability of mixture, it is reasonable to adding anti-rutting agent.

References

1. The ministry of communications highway engineering science institute. JTJ E20-2011.Standard test methods of bitumen and bituminous mixtures for highway engineering [S].
2. Xiaoming,Huang. Asphalt and asphalt mixture [M].
3. The ministry of communications highway engineering science institute. JTG F40-2004. Technical specification for construction of highway asphalt pavement [S].
4. The civil aviation administration of China. JTG F40-2004. Specification for asphalt concrete pavement construction of civil airports [S].
5. The ministry of communications highway engineering science institute. JTG E42-2005. Test methods of aggregate for Highway Engineering [S].

Research on the Design of Disaster Prevention Plan for 22km Ultra-long Highway Tunnel with Multiple Intersections

Xuan Zhang [1], Xuefeng Li [1*], and Xialin Liu [2]

[1] Research Institute of Highway, Ministry of Transport, Beijing 100088, China

[2] CCCC Second Highway Survey, Design and Research Institute Co., Ltd, Wuhan, Hubei 430056, China

Abstract: The Tianshan Shengli Tunnel is currently the longest high-altitude highway tunnel. In order to ensure the safe operation of the tunnel, the SES software is used to investigate the wind speed laws in the tunnel, and the tunnel is divided into 16 disaster prevention sections based on the shaft and critical wind speed. The principles of disaster prevention and rescue of Tianshan Shengli Highway Tunnel are given, and the emergency treatment process of tunnel disaster prevention and rescue is given. The disaster prevention and rescue plan of Tianshan Shengli Tunnel from fire occurrence, personnel and vehicle evacuation, and fire treatment are discussed in detail.

1 Introduction

Highway tunnels have narrow space and limited safe evacuation exits. After a fire occurs, the fire fighting route is single, resulting in the detention of people and vehicles. The toxic gas and smoke generated by the fire will pose a serious threat to the people in the tunnel [1,2]. In order to ensure the orderly development of evacuation and rescue work, it is necessary to establish a scientific and effective disaster prevention plan. The establishment of a disaster prevention plan can reduce the casualties that may be caused by the fire, reduce property losses, and improve the operation and management level of the tunnel [3].

Wang Mingnian et al. [4] studied the disaster prevention and rescue of the Qinling Zhongnanshan Highway Tunnel and found that the fire control in the tunnel includes three parts: system linkage, traffic evacuation, and rescue and fire extinguishing. Li Guohui [5] analyzed the emergency rescue of the Erlangshan Tunnel on the Sichuan-Tibet Highway and obtained the basic theories of the organization and command , force mobilization, combat deployment and disposal procedures of the tunnel emergency rescue. Zhou Qing et al. [6] summarized the basic principles of tunnel disaster prevention and rescue, established a set of tunnel disaster prevention and rescue management system, and analyzed the process of tunnel disaster prevention and rescue system and the factors affecting fire evacuation and rescue.

The Tianshan Shengli Highway Tunnel is located on the southwest side of Urumqi, Xinjiang. It is a two-way four-lane road with a length of more than 22km and an altitude of 2772m. It is an ultra-long high-altitude highway tunnel. The straight-line distance from the entrance to Urumqi is about 92km, and the straight-line distance from the exit to Hejing County is about 76km. The starting and ending pile numbers of the left line of the tunnel are ZK75+660 ~ ZK97+790, the length is 22130m, the maximum buried depth is about 1115m; the starting and ending pile numbers of the right line are YK75+790 ~ YK97+821.7, the length is 22031.7m, the maximum buried depth is about approximately 1122.2m. The tunnel is composed of two lines on the left and right plus an intermediate service tunnel, with 3 vertical, 118 horizontal, and 4 connecting passages for fan rooms, and there are 560 intersections.

Based on the characteristics of the tunnel, the disaster prevention and rescue of the Tianshan Shengli Highway Tunnel is faced with the difficulties of ventilation, smoke control, personnel escape and rescue. Therefore, it is necessary to conduct an in-depth study on the disaster prevention zone and rescue plan of the Tianshan Shengli Tunnel to provide reference for the disaster prevention and rescue design of the tunnel.

2 Disaster prevention and rescue zone division

When a fire occurs, the direction of the wind flow in the tunnel is the same as the direction of the traffic. If the people upstream of the fire source cannot evacuate to the non-fire tunnel through the vehicle passing, they must get off the vehicle and evacuate through crosswalk. Vehicles downstream of the fire source will quickly drive away from the tunnel exit. Therefore, in the event of a fire, the downstream shaft closest to the fire source exhausts air, and the other shafts continuously supply air to the tunnel, so that the toxic gas and smoke in the tunnel can be discharged, ensuring that the fire smoke is controlled between the fire source and the closest exhaust outlet

* Corresponding author: lxfsure@163.com

downstream of the fire source. Guidelines for Ventilation Design of Highway Tunnels stipulates that the minimum fire critical wind speed is 2m/s [7], where the fire critical wind speed for smoke control is 2m/s. The specific schematic diagram of tobacco control is shown in Figure 1.

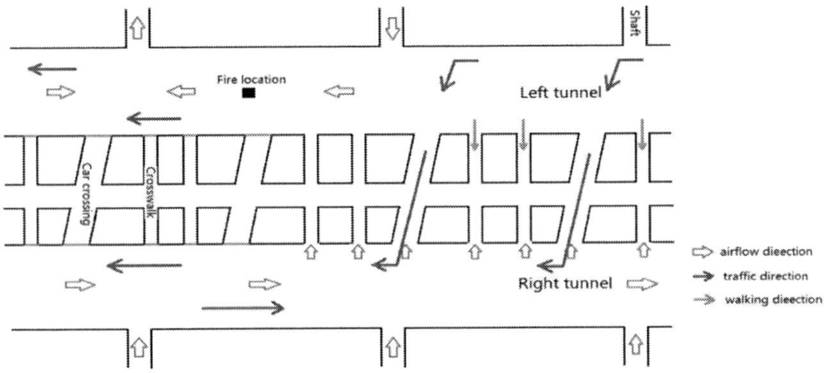

Figure 1. Schematic diagram of tobacco control

Taking the tunnel shaft as the boundary, the left and right tunnels are each divided into 5 pre-zones, a total of 10 pre-zones. Use SES software to simulate the exhaust ventilation under each pre-zone fire. The principle is as follows: Take a wind network composed of 5 edges and 4 nodes as an example, where the arrow indicates the direction of wind flow, and the dashed line is the added natural wind pressure edge. The ventilation schematic diagram and network diagram of the ventilation tunnel are shown in Figure 2 and Figure 3.

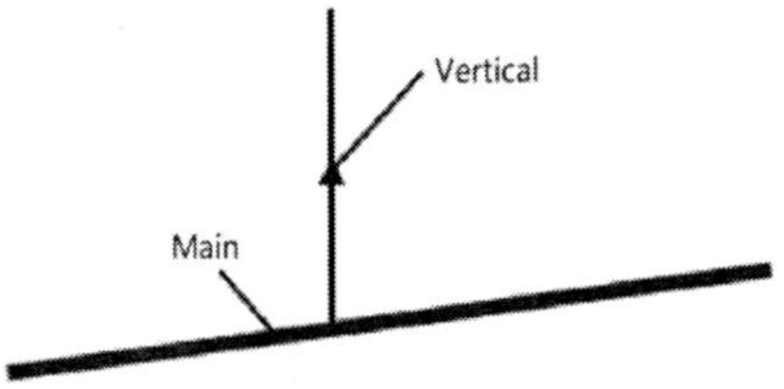

Figure 2. Schematic diagram of ventilation system

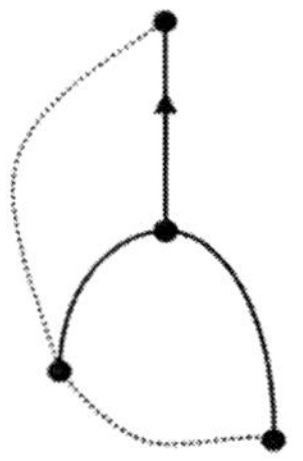

Figure 3. Ventilation network diagram

The basic laws of wind flow in the tunnel ventilation network are as follows:

(1) The law of air volume balance for a node in the ventilation network, according to the continuity equation, the amount of air flowing into the node should be equal to the amount of air flowing out of the node, namely:

$$\sum Q_i = 0 \quad (1)$$

In the formula: Qi ——The amount of air flowing into or out of a node or mesh, flowing into a positive value, and flowing out of a negative value.

(2) The law of wind pressure balance

The algebraic sum of wind pressure at any mesh in the wind network (the wind pressure of the wind flow in the clockwise direction is positive, and the negative value in the counterclockwise direction) should be equal to zero. The so-called non-pressure source mesh refers to a loop where the mesh does not have the effect of natural wind pressure or fan or traffic ventilation. The wind pressure balance law is:

$$\sum \Delta P_i = 0 \quad (2)$$

In the formula: Pi —— the biased wind pressure in the mesh, which is negative and positive, counterclockwise.

The pressure source mesh refers to the function of the mesh natural wind pressure exhaust or fan or traffic ventilation or air supply. The wind pressure algorithm is:

$$\sum\triangle P_i - \left(\sum H_{fan} + \sum H_{self} + \sum\triangle P_t\right) = 0 \quad (3)$$

In the formula: H_{fan} ——wind pressure of the fan in the mesh, clockwise to be positive, counterclockwise to be negative;

H_{self} ——wind pressure in the mesh, the sign of the sign is the same as above;

△Pt ——Traffic ventilation capacity, the sign of the sign is the same as above;

(3) The law of resistance

The wind flow in the tunnel wind path is almost all stable turbulent flow, so the ventilation resistance is proportional to the square of the air volume.

$$\triangle P = RQ_2 \quad (4)$$

In the formula: △P——is the ventilation pressure or ventilation resistance on the wind road, Pa;

R——is the wind resistance on the wind road, kg/m7;

Q——is the air volume passing through the wind road, m3/s.

On the basis of the conservation of air volume at the nodes of the ventilation network, the critical smoke control wind speed of 2m/s is used as the partition boundary to determine the economic opening number of the fans and the detailed partition of the pre-partition. The Tianshan Shengli Tunnel is divided into 16 disaster prevention zones. The overall disaster prevention zone division diagram of the tunnel is shown in Figure 4.

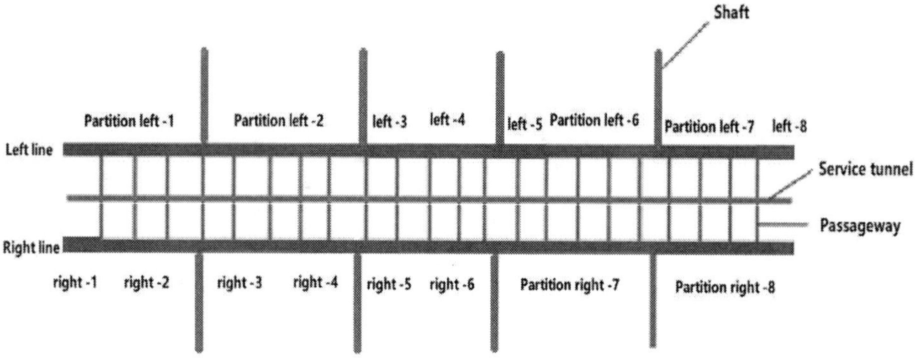

Figure 4. The overall disaster prevention zone of the tunnel

3 Disaster prevention and rescue plan control process

The disaster prevention and rescue plan of the Tianshan Shengli Tunnel should adhere to the basic ideas of prevention, alarm, monitoring, evacuation, rescue and fire fighting, and implement the general principle of "people-oriented, prevention-oriented, and combination of prevention and fire fighting ", and ensure the safety of personnel as the primary principle, and establish a safety precaution system for fire warning, rescue, evacuation of personnel and vehicles, and fire extinguishing. On the basis of zoning, different disaster prevention plans are established, and corresponding disaster prevention measures are initiated according to the different locations of the fire.

When a fire occurs in the tunnel, measures shall be taken at the initial stage of the fire to evacuate people and vehicles in the fire tunnel in time to reduce the further spread of the fire. The control sequence in case of fire is shown in Figure 5.

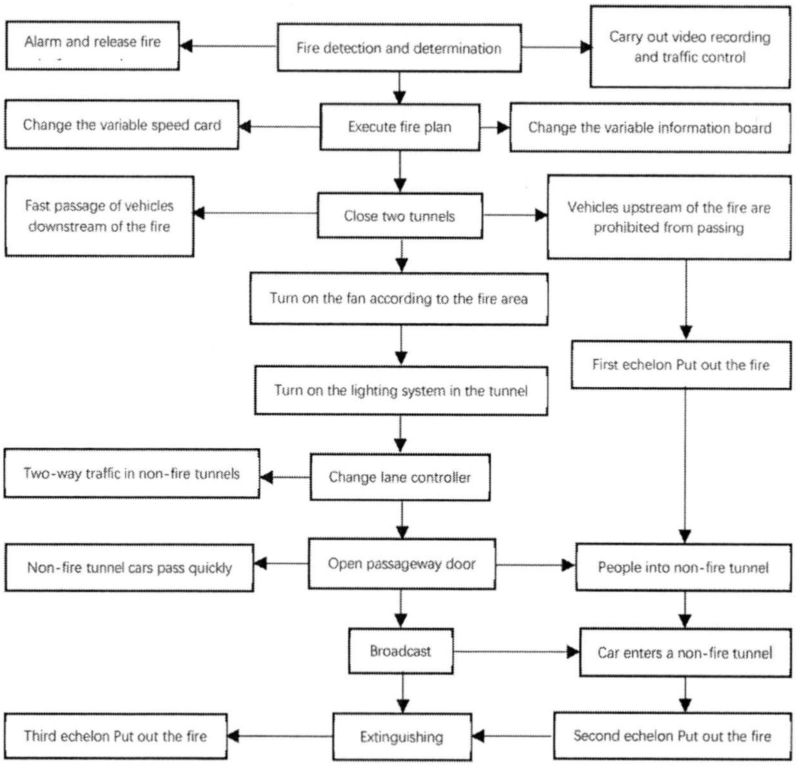

Figure 5. Control sequence in case of fire

The detailed control sequence in case of fire is as follows:

(1) After the fire is confirmed, contact the professional staff of the corresponding unit to rush to the scene for evacuation, rescue and fire fighting, and implement the disaster prevention and rescue plan.

(2) Close the tunnel entrance. The traffic light at the tunnel entrance is red, indicating that entry is prohibited. The prompt message of the variable information board becomes "Tunnel fire, no entry", and the speed of the variable speed board becomes 0. At the same time, the fire information was released, the drivers and passengers were called to prepare to escape through the broadcast in the tunnel, and vehicles outside the tunnel were prohibited from entering the tunnel through the broadcast outside the tunnel.

(3) Determine the number and direction of the fan opening according to the fire location. The direction of the upstream ventilation of the fire is the same as the driving direction, and the wind speed is greater than 2m/s; the smoke generated by the fire is removed through the vertical shaft closest to the fire area to prevent the smoke from flowing back. Turn on the fan in the non-fire tunnel to prevent the smoke from spreading to the non-fire tunnel through the cross passage. Turn on all the lighting systems in the tunnel to facilitate fire fighting and personnel escape.

(4) Open all the cross passage doors upstream of the fire, the broadcast system will broadcast, and the drivers and passengers in the upstream will escape to the non-fire tunnel through the pedestrian and vehicular cross passage. The non-fire tunnel becomes a two-way lane, and the driving speed is limited to 30km/h, and overtaking is strictly prohibited. Vehicles downstream of the fire passed orderly and quickly.

(5) After the evacuation of personnel is completed, organize professionals to put out the fire.

4 Disaster prevention and rescue plans

The smoke generated by the fire is exhausted through the vertical shaft closest to the fire area downstream to prevent the smoke from flowing back. Turn on the fan in the non-fire tunnel to prevent the smoke from spreading to the non-fire tunnel through the cross passage. The drivers and passengers in the upper reaches escaped to the non-fire tunnel through pedestrian and vehicular cross passage. The non-fire tunnel becomes a two-way lane, and vehicles downstream of the fire pass in an orderly and fast manner. Take the Tianshan Shengli Tunnel Zone Left-4 as an example, the schematic diagram of disaster prevention and rescue plan is shown in Figure 6.

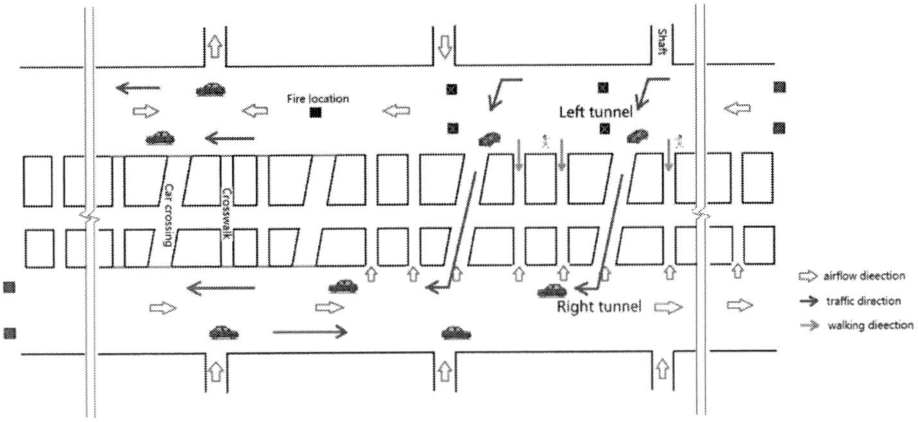

Figure 6. Disaster prevention zone left-4 plan diagram under fire

5 Conclusion

Through the research on the disaster prevention plan of the Tianshan Shengli Tunnel, the following main conclusions are obtained:

(1) Based on the location of the shaft and the critical wind speed, the Tianshan Shengli Tunnel is divided into 16 disaster prevention zones;

(2) Proposed the disaster prevention and rescue process of Tianshan Shengli Tunnel in combination with the mechanism system, wind flow direction and electromechanical system;

(3) Considering the disaster prevention zone and characteristics, the disaster prevention plan design of the Tianshan Shengli Tunnel is given, including the ventilation plan and the driving plan.

References

1. X. Zhou, X. Zhao, Y. Ding, Z. Zheng. Numerical studies on the critical Velocity and smoke distribution affected by tunnel cross-section geometry in tunnel fire. J Guangzhou Univ. 5,54(2006)
2. W. Liu, C. Zeng, Y. Yu, H. Huang. Study on fire preventing technology for subsea tunnel of Xiamen east passage// *Proceedings of the 2006 International Conference on Road Tunnel Operation Management and Safety. Chongqing: Chongqing University Press*,(2006)
3. L. Yu. Research on Ventilation Design and Control Technology of Zhongnanshan Extra-long Road Tunnel in fire mode. SW Jiaotong Univ. (2009)

4. M. Wang, Q. Yang, Q. Zhao, P. Liu. Research on Disaster Prevention Plan of Qinling Zhongnanshan Ultra-long Highway Tunnel. Highway,**11**,87(2000)
5. G.LI, W. YI, B. LIU. Preliminary Analysis on Emergency Rescue in Erlangshan Tunnel of Sichuan-Tibet Road. Fire Sci. Technol. **06**,806(2006)
6. Q. Zhou, X. Zhou. Research on Fire Prevention and Rescue for Yanmenjie Tunnel. Fire Sci.Technol. **28**,599. (2009)
7. JTG/J D70/2-01-2014, Code for Design of Ventilation of Road Tunnel. *China Communications Press*,(2014)

https://doi.org/10.1051/e3sconf/202129302040

Development status and Countermeasures of renewable energy in China

XueBin Hu[1]*

[1]Wuhan University of Science and Technology, Wuhan, China

Abstract. This paper studies the development status of renewable energy in China. In this field, there are some problems in China, such as imperfect legal and policy system, insufficient market operation, weak public awareness of environmental protection, imperfect scientific research system and so on. At the end of the paper, the corresponding improvement measures are put forward.

1 Introduction

Since the deterioration of the earth's environment, renewable energy has been gradually concerned by countries all over the world, especially the healthy development of economy brought by renewable energy and its positive role in maintaining the ecological environment. The latest data showed that China, the United States and the European Union account for more than half of the world's total greenhouse gas emissions, which are the three largest countries in the world. Global greenhouse gas concentrations continue to reach new highs, which will lead to increasingly serious climate change in the future. In view of this phenomenon, based on the development concept of the community of human destiny, it is one of the important tasks for China to carry the construction of ecological civilization forward establish and improve the ecological economic system of green, low-carbon and circular development. Therefore, China urgently needs to develop low-carbon economy, save energy and reduce the emission of environmental hazards. This requires full play to the competitive advantage of renewable energy, giving priority to development and utilization of renewable energy. China has proposed the overall goal of renewable energy development and the development goals of hydraulic power, solar energy, wind energy, ocean energy, etc In order to achieve the above goals, we need to deeply analyze the present condition of China's energy development and the development of renewable energy. Through the understanding of the actual situation, the paper analyzes the problems existing in the development of renewable energy, and puts forward corresponding suggestions.

2 Present condition of energy development in China

In early 2020, COVID-19 broke out in China. This public health emergency has caused a huge impact on China's economy and society. With the gradual improvement of the epidemic situation and the gradual control of the problem, the domestic economy is also improving under the implementation of the macro cycle policy, and the development trend is good. In the meantime, the development of China's energy industry is back on track, coal consumption continues to pick up, oil and gas imports increase, and electric power load continues to reach a new high. The energy structure has been continuously optimized, and all indicators have returned to a positive year-on-year growth.

With the rapid control of COVID-19, China's economic development and society are gradually on the right track, and the demand for energy is also gradually increasing. According to preliminary calculation, China's total annual energy consumption increased by 2.2% over the previous year, reaching 4.97 billion tons of standard coal. The change trend reflects that China's economic recovery is obvious. According to the relevant survey data, the growth of natural gas consumption is the largest, reaching 7.2%, while the growth of coal is the smallest, only 0.6%. Crude oil and electricity consumption increased by 3.3% and 3.1% respectively. The relevant data is shown in Figure 1.

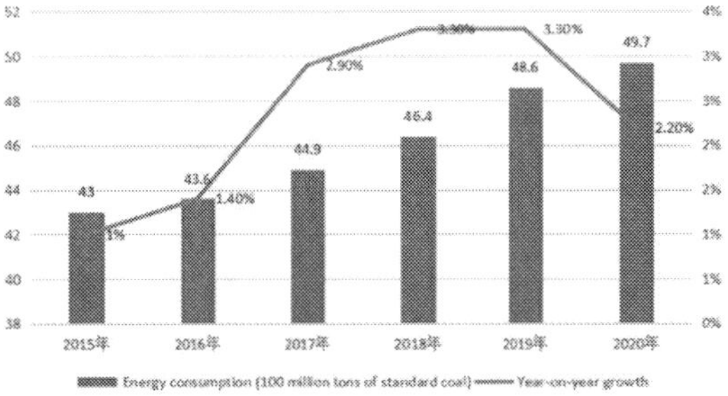

Figure 1. China's total energy consumption and year-on-year growth in 2015-2020

Statistics from the National Energy Administration show that coal consumption accounts for half of the total

* Corresponding author: huxuebin97@126.com

https://doi.org/10.1051/e3sconf/202129302040

energy consumption, up to 56.8%, but it has decreased by 0.9% compared with last year. The total clean energy consumption has increased by 1.0% compared with last year, accounting for 24.4% of the total energy consumption, including renewable energy sources such as hydraulic power and wind power. From 2010 to 2020, the proportion of coal consumption in China's total energy consumption decreased from 69.2% to 56.7%, a decrease of 12.5%; The proportion of clean energy (including renewable energy) consumption in total energy consumption increased from 13.4% to 24.4%, with an increase of 11%. The energy consumption structure continued to change to green and low-carbon. Relevant data are shown in Figure 2. If the proportion of clean energy consumption is to reach the targets of 25% by 2020 and 35% by 2030, efforts from all sectors will need to be redoubled.

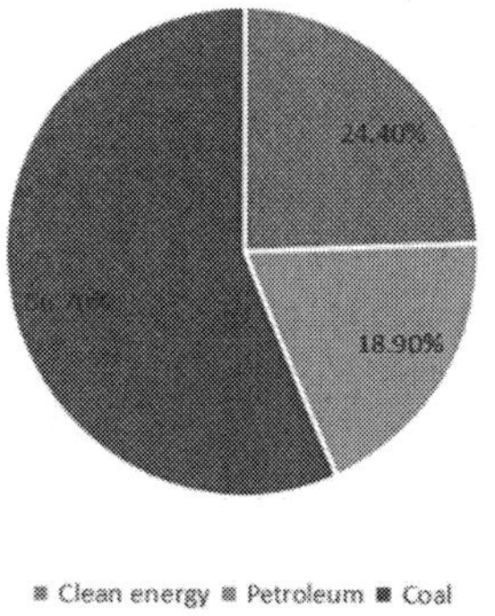

Figure 2. China's energy consumption structure in 2020

3 Development status of renewable energy in China

3.1 China's renewable energy policy orientation

In terms of fiscal policy, China's renewable energy law issued in 2006 has made relevant provisions on the on grid electricity price, strengthened policy coordination, and focused on strengthening policy support for renewable energy land and sea use, finance, taxation, investment and financing, so as to promote sustainable and healthy development and high-quality development of renewable energy.

In terms of market mechanism, China has actively implemented the "green certificate trading mechanism" by learning from other developed countries, which further improves the marketization degree of China's renewable energy industry, improves the full guarantee purchase system of renewable energy, links up with market-oriented trading, and encourages renewable energy to participate in the market orderly, so as to realize full consumption.

On the development goal of renewable energy, the national strategic positioning order of renewable energy is from alternative energy to mainstream energy to leading energy. The strategy can be realized in three time points, that is, renewable energy accounts for 15% of energy consumption in 2020 and about 25% in 2030. By 2050, renewable energy, as the leading energy, should account for about 40% of energy consumption.

Renewable energy policy mainly includes administrative policy and market policy. At present, in the heyday of renewable energy development, China mainly uses administrative means to promote the development of renewable energy, including the implementation of various key projects, financial subsidies and tax policies. China's renewable energy has developed rapidly in recent years, which can be said to achieve leapfrog development.

3.2 Technology orientation of renewable energy in China

In 2020, the energy data report released by the China Federation of electric power enterprises shows that, from the perspective of power structure, renewable energy power generation will reach 2045.54 billion kilowatt hour, with a year-on-year increase of about 176.1 billion kilowatt hour. Of the total power generation, renewable energy generation accounted for 27.7%, up 1% year on year. China's non fossil energy power generation increased by 10.7% over last year, reaching 2393.8 billion kilowatt hour, accounting for 32.6% of China's power generation, an increase of 1.7 percentage points over the previous year, and the power supply capacity of non fossil energy continued to increase. As shown in Figure 3, excluding thermal power, in other words, the main technical paths of renewable energy in China are hydraulic power, wind power and solar photovoltaic.

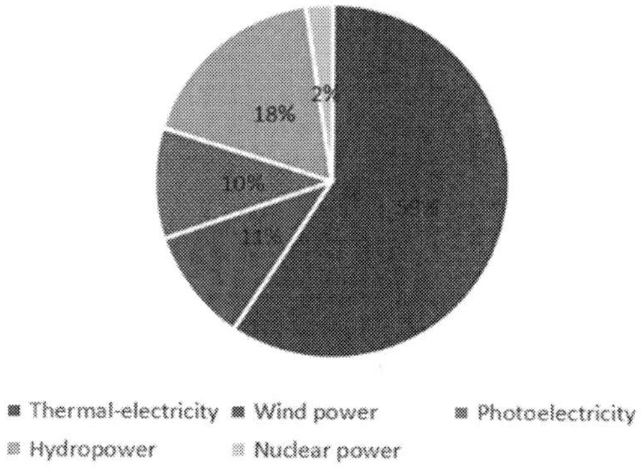

Figure 3. China's power supply structure in 2019

In terms of hydraulic power generation, China's hydraulic power technology has already entered the international advanced ranks, and even exceeded the international level in some aspects. China has become the largest hydraulic power country in the world in 2019, because according to statistics, by the end of 2019, China's annual hydraulic power generation will exceed 1 trillion watt hours, and the total installed capacity of hydraulic power will be about 356 million kilowatts. In fact, as early as 2015, China's total installed hydraulic power capacity has reached 320 million kilowatt hour. In hydraulic power, China mainly uses three-dimensional geological modeling technology, honeycomb hydraulic power technology, hydraulic power automation technology and hydro pneumatic power generation technology. At present, China will continue to develop intelligent hydraulic power technology.

Photothermal power generation and photovoltaic power generation are the main forms of photovoltaic power generation. At present, China has a large scale of photovoltaic power. According to the statistics of the national energy administration, by the end of June 2019, the installed capacity of centralized photovoltaic power generation in China has increased by 6.82gw, a year-on-year increase of 16%, reaching 130.58gw, while the installed capacity of distributed photovoltaic power generation has increased by 4.58gw, a year-on-year increase of 31%, reaching 55.02gw. The total installed capacity of photovoltaic power generation in China has reached 185.59gw.

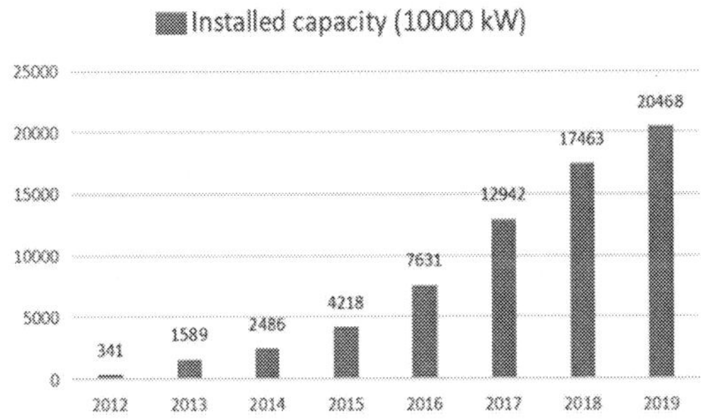

Figure 4. Cumulative installed capacity of solar power generation in China

Compared with photovoltaic power generation, wind power generation has experienced a phased development in China. Since the promulgation of the renewable energy law in 2006, China's wind power has entered the stage of large-scale development and application. However, after several years of development, China has appeared the imagination of abandoning wind power and limiting power. It was not until 2015, when the benchmark price of wind power was lowered, that China entered the wind power industry again, and the scale of new installed capacity was obvious. According to statistics, China's total installed capacity of wind power in 2019 is 21.05 million kilowatts..

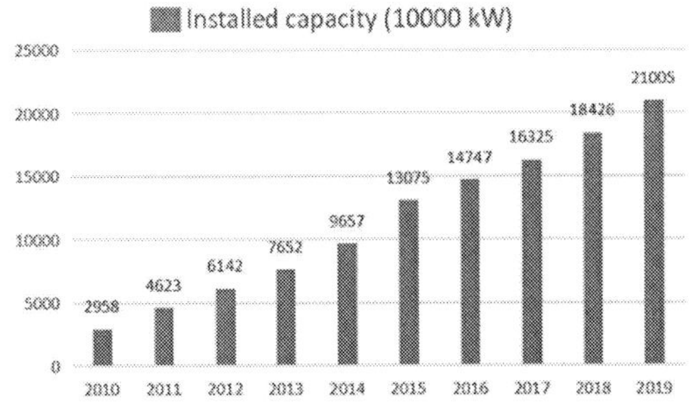

Figure 5. Cumulative installed capacity of wind power in China

4 Problems in the development of renewable energy in China

4.1 The system of renewable energy related laws and policies is not perfect

It was not until 2006 that China promulgated and implemented the renewable energy law, which stipulated the responsibilities and obligations of the Chinese government and society, providing a reliable legal guarantee for the development of renewable energy. However, in terms of content, the bill covers less renewable energy and is too broad. It was only amended in 2009 and lacks comprehensive, in-depth and detailed design of laws and policies. This leads to the low operability of the current law, because the laws and policies are not comprehensive, and the system stipulates the responsibilities and obligations of all stakeholders, which leads to the governments' unwillingness to accept foreign renewable power due to interest problems, and thus leads to the aggravation of power rationing situation. In a word, the incompatibility of relevant laws and policies leads to many contradictions in the development of renewable energy.

4.2 Market operation is insufficient, market allocation is not obvious

At present, the use stage of renewable energy industry in China still does not reflect the commercialization nature, and the market role is not obvious. And due to the lack of stable market security policy to guide the development of renewable energy market in China, the lack of incentives leads to weak innovation ability and slow technological development. Although the state has invested a lot of funds to increase its support for renewable energy, its development in practice is still under the control of other countries, and the main reason is the lack of core technology., It is necessary to strengthen the market-oriented operation of renewable energy in China and play the role of market allocation.

4.3 Public awareness of environmental protection is weak

The majority of the public in China still stay in the understanding of China's vast territory and abundant energy resources, and the awareness of environmental protection is weak, which leads to the low participation of the public in the consumption of renewable energy, and the whole society has not yet formed a comprehensive renewable energy development atmosphere. This requires the government to continuously publicize the current situation of energy shortage and the development advantages of renewable energy, and improve citizens' awareness of consuming such energy products. Only in this way can we expand the application of renewable energy and accelerate the development of renewable energy output.

4.4 The research system of renewable energy is not perfect

Although the state continues to invest a lot of money in the development of renewable energy industry, improve the support for the energy. But from the current situation, the funds are still very scarce. If we want to increase the development potential of renewable energy in China, we must supplement it with efficient financing mechanism and market support. Otherwise, insufficient R & D investment will inevitably lead to poor independent innovation ability and less R & D reserves. The imperfection of renewable energy research system in China also makes it difficult to apply innovative achievements to reality.

5 Suggestions

5.1 Improve the system of renewable energy related laws and policies

We should speed up the pace of Amending China's renewable energy law, establish the legal status of renewable energy development according to the actual needs, improve its utilization proportion in the overall energy structure, and establish administrative authority. According to the development of renewable energy in various regions, problems can be found in the process of continuous development, and relevant policies and regulations can be improved to improve the operability of policies and regulations. Secondly, coordinate the direct relationship between renewable energy laws and policies and other energy development policies, maintain the continuity and consistency of relevant laws and policies, and jointly promote the sustainable development of renewable energy industry.

5.2 Speed up market operation and industrialization

Relevant government departments should strengthen the development planning of renewable energy in China. In other words, it is necessary to strengthen leadership, actively broaden financing channels in accordance with the requirements of industrial development planning, so as to support the rapid development of renewable energy industry. We should establish the corresponding market mechanism to cooperate with the government's planning, increase the development of the renewable energy market, give full play to the main role of the market, and support the renewable energy enterprises to develop the market by means of preferential policies and tax incentives, so as to operate well under the market mechanism. At the same time, we should actively cultivate and standardize the market and intervene selectively.

5.3 Enhance public awareness of environmental protection

The government should strengthen the publicity and information dissemination of environmental protection development and renewable energy. We should make full use of all kinds of publicity and take various publicity forms to improve the public's awareness of renewable energy and make them realize the importance of renewable energy construction. By playing the main role of the people, we can enhance the development atmosphere of renewable energy in China. Secondly, the relevant departments and personnel engaged in the development of renewable energy should be trained regularly to strengthen exchanges. Through the establishment of national and regional renewable energy information websites, the healthy and rapid development of the industry should be guided.

5.4 Increase investment in renewable energy innovation

The first is capital investment, followed by talent investment and international exchanges and cooperation. We should focus on the problems in the development of renewable energy in China and establish a national renewable energy research center. We will strengthen the training of high-end talents. Secondly, we should extensively carry out international exchanges and cooperation, focus on and use China, and purposefully introduce, digest and absorb foreign advanced technologies, processes and key equipment. Enterprises, universities and scientific research institutions are encouraged to carry out the combination of production and research, so as to speed up the related scientific and technological development, accelerate the transformation of its achievements, promote the improvement of the overall technology and equipment level of the industry, and develop China's renewable energy industry better and faster.

References

1. China Electric Power Enterprise Federation. 2020 national electric power industry statistics express [R]. 2021
2. Huang Bo, Xu Wenyong, Wang Quanhui, et al. Reference and Enlightenment of renewable energy development in Germany [J]. China biogas, 2021,39 (1):45-53.Doi:10.3969/j.issn.1000-1166.2021.01.008
3. ZENG MING, LIU XIMEI, LI YULONG, et al. Review of renewable energy investment and financing in China: Status, mode, issues and countermeasures[J]. Renewable & sustainable energy reviews,2014,31(Mar.):23-37.
4. China Coal Industry Association. 2020 annual report on coal industry development [R]. 2021
5. Shi Jingli. Supporting mechanism of international renewable energy participation in power market and Its Enlightenment to China [J]. China energy, 2021,43 (3): 55-58.

Comparison and analysis of conversion methods for self-balancing test of pile bearing capacity

Yonggang *Qiu*[1,*],Chuansheng *Huang*[1],Renhao *Hong*[1],Chunhan *Chen*[1], Jinxiang *Yi*[1], and Weixiong *Li*[1]

1 School of Civil and Architectural Engineering, East China University of Technology, Nanchang, Jiangxi, China

Abstract. The principle and method of self-balancing test are introduced briefly. In this article, through comparison of common Q-S curve of equivalent transformation to transform from the balance of the measured load-displacement curve for the tradition- al method of load - displacement curve, puts forward the advantages and disadvantages of each, through the three bored piles in static load test of pile reaction from balance to detect vertical compressive ultimate bearing capacity of single pile is in line with the requirements and compare the load-displacement curve after conversion. Some problems of the self-balancing method in engineering are pointed out, which is of certain reference value to the popularization and application of the self-balancing test of pile bearing capacity.

1 Introduction

Common single pile bearing capacity testing methods include pile foundation static load test, pile loading test method and self-balancing test method. However, compared with the three methods, self-balancing test method has lower cost, is not limited by the site, and is efficient. There are many methods to transform the traditional static load results. As is known to all, the whole pile t- op load settlement curve is the most reliable data to describe the performance of a single pile. Common conversion methods include equal displacement method, equal load method, simplification method and so on. This paper summarizes and analyses the methods of converting the results of self-balancing test of pile bearing capacity into the results of traditional static load, and makes a comparative analysis of various methods with examples, and points out some problems existing in the engineering of self-balancing method.

2 Self-balancing static load test

2.1 Testing principle

The self-balancing test method is a new static load test method. The ultimate bearing capacity of the pile is tested by burying load cell in the pile body, loading upward and downward along the vertical direction, and using the pile side resistance as the reaction force of the pile end resistance.Used to judge the bearing capacity of pile foundation, pile foundation settlement, pile elastic compression and rock and soil plastic deformation[2].The curve of lateral friction resistance of the pile can be measured at the same time if the reinforcement gauge and other strain measuring instruments are arranged at different elevations of the pile. The ultimate bearing capacity of the whole pile can be calculated by obtaining the ultimate bearing capacity of the upper and lower pile bodies[3-4]. The self-balancing compressive static load test device is shown in Figure 1.

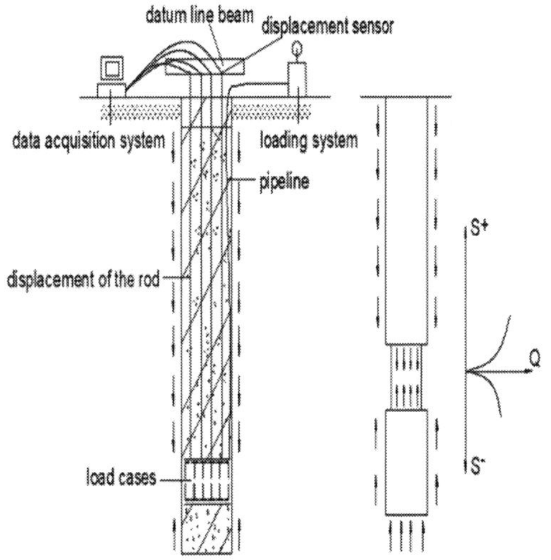

Figure1 self-balancing static load test diagram of foundation pile

2.2 Determination of vertical compressive ultimate bearing capacity of single pile

According to Technical Specification for Static Load Test of Pile Self Reaction Force Balance (DB36/J002-2006) [1], the calculation formula of vertical compressive ultimate

* Corresponding author: 1536421532@qq.com

bearing capacity of a single column tested by the self-balancing method is as follows:

$$Q_{u-up}=Q_{u-down} \quad (1)$$

$$W=\pi\times(D/2)^2\times L_{up}\times\rho \quad (2)$$

$$Q_u=\frac{Q_{u-up}-W}{\gamma}+Q_{u-down} \quad (3)$$

Where: Q_u-- the vertical compressive ultimate bearing capacity of a single pile;

Q_{u-up}-- the ultimate bearing capacity of the upper pile of the load cell;

Q_{u-down}-- the ultimate bearing capacity of the pile under the load cell;

W-- the dead weight of the upper pile of the load cell;

γ-- Correction coefficient of pile lateral friction of the upper pile of load cell.

For clay and silt, γ=0.8; For sandy soil, γ=0.7.

3 Self-balancing conversion method

3.1 Principle of equal displacement method

Equal displacement method is to draw Q_1-S_1 and Q_2-S_2 curves through the load value of each load cell and the displacement of pile. Based on the method of superimposed load by displacement, the following formula can be obtained according to the above principles:

$$Q^i=\frac{Q_1^i}{\lambda_1}+Q_2^j \quad (4)$$

$$S^i=S_1^i=S_2^j \quad (5)$$

Where, Q_1^i, Q_2^j are the upward and downward loads of the load cell at points i and j respectively;S_1^i , S_2^j are the upward and downward displacement of the load cell at points i and j respectively. According to equations (1) and (2), a series of points in the traditional static load test can be obtained, thus obtaining the Q-S curve.

3.2 Principle of equal load method

According to the principle of equal load way. The following formula can be obtained:

$$Q^i=Q_1^i/\lambda_1+Q_2^i \quad (6)$$

$$S^i=\lambda_2\max(S_1^i,\ S_2^i) \quad (7)$$

Where, Q_1^i,Q_2^i are respectively the upward and downward loads of the load cell at point I ; S_1^i,S_2^i are respectively the upward and downward displacement of the load cell at point I ;λ_2 is the displacement correction coefficient, $(S_1^i+S_2^i)/2\max(S_1^i,S_2^i)<\lambda_2<1$, which is desirable$\lambda_2$ =1 in practical engineering.

3.3 Principle of simplification method

The simplified method needs to adopt the theoretical analysis calculation method. For general engineering, we can use the simplified conversion method in the theoretical analysis calculation method. According to the principle of synchronous upward and downward displacement, the simplified conversion method carries out super position of load through displacement, and the following conversion formula can be obtained:

$$\Delta S_1=\frac{Q_{u-down}L}{E_pA_p},\ \Delta S_2=\frac{(Q_{u-up}-W)L/\gamma}{2E_pA_p} \quad (8)$$

$$\Delta S=\Delta S_1+\Delta S_2,\ S=S_{u-down}+\Delta S \quad (9)$$

$$s=s_{u-down}+\frac{[(Q_{u-up}-W)/\gamma+2Q_{u-down}]L}{2E_pA_p} \quad (10)$$

$$Q_u=\frac{Q_{u-up}-W}{\gamma}+Q_{u-down} \quad (11)$$

Where, Q_{u-up} and Q_{u-down} are the ultimate bearing capacity of the upper and lower piles of the load cell; ΔS_1 is the elastic compression deformation of the upper section of the pile under the Q_{u-down} action;ΔS_2 is the elastic compression deformation of the upper section of the pile under compression under the action of lateral friction; W is the dead weight of the upper pile of the load cell; S_{u-down} is the displacement of the bottom position of the load cell; L for the pile length; E_p is elastic modulus of pile; A_p is the cross-sectional area of the pile; γ correction coefficient of lateral friction of upper pile of load cell[5]. The pile top load and pile top displacement corresponding to each stage are calculated, and a series of points are obtained. The equivalent pile top load-displacement curve can be obtained by connecting the points.

3.4 Comparison of characteristics of the three methods

In the equal displacement method, when the difference between S_1^i and S_2^i is small, the fitting Q-S curve is close to the Q-S curve obtained from the conventional static load test. However, in practical work, the displacements of upper and lower sections differ greatly, so it is impossible to find out the corresponding load when the upper and lower displacements are equal, so a complete Q-S curve can not be obtained. Under the condition of equal displacement ($S^i=S_1^i$) with the method of equal load, the value obtained Q^i by Equation (3) is less than that obtained Q^i by Equation (1), so the Q-S curve obtained by this method is lower than that obtained by common methods[6]. The method makes it easier to determine the ultimate bearing capacity in practical engineering, and the obtained value of the ultimate bearing capacity is more safe. The simplified method is applicable in engineering, considering the conversion coefficient and pile compression deformation[7].

4 Engineering Examples

4.1 Engineering test overview

The foundation of a project is designed as reinforced concrete mechanical bored cast-in-place pile, including test pile, 1#2#3#4# building and basement pile foundation.

The pile end bearing layer is medium-weathered mudstone siltstone, and the pile diameter is 900mm and 1000mm. Static load design bearing capacity characteristic value is 4500kN, 5500kN, 6500kN, uplift design bearing capacity characteristic value 1600kN, 2100kN, design pile length is not less than 16.5 meters.

The original landform of the project is the I terrace in the alluvial plain of Gan fu, with an elevation of about 16 ~ 20m. The geotechnical engineering parameters of the site are shown in Table 1. This time, 16# pile, 52# pile of 1# building and 75# pile of 2# building were selected to carry out the pile body reflexively-balanced static load test (slow maintenance load method). The design parameters of self-balanced static load pile and load cell parameters are shown in Table 2.

Table1. The geotechnical engineering parameters of the field.

Name of rock and soil layer	Average thickness (m)	Characteristic value of bearing capacity Fak (kPa)	Cohesion (kPa)	Internal friction Angle (°)	Natural density (g/) cm^3	Recommended value of compression modulus ES of soil (MPa)
Grain filling	4.13	60	8.0	5.0	1.75	3.5
silt	2.28	50	5.8	3.1	1.60	3.15
Silty clay	3.36	100	25.7	13.1	1.86	5.95
Fine sand	3.17	100	-	30.0	1.95	15
Coarse sand	3.66	200	-	35.0	2.00	20
Highly weathered argillaceous siltstone	1.22	300	-	-	-	25
Moderately weathered argillaceous siltstone	3.13	2000	-	-	-	-
Calcareous mudstone	1.98	900	-	-	-	-

Table2. The design parameters of self-balanced static load pile and load cell parameters.

Pile no.	Pile Diameter (mm)	Pile length (m)	Concrete strength grade	Design characteristic value of bearing capacity of single pile (kN)	Expanded low diameter (mm)	Load cell diameter (mm)
16 #	1000	16.85	C35	5500	/	800
52 #	1000	16.54	C35	5500	/	800
75 #	900	16.55	C35	4500	/	700

4.2 Result analysis

The whole test process is normal, the pressure is stable and the curve changes slowly. In this project, 16# pile, 52# pile and 75# pile are all loaded to the maximum test load, and each stage is loaded statically without abnormal phenomenon. Under the maximum load, none of them reaches the termination condition of the specification. Taking 16# pile, 52# pile and 75# pile as examples, the Q-S curves of each pile are shown in figure 2.

As can be seen from figure 2, the upward rebound rate of No.16 pile is 26.9% and the downward rebound rate is 17.7%.The upward and downward rebound rates of No. 52 pile are 24.5% and 21.5% respectively. The upward rebound rate of No.75 pile is 22.2%, and the downward rebound rate is 31.5%.

In this project, the upper part of each test pile load cell is distributed with soil layers, such as moderately weathered argillous siltstone, strongly weathered argillous siltstone, coarse sand, fine sand, silty clay, etc. Considering the burial depth of each soil layer and the pile forming effect of the test pile, the correction coefficient $\gamma=1.0$ is more suitable.No.16 pile diameter D=1000mm, pile length 16.85m, load to 5700kN, settlement at all levels is relatively uniform, corresponding downward settlement is 12.68mm<40mm, the measured vertical compressive ultimate bearing capacity of single pile is 11076kN > 11000kN. No.52 pile diameter D=1000mm, pile length 16.54m, pile loading to 5700kN, settlement at all levels is more uniform, corresponding downward settlement is 13.06mm<40mm, the measured vertical compressive ultimate bearing capacity of single pile is 11082kN > 11000kN.No.75 pile diameter D=900mm, pile length

https://doi.org/10.1051/e3sconf/202129302041

16.55m, pile loading up to 4650kN, settlement at all levels is more uniform, corresponding downward settlement is 15.62mm<40mm, the measured vertical ultimate compressive bearing capacity of single pile is 9042kN > 9000kN.The measured ultimate bearing capacity of the three piles in vertical compression meets the design requirements.

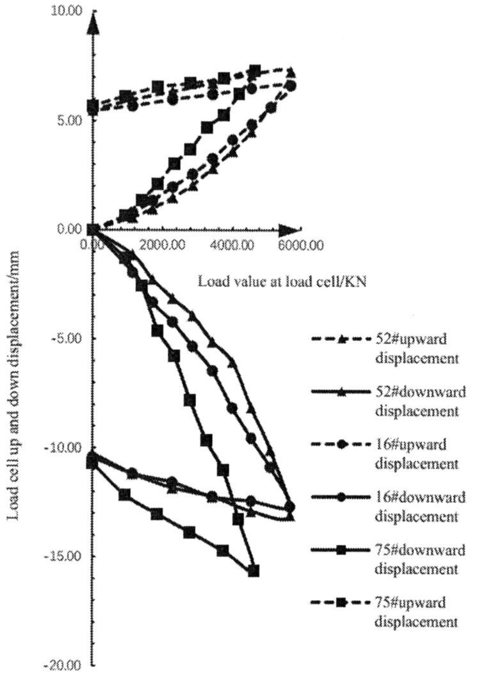

Figure2 16#、52#、75#pile Q-S curve

According to the load method, displacement method and simplified method, such as in the actual project y= 1.0, pulling coefficient λ_1 = 1, displacement correction coefficient λ_2 = 1, which is difficult to choose, such as displacement method due to the same displacement by subsection[8], an equation of the generation of value obtained 16 #, 52 #, 75 # pile under the loading qua- ntity per level corresponding to the settlement of pile top is shown in figure 3, figure 4, shown in figure 5.

From figure 3, figure 4, figure 5 shows 16 #, 52 #, 75 # pile equivalent pile top load displacement curve without a significant downward transition period, shows that bearing capacity limit state has not been achieved, the when S_1 and S_2 the difference is bigger, such as displacement meth- od to get the Q -S curve is not complete, cannot distinguish ultimate bearing capacity, and the lo- ad method and simplified method without the problem, and be safe. The ultimate bearing capacity of 16#, 52# and 75# single pile is obtained by using the method of equal load. The displacement of 19.31mm, 20.33mm and 22.92mm corresponds to the load of 11400kN, 11400kN and 9300kN.Using the simplified method, the ultimate bearing capacity of 16#, 52# and 75# single pile can be calculated as the load corresponding to the displacement of 18.40mm, 18.67mm and 21.28mm, namely, 11082kN, 11088kN and 9047kN. Both methods meet the design requirements. By comparison, it can be seen that the simplified method is safer than the equal-load method, and the equal-load method is safer than the equal-displacement method.

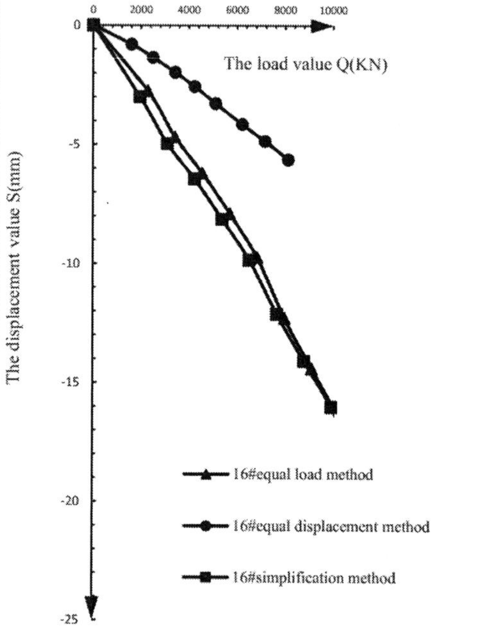

Figure3 Comparison of three conversion methods for 16# pile

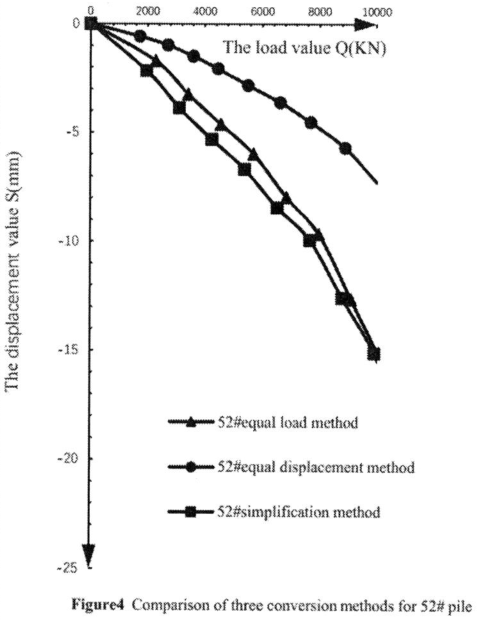

Figure4 Comparison of three conversion methods for 52# pile

https://doi.org/10.1051/e3sconf/202129302041

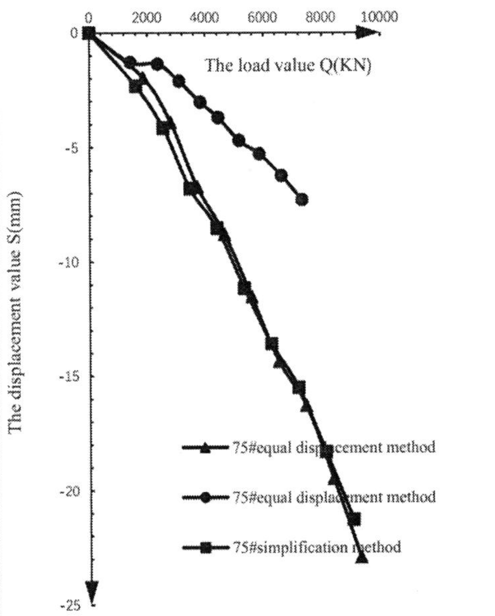

Figure5 Comparison of three conversion methods for 75# pile

5 Conclusion

This paper lists three common methods to comp- are how the Q_1-$S_1$and Q_2-S_2curves measured by the self-balancing method are transformed into the traditional Q-S curve of pile head. The meth- od of reasonable determination of Q-S curve is put forward to meet the requirements of current engineering testing. According to the research status of self-balancing method, the comparison shows that the simplified method is safer than the equal-load method, and the equal-load meth- od is safer than the equal-displacement method.

References

1. DB36/J002-2006, Technical Specification for Static Load Test of Reactive Force Balance of Pile [S].
2. Bo Lu, Shaolei Huo, Yunfeng Ge.Experimental analysis of bearing capacity of pile foundation based on self-balancing method [J]. Railway Construction,2009(12):83-86.
3. Shaolei Huo, Jieqing Zhu, Bo Lu. Application of Self-balancing Pile Measurement Method in Wuhan Building Pile Foundation Engineering [J]. Subgrade Engineering,2009(06):111-113.
4. Dalin Zhang, Weiming Gong, Qinjian Zang. Application Analysis of Self-balancing Test Method [J]. Industrial Building,2006(S1):673-675.
5. Qiyuan Wang. Discussion on Self-balance Test Method in Static Load Test [J]. Shandong Industrial Technology,2017(03):280+231.
6. Xianlin Zhang. Comparison between Self-balancing Pile Test Method and Traditional Anchor Pile Test Method [J].Geotechnical Engineering Technology, 2017,31(06):283-288.
7. ShanShan Lei. Study on bearing capacity mechanism of O-cell test pile [D]. Zhejiang Sci-Tech University,2019. (in Chinese)
8. Sixiong Wu. Application of self-balancing method to bearing capacity test of foundation pile in deep foundation pit [J].Fujian Construction Science and Technology,2016(01):11-13+10.

Discussion on the Construction and Application of Navigation Electronic Map Quality Inspection Sample Database

Jin Bai[1], Jingnan Gao[2,*], Wei Ma[1], Qingqing Yan[1], Xiaodi Wang[1], and Shuai Dong[1]

[1]National Quality Inspection and Testing Center for Surveying and Mapping Products, 100830 Beijing, China

[2]SpaceWill Info. Co., Ltd, 100089 Beijing, China

Abstract. In order to improve the scientificity and efficiency of quality inspection of navigation electronic map, and save the cost of quality inspection, it is necessary to establish the national navigation electronic map quality inspection sample database. In this paper, the construction goal, data acquisition technology process, and database construction method of navigation electronic map quality inspection sample database are described in detail, and the construction content and technical indicators are discussed. The quality test is carried out by combining the in-house verification with the field comparison of the navigation electronic map quality test sample database, and the test results are analyzed and evaluated. The effective application of the sample database in the quality test and evaluation of the internet navigation electronic map in the pilot area is carried out. At the same time, the application of navigation electronic map quality inspection sample database in the research of urban fine management and solid waste recycling economy ecological park planning is prospected.

1 Introduction

With the development of communication, electronics, network technology, and the geographic information industry, map navigation products have been widely used in people's daily life [1-2]. The mobile terminal navigation system and vehicle navigation have become necessary tools for people to travel. As a form of a public map, Internet navigation electronic map has developed rapidly in recent years. According to incomplete statistics, there are more than 40000 Internet map service websites in China. Due to the differences in service fields, service modes, and methods, there are great differences in map basic data, map display, and service functions among various Internet navigation electronic map service websites. However, when users use electronic map navigation and positioning products of different manufacturers, they find that some products often have different degrees of quality problems. Due to a large amount of data, strong timeliness, and complex data structure of navigation electronic map, in the current navigation electronic map quality detection work, for each test sample area, we need to go to the scene for benchmark data collection and field testing. The establishment of national navigation electronic map quality inspection sample database has become a problem to be solved [3]. Therefore, it is very important to develop the construction and application of a sample database for navigation electronic map quality inspection, which is of great significance to improve the scientific and efficiency of the quality detection of navigation electronic map and promote the development of the geographic information industry.

2 Data characteristics of navigation electronic map quality inspection sample database

2.1 Accuracy of point of interest retrieval

As a typical application of navigation system, the accuracy of reaching the specified target can be used as an important parameter to evaluate the navigation system [4]. In common geographic information databases, point of interest（POI） usually represents geographic points and simple related information. But in the construction of navigation electronic map quality detection sample database, users need to retrieve the accurate entrance and exit of the specified target in the process of using the navigation system, so that the system can lay the foundation for users to plan the most accurate route in a short time. Therefore, in the construction of navigation electronic map detection sample database, especially for shopping malls, hospitals and other features with large area and multiple exits, it is necessary to show the routes of entrances and exits in all directions as much as possible.

2.2 High data availability

Vehicle navigation and mobile terminal navigation system are the two most widely used directions of navigation

* Corresponding author: 313219666@qq.com

system at present. When users use the navigation system, whether they are using cars, sharing bicycles or other modes of transportation, they are most concerned about how to reach the destination with the optimal path in a short time. Navigation electronic map is an important basis of navigation system, and the current situation of data will directly affect the accuracy of navigation. The most frequent changes in the process of car driving are traffic restriction information and dynamic traffic information. The actual traffic guidance information and traffic condition information are the two indication conditions that must be followed in the process of car driving. Whether the information in the navigation system is consistent with the field information can be used to judge whether the driving instructions based on the navigation database are effective, Therefore, the current situation of traffic restriction information and dynamic traffic information can directly determine the efficiency of navigation electronic map detection process.

2.3 Reasonable data topological relationship

In the construction of navigation electronic map quality inspection sample database, the sample area data needs to express the geographic location, shape and spatial relationship of the road information and related element data in the target area. Taking road information detection as an example, the connectivity of road network is one of the concerns of navigation electronic map detection personnel, that is, the topological relationship between the arc segment and the node in the road layer, which contains the attributes of traffic capacity. In the process of building the database, when dealing with the topological relationship of Road intersection, especially in the multi-level Road intersection, overpass deck and other places, the topological relationship of the road should be fully reflected, and the combination of the topological relationship and the traffic capacity should be paid attention to at this time. The change of traffic capacity is usually at the node, and the decrease of traffic capacity may affect the traffic degree of topologically connected roads

3 The implementation of sample database for navigation electronic map quality inspection

3.1 Technical process of sample data collection

The sample database of navigation electronic map detection has been collected in 93 cities of 31 provinces, cities and autonomous regions nationwide, and 93 sample areas have been selected. The sample areas include municipalities directly under the central government, provincial capital cities, prefecture level cities and county-level cities. The regional distribution takes into account the East, South, West, North and central regions of China, and the degree of development takes into account the urban, urban-rural integration and rural areas. The sample area size is about 3 km × 3 km. According to the standard data collection scheme of navigation electronic map quality inspection sample database, real-time road network and its ancillary information data are collected in the sample area, including road network data, administrative division data, water system data of grade 6 and above, and basic data of place name data. Collect the road, background and notes of motor vehicles in the city; Collect the data of relevant points of interest in the city, and organize and process the data.

3.2 Collection technical index

3.2.1 POI collection

POI is the point of interest. Whether it is vehicle navigation or mobile terminal navigation, today's navigation systems on the market have their own POI database [5-6]. The number of POI in the navigation map, the accuracy of information and the speed of information update all seriously affect the use of a navigation system. The collection of POI should focus on the important information concerned by users, such as the accurate full name and relevant location information of interest points such as the location of the sample area government, local widely recognized landmark features, hotels, restaurants, shopping malls, transportation hubs, hospitals, gas stations, charging stations, schools, etc. In the process of field collection, it is necessary to collect the name information of POI on the basis of authoritative information, verify the location of POI or collect new POI point information on the spot, take photos on the spot, record the collection time of POI, and record the important attributes of POI, such as scenic spot level and hotel star level. It should be noted that the hotel star rating is subject to the official announcement of the tourism administration, and the Hotels with larger scale equivalent to 4 or 5 stars are classified as the secondary important category. In the process of in-house collation, the POI benchmark data is set up as a separate point layer. Based on the fields including number, name, collection and verification time, photo number and remarks, the POI benchmark data is sorted out, and the fields of classification code, importance and other important attributes are added.

3.2.2 Road information collection

The content of road information collection includes road network information, traffic restriction information, road guidance map information and so on.

The field work of road network information collection is mainly based on the image and the 1:10000 road data in the sample area to carry out on-site verification and mapping, and record the driving track in the whole process to complete the collection of road morphology and structure. In the process of field work, we should focus on collecting the level and code of high-level roads, and refer to the official data for confirmation and on-site verification before the survey. In the process of office work arrangement, it is necessary to establish the line layer of road information separately, specify the direction of road acquisition as the traffic flow direction, and draw all the road elements collected and verified by the field work in the survey area on the image according to the field work track information and 1:10000 road data. In the road information layer, key fields such as reference name, road code, level, one-way restriction, one-way restriction time domain, number of lanes and important one-way line are added.

In navigation electronic map data, structure information, one-way traffic information, and intersection traffic restriction information are combined to express traffic restriction information. It can be divided into important one-way lines and all intersection restriction information, among which important one-way lines, including roundabout, single lane road, often match with traffic restriction information. In the field work, 1:10000 data is used as the base map to collect and identify the single lane, note its time domain, road section range, flow direction and other information, and collect and number the photos related to the single lane information. The basic traffic restriction information includes no left, no right, no U-turn and no straight. In the office work, it is necessary to establish a point layer for traffic restriction information, and add key fields such as road level, restriction information description and vehicle type in the layer.

3.2.3 Background and annotation information collection

The background element information of navigation electronic map data involves administrative division, water system, railway and green space information, and the annotation information involves road name and number, public facilities name and other information. In the construction of navigation database, the background and annotation information need to refer to the basic 1:10000 data acquisition, field verification and photo recording, and the internal industry should build a separate layer and set up the number, name, acquisition / update time, photo number and other fields.

4 Application test of navigation electronic map quality inspection sample database

Organize the quality inspection practitioners to carry out the quality inspection test by combining the in-house verification and field comparison in the navigation electronic map quality inspection sample database as a group, and analyze and evaluate the test results. The evaluation methods and experimental results are as follows:

4.1 Evaluation method

The score of test results is calculated according to the proportion of errors and omissions of different levels, and the score is shown in the formula (1).

$$S_2=100-[a_1*(12/t)+a_2*(4/t)+a_3*(1/t)] \quad (1)$$

In formula (1), S_2 is the score of mass element; a_1 is the number of serious errors; a_2 is the number of large errors; a_3 is the number of general errors; t is the adjustment factor of deduction.

The comprehensive score of elements is calculated by weighted average according to the detection index, and the calculation method is shown in the formula (2).

$$S_1=\sum_{i=1}^{n}(S_{2i}*p_i) \quad (2)$$

In formula (2), S_1 is the score of the corresponding mass element; S_{2i} is the score of the corresponding mass sub elements; P_i is the weight of corresponding quality sub elements, which can be adjusted according to the emphasis of inspection; n is the number of quality sub elements contained in the quality element. For a sample area, the formula for calculating the weighted score of all unit results is:

$$S_I=\sum_{i=1}^{n}(S_{Ii}*p_i) \quad (3)$$

In formula (3), S_I is the data quality score of a sample area; S_{Ii} is the score of each mass element; P_i is the weight of each quality element can be adjusted according to the emphasis of inspection; n is the number of quality elements contained in the unit achievement.

4.2 Analysis of experimental results

Based on the evaluation method of this paper, six influential Internet navigation electronic map service websites are selected to carry out the experiment. The sample area of the experiment is mainly in Prefecture and county-level cities. Six cities are randomly selected from the sample database of navigation electronic map quality detection. In each sample area, according to the complexity of road and poi, the area of not less than 4 square kilometers is cut as the sample area of the actual test.The data in the navigation electronic map quality inspection sample database obtained by field measurement is taken as the reference data of the sample area. Tile map sample data is obtained online and real-time by ordinary users through the official websites of various map service providers. In the experiment, the query function provided by the website is used to query the POI data one by one, and compare with the benchmark reference data to determine the integrity and consistency of the data. The integrity and correctness of the road data are verified by using the line planning function. The test results are shown in Figure 1.

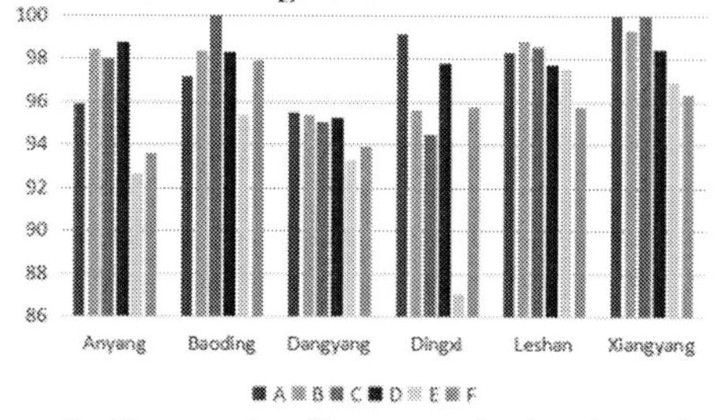

Figure 1. Score results of internet navigation electronic map service test sample area.

The above test results show that the quality evaluation method based on the navigation electronic map quality inspection sample database can not only evaluate the quality of Internet navigation electronic map service well, but also reflect the fine differences in data and services of various Internet map service providers.

5 Prospect of application scenarios

5.1 Application of sample database data in urban fine management

Urban management is the coordinated control behavior of all the activities of the operation and development of the city. Fine management is a management concept and technology that originated from Japanese enterprises in the 1950s, which has been applied to all kinds of social organizations [7]. Urban fine management is to introduce the concept of fine management into urban management. In order to implement fine management in a city, it is necessary to refine the management of underground pipe networks, roads, bridges, buildings and other objects of underground, ground, and above ground space. With the passage of time, the city is constantly changing. The navigation electronic map quality inspection sample database data can more clearly and objectively display and mine the information of urban geospatial-related elements. The timely updated geographic information data can realize the regular observation of large urban areas, and help to grasp the panorama and historical changes of the city. It will greatly promote the application of geographic information data in all aspects of urban fine management. Using high timeliness geospatial information to carry urban management business information, truly restore urban road construction and other information, spatially and sequentially process urban management related business data, and carry it into the urban space-time environment based on geographic information data, so as to provide data support for urban meticulous management.

5.2 Application of sample database data in the planning of solid waste recycling economy ecological park

Industrial Park is an important entry point for the development of circular economy and sustainable development. [8-9]. By using the limited available site space and concentrating all kinds of municipal solid waste treatment facilities, the construction of a circular economy eco-industrial park for municipal solid waste treatment can not only reduce the resistance caused by scattered-site selection, but also improve the joint treatment and disposal level of various types of municipal solid waste, at the same time, it is also conducive to the construction of material, energy and water resources recycling cascade utilization system, which is an effective mode to solve the problem of "garbage siege" in large and medium-sized cities with the land shortage. Using the navigation electronic map quality inspection sample database to obtain the surrounding features information and spatial distribution of the park, to provide basic data support for the planning of the park land layout structure. Combined with the surrounding road planning, congestion information, the process characteristics of each project, and the material supply and demand relationship between projects, the optimal layout of all kinds of land and projects in the park is realized, it provides a scientific and effective plan for the planning and construction of the solid waste park.

6 Summary

As an important foundation of the navigation system, the quality of the navigation electronic map will directly affect the user's evaluation of the navigation system. It is the key to guaranteeing the quality of all kinds of electronic navigation system products to strictly control the quality of navigation electronic maps by adopting the industry technical standards and Technical Specifications issued by state organs. The quality inspection of navigation electronic map in some pilot areas has been confirmed. Through scientific and feasible methods, we can establish a multi-level navigation electronic map quality inspection sample database covering the whole country, and carry out the application of navigation electronic map quality inspection sample database, which can save the inspection cost on the basis of ensuring the authenticity and reliability of navigation electronic map data inspection results Improve inspection efficiency, and continuously improve the quality of navigation electronic map, Internet map service products, and navigation terminal products. At the same time, the establishment of a year-by-year updated navigation electronic map quality testing sample database can not only serve the quality control of Internet map quality testing. It can also be applied to the research of urban fine management, carbon emission monitoring and management, solid waste recycling economy ecological park planning, and so on. It provides a new idea for the above research.

References

1. X.C. Liu, J. Liu, D.L. Li, Bull Surv Map. 33, 5 (2012)
2. Y.H. Cai, W. Zhang, Journal of Navigation and Positioning. 10, 1 (2021)
3. H.Q. Yang, Bull Surv Map. 37, 11 (2014)
4. J. Zhang, N. Jiang, H. Zhou, Geomatics & Spatial Information Technology. 55, 2 (2015)
5. P. Gao, A. H. Wu, Beijing Surv Map. 24, 1 (2013)
6. Q.Q. Yan, X. Ma, H. Li, Geomatics & Spatial Information Technology. 169, 5 (2016)
7. L.Q. Guo, J.Y. Lin, W. Y. Wang, Urban Development Studies. 57, 1 (2012)
8. J.M. Hong, M. Li, Environment Engineering Design. 155, 1 (2020)
9. C.X. Hao, Natural Science. 175, 7 (2019)
10. X.M. Su, J.C. Tan, Y.Z. Chen, Beijing Surv Map. 126, 3 (2016)

https://doi.org/10.1051/e3sconf/202129302043

Mechanical Properties of Gangue Cement Stabilized Road Foundation Mixture

Dongsheng Zhang[1], Naicheng Liu [2], Mingjie Mao [2*], Jiabin Li [1]

[1]Research Group RecyCon, Department of Civil Engineering, KU Leuven, Campus Bruges, 8200, Bruges, Belgium

[2]School of Civil and Hydraulic Engineering, Ningxia University, Yinchuan 750021, China

Abstract. In this study, coal gangue was used to replace the four single-graded natural gravels of 0~5 mm, 5~10 mm, 10~20 mm, and 20~26.5 mm to design the mix ratio and to study the influence in particle size changes of coal gangue on the mechanical properties of the mixture. The research results show that: The maximum dry density of the mixture with coal gangue is lower than the density of the natural crushed stone water-stable mixture, and the optimal moisture content is higher than that of natural crushed stone water-stable mixture; coal gangue aggregate is mixed with cement stabilized in the crushed stone mixture, and the 7d and 28d unconfined compressive strength, splitting strength, and compression rebound modulus values of the mixture are all reduced, which has a degrading effect on the mechanical properties of the mixture; however, both can attain high speed, which is the basic strength requirements for expressways, including first-class highways.

1 Introduction

In China, coal is the major energy source. Coal consumes 70% of its primary energy, which is more than twice the world average[1]. Most of the coal gangue is stacked directly after it is produced. The gangue piles stored in the open air for a long time can easily and spontaneously combust and produce harmful gases, and the weathered coal gangue thus enters the surrounding soil, air, and groundwater, causing pollution. Therefore, recycling of coal gangue is conducive to protecting the environment and restoring the natural ecology[2].

In civil engineering, recycled coal gangue is mostly used as an aggregate for road water-stable layers. First, highway engineering does not have stringent requirements for the type and quality of gangue, and most of the gangue meets engineering construction standards. Second, coal gangue, instead of natural crushed stone aggregate, does not require special technical treatment and only requires a suitable crushing and screening method[2].

Therefore, based on ordinary cement-stabilized crushed stone base materials, referring to the index requirements proposed in the Chinese standard, this study uses coal gangue instead of different grading particle sizes to influence the mechanical properties of the mixture and aims to develop a suitable Ningxia. The coal gangue base material not only alleviates the shortage of natural aggregates but also efficiently uses coal gangue materials. This conforms to the technical principles of localization of road construction materials and converts waste into valuable material.

2 Raw materials and experimental plan

2.1 Raw materials

Cement: Ningxia Saima brand PO42.5 ordinary Portland cement was used; For natural gravel, this study used the natural limestone of Ningxia Taomengou Stone Material Factory and based on the "Highway Pavement Base Construction Technical Regulations" ; four grades of aggregates were used, divided into grades of 0–5 mm, 5–10 mm, 10–20 mm, and 20–26.5 mm. For coal gangue, this test uses coal gangue discharged from coal mines in Ningxia, all of which are crushed and screened using jaw crushers and divided into four categories: 0–5 mm, 5–10 mm, 10–20 mm, and 20–26.5 mm gradation; furthermore, the Yinchuan city tap water was used.

2.2 Experimental program

The cement content was found to be 4%. Based on the screening results, the particle size of natural crushed stone was found to be 20–26.5 mm: 10–20 mm: 5–10 mm: 0–5 mm, and the ratio was 0.18: 0.275: 0.18: 0.365, respectively. The cement-stabilized crushed stone mixture is denoted as M0, the mixture using 0–5 mm coal gangue aggregate particle size is denoted as M(0-5), and that with 5–10 mm coal gangue aggregate particle size is denoted as M(5-10). The mixture with 10–15 mm coal gangue aggregate particle size was recorded as M (10-20), and that with 20–26.5 mm coal gangue aggregate particle size was

* Corresponding author: maomjnxdx@163.com

recorded as M (20-26.5). In the experiment, a heavy-duty compaction test determined the optimal moisture content and dry density index of the material.

The test piece of the mixture was made using a single-horizontal shaft forced mixer to stir the mixture. The mixing sequence was first pour the stabilized material into the mixer and stir for 30s for uniform gradation; subsequently, pre-weighed water was added and stirred for 60 s. Finally, a container was used to slowly add cement and stirred for 120 s to evenly combine the stabilized material and the cement. Based on the static pressure forming method in the "Highway Engineering Inorganic Binder Stabilizing Material Test Regulations" (JTG E51-2009)[3], a WYC-150 II stabilized soil specimen-forming machine was used to form a diameter of 150 mm and a diameter-to-height ratio of 1:1; the compaction coefficient of the cylindrical specimen was 0.98, and the specimen was wrapped tightly in a plastic bag and cured in an environment with a temperature of 20±2 °C and a humidity of ≥95%. The specimen was immersed in water for one day before reaching its age[4]. The unconfined compressive strength, split strength, and compressive resilience modulus tests were conducted after the specimens reached the ages of 7d and 28d.

3 Results and discussion

3.1 Maximum dry density and optimal moisture content

Based on the "Technical Regulations for Highway Pavement Base Construction," the test pieces formed by the heavy compaction method were used to determine the optimum moisture content and maximum dry density of the mixture of coal gangue with different alternative particle sizes. The test results are presented in Table 1.

Table1. The optimal moisture content and maximum dry density of the mixture

Test group	maximum dry density(g/cm^3)	optimal moisture content(%)
M0	2.325	4.90
M(0-5)	2.268	5.13
M（5-10）	2.292	5.01
M（10-20）	2.252	5.23
M（20-26.5）	2.250	5.06

From Table 1, the maximum dry density of the coal gangue cement-stabilized mixture is lower than that of the natural crushed stone water-stable mixture, and the optimal moisture content is higher than that of the natural crushed stone water-stable mixture, which is similar to that of the coal gangue. Density and water absorption are related. The density of coal gangue is lower than that of natural gravel; however, the water absorption is higher. Using 0-5 mm coal gangue as an example, compared with natural gravel, the maximum dry density of the coal gangue cement-stabilized mixture was reduced by 2.45%, and the optimal moisture content was increased by 4.7%. This shows that the incorporation of coal gangue has varying effects on the maximum dry density and optimal moisture content of the mixture. In physical engineering, various technical parameters should be strictly controlled to reduce the risk of cracks owing to the increase in the moisture content of the mixture[5].

3.2 Unconfined compressive strength

The unconfined compressive strength test of the 7d and 28d aged mixture of coal gangue, instead of crushed stones of different single particle size, was conducted, and it was combined with cement stabilized crushed stone mixture benchmark group (M0 group) and cement stabilized coal gangue mixture (M100 group) For comparison, the representative values of unconfined compressive strength and coefficient of variation are shown in Figures 1 and 2, respectively.

From Figure 1, compared with the natural crushed stone mixture benchmark group M0, the unconfined compressive strength of the 7d mixture of coal gangue instead of natural crushed stone of different particle sizes demonstrated different degrees of decline, M(0 -5), M (5-10), M (10-20), and M (20-26.5) groups, and the 7d unconfined compressive strength decreased by 54.2%, 39.0%, 52.3%, and 42.0%, respectively; that is, the substitution of 0-5 mm and 10-20 mm coal gangue coal has a greater impact on the strength of the material. In addition, with the increase in age to 28d , the unconfined compressive strength of each ratio increased, including M(0-5), M(5-10), M(10-20), and M(20- 26.5). Compared with the intensity of 7d, the intensity increased by 16.6%, 9.4%, 20.9%, and 12.5% respectively.

Figure 2 shows that the replacement of natural gravel with coal gangue generally increases the dispersion of the test results of the mixture. At 7d, the coefficient of variation of the M(0-5) mixture was lower than that of the natural crushed stone mixture benchmark group M0, whereas for the other groups, it still improved to varying degrees. At 28d, the coefficients of variation of all coal gangue cement mixtures were much higher than those of the benchmark group, and the M (0-5) group had the smallest increase in mixtures. This shows that replacing the finer natural gravel can slightly weaken the influence of coal gangue on the mechanical properties.

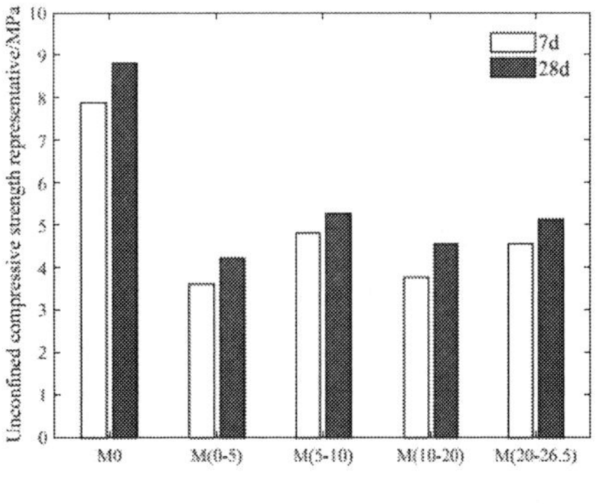

Figure 1. Unconfined compressive strength Representative of cement-stabilized coal gangue mixture

https://doi.org/10.1051/e3sconf/202129302043

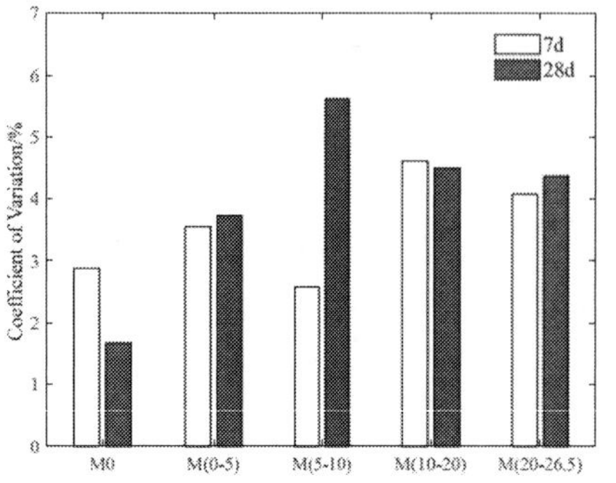

Figure 2. Coefficient of variation of cement stabilized coal gangue mixture

From the perspective of the strength increase of the composite sample with the age and the strength loss of the benchmark group, the strength of the M (5-10) and M (20-26.5) groups was formed quickly, and the strength of the coal gangue was replaced. The loss is relatively low, and the unconfined compressive strength performance is better, which can meet the unconfined compressive strength requirements of the base layer under heavy traffic load conditions of expressways and first-class highways. The substitution of 0-5 mm gravel and 10-20 mm gravel has a greater impact on the unconfined compressive strength of the material[6]. The combination of M (0-5) and M (10-20) can meet the strength requirements under medium and light traffic conditions.

3.3 Splitting strength

The 7d and 28d age splitting strength tests were conducted on a mixture of coal gangue instead of crushed stones of different single-particle sizes and combined with a cement-stabilized crushed stone mixture benchmark group (M0 group), cement stabilized coal gangue mixture (M100 group), and the results of the representative values of splitting strength and the coefficient of variation are shown in Figures 3 and 4.

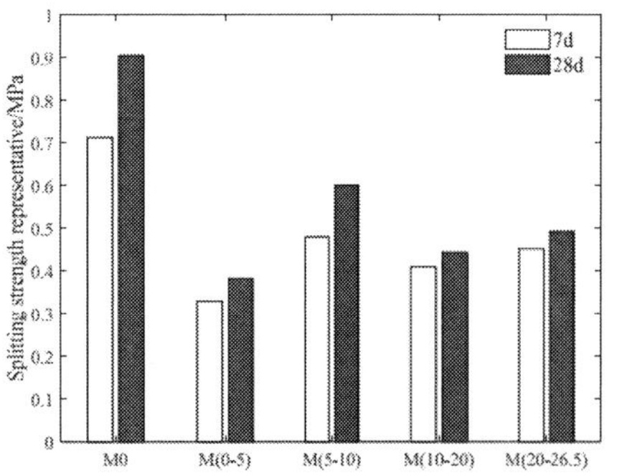

Figure 3. Splitting strength Representative of cement-stabilized coal gangue mixture

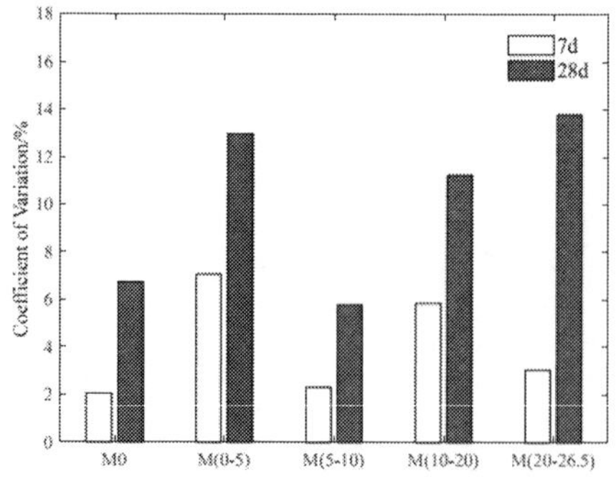

Figure 4. Coefficient of variation of cement-stabilized coal gangue mixture

From Figure 3, compared with the natural crushed stone mixture benchmark group M0, the 7d splitting strength of the mixture of coal gangue instead of natural crushed stone of different single-particle sizes decreases to varying degrees, M(0-5) Compared with the benchmark group M0, the 7-d unconfined compressive strength of the, M(5-10), M (10-20), and M (20-26.5) groups decreased by 53.65%, 32.58%, 42.42%, and 36.52%, respectively. The law of unconfined compressive strength is the same. The substitution of 0–5 mm and 10–20 mm coal gangue has a greater influence on the strength of the material. In addition, with the increase in age, the 28d splitting strength of each ratio increases, including M(0-5), M(5-10), M(10-20), and M(20-26.5). The 7d intensity increased by 16.1%, 25.2%, 8.3%, and 9.1% respectively.

Figure 4 shows that the substitution of coal gangue for natural gravel generally increases the dispersion of the mixture test results. At 7d, the coefficient of variation of the M(0-5) mixture was the lowest except for the benchmark group. Even at 28d, the coefficient of variation of the M(0-5) mixture was lower than that of the natural gravel mixture benchmark group M0. Other groups still have different degrees of improvement. It still shows that replacing the finer natural gravel can slightly weaken the influence of coal gangue on the mechanical properties.

With the combination of the strength increase of the mixture specimen during the age and the strength loss compared with the reference group, the strength of the M (5-10) and M (20-26.5) groups was quickly formed, and that of the coal gangue was replaced. The loss is relatively low, and the unconfined compressive strength performance is better, which can meet the unconfined compressive strength requirements of the base layer under heavy traffic load conditions of expressways and first-class highways; after long-term maintenance, the substitution of 0-5 mm and 10-20 mm coal gangue has a greater impact on the unconfined compressive strength of the mixture. The M(0-5) and M(10-20) groups can meet the strength requirements under medium and light traffic conditions.

In practical applications, the quality standards of coal gangue with particle sizes of 0-5 mm and 10-20 mm should be strictly controlled, or these sizes should not be used to replace natural gravel, and the splitting strength of the mixture should be avoided as much as possible to prevent cracks.

3.4 Compression rebound modulus

The compression rebound modulus representative at 7d and 28d for the mixture of coal gangue instead of crushed stone of different single-particle sizes, the cement-stabilized crushed stone mixture benchmark group (M0 group), and the cement-stabilized coal gangue mixture (M100 group) were compared, and the influence of different particle size coal gangue aggregates into the mixture on its stiffness was discussed and the results are shown in Figure 5.

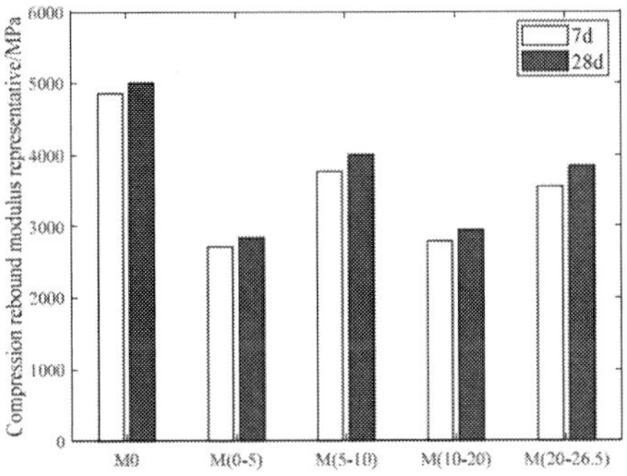

Figure 5. Compression rebound modulus of cement-stabilized coal gangue mixture

From Figure 5, the incorporation of coal gangue reduced the stiffness of the mixture, and the compression rebound modulus of each group was reduced to different degrees, which differs from the previous test results, M(0-5) The compression rebound modulus of the specimens of the group and M(10-20) group decreased to a larger extent, compared with the reference group M0 at the 7d age, they lost 44.1% and 42.6%, respectively, and at 28d , compared with the reference group M0 lost 43.3% and 41.1%, respectively, and the degree of decline is much greater than that of the M(5-10) and M(20-26.50) groups, indicating that 0–5 mm and 10–20 mm coal gangues replace the natural gravel thickness of the mixture. The decrease in stiffness is more obvious. For road sections with higher rigidity requirements, the quality standard of coal gangue of this particle size should be reasonably controlled or not replaced by this particle size during construction to ensure the rigidity of the material, prevent the occurrence of cracks, and affect the construction quality.

4 Conclusions

(1) The maximum dry density of the mixture with coal gangue is lower than the density of the natural crushed stone water-stable mixture, and the optimal moisture content is higher than that of the natural crushed stone water-stable mixture, which is the same as the density of coal gangue. The density of coal gangue is lower than that of natural gravel; however, the water absorption rate is higher.

(2) When coal gangue was mixed with a cement-stabilized crushed stone mixture, the 7d and 28d unconfined compressive strength, splitting strength, and compression rebound modulus of the mixture were all reduced. The mechanical properties showed a deteriorating effect.

(3) All the alternative groups of coal gangue with 4% cement content can meet the strength requirements of expressways and first-class highways.

(4) The change in coal gangue particle size has a significant impact on the mechanical properties of the mixture. After the mixture of 0-5 mm and 10-20 mm coal gangue was replaced, the mixture had unconfined compressive strength, splitting strength, and the compressive resilience modulus decreased significantly. In actual applications, to ensure the quality of the project, the quality standards of coal gangue with particle sizes of 0-5 mm and 10-20 mm should be strictly controlled or the particle size should not be used instead, to avoid the obvious decrease in the mixture's strength and splitting. The strength was too low to cause cracks.

Acknowledgments

This study was supported by the National Natural Science Foundation of China (51768058) and the Key R & D projects of Ningxia Province of China (2021BEE03004). Their results were fundamental to this study.

References

1. Huang Di. Study on research's flow properties and model its heat transfer performance for oxygen-heating CaC_2 production. Master dissertation of Beijing University of Technology. 2014
2. Yang guodong,Wang feng.Experimental Study on Mixture Ratio of Cement Stabilized Base Material with Washed Coal Gangue. Journal of Shandong Jiaotong University. 2018, 26(03):64-70.
3. Fan Wendong.The studies of using about the integrity of the cement stable macadam. Master dissertation of Chang'an University. 2012.
4. Zhou Jie. Performance evaluation of cement stabilized recycled concrete aggregate treated with waste oil and asphalt emulsion. Master dissertation of Hunan University. 2019.
5. Base in Suzhou Area. Journal of China & Foreign Highway. 2012, 32(03):80-85.
6. Fu Luxin. Study on Strengthening Treatment and Application of Cement Stabilized Aggregate Milling Masterial. Shandong University of Technology. 2019.

Multi objective optimization of window to wall ratio of University Gymnasium in severe cold area by coupling natural lighting and energy consumption

ZongJunlin[1] ZhangLongwei[1,*]

[1]School of architecture and planning, Shenyang Jianzhu University, China

Abstract. ***Objective*** To explore the relationship between window opening ratio and natural lighting and energy consumption of University Gymnasiums in severe cold area, and put forward optimization strategies. ***Methods*** Digital simulation technology was used to simulate the energy consumption and natural lighting of University Gymnasium window opening ratio, and the window opening scheme with natural lighting and low energy consumption was obtained. ***Conclusion*** The side window lighting should be used in the window opening scheme of small and medium-sized university gymnasiums. Within the range of experimental data, the North-South lighting is the main lighting mode, and the East-West lighting is the auxiliary.

1 Construction of typical information model

1.1 Establishment of general model of University Gymnasium in cold regions

With the reform and development of China's higher education, the scale of higher education is gradually expanding, and the cultivation of students' sports quality has been widely valued. Gymnasium has gradually become one of the indispensable components of the architectural group in colleges and universities. Due to the large building volume, the energy consumption and the operation and maintenance costs of various facilities in the process of using the stadium are far higher than expected [1-3]. In recent years, with the wide application of digital technology in green design, the single building performance analysis of building energy consumption, such as natural lighting and lighting ventilation, has gradually developed into a multi-objective building performance evaluation. For example, the paper for Zhang on Bin [4] aimed at urban buildings in Beijing, aiming at lighting coefficient, visual comfort and building heat consumption, This paper studies the optimization of the window form of residential buildings in the city. There are also similar simulation experiments that aim at building cluster shape, Da, UDI and various energy consumption, such as , Liu Yao [5], sun Cheng [6], etc. In this paper, grasshopper parametric modeling, ladybug & honeybee energy consumption simulation tools are used. On this basis, the multi-objective optimization tool octopus is used to optimize the window wall ratio of typical university. The natural lighting and energy consumption coupling optimization is carried out, and the window design strategy of University Gymnasiums considering both natural lighting and energy consumption is summarized, It provides a theoretical basis for the natural light environment optimization and low energy consumption design of large space buildings in cold regions.

Gymnasium is a typical large space building, which is obviously different from other general education buildings in space scale, function and structure. In order to get more information about the characteristics of University gymnasiums, the author systematically combs 20 famous University Gymnasiums built in recent years in severe cold areas in China, as shown in Table 1.

Table1. Survey data of University Gymnasium

Region (province)	Building name	Scale	The measure of area (m^2)	height (m)	Number of seats (seats)	Natural lighting	Architectural form	Site size (m)
Liaoning	Shenyang University Gymnasium	Medium-sized	18900	—	5000	Side window	Rectangle	
	Liaoning University Gymnasium	Medium-sized	16500	17.5	4000	Side window	Cylinder	
	Shenyang Jianzhu University	Medium-sized	14000	19.4	1600	Skylights and side	Rectangle	54×32

* Corresponding author: z_lw@sjzu.edu.cn

 https://doi.org/10.1051/e3sconf/202129302044

	Northeast University Gymnasium	Small-scale	11000	17.2	1500	Side window	Ellipse	
	Shenyang University of Technology Gymnasium	Medium-sized	13000	22	4644	Side window	Ellipse	42×55
	Gymnasium of Dalian University of Technology	Medium-sized	17000	25	5800	Side window	Ellipse、rectangle	39×25
Jilin	Jilin University Gymnasium	Medium-sized	36000	24	—	Side window	Circular、rectangle	
	Gymnasium of Changchun Institute of	Medium-sized	11000	17.6	2700	Side window	Rectangle	
	Jilin University of architecture Gymnasium	Small-scale	—	22	2300	Side window	Rectangle	
	Gymnasium of Changchun University of traditional Chinese	Small-scale	—	19	800	Side window	Rectangle	
Heilongjiang	Harbin Institute of Technology Gymnasium	Small-scale	7500		3000	Skylights and side	Irregular rectangle	61×30
	Gymnasium of Harbin University of science and	Medium-sized	33000	22	2700	Skylights and side	Ellipse、circular	
	Gymnasium of Northeast Agricultural University	Small-scale	16000	—	—	Skylights and side	Diamond	
	Heilongjiang University Gymnasium	Small-scale	17000		3000	Side window	Rectangle	

The analysis shows that the basic plane shape of University Gymnasiums in severe cold areas are mostly regular shapes such as rectangle, circle and ellipse, as well as a small number of irregular shapes. Based on the above considerations, this paper constructs a typical information model of University Gymnasium. The scale of University Gymnasium is small and medium-sized, with 3000 seats and 6000 seats respectively. The size of the stadium is about 38m×54m、32m×44m、24m×26m。 In this paper, using grasshopper parametric modeling, the general model is divided into two cases of medium-sized and small university gymnasium, and the size of medium-sized stadium is selected as 38m ×54m, with 3000 seats in the auditorium. According to the seat size (seat width 0.5m, row spacing 0.85m), the plane size is calculated to be 51m×69m, building height 25m; The size of the small playground is 32m ×44m, with 1500 seats in the auditorium, and its plane size is 42m×54m, and the height is set at 15m. According to the design code for sports buildings[7], the starting height of the auditorium stands is about 2m, and the slope angle of the stands is between 28°~ 30°.

1.2 Parameter setting of natural lighting and energy consumption simulation

1.2.1 Setting of natural lighting parameters

Effective natural lighting intensity is an important index to evaluate indoor lighting effect. In this paper, the daylight simulation system module of honeybee tool is used to simulate the natural daylighting experiment. The effective natural daylighting intensity in the year-round dynamic evaluation daylighting simulation is used as the evaluation standard of natural daylighting experiment simulation. The concept is to meet the target space illumination demand in a period of time in the test plane as a percentage of the whole year. The acceptance range of illumination is generally defined as 100lux ~ 2000lux. The main function of University Gymnasium is daily class and training. According to the standard for lighting design of buildings (GB 50034-2013), the required illumination of University Gymnasium in daily training is 300lux, so the effective illumination in this paper needs to be adjusted to the minimum value of 300lux(hereinafter referred to as udi300) during training.

(1) Lighting test point setting

As shown in Figure 2, the test method of light environment in software simulation is the same as the physical test method of light environment. The square grid composed of test points needs to be set in the sports ground part of the model, so the distance between each test point is 2m, and the height of test plane is 1.67m.

(2) Reflectivity setting of enclosure interface

In order to achieve the real indoor natural lighting effect, in the natural lighting simulation experiment, it is necessary to set the reflectivity of transparent materials and materials in the building in honeybee. According to the Architectural Lighting Design Manual, the reflectivity of roof, wall, ground and glass in the general model of University Gymnasium is set as 0.8, 0.5, 0.2 and 0.6 respectively.

(3) Natural lighting simulation experiment process

Finally, the external shape of the building, climate and enclosure interface material information are input into the main simulation engine, and the UDI value data group is obtained through simulation. The data group is the percentage of the illumination time of each lighting test point in a year to meet the daylighting evaluation index, that is, the corresponding UDI value of each test point. Finally, the average value of the data group is taken for comparison.

1.2.2 Building energy consumption parameter setting

The energy consumption of University Gymnasium Building generally comes from the power consumption and building heat consumption of daily classes, training, competitions, air conditioning, fans, lamps and other equipment. The total power consumption includes the power consumption of artificial lighting and the operation of some equipment. Therefore, in this paper, we mainly focus on the energy consumption of cooling in summer, heating in winter, artificial lighting and other equipment operation in cold regions, and count the total energy consumption and energy consumption per unit area as the evaluation index of energy consumption.

(1) Thermal parameter setting

In this paper, Shenyang City is selected as the simulation site, according to its building climate zoning, and with reference to the "design standard for energy efficiency of public buildings" (GB 50189-2015). The general model of this paper is set according to the design standard of Low-E double-layer glass curtain wall, and the relevant setting parameters and simulation settings of external wall, roof and window are shown in Table 2.

Table2. Thermal parameters

Position	Heat transfer coefficient ($W.m^{-2}$)	Solar heat gain coefficient	Visible light transmittance
Exterior wall	0.43	——	——
Roofing	0.35	——	——
Outer window	2.00	0.4	0.45

(2) Indoor environment setting

a. Air conditioning parameter setting

Referring to the relevant parameters of air conditioning design manual, the design temperature of indoor air conditioning in summer and winter of University Gymnasium is set at 26 ℃ and 20 ℃ respectively. According to the climate and temperature characteristics of Shenyang area, the cooling time in summer is set from June 1 to August 31, and the heating time in winter is set from November 1 to March 31 of the next year. During the transitional season, natural ventilation is used for adjustment.

b. Human factor parameter setting

The utilization mode of University Gymnasium is very different from that of conventional gymnasium. In terms of operating time, the university gymnasium can be described as a full-time operation, with very high use efficiency. From Monday to Friday, it is used by university teachers and students in class all day, and provides fitness activities for students on weekends and evenings. In this paper, the maximum indoor personnel density is 0.4 person / m^2, the average metabolic capacity is 120 W / person, the equipment density is 8 w / m^2, and the lighting load density is 15.0 w / m^2.

(3) Energy consumption simulation experiment process

In the same process as the natural lighting simulation experiment, the external shape of the building, the information of climate and enclosure interface materials, and the information of building indoor environment are input into the main simulation engine. Through the simulation, the energy consumption data of heating, cooling, lighting, fan heat pump and other equipment of University Gymnasium is obtained, and the energy consumption data are accumulated together, and the total energy consumption of the whole year and the energy consumption per unit area of the building can be obtained through calculation.

2 Coupling simulation analysis of natural lighting and energy consumption

2.1 Simulation method of natural lighting and energy consumption coupling research

The natural lighting and energy consumption simulation is modeled by grasshopper under rhino platform, and the skylight and side windows of University Gymnasium are simulated. In the "Design Standard for Energy Efficiency of Public Buildings", it is stipulated that the area of transparent part of building roof should not be greater than 20% of the roof area, and the area ratio of single facade window to wall (including transparent curtain wall) of Class A public buildings should not be greater than 0.6. Therefore, the window wall ratio of University Gymnasium side interface is set as 0 ~ 0.6, the growth gradient is 0.05, and the south facade is as shown in the table; The window to wall ratio of skylight is 0 ~ 0.2, and the growth gradient is 0.02.

Then, using the genetic algorithm in the octopus calculator, and the parameter settings are shown in the table below. Through the continuous combination of various parameter variables, similar to the exchange and variation of gene fragments in genetics, the "individual" is a random combination of many parameter variables. Through the selection of building performance evaluation indicators (survival of the fittest), the variable parameter group that ultimately meets the conditions is obtained; After that, the Pareto economics is used to calculate the optimal solution of many parameters that meet the conditions. The parameter settings of octopus are shown in Table 3.

Table3. Octopus parameter setting

Elitism	Mutation Probability	Mutation Rate	Crossover Rate	Population Size
0.5	0.2	0.9	0.8	50

2.2 Simulation experiment of natural lighting and energy consumption coupling research

In the experimental evaluation index, the minimum value of natural lighting is udi300 > 75%, and the energy consumption per unit area is the minimum value by default, that is, the minimum energy consumption. Considering that the main function of University Gymnasium is divided into daily training in class and taking into account the needs of competition, this paper discusses two cases separately.

2.2.1 Simulation experiment of natural lighting and energy consumption coupling research under training mode

(1) Experimental simulation results of gymnasium in medium sized Universities

This iteration has a total of 50 generations. After 125 hours of calculation, the results in the panel tend to be stable, and the results are convergent enough. The solution distribution is shown in Figure 1.

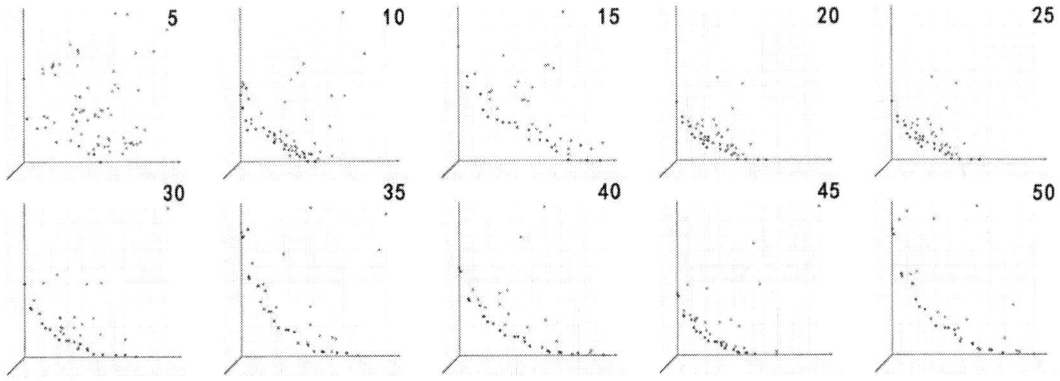

Figure 1. Solving distribution (medium)

(2) Experimental simulation results of small University Gymnasium Similarly. after 125 hours of calculation, the results in the panel tend to be stable, and the solution distribution is shown in Figure 2.

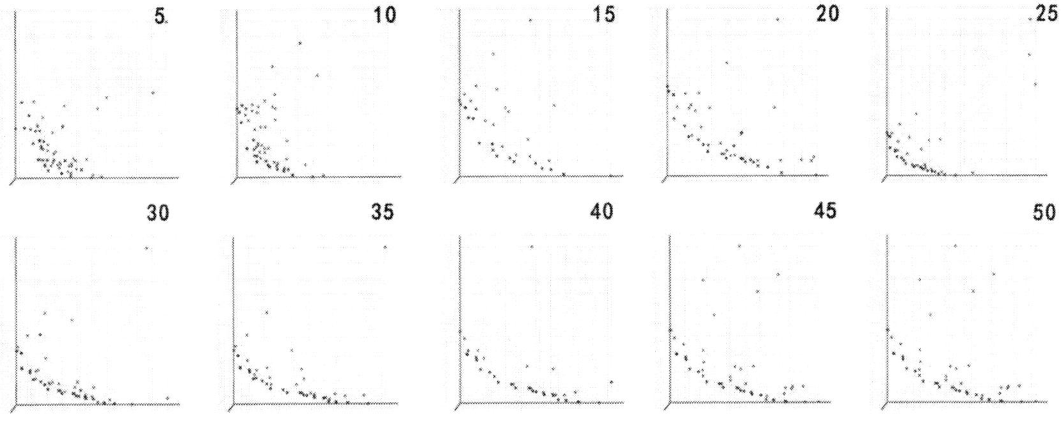

Figure 2. Solving distribution (small)

3 Natural lighting and energy saving design strategy

(1) In the coupling optimization design of natural lighting and energy consumption of medium-sized university gymnasiums, the 50th generation optimal solution analysis shows that the lighting mode should adopt side window lighting, mainly South lighting, or supplemented by less East, West and North lighting, so as to obtain the minimum energy consumption and better effective illumination design values, such as figure 3（a） and （b）; In addition, North lighting can be used as the main lighting, while East, West and South lighting can be used as auxiliary lighting, so as to obtain the optimal effective illumination and less energy consumption design values, such as figure 3 （c） and （d） .

The following are some of the optimal non-dominated solution sets in medium-sized University Gymnasiums:

 https://doi.org/10.1051/e3sconf/202129302044

Table4. Optimal non dominated solution set (medium)

Number	Window to wall ratio (East)	Window to wall ratio (South)	Window to wall ratio (West)	Window to wall ratio (North)	Window to wall ratio of skylight	UDI (300)	Energy consumption per unit area (kWh/m^2)
1	0	0.45	0	0	0	57.3	133.3
2	0	0.25	0.05	0.05	0	69.4	135.1
3	0.05	0.1	0.05	0.2	0	83.9	140.8
4	0.05	0.05	0.05	0.3	0	85.3	145.6

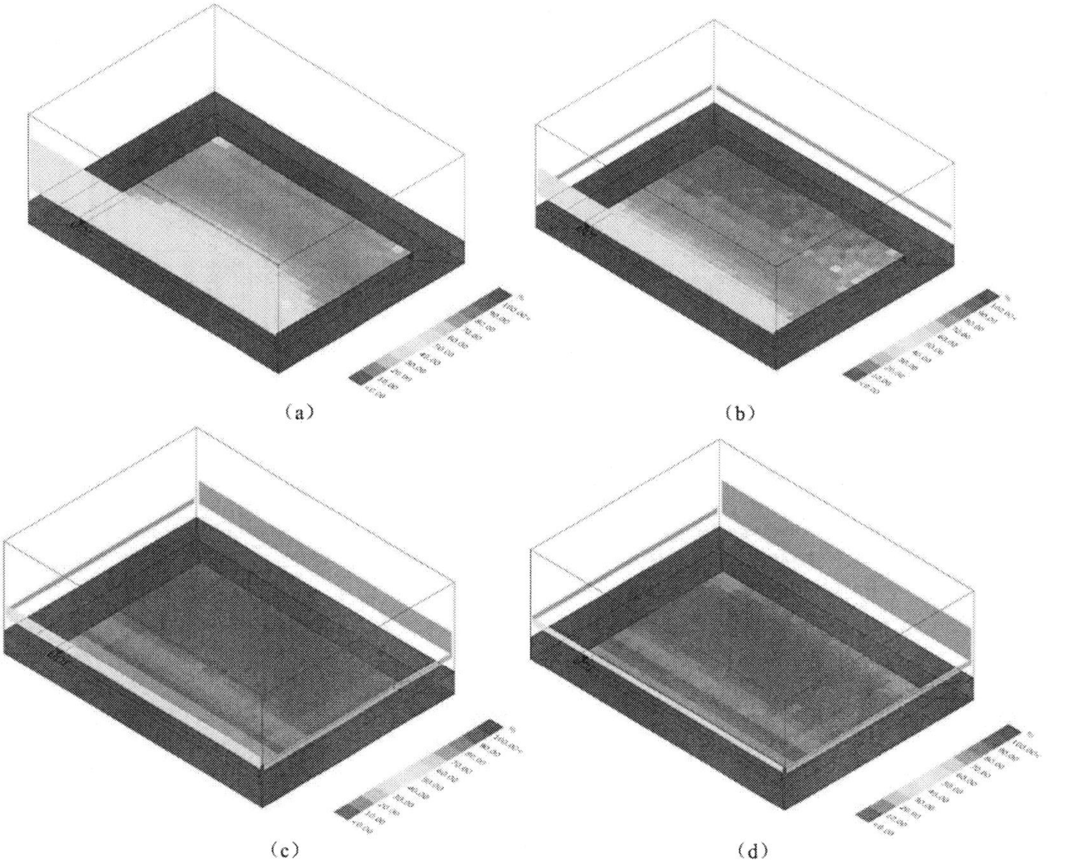

Figure 3. Effective illumination photothermal map (medium)

(2) In the coupling optimization design of natural lighting and energy consumption of small university gymnasium, the 50th generation optimal solution analysis shows that the side window lighting should be adopted in the lighting mode, with the South lighting as the main lighting mode and the East, West and North lighting as the auxiliary. The minimum energy consumption and better effective illumination design values can be obtained, such as figure 4 (a) and (b); It can also be mainly North lighting, supplemented by East, West and South lighting, which can obtain the optimal effective illumination and less energy consumption design values, such as figure 4 (c) and (d).

The following are some of the optimal non-dominated solution sets in small University Gymnasiums:

Table5. Optimal non dominated solution set (small)

Number	Window to wall ratio (East)	Window to wall ratio (South)	Window to wall ratio (West)	Window to wall ratio (North)	Window to wall ratio of skylight	UDI (300)	Energy consumption per unit area (kWh/m^2)
1	0	0.35	0	0.05	0	66.5	118.9
2	0.05	0.2	0.05	0.1	0	78.0	120.8
3	0	0.1	0	0.3	0	85.0	125.2
4	0.05	0.1	0.05	0.35	0	86.8	126.5

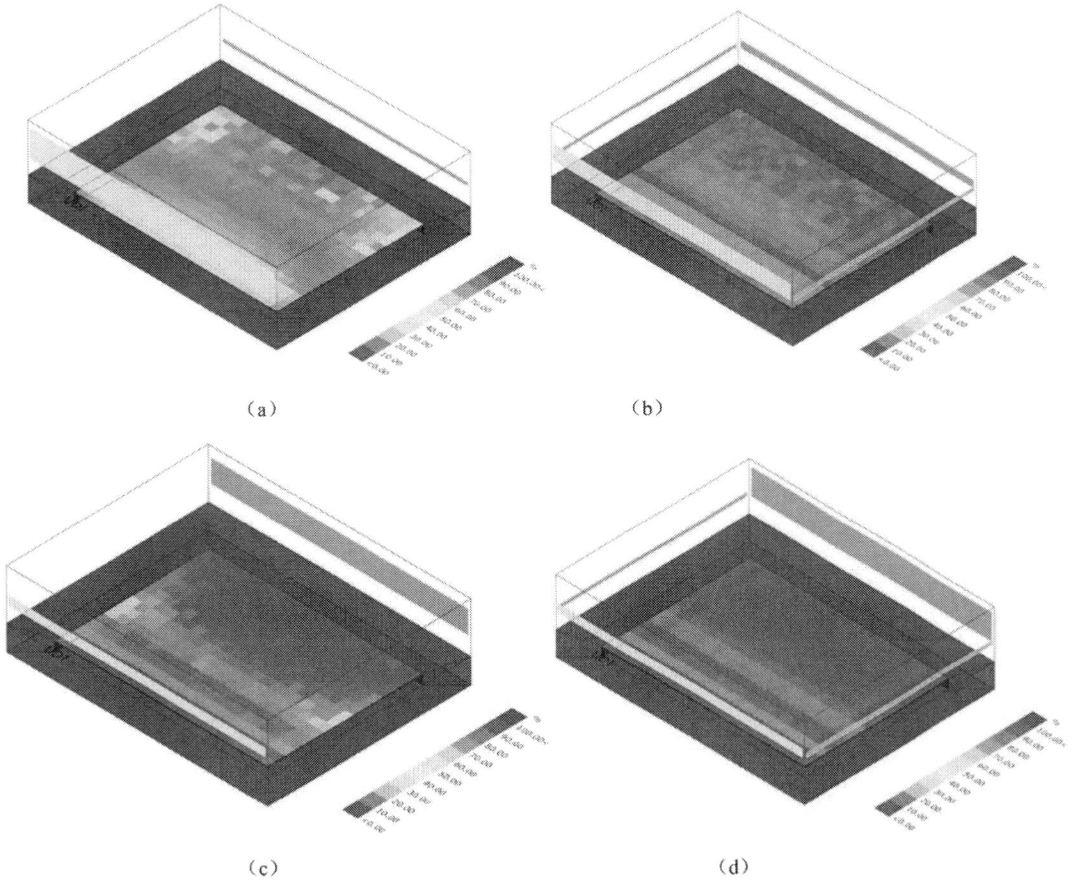

Figure 4. Effective illumination photothermal map (small)

4 Conclusion

Under the background of building energy conservation and emission reduction, the construction of low energy consumption and high comfort University Gymnasium is the general trend. It is more and more important to apply multi-objective optimization design to solve the coupling problem of natural lighting and energy consumption. Based on the sampling survey of University Gymnasium Building, this paper summarizes its basic form, establishes the double objective optimization model of natural lighting and energy consumption by using digital simulation technology, obtains the window opening scheme with consideration of natural lighting and low energy consumption through experiments, reveals its influence law, and summarizes the window opening design strategy of University Gymnasium Building, It provides theoretical basis and design reference for multi-objective performance optimization of large space buildings in cold regions.

Acknowledgments

This paper was completed under the careful guidance of Mr. Zhang Longwei. From the topic selection of the paper, to literature research and data collection, and then to case research, it helped me build up confidence and firm intention to overcome difficulties when I encountered difficulties. I benefited a lot from the profound knowledge and rigorous teaching attitude of Mr. Zhang Longwei.I would like to express my heartfelt thanks to all the teachers, seniors, friends and family members who have helped me.

The authors gratefully acknowledge the support of National Natural Science Foundation of China (Grant 51978418, 51738006) and Scientific Research Plan of the Educational Department of Liaoning Province(Grant lnjc202003).

References

1. Liu Ying. Research on natural light environment design of gymnasium based on value engineering theory [D]. Harbin Institute of technology, 2010.6
2. Bai Xinyu. Preliminary study on light environment evaluation index system of main space of University Gymnasium [D]. Green environmental protection building materials, July 2017
3. Song Zefang, Zhou Yihu. University campus planning and architectural design [M]. China

Construction Industry Press, 2006.9.1[4]Zhang Jian. Study on the influence of building orientation and height on the ventilation behavior of office users [J]. Huazhong architecture, 2011, 29.8

4. Zhang Bin. Study on the optimization design of natural light environment of urban residential buildings in Beijing [d]. Harbin: Harbin University of technology, 2010
5. Liu Yao. Balance judgment of natural lighting and air conditioning energy consumption in public area of large space of railway station [D]. Chengdu: Southwest Jiaotong University, 2015
6. Sun Cheng, Han Yunsong, Zhang Bin, et al. Building of office building environment information model in severe cold area [j]. New building, 2015
7. Ministry of construction of the people's Republic of China. Jgj31-2003 code for design of sports buildings [S]. China Construction Industry Press, 2003.10.01

Influence of Kenyan phonolite manufactured sand on concrete performance

Xuxue Jiao[1,*], Lifeng Liang[1], Xin Li[1], Guodong Fan[1], Xiaojun Chen[1]

[1] China Road and Bridge Co., Beijing 100011, China

Abstract. There is a shortage of river sand and a large number of phonolite deposits around Nairobi, Kenya, but the quality of manufactured sand made from phonolite is poor. In this paper, the effects of phonolite manufactured sand and river sand on the performance of C30 and C50 concrete are studied through the tests of mixture performance, mechanical properties and durability. The results showed that compared with the concrete prepared with river sand, the dosage required of superplasticizer in the preparation of concrete using phonolite manufactured sand was higher, the fluidity was slightly lower, the mechanical properties were basically unchanged, and the durability was slightly better. The above research results preliminarily indicate that it is feasible to use the manufactured sand produced by phonolite around Nairobi, instead of river sand, to prepare concrete. This will reduce the long-distance procurement and transportation of river sand and reduce project construction costs.

1 Introduction

The main line of Nairobi expressway project in Kenya is 27km long, with a design speed of 80km/h, which is a national highway of a grade. The river sand resources around Nairobi are scarce, but the mine resources are rich. The project department considers using the rocks on the mine to produce manufactured sand to prepare high strength concrete. Manufactured sand has been widely used in highway construction in China.

Zheng et al. [1] studied the preparation of C50 prefabricated T-beams with manufactured sand processed from granite in the tunnel, and showed that manufactured sand and gravel aggregates produced from granite can be used to prepare C50 concrete T-beams with excellent workability, mechanical properties and durability, and the stiffness, strength and crack resistance of the concrete T-beam meet the design requirements. Li et al. [2] compared the workability, mechanical properties and durability of concrete prepared by river sand and manufactured sand produced by limestone with a high-stone powder content. The results showed that when the MB value of limestone powder was 0.2 and the content of limestone manufactured sand powder was as high as 15%, it was possible to prepare manufactured sand concrete that had the workability, mechanical properties and durability of RSS (RSS) or exceeded that of RSS, and at the same time, the produce of unilateral concrete could save nearly 40kg of cement. Li et al. [3] studied the application of manufactured sand for the preparation of C50 T-beams, and showed that T-beams did not find cracks and other abnormal phenomena, the force met the design requirements, and was in the elastic working range under the applied load. Gao et al. [4] studied the properties of manufactured sand in 10 bid sections of an expressway, and systematically investigated the preparation technology, workability, mechanical properties and durability of manufactured sand concrete. Zhou [5], Li [6], Li [7] and Yang [8], et al. have studied the influence of different sand-making technology, lithology, and stone powder content on the performance of concrete, all of which can meet the requirements of highway engineering concrete.

The above research provides a certain technical basis for the application of phonolite manufactured sand in highway engineering, but the mines around Nairobi are mainly phonolite mines, which have large crushing value and water absorption, and the research on the application of phonolite manufactured sand in concrete is less. In this paper, a comparative study of phonolite manufactured sand and river sand in Nairobi was conducted to provide technical support for the application of phonolite manufactured sand in concrete of Nairobi expressway project.

2 Material and testing method

2.1 Raw material

(1) Cement: CEM I 42.5 cement whose water consumption of standard consistency is 26%, initial setting is 165min, final setting is 267min, 3d and 28d compressive strength are 27.1MPa and 52.3MPa respectively.

(2) Fine aggregate: the fineness modulus of natural river sand in Kenya is 2.6, and the mud content is 0.7%; the fineness modulus of manufactured sand produced by

* Corresponding author: jiaoxx@crbc.com

moderately weathered phonolite is 2.9, the stone powder content is 6%, the MB value is 1.3, the saturated surface dry water absorption is 2.3%, and the crushing value is 23.7%.

(3) Coarse aggregate: 5mm-25mm continuous graded granite gravel was used.

(4) Fly ash: Grade II fly ash imported from India with a fineness (45 μ m sieve residue) of 14.6%, and a water demand ratio of 98% was used.

(5) The additive is polycarboxylate superplasticizer with solid content of 15.6% and water reducing rate of 21.2%.

2.2 Mix proportion

This experiment referred to the Chinese Standard *General Concrete Mix Proportion Design Regulations (JGJ 55-2011)* to prepare C30 and C50 strength grade concrete respectively. The letter H stood for river sand and J for manufactured sand. The concrete mix proportion of each group was shown in Table 1.

Table 1 Mix proportion of concrete (kg/m^3)

Sample ID	Strength grade	Cement	Fly ash	Fine aggregate	Coarse aggregate	Water	Superplasticizer
C30H	C30	300	70	812	1032	162	3.21
C30J	C30	300	70	812	1032	162	3.54
C50H	C50	440	50	735	1016	160	4.20
C50J	C50	440	50	735	1016	160	4.63

Table 2 Influence of river sand and phonolite manufactured sand on workability of concrete

Sample ID	Strength grade	Superplasticizer (kg/m^3)	Initial slump (mm)	Initial slump-flow (mm)	1h slump (mm)	1h slump-flow (mm)
C30H	C30	3.21	235	575	195	400
C30J	C30	3.54	235	570	170	350
C50P	C50	4.20	240	640	205	460
C50J	C50	4.63	235	625	175	390

2.3 Testing method

The workability of concrete was tested according to the Chinese Standard *Standard for test method of performance on ordinary fresh concrete: GB/T 50080-2016*; the mechanical performance was tested according to the Chinese Standard *Standard for test method of concrete physical and mechanical properties: GB/T 50081-2019*; the durability was tested according to the Chinese standard *Standard for test methods of long-term performance and durability of ordinary concrete: GB/T 50082-2009*.

3 Result and discussion

3.1 Workability

The performance comparison of C30 and C50 concrete mixtures with river sand and phonolite manufactured sand was shown in Table 2. It could be seen from Table 2 that when the initial slump and initial slump-flow of C30 and C50 concrete mixtures of river sand and phonolite manufactured sand were similar, the dosage of superplasticizer for C30 and C50 RSS were 3.21kg/m^3 and 4.20kg/m^3, respectively, and the dosage of superplasticizer for C30 and C50 PMSC (PMSC) were 3.54kg/m^3 and 4.63kg/m^3, respectively, which are nearly 10% higher than that of RSS. This is mainly due to the fact that the dry water absorption of the saturated surface of phonolite manufactured sand was as high as 2.3%, and the phonolite manufactured sand adsorbed a certain amount of superplasticizer and water.

On the other hand, the 1 h slump loss of C30 and C50 RSS were 40 mm and 35 mm respectively, and the 1 h slump-flow loss were 175 mm and 170 mm respectively; the 1 h slump loss of C30 and C50 PMSC were 65 mm and 60 mm respectively, and the 1 h slump-flow loss were 220 mm and 235 mm respectively; the time loss of C30 and C50 PMSC is greater than that of RSS, which was also closely related to the high water absorption of phonolite manufactured sand.

3.2 Mechanical property

The mechanical properties of C30 and C50 concrete mixed with river sand and phonolite manufactured sand at different ages are studied (see Fig. 1). The compressive strength, splitting tensile strength and compressive elastic modulus of C30 and C50 concrete are shown in Fig. 1(a), Fig. 1 (b) and Fig. 1 (c) respectively.

It could be found from Fig. 1(a) that the compressive strength of C30 and C50 PMSC increased steadily with the extension of curing age, which was the same as that of RSS. For C30 concrete, the 3-day, 7-day, 28-day and 56-day compressive strength of PMSC were 107%, 113%, 95% and 96% of RSS respectively, 3d and 7d compressive strength were slightly higher than that of RSS, 28d and 56d compressive strength were slightly lower; For C50 concrete, 3d, 7d, 28d and 56d compressive strength of PMSC were 112%, 102%, 98% and 104% of RSS respectively, meaning that 3d, 7d, 56d compressive strength of PMSC were slightly higher than of RSS, only 28d was slightly lower than that of RSS. Considering C30 and C50 concrete, the compressive strength of PMSC at different curing ages was similar to that of RSS.

Fig. 1 (b) showed that the splitting tensile strength and compressive strength of PMSC had the same development mode, and both of them increased steadily with the extension of curing age. For C30 concrete, the 3d, 7d, 28d and 56d splitting tensile strength of PMSC were 96%, 94%, 102% and 107% of RSS, respectively, the 3d and 7d splitting tensile strength were slightly lower than that of RSS, and the 28d and 56d splitting tensile strength are slightly higher than that of RSS; For C50 concrete, the 3d, 7d, 28d and 56d splitting tensile strength of PMSC were 96%, 94%, 102% and 107% of RSS, respectively. The splitting tensile strength of PMSC was higher than that of river sand concrete in 3d, 7d, 28d and 56d. Considering C30 and C50 concrete, the splitting tensile strength of PMSC was similar to that of RSS.

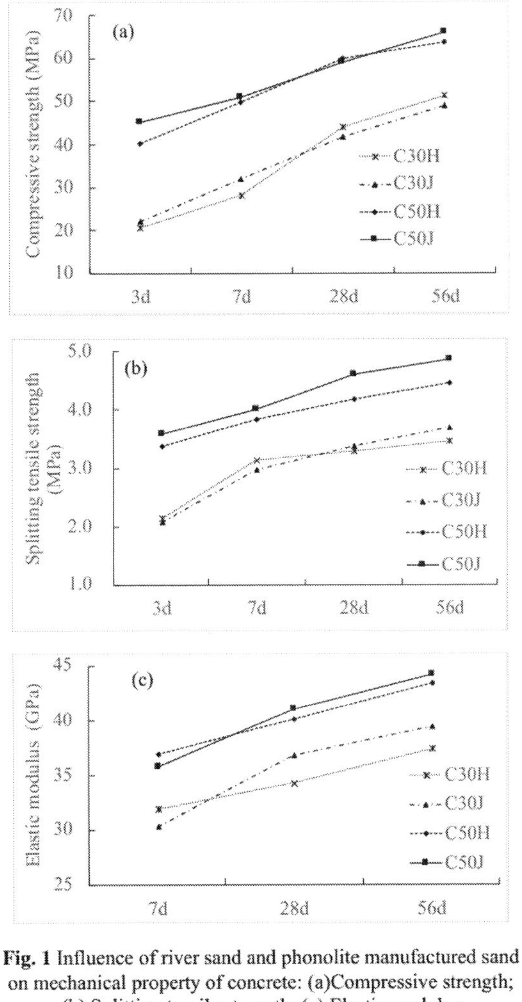

Fig. 1 Influence of river sand and phonolite manufactured sand on mechanical property of concrete: (a)Compressive strength; (b) Splitting tensile strength; (c) Elastic modulus

Fig. 3 (c) illustrated that the elastic modulus and compressive strength of phonolite manufactured sand concrete have the same development mode, and both of them increased steadily with the extension of curing age. For C30 concrete, the 7d, 28d and 56d modulus of elasticity of PMSC are 95%, 107% and 106% of RSS, respectively, the 7d elastic modulus of PMSC was slightly lower than that of RSS, and the 28d and 56d elastic modulus of PMSC were slightly higher than that of RSS; for C50 concrete, the 7d, 28d and 56d elastic modulus of PMSC were 97%, 102% and 1% of RSS, respectively The elastic modulus of PMSC was slightly lower than that of RSS at 7d, and it would not affect the prestressed tension of C50 precast beam. The elastic modulus of PMSC at 28d and 56d were slightly higher than that of RSS. Considering C30 and C50 concrete, elastic modulus of PMSC at 7d was slightly lower than that of RSS, but with the extension of curing age, the growth rate of elastic modulus of PMSC was higher than that of RSS, and the elastic modulus of PMSC at 28d and 56d was higher than that of RSS.

To sum up, although the performance of PMSC was worse than that of RSS, phonolite manufactured sand has more irregular particles, stronger angularity, stronger lap and bite force than river sand. In addition, phonolite manufactured sand contains a certain amount of stone powder, which is conducive to the improvement of mechanical properties [9–11]. Therefore, the compressive strength, splitting tensile strength and elastic modulus of PMSC were similar to that of RSS, indicating that using PMSC to replace RSS would not significantly reduce the bearing capacity of concrete structure.

3.3 Durability

3.3.1 Electric flux

The electric flux of PMSC and RSS in 28d and 56d were shown in Table 3. It could be seen that the electric flux of C30 and C50 PMSC decreased with the extension of curing age, which was the same as that of RSS. For C30 concrete, the 28d and 56d electric fluxes of PMSC were 98% and 94% of RSS, respectively, which were lower than of RSS; For C50 concrete, the 28d and 56d electric fluxes of PMSC are 94% and 93% of RSS, respectively, which are also lower than that of RSS. Considering C30 and C50 concrete, the electric flux of PMSC was slightly lower than that of RSS at 28d and 56d, which was mainly due to the stone powder in manufactured sand promoting the formation speed of cement hydration products [12–14] making the microstructure of concrete more compact.

Table 3 Influence of river sand and phonolite manufactured sand on electric flux of concrete

Sample ID	Eclectic flux(C)	
	28d	56d
C30H	2335	1833
C30J	2283	1714
C50H	1927	1462
C50J	1815	1363

3.3.2 Carbonation resistance

The carbonation resistance of PMSC and RSS was shown in Table 4 For C30 concrete, the carbonation depth of 3d, 7d, 14d and 28d of PMSC were 5%, 6%, 10% and 5% lower than that of RSS; for C50 concrete, the carbonation depth of 3d, 7d, 14d and 28d of PMSC were 10%, 6%, 3% and 7% lower than that of RSS. Considering C30 and C50 concrete, the carbonation depth of PMSC was slightly lower than that of RSS when carbonated for 3d, 7d, 14d and 28d. This was mainly because the manufactured sand contained a certain amount of stone powder, which could optimize the pore structure of hardened paste, reduce the total pore volume of hardened paste, harmful pore volume and multi harmful pore volume [15,16], so it could reduce the carbonation depth of concrete.

Table 4 Influence of river sand and phonolite manufactured sand on electric flux of concrete

Sample ID	Carbonation depth (mm)			
	3d	7d	14d	28d
C30H	4.3	5.4	6.0	6.2
C30J	4.1	5.1	5.4	5.9
C50H	2.1	3.5	3.9	4.6
C50J	1.9	3.3	3.8	4.3

4 Conclusion

(1) The high water absorption of phonolite manufactured sand. On the premise that the fluidity was basically the same, the usage of superplasticizer of phonolite manufactured sand concrete (PMSC) is 10% higher than that of river sand concrete, and the loss of slump with time of PMSC was higher.

(2) The mechanical properties of PMSC were similar to those of river sand concrete due to its large amount of irregular particles, strong angularity and a certain amount of stone powder.

(3) The stone powder in phonolite manufactured sand could promote the formation rate of cement hydration products and optimize the pore structure of hardened paste. Therefore, the durability of PMSC was slightly better than that of river sand concrete.

References

1. X. Zheng, Z. Guan, C. Li, X. Huang, and B. Li, China Concrete Cement Prod. **40** (2020).
2. J. Li, B. Zhang, Y. Wang, C. Gao, Y. Zhou, Y. He, and J. Xia, Bulletin of the Chinese Ceramic Society **37**, 3641 (2018).
3. T. Li, J. Wang, G. Zheng, M. Zhou, and B. Li, Concrete **35** (2009).
4. J. Gao, C. Yu, and G. Li, Highway **155** (2011).
5. H. Zhou, J. Li, R. Li, C. Gao, Y. Zhou, and J. Xia, Build. Sci. **34**, 72 (2018).
6. P. Li, J. Xiong, and S. Wang, Concrete **127** (2012).
7. J. Li, J. Wang, M. Wang, J. Xia, Y. Zhou, C. Gao and H. Zhang, New Build. Mater. **45**, 34 (2018).
8. L. Yang, Y. Tan, Z. Qiu, H. Li, Z. Wang, and Y. Xie, Railway Eng. **119** (2015).
9. M. Guan, G. Wang, Y. Wang, C. Wei, Z. Lai, H. Du, and Z. Liu, Constr. Build. Mater. **269**, 121289 (2021).
10. X. Chen, G. Yuguang, B. Li, M. Zhou, B. Li, Z. Liu, and J. Zhou, Constr. Build. Mater. **240**, 117953 (2020).
11. J. Zhang, D. Li, and Y. Wang, J. Clean. Prod. **258**, 120665 (2020).
12. W. Shen, Y. Liu, Z. Wang, L. Cao, D. Wu, Y. T. Ji, C. Chen, Y. Zhuang, and J. Chen, Constr. Build. Mater. **44**, 422 (2013).
13. W. Shen, Z. Yang, L. Cao, L. Cao, Y. Liu, H. Yang, Z. Lu, and J. Bai, Constr. Build. Mater. **114**, 595 (2016).
14. H. Li, F. Huang, G. Cheng, Y. Xie, Y. Tan, L. Li, and Z. Yi, Constr. Build. Mater. **109**, 41 (2016).
15. B. Li, J. Wang, and M. Zhou, Constr. Build. Mater. **23**, 2846 (2009).
16. X. Ding, C. Li, Y. Xu, F. Li, and S. Zhao, Constr. Build. Mater. **108**, 67 (2016).
17. S. Zhao, X. Ding, M. Zhao, C. Li, and S. Pei, Constr. Build. Mater. **138**, 247 (2017).
18. W. Shen, Y. Liu, Z. Wang, L. Cao, D. Wu, Y. Wang, and X. Ji, Constr. Build. Mater. **172**, 574 (2018).
19. W. Shen, Y. Liu, L. Cao, X. Huo, Z. Yang, C. Zhou, P. He, and Z. Lu, Constr. Build. Mater. **143**, 312 (2017).
20. R. Yang, R. Yu, Z. Shui, C. Guo, S. Wu, X. Gao, and P. Peng, Cement and Concrete Comp. **99**, 203 (2019).
21. Q. Ren, G. D. Schutter, Z. Jiang, and Q. Chen, Constr. Build. Mater. **207**, 218 (2019).
22. H. Yang, D. Liang, Z. Deng, and Y. Qin, Constr. Build. Mater. **188**, 1045 (2018).
23. A. J. Prithika and S. K. Sekar, Constr. Build. Mater. **103**, 1 (2016).
24. B. Li, G. Ke, and M. Zhou, Constr. Build. Mater. **25**, 3849 (2011).
25. P. Nanthagopalan and M. Santhanam, Cement and Concrete Comp. **33**, 353 (2011).

https://doi.org/10.1051/e3sconf/202129302046

Four-Cable-Plane Spatial Cable Stayed Bridge with Two-Amplitude Curved Deck for Canyon-River

Jilin Wang[1]* and JinFa Xu[2]

[1]China Design Group Co., Ltd, Nanjing, China, 210007

[2]Jiangsu traffic engineering consulting and Supervision Co., Ltd, Nanjing, China, 211800

Abstract. On the basis of the original design of Ruck Bridge by Mr. Lin Tongyan, this paper modifies the single-amplitude curved cable-stayed bridge of Ruck Bridge to be an elliptical ring main beam of two-amplitude curved deck, with a duck-egg-shaped arch tower and a spatial cable net with four cable planes, to form a spatial four-cable-plane two-amplitude-curve cable-stayed bridge for the canyon-river topography, so as to improve the structural stress performance of the curved-beam cable-stayed bridge, promote the traffic function and improve the landscape. Combined with the 400m-span New Ruck Cable-stayed Bridge, engineering parameters are designed, a Midas finite element analysis model is established, and the dynamic modal analysis is carried out to verify the structural superiority of this new four-cable-plane spatial cable-stayed bridge with two-amplitude curved deck.

1 Introduction

Under the topographic conditions of canyon rivers, the highway is usually designed as a parallel river. If the curved cable-stayed bridge structure is adopted for the bridge across the river, very favorable conditions can be created for the road line selection, bringing extremely reasonable and economic results [1-2].

As early as 1977, T.Y Lin will provide a design scheme of curved cable-stayed bridge (firm), the firm bridge cable according to the space space hyperboloid anchoring on side slopes, the layout of regular use of the hills on both sides of the superior geological conditions to serve as a powerful anchor structure, in order to balance the load of the bridge, four space on the surface of the cable tension replaces the bridge pier, bear the load of the whole bridge, firm bridge selects the curved cable-stayed bridge was a lot of the stone or tunnel and thus can save millions of dollars, and curved cable-stayed bridge structure itself and the environment to bring out the best in each other beautiful shape [3-4].

The main reasons for the failure of Ruck bridge are as follows: Ruck bridge combines curvilinear main beam with cable-stayed bridge. The structure is complex, and the curvilinear main beam is subjected to the composite action of bending moment, shear force and torque at the same time. The cable design, calculation and analysis are difficult [5-6], and the internal force balance control during construction is very difficult.Ruck bridge cable cable is scattered anchoring on the hillside, the construction is complicated, and the maintenance is quite difficult.Due to the operation effect of the horizontal curved arch structure, the stress of the basic structure is unbalanced[7-8].

River canyon topography of high grade highway, the author of this paper long-span curved cable-stayed bridge across the river, draw lessons from four cable plane cable-stayed bridge of double image design concept, improve firm single curved cable-stayed bridge for elliptic ring type double amplitude curve girder, set the duck eggs arch bridge tower, the arrangement of four network cable plane, space, form a river canyon terrain plane, the space of four double curved cable-stayed bridge, improve the structure performance, improve the traffic function, improve the landscape effect.

Figure 1. Lin Tongyan Ruck bridge design

Combined with the improvement requirements of the original Ruck Bridge design with a span of 400 meters, the structural configuration of the New Ruck Bridge with a four-cable plane and two-amplitude curvilinear cable-stayed Bridge with canyon river terrain was studied,

* Wang Jilin: 37386492@qq.com

engineering parameter design was carried out, and a Midas finite element analysis model was established, so as to verify the superiority of the New structure with a four-cable plane and two-amplitude curvilinear cable-stayed Bridge with a span of four meters.

2 Configuration study

And because the original highway is built along the river and along the mountain, if a straight line bridge is built, the highway on both sides must be changed, will be faced with digging a large number of stone works or tunnel construction, huge investment.

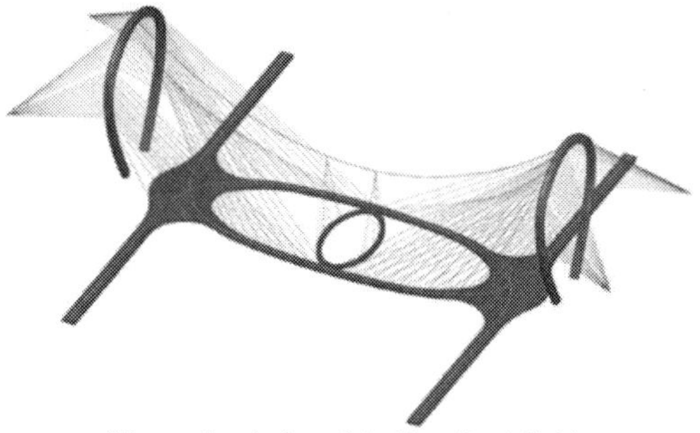

Figure 2. design of the New Ruck Bridge

The design of New Ruck Bridge refers to the design concept of the four-cable-stayed Bridge, and the single-cable-stayed Bridge of the Ruck Bridge is improved to be the main beam of the elliptic-ring double-cable-stayed Bridge. The duck egg arch Bridge tower is set up, and the spatial cable-stayed Bridge of the four-cable-stayed Bridge is arranged to form the double-ring cable-stayed Bridge, so as to improve the traffic function and improve the landscape effect.

Compared with the original Ruck Bridge with a single curved main beam, the New Ruck Bridge adopts the elliptic circular double-curved main beam with symmetric left and right structure, which can greatly reduce the torque in the main beam. The horizontal internal force of the left and right curved main beam is self-balanced, which makes the cantilever construction method feasible indeed, safe, simple and feasible.

Figure 3. Renderings of the New Ruck Bridge

New Ruck Bridge adopts duck egg arch Bridge tower. According to egg-shaped curve equation, steel pipe segments are processed by stewing and bending by fire workers, welded and assembled at the construction site to form egg-shaped steel pipe arch ribs. Egg-shaped steel pipe arch ribs are placed by hoisting egg-shaped steel pipe, and high-strength concrete is poured into the steel pipe to form egg-shaped concrete pipe arch tower.

The duck egg arched bridge tower has excellent mechanical performance and very beautiful shape. The duck egg arched bridge tower has good visual effect and beautiful shape curve. On the hillside, the duck egg arched bridge tower is set up and the hyperboloid space cable cable network with four cable faces is anchored, which is very beautiful and harmonious with the surrounding mountain and canyon environment.

The duck egg arch bridge tower anchors the four-cable plane space cable, and the double-width four-cable plane space curved cable tightly holds the elliptic ring double-curved line main beam, forming the hyperboloid space cable network, which is simple in structure, rich in facade and beautiful in shape.

Elliptic circular double-bent main beam, duck egg arch bridge tower and four-cable space cable are all curvilinear structures, which are cleverly coordinated and graceful, making cable-stayed bridge present a special curvilinear beauty and bringing out the best in each other. The four-cable space cable has a novel and unique shape, which is an ideal combination of structural science and architecture.

Single main cable suspended between two oval arch bridge tower, in the construction phase, a similar cable hoisting method of concrete filled steel tube arch bridge, in the picture around the elliptical ring double curved lines between girders set several temporary connecting steel truss beam, main single cable suspension cable connecting tightly capture steel truss beam, ensure that the elliptical ring double curved girder of the linear segment suspension structure in the process of cantilever construction smoothly, to ensure that the construction of avoid elliptic ring double curved girder structure lifting reverse phenomenon, through the cable hoisting method auxiliary cantilever construction, until the folded cantilever bridge.

Figure 4. Cantilever assembly construction

After a bridge, single main cable suspension cable central set several radial oblique sling, bear the weight of the oval ring beam, safe and reliable to ensure the central oval ring beam stress, the central circle beam as the poles artifacts, limit the lateral deformation of elliptic ring double curved girder, increase the integrity of bridge deck structure and at the same time, the central oval ring beam and a sightseeing platform.

3 Parameter design

The New Ruck Bridge adopts the New technology of cable-stayed Bridge with four cables and two curves in the space of canyon river topography. The design load is highway 2 and the river is 400 meters wide.

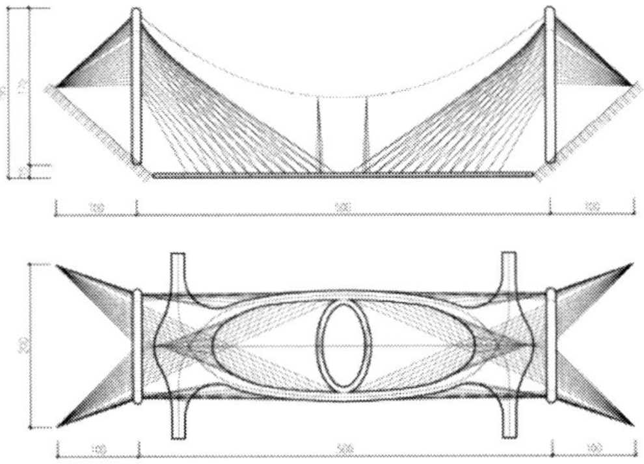

Figure 5. General design of the New Ruck Bridge

The distance between the two arched bridge towers is 500 meters, the height of the arched bridge tower is 170 meters, and the widest part of the bridge tower is 130 meters. The arched bridge tower adopts circular steel tubes with a diameter of 10 meters, the thickness of the steel pipe is 40mm, and C60 concrete is poured inside.

The elliptic circular double-amplitude curved main beam is composed of the elliptic curved deck section, the connecting transition bridge section and the curved cantilever plate. The elliptic curved deck section is the curved composite box girder with single box and double-chamber corrugated steel webs. The beam is 3 meters high and the width of the double-width bridge deck is 10 meters wide.

The connection bridge deck is made of reinforced concrete slab with a thickness of 3 meters.The curved cantilever plate is 0.4 meters thick reinforced concrete slab, and the curved cantilever plate is embedded to support the hillside body.

Elliptic ring type double amplitude curve girder central set elliptic ring beam, oval ring beam supporting elliptic ring double amplitude so that both of elliptical arc of curved girder bridge deck section of the inside of the bridge deck, oval ring beam and a sightseeing platform, oval ring beam for ordinary steel box girder structure, the oval ring Liang Liang is 2.5 meters high, width of 5 meters.

Four cable plane space is divided into four regions anchorage cable on the oval arch bridge tower, four cable plane space network cable from the inside of the cable net and the outer cable network, the inside of the cable net anchorage in the oval curve arch bridge tower inflection point on the top section of the above, the lateral cable network anchor in the oval curve arch bridge tower on top of the inflection point of the following sections, including the inside of the anchor cable network using cross way.

The distance between anchor points on the deck of the four cable planes is 15 meters, and the diameter of the cable is 0.15 ~ 0.20 meters from the end of the beam to the middle of the span. One end of the diagonal cable network is dispersed and anchored on the arch tower of the duck egg, and the other end is divided into two sides and anchored in the anchorage.

The single main cable suspension cable is arranged between two arched bridge towers. The diameter of the single main cable suspension cable is 0.5 meters for the finished cable of high strength steel wire.

After the completion of the bridge, the temporary connecting beam was removed, and the single main cable suspension cable was set with a radial sling. There were a total of 12 high-strength steel cable cables, with each cable having a diameter of 0.1 meters and an elliptic ring beam in the middle of the suspension.

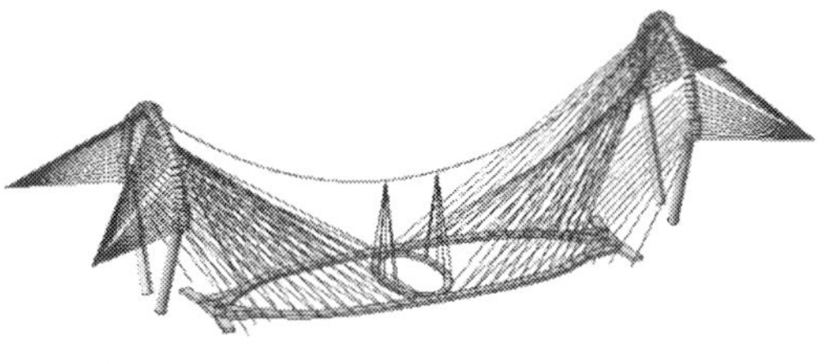

Figure 6. Midas finite element model

4 Calculation results under vertical load

The main span bridge deck is loaded with full load, the additional constant load of the bridge deck adopts the uniformly distributed load standard value of 5 kN/m2, and the live load of the bridge deck adopts the highway second-level load standard value. The internal force of the stiffened beam under direct load is analyzed in the model.

The calculation results are as follows:

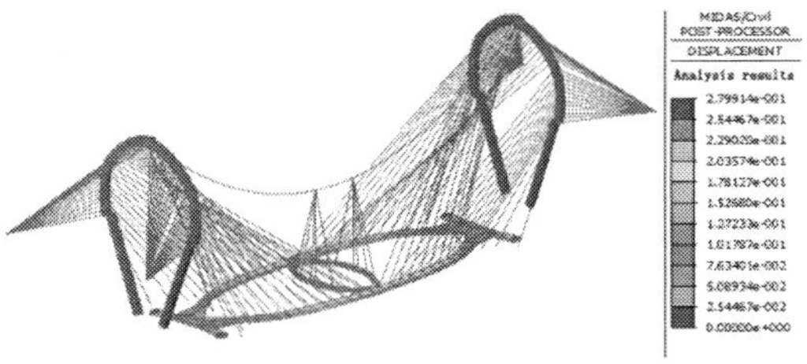

Figure 7. Calculation results of displacement under vertical load（dead load+live load）

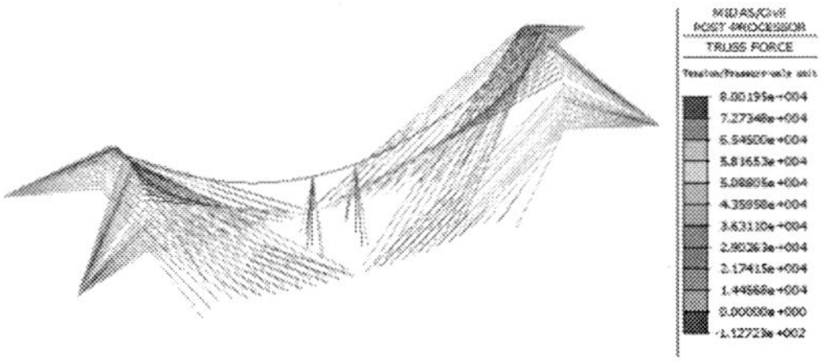

Figure 8. Calculation result of cable internal force（dead load+live load）

 https://doi.org/10.1051/e3sconf/202129302046

Figure 9. Calculation results of cable stress (dead load+live load)

The maximum vertical displacement occurs in the middle of the span, and the maximum displacement is 0.280m, which meets the requirement of 1/500 limit specified in the specification.

The internal force of single main cable suspension is 80019kN, and the stress is 407MPa, which meets the strength requirements.

The maximum internal force of the stay cable is 18758kN, and the maximum stress of the stay cable is 933.0mpa, which meets the strength requirement.

The analysis shows that: under the vertical load, the stress of New Ruck Bridge is symmetrical, and there is no great torsion deformation.

5 Dynamic modal analysis

Cable is the main load-bearing structure of cable-stayed bridge, the influence of gravity stiffness must be considered in the dynamic characteristic analysis, and the internal forces of main cable and sling should be calculated in the form of initial tension force in the modeling.

Based on Midas nonlinear static analysis, stress-rigidity effect and modal analysis function, the natural vibration characteristics of the structure were analyzed. In order not to omit any mode, the subblock method is used to solve the characteristic equation.

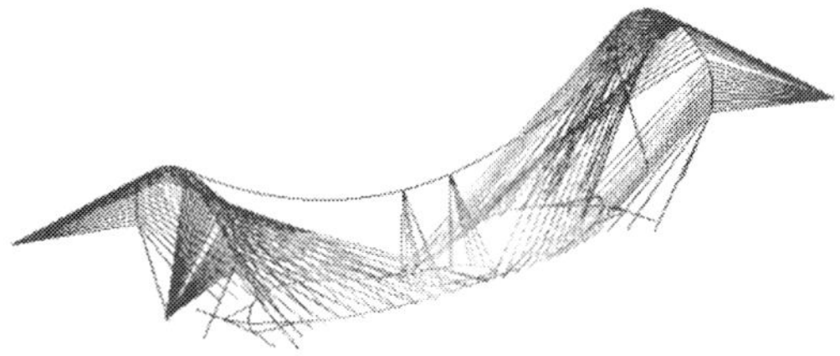

(a) 1 order mode (0.740HZ)
(Positive symmetrical side bend)

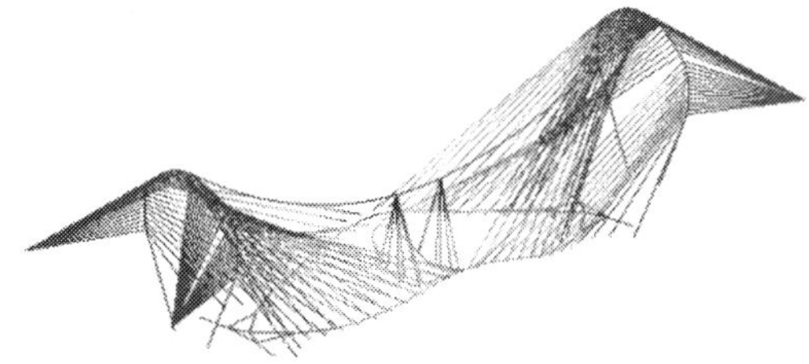

(b) 2 order mode (0.790HZ)
(Positive symmetric vertical bending)

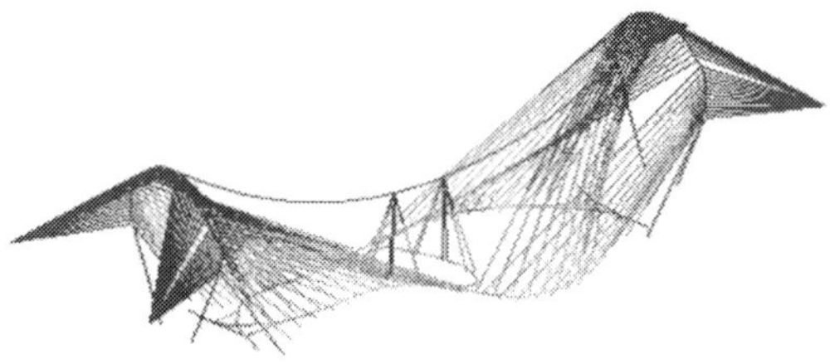

(c) 3 order mode (0.896HZ)
(Antisymmetric side bend)

(d) 4 order mode (0.925HZ)
(Antisymmetric side bend)

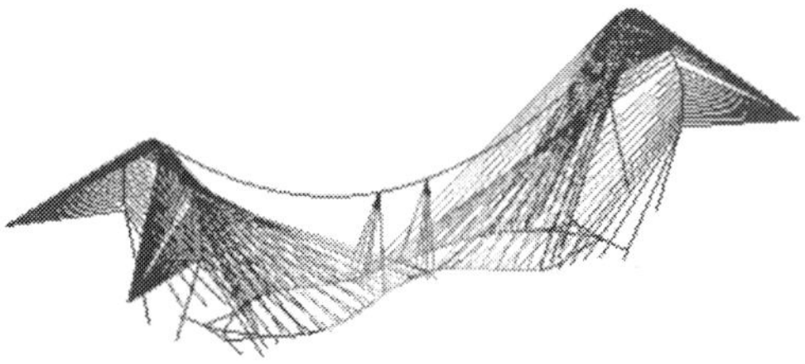

(e) 5 order mode (1.035 HZ)
(Positive symmetric vertical bending)

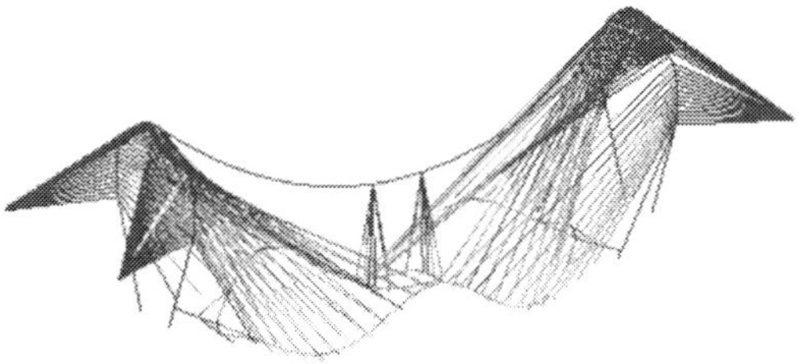

(f) 10 order mode (1.358HZ)
(The deck to reverse)

Figure 10. Typical vibration mode

According to the calculation results in the figure above, the first-order mode is positively symmetrical lateral bending with a high frequency of 0.740Hz.The second mode is the positive symmetrical vertical bend and the frequency is 0.790Hz, indicating that the vertical and lateral stiffness of the New Ruck Bridge structure is relatively large.The 10th order mode is torsional, and the positive symmetrical torsional frequency is 1.358Hz, which indicates that it has good torsional stiffness and strong spatial integrity of the structure.

On the whole, the vibration modes are dense, and there is an obvious pattern grouping phenomenon. The first 10 vibration modes are mainly side bending and vertical bending vibration, and the positive symmetrical torsional vibration mode does not appear until the 10th order. The torsional bending frequency ratio of 1.84 is high, indicating that the structure has a good wind-resistant stability.

6 Conclusion

In this paper, a Midas finite element model was established against the background of the New Ruck Bridge, a canyon river topography with a span of 400 meters. The static analysis was carried out under vertical load, and the dynamic modal analysis was carried out, and the following conclusions were drawn:

(1) to improve the original firm single curved cable-stayed bridge for elliptic ring type double amplitude curve girder, set up the duck eggs arch bridge tower, the four cable plane space network cable arrangement, four cables of cable plane space curved surface has kept a elliptic ring double curved line girders, form the river canyon terrain space plane, four double curved cable-stayed bridge, the curved beam type cable-stayed bridge structure performance is improved and improve the traffic function, improve the landscape effect, construction is convenient, beautiful modelling.

(2) The elliptic circular double-bent main beam, duck egg arch Bridge tower and four-cable surface space cable network of New Ruck Bridge are all curved structure forms, with clever coordination and simple structure. The four-cable surface space cable is novel and unique in modeling, which is an ideal combination of structural science and architecture.

(3) Under the action of vertical constant live load, the maximum vertical displacement occurs in the middle of the span, and the maximum displacement is 0.280m, which meets the requirements of the specification.

The internal force of single main cable suspension is 80019kN, and the stress is 407MPa, which meets the strength requirements.

The maximum internal force of the stay cable is 18758kN, and the maximum stress of the stay cable is 933.0mpa, which meets the strength requirement.

Under the vertical load, the stress of New Ruck Bridge Bridge structure is symmetrical, without great torsion displacement and stress.

(4) the first ten order modes are mainly composed of lateral and vertical bending vibration, the first order vibration mode of symmetric lateral bending, the fundamental frequency of 0.740 Hz is higher, shows that the New firm Bridge, the Bridge structure of the vertical and lateral stiffness is bigger, didn't appear until the 10th order is symmetric torsional vibration model, are symmetric torsional frequency of 1.358 Hz is higher, twisted the frequency ratio of 1.84 is higher, shows that the New firm Bridge, the Bridge structure space integrity is strong, the structure stability of wind resistance is better.

References

1. H. Bangben, Introduction to the Design of THE American Ruck-A-Chuky Bending bridge [J].Foreign Highways,1980(01):58 (in Chinese)
2. Takenouchi, Kyo. Aesthetics of cable-stayed bridges with spatial cable form -conoid ruled surface, along a curved deck[J]. Journal of the International Association for Shell and Spatial Structures, 1997,38,(124),91-98.
3. HUNG, G. M., & STUCCHI, F. R. Conception of cable-stayed curved deck: the effects of unilateral suspension / Concepção de tabuleiros curvos e estaiados: os efeitos da suspensão unilateral[J]. Revista IBRACON de Estruturas e Materiais, 2018,11(4), 686–695.
4. Galante Bardín, D., & Astiz Suárez, M. L. A. (n.d.). Design of a curved cable-stayed bridge deck's directives, by the funicular and antifunicular curves of the concentric load introduced by the cables[J]. nformes de La Construccion,
2017, 69(548), 1–14.
5. Raftoyiannis, I. G., & Michaltsos, G. T. (n.d.). Curved-in-plane cable-stayed bridges: A mathematical model[J]. International Journal of Structural Stability and Dynamics,2012, 12(3).
6. Hoffman, J. J. Analytical and field investigation of horizontally curved girder bridges[D]. Iowa State University,2013.
7. Lin, W., & Yoda, T. Analysis, design and construction of curved composite girder bridges: State-of-the-art. International Journal of Steel Structures,2010, 10(3), 207.
8. Z. Erhua, S. Deshan, Z. Yongtao.Long-span curve girder cable-stayed bridge Π shape analysis method research [J]. World bridge, 2017, (5) : 44-48. (in Chinese)

Building integrated photovoltaic design of nonlinear airport roof based on ruled surface fitting

Jinting Sui[1], Longwei Zhang[1,*]

[1]School of Architecture and Urban Design, Shenyang Jianzhu University, 110168 Shenyang , China

Abstract. Solar energy is a new renewable energy. Using solar energy to generate electricity is an effective means to achieve sustainable development. The huge roof at the top of the airport is a natural interface to receive solar energy, which can be used to supply the energy needed by the airport through building integrated photovoltaic design. However, due to the curved shape of the airport roof, there are limitations in solar energy acquisition and construction of the airport roof. In this paper, the method of "ruled surface fitting-solar radiation analysis-modular unit arrangement" is used to optimize the shape of the non-linear airport roof, and convert the complex roof surface shape into a straight line with good sunlight performance and easy construction. The ruled surface can better integrate BIPV and nonlinear airport, and improve the solar power generation efficiency of the building, thus improving the sustainable performance of the building.

1 Background

With the rapid development of economy, the concept of sustainable development has become increasingly popular in the world. Many countries regard solar energy as an important emerging industry. The scale of global solar energy development and utilization is rapidly expanding, technology keeps improving and cost is significantly reduced, showing a good development prospect. The introduction of building integrated photovoltaic is not only in line with China's strategic planning for sustainable development of energy in the 21st century, but also an important part of developing a circular economic model and the construction of a harmonious society. this is also of great significance to promote the use of solar energy and the development process of the photovoltaic power generation industry.

By the end of June 2020, the number of registered airports in China has reached 296, and the number of airports in China is growing exponentially. With the rapid growth of airport construction, the problem of high energy consumption is becoming more and more prominent. In order to meet the airport energy demand without damaging the ecological environment, the use of solar power generation, especially the large-scale development and application of BIPV is particularly important. The huge building volume and large roof area of the airport are favorable conditions to apply the buildings integration photovoltaic in the airport. First of all, the airport roof area is large, the shape is usually gentle and complete, and the complexity of the form is low, which meets the requirements of photovoltaic integration for uniform and smooth interface receiving light. Secondly, the airport covers a large area, has a low building density in the surrounding environment and an open site. Therefore, the solar radiation receiving surface is less blocked by adjacent buildings, which can effectively improve the irradiation time and irradiation quality of the roof exposed to light interface. Therefore, as to a civil transport airport with vast roof and huge annual energy consumption, it can effectively alleviate the energy consumption pressure of the airport by integrating it with photovoltaic buildings.

This research takes "non-linear airport roof photovoltaic integration" as the research direction, applies photovoltaic integration technology on the airport roof, constructs a new type of nonlinear airport photovoltaic roof system, and further summarizes a set of airport roof BIPV implementation method, which allows the non-linear roof shape to be organically combined with the planar photovoltaic unit to solve the problem of combining BIPV with non-linear airports. This article takes the corridor type of typical airport roof as an example, and elaborates the design method of airport photovoltaic roof in detail.

2 Topology optimization of macroscopic shape

Based on the nonlinear airport roof shape, this paper introduces the ruled surface architecture logic under the premise of meeting its form and function requirements. By adding wires and bus bars, the nonlinear airport roof is transformed into a ruled surface, so that the nonlinear shape is constructed by geometrical logic construction, which lays the foundation for the later arrangement of photovoltaic panels. Secondly, using the solar radiation analysis software Ladybug and the multi-objective genetic algorithm tool Octopus, the topological optimization of

* Corresponding author: z_lw@sjzu.edu.cn

the shape is carried out on the basis of solar radiation, prototype fit, and construction implementation degree [1], and the optimal solution is selected.

2.1 Parameterized geometric logic construction

Most of the non-linear airport roof are free-form surfaces, which are complex, diverse and irregular. Direct application of photovoltaic panels cannot fit free-form surfaces. Therefore, the primary basis for the combination of the two is to introduce geometric logic to reconstruct the nonlinear airport roof, and use ruled surface as the transition between the two. The ruled surface is composed of wires and straight bus bars. In the wire dimension, it is a free curve to control the straight lines. The spatial shape and contour of the curved surface ensure the shape of the free-form surface. The dimension of the bus bar is a straight line with certain regularity. The arrangement method controls the shape of the free-form surface to ensure the reasonable fitting of the photovoltaic unit.(Figure 1).

Taking the typical form of airport roof as an example, the integration process of nonlinear airport roof and photovoltaic building integration is further illustrated as follows. Firstly, based on the form and function of the nonlinear airport roof, the nonlinear morphological features are analyzed, and the upper and lower boundary lines are extracted as the wires of the ruled surface. Secondly, the planned copper indium gallium selenide photovoltaic module size 600*1200 is set as the unit modulus. Based on the grasshopper platform, combined with the airport structure, it is divided horizontally with 600mm as the modulus along the wires through the vertical generatrix, and 1200mm as the modulus for vertical division. Then the free-form surface is divided into several 600*1200 units according to geometric logic. The horizontal and vertical division transforms the initial nonlinear surface into a ruled surface with geometric relations, generates a geometric model with parameterized logic, and transforms the complex and irregular free-form surface into a relatively simple and easy-to-control dynamic model [2].

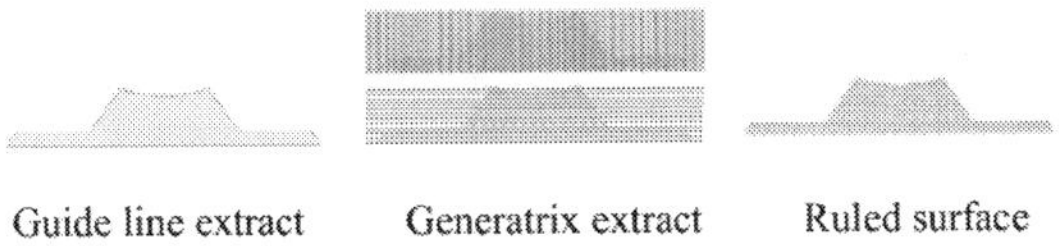

Guide line extract Generatrix extract Ruled surface

Figure 1. Geometric logic construction of ruled surface.

2.2 Solar radiation analysis

The analysis of sunlight acquisition is the premise and foundation of the integration of solar photovoltaic application. Under the premise of ensuring the nonlinear form of the building, the main goal of the combination of BIPV and airport roof is to achieve the maximum solar radiation acquisition. The LadyBug tool based on Grasshopper platform is used to simulate the solar radiation amount of roof [3] and generate the distribution map of annual sunshine access, which provides data support for roof shape optimization to obtain the maximum solar radiation amount [4].

In this essay, the genetic algorithm tool Octopus is used to generate a large number of alternative roof forms with the annual solar radiation as the target, and automatic screening is conducted to obtain three optimal surfaces (Figure 2). On this basis, solar radiation amount, prototype fit degree, construction implementation degree and space comfort degree are taken as the criteria to compare the three:

Scheme 1: The annual solar radiation of this form is 1,753.91kWh /m^2. Although the solar radiation obtained per unit area is relatively high, the agreement between this surface and the initial form of the airport roof is relatively low. In addition, due to the large protrusion on the upper part, the internal space of the terminal building is wasted, and the space utilization rate is low. Meanwhile, the local protrusion is too large, and the continuity of the curved surface is not strong.

Scheme 2: It has a high fit with the original form of the airport, but the annual solar radiation of this form is 1688.49kWh /m^2, and the amount of sunshine obtained is lower than that of other schemes.

Scheme 3: The annual solar radiation of this body is 1691.02 kWh /m^2, the annual solar radiation acquisition is high, and the initial form of the airport roof has a high fit degree, the curved surface has strong continuity, and the space utilization rate is high.

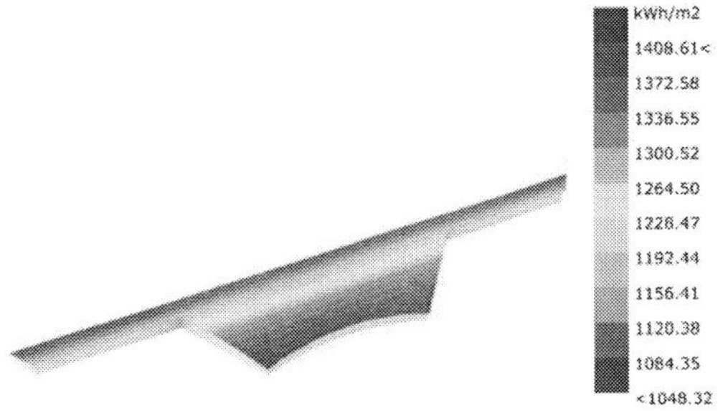

Scheme1 Roof axonometric

Scheme1 Solar radiation data

 https://doi.org/10.1051/e3sconf/202129302047

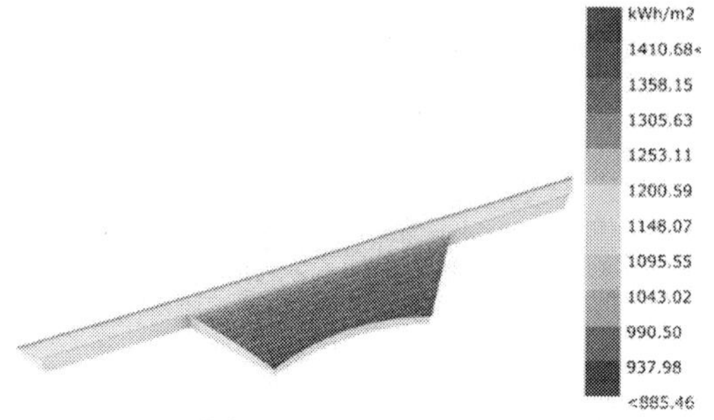

Scheme2 Roof axonometric

Scheme2 Solar radiation data

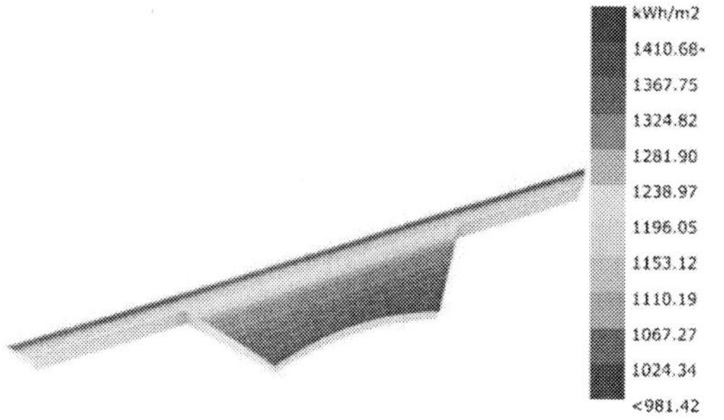

Scheme3 Roof axonometric

Scheme3 Solar radiation data

Figure 2. Typical optimization scheme and solar radiation data

The analysis shows that the scheme has high solar radiation acquisition amount, good morphology fit, full sunlight receiving area and the best comprehensive performance in the three years. It has a good optimization space and is suitable for the foundation of nonlinear airport roof.

3 Mesoscopic shape surface partition

Based on Grasshopper platform, interval processing was carried out on the sunshine analysis data, gradient partition was carried out according to the solar irradiance, and regions were divided on the roof surface [5], which were divided into areas with sufficient sunshine, intermittent sunshine and insufficient sunshine. In order to meet the basic shape and function requirements of the roof, the form of the intermittent sunshine area was partially adjusted to convert it into a full sunshine area, and the laying area of photovoltaic panels was increased.

3.1 Airport roof surface partition

On the one hand, the surface partition of the airport roof determines the functional partition according to the structure and function. According to the function of the corridor type airport, it is divided into three areas, namely, two corridor areas and one terminal area. On the other hand, the photovoltaic installation area is determined through annual solar radiation analysis. According to the sunshine analysis data above, the annual solar radiation acquisition can be roughly divided into three gradients. The red gradient is the area with sufficient sunshine, and the annual solar radiation acquisition is greater than 1000kWh /m². Yellow gradient for intermittent sunshine area, annual solar radiation acquisition in 800-1000kWh / m²; The blue gradient is the insolation lack area, the annual solar radiation acquisition amount is less than 800kWh / m²; The straight grain surface is divided into three regions to form three straight grain surfaces. The rationalized airport roof surface partition is generated from the above two aspects(Figure 3).

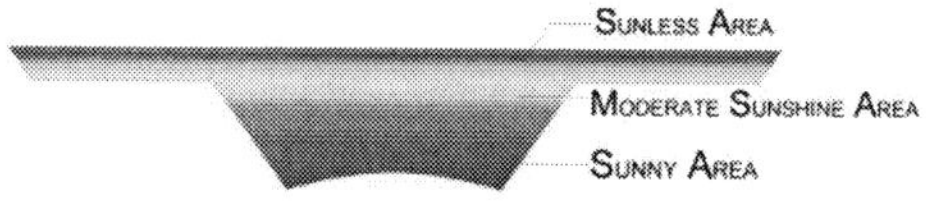

Solar radiation partition

Function partition

Figure 3. Surface partition.

3.2 Shape partition optimization

The red gradient does not need to be adjusted due to high solar radiation. The blue gradient solar radiation is too low, which can only be satisfied if the body changes greatly, and the adjustment meaning is low. On the basis of the local adjustment of the shape, the yellow gradient can be transformed into a full sunshine area, which has great adjustment value. In this paper, genetic algorithm is used to make local adjustment of the surface of intermittent sunshine zone with the objective of maximizing the annual solar radiation acquisition [6]. At the same time, using Kangaroo tool of Grasshopper platform, the intermittent sunshine area surface edge is fixed, to ensure that the body and the full sunshine area and the lack of sunshine smooth connection. Through dynamic feedback model real-time adjustment, the intermittent sunshine area is effectively converted into a full sunshine area, thus improving the overall body's annual solar radiation acquisition. In the process, the bucket effect of photovoltaic should be paid attention to, which refers to the photovoltaic power generation system in general with 6-9 photovoltaic cell to form a set of photovoltaic (pv) set of strings, and the group list of photovoltaic cell within the power generation efficiency should be close. Otherwise, similar to the barrel theory, the whole power generation efficiency is determined by photovoltaic power generation efficiency with the lowest unit, thus forming a greater influence of the uniformity on the photovoltaic efficiency. Based on the above analysis, two variables, annual solar radiation amount and body uniformity, should be taken as optimization objectives when adjusting the body locally, so as to effectively improve the efficiency of photovoltaic power generation.

4 Microscopic shape modularization

Optimized by the above steps, the roof shape is transformed into two large areas -- a sunny area and a lack of sunshine area. Firstly, photovoltaic units are introduced in the area with sufficient sunshine. Based on Grasshopper platform, photovoltaic units are taken as modules to modularize the surface of the body. The sunshine deficiency area is laid with ordinary material units. Secondly, genetic algorithm tools were used to optimize the sunlight incidence angle of each photovoltaic module [7] according to the sun angle to form a scale-type roof arrangement. The arrangement of modules reflects the geometric construction logic of the overall shape of the roof, and reduces the difficulty of construction.

4.1 Standardized module unit

Standardization of module units plays a crucial role in the construction and implementation of the scheme. If the body fit degree is completely pursued, the size of each photovoltaic unit is different, which greatly increases the construction period and difficulty. Therefore, standardization of module units is needed. In the shape topology optimization stage, although the initial horizontal and vertical partitioning was carried out through the 600*1200mm size of photovoltaic cell, the size changed during the fitting process. Therefore, the Kangaroo tool of Grasshopper platform is used to introduce dynamics calculation into GH and achieve interactive simulation, form-finding optimization, constraint solution through physical and mechanical simulation. The tool controls four factors: edge, side length, control point and rotation angle, and standardizes it to 600 * 1200 mm by using the principle of elasticity. After optimization, although the size of each unit fluctuates slightly, the value is within the allowable range of construction error, reaching the standard module unit(Figure 4).

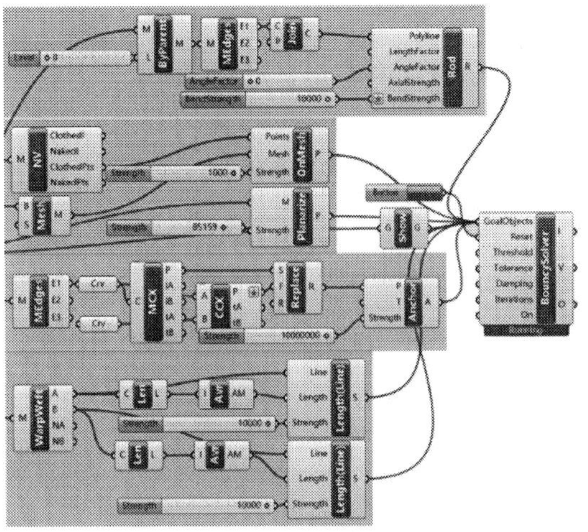

Figure 4. Standardized unit program.

4.2 Photovoltaic cell detail optimization

After the fitting of the photovoltaic cells and the nonlinear airport roof, the shading between the photovoltaic cells and the rotation angle of the photovoltaic panels have a crucial effect on the efficiency of photovoltaic power generation. In order to achieve the maximum solar radiation acquisition, each photovoltaic unit needs to rotate at a certain angle to receive sunshine. At the same time, in order to avoid the shadow blocking between photovoltaic panels affecting the efficiency of photovoltaic power generation, a shadow transition zone is introduced between photovoltaic panels. Under the premise of keeping the large structure of the roof shape unchanged, the Octopus tool was used for multi-objective optimization. The optimization variables are the rotation angle and the width of the shadow transition band. Considering the uniformity and practicability of the roof shape, the rotation Angle range of each photovoltaic module is controlled within $\pm 5°$ compared with the original surface, and the width of the shadow transition band is controlled within 100mm. By adjusting the rotation angle of the photovoltaic unit and the width of the shadow transition zone, the material consumption is minimized on the premise of guaranteeing the annual solar radiation acquisition, and the economic efficiency and realizability of construction are increased. After several iterations through the Octopus tool, the final optimal shape of the airport roof is formed, and the organic combination of photovoltaic integration and nonlinear airport roof is realized(Figure 5).

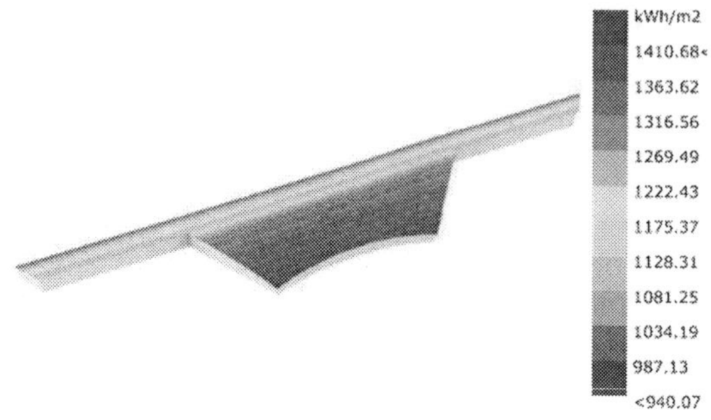

Roof axonometric

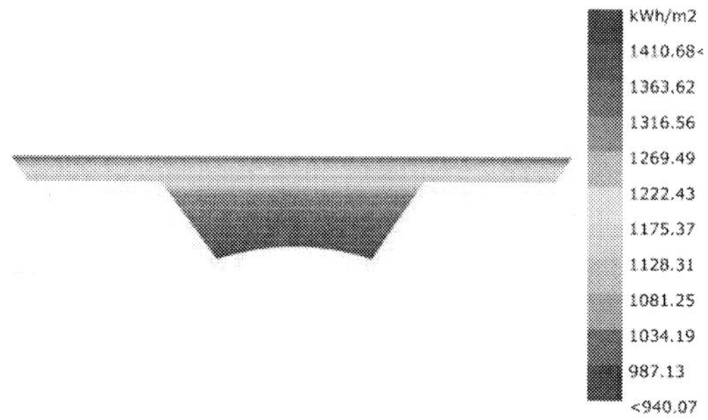

Solar radiation data

Figure 5. Final optimization scheme.

5 Summary

This essay proposes a nonlinear airport roof optimization method based on ruled surface fitting methods, which can transform complex and irregular roof free-form surfaces into ruled surfaces with good sunlight performance and easy construction, making better integration of BIPV and nonlinear airport.

(1)Establish geometric logic, and implement the horizontal and vertical fitting of the airport roof by introducing a 600*1200 modulus to realize the preliminary standardization of the roof shape. At the same time, the dynamic parameter frame is used to realize the data linkage from the whole to the part of the roof shape, which lays the foundation for the optimization of the genetic algorithm based on the grasshopper platform, and realizes the shape-level fitting of the nonlinear roof shape.

(2) The sunshine data of the airport roof shape is analyzed by the Ladybug tool, and by gradient partitioning, it is divided into sunshine sufficient area, sunshine intermittent area and sunshine insufficient area, which provides quantitative basis for further optimizing the shape. Through zone fitting and local adjustment of the shapes of intermittent sunshine areas and low sunshine areas, group-level fitting of nonlinear roof shapes is realized, which effectively increases the photovoltaic paving area and improves the efficiency of photovoltaic power generation.

(3)The standardization of photovoltaic module units is an important factor affecting the feasibility of construction. Through the Kangaroo tool, the principle of elasticity is used to standardize it, and at the same time, based on the Octopus tool, with the annual solar radiation acquisition as the optimization goal, the optimal rotation angle and transition zone width of the photovoltaic unit are obtained. From macro to micro, it realizes the structural-level fitting of nonlinear roof shapes, and transforms complex free-form surfaces into a highly feasible nonlinear integrated photovoltaic airport.

Non-linear airport roofs provide us with a new way of efficiently using solar energy from the perspective of renewable energy utilization. Because of its huge solar receiving surface, it has a huge advantage in the application of photovoltaic power generation. Whether it is the reconstruction of existing airports or the construction of new airports, it is easy to form a cluster effect, and does not occupy land, which greatly reduces the investment cost of photovoltaic facilities and alleviates the problem of high energy consumption in the airport. It has a large application prospect and development potential, and complies with China's Energy Strategic Plan for Sustainable Development in the 21st Century.

Acknowledgments

The authors gratefully acknowledge the support of National Natural Science Foundation of China (Grant 51978418, 51738006) and Scientific Research Plan of the Educational Department of Liaoning Province(Grant lnjc202003).

References

1. J. Harding, S. Joyce, P. Shepherd, C. Williams, Thinking topologically at early stage parametric design, in: Adv. Archit. Geom., Springer, Vienna, 2012, pp. 67–76.
2. Negendahl, Building performance simulation in the early design stage: an introduction to integrated dynamic models, Autom. Constr. 54 (2015) 39–53.
3. Qingsong M, Fukuda H. Parametric Office Building for Daylight and Energy Analysis in the Early Design Stages. Procedia - Soc Behav Sci 2016;216:818–28.
4. Longwei Zhang. Lingling Zhang , Yuetao Wang, Shape optimization of free-form buildings based on solar radiation gain and space efficiency using a multi-objective genetic algorithm in the severe cold zones of China, Solar Energy 132 (2016) 38–50.
5. Y.G. Yi, A.M. Malkawi., 2009. Optimizing building form for energy performance based on hierarchical geometry relation, Automation in Construction. 18. 825–833.
6. Jin, J.T., Jeong, J.W., 2014. Optimization of a free-form building shape to minimize external thermal load using genetic algorithm. Energy Build. 85, 473–482.
7. Ka¨mpf, J.H., Montavon, M., Bunyesc, J., Bolliger, R., Robinson, D., 2010. Optimisation of building's solar irradiation availability. Sol. Energy 84, 596–603.

Research on safety risk early warning of tunnel construction based on BIM and RFID Technology

Chen Yong [1], He Xudong [1], Lin Guojun [2], Wu Ping [3*]

[1]China Railway Construction Group Co., Ltd., 100040 Beijing, China

[2]Ninghai Transportation Bureau,315600 Ningbo, China

[3]School of Civil Engineering and Architecture, NingboTech University, 315100 Ningbo, China

Abstract. Due to the frequent occurrence of safety accidents in tunnel construction, it poses a great threat to property safety and the life safety of construction personnel. According to the classification of accident types, this paper standardizes the data of influential risk factors, delimits the boundary of dangerous areas, and proposes the specific steps of inputting field information into BIM database. The results show that the combination of BIM and RFID technology can monitor the conditions of the construction site in real time, and realize the automatic early warning of the dangerous sources in the tunnel construction.

1 Introduction

Underground engineering construction is a high-risk construction project with the characteristics of large investment, long cycle, complex technology, unpredictable risk factors and large impact on social environment[1]. Common tunnel accidents can be summarized into eight categories: fire and explosion accidents, tunnel collapse accidents, implement injury accidents, electric shock accidents, poisoning and asphyxiation accidents, falling accidents from high places and object striking accidents.

Building Information Modelling and Internet of Things are information technology developed by the country in recent years. BIM technology has the characteristics of integration and visualization[2], which is of great significance for improving engineering quality and efficiency, reducing cost, and realizing resource dynamics and sharing. Internet of Things technology can effectively monitor the construction site through wireless sensing, which can provide reliable data. Therefore, the combination of BIM and the Internet of Things can achieve the purpose of remote automatic monitoring and real-time early warning, hence reducing construction safety risks, reducing the occurrence of accidents, ensuring the safety of construction personnel and reducing economic losses.

By establishing tunnel construction safety risk early warning database, this study finally formed and constructed a tunnel construction safety early warning platform based on BIM and RFID.

2 warning database for tunnel construction based on BIM and RFID

RFID technology is used to mark the location of construction site and locate the dynamic location of operators. Through the real-time interaction between the collected data and BIM model data, the safety status information of personnel is dynamically analysed and early warning of dangerous status is provided. The early warning model needs to collect, process, filter, refine and judge the obtained information to determine whether the data value monitored by the security early warning platform is in a safe state. According to the types of tunnel accidents in 8 categories, the boundaries of dangerous areas are delineated for all kinds of dangerous sources in the construction of tunnel projects, which is the basis for early warning of safety risks.

2.1 Establishment of Risk Source Database

2.1.1 Flammable gas concentration monitoring

The accident caused by gas is the most common accident in tunnel construction explosion. Explosions must meet three basic conditions at the same time, one is the gas concentration, the other is an ignition source with sufficient energy, and the third is sufficient oxygen content. The critical gas concentration value in the tunnel is shown in Table 1.

* Corresponding author: wuping@nbt.edu.cn

Table1. Critical value of gas concentration in tunnel.

Concentration	Impact
<5%	No explosion in case of fire
5%-16%	Gas in the air can cause explosion when it meets fire
>16%	No explosion in case of fire, but in the air fire will still burn.

2.1.2 Special geotechnical monitoring

According to the statistical analysis of the accident, the tunnel collapse is the highest incidence and the more serious casualties of the accident, so the cause of the tunnel collapse needs to be paid attention to.

(1) Rock explosion

There are three conditions for rock explosion: ①hard and complete rock; ②highland stress; ③rock of high fragility. The integrity of the rock directly determines whether a rock explosion will occur. The rock integrity coefficient Kv can be used as the criterion for determining the occurrence of rock explosion. In the initial ground stress grading of a rock body, the strength stress ratio S refers to the ratio of the rock's pressure strength to the ground stress[3]. The brittle coefficient B of rock can describe the brittleness of rock very well, reflect its tendency to occur rock explosion. The three conditions under which the integrated rock explosion occurred are summarized as shown in Table 2.

Table2. Early Warning for Rock Explosion.

Condition	Outcome	Early warning
$0.55<Q<0.75$	Medium to low intensity	When any two of these values meet the criteria. Yellow alert.
$2.5\le S\le 5$		
$3<B<5$		
$Kv>0.75$	Severe intensity	When any two of these values meet the criteria. Red alert.
$S<2.5$		
$B\ge 5$		

(2) Surrounding rock

Expansion, water-rich weak fracture, extrusion and other surrounding rocks are very easy to cause tunnel collapse, and the mechanics is the most affecting factors of rock stability, which refers to the surrounding rock resistance deformation and destruction performance. The level of surrounding rock is shown in Table 3. There are six levels of surrounding rock, starting from IV level, surrounding rock may lose stability, resulting in a larger collapse, and after the accident, it was found that some tunnel collapse accidents, is due to the sudden change of surrounding rock from level III to V, resulting in collapse. Therefore, with class III perimeter rock as the boundary, when the elastic vertical wave velocity of the surrounding rock >3km/s, it is considered a safe state, when velocity tends to be 3km/s, it is considered unsafe.

Table3. Surrounding Rock Grading.

Level	Stable state after excavation	Kv (km/s)
I	Stable.	>4.5
II	When exposed for a long time, local slight collapse may occur. The side wall is stable.	3.5-4.5
III	When arch is not supported, slight collapse may occur. The side wall is basically stable. When the blasting vibration is too large, collapse is easy to occur.	2.5-4.0
IV	When arch is not supported, serious collapse may occur. The side walls are sometimes unstable	1.5-3.0
V	Easy to collapse.	1.0-2.0
VI	Easy to collapse and deform. In the presence of water, soil and sand often gush out together with water.	<1.0

2.1.3 Toxic gas monitoring

Toxic gas in tunnels has always been the main source of danger in the construction process,which is easy to cause poisoning or asphyxiation of major safety accidents. The main toxic gases are CH_4, CO, CO_2, H_2S and NH_3. The permissible concentration values of toxic gases commonly found in tunnels are shown in Table 4.

Table4. Allowed concentrations of common toxic gases in tunnels.

serial number	The name of the harmful gas	Limit concentration
1	CH_4	1%
2	CO	0.0024%
3	CO_2	1.5%
4	H_2S	0.00066%
5	SO_2	0.0005%
6	NH_3	0.0004%
7	NO_2	0.0025%

2.2 RFID requirements analysis

2.2.1 People related information collection

Information about a person needs to be reflected by the worker wearing an RFID card with personal information. Its main information includes person identity information, personnel location information, and personnel status information.

Firstly, person identification information, including worker's name, blood type, emergency contact name and contact information. In addition, in order to ensure that the quality of workers meet the standards, RFID card should also record the safety of workers on-the-job production education and training and obtain the corresponding qualifications of induction certificate information. Person identification information is used both for clocking in at work and for personnel identification.

Secondly, personnel location information, construction personnel close to hazardous areas at different distances of the early warning results as shown in Table 5.

Table5. Early warning results for construction personnel near hazardous areas at different distances.

The location of the person	Early warning information
2.5m from the hazardous area	The BIM module appears "yellow" at the appropriate location and the person icon flashes
1.5m from the hazardous area	"Orange" appears at the appropriate location of the BIM module, and the person icon flashes and alarms

Common tunnel accidents also include object blows, implement injuries, electric shocks, etc. According to the classification criteria, dangerous areas are defined. Take the collision zone and the electric shock zone as an example[4].

Collision zone. Unsafe factors in its construction may include mechanical equipment, etc. The mechanical equipment here refers to the construction machinery and transport vehicles at the construction site, etc., and when such machinery and equipment is in working condition, it is within 2m of its outline as a collision zone

Electric shock zone. Unsafe factors in its construction may include wires or cables, etc. Therefore, the ribbon area of 1m next to a wire or cable is defined as a electrocution zone.

Using the simulation property of BIM, a circular danger area with the location of the "object" as the center of the circle and the safety distance as the radius, namely the collision area and the electric shock area, is simulated, and the red color is displayed in the BIM platform. In addition, the radius of 1.5m+ safety distance /2.5m+ safety distance in Table 6 above should be used to draw circular areas respectively and distinguish them with orange/yellow[5].

In addition, in addition to collisions between people and objects, there are collisions between objects and objects in the definition of collision zones, the mechanical installation of a cluster of travel, weight, inclination sensors, LCD hosts and video cameras equal to a cluster of tree-shaped wireless sensing network, for real-time dynamic monitoring[6].

Finally, personnel status information refers to the length and length of construction workers working hours, if the construction personnel excessive fatigue is very easy to cause safety accidents, through wearing RFID card to monitor workers working hours, timely adjustment.

2.2.2 Information collection of objects

"Objects" mainly refer to materials and machinery. In the tunnel construction process, the unsafe state of the material is also very easy to cause the occurrence of safety accidents. The security risk warning platform monitors the status of the relevant object equipment in real time, enters the relevant information of the "thing" into the RFID card, and guarantees that each "object" corresponds to an RFID card. Input information includes parameter information about materials and machinery, such as model number, date of manufacture, manufacturer, etc., as well as its location information and trajectory.

3 Security management system construction

Through the above research, an early warning platform for tunnel construction safety risks based on BIM technology and RFID technology is established. Including data collection, data processing and security alert three parts.

(1) Data collection

Data collection is used to obtain various unsafe data during construction. Various sources of hazards can be monitored and located using radio frequency identification (RFID) technology. In addition, the location information of personnel and mechanical targets can be dynamically monitored and the information of emergencies can be processed and analysed. Self-organizing and information collaboration with sensor networks enables close-up transmission of data from sensors, RFIDs, and other data collection devices. Environmentally aware information (stress, temperature, gas concentration, etc.) and the collected data are submitted to the monitoring host via a gateway in the data transmission subsystem.

(2) Data processing and transmission

Data processing is conducted based on LAN network and PAN network. After RFID tag scanning information is automatically transmitted to the BIM model through the network, the collected data is transmitted, filtered, checked, grouped, and the data is systematically managed. The risk data is interacted with the BIM model that passes the IFC standard, and the security status of the target location, surrounding environment, and stress level is finally presented through comparative analysis.

(3) Security alert

Safety early warning is used to collect various safety data on the construction site in order to realize the various functions of the platform. It is divided into analysis stage and monitoring stage, including BIM dynamic simulation, target dynamic monitoring, real-time safety analysis and early warning of construction process. Firstly, according to the results of the analysis stage, the monitoring object is set up, the tag type is defined, and the equipment and tags are set up. Secondly, during the construction process, the BIM model is used to track and locate the target visually; Finally, the monitoring process information is updated automatically to form BIM security information database. The project safety managers can obtain the safety information of the construction site in real time. The

https://doi.org/10.1051/e3sconf/202129302048

accuracy and real-time of safety early warning can be ensured by adding and correcting the safety information in time. The early warning system can also sends out early warning signals to safety managers and proposes safety solutions.

4 Conclusion

This paper uses information technology to realize real-time monitoring and early warning of safety risks in tunnel construction. A tunnel engineering security risk warning platform based on BIM and RFID technology is established, the demand data of the early warning platform is obtained according to the analysis of risk sources. The corresponding data standardization are formulated and embedded into the database. The platform is divided into data collection, processing transmission and security early warning subsystem, in which case, the three subsystems coordinate the work of the construction process unsafe people, objects, geological environment data collection and real-time monitoring. By transmitting the data to the BIM model in a timely manner and performing visual analysis, early warnings can be effectively and accurately issued to notify relevant personnel to take appropriate measures to ensure the safety of construction personnel.

Acknowledgement:

China Railway Construction Group Co., Ltd. Science and Technology Research and Development Project (2020-12), Ningbo Municipal Transportation Bureau Science and Technology Project (202007).

Reference

1. Q. H. Qian, R. X. Li, Journal of rock mechanics and engineering, 649-655(2008).
2. H. Q. Lv, Sichuan Cement, 120(2019).
3. L. D. Zhu, Shanxi architecture, **42**, 176-177 (2016).
4. Y. L. Liu, Y. H. Chen, Y. L. Gao, Shanxi architecture, **45**, 188-189(2019).
5. X. W. Zou, D. Zhang, H. Ma, Journal of engineering management, **33**, 124-129(2019).

E3S Web of Conferences 293, 02049 (2021)
GCEECE 2021
https://doi.org/10.1051/e3sconf/202129302049

Research on load bearing performance recovery method of a RC simply-supported beam bridge

Dejun Mao[1,*], Zongxing Zhang[2], Xun Jiang[3], Liu Jinlin[4], Lan Shuwei[1]

[1]College of Architecture and Civil Engineering, Kunming University, Kunming 650214, China

[2]Kunming Municipal Engineering Design & Science Research Institute Co.,Ltd , Kunming 650228, China

[3]Yunnan Communications Investment Group Highway Construction Co., Ltd., Kunming 650034, China

[4]Zhaotong Expressway Construction and Maintenance Engineering Co., Ltd, Zhaotong 657000, China

Abstract: The load-bearing performance of a RC simply supported beam bridge was comprehensively detected and evaluated. The appearance detection results show that there is a large area of damaged and exposed reinforcement in the main girder and diaphragm plate of the bridge. The detecting results of material condition and state parameters show that the durability of the bridge is insufficient, the risk of steel corrosion is large, and the strength condition of concrete material is poor. The load test results show that the bearing capacity of the bridge does not meet the requirements. In view of the current situation of the bridge, a comprehensive reinforcement treatment scheme was proposed, in which modified epoxy mortar is used to repair the main beam and diaphragm of the whole bridge, steel plate is pasted to reinforce the main beam of the whole bridge, and anticorrosive coating is applied on the main beam and diaphragm.

1 Introduction

In actual engineering, due to various reasons, such as long-term use of functional degradation, earthquakes and other natural disasters, changes in structural use functions, improper design, poor construction quality, the bridge's bearing capacity is insufficient, safety and reliability are reduced, and reinforcements are needed. Reinforced concrete (RC) simply supported beam bridges have the characteristics of simple force, convenient design and calculation, and low construction cost, and have been widely used worldwide. This article takes an RC simply supported beam bridge as the background, and according to the detailed inspection and evaluation results of the bridge, a comprehensive reinforcement treatment plan was formulated to repair the main girder and diaphragm of the full bridge with modified epoxy mortar, reinforce the main girder of the bridge with glued steel plates, and apply anticorrosive coating to the main girder and diaphragm of the entire bridge, so as to restore and improve the bearing performance of the bridge. The reinforcement treatment plan in this article can provide reference for similar bridge treatment.

2 Project profile

An RC simply supported beam bridge in Guangdong Province, China, was built in 2002 with a span of 1×13.0m. The full bridge consists of two left and right parallels, each with a width of 10.2 m, and the total width of the bridge is 20.4m. Both the left and right parallels are arranged with 6 pieces of reinforced concrete simply supported T beams, 4 transverse partitions are arranged at equal intervals along the longitudinal direction of the T beams, the support is plate rubber bearing, the abutment is gravity abutment, and the deck pavement for asphalt concrete. The design load is car -20 class, trailer -100, and crowd load is 3.5kN/m^2. The cross-section layout of the bridge is shown in Figure 1, and the cross-section size of the main beam is shown in Figure 2.

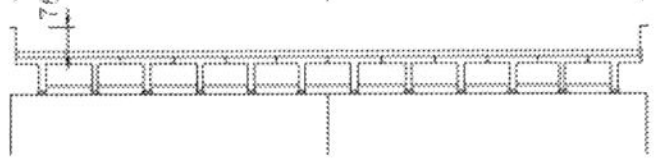

Figure 1. Cross-section layout of the bridge (Unit: cm).

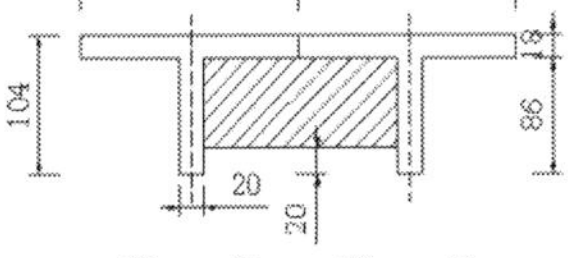

Figure 2. Cross-section size of the main beam (Unit: cm).

*541089451@qq.com

 https://doi.org/10.1051/e3sconf/202129302049

3 Bridge detection results

3.1 Appearance detection

According to the method of the Chinese standard "Specification for Inspection and Evaluation of Load-bearing Capacity of Highway Bridges" JTG/T J21-2011, the bridge has been visually detected. The main results are as follows:

- Bridge deck system: There are 2 traces of transverse crack repair on the bridge deck, which are good after repair; 1 trace of longitudinal crack repair, which cracks after repair; there is 1 loose and dislocation of the curb in the green belt on the right.
- Superstructure: The main beams and diaphragms of the full bridge have large areas of damaged and exposed bars. The exposed bars of some main beams and diaphragms are shown in Figures 3 to 4; the overall condition of the full bridge supports is good.
- substructure: The abutment is slightly scoured with pockmarked surface and attached with oil.
- Underwater foundation: There is no erosion of the foundation.

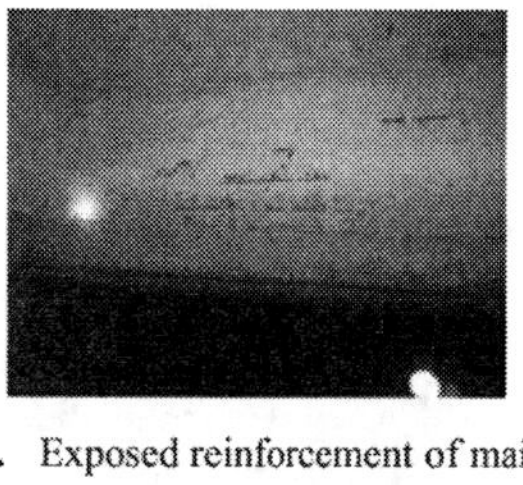

Figure 3. Exposed reinforcement of main girder.

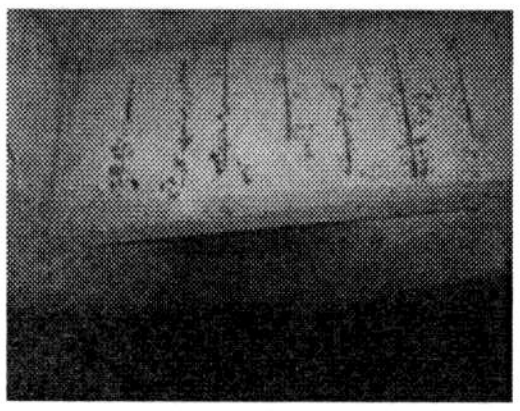

Figure 4. Exposed reinforcement of diaphragm plate.

The appearance detection results show that the main diseases of the bridge are: the main girder and the diaphragm have large areas of damaged and exposed bars. The reason is that the thickness of the concrete protective layer is insufficient, and the corrosion and expansion of the steel bars cause the protective layer to peel off.

3.2 Material condition and state parameter detection

According to the method of the Chinese standard " Standards for Technical Condition Evaluation of Highway Bridges" JTG/T H21-2011, the material condition and state parameters of the bridge were detected and evaluated. The main results are as follows:

- The test results of steel corrosion potential show that the lowest value of steel corrosion potential of each test component is (-300, -400] mV, which indicates that the bridge has corrosion activity and the probability of corrosion is greater than 90%. The evaluation scale is 3.
- The concrete resistivity test results show that: the resistivity of each test member is $10000 \leq \rho < 15000$, and the possible corrosion rate of the steel bar is general, and the evaluation scale is 3.
- The test results of the thickness of the steel bar protection layer show that the thickness of the steel bar protection layer D_{ne}/D_{nd} of each test component is less than or equal to 0.55, the thickness of the steel bar protection layer is poorly controlled, and the steel bars are easy to lose alkaline protection and rust. The evaluation scale is 5.
- The test results of concrete carbonation depth show that the concrete carbonation depth Kc of each test member is less than 0.5, which has a small effect on the corrosion of steel bars, and the evaluation scale is 1.
- The test results of concrete strength show that the concrete strength of each test member is $0.80 \leq K_{bt} < 0.90$, and the overall strength of the bridge material is poor, and the evaluation scale is 3.
- The test results of the chloride ion content show that the chloride ion content of each test component (percentage of the cement content) $0.40 \leq § < 0.70$， may induce corrosion of steel bars. The evaluation scale is 3.

The detecting results of material conditions and state parameters show that the bridge has insufficient durability, high risk of steel corrosion, and poor strength of concrete materials. The reason is that the location of the bridge is in a typical atmospheric ocean climate environment, with high ambient temperature, high air humidity, and high content of corrosive ions in the air. The construction quality of the concrete was not well controlled during the construction of the bridge.

3.3 Load test

According to the method of the Chinese standard " Load Test Methods for Highway Bridge" JTG/T J21-01-2015, the static load test and dynamic load test were carried out on the bridge, and the main results are as follows:

- Static load test: The strain and deflection calibration coefficients of the key measuring points in the left and right parallels main beams are greater than 1, indicating that the bearing capacity does not meet the requirements; the relative residual deformation of the beam after unloading is less than 20%.
- Dynamic load test: The measured pulsation, sports car, brake, and car jump fundamental frequencies of the left and right parallels main beams are all greater than the theoretically calculated values.

The load test results show that the load-bearing capacity of the bridge does not meet the requirements due to the large-area damage of the main girder and the diaphragm plate, the exposed reinforcement, the corrosion of the steel bar, and the poor concrete strength.

4 Treatment plan

The bearing capacity of the bridge does not meet the requirements of use, and it should be repaired and reinforced. According to the disease condition of the bridge, the following comprehensive reinforcement treatment plan has been formulated.

4.1 Repair damage and exposed reinforcement

Use modified epoxy mortar to repair the damaged and exposed steel reinforcement of the main girder and diaphragm plate of the full bridge. The repair process is shown in Figure 5.

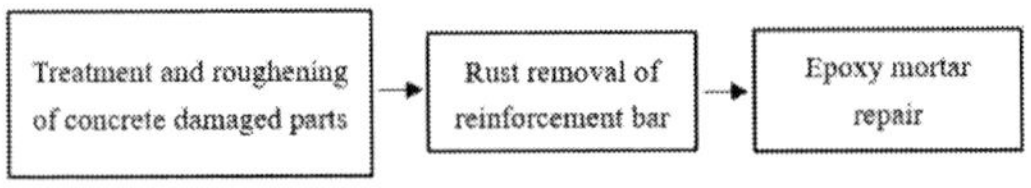

Figure 5. Repair process.

The purpose of processing and chipping the damaged concrete is to clean the loose part thoroughly and to make the concrete surface have a certain degree of roughness. During the construction process, it is forbidden to use equipment with high impact force. Manual methods can be used to ensure that the steel bars and adjacent concrete are not damaged. After the treatment is completed, compressed air is used to remove dust. When the concrete base surface is wet, the moisture fills the cracks, causing the epoxy mortar slurry to not penetrate into the concrete matrix, which will significantly reduce the bonding effect [1]. Therefore, the concrete surface should be kept dry during the treatment process, and it is strictly forbidden to wash it with water.

After concrete treatment and chiseling, the steel bars should be derusted. After the steel bar is derusted, apply rust inhibitor on its surface. The rust inhibitor can improve the passive film on the surface of the steel bar, improve the anti-corrosion ability of the passive film, and thus increase the service life of the reinforced concrete structure. According to different classification methods, steel bar rust inhibitors can be divided into organic rust inhibitors, inorganic rust inhibitors, blended rust inhibitors, migration type rust inhibitors, etc. [2]. There are many kinds of steel bar rust inhibitors on the market, and the effect is uneven [3]. The RI-1 series of rust inhibitors are the earliest developed and most widely used products. For example, RI-1C2 type steel bar rust inhibitors are used in Guangdong Foshan Expressway Bridge, Yuehai Railway Hub Overpass, and Hangzhou Bay Bridge [4]. RI-1C2 type rust inhibitor is proposed to be used for the protection of the bridge steel bars.

Before repairing with epoxy mortar, a layer of modified epoxy-based liquid should be applied to the surface of the concrete to keep the concrete moist to ensure reliable bonding between the epoxy mortar and the concrete. Epoxy mortar is made of ordinary cement mortar and epoxy resin modified materials according to a certain proportion. The amount of epoxy resin, curing agent, polymer cement ratio, polymer sand ratio and aggregate gradation have obvious effects on its strength and bonding performance. The epoxy mortar mix ratio used in the bridge is: epoxy resin: thinner: curing agent mass ratio 10:4:3, polymer-cement ratio is 1.5, and polymer-sand ratio is 1:6. Studies have shown that the epoxy mortar material with this ratio has high strength and good bonding performance [5].

4.2 Bonding steel plate reinforcement

The main girders of the whole bridge are reinforced with glued steel plates to restore and improve its bearing capacity. The reinforcement scheme is designed as follows:

A 5mm thick, 18cm wide Q345qE steel plate strip is pasted along the bottom of each main beam within 6m on both sides of the middle span; U-shaped hoop plates are added to strengthen the anchorage and improve the end shear capacity, and the main beam is glued to steel the reinforcement design is shown in Figure 6, and the specifications and dimensions of the steel plates required for the reinforcement of each main beam are shown in Table 1. Before sticking the steel plate, chemically inject the cracks with a width greater than 0.15mm.

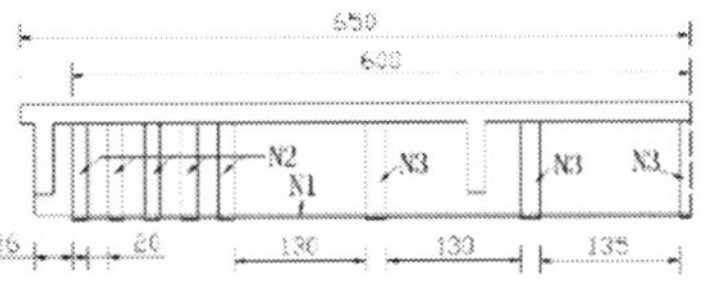

Figure 6. Main beam reinforcement design drawing (Unit: cm).

Table1. Specification and size of steel plate required for reinforcement of main beam.

Type number	Length /cm	Width /cm	Thickness /mm	Quantity
N1	1200	20	5	1
N2	190	15	5	10
N3	190	20	5	5

The construction process of sticking steel plate reinforcement is shown in Figure 7.

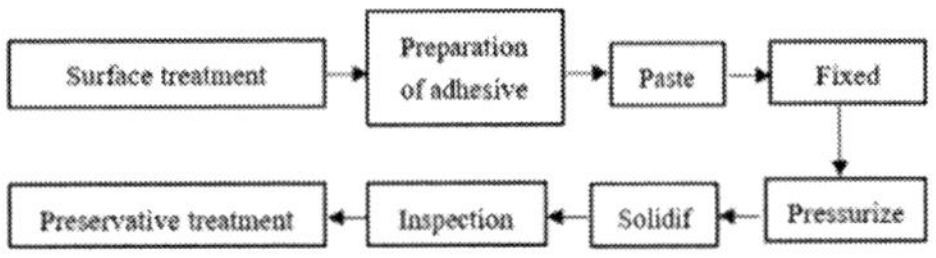

Figure 7. Construction process of sticking steel plate reinforcement.

The construction should be carried out by a skilled construction team to strictly control the construction quality. The method of sticking with steel plate reinforcement should be unloaded as much as possible. Considering that the bridge is located in an urban area, where the flow of people and vehicles is intensive, it is impossible to completely close the traffic, so the framing reinforcement is adopted.

4.3 To coat anti-corrosion coating

The main beam and diaphragm of the whole bridge are coated with anti-corrosion coating to solve the problems of steel corrosion and insufficient durability. The anti-corrosion coating construction shall be carried out after the steel plate is pasted. A large number of engineering practices have shown that coating anticorrosive coatings on RC bridge structures can effectively block the corrosive ion erosion in the atmosphere, delay the corrosion of steel bars, and carbonization and cracking of component concrete [6]. At present, anti-corrosion coatings have been widely used in bridge engineering, especially for river-crossing and river-crossing bridges in offshore areas [7].

Considering the environmental temperature, humidity and other factors of the bridge, the material of anti-corrosion coating is selected, and E26 anti-corrosion coating material which can effectively resist external factors and invasion erosion and meet the requirements of relevant standards is adopted [8]. The material information of E26 anti-corrosion coating is shown in Table 2, and its performance can meet the anti-corrosion requirements of the bridge. The anti-corrosion coating construction includes three layers, namely bottom layer, middle layer and surface layer.

Table2. E26 anti-corrosion coating material information.

Coating name	Coatings	Materials	Mechanism of action
E26	Bottom layer	H06-10 primers	Infiltration into concrete
	Intermediate layer	H45-ZF101	Film forming on concrete surface
	Surface	PV-50 topcoat	Film forming on concrete surface

5 concluding remarks

This article takes an RC simply supported beam bridge as the background, the bearing capacity of the bridge is comprehensively evaluated by bridge appearance inspection, bridge material condition inspection and state parameter inspection, static load test and dynamic load test. According to the results of bearing capacity evaluation, a comprehensive reinforcement treatment plan was formulated to repair the main girder and diaphragm of the full bridge with modified epoxy mortar, reinforce the main girder of the bridge with glued steel plates, and apply anticorrosive coating to the main girder and diaphragm of the entire bridge, so as to restore and improve the bearing performance of the bridge. The reinforcement treatment plan in this article can provide reference for similar bridge treatment.

Acknowledgments

Authors wishing to acknowledge assistance from the Youth Project of the Joint Special Fund for Basic Research of Local colleges and universities in Yunnan Province (2019FH001(-100)).

References

1. Fang Liu, Lingling Xu, Mingshu Tang. Effects of interfacial roughness and wetness on repair effectiveness of epoxy resin based repairing mortar. Journal of Nanjing University of Technology (Natural Science Edition), 35: 36-41, (2013)
2. Yinbo Gao, Juan Hu, Qing Liu, et al. Applications and prospects of corrosion inhibitors for reinforcing steel. Journal of Xiamen University (Natural Science), 54: 713-720, (2015)
3. Xiaotong Wang, Congtao Sun, Qing Liu, et al. Inhibition mechanism of amino alcohol corrosion inhibitors. Bulletin of the Chinese Ceramic Society, 36: 84-88, (2017)
4. Naifeng Hong. Development and use of steel bar inhibitor. Industrial Construction, 35: 68-70+79, (2005)
5. Chanchan Li, Dejin Cheng, Cheng Lian, et al. Study of factors influencing on epoxy mortar strength. Adhesion, 37: 58-60+75, (2016)
6. Yaoqing Wen, Wenchun Wen, Zuxin Sun. The anticorrosive maintenance painting program for surface of anchored reinforced concrete structure of Haicang Bridge in Xiamen. Shanghai Coatings, 54: 49-51, (2016)
7. Sike Shen, Shuya Bai, Yunsheng Tian, et al. Discussion on the use of industrial protection coatings in bridge engineering. China Coatings, 25: 55-58, (2010)
8. Ming Ji, Yueliang Gan, Chaohai Ding, et al. Application of high performance concrete and coating of protective paint on a port in Zhejiang province. Journal of Civil and Environmental Engineering, 37: 23-27, (2015)

https://doi.org/10.1051/e3sconf/202129302050

Experimental investigation on pump efficiency and cavitation performance of Huang-jin-xia hydro-junction

Shaojiang Luo [1], Tao Dong [1], Yong Mou [1], and Chenhao Li[2*]

[1] HanJiang-To-WeiHe River Valley Water Diversion Project Construction Co.Ltd., Shaanxi Province , Xi'an, 710100, China

[2] State Key Laboratory of Eco-hydraulics in Northwest Arid Reign of China， Xi'an University of Technology， Xi'an, 710048, China

Abstract. In this paper, the performance of A1515 centrifugal pump was studied on a high parameters and precision large general test platform in combination with the Huang-jin-xia hydro-junction in Hanjiang-to-Weihe River valley water diversion project. The efficiency characteristics and cavitation performance of the pump were tested and measured respectively. All measured parameters were collected and recorded through the acquisition system. The experimental results show that the optimal efficiency and weighted average efficiency of the pump is 92.46% and 92.18%, and the design point efficiency is 92.39%, all of which meet the design requirements. After conversion to the prototype pump, the efficiency value at each working point is higher than the guaranteed value. The pump hump margin is 3.03%, which is greater than 2% of the required value and meets the requirements. In the whole normal operating range, the cavitation performance of the model pump meets the design and test requirements. The comprehensive error of model test is less than ±0.2%. The research results provide reference for the design of large centrifugal pump.

1 Introduction

Centrifugal pump is widely used in industrial production because of its advantages of high head. Because the large pump technology is not mature interiorly, there is no relevant data reference, it is necessary to research the pump by model experiment[1].The energy and efficiency performance of centrifugal pump unit is not only related to the degree of energy consumption, but also affects the safety of pump unit. It is of great significance to carry out the pump performance under different working conditions[2-4]. Secondly, cavitation is a physical phenomenon in the operation of pump, and the cavitation process will produce large pressure pulsation. And cavitation collapse will produce a lot of impact and noise, resulting in the unit performance decline; therefore, it is very necessary to study the cavitation performance through experiments.

At present, researchers have done a lot of research on pumps' cavitation. Lin[5] conducted numerical calculation on LES turbulence model and cavitation model for the whole flow passage, and concluded that the draft tube was affected by the runner inlet cavitation, and the amplitude of pulsation frequency increased. Zhao et al.[6] studied the pressure pulsation law and the change law of cavitation volume in a low specific speed centrifugal pump during cavitation, the results showed that the bubble volume increased non-linearly in the cavitation process. Luo et al.[7] studied the characteristics of the centrifugal pump's transient speed under different conditions by experiments. When cavitation just happened, the pump's speed varied in the largest range, and the degree of rotation speed fluctuation under severe cavitation is higher than that without cavitation. Wang et al.[8] simulated the transient cavitation flow in a centrifugal pump based on the RNG k-ε turbulence model and the improved transport cavitation model. Zhao[9] studied the positive and negative problems of cavitation control by placing obstacles on the blade surface of a centrifugal pump through experiments and numerical simulation. The results show that the head value of pump can be increased by 30% when the blade pressure side is placed with obstacles.

This paper takes the centrifugal pump of Huangjinxia hydro-junction as the research object and carries out the model pump experiment. The cavitation performance of pump is studied by visualization, and the energy and efficiency characteristics of pump under different speed and flow conditions was carried out, finally, the characteristic curve of the pump was drawn, the cavitation performance of pump with different cavitation numbers was analysed.

2 Pump model and experimental apparatus

Model pumps including inlet pipe, volute, stay ring, head cover, impeller et al. The model device is made of metal materials except for the straight conical section of the inlet pipe which is transparent Perspex, the rest are metal materials. The whole flow channel of the pump from inlet pipe to volute outlet is geometrically similar to the prototype pump, the scale coefficient of the prototype and

* Corresponding author: drleech_xaut@126.com

the model is D1p/D1m= 4.551. Fig. 1 shows a schematic diagram of the pump model device. Table 1 shows the basic parameters of the test model pump. All test parameters meet the requirements of relevant standards of GB15613-2008 and IEC60193-1999. The measurement accuracy and operation stability of all test parameters are required. The comprehensive test error of the model efficiency of the test bench is less than ±0.20%.

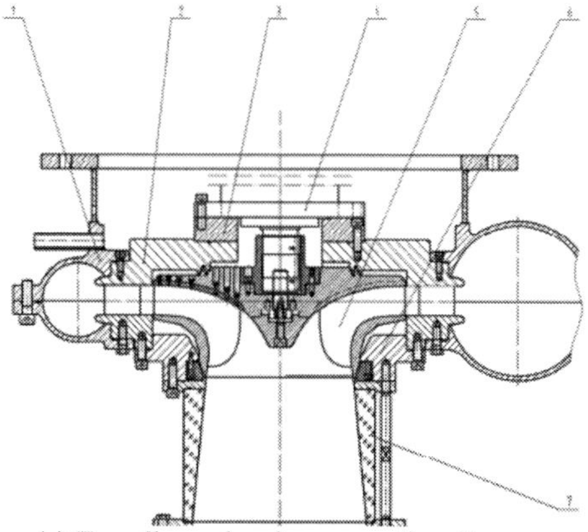

(a) Two dimensional cross-section diagram

(b) Experimental device diagram

Figure 1. Caption of the Figure 1. Below the figure. 1-volute; 2-head cover; 3-sub-plate; 4-binder-plate; 5-impeller; 6- stay ring; 7-inlet taper pipe;

Table1. Main parameters of model pump

Designations	Parameters
Pump type	A1515
Pump inlet diameter	D_{2m}=0.307m
Pump outlet diameter	D_{1m}=0.529m
Blade number	9
Guide vane number	Z_s=13
Design flow rate	Q_d=0.348m3/s
Design head	H_d=37.93m
Rated speed	n_m=1000 r/min

3 Results and analysis

3.1 Analysis of efficiency experimental results

Efficiency test is to measure the efficiency, flow rate, head, power and other parameters of the pump under the water condition. Including the efficiency characteristics of the normal frequency variation range 49.8~50.5Hz and the head range. In the test, set the dynamometer speed as n_m=1000r/min.When drawing model pump characteristic curve, horizontal axis represents flow rate vertical axis represents head, pump efficiency and shaft power, andthe actual operating area of prototype pump should be marked on the pump characteristic curve. The formula for calculating the weighted average efficiency of pump is as follows:

$$\eta_{ave} = \sum_{j=1}^{n} \omega_j \cdot \eta_j / 100 \quad (1)$$

Where, ω_j and η_j are efficiency weighting factor and efficiency, respectively.

According to the test requirements, when the model pump speed is nm=1000r/min and the power frequency is 50Hz, the efficiency tests are carried out on the optimal and weighted average points of the pump under 6 groups of different head. The efficiency test results are shown in Table 2. In addition, the flow-head curve test was carried out in two ways of flow increasing and flow decreasing under the same speed condition. Finally, in the range of 100% N_r ~ 92% N_r, the pump efficiency test at different speeds is carried out according to 1%N_r interval. The results are shown in Fig. 2 and Fig. 3.

Table2. Weighted average efficiency of model A1515

Parameters	Head (m)					
	101.7	104.5	106.5	108.5	110	116.5
Weighting coefficient	10	31	38	14	3	4
Model head (m)	34.9	35.86	36.52	37.26	37.73	39.92
Mode flow rate (m^3/s)	0.3868	0.3745	0.3658	0.357	0.3504	0.3214
Model efficiency (%)	91.74	92.08	92.25	92.39	92.44	92.37
Prototype efficiency (%)	92.74	93.08	93.25	93.39	93.44	93.37
Prototype head (m)	101.73	104.57	106.53	108.58	110.01	116.51
Prototype flow rate (m^3/s)	13.66	13.23	12.93	12.61	12.38	11.36
Input power (MW)	14.66	14.54	14.45	14.34	14.26	13.87
	Experimental value			**Guarantee value**		
Model Weighted average efficiency (%)	92.18			-		
Prototype Weighted average efficiency (%)	93.18			≥93.14		

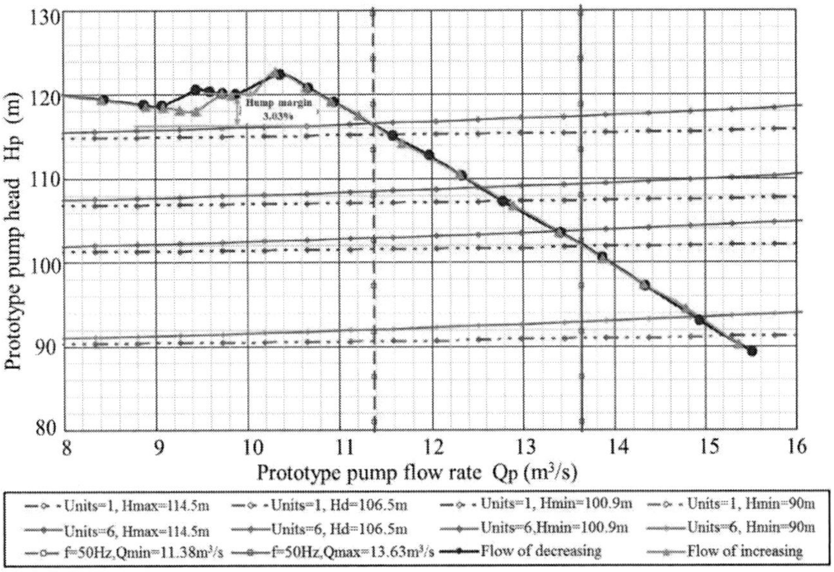

Figure 2. A1515 prototype pump energy characteristic curve (n_{rp}=375r/min f_0 =50Hz D_p=1.397m)

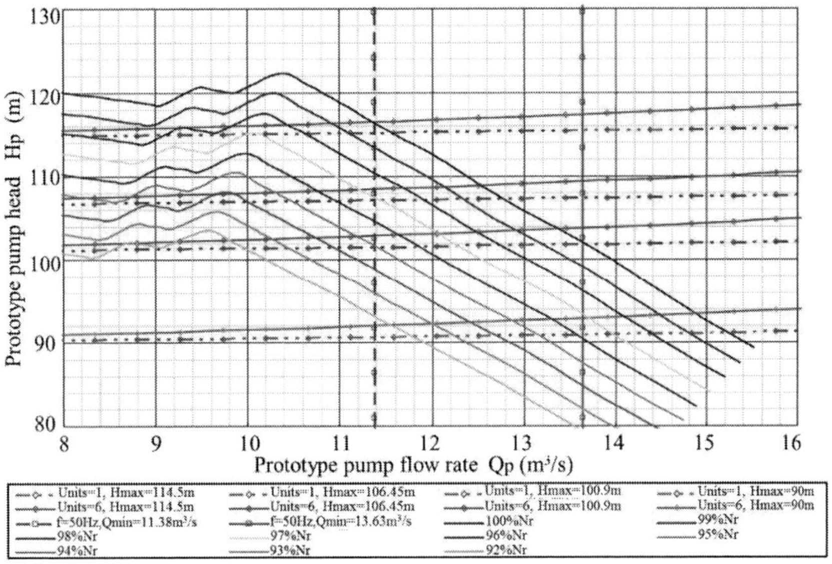

Figure 3. Energy characteristic curve of prototype pump at different rotating speeds

3.2 Cavitation performance experimental results analysis

According to the test requirements, the cavitation performance of the pump was carried out, and the results were recorded and analysed. The pump condition cavitation test is carried out under the specified head, input power and frequency (including abnormal frequency). The Cavitation tests were conducted under different heads with cavitation numbers, including critical cavitation tests and determination of primary cavitation coefficients.

The cavitation test of the model pump needs to determine the primary cavitation $NPSH_i$ and critical cavitation $NPSH_{1.0\%}$ at different conditions, and indicate the $NPSH_P$ of the pump station device, which met the normal operation conditions of the pump station, the pump station cavitation $NPSH_P$ is greater than the primary cavitation NPSHi. The critical cavitation $NPSH_{1.0\%}$ value is defined as the cavitation NPSH value when the suction height decreases and the efficiency is lower than the efficiency of 1.0% in the no-cavitation condition. Primary cavitation $NPSH_i$ value is defined as the cavitation NPSH value when stable bubbles were attached on two blades of the impeller by visual observation. Pumping unit cavitation coefficient is $\sigma_P = NPSH_P/H_P$; primary cavitation coefficient is $\sigma_i = NPSH_i/H_P$; critical cavitation coefficient is$\sigma_{1.0\%} = NPSH_{1.0\%}/H_P$.

The test results in Fig. 4 and Fig. 5 show that within the normal operation range of the pump, the pump device's net positive suction head station $NPSH_P$ is greater than the primary cavitation $NPSH_i$, and the pump station's cavitation safety factors $K_{NPSH_i} \geq 1.2$, $K_{NPSH_{1.0\%}} \geq 1.6$, both meet the requirements.

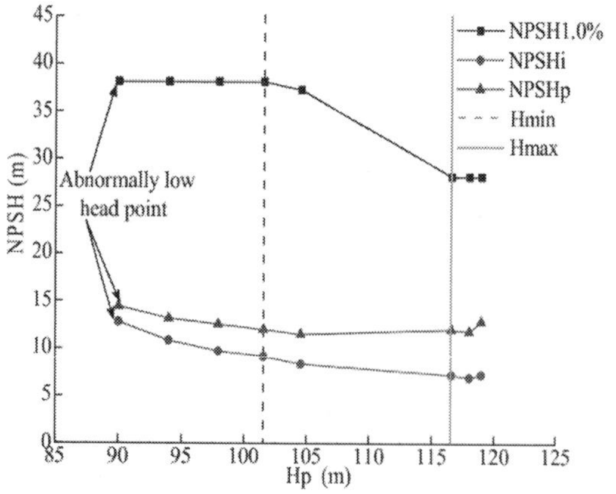

Figure 4. Cavitation experimental results of A1515 prototype pump

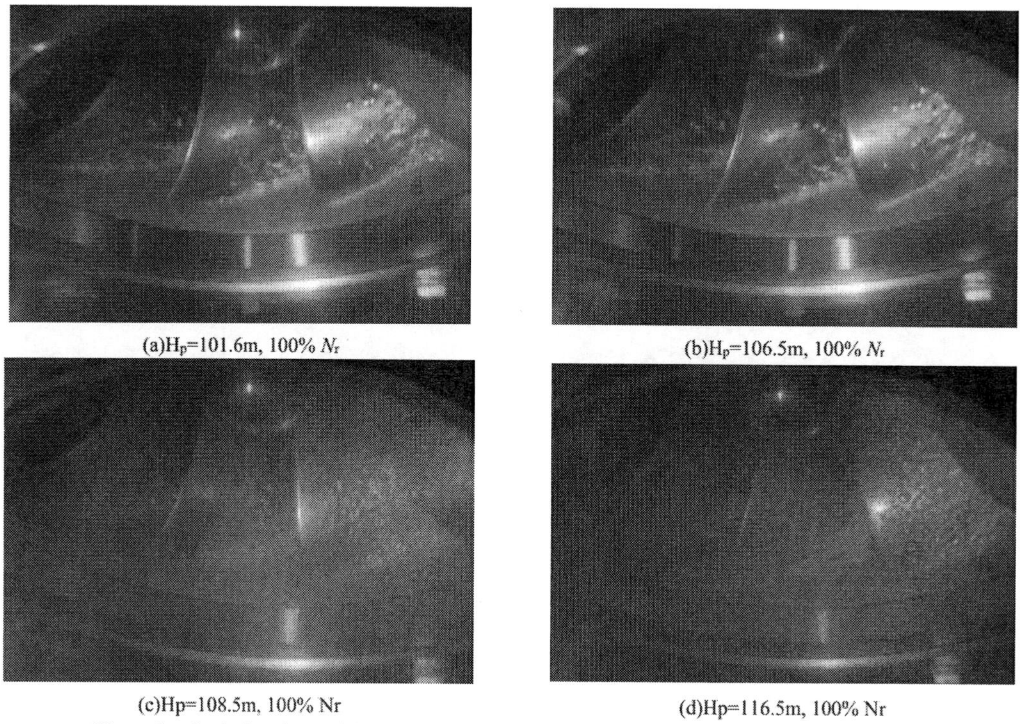

(a)H_P=101.6m, 100% N_r
(b)H_P=106.5m, 100% N_r
(c)Hp=108.5m, 100% Nr
(d)Hp=116.5m, 100% Nr

Figure 5. Cavitation shape of the prototype pump at critical cavitation (σ 1.0%) under different heads

4 Conclusions

Based on the high-precision and high-parameters pump test platform, this paper conducts experimental research on the efficiency characteristics and cavitation performance of A1515 model pump under different working conditions through fine test methods, and analyses the efficiency and cavitation characteristics of pump under different working conditions. The main conclusions are as follows:

(1) The optimal efficiency, weighted average efficiency and design efficiency of the model pump are 92.46%, 92.18% and 92.39% respectively, all of which meet the design requirements. After the conversion to the prototype pump, the efficiency values at each condition are higher than the guaranteed value, which indicates that the design indexes of the pump meet the contract requirements.

(2) A1515 model pump can meet the contract requirements at the minimum flow rate and design head flow rate, and the design flow rate is close to the optimal condition. Pump hump margin is 3.03%, which is greater than the required 2% and meet the requirements.

(3) Within the whole normal operating range, the cavitation performance of the A1515 model pump with different cavitation numbers meets the design and test requirements.

Acknowledgements

This study was supported by the Science and Technology Innovation Fund Project of Hanjiang-to-Weihe River Valley Water Diversion Project Construction Co. Ltd. "Study on the outage and escape transition process of centrifugal pump under the influence of long water pipeline" (NO: 2020203) and Open project of State Key Laboratory of Hydroelectric Equipment (NO: SKLHE-ORF-202004)

References

1. Y Zhang, D Zhang, Y Liu, et al. Fluid Machinery, **31** 1 (2003),.
2. B Chen,B Zhen, H Zhang, et al. China Water & Wastewater, **35**,91-95(2019).
3. H Xie,Z LI, D Liu, S Huang. China Rural Water and Hydropower, **9**1,29 (2019).
4. R Tao, R Xiao, W Yang. Journal of Drainage and Irrigation Machinery Engineering, **36,**573(2018).
5. S Lin. Study on the influence of cavitation on the hump characteristics of pump turbine, Harbin Institute of Technology,(2019).
6. W Zhao, G Wang, G Zhao, et al. Journal of Lanzhou University of Technology, **46,**50 (2020).
7. Y Luo, J Dong, Y Han, et al. Journal of vibration and shock, **40**,23 (2021).
8. W Wang, Q Meng, H Lu, et al. China Rural Water and Hydropower, **92**,202 (2017).
9. G Zhao. Investigation of Cavitation instabilities and it's Control in a Centrifugal pump, Lanzhou University of Technology, (2018).

Analysis on spatial distribution pattern of football fields: A case study in Wuhan, China

Yuwei Wang[1,*], Kaicheng Wu[1], and Wang Gao[1]

[1]School of Artificial Intelligence, Jianghan University, Wuhan 430056, China

Abstract. In order to promote the sustainable development of sports facilities, the statistical data of football fields in 13 districts of Wuhan city were selected. Based on geographic information system (GIS) and exploratory spatial data analysis (ESDA) technology, this paper analyzed the global and local autocorrelation of the number of football fields per capita in Wuhan city. The results indicate that: (1) Comparing to the new districts in Wuhan, the number of football fields per capita in the central districts was relatively few. (2) In the global scale, the number of football fields per capita in Wuhan presented a spatial aggregation pattern. (3) In the local scale, the regional discrepancy and spatial clusters were confirmed in the local spatial autocorrelation analysis for football fields in Wuhan. At the same time, the research results could provide reasonable references for the planning of football field facilities.

1 Introduction

Football fields are the material basis for the development of football and are important parts of the national activity venues [1,2]. In recent years, people's enthusiasm for participating in football has gradually increased. Consequently, improving the degree of participation in football and the number of people involved are important approaches to build China into a world sports power. But at present, our country is facing the problems of insufficient supply and unreasonable spatial distribution for football field facilities, which restricts the sustainable development of football [3,4].

With the help of GIS and ESDA technology, the spatial distribution of different sports venues can be mapped, and the spatial dependence and heterogeneity distribution characteristics of sports venues can be defined, quantified and explained [5–7]. Nevertheless, few studies have discussed the distribution patterns of football fields, especially at county level.

Wuhan is one of the first batches of China's football reform pilot cities. Therefore, it is of great value to explore the spatial distribution pattern of football fields in Wuhan. This study applied ESDA method and Moran's I index to measure the spatial association patterns of football field distribution in 13 districts of Wuhan city. Firstly, from a global perspective, the Moran's I index was used to detect the spatial distribution pattern of the number of football fields in Wuhan. Secondly, from the local point of view, the local Moran's I analysis were exerted to identify different local clustering patterns in the study area.

2 Study area, data and methodology

2.1 Study Area

Wuhan is located in central China, between 29°58′N-31°22′N and 113°41′E -115°05′E. The total area is about 8569.15 km^2, including 13 administrative districts. Among them, Jiang'an District (Marked as 1 in Figure 1, the same below), Jianghan District(2), Qiaokou District(3), Hanyang District(4), Hongshan District(5), Wuchang District(6) and Qingshan District(7) are the central districts, while Jiangxia District(8), Hannan District(9), Caidian District(10), Dongxihu District(11), Huangpi District(12) and Xinzhou District(13) are the new districts [8]. With the rise of national fitness as national strategy, the number of football fields has reached a certain extent in Wuhan, the spatial distribution pattern of football fields needs to be further examined.

2.2 Data

The data used in this study includes three sections, namely, Wuhan football field survey data, Wuhan administrative data and demographic data. (1) Wuhan football field survey data were originated from field census (10 years as a census cycle). The data were based on the 5th national general survey of sports fields (2004), the 6th national general survey of sports fields (2014), and general survey of football fields in Wuhan (2018). The data contain the detailed name, location, and year of completion for each football field. (2) The data of Wuhan administrative division was from the vector map of Wuhan City (scale 1:10000) drawn by Wuhan Planning Institute, and the data

* Corresponding author: weberwang@jhun.edu.cn

format is Shapefile. It includes 13 regions under the jurisdiction of Wuhan. (3) The demographic data was obtained from Hubei statistical yearbook 2019, including the number of permanent residents in 13 districts of Wuhan in 2018.

In this study, the number of football fields per capita was applied as the index to explore the spatial distribution patterns of football fields in Wuhan. The number of football fields per capita is calculated by dividing the number of football fields per district by the permanent residents (Unit: Number/10,000 person, the same below).

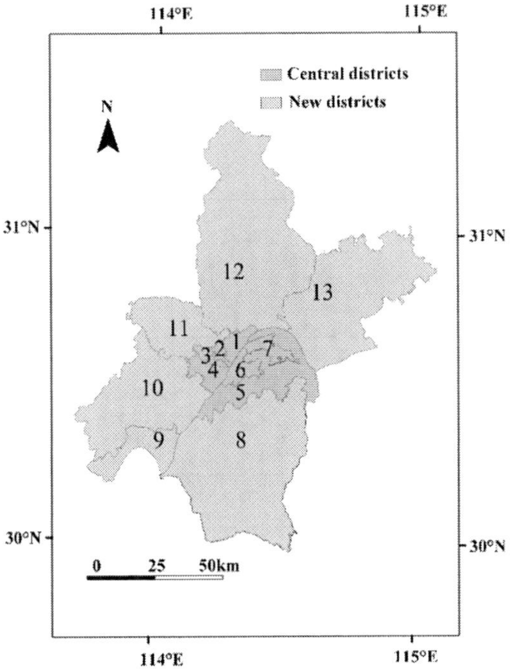

Figure 1. The administrative division of the study area.

2.3 Methodology

2.3.1 The analysis on global spatial autocorrelation

The first law of geography is the basis of spatial data analysis, which points out that everything is related to other things, but the similar things are more closely related. Spatial autocorrelation is an important index to study the degree of spatial data association, which is used to measure the similarity between the same spatial attributes. Spatial autocorrelation can reveal three distribution associations of spatial data, namely positive association (clustered distribution), negative association (distributed distribution), non-association (random distribution) [9,10].

Two metrics, i.e., global Moran's I index and local Moran's I index, were adopted to estimate the global and local spatial autocorrelation for football fields respectively. Global Moran's I index is used to measure the overall distribution for a certain spatial attribute in the whole research area. The calculation formula is as follows:

$$I = \frac{n}{s_0} \cdot \frac{\sum_{i=1}^{n} \sum_{j=1}^{n} w_{ij} (x_i - \bar{x})(x_j - \bar{x})}{\sum_{i=1}^{n} (x_i - \bar{x})^2} \quad (1)$$

Where n represents the number of spatial units in the study area; x_i (x_j) represents the observation value of the i(j)th space unit in the study area; w_{ij} is the spatial weight matrix, which represents the adjacency of spatial units; $\bar{x}$ is the average value; s_0 is the sum of the spatial units of the spatial weight matrix [11].

The statistical test of global Moran's I usually adopts two-sided test, and the calculation formula of Z value is as follows:

$$Z = \frac{I - E(I)}{\sqrt{V(I)}} \quad (2)$$

Where E(I) and V(I) denote the expected values and variances of global Moran's I index.

Global Moran's I value generally ranges from -1 to 1, and greater than 0 indicates that there is a positive spatial association in the distribution of football fields in the study area. On the contrary, less than 0 indicates that there is a negative spatial association in the distribution of football fields. The closer the value is to 1, the higher the spatial positive association degree is, and revealing the distribution of football fields tends to be concentrated. Particularly, when the value reaches 0, the football fields are randomly distributed in the study area.

2.3.2 The analysis on local spatial autocorrelation

Global spatial autocorrelation is essentially a global average measure in space, which can't reflect the spatial association among local regions. Local spatial autocorrelation is often employed to access the spatial association of every spatial unit in the study area. The measurement of local spatial autocorrelation usually depends on local indicators of spatial association (LISA). LISA transforms global spatial association into local form. The idea of LISA is introduced into local Moran's I index, which can define the local Moran's I index for any spatial observation i in the study area [12]. For spatial observation i, the LISA definition for Moran's I is as follows:

$$I_i = z_i \sum_j w_{ij} z_j \quad (3)$$

z_i and z_j record the deviation between the observed values and the mean values of all observed values. The addition operation of j targets at all the observed values in the neighborhood set j_i of i, and the definition of spatial neighborhood is expressed by the spatial weight matrix w_{ij}.

In literature, Moran scatter plot and LISA cluster map are often used in local spatial autocorrelation analysis. The combination of the two methods can reveal the spatial association characteristics of the spatial objects from different perspectives.

Moran scatter plot depends on Cartesian coordinate system, where the X-axis is the normalized attribute value of each spatial unit, Y-axis is the spatial lag value. Moran scatter plot contains four quadrants, representing four local spatial clustering patterns between observation i and its neighbourhood observations, namely, first quadrant: local high-high clustering pattern, high value observation is encompassed by high value neighbors; second quadrant: local high-low clustering pattern, high value observation is encompassed by low value neighbors; third quadrant: local low-low clustering pattern, low value observation is encompassed by low value neighbors; fourth quadrant: local low-high clustering pattern, low value observation is encompassed by high value neighbors.

In addtion, through the LISA cluster map, the statistical significance of four local spatial clustering patterns can be inspected, and the statistically significant autocorrelation patterns can be highlighted and visualized on the LISA cluster map. Generally, the statistical significance level is set at α= 0.05.

3 Result

3.1 The spatial distribution of football fields in each district in Wuhan

By cleaning and integrating the census data of sports fields in three years (2004, 2014 and 2018), a total of 1022 football fields in Wuhan were obtained in this study. Table 1 summarizes the number of football fields and population of 13 districts in Wuhan.

Table1. The number of football fields and population of 13 districts in Wuhan.

District	The number of football fields	Population (Unit:10,000)
Jiang'an	77	96.27
Jianghan	64	72.97
Qiaokou	53	86.87
Hanyang	43	66.42
Wuchang	77	128.28
Hongshan	128	167.73
Qingshan	13	52.89
Dongxihu	66	58.48
Caidian	72	76.16
Jiangxia	201	96.20
Huangpi	70	101.19
Xinzhou	88	91.06
Hannan	70	13.58

Figure 2 is a hierarchical map to display the number of football fields per capita in 13 districts in Wuhan. The number of football fields was classified into 5 categories. As is displayed in the map, the spatial distribution of football fields in different districts shows regional variations. The areas with less football fields were mainly concentrated in central districts, including Jiang'an District, Qiaokou District, Hanyang District, Wuchang District, Hongshan District and Qingshan District. Correspondingly, the new districts occupied more football fields, including Dongxihu District, Caidian District, Jiangxia District, Xinzhou District and Hannan District.

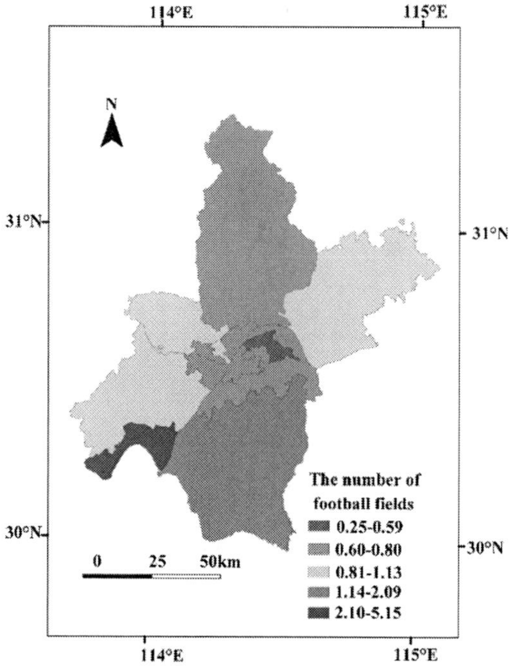

Figure 2. The hierarchical map of the number of football fields per capita in Wuhan.

3.2 Global spatial autocorrelation analysis for number of football fields per capita

The global Moran's I index and its z-score for the number of football fields per capita were calculated by using the GeoDa software (Version 1.18.0), and the pseudo p-value was obtained by using 99, 199, 499, 999 and 9999 random permutations respectively. In the calculation, the queen contiguity matrix was adopted as the spatial matrix in this study. The Global Moran's I index, z-value and p-value for the number of football fields per capita are listed in Table 2.

Table2. Estimated value of Morans' I Index of football fields in Wuhan.

The number of football fields per capita	Global Moran's I index	z-value	p-value
	0.206	3.32	0.006**

**significant at 1% level

The results indicate that the global Moran's I index of the number of football fields per capita in Wuhan is significant at the level of 0.01. The global Moran's I index

https://doi.org/10.1051/e3sconf/202129302051

is 0.206, which implies that the number of football fields per capita is positively aggregated in space. With the proximity of geographical positions, the spatial association of the football fields is more significant. That is, with the aggregation of spatial position, the spatial association becomes more significant. High values tend to be surrounded with high values and low values tend to be surrounded with low values.

3.3 Local spatial autocorrelation analysis for number of football fields per capita

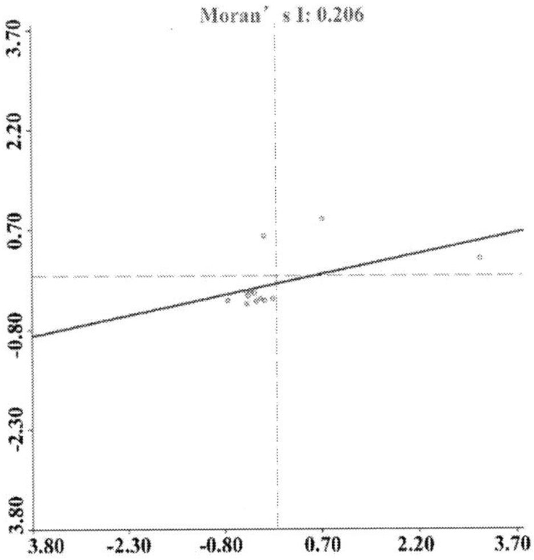

Figure 3. The Moran scatter plot of football fields in Wuhan.

The Moran scatter plot (Figure 3) displays the local spatial distribution of the number of football fields per capita in Wuhan. Two districts were distributed in the first quadrant, including Jiangxia District and Hannan District, indicating a high-high clustering pattern for the number of football fields per capita. At the same time, most of the districts in Wuhan were distributed in the third quadrant, i.e., Jiang'an District, Jianghan District, Qiaokou District, Hanyang District, Wuchang District, Hongshan District, Qingshan District, Dongxihu District, Huangpi District and Xinzhou District, indicating a low-low clustering pattern for the number of football fields per capita. In comparsion, the Caidian District was the only region located in the fourth quadrant. It could be derived that the number of football fields per capita in Caidian District was lower than the surrounding areas.

Figure 4 shows the univariate LISA cluster map of football fields in Wuhan. The map depicts the clustering characteristics of the number of football fields per capita in Wuhan in terms of local scale. The red part (Wuchang District) in Figure 4 represents that the spatial association mode of football fields for the region and its surrounding areas is a local low-low value clustering pattern, that is, the number of football fields per capita in this region and its neighborhood are both low, and the low value are encompassed by the low values. The green part is a low-high value clustering pattern, namely, the number of football fields in the region is low, while the number of football fields in the neighborhood area is high, the low value are encompassed by the high values. The area showing low-high pattern is Caidian District.

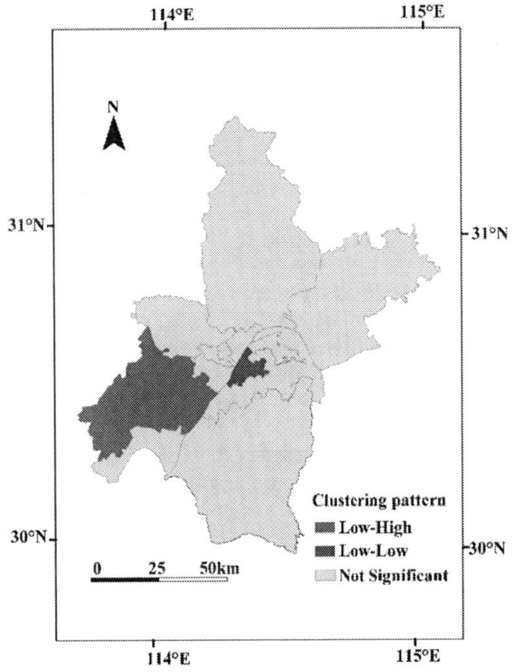

Figure 4. The univariate LISA cluster map of football fields in Wuhan.

4 Discussion

This paper tried to analyze spatial distribution pattern of football fields in Wuhan through ESDA and GIS technology. ESDA technology has been proven to be an effective tool to analyze spatial data, with spatial association measurement as the core, which can reveal the spatial dependence and spatial heterogeneity of data. GIS provides technical support for ESDA analysis. Based on GIS technology, spatial data can be visualized, stored, simulated and predicted. By adding spatial location information into football field data, the field data can be regarded as spatial data.

First of all, the hierarchical map for the number of football fields per capita in Wuhan provides evidence for the uneven distribution of football fields. We have discovered that central districts in Wuhan tend to own less football fields. On the contrary, the number of football fields per capita in new districts increases compared to the central districts. With the acceleration process of urbanization and population migration, the construction areas are expanding obviously [13,14]. In the central areas of Wuhan, most of the lands are occupied by urban areas, the areas planned for sports facilities are relatively rare. Correspondingly, in the new districts of Wuhan, the level of urbanization is comparatively low, and therefore more land has been planned for the construction of football fields.

Secondly, the results of global spatial autocorrelation, i.e., calculation of global Moran's I index, have confirmed the positive spatial association of football fields in whole area of Wuhan city. It means that the football field allocation level of Wuhan city is in a state of clustering distribution among neighboring districts. In the vast majority of cases, when the per capita number of football fields in a region is high, the number of football fields in the surrounding region is also high [15,16]. In line with the high-high clustering pattern, when the number of football fields per capita in a certain area is low, the number of neighboring areas is also low. These two clustering patterns are further confirmed in local spatial autocorrelation subsequently.

Last but not least, the Moran scatter plot and univariate LISA cluster map of football fields in Wuhan provide contribution to the study of local spatial autocorrelation. The local Moran's I index in most districts mainly falls into the third quadrant (low-low clustering pattern). This provides some insights that there exists a clustering trend for the districts with lower number of football fields [17]. When the number of football fields per capita in a region is small, the number of fields in the surrounding regions tends to be small with high probabilities.

Under the statistical test level α of 0.05, only two districts were displayed on the LISA cluster map. The number of football fields per capita in Caidian District and its surrounding districts showed a low-high spatial clustering pattern. This reminds the policymaker to pay much attention to the unbalanced development of football fields in Wuhan. To cope with this, for the areas with less football fields, the number of football fields per capita should be expanded. During the implementation of sports venues planning, these areas need to be given priority during the development process. Meanwhile, the significant low-low spatial distribution pattern for Wuchang District and its surrounding districts was found. This could be explained by the spatial diffusion effect between Wuchang District and its surrounding areas, the spatial discrepancies between these regions were decreasing. It reflects the lack of implementation of the planning policies, indicating that the government's management and planning of sports facilities need to be strengthened.

This study only targets to analyse the global and local spatial distribution patterns of football fields in Wuhan city. We call for future research to study the driving factors of these distribution patterns, including socio-economic factors, natural factors and policy factors.

5 Conclusion

Sports facilities play an important role on physical fitness and health. Our research found that comparing to the new districts in Wuhan, the number of football fields per capita in the central districts is relatively less. To address the uneven distribution of football fields in Wuhan city, the spatial distribution pattern of football fields should be captured in global and local scales.

The results of global spatial autocorrelation for football fields indicate that the number of football fields per capita tended to aggregate at district level. The spatial positions of football fields have potential mutual influence, the football fields tend to gather in adjacent areas. In addition, the results of local spatial analysis indicate that there are regional disparities for the number of football fields per capita in Wuhan.

Therefore, the land use planning policy for football fields should be initiated. For instance, more football fields should be planned and built in the central areas of Wuhan.

References

1. C. Vater, S. Luginbühl, and L. Magnaguagno, J. Sports Sci. **37**, 2789 (2019)
2. Y. Hosokawa, A. J. Grundstein, and D. J. Casa, J. Athl. Train. **53**, 860 (2018)
3. J. Connell, Geogr. Res. **56**, 5 (2018)
4. R. Izzo and E. Belpassi, Risk **178**, 2016 (2018)
5. E. C. A. Nordbø, H. Nordh, R. K. Raanaas, and G. Aamodt, Landsc. Urban Plan. **177**, 19 (2018)
6. F. He, Arab. J. Geosci. **14**, 1 (2021)
7. G. Rybarczyk, S. Banerjee, M. D. Starking-Szymanski, and R. R. Shaker, J. Locat. Based Serv. **12**, 40 (2018)
8. Y. Liu, T. Luo, Z. Liu, X. Kong, J. Li, and R. Tan, Habitat Int. **47**, 113 (2015)
9. G. Geniaux and D. Martinetti, Reg. Sci. Urban Econ. **72**, 74 (2018)
10. M. Ceci, R. Corizzo, D. Malerba, and A. Rashkovska, Data Min. Knowl. Discov. **33**, 698 (2019)
11. H. Pu, K. Luo, P. Wang, S. Wang, and S. Kang, Environ. Sci. Pollut. Res. **24**, 4457 (2017)
12. R. S. Bivand and D. W. S. Wong, TEST (2018)
13. H. Li, Y. Liu, Q. He, X. Peng, and C. Yin, J. Urban Plan. Dev. **144**, 5018002 (2018)
14. T. Luo, R. Tan, X. Kong, and J. Zhou, Sustainability **11**, 2207 (2019)
15. L. Anselin, Geogr. Anal. **51**, 133 (2019)
16. L. Anselin, Geogr. Anal. **27**, 93 (1995)
17. T. N. TİEN, Maden Tetk. ve Aram. Derg. **156**, 179 (2018)

https://doi.org/10.1051/e3sconf/202129302052

Analysis on Space accessibility of Medical Service Facilities in Central Wuhan Based on Improvement of 2SFCA

Yi Wang[1,a], Jun Shao[2*], Miao Xu[3], Changyou Wu[4] and Yibang Li[5]

[1] Urban and Rural Planning, Wuhan University of Science and Technology, Wuhan, Hubei , 430000, China

[2] Urban and Rural Planning, Wuhan University of Science and Technology, Wuhan, Hubei , 430000, China

[3] Urban and Rural Planning, Wuhan University of Science and Technology, Wuhan, Hubei , 430000, China

[4] Urban and Rural Planning, Wuhan University of Science and Technology, Wuhan, Hubei , 430000, China

[5]Urban and Rural Planning, Wuhan University of Science and Technology, Wuhan, Hubei , 430000, China

Abstract. Aiming at the current imperfect hierarchical diagnosis and treatment system and the unreasonable spatial distribution of medical facilities. This article introduces 2SFCA model improved by the Gaussian distance decay function and multiple stage radius to evaluate the accessibility of medical facilities in Wuhan. The results indicate that: firstly, compared with the traditional 2SFCA model, the improved 2SFCA model is more sensitive to the identification of spatial accessibility, and the transition effect of the identification results is more obvious, which can accurately reflect the spatial distribution characteristics of the accessibility of medical facilities in Wuhan. Secondly, the distribution of medical facilities in Wuhan is obvious and the main factors that affect the accessibility value in different urban areas are quite different. Thirdly, the high value areas of medical facilities in the main urban area of Wuhan are mainly concentrated within the second ring line, the transition effect between the second and third ring lines is obvious, and the low value areas are mainly concentrated outside the third ring line. In the end, the main reasons for the low accessibility of the area outside the Third Ring Road are the low service quantity and low service ability of the community health service institutions and the high traffic resistance. The research results can provide a reference for the layout and planning of medical institutions in the main urban area of Wuhan.

1 Introduction

Medical and health services have been an important part of social public service facilities for a long time, and the rationality and accessibility of medical facilities layout is one of the key factors of public service. With the outbreak of COVID - 19 and the further improvement of the hierarchical diagnosis and treatment system in Wuhan, residents have put forward higher requirements for the improvement of the medical service system. But at present, large hospitals are overcrowded and community hospitals are deserted, which is a true picture of patients in many big cities. 80% of the high-quality medical resources of the whole society are concentrated in big cities, and 80% of the medical resources are concentrated in several high-level hospitals. The high concentration of medical resources is an important reason for the difficulty of medical treatment[1]. Investigative studies have shown that at least more than half of the outpatients and inpatients can divert downward on the premise of ensuring medical quality[2]. The phenomenon of patients blindly crowding in large Grade 3-A hospitals is caused by the unbalanced hierarchical allocation of medical resources in each region. Optimizing the allocation of medical resources and building a complete hierarchical medical and treatment system have become an urgent proposition for diagnosis and treatment reform.

Therefore, this paper intends to analyse the accessibility of medical facilities in the main urban area of Wuhan by using the improved 2SFCA model. Taking residential areas as the research unit, aiming at medical facilities of different levels, the improved 2SFCA model is adopted to combine the traffic network data of Wuhan and the multi-level radius data obtained from the survey. Then, apply ARCGIS software for statistical computing and visualization analysis to scientifically measure the spatial distribution of medical facilities in Wuhan city, and to provide reference for the layout and planning of medical institutions in Wuhan city.

*Jun Shao: shaojun@wust.edu.cn

2 Study area and data

2.1 Study area

Wuhan is an important city in central China and the capital city of Hubei Province. By the end of 2020, Wuhan had 13 administrative districts under its jurisdiction. This study takes the main urban area of Wuhan City as the research scope, with a total area of 955.15km^2. There are a total of 90 street towns, including Jiangan District, Qiaokou District, Jianghan District, Hanyang District, Wuchang District, Hongshan District and Qingshan District. The general situation of the study areas is shown in Figure 1.

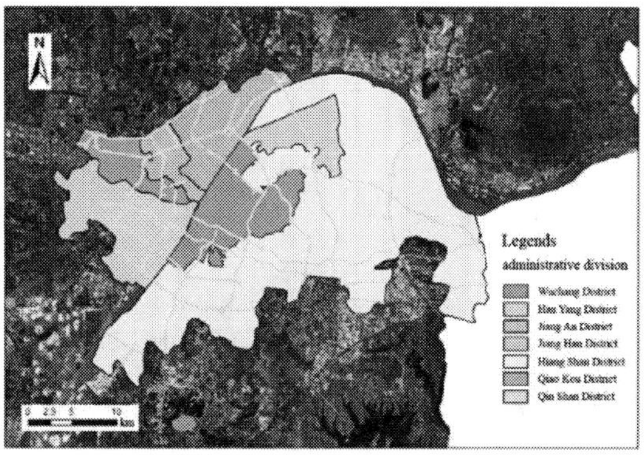

Figure 1. Research Area Overview

2.2 Data Sources

Taking smaller units as the research subjects can better reflect the differences in the medical resources acquired within the region. Therefore, this study takes the residential community as the minimum spatial scale to study the accessibility of medical facilities in the main urban area of Wuhan. The data mainly include:

(1) Information of medical institutions in the main urban area of Wuhan. Among the 421 medical institutions selected in this paper, there are 286 community health service institutions, 59 secondary hospitals and 76 tertiary hospitals. The data mainly came from the official website of Wuhan Health Commission Hospital and Baidu POI coordinate picking tool (Figure 2).

(2) Population data. In order to determine the demand points more accurately, the demand points here are 4934 residential areas in the main urban area of Wuhan obtained in December 2018. The geometric particle of the community is taken as the centre of mass of population distribution, and the number of households in the community is used to replace the number of population for analysis (Figure 3).

(3) **Road traffic data.** The road network data of the main urban area was derived from the electronic map of Wuhan in 2018, including different grades of roads such as urban expressways, expressways, national roads, provincial roads, county roads and township streets in the main urban area, and their speed attributes were assigned according to the road classification. Road speed according to Highway Engineering Design Standards (JTGB01-2014)) [3], Code for Design of Urban Road Engineering [4], and combined with the actual traffic conditions, to determine the speed of different levels of roads (Figure 4).

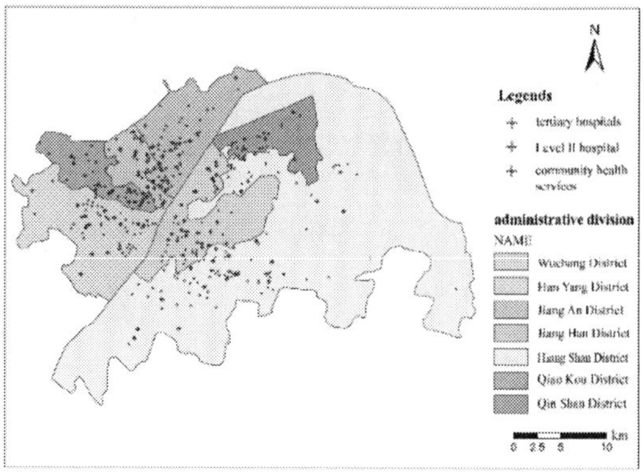

Figure 2. Distribution of Medical Points in main urban area

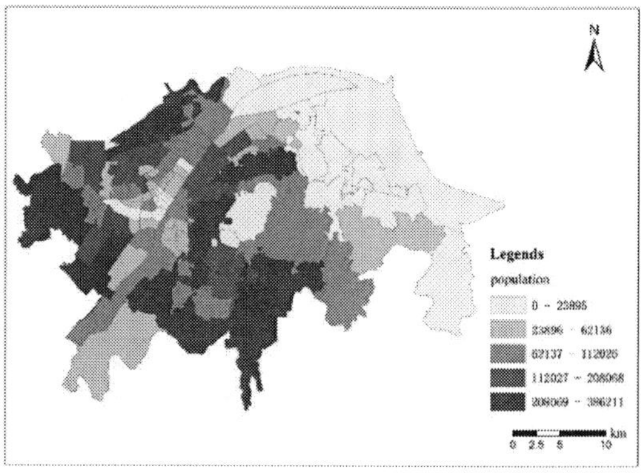

Figure 3. Main urban block population distribution

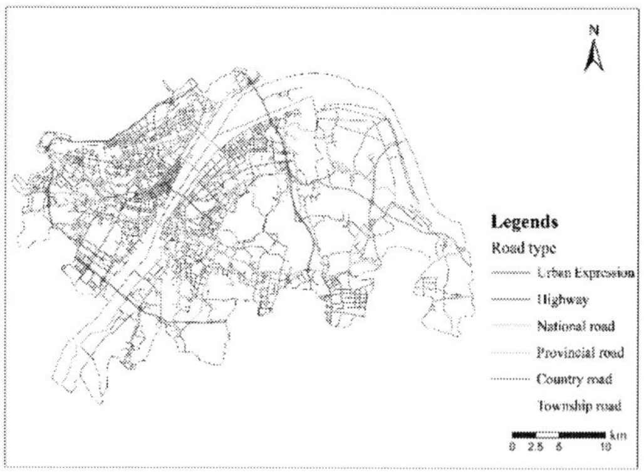

Figure 4. Main urban road classification

2.3 Research Methods

2.3.1 2SFCA

1) Taking hospital j as the centre, determine the time threshold and search all the plot points k within the search radius. Then calculate the ratio of supply and demand, as shown in formula (1)

$$R_j = \frac{S_j}{\sum_{k\in\{d_{kj}\le d_0\}} D_k} \quad (1)$$

In the formula： R_j is the ratio of medical service supply and demand at point j; S_j is supply scale at point j which is expressed by the number of hospital beds in this paper. D_k represents the demand scale of K point within the search scope, which is represented by the number of residential households in the community; d_{kj} refers to the cost of obtaining medical services between point K and point J, which is usually expressed by distance or time, time is used in this paper; d_0 is the search radius.

2) From the centre of cell point i, search for all medical points j within the search radius, then the accessibility A_i^F at Settlement i is obtained by summing up the services R provided by all medical points, as shown in formula (2)

$$A_i^F = \sum_{j\in\{d_{ij}\le d_0\}} R_j = \sum_{j\in\{d_{ij}\le d_0\}} \frac{S_j}{\sum_{k\in\{d_{kj}\le d_0\}} D_k} \quad (2)$$

In the formula： A_i^F is the accessibility of medical services at cell point i; d_{ij} is the time cost of obtaining medical services between points i and j.

2.3.2 Improved 2SFCA

The traditional 2SFCA fully considers the balance between the supply and demand points and the resident competition for the limited medical resources, but ignores the impact of the hierarchical scale of the medical points on the residents' medical care conduct. Generally speaking, the larger the grade scale, the more perfect the medical facilities, the farther they can attract residents; however, the smaller the scale, the weaker the technology, and the less likely it is to attract distant residents. In addition, within the search radius, with the increase of distance, the residents' travel willingness will also be reduced, and accessibility has a law of attenuation with the distance. Based on this, Many scholars improve the 2SFCA model by extending the search radius and the introduction of the distance attenuation function. Wenli Yang[5]al. improved the 2SFCA model by referring to the shortest time of Baidu API planning service to set the multi-level radius and introducing the power function attenuation function in the accessibility study of downtown Shanghai. Li Fu[6] et al. adopted the method of field questionnaire survey to set multi-level radius and used Gaussian distance attenuation function to measure the accessibility of medical facilities in Beipei District of Chongqing. Jiangxia Yu et al. calculated the accessibility of medical facilities in the main urban area of Xi 'an by taking the number of medical service population as the basis and setting dynamic radius based on programming program. These extended forms make up for the deficiency of traditional single search radius to some extent, but they are also inevitably subjective and need more research and continuous improvement. For the distance attenuation, most of the extension forms are based on the distance attenuation function, and the general form is shown in formula (3) and (4). The specific extended form varies with the form of the function G (d_{ij}), mainly including the piecemeal attenuation form of gravity model kernel density form IV and Gaussian form, etc., which makes up for the shortage of treating G (d_{ij})) as a constant in the traditional 2SFCA model.

This paper comprehensively considers the service ability of medical points, the attraction ability of different hospital scale on the residents, and the influence of distance attenuation on their travel willingness. The improved 2SFCA method adopted in this paper has three improvements compared with the traditional method. Firstly, it introduces the Gaussian distance attenuation function; secondly, using the road network data to calculate the shortest time required between the starting and ending points; in the end, through the setting of the multi-level radius and combining the actual situation of the area. The results indicate that:

For community health services, 44 % of residents think that medical travel time is acceptable within 10 ~ 15 min, and 33% think that medical travel time is acceptable within 15 ~ 20 min. For secondary medical sites, 62 % of residents consider medical travel time acceptable within 20 ~ 30min, and 25 % believe that medical travel time is acceptable within 30 ~ 40 min. For tertiary medical sites, 45 % of residents believe that travel time is acceptable within 40 ~ 60 min, and 20 % believe that travel time is acceptable within 60 ~ 80 min. Therefore, the maximum travel time of community health service institutions is set to the maximum travel time of 15 min, secondary hospital to the maximum travel time of 30 min, tertiary hospital to 60 min.

$$A_i^F = \sum_{j\in\{d_{kj}\le d_0\}} \frac{S_j \times f(d_{ij})}{\sum_{k\in\{d_{kj}\le d_0\}} D_k f(d_{ij})} \quad (3)$$

$$f(d_{ij}) = \begin{cases} g(d_{ij}) & d_{ij} \le d_0 \\ 0 & d_{ij} > d_0 \end{cases} \quad (4)$$

$$g(d_{ij}) = \frac{e^{-1/2\times(d_{ij}/d_0)^2} - e^{-1/2}}{1 - e^{-1/2}} \quad (5)$$

In the formula: f (d_{ij}) is the general form of the distance decay function; g (d_{ij}) is the distance decay function within the search radius do, and the Gaussian distance decay function in this paper.

Table1. Improve the 2SFCA Model Search Thresholds

Medical facilities	Search range
Third Hospital	60
Secondary Hospital	30
Community Health Services Agency	15

2.3.3 Space accessibility calculation of medical facilities

Because the straight-line distance doesn't represent the actual distance between the two points, and the traffic speed of the road of different grades is also very different. In order to make the research more suitable for the actual situation, this paper takes time to characterize the travel impedance of the residents.

1) Based on the spatial location information of the residential areas and medical points, combine it with the road traffic data in ARCGIS to establish a spatial road network data set. Give different levels of roads to different speed attribute values according to Table 1, establish the OD cost matrix, and calculate the travel time of each resident to all medical points.

2) According to the established OD cost matrix, the shortest travel time of each resident to the medical point is calculated in ARCGIS, and the average of the shortest medical time consumption of each resident is 32.23min, so this paper sets the fixed threshold for the unimproved 2SFCA medical points to 32.23min. For the improved 2SFCA model, different search radii are set according to the grade scale of the medical points and the data of the field survey questionnaire. The Community Health Services are established as the 15min, the secondary hospitals are set to 30min, tertiary hospitals are set to 60 minutes.

Table2. Road speeds of different grades in Wuhan main urban area

Road Type	Speed(km/h)
Highway	80
Urban Expressway	80
National road	60
Provincial road	60
County Road	40
Township streets	20

3) Taking the medical points as the starting point, all the community points falling within the fixed search radius of 32.23min are obtained, and the supply and demand ratio of each medical point is calculated according to formula (1); for improvement 2SFCA, all cell points under the corresponding service threshold shall be searched according on the level of different medical points, the distance weight of the community to each medical point is calculated according to formula (4) and formula (5), and the service supply and demand ratio of each medical point is calculated.

4) Taking the community point as the starting point, calculate the sum of the supply and demand ratio of all medical points within the search threshold range; improvement 2FSCA obtains distance weight according to travel time basis formula (4) and formula (5), then calculating the weighted sum of the supply-demand ratio of the residents is worth reaching the spatial accessibility of the medical facilities of each settlement.

5) The reachability of the calculation results is assigned to each cell point and rendered in ArcGIS to obtain the reachability distribution diagram.

3 Results and Analysis

The spatial accessibility results of central urban medical facilities in Wuhan with traditional 2SFCA and improved 2SFCA were interpolated, and the results are shown in Figure 5 and Figure 6.

3.1 Comparison of Traditional 2SFCA Model and improved 2SFCA Model

Comparing Figure 5 and Figure 6, using traditional 2SFCA and improved 2SFCA, there are roughly features decreasing from east to west, from center to four sides, but there are also two differences.

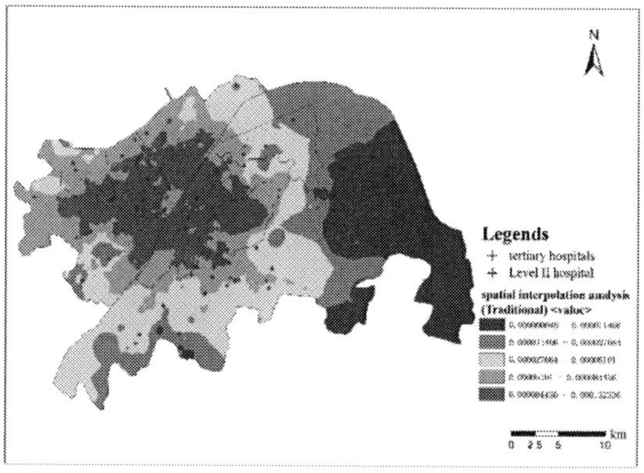

Figure 5. Traditional 2SFCA

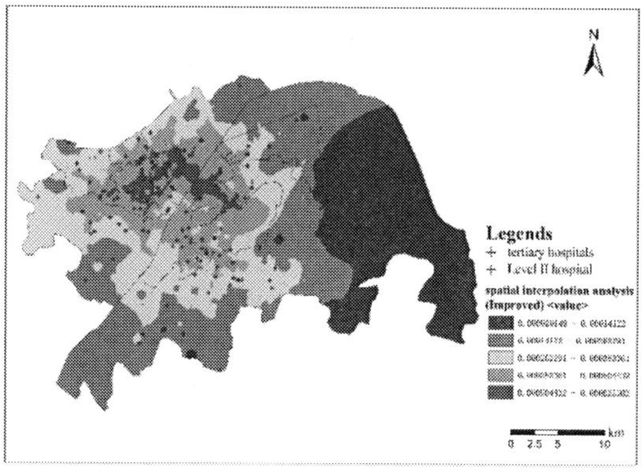

Figure 6. Improved 2SFCA

From the spatial distribution perspective, the results of traditional 2SFCA show the characteristics of two-level distribution, the circle transition result of improved 2SFCA is more significant, which are distributed at both ends of high accessibility and low accessibility. However, the circle layer transition result of improved 2SFCA is more significant, which further classifies the areas of high accessibility judged by traditional 2SFCA in a more detailed way. It is shown that the improved 2SFCA has advantages in identifying internal differences in high reachable value region. The improved 2SFCA also identifies the low accessibility of marginal areas such as Zhangjiawan Street, Qingling Town, Tianxing Town, Hongshan District, Qingjiajie Street, Jiangan District,

https://doi.org/10.1051/e3sconf/202129302052

Changqian Street, Gongmingcun Street, Qingshan District, which better reflects the differences in accessibility within the study area. In addition, the distribution area with the highest accessibility distribution of traditional 2SFCA is too large, some areas far from hospitals and without intensive distribution of medical points are also divided into high accessibility areas, such as Zhoutou Street in Hanyang District and Huashan Town in Donghu Development Zone ZuoLing town streets and the kanto. However, the improved 2SFCA takes into account the population scale competition for medical resources and distance attenuation factors, and the areas with the highest accessibility value only appear in the north of Wuchang District, the south of Jiangan District, the middle of Jianghan District and part of Qiaokou District, which is run through by Jianshe Avenue and Xudong Avenue. This calculation result is more consistent with the actual situation.

According to the calculation results, there is a big difference between the maximum and minimum values of the two methods, indicating that there is a big difference in the accessibility of medical facilities in the central urban area, and the distribution of medical resources is uneven. In addition, after the normalization of the calculation results of the two methods, statistical features such as the mean median standard difference are obtained, as shown in Table 3. It is found that the calculation results of the improved 2SFCA are slightly smaller than those of the traditional 2SFCA. This is because the improved 2 SFCA considering the size of the medical level and the distance between the point of supply and demand factors on the impact of accessibility value, using different search radius and add gaussian distance decay function etc, resulting in part under the traditional algorithm can reach the quantities under the improved algorithm and inaccessible or lower accessibility values, lower part of the grade Remote medical sites are excluded. It is more in line with the actual medical treatment situation, so the overall calculation result is smaller.

Table3. Comparison of the two methods

	Traditional 2SFCA	Improved 2SFCA
Minimum value	0.00002	0.00002
Maximum value	0.000726	0.000626
Average value	0.0004675	0.000409
Median number	0.000421	0.000495
Standard difference	0.00011	0.00009

To sum up, the improved 2SFCA takes into account the range attenuation factor, the grade scale factor of the medical point, and uses the Gaussian range attenuation function combined with multi-stage radius to make up for the shortcomings of the traditional 2SFCA. The calculation results are more in line with the actual situation in all aspects and are better than the traditional 2SFCA. Therefore, the following article will focus on the analysis of the calculation results of the improved 2SFCA.

3.2 Spatial distribution of accessibility of medical facilities in the main urban area of Wuhan

On the whole, the accessibility of medical facilities in Wuhan downtown is high, the results present the circle distribution and east-west imbalance. From the east to the west and from the centre to the periphery, the accessibility value decreases gradually, especially in the east of Hongshan District, which is the lowest area in the main urban area, while the accessibility value in the central Jianghan District of Wuchang District is higher. The distribution of medical facilities in Wuhan's main urban area can be divided into 5 circles:

1) First-class area is the highest accessibility circle, the north of Wuchang District, south of Jiangan District, central Jianghan District and Qiaokou District have the highest accessibility, belonging to the first-class area. It has obvious advantages over the surrounding area.

The perfect traffic network, good connectivity between medical sites and residents, the perfect hierarchical construction of medical institutions, reasonable spatial layout and rich overall medical resources lead the peak of medical accessibility value in this region in the overall central urban area.

2) The second-class area is the circle layer with high accessibility. Most areas in the south of Qiaokou District of Jianghan District of Wuchang District and the north of Hanyang District have better accessibility and belong to the second-class area. This area basically surrounds inside and outside the Second Ring Road of Wuhan City and occupies a large area.

However, compared with the first-class area, the population density of most blocks in this area is very high, the supply-demand balance relationship of medical resources is slightly less than that of the first-class area, but it can better meet the daily medical needs of residents.

3) The third-class area is generally accessible circle. This area is basically located between the second and the third rings in the central urban area. The regional area is large, and the population density of most blocks in the area is very high. The main problem in the three types of regions is that, on the one hand, there is a lack of hospitals in highly densely populated areas in the region, which cannot be supported by community service agencies alone; on the other hand, some areas with more hospitals in the region have the problem of unreasonable spatial distribution of supply and demand mismatch.

4) The fourth type of area is the circle layer with low accessibility. Most blocks in this area have low residential density. There are 6 level-III hospitals,8 level-II hospitals and 26 community health service institutions in this area. The regional population density is not high, and the main reasons for the low accessibility value are the high travel impedance caused by the low road network density and the inconvenience of nearby medical treatment caused by too few community health service institutions.

5) The five types of areas are low accessibility zones, and the areas with the worst accessibility are the eastern part of Hongshan District and a small part of Qingshan District, which are completely outside the Third Ring Road of Wuhan. Compared with other areas, the area is larger, the location is more remote, the population density is very low, and the distribution of medical points is more dispersed.

4 Conclusion and Discussion

Based on the basis of traditional 2SFCA, considering the influence of hospital scale and distance attenuation factors, the multi-level radius and Gaussian distance attenuation function are introduced to improve the 2SFCA model, based on GIS space analysis technology and the minimum research unit to study the accessibility of medical facilities in Wuhan, making the research results closer to reality. The research results indicate that:

1) The improved two-step mobile search method presents a more significant layer transition feature in the spatial distribution, with high sensitivity in identifying the internal differences between high and low reachability regions. In statistical features, the average and standard difference are slightly less than the traditional 2SFCA, the more uniform distribution characteristics, and the analysis found that low accessibility and general accessibility of improved 2SFCA recognition, such as Baishazhou Street, Wuchang District and Zhangjiawan Street, Hongshan District, The high accessibility area identified by traditional 2SFCA, after in-depth analysis, found that there were no secondary hospitals and tertiary hospitals in the two blocks, and the actual medical accessibility value is not high. The results of improving 2SFCA more accurately reflect the spatial distribution characteristics of accessibility of medical facilities in the central urban area of Wuhan.

2) The overall medical facilities in the central urban area of Wuhan are high, showing a trend of decreasing from center to around and from west to east.

3) The domestic research, such as Li Fu, improved the two-step mobile search method by introducing Gaussian distance decay function and multi-level service radius, and also compared the improved 2SFCA with the traditional 2SFCA results, and found that the improved statistical eigenvalues are less than the pre-improvement results, in addition, the characteristics of the spatial distribution characteristics are also inconsistent. Through comprehensive analysis, the improved results are more consistent with the actual situation and are consistent with the conclusions of this paper, it shows that the improved two-step movements can consider various factors comprehensively, and the results can more relatively accurately describe the accessibility spatial distribution of medical facilities in the study area.

The results of this paper help to understand the spatial accessibility of medical facilities in Wuhan and provide reference for the rational layout planning of medical institutions, but there are many deficiencies: firstly, still sets the maximum time threshold for residents at different levels of medical points. In the next step, the study can consider setting the sensitivity analysis of different travel time impedance to determine the threshold d_0. Secondly, the study is based on the assumption that residents only seek medical treatment in the central area of Wuhan and the internal medical resources are only for residents in the study area, without considering the behavior of the surrounding area and the accessibility of the region.

References:

1. Hu X.R., Xiong W., Ren F., Mao Y., Wang G.C. (2017) Analysis of medical service capability by road network—— Taking Wuhan as an example. Surveying and GIS, 40 (11): 67 – 70.
2. Yu J.X., Zhang X.Q. (2017) Study on Medical Facilities Based on Service Radius —— Taking the main urban area of Xi ' an as an example. Journal of Shaanxi Normal University (Natural Science Edition), 45 (05): 78 – 84.
3. Yang W.L., Chen N., Wang H.M., Yang W., Xia C.L. (2020) Study on Reachability of Medical Facilities in 2SFCA Shanghai Central Urban Area. Geographic Information World, 27 (06): 75 – 79.
4. Fu L., Wang Y., Zeng B., Mao Y., Gao M. (2019) Space Reachability Analysis of Beibei Medical Facilities Based on Improving the Two - step Mobile Search Method. Journal of Earth Information Sciences, 21 (10): 1565 – 1575.
5. Guo J.X., Liu T., Wang K.Y., Xie S.J., Zhang Y.S. (2020) Educational Resources Based on Improved Two - step Mobile Search Method —— takes the main urban area of Hangzhou as an example. Journal of Hangzhou Normal University (Natural Science Edition), 19 (03): 329 – 336.
6. Wang Q. (2019) Reachability in Shanghai Based on Improved Two - step Mobile Search. Wuhan University.
7. Liu C.C. (2019) Reachability Analysis and Layout Optimization of Medical Facilities Based on Improved 2SFCA Algorithm. Chengdu University of Technology.

 https://doi.org/10.1051/e3sconf/202129302053

Measurement and analysis of bridge expansion joint noise

Dingtao Mao[1], and Yong Ding[1,*]

[1]Department of Civil Engineering, 315211 Ningbo, China

Abstract. The structure-borne noise while the vehicle passing across the bridge expansion joint is the main source of urban bridge noise. In order to control this noise pollution, 20 bridges including three types of typical expansion joints in Ningbo City were selected, and the noises were measured while vehicle passed across the bridge expansion joints. The measured results are expressed by the Z-weighted sound pressure level, which kept the effect of the low-frequency noise. Then the influence factors of this noise are discussed. The results show that: (1) The sound pressure while vehicle on the bridge expansion joints is much greater than that on the normal road or mid-span of the bridges, which results in significant environmental noise pollution; (2) The wider the gap of the bridge expansion joints, the greater the noise level; (3) The noises produced by the modular expansion joints and comb-plate expansion joints are greater than that from the single-gap expansion joints.

1 Introduction

With the development of urban modernization, the number of viaduct roads has increased significantly. Ghimire [1] and Sexton [2] pointed out that the noise generated by vehicles passing bridges, especially the bridge expansion joints, has become a kind of environmental pollution. Therefore, it is very important to analyse the noise of bridge expansion joints. There are many work on the noise of the main structure of the bridge, but only a few researches on the noise of the bridge expansion joints. Wakefield [3] measured the sound pressure level of modular bridge expansion joints; Ghimire [4-5] used the boundary element method to simulate the structure noise of bridge expansion joints; Ding [6] found that the traffic load has a great impact on the bridge expansion joint structure, and will produce large noise radiation; Glaeser [7] proposed that the noise of bridge expansion joints is the result of vehicle tire impacting the discontinuous surface profile of expansion joints.

Up to now, the work of using the measured data to analyse the noise of the bridge expansion joint is still lacking. Therefore, more than 20 bridges in Ningbo are selected in this paper, the sound pressure of vehicles passing through the bridge expansion joints are measured, the sound pressure level are calculated, and the influence factors of bridge expansion joint noise are discussed.

2 Measurement process

2.1 Test method

The sound pressure is measured by using the acoustic sensor of BK4190, the effective frequency range of which is 3.15-20000Hz. Four-channel acquisition instrument from Beijing BSWA company is used for date collection.

As shown in Figure 1, three test points on the bridge are taken. According to the acoustic test specification [8], each measuring point is 1.2 meters above the bridge deck and 0.2 meters from the inner edge of the pavement. Each test takes about twenty seconds, while the vehicle passes through the bridge. The test date is from June 2016 to October 2016

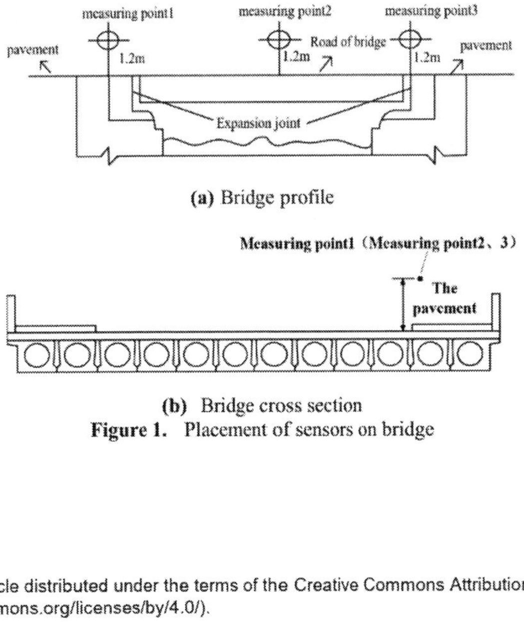

(a) Bridge profile

(b) Bridge cross section

Figure 1. Placement of sensors on bridge

* Yong Ding: dingyong@nbu.edu.cn

2.2 Test objects

It is necessary to measure and analyse the bridge noise by using the Z-weighted sound pressure level, which kept the effect of the low-frequency noise [9]. Therefore, the Z-weighted sound pressure is adopted in the measurement of the noise generated by the vehicles passing across the bridge expansion joints. Considering that the test objects should include various types of bridges and expansion joints, after on-site investigation of more than 100 bridges and their expansion joints in Ningbo, 20 of them were selected for field measurement. The selection was based on the fact that there is no large sound interference sources nearby. The bridges are shown in Table 1 and two of them are shown in Figure 2.

(a) Yuanshi Bridge

(b) Wantou Bridge
Figure 2. Typical bridges

The types of the bridge expansion joints are shown in Table 2, and several typical expansion joints are shown in Figure 3.

(a) Single-gap expansion joint (Cailian Bridge)

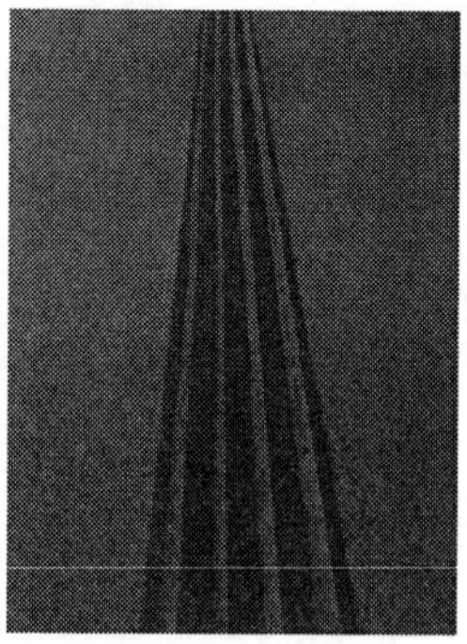

(b) Modular expansion joint (Mingzhou Bridge)

(c) Comb-plate expansion joint (Yongjiang Bridge)
Figure 3. Various types of typical expansion joints

2.3 Data process

During the test, it is necessary to reduce the influence of surrounding noises on the bridge expansion joint noise. Therefore, the noise is measured in the mid-night, and the typical sound pressure is shown in Figure 4.

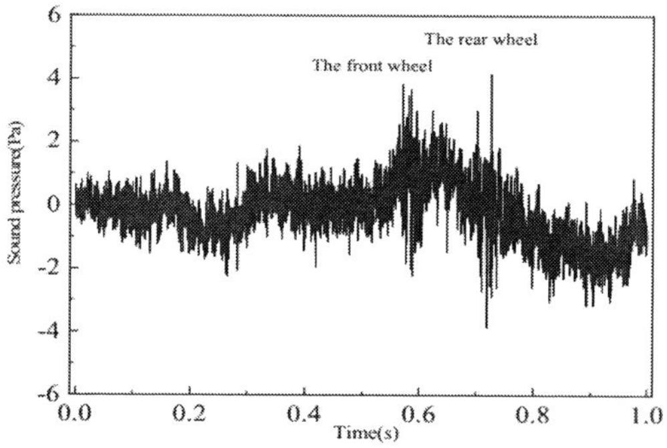

Figure 4. The time-history sound pressure of the vehicle passing across the expansion joint (Cailian Bridge)

When the vehicle passing the normal road or mid-span of the bridge, the main sources of the noise are the vehicle noise and the friction noise between tire and road. When the front and rear wheels of the vehicle are on the bridge expansion joint, it generates structural noise [6-7], and it is observed that the vehicle jumping on the expansion joints, that causes the low-frequency noise because of the vibration of the bridge structures [9-10].

https://doi.org/10.1051/e3sconf/202129302053

Table1. Statistics of bridge types

Bridge type	Simple beam bridge	Continuous beam bridge	Arch bridge	Cable-stayed bridge	Suspension bridge
Amount	8	6	3	2	1

Table2. Statistics of expansion joints

Expansion joint type	Single-gap expansion joint	Modular expansion joint	Comb-plate expansion joint
Amount	15	2	5

In Figure 4, the peak value of the sound pressure occurs when the wheels passing across the expansion joint, and the low-frequency fluctuation of the sound pressure is caused by the influence of wind pressure on the acoustic sensor [11].

According to the measured sound pressure, the sound pressure level is calculated by the following formula

$$L_P = 20 \times \frac{\lg P}{\lg P_0} \tag{1}$$

Where L_P is the sound pressure level (dB), P is the measured sound pressure, and P_0 is the standard sound pressure, which is taken as $2\times10^{-5}Pa$.

3 Measured results and analysis of influence factors

3.1 Bridge expansion joint noise

From the time-history of expansion joint noise shown in Figure 4, it can be seen that two peaks of the sound pressure appear when the front and rear wheels of the vehicle passing across the bridge expansion joint, respectively. And these two sound pressure are much greater than the others Therefore, the bridge expansion joint noise is an important source of the traffic noise of bridge structure.

In order to verify this conclusion, a series of measurement of sound pressure are performed on the single-gap expansion joints of Yuanshi Bridge and Wantou Bridge (Figure 2) while vehicles passing across.

The measured sound pressure when the vehicle on the mid-span of the bridge are compared with that when the vehicle on the bridge expansion joint.

The measuring point 1 and 3 are beside the bridge expansion joints, and the measuring point 2 are beside the mid-span of the bridge.

The average sound pressure levels are calculated after multiple measurements, and they are shown in Tables 3 and 4. It indicates thatthe noise of bridge expansion joint is much greater than that of the bridge structure.

The environmental standard [8] suggests that the acceptable sound pressure level for human body is form 60 dB to 70 dB. However, the sound pressure level of the bridge expansion joints when vehicle passing across reaches 80-100 dB. This phenomenon needs to be paid enough attention.

3.2 Analysis of influence factors of expansion joint noise

3.2.1 Width of the gaps in expansion joint

It can be seen from Table 3 and Table 4 that the average noise of bridge expansion joints generated by vehicles on the Yuanshi Bridge is greater than that on the Wantou Bridge.

During the test, it was found that the width of the single-gap expansion joints on the Yuanshi Bridge is 2-3mm larger than that on the Wantou Bridge, which may be one of the factors causing the differences of the noise of bridge expansion joints.

In order to confirm the conclusion, the cement truck passing the single-gap expansion joint is taken as an example. A series of noise measurements are performed for the bridges on the Huizhan Road and Chuangyuan Road. The maximum sound pressure when the vehicles passing across the bridge expansion joints are obtained, and the relationship between the maximum sound pressure level and the width of the gaps in bridge expansion joints are shown in Figure 5.

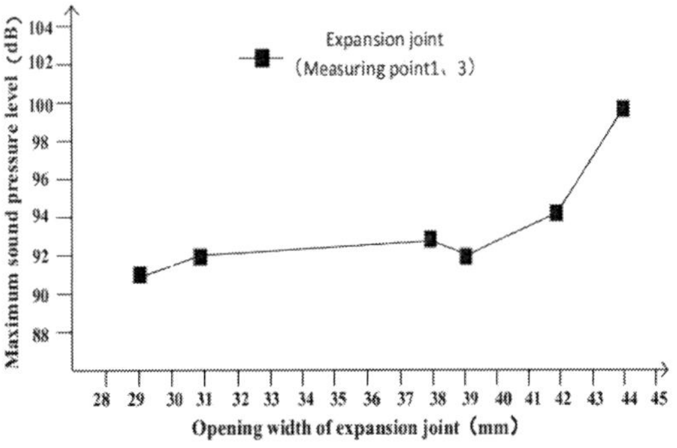

Figure 5. The maximum sound pressure level varies with the width of the gap in bridge expansion joints

Figure 5 shows that as the width of the gap of single-gap expansion joint increases, the noise generally rises. The reason is that the wider the gap of the bridge expansion joints, the more serious of vehicle jumping on the expansion joints, and the greater the low-frequency noise because of the vibration of the bridge structures.

Table3. Sound pressure level of the expansion joint of Yuanshi bridge

Vehicle Type	Point 1 beside the expansion joint	Point 2 beiside the mid-span of bridge	Point 3 beside the expansion joint
Cement truck	88.9	59.7	86.6
Bus	83.1	56.7	82.3
Mini bus	81.7	58.0	82.7

Table4. Sound pressure level of the expansion joint of Wantou Bridge

Vehicle Type	Point 1 beside the expansion joint	Point 2 beiside the mid-span of bridge	Point 3 beside the expansion joint
Cement truck	83.8	53.8	83.6
Bus	76.7	53.6	79.4
Mini bus	75.2	53.3	76.0

Table5. Maximum sound pressure level of different types of bridge expansion joints

Number	Type of the expansion joint	Maximum sound pressure level
1	Single-gap expansion joint	91
2	Modular expansion joint	93
3	Comb-plate expansion joint	93

3.2.2 Type of expansion joint

During the test, the noise of three different types of bridge expansion joints (single-gap expansion joint, modular expansion joint and comb-plate expansion joint) were measured. In order to compare the differences in noise among three different types of expansion joints, the sound pressure level are measured when the cement trucks passing across the expansion joints on Mingzhou bridge, and the results are shown in table 5.

As shown in Table 5, there is little difference between the noise generated by the modular expansion joint and the comb-plate expansion joint, and both of them are 2-3 dB higher than the noise generated by the single-gap expansion joint. The reason is that the modular expansion joint and the comb-plate expansion joint have more complex structures and wider gaps, which produce greater sound pressure [4-5, 12-13].

4 Conclusion

Based on the measured sound pressure of the expansion joints of 20 bridges in Ningbo, the sound pressure level is calculated, and the influence factors of bridge expansion joint noise are discussed. The conclusions are as follows.

(1) The sound pressure while vehicle on the bridge expansion joints is much greater than that on the normal road or mid-span of the bridges, which results in obvious noise pollution, that needs to be paid enough attention;

(2) The wider the gap of the bridge expansion joints, the more obvious the vehicles jumping on the expansion joints, and the greater the low-frequency noise because of the vibration of the bridge structures. Therefore, the noise can be controlled by reducing the width of the gap in expansion joints;

(3) The noise generated by the modular expansion joints and the comb-plate expansion joints are greater than that generated by the single-gap expansion joints.

Acknowledgements

This work has been supported by the Natural Science Foundation of Zhejiang Province (LY19E080009), the Comprehensive Insurance for Country Road in Haishu District of Ningbo City, and Ningbo Transportation Science and Technology Project (202104).

References

1. Ghimire J P , Matsumoto Y, Yamaguchi H, Noise Control Eng J, **56** (2008)
2. Sexton T, INCEJ and ASJ, 734-739 (2011)
3. Wakefield C W, Marriner, Duane E, Can Acoust, **39** (2011)
4. Ghimire J P, Matsumoto Y, Yamaguchi H, J. Sound Vib., **328** (2009)
5. Ghimire J P, Matsumoto Y, Yamaguchi H, Noise Control Eng J, **56**:442-450(9) (2008)
6. Ding Y, Zhang W, Francis T.K. Au , Eng. Struct., **127** (2016)
7. Glaeser K P, Schwalbe G, Zöller M, INCEJ and ASJ 753-762 (2011)
8. AQSIQ, *Environmental quality standard for noise(GB 3096-2008)* (2008)
9. Zhang W, Jin T, Ding Y, UEUE, 42-46 (2015)
10. Xie X, Wu D Y, Wang J F, JZU, **43**(2009)
11. Ravshanovich K A, Yamaguchi H, Matsumoto Y, Eng. Struct., **29** (2007)
12. Mao W S, JURBFC, **7** (2013)
13. Wang C K, JCT, 43 (2016)

Grey prediction analysis of cutoff wall affected by construction effect

YUAN Hong-hu[1a], Dai Yao[2], Wang Yu[2*]

[1] Beijing Institute of water, Beijing, 100048, China

[2] School of Mechanical Engineering, Xihua University

Abstract: The long-term stability analysis of cutoff wall affected by construction effect is crucial. The long-term stability of the cutoff wall of Daning Reservoir is mainly affected by the combined action of highway vehicle load, extreme highway load and water load under different water storage loads. In this paper, the prediction model and method of grey theory are used to predict the horizontal displacement at the top of the cutoff wall, the settlement at the top, the maximum principal stress of the reinforced concrete wall, and the maximum principal stress of the plastic concrete wall based on grey analysis with each storage and drainage cycle as the time series point combined with the actual situation of the project. The long-term stability of the cutoff wall under load is predicted and evaluated. The results show that the maximum stress of the monitoring 2 – 2 section cutoff wall gradually tend to be stable during the whole storage and drainage cycle. The limit maximum compressive stress of reinforced concrete wall is 4.320 MPa, and the limit maximum compressive stress of plastic concrete wall is 1.758 MPa. Judging from the material strength, the maximum principal stress is within the design strength of the material, and the plastic failure of the wall does not occur, indicating that the combined anti-seepage system of rigid cutoff wall and plastic cutoff wall adopted by the Daning Reservoir is stable and safe in the later normal operation period.

1 Introduction

Some sections of the cutoff wall of Daning Reservoir are located in a large regional terrain difference. The upper wall is built first and then buried. There is a large amount of backfill soil on both sides of some walls. At the same time, a highway is planned to be built above the backfill area. The engineering disturbance is serious, and the safety of the cutoff wall needs to be monitored and evaluated during the construction period. After backfill construction, the main load form on both sides of the cutoff wall is the vehicle load of Jingshi highway. After the reservoir impoundment, it is necessary to consider the influence of water pressure on the cutoff wall and the surrounding soil to predict and evaluate the long-term stability of the cutoff wall.

Grey theory method is widely used in deformation prediction because it only needs a small number of data samples to effectively predict deformation. Zhang Jin, et al[1] proposed a tunnel deformation prediction model using improved genetic algorithm to optimize grey neural network. Weng Zhijian, et al[2] Aiming at the shortcomings of the grey GM (1,1) model in predicting the settlement data sequence with large random fluctuation, combined with the grey theory model and Markov chain theory, a new dimension GM (1,1) settlement prediction model based on Markov correction is established. Zhao Zilong[3] proposed a differential grey neural network-AR model based on wavelet decomposition. Tang Chao, et al[4] In view of the fact that the traditional GM (1,1) model cannot consider the influence of the external environment change on the system change trend and the shortcomings of the GM (1, N) multivariate model in the model structure, the OGM (1, N) model is introduced and the adaptive mutation particle swarm optimization algorithm is used to optimize the background value for the error of the background value. Zheng Xudong et al[5] reasonably combined the model based on the principle of maximum entropy and established a hybrid prediction model based on the principle of maximum entropy. Qi Lan, et al[6] introduced the idea of iterative weight and genetic variation to improve the FOA algorithm of Drosophila optimization to improve its search ability and speed. Zhang Yinghao, et al[7] introduced iterative data grouping technology to further improve the accuracy of grey Markov model for deformation prediction. Wang Xinsheng, et al[8] used conventional GM (1,1) model, homogeneous exponential function grey model and non-homogeneous exponential function grey model to predict and compare the deformation of surrounding rock of tunnel. Shen Zhehui, et al[9] established a non-equidistance grey-Markov model to predict the internal deformation of the dam. Guo Haiqing, et al[10] combined with time-shift wavelet denoising and grey theory, studied the monitoring data of Jinping I hydropower station slope displacement,

Corresponding author: Wang Yu, xhwangyu@yeah.net
[a]E-mail: bjyuanhonghu_@126.com

https://doi.org/10.1051/e3sconf/202129302054

proposed time-shift wavelet coefficient correlation denoising and wavelet-MGM (1, n) prediction model. Wang Feng, et al[11] used dynamic multivariate grey model to predict the deformation of dangerous rock.

2 Engineering background

The Daning Reservoir is located in the north of Daning Village, Fangshan District, Beijing. The total storage capacity of the reservoir is 46.11 million m^3, and the highest water level is 58.5m. It is surrounded by a plastic concrete vertical cut-off wall with a thickness of 0.60m and a total length of 7.84km[12]. Among them, the impervious wall of dujiakan section of Xidi passes under the Beijing Shijiazhuang Expressway to be expanded. The impervious wall of this section is composed of reinforced concrete impervious wall above ground and plastic concrete impervious wall below ground. The subgrade of Beijing Shijiazhuang expressway is composed of gravel soil backfilled and rolled on both sides of reinforced concrete wall.

Due to the reconstruction project of Beijing Shijiazhuang Expressway, dujiakan toll station is extended to Daning reservoir area. The extension scope includes the built impervious wall of the reservoir, pile number xd0 + 697.2 ~ xd0 + 835.9, with a length of 138.7m. The cross influences position of highway and cutoff wall is shown in Figure 1[12].

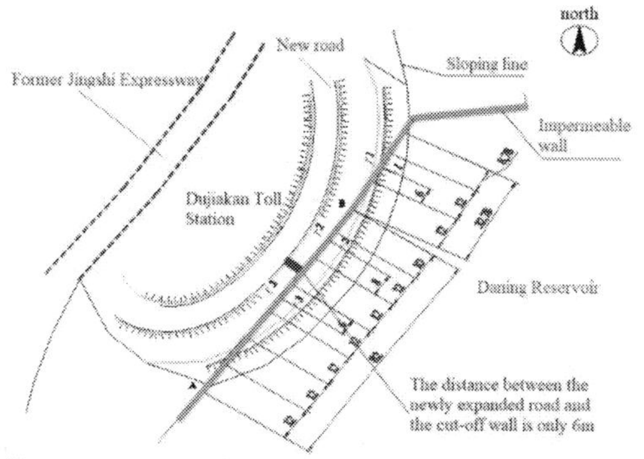

Figure 1. Schematic diagram of the Beijing-Shijiazhuang Expressway and the impervious wall of the reservoir

After the filling construction is completed, the construction near the cut-off wall is mainly the construction of the Dujiakan toll station on the Beijing-Shijiazhuang Expressway. The construction characteristics of this period are the small amount of construction earthwork and the long distance from the cut-off wall, which has very limited impact on the cut-off wall, so it is not considered in the long-term stability calculation; After the reconstruction project of Dujiakan toll station is completed and opened to traffic, the main load form on both sides of the cut-off wall is the vehicle load of Jingshi Expressway; After reservoir impoundment, the influence of water pressure on cutoff wall and surrounding soil should be considered at the same time.

According to the characteristics of the project, the long-term stability of the cut-off wall is mainly affected by road vehicle loads, different water storage loads, extreme road loads and water loads. This article mainly analyzes the stability of the cut-off wall under the above-mentioned load, and predicts the long-term stability. The analysis model is shown in Figure 2.

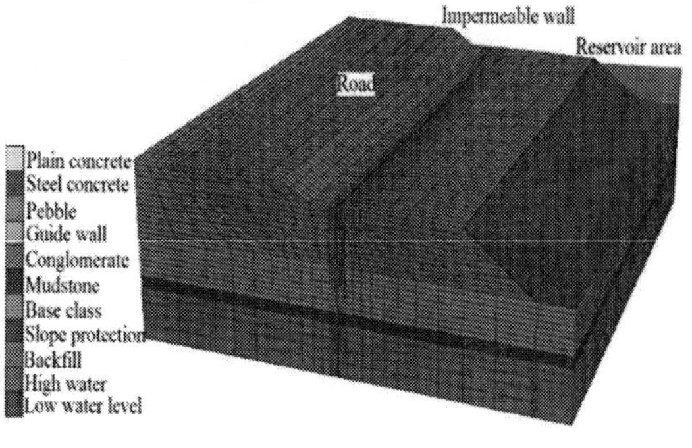

Figure 2. Long-term stability analysis model of cutoff wall

3 GM (1, 1) gray model

The GM(1,1) model establishes an approximate differential equation by performing gray transformation processing on univariate data, converts it into a matrix form, and solves the matrix to predict the value of the variable at the next moment. The prediction process is:

1)Select the initial data series with equal time intervals, This sequence is the data sequence to be predicted

$$X^{(0)}=\left(x^{(0)}(1),x^{(0)}(2),\cdots,x^{(0)}(n)\right) \quad (1)$$

Where $x^{(0)}(n)$ represents the time series data at time n。

2) Perform an accumulation generation on the initial data sequence $X^{(0)}$ (1-AGO, AGO is the accumulation symbol), and generate a new sequence $X^{(1)}$:

$$X^{(1)}=\left(x^{(1)}(1),x^{(1)}(2),\cdots,x^{(1)}(k),\cdots,x^{(1)}(n)\right) \quad (2)$$

其中 $x^{(1)}(k)$ 为:

$$x^{(1)}(k)=\sum_{i=0}^{k}x^{(0)}(i), \quad k=1,2,\cdots,n \quad (3)$$

3) Establish the GM(1,1) prediction model, which is the first-order linear differential equation of $X^{(1)}$:

$$\frac{dX(t)^{(1)}}{dt}+aX(t)^{(1)}=u \qquad t\in[0,\infty] \quad (4)$$

4) Solve differential equations, use differential equations to make predictions

For a sequence of accumulated numbers, when $x^{(1)}(k)$ is known, the prediction function $x^{(1)}(k+1)$ of $x^{(1)}$ is:

$$x'^{(1)}(k+1)=\left(x^{(0)}(1)-\frac{u}{a}\right)e^{-ak}+\frac{u}{a} \quad (5)$$

Thus the predicted value of the original sequence $x^{(0)}$ is:

$$\begin{cases} x'^{(0)}(1) = x^{(0)}(1) \\ x'^{(0)}(k+1) = x'^{(1)}(k+1) - x'^{(1)}(k) = (1-e^{a})\left(x^{(0)}(1) - \frac{u}{a}\right)e^{-ak} \quad k = 1,2,\cdots,n \end{cases} \quad (6).$$

Based on the above prediction model and method of grey theory, combined with the actual situation of the project, taking each storage and drainage cycle as the time series point, based on the grey analysis. The horizontal displacement of the top of the cutoff wall, the settlement of the top, the maximum principal stress of the reinforced concrete wall and the maximum principal stress of the plastic concrete wall are predicted respectively.

4 Prediction of Deformation Law of Cutoff Wall

4.1 Predict sample parameters

The main source of the predicted sample parameters is the model parameters in the process of multiple reservoir storage and drainage. There are 4 parameters to be predicted, and their values are shown in Table 1. From the calculated parameter table, it can be seen that in each storage and drainage cycle, the difference between the parameters of water storage and drainage is small, so in the gray analysis, each adjacent water storage and drainage process is regarded as a time series point. The value is the average of the two, and the final data sequence table for gray analysis is shown in Table 2.

Table1. Table of calculation results of parameters to be predicted

Time schedule	Horizontal displacement of wall top (mm)	Wall settlem ent (mm)	Maximum principal stress of steel concrete (MPa)	The maximum principal stress of plain concrete (MPa)
After the construction is over	17.57	5.043	1.069	0.876
Water storage for the first time	-42.52	4.666	2.793	1.435
First drain	-42.74	4.523	3.803	1.423
Second impoundment	-61.86	4.896	5.36	1.704
Second drainage	-62.23	4.553	5.987	1.463
Third impoundment	-63.3	4.591	6.116	1.535
Third drainage	-61.81	4.725	5.023	1.46
Fourth impoundment	-63.93	4.724	5.049	1.532
Fourth drainage	-64.14	4.849	5.032	1.542
Fifth impoundment	-65.7	4.751	5.161	1.565
Fifth drainage	-63.62	4.76	5.122	1.557

Table2. Parameter table used in grey analysis

Time series	Horizontal displacement of wall top (mm)	Wall settlement (mm)	Maximum principal stress of steel concrete (MPa)	The maximum principal stress of plain concrete (MPa)
1	-62.045	4.7245	5.6735	1.5835
2	-62.555	4.658	5.5695	1.4975
3	-64.035	4.7865	5.0405	1.537
4	-64.66	4.751	5.161	1.565

According to the prediction results, the maximum stress of the impervious wall of the section 2-2 during the entire storage and drainage cycle gradually stabilized. The ultimate maximum compressive stress of the reinforced concrete wall is 4.320MPa, and the ultimate maximum compressive stress of the plastic concrete wall is 1.758MPa.

5 conclusion

The results show that from the point of view of material strength, the maximum principal stress is within the range of material design strength, and the wall has no plastic failure, indicating that Daning adopts a combined anti-seepage system of rigid anti-seepage wall and plastic anti-seepage wall. The reservoir is stable and safe in the later period of normal operation.

Author

Yuan Honghu, male, 1984-, PhD, senior engineer, deputy director, mainly engaged in survey, design and scientific research in water conservancy engineering and geotechnical engineering.

References

1. Zhang Jin, Chen Lin, Lai Zulong. Improved genetic algorithm to optimize gray neural network tunnel deformation prediction[J]. Science of Surveying and Mapping, 2021, 46(02): 55-61+77.
2. Weng Zhijian, Qiu Chenjie, Qiu Fuxiang, Yang Yunhong, Lu Rufa, He Shenglong. Grey GM (1,1) settlement prediction model based on Markov optimization and its application[J]. Science Technology and Engineering, 2020, 20(29): 12065 - 12070.
3. Zhao Zilong. Differential gray neural network-AR model based on wavelet decomposition and its application in subway tunnel settlement prediction[J]. Bulletin of Surveying and Mapping, 2020(S1): 99-103.
4. Tang Chao, Chen Yanying, Li Shulin, Liu Yinchi, Hu Jingyun, Peng Fuhua. Improved OGM(1,N) based on adaptive mutation particle swarm optimization algorithm and its application in dumping yard deformation prediction[J]. Chinese Journal of Rock Mechanics and Engineering, 2020, 39(S1): 3197-3205.

5. Zheng Xudong, Chen Tianwei, Deng Jieli, Duan Qingda, Gan Ruo, Wang Lei. Application of entropy weight method parallel combination model in dam deformation monitoring[J]. Hydropower, 2019, 45(03): 115-119.
6. Qi Lan, Li Qixian. Box culvert settlement prediction based on improved FOA-GM(1,1) gray prediction model[J]. Hydropower Energy Science, 2018, 36(06): 129-132.
7. Zhang Yinghao, Liu Xiaoqing, Ning Xinyang. Application of gray-Markov model in dam deformation prediction of South Russia 5 Hydropower Station[J]. Water Resources and Hydropower Technology, 2018, 49(02): 139-143.
8. Wang Xinsheng, Zuo Weiqin, Zhou Chengtao, Chen Yu, Hou Yabin. Tunnel surrounding rock deformation prediction based on inhomogeneous exponential function gray model[J]. Modern Tunnelling Technology, 2017, 54(01): 105-109.
9. Shen Zhehui, Huang Teng, Tang Youhui. The application of gray-Markov model in dam internal deformation prediction[J]. Engineering of Surveying and Mapping, 2015, 24(02): 69-74.
10. Guo Haiqing, Zhang Min, Huang Tao, Ai Chunbin. Research on slope displacement prediction model based on time-lapse wavelet-gray theory[J]. Journal of Yangtze River Scientific Research Institute, 2015, 32(09): 146-152.
11. Wang Gaofeng, Sun Xiujuan, Sun Xiangdong, Gao Youlong, Wang Hongde, Le Qilang, Shi Xuelei. Application of dynamic multivariate grey model in the prediction of dangerous rock deformation[J]. Journal of Hohai University (Natural Science Edition), 2014 , 42(06): 508-512.
12. Beijing Institute of Water Resources Planning and Design. On-site monitoring and analysis of the impact of construction on the seepage wall of the west embankment of Daning Reservoir in Beijing [R]. Beijing: Beijing Institute of Water Resources Planning and Design, 2012.

Study on water stability of asphalt binder with medium weathered igneous rock

Changgui Li[1], Yuanhong Tao[1,*], Jingliang Xia[2,3,4]

[1] China Road and Bridge Co., Beijing 100011, China

[2] China Academy of Building Research, Beijing 100013, China

[3] National Engineering Research Center of Building Technology, Beijing 100013, China

[4] State Key Laboratory of Building Safety and Built Environment, Beijing 100013, China

Abstract. Aiming at the problems of weak acidity of medium weathered igneous rock around Nairobi, Kenya, poor adhesion with asphalt and poor water stability of asphalt binder, the article studied the use of anti-stripping agent, cement, hydrated Lime and other technical measures to improve the water stability of asphalt binder with medium weathered igneous rock. The results showed that the 48h Marshall residual stability of the benchmark asphalt binder without any measures was 78.5%, which did not meet the standard requirements. The Marshall residual stability of medium weathered igneous rock can be significantly improved by adding anti stripping agent, cement and hydrated Lime. After freeze-thaw cycles, the splitting tensile strength of the asphalt binder with medium weathered igneous rock decreased obviously, and the TSR values of the asphalt binder with anti-spalling measures from small to large were K-4, K-3, K-2, K-6 and K-5. The water stability of the medium weathered igneous rock asphalt binder mixed with anti-stripping agent alone had relatively poor durability, and the medium weathered igneous rock asphalt binder mixed with cement and anti-stripping agent had the strongest ability to resist deformation when immersed in water.

1 Introduction

China is pursuing the "Belt and Road" construction. The Nairobi Expressway in Kenya is one of the important infrastructure projects in East Africa. The main line of the expressway has a total length of 27km and a design speed of 80 kilometers per hour, which is a national highway A-class highway. The rocks in Nairobi and surrounding areas are mainly medium weathered igneous rock. After investigation, the medium weathered igneous rock available along the project are weakly acidic aggregates, and other rock quarries are far away and costly. The project considers using medium weathered igneous rock along the highway for engineering construction.

Since medium weathered igneous rock is weakly acidic aggregate, and asphalt is also acidic, the adhesion of asphalt and aggregate needs to be studied. In the application of acidic aggregates, Dong et al. [1] studied the adhesion of lime to asphalt and granite aggregates and found that lime could increase the surface energy of asphalt and enhance the adhesion work between asphalt and granite aggregates, thereby improving adhesion; the adhesion of asphalt and granite aggregate increased with the increase of lime fineness; when the lime content was 10% of the asphalt mass, the adhesion between asphalt and granite aggregate was the best. Tan et al. [2–7] studied to improve the adhesion of asphalt and acidic rocks by adding anti-stripping agent, cement, etc., thereby improving the water stability performance. The research results also showed that different technical measures have improved the adhesion of asphalt and aggregate to a certain extent.

However, the investigation of domestic and foreign related research showed that there were few reports on the adhesion of medium weathered igneous rock and asphalt, especially the water stability. In order to promote the application of medium weathered igneous rock in the Nairobi Expressway in Kenya, this research used the test methods such as the immersion Marshall test, the split strength test after freezing and thawing, and the water rutting test to investigate the improvement of anti-stripping agent, cement, hydrated Lime and other materials on the water stability of asphalt binder with medium weathered igneous rock.

2 Material and testing method

2.1 Raw material

(1) Aggregate: The aggregate used in this experiment was medium weathered igneous rock in Nairobi and surrounding areas. The rock was processed into coarse and fine aggregates using a laboratory small jaw crusher.

* Corresponding author: taoyh@crbc.com

For stone powder with a particle size of 0~0.075mm, chemical titration was used for chemical composition analysis. According to Chinese Standard *the Methods of Aggregate for Highway Engineering* (JTG E42-2005), the compressive strength, apparent density, firmness, and crush value of aggregates with a particle size of 5-25mm were tested, and the test results were shown in Table 1:

Table 1 Technical indicators of aggregate

Oxide composition/%	SiO_2	65.39
	Al_2O_3	13.18
	Fe_2O_3	12.27
	CaO	2.66
	MgO	0.82
Compressive strength (MPa)		95.2
Apparent density ($kg \cdot m^{-3}$)		2672
Crush value (%)		10.15
Robustness (%)		3.2

(2) Asphalt: The asphalt used in this test was road petroleum asphalt 90#, and its technical indicators were shown in Table 2:

Table 2 Technical indicators of asphalt

Penetration at 25°C (0.1mm)	Softening point (°C)	Ductility (15°C, 5cm/min)
95.6	47.2	145.1

(3) Cement: CEM I 42.5 cement produced locally in Kenya was used. The setting time is 165min for initial setting and 267min for final setting. Water consumption for standard consistency is 26%, 3d compressive strength is 27.1MPa, 28d compressive strength is 52.3MPa.

(4) Hydrated Lime: hydrated Lime produced locally in Kenya with an effective CaO and MgO content of 68.2% was used.

(5) Amine anti-stripping agent: liquid amine anti-stripping agent was used, whose appearance is milky white thick liquid, pH value is between 10 and 12, and freezing point is less than 0 °C. And its content is 0.4% of the asphalt mass.

2.2 Testing Proportion

In order to study and improve the road performance of medium weathered igneous rock asphalt binder, anti-stripping agent, cement, hydrated Lime, and Compound of anti-stripping agent, cement and hydrated Lime were used as admixtures to study their influence on the water stability of asphalt binder. The amount of slag powder accounts for 2% of the total amount of ore; the blending amounts of cement and hydrated Lime account for 50% of the total amount of slag powder. When used alone, the amount of anti-stripping agent was 0.3% of the asphalt content; when it was compounded with cement and hydrated Lime, the amount of anti-stripping agent was 0.2% of the amount of asphalt, and cement and hydrated Lime respectively account for 50% of the amount of slag powder.

Table 3 Proportion of ore materials

Sample number	Slag powder (%)	Anti-stripping agent (%)	Cement (%)	Hydrated Lime (%)
K-1	2	—	—	—
K-2	2	0.3	—	—
K-3	1	—	1	—
K-4	1	—	—	1
K-5	1	0.2	1	—
K-6	1	0.2	—	1

2.3 Testing method

At present, the evaluation methods for the water stability of asphalt binder mainly include the immersion Marshall test, the split strength test after freezing and thawing, and the water immersion rutting test. The split strength test after freezing and thawing and the water immersion rutting test are in accordance with Chinese Standard Test Method of Bitumen and Bituminous Mixtures for Highway Engineering (JTU E20-2011). Based on this standard, the immersion Marshall test was adjusted. Specifically, on the basis of the conventional Marshall stability test, the immersion time was extended to 0.5h, 48h, 72h, 96h, respectively, to test the Marshall stability after immersion, and calculate the residual stability of different immersion times.

3 Result and discussion

3.1 The immersion Marshall test

The main reason of water damage and poor water stability of asphalt binder pavement structure is the decrease of adhesion between asphalt and aggregate. The reason for the decrease of adhesion is that water enters between asphalt and aggregate, which makes the adhesion between asphalt and aggregate decrease until asphalt is separated from aggregate surface. On the other hand, the binding force of water and aggregate is greater than that of asphalt and aggregate, so it can break off the bonding between asphalt and aggregate, and make asphalt separate from aggregate surface. Therefore, the immersion Marshall test was used to evaluate the water stability of the asphalt binder.

The test results of Marshall stability and residual stability of different test schemes for improving the bonding performance of asphalt and aggregate were shown in Figure 1 and Figure 2. It could be seen that the Marshall stability of the different test groups gradually decreased with the increase of the immersion time of the test piece. The 48h Marshall residual stability of the K-1 test group without any measures was 78.5%, which did not meet the requirements of Technical Specifications for Construction of Highway Asphalt Pavements JTG F40-2017 to be higher than 80%, and the Marshall stability cannot be measured after the test piece is immersed in water 96h. Therefore, it was clear that for medium weathered igneous rock, appropriate anti-stripping measures need to be taken when used in asphalt binders.

Combining Figure 1 and Figure 2, it could be seen that under different immersion times, the Marshall stability of the K-2 to K-6 groups were significantly higher than that of the K-1 group, indicating that different technical methods had a certain effect on improving the water stability of the medium weathered igneous rock asphalt binder. Comparing different technical measures, it could be found that the Marshall stability of the asphalt binder mixed with the anti-stripping agent alone was not much different from other technical measures at the immersion time of 48h, but when the immersion time reached 96h, the Marshall stability of K-2 has a significant decrease. The main reason was that the anti-stripping agent is a liquid amine, which is organic, and its anti-aging performance is relatively poor, and the anti-stripping effect becomes weak during long-term immersion. Hence, it was recommended to use cement or hydrated Lime, or a combination of cement, hydrated Lime and anti-stripping agent, for the medium weathered igneous rock asphalt binder, which can improve the water stability of the asphalt binder for a long time.

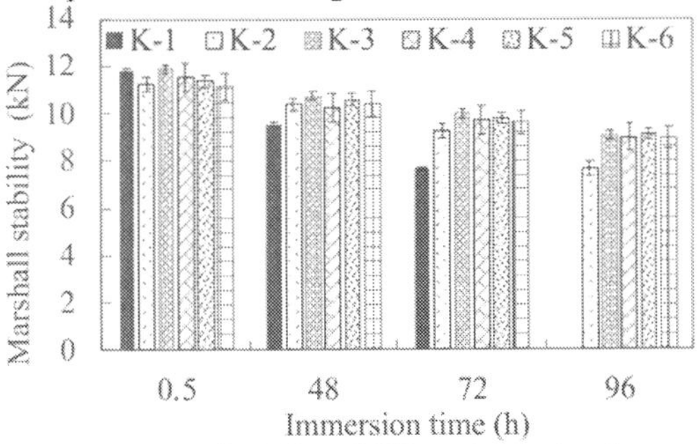

Fig. 1 Marshall stability of different immersion time

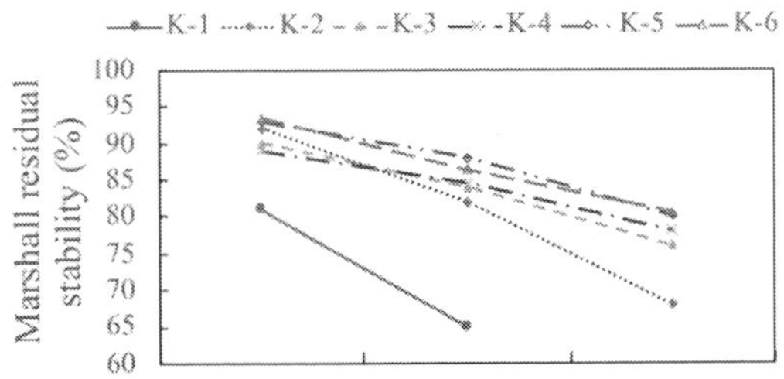

Fig. 2 Marshall Residual Stability of Different Immersion time

3.2 Freeze-thaw test

The main reason for the water damage of asphalt binder pavement structure and poor water stability of asphalt binder is not only the decrease of the adhesion between asphalt and aggregate, but also the existence of the water between asphalt and aggregate, which is prone to hydrodynamic pressure. This repeatedly acts on the interface between asphalt and aggregate, causing aggregate detachment in the asphalt-bound material pavement structure, forming potholes and other pavement diseases. For this reason, the study uses the splitting tensile strength of the asphalt binder after freezing and thawing to compare the splitting tensile strength of the asphalt binder without freezing and thawing to evaluate the water stability of the weathered igneous rock asphalt binder.

The test results of freeze-thaw splitting tensile strength and splitting tensile strength ratio (TSR) of different test schemes to improve the binding performance of asphalt and aggregate were shown in Fig. 3 and Fig. 4. It could be seen that the splitting tensile strength of medium weathered igneous rock asphalt binder decreased significantly after freeze-thaw cycle, and the TSR of the benchmark test group k-1 without any measures was only 67.1%, which did not meet the requirement of more than 75% in JTG F40-2017. Therefore, it was necessary to take corresponding technical measures to improve the water stability when the medium weathered igneous rock is applied in asphalt binder.

It could be seen from Figure 4 that the TSR values of asphalt binders with anti-spalling technical measures from small to large were K-4 (hydrated Lime accounts for 50% of mineral powder), K-3 (cement accounts for 50% of mineral powder), K-2 (0.3% of asphalt), K-6 (0.2% of anti-spalling agent combined with hydrated Lime), K-5 (0.2% of anti-spalling agent combined with cement). The TSR values of asphalt binders of K-2 to K-6 could meet the requirements of JTG F40-2017, in which the minimum TSR value of asphalt binders with single hydrated Lime were 79.2%, the maximum TSR value of asphalt binders with anti-stripping agent 0.2% and cement compound technical measures was 87.3%. In general, the single addition of anti-stripping agent, cement or hydrated Lime can greatly improve the TSR of medium weathered igneous rock asphalt binder. The single addition of anti-stripping agent had the best effect, and the single addition of cement was better than hydrated Lime. In addition, through the combination of anti-stripping agent and cement or hydrated Lime, the water stability of weathered igneous rock asphalt binder was improved most significantly.

https://doi.org/10.1051/e3sconf/202129302055

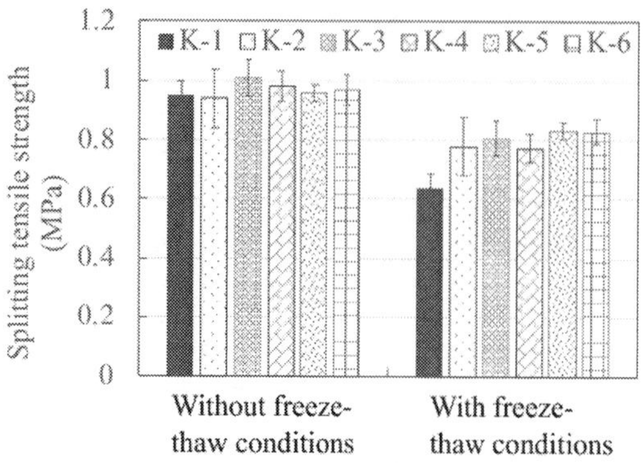

Fig. 3 The effect of freezing and thawing on split tensile strength

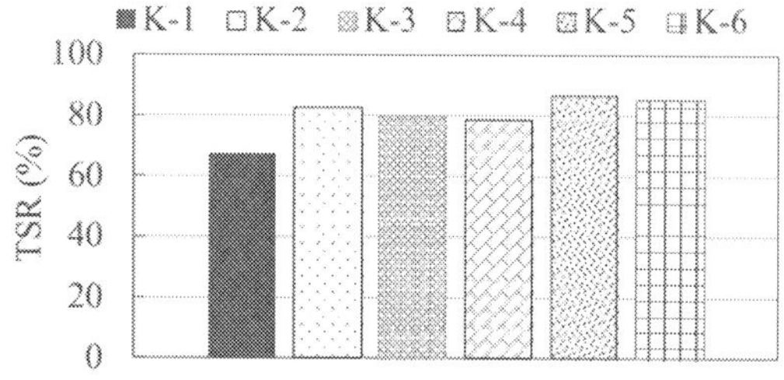

(b) C40 concrete
Fig. 4 Splitting tensile strength ratio (TSR) of different test samples

3.3 Immersion rutting test

Traffic load is also one of the main reasons for the deterioration of water stability of asphalt binder. From the perspective of macro mechanics, asphalt pavement structure needs to face the repeated action of traffic load in the service process; at this time, the interface between asphalt and aggregate will have repeated shear action, and different aggregate particles will also have repeated shear action. If the interface between asphalt and aggregate has shear failure, the water can quickly enter between the interface of asphalt and aggregate through the point of shear failure, resulting in the deterioration of water stability of asphalt binder. On the other hand, when the water enters the asphalt binder, under the repeated load, the ordinary water becomes dynamic water, which further accelerates the deterioration of water stability of asphalt binder. Therefore, the immersion rutting test was used to further study the water stability of the medium weathered igneous rock asphalt binder under load.

The results of immersion rutting test under different test schemes to improve the binding performance of asphalt and aggregate were shown in Figure 5.It could be seen that the rutting depth curve of immersion rutting test showed an obvious slope change at 10 min. In the first 10 min of immersion rutting test, the rutting depth of each test group increased greatly, and then the rutting depth of each group showed a linear relationship with the extension of immersion time, with a small increase.

For the K-1 group without anti-stripping measures, the rutting depth was the largest in the rutting test of each immersion time, which indicates that the application of medium weathered igneous rock in asphalt binder needs to take anti-stripping measures to improve its water stability. The rutting depth of each test group with different technical measures within 10 minutes of immersion was basically similar, and then with the extension of immersion time, the rutting depth of K-2 group with anti-stripping agent was deeper than that of other groups, which indicated that the durability of water stability of medium weathered igneous rock asphalt binder with anti-stripping agent was relatively poor. In the 60 min immersion rutting test, the K-5 test group with cement and anti-stripping agent mixed had the shallowest rutting depth, which indicated that the medium weathered igneous rock asphalt binder with cement and anti-stripping agent mixed had the strongest resistance to deformation.

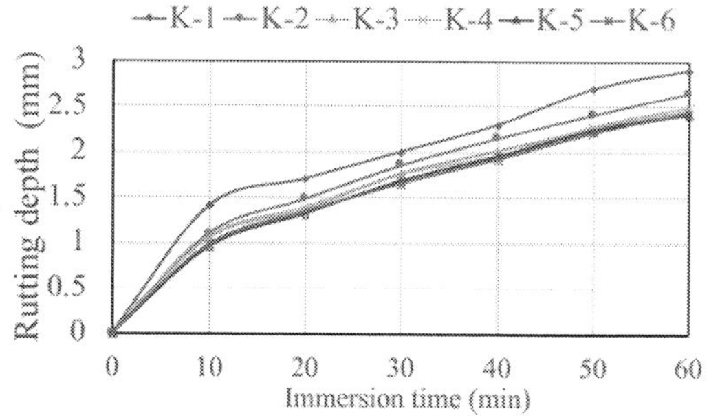

Fig. 5 Result of Immersion rutting test

4 Conclusion

(1) The 48h Marshall residual stability of the benchmark asphalt binder without any measures was 78.5%, which did not meet the standard requirements. The Marshall residual stability of medium weathered igneous rock could be significantly improved by adding anti-stripping agent, cement and hydrated Lime.

(2) After the freezing and thawing cycles of the medium weathered igneous rock asphalt binder, the split tensile strength decreased significantly. The TSR value of the asphalt binder with anti-stripping technical measures was K-4, K-3, K-2, K-6 and K-5 in order from small to large.

(3) Single mixing of anti-stripping agent, cement and hydrated Lime could greatly increase the TSR of medium weathered igneous rock asphalt binder. Single mixing of anti-stripping agent has the best effect, and single mixing of cement had better effect than hydrated Lime. Through compounding anti-stripping agent and with cement or hydrated Lime composite, the water stability of the medium weathered igneous rock asphalt binder was most significantly improved.

(4) The durability of water stability performance of medium weathered igneous rock asphalt binder with anti-stripping agent was relatively poor, while that of medium weathered igneous rock asphalt binder with cement and anti-stripping agent was the strongest.

Acknowledgements

This research was funded by National Natural Science Foundation of China, grant number 51978408, and National Key R&D Program of China, grant number 2016YFC0701000.

References

[1] S. Dong, S. Han, Y. Yi, S. Wu, and Q. Zhang, J. Chongqing Jiaotong Univ. Sci. **40**, 89 (2021).

[2] J. Tan, West. China Commun. Sci. Technol. 21 (2018).

[3] Z. Wei, Transp. Sci. Technol. 132 (2017).

[4] Y. Wu, F. Wang, and Y. Zhang, J. China Foreign Highw. **36**, 289 (2016).

[5] D. Wu, The Research of Performance on Cement and Amine Anti-Stripping Material's Asphalt Mixture Pavement, Dissertation, Chongqing Jiaotong University, 2016.

[6] S. Du and X. Suo, J. Xi'an Technol. Univ. **33**, 982 (2013).

[7] X. Zheng, Q. Yang, and W. Lv, J. Build. Mater. 480 (2005).

 https://doi.org/10.1051/e3sconf/202129302056

Study on Diagnosis and Treatment Technology of Cracks in Consolidated Continuous Box Girders of Middle Piers during Construction

Guan Liangyong[1], Dong Shuiying[2]

[1.] Jiaxing Transportation Investment Group Co., Ltd., Zhejiang, Jiaxing 3140501

[2.] Beijing Gongke Guqiao Technology Co., Ltd., Beijing 10088

Abstract: In this paper, taking the cast-in-place reinforced concrete continuous box girders of a bridge as the research object, we studied the cracking type of the box girders and their construction. Combined with the calculation and analysis of the construction, the causes and effects of the cracks of box girders during the construction were further studied. Then the maintenance and strengthening scheme that UT10 super-high-performance concrete should be used on the bridge floor was proposed, which will provide reference for the diagnosis and treatment of similar bridge structures.

1 Introduction

In the process of bridge construction and use, the cracks of concrete structure often perplex bridge engineers and technicians, because they will affect the whole acceptance check. Sometimes even a do-over would happen, therefore, it's really crucial and necessary to find out the cause of cracks and work out reasonable plans, which can also provide reference for similar projects in the future. In this paper, taking the transverse cracks in the top section of the box girders and the flange plates as well as the L-type cracks in the bottom plates and web plates of the mid-span piers as the samples, we analyze the causes of the cracks in the box girders, and then put forward the measures for repairing these cracks to achieve the simple and effective repair for the box girders.

2 Overview of the Bridge

The superstructure of the bridge is made up of cast-in-place reinforced concrete continuous box girders with a combined span of 2 × 18 m, and the length of the bridge is 50 m. The box girder is made of C40 concrete with single box and single chamber, and the height of the beam is 130 cm. The substructure of the bridge is made up of column piers, consolidated by piers and beams. The pier columns, abutment caps, back walls, abutment bodies and retaining blocks are made of C30 concrete. The abutment is arranged radially, and U-type pile foundation is adopted. And the length of the slab behind the abutment cap is 8 m. The plane of the bridge is located in the right circular curve of radius, 67 m, and the longitudinal slope of the longitudinal section is 3.814%. The bridge deck has a net width of 7.5 ~ 8.94 m. The leveling layer is made of 8 cm C40 concrete while the bridge deck is made of 10cm bituminous concrete. The waterproof layer is made of SBR (modified emulsified asphalt water seal layer); GPZ(II)-2.5-DX basin bearing is used. Model D60 is used as reference for the expansion device for the bridge. Vehicle load class of the bridge is highway-I. The elevation and I-I section of the bridge are shown in Figure1 and 2.

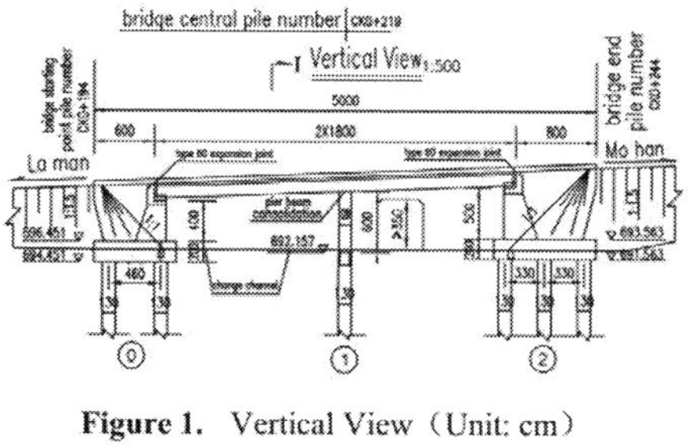

Figure 1. Vertical View (Unit: cm)

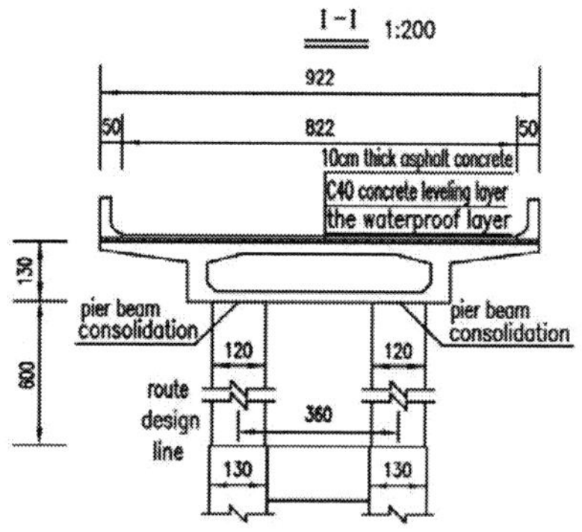

Figure 2. Cross-section Diagram (Unit: cm)

corresponding author: 1161835210@qq.com

https://doi.org/10.1051/e3sconf/202129302056

3 Construction Procedures and Cracking Status of Box Girders

3.1 Construction Procedures

The bridge is constructed with full supports, and the detailed process is as follow: pouring the ground where the bridge is located (6 days) → setting up the socket-type steel pipe supports (5 days) → accepting the supports (1 day) → pouring the concrete of the bottom plates and the web plates (6 days) → pouring the concrete of the wing plates and the roof plates → removing the box girder formwork (28 days) → removing the socket-type steel pipe supports (30 days)

3.2 Cracking Status of Box Girders

The cracks appeared in the box girders when the socket-type plate-and-buckle steel pipe supports were removed. The transverse cracks mainly appeared at the top of the piers and in the flange plates. Besides, the transverse cracks also appeared in the bottom plates near the midspans while the vertical cracks appeared in the web plates. The view of cracks can be seen as Figure 3. There are six transverse cracks at the top plate of the box girder on the top of No. 1 pier, one of which is 5 meters long, the other five 9.2 meters long. Moreover, there're seven transverse cracks in the left flange plate and nine cracks in the right flange plate, the length of which is between 1.0 meter and 1.7 meters. There are nine cracks in the middle web of the two midspans, which extended to the bottom plate with the L shape. The width of the maximum crack is 0.20mm, while the depth of the cracks is between 14mm and 48mm. The photos are shown in Figure 4.

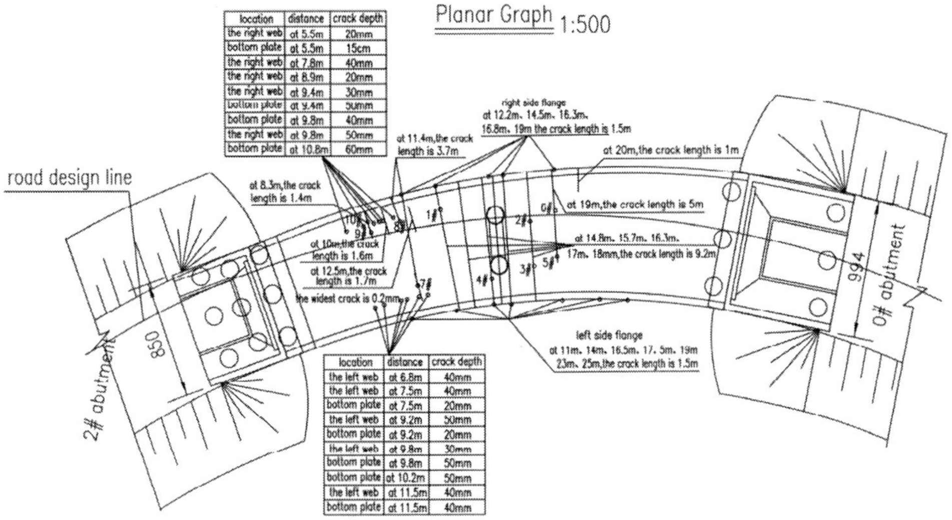

Figure 3. Diagram of the Cracks

Figure 4. Diagram of the Crack Depth of No. 5 Hole

4 Back Analysis of the Bridge Construction

According to the calculation, the stress during the construction can meet the requirements. The bending and shear capacity of the box girders which were originally designed can meet the requirements of the design code. The crack width and the long-term deflection of the main girders can meet the requirements of the code when the bridge is used in limiting cases[1,2] (frequency combination, quasi-permanent effect combination and standard combination). A numerical calculation model of the bridge is established to analyze the causes of cracking of box girders during construction. Taking the settlement of the supports into account, we conduct the back analysis of the construction.

https://doi.org/10.1051/e3sconf/202129302056

(1) Parameters of the calculation of the model: C40 concrete is used for the concrete box girders as well as the piers. Common steel bars are HPB300 and HRB400 ones. Support settlement (construction record): the settlement of 1/4 of the span supports is 3mm, 1/2 of them 5.8 mm, and 3/4 of them 3.8mm. The temperature is considered as: temperature rises 20°C or it drops 20 °C; The temperature gradient: according to Article 4.3.12 of the specification (JTG D60-2015), the temperature rise of the top of the beam is T1 = 25 °C and T2 = 6.7 °C, and the inverse temperature difference of vertical sunlight equals to it multiplying -0.5. The stiffness of the supports is 1×10^7kN/m. The numerical model is shown in Figure

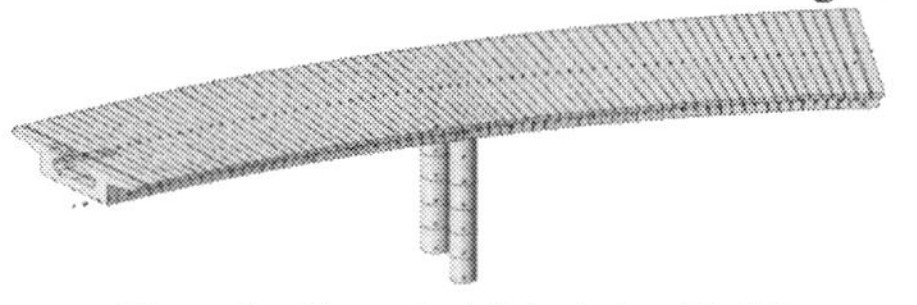

Figure 5. Numerical Calculation Model

(2) Load combination: Stress calculation of the bridge components is carried out according to transient conditions in the construction stages such as fabrication, transportation and installation. And it's adopted according to "The Code for Design of Reinforced Concrete and Pre-Stressed Concrete Bridges and Culverts" (JTG 3362-2018).

(3) Construction stages: There are four stages of the construction, which are: building pier columns → setting up superstructure of the full supports → support and beam settlement → removal of the supports.

(4) Calculation results: The calculation results of removing the supports are shown in Figure 6 and 7. It can be seen from the figures that the maximum positive stress of the upper section and the maximum stress of the lower section of the box girder (at the top of No. 1pier) are 1.69MPa and 5.23MPa respectively when the supports are removed in the fourth construction stage near the mid-span.

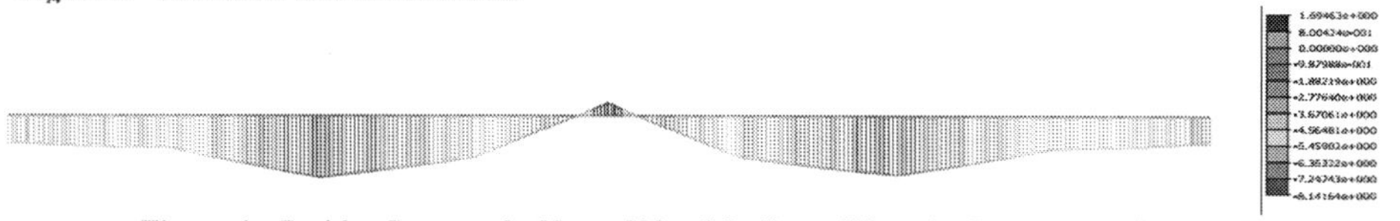

Figure 6. Positive Stress at the Upper Side of the Beam When the Supports Are Removed

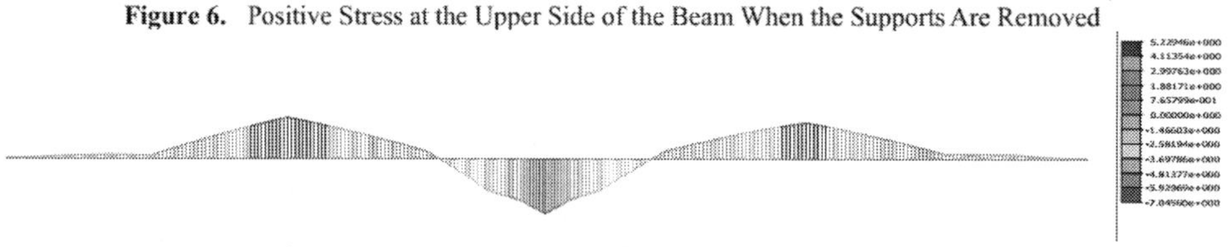

Figure 7. Positive Stress at the Lower Side of the Beam When the Supports Are Removed

According to the combination of 1.0 permanent action (without the second-phase load) that the temperature gradient is +0.8 and the whole temperature will be increased and decreased, how the cracks will be in construction stage is checked and calculated. The results show that the maximum crack width of the upper edge of the structure is 0.26 mm, at the top of No.1 pier. And the maximum one at the lower edge is 0.27 mm, near the mid-span, which is 0.20 mm larger than the crack width limit[3].

5 Analysis on the causes of the cracks and Their Lingering Effects[4,5,6]

5.1 Analysis on the Causes of the Cracks

The location and trend of the cracks in the box girders are obviously related to the disadvantageous position of bending force under dead load, which indicates that the cracks are bending force cracks. When bending force cracks are produced in the structure, it is always related to load effect and structural resistance.

(1) As for load effect: The dead weight of the structure has not changed, but the boundary conditions have changed during the construction of the structure. No cracks have been found during the course of removing the formwork of the bridge. However, cracks have been found during the removal of the steel pipe supports. The inversion results of the construction show that the structure has a good stress state when there're full supports, but the stress state is unfavorable when the structure system is changed. The internal force is redistributed when the structure is changed into a continuous system to bear dead load, which results in stress cracks in the structure.

(2) As for structural resistance: The superstructure of the bridge is poured in two times. It was poured for the first time on January 28th, 2018, the second time on February 4th, 2018. The time interval between the first and second pouring is long, so there is a time difference. Due to that, unsynchronized shrinkage of concrete occurred, which might affect the integrity of the structure. Furthermore, the reduced integrity will affect the stiffness and bearing capacity of the structure.

5.2 Lingering Effects of the Cracks

The analysis on crack causes shows that the typical cracks of the bridge are stress ones which occur when the structure bends under dead load. The occurrence of these stress cracks indicates that the stress of the concrete has exceeded its tensile strength. When the steel bars share the stress, it will have the following effect on the structure:

 https://doi.org/10.1051/e3sconf/202129302056

(1) Since the tensile strength of steel bars is obviously higher than that of concrete, the possibility of further development of the width of the cracks is small.

(2) The cracking of reinforced concrete structure will reduce the bonding between steel bars and concrete. When the bonding is reduced, the number of cracks will increase more easily under the same load. Therefore, the number of cracks may further increase when the bridge is under the second-phase construction with dead load.

(3) The structural stress exceeds the tensile strength of concrete while the reinforcement stress increases, which indicates that the safety of the structure decreases.

(4) When cracks appear, the steel bars wrapped up the concrete are prone to contact with air, and they will corrode more easily, leading to the loss of the durability of the structure.

6 Maintenance and Reinforcement Plans

Based on the analysis on the causes of the cracks in box girders and their lingering effects, the repair and reinforcement plans are carried out from three aspects: the repair of concrete cracks, structural durability and structural safety. The detailed construction process and its core points are shown in Figure 8.

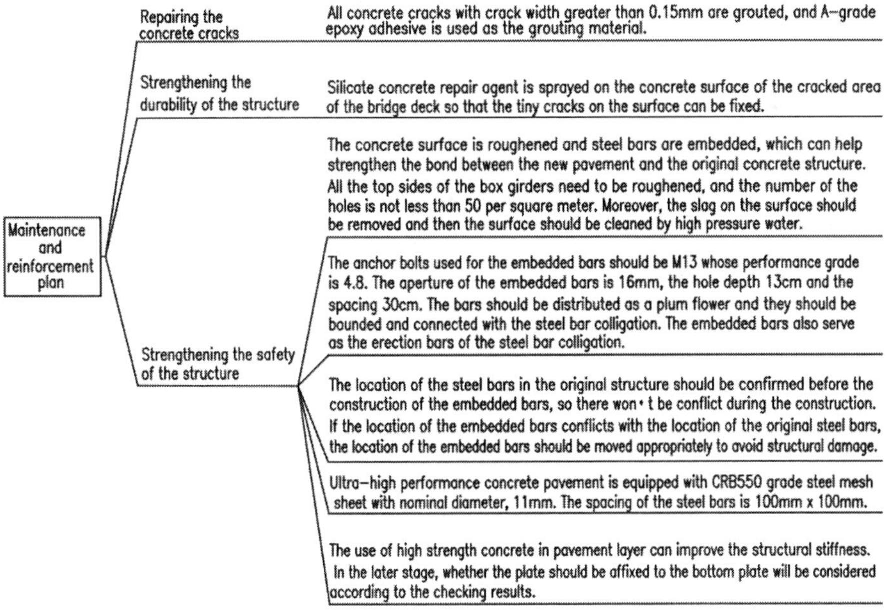

Figure 8. Flow Chart of the Plan and the Key of the Construction

7 Conclusion

Based on the investigation of the cracking status of the reinforced concrete box girders of the bridge, construction data and numerical simulation, the causes of the cracks in the box girders and their lingering effect were analyzed. Besides, the maintenance and reinforcement plan that UT10 ultra-high performance concrete should be used as the pavement material of the bridge deck was put forward, which realizes the simple and effective reinforcement and transformation on the cracks in the box girders. The technical idea and the strengthening methods mentioned in this paper can provide reference for the diagnosis and improvement of bridges with similar problems.

Reference

1. "General Code for Design of Highway Bridges and Culverts" (JTG D60-2015).
2. "Code for Design of Highway Reinforced Concrete and Pre-stressed Concrete Bridges and Culverts" (JTG3362-2018).
3. "Technical Code for Construction of Highway Bridges and Culverts" (JTG/T F50-2011).
4. Du Lingjun, Tao Haihong. Cause Analysis and Prevention and Control of Cracks in Pre-stressed Concrete Box Girder of Continuous Rigid Frame Bridge. Henan Building Materials, 2012(04): 42-44.
5. Xu Zhen, Zhang Xuesong, Xu Junlan. Influence Analysis of Parameters on Mid-Span Deflection of Continuous Rigid Frame. Journal of Chongqing Jiaotong University: Natural Science Edition, 2007,26(4):9-13.
6. Zhao Dianjun, Liu Congfeng. Analysis and Control of Construction Crack in Reinforced Concrete Box Girder. Technology News, 2019(15):680.

 https://doi.org/10.1051/e3sconf/202129302057

Underestimated angle of submarine slope at failure: A short discussion

Bo Chen[1], Chaoqi Zhu[2,1,3,4*], Yasheng Feng[1], Xiaohui Han[1], Weite Zeng[1], Congcong Xing[5], Shili Lin[1], and Gang Liu[1]

[1]Hainan Key Laboratory of Marine Geological Resources and Environment, Hainan Marine Geological Survey Academy, 570206, Haikou, China

[2]Laboratory for Marine Geology, Qingdao National Laboratory for Marine Science and Technology, 266061, Qingdao, China

[3]Shandong Provincial Key Laboratory of Marine Environment and Geological Engineering, Ocean University of China, 266100, Qingdao, China

[4]College of Earth Science and Engineering, Shandong University of Science and Technology, 266590, Qingdao, China

[5]North China Sea Environmental Monitoring Center, Ministry of Natural Resources, Qingdao, China

Abstract. Submarine landslides always occur along gently inclined continental slopes, but the reasons for such failure of low-angle submarine slopes are unclear and contentious. A short discussion is presented here to some recent published papers including some inspirations about the low-angle failure of submarine slopes. Because fluid overpressure could lead to steepening slopes before landslide, the inclination of the slopes at failure may be underestimated. We offer a new perspective on low-angle failure of low-permeability slopes.

1 Introduction

Hance (2003) [1] developed a database of submarine landslides on published literature and 399 submarine landslides in this database are cataloged with slope angle data. The results (Figure 1) show that the 3 to 4 degree slope angle interval has the highest frequency of submarine landslide and that the minimum is 0.22 degree. About 85 percent of these submarine landslides occurred on seafloor slopes flatter than about 10 degrees (Figure 2). Such flat slope angles are much less than for most subaerial landslides. Submarine landslides always occur along gently inclined continental slopes, but the reasons for such failure of low-angle submarine slopes are unclear and contentious.

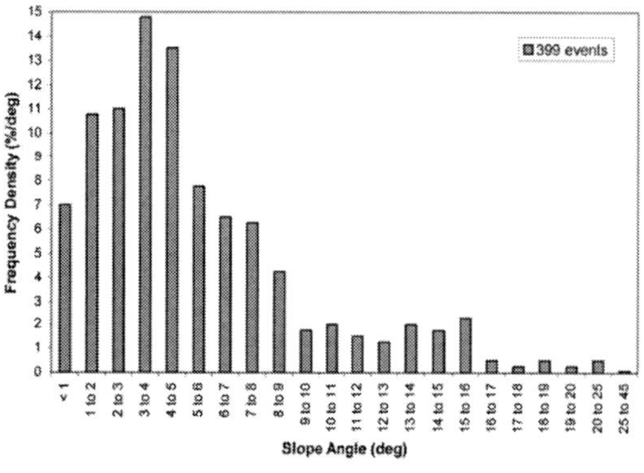

Fig. 1. Frequency density distribution of the average angle of the slope at failure for the seafloor slope failures[1].

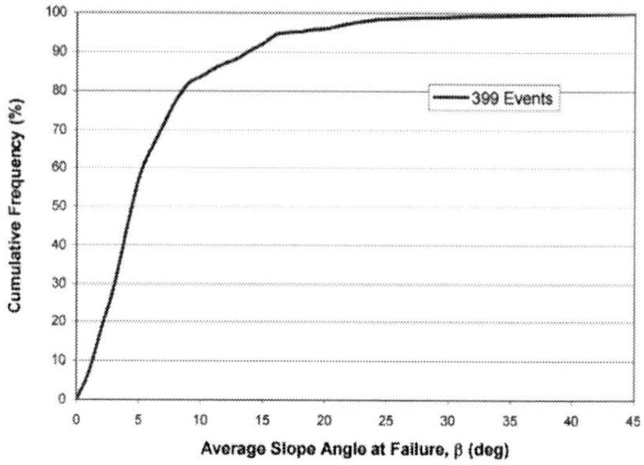

Fig.2 Cumulative frequency distribution of the average angle of the slope at failure for the seafloor slope failures[1].

2 Discussions

Submarine slope failures are usually occur on very gentle slopes. The slope angle at failure is usually obtained by determining the average angle of the unfailed adjacent seafloor slope. Without consideration of seafloor evolution before failure, this method may lead to slope angle underestimation.

Yan et al. (2020) [2] have designed a series of new physical model tests to investigate the formation mechanism of pockmark. The gas produced by natural gas hydrate dissociation was simulated by aeration in their experiments. Based on the experimental

* Corresponding author: george-zhu@foxmail.com

phenomena, they found some interesting results for the pockmark formation. Though their physical model tests were specifically designed for use in pockmarks, the experimental phenomena (Figure 3) would help to solve unsettled questions about the failure of low-angle submarine slopes. Inspired by Yan et al. (2020) [2] and some recent results, we present the possible reasons for such failure of low-angle submarine slopes.

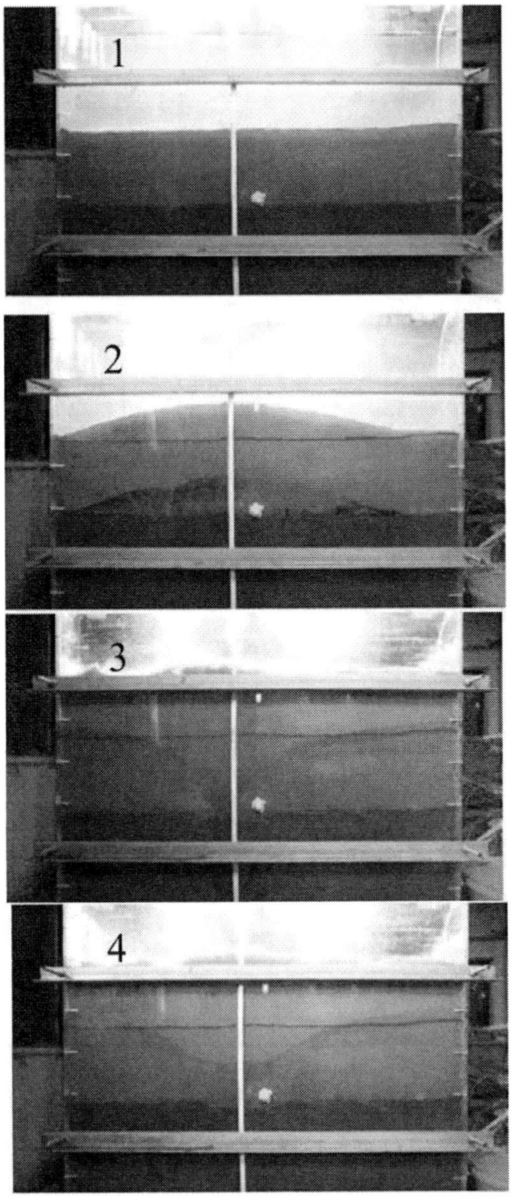

Fig. 3. The experimental phenomena in Yan et al. (2020)

Figure 3 shows that a cavity forms along the interface between the overlying and underlying layers due to aeration. Meanwhile, the clay overlying layer gradually arches upward. Then, pockmark forms after gas eruptions. Such seafloor deformation evolution has been proven by the founding of a cluster of kilometer-wide pockmarks and domes, as well as methane expulsion, in the Barents Sea floor[3]. In Figure 4, Andreassen et al. (2017) [4] proposed a conceptual model that links the domes and pockmarks. Build up of overpressurized gas below the gas hydrate zone leads to formation of a gas hydrate pingo (dome). After gas eruptions, the craters (pockmark) forms. Both the experimental and field results show fluid overpressure leads to steepening slopes before pockmark formation.

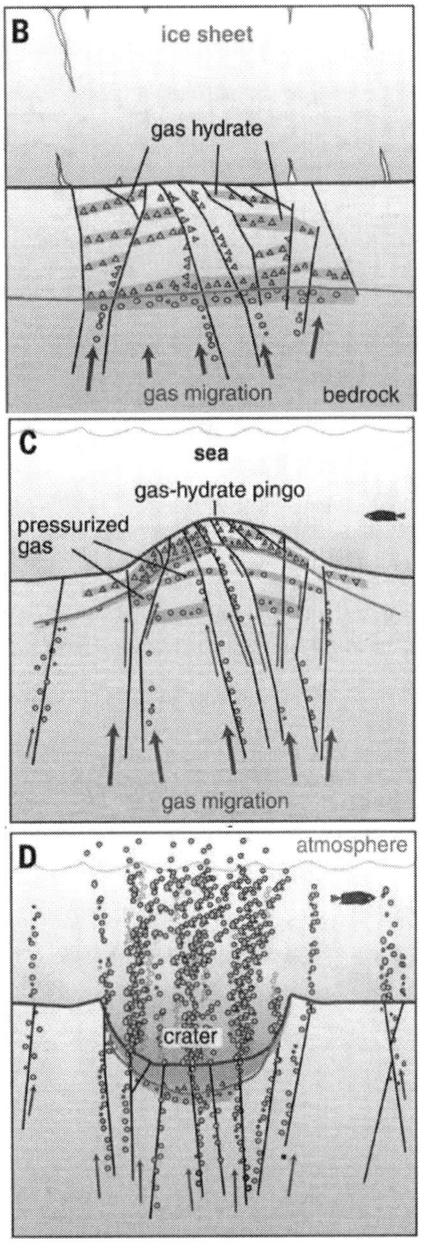

https://doi.org/10.1051/e3sconf/202129302057

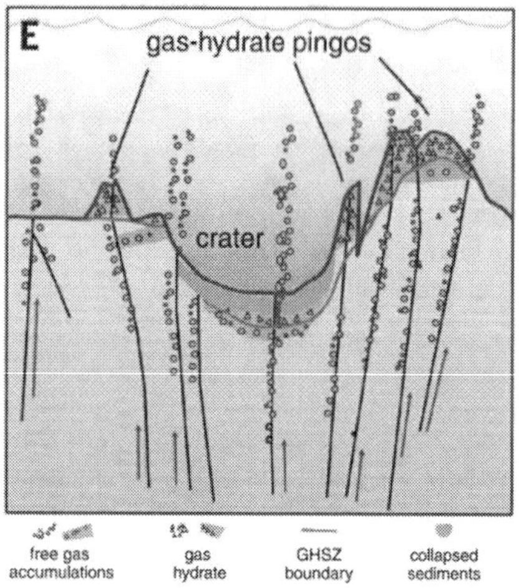

Fig. 4. The conceptual model for pockmark formation in the northern Barents [4]

If the original seafloor in Figure 3 or Figure 4 was inclined at an angle (e.g. 5°), the final negative landform could be regarded as a symbol of submarine landslide in a certain sense. Then the angle of submarine slope at failure is estimated as 5° based on the current method. Because overpressure leads to steepening slopes before landslide, the angle at failure is bigger than 5°. Once the landslide is initiated, the trapped fluid leaks into seawater and the overpressure dissipates. As a result, the steepening slope forms negative landform. The subsequent field survey cannot identify previously tumefied seafloor and steepening slopes. Finally, the inclination of the slopes at failure may be underestimated. Just like Yan et al. (2020) [2], the most recent physical model tests also show the steepening slope due to overpressure [5-6].

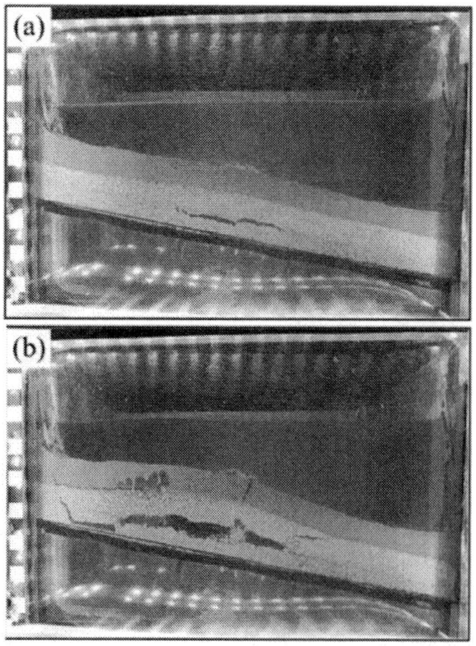

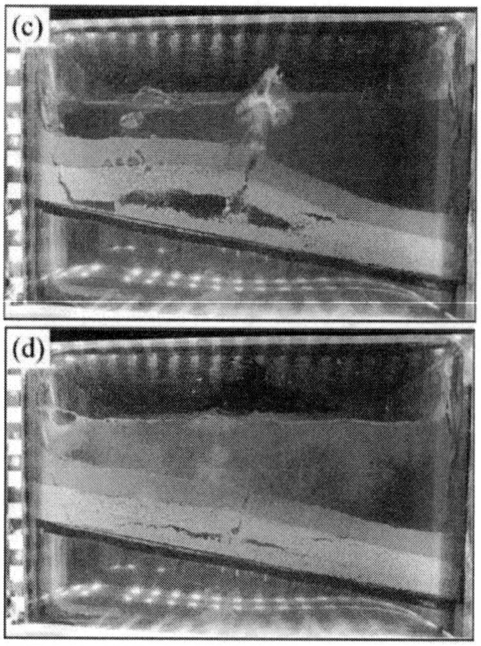

Fig. 5. The experimental phenomena in Nian et al. (2020) [5]

As is shown in Figure 6 or Figure 7, seafloor slope angles evolve over time, and the corresponding landforms become more complex. If ignoring the seafloor deformation evolution before slope failure, the inclination of these slopes at failure may be underestimated.

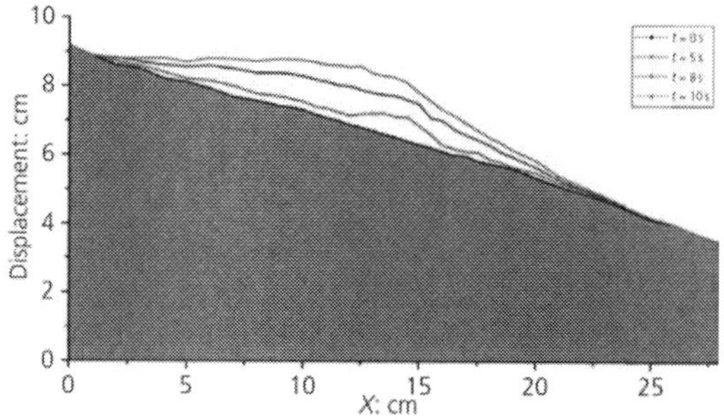

Fig. 6. The slope angle evolution in Nian et al. (2020) [5]

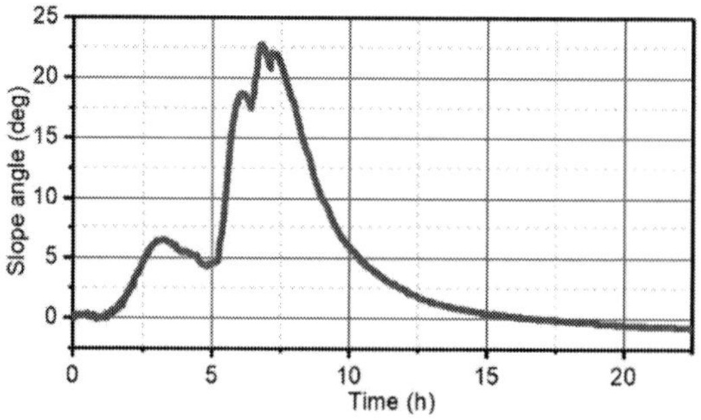

Fig. 7. The slope angle evolution in Zhu et al. (2020) [6]

3 Conclusions

Based on Yan et al. (2020) [2] and some recent results[4-6], we propose a new perspective on such low-angle failure of submarine slopes (Figure 8, taking hydrate dissociation for example). If the sediments are fine-grained and have a low permeability, excess pore pressures would be generated and act as a loading mechanism on the overlying soil slope. The angle of submarine slope at failure would be underestimated, especially among those with overpressure under low-permeability overlying layer.

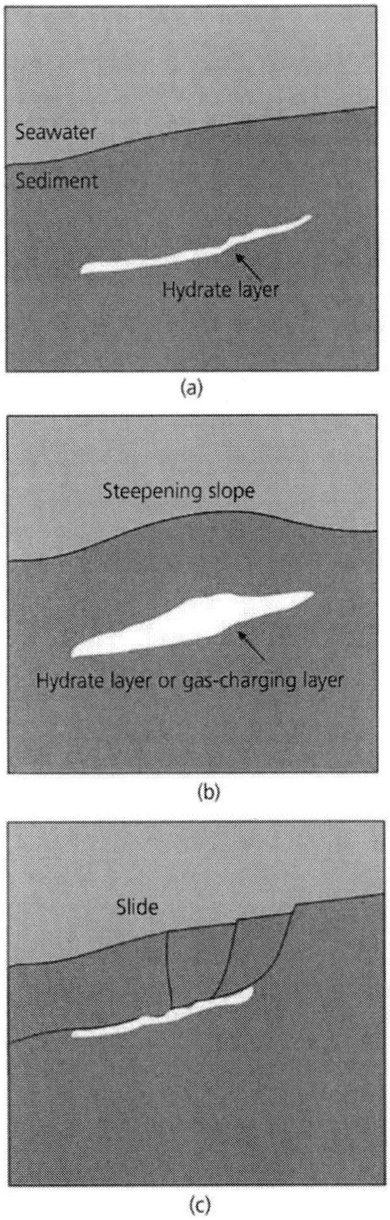

Fig. 8. Overpressure, steepening slope and submarine landslide during hydrate dissociation (Zhu et al., 2020)[6]

Not all the inclinations of the slopes at failure are underestimated; however, the possibility of angle underestimation cannot be ruled out. Because fluid overpressure leads to steepening slopes before landslide, the inclination of the slopes at failure may be underestimated. We offer a new perspective on low-angle failure of low-permeability slopes.

Acknowledgements:

This study was sponsored by the Hainan Key Laboratory of Marine Geological Resources and Environment (Grant No. HNHYDZZYHJKF008) and the National Natural Science Foundation of China (Grant No. 41776068).

References

1. Hance, J. J. (2003). Development of a Database and Assessment of Seafloor Slope Stability Based on Published Literature, The University of Texas at Austin.
2. Yan, X., H. Sun, Z. Chen, F. Shuai, Z. Wei, Y. Xu. 2020. Physical experimental study on the formation mechanism of pockmark by aeration. Marine Georesources & Geotechnology, 38(3), 322-331. doi: 10.1080/1064119X.2019.1571539
3. Serov, P., S. Vadakkepuliyambatta, J. Mienert, H. Patton, A. Portnov, A. Silyakova, G. Panieri, M.L. Carroll, J. Carroll, K. Andreassen, A. Hubbard. 2017. Postglacial response of Arctic Ocean gas hydrates to climatic amelioration. Proceedings of the National Academy of Sciences, 114(24), 6215-6220. doi: 10.1073/pnas.1619288114
4. Andreassen, K., A. Hubbard, M. Winsborrow, H. Patton, S. Vadakkepuliyambatta, A. Plaza-Faverola, E. Gudlaugsson, P. Serov, A. Deryabin, R. Mattingsdal, J. Mienert, S. Bunz. 2017. Massive blow-out craters formed by hydrate-controlled methane expulsion from the Arctic seafloor. Science, 356(6341), 948-952. doi: 10.1126/science.aal4500
5. Nian, T., X. Song, W. Zhao, H. Jiao, X. Guo. 2020. Submarine Slope Failure Due to Overpressure Fluid Associated with Gas Hydrate Dissociation. Environmental Geotechnics, 2020, doi: 10.1680/jenge.19.00070
6. Zhu, C., X. Jiao, S. Cheng, Q. Li, K. Liu, H. Shan, C. Li, Y. Jia. 2020. Visualizing fluid migration due to hydrate dissociation: Implications for submarine slides. Environmental Geotechnics. doi: 10.1680/jenge.19.00068

Discussion of 800m Composite Bridge with CFST Flying Swallow Arch and Self-anchored Suspension Cable

ZhaoYang Chen[1*] and WenPing Xu[2]

[1]CSCEC AECOM CONSULTANTS CO.， LTD, LanZhou China,730000

[2]College of Civil Engineering, Southeast University, NanJing China,210096

Abstract. In view of the demand of 800 meters Super Long Span CFST arch bridge, the composite bridge of CFST flying swallow arch and self-anchored suspension cable is proposed. The thrust of flying-bird CFST arch bridge and the tension of self-anchored suspension bridge are balanced, forming a self-balanced structure system. The arch rib structure is mainly stressed, supplemented by the self-anchored suspension system, which works together and has complementary advantages. Using the single leaf hyperboloid variable section steel tube four limb space truss arch rib structure, the self-weight of the mid span arch rib section structure is reduced, the risk of construction and hoisting of the mid span section is reduced, the section size at the arch foot is increased, the mass center and stiffness center of the arch bridge structure are effectively reduced, and the stability of the super long span concrete-filled steel tube arch bridge is increased. Combined with the actual project, the parameters are designed, the Midas finite element model is established, the internal force analysis and calculation, modal analysis and buckling analysis are carried out, and the superiority of the structural technical measures is verified.

1 Introduction

Flying-swan arch bridge is a self-balanced structure system, which is deeply loved by people for its advantages such as beautiful shape, large span capacity, good economy, convenient construction, large structural stiffness and good durability [1-2].

With the increase of span swallow type of concrete-filled steel tube arch bridge, arch bridge weight increasing, structural internal force increase, segmental hoisting construction difficulties, a significant reduction in the arch bridge structure stability, can lead to the development of swallow type of concrete-filled steel tube arch bridge in technical bottleneck [3-4], build more long-span concrete-filled steel tube flitting models type arch bridge will have a greater risk of structural technology.

At present, under construction in span length of 507 meters mountain Yangtz river highway bridge is the world's largest span concrete filled steel tube swallow type arch bridge, for the sake of giving full play to the advantages of swallow type steel tube concrete arch bridge of the economy, to carry out the swallow type steel tube concrete arch bridge design and construction technical improvement of level to build 800 meters of large span concrete filled steel tube swallow type arch bridge will be a big challenge problem of the civil engineer [5-6].

It is impossible to build an 800-meter arch bridge only relying on a single structure of concrete filled steel tube Flying-Swallow arch bridge. The main problem lies in ： As the span of flying-Swallow arch bridge increases, the internal force of arch foot increases, and the section size of arch foot increases, which leads to the problem of economic rationality. With the increase of flying-bird arch span, the structural stability of the arch bridge decreases significantly, which leads to the problem of structural safety. With the increase of span of flying-bird arch bridge, the lifting weight of mid-span segment increases, which leads to the difficulty of lifting construction of mid-span segment. As swallow the arch span increases, the variable load will be large arch foot base imbalance caused by internal force (half through bridge swallow type arc bridge under dead load only can balance arch foot base internal force), will lead to the arch foot base to produce larger deformation displacement, will appear the arch foot based low reversed cyclic stress fatigue damage problem [7-8].

As a single bridge structure has its own advantages and disadvantages, it is often difficult to meet the carrying capacity demand of super-large span Bridges. Therefore, it is necessary to consider the combined use of multiple bridge structure forms at the same time, so as to form a composite bridge structure system, complement each other's advantages and work together, so as to give full play to the advantages of various bridge structures [9-10].

The main cable of self-anchored suspension bridge is anchored at the end of the beam, which transfers the horizontal force to the main beam, and the main beam is under compression. Self-anchored suspension bridge can avoid the construction of large volume anchoring, and its economic index is reasonable. The disadvantages of self-

*ZhaoYang Chen：807105720@qq.com

anchored suspension bridge are that the construction method of "first beam then arch" must be adopted, which is complicated in construction.

The thrust of concrete filled steel tube arch bridge exerts huge thrust on the foundation, while the self-anchored suspension bridge exerts huge pull on the foundation. Therefore, the two types of bridge types can be considered. The thrust of the Flying-type concrete filled steel tube arch bridge and the pull of the self-anchored suspension bridge balance each other, so as to improve the working performance of the flying-type concrete filled steel tube arch bridge and form a new bridge type structure of the super-large span concrete filled steel tube arch bridge and self-anchored suspension bridge.

With the development of transportation, it will be necessary to build a large number of Bridges with 800 meters span on large rivers. The traditional flying-type concrete filled steel tube arch bridge will be faced with great structural technical risks. If the suspension bridge and cable-stayed bridge are adopted, the price is expensive and the construction is difficult. The combination bridge with 800m steel tube flying-type arch and self-anchored suspension cable has the advantages of convenient construction, large span capacity, low cost, large stiffness, high bearing capacity and good durability, etc., and its structural technical advantages are very obvious.

Combining with the design requirements of Chongqing in the planning of the Yangtze river highway bridge, for 800 meters level of large span concrete filled steel tube swallow type arch combined with the anchor cable bridge geometric configuration research, engineering parameter design, and establish the Midas finite element model, internal force analysis, modal analysis and buckling analysis research, in order to verify the structural advantages of technical measures.

2 Configuration research

At present, the traditional technology of concrete filled steel tube flying-type arch bridge cannot be used to build the super-large span bridge with 800 meters of span, so technical improvement should be carried out. There are two aspects of technical improvement measures:

Firstly, steel tubes with single leaf hyperboloid shaped variable section form the arch ribs of basket arch limbs space truss, which can reduce the dead weight of middle span segment, improve the internal force of arch bridge structure, reduce the difficulty of hoisting construction of middle span segment, reduce the structure gravity center and improve the stability.

Second, a self-anchoring suspension cable system is added to strengthen the structural system of the flying-type arch bridge of concrete filled steel tube, forming a super-large span flying-type arch bridge of concrete filled steel tube and a self-anchoring suspension bridge. The two sets of structural systems work together.

Flying-arch and self-anchored suspension cable composite bridge is an organic combination of modern suspension bridge and classical arch bridge. It is a composite structure system composed of flying-arch ribs, double slings, bridge towers, self-anchored suspension cable and deck system. The arch structure is mainly stressed by the suspension cable.

Swallow arch combined with the anchor cable bridge shows the arch bridge and suspension bridge, the characteristics of both the advantages of two kinds of bridge is again complement each other, the existence of the main tower, is the body of the cable attachment, also in construction as the guy in the tower and cable hoisting tower, cable to assist the main arch stress, can rise to adjust pressure line, the improvement of arch rib structure stiffness, and reduce the role of the main arch thrust.

The 800 meters flying-type arch and self-anchored suspension cable composite bridge adopts four single-blade hyperboloid variable section steel tubes to form the space truss arch ribs of the base-shaped arch limbs. The variable section steel tube arch rib reduces the section size at the arch top, increases the section size at the arch foot, reduces the lifting construction weight of the mid-span segment, and increases the stability of the super-large span Flying-type concrete filled steel tube arch bridge.

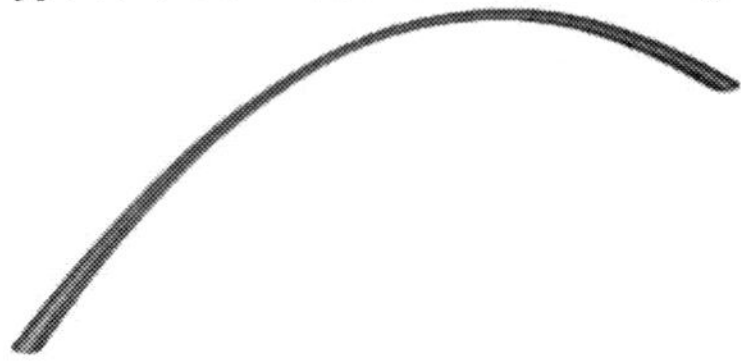

Figure 1. Single - lobed Hyperboloid Variable Section Steel Tube

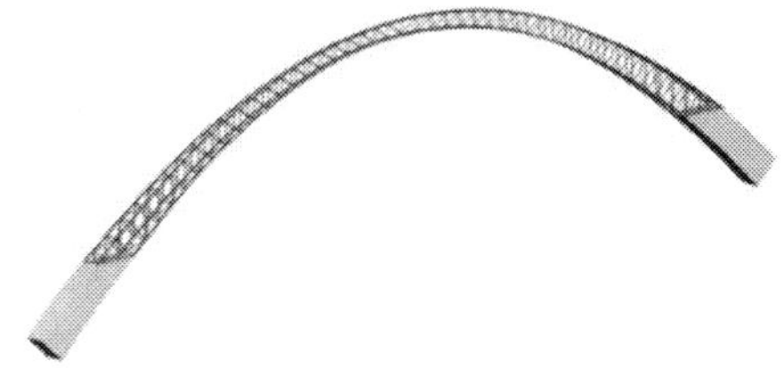

Figure 2. Four - limb Truss Arch Ribs with Hyperboloid Variable Section

Drawing on the concept of external prestressed concrete structure reinforcement, the two sets of suspender system work together with the flying-type concrete filled steel tube arch bridge as the main body and the self-anchored suspension bridge as the auxiliary, complementing each other's advantages. The diagonal back cable of the self-anchored suspension bridge is anchored in the rear beam of the Flying-type CFST arch bridge. The tension of the self-anchored suspension cable system on the foundation is balanced with the thrust of the flying-type CFST arch bridge on the foundation.

Swallow type of concrete-filled steel tube arch bridge self-anchored suspension bridge and share the same base, almost no increase infrastructure investment cost of the bridge, self-anchored suspension bridge pylon is both swallow type of concrete-filled steel tube arch bridge without stents cable hoisting temporary pillar structure, and permanent pillar structure of self-anchored suspension bridge structure, kill two birds with one stone, no increase bridge tower construction investment, only added a small amount of suspension cable, new models combined hybrid bridge has a good economic performance.

Figure 3. Effect Diagram of Combined Bridge with Swallow Arch and Self-anchoring Suspension Cable

Due to swallow type of concrete-filled steel tube arch bridge thrust to balance the tension of the self-anchored suspension bridge structure on the basis, therefore, self-anchored suspension bridge structure system can realize the construction scheme of "cable back rest is first", to avoid the self-anchored suspension bridge in the temporary pier, avoiding the complex system transformation from a mooring line cable structure construction process, save cost, greatly shortens the self-anchored suspension bridge in construction period.

Due to fly into a bridge type horizontal tie bar tensioning prestressed concrete filled steel tube arch bridge is constant, therefore, in the variable of vehicle under live load, swallow type of concrete-filled steel tube arch bridge foundation models will produce certain thrust, variable of vehicle of live load imbalance base internal force will affect the safety of the infrastructure. As the self-anchored suspension cable system participates in the stress, the self-anchored suspension cable system generates tension on the foundation, and the thrust and tension of the two cancel, which greatly reduces the unbalanced horizontal force of the bridge foundation under the action of variable vehicle live load, reduces the displacement of the foundation structure, and ensures the safety of pile foundation structure.

Traditional self-anchored suspension bridge's main cable can be replaced, the wind blows the rain has led to the damage of suspension bridge cable corrosion, self-anchored suspension bridge after using for a long time, the main cable of suspension bridge structure has certain security hidden danger, the swallow type of concrete-filled steel tube arch bridge self-anchored suspension bridge and two sets of bearing structural system, therefore, composite bridge models can easily replace the self-anchored suspension bridge cable.

In order to swallow type of concrete-filled steel tube arch bridge, the use of suspension bridge tower, draw lessons from in vitro pre-stressed reinforcement concept, using the as external prestressed strengthening cable anchor cable system, the integrated use of swallow type of concrete-filled steel tube arch bridge, and technical advantage of self-anchored suspension bridge, and work together, complementary advantages, one of the new hybrid bridge formed two bridge, concrete-filled steel tube arch bridge, seventy percent thirty percent of the self-anchored suspension bridge, spanning ability, good efficiency, structure rigidity, high bearing capacity, the advantages of beautiful appearance and good durability.

3 The establishment of finite element model

Chongqing Changjiang river highway bridge, in the planning of river width of 1400 meters, the need to build a 800 meters order to swallow type steel tube concrete arch bridge, with consideration of requirements, durability and economy to choose concrete-filled steel tube swallow arch and structure form of the combination of anchor cable bridge, expected to break through the existing structure of concrete filled steel tube arch bridge structure technology bottleneck.

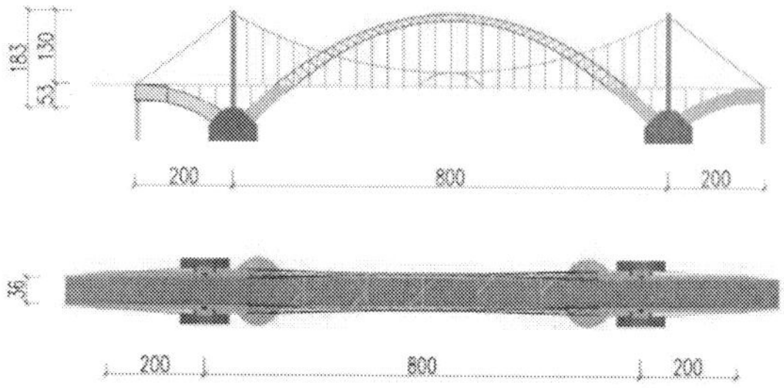

Figure 4. General Layout

Bridge span arrangement for (200 meters + 800 meters + 200 meters), the design load level for road, bridge deck width 32 meters (left the pavement + driveway 28 m + 2 m right sidewalk 2 meters), derrick spacing of 25 m, arch rib rise-span ratio of 1/5, the arch vector is 160 meters high, arch axis coefficient is 1.45, from the door of the anchor cable system bridge tower height is 173 meters, the distance from the top of the base to the bridge panel for 53 meters, bridge panel to the top of the tower height of 120 meters.

Figure 5. Elevation Drawing

Left and right side of the large span concrete filled steel tubular arch bridge hyperboloid of one sheet each set of variable cross-section tube limbs space truss arch rib, limbs space truss arch rib by two cross lean close to each other into a basket of arch truss arch, limbs space truss arch rib arch section 12 meters in height, the arch foot section height of 25 meters, limbs space truss arch rib in the vault after two upper chord close distance of 5 meters, two in two upper chord spacing is 20 meters.

Four limbs space truss arch rib using simple hyperboloid shape cross-section of steel tube arch arch limb, lower chord on the dome. Steel pipe diameter is 1500 mm, on the arch foot bottom chord steel pipe diameter is 3000 mm, the rest of the section, the lower chord variable cross-section steel pipe diameter determined by hyperboloid of one sheet equation, hyperboloid of one sheet of variable cross-section arch steel tube wall

https://doi.org/10.1051/e3sconf/202129302058

thickness 25 mm from the vault on the basis of hyperboloid of one sheet equation change to the arch foot 36 mm.

From the arch foot to the deck section, steel plates are wrapped around the outer ribs of the space truss limbs. The thickness of the steel plates changes from 14mm at the deck to 18mm at the foot of the arch.

The wall thickness of inclined ventral steel tube of space truss arch rib of four limbs changed from the wall thickness of 500mm in diameter at the arch top to the wall thickness of 800mm in diameter at the bridge deck.

The no brace cable lifting scheme was adopted. Starting from the arch foot foundation, the space truss rib segment of the steel tubular limbs with hyperboloid shape variable section was piecewise hoisted and splice until the arch bridge was closed.

During the cable hoisting of the arch rib segment of the space truss for four limbs, the wind-resistant K-type supports are installed synchronously. The K-type supports are made of steel tubes. The diameter of the K-type supports varies linearly from the wall thickness of 600mm at the arch top to the wall thickness of 1200mm at the arch foot to 16mm.

The vacuum assisted continuous filling tube concrete technology and the single blade hyperboloid variable section arch steel tube were filled with C60 high performance concrete by jacking method.

The semi-span auxiliary arch rib adopts the solid arch of steel concrete with variable section, and adopts the form of rectangular section, which changes from the linearization of 20×25 meters at the base of arch to 12×15 meters at the deck.

The sling spacing of flying-bird arch bridge is 25 meters, and the derrick is made of 1670MPa high-strength steel wire with a diameter of 0.16 meters.

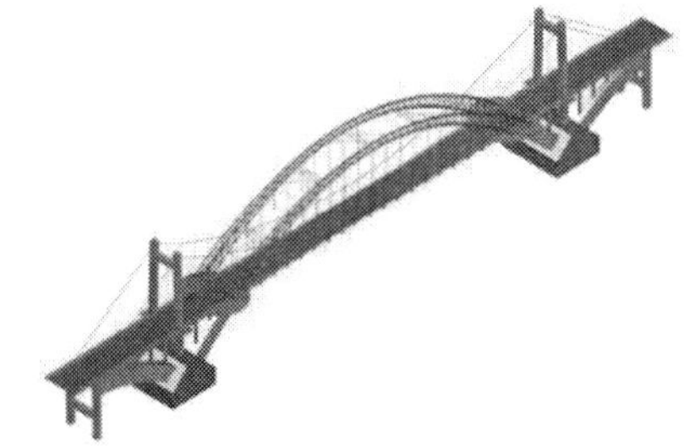

Figure 6. 3D Design Drawing of 800m Concrete Filled Steel Tube Arch Bridge

Each side arch rib offlying-bird arch bridge is provided with 8 finished arch bridge tie cable, which is arranged under the pavement plate. The single-strand tie cable is made of high-strength steel wire cable with a diameter of 0.4 meters, the strength of which is 1860MPa, and the high-strength steel wire cable is covered with PE protective sleeve.

The self-anchoring suspension cable system tower column is the structure of the rigid concrete hollow bridge tower, with the external dimension of 12×12 meters and the wall thickness of 1.2 meters. C70 high-performance concrete is used for pouring, and the steel framework is set inside.

The cable of self-anchoring suspension system adopts 1860MPa high-strength wire rope with diameter of 0.65 meters, the span ratio of self-anchoring suspension cable is 1/8, the spacing of self-anchoring suspension cable system is 25 meters, and the self-anchoring suspension system adopts 1670MPa high-strength wire rope with diameter of 0.1 meters.

The deck stiffened girder is made of concrete girder grid system. The stiffened girder is suspended under the sling of arch bridge and self-anchored suspension bridge, and the arch structure is mainly stressed, supplemented by the suspension.

The central arch buckle is installed in the middle span to adjust the tension of the suspender of the arch bridge and the suspension cable of the self-anchored suspension bridge, the internal forces of the flying-type concrete filled steel tube arch bridge system and the self-anchored suspension bridge system are adjusted, asphalt concrete pavement is installed, and balustrades are installed to form an 800-meter composite bridge of the flying-type concrete filled steel tube arch and the self-anchored suspension cable.

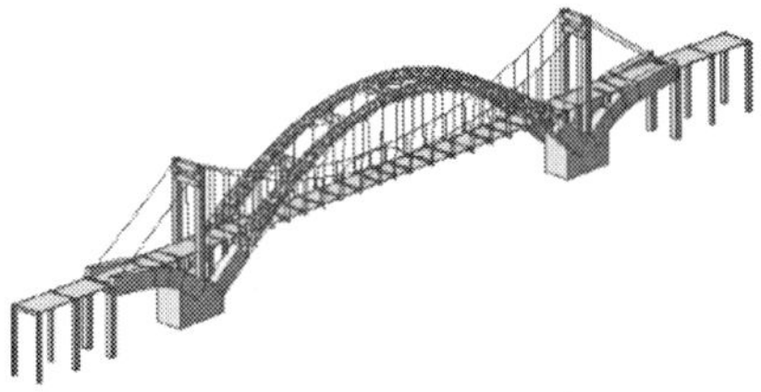

Figure 7. Midas Model

4 Calculation results under vertical load

Full load is applied to the main bridge deck, uniform load standard value is adopted for the additional constant load of the bridge deck, 5KN/m2 is adopted for the standard value of road first-grade load is adopted for the live load of the bridge deck, and the calculation results are as follows:

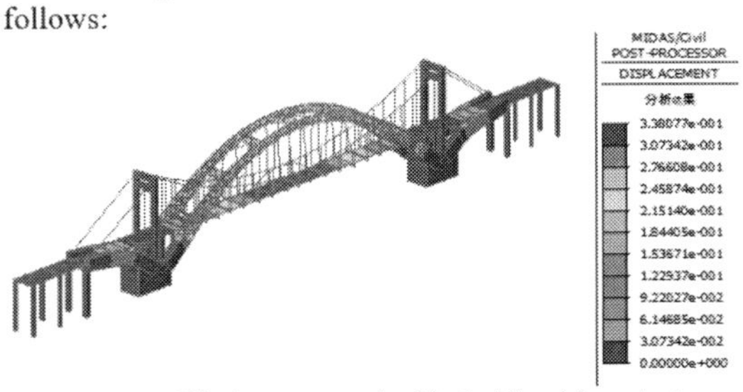

Figure 8. Displacement under Vertical Load (constant + active)

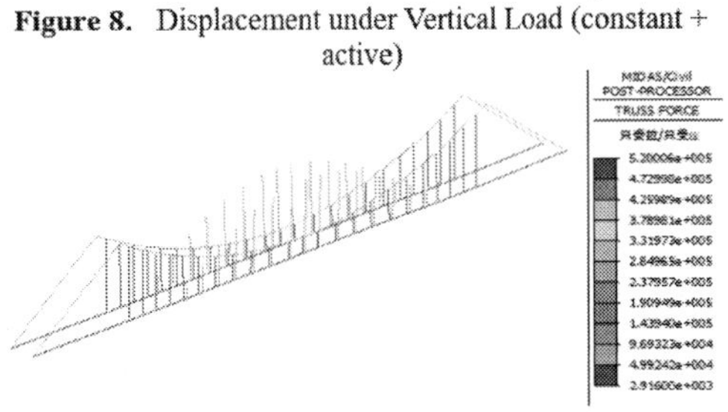

(a) Internal force of cable element

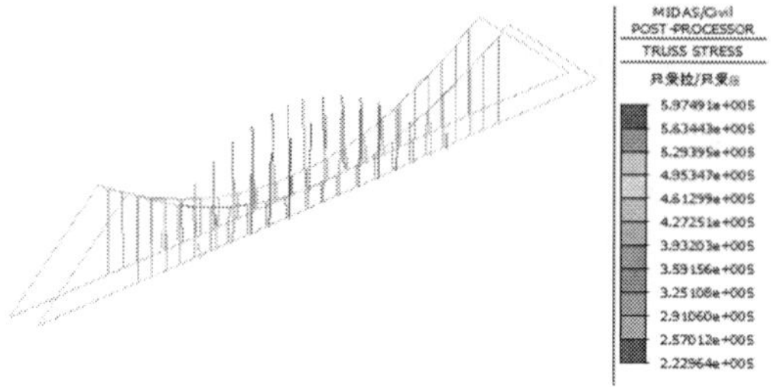

(b) Internal force of cable element

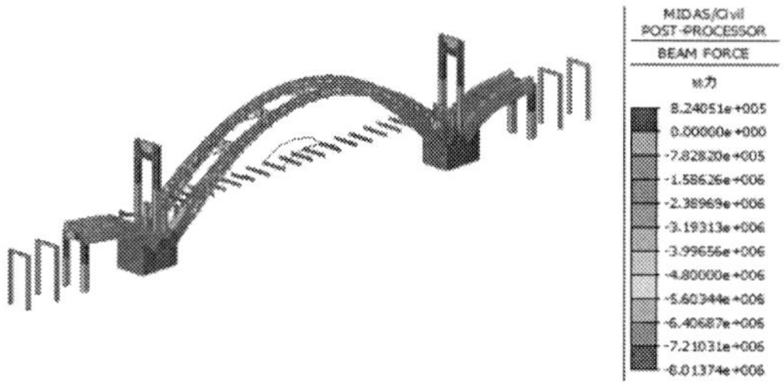

(c) Internal force of beam element

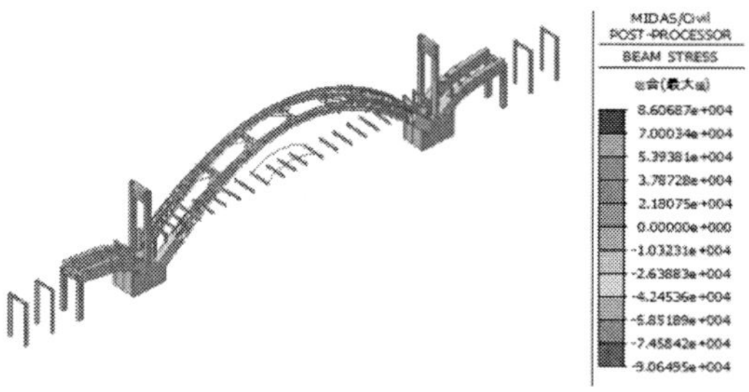

(d) Internal force of beam element

Figure 9. Calculation Results of Internal Forces and Stresses under Load

Under the action of vertical live load, the maximum vertical displacement occurs in the middle of the span, with the maximum displacement of 0.338m, which meets the requirement of 1/500 limit specified in the code. The combined structure of flying-arch and self-anchored suspension cable has large stiffness, which is conducive to reducing the resonance effect of vehicle and bridge and improving the driving comfort.

The maximum internal force of the suspension cable is 185,034.5 KN, and the maximum stress of the suspension cable is 557.6MPa, meeting the strength requirement.

The maximum internal force of arch bridge sling is 18767.8kN, and the maximum stress of arch bridge sling is 597.4mpa, which meets the strength requirement.

The maximum internal force of suspension system sling is 3206.8kN, and the maximum stress of suspension system sling is 408.3mpa, meeting the strength requirement.

The maximum internal force of the arch cable is 520006KN, and the maximum stress of the arch cable is 459.8mpa, which meets the strength requirement.

The maximum internal force of the truss arch rib (single limb) is 1057628KN, and the maximum stress of the truss arch rib is 90.6mpa, which meets the strength requirement.

The maximum internal force of half-span auxiliary arch rib is 4259536KN, and the maximum stress of half-span auxiliary arch rib is 29.1mpa, which meets the strength requirement.

The analysis shows that the horizontal thrust on the left and right sides of the arch foot is basically balanced under the action of constant + live load, which is beneficial to improve the stress condition of the foundation of the flying-type arch and self-anchored suspension cable composite bridge. The self-balancing cable of the flying-arch realizes the horizontal thrust balance of the foundation under the action of constant load. Under the action of live load, the self-anchored suspension cable system exerts tension on the foundation, and the horizontal thrust of the arch bridge and the tension of the self-anchored suspension cable are basically balanced.

5 Modal Calculation Results

Midas structural analysis software is used for dynamic modal analysis. In order not to omit any vibration mode, the sub block method is used to solve the characteristic equation.

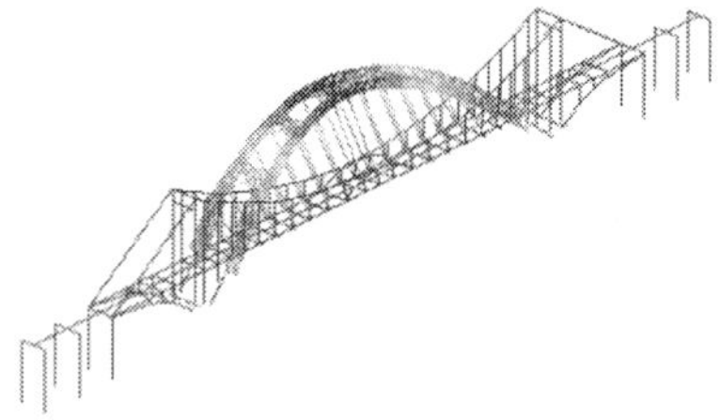

(a) First-order mode of vibration (positive symmetrical arch lateral bending) (0.236Hz)

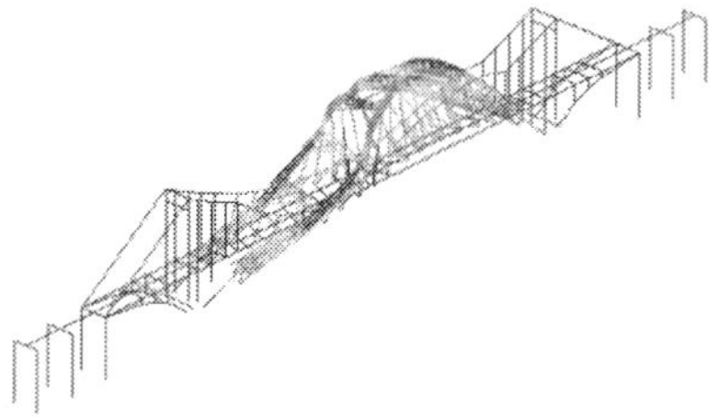

(b) Second-order mode (arch anti-symmetric lateral bending) (0.415Hz)

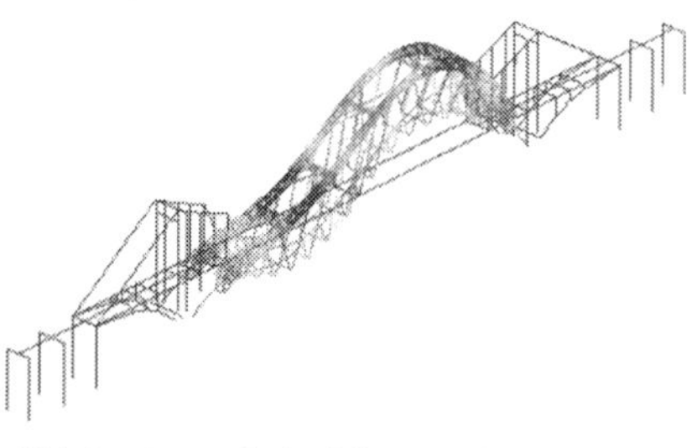

(c) Third-order mode (anti-Symmetric vertical bend) (0.438Hz)

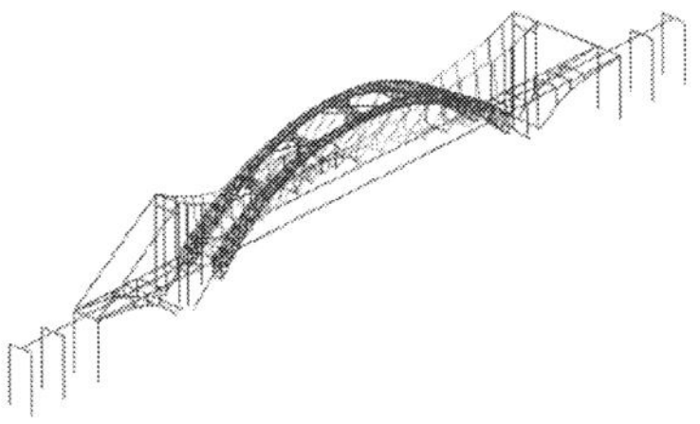

(d) Mode of vibration (vertical bending with positive symmetry of bridge deck) (0.454Hz)

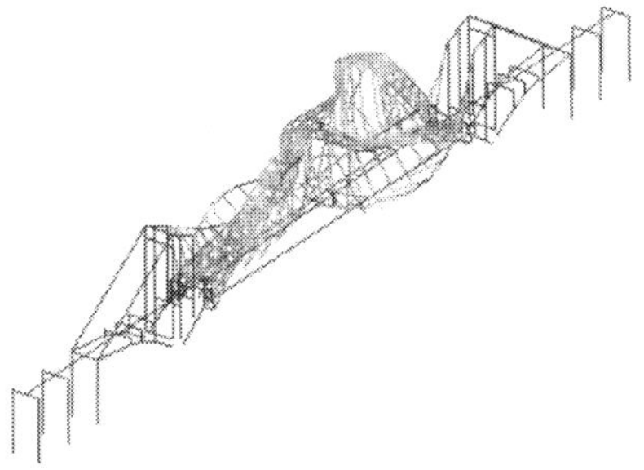

(e) 7 Mode of Vibration (torsion) (0.775Hz)

Figure 10. Typical Modes

From the calculation results in the figure above, it can be seen that the first-order mode of vibration is arch positive symmetric lateral bending, and the frequency is 0.236Hz. The second mode is the positive symmetrical lateral bending of the bridge deck, and the frequency is 0.415Hz. The 7th mode is a torsion mode with a frequency of 0.775 Hz.

On the whole, the vibration modes are dense and obvious grouping phenomenon appears. The first six modes are mainly side bending and vertical bending vibration, and the torsional vibration mode does not appear until the seventh order. The torsional frequency ratio is 1.87, which is high, and the structure has good wind-resistant stability.

Vibration shape in order is the first level of lateral vibration shape vertical vibration form before they occur, the analysis shows that the increase in the auxiliary system of the mooring line cable obviously improves the vertical frequency of vertical stiffness has a larger increase, but its level of lateral frequency is low, the future remains to be further order to swallow the level of the arch combined with the anchor cable bridge to the research of structure stiffness of technical improvement measures.

6 Stability Analysis

Arch bridge is bridge is a kind of compression member, when the structure is unstable, there is no obvious sign, it is very sudden and destructive. Therefore, it is very necessary to analyze the stability of the structure based on the static and dynamic analysis, so as to avoid the failure of the structure.

Midas software was used to analyze the buckling of the designed bridge, and the structural stability coefficient was obtained. In general, the structure with the minimum stability coefficient has the greatest possibility of instability. The first 10 stability coefficients are listed in the table below.

The modal	The eigenvalue	Allowable error
1	10.010	1.4846e-018
2	10.657	8.6848e-013
3	11.484	3.1153e-010
4	12.410	1.3817e-010
5	14.003	6.6927e-009
6	14.401	5.8696e-009
7	16.103	1.2716e-009
8	16.234	4.2738e-009
9	16.318	2.5674e-009
10	17.238	1.1442e-009

It can be seen from the above table that the stability and safety coefficient of the tied arch bridge is 10.010, which meets the requirement of GB 50923-2013 technical Specification for Concrete-filled Steel Tube Arch Bridge greater than 4.0.

Analysis shows that the level of 800 meters swallows of concrete filled steel tubular arch bridge adopts hyperboloid of one sheet shape cross-section of concrete filled steel tube space truss arch rib limbs and swallow arch combined with suspension cable bridge two new technology, greatly improved the swallow type of concrete-filled steel tube arch bridge structural stability models, successfully solved the level of 800 meters of concrete filled steel tubular arch bridge, the stability of the key problems to break the swallow type of concrete-filled steel tube arch bridge structure of the development of the technology bottleneck.

7 Conclusion

Based on the engineering background of a concrete filled steel tube arch bridge of 800 meters, this paper studies the configuration of concrete filled steel tube flying-type arch bridge and self-anchored suspension cable composite bridge, establishes Midas finite element model, conducts internal force analysis under vertical load, conducts modal analysis and buckling analysis, and draws the following conclusions:

(1) The thrust of concrete filled steel tubular arch bridge and the tension of the self-anchored suspension bridge with both balance, form self-balancing structure of the system models, dropped sharply in variable based imbalance internal force under live load, concrete filled steel tube arch rib structure is given priority to, complementary with the force of anchor cable system, two sets of structural system work together, complementary advantages.

(2) Reform the traditional section steel tube arch arch body structure, adopts the hyperboloid of one sheet of variable cross-section tube limbs space truss arch rib structure, reduce the cross section in the weight of lifting, lowering the risk of construction hoisting difficulty,

increased the arch foot section dimension, the arch structure of mass center and stiffness center effectively reduce, increase the stability of the large span concrete filled steel tube arch bridge.

(3) The combination of flying-type arch and self-anchored suspension cable of concrete filled steel tube greatly improves the vertical stiffness of the structure. Under the action of vertical variable live load, the maximum displacement of mid-span position is 0.338m, which meets the requirement of 1/500 limit limit of the specification.

(4) The whole, modal density, apparent modal group phenomenon, the first six order vibration mode are mainly composed of lateral and vertical bending vibration, until the seventh order torsional vibration mode, vibration shape in order is the first level of lateral vibration shape vertical vibration form before they occur, analysis shows that the technical measures to increase the frequency of the vertical bridge, but the horizontal lateral frequency is low, the future remains to be further developed on the level of the combination arch bridge models to the structural stiffness of technical improvement measures.

(5) Buckling analysis calculation, the bridge is 800 meters swallow the arch models of minimum stability safety coefficient is 10.01, the analysis shows that the hyperboloid of one sheet shape cross-section of concrete filled steel tube space truss arch rib limbs and swallow arch combined with suspension cable bridge two new technology, successfully solved the level of 800 meters of concrete-filled steel tube arch bridge stability is the key problem.

References

1. Bouras Y, Vrcelj Z. Thermal in-plane stability of concrete-filled steel tubular arches[J]. International Journal Of Mechanical Sciences.,2019,163.
2. Zhang Zhenwei, Zhang Wei. Analysis of static and dynamic characteristics of Flying-Type concrete-filled steel tubular tied arch bridge [J]. The Chinese and foreign road,2019,39(01):113-116. (In Chinese)
3. Qin Yuyi. Analysis of ultimate Bearing capacity of 500m concrete filled steel tube arch bridge [D].Chongqing: Chongqing Jiaotong University, 2018.(In Chinese)
4. Guo-fu Sun, Shu-cai Li, Xue-jun Zhou,et al. Simulation Calculation Research of Long Span CFST Arch Bridge Based on Optimization Algorithm[J]. Second International Conference on Intelligent Computation Technology and Automation, 2009,2:216-219.
5. Zheng Jielian, Wang Jianjun, Mou Tingmin, etc. Feasibility study on design and construction of 700 M class Concrete filled steel Tube arch Bridge [J]. Chinese Engineering Science,2014(8):33-37.(In Chinese)
6. Morcous, G., Hanna, K., Deng, Y., & Tadros, M. K. . Concrete-Filled Steel Tubular Tied Arch Bridge System: Application to Columbus Viaduct[J]. Journal Of Bridge Engineering, 2012,1, 107.
7. Yuan Yi, Yi Lunxiong. Wuhan Gutian Bridge -- Design and key technology of self-anchored suspension bridge [J]. Bridge Construction, 2019, 49(02):80-85.(In Chinese)
8. Shun-ichi Nakamura, Hiroyasu Tanaka, Kazutoshi Kato. Static analysis of cable-stayed bridge with CFT arch ribs[J]. Journal of Constructional Steel Research.2009,5,776-783.
9. Qiu Jing, Li Guangfeng, Shen Ruili. Study on the stability of self-anchored suspension bridge and upper - supported arch Bridge[J]. Journal of Chongqing Jiaotong University (Natural Science edition),2012,31(01):1-5+10.(In Chinese)
10. Jielian Zheng, Jianjun wang. Concrete-Filled Steel Tube Arch Bridges in China[J]. Engineerin, 2018,4(1):143-155.

Study on the compensating calcium ion method for existing concrete based on crystal infiltration waterproof materials

Yang Dishan [1] , Yan Tongyu [1] , Lin Yubin [1] , Li Jiyu [1] , Ruan Xiaofei [1] , Zhou Hong [2] , Shi Jianguang [2]

[1]National Grid Fujian Electric Power Co., Ltd. Institute of Economics and Technology, Fuzhou 350012, China

[2]School of Architecture and Civil Engineering, Xiamen University, Fujian 361005, China

Abstract: The quality of the the existing concrete surface in its repairing and strengthening is the key factor affecting the together working performance. The effective method to solve this problem is to use permeable crystal waterproof material to strengthen the existing concrete surface. In view of the existing old concrete which lacks free Ca2+ in the interior, the method of compensating calcium ion strengthening is proposed based on the action mechanism of permeable crystalline waterproof material. On the basis of DPS, calcium ion compensating agent Ca $(OH)_2$, Na_2SiO_3 (sodium silicate) and Na_2CO_3 (potassium carbonate) solutes are added to prepare composite reinforcement solution for impregnation strengthening of concrete. Scanning electron microscopy (SEM) was used to observe the microstructure of the concrete before and after strengthening. It showed that after the entry of silicate ions, C-S-H colloid was formed by reaction with Ca $(OH)_2$, and C-S-H gel blocked some pores, which made the inner microstructure more compact. The results of compressive strength before and after concrete strengthening show that the strengthening effect of calcium ion compensation method for low-strength concrete is better than that of permeable crystallization material strengthening method. The strength of C5 and C15 grade concrete is increased by 36.1% and 6% respectively, and the surface strength of 13.7 MPa concrete is increased by 16.7%.

1 Introduction

Existing concrete is required to have a certain strength in the repair and reinforcement of the existing concrete base.For example, the various reinforcement methods in the GB 50367-2013 <Code for Design of Reinforcement of Concrete Structures> require that the positive tensile bonding strength of the substrate surface of the concrete is not less than 1.5MPa and the strength is above C10 or C20 when the concrete is reinforced.JGJ/T259-2012 <Technical Specification for Durability Repair and Protection of Concrete Structures> divides the repair construction into four processes: base treatment, interface treatment, repair treatment, and surface treatment. base treatment and interface treatment are important measures to ensure the bonding effect between the base concrete and the repair material.The base treatment is mainly to solve the existing concrete foundation hollowing, cracking, loosening, steel corrosion and other problems that affect the repair treatment and durability in the later use, and create good base conditions for the restoration.Whether it is the repair or reinforcement of the concrete structure, the quality of the existing concrete base and the treatment method of the bonding surface are the key factors that affect the bonding of the existing concrete and the repair or reinforcement materials to form a common force[1].In order to ensure the quality of the existing concrete base, surface treatment methods such as flower hammers, grinders or high-pressure water jets are required to remove loose aggregates, gravel, scum, leaked cracks and dirt or damaged concrete, surface chiseling, surface cleaning, etc[2]. The surface roughness after base treatment directly affects the bond strength of existing concrete and repair and reinforcement materials[3].The more the existing concrete base layer is removed and chipped, the more difficult the construction, the greater the impact on the existing concrete, and the more environmental problems such as dust.Especially for the concrete components with serious carbonization and corrosion of the base layer and the bare leakage of external high-altitude steel bars, the difficulty of the base layer treatment is higher.Taking base reinforcement treatment is an effective technical way to solve these problems. After the surface is cleaned, the existing concrete base can be directly reinforced, so as to improve the quality and strength of the concrete at the base, and create better conditions for subsequent interface treatment, repair or reinforcement. In order to explore the measures of this technical approach, on the basis of summarizing the research results of concrete infiltration crystalline material reinforcement, the method of compensating calcium ion reinforcement was researched and developed.

* Corresponding author: 550818454@qq.com

https://doi.org/10.1051/e3sconf/202129302059

2 Existing concrete base layer infiltration crystalline material reinforcement method

Permeable crystalline materials are divided into water-based infiltrated crystalline materials and cement-based infiltrated crystalline materials.The water-based osmotic crystallization material is the most famous DPS (Deep Penetrating Sealer) permanent condensate, and it is a kind of chemical permeable waterproof agent. The building materials industry of our country also issued the JCT 1018-2006 <water-based permeable inorganic waterproof agent>.Jiang Qian [4] applied water-based permeable crystalline material (DPS) on the surface of concrete with different strength grades, the compressive strength and chloride ion penetration resistance of concrete were significantly improved, and the improvement degree decreased with the increase of concrete strength grade.Gong Yunli[5] used three kinds of water-based osmotic crystallization (seal-100, karst, DPS) materials to do the test of improving the strength of concrete. The three kinds of waterproof agents showed slightly different effect of improving the compressive strength for different water cement ratio. When the water cement ratio was 0.5, the improvement effect was more obvious than that when the water cement ratio was 0.4.Weng [6] studied the effect of DPS on the surface pore structure, mechanical properties and durability of cement-based composites. Concrete specimens with different water cement ratios ($W/C = 0.35$, 0.45, 0.55) were treated with different amounts of DPS. The results show that DPS improves the compressive strength and chloride ion resistance of concrete,and the penetration depth of DPS is about 2 cm.Zhang Yongchun [7] sprayed concrete with permeable crystalline material for 72h. The results showed that the maximum penetration depth was 195mm, the maximum penetration depth of side wall was 30mm, and the surface strength and hardness were significantly improved.Zong Shizhen [8] used the water-based permeable crystalline material in the protection of ancient relics, and found that DPS has strong infiltration ability on the surface of dense (gray pottery) and porous (red pottery). After the treatment test of 10% phosphoric acid and 10% sulphurous acid, the surface has no change, indicating that DPS has strong resistance to chemical corrosion and air invasion.

Cementitious permeable crystalline waterproof material (CCCW) is a new type of waterproof material based on Portland cement, quartz sand and other basic materials mixed with active chemicals.The general use method is to mix the cement-based permeable crystalline material with water and then brush or spray it on the concrete surface.Cement based permeable crystalline materials have developed from VANDEX in Germany to XYPEX in Canada, VANDEX in Switzerland, FORMDEX in Germany and PENETRON in the United States.China has also issued standards for cementitious permeable crystalline materials, such as GB 18445-2001<cementitious permeable crystalline waterproof materials> .

Chen Guangyao[9] proposed the crystallization mechanism of cementitious osmotic crystallization materials: precipitation reaction crystallization mechanism, complex precipitation reaction crystallization mechanism.The active chemical substances in the permeable crystalline materials react with the free alkali ions dissolved in the capillary pore solution to form crystals insoluble in water, which can effectively seal the pores and micro cracks in the concrete, thus playing the role of water resistance and waterproof.The active chemical substances entering into the concrete react with Ca^{2+} to form a water-soluble but unstable calcium complex.The calcium complex dissolved in water and diffused with the water in the coagulation. When the calcium complex met the highly active unhydrated cement gel, the active chemicals were replaced by silicates and aluminates, which had more stable chemical properties. The precipitation reaction took place. $Ca(OH)_2$ was then transformed into a certain strength crystal, filling the micro cracks and capillary pores in the concrete. So as to enhance the concrete strength and waterproof effect.

Xu Minggen [10] applied cementitious permeable crystalline materials to xiazixi tunnel of Jiaoliu line,SEM combined with energy spectrum analysis shows that the penetration depth increases with the increase of age. The scanning electron microscope shows that the penetration depth of the material in the mortar can reach 29.1 mm after 28 days of curing in the standard state after brushing.Cheng Wei [11] according to the phenomenon that the concrete surface is uneven and loose, the tunnel wall is wet in a large area, and the local external pressure seepage is serious in the transition section of the downstream impedance surge shaft of No. 3 diversion tunnel of Tianshengqiao II hydropower station,After consolidation grouting, the pores are coated with XYPEX plugging agent. After treatment, the loose pitted surface, pores and local honeycomb are solved. The surface is smooth, the water seepage phenomenon disappears, the surface is gradually dry, the handle is hard, and the strength is improved.

Cement based permeable crystalline materials have their limitations. Now, different scholars do not have the same views on how deep the penetration depth of CCCW can be. Xue Shaozu [12] observed by scanning electron microscope that the penetration depth of active substances in the base course is about 5cm within 28 days. However, some researchers have shown that the penetration depth of the substance can reach 1.2m, and the active ingredients will continue to penetrate with the passage of time [13].

These studies and engineering practices have proved that both water-based permeable crystalline materials and cement-based permeable crystalline materials can improve the compressive strength and durability of concrete on the basis of increasing the waterproof capacity of concrete base. In order to improve the reinforcing effect of water-based permeable crystalline material DPS, the method of compensating calcium ion was formed by adding calcium ion compensating agent.

3 Research on strengthening technology of existing concrete base with permeable crystalline material

3.1 selection of reinforcement materials and preparation of reinforcement solution

The reinforcement material of permeable crystalline material is made of Keluo permanent condensate DPS produced by Keluo waterproof technology (Shenzhen) Co., Ltd. the basic properties of the product are shown in table 2-1.

Table 2.1 basic properties and applicability of DPS

Basic attribute name	Attribute characteristics
Environmental protection	Pure environmental protection materials
Acid and alkali resistance, gasoline and freeze thaw resistance	have
Improvement of water absorption of concrete	83% reduction, with secondary impermeability
Strengthening effect on concrete	Surface strength increased by 20% - 25%
Anti cracking (automatic crack sealing)	seal 0.3mm crack automatically
Requirements for drying degree of base course	Dry and wet base surface can be constructed normally
Requirements for flatness and regularity of base surface	No
Construction environment requirements	All weather

DPS liquid is colorless and transparent. In order to determine which anions are contained in DPS, $CaCl_2$ solution is added to it and a large number of white flocculent substances are found. Therefore, it can be concluded that there may be sio_3^{2-}, CO_3^{2-} and OH^- in DPS, as shown in Figure 2.1. The pH test showed that the test paper was blue purple and the liquid was alkaline. When sodium silicate solution was added, a very small amount of precipitation was formed, as shown in Figure 2.2. There was a small amount of Ca $(OH)_2$ in DPS. It can be judged that the internal "active substances" are silicate, carbonate and a small amount of Ca^{2+}.

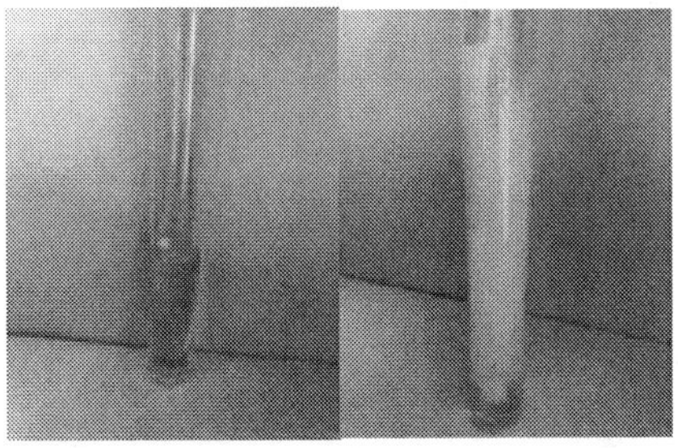

Figure2.1 DPS permanent condensate and massive precipitation

Figure2.2 Alkaline pH test paper and white precipitation

Because the content of Ca^{2+} in DPS is very limited, in order to obtain better reinforcement effect, more Ca^{2+} must be added as compensation.Due to the age of the old concrete is too long, the free Ca^{2+} in the concrete has been basically consumed by the influence of carbonation and other conditions. At this time, it is more necessary to compensate more Ca^{2+} in the concrete, so that the reinforcement liquid can react with Ca^{2+} after penetration to play a reinforcing role.For the choice of Ca^{2+} compensation agent, the solubility and acidity and basicity are mainly considered. Because Ca^{2+} needs to penetrate into the interior of the concrete substrate in order to have a better reinforcement effect, the calcium ion compensation agent needs to dissolve into the water and immerse into the concrete substrate with the water, which requires the calcium ion compensation agent to have good solubility.The solubility of common compounds containing Ca^{2+} is shown in table 2.2.

Table 2.2 solubility of Ca2+ compounds (unit: g)

Ca^{2+} containing compounds	Solubility of 100g water at 20℃
$Ca(OH)_2$	0.17
$CaCl_2$	74.5
$Ca(NO_3)_2$	129
$CaSO_4$	0.25

It can be seen from the table that the solubility of Ca $(NO_3)_2$ and $CaCl_2$ is very high, while the solubility of Ca $(OH)_2$ and $CaSO_4$ is small.In theory, the effect of using Ca $(NO_3)_2$ and $CaCl_2$ as calcium ion compensators is very good, but considering that the reinforcement inside the concrete can not be eroded, the acidity and alkalinity of calcium ion compensators should also be considered.Because the acid solution will not only react with the hydration products of cement and the stones in concrete, but also destroy the alkaline environment in concrete, which has a very negative impact on the internal reinforcement, the calcium ion compensation agent must be alkaline solution.It is reasonable to choose $Ca(OH)_2$ as calcium ion compensation agent, the solubility of $Ca(OH)_2$ can meet the requirements, and the pH value of saturated $Ca(OH)_2$ solution is 12 ~ 13, which has a positive effect on the protection of internal reinforcement of concrete.

Because the strength of concrete mainly comes from the high stability hydrated calcium silicate compound (C-S-H) generated by cement clinker such as tricalcium silicate and dicalcium silicate under the action of water, and the internal concrete is still carrying out C-S-H hydration reaction slowly after its initial setting, in addition, the $CaCO_3$ in concrete also contributes to the strength.Therefore, three principles should be followed in the preparation of reinforcement solution: it is easy to dissolve in water and can penetrate into concrete with water; It can react with Ca^{2+} to form insoluble silicate or carbonate which contribute to the strength; It does not react with other substances in the concrete.Na_2SiO_3 (sodium silicate) and Na_2CO_3 (potassium carbonate) are selected to form a composite reinforcement solution. $CaSiO_3$ and $CaCO_3$ are insoluble, which can effectively seal the internal pores of concrete and enhance the strength of concrete.

The calcium hydroxide used is produced by Tianjin Damao chemical reagent factory, the molecular formula is $Ca(OH)_2$, the purity is 95%, the sodium silicate is produced by Wuxi Yatai United Chemical Co., Ltd., the molecular formula is $Na_2SiO_3{\cdot}9H_2O$, the molecular weight is 284, the potassium carbonate is produced by Tianjin Damao chemical reagent factory, the molecular formula is K_2CO_3, the molecular weight is 138.21. The purity is 99%.

Through the preparation of composite reinforcement solution, the calcium ion compensation reinforcement method based on the osmotic crystallization material reinforcement method is formed.

3.2 preparation of concrete test block and impregnation of reinforcement solution

The main materials of concrete test block are Portland cement, water, coarse aggregate and fine aggregate.The cement is P·O 32.5 ordinary portland cement, which conforms to the national standard GB 175-1999 <Portland cement, ordinary portland cement> . Its 3-day and 28 day flexural strength is 4.0Mpa and 7.6Mpa respectively, and its 3 day and 28 day compressive strength is 18.8Mpa and 37.4Mpa respectively. The fine aggregate used is machine-made sand from Quanzhou, Fujian, with an apparent density of 2680kg/m^3. According to GB/T 14684-2011 <the requirements of sand for building>, the percentage of sieve residue meets the requirements. The coarse aggregate is crushed stone from Xiamen, Fujian Province, with a continuous grading of 5-25mm. According to GB/T 14685-2011 <the requirements of pebbles and crushed stones for construction>, the cumulative sieve residue percentage of crushed stone aggregate meets <the requirements of the specification>. The water is tap water.

In order to explore the reinforcement effect of low-strength concrete, concrete test blocks with C5, C10 and C15 different mix proportions are designed. According to the selection table of concrete sand ratio in JGJ55-2011 <code for design of common concrete mix proportion>, the sand ratio is determined to be 45%, and then the amount of coarse and fine aggregate is determined by absolute volume method. The mix proportion of three kinds of concrete is shown in table 2.3.

Table 2.3 mix proportion of three concrete test blocks

Concrete grade	Material consumption（kg/m^3）			
	Cement	Water	Sand	Gravel
C5	156	195	915	1118
C10	207	195	894	1092
C15	256	195	875	1069

In order to make the reinforcement liquid penetrate deeply in the concrete, the concrete test block is infiltrated by immersion method. Three concrete test blocks of different strength are put into impermeable PE plastic bags, and then the reinforcement liquid is added. The closed opening ensures that the reinforcing liquid is not evaporated, and it is soaked in the ambient temperature of 20 °C and the relative humidity of 70% for 7 days.

3.3 influence of penetration reinforcement materials on concrete strength

According to GB 50081-2002 <standard for test methods of mechanical properties of ordinary concrete>,the compressive strength of concrete reinforced by permeable crystalline material and compensated calcium ion is shown in table 2.4, and the comparison of compressive strength of reinforced concrete is shown in Figure 2.3.It can be seen that the permeable crystalline material DPS has a reinforcing effect on low strength concrete. The lower the strength of concrete is, the better the reinforcing effect is. The strength of C5 concrete is increased by 33.4%, and that of C15 concrete is only increased by 6%.Low strength concrete has larger water cement ratio, more internal pores and poor anti permeability, which makes it easier for DPS to penetrate. The deeper the penetration depth is, the better the reinforcement effect is.

Table 2.4 comparison of compressive strength improvement of concrete reinforced by permeable crystalline material and compensated calcium ion

Group	Concrete grade	Compressive strength（MPa）			Average（MPa）
		1	2	3	
Control group	C5	4.8	4.8	4.6	4.7
	C10	8.6	8.1	8.6	8.4
	C15	14.7	13.8	14.0	14.2
Strengthening method of osmotic crystalline materials	C5	6.3	6.2	6.4	6.3
	C10	9.8	9.5	10.0	9.8
	C15	15.9	14.2	15.2	15.1
Compensation calcium ion Reinforcement method	C5	6.3	6.4	6.5	6.4
	C10	10.0	10.3	9.8	10.0
	C15	15.9	14.3	15.2	15.1

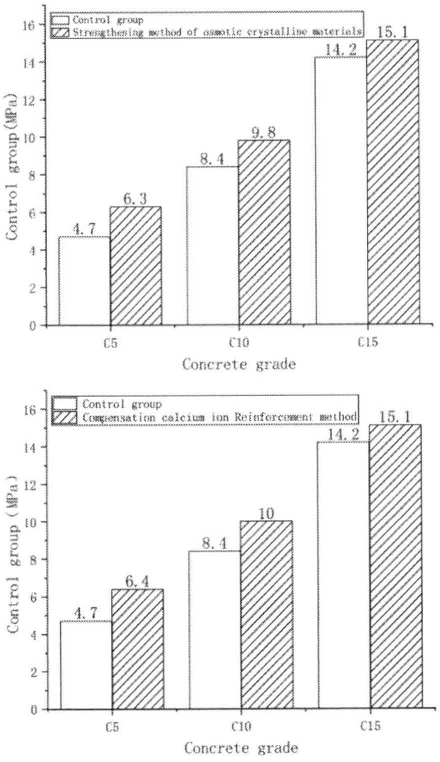

Fig. 2.3 comparison of compressive strength of concrete reinforced by permeable crystalline material and compensated calcium ion

The results show that the reinforcing effect of calcium ion compensation method for low strength concrete is similar to that of osmotic crystallization material. The strength of C5 concrete is increased by 36.1%, and that of C15 concrete is only increased by 6%.

3.4 influence of penetration reinforcement materials on concrete surface strength

According to JGJT23-2011 <the technical specification for testing concrete compressive strength by rebound method>, the surface rebound strength of concrete before and after reinforcement is tested. After the surface rebound strength and carbonation depth are obtained, the compressive strength of concrete can be obtained (see table 2-5 and table 2-6). It can be seen that the surface strength of concrete with 14MPa strength is increased by 2.5MPa, i.e. 17.8%, and the surface strength of concrete with 13.7mpa strength is increased by 2.3mpa, i.e. 16.7%, after using the calcium ion compensation reinforcement method.

Table 2.5 surface strength of concrete before and after reinforcement with permeable crystalline materials

Group	Rebound value Rm				Average	Strength conversion (MPa)
Before reinforcement of infiltration crystalline material	22	22	28	21	23.3	14.0
	23	24	28	24		
	26	28	32	22		
	28	28	26	25		
After the reinforcement of infiltration crystalline material	21	25	24	25	25.2	16.5
	25	25	25	26		
	24	31	35	21		
	25	32	25	27		

Table 2.6 concrete surface strength before and after calcium ion reinforcement

Group	Rebound value Rm				Average	Strength conversion (MPa)
Before compensating calcium ion reinforcement	23	26	30	24	23.0	13.7
	28	26	23	22		
	20	22	21	22		
	22	23	22	20		
After calcium ion compensation	25	28	28	26	24.8	16.0
	28	28	25	24		
	22	24	23	24		
	24	25	22	22		

3.5 microstructure of concrete after penetration reinforcement

After being permeated with crystalline material, the concrete inside is covered by wormlike C-S-H gel. According to the mechanism of precipitation reaction and crystallization, after infiltration of crystalline material into concrete, its active substances react with a large number of alkaline substances such as $Ca(OH)_2$ and free Ca^{2+} in concrete, forming insoluble calcium silicate gel (Fig. 2.4).

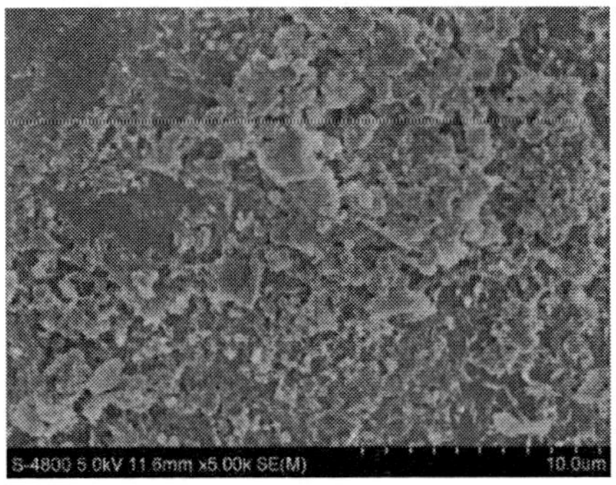

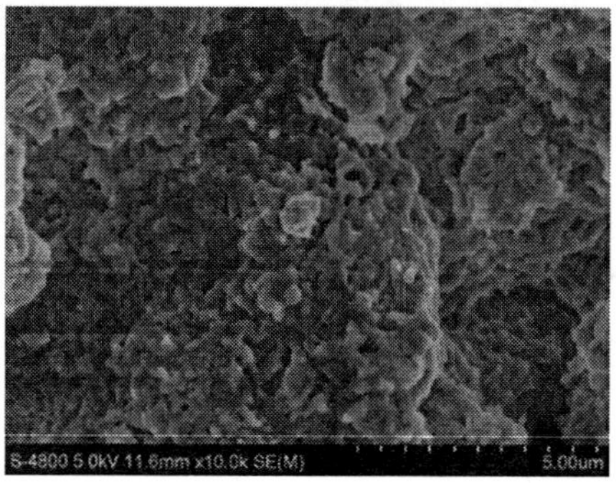

Figure. 2.4 The internal microstructure of concrete after reinforcement by infiltrating crystalline material (5000 times, 10000 times)

The microstructure of the concrete after calcium compensation is similar to that of the concrete after strengthening by infiltration and crystallization. (Fig. 2.5), there are hardly any six corner lamellar $Ca(OH)_2$ on the surface. All of them are wrapped by C-S-H. After considering the entry of silicate ions, C-S-H colloids react with $Ca(OH)_2$, and C-S-H gel blocks some pores. So that the internal microstructure is more compact.

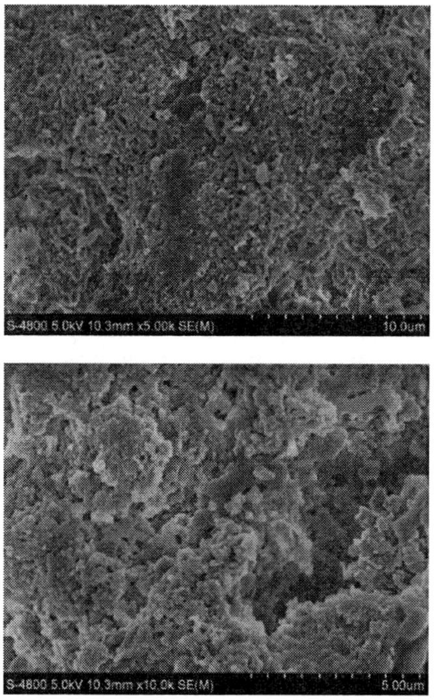

Figure. 2.5. The internal microstructure of concrete after calcium ion compensation (5000 times, 10000 times)

4 Summary

There are two kinds of strengthening materials for existing concrete base: water-based permeable crystalline materials and cement-based permeable crystalline materials. Different strengthening materials provide different technical measures for various engineering needs. In order to improve the reinforcing effect of compressive strength, the composite reinforcing solution was prepared by adding calcium ion compensating agent $Ca(OH)_2$, Na_2SiO_3 (sodium silicate) and Na_2CO_3 (potassium carbonate) on the basis of permeating crystallization reinforcing material DPS. Through the comparison of concrete immersion tests, it is found that:

(1) The permeable crystalline material DPS can reinforce the low strength concrete. The lower the strength of concrete, the better the reinforcement effect. The strength of C5 concrete is increased by 33.4%, and that of C15 concrete is only increased by 6%.

(2) The results show that the compensation calcium ion reinforcement method is better for low strength concrete, the strength of C5 concrete is increased by 36.1%, and the strength of C15 concrete is increased by 6%. It can be determined that calcium ion and silicate ion play a major role in the reinforcement of permeable crystalline materials

(3) The results show that the surface strength of 14MPa concrete is increased by 17.8%. The surface strength of the concrete with 13.7mpa strength increased by 16.7% after being strengthened by the calcium ion compensation method.

(4) The internal microstructure of concrete reinforced by calcium ion compensation and the microstructure of concrete reinforced by osmotic crystallization method show that,after the entry of silicates, C-S-H colloid was formed by reaction with $Ca(OH)_2$, and C-S-H gel blocked some pores, which made the inner microstructure more compact.

The results show that the method of compensating calcium ion is proposed based on the mechanism of osmotic crystallization waterproof material, which can improve the old concrete which lacks free Ca^{2+} in the interior.

Reference:

1. Tian Wenling,Zhao Zhifang,Zhao Guofan,et al.Review on bond mechanism and test methods of new and old concrete[J]. *Journal of Hebei institute of technology*, **1**(1998):84-94.
2. GB 50550-2010, building structure reinforcement engineering construction quality acceptance specification. [3] Ao Jintao,Chen Zhaoyuan. Main factors affecting the bond strength of new and old concrete[J].*Engineering mechanics*, A02(1997):328-333
3. Jiang Qian,Mu Song,Liu Jianzhong,Shi Liang.Research and Mechanism Analysis on performance improvement of concrete with water-based permeable crystalline materials[J].*New building materials*, **43**, 03(2016):49-52,89.
4. Gong Yunli. Study on the influence of water-borne permeable inorganic waterproof agent on the performance of waterproof concrete[D].Central South University,(2012).
5. Tasi-Lung Weng. Effect of Surface Penetrating Sealer on the Microstructure Properties of Cement-based Composites[A]. Information Engineering Research

https://doi.org/10.1051/e3sconf/202129302059

Institute, USA.Proceedings of 2012 International Conference on Engineering Materials (ICEM 2012 V651)[C].Information Engineering Research Institute, USA:Institute of intelligent information technology applications,2012:6.

6. Zhang Yongchun.Application of permeable crystalline waterproof material in construction[J]. *Scientific management of Railway Materials*, **04**(2003):25-26.
7. Zong Shizhen.Application of DPS permanent coagulant in the protection of ceramic cultural relics[J].Restoration and research of cultural relics, 2003(00):176-179.
8. Chen Guangyao,Wu Xiaomei, Fan Yueming. Mechanism analysis of cementitious permeable crystalline waterproof material[J].*New building materials*,**36**, 08(2009):68-71.
9. Xu Minggen, SEM and EDS analysis of penetration depth of cementitious permeable crystalline waterproofing materials[J].*Journal of Railway Science and Engineering*,03(2005):67-70.
10. [11] Cheng Wei, Action mechanism of XYPEX material and its application in Tianshengqiao secondary structure.*Construction organization design*, **2**(2001):21-24
11. Development status and market of cementitious permeable crystalline waterproofing materials, Waterproofing engineering and materials "conference news" No.6 (total 101) compilation of special materials of waterproofing technology committee of construction and building materials branch of Chinese Architectural Society
12. Shang Xiaohua,Jing Denghu.Experimental study on CCCW improving performance of low strength concrete[J]. Journal of water conservancy and Construction Engineering, **78**,02(2018),:42-47.

 https://doi.org/10.1051/e3sconf/202129302060

Test Study on Composite Strengthened Beam with CFRP Board, Angle Steel and External Prestressed Steel Strand

Tao Jiang[1,*], BiQing Li[1] and kai Chen [2]

[1.] Jiangsu Southeas Special Technical Engineering Co., Ltd, Nanjing, 210008, China

[2.] School of Civil Engineering, Southeast University, Jiangsu Nanjing, 210096, China

Abstract: The composite strengthening beam combined with a variety of reinforcement technology, complementary advantages.In order to study the mechanical behavior of T-beam strengthened with CFRP board, external angle steel and external prestressed steel strand, six reinforced test beams were fabricated to compare and analyze the structural characteristics of bending stiffness, crack development and bearing capacity. The research shows that the beam strengthened by CFRP board, external angle steel and external prestressed steel strand has good working performance. The rebound force of the bending deformation of the external angle steel and the upward force of the external prestressed steel strand reliably anchor the carbon fiber board, avoiding the bond failure and slippage of the CFRP.The three kinds of reinforcing materials work together and complement each other, which can give full play to their respective strength. The composite strengthened beam with CFRP board, external angle steel and external prestressed steel strand can greatly improve the cracking load, structural stiffness and ultimate bearing capacity of the original beam, and the reinforcement effect is optima.

1 Introduction

A single reinforcement method has its own advantages and disadvantages, and it is often unable to complete the reinforcement beam with a large bearing capacity demand. Therefore, it is necessary to consider using multiple reinforcement methods to comprehensively strengthen the structural beam at the same time, so as to form a comprehensive reinforcement beam. Comprehensive reinforcement beams complement each other's advantages and work together to give full play to the performance of various reinforcement materials [1-3].

Sticking carbon fiber board, outsourcing Angle steel and external prestressing are the most commonly used reinforcement methods for reinforced concrete beams. There are many successful reinforcement projects. The three reinforcement methods have their own advantages and disadvantages. When a large strengthening bearing capacity is needed, if carbon fiber plate is used for reinforcement, interface peeling failure and anchoring failure will occur due to insufficient bond strength between carbon fiber plate and concrete, so the strength of carbon fiber material cannot be fully utilized [4-7]. If the outsourcing Angle steel reinforcement with larger geometry size is adopted, the construction is difficult and the cost of reinforcement is high. If a large number of external prestressing tendons are used for reinforcement, there will be adverse phenomena such as overarching and upper flange being pulled apart. Therefore, if the three reinforcement methods of sticking carbon fiber plate, outsourcing Angle steel and external prestress are applied comprehensively, the composite reinforcement beam of sticking carbon fiber plate-outsourcing Angle steel and external prestress can be formed, so as to achieve the optimal comprehensive effect of strengthening beam [8-10]..

Pasting carbon fiber board - outsourcing Angle - externally prestressed composite beam is the use of Angle steel bending deformation of rebound and prestressed reinforcement method is lifted up to reliable anchorage of the carbon fiber plate, can reduce the carbon fiber board bond-slip, ensure that the carbon fiber plate material strength, give full play to the three kinds of reinforcement materials to work together, the comprehensive reinforcement beam can greatly improve the cracking load and ultimate load, significantly improve the performance of the concrete beam.

2 Design of reinforced beam

Six reinforced beams were made, and comparative experiments were carried out to analyze and study the bearing capacity, deformation capacity and fracture development of the beams with CFRP - Angle steel - external prestressed steel strand composite reinforcement. Reinforcement method and beam number of test beam are shown in Table 1.

* Corresponding author: Tao Jiang, 13815419890@163.com

Table1. Test beam number and reinforcement method

Test beam no.	Reinforcement description of beams
L1	Unreinforced (original beam)
L2	Paste carbon fiber sheet reinforcement
L3	External Angle reinforcement
L4	Paste carbon fiber board - external Angle steel reinforcement
L5	External prestressed reinforcement
L6	Carbon fiber plate - external Angle steel - external pre - stress reinforcement

(1) Test the original beam L1

Test the original beam of reinforced concrete simply supported T beam (as shown in figure 1), test beam total length of 6000 mm, calculation of span is 5700 mm, high beam of 500 mm, the bottom of the beam width of 200 mm, HRB400 reinforced by using the bottom longitudinal steel beams, with two phi 16 steel bar, steel section reinforcement ratio of 0.335%, reinforced HPB235 level by using the longitudinal compression steel roof beams, with 4 phi 10 steel bar, stirrup used reinforced HPB235 level, stirrup for phi 8 @ 150, the strength grade of concrete design for C25.

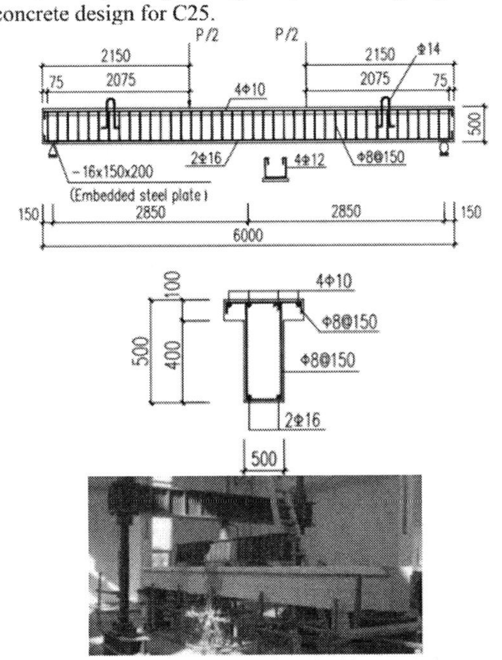

Figure 1. Schematic diagram of original beam reinforcement and test site diagram

(2) beams strengthened with carbon fiber plates L2

A 1.4mm thick 200mm wide 6m overlaid carbon fiber plate is pasted at the bottom of L2 beam. The carbon fiber boardf_{ptk} = 3000MPa.The carbon fiber plate is provided with special clamping steel plate with tooth groove at the end of the beam to anchor the carbon fiber plate. Three 100mmU type steel rings are set at both ends, and four 100mm wide U-type steel rings are pasted in the middle.Fig.2 shows the test site of L2, a beam strengthened with carbon fiber plates.

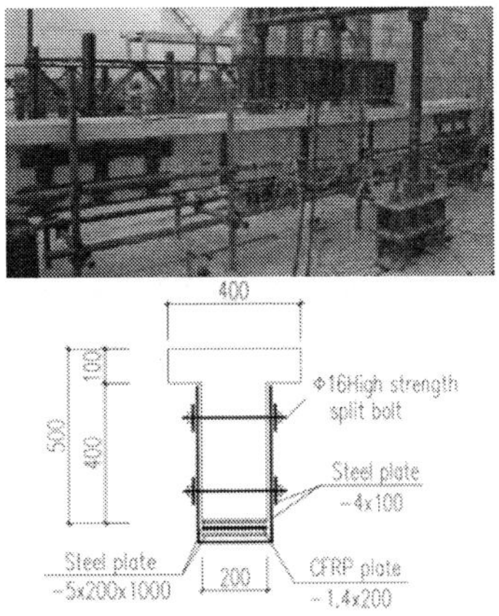

Figure 2. Test site diagram of reinforced beam L2

(3) Reinforced beam L3 with external Angle steel

Both sides of the bottom of L3 beam are supported by Angle steel. Angle steel of Model L50×50×5 is made of Q235B, which is symmetrically bonded to both sides of the beam bottom and anchored with steel hoop plate.

The two ends of the beam are anchored by two U-shaped steel hoop with a spacing of 250mm and a width of 80mm. In the beam, four steel hoop plates with a width of 80mm are set in corresponding positions.

The test site of L3 reinforced beam with outsourcing Angle steel is shown in Figure 3.

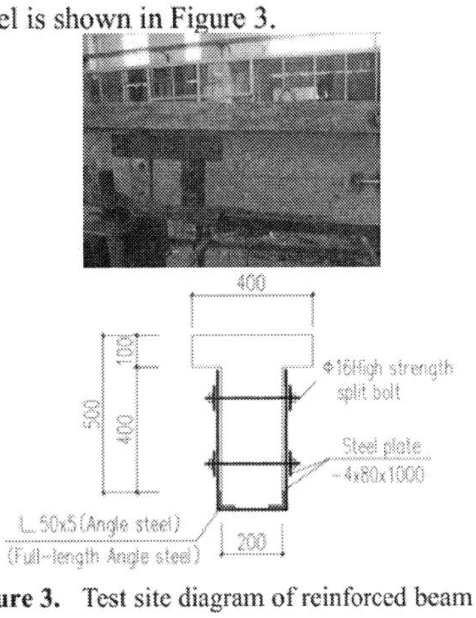

Figure 3. Test site diagram of reinforced beam L3

(4)CFRP - reinforced beam L4 with external Angle steel

This beam is a composite reinforcement beam with carbon fiber bonded to external Angle steel. The Angle steel normal force is used to anchor the carbon fiber plate reliably and work together. Paste carbon fiber plate according to the production method of L3, paste steel pressure plate with thickness of 5mm, length of 170mm and width of 50mm at every interval of 250mm under the carbon fiber plate, then paste Angle steel on both sides of the bottom of the beam for support, and finally set up three U-shaped steel band plates at the end of the beam, and set up four U-shaped steel band plates at the middle of the beam. FIG. 4 shows the test site of the composite reinforcement beam L4 with carbon fiber plate-outsourcing Angle steel.

 https://doi.org/10.1051/e3sconf/202129302060

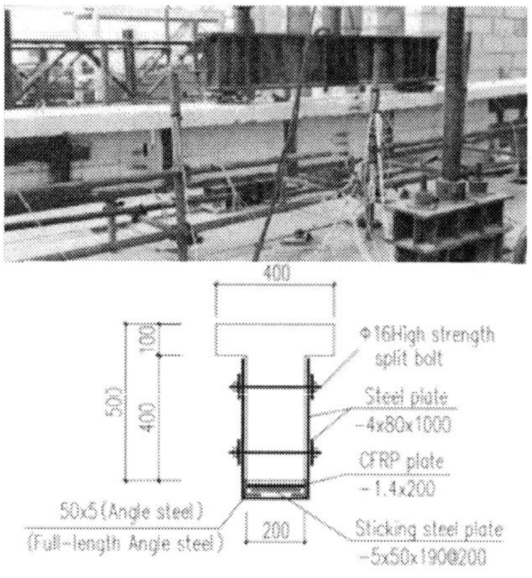

Figure 4. Test site diagram of reinforced beam L4

(5) Externally prestressed reinforced beam L5

The external prestressed reinforcement beam L5 is the external prestressed reinforcement beam. The external prestressed reinforcement bars are $2\varphi^2 15.2$ unbonded tendons and the external prestressed steel strand is in the form of double broken lines. The prestressing tendons are arranged symmetrically, the area of each beam section $A_p = 139mm^2$, the standard value of tensile strength $f_{ptk} = 1860MPa$, and the tension control stress are:$\sigma_{con} = 1023MPa$. The effective stress readings of the prestressing tendons obtained by the pressure sensors on both sides are approximately the same, and the actual effective prestress is about 900MPa. The test site of externally prestressed reinforced beam L5 is shown in Figure 5.

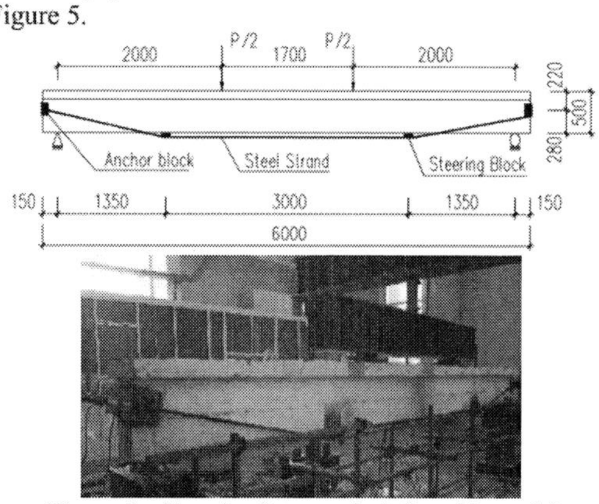

Figure 5. Test site diagram of reinforced beam L5

(6) Carbon fiber plate - external Angle steel - external prestressed reinforcement beam L6

The reinforcement beam L6 is made according to the method of L4 (carbon fiber plate-angle steel composite reinforcement beam), and then the external prestressing is applied according to the construction process of L5 (external prestressing reinforcement beam). By using Angle steel rebound force and normal force of prestressing steel to anchor carbon fiber plate reliably, the bonding slip of carbon fiber plate can be reduced and the strength of carbon fiber plate can be fully exerted. The three reinforcement materials work together and complement each other. FIG. 6 shows the test site of externally prestressed reinforced beam L6 with CFRP plates and Angle steel.

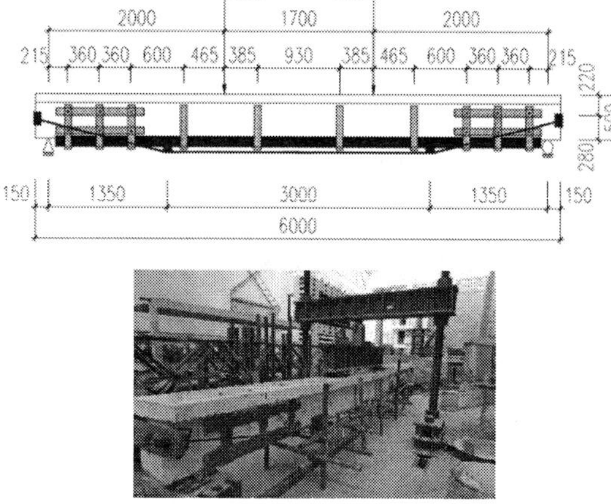

Figure 6. Test site diagram of reinforced beam L6

3 The test phenomenon

(1) Unreinforced beam (L1)

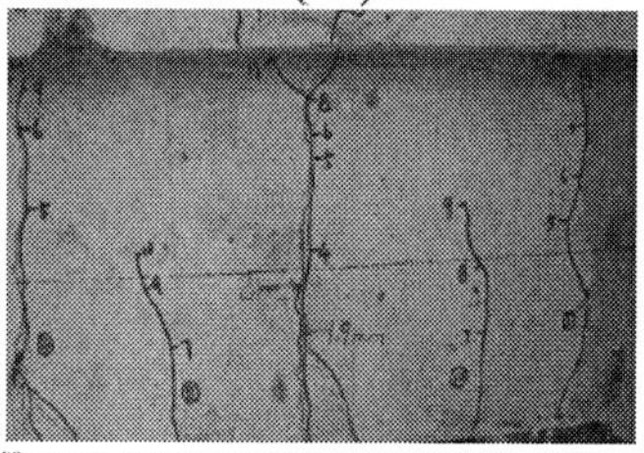

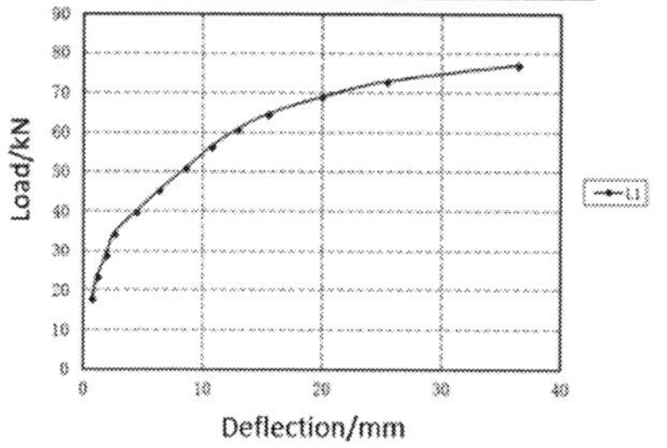

Figure 7. Load - deflection curve of L1 midspan

When the load was applied to 34.33kN, the first subtle vertical vertical crack was observed near the mid-span of the test beam, and the measured vertical deflection of the mid-span of the beam was 2.51mm. With the increase of the load, new cracks appear in the section near the loading point and the pure bending section of the middle span. The original cracks extend upward on the basis of the original cracks, the crack width increases, and the deflection of the beam increases faster than that before the cracking. When loading to 56.32kN, cracks occur at 1/4 span away from the support, and the longest crack in the middle span has developed to the bottom of the flange of the T-beam. When the load is increased to 68.86kN, the tensile reinforcement in the beam yields, the deflection of the beam increases sharply, the cracks in the pure bending section of the span are uniformly distributed, and most of the cracks converge at the bottom of the beam. When the load increases to 77.15kN, the reinforcement strain increases sharply, the crack width expands rapidly, and the deflection development speeds up. Finally, the maximum crack width of the beam reaches 2.1mm, the measured deflection value is 36.43mm, and the loading is terminated.

(2) Bonded carbon fiber reinforced beam (L2)

When the load was added to 36.09kN, the first crack appeared. After the crack appeared, the strain of the carbon fiber plate suddenly increased. As the load increases, bending cracks develop rapidly and longitudinal cracks appear on the carbon fiber plates.

When the load reaches 118.11kN, the tensile reinforcement reaches its yield, and the "crackle" sound is obviously heard at this time, and the carbon fiber plate slides locally. The load continues to increase, and when it reaches 145.61kN, the crack width reaches 2.0mm, which is judged to be the ultimate state of the beam. The crack load of the carbon fiber reinforced beam L2 is slightly increased, and with the development of cracks, the bonding and sliding phenomenon occurs, and finally the arch tie system is formed (the anchor-fixture is reliable). After cracking, the crack develops rapidly and the cracking load decreases rapidly, and the beam stiffness decreases rapidly, and the ultimate bearing capacity is improved compared with the original beam L1.

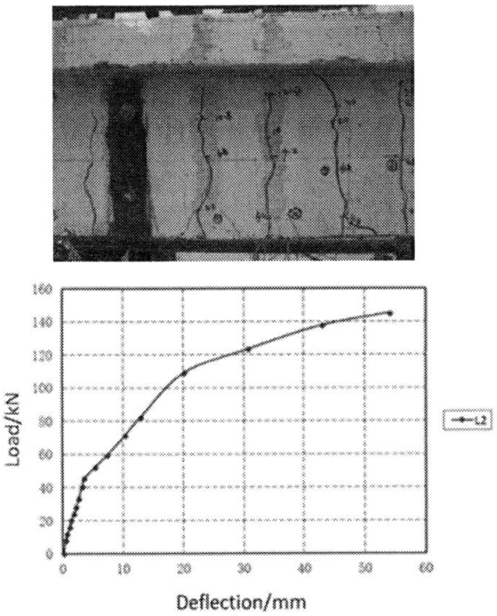

Figure 8. L2 shows the load-deflection curve of the mid-span

(3) External Angle reinforced beam (L3)

At the beginning, the load-deflection curve of the beam was basically linear. When the beam was loaded to 35kN, cracks began to appear in the beam body, with fewer and thinner cracks, the width of which was 0.02mm and the deflection was 2.26mm. When loading continued, the deflection increased, but the load-deflection curve remained uniform, and the cracks were still relatively small. When loading reached 172.2kN, the Angle steel began to yield, and the cracks suddenly increased to 0.72mm, with more cracks appearing, and the mid-span deflection was 20.2mm. When loading to 217.1kN, the crack width has reached 2.0mm, the deflection has increased to 43.24mm, and the test is terminated. The test shows that the outsourced Angle steel reinforced beam L3 has good working performance.

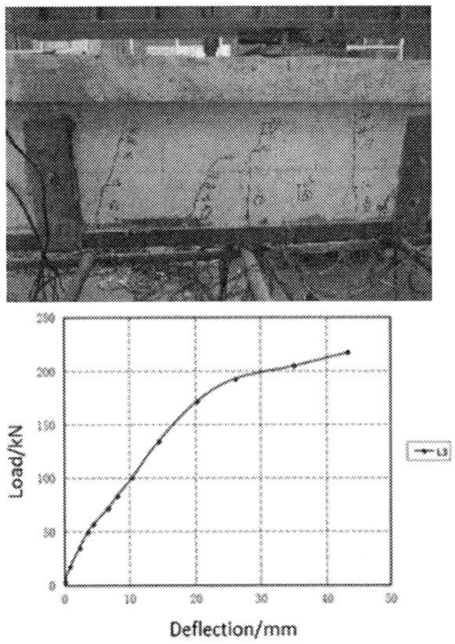

Figure 9. L3 Load - deflection curve of mid-span

(4) CFRP reinforced beam with external Angle steel (L4)

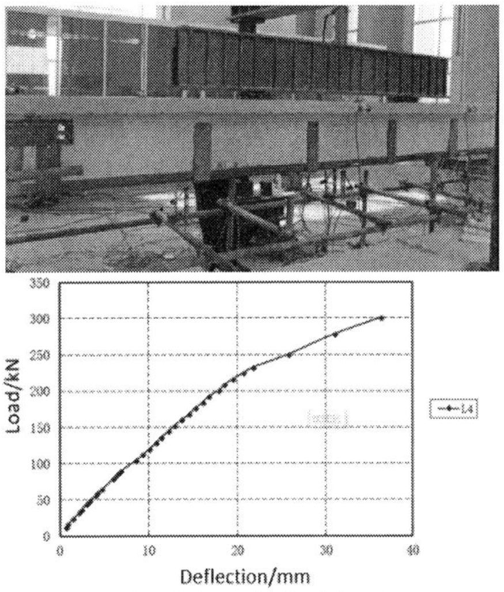

Figure 10. L4 Mid-span load - deflection curve

At the beginning of loading, the deflection of the beam changes uniformly with the increase of load. When the beam is loaded to 47.67kN, the first crack begins to appear, and the deflection of the beam is 3.25mm. When the loading continues, the increment of deflection increases, but the load-deflection curve still changes uniformly, and the cracks are still fine. When the load is added to 231.67kN, the Angle steel begins to yield, and the cracks suddenly increase to 1.36mm, resulting in bending, shear and oblique cracks, and the deflection of the beam body is

https://doi.org/10.1051/e3sconf/202129302060

31mm. When loading continues to 298.42kN, the crack width has reached 1.8mm, and the deflection has reached 36.4mm. After stopping loading, the crack is basically closed after unloading. Test beam stiffness L4 stiffness degradation is not obvious in the whole failure process, crack development overall small and concentrated, as a result of the existence of Angle steel, Angle steel rebound method to force to reliable anchorage of the carbon fiber plate, can reduce the carbon fiber plate bond-slip, ensure that the carbon fiber plate material strength, give full play to the two types of reinforcement materials to work together, the complementary advantages. The carbon fiber board bonded well to the beam body, and there was no obvious bond slip.

(5) Externally prestressed reinforced beam (L5)

At the initial stage of loading, 20kN is applied step by step, and the deflection, stress of non-prestressed reinforcement and stress of steel strand of the beam body increase slowly, which is basically linearly related to the load growth. When the load increased to 100kN, 10kN was loaded step by step until 118.1kN, vertical cracks appeared near the loading point, and the deflection value in the span was 6.54mm. When the load increased to 163.1kN, the ordinary steel bars began to yield, the deflection of the beam increased rapidly, new vertical cracks appeared in the pure bending section of the span, and the original cracks began to extend up and down.
When the load increases to 190.1kN, the crack extends up to the lower edge of the wing plate, the maximum crack width has reached 1.5mm, the mid-span deflection value is 23.46mm, and the loading stops.

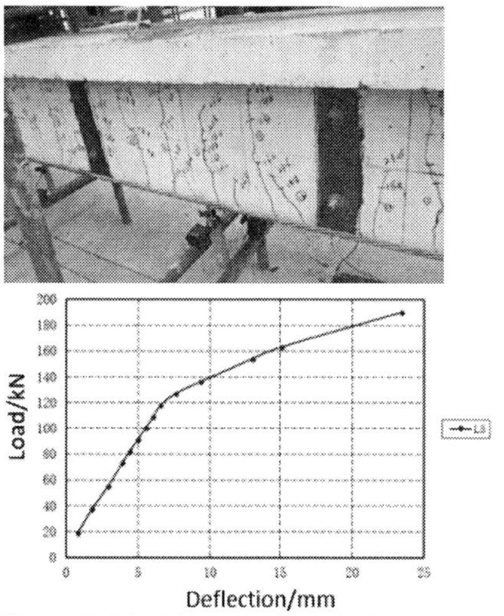

Figure 11. L5 Mid-span load - deflection curve

(6) Carbon fiber plate - external Angle steel - External prestressed reinforcement beam (L6)

Began to load the girder deflection of uniform change appears as load increases, the load of 181.30 kN beam body began to appear when the first crack, deflection of beam body is 8.92 mm, load - deflection curve of a turning point, with the load increasing, pure bending section and cross section near mount point across in the emergence of new cracks, the original cracks on the basis of the extended upward, crack width increase; When loaded to 252.13kN, bending, shear and oblique cracks began to appear in the beam body. When the load is applied to 341.60kN, the cracks are basically complete and the maximum crack width is 0.3mm. When loading to 379.32kN, the reinforcement yielded and the load-deflection curve began to show turning points. When loading continues to 455.56kN, the crack width reaches 1.6mm, and the test stops. By using Angle steel rebound force and normal force of prestressing steel to anchor carbon fiber plate reliably, the bonding slip of carbon fiber plate can be reduced and the strength of carbon fiber plate can be fully exerted. The three reinforcement materials work together and complement each other.

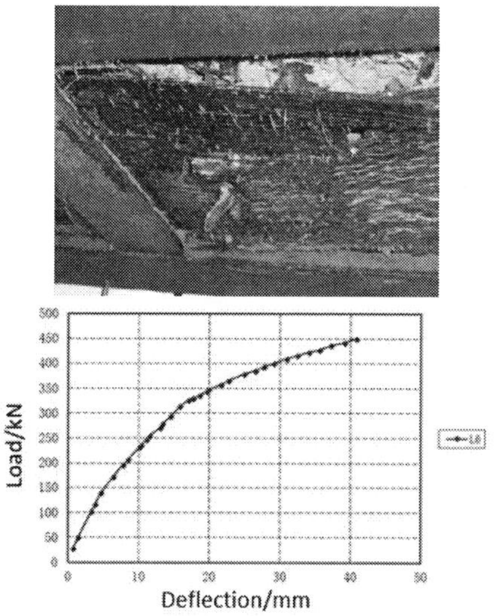

Figure 12. L6 Mid-span load - deflection curve

4 Comparative analysis of test results

The cracking load, yield load and ultimate load of the test beam obtained by load test are summarized in Table 2.

Table2. L1~L6 Summary table of test result data

Beam no.	*P*(kN)		
	P_{cr}	P_y	P_u
L1	34.33	68.86	77.15
L2	36.09	118.11	145.61
L3	35.00	193.10	217.10
L4	47.67	230.39	298.42
L5	118.14	163.10	190.22
L6	127.26	379.32	455.56

4.1 Ultimate load

The analysis shows that the three strengthening methods, including sticking carbon fiber plate, outsourcing Angle steel and external prestressing, can effectively increase the ultimate load of the original beam. The comprehensive strengthening beam has complementary advantages in

https://doi.org/10.1051/e3sconf/202129302060

collaborative work, and the comprehensive strengthening beam can greatly increase the ultimate load of the original beam. The column diagram of ultimate load of the strengthened test beam is shown in Figure 14.

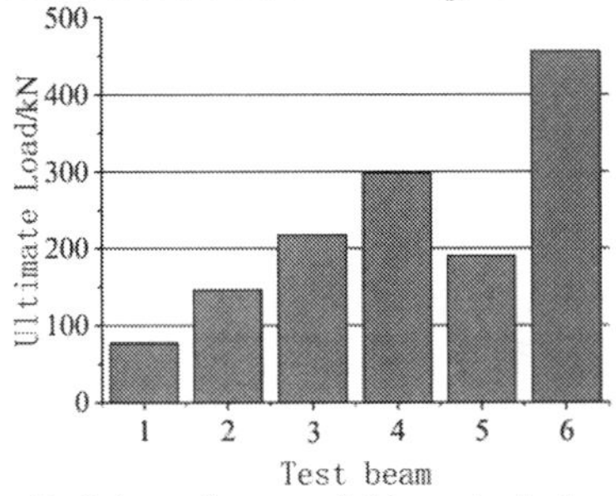

Figure 13. Column diagram of ultimate load of test beam

L2 (CFRP reinforced beam) is increased by (68.46kN) 88.7%. CFRP strengthened with CFRP has little effect on the ultimate bearing capacity of concrete beam, because CFRP sheets fail to give full play to the material strength due to the bonding slip of CFRP sheets in the mid-span area of beam in the test. Although the anchoring fixture was firm, the arching rod system was formed in the later stage of the test beam, resulting in the rapid development of cracks in the test beam, which reached 2.0mm and the test was terminated.

The ultimate loads of L3 (outfitted Angle reinforced beam) and L5 (externally prestressed reinforced beam) were increased by (139.95kN) 181.4% and (113.07kN) 146.6%, respectively, compared with the original L1 (77.15kN). Therefore, the two reinforcement methods (external Angle steel reinforcement and external prestress reinforcement) have a better effect on increasing the ultimate bearing capacity of concrete beams.

L4 (pasting carbon fiber plate - outsourcing Angle steel reinforcement beam) composite reinforcement beam the limit load of 221.27 kN (286.8%) increased, respectively, compared with the L1, and separately use pasting carbon fiber board reinforcement beam L2, paste the Angle steel reinforcement beam L3 limit load by 68.46 kN (88.7%), 139.95 kN (181.4%), and the improvement of L4 limit load capacity than the L2, L3 the improvement of the larger sum, namely the "1 + 1>2 "reinforcement effect.

L6 (Carbon fiber plate - external Angle steel - external prestressed reinforcement beam) composite beam reinforcement of ultimate load increased 378.41 kN (490.5%), and single beam strengthened with pasting carbon fiber plate L2, paste the Angle steel reinforcement beam L3 and L5 externally prestressed reinforcement beam limit load 68.46 kN increased respectively (88.7%), 139.95 kN (181.4%), and 113.07 kN (146.6%), increase the amount of L6 limit load than L2, L3 and L5 both increase the sum of the amount, which produced the "1 + 1 + 1 >3 "reinforcement effect.

From the test results of the above six reinforced beams, the reinforcement effect of L4 and L6 is the most obvious, which fully demonstrates the superiority of the comprehensive reinforcement in improving the ultimate bearing capacity.

4.2 Correlation curve of load and displacement

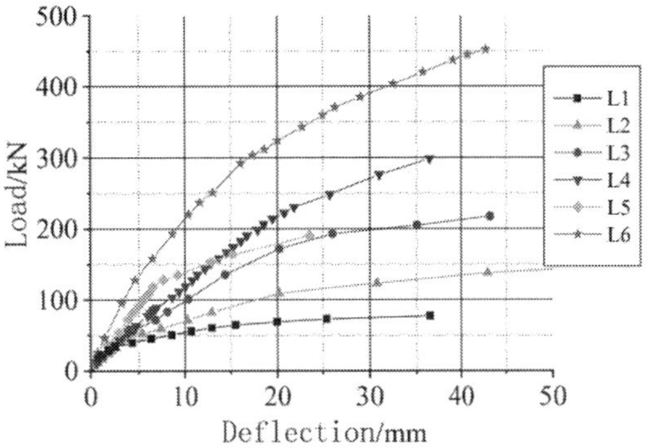

Figure 14. The correlation curve between load and displacement in mid-span position

As can be seen from Figure 15 the stiffness of the original beam L1 decreases rapidly after cracking, and the ultimate bearing capacity is very low. The cracking load of the strengthened beam L2 with carbon fiber plate is slightly increased, but with the development of cracks, the phenomenon of bonding and slippage occurs, and eventually the arching rod system is formed (the anchor-clamp is reliable), resulting in the rapid decrease of the stiffness of the strengthened beam L2 and a slight increase in the ultimate bearing capacity compared with the original beam L1. The cracking load of Angle steel strengthened beam L3 is increased to some extent, and the ultimate bearing capacity is increased to some extent compared with the original beam L1 as the stiffness of crack development decreases slowly, which indicates that Angle steel strengthened beam L3 has good working performance. The cracking load of beam L4 reinforced with carbon fiber plate-outsourcing Angle steel is greatly increased, and the ultimate bearing capacity is greatly increased compared with the original beam L1 as the stiffness decreases slowly with the development of cracks, indicating that beam L4 reinforced with carbon fiber plate-outsourcing Angle steel has good working performance, and the two reinforcement materials have complementary advantages in cooperative work. The cracking load of externally prestressed reinforced beam L5 is greatly increased. After the cracking of beam body, the crack develops rapidly. With the rapid decline in the stiffness of crack development, the ultimate bearing capacity is improved compared with the original beam L1. Stick carbon fiber plate Angle -- - outsourcing externally prestressed reinforcement beam L6, cracking load is greatly improved, with the development of fracture stiffness fell very slowly, ultimate bearing capacity than the original beam L1 greatly improved, show that the carbon fiber plate - outsourcing Angle - in vitro prestressed composite beam L6 has good work performance, work with three kinds of reinforced materials have complementary advantages, the reinforcement effect optimal.

4.3 Correlation curve of load and crack width

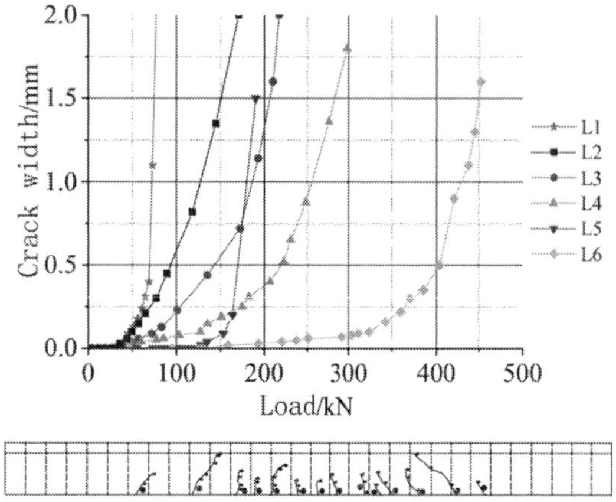

Figure 15. The correlation curve between load and crack width in mid-span position

It can be seen from the development curve of crack width of the six reinforced beams that the crack load of original Beam L1 is low, the crack spacing after cracking is large, and the crack development is very rapid. The cracking load of L2 beam strengthened by carbon fiber plate is slightly increased, but with the development of cracks, the phenomenon of bonding and slippage occurs, and eventually the arching rod system is formed (the anchoring fixture is reliable). After cracking, the crack spacing decreases to some extent and the crack develops rapidly. The crack load is increased, the crack spacing is small, and the crack development is slow. With the slow development of cracks, the crack spacing is relatively dense, which indicates that the CFRP composite reinforced beam L4 has good working performance, and the two reinforcement materials have complementary advantages in cooperative work. The crack load of externally prestressed reinforced beam L5 is greatly increased, and the crack spacing is large and develops rapidly after the crack of the beam body. Carbon fiber board - outsourcing Angle - externally prestressed composite beam L6, has greatly improve the cracking load, crack spacing is very dense, crack development is very slow, show that the carbon fiber plate - outsourcing Angle - externally prestressed composite beam L6 has good work performance, work with three kinds of reinforced materials have complementary advantages, the reinforcement effect optimal.

5 conclusions

1) The carbon fiber plate-external Angle steel-external prestressed composite reinforcement beam has good mechanical performance. The three reinforcement materials work together to enhance strengths and avoid weaknesses, complement each other's advantages and achieve the best reinforcement effect.

2) The reflexion of the bending deformation of the external Angle steel and the upward lifting force of the prestressed reinforcement method reliably anchor the carbon fiber plate, avoid the bonding and sliding phenomenon of the carbon fiber plate, and give full play to the strength of the three reinforcement materials.

3) The cracking load, structural stiffness and ultimate bearing capacity of the original beam can be greatly improved simultaneously by the composite reinforcement beam of carbon fiber plate-angle steel outsource-external prestress.

4) The increment of cracking load and ultimate load of the composite beam strengthened with carbon fiber plate-external Angle steel and external prestress is greater than the sum of the increment of the beam strengthened with carbon fiber sticking, external Angle steel and external prestress alone, which is 1+1+1>3. Reinforcement effect.

References

1. Cao Fubo, Yin Runping, Wang Chenxia. Flexural performance test of composite reinforced concrete beams [J]. China Journal of Chongqing University, 2017, (07): 9-18.
2. Long Qin, Chu long CE. Application of externally prestressed CFRP in strengthening beam bridges [J]. Chinese foreign highway, 2020, (03): 157-159.
3. Karam E C, Hawileh R A, El M T, et al. Experimental investigations of repair of predamaged steelconcrete composite beams using CFRP laminates and mechanical anchors[J].Thinwalled Structures,2017,112:107-117.
4. Correia L, Teixeira T, Michels J, et al. Flexural behaviour of RC slabs strengthened with prestressed CFRP strips using different anchorage systems[J].Composites Part B:Engineering,2015,81:158-170.
5. Fu Zhiqiang, Liu Guohui, Xiao Chengzhi. Experimental study on strengthening effect of prestressed concrete hollow slab beam with steel plate and CFRP [J]. Journal of Hebei University of water resources and electric power, 2020, (03): 22-27.
6. Macdougall C, Green M, Amato L. CFRP tendons for the repair of post-tensioned, unbonded concrete buildings[J].Journal of Performance of Constructed Facilities,2010,25(3):149-157.
7. Grelle S V, Sneed L H. Review of anchorage systems for externally bonded FRP laminates[J]. International Journal of Concrete Structures and Materials, 2013, 7 (1) :17-33.
8. Shan Cong. Finite element analysis of fire resistance of reinforced concrete beams strengthened with angle steel [J]. Jiangsu architecture, 2017, (04): 19-21.
9. Zhang Yunqing, Zhao Qingyun, Meng Tao. Study on external prestressing reinforcement technology of concrete continuous box girder [J]. Chinese foreign highway, 2020, (03): 164-167.
10. Nie Shuhua, sun Jufeng, Liu fjiang, et al. Crack development and failure analysis of composite strengthened beams [J].China Journal of Shandong Agricultural University, 2010, (01): 129-136.

Establishment and prediction of energy demand model in Yunnan Province based on spatial nodes

Liu Siyang[1], Wei Zirui[2], Qian Wen[3], Chen Yu[3], Liu Qian[2], Long Zhen[2], Yang Junwen[3*]

[1] Electric Power Research Institute, Yunnan Power Grid Co. Ltd, Kunming, China
[2] Kev Laboratory in Software Engineering of Yunnan Province, School of Software, Yunnan University, China
[3] Information Center Yunnan Power Grid Co. Ltd Kunming China

Abstract. Energy demand is closely related to energy price, GDP and population. By using the shortest path algorithm and K-means clustering, we set up the spatial nodes, and carried out the model simulation to predict the energy demand of Yunnan Province. The results show that the total energy consumption of Yunnan Province will still show an upward trend from 2020 to 2015; hydropower silicon integration projects in Yunnan Province, the power supply and demand situation in Yunnan Province will change from oversupply to basic balance between supply and demand, and the role of thermal power in dry season will be played to make the decline of coal consumption tend to be smooth; from 2020 to 2025, Yunnan's electricity consumption will increase by about 8.02% year-on-year. However, according to the commissioning of some projects, the total electricity consumption in the province will be about 192.9 billion kwh in 2020, with a year-on-year increase of 12.3%.

1 Introduction

Energy demand is closely related to energy price, GDP and population. The energy demand forecast of Yunnan Province should help to achieve self-sufficiency and cost efficiency, provide some useful references for improving energy utilization efficiency[2] and realize sustainable economic development. Si Wei et al [3] research, the change of economic structure in Yunnan Province has played a certain role in promoting the change of energy intensity at the industrial level. It can also be seen that Yunnan has made some achievements in reducing energy intensity through economic structure adjustment. According to our prediction, this direction still has potential to be tapped. Based on the spatial nodes, we model the energy demand of Yunnan Province, taking into account the influencing factors of energy demand in Yunnan Province[4], and make a forecast, including the total energy consumption, hydropower production and solar power generation. The energy demand forecasting model uses microeconomic variables to forecast energy demand, which helps to plan and draft energy demand management policies.

2 Research objective

The energy demand forecast should contribute to the following aspects:

(1) Planning future needs and determining protection measures;

(2) Identify and rank energy resources, optimize energy utilization, and formulate energy efficiency development strategies;

(3) Policy decision framework;

(4) Energy conservation and emission reduction policies.

3 Application analysis

Different regions have different energy structure, and the demand for energy will be different. Weather factors such as temperature will directly or indirectly affect the energy use of residents. The trend of economic growth in different regions will also have a periodic impact on energy consumption[6]. The distribution of major industries also reflects the consumption of energy[5]. According to the consumption data of electric energy, gas energy and fossil energy in various regions of Yunnan Province, the multi-dimensional energy consumption data of Yunnan Province is constructed based on the geographical location data, weather data and time data. The data preprocessing of these raw data includes data cleaning, data integration, data specification, etc., and then carry out feature engineering to make the data better represent the actual problems of prediction model processing. Using the algorithm model of deep learning and machine learning, the paper forecasts the demand of various energy sources in Yunnan Province in the long term, medium term, short term and ultra short term.

Different energy sources have different trends in demand. For example, electricity is a kind of energy with strong real-time performance. At different time points in a

* Corresponding author: junwenyang_1@163.com

day, the demand for electric energy will also be very different. The demand for electric energy is also a energy with seasonal change. Therefore, the demand prediction of electric energy is suitable for studying the demand change in a short time, while natural gas Fossil energy such as oil is more suitable for studying the change of demand for a long time.

4 Modeling based on spatial nodes

4.1 Modeling

Similar energy demand is the same in different regions the first mock exam model is the most similar area with the same energy demand.

(1) Region division based on shortest path algorithm

In the prediction of regional division needs to be as fine as possible, according to the regional population distribution, regional development, regional characteristic industry to judge the regional demand for new energy. In addition to Kunming, the capital of Yunnan Province, there are Qujing, Zhaotong, Pu'er, Baoshan, Lincang and other large cities with a population of more than 100W. In addition, the tourism industry in Yunnan Province is developed. Although the resident population in Lijiang, Dali, Xishuangbanna and Diqing is small, the floating population is large, and the demand for transportation is no less than that in relatively large cities.

Considering the circulation problem of Yunnan Province, the demand is forecasted according to the communication lines of each city. First of all, Yunnan Province takes the city as the unit, according to the time series model, carries on the short-term, medium-term and long-term energy demand forecast to the entity. Then, Dijkstra algorithm is used to solve the optimal path covering all the cities in Yunnan Province. In Dijkstra algorithm, all municipal units in Yunnan Province are regarded as an independent point in the algorithm. The model figure G is formed by the inverse scaling weight of the demand forecast quantity of each city. V is the set of all vertices in G, s = {V}. D [x] is the known path from the source point to X. The Initial D [v] is 0, and the rest is infinite.

Then we run one-step breadth first algorithm from the source point v to find its adjacent points. Starting from source point 0, the visible point found. Then calculate the path length from the visible point to the source point v, and update d [x]. Then the paths are sorted, the shortest path is selected as the shortest path, and its end point is added to s. Select the newly added points from s, run the breadth first algorithm to find their adjacent points, and repeat the previous step until all points have joined s or no new visible points can be found (there are unconnected points in the graph, and s < V) to terminate the algorithm. At this time, the spanning tree in the model graph is the path with the smallest sum of weights and the largest energy demand. This route covers all the energy demand among the cities in Yunnan Province, and plans the route with the largest demand. Then, the short-term, medium-term and long-term demand forecasting based on time series is carried out independently for the lines on the spanning tree.

(2) Region division based on K-means clustering

K-means clustering algorithm divides many feature points into k categories by minimizing the average distance of the same class feature points, which is an unsupervised learning method[1]. When K-means is applied to regional division, the comprehensive characteristics of various types of energy consumption historical data, GDP proportion and car ownership of each region are used to normalize the characteristics and send them into K-means model for clustering.

Step 1: in the feature space constructed above, k points are randomly initialized as the initial clustering center; Step 2: calculate the cluster of each point to K cluster centers, and then divide the point to the nearest cluster center to get K clusters; Step 3: recalculate the centroid of each cluster (cluster center); Step4: repeat Step2 to Step4 until the centroid position does not change, or only changes in a small range, and get K clusters, that is, the region is divided into K clusters. In the iterative process, K-means algorithm takes the objective function as the nearest clustering center. After several iterations, the objective function J converges to a small range to ensure the average optimal distance between the feature points in the positive feature space and the clustering center[7][8].

4.2 Model simulation

(1) Total energy consumption of Yunnan Province, time period: 2011-2017

Simulation data: energy consumption of Yunnan Province, as shown in table 1;

Application model: energy demand forecasting model based on seq2seq

Related data: GDP, population and output value of main industrial products of Yunnan Province

Table 1 total energy consumption of Yunnan Province from 2011 to 2017 (data source: Yunnan Statistical Yearbook)

	Energy consumption (10000 tons of standard coal)	GDP (100 million yuan)	Population (10000 people)		Output of main industrial products								
			total	Urban population	Raw coal (10000 tons)	Crude steel (10000 tons)	Steel (10000 tons)	Power generation (billion KWH)	Finished sugar (10000 tons)	Cigarettes (100 million)	Cement (10000 tons)	Cloth (10000 meters)	Machine made paper and paperboard (10000 tons)
2011	9540	8893.12	4631	1704.2	9957.41	1323.23	1351.85	1555.13	173.51	3649.91	6788.88	449.58	49.12
2012	10434	10309.47	4659	1831.453	7610.37	1526.69	1600.04	1533.94	205.93	3841.16	7793.66	419	51.46
2013	10072	11832.31	4686.6	1897.136	8185.53	1884.8	2053.92	1954.62	236.52	3787.75	9009.16	298	41.43

 https://doi.org/10.1051/e3sconf/202129302061

2014	10455	12814.59	4713.9	1967.11	4013.656997	1689.07	1935.05	2347.206114	249.68	3848.06	9492.64	156	46.58
2015	10357	13619.17	4741.8	2054.622	4590.14	1418.08	1695.37	2352.4	249.58	3903.63	9305.31	118	58.89
2016	10656	14788.42	4770.5	2148.156	4251.82	1417.33	1654.65	2469.5	220.77	3736.89	10963.53	104	78.46
2017	11091	16376.34	4800.5	2241.353	4392.91	1517.5	1607.38	2730.09	227.13	3589.36	11292.89	93	89.55

The simulation results are shown in figure 1:

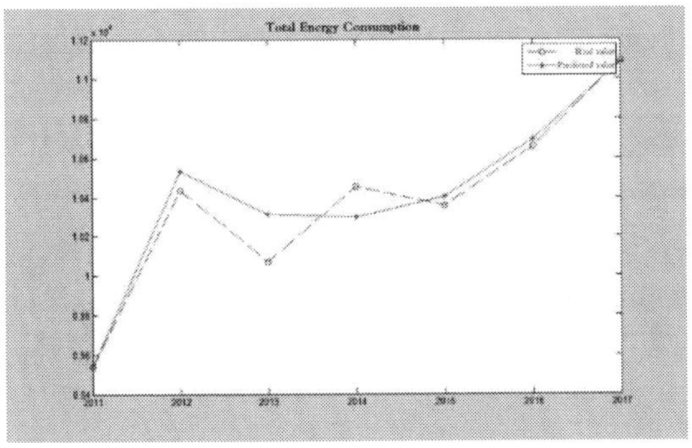

Figure 1 simulation results of energy consumption in Yunnan Province from 2011 to 2017

(2) Simulation of hydropower production in Yunnan Province, time period: 2013-2018

Data: hydropower production in Yunnan Province, as shown in table 2.

Application model: energy demand forecasting model based on seq2seq

Associated data: precipitation in Yunnan Province, as shown in table 2.

Table 2 regional rainfall of Yunnan Province from 2013 to 2018 (data source: Yunnan Statistical Yearbook)

	2013	2014	2015	2016	2017	2018
State and city	Annual precipitation (mm)	Annual precipitation (mm)	Annual precipitation (mm)	Annual precipitation (mm)	Annual precipitation (mm)	Annual precipitation (mm)
whole province	835	981	1 106.6	1 154.6	1 178.7	1 117.3
Kunming City	604	966	1 086.3	956.1	1 042.6	902.8
Wuhua District	551	906	1 190.7	1 150.2	1 186.4	1 084.8
Dongchuan District	566	1 078	810.8	729.5	830.9	620.6
Chenggong District	428	763	968.8	885.5	943.3	868.1
Jinning District	608	904	925.7	916.5	1 146.2	917.4
Fumin County	558	851	999.6	1 071.6	982.1	780.8
Yiliang County	531	1135	1 274.7	625.1	1 055.5	967.3
Shilin County	570	1008	1 069.4	893.3	1 008.4	861.6
Songming County	672	1007	1 176.9	1 104.4	1 150.1	1 145.0
Luquan County	622	963	946.7	1 024.8	810.2	962.1
Xundian County	827	969	1 253.9	1 044.9	1 290.0	921.1
Anning City	824	885	998.3	902.1	1 065.2	802.2
Qujing City	740	1054	1 248.3	1 047.2	1 093.6	1 019.0
Qilin District	679	880	1 231.0	1 043.3	1 107.3	928.5
Zhanyi District	636	877	1 257.0	887.8	1 149.2	875.5
Malong District	650	795	971.5	938.3	900.3	1 060.3
Luliang County	629	1199	1 324.2	882.3	866.7	798.1
Shizong County	1 043	1577	1 982.0	1 218.7	1 389.1	1 237.9
Luoping County	784	1431	1 342.6	1 530.0	1 598.5	1 521.9
Fuyuan County	623	676	909.1	1 104.8	1 047.9	936.6
Huize County	758	939	1 219.7	822.3	848.1	750.7
Xuanwei City	856	1115	997.2	997.3	935.5	1 061.3
YuXi City	673	872	938.7	932.7	1 050.5	903.9
Hongta Distric	628	886	941.1	988.1	1 184.3	837.9
Jiangchuan District	720	811	1 032.7	865.3	1 009.5	909.1
Chengjiang County	691	1023	1 111.5	955.2	1 151.1	918.8
Tonghai County	591	1010	1 017.7	998.9	1 229.6	903.7
Huaning County	838	957	1 034.2	986.5	1 023.9	908.3
Yimen County	565	767	967.9	959.5	981.3	830.4
Eshan County	609	760	782.5	935.6	922.5	946.3
Xinping County	736	674	895.0	1 008.4	1 031.2	1 046.9
Yuanjiang County	677	965	665.7	697.2	921.1	833.4

https://doi.org/10.1051/e3sconf/202129302061

The simulation results are shown in figure 2:

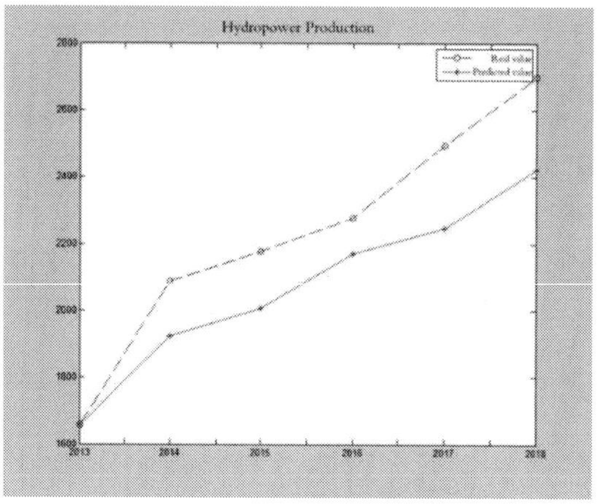

Figure 2 simulation results of hydropower production in Yunnan Province from 2013 to 2018

(3) Simulation of solar power generation in Yunnan Province, time period: 2015-2018

Data: solar power generation in Yunnan Province is shown in table 3.

Application model: Based on deep learning and ridge regression coupling prediction model

Associated data: temperature in Yunnan Province, as shown in table 3.

Table 3 temperature of various cities in Yunnan Province from 2013 to 2018 (data source: Yunnan Statistical Yearbook)

	2013	2014	2015	2016	2017	2018
State and city	Annual average temperature	Annual average temperature	Annual average temperature	Annual average temperature	Annual average temperature	Annual average temperature
	(°C)	(°C)	(°C)	(°C)	(°C)	(°C)
whole province	**17.4**	**16.9**	**17.4**	**17.2**	**17.1**	**17.1**
Kunming City	**16.8**	**16.3**	**16.5**	**16.1**	**16.3**	**16.3**
Wuhua District	16.4	15.8	16.2	15.8	15.7	15.7
Dongchuan District	16.6	16.1	20.6	20.4	20.3	20.8
Chenggong District	20.7	20.3	16.1	15.7	15.7	15.7
Jinning District	15.8	16.0	16.2	15.9	15.4	15.3
Fumin County	16.8	16.9	17.1	16.5	16.4	16.3
Yiliang County	17.7	17.5	17.7	17.5	17.4	17.2
Shilin County	17.5	16.9	17.2	16.8	16.7	16.8
Songming County	15.6	14.9	15.1	14.9	14.9	14.9
Luquan County	16.4	16.4	16.3	16.1	16.2	16.0
Xundian County	16.0	15.8	15.9	15.5	14.9	14.9
Anning City	16.3	16.3	16.4	15.9	15.9	15.7
Qujing City	**15.1**	**14.6**	**15.1**	**14.8**	**14.6**	**14.7**
Qilin District	16.0	15.8	16.2	15.8	15.6	15.8
Zhanyi District	14.7	14.3	14.5	15.4	15.3	15.4

Malong District	15.7	15.3	15.6	14.1	14.1	14.1
Luliang County	15.0	14.2	14.6	15.6	15.4	15.3
Shizong County	15.9	15.8	16.4	14.3	14.1	14.2
Luoping County	15.1	13.9	14.4	16.3	16.1	16.0
Fuyuan County	14.1	12.9	13.4	14.1	13.9	14.1
Huize County	15.4	15.3	15.7	12.9	12.9	13.1
Xuanwei City	14.4	14.2	14.9	14.4	14.4	14.4
YuXi City	**18.0**	**17.7**	**17.9**	**17.6**	**17.6**	**17.4**
Hongta Distric	16.9	17.0	17.0	16.7	16.7	16.4
Jiangchuan District	17.0	17.0	17.2	17.0	16.9	16.7
Chengjiang County	17.2	16.8	17.0	16.6	16.6	16.6
Tonghai County	16.5	16.4	16.5	16.2	16.2	16.2
Huaning County	17.5	16.5	16.7	16.3	16.3	16.3
Yimen County	17.4	16.9	17.1	16.8	16.9	16.6
Eshan County	17.1	16.7	16.9	16.6	16.5	16.3
Xinping County	18.1	17.8	17.9	17.6	17.6	17.5
Yuanjiang County	24.6	24.3	24.8	24.4	24.3	24.2

The simulation results are shown in figure 3:

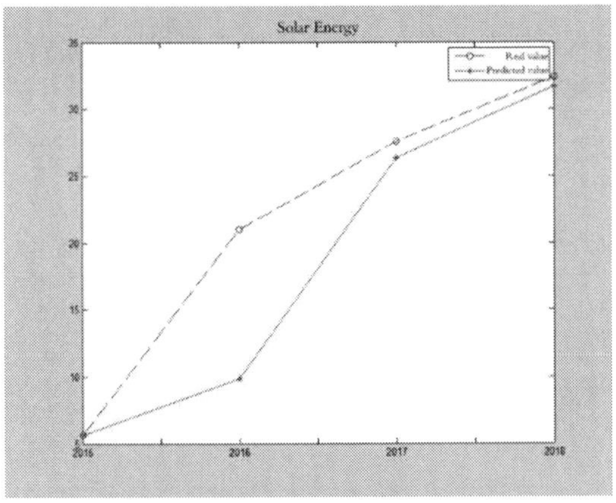

Figure 3 simulation results of solar power generation in Yunnan Province from 2015 to 2018

5 Discussion

Based on the above spatial node model and simulation results, our energy demand forecast for Yunnan Province is as follows:

（1） Total energy consumption forecast

According to China's National Statistical Yearbook, Yunnan's GDP has maintained a rapid growth, which is higher than the national level. From 2011 to 2017, the GDP of Yunnan Province increased from 0.88 trillion yuan to 1.8 trillion yuan.

According to the data simulation results of Yunnan Province from 2011 to 2017, the total energy consumption of Yunnan Province from 2020 to 2025 will still show an upward trend, as shown in table 4.

Table 4 total energy consumption forecast of Yunnan Province

Year	Total energy consumption forecast (10000 tons of standard coal)
2020	12663.08
2021	13211.21
2022	13771.29
2023	14353.93
2024	14958.37
2025	15554.95

（2） Coal production and consumption forecast

According to the gray prediction model, the raw coal output of Yunnan Province in 2020 can reach 50 million tons. According to the policy forecast, Yunnan's raw coal output will reach 55 million tons from 2021 to 2025, basically meeting the coal consumption demand in the province.

According to the data simulation results of Yunnan Province from 2012 to 2018, the coal consumption of Yunnan Province will continue to decline from 2020 to 2025. However, combined with the promotion of hydropower aluminum, hydropower silicon integration

https://doi.org/10.1051/e3sconf/202129302061

projects in Yunnan Province, the power supply and demand situation in Yunnan Province will change from oversupply to basic balance of supply and demand. In dry season, there will be power shortage, and the role of thermal power in maintaining dry season will be brought into play, so that the downward trend of coal consumption tends to be flat. Therefore, the coal consumption forecast of Yunnan Province from 2020 to 2025 is shown in table 5.

Table 5 raw coal consumption forecast of Yunnan Province

Year	Raw coal (unit: 100 million tons)
2020	5921.596
2021	5832.5
2022	5744.744
2023	5658.308
2024	5573.173
2025	5489.319

（3） Electricity forecast

According to the model prediction, the power consumption of Yunnan Province will increase by about 8.02% year-on-year in 2020-2025. However, due to the estimated increase of large industrial load of 3.4 million kW in 2020, under the condition of timely production, it is estimated that the new power consumption is about 15billion kwh, and the power load is respectively Matang hydropower aluminum phase I (820000 kW), Funing Shenhuo Group hydropower aluminum project phase I (750000kw), and Its sub electrolytic aluminum (54000kw), Heqing Yixin Aluminum Co., Ltd. phase II (350000kw), Lianghe Yunju Technology Co., Ltd. (200000 kW), Yunnan Yonggang iron and Steel Group giant Lida Steel Co., Ltd. (100000 kW), Dehong Hebao Technology Co., Ltd. (70000 kW), Lijiang Longji silicon material Co., Ltd. phase II (65000kw), and Baoshan Longji silicon material company phase III (62000 kW) and so on. In comprehensive consideration, according to the above-mentioned production situation, the total power consumption in 2020 is about 1929 billion kwh, an increase of 12.3% year-on-year.

Without the construction of large projects, it is predicted that the power consumption of Yunnan Province will increase by about 8.02% year on year in 2021-2025.

（4） Natural gas forecast

Because of the lack of oil and gas resources in Yunnan, it is mainly transported by China Myanmar oil and gas pipelines. In 2020, the implementation opinions of Yunnan Provincial People's Government on further promoting the coordinated and stable development of natural gas were issued, which put forward that: "to build a multi-element gas supply pattern, supplement the short board of natural gas infrastructure construction, accelerate the realization of the strategic goal of" gasification Yunnan ", strive to achieve 2.8 billion cubic meters of natural gas consumption in 2020 and 6 billion cubic meters by 2025; Ensure the safety and efficiency of the facilities, balance of natural gas supply and demand, and strong guarantee for people's livelihood. "

It is predicted that the demand for natural gas in Yunnan Province will increase from 2.8 billion cubic meters to 6 billion cubic meters in 2020-2025.

Acknowledgments

This work has been supported by the National Key Research and Development Program No.2018YFB2100100, Postdoctoral Science Foundation of Yunnan Province, Science Foundation of Yunnan Province No.202101AT070167, Project of the Yunnan Provincial Department of Education scientific research fund No. 2019J0010, DongLu Young and Middle-aged backbone Teachers Project of Yunnan University, Open Foundation of Key Laboratory in Software Engineering of Yunnan Province under Grant No.2020SE311, Science and Technology Project of Yunnan Power Grid Co., Ltd No.YNKJXM20180092 and YNKJXM20190639.

References

1. Yang Junchuang, Zhao Chao. Survey on K- Means clustering algorithm. Computer Engineering and Applications, 2019, 55 (23) : 7-14
2. Shi Hongliang, Yang Xianming. Decomposition model of energy efficiency change and Its Empirical Study -- Based on six major cities in Yunnan Province, Technology Economics, 2016
3. Si Wei, Guan Bo. Influence of economic structure change on energy intensity in Yunnan Province, Market modernization, 2010
4. Qin Guozhen. Analysis and forecast of influencing factors of energy demand in Yunnan Province, 2012
5. Li Yinlong, Index system construction and comprehensive evaluation of economic development in Yunnan Province -- from the perspective of pillar industries, 2015
6. Wu Bing. Econometric analysis of energy consumption and economic growth in Yunnan Province, 2011
7. Li Jintao, Ai Ping, Yue Zhaoxin, Ma Mengmeng, Bian Shizhe. Improvement of K-means clustering algorithm, Theory and method, 2017
8. Li Huirao. Improvement and application of K-means clustering method, 2014

Online Dispatch of Shared Energy Storage with Remote Renewable Generations

Shenghe Wang[1], Jinzhong Li[2], Yuguang Xie[2], Jiayu Bai[3,*], Bo Gao[2], and Shengwei Mei[3]

[1]State Grid Anhui Electric Power CO., LTD., Hefei 230022, China
[2]State Grid Anhui Electric Power Research Institute, 230601 Hefei, China
[3]State Key Lab of Control and Simulation of Power Systems and Generation Equipments (Dept. of Electrical Engineering), Tsinghua University, 100084 Beijing, China

Abstract. The sustained penetration of wind and solar generation is conducive to alleviating the energy crisis and environmental pollution. However, the finite capacity of transmission corridor limits the delivery of remote renewables and leads to curtailment. Energy storage plays an important role in renewables accommodation and improving equipment utilization, and shared energy storage can magnify the benefits through a temporal and spatial complementary. This paper proposes an online dispatch approach of energy storage shared by multiple renewable plants. The uncertainty and non-anticipativity issue are addressed via a two-stage decision-making process. The day-ahead scheduling determines the allowable energy and power levels of shared energy storage; the real-time online dispatch decides the wait-and-see strategies of charging and discharging power based on a single-period optimization model constrained by the pre-scheduled bounds. The case study verifies the effectiveness of the proposed method.

1 Instruction

In response to the worldwide challenges of energy crisis and environmental pollution, renewable energy represented by wind and photovoltaic (PV) power is developing rapidly in recent years. In China, the installed capacity of renewable generation has surged to 4000 GW by 2020 and is projected to be increased to 12 billion kW by 2030. Although the penetration of renewable energy accelerates grid decarbonization and promotes society's sustainable development, the inherent volatility and stochasticity of renewables bring huge challenges to power system dispatch. Moreover, the unbalanced geographical distribution and the insufficient capacity of supporting transmission channels result in massive resource abandonment. Energy storage is a flexible resource to compensate for the mismatch of renewables and load demand and mitigate the curtailment [1].

Motivated by the success of sharing economy, shared energy storage is attracting increasing attention. Considering the high cost and low utilization ratio of energy storage in individual frameworks, shared energy storage exploits the temporal and spatial demand complementary of different users and brings about cost reduction [2]. The shared energy storage can be implemented on the demand-side or generation side. The day-ahead operation of shared energy storage with PV power producers is studied in [3], and the sensitivity-based allocation ratio is adopted for profit distribution. In [4], a combinatorial auction mechanism is induced to the allocation of residential shared energy storage, wherein the capacity, energy storage level, charging and discharging power are taken as multi-sources to be auctioned. [5] pays attention to the electricity losses during the sharing process and develops a hierarchical capacity planning approach for community-level shared batteries. Taking into account the price uncertainty, [6] proposes a sharing strategy of energy storage with two users based on a robust optimization model. Based on the equilibrium approach, [7] presents a pricing and dispatch strategy of a residential system with shared energy storage and multiple consumers.

Considering the randomness and limited predictability of renewables, numerous methods have been proposed for the optimal dispatch of energy storage with renewable energy. Stochastic optimization [8] and robust optimization [9] are classical approaches to address the uncertainty issues, and both methods are formulated based on offline data, such as forecast and historical data while ignoring the non-anticipativity effect. However, the uncertainty is observed period by period and may not be realized as anticipated, thus the anticipative schedules generated by conventional two-stage stochastic/robust optimization may be infeasible in practical operation [10]. To guarantee the causality of decision-making, the scheduling strategy is supposed to be independent of future information.

To resolve the uncertainty and non-anticipatively issues of shared energy storage dispatching with renewable generations, this paper proposes an online dispatch method, which is composed of two stages of decision-making. The day-ahead dispatch pre-schedules

* Corresponding author: Email: bai_jiayu01@163.com, Tel: +86 18200542600.

the allowable state-of-charge (SoC) ranges of shared energy storage; the real-time scheduling determines the wait-and-see decisions of charging and discharging power based on a single-period optimization model constrained by the pre-prespecified bounds without any future information.

The remainder of the paper is organized as follows. Section 2 presents the online dispatch framework of the shared energy storage with renewable plants and establishes the models of non-anticipative day-ahead scheduling and online dispatch. The case study is analyzed in Section 4 and conclusions are drawn in Section 5.

2 Problem Formulation

2.1 Online dispatch framework

The structure of an integrated system with renewables and shared energy storage is shown in Fig. 1. The renewables, including wind power and photovoltaic power, are sold to the grid through the transmission line. Considering the fluctuation of renewable outputs, curtailment takes place in abundant scenarios due to line capacity limitation, and the under-utilization of devices results from insufficient outputs. The energy storage is shared by the solar plants and wind plants to cope with uncertainty and smooth the output.

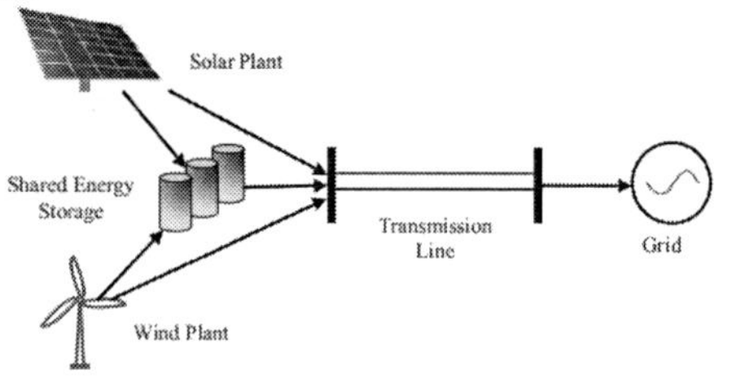

Fig.1 System structure.

In the day-ahead stage, the allowable SoC ranges and the permitted charging/discharging level of shared energy storage is determined based on a rough prediction of the renewable outputs; in the real-time stage, the scheduling moves on hour by hour, and the strategy is determined by a single-period optimization problem with the pre-specified bounds after observing the real output of renewables. The online dispatch framework is displayed as follows.

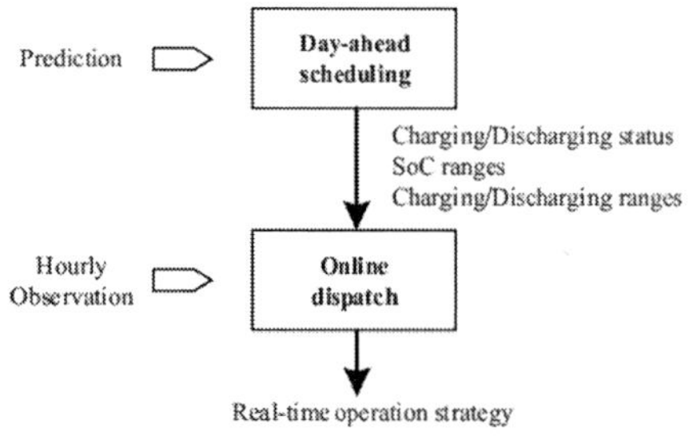

Fig.2 Online dispatch framework.

2.2 Non-anticipative day-ahead scheduling

The uncertainty of accessible renewable outputs is described via the following box set.

$$U^{ra} = \left\{ p_{i,t}^{ra} \mid \underline{p}_{i,t}^{ra} \le p_{i,t}^{ra} \le \overline{p}_{i,t}^{ra}; \forall i,t \right\} \tag{1}$$

where the subscript i is the index of renewable plant, and subscript t is the index of time slot. $p_{i,t}^{ra}$ is the uncertain output of each renewable plant, $\underline{p}_{i,t}^{ra}$ and $\overline{p}_{i,t}^{ra}$ are the predicted lower and upper bounds. The minimal possible total output $\underline{p}_{i,t}^{r\Sigma}$ can be calculated by $\sum_i^I \underline{p}_{i,t}^{r}$, and the maximal possible total output $\overline{p}_{i,t}^{r\Sigma}$ equals $\sum_i^I \overline{p}_{i,t}^{r}$, where I is the number of renewable plants. Then, the uncertainty of total renewablw output $p_t^{ra\Sigma}$ can be judged by

$$U^{ra\Sigma} = \left\{ p_t^{ra\Sigma} \mid \underline{p}_t^{ra\Sigma} \le p_t^{ra\Sigma} \le \overline{p}_t^{ra\Sigma}; \forall t \right\} \tag{2}$$

Consider the curtailment of renewables, the uncertainty set of real accommodated renewables is adjusted by

$$U^{r} = \left\{ p_t^{r\Sigma} \mid \min\left\{ \overline{p}_t^{ra\Sigma} - h_t, \underline{p}_t^{ra\Sigma} \right\} \le p_t^{r\Sigma} \le \overline{p}_t^{ra\Sigma} - h_t; \forall t \right\} \tag{3}$$

where h_t is the curtailment level, and is determined in the day-ahead scheduling. Two vertex scenarios denoted by $S1$ and $S2$ are picked from the uncertainty realizations:

$$p_{t,S1}^{r\Sigma} = \overline{p}_t^{ra\Sigma} - h_t; \forall t \tag{4}$$

which represents the maximal renewable output ;

$$p_{t,S2}^{r\Sigma} = \min\left\{ \overline{p}_t^{ra\Sigma} - h_t, \underline{p}_t^{ra\Sigma} \right\}; \forall t \tag{5}$$

which represents the minimal renewable output. The day-ahead dispatch problem is formulated as follows.

$$\max_{\substack{\mu_t^c, \mu_t^d, h_t, \tilde{p}_{t,s}^{r\Sigma} \\ E_t^{\min}, E_t^{\max}, \\ p_t^{c,\min}, p_t^{c,\max}, p_{t,s}^c \\ p_t^{d,\min}, p_t^{d,\max}, p_{t,s}^d}} \sum_{t=1}^{T} C_t \sum_{s=1}^{S} \omega_s \left(\tilde{p}_{t,s}^{r\Sigma} - p_{t,s}^c + p_{t,s}^d \right) \tag{6a}$$

$$\text{s.t. } \alpha^L p^L \le \tilde{p}_{t,s}^{r\Sigma} - p_{t,s}^c + p_{t,s}^d \le p^L; \forall t,s \tag{6b}$$

$$\tilde{p}_{t,s}^{r\Sigma} = \min\left\{ \overline{p}_t^{ra\Sigma} - h_t, p_{t,s}^{r\Sigma} \right\}; \forall t,s \tag{6c}$$

$$0 \le h_t \le \overline{p}_t^{ra\Sigma} \tag{6d}$$

$$p_t^{c,\min} \le p_{t,s}^c \le p_t^{c,\max}; \forall t,s \tag{6e}$$

$$p_t^{d,\min} \le p_{t,s}^d \le p_t^{d,\max}; \forall t,s \tag{6f}$$

$$E_t^{\min} = E_{t-1}^{\min} - p_t^{d,\max} / \eta^d + \eta^c p_t^{c,\min}; \forall t \tag{6g}$$

$$E_t^{\max} = E_{t-1}^{\max} - p_t^{d,\min} / \eta^d + \eta^c p_t^{c,\max}; \forall t \tag{6h}$$

$$\underline{E} \le E_t^{\min} \le E_t^{\max} \le \overline{E}; \forall t \tag{6i}$$

$$\mu_t^c \underline{p}^c \le p_t^{c,\min} \le p_t^{c,\max} \le \mu_t^c \overline{p}^c; \forall t \tag{6j}$$

$$\mu_t^d \underline{p}^d \le p_t^{d,\min} \le p_t^{d,\max} \le \mu_t^d \overline{p}^d; \forall t \tag{6k}$$

$$\mu_t^d, \mu_t^c \in \{0,1\}, \mu_t^d + \mu_t^c \le 1; \forall t \tag{6l}$$

where C_t is the electricity price, ω_s is the weighting factor of the selected scenario and the subscript $s \in \{S1, S2\}$ represents the index of scenario, $\tilde{p}_{t,s}^{r\Sigma}$ is the actual total output of renewable plants, $p_{t,s}^c$ and $p_{t,s}^d$ are the charging and discharging power of shared energy storage. Binary variables μ_t^c and μ_t^d denote the charging and discharging status of share energy storage. The auxiliary variables $p_t^{c,\min}$ and $p_t^{c,\max}$ are the allowable lower and upper bounds of charging power, $p_t^{d,\min}$ and $p_t^{d,\max}$ denote the permitted discharging level, $E_t^{\min}$ and $E_t^{\max}$ give the safe ranges of SoC. The parameter p^L is the transmission line capacity and α^L is the minimal load rate , η^c and η^d are the charging and discharging efficiencies, $\underline{E}$ and $\overline{E}$ are the physical lower and upper bounds of SoC, $\underline{p}^c$ and $\overline{p}^c$, $\underline{p}^d$ and $\overline{p}^d$ give the physical ranges of charging and discharging power, respectively. The objective (6a) is to maximize the weighted sum of generating income. Constraint (6b) gives the transmission line capacity constraint and the minimal load rate restrictions. Constraint (6c)-(6d) describe the real renewable output under each selected scenario considering the curtailment level. Expressions (6e)-(6k) are non-anticipative constraints of shared energy storage that enforce the all-scenario feasible of the charging and discharging strategies. Constraint (6l) forbids simultaneous charging and discharging. The minimal value expression in (6c) can be transmitted to an equivalent mixed-linear formulation as follows.

$$
\begin{aligned}
&\tilde{p}_{t,S1}^{r\Sigma} = \overline{p}_t^{ra\Sigma} - h_t;\forall t \\
&\tilde{p}_{t,S2}^{r\Sigma} \le \overline{p}_t^{ra\Sigma} - h_t;\forall t \\
&\tilde{p}_{t,S2}^{r\Sigma} \le \underline{p}_t^{ra\Sigma};\forall t \\
&\tilde{p}_{t,S2}^{r\Sigma} \ge \overline{p}_t^{ra\Sigma} - h_t - \overline{p}_t^{ra\Sigma}\left(1-z_t^1\right);\forall t \\
&\tilde{p}_{t,S2}^{r\Sigma} \ge \underline{p}_t^{ra\Sigma} - \underline{p}_t^{ra\Sigma}\left(1-z_t^2\right);\forall t \\
&z_t^1, z_t^2 \in \{0,1\}, z_t^1 + z_t^2 = 1;\forall t
\end{aligned}
\tag{7}
$$

where z_t^1 and z_t^2 are induced binary variables.

2.3 Online dispatch

In the real-time dispatch stage, after observing the real accessible total output of renewables $p_t^{r\Sigma}$, the scheduling of shared energy storage is determined based on the following model.

$$\max_{\tilde{p}_t^r, p_t^c, p_t^d, h_t} C_t\left(\tilde{p}_t^{r\Sigma} - p_t^c + p_t^d\right) \tag{8a}$$

$$\text{s.t. } \alpha p^L \le \tilde{p}_t^{r\Sigma} - p_t^c + p_t^d \le p^L \tag{8b}$$

$$\tilde{p}_t^{r\Sigma} = \min\left\{\overline{p}_t^{ra\Sigma} - h_t, p_t^{r\Sigma}\right\} \tag{8c}$$

$$0 \le h_t \le \overline{p}_t^{ra\Sigma} \tag{8d}$$

$$p_t^{c,\min *} \le p_t^c \le p_t^{c,\max *} \tag{8e}$$

$$p_t^{d,\min *} \le p_t^d \le p_t^{d,\max *} \tag{8f}$$

where $p_t^{c,\min *}$ and $p_t^{c,\max *}$ represent the permitted charging ranges determined in the day-ahead stage, $p_t^{d,\min *}$ and $p_t^{d,\max *}$ are the pre-specified allowable discharging range, and they are the optimal solution to problem (6). It can be proved that if the non-anticipative day-ahead scheduling problem (6) has feasible solutions, then there must be at least one feasible scheduling strategy for the real-time dispatch under all the uncertainty realizations that reside in the pre-specified sets (1) [10-11]. Different from the models of multi-period dispatch with energy storage, the time-couping SoC constraints are relaxed, but the bounds of SoC will not be violated under any uncertainty realizations due to the pre-defined bounds of charging/discharging power and SoC level. Moreover, the real-time decision-making only depends on the current observations and requires no predictions, which meets online dispatch requirements.

3 Case study

Case studies are carried out on an integrated system with 2 wind plants whose capacities are both 400 MW, a 400 MW solar plant, and a shared energy storage with the capacity of 200 MW/600 MWh. Other system parameters are listed in Table. 1. and the predicted renewable output is plotted in Fig.3. All the tests are implemented via Matlab 2016a and Gurobi 9.1.1.

Table1. System parameters.

Parameters	Value (unit)	Parameters	Value (unit)
α^L	0.5	p^L	600 MW
η^c	0.95	η^d	0.95
$\underline{E}$	30 MWh	$\overline{E}$	600 MWh
$\underline{p}^c$	50 MW	$\overline{p}^c$	200 MW
$\underline{p}^d$	50 MW	$\overline{p}^d$	200 MW
C_t	0.8 yuan/MWh	E_0	180 MWh

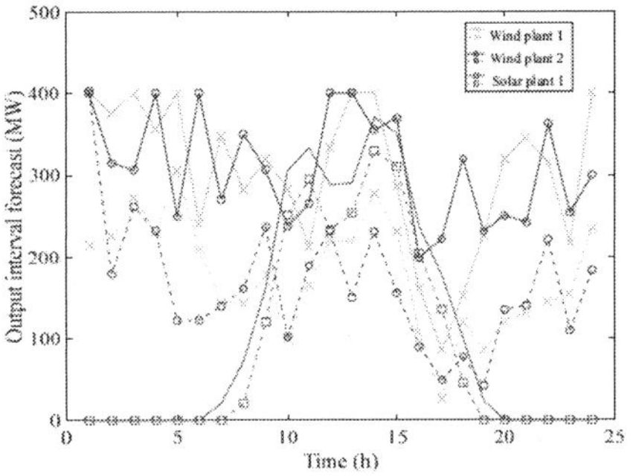

Fig.3 Prediction of renewable output intervals.

The day-ahead scheduling results are shown in Fig. 4-5. Fig. 4 displays the safe ranges of charging and discharging power; Fig. 5 displays the permitted variations of SoC, which are kept within the physical operation ranges of energy storage, indicating that the adjustment of energy storage can guarantee the minimal load rate constraints and transmission capacity limitation under all the uncertainty realizations.

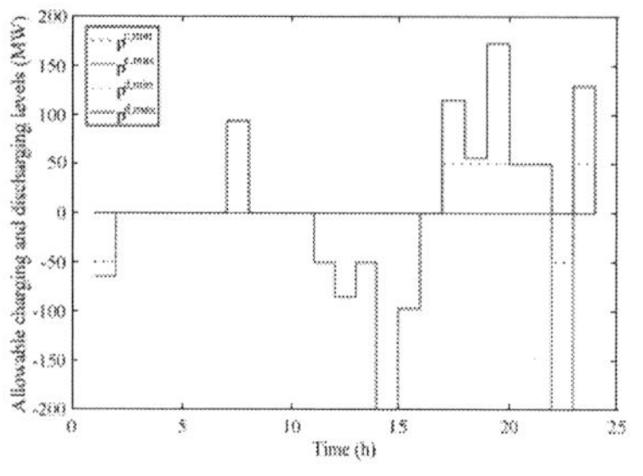

Fig.4 Allowable charging and discharging ranges.

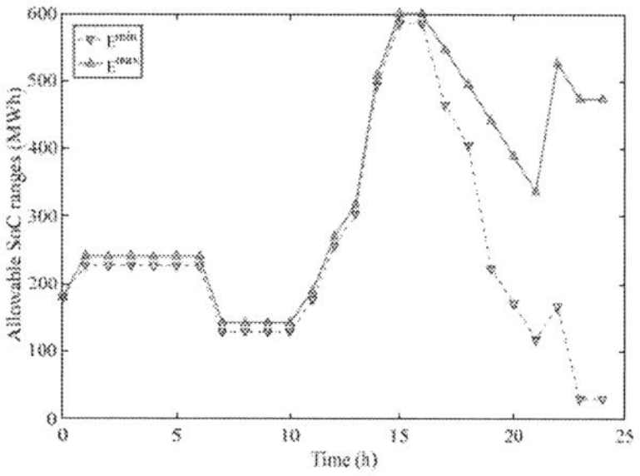

Fig.5 Allowable SoC ranges.

The observed renewable output and the power delivery through the transmission line in the real-time stage are shown in Fig. 6. In periods 11-15, the accessible renewable output exceeds the transmission line capacity, thus part of the surplus is used to charge the shared energy storage and the residual energy is abandoned which is represented by the shadow area in Fig. 6, leading to 257.3 MWh renewable curtailment. In periods 17-21, the actual total renewable output below the required minimum load, and the vacancy is compensated by the discharging power.

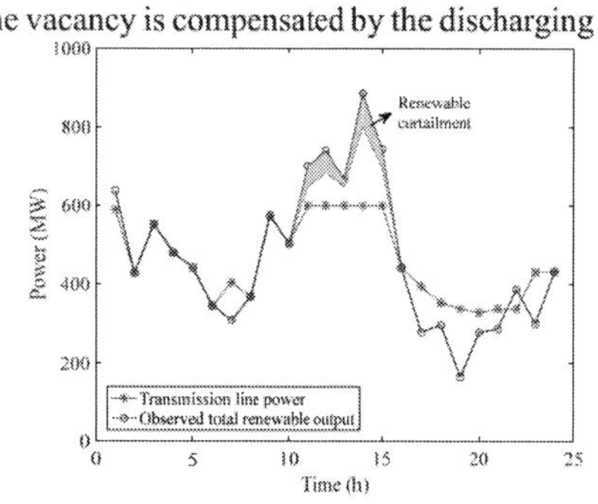

Fig.6 Online dispatch results.

The weighted sum of generating income and total curtailment scheduled in the day-ahead stage with different line capacity settings is plotted in Fig. 7. As can be observed, with the increase of the transmission line capacity, the generation income grows up and the total curtailment decreases. However, considering the minimal load ratio constraints, when transmission line capacity is too large, i.e., more than 630 MW, the discharging of energy storage can't fully compensate for the insufficiency of renewables output, leading to infeasible day-ahead solutions.

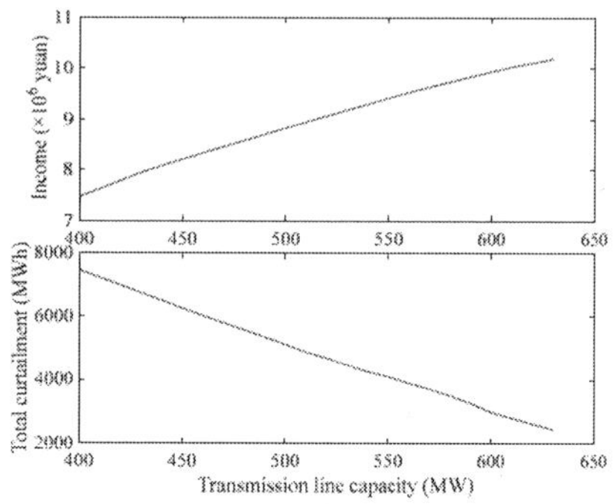

Fig.7 Influence of transmission line capacity.

The weighted sum of generating income and total curtailment scheduled in the day-ahead stage with different minimal load ratios is plotted in Fig. 8. With the increase of minimal load ratio, the generating income decreases because the more renewable output is consumed to charge the energy storage to support the discharging during the periods with inadequate renewable outputs and less renewable energy can be sold; when the minimal load ratio reaches 0.53, the vacancy cannot be fully compensated by energy storage due to its maximal charging power. When the minimal load ratio is set below 0.45, the generation income cannot be elevated by the decrease of load ratio; when less than 0.42, it has little impact on the total curtailment.

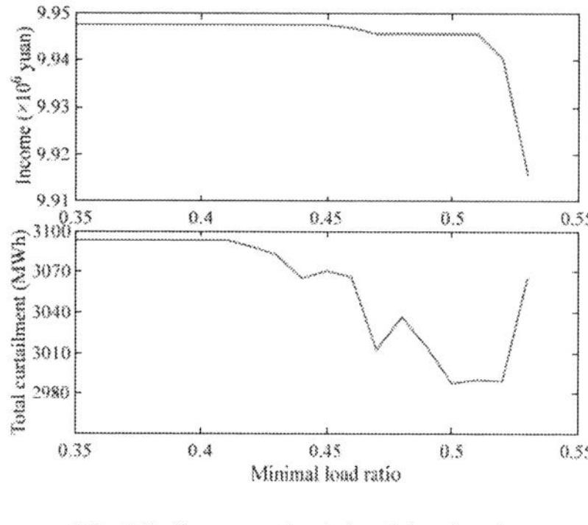

Fig.8 Influence of minimal load ratio.

4 Conclusion

This paper explores the online dispatch of a renewable-storage integrated system where a central energy storage is shared by multiple renewable plants. The proposed method guarantees that the total power exported via the transmission line satisfies the requirement on the utilization rate of transmission capacity, regardless of the hourly unfolding real renewable output. Case study shows that the SoC of the shared energy storage is constrained by two additional bounds, implying that the SoC can vary within a subrange of its physical ranges.

Acknowledgment

This paper is supported by State Grid Anhui Electric Power Co., Ltd. Science and Technology Project (52120520004M).

References

1. J. Jo, J. Park, 11, IEEE T. Smart Grid **11**, 4466-4476 (2020)
2. R. Dai, R. Esmaeilbeigi, H. Charkhgard, IEEE T. Smart Grid (2021)
3. D. Kim, H. Kim, D. Won, IEEE Access **8**, 197097-197110 (2020)
4. W. Zhong, K. Xie, Y. Liu, C. Yang, S. Xie, IEEE T. Smart Grid **11**, 4105-4115 (2020)
5. P. Huang, Y. Sun, M. Lovati, X. Zhang, Energy **222**, 119931 (2021)
6. R. Dai, H. Charkhgard, F. Rigterink, Int. T. Oper. Res. (2020)
7. W. Zhang, W. Wei, L. Chen, B. Zheng, S. Mei, Energy **202**, 117543 (2020)
8. M.A. Mirzaei, A.S. Yazdankhah, B. Mohammadi-Ivatloo, M. Marzband, M. Shafie-khah, J.P. Catalão, J. Clean Prod. **223**, 747-758 (2019)
9. B. Dey, B. Bhattacharyya, S. Sharma, International Journal of Energy Optimization and Engineering **8**, 67-87 (2019)
10. Y. Zhou, Q. Zhai, M. Zhou, X. Li, IEEE T. Sustain. Energy **11**, 758-770 (2019)
11. Y. Zhou, Q. Zhai, L. Wu, IEEE T. Sustain. Energy (2020)

 https://doi.org/10.1051/e3sconf/202129302063

Energy Demand Forecast in Yunnan Province Based on Seq2Seq Model

Anrui Li[2], Shi Su[1], Tong Han[1], Chunlin Yin[1], Jie Li[1], Linjie Chen[2], Xinkai Wang[2], Shichen Wang[2] and Siyang Liu[1, *]

[1] Electric Power Research Institute, Yunnan Power Grid Co. Ltd, Kunming, China
[2] Key Laboratory in Software Engineering of Yunnan Province, School of Software, Yunnan University, China

Abstract. Energy demand forecast has an important practical significance for the sustainable development of the national economy, the reasonable allocation of resources, and the construction of modernization goals. This study is based on the analysis of coal, electricity, natural gas, and other energy data in Yunnan Province from 2011 to 2018 and uses long short-term memory, sequence to sequence, deep learning, and ridge regression coupling methods to construct an energy demand forecast model in Yunnan Province. Forecast results show the following. The total energy consumption of Yunnan Province from 2021 to 2025 will continue to increase. Moreover, the coal consumption of Yunnan Province will continue to decline from 2021 to 2025. Furthermore, the electricity consumption of Yunnan will increase by about 8.02% year-on-year from 2021 to 2025. The experiment proves that the forecasting effect of the energy demand forecast model proposed in this study is excellent.

1 Introduction

The impact of the construction of the Energy Internet on the traditional energy industry has revolutionary challenges and huge opportunities. The traditional energy industry should combine its own characteristics and rely on Internet technology and Internet concepts to make a reasonable change in its positioning to adapt to the development trend of the Energy Internet. Therefore, predicting the relevant demand of the energy industry is particularly important to determine the direction of the development of traditional industries.

In recent years, many scholars have conducted considerable research on energy demand forecasting. Common forecasting methods can be divided into trend extrapolation and scenario analysis. The trend extrapolation method is mainly based on past and current development trends to infer possible future states, such as time series method [1], energy elasticity coefficient method [2], and input–output method [3-5]. The scenario analysis method mainly identifies the external factors affecting the development of the main body of the research through the study of the environment and simulates various cross scenarios that may occur in the external factors to analyze and predict various possible prospects [6-9].

On the basis of the situation of Yunnan Province, this study selects and obtains comprehensive data, including energy, electric power, coal, wind power, natural gas, and non-fossil energy the from National Bureau of Statistics, power companies, and other data source units through national and local statistical channels, energy companies, and industry associations. On the basis of the energy data of Yunnan Province, the energy demand forecast model of Yunnan Province is established through a comprehensive analysis of the energy supply and demand relationship and energy structure. By using this model, the energy demand in Yunnan Province from 2021 to 2025 is predicted, and the energy development trend is also analyzed.

2 Models and forecasting methods

Long short-term memory (LSTM), sequence to sequence (Seq2Seq), deep learning, and ridge regression coupling prediction model are used to predict the total energy, coal, electricity, and natural gas consumption in Yunnan Province. LSTM performs well on unstable time series with fixed components, and deep learning performs accurately in model prediction.

2.1 Energy demand forecasting model based on LSTM

LSTM is a special RNN that selects long-term information through the "gate" design, thereby avoiding the problem of long-term dependence. The gates of LSTM at each sequence index position t generally include forget, input, and output gates. The forget gate controls whether information is forgotten, the input gate processes the input of the current sequence position and determines what new information can be updated in the cell state, and the output gate is responsible for processing the output information.

The calculations corresponding to the forget, input, and output gates are as follows:

* Corresponding author: lsyfd27@sina.com

$$f_t=\sigma\left(W_f\cdot[h_{t-1},x_t]+b_f\right) \quad (1)$$

$$i_t=\sigma\left(W_i\cdot[h_{t-1},x_t]+b_i\right) \quad (2)$$

$$C_t=f_t*C_{t-1}+i_t*tanh\left(W_C\cdot[h_{t-1},x_t]+b_C\right) \quad (3)$$

$$h_t=\sigma(W_o[h_{t-1},x_t])*tanh(C_t) \quad (4)$$

When we use 7-day data to predict the energy demand for the next day, the input of the model is $[x_0, x_1, ..., x_6]$. For example, when calculating the expression of x_1, the parameter learns through the forget gate W_f. The expressions of x_0 (i.e., h_0) and x_1 are used to calculate the proportion that needs to be forgotten f_1. The input proportion i_1 is calculated by the parameter learned through the input gate W_i, the expression of x_0 (i.e., h_0), and x_1. The state of x_0C_0 at the previous moment is controlled through the calculation results of the forget and input gates. Through the forgotten part of information and the passing ratio of the input $tanh(W_C[h_{t-1},x_t]+b_C)$, the state of x_1 C_1 can be obtained. Finally, the output gate parameter W_o can obtain the following.

2.2 Energy demand forecasting model based on Seq2Seq

When the Seq2seq model maps a sequence as an input to another output sequence, it mainly completes the two basic links of the encoder input and the decoder output. The encoding end encodes the input sequence data and outputs the encoding status on the network end. The decoder receives the final state of the encoder as the initial state, and the output value of each step after decoding is the input value of the next step.

On the basis of the energy demand data of different time scales, a chronological data sequence is obtained, such that each piece of data contains at least two parts, namely, the time sequence and the corresponding energy consumption. The time sequence is used as a feature to input into Seq2Seq, and the corresponding energy consumption as forecast content. For the energy demand forecasting task of the time scale of day, week, and month, weather characteristics can be added. The weather forecast data in the future can be used to form the comprehensive features with the time series as the input of the Seq2Seq model.

2.3 Deep learning and ridge regression coupling forecast model

In comparison with traditional machine learning methods, the deep learning model eliminates the feature selection process and can learn the feature selection strategy by itself. Ridge regression is a biased estimation regression method for multicollinearity data analysis. It is an improvement on the basis of the least squares method. It does not consider the unbiased of the least square method and loses part of the hidden information of the characteristic factors. Thus, it establishes a realistic regression model. The regression coefficients of the ridge regression model can reflect the overall trend and avoid the overfitting and underfitting problems of the least squares linear regression, making the model practical. Coupling weather forecast data based on the ridge regression method and building an energy demand model and making predictions can stabilize forecast results and improve forecast accuracy.

An energy demand forecast model with a forecast period of 1 month is established. First, the monthly weather forecast data in this area of Yunnan Province is obtained based on the inverse distance weighting method, and the energy demand forecast model based on the LSTM network is obtained by using data training, with the forecast data obtained. Then, through monthly energy consumption rolling forecasts, on the basis of the measured data, the sliding load is carried forward time by time to revise the forecasts of energy demand for subsequent months. Finally, a ridge regression model is established, with the independent variables being the forecast data of weather factors and the energy consumption of next month predicted by the LSTM network model.

When collecting weather data, weather forecast data are collected for the next month to construct a mixed feature x = [LSTM-out, temperature, humidity, weather...], normalize x, and send it to the ridge regression model, which obtains the hidden layer output $h_i=Wx_i$, and finally obtains the prediction result through the output layer. The ridge regression model uses $J(\theta)=\frac{1}{2m}\sum_{i=1}^{m}\left(h_\theta\left(x^{(i)}\right)-y^{(i)}\right)^2+\lambda\sum_{j=1}^{n}\theta_j^2$ as the loss function. During training, the mean square error between the predicted value h and the true value y is calculated, and a regular term is added to alleviate the problem of model overfitting and underfitting.

3 Model building

On the basis of the different years and types of data collected, LSTM, Seq2Seq, deep learning, and ridge regression coupling prediction models are used to construct prediction models for total energy and each energy component. Particularly, the total energy consumption is subject to macro-control; hence, the Seq2Seq model is used. For the construction of a single energy demand model, the LSTM model and the coupling prediction model based on deep learning and ridge regression are used to build the prediction model.

3.1 Forecast model of total energy consumption in Yunnan Province

Experimental data: Energy consumption in Yunnan Province, as shown in Figure. 1.

Application model: Energy demand forecasting model based on Seq2Seq

Associated data: Yunnan's GDP, Yunnan's population, and output value of main industrial products

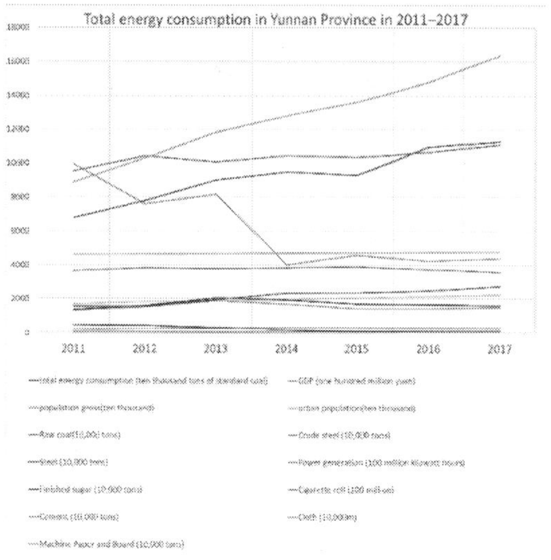

Figure 1. Total energy consumption in Yunnan Province in 2011–2017 (Data source: Yunnan Statistical Yearbook)

The fitting result is shown in Figure. 2.

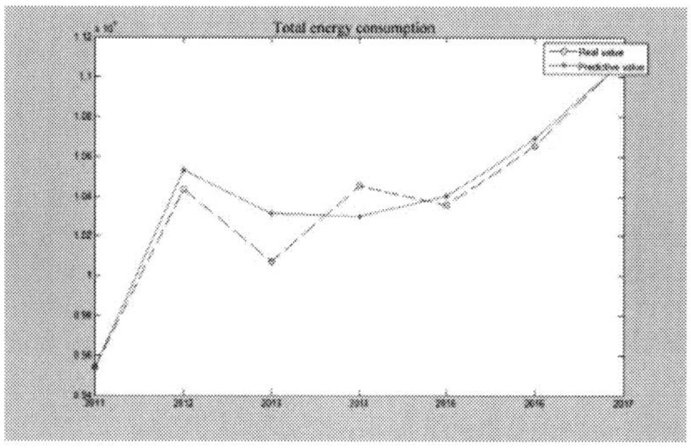

Figure 2. Energy consumption forecasting model of Yunnan Province based on Seq2Seq model

3.2 Thermal power production forecast model in Yunnan Province

Experimental data: Thermal power production in Yunnan Province, as shown in Table 1, and hydropower production in Yunnan Province

Application model: Energy demand forecasting model based on LSTM

Associated data: Hydroelectric power generation and precipitation in Yunnan Province

Table1. Thermal power production in Yunnan Province

Year	Thermal power generation (hundred million kWh)	Annual precipitation (mm)
2013	479.3	835.271172
2014	391.41	980.9
2015	266.47	1106.60238
2016	238.04	1154.6

The fitting result is shown in Figure. 3.

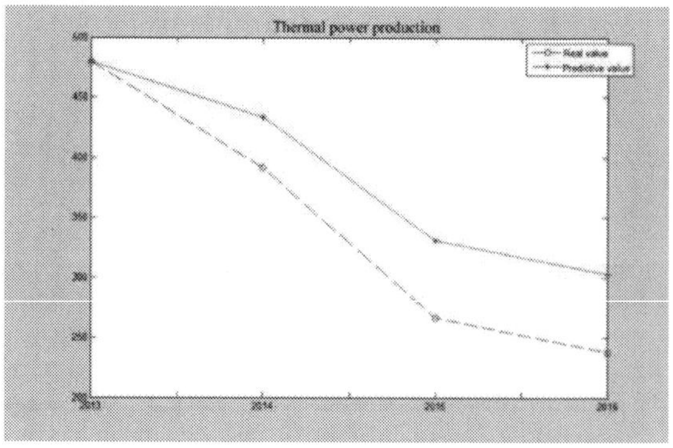

Figure 3. Thermal power production forecast model in Yunnan Province based on LSTM model

3.3 Simulation of wind power production in Yunnan Province

Data: Wind power production in Yunnan Province, as shown in Table 2., and temperature in Yunnan Province (Yunnan Statistical Yearbook)

Application model: Coupling prediction model based on deep learning and ridge regression

Associated data: Weather in Yunnan Province

Table2. Wind power generation in Yunnan Province from 2014 to 2018 (hundred million kWh)

	Wind power generation (hundred million kWh)
2018	230.7
2017	192.9
2016	141.7
2015	80.3
2014	55

The fitting result is shown in Figure. 4.

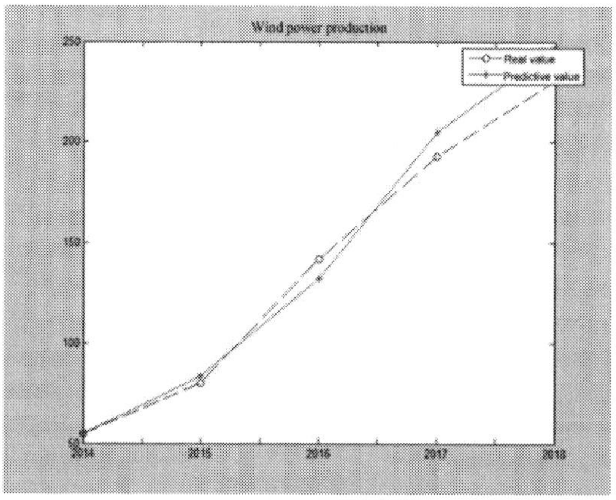

Figure 4. Forecast model of wind power production in Yunnan Province based on the coupling model of deep learning and ridge regression

https://doi.org/10.1051/e3sconf/202129302063

4 Model accuracy analysis

The comparison between the predicted value of the model simulation part and the true value curve indicates that the prediction simulation results are quite impressive. Thus, the total energy, raw coal, electricity, and natural gas in Yunnan Province of China are predicted by this prediction model. Some of the data in 2019 in the trained model are compared with the corresponding data in 2019 in the 2020 Statistical Yearbook of Yunnan Province to analyze the accuracy of the model.

Table3. Part of the energy demand forecast tables in Yunnan Province

Category	Forecast of 2019 (ten thousand tons of standard coal)	Official data for 2019(ten thousand tons of standard coal)
Total energy consumption	12663.08	12157.65
Raw coal consumption	5921.596	6759.46
Power consumption	1256.4	1112.40

According to the analysis of the forecast table, the forecast of total energy consumption and electricity consumption is accurate, but the forecast value of raw coal consumption is slightly lower than the data published in the Yunnan Statistical Yearbook. The main reason for the difference of raw coal consumption may be that in 2019, the government increases the input of raw coal consumption to support the corresponding industrial development, but the prediction results of other energy consumption structure are basically consistent with the actual situation, and the accuracy can meet the prediction demand.

5 Model prediction

5.1 Prediction of total energy consumption

On the basis of the National Statistical Yearbook of China, the regional GDP of Yunnan Province has maintained a rapid growth, which is higher than the national level. From 2011 to 2017, Yunnan's GDP rose from 0.88 trillion yuan to 1.8 trillion yuan.

The simulation results of the 2011–2017 data of Yunnan Province based on the forecast model indicate that the total energy consumption of Yunnan Province from 2021 to 2025 will still show an upward trend, as shown in Table 4.

Table4. Forecast of total energy consumption in Yunnan Province

Year	Forecast value of total energy consumption (ten thousand tons of standard coal)
2021	13211.21
2022	13771.29
2023	14353.93
2024	14958.37
2025	15554.95

5.2 Raw coal consumption forecast

On the basis of the gray prediction model, the output of raw coal in Yunnan Province in 2021 can reach 50 million tons. According to policy forecasts, from 2021 to 2025, Yunnan's raw coal output can reach 55 million tons, which basically meets the coal consumption demand in the province.

The simulation results of the 2012–2018 data of Yunnan Province based on the forecast model indicate that the coal consumption of Yunnan Province from 2021 to 2025 will show a downward trend.

However, combined with the advancement of hydropower, aluminum, hydropower, and silicon integration projects in Yunnan Province, the power supply and demand in Yunnan Province will change from oversupply to a basic balance between supply and demand. During the dry period, power shortage will be experienced, the role of thermal power will be played out, and the decline trend of coal consumption tends to be flat. Therefore, the forecast of coal consumption in Yunnan Province from 2021 to 2025 is shown in Table 5.

Table5. Forecast of raw coal consumption in Yunnan Province

Year	Raw coal (unit: hundred million tons)
2021	5832.5
2022	5744.744
2023	5658.308
2024	5573.173
2025	5489.319

5.3 Electricity load forecasting

On the basis of model prediction, the electricity consumption in Yunnan Province will increase by about 8.02% year-on-year from 2021 to 2025. However, as the new large industrial load in Yunnan Province is expected to be 3.4 million kW in 2020, under the condition that production is put into operation on time, it is estimated that the new electricity will be about 15 billion kWh, and the largest electricity load includes Matang Hydropower Aluminum Phase I (820000 kW), Funing Shenhuo Group Hydropower Aluminum Project Phase I (750000 kW), Qiya Electrolytic Aluminum (540000 kW), Heqing Yixin Aluminum Co., Ltd. Phase II (350000 kW), Lianghe County Yunju Technology Co., Ltd. (200000 kW), Julida Iron and Steel Co., Ltd. of Yunnan Yonggang Iron and Steel Group (100000 kW), Dehonghebao Technology Co., Ltd. (70000 kW), Lijiang Longji Silicon Material Company Phase II (65,000 kW), and Baoshan Longji Silicon Material Company Phase III (62000 kW). Through a comprehensive consideration, on the basis of the abovementioned production situation, the total electricity consumption in the province in 2020 is about 192.9 billion kWh, with a year-on-year increase of 12.3%.

Without the construction of megaprojects, the electricity consumption in Yunnan Province is predicted to increase by about 8.02% year-on-year from 2021 to 2025.

6 Discussion

On the basis of the prediction results of various future energy consumption indicators by our trained prediction model, our recommendations are as follows.

(1) Yunnan Province should develop clean coal power actively and promote deep peak shaving of thermal power units.

With the rapid growth of power consumption in and outside the province and the slowdown in power production in the province, the scale of surplus hydropower in Yunnan Province during the flood season has shown a trend of shrinking year by year, and the contradiction of power supply balance in the whole province has gradually changed from promoting the full consumption of clean energy to the transformation of power supply in the dry period. The supporting role of thermal power supply in dry period has become prominent. With the profound development of the Energy Internet, the power market compensation mechanism is gradually improving, and the thermal power in Yunnan Province can be fully positioned as a peak-shaving power source and a backup power source. In future energy planning, thermal power location, scale, and access system should be discussed according to the above positioning.

(2) Yunnan Province should further explore commercial innovations to promote the transformation of traditional energy companies into integrated energy service providers.

With the development of distributed power generation and regional energy networks, integrated energy companies will exist in the future, which can centrally supply water, electricity, gas, cooling, heat, and other resources and realize the "water and electricity centralized reading" of intelligent meter, intelligent water meter, and intelligent gas meter through power fiber to the home. The two-way interactive demand of users in the era of mobile Internet will prompt enterprises to continue to extend their service areas and tap the added value of energy products and services. Traditional energy enterprises in Yunnan Province should promote the construction of intelligent power service system actively, deploy intelligent power interactive terminals, improve residents' awareness of intelligent power consumption, build demonstration intelligent communities, improve the intelligent level of charging, support the innovation of charging business model, and transform from a single energy supplier to a comprehensive energy service provider.

(3) Yunnan Province should speed up the replacement of electric energy to build a green Yunnan.

Yunnan shall accelerate the replacement of electric energy, use secondary electric energy to replace traditional primary fossil energy consumption, such as electric vehicles/hydrogen energy vehicles instead of fuel vehicles, and promote the combination of the advantages of clean energy of hydropower, silicon and aluminum resources, and industrial basis, push forward the integrated development of hydropower aluminum and hydropower silicon industry, and accelerate the full consumption of clean energy in Yunnan Province.

Acknowledgments

This work has been supported by the National Key Research and Development Program No.2018YFB2100100, Postdoctoral Science Foundation of Yunnan Province, Science Foundation of Yunnan Province No.202101AT070167, Project of the Yunnan Provincial Department of Education scientific research fund No. 2019J0010, DongLu Young and Middle-aged backbone Teachers Project of Yunnan University, Open Foundation of Key Laboratory in Software Engineering of Yunnan Province under Grant No.2020SE311, Science and Technology Project of Yunnan Power Grid Co., Ltd No.YNKJXM20180092 and YNKJXM20190639.

References

1. T. Wang, H. Leung, J. Zhao and W. Wang, "Multiseries Featural LSTM for Partial Periodic Time-Series Prediction: A Case Study for Steel Industry," in *IEEE Trans. Instrum. Meas.*, **69**, no. 9, pp. 5994-6003 (2020)
2. H. Yang, J. Zhang, J. Qiu, S. Zhang, M. Lai and Z. Y. Dong, "A Practical Pricing Approach to Smart Grid Demand Response Based on Load Classification," in *IEEE Trans. Smart Grid*, **9**, no. 1, pp. 179-190 (2018)
3. J. Ke et al., "Comprehensive Economic Benefit Assessment Method and Example of Energy Storage Based on Power Grid," *iSPEC - IEEE Sustain. Power Energy Conf.: Grid Mod. Energy Revolut., Proc.*, pp. 2287-2292 (2019)
4. A. Enescu, H. Andrei, E. Diaconu and V. Ion, "Numerical method for modeling the input-output characteristic in a cogeneration power plant," *Proc. Int. Conf. Electron., Comput. Artif. Intell., ECAI*, pp. 1-6 (2019)
5. M. Yan, N. Zhang, X. Ai, M. Shahidehpour, C. Kang and J. Wen, "Robust Two-Stage Regional-District Scheduling of Multi-carrier Energy Systems With a Large Penetration of Wind Power," in *IEEE Trans. Sustainable Energy*, **10**, no. 3, pp. 1227-1239 (2019)
6. F. Zhang et al., "Understanding WeChat User Preferences and "Wow" Diffusion," in *IEEE Trans Knowl Data Eng* (2021)
7. Z. Zhao, H. Zhou, L. Qi, L. Chang and M. Zhou, "Inductive Representation Learning via CNN for Partially-Unseen Attributed Networks," in *IEEE Trans. Netw. Sci. Eng.*, **8**, no. 1, pp. 695-706 (2021)
8. Z. Zhao, H. Zhou, L. Qi, L. Chang and M. Zhou, "Inductive Representation Learning via CNN for Partially-Unseen Attributed Networks," in *IEEE Trans. Netw. Sci. Eng.*, **8**, no. 1, pp. 695-706 (2021)
9. W. Ahmed et al., "Machine Learning Based Energy Management Model for Smart Grid and Renewable

Energy Districts," in *IEEE Access*, **8**, pp. 185059-185078 (2020)

Analysis method of cluster perforating tubing strings trafficability in horizontal wells for unconventional reservoirs

Xiaoqiang Guo[1*], Jun Liu[1, 2*], Xiang Huang[1], Zhigang Du[1] and Xiao Li[1]

[1]School of mechanical and electrical engineering, Southwest Petroleum University, Chengdu Sichuan 610500, China
[2]School of Mechanical Engineering, Chengdu University, Chengdu, 610106, China.

Abstract. The feasibility of a normal perforating operation depends on the perforating tubing string, which is the main component of the cluster perforating tool for the exploration and development of unconventional oil&gas reservoirs. To improve the efficiency of cluster perforation in horizontal wells, a passing capability analysis model for the tubing string in a wellbore was established, thereby generating complex equations with coefficients that are solved via geometric analyses and the beam-column theory. This model was based on a comprehensive consideration of the friction between the downhole tools and the borehole wall, fluid resistance, wellbore geometric constraint, pump thrust, axial tension, tool variable cross-section, tension of cable head, and tool elastic deformation. Based on field test data, the pass-through capacity of "1 bridge plug + 4 cluster perforating gun string" in the X206 well and the main factors affecting pass-through ability were analyzed to test the validity of the model. The model presented in this paper can provide an effective tool for analyzing the design and operation of a cluster perforating string.

1 Introduction

In recent years, cluster perforations 1 have resulted in revolutionary developments in perforation technologies 2. During a cluster perforating operation, the tool string is conveyed to the downhole target stratum via a cable; this is followed by bridge plug setting and multi-cluster perforation 3. As the main component of perforating tools, the cluster perforating string plays an important role during perforating operations. Generally, the length of the tool string can be excessively large. A large tool string can easily get stuck, especially under specific well conditions such as irregular well trajectories and a high dogleg degree. This can result in a sudden rise in the pumping pressure, thereby leading to accidents such as premature setting of a composite bridge plug, deformation of perforated tubing strings, wear failure of tubing-casing and strength failure 5. Hence, more accurate safety evaluation method of cluster perforating tubing strings is urgently needed to ensure the safety of field operation.

There are a wide range of geometric and mechanical methods that can be applied for research on downhole tool pass-through capacity. In the early years, a calculation method for downhole tool pass-through capacity under rigid conditions was proposed 6. Later, the rigid pass-through capacity model was perfected and a mechanical model for solving the pass-through capacity of downhole tools was established, using the beam-column theory under flexible conditions 7. Based on this, Di et al. 8 considered the influence of a centralizer on the pass-through capacity of the drilling tool, however, the influence of axial force was not be taken into account. Therefore, it is impossible to accurately predict the magnitude of pump thrust at different positions. In 1997, He et al. 9 established a friction calculation model for casing in horizontal wells through mechanical analyses. Subsequently, for the analysis of casing friction, Sui et al. 10 considered the effect of rotating down. In 2008, Wang 11 elaborated on a mechanical analysis method for the downhole tool pass-through capacity. In 2013, Zhu et al. 12 established the control equation for the length of the tool string under rigid conditions. However, as the deformation of the string was not considered, the calculation result was conservative. In 2016, Feng et al. 13 carried out a study on the pass-through capacity of the string under the assumption that the position of the maximum curvature of the borehole trajectory was the sticking point of the multi-layer splitter string, which was an inaccurate assumption.

Currently, for the calculation of downhole tool pass-through capacity, a majority of relevant studies 14 consider the drill tool and pipe as examples and simplify the downhole tool string to an analysis model of the beam with an equal cross section under the transverse component of the net weight and axial pressure. Existing studies also assume that the tool string is stuck at a position of maximum dogleg degree. These models are appropriate for rigid tool strings; however, for tool strings with greater flexibility, including cluster perforating strings, such simplifications and assumptions are significantly different from the actual situation, resulting in a conservative prediction.

* Corresponding authors: 786526101@qq.com (Xiaoqiang Guo, PhD)
and 201031010081@swpu.edu.cn (Jun Liu, PhD, Professor)

In this study, a pass-through capacity analysis model for a cable pumping cluster perforating string in a wellbore was established, and the field test data was used to verify the effectiveness of the model. The model was based on the comprehensive consideration of factors such as the friction force between the downhole tools and wellbore, wellbore geometry limitation, pump thrust, axial tension, string variable cross section, tension of cable head and tool elastic deformation.

2 Pass-through capacity analysis model

2.1 Determination of well trajectory curvature radius

The borehole curvature, which describes the degree of curvature of the well trajectory, directly affects the pass-through capacity of downhole tools. The pass-through capacity of downhole tools decreases as the radius of the borehole curvature decreases. Therefore, the method to determine the borehole curvature radius is provided in this study. A section of the well trajectory for any two adjacent measuring points is set as shown in Fig. 1 17.

It is assumed that the well trajectory has N measuring points and $N+1$ sections. Concurrently, assuming that the sounding increment of the adjacent measuring points section i is ΔS, the increment of the deviation angle is $\Delta\alpha$, and the increment of the azimuth angle is $\Delta\varphi$. Thus, the overall angle change rate (dogleg degree) in section i is K_i, which is expressed as

$$K_i = \sqrt{\left(\frac{\Delta\alpha}{\Delta S}\right)^2 + \left(\frac{\Delta\varphi}{\Delta S}\right)^2 \times \sin^2\left(\frac{\alpha_{i+1}+\alpha_i}{2}\right)} \quad (1)$$

The hole curvature radius in section i is defined as

$$R_i = \frac{1}{K_i} \quad (2)$$

Assuming that the dogleg degree G_i (°/30 m) in well section i is known, the radius of curvature R_i can be expressed as follows:

$$R_i = \frac{30\times 180}{G_i \times \pi} \quad (3)$$

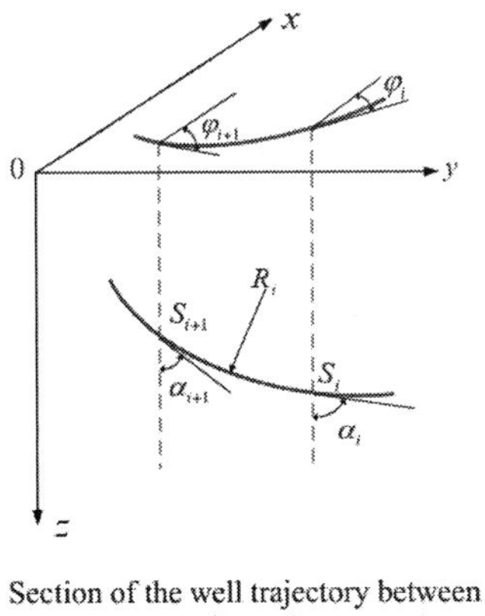

Figure 1. Section of the well trajectory between adjacent measuring points

2.2 Conditions of string strings passing through wellbore

As shown in Fig. 4, the hole curvature radius is R, wellbore diameter is d_b, outside diameter at the midpoint of the perforating string is d_z, outside diameter of the bridge plug is d_q, total length is L, and bending deformation caused by the wellbore constraint in the middle of the string is y_c.

According to the geometric relationship shown in Fig. 2, the following can be obtained:

$$y_c = R - \frac{1}{2}d_b + d_z - \sqrt{\left(R + \frac{1}{2}d_b\right)^2 - \left(L^2/4\right)} - \frac{1}{2}\left(d_q - d_z\right) \quad (4)$$

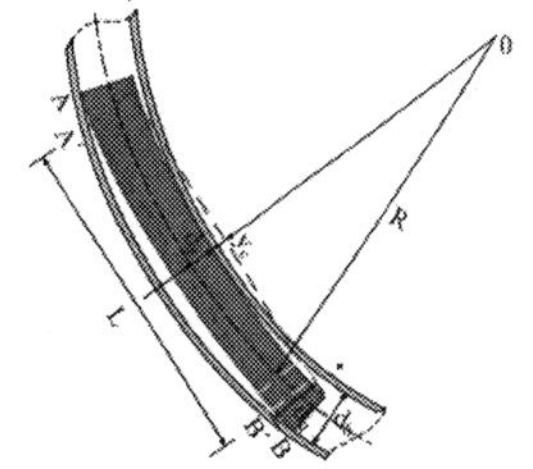

Figure 2. Analysis diagram of pass-through capacity for the cluster perforating tubing strings

The field test indicates that the maximum deflection when running a clustered perforating string occurs near the midpoint of the string. For convenience in the calculation, the midpoint of the tubing string is considered to be the maximum deflection position. The pass-through capacity of the string can be described by the following equations:

Through: $y_{max} \geq y_c$ (5)

Not through: $y_{max} < y_c$ (6)

2.3 Force and deformation analysis

According to Eq. (5), and (6), the pass-through capacity of the tubing string is related to the maximum deflection when running, which can be obtained by analyzing the force and deformation of the tubing string.

(1) Force analysis

For horizontal wells, the friction between the tubing string and inner wall of the wellbore increases as the deviation angle increases. When the tubing string cannot run on its own under the influence of gravity, fluid is generally pumped at the wellhead, forming a differential pressure thrust (pump thrust). This enables the tubing string to smoothly reach the bottom of the well. While running the cluster perforating tool, the axial and transverse force components of the tubing string, total resistance, friction force of the cable, and pump thrust, can be expressed as follows:

$$\begin{cases} F_p = WL\cos\alpha;\ \ F_n = WL\sin\alpha \\ F_f = WL\sin\alpha\cdot f + \frac{1}{2}C_d\rho_m v_r^2 A \\ F_d = W'L'\sin\alpha\cdot f \\ F_b = WL\sin\alpha\cdot f - WL\cos\alpha + W'L'\sin\alpha\cdot f \end{cases} \tag{7}$$

(2) Deformation analysis

In order to analyze the pass-through capacity of the cluster perforating string in the wellbore, the deformation analysis of the string under various forces in Eq. (7) should be performed.

The concrete cluster perforating tubing string structure is composed of several tools with different lengths and outer diameters. Hence, the calculation of the pass-through capacity is significantly complicated. Therefore, the specific string structure must be simplified to some extent. In this study, part of the string with the same outer diameter is regarded as a section of beam, and the entire cluster perforating string is connected using several sections of the beam with different outer diameters, as shown in Fig. 3. In the process of running, the tubing string is carried by the cable and runs at a relatively stable speed. In this case, the string is affected by the net weight, cable tension, wellbore friction and fluid resistance. This is equivalent to a variable cross section of a simply supported beam under axial tension and transverse distributed load, as shown in Fig. 4.

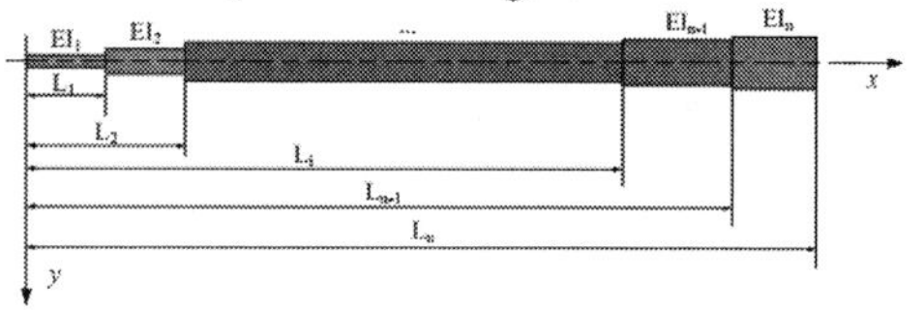

Figure 3. Schematic diagram of the of cluster perforating string structure

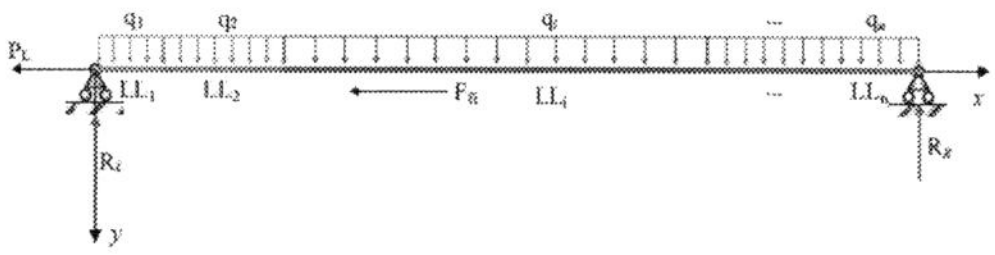

Figure 4. Force analysis diagram of the cluster perforating string

① According to the static equilibrium relationship, the reaction force of the two ends of the beam can be obtained, where $q_i = W\sin\alpha$, $L_0 = 0$:

$$\begin{cases} R_L = \dfrac{\sum_{i=1}^{n} q_i LL_i (L_5 - L_{i-1} - \frac{1}{2}LL_i)}{L_5} \\ R_R = \dfrac{\sum_{i=1}^{n} q_i LL_i \left(L_{i-1} + \frac{1}{2}LL_i\right)}{L_n} \end{cases} \tag{8}$$

② Bending moment at any point of the beam can be expressed by:

$$\begin{cases} M_1(x) = R_L x + P_1 y - \frac{1}{2}q_1 x^2 & (0 \le x \le L_1) \\ M_2(x) = R_L x + P_2 y - \frac{1}{2}q_2 (x - L_1)^2 \\ \qquad - q_1 LL_1\left(x - \frac{1}{2}LL_1\right) & (L_1 \le x \le L_2) \\ \vdots \\ M_n(x) = R_L x + P_n y - \frac{1}{2}q_n (x - L_{n-1})^2 \\ \qquad - \sum_{i=1}^{n-1} q_i LL_i \left(x - L_{i-1} - \frac{1}{2}LL_i\right) & (L_{n-1} \le x \le L_n) \end{cases} \tag{9}$$

where $P_i (i = 1, n)$ is the axial tension in section i, and $P_i < 0$ is

$$\begin{cases} P_i = -\frac{1}{2}q_i LL_i \cos\alpha - \sum_{j=i+1}^{n} q_j LL_j \cos\alpha \\ \qquad + q_i LL_i \sin\alpha\cdot f + \frac{LL_i}{2L_i}C_d\rho_m v_r^2 A & (i = 1, n-1) \\ P_n = -\frac{1}{2}q_n LL_n \cos\alpha + q_n LL_n \sin\alpha\cdot f \\ \qquad + \frac{LL_n}{2L_n}C_d\rho_m v_r^2 A & (i = n) \end{cases} \tag{10}$$

③ The approximate differential equation for the deflection curve of the beam is

$$EI_i y_i'' = -M_i(x) \quad (L_{i-1} \le x \le L_i) \tag{11}$$

④ The general solution of the differential equation is as follows:

$$\begin{cases} y_i = A_i \sin k_i x + B_i \cos k_i x - \frac{N_i(x)}{P_i} - \frac{EI_i q_i}{P_i^2} & (L_{i-1} \le x \le L_i) \\ y_i' = A_i k_i \cos k_i x - B_i k_i \sin k_i x - \frac{N_i'(x)}{P_i} & (L_{i-1} \le x \le L_i) \end{cases} \tag{12}$$

where

$$\begin{cases} k_i = j\sqrt{\frac{-P_i}{EI_i}} = jk_i';\sin(jx) = jsh(x);\cos(jx) = ch(x); j = \sqrt{-1} \\ N_i(x) = R_L x - \frac{1}{2}q_i (x - L_{i-1})^2 - \sum_{j=1}^{i-1} q_j LL_j \left(x - L_{j-1} - \frac{1}{2}LL_j\right) \\ N_i'(x) = R_L - q_i (x - L_{i-1}) - \sum_{j=1}^{i-1} q_j LL_j \end{cases} \tag{13}$$

The deflection and slope equations of the beam determined using Eq. (12) can be further expressed as

$$\begin{cases} y_i = jA_i \mathrm{sh}(k_i'x) + B_i \mathrm{ch}(k_i'x) - \frac{N_i(x)}{P_i} - \frac{EI_i q_i}{P_i^2} (L_{i-1} \le x \le L_i) \\ y_i' = jA_i k_i' \mathrm{ch}(k_i'x) + B_i k_i' \mathrm{sh}(k_i'x) - \frac{N_i'(x)}{P_i} \quad (L_{i-1} \le x \le L_i) \end{cases} \tag{14}$$

To solve Eq.(11), the corresponding boundary conditions and continuous conditions must also be known.

The boundary conditions are as follows:

$$y_1(0)=0,\ \ y_n(L_n)=0 \tag{15}$$

In the continuous condition, the deflection and slope at the variable cross section
are equal:

$$\begin{cases} y_1(L_1)=y_2(L_1) \\ y_2(L_2)=y_3(L_2) \\ \cdots \\ y_{n-1}(L_{n-1})=y_n(L_{n-1}) \end{cases},\quad \begin{cases} y_1{}'(L_1)=y_2{}'(L_1) \\ y_2{}'(L_2)=y_3{}'(L_2) \\ \cdots \\ y_{n-1}{}'(L_{n-1})=y_n{}'(L_{n-1}) \end{cases} \tag{16}$$

2.4 Solution of the model

On substituting Eq. (15) and (16) into Eq. (14), the equations of deflection and slope in the matrix form can be expressed as

$$\boldsymbol{HX}=\boldsymbol{b} \tag{17}$$

where $\boldsymbol{H}=\boldsymbol{AR}+\boldsymbol{AI}i$ and $\boldsymbol{b}=\boldsymbol{BR}+\boldsymbol{BI}i$ are both complex numbers.

Let $\mathrm{sh}(k_i'L_j)=E_{ij}'$, $\mathrm{ch}(k_i'L_j)=F_{ij}'$ $(i,j=1,\mathrm{n})$; the terms in Eq. (18) are as follows:

$$\boldsymbol{AR}_{2n\times 2n}=\begin{pmatrix}
0 & 1 & 0 & 0 & 0 & 0 & \cdots & 0 & 0 & 0 & 0 & 0 \\
0 & 0 & 0 & 0 & 0 & 0 & \cdots & 0 & 0 & 0 & 0 & F'_{nn} \\
0 & F'_{11} & 0 & -F'_{21} & 0 & 0 & \cdots & 0 & 0 & 0 & 0 & 0 \\
0 & 0 & 0 & F'_{22} & 0 & -F'_{32} & \cdots & 0 & 0 & 0 & 0 & 0 \\
\vdots & \vdots & \vdots & \vdots & \vdots & \vdots & & \vdots & \vdots & \vdots & \vdots & \vdots \\
0 & 0 & 0 & 0 & 0 & 0 & \cdots & F'_{n-2,n-2} & 0 & -F'_{n-1,n-2} & 0 & 0 \\
0 & 0 & 0 & 0 & 0 & 0 & \cdots & 0 & 0 & F'_{n-1,n-1} & 0 & -F'_{n,n-1} \\
0 & k_1'E'_{11} & 0 & -k_2'E'_{21} & 0 & 0 & \cdots & 0 & 0 & 0 & 0 & 0 \\
0 & 0 & 0 & k_2'E'_{22} & 0 & -k_3'E'_{32} & \cdots & 0 & 0 & 0 & 0 & 0 \\
\vdots & \vdots & \vdots & \vdots & \vdots & \vdots & & \vdots & \vdots & \vdots & \vdots & \vdots \\
0 & 0 & 0 & 0 & 0 & 0 & \cdots & k'_{n-2}E'_{n-2,n-2} & 0 & -k'_{n-1}E'_{n-1,n-2} & 0 & 0 \\
0 & 0 & 0 & 0 & 0 & 0 & \cdots & 0 & 0 & k'_{n-1}E'_{n-1,n-1} & 0 & -k'_nE'_{n,n-1}
\end{pmatrix} \tag{18}$$

$$\boldsymbol{AI}_{2n\times 2n}=\begin{pmatrix}
0 & 0 & 0 & 0 & 0 & \cdots & 0 & 0 & 0 & 0 & 0 & 0 \\
0 & 0 & 0 & 0 & 0 & \cdots & 0 & 0 & 0 & 0 & E'_{n,n} & 0 \\
E'_{11} & 0 & -E'_{21} & 0 & 0 & \cdots & 0 & 0 & 0 & 0 & 0 & 0 \\
0 & 0 & E'_{22} & 0 & -E'_{32} & \cdots & 0 & 0 & 0 & 0 & 0 & 0 \\
\vdots & \vdots & \vdots & \vdots & \vdots & & \vdots & \vdots & \vdots & \vdots & \vdots & \vdots \\
0 & 0 & 0 & 0 & 0 & \cdots & E'_{n-2,n-2} & 0 & -E'_{n-1,n-2} & 0 & 0 & 0 \\
0 & 0 & 0 & 0 & 0 & \cdots & 0 & 0 & E'_{n-1,n-1} & 0 & -E'_{n,n-1} & 0 \\
k_1'F'_{11} & 0 & -k_2'F'_{21} & 0 & 0 & \cdots & 0 & 0 & 0 & 0 & 0 & 0 \\
0 & 0 & k_2'F'_{22} & 0 & -k_3'F'_{32} & \cdots & 0 & 0 & 0 & 0 & 0 & 0 \\
\vdots & \vdots & \vdots & \vdots & \vdots & & \vdots & \vdots & \vdots & \vdots & \vdots & \vdots \\
0 & 0 & 0 & 0 & 0 & \cdots & k'_{n-2}F'_{n-2,n-2} & 0 & -k'_{n-1}F'_{n-1,n-2} & 0 & 0 & 0 \\
0 & 0 & 0 & 0 & 0 & \cdots & 0 & 0 & k'_{n-1}F'_{n-1,n-1} & 0 & -k'_nF'_{n,n-1} & 0
\end{pmatrix} \tag{19}$$

$$\boldsymbol{BR}_{2n\times 1}=\begin{pmatrix} b_1 & b_2 & \cdots & b_i & \cdots & b_{2n-1} & b_{2n} \end{pmatrix}^{\mathrm{T}} \tag{20}$$

$$\boldsymbol{BI}_{2n\times 1}=0 \tag{21}$$

$$\boldsymbol{X}_{2n\times 1}=\begin{pmatrix} A_1 & B_1 & \cdots & A_i & B_i & \cdots & A_n & B_n \end{pmatrix} \tag{22}$$

In Eq. (20),

$$\begin{cases} b_1=\dfrac{N_1(0)}{P_1}+\dfrac{EI_1q_1}{P_1^2},\ \ b_2=\dfrac{N_n(L_n)}{P_n}+\dfrac{EI_nq_n}{P_n^{\ 2}} \\ \vdots \\ b_{n+1}=\dfrac{N_{n-1}(L_{n-1})}{P_{n-1}}-\dfrac{N_n(L_{n-1})}{P_n}+\dfrac{EI_{n-1}q_{n-1}}{P_{n-1}^{\ \ 2}}-\dfrac{EI_nq_n}{P_n^{\ 2}} \\ \vdots \\ b_{2n}=\dfrac{N_{n-1}{}'(L_{n-1})}{P_{n-1}}-\dfrac{N_n{}'(L_{n-1})}{P_n} \end{cases} \tag{23}$$

The Eq. (17) were solved by the complex coefficient all-selected principal element Gaussian elimination method, and then the Eq. (22) was substituted into the Eq. (14) to obtain the deflection and corner equation of the variable-section simply supported beam. On this basis, the maximum deflection $y'_{\max}$ of the string under the situation of no borehole wall constraint can be obtained. In the actual situation of the wall constraint, when $y_c>0$ and $y'_{\max}\geq y_c+(d_b-d_w)$, the maximum deflection $y_{\max}=y_c+(d_b-d_w)$.

The abovementioned derivation is for the axial force $P_i<0$, that is, tension. For horizontal wells, as the string runs, the deviation angle increases and the axial component of the net weight decreases. In this case, the string may be subjected to axial pressure, that is $P_i>0$, due to resistance or pump thrust. In this case, k_i is replaced 18 by Eq. (24) and substituted into the abovementioned equation to obtain the deflection and slope equation under pressure.

$$k_i=\sqrt{\frac{P_i}{EI_i}} \tag{24}$$

The calculation code based on the abovementioned theoretical model is written using the Fortran language, and the procedure shown in Fig. 5.

https://doi.org/10.1051/e3sconf/202129302064

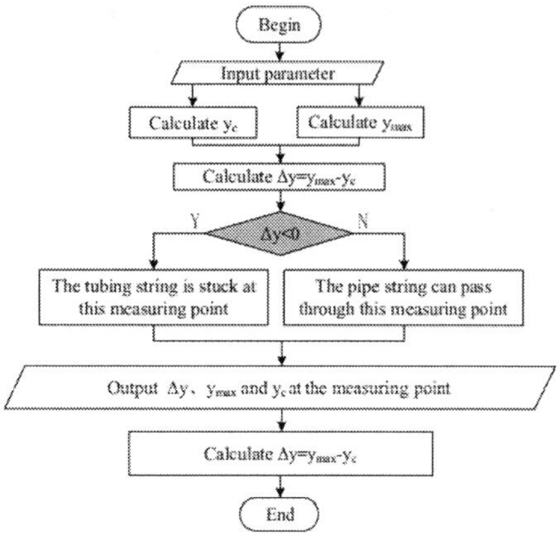

Figure 5. Flow chart of solving model

3 Results and discussions

Considering the running of a "1 bridge plug + 4 cluster perforating gun string" in well X206 as an example, a field test of the pass-through capacity of a cable pumping cluster perforating string in a wellbore was performed.

As the measured borehole trajectory parameter is a discontinuous point, the least square method is adopted to fit the actual borehole trajectory data through cubic spline interpolation, to facilitate observation. Starting from the well depth of 3800 m, the measuring points of the borehole trajectory of X206 was fitted. The fitting trajectory is shown in Fig. 6, and the well's inner diameter is 114.3 mm. As can be seen from the borehole trajectory of well X206, it is a typical horizontal well. Moreover, the well depth corresponding to the measuring points is shown in Fig. 7.

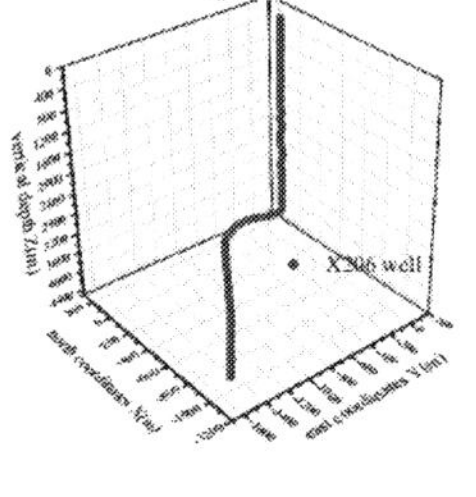

Figure 6. X206 hole trajectory diagram

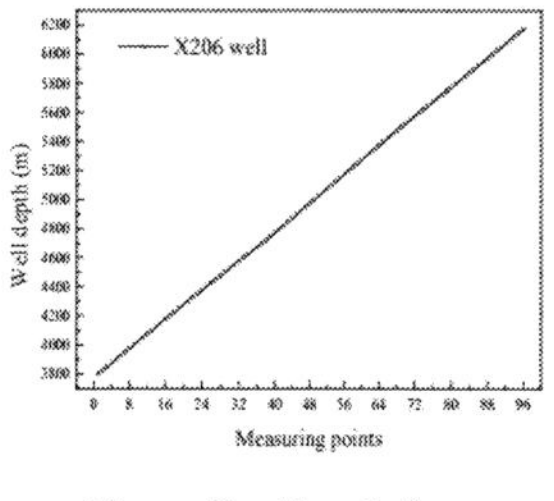

Figure 7. Correlation between measuring points and well depth

The "1 bridge plug + 4 cluster perforating string" run in the X206 well and the perforating tool with the same outer diameter were segmented. The cluster perforating string was divided into 8 sections. The size parameters are shown in Table 1. The relevant calculation parameters are listed in Table 2.

In the field test, the string ran at a relatively stable speed. When the string ran into a well depth of 3800-3900 m, the string exhibited signs of resistance and deceleration, because the section of the well transitioned from the vertical well to the build-up well. The well depth of 3900-4500 m is termed as the "build-up" section, and the well depth exceeding 4500 m is termed as the "horizontal" section. When the string runs to a depth of approximately 4500 m, it is difficult for the string to run under the influence of gravity alone. At this point, the string is run to the bottom of the well by applying pumping thrust.

Subsequently, the pass-through capacity analysis model for the cable pumping cluster perforating string was used to analyze the downhole performance of the "1 bridge plug + 4 cluster perforating gun string" in well X206 and compared with the field test results.

Table1. Size parameters of the cluster perforating string

Sequence number	Outer diameter/m	Volume / m³	Length/ m
1	0.043	0.00078	0.58
2	0.073	0.00043	0.12
3	0.083	0.01335	2.52
4	0.089	0.01374	2.25
5	0.073	0.00537	1.367
6	0.089	0.02246	7.104
7	0.099	0.00756	2.119
8	0.103	0.00504	0.98

Table2. Related parameters of calculation

Parameters	Value
String elastic modulus	206GPa
String density	7850kg/m³
Cable linear density in well fluid	230kg/km
Well fluid density	1100 kg/m³
String down speed	1.45m/s
Friction coefficient	0.25
Fluid resistance coefficient	165

3.1 Situation analysis of string stuck

During the running process of the string, the net weight and axial tension cause a certain bending deformation. When the maximum deflection y_{max} is less than the y_c value of the corresponding measuring points, that is, $y_{max} < y_c$, the string will get stuck. Fig. 8 is a schematic diagram of y_{max} and y_c; it presents the changes in the measuring points when string is run into the X206 well.

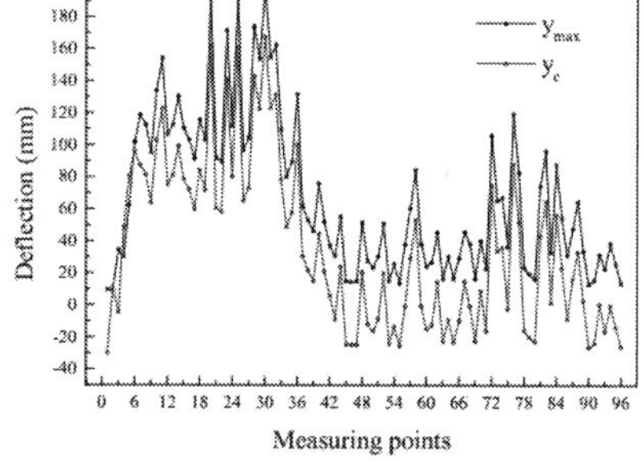

Figure 8. Variation in y_{max} and y_c at different measuring points

In Fig. 8, for the first 6 measuring points, $y_{max} < y_c$ for three measuring points, and the string was blocked at these positions. This was attributed to the large dogleg degree and small deviation angle of the well segment. In the measuring points between 6 and 32, the dogleg degree is large, and the deviation angle increases gradually. Hence, y_{max} as well as y_c are large. In this case, $y_{max} > y_c$, and the string will not be blocked. After the string runs through measuring point 32, the change in y_{max} and y_c is consistent, and y_c is always less than y_{max} because the string is close to the inner wall of the wellbore due to net weight deformation in the horizontal section. These observations are consistent with the actual phenomena observed in the field.

3.2 Analysis of downhole mechanical analysis

As the string runs, the deviation angle and the resistance to the string increases. When the resistance is greater than the axial component of the net weight of the string, the string will stop running. At this point, the pump thrust is needed to push the string down to the bottom of the well. According to Eq. (7), the string's friction, axial forces and the cable friction force vary with the measuring points while applying the pump thrust, as shown in Fig. 9.

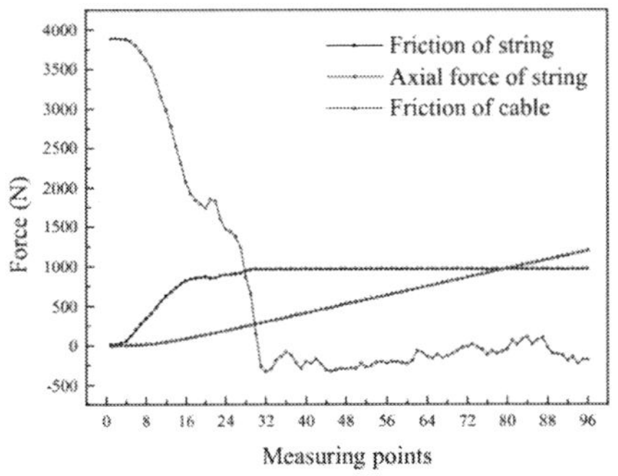

Figure 9. Change in friction and axial force of the string

Observing Fig. 9, the build-up section is between measuring points 1 and 30, and the deviation angle increases gradually, so the axial component of the string decreases, and the friction of the string increases. The horizontal section of the well is after the measuring point of 30, and the deviation angle fluctuates around 90°. Therefore, the friction of the string changes steadily, and the axial force fluctuates around 0 N. As the downhole cable gradually increases with the downhole string, the friction force generated by the net weight component of the cable increases gradually in the build-up section and the horizontal section. With the combination of Eq. (7) and Fig. 9, the pump thrust can be calculated, as shown in Fig. 10.

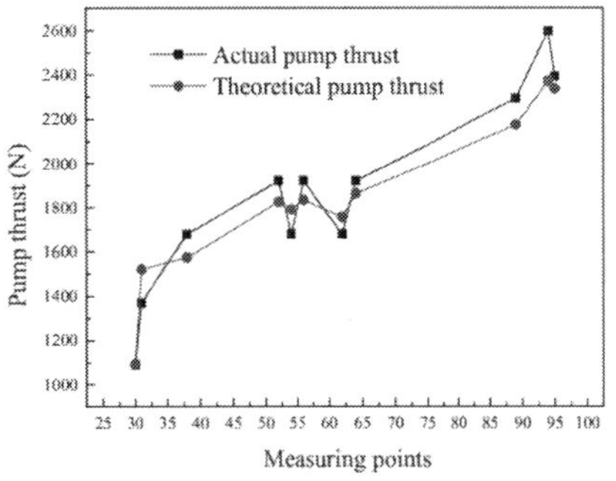

Figure 10. Schematic diagram of the change in pump thrust

Fig. 10 shows that the difference between the calculated and measured pump thrust is small in quantity, and consistent throughout. The main reasons for the difference are due to the fact that the calculation of the pump thrust 19 is relatively complicated, it is not only related to the above various resistances, but also affected by string structure, pumping fluid impact force, the fluid velocity calculation method, and other factors [20].

For general working conditions, if the string can run to the bottom of the well, it can also pass through the wellbore smoothly when pulled up. The difference is that the cable head may be pulled off due to excessive tension, so the cable head should be checked before designing the string. During the process of pulling the string at a relatively stable speed, the tension exerted by the cable head can be expressed as

$$F_L = WL\cos\alpha + WL\sin\alpha \cdot f + \frac{1}{2}C_d\rho_m v_r^2 A \qquad (25)$$

Combined Eq. (25) with field data, the change of tension of cable head in the pulling can be obtained

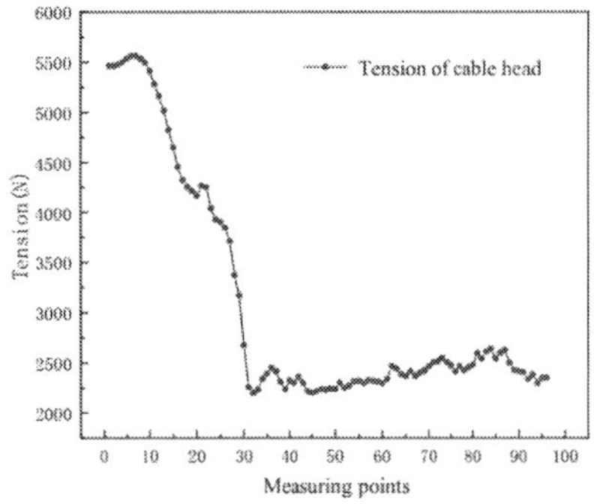

Figure 11. The change of tension of cable head in the pulling

Fig. 11 shows that during the pulling process, the cable head tension was about 2.5 kN in the horizontal section. With the pulling of the string, the deviation angle decreases, the axial component of the string increases, and the cable head tension increases. When the string reached the measuring point 5~8, the cable head tension reached the peak value of 5.56 kN. In the field test, the peak tension of cable head was between 5~5.5 kN, and the allowable tension was 12 kN, so the calculation of cable head tension in this model is close to the reality.

4 Conclusions

(1) Based on a comprehensive consideration of factors such as the friction force between the downhole tools and borehole wall, fluid resistance, wellbore geometric limitation, pump thrust, axial tension, string variable cross section, and tool elastic deformation, a pass-through capability analysis model for a cable pumping cluster perforating string in a wellbore was established. For this model, the geometric analysis method and the beam-column theory are mainly used to establish the complex coefficient equations to be solved. According to the continuous and boundary conditions, the pass-through capacity of downhole tools in wellbores with different structures can be solved using the mathematical model and proposed solution method.

(2) On the basis of the field test data, the pass-through capacity of a cluster perforating string and the main factors affecting the pass-through capacity were analyzed. The field data were compared with the model prediction results to verify the effectiveness of the model. Prior to designing and running the string, the analysis model can be used to predict the sticking points of the string and the required pump thrust.

(3) When the string runs through the sticking points, a relaxing cable tension or an acceptable pump thrust is not sufficient to run the string, indicating that the string cannot pass through the well section. Therefore, it should be replaced with a string of smaller outer diameter or a shorter length of string for perforating. In this study, for the calculation of fluid resistance, the values of drag coefficient and the string running velocity were within a certain range. To obtain a more accurate relative velocity and fluid resistance coefficient when running the cluster perforating string, the drag coefficient should be determined through several calibration tests, and the running velocity of the string should be measured at each measuring point

Acknowledgments

This work has been supported in part by National Natural Science Foundation of China (Grant No. 51875489) and the China Scholarship Council (CSC, File No.201908510191).

Nomenclature

ΔS	Sounding increment of the adjacent measuring points section i, m
$\Delta\alpha$	Increment of the deviation angle, rad
$\Delta\varphi$	Increment of the azimuth angle, rad
K_i	Overall angle change rate in section
R_i	Hole curvature radius in section i, m
G_i	Dogleg degree, °/30 m
d_b	Wellbore diameter, m
d_z	Outside diameter at the midpoint of the perforating string, m
d_q	Outside diameter of the bridge plug, m
L	Total length of tubing string, m
y_c	bending deformation caused by the wellbore constraint in the middle of the string, m
y_{max}	Maximum deflection of pipe string, m
F_p	Axial component of the string, N
F_n	Transverse component of the string, N
F_f	Total resistance of the string, N
F_d	Friction force suffered by the cable, N
F_b	Pump thrust, N
W	String and cable weight per unit length in well fluid, N/m
W'	Cable weight per unit length in well fluid, N/m;
L'	Length of cable in deviation section and horizontal section, m
α	Deviation angle, (°)
f	Friction coefficient between pipe string, cable and wellbore
C_d	Fluid resistance coefficient
V_r	Corresponding velocity of the string and the well liquid, m/s
ρ_m	Well fluid density, kg/m^3
A	Maximum cross-sectional area of the string, m^2
y_i	Deformation deflection in section i, m
y_i'	Corner in section i, rad
R_L	Reaction force of the left end of the simply-supported beam, N
R_R	Reaction force of the right end of the simply-supported beam, N
LL_i	Distance from the variable cross section to the origin of coordinates, m
q_i	Distributed load of section i of beam, N/m
L_i	Length of section i of beam, m
P_L	Cable tension, N
F_{fi}	Friction force on each beam, N
E	Elastic modulus of perforating string, Pa
I_i	Moment of inertia of section i of beam, m^4
P_i	Axial tension at section i of perforating string, N
F_L	Cable head tension, N

References

1. T.X. Jiang, C.G. Jia, H.T. Wang. Shale gas Horizontal Well Sru Fracturing Technology. Beijing: Science Press, 262-265 (2017)
2. H.B. Chen, K. Tang, F. Chen, J.B. Chen, B.C. Li, G.H. Ren. Oriented cluster perforating technology and its application in horizontal wells. Natural Gas Industry, 36(07):33-39 (2016)
3. D. Smith, P. Starr. Method to pump bridge/frac plugs at reduced fluid rate//SPE, 13-15 (2008)

4. C.L. Cipolla, N.R. Warpinskin, M.J. Mayerhofer, E. Lolon, M.C. Vincent. The relationship between fracture complexity, reservoir properties, and fracture treatment design//SPE, 21-24 (2008)
5. D. Vaucher, C. Parrish, R.T. Brooks. Challenges and solutions to running tools through older well completions containing significant paraffin and asphaltene buildup//SPE, 4-5 (2009).
6. J. Ran. Downhole tools in highly deviated fracturing and acidizing wells. Oil Drilling and Production Technology, (02):70-73 (1990)
7. J.P. Zhao, Y.N Su. Drilling tool Assembly passage capacity model and its analysis. Oil Drilling and Production Technology, (05):1-6 (1993)
8. Q.F. Di, Z.Q. Yu. The formulas and analysis of passing-through ability of downhole motor (PTAM). Fault-block Oil & Gas Field, (04):40-43,48 (1996)
9. S.M. He, X.Y. Guo, Z.H. Wang. Calculation of frictional drag of casing running in horizontal wells. Journal of Southwest Petroleum Institute, (02):29-34,4 (1997)
10. M.C. Sui, K.L. Meng, L Du. Analysis of friction on drill string in horizontal drilling. China Petroleum Machinery, (02):9-12,1 (1999)
11. Y.H. Wang. Study on the Tripability of Completion String in Horizontal Well. China University of Petroleum. (2008)
12. X.X. Zhu, S.F. Xue, X.H. Tong. Parameter control methods for the pumping toolstring composed of perforating gun and fracturing plug in a horizontal well. Petroleum Exploration and Development, 40(03):371-376 (2013)
13. D. Feng, L. Shi, C.Y. Xia, H. Zhang, Y.L. Tu. The pass-through capacity analysis for multi-layer injection string in directional wells. Journal of Southwest Petroleum University: Science & Technology Edition, 38(03):162-169 (2016)
14. Z.X. Chen, X.P. Tang, Z.Q. Gao. Passing ability of medium/short-radius whipstocking screw drill tools inside casing. Oil Drilling and Production Technology, (06):7-10 (1999)
15. Z.J. Wei, J.H. Fu, Y.H. Liu, G. W Zeng. The analysis of PDM with double bent angle running into casing. Fault-block Oil & Gas Field, (01):68-70,93 (2005)
16. X.H. Zhu, Y. Gao, S.H. Liu, J.M. Xu. Feasibility analysis of single bend screw tripping. Oil Drilling and Production Technology, 33(03):106-108 (2011)
17. C. Qin, X.B. Hu, D.L. Gao. Three-dimensional calculation of sucker rod centralizer spacing in deviated well. China Petroleum Machinery, (05):45-48,61 (1997)
18. X.P Tang. Timoshenko beam of variable cross-section (Variablestiffness). Mechanics and Engineering, (04):47-50 (1999)
19. F. Chen, D.B. Yang, X.W. Guo, K. Tang, G.H Ren, Q.B. Zhang. Analysis of pumping process and study of displacement control for static start-up pumping in horizontal section. Well Logging Technology, 42(06):720-725 (2018)
20. F. Chen, D.B.Yang, K. Tang, G.H Ren, Q.B. Zhang, B.C. Li. Pumping multi-cluster perforation and bridge plug joint operation technology for upward inclined wells. Well Logging Technology, 42(01):117-121 (2018)

Comprehensive Evaluation of Cultivated Land Resources in China Based on Three Dimensions: Quantity, Quality and Ecology Danger

Renyi Yang[1], Zisheng Yang [1,*], and Fenglian Liu[1]

[1] Institute of Land & Resources and Sustainable Development, Yunnan University of Finance and Economics, Kunming 650221, China

Abstract. Based on TOPSIS-Grey Relational Analysis Method, the three dimensions of cultivated land quantity, quality and ecological danger are organically integrated to build a comprehensive evaluation indicator system of practical and national cultivated land resources and quantitative calculation method by taking the province-level administrative region as the evaluation unit. The index of cultivated land quantity, quality, and ecological danger and the comprehensive index of cultivated land resources are calculated quantitatively in 31 province-level administrative regions of China. The comprehensive endowment and superiority of cultivated land resources in each province are also analysed systematically. The results show that there are 10 provinces (municipalities directly under the Central Government and autonomous regions) classified as "high superiority" provinces in terms of cultivated land resources. There are 9 provinces classified as "relatively high superiority" provinces. The dominant types of cultivated land resources in China can be classified into six types: superiority type of quantity, superiority type of quantity and ecology, superiority type of quality and ecology, superiority type of ecology, comprehensive superiority type, and unclear superiority type. Most provinces have a certain "single superiority", "double superiority", or "comprehensive superiority".

1 Introduction

The ancient Chinese sayings "all things are born in the soil" and "where there is soil, there is grain", which profoundly reveal the close relationship between cultivated land resources and food security. The quantity, quality and utilization of cultivated land resources in a country and region fundamentally determine the effective food supply and food security plan [1]. Internationally, the United States, the United Kingdom, Germany, France, the Netherlands and other countries have early started the research on the investigation, evaluation and protection of land resources, including cultivated land resources. Among them, the most representative is the land capability classification system issued by the United States Department of agriculture [2] (1961) for servicing land protection, land suitability evaluation system formulated by FAO [3] (1976) and modern information system technology was used to predict crop yield in large regions by FAO [4] and dumanski [5], which are widely used in land resource investigation, land suitability evaluation, cultivated land quality classifying and grading in the world. China is a country with a large population and a small amount of cultivated land. The per capita cultivated land area is less than 1/2 of the world's per capita level. For a long time, China has been feeding 20%~22% of the world's population only by using about 7%~10% of the world's cultivated land. Therefore, the research on the quantity change, quality evaluation and ecological security condition of cultivated land resources has been highly paid attention by governments at all levels and academic circles. In recent decades, research results have been emerging constantly. In terms of cultivated land quantity change, Bi Yuyun et al. [6] (2000) and Feng Zhiming et al. [7] (2005) have analyzed the dynamic changes of cultivated land resources of China since 1949. Yang Zisheng [8] (2008) has discussed the necessity of adhering to the cultivated land red line of 18 million mu on the basis of analyzing cultivated land area change of China from 1996 to 2006. Cao Xue et al. [9] (2014) have reconstructed data sets of cultivated land in China in recent 300 years and have analyzed changes of cultivated land of China. Tan Yongzhong et al. [10] (2017) have analyzed the spatial pattern of cultivated land area change of China before and after the second national land survey. GAO et al. [11] (2019) have analyzed the change of cultivated land spatial and temporal characteristics of China from 1990 to 2015 based on the geomorphic zoning. LAI et al. [12] (2020) analyzed the change of cultivated land use of China since the reform and opening up period. In addition, there were plenty of the studies of cultivated land quantity change. Wu et al. [13] (2017) studied cultivated land protection policy of China based on the dynamic balance of total cultivated land and basic cultivated land zoning. Deng et al. [14] (2015), Li et al. [15] (2015) and Cui Xufeng [16] (2018) studied cultivated

* Corresponding author: yangzisheng@126.com

land non-agriculturalization of China, etc. From the perspective of national cultivated land survey, the first national detailed survey of land resources [17] and the second national land survey [18] were completed in 1996 and 2009 respectively. The national cultivated land and other types of land area and vector data were summarized. At present, the third national land survey will nearly complete [19]. In the aspect of cultivated land quality evaluation, the most representative was 1:1 million land resources map of China compiled in 1986, which was the first time to assess the suitability grades of land in China, including the grades of arable land. Since 1999, the Ministry of land and resources (renamed the Ministry of natural resources now) has deployed and carried out the nationwide cultivated land quality grading survey, and completed national cultivated land quality grade achievements based on the first detailed survey of land resources in December 2009. Since 2011, based on the results of the second land survey and its change survey results, the Ministry of land and resources has conducted a supplementary survey on the national cultivated land quality grade achievements. Since 2014, it has carried out annual updating evaluation and monitoring of cultivated land quality grade. In December 2017, the Ministry of land and resources published the 2016 national cultivated land quality grade update evaluation results [21-22], which divided the cultivated land into 15 grades. At the same time, since 2002, the Ministry of agriculture has also launched a nationwide capability investigation and quality evaluation of cultivated land, and issued the national Bulletin of cultivated land quality grade [23] in December 2014. In May 2020, the Ministry of agriculture and rural areas issued the bulletin on the national cultivated land quality grade in 2019 [24], which divided the cultivated land into 10 grades. In addition, Wang Hongbo et al. [25] (2011) have analyzed the differentiation characteristics of cultivated land in China and its impact on cultivated land protection. Zou Jinlang et al. [26] (2015) analyzed the structural characteristics of cultivated land output of China. Xu et al. [27] (2017), Wang et al. [28] (2019), Gao Chan et al. [29] (2020) have analyzed and evaluated cultivated land of productivity in China. In terms of cultivated land ecological evaluation, it mainly involved cultivated land ecological danger [30] (including cultivated land pollution [31-33], cultivated land soil erosion [34], cultivated land natural disasters [35-38], etc.), cultivated land ecological security [39-41] and cultivated land system security evaluation [42], and cultivated land restoration in ecological damaged areas [43], etc.

The practice of cultivated land quantity survey and evaluation, cultivated land quality evaluation and cultivated land ecological evaluation has greatly promoted the development of cultivated land resources evaluation theory and method, and provided an important basis for a country to formulate cultivated land utilization and remediation planning, cultivated land protection policy, and national land space management decision-making. However, on the whole, the quantitative index, quality index and ecological index of cultivated land evaluation at present basically belong to "single item" index. It is urgent to establish a comprehensive evaluation indicator system and quantitative measurement method of cultivated land resources, which can integrate quantitative evaluation, quality evaluation and ecological danger evaluation, so as to evaluate, measure and compare differences of comprehensive endowment and superiority and inferiority of cultivated land resources among all provinces (municipality directly under the Central Government, Autonomous Region, hereinafter referred to as the province) to formulate the strategic decision-making and scientific planning of comprehensive utilization, protection and remediation of cultivated land resources at the national level. Therefore, this paper plans to take the province-level administrative region as the analysis and evaluation unit, organically integrate the three dimensions of cultivated land: quantity evaluation, quality evaluation and ecological danger evaluation, and build a set of practical and feasible comprehensive evaluation indicator system and quantitative calculation method of cultivated land resources, so as to provide technical support for scientific analysis of comprehensive endowment and development and utilization superiority of cultivated land resources in all provinces of China.

2 Research Method

2.1 Research Ideas

The comprehensive evaluation of cultivated land resources in this paper takes the province-level administrative region as the evaluation unit, organically integrates the quantity of cultivated land resources, the quality of cultivated land resources and the ecological danger of cultivated land utilization, and comprehensively evaluates the comprehensive endowment and superiority and inferiority of provincial cultivated land resources. At the national level, the evaluation research results are not only the important basis for formulating the planning of cultivated land utilization, protection and remediation, but also the important basis for land spatial management decision-making and management effect evaluation. Therefore, it is necessary to study and establish a set of specific, feasible and practical comprehensive evaluation indicator system of cultivated land resources. In the past, most researchers used to divide the indicator system into three levels (i.e., target level, criterion level and indicator level) before hierarchical processing. The commonly used evaluation methods mainly include analytic hierarchy process (AHP), entropy weight method, comprehensive analysis method, etc., which are relatively mature in application, but each has its superiority and inferiority [44]. Based on the development of similar evaluation research field in recent years, this paper combines TOPSIS (approximate ideal solution sorting method) and grey correlation analysis method, and calculates the quantity index, quality index and ecological danger index of cultivated land resources in each province on the basis of constructing a series of indicator systems of cultivated land resources quantity, quality and ecological danger. Then the three indexes are organically integrated to calculate the province-level comprehensive evaluation index of cultivated land resources, so as to get the overall evaluation and cognition of the cultivated land utilization system of each province, aiming at comprehensively measuring and comparing the comprehensive endowment and superiority and inferiority of the cultivated land resources in each province.

2.2 Comprehensive Evaluation Method of Cultivated Land Resources Based On TOPSIS-Grey Correlation Analysis Method

Technique for Order Preference by Similarity to an Ideal Solution is a common decision-making technique for multi-objective decision-making analysis of finite schemes in systems engineering [45]. This method judges the superiority and inferiority of a system by calculating the weighted Euclidean distance between the system and the positive and negative ideal solutions. It has a concise thought and convenient calculation. In recent years, it has also been used in the comprehensive evaluation of multiple indexes. However, there are still some problems in the applications of this method to the comprehensive evaluation. It reflects the relative position relationship of the data curve, and cannot reflect the dynamic change of the system, so it is difficult to obtain a reasonable evaluation result [46]. Grey relational analysis method is a branch of grey system theory. One of the characteristics different from traditional statistical methods is that it can use relatively small amount of data to evaluate the quantitative and qualitative relationship between factors and variables [47], and has the superiorities of low sample size, simple calculation and concise regularity [48]. The combination of TOPSIS method and grey correlation analysis method is used to improve the TOPSIS method, which can more accurately express the degree of fit between alternative schemes and ideal schemes [49], so as to improve the accuracy of the evaluation result [50]. In this paper, the comprehensive evaluation method of cultivated land resources based on TOPSIS-Grey Relational Analysis Method is as follows:

1. Constructing Weighted Decision Matrix R.

$$\boldsymbol{R}=[R_{ij}]_{m\times n}=[I_{ij}\times W_j]_{m\times n} \quad (1)$$

In formula (1), m is the total number of province-level administrative regions; n is the total number of indicators; I=1,2,...,m is the province-level administrative region; J=1,2,...,n is the indicator; I_{ij} is the attribute value of indicator J of province I after dimensionless treatment; W_J is the indicator weight determined by entropy weight method.

The specific method of calculating indicator weight by entropy weight method is as follows:

(1) The indicators with positive and negative attributes were dimensionless respectively. The specific method is shown below.

(2) Calculate the information entropy of each indicator.

$$E_j=-\frac{\sum_{i=1}^{m}P_{ij}\ln(P_{ij})}{\ln(m)},\quad P_{ij}=\frac{I_{ij}}{\sum_{i=1}^{m}I_{ij}} \quad (2)$$

In formula (1), if P_{ij}=0, then we will definition $P_{ij}\ln(P_{ij})$=0.

(3) According to the information entropy (E_1, E_2,..., E_n), The weight of each indicator is calculated as:

$$w_j=\frac{1-E_j}{n-\sum_{j=1}^{n}E_j} \quad (3)$$

2. Calculating the Euclidean Distance of Positive and Negative Ideal Solutions (D^+, D_i^-)

$$D_i^+=\sqrt{\sum_{j=1}^{n}\left(R_{ij}-Q_j^+\right)^2},D_i^-=\sqrt{\sum_{j=1}^{n}\left(R_{ij}-Q_j^-\right)^2} \quad (4)$$

In formula (4), $\begin{cases}Q_j^+=\max(R_{ij}),i=1,2,\cdots,m\\Q_j^-=\min(R_{ij}),i=1,2,\cdots,m\end{cases}$ indicate positive and negative ideal solutions of each province-level administrative region respectively.

3. Calculating the Grey Correlation Degree of Positive and Negative Ideal Solutions (P_i^+, P_i^-)

The grey correlation degree of positive and negative ideal solutions of province I can be expressed as follows:

$$P_i^+=\frac{\sum_{j=1}^{n}\rho_{ij}^+}{n},P_i^-=\frac{\sum_{j=1}^{n}\rho_{ij}^-}{n} \quad (5)$$

We assume that $\mu\in[0,1]$ is the resolution coefficient (generally taken as 0.5), Then the grey correlation coefficient in formula (6) can be expressed as follows:

$$\rho_{ij}^+=\frac{\min\limits_{1\le j\le n}\min\limits_{1\le i\le m}\left|Q_j^+-R_{ij}\right|+\mu\max\limits_{1\le j\le n}\max\limits_{1\le i\le m}\left|Q_j^+-R_{ij}\right|}{\left|Q_j^+-R_{ij}\right|+\mu\max\limits_{1\le j\le n}\max\limits_{1\le i\le m}\left|Q_j^+-R_{ij}\right|},$$

$$\rho_{ij}^-=\frac{\min\limits_{1\le j\le n}\min\limits_{1\le i\le m}\left|Q_j^--R_{ij}\right|+\mu\max\limits_{1\le j\le n}\max\limits_{1\le i\le m}\left|Q_j^--R_{ij}\right|}{\left|Q_j^--R_{ij}\right|+\mu\max\limits_{1\le j\le n}\max\limits_{1\le i\le m}\left|Q_j^--R_{ij}\right|} \quad (6)$$

4. Calculate Closeness Degree (C_i^+)

$$C_i^+=\frac{e_1d_i^-+e_2p_i^+}{e_1\left(d_i^-+d_i^+\right)+e_2\left(p_i^-+p_i^+\right)} \quad (7)$$

In formula (7), d_i^-, d_i^+, p_i^-, p_i^+ respectively indicate dimensionless values of D_i^-, D_i^+, P_i^-, P_i^+. e_1, e_2 is the degree of decision preference. Normally, $e_1+e_2=1$. For the sake of simplification, we can define $e_1=e_2=0.5$. The specific dimensionless method is as follows:

$$\begin{cases}d_i^-=D_i^-/\max(D_i^-),\ d_i^+=D_i^+/\max(D_i^+)\\p_i^-=P_i^-/\max(P_i^-),\ p_i^+=P_i^+/\max(P_i^+)\end{cases} \quad (8)$$

5. Calculating the Index of Farmland Quantity, Index of Farmland Quality and Index of Ecological Danger

According to the final calculated closeness degree, the quantity, quality and ecological danger of cultivated land in each province-level administrative region of China will be evaluated. The value range of closeness degree is [0,1]. The higher value of closeness degree is, the richer the cultivated land resource endowment will have, the better quality of cultivated land or the lower ecological danger degree will have and vice versa. According to it, the index of farmland quantity (I_{FQ1}), index of farmland quality (I_{FQ2}) and index of ecological danger (I_{ED}) can be calculated by using the closeness degree to represent the cultivated land conditions of each province-level administrative region in China.

$$\begin{cases} I_{FQ1} = FQ_1C_i^+ \times 100 \\ I_{FQ2} = FQ_2C_i^+ \times 100 \\ I_{ED} = EDC_i^+ \times 100 \end{cases} \quad (9)$$

In formula (9), $FQ_1C_i^+$, $FQ_2C_i^+$ and EDC_i^+ respectively indicate the closeness degree of comprehensive quantity, quality and ecological danger index of farmland. The reason why the formula is multiplied by 100 is to ensure that the value range of the three indexes is [0,100].

6. Calculating the Comprehensive Index of Farmland Resources

Comprehensive Index of Farmland Resources (CI_{FR}) is a quantitative evaluation index for the comprehensive condition of cultivated land resources. It is an organic synthesis and integration of I_{FQ1}, I_{FQ2} and I_{ED}.

The total output of cultivated land is affected by a lot of conditions. Among them, the quantity, quality and ecological danger of cultivated land are very important factors. However, these factors are not isolated, they are an organic whole which is related to each other. It is similar to the "barrel law". In this whole, if a certain index is low, it will critically affect the level of the whole. According to this, we can refer to the calculation formula of human development index (HDI), and use the geometric average of I_{FQ1}, I_{FQ2} and I_{ED} to reflect the comprehensive endowment conditions of cultivated land resources. The calculation formula of CI_{FR} is set as follows:

$$CI_{FR} = \sqrt[3]{I_{FQ1} \cdot I_{FQ2} \cdot I_{ED}} = I_{FQ1}^{\frac{1}{3}} \cdot I_{FQ2}^{\frac{1}{3}} \cdot I_{ED}^{\frac{1}{3}} \quad (10)$$

The comprehensive endowment of cultivated land resources in a province is affected by the quantity, quality and ecological danger of cultivated land, but the degree of influence will be quite different. However, formula (10) takes the product of the 1/3 power of each index as the comprehensive endowment index of cultivated land resources. In other words, its weight setting value is the same, which is obviously unreasonable. So the comprehensive endowment of cultivated land resources calculated by formula (10) may deviate from the actual situation. Therefore, formula (10) needs to be improved properly. The improved calculation method of comprehensive index of farmland resources is as follows:

$$CI_{FR} = I_{FQ1}^{\ \alpha} \cdot I_{FQ2}^{\ \beta} \cdot I_{ED}^{\ \gamma} \quad (11)$$

In formula (11), α、β、γ respectively indicate weight value of I_{FQ1}, I_{FQ2} and I_{ED}. They satisfy the following relations: $\alpha+\beta+\gamma=1$. Because I_{FQ1}, I_{FQ2} and I_{ED} belong to three different categories and different dimensions of comprehensive index, and entropy weight method determines the index weight according to the information provided by the observation value of each index, ignoring the importance of the index itself. Therefore, Delphi method (also called expert consultation method) is properly adopted here. In October 2020, 18 experts and scholars were organized to assign the weights of the above-mentioned I_{FQ1}, I_{FQ2} and I_{ED}. After treatment, the weight values of the three indexes were obtained: $\alpha=0.58$, $\beta=0.24$, $\gamma=0.18$.

The higher value of CI_{FR} is, The better comprehensive endowment of cultivated land resources will have and the more benefit of superiority and conditions of cultivated land utilization will show.

2.3 Construction of Evaluation Indicator System

As a form of policy guidance, the evaluation indicator system should not only reflect the leading idea of coordinated development of many aspects with the owner of cultivated land, but also make each evaluation indicator the most sensitive, convenient, and abundant leading indicator among the numerous indicators reflecting regional cultivated land use system, so that the evaluation indicator system can best describe the state change trend of regional cultivated land use In view of this, the following principles should be followed in the construction of comprehensive evaluation indicator system of cultivated land resources: (1) To reflect the superiority and inferiority of regional comprehensive endowment and development and utilisation of cultivated land resources. (2) To embody the practicability and pertinence of "province region" — a large regional scale feature. (3) Strengthen and pay attention to the scientific nature, operability, and relative completeness of indicator selection. (4) Fully consider the availability, sensitivity, and quantifiable of the basic data enshrined in such indicators.

A cultivated land use system is a complex ecological economic and social composite system, which involves a series of related factors and several coordinated development conditions. It endows the comprehensive evaluation of cultivated land resources with characteristics of system complexity, multi-factor relevance, multiple realisation mechanisms, regional differences and particularities, *etc*. Therefore, the construction of provincial comprehensive evaluation indicator system of cultivated land resources should focus on the characteristics of provincial ecological environment and cultivated land utilisation, taking the comprehensive evaluation of quantity, quality, and ecological danger of cultivated land resources as its core, while highlighting the impact of comprehensive endowment conditions of cultivated land resources on ecological, economic, and social progress. In view of this, in setting the evaluation indicators, we should select typical, sensitive indicators which are independent and can reflect the characteristics of various aspects on the basis of fully understanding the structure, function, characteristics, and utilisation objectives of provincial land use systems to establish an indicator system framework of comprehensive evaluation of cultivated land resources and we divided the comprehensive evaluation indicator system of cultivated land resources into three levels: indicator category, evaluation indicator, and element indicator. We constructed a concise, practical, and feasible provincial cultivated land resource evaluation indicator system (Table 1).

Table1. Comprehensive evaluation indicator system and data acquisition modes of cultivated land resources at provincial level.

Evaluation indicator categories	Evaluation indicators				Element indicator	Data source
	First level		Second level			
	Indicator	Weight	Indicator	Weight		
Indicator categories of cultivated land quantity	Index of Total Area	0.5113	—	—	Total area of each province	Statistical Yearbook of Land and Resources of China, China Statistical Yearbook
					Total area of China	
	Index of Area Per Capita	0.4887	—	—	Total population of each province	Statistical Yearbook of Land and Resources of China, China Statistical Yearbook
					Total cultivated land area of each province	
Indicator categories of cultivated land quality	Index of Quality Grade	0.4423	—	—	Cultivated land quality grade of each province	Special Investigation and Evaluation of the Ministry of Land and Resources
					Cultivated land quality grade of China	
	Index of Annual Grain Yield Per Unit Area of cultivated land	0.5577	—	—	Annual grain yield per unit area of cultivated land of each province	China Statistical Yearbook
					Annual grain yield per unit area of cultivated land of China	
Indicator categories of cultivated land ecological danger	Index of Soil Loss	0.5492	Index of Soil Loss Area	0.4204	Area and proportion of Soil Loss	Special Investigation and Monitoring of Soil and Water Loss of the Ministry of Water Resources
			Index of Soil Loss Intensity	0.5796	Soil and water loss area of each intensity level	
	Index of Natural Disaster	0.4508	Index of Disaster-affected Crops Area	0.6771	Area of disaster-affected Crops	China Statistical Yearbook
					Total sown area of crops	
			Index of Grain Production Reduced by Disaster	0.3229	Area of disaster-affected crops and disaster-destroyed crops	China Statistical Yearbook
					Total sown area and grain crop area	
					Total output of grain	

As the first level of comprehensive evaluation index system, indicator category depends on the basic content of evaluation. As mentioned above, the basic content of comprehensive evaluation of cultivated land resources includes three parts: first, the quantity of cultivated land resource (reflecting the status of cultivated land in the aspect of "quantity"), the second is the quality of cultivated land (reflecting the status of cultivated land in the aspect of "quality"), and the third is the ecological danger to cultivated land (reflecting the status of cultivated land in the aspect of "environment"). These three constitute a whole evaluation system of cultivated land resources. The formation, cultivation, and development of sustainable utilisation capacity of cultivated land in any region are the result of the comprehensive action of all parts of the whole system, rather than the any single part and its content. According to this, the comprehensive evaluation index of cultivated land resources can be divided into three categories: quantity, quality, and ecological danger indicator category of cultivated land.

The evaluation indicator is a quantitative analysis index under the indicator category, selected to reflect each evaluation content of cultivated land use system. Generally, the basic content of each evaluation needs several indices to be expressed specifically and quantitatively, so each indicator category is often composed of specific evaluation indicators with different numbers. In view of the complexity of cultivated land ecological danger evaluation indicators, it can be further divided into two levels of indicators (Table 1). Each evaluation indicator is specifically expressed by several element indicators.

Element indicator is the basic indicator used to represent the evaluation index. It is the lowest level in the whole evaluation index system and belongs to the smallest unit [51]. It is a specific measure used to evaluate the regional cultivated land resources. According to the actual situation of the province, we should highlight the key points and select measurable, comparable, and available quantitative indicators or groups of indicators to express directly the quantitative performance, intensity performance, and speed performance of evaluation indicators.

2.4 The Method of Dimensionless Index

In the general sense of cultivated land resource evaluation research, we usually need to select a large number of indicators for evaluation, but because the nature of these indicators differs, their ranges also differ significantly, so they cannot directly comparable, and in particular they cannot be directly combined in a comprehensive evaluation. Therefore, it is necessary to find a method so that all indicators can be converted into values that can be evaluated uniformly [52]. In the design of this calculation method, the range of each evaluation indicator is set to [0,100], and then the calculated values of all indicators are converted through the system conversion index, where reaching 100 means that the cultivated land resource endowment is high and the condition is excellent, and reaching 0 means that the cultivated land resource endowment is low and the condition is poor. In this way, weighted summation can be conducted to obtain the final comprehensive index value for the overall evaluation. In the process of conversion, the main mathematical methods that can be used are power conversion, satisfaction conversion, grading, a reciprocal method, *etc*., which can be selected according to need.

1. Evaluation indicators of cultivated land quantity

The cultivated land quantity index is the basic index in the whole cultivated land use system. It can be divided into absolute quantity indicator and relative quantity indicator. The former refers to the total area of cultivated land, which reflects the richness of the overall cultivated land resource in the province. The latter mainly refers to the *per capita* cultivated land area, which reflects the amount of cultivated land *per capita* in the province.

(1) Index of total area (I_{TA})

The total area of cultivated land plays an important role in the total grain output and food security of a country and region. At present, the latest cultivated land data available in China are the survey data pertaining to land use status change at the end of 2017 [53], *i.e.* 134,881,218.29 ha. The cultivated land quantity of 31 provinces in China varies greatly, with the largest difference of 1584.57 × 10^4 ha in Heilongjiang Province and 19.16 × 10^4 ha in Shanghai at the minimum, with a difference of 82.7 times between them. To reflect the total amount of cultivated land resources in the province and the actual situation in China, the following formula is used to calculate I_{TA}:

$$I_{TA} = \frac{TA_i}{TA_{\max}} \times 100 \qquad (12)$$

In Eq. (12), TA_i is the total area of cultivated land in province i. Here we take the average value from the year of completion of the second national land survey (2009) to that in 2017. TA_{max} is the maximum value of the total cultivated land area of provinces in China, and the average value from 2009 to 2017 is also taken.

(2) Index of area *per capita* (I_{APC})

Considering the obvious differences in population, total land area, and cultivated land area among the province-level administrative regions in China, it is not enough to consider only the absolute abundance indicator. Therefore, we introduce a relative indicator: cultivated land area *per capita*:

$$I_{APC} = \frac{APC_i}{APC_{\max}} \times 100 \qquad (13)$$

In Eq. (13), APC_i is the *per capita* cultivated land area of province i, APC_{max} is the maximum *per capita* cultivated land area of provinces in China. These two indicators are taken as the average value from 2009 to 2017.

2. Evaluation indicators of cultivated land quality

The quality of cultivated land has a fundamental impact on the per unit yield of cultivated land. To a large extent, the unit yield level of cultivated land reflects the quality of cultivated land. From the perspective of the availability of basic data for this evaluation, in recent years, the Ministry of Land and Resources and the Ministry of Agriculture and Rural Areas have published the National Cultivated Land Quality Survey and Evaluation results [21-24], but in the *Announcement of National Arable Land Quality Grading in 2019* issued by the Ministry of Agriculture and Rural Areas only national data, no provincial data, are used. While the *Announcement on the Release of the Main Data Results of the 2016 National Cultivated Land Quality Grades Updating Evaluation* released by the Ministry of Land and Resources in December 2017 [22] (the Appendix thereto) the cultivated land quality status of 31 provinces in China (*i.e.*, areas of 15 cultivated land grades in each province) is published, so it can be used for the comprehensive evaluation of cultivated land resources. However, it should be noted that the evaluation of 15 grades of cultivated land quality established by the Ministry of Land and Resources is the evaluation standard for determining the quality of cultivated land by comprehensively considering the annual yield thereof. The 15th grade consists of cultivated land with the lowest yield per unit area of grain, and its annual production capacity is about 1500 kg/ha, and so on in increments 1500 kg/ha. The annual production capacity of first-grade land is 22,500 kg/ha [21]. According to this standard, the annual production capacity of the province is not consistent with the actual yield: for example, according to the *Announcement on the Release of the Main Data Results of the 2016 National Cultivated Land Quality Grades Updating Evaluation* by the Ministry of Land and resources, the average cultivated land quality grade of Inner Mongolia is 14.02 when using the area-weighting method, so the annual production capacity is about 3000 kg/ha, however, the average grain yield in Inner Mongolia from 2009 to 2019 has reached 4697.8 kg/ha, which is 1.57 times its annual production capacity, which is illogical. Other provinces also have the same problem. To make up for this deficiency, we not only used the cultivated land quality grade survey and evaluation the data released by the Ministry of Land and Natural Resources, but also considered the actual yield of each province, and combined both to evaluate the provincial cultivated land quality.

(1) Index of quality grade (I_{QG})

Firstly, the average quality grade (V_{AQGi}) of each province-level administrative region is calculated by an area-weighting method, and then I_{QG} is calculated by maximum method. The calculation is determined as follows:

$$V_{AQG_i} = 100 \times \sum_{j=1}^{15} \left(\frac{TA_{ij}}{TA_i} \times \frac{16-j}{15} \right) \quad (14)$$

$$I_{QG} = \frac{V_{AQGi}}{V_{AQG_{\max}}} \times 100 \quad (15)$$

In Eqs. (14) and (15), TA_{ij} is the cultivated land area of cultivated land quality grade j in province i. TA_i is the total area of cultivated land in province i. j is the quality grade of cultivated land (15 grades in total). For the first grade $j = 1$, second grade $j = 2$, and so on. V_{AQGmax} is the maximum value of the average grade of the provincial cultivated land quality in China.

(2) Index of annual grain yield per unit area of cultivated land (I_{AGY})

Referring to the above indicators, the maximum method (*i.e.* the comparison between the maximum value of annual grain yield per unit of cultivated land in a province and the maximum in the whole country) is used to calculate I_{AGY}. The calculation is determined as follows:

$$I_{AGY} = \frac{V_{AGYi}}{V_{AGY\max}} \times 100 \quad (16)$$

3. Evaluation indicators of cultivated land ecological danger

The evaluation index of cultivated land ecological danger covers a wide range, mainly including soil erosion, natural disasters, soil pollution, *etc*. From the perspective of the availability of basic data, although the Ministry of Environmental Protection and the Ministry of Land and Natural Resources issued the *Bulletin of the National Survey of Soil Pollution* in 2014, the first survey obtained the proportion of soil pollution points in different land grades such as cultivated land in China, as well as the proportion of slight, mild, moderate, and severe pollution points [31], but the survey data of each province-level administrative region were not published. In another development, all provinces have completed the detailed survey of soil pollution by the end of 2018, but the specific data have not been published. Therefore, the evaluation does not consider the cultivated land soil pollution indicators.

(1) Index of soil loss (I_{SL})

Soil loss is the product of many unreasonable development and utilisation activities of land resources, such as destroying forest and grass to open up wasteland, steep slope farming, *etc*., which is known as the principal ecological environment problem in recent years. The soil loss index is an indicator to reflect the scale and intensity of soil loss in a province. It is characterised by soil loss area and loss intensity indices. It should be noted that, since, at national level, there has not yet been a special soil loss survey on cultivated land, the latest data that can be collected at present are those pertaining to monitoring data in a soil loss survey commissioned by the Ministry of Water Resources in 2018, including the total area of soil loss and the areas of various intensity levels of soil loss in 31 provinces. However, it should be noted that cultivated land is subject to severe human disturbance due to the whole process of crop management and planting (such as soil ploughing, soil preparation, sowing, loosening, weeding, fertilisation, and harvesting) that is readily causes severe soil loss in many places. Taking Yunnan Province as an example, according to Yang Zisheng *et al.* [34], the total amount of cultivated land soil loss in the whole province accounts for about 70% to 75% of the total annual average soil loss from the land. In addition, soil loss from other types of land often become an important ecological and environmental factor, affecting the use of cultivated land. Therefore, in the current situation of a lack of data relating to soil loss from cultivated land in each province-level administrative region, we temporarily used the monitoring data from the soil and water loss survey commissioned by the Ministry of Water Resources in 2018 for analysis and evaluation.

① Index of soil loss area (I_{SLA})

The common index reflecting the scale of regional soil loss is the rate of soil loss area (R_{SLA}), which refers to the percentage ratio of soil loss area to total land area. According to the *Standard for Classification and Gradation of Soil Erosion* issued by the Ministry of Water Resources (1997), the area of soil erosion here refers to the area subject to a minimum of mild erosion. The calculation is determined as follows:

$$R_{LSA} = \frac{A_{SL}}{A_{TL}} \times 100\% \quad (17)$$

In Eq. (17), A_{SL} indicates the area of soil loss; A_{TL} indicates the total land area.

The higher the value of R_{SLA}, the more scale (range) of soil erosion suffered and the ecological environment degrades. According to the above method, the calculated values of R_{SLA} need to be converted to ensure that the result is within [0,100]. The converted index value is called the index of soil loss area (I_{SLA}):

$$I_{SLA} = 100 - (R_{SLA} \times 100) \quad (18)$$

The greater the value of the I_{SLA}, the lower the proportion of soil loss in a certain area, and the better the ecological environment.

② Index of soil loss intensity (I_{SLI})

In the current investigation and study of soil loss in China, the intensity grade of soil erosion is generally classified by soil erosion modulus. Soil erosion modulus (*SEM*) refers to the average annual soil erosion amount per unit land area in a certain region (or watershed), and its unit is t/km^2·y. It is the main index used to characterise the intensity of soil loss. According to the *Standard for Classification and Gradation of Soil Erosion* [54] issued by the Ministry of Water Resources in January 2008, soil erosion grades are divided into slight erosion, mild erosion, moderate erosion, intensity erosion, extreme intensity erosion, and severe erosion. The SEM values are $<$ 500 t/km^2·y (north-east black soil area, northern rock mountain area and wind erosion area are $<$ 200 t/km^2·y; north-west loess plateau area is $<$ 1000 t/km^2·y), $\geq$ 500

(200, 1000) but < 2500 t/km^2·y, ≥ 2500 but < 5000 t/km^2·y, ≥ 5000 but < 8000 t/km^2·y, ≥ 8000 but < 15000 t/km^2·y, ≥ 15000 t/km^2·y. The soil loss grades in the soil loss survey and monitoring the data published by the Ministry of Water Resources in 2018 include mild erosion, moderate erosion, intense erosion, and worse. According to this, when calculating the average soil erosion modulus of each province, the SEM of each erosion grade is as follows: 200 t/km^2·y for slight erosion, 1500 t/km^2·y for mild erosion, 3750 t/km^2·y for moderate erosion, and 12,000 t/km^2·y for intense erosion and worse. On this basis, the total annual soil loss and average soil erosion modulus of each province can be calculated, and then I_{SLI} can be calculated according to the following formula:

$$I_{SLA} = 100 - (\frac{SEM_i}{SEM_{NA}} \times 100) \quad (19)$$

In Eq. (19), SEM_i indicates the soil erosion modulus for province i; SEM_{NA} indicates the national average soil erosion modulus of soil-eroded land in China. It also is the average amount of soil erosion per unit soil loss area (the sum of the areas of mild erosion, moderate erosion, intense erosion, and worse). According to the survey and monitoring results published by the Ministry of Water Resources in 2018, the SEM_{NA} value is 4128.64 t/km^2·y. The higher the value of I_{SLI}, the lower the intensity of soil loss, and the better the ecological environment.

(2) Index of natural disaster (I_{ND})

Natural disasters are an important factor affecting the output and efficiency of cultivated land use. At present, the annual *China Statistical Yearbook* has included a comprehensive range of area of disaster-affected crops and area of disaster-destroyed crops caused by drought, floods, landslides, debris flows, typhoons, hailstorms, and ice and snow disasters, which provides a basis for this evaluation. Here, the disaster area and disaster intensity are also used to characterise I_{ND}.

① Index of disaster-affected crops area (I_{DCA})

Referring to the calculation method of I_{SLA}, the following formula is adopted for calculation:

$$I_{DCA} = 100 - (\frac{A_{DC}}{A_{TS}} \times 100) \quad (20)$$

In Eq. (20), A_{DC} indicates area of disaster-affected crops; A_{TS} indicates the total sown area of crops.

The higher value of I_{SLA}, the lower the areal proportion of disaster-affected crop, and the better the ecological environment.

② Index of grain production reduced by disaster (I_{GPD})

According to the *Natural Disaster Information Statistics* [55], in the agricultural disaster statistics, it is necessary to count areas of disaster-affected crops, area of disaster-damaged crops, and area of disaster-destroyed crops. These three indicators can reflect the difference in severity of a disaster. They respectively refer to the planting area of crops whose crop yield is reduced by 10% or more, 30% or more, and 80% or more than in a normal year due to natural disasters. According to this, we can calculate the amount of grain reduced due to each disaster and reflect the intensity of the disaster therein. In recent years, the *China Statistical Yearbook* has included areas of disaster-affected crops and area of disaster-destroyed crops in each province-level administrative region, but omitted the intermediate-level area. Therefore, the following method is used to calculate the grain production as reduced by disaster (GP_D):

$$GP_D = [C_1 \times (A_{AC} - A_{DC}) + C_2 \times A_{DC}] \times AY_{UA} \quad (21)$$

In Eq. (21), A_{AC} indicates area of disaster-affected crops, taking the average value from 2009 to 2019. A_{DC} indicates area of disaster-destroyed crops, taking the average value from 2009 to 2019. AY_{UA} indicates average yield per unit area of grain crops, also taking the average value from 2009 to 2019; C_1 and C_2 respectively indicate production reduction coefficients of disaster-affected areas and disaster-destroyed areas, with average values of 0.375 and 0.900.

The index of grain production reduced by disaster (I_{GPD}) indicates the comparative relationship between the amount of grain reduced due to disasters and the total grain output as calculated by:

$$I_{GPD} = 100 - (\frac{GP_D}{AO_G} \times 100) \quad (22)$$

In Eq. (22), GP_D indicates the grain production as reduced by disaster; TO_G indicates the total output of grain. The higher the value of I_{GPD}, the lower the degree of grain production reduced by disaster, and the better the ecological environment.

3 Results and Analysis

3.1 Quantitative Comprehensive Evaluation Results of Cultivated Land Resources

According to the above comprehensive evaluation method of cultivated land resources based on the TOPSIS-Grey Relational Analysis Method, the index of farmland quantity (I_{FQ1}), index of farmland quality (I_{FQ2}), index of ecological danger (I_{ED}), and comprehensive index of farmland resource (CI_{FR}) of 31 provinces are calculated (Table 2). According to the calculated results, the classification and spatial distribution of each index can be analysed, and the comprehensive endowment and superiority of cultivated land resources in each province can be analysed and ranked, respectively.

Table2. Cultivated land quantity index, quality index, ecological danger index, and comprehensive index of each province in China.

Region	I_{FQ1}	I_{FQ2}	I_{ED}	CI_{FR}	Comprehensive ranking of CI_{FR}
Beijing	14.78	45.46	69.30	25.57	30
Tianjin	16.92	37.95	81.20	27.24	29
Hebei	35.98	40.97	61.98	40.94	9
Shanxi	31.70	24.19	31.98	29.75	28
Inner Mongolia	65.93	18.22	15.72	37.41	15
Liaoning	34.11	32.89	48.45	36.02	16
Jilin	51.37	36.14	57.09	48.12	3

 https://doi.org/10.1051/e3sconf/202129303001

Region	I_{FQ1}	I_{FQ2}	I_{ED}	CI_{FR}	Comprehensive ranking of CI_{FR}
Heilongjiang	85.50	27.45	58.14	60.73	1
Shanghai	14.50	82.73	84.28	30.23	27
Jiangsu	28.68	72.97	79.29	43.09	6
Zhejiang	20.63	50.65	63.19	31.31	25
Anhui	34.88	52.85	70.19	43.71	5
Fujian	19.07	58.42	72.11	31.70	22
Jiangxi	25.71	67.30	63.78	38.14	14
Shandong	37.86	58.66	67.00	46.61	4
Henan	39.72	66.66	72.16	50.08	2
Hubei	32.92	67.35	53.62	42.68	8
Hunan	27.87	71.43	60.62	40.18	10
Guangdong	21.46	69.45	68.44	35.05	19
Guangxi Zhuang	30.91	48.40	62.78	39.10	12
Hainan	22.06	43.48	58.59	30.95	26
Chongqing	25.33	46.31	56.52	33.83	20
Sichuan	35.93	49.07	61.75	42.68	7
Guizhou	34.31	29.22	50.17	35.35	17
Yunnan	38.70	32.91	47.29	38.59	13
Tibet	27.17	23.56	74.65	31.49	23
Shaanxi	30.97	28.50	46.12	32.62	21
Gansu	43.32	18.62	22.58	31.46	24
Qinghai	23.78	20.61	40.29	25.27	31
Ningxia Hui	33.61	30.52	48.64	35.10	18
Xinjiang Uygur	44.42	36.07	31.96	39.82	11

3.2 Quantity, Quality, and Ecological Danger of Cultivated Land

1. Quantity of cultivated land

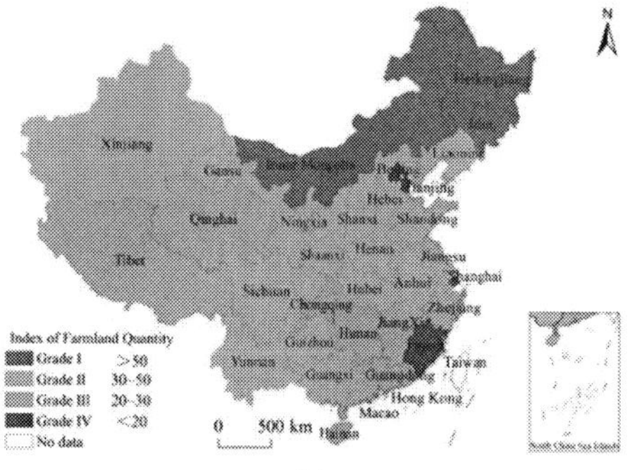

(*a*)

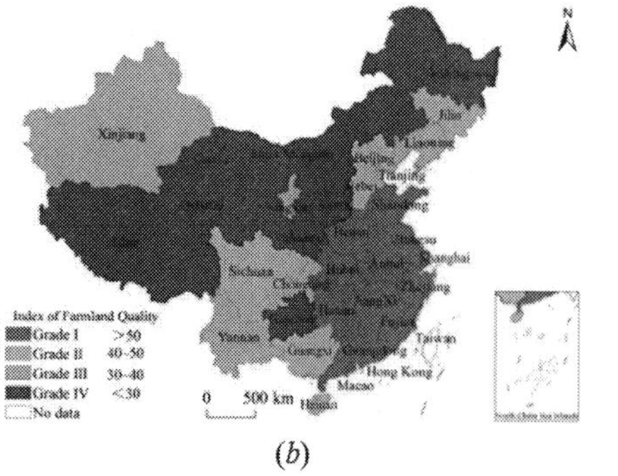

(*b*)

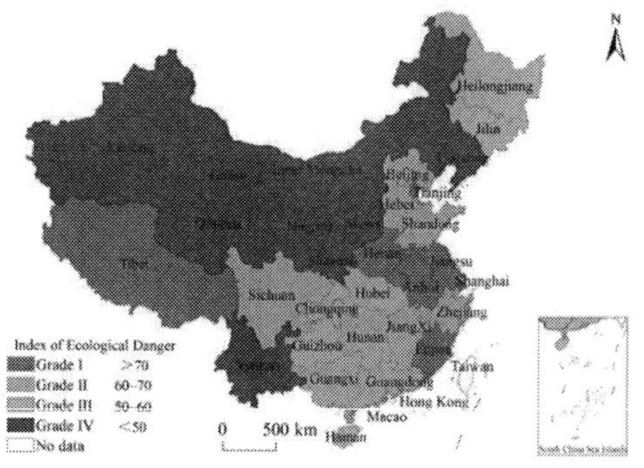

(*c*)

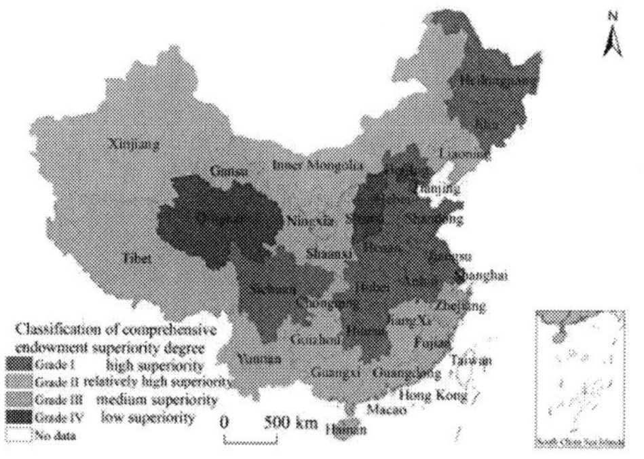

(*d*)

Figure 1. Classification map of cultivated land quantity, quality, ecological danger, and comprehensive endowment dominance in China. (**a**) Classification of index of farmland quantity; (**b**) Classification of index of farmland quality. (**c**) Classification of index of farmland ecological danger; (**d**) Classification of comprehensive endowment superiority degree of farmland resources.

In terms of the quantity of cultivated land, I_{FQ1}, which is determined by the total area of cultivated land and the *per capita* cultivated land area, have obvious differences among provinces. The I_{FQ1} values of Heilongjiang Province, Inner Mongolia Autonomous Region, and Jilin Province are 85.50, 65.93, and 51.37 respectively. Among them, the I_{FQ1} value of Heilongjiang Province is 5.9 times higher than that of Shanghai (I_{FQ1} value = 14.50) which is the lowest in China. These are followed by Xinjiang Uygur Autonomous Region, Gansu Province, Henan Province, Yunnan Province, Shandong, Hebei Province, and Sichuan Province. The I_{FQ1} values of these seven provinces are about 35 to 45. The I_{FQ1} values of Anhui Province, Guizhou Province, Liaoning Province, Ningxia Hui Autonomous Region, Hubei Province, Shanxi Province, Shaanxi Province, and Guangxi Zhuang Autonomous Region are about 30 to 35. The I_{FQ1} values of Jiangsu Province, Hunan Province, Tibet Autonomous Region, Jiangxi Province, Chongqing, Qinghai Province, Hainan Province, Guangdong Province, and Zhejiang Province are relatively low, at about 20 to 30. Fujian Province, Tianjin, Beijing, and Shanghai have the lowest endowment of cultivated land, the I_{FQ1} values of the four provinces are all below 20. According to this, the I_{FQ1} of each province can be divided into four levels (Fig. 1*a*). Among them, provinces with an I_{FQ1} value above 30 (*i.e.*

belonging to Grade I or II) mean that their cultivated land quantity has obvious superiority.

2. Quality of cultivated land

In terms of the quality of cultivated land, the I_{FQ2} determined by the quality grade of cultivated land and the annual grain yield per unit area of cultivated land in each province differ significantly. The I_{FQ2} values of Shanghai, Jiangsu Province, and Hunan Province are 82.73, 72.97, and 71.43 respectively. Among them, the I_{FQ2} value of Shanghai is 4.5 times higher than that of Inner Mongolia Autonomous Region (I_{FQ2} value = 18.22) which is the lowest in China. In Guangdong Province, Hubei Province, Jiangxi Province, Henan Province, Shandong Province, Fujian Province, Anhui Province, and Zhejiang Province, the I_{FQ2} values of these eight provinces are about 50 to 70 followed by Sichuan Province, Guangxi Zhuang Autonomous Region, Chongqing, Beijing, Hainan Province, and Hebei Province. The I_{FQ2} values of the six provinces are about 40 to 50. The I_{FQ2} values of Tianjin, Jilin Province, Xinjiang Uygur Autonomous Region, Yunnan Province, Liaoning Province, and Ningxia Hui Autonomous Region are relatively low, at about 30 to 40. The lowest I_{FQ2} values are in Guizhou Province, Shaanxi Province, Heilongjiang Province, Shanxi Province, Tibet Autonomous Region, Qinghai Province, Gansu Province, and Inner Mongolia Autonomous Region. The I_{FQ2} values of the eight provinces are all below 30, and the I_{FQ2} values of Gansu Province and Inner Mongolia Autonomous Region are both below 20. According to this, the I_{FQ2} of each province is divided into four levels (Fig. 1*b*). Provinces with I_{FQ2} values of above 40 (*i.e.* belonging to Grades I or II) mean that their cultivated land quality conditions have obvious superiority.

3. Ecological danger of cultivated land

In terms of cultivated land ecological danger situation, the differences of index of farmland ecological danger (I_{ED}) in different provinces determined by soil loss status (soil loss area ratio and soil loss intensity) and cultivated land natural disaster status (area proportion of disaster-affected crops and grain production reduced by disaster) exhibit relatively significant differences. The lowest ecological danger to cultivated land (meaning the highest value of I_{ED}) is in Shanghai and Tianjin, where the I_{ED} values are 84.28 and 81.20 respectively. Among them, I_{ED} value of Shanghai is 5.4 times higher than that of Inner Mongolia Autonomous Region (I_{ED} value = 15.72) which is the lowest in China followed by Jiangsu Province, Tibet Autonomous Region, Henan Province, Fujian Province, and Anhui Province, the I_{ED} value of these five provinces is about 70 to 80. The third tier of I_{ED} values occur in Beijing, Guangdong Province, Shandong, Jiangxi Province, Zhejiang Province, Guangxi Zhuang Autonomous Region, Hebei Province, Sichuan Province, and Hunan Province. The I_{ED} values of these nine provinces are about 60 to 70. The I_{ED} values of Hainan Province, Heilongjiang Province, Jilin Province, Chongqing, Hubei Province, and Guizhou Province are about 50 to 60. The relatively low values of I_{ED} occur in Ningxia Hui Autonomous Region, Liaoning Province, Yunnan Province, Shaanxi Province, Qinghai Province, Shanxi Province, Xinjiang Uygur Autonomous Region, Gansu Province, and Inner Mongolia. The I_{ED} values of the nine provinces are all below 50. Among them, the I_{ED} values of Gansu Province and Inner Mongolia Autonomous Region are all below 23. It can be seen that indices of ecological danger in most provinces are relatively high. Here, the index of ecological danger in each province is divided into four levels (Figure 1*c*). Among them, provinces with an I_{FQ1} value of above 50 (*i.e.*, belonging to Grades I, II, or III) mean that the ecological danger of cultivated land is relatively slight and the ecological environments remain in good condition.

3.3 Classification of Comprehensive Endowment Superiority of Cultivated land Resources

Table 2 shows that the comprehensive index of farmland resources (CI_{FR}) determined by I_{FQ1}, I_{FQ2} and I_{ED} differs significantly. Heilongjiang Province has the best comprehensive endowment condition of cultivated land resources, and its CI_{FR} value is 60.73, which is 2.4 times that of Qinghai Province (CI_{FR} value is 25.27) which is the smallest in China.

The further to analyse the difference of comprehensive degree of endowment of farmland resources in 31 provinces in China, the natural breakpoint method provided by ArcGIS software is used here based on the natural classification of CI_{FR} with appropriate adjustments. Finally, the comprehensive endowment superiority of cultivated land resources of 31 province-level administrative regions in China is divided into four grades: high superiority, relatively high superiority, intermediate superiority, and low superiority.

The results (Fig. 1*d*) show that the CI_{FR} values of Heilongjiang Province, Henan Province, Jilin Province, Shandong Province, Anhui Province, Jiangsu Province, Sichuan Province, Hubei Province, Hebei Province, and Hunan Province are all above 40: these are the 10 provinces with the highest comprehensive degree of endowment superiority of cultivated land resources in China, belonging to the first level — "high superiority". The CI_{FR} values of Xinjiang Uygur Autonomous Region, Guangxi Zhuang Autonomous Region, Yunnan Province, Jiangxi Province, Inner Mongolia Autonomous Region, Liaoning Province, Guizhou Province, Ningxia Hui Autonomous Region, and Guangdong Province are about 35 to 40. These are the nine provinces with relatively high comprehensive endowment superiority of cultivated land resources in China, belonging to the second level — "relatively high superiority". The CI_{FR} values of Chongqing, Shaanxi Province, Fujian Province, Tibet Autonomous Region, Gansu Province, Zhejiang Province, and Hainan Province are between 31 and 34. From the national level, the comprehensive endowment statuses of cultivated land resources in these seven provinces are at an intermediate level, belonging to the third level — "intermediate superiority". The CI_{FR} values of Shanghai, Shanxi Province, Tianjin, Beijing, and Qinghai Province are relatively low, ranging from 25 to 30. From the national level, the comprehensive endowment of cultivated land resources in these five provinces is at a low level, belonging to the fourth level — "low superiority".

https://doi.org/10.1051/e3sconf/202129303001

It is worth noting that the 10 provinces with relatively high comprehensive endowment superiority and nine provinces with relatively high dominance degree of cultivated land resources are usually those provinces with relatively large total grain output. According to *China Statistical Yearbook*, among the 13 provinces with an average annual grain output of more than 20 million tons from 2009 to 2019, there are 10 provinces (*i.e.*, Heilongjiang Province, Henan Province, Shandong Province, Anhui Province, Jilin Province, Jiangsu Province, Hebei Province, Sichuan Province, Hunan Province, and Hubei Province) classified as "high superiority" provinces with a high comprehensive endowment superiority of cultivated land resources in China and there are three provinces (*i.e.*, Inner Mongolia Autonomous Region, Jiangxi Province, and Liaoning Province) classified as "relatively high superiority" provinces with relatively high comprehensive endowment of cultivated land resources.

3.4 Classification of the Types of Superiority of Cultivated Land Resources in Different Provinces

According to the classification of I_{FQ1}, I_{FQ2}, and I_{ED}, 31 provinces in China can be classified into six categories (Fig. 2): ① superiority in quantitative terms; ② superiority in quantity and ecology terms; ③superiority in quality and ecology terms; ④superiority in ecology terms; ⑤comprehensive superiority; ⑥unclear.

Among the 19 "high superiority" and "relatively high superiority" provinces, the superiority categories of each province differ and can be roughly divided into four categories. The first is superiority in quantitative terms, including Xinjiang Uygur Autonomous Region, Inner Mongolia Autonomous Region, Yunnan Province, Liaoning Province, and Ningxia Hui Autonomous Region. The second is superiority in terms of quantity and ecology, including Heilongjiang Province, Jilin Province, and Guizhou Province. The third is superiority in terms of quality and ecology, including Jiangsu Province, Hunan Province, Jiangxi Province, and Guangdong Province. The fourth is comprehensive superiority, including Henan Province, Shandong Province, Anhui Province, Hebei Province, Sichuan Province, Hubei Province, and Guangxi Zhuang Autonomous Region.

Even among the 14 provinces with "medium superiority" and "low superiority" whose comprehensive endowment superiority of cultivated land resources is not prominent, they will have obvious superiority in terms of quantity, quality, or ecology (although their classification is below that of "high superiority" and "relatively high superiority" provinces), which can be divided into four categories. The first is superiority in quantitative terms, including Shaanxi Province, Gansu Province, and Shanxi Province. The second is superiority in terms of quality and ecology, including Chongqing, Fujian Province, Zhejiang Province, Shanghai, Beijing, and Hainan Province. The third is superiority in terms of ecology, including Tianjin and Tibet Autonomous Region. The fourth is unclear, such as Qinghai Province.

The majority of provinces have a "single" superiority, a "double" superiority, or comprehensive superiority, which provides a direction and path for the provinces to develop, utilise, and protect their limited cultivated land resources in a rational manner.

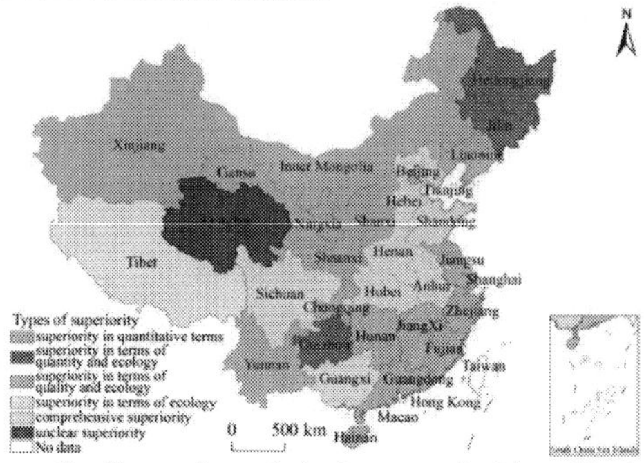

Figure 2. Types of superiority in terms of cultivated land resources in each province of China.

4 Conclusions and Discussions

4.1 Main Conclusions

The evaluation of cultivated land resources at the national level is an important basis for the country to formulate the plan of cultivated land utilisation and renovation, the policy of cultivated land protection and decision-making around land space management. It is necessary to establish a comprehensive evaluation indicator system and quantitative calculation method of cultivated land resources, which integrates quantitative evaluation, quality evaluation, and ecological danger assessment. In this paper, based on TOPSIS-Grey Relational Analysis Method, the province-level administrative region is taken as the analysis and evaluation unit, and the three dimensions of cultivated land: quantity, quality, and ecological danger are combined to construct a practical and feasible comprehensive evaluation indicator system and quantitative calculation method of cultivated land resources in China, to measure and compare the superiority, inferiority, and comprehensive endowment of cultivated land resources in various provinces.

The results show that Heilongjiang Province, Henan Province, Jilin Province, Shandong Province, Anhui Province, Jiangsu Province, Sichuan Province, Hubei Province, Hebei Province, and Hunan Province are ten "high superiority" provinces with the high comprehensive endowment of cultivated land resources in China. Xinjiang Uygur Autonomous Region, Guangxi Zhuang Autonomous Region, Yunnan Province, Jiangxi Province, Inner Mongolia Autonomous Region, Liaoning Province, Guizhou Province, Ningxia Hui Autonomous Region, and Guangdong Province are nine "relatively high superiority" provinces with relatively high comprehensive endowment of cultivated land resources in China. These provinces are also those with more total grain output in China. There are 13 provinces with an average total grain output of more than 20 million tons from 2009 to 2019 belonging to the scope of the above-mentioned "high superiority" provinces or "relatively high superiority" provinces.

From the "single" superiority degree of cultivated land quantity, quality, and ecological danger, Heilongjiang Province, Inner Mongolia Autonomous Region, and Jilin Province are the more prominent in the field of quality, whose I_{FQ1} values exceed 50. Next are 15 provinces: Xinjiang Uygur Autonomous Region, Gansu Province, Henan Province, Yunnan Province, Shandong Province, Hebei Province, Sichuan Province, Anhui Province, Guizhou Province, Liaoning Province, Ningxia Hui Autonomous Region, Hubei Province, Shanxi Province, Shaanxi Province, and Guangxi Zhuang Autonomous Region with relatively high I_{FQ1} values of between 30 and 50. The cultivated land qualities of Shanghai, Jiangsu Province, Hunan Province, Guangdong Province, Hubei Province, Jiangxi Province, Henan Province, Shandong Province, Fujian Province, Anhui Province, and Zhejiang Province in central and eastern regions are better. The I_{FQ2} values of these 11 provinces exceed 50. These are followed by Sichuan Province, Guangxi Zhuang Autonomous Region, Chongqing, Beijing, Hainan Province, and Hebei Province, and the I_{FQ2} values of these six provinces are between 40 and 50. In terms of cultivated land ecological danger, Shanghai, Tianjin, Jiangsu Province, Tibet Autonomous Region, Henan Province, Fujian Province, and Anhui Province have greater superiority, and the I_{ED} values of these seven provinces exceed 70. These are followed by Beijing, Guangdong Province, Shandong Province, Jiangxi Province, Zhejiang Province, Guangxi Province, Hebei Province, Sichuan Province, Hunan Province, Hainan Province, Heilongjiang Province, Jilin Province, Chongqing, Hubei Province, and Guizhou Province. The I_{ED} values of these provinces are between 50 and 60.

The 31 provinces are classified into six categories: ①superiority in quantitative terms; ② superiority in terms of quantity and ecology; ③superiority in terms of quality and ecology; ④superiority in terms of ecology; ⑤comprehensive superiority; ⑥unclear superiority. The cultivated land resources in most provinces have a "single" superiority, "double" superiority, or comprehensive superiority.

4.2 Discussions

In the evaluation of cultivated land resources at the national level, it is not enough to evaluate the single index of quantity, quality, or ecology of cultivated land, which is not conducive to objectively and scientifically measuring and comparing the differences in superiority, inferiority, and comprehensive endowment of cultivated land resources in different provinces. In this paper, a quantitative index system and its calculation method for the comprehensive evaluation of cultivated land resources are established to measure systematically the comprehensive condition of cultivated land quantity, quality, and ecological danger in each province, which promotes the development of the traditional land resources evaluation system, and helps to objectively reveal the endowment status and limiting factors of cultivated land resources in each province, and provides the basis for the development, utilisation, and protection strategy decision, comprehensive renovation planning, and management of cultivated land resources according to local conditions. However, this does not mean that the quantitative index system and its calculation method of comprehensive evaluation of cultivated land resources constructed in this paper are only applicable to the evaluation of province-level administrative regions at the national level. This index system and its calculation method are also applicable to the analysis and research across different regional scales, such as city level, county level, and township level, and can even be used as a reference in the analysis and evaluation of the comprehensive status of cultivated land resources between countries.

In view of the complexity of this comprehensive evaluation work and the difficulty of obtaining some indicator data, the evaluation needs to be further improved and perfected. For example, cultivated land soil pollution is an important indicator of ecological danger, however, due to the lack of provincial survey data at the national level, only a few provinces such as Sichuan Province have published provincial survey data at the provincial level, so we could not measure the index. Although the author has attempted to refer to the map of the distribution of heavy-metal pollution in surface soil samples in China drawn by Duan Qiannan *et al.* [33] and relevant reports pertaining to various provinces, and referring to the statistical data of major pollutant emissions, fertiliser application, population density, and other statistical data in the *China Statistical Yearbook*, and comprehensively judged and roughly calculated the cultivated land soil pollution status index of each province, the results are not perfect. We supplemented this research after the national-level issue of survey data relating to cultivated land soil pollution in each province.

Due to the significant differences in natural and social economic conditions such as landscape, topography, climate, soil, hydrology, population, economy, and other natural and socio-economic conditions of 31 provinces, the quantity, quality, and ecological environment of cultivated land resources in each province also differ to a significant extent, therefore, the spatial distributions of provinces with quantity superiority, quality superiority, and ecological environment superiority are not identical. For example, Heilongjiang is the province with the most prominent quantitative superiority of cultivated land resources in China. Its index of farmland quantity reaches 85.50, ranking first among first level (high superiority) provinces, but its index of farmland quality is only 27.45, classifying it as a fourth level province. Another example is that the index of farmland quantity of Jiangxi Province is only 25.71, which belongs to the third level, but its index of farmland quality and ecological danger reach 67.30 and 63.78 respectively, which belong to the first and second levels, respectively. In other words, it has superiority in cultivated land quality and ecological environment, therefore, the comprehensive index of farmland resources in Jiangxi Province reaches 38.14, and the comprehensive endowment superiority is classified at the second level, as it has become one of the major grain-producing provinces in China.

Based on this, when evaluating the comprehensive endowment of cultivated land resources in each province, we should have such a basic understanding such that: a province has abundant quantity, high quality, and light (or unapparent) ecological danger, which indicates comprehensive superiority of cultivated land resources. However, it cannot be denied that even if the quality of cultivated land is low, even there is a certain ecological crisis, if there is a relatively rich amount of cultivated land, it will also exhibit superiority of cultivated land resources, and remains an important pillar of national food security. Therefore, those provinces with the worst comprehensive endowment and inferiority of cultivated land resources mainly refer to the provinces whose cultivated land quantity is inferior, and the cultivated land quality and ecological conditions are not dominant. According to this understanding, most provinces in China have a certain "single" superiority, a "double" superiority, or comprehensive superiority, which provides an important direction and path for those provinces to develop, utilise, and protect their limited cultivated land resources in a rational manner.

References

1 Yang Zisheng, Liu Yansui, Zhao Qiaogui, et al. Principles and Methods Regional Food Security Assessment Based on cultivated land Utilization and Its Practice in Yunnan Province [M] . Beijing: China Science and Technology Press, 2017: 1-180.

2 Klingebiel A. A., Montgomery P. H. Land Capability Classification [N]. Washington D.C.: US Department of Agriculture Handbook,1961, (210): 1-24.

3 FAO. A framework for land evaluation [N]. Rome: FAO soil bulletin, 1976, (32): 1-25.

4 FAO. Report on the Agr o -ecological Zones Project [N]. Rome: World Soil Resources Reports, 1982, (48):1-4.

5 Dumanski J, Onofrei C. Techniques of crop yield assessment for agricultural land evaluation [J]. Soil Use and Management, 1989, 5(1): 9-15.

6 Bi Yuyun, Zheng Zhenyuan. The actual changes of cultivated area since the founding of new China [J] . Resource Science, 2000

7 FENG Zhiming, LIU Baoqin, YANG Yanzhao. A study of the changing trend of Chinese cultivated land amount and data reconstructing: 1949-2003. Journal of Natural Resources, 2005, 20(1): 35-43.

8 YANG Zisheng. On the necessity of keeping the red line of cultivated land area about 120 million hectares in china and its countermeasures system [C]//LIU Yansui. Research on Sustainable Use of Land Resources and New Countryside Construction in China. Chongqing: Southwest China Normal University press, 2008: 20-26.

9 Xue CAO, Xiaobin JIN, Jinshuo WANG, Lijuan MIAO, Yinkang ZHOU. Reconstruction and change analysis of cropland data of China in recent 300 years[J]. Acta Geographica Sinica, 2014, 69(7): 896-906.

10 TAN Yongzhong, HE Ju, YUE Wenze, et al. Spatial pattern change of the cultivated land before and after the Second National Land Survey in China. Journal of Natural Resources, 2017, 32(2): 186-197.

11 GAO Xiaoyu, CHENG Weiming, WANG Nan,et al. Spatio-temporal distribution and transformation of cropland in geomorphologic regions of China during 1990-2015 [J]. Journal of Geographical Sciences, 2019, 29(2): 180-196

12 Zhaohao Lai, Meiqiu Chen,Taoju Liu. Changes in and prospects for cultivated land use since the reform and opening up in China [J]. Land Use Policy, 2020, 97: 104781

13 Yuzhe Wu, Liping Shan, Zhen Guo,Yi Peng. cultivated land protection policies in China facing 2030: Dynamic balance system versus basic cultivated land zoning [J]. Habitat International, 2017, 69: 126-138

14 Xiangzheng Deng, Jikun Huang, Scott Rozelle, Jipeng Zhang, Zhihui Li. Impact of urbanization on cultivated land changes in China [J]. Land Use Policy, 2015, 45: 1-7

15 Yuheng Li,Yurui Li,Hans Westlund,Yansui Liu. Urban–rural transformation in relation to cultivated land conversion in China: Implications for optimizing land use and balanced regional development [J]. Land Use Policy, 2015, 47: 218-224

16 CUI XuFeng, MA YunMeng, ZHANG GuangHong. The factors of cultivated land conversion and its temporal and spatial characteristics: an integrated model [J]. Scientia Agricultura Sinica, 2018, 51(22): 4316-4327.

17 LI Yuan. Land Resources of China [M] . Beijing: China Land Press, 2000. 1-235.

18 Ministry of Land and Resources of the People's Republic of China, National Bureau of Statistics of the People's Republic of China, Office of the Second National Land Survey Leading Group of the State Council. Communique on the key data results of the Second National Land Survey [EB/OL]. (2013-12-30) http://www.china.com.cn/zhibo/zhuanti/ch-xinwen/2013-12/30/content_31040885.htm.

19 WANG Hao. The Third Land Survey to be completed by 2020[N]. People's Daily, 2019-08-17(2).

20 SHI Yulin. Land Resources Data Set of China's 1:1 Million Land Resources Map. Beijing: China Renmin University Press, 1991: 1-45.

21 Ministry of Land and Resources of the People's Republic of China. Press conference on the results of the 2016 national cultivated land quality grades updating evaluation [EB/OL]. (2017-12-26) http://www.scio.gov.cn/xwfbh/gbwxwfbh/xwfbh/gtzyb/Document/1614366/1614366.htm.

22 The Ministry of Land and Resources of the People's Republic of China. Announcement on the release of

the main data results of the 2016 national cultivated land quality grades updating evaluation [EB/OL]. (2017-12-21) http://g.mnr.gov.cn/201712/t20171226_1711147.html.

23 Department of Agriculture of the People's Republic of China. Bulletin on the quality grades of cultivated land throughout the country [R] . Farmers Daily, 2014-12-18 (7).

24 The Ministry of Agriculture and Rural Affairs of the People' s Republic of China. Announcement of national arable land quality grading in 2019[R]. Gazette of the Ministry of Agriculture and Rural Affairs of the People' s Republic of China, 2020, (4): 113-120.

25 WANG Hongbo, CHENG Feng, ZHANG Zhongfan, et al. Differential characteristics of cultivated land grade and its effect on cultivated land protection in China [J]. Transactions of the CSAE, 2011, 27(11): 1—8.

26 ZOU Jinlang, YANG Zisheng, WU Qun. Structural feature of cultivated land use output in China. Journal of Natural Resources, 2015, 30(8): 1267-1277.

27 XU Xinliang, WANG Liang, CAI Hongyan, et al . The influences of spatiotemporal change of cultivated land on food crop production potential in China [J]. Food Security, 2017, 9(3): 485-495

28 WANG Shichao, WANG Jinzhou, ZHAO Ya-wen, et al. Assessment of the contribution percentage of inherent soil productivity of cultivated land in China [J]. Journal of Integrative Agriculture, 2019, 18(11): 2619-2627.

29 GAO Chan, ZHANG Bangbang, ZHAO Minjuan, et al. Grain Productivity potential of cultivated land and yield gap analysis in China [J] . Journal of China Agricultural University, 2020, 25(1): 10-18.

30 YANG Shiqin, YANG Zisheng. Problems and strategies for arable land eco-conservation in China [J]. Agricultural Science & Technology, 2017, 18(7): 1355-1357+1360.

31 Department of Environmental Protection, Ministry of Land and Resources of the People's Republic of China. Bulletin of the National Survey of Soil Pollution [R]. China Land and Resources Daily, 2014-04-18(2).

32 CHEN Nengchang, ZHENG Yuji, HE Xiaofeng, et al. Analysis of the Report on the national general survey of soil contamination [J]. Journal of Agro-Environment Science, 2017, 36(9): 1689-1692.

33 DUAN Qiannan, LEE Jianchao, LIU Yansong, et al. Distribution of heavy metal pollution in surface soil samples in China: a graphical review [J]. Bulletin of Environmental Contamination and Toxicology, 2016, 97(3): 303-309.

34 YANG Zisheng, HE Yimei. Soil erosion of cultivated land in mountainous areas at China's southwest Border: a Case Study in Yunnan Province [J]. Research of Soil and Water Conservation, 2009, 16(1): 1-7.

35 WANG Ping, SHI peijun. Comprehensive regionalization of agricultural natural disaster in China [J]. Journal of Natural Disasters, 2000, 9(4):16-23.

36 YANG Zisheng, LIU Yansui. Study on drought disaster regionalization in China [C]/YANG Zisheng. Research on Water Governance and Sustainable Development of China. Beijing: Social Sciences Academic Press (China), 2012. 169-180.

37 YANG Zisheng, LIU Yansui. Study on flood disaster regionalization in China [C]/YANG Zisheng. Research on Water Governance and Sustainable Development of China. Beijing: Social Sciences Academic Press (China), 2012. 192-201.

38 YANG Zisheng, HE Yimei. Study on grain yield reduction of China due to flood and drought disasters from 1950 to 2010 [C]/YANG Zisheng. Research on Water Governance and Sustainable Development of China. Beijing: Social Sciences Academic Press (China), 2012. 202-213.

39 ZHU Hongbo, ZHANG Anlu. Analysis on spatial-temporal difference in ecological security of cultivated land resources in China [J]. Resources and Environment in the Yangtza Basin, 2007, 16(6): 754-754.

40 Wang Qian, Jin Xiaobin, Zhou Yinkang. cultivated land ecological security and spatial aggregation pattern n in Hebei Province [J]. Transactions of the Chinese Society of Agricultural Engineering, 2011, 27(8): 338-344.

41 ZHANG Rui, LIU Youzhao. Evaluation on cultivated land ecological security based on the PSR model and diagnosis of its obstacle indicators in China [J]. Resources and Environment in the Yangtza Basin, 2013, 22(7): 945-951.

42 KUANG Lihua, YE Yingcong, ZHAO Xiaomin, et al. Evaluation and obstacle factor diagnosis of cultivated land system security in Yingtan City based on the improved TOPSIS Method. Journal of Natural Resources, 2018, 33(9): 1627-1641.

43 LU Hua, XIE Hualin, LV Tiangui, et al. Determinants of cultivated land recuperation in ecologically damaged areas in China [J]. Land Use Policy, 2019, 81: 160-166.

44 LI Hong, ZHI Shuonan. Dynamic Study on China's Energy Security under the New Normal: Based on Gray Correlation TOPSIS Model [J]. Ecological Economy, 2020, 36(8): 57-62+87.

45 XIA Shaowei, YANG JiaBen, YANG Zhenbin. Introduction to Systems Engineering [M]. Beijing: Tsinghua University Press, 1995. 30-210.

46 HU Yonghong. The Improved Method for TOPSIS in Comprehensive Evaluation [J]. Mathematics in Practice and Theory, 2002, 32(4): 572-575.

47 María Elena Arce, Ángeles Saavedry, José L.Míguez, Enrique Granada. (Arce M E, Saavedra Á, Míguez J L, et al.) The use of grey-based methods in multi-criteria decision analysis for the evaluation of sustainable energy systems: A review [J]. Renewable & Sustainable Energy Reviews, 2015, 47: 924-932.

48 HE Bowen, LI Ding, LIU Xiaojie. Regional differentiation characteristics and the influencing factors of spatial poverty in rural areas of Gansu Province [J]. Research of Agricultural Modernization, 2019, 40(5): 819-829.

49 ZHOU Lei, HUANG Qiuhao. Efficiency Evaluation on Urban Land Use Based on the Gray Correlation TOPSIS Method and Diagnosis of Its Obstacle Indicators [J]. Research of soil and water conservation, 2014, 21(4): 39-44.

50 ZHOU Honghao, CHEN Xiaohong. Spatio-temporal evolution of sustainable livelihood security and study of its obstacle indicators in northeast China[J]. Scientia Geographica Sinica, 2018, 38(11): 1864-1874.

51 ZHANG Fengrong, Wang Jing, CHEN Baiming, et al. Index System and Evaluation Method of Sustainable Land Use [M]. Beijing: China Agricultural Publishing House, 2003.18-56.

52 CHEN Baiming. Design and evaluation of indicator system of regional land for sustainable use [J]. Progress in Geography, 2002, 21(3): 204-215.

53 Ministry of Natural Resources of the People's Republic of China. Statistical Yearbook of Land and Resources in China [M]. Beijing: Geological Publishing House, 2019. 3-12.

54 Ministry of Water Resources of the People's Republic of China. Water Conservancy Industry Standard of the People's Republic of China: SL 190-2007, Standards for classification and gradation of soil erosion [S]. Beijing: China Water Conservancy and hydropower press, 2008. 8-12.

55 General Administration of Quality Supervision, Inspection and Quarantine of the People's Republic of China, Standardization Administration of the People's Republic of China. National Standard of the People's Republic of China: GB/T2438.1-2009, Natural Disaster Information Statistics-Part1: Basic Indicators [S]. Beijing: China Standard Press, 2009. 1-5.

E3S Web of Conferences **293**, 03002 (2021) https://doi.org/10.1051/e3sconf/202129303002
GCEECE 2021

The Sustainable Development of Smart Cities: A Case Study of the Design of driverless bus stops in Zhengzhou

Xiaoguang Chen[1], Jinting Sui[1,*], Xufeng Tai[1], and Heqi Li[1]

[1]Institute of Fourth, China Northeast Architectural Design & Research Institute Co., Ltd., Shenyang 110168, China

Abstract. With the arrival of the "5G" era and the digital economy era, the development of the city has permeated all aspects of the city, and with the emergence of "smart city", it is likely to have "smart" everywhere in the future city. At the same time, "Future Smart Transportation" was proposed as a discussion topic at the "2019 (8th) International Smart City Summit and Smart Ecology Expo". The research and development of the driverless bus stops will spur the driverless bus stops to take on the past with a brand-new look, welcome the future, and contribute to the sustainable development of the smart city. Taking the driverless bus station in Zhengdong New District as an example, this paper explores the sustainable development of a smart city through the creation of manned (open)-shared (transition space)-driverless (semi-enclosed space) space and the introduction of modular design approach. To connect people's work, life and transportation, and derive a new public transportation environment based on manned bus stops, namely, intelligence, sustainability, environmental friendliness and safety, thus create a smart transportation pilot to embrace the bright future of sustainable smart cities.

1 Background

The project is located in Zhengdong New District of Zhengzhou City (Figure 1), which adheres to the concept of green and sustainable development and strives to become an international regional financial center as well as an ecologically livable city up to international standard. Longhu Central Road is the most important urban trunk road in longhu district, where relying on the road shape and important traffic functions throughout the region and in combination with the fast-developing unmanned technology, it will be built as an unmanned experimental demonstration area. On this basis, relevant ancillary facilities along the road will be improved, and efforts will be made to build a comprehensive intelligent road.

Figure 1. Aerial view of Zhengdong New District.

The project currently includes 35 general bus stops (Figure 2), with completed platforms but no bus shelters yet and some of the manned buses are in operation. The government decided to build an unmanned bus stop base on this situation. Questions are: First, how to make use of the existing platform without dismantling and damaging it using sustainable development design method. Secondly, how to guarantee the normal operation of the manned bus stops before the driverless bus stops are completed and put into use, and the solve problem of functional transformation and transition between the two in the future.

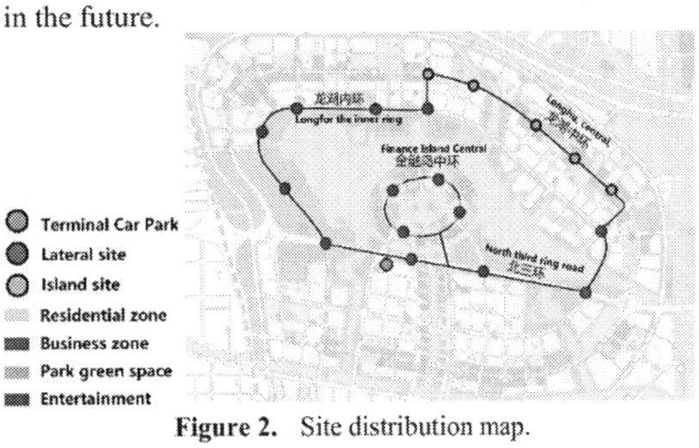

Figure 2. Site distribution map.

Based on the analysis of the site status, there are three types according to different road conditions (Figure 3).

Figure 3. Island platform (left) and side platform (right).

* Corresponding author: 1402874257@qq.com

2 Architectural design

Considering the present situation of bus stops at the beginning of the design (Figure 4), three problems are summarized: First, the functional connotation of urban bus stops is simple, which can only provide functions such as stops, seats, maps and so on. Second, the space is simple with structures that consists of columns and ceilings, which are completely open and disordered thus cannot form an effective crowd organization. Third, one-time investment leads to a short service life and no reserve conditions for upgrading. To sum up, the existing bus stations can no longer meet the future multi-functional needs, and there is no possibility of upgrading. The existing spatial structure conditions cannot effectively dock self-driving buses.

Figure 4. Present situation map of bus station.

2.1 Sustainable development strategy

By studying the unmanned bus and its carrying stations, the unmanned public transport platform is considered as the carrier to meet the current and future public transport system. In the practical stage, it should have the ability to adapt to the changes of the present and the future. Through one-time investment in the construction, it should meet the functional requirements of the manned bus station at the present stage, and meet the requirements of the unmanned bus station through upgrading and transformation in the future. In this way, it is possible to avoid dismantling the bus station on duty, but upgrade it in the future to cope with the era of popularization of unmanned driving, thus saving a lot of social resources, investment and time cost. This is the design of sustainable development.

2.2 Platform space strategy

The project itself takes into account the coexistence of the present manned and the future unmanned. We put forward the concept of manned (open) - shared (transitional space) - unmanned (semi-closed space), which makes the current manned bus platform a transition from the traditional platform to unmanned platform (Figure 5). At the same time, a shared space is added between manned and unmanned driving, which is used as a transitional space to effectively connect these three parts. And through later modularity, it can be combined freely according to functions and requirements [1-2].

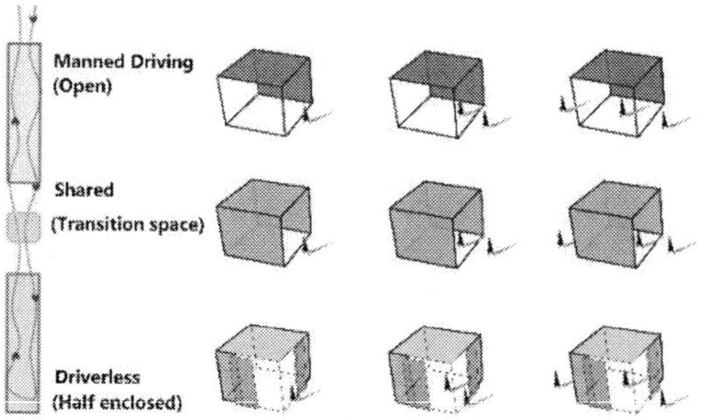

Figure 5. Schematic diagram of platform space.

3 Modularization

3.1 Constructional modularization

As a sustainable design and production method, the modularization strategy can achieve rapid construction, in line with the development of smart cities. While the idea of integrating multiple functions and demands by means of splitting and combining in the design embodies the concept of sustainability.

Inspired by the space station (Figure 6), the design takes a "space capsule" as the basic module unit for unit connection, which can realize the growth from small to large and from simple to complex, thus realizing the growth transition and transformation from manned station (with simple requirements) to driverless station (with diverse requirements).

Figure 6. Image of space station.

In the design of the "space capsule" (Figure 7), the module is divided into five plate-type modules and two structural modules by adopting the modular design concept (Figure 8), wherein the five plate-type modules are maintenance modules. The basic combination is an external decorative panel plus a plate supporting structure system, in which the former can be arbitrarily switched among solar panels, electronic screens, perforated aluminum plates and toughened glass, so as to realize different functional requirements of the station.

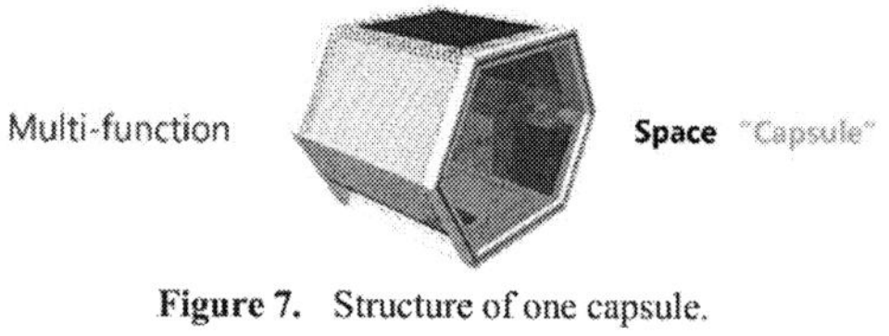

Figure 7. Structure of one capsule.

 https://doi.org/10.1051/e3sconf/202129303002

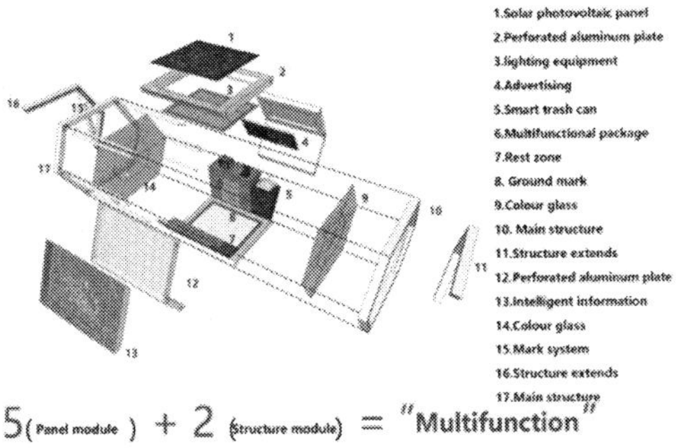

Figure 8. Explosive view of one capsule.

The structural modules on both sides are hexagonal or C-shaped structural support modules, which are welded and formed by fixed-length box steel beams in the factory and transported to the site for assembly and connection with the plate modules (Figure 9). The production mode of modular construction introduces factory prefabrication production into the construction process. Through factory prefabrication, the site construction time and difficulty are greatly reduced, while the construction efficiency is improved. Meanwhile, the modular construction strategy can effectively improve the building quality. In case of partial damage, the module can be replaced separately, thus prolonging the service life of the building, which is of sustainable development significance [3].

Figure 9. Real scene of stations[1].

3.2 Functional modularization

Building modularity is connected with function modularity. Function is the basic attribute of the building itself. Therefore, in the modular building, functional zoning is mostly used as the main basis of modular divisions. The sustainable meaning of using function-based modular division lies in that a specific module can correspond to a specific function. After summarizing and analyzing the functions of future unmanned bus stations, eight basic functional modules are designed (Figure 10), including multifunctional modules, intelligent modules and barrier-free modules[4].

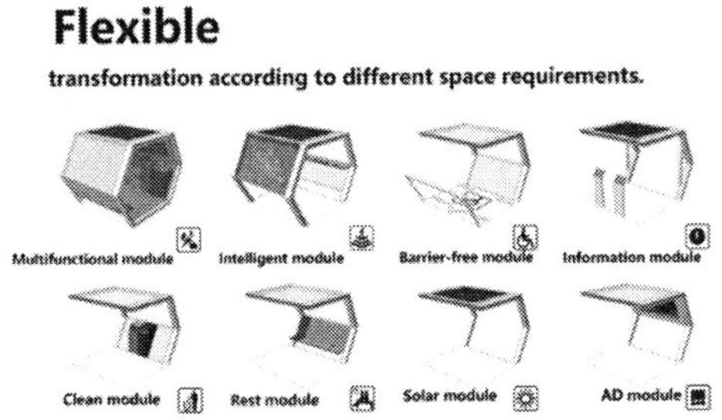

Figure 10. Diagram of one unit module.

On the basis of basic modules, different module is spliced and combined, and then auxiliary equipment is integrated to realize different combinations of multifunctional modules, including advertisement module, cleaning module, barrier-free module, electronic information module, intelligent module, etc. These modules can be flexibly combined according to different fields and requirements, and can be quickly produced and installed (Figure 11). Figures 1, 2 and 3 are relatively simple bus stops, including basic bus modules such as rest module, cleaning module and barrier-free module, which are suitable for existing platforms. 4, 5 and 6 have added solar modules, safety modules and multi-functional modules, which are suitable for the needs of smart city. 7 and 8 are the most complex ones, adding intelligent modules and shell modules to meet the needs of unmanned driving. On this basis, we can realize the transition from simple function to complex function, and upgrade from manned stations to unmanned stations, so as to meet the current needs and foreseeing the future.

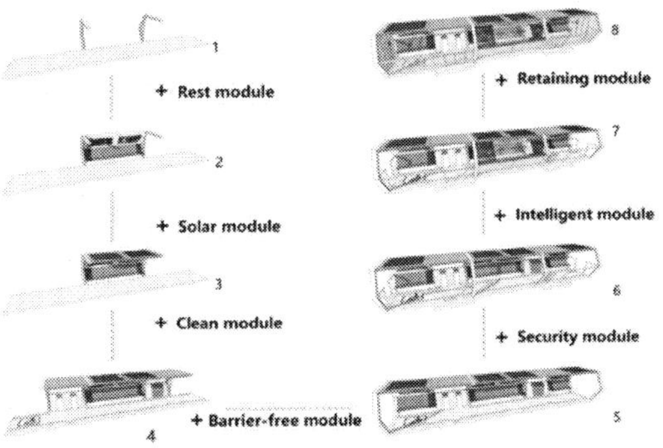

Figure 11. Growth of a bus stop.

4 Intelligent Network: Connection of Intelligent Technology

The new unmanned bus station is guided by the unmanned bus station and passengers' demand by introducing intelligent technology. Through the research of related new technologies, it connects people, people and cities, and matches the most advanced and suitable technology with the unmanned bus stations, providing passengers with a brand-new bus environment (Figure 12), so that every citizen can feel the coming of smart city in advance through infrastructure [5].

Figure 12. Real scene of stations[1].

[1] Source: https://www.vcg.com/

- Urban network: micro 5G base station, providing urban WIFI hot spots for citizens to connect (Figure 13).
- Photovoltaic applications: PV panels mounted on the top of the kiosks can provide green energy for bus stops and provide LED inductive lighting systems at night [6].

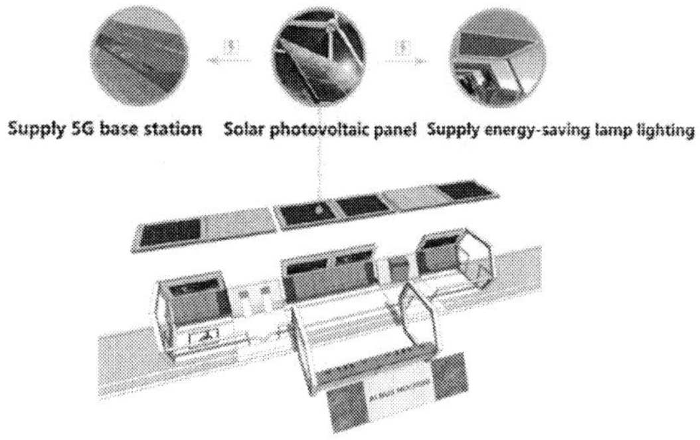

Figure 13. Analysis diagram of intelligence.

- Environmental monitoring: real-time detection of urban road environment, air quality, urban temperature, etc., and aggregation of real-time urban data.
- Shared platform: With the help of a third-party platform, it provides functions such as charging mobile phones, debiting from different platforms, and vending machines.
- Visualization: LCD high-definition screen, playing city cultural propaganda video, displaying city image, advertising and other video information.
- Voice station-reporting function: timely broadcasting the bus arrival and departure time and updating the time for passengers in real time (Figure 14).

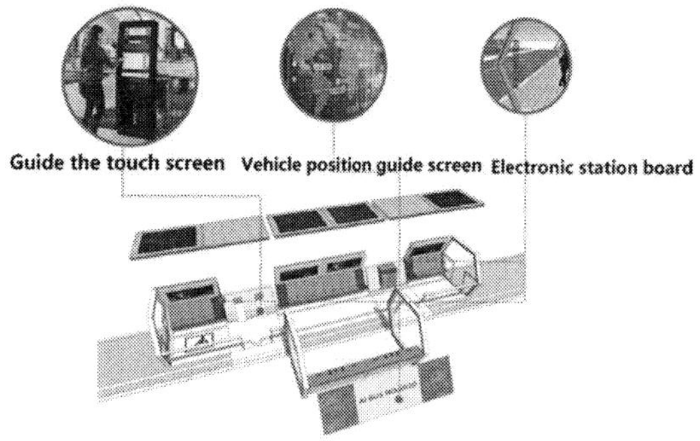

Figure 14. Analysis diagram of intelligence.

- Humanization design: Zhengzhou, as a northern city, it is difficult to sit for a long time in winter, so the waiting seats will be equipped with a seat heating function.
- Smart city: LCD touch screen, that can show the real-time operation status of public transportation and information about stops, routes and transfers, as well as searching for ticketing information of surrounding scenic spots and life service information such as weather forecast and map navigation.

5 Summary

This paper expounds the design strategy of future "intelligent transportation" based on the sustainable development of the intelligent city through the thinking of the design of Zhengzhou driverless bus stops.

(1) by creating manned (open) - shared (transitional space) - unmanned (semi-enclosed space) space, adding transitional space and unmanned mode to the existing platform, it can adapt to the changes now and in the future, that is, meet the functional requirements of manned bus stations at the present and unmanned bus stations in the future. To avoid dismantling manned bus stations in an all-round way but upgrade and optimize, it can save a lot of social resources, investment and time cost, and echo the urban sustainable development strategy in intelligent transportation.

(2) Modular design is an important application method of the sustainable concept in the field of architectural design. Unify the image of the city through modularization, and increase bus stop services through diversification, so as to enhance the city. The application of modular design strategy in the field of architectural design can not only save architectural resources effectively, but also maintain the sustainable vitality of architectural space. However, modular construction can continue to develop and flourish only if it meets the actual functional requirements, conforms to the corresponding social and cultural forms and adapts to the means of material construction.

References

1. General Rules for Design of Civil Buildings, GB50352-2019, (2019)
2. Code for Fire Protection Design of Buildings, GB50016-2014, (2018)
3. . Abe, Introducing autonomous buses and taxis: Quantifying the potential benefits in Japanese transportation systems. Transportation research part A-policy and practice, 126， 94-113 (2019)
4. A. Chehri, H.T. Mouftah, Autonomous vehicles in the sustainable cities, the beginning of a green adventure. Transportation Research Record, 37-46 (2016)
5. X.X. Dong, M. Di Scenna, E. Guerra, Transit user perceptions of driverless buses. Transportation, 46, 35-50 (2019)
6. L.M. Eugenia, A. Alonso, The Driverless Bus: An Analysis of Public Perceptions and Acceptability. Sustainability, **11**, 4986 (2019)

Research on the application of GIS technology in the spatial status grooming of village areas

Song hao[1]

[1]Jianghan University, Wuhan, China

Abstract. The preparation of five levels and three categories of territorial spatial planning includes many aspects, and this article mainly deals with one of the levels and categories: detailed territorial spatial planning of village areas; it focuses on the preliminary work of planning preparation: the current situation combing of village territorial spatial areas. The article firstly outlines the basic application of GIS technology in the current situation combing; then introduces the relevant contents of the Third National Land Survey.Finally, with actual cases, we focus on the significant role played by GIS technology in the integration of land use data in the base period of village territorial space, the location of permanent basic farmland indicators, the location of ecological protection red line, and the statistics of conflicting patches of ecological environmental protection and basic farmland protection.

1 Introduction

The practice of five levels and three categories of territorial spatial planning is now in full swing in China, which contains a lot of work, so I will not repeat it here. Geddes, a Western humanist planning master, creatively proposed the general method of regional planning: "investigation-analysis-planning", which was later regarded as the general procedure of planning;The author believes that this point coincides with the words of a great Chinese man, Mao Zedong: "Without investigation, there is no right to speak". The author is convinced that without investigation, one cannot even speak, let alone carry out subsequent planning work. Any level of planning work can not engage in one-size-fits-all, not just to meet the laws and regulations and technical specifications can be; also cannot be because of time constraints, heavy tasks will only focus on the final planning results, while ignoring the decision to plan the results can be achieved "tailored", "into the countryside with the customs The key part of "tailor-made" and "customizable" planning results: the overall grasp of the current situation of the project and resource conditions. This paper mainly focuses on the use of GIS technology in the completion of the current situation of land use in the village, the location of permanent basic farmland indicators, the ecological protection red line, the statistics of conflicting patches, etc. In this paper, we would like to make some limited discussion on the use of GIS technology in the village area.

2 GIS and current status combing work

2.1 The concept of GIS

GIS is an evolving concept, and the father of GIS, Roger Tomlinson (1966), first proposed GIS as a digital system for analysing and manipulating geographic data in a comprehensive manner. At present, experts and scholars worldwide prefer the definition of GIS by the Federal Institute for the Coordination of Digital Maps (FICCDC): "a system consisting of computer hardware, software, and different methods designed to support the acquisition, management, processing, analysis, modelling, and display of spatial data in order to solve complex planning and management problems. " .

2.2 Specific uses of GIS in status quo research and analysis

(1) Status quo research stage: GIS can be used to manage the status quo data (e.g., land use status quo data, road data, municipal facilities data, etc.). Use handheld GIS equipment to assist in site investigation. Handheld devices integrating GPS, RS and GIS can tell the planner the location and surrounding geographic environment, as well as relevant geographic data, so that the planner can grasp the site situation faster and more accurately.

(2) Current situation analysis stage: GIS overlay analysis function can be used to evaluate the suitability of the site; make various types of current situation drawings; use the spatial statistics function to explore the spatial distribution pattern of geographic things; analyse the spatial structure; simulate three-dimensional terrain and

* Corresponding author: balok@126.com

https://doi.org/10.1051/e3sconf/202129303003

landscape, virtual city scenes; analyse the landscape view domain to produce urban evolution animation, etc.

3 The Third National Land Survey

Land resources are the important material basis for national economic construction and the fundamental of land spatial planning. The basic survey and special survey of national land resources is the basis for the evaluation of the bearing capacity of resources and environment and is of great significance to the good spatial planning of the national land. The main task of the survey of national land resources is to find out the distribution and scope of various land use patterns projected on the surface as well as the basic situation of development, utilization and protection, and to grasp the most basic national background situation of national land resources and common features.

3.1 Overview of the "Three Surveys"

As a major national survey, the Third National Land Survey (NLS-3) aims to comprehensively refine and improve the national basic land use data based on the results of the Second National Land Survey (NLS-2). The purpose is to comprehensively refine and improve the basic data of national land use based on the results of the Second National Land Survey (referred to as "Second Survey"). The "Three Surveys" is an extremely important basic national survey of China's development into a new era, which is related to the overall situation of ecological civilization construction; it is related to a series of the most basic natural resource conditions, national conditions and national strength after the first century goal is achieved and towards the second century goal of modernization; it is also the most complex and important survey of the entire natural resource system. It is also the most complex and important basic work of the whole natural resource system [1].

3.2 "Three tuning" important features

(1) The results of the "three surveys" are more realistic and reliable. Since the launch of the "three surveys" in 2017, the initial survey results were formed at the end of 2019, and the "three surveys" results were finally formed with December 31, 2019 as the unified point of time through the unified point of time update work. The whole survey process makes full use of satellite remote sensing technology and "Internet evidence" technology, adopts the integration of internal and external database construction technology, takes counties (districts) as the survey unit, and the results pass county-level self-inspection, provincial pre-inspection, national verification and multiple rounds of inspection, so the results are true, accurate and reliable.

(2) The survey results cover the whole area: the classification of land use status quo is a classification of land use types from the perspective of resource development and utilization, which is a classification with relatively wide coverage and the largest number of resource types. The "three surveys" work classification is in line with the current land use classification system.

The working classification is based on GB /T 21010-2017 "Land Use Status Classification", and some of the land types have been refined and consolidated, with 12 primary and 73 secondary classifications. The primary classification includes wetland, arable land, plantation land, forest land, grassland, commercial and service land, industrial and mining land, residential land, public administration and public service land, special land, transportation land, water and water conservancy facilities land, and other land. The survey results formed based on this cover the whole area and all elements.

(3) More detailed survey content: the Second National Land Survey (abbreviated as "Second Survey") and the change surveys in the past years did not carry out the detailed work of land use within the urban and village areas , in order to improve the level of natural resource management and achieve accurate management, the Third National Land Survey (abbreviated as In order to improve natural resources management and achieve accurate management, the Third National Land Survey (the "Survey") explicitly proposes to conduct "survey on the current status of land use in urban and village areas", and to find out the land use status of commercial services, industry, storage and other land types in urban and village areas. In addition, the "three surveys" labelled some land attributes, such as cultivated land with planting attributes and plantation attributes [2].

(4) The importance of agriculture and ecosystems: the "three surveys" pay more attention to the ecosystem, adding a class of wetlands; and improve the accuracy of agricultural applications of the survey. The "three surveys" are oriented to the fine management of natural resources and the evaluation of land conservation and intensification, and expand the content of special surveys, including the detailed survey of arable land, the survey and evaluation of the quality level of arable land and the evaluation survey of the grading of arable land, so as to provide support for the "trinity" of arable land quantity, quality and ecology It provides support for the "three-in-one" protection and management of arable land in terms of quantity, quality and ecology. This is especially important for the in-depth development of village territorial spatial planning to lay an important foundation for the current situation.

GIS system has natural advantages in data collection, processing, analysis and visualization expression, which becomes an important tool for the current state of the land resources investigation. In the initial stage of the current situation investigation of national land resources, the powerful data organization and management capability of GIS system can be used to establish the spatial database of national land resources and provide data basis for national land spatial planning.

4 Sorting out the current situation of land and space utilization in the village area

The main vector data covered by the territorial spatial planning include the current status data of land use and planning data. The specific data are as follows: (1) the data of the Third National Land Survey; (2) ecological red line; (3) permanent basic farmland; (4) arable land use, etc.; (5) forestry small class surface; (6) land use control; (7) land change database; (8) land supply data; (9) land use planning data; (10) urban general planning data; (11) urban control planning data; (12) urban elastic development zone; (13) mineral resources distribution data; (14) other special planning data, etc. This paper mainly deals with the use and analysis of the first three status quo databases [3].

The work of sorting out the current situation of land and space utilization at the village level includes the following contents: integration of land and space data of the base period of village land and space, location of permanent basic farmland indicators, location of ecological protection red line, statistics of conflicting patches of ecological protection and basic farmland protection, etc. And the completion of these elements needs to take advantage of GIS technology to achieve. The author will take Feng Sheng Village in Xiao chang County, Hubei Province and Pan Long Village in Shen nong jia Forestry District, Hubei Province as examples, to start specific discussions respectively.

4.1 Village land use in the base period

In order to sort out the land use categories and scale of the village area, it is necessary to retrieve and integrate the "three surveys" data on the GIS platform: since the land use classification used in the "three surveys" data is the status quo land use classification system (12 primary classifications, 73 secondary classifications), and the land use classification of land spatial planning (24 primary classes, 106 secondary classes, 39 tertiary classes), it is impossible to correspond to all the data. (12 primary, 73 secondary), while the spatial planning of land use classification (24 primary, 106 secondary, 39 tertiary), it is impossible to achieve one-to-one correspondence of all data; therefore, after importing the "three investigations" data on the GIS platform, it is necessary to first apply the base conversion rules: (1) direct conversion (one-to-one or multi-to-one type); (2) refinement of the survey data; and (3) the integration of the "three investigations" data on the GIS platform.); (2) refinement survey (one-to-many) (3) no correspondence (need to refine the survey), the land information can be directly converted directly, and then combined with the site field survey process, the other two cases of land information for conversion, and finally on the basis of the "three investigations" data to form the village area land Based on the data of "three surveys", the base period land data required for spatial planning can be formed, and then the statistical table and the land status map can be completed by using GIS platform to classify the current land status of village land space [4].

Table1. Statistics of land use classification in Fong sheng Village in the base period of the national land space

Land type			Base year: Area (ha)	Base year: %
Total land area			405	100
Agricultural and forestry land	Arable land	Water Field	101	25
		Dry land	94.5	23.3
		Subtotal	195	48.3
	Garden	Orchard	38	9.27
		Other gardens	9.3	2.3
	Wood land	Tree woodland	12.7	3.12
		Other forest land	24.5	6.04
		Rural Roads	17.27	4.26
	Other agricultural land	Pond water surface	46.2	11.4
		Ditch	11.31	2.79
		Agricultural land for facilities	3.9	0.98
	Subtotal		359	88.5
Construction Land	Rural Residential Land	Rural residential land	30.5	7.53
	Public Facility Land	Land for Office, Business Service, Science, Education, Culture and Health	4.64	0.65
	Industrial Land	Industrial land	2.81	0.7
	Special Sites	Highways	4.55	1.13
	Subtotal		40.6	10.0
Nature Conservation and Preservation	Terrestrial waters	River waters	6.14	1.51
	Subtotal		6.14	1.51

4.2 Permanent basic farmland

After completing the integration of the village land space base period land data of Feng Sheng village, then retrieve the village permanent basic farmland index data from the upper planning database, we can know that the total amount of basic farmland control in Feng Sheng village is 192.29 hectares; then import the GIS drawings of permanent basic farmland in Feng Sheng village into the village land space base period land map for overlay, if the land space base period land and permanent basic farmland land overlap If the location of the base period land and permanent basic farmland overlap is not arable land type, it is the land conflict patch, and these patches are also the place where the permanent basic farmland is occupied, and finally the conflict patches can be classified and counted by GIS platform to get the classification statistics of occupied permanent basic farmland in Feng Sheng Village. As shown in Table 2, it can be clearly seen that about 20.33 hectares of permanent basic farmland in Feng sheng Village, which was designated in the above plan, were occupied (10.57% of the total), and the most occupied were garden land and forest land (about 90%), i.e. non-construction land; according to the basic principle of "the total amount of basic farmland remains unchanged", in the later planning In accordance with the basic principle of "the total amount of basic farmland remains unchanged", direct reclamation to basic farmland can be considered at the later planning stage[5]. Others are rural homesteads and other construction sites that occupy basic farmland. For these land indicators, in the later planning stage, we can consider moving them out of the basic farmland indicators and select other sites to be allocated to basic farmland, to finally achieve the "balance" of basic farmland.

Table2. Classification statistics of basic farmland occupation in Feng sheng Village

Land type	Area of occupied basic farmland (m^2)	Percent (%)
Fruit Garden	21037	184732 m^2 (91%)
Tea Plantation	1016	
Other gardens	31554	
Other forest land	104822	
Tree woodland	15684	
Other Grassland	611	
Farming ponds	537	
Agricultural land for facilities	2803	
Special Sites	6578	
Rural residential land	8820	18546 m^2 (9%)
Highway land	9229	
Urban and village road land	587	
Total	203278	100%

4.3 Ecological Protection Red Line

After completing the integration of the village territorial space base period site data of Pan Long Village, and then retrieving the village ecological protection red line data from the upper planning database, we can learn that the total area of Pan Long Village ecological protection zone is 12022.65 hectares. The GIS drawing of the ecological protection red line of Pan Long Village can then be imported into the village area of the spatial base period site map overlay, if the location of the spatial base period site and the ecological protection red line overlap is the type of dry land, rural residential land and mining land, that is, land conflict patches, these patches are also the ecological environment protection zone is occupied, and finally the conflict patches can be classified through the GIS platform statistics Finally, we can get the classification statistics of the occupied ecological protection red line in Pan Long Village through GIS platform. As shown in Table 3, it can be clearly seen that about 90616 square meters in the ecological protection zone of Pan Long Village designated in the above plan are occupied (0.75% of the total); according to the basic principle of "the ecological protection red line will only increase but not decrease", all these sites should be withdrawn in the later planning stage.

Table3. Classification statistics of ecological protection red line occupation in Pan Long Village

Land type	Area occupied by ecological protection red line (m^2)
Dry land	70327
Rural residential land	10509
Mining land	9780
Total	90616

5 Conclusion

Village territorial spatial planning involves rural revitalization and the personal interests of the majority of farmers, so to do a good job of planning, it is the biggest prerequisite to get a clear picture of the family background; the current village planning work is different from the past, with more emphasis on "multi-planning", which involves the entire village territorial space and needs to grasp its current situation.

In order to do a good job of sorting out the current state of the village land space, it is necessary to fully combine GIS technology to read the data of the three surveys, the index requirements of the basic farmland and the index requirements of the ecological protection red line; and then obtain the location and area of the basic farmland and the ecological protection red line occupied in the base year through the technology of GIS platform overlay processing data, and finally obtain the most accurate and detailed working base map, so as to better lay an important basic role for the next step of the village land space planning work.

References

1. Ministry of Natural Resources of the People's Republic of China. TD/T 1055-2019 Technical Regulations for the Third National Land Survey [S]. Beijing: *Ministry of Natural Resources of the People's Republic of China*, 2019.
2. Natural Resources Survey and Monitoring Division, Guangxi Natural Resources Department. What is new in the third survey --- A chart to read the new changes of the Third National Land Survey [J]. *Southern Land and Resources*, **6**(2019): 18-19.
3. Deng，F. The Problems Facing the Comprehensive Classification of Natural Resources and Its Suggestions for Improvement [J]. *China Land*, **5**(2020): 20-22

4. Sui， Y.T. The policy orientation of national land spatial governance through the technical adjustment of "three investigations" [J]. *China Land,* **12**(2019): 24-27.
5. Feng， D.B. Application Analysis of Geographic Information System in Territorial Spatial Planning [J]. *Intelligent Planning*, **08**(2020):145-146.

 https://doi.org/10.1051/e3sconf/202129303004

Research on landscape design of rural linear space——Taking the design of Haiou Island Rural pedestrian street as an example

ZhongWeiWang[1,*], ZhaomingDu[2], PengnanLi[3]

[1] School of art and design, Guangdong University of Finance & Economics, Guangzhou, China
[2] School of art and design, Guangdong University of Finance & Economics, Guangzhou, China
[3] School of art and design, Guangdong University of Finance & Economics, Guangzhou, China

Abstract. Based on the rural regional characteristics, this paper analyzes the types and characteristics of linear space, and discusses the landscape design strategies of rural linear space in the context of territoriality. Specifically, it explores the landscape design path of rural linear space through spatial sequence fabric, detail context narrative, local materials and technology vocabulary application, and takes the landscape design of Haiou Island Rural pedestrian street as an example to do a further practical demonstration.

1 Overview of rural linear space

In the field of place space, linear space refers to the narrow and long banded space form formed by the strong directivity of space in a certain direction. Any space with obvious unidirectional dominance can be called linear space. In the rural space, there are three types of linear characteristics: street space, shoreline space and corridor space. According to the changes of the location and scale of street and shoreline, different structural relationships are formed with the main body of the building (table 1), the street includes several levels of main street, alley and lane. Rural linear space connects the village square, street node, construction group and courtyard. By creating its boundary transition space and improving its publicity, the integration of functional units under the original order of settlement can be realized. The landscape design of linear space helps to strengthen the regional characteristics and landscape features of settlement space.

Table1. Analysis of the relationship between buildings, streets, rivers and roads

Name	type	Plane form
Outer space Street	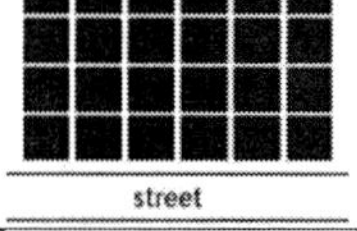	
Inner space Street	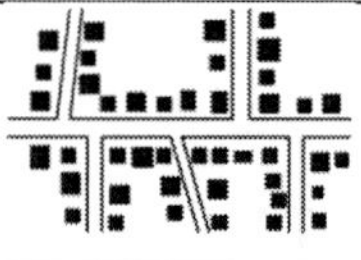	
Street Lane and shoreline space		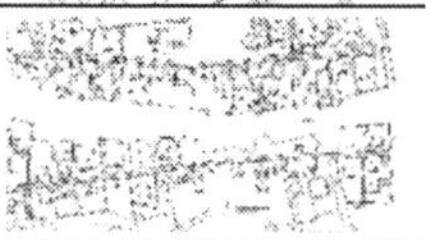

2 The construction strategy of rural linear space

2.1 The integrity and variability of linear space sequence

Linear space is the pattern of group consciousness generated by imitating and adjusting the paradigm under the self-organization state of the village subject. Therefore, the linear space design should pay attention to the balance of its integrity and variability, so as to effectively weaken the disadvantages of monotonous space. The design applies the construction method of paragraph sequence. Firstly, set the overall style orientation and construction mode, and then segment the suitability of the zonal space. According to the differences between the site landscape and the architectural interface, the corresponding theme is established, and the characteristics of each section under the overall unity are shaped, so as to eliminate the boredom of the long zonal space. For example, group construction of important common nodes in linear space belt can form multiple spatial attraction points; Or take the old buildings as the center, together with the surrounding houses and courtyards to construct the landscape group with regional characteristics, and combine the corridors to connect multiple groups, with the radiation force and attraction of spatial clusters, the integration of functional units and style coordination under the original order of the place can be realized (Fig. 1), thus enhancing the features and sequence changes of the stage space.

* Corresponding author e-mail: 20191059@gdufe.edu.cn

2.2 Narrative expression of memory context in linear space

There are a lot of memory places in the settlement space. In the construction of linear space, we should actively explore the cultural information of historical memory, and form the context narrative expression and aesthetic experience of homesickness of each space point. For example, the village entrance space in the linear space paragraph, the intermediary space between houses, which contains a large number of historical sites and story settings, through the ritual narrative writing of these point space, strengthen its identifiability, explain the life memory information. On the one hand, it can continue the derivative relationship between history and modern through the renewal of structures or landscape heritage; on the other hand, it can generate local cultural images through the patterned construction of local detail materials, and gradually establish the narrative expression with nostalgia aesthetic feelings.

2.3 Ruralization and ecological landscape construction

" Ruralization " is an important feature of rural aesthetics which is different from urban aesthetics. It is reflected in the low cost and saving of rural landscape and the shaping of local characteristics. Local construction focuses on the use of local materials, low-tech construction methods and local design combined with local technology, and the design process of integrating materials, structure and construction; In addition to the emphasis on the protection of local environment and the application of local materials, "ecologization" actively introduces modern ecological technology to promote the formation of new forms and new functional places of local landscape. The upgrading of vernacular landscape first presents the characteristics of "ecological adaptation, repairing the old from the abandoned". Drawing lessons from the grassroots and spontaneous characteristics of rural construction, it pays attention to the reuse of vernacular materials and wastes. On the other hand, it pays attention to the combination of process standardization and localization. The effective application of modern technology can make up for the shortcomings of traditional technology and improve the design effect and efficiency, It provides support for the framework and construction method of open and elastic space.

3 Landscape planning and design of Haiou Island Rural pedestrian street

3.1 location background and current situation analysis

Haiou island is located in the eastern suburb of Panyu District, Guangzhou, and the east of Shilou town. It is located at the estuary of the Pearl River and faces the lion ocean. It is a typical inland river island which is surrounded by the main channel of the Pearl River and Lianhuashan road.formed by the alluvial rivers of the Pearl River Delta. The 13 kilometer seagull road runs through the island from north to south. The island is mainly engaged in planting and breeding, and the main crops are rice, bananas, flowers and so on. It is a unique green ecological island in the urban community of the Pearl River Delta. The project is located in the east side of Haixin section of Haiou Island Haiou Road, Shilou Town, Panyu District, Guangzhou City, reaching the lower bridge site of Haiou bridge in the north and the section at the junction of four lanes and two lanes of Haiou road in the south, with a distance of about 1.3km. Haixin road section is located at the foot of Haiou bridge. At present, it is the only way to get in and out of Haiou Island, with vegetable market, fruit farm and Qifeng farm on both sides. Therefore, the government has built this road section into a rural commercial pedestrian street, so as to further improve the environmental appearance of Haiou Island entrance and exit, and promote the ecological environment construction and economic and social development of Haiou island.

The reconstruction site is the East and west sides of a highway, belonging to the peripheral part of Haixin village of Haiou island. Haixin village belongs to DanJia cultural village. The early buildings include red brick houses, gray sand houses, etc. the buildings are mostly decorated with concrete flowers and glazed flowers. Nowadays, in the context of rapid urbanization, the buildings on both sides of the street have become 3 to 4 floors tile houses built by residents themselves. The first floor is usually used as shops, with a more modern architectural style, and the surrounding environment is in a state of disordered spontaneous construction. The whole building interface forms an open space distance of about 20 meters from the highway, and forms a terrain elevation difference of about 1 meter from the building to the highway. The original open space is cement ground, the boundary between the site and the road is not clear, the parking on the site is chaotic, and there are potential safety hazards for vehicles entering and leaving (Fig.1). Secondly, all kinds of plants are planted disorderly in the site, and there are all kinds of temporary structures in some parts, which seriously affect the overall landscape on both sides of the road. Third, the information of shops along the street is messy, which makes it difficult to form an overall intuitive main interface. Therefore, this design is mainly faced with the problem of site function rectification, and how to realize the construction of the local style sequence of the landscape pedestrian road on both sides.

Figure 1. Current situation of original site

3.2 Planning and design

3.2.1 Planning concept

In order to reflect the local characteristics of the region, effectively rectify the disorder of the original place and the lack of obvious style and features as the goal, combined with peripheral business fundamentals, build a commercial business area with rural vitality and characteristics. Based on the linear spatial characteristics of the site, the planning divides the site into five sections according to the location of streets and alleys in the way of paragraph construction (Fig.2), and each section makes characteristic theme changes in the aspects of flower pool modeling, detail decoration, road pavement and planting, so as to enrich the spatial content in a single pattern and form the characteristic changes among each section under the overall style.

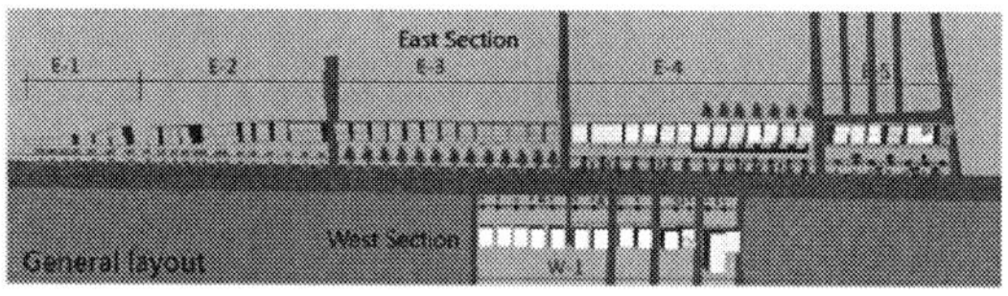

Figure 2. General layout (source: drawing by author)

3.2.2 Function planning and diversion of people and vehicles

Using the height difference of nearly one meter at the beginning of the original site, the design steps divide the site into three functional levels: commercial recreation area, parking area and peripheral footpath area. The three areas are divided by the red brick pool as the boundary, supplemented by steps to connect, forming the separation of people flow and traffic flow of Commercial Recreation (Fig.3). The commercial recreation is divided into two parts: corridor footpath and peripheral footpath, forming the inner and outer space changes of streets. At the same time, it is clear that the original disordered green planting, corresponding to the space of about 1 meter between each house, is planted with local trees, forming the counterpoint relationship of different tree species in front of the house, which can not only create a green atmosphere of the site, but also strengthen the sense of order in the space sequence of the site (fig. 4).

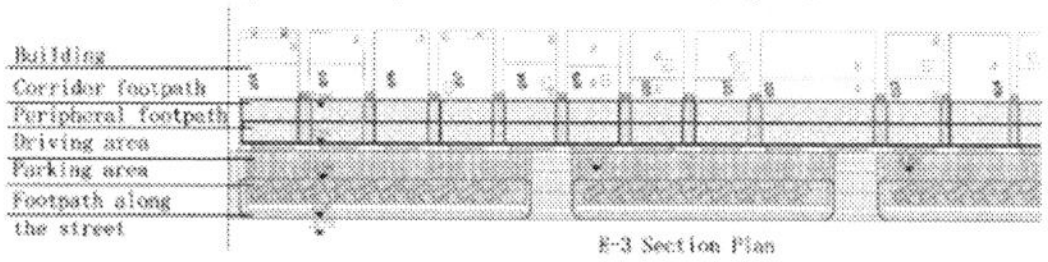

Figure 3. local plan (source: drawing by author)

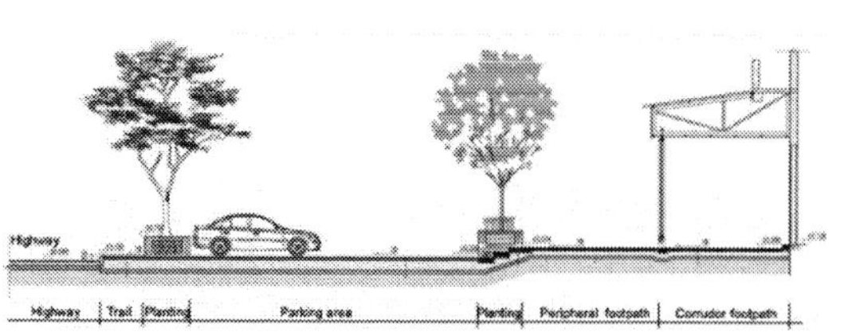

Figure 4. Section (source: drawing by author)

3.2.3 Walking space and overall image

In view of the strong light, hot and humid climate in Lingnan area, residents like to be ventilated and cool. Referring to the image characteristics of Guangzhou style arcade commercial street, a steel structure scaffolding with a span of 4 meters is overhanging according to the building. The front end is supported by a galvanized cross steel frame, the outside is embedded with 4 log squares (bolted), and the foundation is made of precast concrete piles (Figures 7 and 8). The "purlin" and "rafter" of the framework are laid with galvanized steel pipe, and the steel pipe is used as the transverse connection "beam" between the columns to make the overall structure stable. The scaffolding is decorated with anti-corrosion wood, the cornice is closed with aluminum alloy edging, and the metal with local pattern is hung under the external cornice. The inside of the eaves buckle is tested and designed with stretchable bamboo curtain to shade (FIG. 10, 11). The roof is made of aluminum alloy roof panel, creating a commercial space with local characteristics. The integral design of the veranda constructs a unified spatial sequence for the disordered space; On the other hand, the corridor sequence is based on the forward extension of the building, and the pedestrian's line of sight is attracted by it. To a certain extent, it obscures the "modernization" of the ceramic tile house, forming the iconic style keynote of the rural pedestrian street.

Figure 5. arcade in Guangzhou

Figure 6. pedestrian street

Figure 7. space outside pedestrian street Figure

Figure 8. gray space of pedestrian street

3.3 Narrative expression of local memory

Local context narrative is mainly reflected in the application of local element symbols, local materials and local crafts. For example, different sections have different decorative symbol types, hang mosaic symbols on the trellises of different sections and the difference between the flower beds and the ornaments in different sections is made. The facade of the flower pond is organically combined with the elements of glazed flower, bamboo, concrete, and concrete board, etc. Create visual content and detail changes in different areas (Table2).

https://doi.org/10.1051/e3sconf/202129303004

Table2. flowerpot style sheet (source: photo by author)

Red sandstone coping + concrete relief inlay	Red sandstone coping + glazed decoration

Red sandstone coping + glazed tube bamboo inlay	Red sandstone coping + glazed decoration

The paving materials of rural pedestrian street are mainly hexagonal bricks, with red, green and yellow colored hexagonal bricks as the central pattern , and the red and colored hexagonal bricks as the edge of the area (FIG. 9). Different color and material combinations are applied in different sections.

In the paving of red brick footpath, the method of absorbing rural soil is adopted. The red brick is built vertically without cement mortar in the middle. Only the red brick is compacted to form a plane, and the surface is sprinkled with fine sand (Fig.10). This method can make use of the gap between red bricks to form surface water infiltration, which is very conducive to pavement drainage. In terms of planting, longchuanhua, Forsythia suspensa, Rhododendron pubescens and other shrubs were planted in different sections of the inner space flower pool to strengthen the characteristics of different sections. The trees are planted with longan, a local tree species. Longan is short and will not block the store. Beautiful kapok is planted near the periphery of the road. It is tall and straight, and belongs to flowering plants. When visitors drive down the bridge, they will form their first impression of Haiou island(Fig.11-12).

Figure 9. Floor paving with hexagonal brick

Figure 10. Hexagonal brick and red brick

Figure 11. Paving details

Figure 12. Paving details

4 Summary

Rural linear space is the main space and Interface of the countryside, which Display the most distinctive rural style.From a design point of view, we should not only consider the variability of a single linear space, but also consider the landscape expression of the aesthetic characteristics of rural linear space. A large number of material and social cultural contents in the village are comprehensive and integral. Compared with the settlement as a whole, architecture is also a part of the settlement landscape. The boundary between the two forms is vague and their functions are highly integrated. As Dr. Luo Hui said: the rural landscape environment is a combination of weak artificial and the strong nature relatively, such as the agricultural landscape of fields, fish ponds and vegetable gardens, behind them is human agriculture. Therefore, the settlement linear space design should emphasize the form interaction and function translation between architecture and landscape, and explore the symbiotic path of new rural landscape and architecture in an integrated and mixed way.

At the same time, the landscape construction integrates the standardized and modular modern technology to make up for the shortcomings of traditional technology and improve the design effect and efficiency. At the same time, the penetration of modern technology expands the openness and plasticity of the space. On the other hand, it involves the refining and innovation of regional decorative symbols, and the application of local materials and techniques. In the process of implementation, architects do not need to be strict with the control of all design details, and it is best to hand over some of the design to craftsmen or villagers: 60% - 70% of the architects determine the design by themselves, 30% of them are left on the site to determine according to the opinions of craftsmen and villagers, and the construction drawings will be supplemented after the construction is completed. This mode can not only ensure that the whole architectural scheme is carried out according to the architect's ideas, ensure the basic degree of completion and style, but also set aside the possibility of "surprise" for the project according to local conditions. Because many practices come from the understanding of local craftsmen, the aesthetic perspective is quite different from the academic perspective. On the contrary, it can break through the conventional thinking and create boundaries, highlight the local aesthetic appeal.

Acknowledgments

This Paper was supported by the following Research Funds: Characteristic innovation project of Guangdong Universities in 2020:"Research on the Construction and application of landscape schema language in traditional villages of Guangfu area"（No.2020WTSCX026）;

Key Humanities and Social Sciences projects in Guangdong universities from Education Department in 2019:"Research on Inheritance and innovation of ecological wisdom of traditional human settlements in Lingnan"(No.2019WZDXM007）;

Social science project in Foshan: Research on the construction paradigm of new rural communities in Foshan (No. 2021-GJ081);

https://doi.org/10.1051/e3sconf/202129303004

References

1. Y. Fu, Z.C. Yu, A.C **9**,100(2015)
2. A. Rapoport, *House Form and culture*, (China Architecture & Building Press, Beijing, 2007)
3. ZW.Wang,Research on rural landscape construction strategies from the perspective of rural aesthetics — Take villages in the Pearl River Delta area as an example,189,(2019)
4. H. Luo, C. Zhao, N.B **9**, 1 (2015)

Experimental Study on Vegetation Restoration of Abandoned Land in Desert Steppe Nonferrous Metal Mine

Rong Hao[1,*], Shan Dan[1], Zhang Tiegang[1], Wang Xiaoli[2], Wu Xunan[2]

[1] Institute of Water Resources for Pastoral Area, Ministry of Water Resources of China, Hohhot, Inner Mongolia

[2] Sunite Jinxi Gold Mining Co. Ltd, Suniteyou Banner, Inner Mongolia

Abstract. Taking the mine dump in nonferrous metal mine as the research object in desert steppe Inner Mongolia, according to the artificial remodeling of the landforms and the characteristics of water and soil erosion, the vegetation restoration model tests on the dumping site were carried out. The ecological effects of four vegetation restoration measures on soil and water storage were analyzed which including shrubs, shrub-grass, grass and natural restoration of vegetation. The research results showed that compared with the natural restoration plant community, the soil surface water content of the three planting restoration measures increased obviously, the soil organic matter content increased but the change difference was not obvious. The vegetation coverage of shrub-grass was 67%, and the aboveground biomass was 132 g/m2. The vegetation restoration measure of shrub-grass had stronger wind-proof and soil-fixing ability than other measures, the vegetation coverage reaches 67% and the above-ground biomass is 132 g/m2. Efficacy evaluation was carried out by the grey correlation method, and Comprehensive benefit of shrubs, shrub-grass and grass all reached higher value. The most comprehensive benefit was the shrub-grass measure, which reached 0.68, and the smallest was shrubs measure also reached 0.60. Three measures cloud be applied to the vegetation restoration of desert steppe metal mining areas, and shrub-grass measure was the best.

1 Introduction

The desert steppe in Inner Mongolia was located in the northern sand prevention belt of the national ecological security pattern of "two screens and three belts". It was the key construction area of the belt and road initiative area and also was an important natural ecological barrier in China. The desert steppe was characterized by drought climate, strong wind, low rainfall, serious vegetation degradation and it was extremely fragile ecosystem. With the increasing of the scale and intensity of mineral resources construction in desert steppe, the area of loose accumulation abandoned land formed by mining construction and production was increasing continuously. Due to the poor site conditions of the abandoned land, vegetation was lack of effective supply of nutrients and water, and it was difficult to restore vegetation[1,2]. The ecological and environmental problems in desert steppe mining areas were becoming more and more serious, especially in nonferrous metal mine, and had adverse effects on the green mine construction.

Scholars had carried out a large number of researches on the restoration of vegetation in mines around the world. The research results mainly included the influence of the conditions of mining opposite sites, the ecology and comprehensive treatment of abandoned lands, the physical restoration, biological reclamation, phytoremediation technology, erosion resistance and reclamation technology, etc. The United States, Canada, Australia, France and other developed countries had basically realized the comprehensive restoration of land, environment and ecology. The United States mainly focused on the restoration of water bodies and woodlands, Australia focused on mine management and restoration, and France and other European countries focused on the restoration of abandoned land[3-7]. At present, there was a lack of scientific and effective technical means and governance model for ecological restoration in metal mine of desert steppe in China, and the scientific and technological foundation was very weak. In this study, the mine dump in nonferrous metal mine of desert steppe was selected as the research object. According to the artificial remodeling landform characteristics and soil and water loss characteristics of the abandoned land, the vegetation restoration test was carried out. The ecological adaptability and the effect of water storage and soil conservation of different vegetation restoration measures were analysed from the aspects of soil moisture conservation, wind erosion resistance and vegetation restoration. The grey relational degree method was used to quantitatively optimize the comparison and selection. The results would provide a scientific basis for vegetation restoration in abandoned land of nonferrous metal mine.

* Corresponding author: mksrh@126.com

2 Materials and methods

2.1 Study area

The study area was located at the nonferrous metal mine dump of Sunite Jinxi Gold Mining Co. Ltd in Xilinguole League, Inner Mongolia. The nonferrous metal mine dump was located in the south of the mining area and covered an area of 40 hm2. It was a stepped landform with platforms and slopes distributed and the relative height is 100m. The height of each step is about 20m. The metal slag was mixed with soil and then discharged to the dump and then covered with soil 50 cm ~ 80 cm. The soil quality was poor and the nutrient content was low according to the test of soil physical and chemical properties. The test area was located in the middle of the Inner Mongolia Plateau and it belonged to the arid and semi-arid climate in the mid-temperate zone. The average annual rainfall is 209 mm, which is concentrated in June to September. The native soil in the test area was sandy soil, and the original vegetation belongs to desert steppe vegetation.

2.2 Experimental design

Vegetation restoration measures adopted in mine dump treatment mainly had three forms, which are shrubs(I), shrub-grass (II), grass (III), and natural restoration plant community without taking measures. The experiment sample plots were set in each vegetation restoration measure, and the windproof soil consolidation and vegetation growth status of each experiment sample plots were analysed, and the effects of soil and water conservation were evaluated.

Table1. Main situation of vegetation restoration measures sample plots.

No.	Measure	Place	The main technical points
I	shrubs	Dump platform	Planting *Caragana korshinskii*, plant row spacing was 2.0m×2.0m.
II	shrub-grass	Dump platform	Planting *Caragana korshinskii*, plant row spacing was 2.0m×2.0m and Sow *Melilotus officinalis*, *Artemisia sphaerocephala* and *Avena sativa* in open space
III	grass	Dump platform	Sow *Melilotus officinalis*, *Artemisia sphaerocephala* and *Avena sativa* in open space
CK	natural restoration	Dump platform	Natural recovery state

2.3 Determination method

Surface soil (0-20cm) was randomly selected from each vegetation restoration sample plot to measure soil water content, Soil moisture content was determined by weighing method. Soil moisture content were tested once a month during the growing season from 2019 to 2020 (June to September). Soil samples were collected in the beginning of September 2020, and soil organic matter content was determined by potassium dichromate volumetric method[8]. The total coverage of vegetation was measured by a self-made 1m2 aluminium alloy frame in the peak growth period of 2020, which was divided into 100 small grids with an area of 1cm2 on average. The aboveground biomass of the vegetation was measured by cutting method. All plants were cut on the ground in 1m2 quadrat and weighed after drying at 80°C. Each measurement index was repeated for 9 times in each sample plot.

Soil wind erosion was measured by chain pin method, and soil wind erosion monitoring areas were set up in each sample. Nine chain pins were buried in the monitoring area, and three rows were laid in the vertical main wind direction, three branches in each row, with an interval of 1m×1m. The height changed between the chain pins and the ground were monitored regularly, and the wind erosion amount was analyzed. The test was conducted every three months from late October 2018 to late November 2020.

2.4 Data analysis

Analytic Hierarchy Process and Grey correlation method were used to determine the comprehensive benefit evaluation index system of vegetation restoration measures, and benefit evaluation and adaptability evaluation were conducted. Analytic hierarchy process can be carried out in the following four steps: (1) establish the hierarchical structure model; (2) construct all judgment matrices in each level; (3) hierarchical single sorting and consistency test; (4) hierarchical total sorting and consistency test[9]. Grey correlation method is a quantitative comparative analysis method. It analyses the correlation degree among the main factors in the system and determines the factor with the largest correlation degree according to the comparability and nearness of the sequence[10]. The basic principle of grey relational degree method is to judge whether the connection is close according to the similarity degree of geometric shapes of sequence curves. If a group of geometric curves are more similar in shape, the greater the correlation. According to this principle, we can find out the primary and secondary factors in the complex system, which can provide information reference for the comprehensive decision making of the system and the improvement of comprehensive benefit.

3 Results and analysis

3.1 Change of vegetation

The plant community of the nonferrous metal mine dump had certain difference after implementing different vegetation restoration measures. the vegetation coverage of the natural restoration plot was only 12%, and a large area of the ground was exposed according to Table 2. After densely planting shrubs in the nonferrous metal mine dump, the survival rate and preservation rate of shrubs were higher, the vegetation coverage reached 47%, and the aboveground biomass was 101 g/m^2. The shrub preservation rate of shrub-grass plots reached 90.2%, the vegetation coverage was 67%, and the above-ground biomass was 132 g/m^2, the plant growth was relatively good. The vegetation coverage of the grass plot was 54% and the aboveground biomass was 84 g/m^2.

Table2. Change of vegetation under different vegetation restoration measures.

Vegetation	I	II	III	CK
vegetation coverage（%）	47	77	54	12
vegetation height（m）	13	17	16	6
above-ground biomass（g/m^2）	84	132	101	18
Shrub retention rate（%）	79.5	90.2		
Shrub survival rate（%）	85.2	96		

3.2 Change of soil properties

Different vegetation restoration measures had a certain influence on the surface soil moisture as shown in the fig. 1, and the surface soil moisture of the natural restoration plots was lower than that of the plots after the implementation of vegetation restoration measures. The average variation of soil water content in each test plot was as follows: shrub-grass (II) > shrubs (I) > grass (III) > natural restoration (CK). The difference of soil water content between three vegetation restoration measures plots and the natural restoration plot was significant ($p<0.05$). The soil water content of shrub-grass plot was the highest, which was 12.5% higher than that of grass plot, and the difference of soil water content between the shrub-grass plot and grass plot was significant ($p<0.05$).

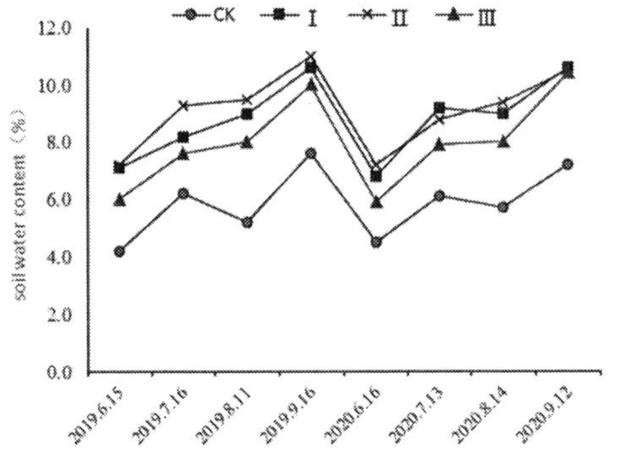

Figure 1. Change of soil water content on different measures.

According to the change of soil organic matter content (Fig. 2), the soil organic matter content of the natural restoration plot was the lowest (4.8 g/kg), and the soil organic matter content of the shrub-grass plot was the highest (5.8 g/kg), which was 17%, 12% and 9% higher than that of the natural restoration plot, shrubs plot and grass plot respectively. The results showed that the vegetation restoration measures had a certain promotion effect on the accumulation of soil organic matter in nonferrous metal mine dump, and provided the basic conditions for the restoration and reconstruction of vegetation. However, the soil organic matter content did not increase significantly after the implementation of vegetation restoration measures, and the soil conditions in nonferrous metal mine dump were poor through comparative analysis. The plant community of the nonferrous metal mine dump had certain difference after implementing different vegetation restoration measures.

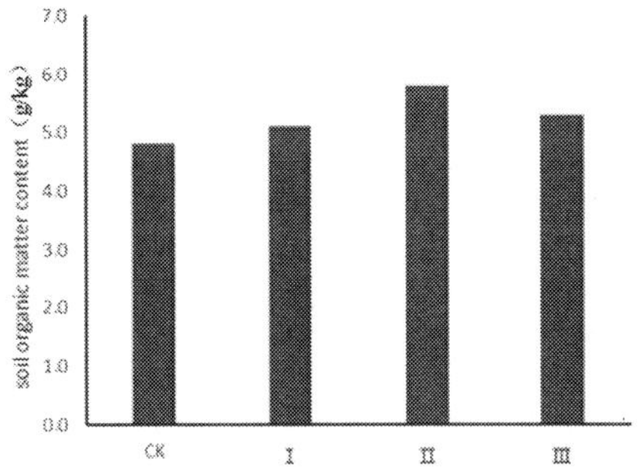

Figure 2. Change of soil organic matter content

3.3 Change of Soil wind erosion

Soil wind erosion resistance of different vegetation restoration measures varied to some extent, and the annual average soil wind erosion thickness changed greatly among different plots. four plots are arranged was as follows according to the annual wind erosion amount: natural restoration (CK) > shrubs (I) > grass (III) >.shrub-grass (II) by the observation results of soil wind erosion (Table 3). Variance analysis showed that the annual wind erosion of natural restoration plot (CK) was significantly different from that of shrubs plot (I), grass plot (III) and shrub-grass plot (II) ($p<0.05$). Wind erosion resistance abilities of different vegetation restoration measures were also different, the soil wind erosion of shrub-grass plot (II) was 53% and 30% lower than that of shrubs plot (I) and grass plot (III) respectively, which indicated that the windproof and soil consolidation ability of shrub-grass measures was better than other vegetation restoration measures.

Table3. Wind erosion index observation on different vegetation restoration measures.

Wind erosion index	I	II	III	CK
average annual wind erosion thickness（mm）	0.6	0.4	0.5	1.9
wind erosion amount（t/hm^2）	8.7	5.7	7.4	30.8

3.4 Change of Soil wind erosion

3.4.1 Comprehensive benefit calculation

The comprehensive benefit evaluation of vegetation restoration measures in the mine dump of nonferrous metal mine could reflected whether the technical measures were suitable for the ecological restoration requirements of local mining areas, so as to achieve the purpose of priority to the application of appropriate technical measures. Evaluation system of weights and calculation model were established by application of analytic hierarchy process (AHP), the evaluation index included soil and water conservation benefit index (soil wind erosion, soil water content, soil organic matter content), ecological index (vegetation coverage, aboveground biomass) and economic benefit index (investment cost and maintenance cost) which were three class a total of seven indicators. Benefit index values of different vegetation restoration measures were determined by actual measured data. The membership function of ascending half trapezoidal distribution and descending half trapezoidal membership function were used to standardize the values of different measures respectively. The standardized values were shown in Table 4. Grey correlation method was adopted to evaluate the comprehensive benefit, the correlation of sequence and measured values were calculated, the evaluation standard size, measured sequences with the reference sequence proximity to assess the testing sequence, and different measures the correlation coefficient of each index was obtained by calculation, and determine the comprehensive benefit correlation (table 5).

Table4. Standardized evaluation index values.

Evaluation index	I	II	III	CK
soil wind erosion	0.71	0.82	0.69	0.31
soil water content	0.43	0.52	0.45	0.18
soil organic matter content	0.24	0.27	0.24	0.13
vegetation coverage	0.63	0.95	0.68	0.21
aboveground biomass	0.69	0.76	0.62	0.4
investment cost	0.36	0.30	0.50	0.95
maintenance cost	0.25	0.32	0.85	0.98

Table5. Comprehensive correlation degree of different measures.

Evaluation index	I	II	III	CK
soil and water conservation benefit index	0.62	0.72	0.61	0.26
ecological benefit iindex	0.63	0.82	0.64	0.35
economic benefit index	0.52	0.49	0.58	0.99
comprehensive benefit index	0.6	0.68	0.61	0.53
Comprehensive benefit ranking	3	1	2	4

3.4.2 Adaptive analysis

Based on the comprehensive benefit analysis, although the economic benefit of natural restoration was high and the ecological management cost can be saved, the water and soil conservation ability of natural restoration vegetation was very low. The vegetation restoration cycle was longer and it was easy to cause serious soil and water loss in the process of ecological restoration. comprehensive benefit of shrubs, grass and shrub-grass vegetation restoration measures all achieved high value, and the differences between each other was lesser, the biggest comprehensive benefit of shrub-grass measures were up to 0.68, the smallest shrubs measure was 0.60, the result showed that three measures were applicable to vegetation restoration in the nonferrous metal mine dump, and had more apparent effect of ecological restoration. Among the three measures, soil and water conservation benefit and ecological benefit of shrub-grass measures was the best, and the economic benefit was slightly lower than the other two measures, the shrub-grass measure should be adopted first in the vegetation restoration of the nonferrous metal mine dump if the economic conditions allow.

4 Conclusion

4.1 Vegetation restoration measures have influence on the soil moisture content and organic matter content in the nonferrous metal mine dump of desert steppe, which enhanced the soil water holding capacity and provided basic conditions for vegetation restoration. However, the soil conditions in mine dump were not conducive to the growth of plants.

4.2 The comprehensive benefit of shrub-grass measures was the highest, up to 0.68, and the soil and water conservation and ecological benefits were obvious through the grey relational degree method. Therefore, it was suitable to be applied in the vegetation restoration and reconstruction of the nonferrous metal mine dump with poor natural conditions and difficult vegetation restoration.

4.3 Due to different natural environmental conditions in different areas, vegetation restoration inevitably presented certain differences. Therefore, vegetation restoration measures need to be modified according to the objective reality, and the specific configuration mode should be determined according to the plant ecological characteristics and local natural conditions.

Acknowledgments

This work was supported by Major Special Science and Technology Project of Inner Mongolia Autonomous Region (zdzx2018058) and Inner Mongolia Autonomous Region Science and Technology Plan Project (2019GG023).

References

1. Wei Z Y, Bai, Z K. The Concept and Measures of Runoff-dispersing on Water Erosion Control in the Large Dump of Opencast Mine [J]. Journal of China Coal Society, 2003, 28(5):486-490.
2. Lv G, Liu Y Z, L Y X et al. Evaluation of Soil Hydrology Effects of the Vegetation Restortion in Haizhou Open-pit Coal Mine Dump [J]. Ecology and Environment Sciences, 2017, 26(1):67-72
3. Carter C T , Ungar I A. Aboveground Vegetation, Seed Bank and Soil Analysis of a 31-Year-Old Forest Restoration on Coal Mine Spoil in Southeastern Ohio[J]. American Midland Naturalist, 2002, 147:44-59.

4. Stewart K M E, Penlidis A. Designing polymeric sensing materials for analyte detection and related mechanisms[J]. Macromolecular Symposia, 2016, 36(1): 123-132.
5. Nussbaumer Y, Cole M A, Offler C E, et al. Identifying and ameliorating nutrient limitations to reconstructing a forest ecosystem on mined land[J]. Restoration Ecology, 2016, 24(2): 202-211.
6. Sizmur T, Martin E, Wagner K, et al. Milled cereal straw accelerates earthworm (Lumbricus terrestris) growth more than selected organic amendments[J]. Applied Soil Ecology, 2017, 113- 166.
7. Zhang S L, Mi J X, Hou H P et al. Research Progress on Mining Ecological Restoration: Based on the Report of Three Consecutive World Ecological Restoration Conferences [J]. Acta Ecologica Sinica, 2018, 38(15):345-353.
8. Bao S D. Soil Agrochemical Analysis [M]. Beijing: China Agriculture Press, 2010.
9. Saaty T L. Highlights and critical points in the theory and application of the Analytic Hierarchy Process[J]. European Journal of Operational Research, 2007, 74(3):426-447.
10. Li J , Zhou X , Yan J , et al. Effects of Regenerating Vegetation on Soil Enzyme Activity and Microbial Structure in Reclaimed Soils on a Surface Coal Mine Site [J]. Applied Soil Ecology, 2015, 87:56-62.

Deformation prediction and anti floating analysis of a raft foundation

Binbin Xu[1,2,3,4*]

[1]Tianjin Port Engineering Institute Co. Ltd. of CCCC First Harbor Engineering Co. Ltd., Tianjin, 300222, China

[2]CCCC First Harbor Engineering Co. Ltd., Tianjin, 300461, China

[3]Key Lab. of Geotechnical Engineering of Tianjin, Tianjin, 300222, China

[4]Key Lab. of Geotechnical Engineering, Ministry of Communication, Tianjin, 300222, China

Abstract. This paper mainly studies the deformation and stress of the concrete plat under the uniform load considering the thickness of the compression layer of layered ground. Firstly, the load characteristics and the ground strategy are discussed in detail to clarify the boundary condition. Then the deformation of the ground in each layer is calculated by GEO-Cal program and the final settlement and uneven deformation is also predicted using consolidation theory. Ground stress and settlement is also calculated by FEM to verify the previous calculation.

1 Project overview

According to the general layout plan and borehole layout plan, the bottom plate of the reservoir is a large area concrete structure. Under the water load and model load, the foundation will produce obvious settlement. In addition, according to the geological survey report, the foundation compression layer within the scope of the reservoir is large and uneven in thickness, which is easy to produce uneven settlement[1-4]. In order to reduce the influence of uneven settlement on the concrete floor, the designer proposes to use pile foundation[5-6].

In this paper, considering the influence of the different thickness of the compression layer of layered foundation, the settlement deformation of the reservoir foundation under the action of service load is studied, and the causes and changes of the uneven settlement are mainly analyzed. The research results can provide the basis for the later foundation selection.

2 Load, soil layer distribution and soil property analysis

2.1 Load analysis

The section and elevation of the reservoir are shown in Figure 1. It can be seen from Figure 4 that according to the converted Xingang elevation, the average elevation of the borehole within the scope of the reservoir is +3.9m, while the designed outdoor floor elevation is +4.9m, which indicates that the reservoir needs to be filled with 1m high fill after the excavation. If the elevation of concrete floor is taken as the analysis object, the load acting on the surface before and after construction is shown in Figure 2, and the specific explanation is as follows:

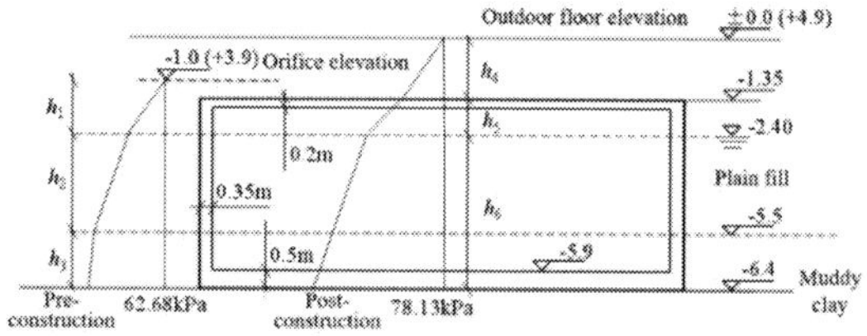

Figure 1. Elevation, size and load diagram of reservoir

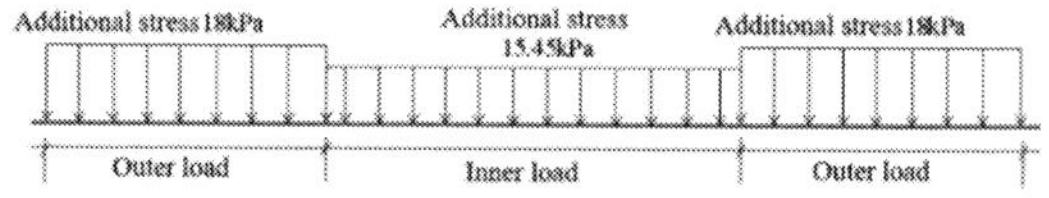

Figure 2. Schematic diagram of foundation action load on the bottom of reservoir

a) Pre-construction load within the scope of reservoir

The pre construction load is the self weight stress of the soil at this depth. It should be noted that the relative elevation of the ground surface before construction is -1.0m, then the self weight stress is:

$$\gamma_1 h_1+(\gamma_1-\gamma_w)h_2+(\gamma_2-\gamma_w)h_3$$
$$=19.2*1.4+9.4*3.1+7.4*0.9=62.68\text{kPa} \quad (1)$$

b) Post-construction load within the scope of reservoir

* Corresponding author: 109930595@qq.com

The post construction load includes the concrete weight acting on the bottom, which can be divided into the above water part and the underwater part (the groundwater level is taken as the relative elevation of -2.4m); the water weight when the reservoir is full of water and the soil weight on the reservoir. After calculation, the load of concrete acting on the foundation is 17.5kpa, the water weight is 42.63kpa, the overlying soil weight is 18kpa, and the total load after construction is 78.13kpa. Therefore, the additional stress acting on the foundation of the reservoir at the relative elevation of - 6.4m is 15.45kpa, which can also be regarded as a large area uniform load.

c) Additional stress on both sides of reservoir

When the additional stress of the foundation on both sides of the reservoir is 1m at the relative elevation of -6.4m, the additional load is 18kpa.

2.2 Soil distribution

According to the drilling data, it can be divided into ten layers from top to bottom, mainly including plain fill 1, muddy clay 2a, silty clay 2B, muddy clay 2c, silty clay 2d, silty clay 3, silty clay 4a, silty clay 4b, silty clay 4c and clay 5a. The thickness of soft soil mucky clay 2c and silty clay 2d in the hydraulic hall is the most uneven, while the thickness of soft soil mucky clay 2a, silty clay 2b and mucky clay 2c in the reservoir is the most uneven.

2.3 Soil parameters

The indoor test results of 875 undisturbed soil samples and 88 disturbed soil samples collected in this survey are stratified according to the engineering geology (from top to bottom are plain fill, muddy clay, silty clay, silt, silt, silty clay, silt, etc.), and the statistical results are listed in Table 1.

Table1. Model size

Layer	γ(kN/m^3)	E(Mpa)	k	c	φ (°)
plain fill	19.2	4.3	7.9e-8	11.02	6.02
muddy clay	17.2	2.6	4.6e-8	6.74	1.86
silty clay	19.6	6.0	3.0e-8	12.08	7.89
silt	17.9	2.8	2.9e-8	10.58	2.30
silty clay	20.2	9.1	2.8e-8	2.43	27.25
silty clay	19.8	5.1	3.5e-8	13.49	5.75
silt	20.1	5.8	7.0e-8	17.16	11.83
silty clay	19.8	10.6	3.0e-8	1.78	29.43
clay	19.2	4.8	6.8e-8	19.24	8.16

3 Settlement calculation

3.1 Calculation conditions

According to the general planning plan and the layout plan of boreholes, the B-B geological section in Figure 1 is determined as the calculation section. The boreholes included in the scope of pool foundation are zk46, zk47, zk48 and zk49.

The length of the reservoir is about 70m and the width is about 15.9m. The bottom surface of the bottom plate of the reservoir is taken as the ground surface. The load acting on both sides of the reservoir is detailed in the chapter of load analysis. Figure 3 shows the stratigraphic profile along the length of the pool. In order to calculate the settlement value of the surface and stratum in the depth direction, the coordinate origin (0,0) is set at the midpoint of the surface pool length, which is the x axis along the surface and the z axis in the depth direction.

In addition, in order to consider the consolidation degree of foundation at different time, the consolidation degrees of 200d, 300d, 10y, 20y, 30y, 40y and 50y are calculated respectively. Considering the influence of layered soil layer, the consolidation coefficient of foundation is calculated as $Cv=(\sum Hi)/(\sum Hi/Cvi)$.

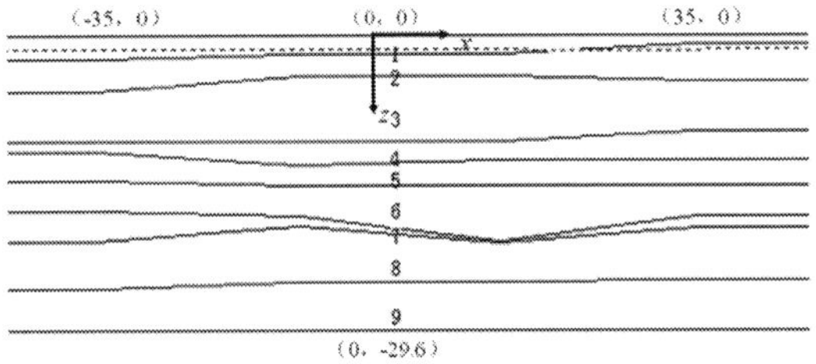

Figure 3. Calculation of soil layer and "block" division

3.2 Calculation results and analysis

Table 2 lists the corresponding soil layer compression at different positions of the reservoir bottom foundation, the soil layer with the largest compression, the total surface settlement and the uneven surface settlement, in which the empirical coefficient of foundation settlement is 1.3. Figure 4 shows the compression of each layer of soil at different coordinates, and the uneven compression of silt layer and muddy clay layer is the most obvious. Figures 5 and 6 show the variation of the total settlement and the average degree of consolidation of the foundation at the bottom of the reservoir at different coordinates at different times. Due to the drainage conditions and soil permeability, the foundation consolidation is slow, and it takes a long time to complete the whole consolidation process. Table 3 shows the magnitude of the differential settlement at the bottom of the reservoir at different consolidation stages.

Table2. Compression of each layer and total surface settlement at different coordinates under the reservoir

	-35	-25	-15	-5	5	15	25	35
1	**0.079**	0.076	0.070	0.067	0.067	0.066	0.060	0.051
2	0.026	0.024	0.022	0.020	0.020	0.021	0.024	0.029
3	0.071	**0.081**	**0.091**	**0.098**	**0.098**	**0.098**	**0.091**	**0.082**
4	0.011	0.016	0.021	0.023	0.022	0.020	0.022	0.026
5	0.016	0.014	0.012	0.011	0.012	0.013	0.013	0.013
6	0.019	0.019	0.020	0.022	0.027	0.032	0.030	0.024
7	0.014	0.011	0.007	0.004	0.003	0.001	0.002	0.004
8	0.021	0.023	0.024	0.025	0.021	0.018	0.019	0.021
9	0.022	0.023	0.025	0.025	0.025	0.025	0.025	0.026
Max.	1	3	3	3	3	3	3	3

 https://doi.org/10.1051/e3sconf/202129303006

Total	0.278	0.286	0.291	0.296	0.295	0.294	0.287	0.275
	Maximum settlement 0.296m is located at -5m, and minimum 0.275m is located at 35m. Uneven settlement is 2.1cm.							

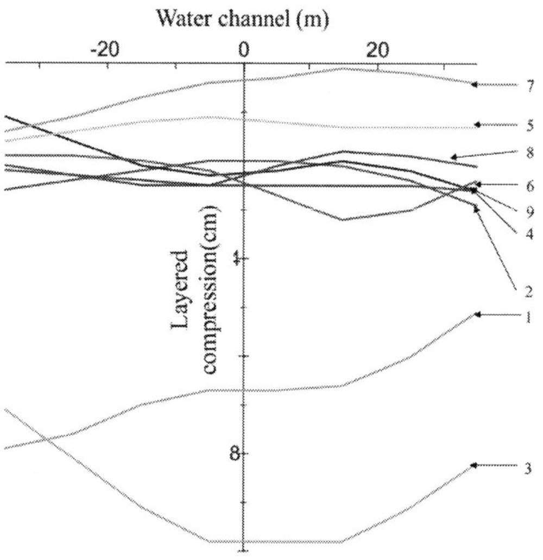

Figure 4. Calculation of soil layer and "block" division

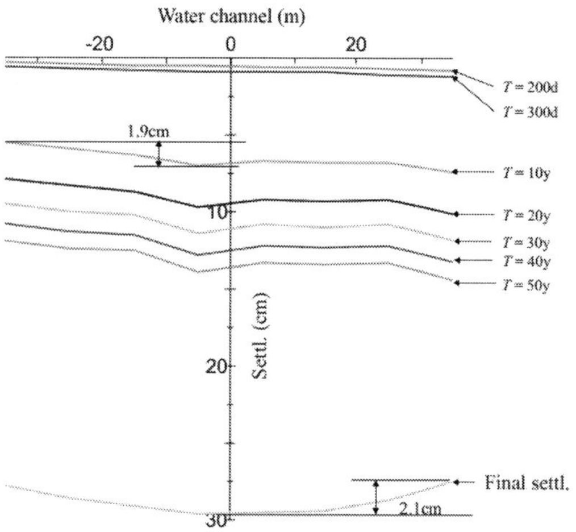

Figure 5. Total settlement at different coordinates at different time of reservoir bottom

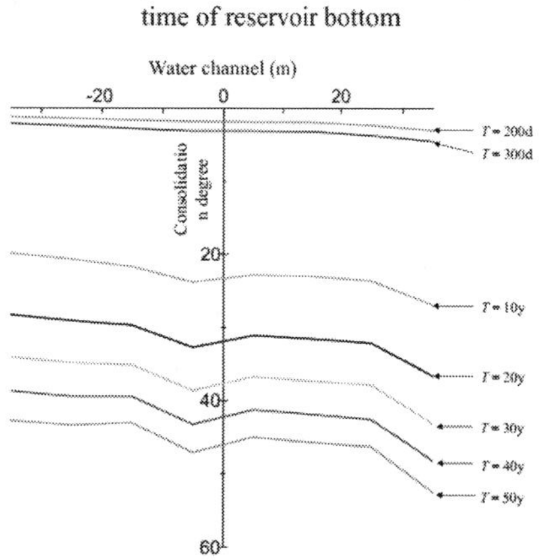

Figure 6. Average consolidation degree of foundation at different coordinates at different times at the bottom of reservoir

3.3 Finite element results

In order to ensure the accuracy of the calculation results, PLAXIS 3D finite element software is used to calculate the foundation settlement of the reservoir. In the calculation, the soil parameters of ZK46 ~ ZK49 drilling wells are selected. The physical and mechanical indexes are shown in Table 1 and the calculation conditions are the same as those in 3.2. The numerical model of PLAXIS after meshing is shown in Figure 7.

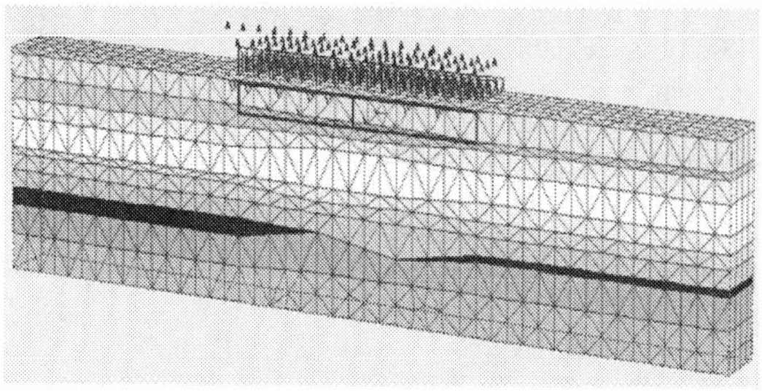

Figure 7. Numerical model after meshing

Because of the discontinuity of the soil layer in this case, in order to ensure the accuracy of the calculation, the location of the discontinuous soil layer is locally optimized in grid time sharing. The distribution of groundwater in the calculation is the same as that in 3.2, and the hydrostatic pressure calculated by PLAXIS 3D finite element is shown in Figure 8. It can be seen from figure 30 that the maximum hydrostatic pressure generated by the foundation is 356kPa.

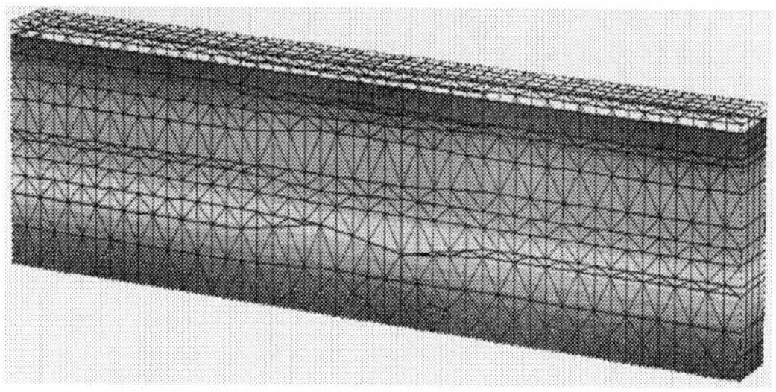

Figure 8. Hydrostatic pressure (kPa)

It can be seen from figure 9 that the settlement at different locations of the reservoir foundation is different. The maximum settlement occurs at x=-5m, and the maximum settlement is 23.2 cm. The effective stress of foundation soil is shown in Figure 10.

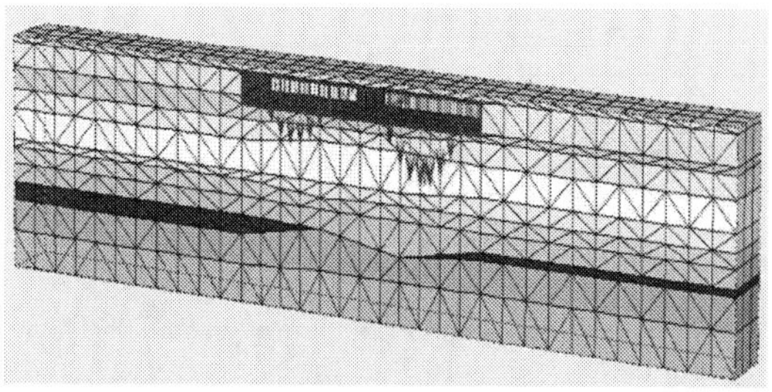

Figure 9. Vector diagram of foundation settlement (m)

 https://doi.org/10.1051/e3sconf/202129303006

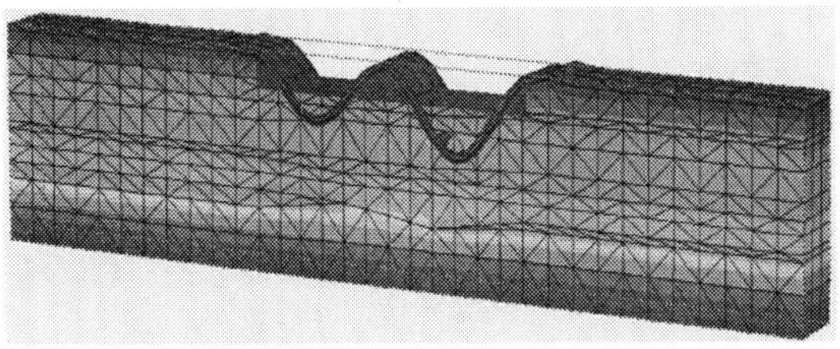

Figure 10. Effective stress of foundation (kPa)

It can be seen from Figure 10 that the maximum effective stress produced by the foundation is compressive stress, which is 500 kPa. The calculated stress of the reservoir is shown in Figure 9.

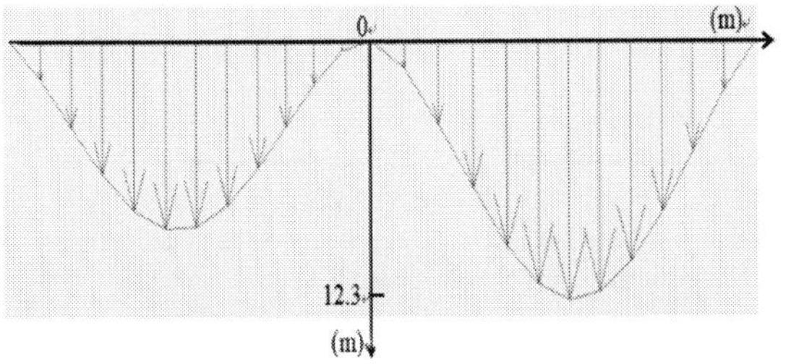

(a) Vertical displacement of reservoir (m)

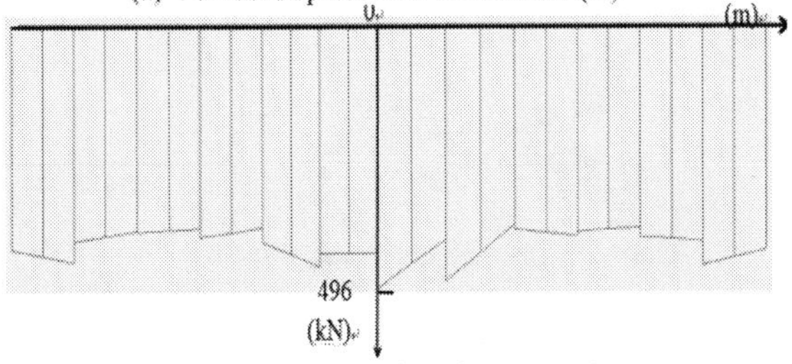

(b) Axial force distribution of reservoir (kN)

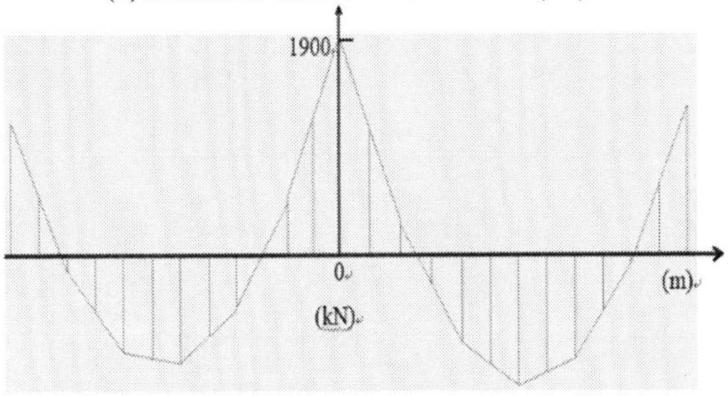

(c) Moment distribution diagram of reservoir (kNm)
Figure 11. Stress of reservoir

It can be seen from Figure 11 that the maximum vertical displacement of the reservoir is 12.3cm, the axial force is compressive stress, the maximum value is 496kN, and the maximum bending moment is 1900kNm. The maximum compressive stress of the reservoir is 3.6MPa calculated from the most unfavorable load combination of maximum pressure and maximum bending moment.

4 Conclusions

The settlement deformation and anti floating of the foundation under the concrete floor of the reservoir are analyzed in detail. The main conclusions are as follows:

(1) The results show that the uneven compression of silt layer and muddy clay layer is the most obvious, and the final uneven settlement is only 2.1cm.

(2) Because the empirical coefficient of 1.3 is selected in the calculation of settlement by the port engineering foundation calculation system, the calculation results of the finite element method are expanded by 1.3 times. Compared with the results of the port engineering foundation calculation system, when the service load is 2t, the maximum difference between the two methods is less than 0.7%, and the maximum settlement of the reservoir foundation calculated by the two methods is less than 0.7% The maximum difference is less than 2%.

(3) In the finite element calculation, the surface uneven settlement is 4.7cm without raft foundation, 2.3 cm with raft foundation, the maximum tensile stress is 48.7kN, the maximum bending moment is 790.3 kNm, and the maximum compressive stress is 0.26MPa.

References

1. F. Han, H.M. Han, L.Q. Zou, ShangDong Land and Resources **35**, 11 (2019)
2. H.W. Zhang, Transportation Standardization **42**, 8 (2014)
3. B.X. Li, Transpo World **1**, 3 (2016)
4. Y.L. Qiao, G.H. Yang, Y.C. Zhang, Guangdong Water Resources and Hydropower 7 (2013)
5. Q.J. Jiang, Architecture & Culture，**10**（2013）
6. W.X. Cui, Journal of railway engineering society, **4** (2009)
7. G.Y. Wang, J. Yu, S.G. Wu, J.Q. Wu, Geology and exploration **45**, (5) (2009)
8. R.L. Hua, L.C. Wang, Z.Q. Yue, Engineering Geology **76**, 1 (2004)

Incremental dynamic analysis of underground subway station structure

Tong Liu[1,*], Fuwang Xu[2], Qinghe Wang[3], Huaiyu Xu[3]

[1]School of Management, Shenyang Jianzhu University, Shenyang 110168, China.

[2]China Communications Tunnel Engineering Company Limited, Fujian 353100, China

[3]School of Civil Engineering, Shenyang Jianzhu University, Shenyang 110168, China.

Abstract. Incremental dynamic analysis constitutes the basis of seismic performance evaluation and seismic vulnerability assessment. In this paper, a typical two-story three-span subway station structure is selected as an example structure and its incremental dynamic analysis procedure are presented, including selection of ground motion, intensity measure and limit state determination. The incremental dynamic analysis procedure provided in this study can be a basis for further study on seismic design for underground structural systems.

1 Introduction

With the continuous development and utilization of urban underground space, the safety of underground structure is of great importance, especially the structural safety under seismic load. Up to now, many researchers focused on the seismic performance of underground structure [1-5]. Among these studies, seismic fragility analysis based on incremental dynamic analysis (IDA) has been applied to many different kinds of underground structures, such as subway station structure[2], underground cavities[4], tunnel[5]. Through probabilistic approach, the full range of structural dynamic behavior, from elastic to elastic-plastic, until collapse can be considered. However, the implement of IDA for underground structure is still based on surface structure proposed by Vamvatsikos and Cornell in 2002[6] and few study focused on the detail of IDA for underground structure, such as how to choose effective intensity measure, ground motions.

For this purpose, a two-story three-span subway station structure is taken as a typical example to illustrate the detail of IDA, mainly including selection of ground motion, intensity measure and limit state determination. The results obtained herein can be the basis for further study of seismic performance for underground structure.

2 Case study

2.1 Typical subway station structure

The rectangular reinforced concrete box is of 20.9 m wide and 12.37 m high, as shown in Figure 1. The first floor of the station is the lobby floor while the second floor is an island platform. The cross section of all the columns is 0.6m by 1.0m. All column spacing is 8 m. The roof depth of the example structure is 2.9m, the soil condition from the geological investigation report is shown in Table 1.

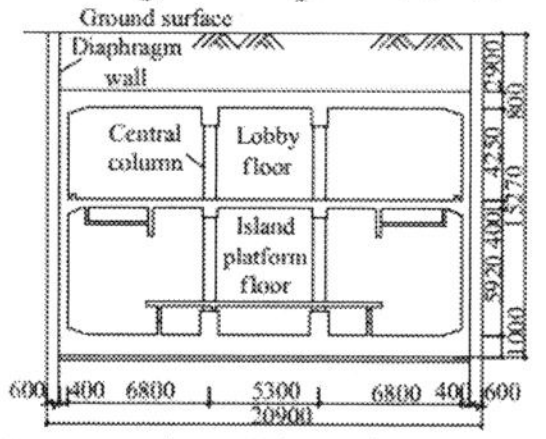

Figure 1. Cross section of the subway station structure.

Table1. Soil parameters of the example structure.

Sequence	Name	Depth (m)	Unit Weight (kN/m^3)	Elastic Modulus (MPa)	Poisson's Ratio	Friction Angle (°)	Cohesion (kPa)
1	Man-made fill	0-1.3	19.0	20.34	0.32	15.0	20.0
2	Isabelline silty clay	1.3-2.4	19.2	20.34	0.32	31.3	9.5
3	Isabelline silty clay	2.4-3.3	18.0	14.00	0.34	33.8	15.1
4	Grey mucky soil	3.3-6.9	17.4	10.85	0.38	28.3	5.3
5	Grey mucky soil	6.9-14.8	16.7	7.39	0.40	24.9	7.2

* Corresponding author: 262liut@tongji.edu.cn

6	Grey clay	14.8-16.7	17.4	11.55	0.35	29.7	10.0
7	Dark green clay	16.7-21.1	19.5	24.85	0.29	29.1	31.3
8	Grey silt	21.1-28.0	18.2	32.20	0.29	31.1	2.0
9	Grey clay	28.0-43.0	17.7	15.09	0.33	32.5	8.1
10	Grey clay & Silty sand	43.0-60	18.4	28.70	0.32	28.1	8.0

Table2. Material Parameters of the subway station.

Component	Material					
	Concrete				Steel	
	Elastic Modulus (GPa)	Axial Tensile Strength (MPa)	Axial Compression Strength (MPa)	Possion's Ratio	Elastic Modulus (GPa)	Yield strength (MPa)
Central column	33.5	2.51	29.6	0.2	200	400
Others	31.5	2.20	23.4	0.2	200	400

2.2 Numerical model

The two dimensional finite element model with both the surrounding soil and subway station structure is established according to the general profile mentioned above through the finite element software ABAQUS[7], as shown in Figure 2. The subway station structural model is 1000 m long and 60 m high. To enhance the calculation efficiency, beam element (B21) is adopted for the reinforced concrete structural members here. The material properties of both concrete and steel in the subway station structures are shown in Table 2, which are obtained from the engineering project at Shanghai.

The concrete plastic damage model proposed by Lubliner et al.[8] and Lee and Fenves[9] was adopted. The reinforcement of the two dimensional frame is attained through the rebar command.

The 4-nodes plane strain element (CPE4R) and the quadrilateral plane strain infinite element (CINPE4) were adopted for soil element and the Mohr-Coulomb model was used to simulate the soil's constitutive characteristics. The soil parameters are shown in Table 1. The interface between the soil and the structure is modeled as a frictional surface with a coefficient of friction μ of 0.4 and a friction angle of 22°. There is no cohesion between the structure and the soil.

The boundary conditions of the model are as follows: the horizontal and vertical displacements are fixed at the bottom surface while the top of the structure is free. Infinite elements were applied at the lateral boundaries and the ground motions are imposed at the bottom of the model.

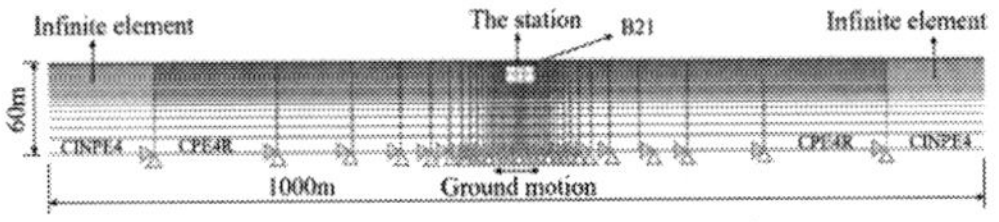

Figure 2. Finite element model.

3 Incremental dynamic analysisprocedure

IDA can take the randomness of seismic ground motion as well as the seismic response under strong ground motion intensity into consideration [6]. In this section, the IDA procedures are discussed detail by detail.

3.1 Selection of ground motion

Generally, 10-20 records are enough to provide sufficient accuracy in the estimation of seismic demands of buildings. Besides, the selected ground motion records should be able to represent the structure's site condition. According to the NEHRP(National Earthquake Hazards Reduction Program),the site condition can be divided into six groups, denoted as A, B, C, D, E, F[10]. The character of the site in this study refers to E group based on the shear wave velocity. Therefore, for IDA of underground structure, a series of twelve ground motion records that belong to a bin of relatively large magnitudes of 7.5-8.0 and near-fault are selected from the Pacific Earthquake Engineering Research Center [11], as illustrated in Table 3.

3.2 Selection of intensity measure

Intensity measure (IM) represents the intensity of ground motion, which plays an important role for the accuracy of IDA. In this section, four indexes are selected as IM, including Peak Acceleration (PA), Peak Velocity (PV) at the basement rock and Peak Base Acceleration (PBA), Peak Base Velocity (PBV) at the bottom of subway station structure. The ground motions are scaled with PV as 0.03, 0.05, 0.1, 0.2, 0.3, 0.4, 0.5, 0.6, 0.7, 0.8m/s and the maximum inter-story drift angle (θ_{max}) is chosen as damage measure of IDA. Then, with nonlinear dynamic time-history analysis, the IDA curves can be obtained.

 https://doi.org/10.1051/e3sconf/202129303007

Table3. Twelve ground motion records adopted.

No.	Event	Station	Component	PGA(g)	PGV(cm/s)
M1	Imperial Valley-06, 1979	El Centro Array #3	E03140	0.267	47.97
M2	Imperial Valley-06, 1979	El Centro Array #3	E03230	0.223	43.29
M3	Loma Prieta, 1989	APEEL 2 - Redwood City	A02043	0.274	53.65
M4	Loma Prieta, 1989	APEEL 2 - Redwood City	A02133	0.220	34.12
M5	Loma Prieta, 1989	Foster City - Menhaden Court	MEN270	0.110	21.98
M6	Loma Prieta, 1989	Foster City - Menhaden Court	MEN360	0.119	20.93
M7	Loma Prieta, 1989	Treasure Island	TRI000	0.100	15.59
M8	Loma Prieta, 1989	Treasure Island	TRI090	0.160	33.20
M9	Superstition Hills-02, 1987	Imperial Valley Wildlife Liquefaction Array	IVW090	0.179	31.67
M10	Superstition Hills-02, 1987	Imperial Valley Wildlife Liquefaction Array	IVW360	0.208	36.21
M11	Northridge-01, 1994	Carson - Water St	WAT180	0.091	6.33
M12	Northridge-01, 1994	Carson - Water St	WAT270	0.088	8.32

Generally, the basic principle of IM selection is to make the discrepancy of DM under different ground motions as small as possible. The IDA curves with different IMs are shown in Figure 3. To better illustrate the dispersion of curves, we can compare the average values of ln(DM)'s standard deviation, among which the smaller one means lower dispersion and the corresponding IM is more suitable for IDA. Table 4 shows the average values of ln(DM)'s standard deviation of IDA curves with different IMs, respectively. From the statistic in the Table, it can be found that the $\sigma_{\ln(DM)}$'s average value of PBA is smaller than that of the other IMs, while PBV has the largest value among four IMs. Consequently, for seismic performance evaluation of the example subway station structure, PBA is a better IM candidate than the others.

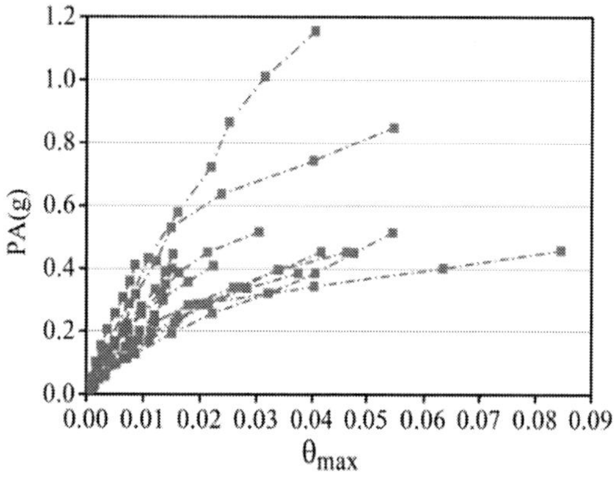

(a) PA as IM

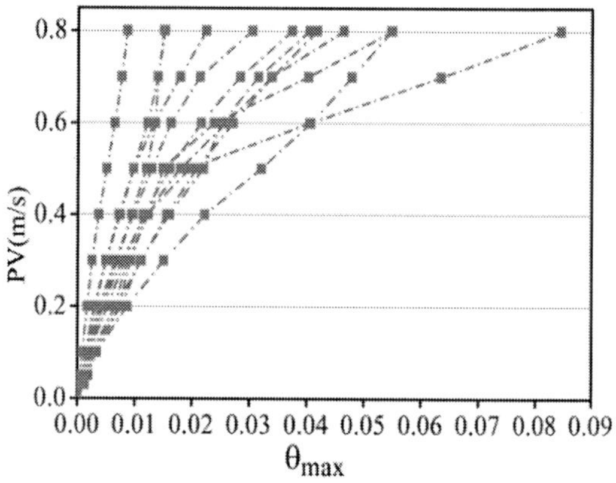

(b) PV as IM

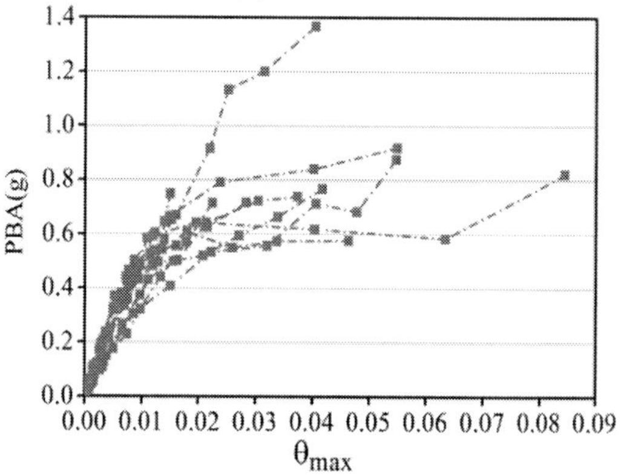

(c) PBA as IM

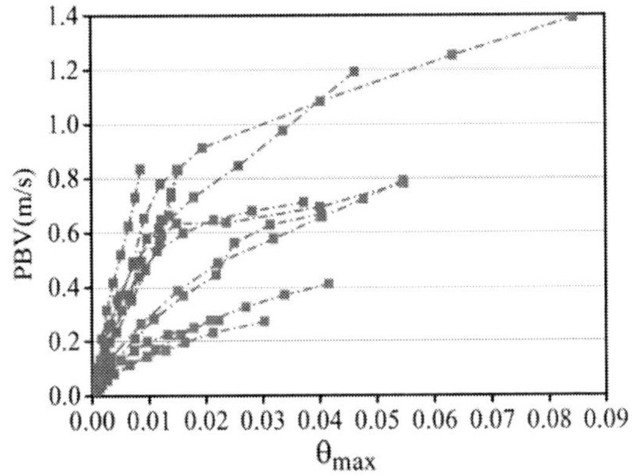

(d) PBV as IM

Figure 3. IDA curves with different Ims.

Table4. Average value of ln(DM)'s standard deviation.

IM	PA	PV	PBA	PBV
Average value of $\sigma_{\ln(DM)}$	0.485	0.47	0.27	0.505
Ratio of average value to PV	1.03	1	0.57	1.07

3.3 Limit state determination

For surface structure, limit states are divided into two levels: Immediate Occupancy and Collapse Prevention (CP), among which the CP state is defined when the slope of IDA curves decreased to 20% of elastic slope (Ke)[12]. However, when it comes to the IDA of underground structure, the property of IDA curve has great discrepancy due to the restraint of the surrounding soil. To be more clearly, we choose M3 ground motion record, among which Ki means the slope between point i-1 and point i of IDA curve. From the statistics in Table 5, it is shown that the decline of slope is limited and larger than 20% of elastic slope (0.2Ke) all the time.

The result shows that it is not proper for the example subway station structure to define limit state based on the slope decrease of IDA curve. Two approaches are suggested herein. Firstly, define limit state quantitatively based on the design code. Secondly, define limit state quantitatively based on the structural damage distribution under seismic load. These two suggestions have already applied to different kinds of underground structures[2,6,13], but still need further study to make the process of limit state determination more clearly.

Table5. Slope decrease of IDA curve when scale M3.

Slope/Vertical axis	PA	PV
K1(Ke)	30.82091	6027.558
K2	32.63887	6381.01
K3	33.27021	6503.168
K4	32.57325	6369.427
K5	27.06349	5291.005
K6	21.39749	4184.1
K7	22.14286	4329.004
K8	14.20833	2777.778
K9	11.23956	2197.802
K10	11.26652	2202.643
0.2Ke	6.780601	1326.063

4 Conclusion

To conduct IDA of underground structure, effective procedures are provided as follows.

(1) Ground motions for IDA should be able to represent the structure's site condition and 10-20 records are usually enough.

(2) PBA can be selected as the IM for IDA. When the structure type change, its applicability should be verified.

(3) As for the limit state determination, two approaches are suggested and further study is still needed.

Acknowledgements

The research work in this paper was supported by LiaoNing Revitalization Talents Program, China (XLYC1902027), by Doctoral Scientific Research Foundation, China (2019 - BS- 193), by Science and Technology Project of MHRUD, China (2019 - K - 054), by Shenyang Science and Technology Project, China (RC200143) and Tianjin Science and Technology Project, China (18ZXGDGX00030).

Reference

1. Huo H, Bobet A, Fernández G, et al. Load Transfer Mechanisms between Underground Structure and Surrounding Ground: Evaluation of the Failure of the Daikai Station[J]. Journal of Geotechnical & Geoenvironmental Engineering. 2005, **131**(12): 1522-1533.
2. Liu T, Chen Z Y, Yuan Y, et al. Fragility analysis of a subway station structure by incremental dynamic analysis[J]. Advances in Structural Engineering, 2016, **20**(7): 1111-1124.
3. Liu J B, Wang W H, Dasgupta G. Pushover analysis of underground structures: method and application[J]. Science China Technological Sciences, 2014, **57**(2): 423–437.
4. Cui Z, Shen Q, Leng X L, et al. Performance-based seismic stability assessment of large underground cavern group with incremental dynamic analysis[J].

Chinese Journal of Rock Mechanics and Engineering, 2012, **31**(4): 703-712.

5. Argyroudis S, Pitilakis K. Seismic fragility curves of shallow tunnels in alluvial deposits[J]. Soil Dynamics and Earthquake Engineering, 2012, **35**: 1-12.
6. Vamvatsikos D, Cornell C A. Incremental dynamic analysis[J]. Earthquake Engineering and Structural Dynamics, 2002, **31**(3): 491-514.
7. ABAQUS, Users Manual V. 6.10-1. 2010. Dassault Systemes Simulia Corp., Providence, RI.
8. Lubliner J, Oliver J, Oller S, et al. A plastic-damage model for concrete[J]. International Journal of Solids and Structures, 1989, **25**(3): 299-326.
9. Lee J, Fenves G L. Plastic-damage model for cyclic loading of concrete structures[J]. Journal of Engineering Mechanics, ASCE, 1998, **124**: 892-900.
10. BSSC. "2003 edition NEHRP Recommended Provisions for Seismic Regulations for New Buildings and Other Structures", FEMA 450. Part 1: Provisions and Part 2: Commentary. Washington, D. C. 2004.
11. Pacific Earthquake Engineering Research Center (PEER). PEER strong motion database. Berkeley: University of California, 2000.
12. FEMA. Recommended seismic evaluation and upgrade criteria for existing welded steel moment-frame buildings[R]. Report No. FEMA-351, SAC Joint Venture, Federal Emergency Management Agency, Washington DC, 2000.
13. Alembagheri M, Ghaemian M. Damage assessment of a concrete arch dam through nonlinear incremental dynamic analysis[J]. Soil Dynamics & Earthquake Engineering. 2013, **44**(1): 127-137.

https://doi.org/10.1051/e3sconf/202129303008

In-situ rock tests for fault gouge zone: A case in Fengman hydropower station, China

Xinchuan Xu[1,4], Zhaoyue Yu[2,*], Fangfang Xue[3] and Xiaogang Long[3], Xinyu Mao[2], Changyan Ke[2], Yuefei Wang[2], Huanyu Meng[2]

[1]China Water Northeastern Investigation, Design & Research Co. Ltd, Changchun 130061, Jilin, China

[2]School of Geological Engineering and Geomatics, Chang'an University, Xi'an 710054, Shaanxi, China

[3]Shaanxi Zhenan Pumped Storage Co. Ltd, Xi'an 710061, Shaanxi, China

[4]Research Center on Cold Region Engineering Ministry of Water Resources, Changchun 130000, Jilin, China

Abstract. The existence of faults in the dam site area threatens the stability and safety of large-scale hydropower projects in China. The fault argillaceous zone is the worst kind of fault fracture zone, and the determination of its deformation and strength parameters is the key point of rock engineering investigation. In this study, the in-situ bearing plate test and direct shear test were carried out on the gouge zone of F67 fault in the dam site of Fengman Hydropower Station. The test results show that the deformation and shear law of each test point is good, which is basically consistent with the actual condition of the measured rock mass. However, due to the limited number of measurements, the results are limited in terms of macroscopic representation. The experimental results provide scientific basis for subsequent engineering design and further enhance the understanding of mechanical properties of fault gouges.

1 Introduction

With the rapid development of China's economy, the pace of various types of energy and infrastructure construction is accelerating. Regional stability in the fields of open-pit mining, energy, transportation and water conservancy and hydropower engineering often plays a controlling role in the safety, feasibility and economy of the whole project [1, 2]. As an important factor affecting the fault engineering characteristics, the study on the mechanical properties of fault gouges is helpful to explain and predict the disaster risk in the fault crossing area [3].

From the perspective of engineering geology, the fault argillaceous zone is a kind of weak zone compared with surrounding rock, and it is the boundary condition for many geological disasters. From the perspective of rock mechanics, the fault argillaceous zone is the key research object of rock engineering due to its poor mechanical properties, low strength and wide distribution [4]. In the long-term engineering practice, the physical and mechanical properties of rock mass in fault zone have been extensively studied, and abundant test data have been accumulated. Many scholars have studied the physical and mechanical properties of fault fracture zones in depth [5-9]. In terms of the physical and mechanical research of fault gouges, Geng et al. conducted systematic experimental studies on the mechanical properties of fault gouge in Honghe fault, Yimu fault, Haiyuan fault, Fuyun fault and Xianshuihe-Xiaojiang fault, which made scholars have a preliminary understanding of the mechanical properties of fault gouge in China [10]. Han made a detailed study and analysis of the physical and mechanical parameters of fault gouges at the Daliushu dam site of the Black Three Gorges, and connected it with its formative environment, which provided a reliable basis for demonstrating the quality of loose rock mass in the Black Three Gorges [11]. Klima conducted shear tests on clay-rich fault gouge samples and established a probabilistic prediction model for the shear strength performance [12].

The strength and integrity of dam foundation rock mass is one of the key factors affecting the dam safety [13]. The reasonable determination of mechanical parameters of dam foundation rock masses is the basis of stability analysis and safety evaluation of dam engineering. In-situ rock test is the main method to study the mechanism of rock mass deformation and failure, evaluate the mechanical properties of engineering rock mass and solve complex rock engineering problems [14].

The F67 fault is the largest developed fault zone with the worst engineering geological conditions in the dam site of the Fengman Hydropower Station. It has a great influence on the dam foundation deformation, anti-sliding and seepage stability. The purpose of this study is to determine the shear and deformation properties of the fault argillaceous zone of the F67 fault on the foundation plane of the dam foundation. It is proposed to use fine in-

* Corresponding author: 2020226111@chd.edu.cn

 https://doi.org/10.1051/e3sconf/202129303008

situ rock testing to provide scientific basis for subsequent engineering design and further enhance the understanding of mechanical properties of fault gouges.

2 Study area

2.1 Project overview

Fengman Hydropower Station is a large-scale hydroelectric power station built on the Songhua River in the Jilin City, Jilin Province. It mainly generates electricity, and also takes into account flood control, water supply, irrigation, ecological environment protection and tourism (Fig. 1). The project is 210 km away from the upstream Baishan Hydropower Station, and 16 km away from the downstream Jilin City. Chang-Ji Expressway is directly connected with the project area through National Road 202, with convenient transportation.

The Fengman Hydropower Station project is a renovation and upgrading project, and the project grade is first class. The dam, the spillway, diversion tunnel and the powerhouse are first-class buildings. The dam site is arranged 120 m downstream of the original dam site. The dam type is selected as a roller compacted concrete gravity dam. The dam crest is 269.9 m high and 1068.0 m long. The maximum dam height is 94.5 m.

Fig. 1 anorama of the Fengman Hydropower Station.

2.2 Test section description

The F67 fault is the largest developed fault fracture zone with the worst engineering geological conditions in the dam site area. It passes through the dam site area along the river at the terrace on the right bank and is nearly orthogonal to the dam axis. The fault fracture zone is composed of multiple faults such as F67-1, F67-2, F67-3, F67-4 and F67-5. The overall trend of faults is N15°~50°W, with dip direction of SW and dip angle of 60°~85°. Due to the multi-period activities of the F67 fault, the fracture zone has a large number of faults and their distances are relatively close. In addition, under the influence of weathering, the faults have different widths and different characteristics. However, the width of the fracture zone gradually decreases from top to bottom, and the degree of fracture decreases with the increase of the depth. The argillaceous zone in the central zone of F67-1, F67-2 and F67-3 faults accounts for about 2.3% of the total width of the fault fracture zone. The F67-1 fault argillaceous zone has a width of about 0.5~0.7 m, and can be up to 2.5 m locally (revealed by drilling FZK138). As shown in Fig. 2, the fault gouge is composed of grey to grayish yellow mud and rock debris, which is plastic when wet. The widths of the argillaceous zone of F67-2 fault and F67-3 fault are 0.4~0.6 m and 0.05~0.1 m, respectively, and both of them are composed of mud and rock debris.

Fig. 2 The fault gouge of the F67 fault.

3 Methodology

3.1 Instrument and equipment

The main instrument and equipment used in this study include: the specimen processing equipment, reaction force device, normal load system, shear load system, measurement system and calibration system (Fig. 3a). Among them, the reaction force device uses anchor rods to provide the reaction load (Fig. 3b).

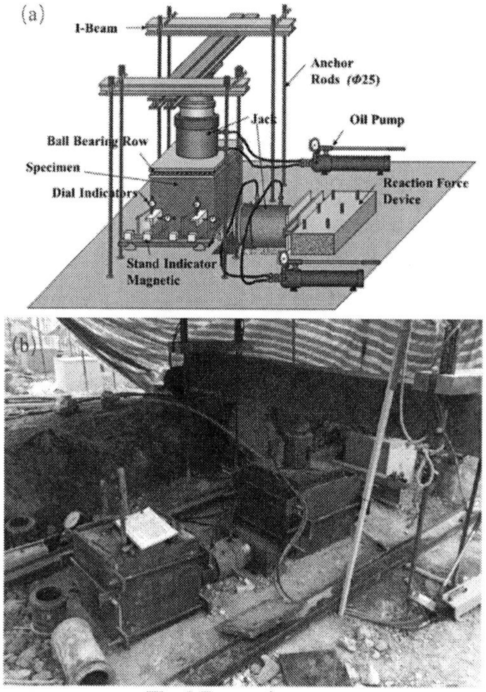

Fig. 3 Test equipments.

(a) Diagram of direct shear test. (b) The reaction force.

https://doi.org/10.1051/e3sconf/202129303008

3.2 Processing of test points

In order to accurately select representative test sections and test points, it is necessary to clean up the gravel left by the excavation of the test site. As shown in Fig. 4, the cleaning method adopts waterwheel scouring and manual cleaning.

Fig. 4 Test site cleaning.

(a) Waterwheel scouring. (b) Manual cleaning.

3.2.1 Deformation test points processing

According to the requirements of "Rock Test Regulations for Hydropower and Hydraulic Engineering" (DL/T 5368-2007), the bearing plate method was used in the deformation test in this study. According to the requirements of test point processing, the steel drill and hand hammer were used to remove loose rock mass at first, and then flatten the Φ200 cm area centered on the test point, with the undulation difference controlled within 3~5 cm. A total of 6 test sites were set up. For specific operation, refer to the "Rock Test Regulations for Hydropower and Water Conservancy Engineering" (DL/T 5368-2007). The on-site work photos are shown in Fig. 5.

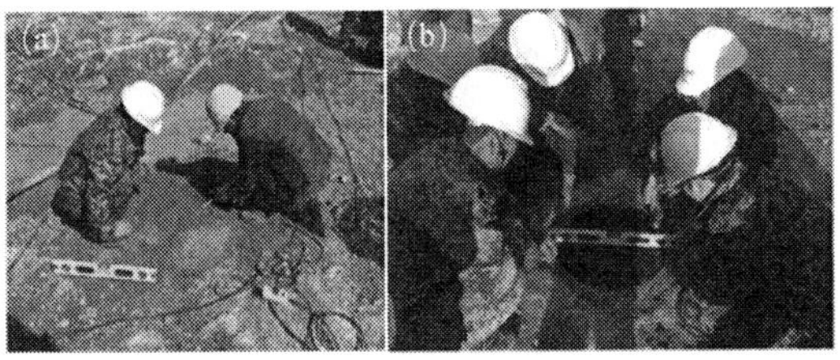

Fig. 5 Processing of on-site deformation test points.

3.2.2 Direct shear test pilot processing

The horizontal push method was used in the direct shear test of rock mass. During the test, the normal load direction applied to the sample was perpendicular to the shear plane, and the thrust direction of the sample was consistent with the predetermined shear direction. The rock surface around the sample was trimmed flat and in the same plane as the predetermined shear plane. According to the processing requirements of the specimen and to reduce the disturbance of the rock specimen, the method of step-by-step processing was used in this study. A total of 6 test sites were set up. The specific processing method refers to the "Rock Test Regulations for Hydropower and Water Conservancy Engineering" (DL/T 5368-2007). The photos of on-site work are shown in Fig. 6.

Fig. 6 Processing of on-site direct shear test points.

(a) Construction of test sites. (b) Encapsulation of test sites.

3.3 Rock mass deformation test

The principle of rock mass deformation test is based on the elastic theory. It is considered that rock mass is ideal uniform isotropic elastomer. The elastic modulus and deformation modulus of rock mass are calculated according to the Boussinesque formula of the elastic theory of half-space infinite body. The test deformation parameters of the rigid bearing plate method are calculated as follows:

$$E = \frac{\pi}{4} \cdot \frac{(1-\mu^2)pD}{W} \tag{1}$$

where E is the deformation modulus or elastic modulus; μ is Poisson's ratio of rock mass; p is the pressure calculated according to the unit area of the bearing surface; D is the diameter of the pressure plate; and W is the deformation of rock mass surface.

The deformation test of rock mass was carried out by rigid bearing plate method under the condition of artificial water immersion. The loading method adopts the step-by-step cyclic method, and the test pressure load is divided into five stages. The maximum engineering design load is 2.0 MPa, which is divided into five levels, i.e., the difference is 0.4 MPa. Therefore, it is preliminary determined that the stresses at all levels are 0.4 MPa, 0.8 MPa, 1.2 MPa, 1.6 MPa and 2.0 Ma, respectively. The field test photos are shown in Fig. 7.

Fig. 7 Field photos of in-situ deformation tests.

According to the "Rock Test Regulations for Hydropower and Water Conservancy Engineering" (DL/T 5368-2007), the test results were comprehensively sorted, and the corresponding curves and charts were drawn and listed.

3.4 Rock mass direct shear test

In the direct shear test of rock mass, the shear failure of rock mass follows the Coulomb criterion, and its expression is as follows:

$$\tau = \sigma tg\varphi + C \tag{2}$$

 https://doi.org/10.1051/e3sconf/202129303008

where τ is the shear stress applied on the shear surface; σ is the normal stress applied on the shear surface; φ is the internal friction angle of the material; C is the cohesive force of the material.

The direct shear test of rock mass was carried out by horizontal pushing method under the condition of artificial water immersion. Significantly, the direction of shear load was consistent with the direction of engineering water thrust. According to the collected data, the foundation stress of Fengman Hydropower Station project is 2.0 MPa. Therefore, five levels with a difference of 0.4 MPa were adopted in this study. it is preliminary determined that the stresses at all levels are 0.4 MPa, 0.8 Mpa, 1.2 Mpa, 1.6 Mpa and 2.0 Mpa. The actual normal load varies with the area of the specimen, but it does not affect the final test results (reference). The corresponding shear stress was obtained by tests under various normal stresses, and then the mechanical parameters of shear strength were obtained. The field test photos are shown in Fig. 8.

Fig. 8 In-situ direct shear test.

(a) Artificial water immersion. (b) Test in progress.

According to the "Rock Test Regulations for Hydropower and Water Conservancy Engineering" (DL/T 5368-2007), the test results were comprehensively sorted, and the corresponding curve charts were drawn and listed.

4 Results

4.1 Deformation test results

According to the summary table of test results and the relationship curve between pressure and deformation at each test point of the argillaceous zone (Fig. 9), the relationship between the pressure and deformation (or elastic) modulus of the argillaceous zone can be obtained (Figs. 10 and 11).

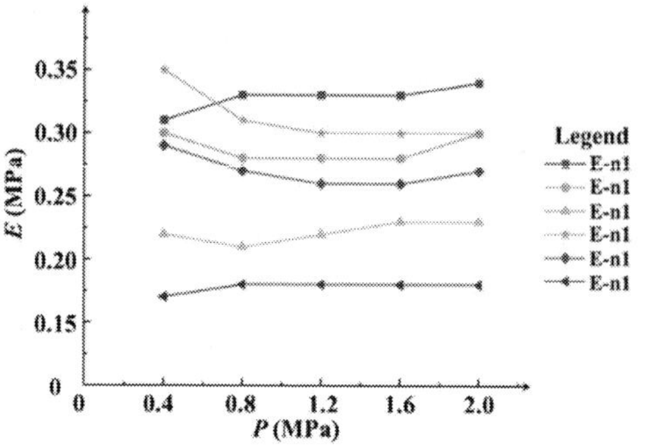

Fig. 10 Relation curve of deformation modulus for the F67 fault argillaceous zone.

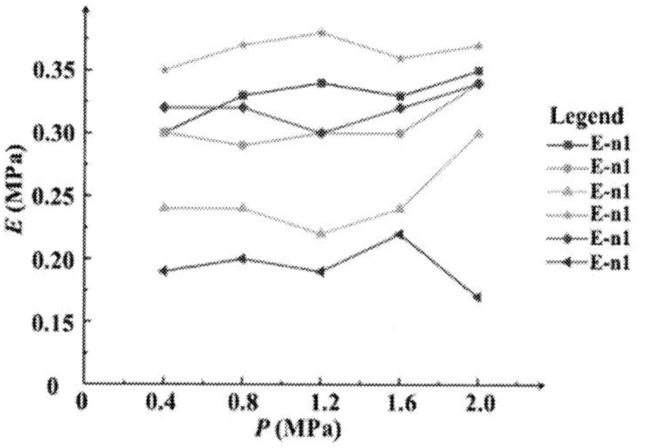

Fig. 11 Relation curve of elastic modulus for the F67 fault argillaceous zone.

According to the "Rock Test Regulations for Hydropower and Water Conservancy Engineering" (DL/T 5368-2007), the relationship curve between pressure and deformation of each test point can be determined. The results are shown in Table 1.

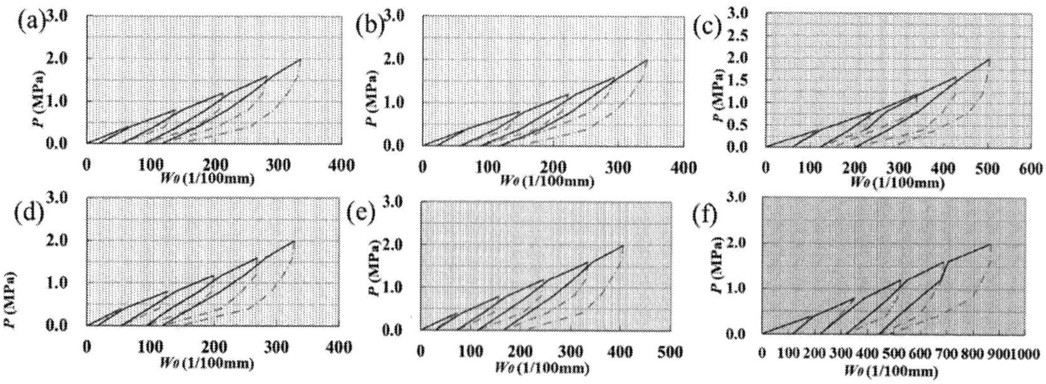

Fig. 9 Pressure-deformation curve. (a-f) Test point n1-n6.

Table1. Summary of deformation test results for the fault argillaceous zone.

Lithology	Loading direction	Project	Test results (GPa)					
Argillaceous zone	Vertical	Test point	E-n1	E-n2	E-n3	E-n4	E-n5	E-n6
		Deformation modulus	0.24	0.21	0.13	0.20	0.17	0.08
		Elastic modulus	0.35	0.38	0.30	0.37	0.34	0.18

From the analysis of Fig. 9, Fig. 10, Fig. 11 and Table 1, it can be seen that the test results have a certain degree of discreteness, but they are all in the same order of magnitude, with little difference. The loading curve of the test rock mass is gentle, indicating that the stiffness of the rock mass is low. After unloading, the deformation of the rock mass can only partially recover, and there are obvious plastic deformation and slow hysteresis. It indicates that the structure of the test rock mass is loose and broken, but its composition and structure are relatively uniform. The measured values of deformation modulus at each test point ranged from 0.08 to 0.24 GPa with an average of 0.17 GPa, and the elastic modulus ranged from 0.18 to 0.38 GPa with an average of 0.32 GPa.

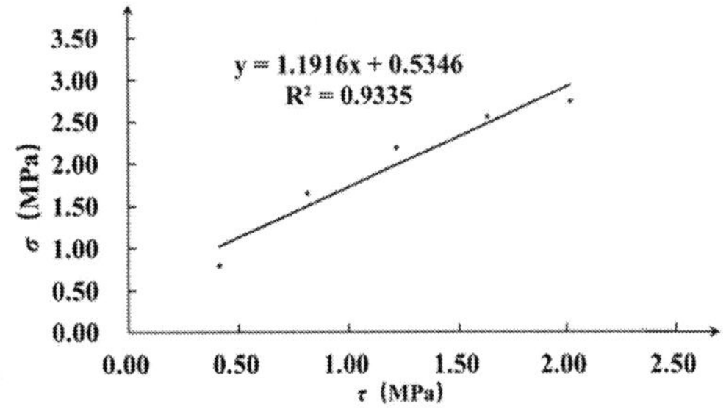

Fig. 13 Normal stress-shear stress curve for shear strength of rock for the fault argillaceous zone.

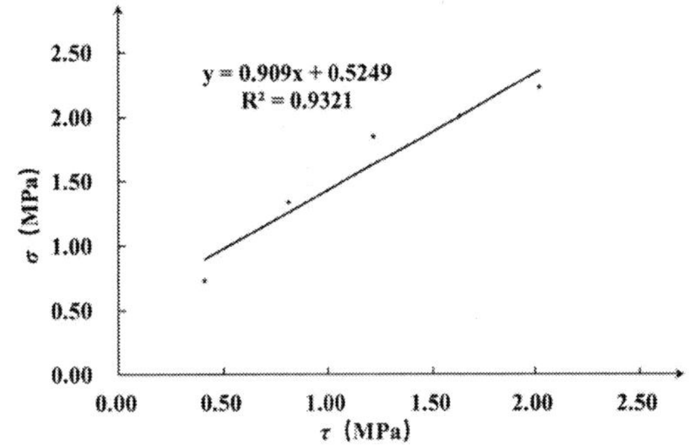

Fig. 14 Normal stress-shear stress curve for shear strength of rock joint for the fault argillaceous zone.

4.2 Direct shear test results

6 test points, numbered τ-n1 ~ τ-n6, were set up in the direct shear test of the argillaceous zone. The relation curves between shear stress and shear displacement for each point are shown in Fig. 12. In the process of sorting out the test results, based on the comprehensive analysis of the shear surface characteristics, this study used the least square method to determine the shear of rock (rock joint) peak and the corresponding strength parameters. The test results are shown in Table 2, and the fitting curves of normal stress-shear stress are shown in Figs. 13 and 14.

From the test results in Fig. 13, Fig. 14 and Table 2, it can be seen that the results of this group are relatively regular, which is consistent with the geological situation.

As can be seen from the stress-displacement curve in Fig. 12, the specimen basically belongs to plastic destruction type.

Table2. Summary of shear test results for the fault argillaceous zone.

Test point	Pile number			σ(MPa)	τ(MPa)		Shear strength of rock		Shear strength of rock joint	
	X	Y	H		Shear of rock	Shear of rock joint	tg φ'	C' (MPa)	tg φ	C' (MPa)
τ-n4	611.915	-35.963	178.101	0.41	0.79	0.73	1.191	0.534	0.909	0.524
τ-n2	616.862	-37.573	178.309	/	/	/				
τ-n3	616.398	-39.881	178.642	0.81	1.65	1.34				
τ-n1	616.429	-34.892	178.439	1.22	2.19	1.85				
τ-n6	611.358	-41.431	178.290	1.63	2.56	2.01				
τ-n5	611.634	-38.520	178.237	2.02	2.74	2.23				

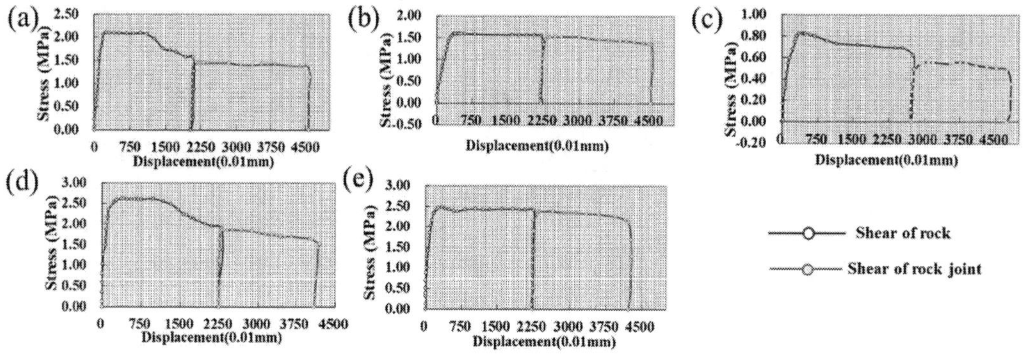

Fig. 12 Shear stress-displacement curve. (a-e) Test point n1and n3-n6.

5 Discussion

The F67 fault at the dam foundation is the largest developed fault zone with the worst engineering geological conditions in the dam site area. It is an important research object in the field of engineering geology. Its strength parameters are particularly important, and are affected by complex factors such as lithology, genesis, and occurrence environment. Therefore, in the investigation of dam construction area, the testing and survey of faults are put in the first place. At present, there are two main methods for selecting rock mass deformation and shear strength parameters in fault fracture zones. One of them is the engineering analogy method, which is generally used in the planning and feasibility study stage for small and medium-sized projects or large-scale projects with insufficient data. The other is the field investigation method, which is mainly used for large-scale projects and is carried out in geological exploration tunnels or on the foundation surface of dam foundations.

The rock mass deformation tests (6 test points) and the rock mass direct shear tests (6 test points) were carried out in the argillaceous zone of the F67 fault. On the whole, the deformation law of each test point is good, which is basically consistent with the actual condition of the measured rock mass. Besides, the deformation parameters obtained are reasonable in order of magnitude. However, due to the influence of site construction, geological structure, lithology distribution, etc., the selection of test sections and the arrangement of test points are limited to some extent. Therefore, the number of measurements is limited at the dam site. As a result, the macroscopic representation of test results in the whole project area has some limitations.

The results of this test are different from the in-situ test results of rock mass of the F67 fault zone in the exploration tunnel 189 and the foundation of the lower dam site in 1973, 2008, and 2009 [15]. This is mainly due to the different excavation mode, different elevation, different overlying mountain, and the existence of fracture zone, weathering zone and unloading zone in the test area. Moreover, the hydrological environment is also quite different, especially in the deformation test area. The elevation of this test section is 174 m, which is the lowest point of several nearby dam sections. Therefore, both groundwater and surface water gather here. Considering the effect of artificial drainage and natural evaporation, the test rock mass gradually changes from relatively dense structure to loose irregular structure after the repeated soaking and drying cycles [16, 17].

Due to the incomplete grasp of the geological data of the entire project, the current research is difficult to in-depth. Therefore, the above is mainly from an experimental point of view. The design and calculation indexes of different rock masses in the dam site area are suggested to be selected on the basis of multiple test results and in combination with other relevant data of the project and engineering analogy.

6 Conclusion

It is very important to understand the shear and deformation properties of the fault gouges in hydroelectric engineering. In this rock mass deformation test, a group (6 test points) of rock mass deformation tests were carried out on the fault gouges for the F67 fault. On the whole, the deformation law of each test point is good, which is basically consistent with the actual situation of the measured rock mass, and the deformation parameters are reasonable in the general order of magnitude. The deformation modulus of fault gouges ranged from 0.08 to 0.24 GPa with an average of 0.17 GPa, and the elastic modulus ranged from 0.18 to 0.38 GPa with an average of 0.32 GPa. A group (6 test points) of rock mass direct shear tests have been carried out on the fault gouges for the F67 fault. In general, the regularity of the test results is good. The peak strength of the fault gouges is $tg\varphi' = 1.191$ and $C = 0.534$ MPa, with correlation coefficient $R^2 = 0.933$. The peak intensity $tg\varphi = 0.909$ and $C = 0.524$ MPa, with correlation coefficient $R^2 = 0.932$.

However, due to the geological structure and lithology distribution and other reasons, the selection of test sections and the arrangement of test points are limited. Therefore, the macroscopic representation of test results in the whole project area has some limitations.

Acknowledgements

This research is supported by the Scientific Innovation Practice Project of Postgraduates of Chang'an University (No. 300103714049).

References

1. Y. Chahra, B. Riad, T. Mustapha. Application of classification systems for the assessment of rock mass stability—case of national road 43, Jijel, Algeria. Arab. J. Geosci. **14**, 3. (2021)
2. S. Tao, Y. Baoquan, C. Yuan, et al. Research on the Stability of Dam Foundation of a Sluice Dam Project in the Pearl River. IOP Conference Series: Earth and Environmental Science. **768**, 1. (2021)
3. C.H. Scholz Wear and gouge formation in brittle faulting. Geology. **15**, 6, 493-495. (1987)
4. W.W. Chen, W.F. Han, X.G. Zhang. Engineering Geology of the Fault Rock. Beijing, China. (2008) (In Chinese)
5. J.T. Engelder. Cataclasis and the Generation of Fault Gouge. GSA. Bulletin. **85**, 10, 1515-1522. (1974)
6. R.H. Sibson Fault rocks and fault mechanisms. J. Geol. Soc. **133**, 3, 191-213. (1977)
7. E.H. Rutter, R.H. Maddock, S.H. Hall, et al. Comparative microstructures of natural and experimentally produced clay-bearing fault gouges. Pure. Appl. Geophys. **124**, 1, 3-30. (1986)
8. F.M. Chester, J.P. Evans, R.L. Biegel. Internal structure and weakening mechanisms of the San Andreas Fault. J. Geophys. Res-Sol. Ea. **98**, B1. (1993)
9. C. Marone, C.H. Scholz. Particle-size distribution and microstructures within simulated fault gouge. Int. J. Rock. Mech. Min. **27**, 3, A142. (1990)
10. N.G. Geng, X.X. Yao, Y. Chen. Preliminary study on mechanical properties of the fault gouge of five faults in China. ERC. **1**, 04, 62-67. 1985 (In Chinese)
11. W.F. Han. Research on major engineering geological problems in the development of the Heishangxia Reach of the Yellow River. Research on major engineering geological problems in the development of the Heishangxia Reach of the Yellow River. (2004) (In Chinese)
12. Q. Klima. Investigation for probabilistic prediction of shear strength properties of clay-rich fault gouge in the Austrian Alps. Eng. Geol. **94**, 1-2, 103-121. (2007)
13. P. Lin, R.K. Wang, S.Z. Kang, et al. Research on the Key Problems of Damage, Reinforcement and stability of extremely high arch dam foundation. J. Rock. Mech. Eng. **30**, 10, 1945-1958. (2011) (In Chinese)
14. H.M. Zhou, Y.H. Zhang A review of the application of new technology of in-situ test of rock mass in hydraulic complex rock mass engineering. J Yr Sci Res Inst. **37**, 04, 1-6. (2020) (In Chinese)
15. H.Y. Qi, Y.B. Cai, Z.F. Liu, et al. Experimental study on deformation modulus of borehole for reconstruction of metamorphic conglomerate in dam foundation of Fengman Hydropower Station. Ne. Water. Resour. H-p. **28**, 11, 50-53. (2010) (In Chinese)
16. B. Xiao, D.N. Zhou, Z.H. Min. Relaxation Analysis of the Arch Dam Foundation Rock Masses of Jinping I Hydropower Project. Appl. Mech. Mater. **580-583**, 2108-2111. (2014)
17. E.F. Zhao, Y.F. Jiang, L. Foong. Seepage Evolution Model of the Fractured Rock Mass under High Seepage Pressure in Dam Foundation. Adv. Civ. Eng. **2021**. (2021)

Study on regional characteristics of plant landscape in yao village of longsheng county

Hao Xin Li[1,*], and Zhen Ran Liang[1]

[1]School of Tourism and Landscape Architecture, Guilin University of Technology, 541006 Guilin, Guangxi, China

Abstract. Taking the plant landscape of Baimian Yao Village and Huangluo Yao Village in the regional environment of Longsheng Autonomous County as the research object, combing and analyzing the status quo of their plant species and plant communities, and combining the folk culture and living area of the minority ethnic group in Guangxi—the Yao nationality,it summarizes the characteristics of plant landscape such as the application of native plant varieties, the level of plant vertical landscape space, and the combination of plant application and folk culture. In order to provide theoretical and practical guidance for the future plant landscape construction in this area.

1 Introduction

Regionality refers to the comprehensive characteristics of a certain area under the influence of the natural environment, things and activities, and historical and social context[1].The time category of the concept of regionalism is variable,it not only refers to the comprehensive characteristics of a region in a specific space-time background, but also refers to the regional comprehensive characteristics presented by a region in the process of time, which has the characteristics of changing with time[2].With the development of globalization and the improvement of economic foundation, people's requirements for the quality of plant landscape are increasingly high, the highly distinctive garden plant landscape has become the aesthetic object that people expect, there is also mutual learning and reference in the construction of plant landscape between different regions. In this process, problems such as "one thousand cities, one side of the same landscape in different regions" appeared in the rural landscape design, so it is very necessary to develop and protect the regional landscape of garden plants [3].

As one of the basic types of traditional settlements, the landscape features of ancient villages have been enriched with the historical process of human settlement activities[4].In the context of ecological civilization, the economy has driven the rapid development of tourism, due to the long-term effect of modern elements on the traditional ethnic minorities, the plant landscape in the traditional ethnic minority villages is affected by the urban and rural greening pattern, and deviates from the traditional construction method of ethnic local plant landscape.Minority folk villages contain rich ethnic history and culture as well as distinctive ethnic local features and features, they often form ethnic minority villages based on the characteristic natural landforms and combined with their own living customs and cultural traditions. Therefore, to study the regional characteristics of plant landscape in a few folk villages is not only conducive to coordinating the development, protection and coordinated development of traditional ethnic villages, but also helpful to provide basic information and design ideas for local rural landscape planning and urban garden design.

2 Materials and methods

2.1 Overview of the study area

Longsheng County, located in the northeast of Guangxi, is a south subtropical monsoon climate zone [5].There is plenty of sunshine and rainfall, and the average temperature is 18.1 degrees Celsius. There are four types of soil: yellow brown soil, yellow red soil, red soil and yellow soil. Its zonal vegetation is dominated by subtropical monsoon evergreen and deciduous mixed forest, with diverse plant habitats and rich species [6].The forest coverage rate is as high as 76.9%, including woodland, shrubbery, forest and suitable woodland.Among them, economic forest is about 134,000 square hectares, accounting for 65.1%. The other ecological forests accounted for 34.9% of the woodland area. Due to the influence of human activities, the forest native vegetation stock is small, mainly composed of Moraceae, Camelaceae, Hamamelisaceae, Lauraceae, Araceae and other species of plants.

2.2 Survey contents and methods

Plant landscape is an indispensable part of environmental construction, and shows the overall plant style and

* Corresponding author: 3038674536@qq.com

https://doi.org/10.1051/e3sconf/202129303009

landscape construction level. Taking into account the history, construction characteristics, conservation and management level of a few folk villages and other factors comprehensively, two typical plant communities with good effects in two folk villages were selected to take photos and draw the plan. The plant name, number (tree record, shrub and grass record tree/cluster number), plant community structure, DBH, crown width or tree degree, crown height and growth and development information in the survey area were recorded.

3 Results and analysis

3.1 The regional characteristics of plant species of Yao nationality in Longsheng

A field survey of plant species in two typical ethnic minority villages in Longsheng Autonomous County was carried out,a total of 66 species of garden plants, belonging to 49 families and 62 genera, were counted. According to the distribution of families and genera in the region, the plants in the survey area are mainly subtropical species, mainly distributed in Moraceae, Camelliaceae, Araceae, Rutaceae.According to "Flora of Guangxi", the native plants and exotic plants in the tree layer accounted for 76% and 24%, respectively, according to the statistics of plant sources in the region;the proportion of native plants and exotic plants in shrub layer was 80% and 20%, respectively, the proportion of native and exotic plants in herbaceous layer was 88% and 12%, respectively. Therefore, in the plant landscape communities of Baimian Yao Village and Huangluo Yao Village, there are abundant and diverse native plants planted in the arboreal layer, and the proportion of exotic plants in the arboreal layer is large. The main native tree, shrub and herb species with high planting amount are shown in Table 1.

Table1. List of major native plant species.

A scientific name	Family name	Generic name	Life attribute
Eucalyptus	Myrtaceae	Eucalyptus	Evergreen trees
Osmanthus	Oleaceae	Osmanthus	Evergreen trees
Taxus chinensis	Taxaceae	Pacltitaxel	Evergreen trees
Syzygium jambos	Myrtaceae	Syzygium	Evergreen trees
FicusbenjaminaL.	Moraceae	Ficus	Evergreen trees
Cinnamomum pedunculatum.	Lauraceae	Cinnamomum	Evergreen trees
Ficus microcarpa	Moraceae	Ficus	Evergreen trees
Cycas revoluta	Cycadaceae	Cycas	Evergreen trees
Trachycarpus fortunei.	Palmae	Trachycarpus	Evergreen trees
Lithocarpus corneus	fagaceae	Lithocarpus	Evergreen trees
Ailanthus altissima	simaroubaceae	ailanthus	Deciduous trees
Ovate Catalpa,	Bignoniaceae	Catalpa genera	Deciduous trees
Camptotheca Acuminata	Nyssaceae	Camptotheca	Deciduous trees
Hovenia acerba Lindl.	Rhamnaceae	Hovenia	Deciduous trees
Pterocarya stenoptera	Juglandaceae	Pterocarya	Deciduous trees
Magnolia denudata Desr.	magnoliaceae	magnolia	Deciduous trees
Liquidambar formosana	hamamelidaceae	sweetgum	Deciduous trees
Larix gmelinii	Pinaceae	Iarix	Deciduous trees
Diospyros kaki	Ebenaceae	Diospyros L	Deciduous trees
Ficus virens Ait.	Moraceae	Ficus	Deciduous trees
Glycosmis pentaphylla	Rutaceae	Mountain small orange	Small evergreen trees
Zanthoxylum bungeanum	Rutaceae	Zanthoxylum L.	Deciduous small trees
Aegiceras corniculatum Linn.	Myrsinaceae	Parmentiera	Evergreen shrubs or small trees
Ficus variolosa	Moraceae	Ficus	Evergreen shrubs or small trees
Ilex cornuta Lindl	Aquifoliaceae	Ilex	Evergreen shrubs or small trees
Lagerstroemia indica L.	Lythraceae	Lagerstroemia	Deciduous shrubs or small trees
Mallotus japonicus	Euphorbiaceae	Mallotus	Deciduous shrubs or small trees
Ficus pumila Linn.	Moraceae	Ficus	Evergreen shrubs
Camellia japonica L.	theaceae	Camellia	Evergreen shrubs
Ardisia crenata Sims	Myrsinaceae	Ardisia Swartz	Evergreen shrubs
Oreocnide frutescens	Urticaceae	Oreocnide	Evergreen shrubs
Gardenia jasminoides	rubiaceae	gardenia ellis	Evergreen shrubs
Loropetalum chinense	Hamamelidaceae	Loropia	Evergreen shrubs
Malvaviscus arboreus	Malvaceae	Platanus	Small evergreen shrubs
Indocalamus tessellatus	The Grass Family	Indocalamus	Small evergreen shrubs
Ligustrum vicaryi	Oleaceae	Ligustrum L.	Semi-deciduous shrub
Celastrus orbiculatus	Celastraceae	Celastrus L.	Vines shrubs
Palhinhaea cernua	Rosaceae	Cotoneaster	Evergreen, deciduous shrubs
Desmos chinensis Lour.	custard apple familyn	Desmos Lour	Erect or climb shrubs
Dendranthema morifolium	compositae	Chrysanthemum	Herbs perennial
Rumex acetosa	Polygonaceae	Rumex	Herbs perennial

Typhonium giganteum	Araceae	Typhonium Schott	Herbs perennial
womb doix miev	Verbenaceae	Caryopteris	Herbs perennial
Baphicacanthus cusia	Acanthaceae	Baphicacanthus	Herbs perennial
Bothriochloa ischaemum	The Grass Family	Bothriochloa	Herbs perennial
Chrysosplenium grayanum	Saxifragaceae	Chrysanthemum	Herbs perennial
Pistia stratiotes	Acanthaceae	Pistia Linn.	Floating perennial herbs
Phragmites australis	The Grass Family	Phragmites	Herbs aquatic or wet for many years
Siraitia grosvenorii	Cucurbitaceae	Momordica	Perennial climbing herb
Senecio scandens	Composite	Senecio	Perennial climbing herb
Commelina communis	Commelinaceae	Commelina	Herbs of annual spreading
Lablab purpureus	Leguminous	Leguminosae	Annual twining grass lianas
Urena lobata Linn.	Malvaceae	urena	Erect subshrubby herbs
Eichhornia crassipes	Pontederiaceae	Eichhornia Kunth	Floating herb
Passiflora edulis Sims	Passifloraceae	passiflora	Careless qualitative fujimoto

It can be seen from Table 1 that the local plant resources in the counties and autonomous regions of Longsheng are not only rich in application, but also highly adaptable and of high ornamental value, which provides guarantee for the biodiversity and ecosystem stability of a few folk villages. While planting local plants in a reasonable way, the planting of exotic plants in a few folk villages also takes up a certain proportion.The exotic plants have been cultivated and domesticated for many days and months, and have been adapted to the local environmental conditions and shown a good growth state,they have played a positive role in enriching the biodiversity of a few folk villages and constructing the landscape of plant communities. The species of exotic trees, shrubs and grasses with high planting amount are shown in Table 2.

Table2. List of major exotic plant species.

A scientific name	Family name	Generic name	Life attribute
Casuarina equisetifolia Forst.	Casuarinaceae	Casuarina L.	Evergreen trees
Grevillea robusta	Proteaceae	Grevillea	Evergreen trees
Punica granatum	Punicaceae	Punica L	Deciduous trees
Prunus cerasifera	Rosaceae	Prunus	Deciduous small trees
Schefflera octophylla	Araliaceae	Schefflera	Evergreen shrubs
Bougainvillea spectabilis	Nyctaginaceae	bougainvillea	Vines shrubs
Fatsia japonica	Araliaceae	Illicium	Evergreen shrubs or small trees
Epipremnum aureum	Araceae	Epipremnum	Evergreen vine
Ricinus communis	Euphorbiaceae	Ricinus L	Herbs annual or perennial
Hymenocallis littoralis	Amaryllidaceae	Hymenocallis	Herbs of perennial bulbs
Crinum asiaticum	Amaryllidaceae	Crinum	perennial herbs

3.2 The regional characteristics of the plant community of Yao nationality in Longsheng

Plant community refers to the combination of all plants living in a certain area, which is formed by mutual influence of each individual plant through reciprocity and competition, and is the result of their adaptation to the common living environment. Each relatively stable plant community has certain species composition and structure [7].In this study, the plant planting collocation in the two selected typical minority folk villages was investigated and statistics were made, and the community structure and ornamental characteristics were analyzed, and 8 plant communities with rich structure and high ornamental value were obtained (Table 3).As shown in Table 3, the plant communities in a few folk villages take native plants as the backbone tree species and exotic plants as the ornament, so that the composition of the plant communities can most intuitively express the natural plant landscape with regional characteristics.

Table3. 8 typical plant communities.

Main plant communities	Community structure	Main viewing season
liquidambar+Loro petalum safflower+ Schefflera+indocalamus+ castor+dayflower	Large tree + small tree + large shrub + small shrub + herb	Spring, summer, autumn, winter
Ficus lacor+Lith-ocarpus corneus+ camellia+Gold waist+ Bothriochloa ischaemum	Large tree + small tree + herbs	Spring, summer, autumn
eucalyptus+osmanthus+ Sichuan Pepper + Xun son+Fatsia japonica	Large tree + small tree + small shrub	Spring, summer, autumn, winter
Cinnamomum japonicum+Aegiceras corniculatum+Mallotus+Urena lobata	Large tree + large shrub + small shrub	Summer, autumn

yulan+horsetail beefwood+Ilex pern -yi+Ficus variolosa+ HYmenocAllisSpciosa	Large trees + small shrubs + herbs	Spring, summer, autumn, winter
Purple Leaf Plum+ Ligustrum vicaryi+ gardenia+Malvaiscus arboreus+sen-ecio	Small trees + large shrubs + small shrubs + herbs	Spring, summer, autumn
Grevillearobusta+Mal-vaiscus arboreus+ Chrysanthemum	Large trees + small shrubs + herbs	Spring, summer, autumn
Syzygium jambos+ Cycas revoluta+Ardisia crenata+Celastrus orbiculatus+Oreocnide frutescens	Large trees + small shrubs	Spring, summer, autumn

3.3 Summary of regional characteristics of Longsheng Yao plants

Both Baimian Yao Village and Huangluo Yao Village have remarkable characteristics of native plants. In the application of native plants in the landscape, the two Yao villages have their own application principles and aesthetic characteristics.Is common, there are some regional plant application such as *osmanthus*, *liquidambar*, *Chinese beech*, *eucalyptus*, *fields* such as strong adaptability, growth in good condition of native plant species as the backbone, the plant landscape of the whole ecological function, decoration, space, define functions are rev to the leading role, showing LongSheng County living style and features.The application of some regional plants is unique and has certain aesthetic value and ecological significance, such as the application of fruiting plants such as *Orangeberry* and *Ardisia crenata*. For example, the preservation of wild vegetation, such as *Polygonum capitatum*, *climbing groundsel herb*, and *Cibotium barometz,* not only makes the villages' aesthetic features remarkable, but also shows the unique features of the mountainous areas of ethnic minorities. In general, the plant landscape inside the Yao Village in Longsheng County is different from the conventional plant landscape configuration in the general urban park and green space. The plant species selection is rich, the form is characteristic, and the scale and distribution are suitable for the internal space of the village.

4 Plant landscape characteristics of Yao Village in Longsheng County

Baimian Yao Village and Huangluo Yao Village are both villages with the characteristics of Yao style plant landscape, and are related to the folk culture and living area of Yao minority.Yao nationality is a minority nationality with profound historical and cultural deposits in China. The regional characteristics of its main living environment have created a variety of local plant species, and also increased people's application and love of plants.

4.1 Application of native plant varieties

Local plants are plants that are native to the region or have been introduced, cultivated and propagated for a long time, and have many advantages such as adapting to the local natural environment, maintaining the local ecological stability, strong growth adaptability and low maintenance and management cost [8]. Moreover, some local plants are closely related to local legends, allusion and customs.In the greening of traditional villages, the application of native plant varieties is not only conducive to improving the local greening environment, but also can well reflect the regional natural characteristics and traditional local culture of a village.At the same time, because native plants have low requirements for living environment and are easy to survive, the cost of planting and maintenance is relatively low.Therefore, the construction of local plants as the main plant landscape has become one of the important development directions of gardens.

4.2 Plant vertical landscape space level

The rational collocation of plants not only makes the landscape spatial hierarchy diverse[9], but also has different visual senses.Therefore, the plant landscape can be divided into four landscape levels, namely, large tree -- small tree, large shrub -- small shrub, erect herbaceous -- ground covering herbaceous. The top plants are mainly Ficus in Moraceae and Phyllostachys graminae, which are mostly planted at the border of the village to construct the green skyline of Yaozhai. The second layer is dominated by small trees and large shrubs, mostly planted with evergreen or color shrubs and crape myrtle to highlight the seasonal image. The third layer is small shrubs and perennial, erect herbs, the plants are mainly ornamental (flower, fruit), medicinal type;the lowest layer for the floor of the herb can be eaten.Among them, the lianas are shade-resistant and climbing with other plants, while the lianas are sunny and climbing on the bamboo fence, so that the whole plant landscape is connected in series, presenting the stylized style of Yao Village.

4.3 The combination of plant application and folk culture

The application of plants in Longsheng Autonomous County is closely related to the culture and folk custom of Yao Nationality.They believe that the tree deity can protect the peace of the village, help the Yao people eliminate disasters and diseases, and bless the Yao people, so the Yao people have the traditional idea of worships trees.Especially for maple trees, Yao people think that maple trees are the incarnation of their ancestor Chiyou and their protectant god [10].In one of the main courtyars of the village, there is a sacred monument to be worshipped by people.Behind the sacred monument are towering ancient maple trees, which can not be damaged unless they are pruned by special personnel on festivals and festivals.In addition, the residents of Huangluo Yao Village pick wild plants with decorative effects in the garden in front and back of the house, forming a landscape with Yao characteristics together with other landscape elements in the village, such as residential buildings, castellan rules and steles. The main plants such as *Cinnamomum camphora* and *eucalyptus* are known as "feng shui trees", which signify protection from evil spirits, longevity and good luck. There are also *Cedar* and *Ficus virens*, which symbolize good life and evergreen life.*Maple,Taxus chinensis,Lindera glauca,Musa acuminata,Momordica Grosvenori* and *Crassocephalum crepidioides* are all plants of Yao culture and folk custom in Yao village of Longsheng Autonomous County.The application of plants in yao village in longsheng autonomous county reflects the combination with folk culture and forms the unique plant culture and plant landscape in yao village.

5 Conclusion

However, due to the development of tourism, the distinctive style of Yaozhai village has gradually disappeared, first, in the construction of public green space, the planting, variety and pruning methods of plants are the most common and most used in the greening of surrounding towns in these two areas, which destroys the original plant landscape ecology to a certain extent;second, with regard to the greening of the small courtyards in front of and behind the villagers' houses, since Longsheng Yaozhai itself is located in a mountainous region, the available construction area is limited, so the area of courtyards for residents is even more limited.The relevant local departments should strengthen the protection and conservation management of plants to avoid the encroachment of green space or plant space by infrastructure construction, especially in some plants and landscapes with characteristics, such as maple trees.In this perfect the construction of some characteristic regional landscape, and increase the application of native plant varieties. At the same time, combined with the garden characteristics of Yao people, the space of Yao village is decorated to interpret and show the Yao culture.

References

1. S.K.Xu, S.Q.Fang, Y.P.Gan. Tourism Destination Image Perception in Poor Counties: A Case Study of Longsheng Ethnic Autonomous County in Guangxi. Journal of Nanning Normal University (Natural Science Edition), 36(04), 94-101(2019).
2. D.N.Zhang, L.W.Zhou, C.Y.Duan, X.X.Chen, W.He, Q.M.Teng, Y.J. Sun, Z.F.Zhang. Heavy metal content and potential ecological damage in soil and fruit of Monohan in Longsheng County.Plant in guangxi, 9 (02) , 243-250, (2021).
3. X.L.Sun, X.X.Huang. The expression of regional characteristics of waterfront plant landscape in Suzhou. China Urban Forestry,17(03), 24-28(2019).
4. L.Wang,B. Gong, J.D.Zhao. Investigation and analysis of plant landscape in ancient village of Dangjiacun in Shaanxi Province. Journal of Northwest Forestry University, 30(06),284-288(2015).
5. C.L.Qin, C.F.Qin. Reflection on the construction of greening plant landscape in Baimian Yao Village of Longsheng County.Heilongjiang Agricultural Sciences,no.09,84-87(2018).
6. L.Zhou,L.Dong. Research on plant landscape characteristics of Xihua Garden in Kunming.Guizhou Agricultural Sciences,39(02),188-191(2011).
7. Q.Wang, B.Gong,X.Wang,J.Z.Sun. Research on regional expression of plant landscape in waterfront green space: A case study of Fenhe waterfront green space in Linfen City.Journal of Northwest Forestry University, 29(05), 227-231+242(2014).
8. S.H.Li.Construction of Plant Landscape Architecture Theory System Based on the "Three Talents of Heaven and Earth"--Research on Landscape Architecture Design Theory Based on the "Three Talents of Heaven and Earth and Earth" (II). Chinese Landscape Architecture, 27(07), 51-56(2011).
9. J.F.Zhou,Y.Ou,L. Zhou,X.C. Liu,R. Bao. Study on plant landscape and application in traditional villages of Dai nationality in Dehong, Yunnan.Guangxi Botany,39(03),346-358(2019).
10. J.C.Yang,J.K. Li. Long Shenghongyao. Guangxi CPPCC News,no.03(2010-06-05).

 https://doi.org/10.1051/e3sconf/202129303010

Experimental study on bending mechanical properties of moso bamboo

Yousheng Deng[1*], Chengpu Peng[1]

[1]Research Center of Plie-supported Structures, Xi'an University of Science and Technology, Xi'an China

Abstract. Moso bamboo, as an environmentally friendly material, has become a research hotspot in civil engineering circles, and its good mechanical properties make it have great potential in civil engineering applications. To take full advantage of moso bamboo, a series of bending tests were carried out on bamboo pipes (RBP), hollow bamboo pipes (HBP, without bamboo membrane) and bamboo pipes with concrete(BPC), and analyzed such as bending strength, flexural bearing capacity, deformation law, failure characteristics and other test results. The results showed that the smaller the diameter of moso bamboo is, the greater the bending strength and elastic modulus are; the maximum strain of BPC is 2.5 times the HBP, the distribution of strain along cross section is linear; RBP, HBP, and BPC show different failure pattern; there is a little difference between RBP and HBP that come from the same moso bamboo and have a similar diameter in limit capacity, bending strength, and elastic modulus; the flexural bearing capacity of BPC is 2.4 times the HBP, the bending strength is 1.7 times, the composite structure is effectively enhanced by concrete. The test results can provide references for the application of bamboo in landslide treatment, the supporting structure of the foundation, and other engineering.

1 Introduction

Bamboo is of an extremely important position in the global forest resources, known as the second forest. China is the richest country in bamboo resources with the largest distribution area of bamboo forest in the world. Bamboo is a renewable and degradable material[1], has low cost and a short growth cycle, which can be useful within 3-5 years. Moso bamboo is also a good structure material in civil engineering with low density, great rigidity, and strength[2,3], used in civil engineering for thousands of years. The application research of bamboo in civil engineering has attracted extensive attention of scholars. Some scholars have examined the mechanical properties of bamboo. Because of lacking relevant theoretical systems such as material mechanics test standards, structural design methods, modern construction technical specifications, which are similar to the standards used in steel structure, concrete structure, wood structure, and other materials, moso bamboo is still not fully utilized in modern structure[4]. Nogata Fumin and Tan et al.[5-6] regarded bamboo as a composite material composed of vascular bundle and matrix, and proposed that the mechanical properties of bamboo bundle depend on the density of vascular bundle. The vascular bundle distribution in the longitudinal and radial direction is different, which leads to different mechanical properties of materials in two directions. Chung et al.[7] developed a range of experiments about the mechanical properties and buckling failure of moso bamboo and bambusa pervariabilis mcclure, stated two types of standard values of basic mechanical properties and calculation method of bearing capacity of the compressive bamboo structure. Tommy et al.[8] tested the influence of age, diameter, and other factors on the compressive strength of bamboo, obtained the general range of compressive strength, the results showed that compressive strength is reduced evidently with the increase of outer diameter. Jian et al.[9] analyzed the mechanical properties of the bamboo reinforced concrete beam, the results proved that bamboo possesses good bending strength and can be used as reinforcement for low-cost building concrete structure. Wei et al.[10] carried out a bending test of large-scale rectangular glued bamboo beam, pointed out four typical failure characteristics of glued bamboo beams, the section stiffness is the key to bearing capacity. Zhang et al.[11] tested various mechanical properties of moso bamboo pipe compared with bamboo pieces, it was pointed that there is a little difference in compressive and shear strength between the two specimens, there is a great difference in bending strength, but their variation regularity of mechanical properties is similar. Si et al[12]. analyzed the physical and mechanical properties of different sorts of bamboo, the results showed that there is a linear relationship between the dry density and compressive strength and bending strength parallel to the grain of bamboo species, pointed out a regression equation that can be used to estimate the relationship between the two species. Cheng[13] utilized the moso bamboo in the support of a shallow foundation, the supporting system composed of moso bamboo pipe pile shows better ability

* Corresponding author: dengys2009@126.com

https://doi.org/10.1051/e3sconf/202129303010

in constraining soil displacement and stability. The aforementioned studies examined the bamboo has great potential as an eco-friendly material. But its mechanical properties are related to the position and moisture content of moso bamboo and whether it contains bamboo joint. To make full use of moso bamboo, it is necessary to test its physical and mechanical properties. In the paper, a series of bending tests were carried out to analyze the bending strength, deformation law, and failure pattern of RBP and BPC.

2 Experimental materials

The row materials were 4 years old, selected from a bamboo forest in Xianning, Hubei. There were representatives and no defects. The specimens were placed in the room until the moisture content was 12 ~ 15 %. One meter long bamboo pipe was cut from the top, middle, and lower parts of the whole moso bamboo as specimens for bending test, numbered T1-1~T4-1， M1-1~M4-1, D1-1~4-1. The size of the specimens was 1000 mm long, d mm in diameter, d was 80~120 mm, as shown in Fig. 1.

Figure 1. Specimens for bending test

3 Bending test for moso bamboo

3.1 Bending strength and elastic modulus

(1) Experiment scheme

Both ends of the specimen were simply supported, tested through the third-point bending method, shown in Fig. 2. Five strain gauges were attached on the cross section at the middle span to measure the strain variation. Three displacement meters were placed in the middle span and points of trisection to measure the deflection of the specimen, the data was collected by a static strain test system. According to the *Determination of physical and mechanical properties of bamboo culm* (LY/T2564-2015), the preload was not more than 100 N, applied with uniform speed. The specimens were damaged in 1~2 min. The limit load and deflection at the middle span were recorded.

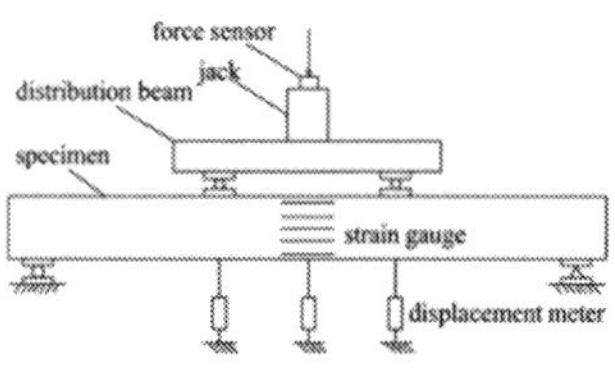

(a) Schematic diagram of bending test

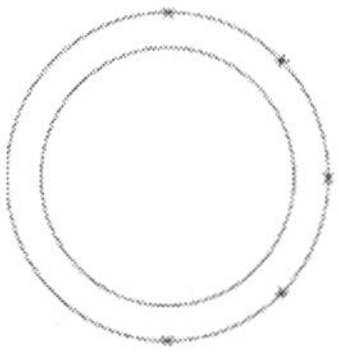

(b) The distribution of strain gauge

(c) Loading diagram

Figure 2. Equipment for bending test of moso bamboo

(2) Results and analysis

From the tested data, the bending strength(σ) and elastic modulus(E) can be calculated by Eq. 1 and 2.

$$\sigma = \frac{F_{\max} \times L \times D_{\min}}{12I} \tag{1}$$

$$E = \frac{23 \times \Delta F \times L^3}{1296 \times \Delta\delta \times I} \tag{2}$$

Where F_{max} is limit load, L is the length of the specimen, D_{min} is the minimum diameter at the loading points, I is the second moment of area of the beam cross-section about the neutral axis, ΔF is the difference between the upper and lower limit loads, $\Delta\delta$ is the deflection change at the middle span between the upper and lower limit load.

At the initial stage of loading, the specimens were in the elastic stage, the development trend of load-displacement curves is basically the same, showing a linear relationship, indicated in Fig. 3. With the continuous increase of load, the plastic deformation of the specimens occurred, the bending specimens located from different parts showed different bearing characteristics. The cracks firstly appeared at the support points of the specimen without bamboo joint, accompanied by a small tearing sound. When the load was up to the limit, it made a huge noise and cracks were obvious. The cracking load of the bending specimen was about 90% of the ultimate, the failure mode is brittle. The typical failure mode of specimens is shown in Fig. 4.

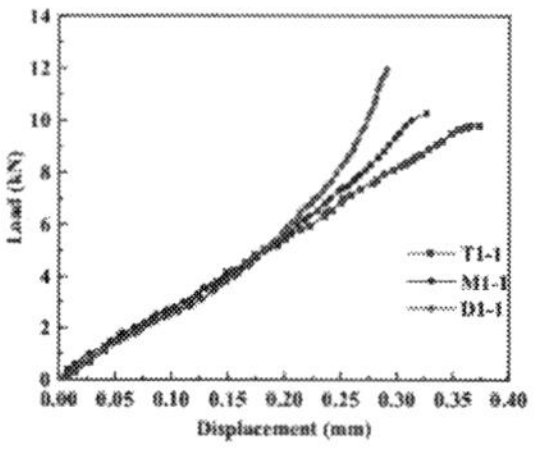

Figure 3. Load-displacement curves at the middle span

 https://doi.org/10.1051/e3sconf/202129303010

The typical failure characteristics of bamboo bending specimens were as follows: there were two symmetrical cracks with the same length in the tensile zone of the section with smaller diameter or the support section without bamboo joint, the angle between the two cracks was 120°, a crack appeared at the bottom the tensile zone of some specimens, the bamboo was damaged by tension. Most of the cracks in the tensile zone were symmetrically distributed, the crack length was approximately the same. It was indicated that the failure of the end section is an important factor influencing the whole bearing capacity of bamboo pipe, and there is an unstable phenomenon at the bamboo joint.

The parameters of specimens and test results are shown in Table 1. The results indicated that the smaller the diameter is, the greater the flexural strength and elastic modulus are. The flexural bearing capacity was up to 12.1 kN.

3.2 Flexural bearing capacity and bending strain

The test load was applied step by step, the preload was 1 kN. At each loading step, the load was 3 kN, the period was 10 min.

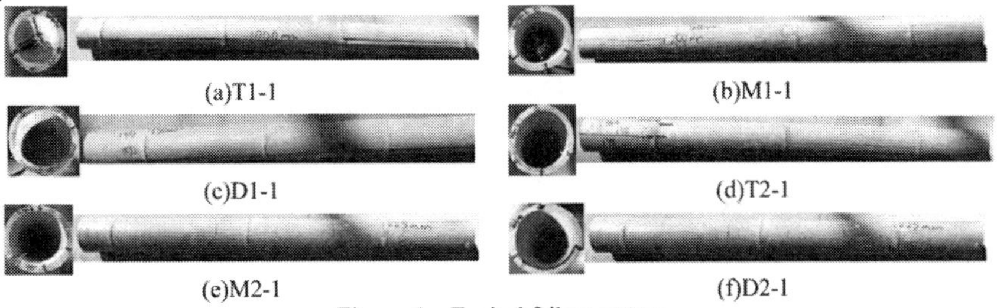

Figure 4. Typical failure pattern

Table1. Parameters and experiment results

Number	D_{min}/mm	L/mm	F_{max}/kN	I/mm4	σ/MPa	ΔF/kN	$\Delta\delta$/mm	E/GPa
T1-1	75.55	900	9.8	971727	57.14	1.9	0.066	381
M1-1	84.57	900	10.3	1381082	47.30	2.1	.094	209
D1-1	88.37	900	12.1	1712128	46.84	2.2	0.101	165
T2-1	78.46	900	6.9	982808	41.31	1.4	0.073	251
T2-1	81.42	900	9.5	1238366	46.85	1.5	0.084	187
D2-1	88.37	900	12.1	1712128	46.84	2.2	0.101	165

Table2. Test results

Number	F_{max}/kN	D/mm	$\varepsilon_c/10^{-6}$	$\varepsilon_t/10^{-6}$	f_t/mm
T3-1	7.3	78.60	1677	3002	0.302
M3-1	6.1	90.14	1515	1911	0.192
D3-1	6.9	98.73	1703	2166	0.206
T4-1	5.0	86.14	1184	1522	0.180
M4-1	7.2	90.58	2159	2378	0.281
D4-1	8.9	104.61	1369	1994	0.179

Where F_{max} is the limit load, D is the diameter of specimens, ε_c, ε_t is the compressive and tensile strain corresponding to the ultimate load, f_t is the deflection at the middle point corresponding to the limit load.

As shown in Table 2, the diameter of specimens is 70~110 mm, the maximum compressive strain is 1100~2200 με , the maximum tensile strain is 1500~3100 με , the maximum tensile strain of the specimens is generally greater than the maximum compressive strain. The load-strain curves of each test point at the section of the middle span are shown in Fig. 5, the negative value is compressive strain, the positive is tensile strain. At the whole loading stage, the load-strain relationship is basically linear in both tension and compression areas.

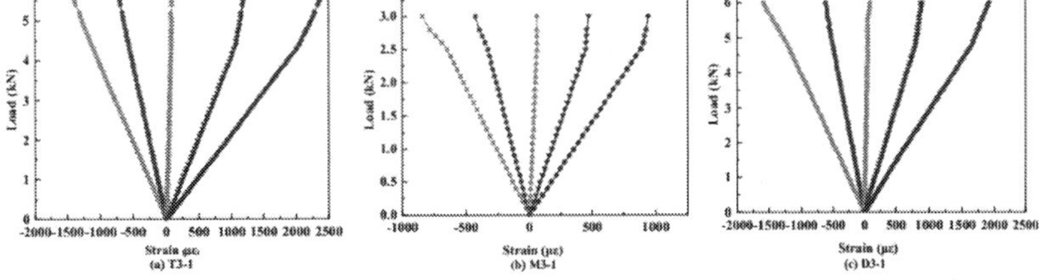

Figure 5. Load-strain curves of the section at middle span

The strain variation along the height of the cross section at the middle span is shown in Fig. 6. It is evident that the development trend of the curves is linear. The phenomenon conforms to the plane section assumption.

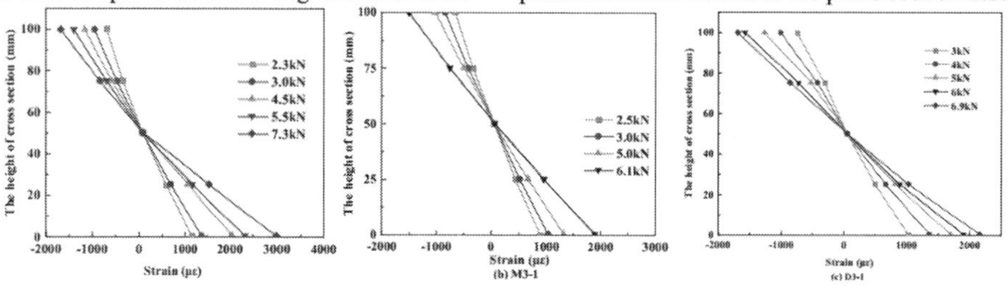

Figure 6. The height-strain curves

4 BPC bending test

4.1 Bending strength and elastic modulus

(1) test scheme

Specimens were six, divided into two groups according to RBP(Y), HBP(K), BPC(KH). The first group was HBP, numbered K-1, K-2, K-3. The second group was numbered Y-1, Y-2, KH-1. According to the *Determination of physical and mechanical properties of bamboo culm* (LY/T2564-2015), the preload, not more than 100 N, was applied with uniform speed, the specimens were damaged in 1~2 min. The limit load and deflection of the middle span were recorded.

(2) Results and analysis

There were different failure modes between RBP, HBP, and BPC. There is a linear relationship between load and displacement at the middle span. When the bamboo pipe was destroyed, there were three cracks at the tensile zone, the one at the bottom, the angle between the other two cracks was 120°. The failure characteristics of RBP and HBP were the same. A few specimens with large diameters were damaged by compression, there were evident cracks at the compressive area. First, the concrete at the tensile area was destroyed. Second, the cracks appeared evidently at the bottom of bamboo pipe. Finally, the bamboo at the compressive area was damaged. There were apparent cracks. The load-displacement curves at the middle span are shown in Fig. 7.

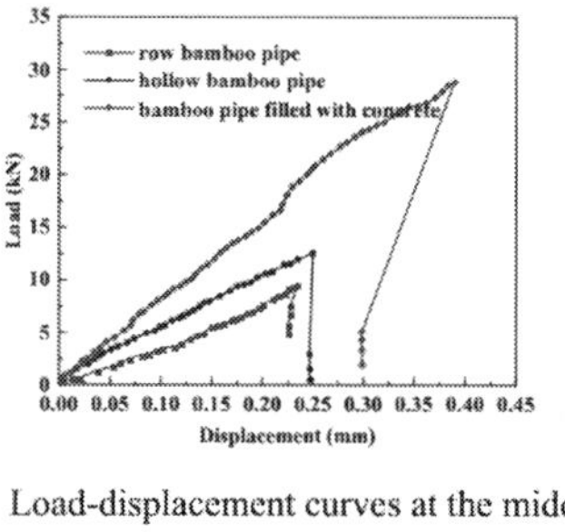

Figure 7. Load-displacement curves at the middle span

The results indicated that the difference between RBP and HBP is little in limit load, bending strength, and elastic modulus, the specimens were from the same bamboo and with the same diameter. The HBP showed greater bearing properties. Therefore, the bamboo joint has little influence on the flexural bearing capacity of bamboo. The flexural bearing capacity of BPC is 2.4 times the HBP with the same diameter, and the bending strength is 1.7 times. The bearing capacity of the composite structure is increased by concrete.

Table3. Parameters and test results

Number	D_{min}/mm	L/mm	F_{max}/kN	I/mm4	σ/MPa	ΔF/kN	$\Delta\delta$/mm	E/GPa
K-1	90.58	900	7.2	1934133	25.29	1.4	0.040	236
Y-1	90.14	900	6.1	1804873	22.85	1.3	0.041	229
K-2	92.71	900	12.4	2271301	37.96	2.4	0.063	218
Y-2	93.52	900	9.3	2258491	28.88	1.9	0.050	216
K-3	115.61	900	12	5800199	17.94	2.5	0.038	146
KH-1	112.09	900	28.7	7743542	31.16	6	0.082	123

4.2 Flexural bearing capacity and bending strain

(1) Test scheme

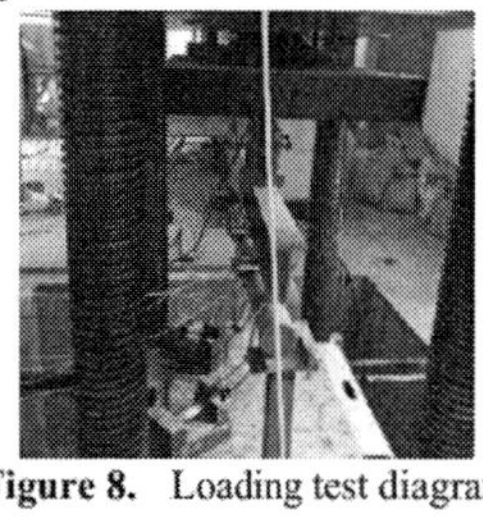

Figure 8. Loading test diagram

Specimens were six, divided into three groups. The first group was RBP, numbered Y-4, Y-5, the second group was HBP, numbered K-4, K-5, the third group was BPC, numbered KH-4, KH-5. The load was applied step by step. The preload was 1 kN, the load was 3 kN in each loading step. The period of each step was 10 min. the loading diagram is shown in Fig. 8.

(2) Test results and analysis

The test results are shown in Table 4. The diameter of specimens is 90~120 mm. For HBP bending specimens, the maximum compressive strain is 1100~2200 με, the maximum tensile strain is 2000~3400 με. For BPC, the maximum compressive strain is 2600~3400 με , the maximum tensile strain is 4500~6100 με. The maximum tensile strain of bending specimens is generally greater than the compressive, and the maximum tensile strain of BPC is 1.8 times the compressive. The maximum strain of BPC specimens is 2.5 times the HPC.

The load-strain curves of each test point at the section of the middle span are shown in Fig. 9 and 10. It is indicated that the strain is linearly related to the load in both compressive and tensile areas at the whole loading stage

Table4. Test results

Number	F_{max}/kN	D/mm	$\varepsilon_c/10^{-6}$	$\varepsilon_t/10^{-6}$	f_t/mm
Y-4	6.1	90.14	-1515	1911	0.192
Y-5	6.9	98.73	-1089	2166	0.206
K-4	7.2	90.58	-2519	2378	0.281
K-5	8.9	109.88	-1074	1994	0.179
KH-4	21.4	110.40	-2627	4554	0.294
KH-5	27.1	110.80	-3388	6031	0.267

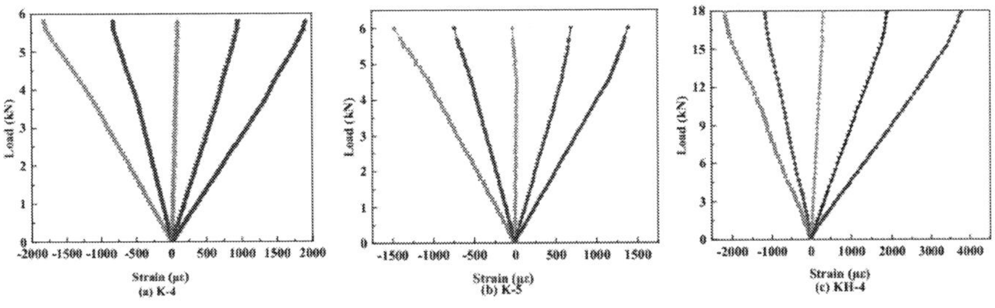

Figure 9. Load-strain curves of the bamboo section at middle span

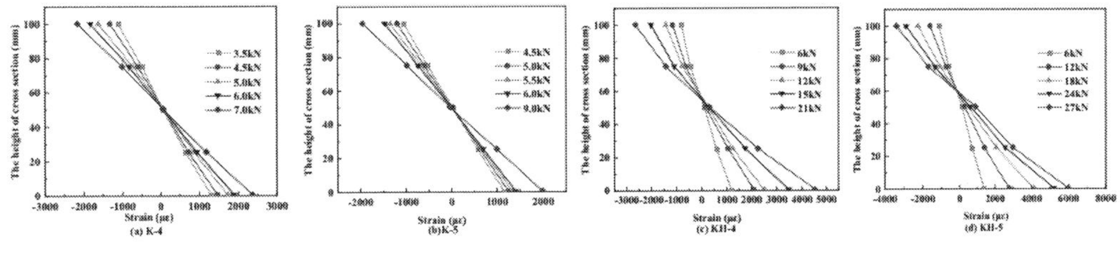

Figure 10. The height of cross section-strain curves

5 conclusion

(1) The diameter of the bamboo pipe was 70~100 mm, the smaller the diameter was, the greater the bending strength and elastic modulus were. The flexural bearing capacity of RBP was up to 12.1 kN.

(2) The maximum strain of the BPC bending specimen was 2.5 times the the HBP specimen. The load-strain relationship of HBP and BPC at each test point was basically linear, whether in the tensile or compressive zone. The relationship conformed to the plane section assumption.

(3) The bending failure mode of RBP, HBP, and BPC was different. The difference of RBP and HBP from the same bamboo was a little in limit load, bending strength, and elastic modulus. The flexural bearing capacity of BPC was 2.4 times the HBP with the same diameter, and the bending strength was 1.7 times.

References

1. LI Hui, LI Xiaohua, LI Xingong, et al. Development Status of Standards on Bamboo Structure Engineering Material[J]. China Standardization, 2012, (10): 89-92.
2. ZHANG Qisheng. Industrial Utilization of Bamboo in China[M]. Beijing: China Forestry Publishing House, 1995:1-12.
3. ZHANG Qisheng. Attaching Importance to Science and Innovation in the Processing and Utilization of Bamboo Timber in China[J]. Journal of Zhejiang A&F University, 2003, 20(1): 1-4.
4. XIAO Yan, SHAN Bo. GluBam Structures[M]. Beijing: China Architecture & Building Press, 2013, 8.
5. Nogata Fumio. Intelligent Functionally Graded Material: Bamboo[J]. Composites Engineering, 1995,5(7):743-751.
6. Tan T. Mechanical Properties of Functionally Graded Hierarchal Bamboo Structures[OL]. Acta Biomater, 2011.

https://doi.org/10.1051/e3sconf/202129303010

7. K.F. Chung, W.K. Yu. Mechanical Properties of Structural Bamboo for Bamboo Scaffoldings[J]. Engineering Structures, 2002, 24: 2-10.
8. Y. Lo Tommy, H. Z Cui, et al. The Effect of Fiber Density on Strength Capacity of Bamboo[J]. Science Direct, 2004, 58(21): 2595-2598.
9. Jain SK, Kurhekar SP, Wadekar C. Strength Analysis of Bamboo and Steel Reinforced Concrete Beam[J]. Engineering& Technology in India , 2015, 6(1): 41-45.
10. WEI Yang, JIANG Shenxue, ZHANG Qisheng, et al. Experimental Study on Flexural Performance of Bamboo Beam[J]. Building Structure, 2010(1): 88-91.
11. ZHANG Dan, WANG Ge, ZHANG Wenfu, et al. Mechanical Properties of Phyllostachys Pubescens[J]. Journal of Central South University of Forestry&Technology, 2012, 32(7): 119-123.
12. SITU Chunnan, DU Wenjun, WANG Jian, et al. The Relationship between Air Dry Density and Bamboo Compressive and Flexural Strength of the Main Bamboo Species in Chishui City, Guizhou Provience[J]. Journal of Bamboo Research, 2012, (02): 16-19.
13. CHENG Zhihe. Study on Bearing Characteristics of Dentate Row Bamboo Pile-Composite Soil Nailing System[D]. Xi'an University of Science and Technology, 2020.
14. Y/T2564-2015, Determination of physical and mechanical properties of bamboo culm[S]. Beijing: Standards Press of China, 2015.

Validation and Optimization of Heat and Mass Transfer Model for Mushroom Convection Drying

Huimin Shan[1], and Kongqing Li[1]

[1]School of Civil Engineering, Hunan University of Science and Technology, Xiangtan Hunan 411201, China

Abstract: In order to solve the problems of time consuming, energy consumption and low simulation accuracy in the hot air-drying system of food drying. Using computational fluid dynamics (CFD) to simulate the drying process of mushrooms can provide a reference for its technology research and development (R&D). The porous media approach was used to model the flow resistance offered by the mushroom. The resistance coefficient and porosity were determined through the experiments. Different air supply ways (wind speed, temperature, fresh air volume, reverse air supply period) were simulated and two optimize ways were suggested according the shortest possible drying time. The simulation results are in good agreement with the experimental results. The air supply way of periodic with reverse (SMPR), which means timed alternate the direction of air flow, or the way of mix in fresh air intermittently can effectively shorten the drying time. Considering comprehensively, the optimal air supply way was the mix in fresh air intermittently with fresh air accounts for 30% or SMPR with the period of 2h under the condition of the hot air temperature of 55°C and wind velocity of 0.6m/s.

1 Introduction

Drying is one of the most energy-consuming and complex devices. There are still many deficiencies in hot air-drying technology for food, such as long process time, large energy consumption, poor uniformity, etc. Modeling technology has great potential to develop drying technology. It can optimize the current drying equipment design and drying process. Moreover, the numerical model has cost advantages in industry, such as no additional lab space, maintenance and technical personnel operating equipment, reducing measurement and lab full-scale testing. Drying model plays an important role in drying technology R&D [1].

The accuracy of dry simulation based on CFD is limited, which is mainly due to the modeling method and technology [2-4]. The drying process is related to the knowledge of material properties, which are not well known. These have a great influence on the accuracy of the drying model results [5-7]. Increasing the reliability and confidence of numerical models can promote the development of drying technology. For specific drying materials, drying ways has to be guided by experience and validation studies.

2 Drying experiment

2.1 Properties experiment

The drying effect varies with the material properties. Mushroom material properties: heat capacity at constant pressure is 1800J/(kg·K); The diffusion coefficient takes an average of $10^{-9}m^2/s$; Density is $860kg/m^3$; Thermal conductivity is 0.515W/(m·K)[8].

2.1.1 Materials and instruments

Fresh mushrooms samples were procured from local vegetable market and had a moisture content of about 80%. The air heated by an electric heater (PTC) was fed into drying box with a volume of $14\times14\times14cm^3$ by an axial flow fan (FD3-2). The inlet and outlet were both $14\times14cm^2$ with opposites. During the experimental process, a thermal anemometer (TSI-9565, USA) was used to measure the wind speed at the inlet of the drying box. A compensated micromanometer (YJB-1500) was used to measure the pressure difference between the inlet and outlet of the drying box, and a thermocouple temperature control box (1A1-K) was used to control the temperature of the supply air. The moisture loss of the sample was recorded at every 1h interval by a precision electronic balance weighing (BSM-220.4).

2.1.2 Experimental scheme

2.1.2.1 Resistance experiment

The porous media approach was used to model the flow resistance offered by the mushroom. The resistance is derived from experimental data of pressure drop at

Kongqing Li: likongqing@qq.com

different velocity of air passing by. The relationship between pressure drops and wind speed was shown into Fig. 1. Fitting result and the goodness are also presented in it. The correlation coefficient of the fit R^2 is 0.9691.

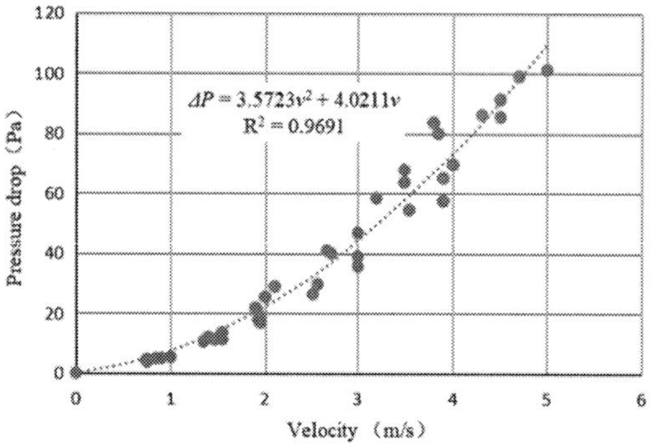

Figure 1. Air Supply Speed at the Inlet and Pressure Drop across the Drying Box

The influence of porous media is modelled by specifying a momentum source term in the standard fluid flow Equation (1) as:

$$S_i = -\left(\frac{\mu}{\alpha} v_i + \frac{1}{2} C_2 \rho |v| v_i\right) \quad (1)$$

Where, v is the velocity of air, m/s; μ is the viscosity, Pa·s; ρ is the density of air, kg/m^3; ΔP is the pressure drop, Pa. where α is the permeability ($1/\alpha$=1604589m^{-2}) and C_2 is the inertial resistance factor. (C_2=41.6m^{-1}). The viscosity resistance drag coefficients and inertial drag coefficients are obtained by sorting out formula (1) and fitting results.

2.1.2.2 Porosity experiments

Since the whole box space were regarded as a porous medium in the following simulations. The procedure of porosity measurement was as follows: the initial apparent volume of the mushroom in its natural state was V which can be approximately considered as the volume of the box. The actual volume of mushrooms was measured by drainage method: the mushrooms were reasonably laid in a thin plastic bag, and the air in the bag was discharged with the pressure of water, then the actual volume of mushrooms Vi was measured. The porosity of can be calculated by:

$$P = \frac{V - V_i}{V} \quad (2)$$

The porosity P is divided into four stages according to the drying time, as shown in Table 1.

Table1. The porosity of mushrooms varies with time

Drying time / (h)	Total volume of the box / (m^3)	Actual volume of mushroom / (m^3)	porosity
0~1	2.74×10^{-3}	1.37×10^{-3}~1.26×10^{-3}	0.5
1~4	2.74×10^{-3}	1.23×10^{-3}~ 1.13×10^{-3}	0.55
4~6	2.74×10^{-3}	1.10×10^{-3}~9.01×10^{-4}	0.6
>6	2.74×10^{-3}	8.51×10^{-4}~8.11×10^{-4}	0.69

2.2 Drying test and model

2.2.1 Drying parameters

The experiment was carried out in winter, and the moisture content was not controlled due to the small difference in humidity in the air. Experimental conditions are shown in Table 2:

Table2. Mushroom drying experimental conditions

NO.	Dry-bulb temp./(°C)	Wet-bulb temp./(°C)	Relative humidity/(%)	Air supply velocity/(m/s)
1	45	20	7.36	1
2	55	23	4.6	1
3	65	26	3.31	1

Table 4. Modified formula of drying kinetics model at different temperatures

Temp. of supply air /(°C)	Formula of drying kinetics model	R^2
45	M_t=3.27exp(-(7.83E-07)×t$^{1.47}$)+1.59exp(-(9.2E-06)×t$^{1.47}$)	0.99995
55	M_t=3.71exp(-(3.71E-06)×t$^{1.26}$)+1.15exp(-(9.13E-05)×t$^{1.26}$)	0.99989
65	M_t=-39.39exp(-(2.18E-04)×t$^{0.91}$)+44.24exp(-(2.29E-04)×t$^{0.91}$)	0.99971

2.2.2 Selection and calculation of hot-air drying models

Calculation formula of dry base moisture content (M_t):

$$M_t = \frac{W_t - G}{G} \quad (3)$$

Where, W_t is the material mass at any time point in the drying process, g; G is the mass of the material when it is completely dried (that is the dry base mass).

Match the experimental results with the commonly used drying model[9-17] (as shown in Table 3). Finally, it was found that Hii model was the most consistent with the law of mushroom drying. Fig. 2 and Table 4 show the model fitting results of M_t changes with time under different drying temperatures.

https://doi.org/10.1051/e3sconf/202129303011

Table 3. Kinetic model of mushroom drying

NO.	The model name	Model formulation
1	Newton	M_t= exp(-k t)
2	Henderson and Pabis	M_t = a exp(-k t)
3	Page	M_t= exp(-k t^n)
4	Logarithmic	M_t= a exp(-k t)+c
5	Two term model	M_t = a exp(-k t) + b exp(-g t)
6	Verma et al.	M_t= a exp(-k t) + (1-a) exp(-g t)
7	Midili-kucuk	M_t = a exp(-k t^n) + b t
8	Wang and smith	M_t = 1+a t+b t^2
9	Hii	M_t = a exp(-k t^n)+b exp(-g t^n)

Where, t is time, M_t is dry base moisture content, others are constants.

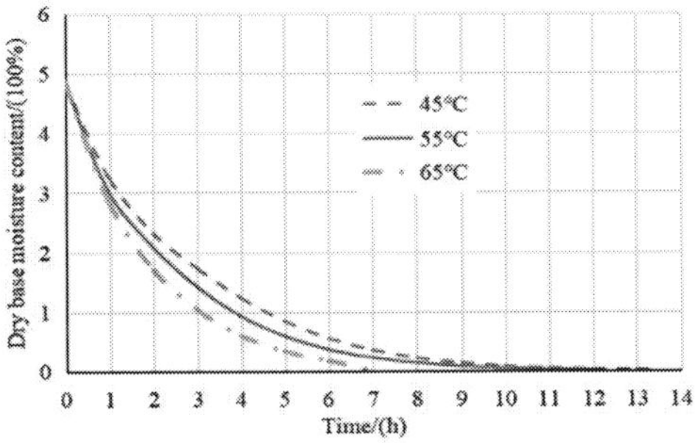

Figure 2. Relationship between Dry Base Moisture Content and Time at Different Temperatures

3 Model establishment

3.1 Numerical model

The turbulence model used in the air flow simulation is the k-Ɛ model [18]. In the dynamic equilibrium state of water absorption and resolution of mushroom, the main factors that affect the amount of water released or absorbed by mushroom include temperature and the concentration of water vapor in the surrounding air. The concentration of water vapor in the air is equal to the sum of the concentration of water vapor in the air and the water vapor evaporated from the mushroom during the drying process. Vapor concentration equation is as follow:

$$\frac{\partial c}{\partial T} = D_A\left(\frac{\partial^2 c}{\partial x^2}+\frac{\partial^2 c}{\partial y^2}+\frac{\partial^2 c}{\partial z^2}\right)-\rho_s\frac{1-V_k}{V_k}\frac{\partial M}{\partial t} \quad (4)$$

The basis for establishing the temperature equation of dry-base matter is that the temperature change of mushrooms is composed of two parts: one is the heat exchange between mushrooms and hot air, the other is the heat consumed by the evaporation of water vapor in mushrooms. Its equation is as follow:

$$\frac{\partial T}{\partial t} = \frac{k}{\rho c}\left(\frac{\partial^2 T}{\partial x^2}+\frac{\partial^2 T}{\partial y^2}+\frac{\partial^2 T}{\partial z^2}\right)+\frac{q}{\rho}\frac{\partial M}{\partial t} \quad (5)$$

The calculation formula of mass transfer rate is:

$$\frac{\partial M}{\partial t} \approx \frac{\Delta M}{\Delta t} \quad (6)$$

The mass of water vapor escaping is the product of the change in dry base moisture content and the dry base mass:

$$\Delta M = \Delta M_t \times G \quad (7)$$

In the simulated drying process, the parameters of each cell are different. Therefore, when calculating the dry basis moisture content of each cell, interpolation should be used to calculate the dry basis moisture content formula with the known temperatures:

$$M_t = M_{t_1} - \frac{M_{t_1}-M_{t2}}{t_1-t_2}(t_1-t_0) \quad (8)$$

Where, D_A is the diffusion coefficient of water vapor in the air, m^2/s; V_k is the volume occupied by air in the porous medium, m^3; T is the temperature of the mushroom, K; c *is* water vapor concentration, kg/m^3; t is drying time, s; ρ_s is wet density of mushroom, kg/m^3; ρ is the apparent density of mushrooms, kg/m^3; k is thermal conductivity, W/(m·K); ΔM is the amount of water vapor escaped, kg; M_{t1},M_{t2} are the moisture content of the dry basis when the temperature is t_1 and t_2, respectively, where t_1 and t_2 are the two temperature values closest to 45°C, 55°C and 65°C for t_0.

3.2 Physical model

The simulation target was a three-dimensional model of the drying box and transient simulations were carried out. The dimension of the box was 3m length, 2m width by 2m high (along x-axis, z-axis and y-axis, respectively). 0m~3m, z-axis: -1m~1m; y-axis: 0m~2m. The entrance was located in the yz plane with x=0m, and the exit was located in the yz plane with x=3m. Three sets of simulation experiments with the initial temperature of 17°C were performed.

The air supply parameters set in the simulations are consistent with those in the experiments. In order to reduce the drying time, energy consumption, and avoid excessive drying, the relative humidity at the end of the drying box needs to be reduced by PRAS or mix in fresh air. Three kinds of simulations have been performed in this paper.

- The first kind one was conducted under three set temperatures of 45°C, 55°Cand 65°C and three wind velocities of 0.4m/s, 0.6m/s and 0.8m/s, in which temperature and velocity of hot drying air at the inlet were kept constant in the whole process of drying.
- The second kind was to add a way to mix in fresh air on the basis of the first kind , it was conducted under three fresh air ratio of 20%, 30%, 40%, in which the mixed fresh air temperature is 5°C, and moisture content is 4.1g/kg, and mixed air supply 10 minutes for every 1 hour.
- The third kind was conducted under three set temperatures of 45°C, 55°C and 65°C, three wind velocities of 0.4m/s, 0.6m/s and 0.8m/s, and three period of 2h, 3h and 4h, in which air supply direction changed every half period.

Reduce unnecessary workload through orthogonal experiment, 9 groups of simulation experiments were needed for each case.

 https://doi.org/10.1051/e3sconf/202129303011

4 Result and analysis

4.1 Accuracy analysis of simulation

As shown in Figure 3. When the temperature was 45 °C, 55 °C and 65°C, the maximum relative errors between the experiment and simulation were 0.52%, 3.35% and 1.36%, respectively, in the drying process. The error tends to increase over drying time. The reasons may be caused by iterative convergence error and discrete error in the simulation process. In addition, the error increases as temperature increasing. This was because the increase of temperature in the actual drying process leads to a large amount of shrinkage of the mushroom, and the drying rule of the different moisture between surface and interior causes a large change in the diffusion rate of internal moisture to the surface. And the diffusion coefficient of this model was averaged. The simulated drying time at 45 °C, 55 °C and 65°C was 13.4h, 12h and 6.8h, respectively, which was similar to the experimental results.

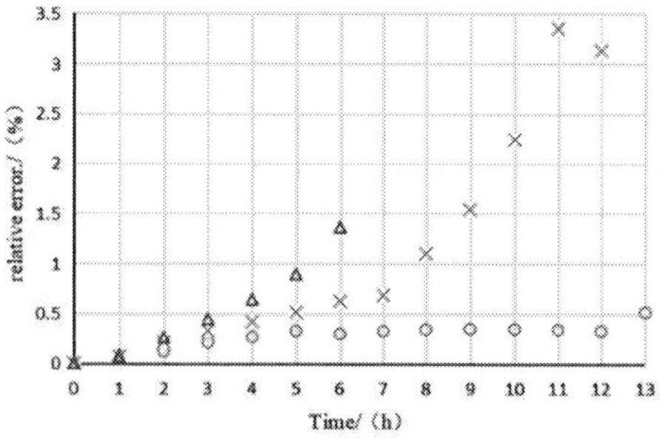

Figure 3. Relative Error between Experimental and Simulated Values of Dry Base Moisture Content of Mushrooms Per Hour at 45、55、65°C

4.2 Optimization of air supply scheme

The results of the two simulation methods are used as typical examples: one was conducted under temperature of 55°C, wind speed of 0.8m/s, fresh air ratio of 30%, and the other was under temperature of 45°C, wind speed of 0.4m/s, period of 2h (other similar figures were omitted due to space limitation). The drying box were divided into 9 parts along the direction of air supply. The dry basis moisture content of each part at different time is shown in Fig. 4 and Fig. 5. The initial decrease rate of dry basis moisture content is fast, but with the process of drying, the time needed gradually becomes longer when the dry basis moisture content decreases by 1. This is due to lower internal moisture content and a slower rate of water diffusion from interior region.

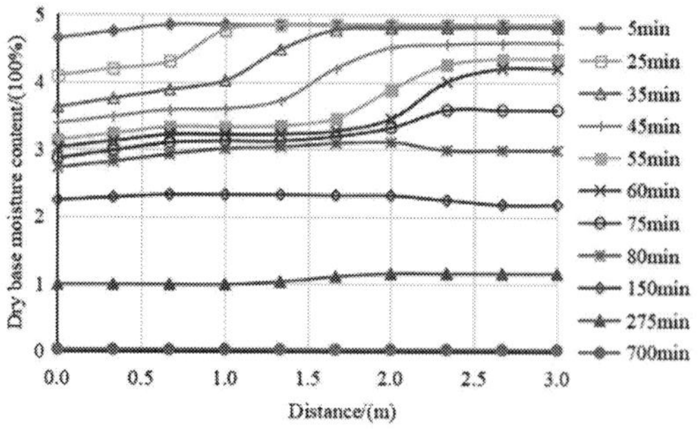

Figure 4. Dry Basis Moisture Content (Mt) at Different Time Points at method of periodic reverse the direction of air, in which was conducted under periodic of 2 hour, temperature of 45°C and wind velocity of 0.4m/s

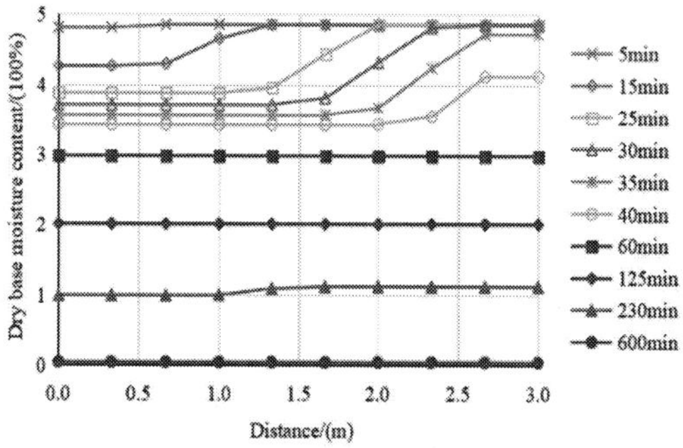

Figure 5. Dry Base Moisture Content (Mt) at Different Time Points at mix in Fresh Air, in which was conducted under Fresh Air Ratio of 30%, temperature of 45°C and wind velocity of 0.4m/s

For the direct air supply, the closer to the outlet, the higher dry basis moisture content (significantly at the initial drying stage), because the relative humidity of air is one of the main factors affecting the drying rate. Simulated physical model according to the industrial production needs of the stacking volume of mushroom. Mushrooms near the entrance have the priority of convective heat and humidity exchange. In this process the increase of moisture content in the air decreases their drying ability. Therefore, the method of periodic reverse air supply and introduce the fresh air with low humidity in time can not only ensure the uniformity of drying, but also reduce drying time and energy consumption.

The total drying time under different working condition is shown in Table 5.

Table 5. Summary of drying time under various conditions

Air supply with constant		Mix in fresh air		SMPR	
Temp-wind speed /(°C-m/s)	Drying time /(min)	Temp-wind speed-fresh air ratio /(°C-m/s-%)	Drying time /(min)	Temp-wind speed-period /(°C-m/s-h)	Drying time /(min)
45-0.4	730	45-0.4-20	695	45-0.4-2	700
45-0.6	720	45-0.6-30	720	45-0.6-4	700
45-0.8	700	45-0.8-40	700	45-0.8-3	690
55-0.4	690	55-0.4-20	650	55-0.4-3	640

55-0.6	670	55-0.6-30	600	55-0.6-2	600
55-0.8	670	55-0.8-40	600	55-0.8-4	600
65-0.4	375	65-0.4-20	370	65-0.4-4	365
65-0.6	370	65-0.6-30	370	65-0.6-3	365
65-0.8	365	65-0.8-40	365	65-0.8-2	365

The method of periodic reverse the direction of air can solve the situation that the relative humidity of the air in the exit section is higher than that in the entrance section, as shown in Fig. 4 after 80mins later, which will lead to low drying speed at the exit section and excessive drying at the entrance section. The method of introducing fresh air can reduce the relative humidity of air from the source, increase the concentration difference and water vapor partial pressure difference, promote the overflow of water, and reduce the drying time.

Temperature plays a more important role on drying rate than wind speed does. At the same temperature, the higher wind speed, the shorter drying time, but it is not obvious. At the same wind speed, the higher the temperature, the faster the drying rate is, and it is more significant. The main reason is that the temperature and enthalpy of mushrooms increase during the drying process, which can promote the evaporation of water. However, temperature will affect the shrinkage degree of the material, and the higher temperature, the greater shrinkage, as shown in Figure 6. The shrinkage of 65°C is obvious, and the difference between 45°C and 55°C is small. Therefore, the optimal working condition is temperature of 55°C, wind speed of 0.6m/s, fresh air ratio of 30% and reverse air supply with a period of 2 hours, both of which are shortened by 70min.

Figure 6. The Original State of Mushrooms and the State at Each Temperature When Drying Completely

5 Conclusions

Based on the mechanism of heat and moisture transfer and drying characteristics of mushrooms, a drying model of mushrooms with high accuracy was established. The simulation results can guide mushroom production practice. This simulation method provides a new idea for other food drying industry production and has a reference significance. The contents are as follows:

- Measure and calculate mushroom physical parameters: resistance coefficient, porosity.
- Drying kinetics model: according to the experimental results and the commonly used drying kinetics model, the Hii model has the highest fitting degree, $R^2 > 0.999$.
- A new mathematical calculation model was proposed, and the UDF program was coupled to the CFD. The maximum relative error between the experimental and simulation results was 3.35%.
- Three air supply ways are simulated: air supply with constant, mix in fresh air and SMPR. Drying time, uniformity, appearance shrinkage, energy saving and other factors should be considered. Analytically there are two optimal dry conditions. One is mix in fresh air, in which is conducted under temperature of 55°C, wind speed of 0.6m/s and fresh air ratio of 30%. The other is SMPR. It is conducted under temperature of 55°C, wind speed of 0.6m/s and period of 2 hours. Both of them can shorten the drying time by 70 minutes.

References

1. T. Defraeye, Appl. Energy 131, 323-344 (2014)
2. T. Norton, D.W.Sun, Trends in Food ence & Technology 17,600-620 (2006)
3. M. Casey, T. Wintergerste, Best Practice Guidelines, (2000).
4. D. Lacasse, R. Turgeon, D. Pelletier, Int. J. Therm. Sci. 43, 925-938 (2004)
5. A.K. Datta, J. Food Eng. 80, 96-110 (2007)
6. A.K. Datta, Compr. Rev. Food. Sci. Food Saf. 7, 121-129 (2008)
7. T. Gulati , A.K. Datta, J. Food Eng. 116, 483-504 (2013)
8. Z.Y.Zhang, Y.Q. Li, C.H. Liu,et al. Food Sci. 623, 239-246 (2020)
9. A. Ayensu, Sol. Energy 59, 121-126 (1997)
10. M. Kashaninejad, A. Mortazavi, A. Safekordi, et al, J. Food Eng. 78, 98-108 (2007)
11. V.T.Karathanos, V.G. Belessiotis, J. Agr. Eng. Res. 74, 355-361 (1999)
12. I. Doymaz, J. Food Eng. 65, 205-209 (2004)
13. Z. Wang, J. Sun, X. Liao, et al, Food Res. Int. 40, 39-46 (2007)
14. X.H. Guo, C.Y. Xia, Y.R Tan, et al, J. Integr. Agric. 13, 207-216 (2014)
15. A. Midilli, H. Kucuk, Energy Conv. Manag. 44, 1111-1122 (2003)
16. C.Y. Wang, R.P. Singh, (1978)
17. C.L. Hii, C.L. Law, M. Cloke, J. Eng. Sci. Tech. 3 (2008)
18. X.T. Li, B. Zhao, Numerical simulation of indoor air flow(CMP, Beijing, 2009)

A study on the Health Warning of Provincial Capital Cities in China based on the Normalized Model

Guoquan Qian[1,2*], Jinxiang Wang[1,2], Xiaoying Nie[1], Mingtao Li[1] and Tianpeng Gao[2,3*]

[1]College of Geography and Environmental engineering, Lanzhou City University, Lanzhou 730070, China

[2]Engineering Research Center of Mining Pollution Treatment and Ecological Restoration of Gansu Province, Lanzhou City University, Lanzhou 730070, China

[3]School of biology and environmental engineering, Xi'an University, Xi'an 710065, China.

Abstract: Along with the rapid advance of urbanization, urban disease appear. For detection city disease earlier, the ecological city is put in this article. The ecological city is regarded as the human body to plan, construct and manage city with the thought of "law in the human body". The city's population, land, water, air, transportation, culture, are considered as the body's nerves, muscles, blood, breath, limbs and the vigor, respectively. And a quantitative standard is determined for each object of study. If the quantitative value is abnormally large or small, it indicates that the city is unhealthy, the early warning system for city will put forward, the causes will lookup and the treatment will give.

1 Introduction

China's urbanization is advancing rapidly, with the urbanization rate of permanent residents rising from 17.9 percent in 1978 to 60.6 percent in 2019. In 2019, there will be 684 cities, 21,013 towns, and 60,300 square kilometers of built-up urban areas. With the rapid growth of the number of cities, "urban diseases" also appear, such as population expansion, housing shortage, employment difficulties, traffic congestion, environmental degradation and frequent public security incidents etc. It not only hinders the healthy and orderly development of the city, but also affects the quality of life of residents [1]. Under this background, it is proposed to establish urban early warning system and scientifically govern the city.

2 The theoretical basis of the establishment of urban early warning system

In 1933, the Athens Charter put forward the "organic city" for the first time, which believed that the various functions of a city should be laid out reasonably and run smoothly. In 1977, the Charter of Machu Picchu proposed that as a living organism, the inner order of a city is formed through the self-organizing behavior caused by the complex interaction of various components in the system, and maintains the balance of "life" in the overall association. In 1943, Finnish scholar Eliel Saarinen gave a systematic exposition of the theory of urban organic evacuation in his book The City, Its Growth, Decline and Future. He thought tends to decline of city today, need to have a on the basis of the principle of reasonable urban planning, the revolutionary changes in the structure of the city has a good, for the healthy development of heavy industry layout and light industry should be evacuated from the city center, only the enterprise administrative department and urban layout in in the center of the city, Large areas of land left vacant by industrial relocation in central urban areas should be used to increase green space and residential land for technicians, administrative personnel and business people who must work in central urban areas so that they can enjoy family life close to them. He believed that such a structure should not only accord with the nature of human habitation, but also facilitate people to live a common social life and feel the pulse of the city, while not divorced from the nature [2].

In 2019, Professor Yang Zhi [3], dean of the School of Marxism at Harbin Institute of Technology, took charge of the major research project of philosophy and social sciences of the Ministry of Education, "Theoretical Research on Socialist Urban Organism with Chinese Characteristics". He also proposed the idea that cities are living organisms.

Based on the above understanding of the city as a living entity[4], we believe that the urban construction with Chinese characteristics can be planned, constructed and managed with the idea of "law in human body". The core of "Law on Human Body" is to treat the city as an organism like human being, in which the population is like human's nerves to the city. Land to the city is like human muscle; Water to a city is like human blood, and air is like human breath to a city. Traffic to the city, like human limbs; Culture is as vital to a city as it is to a person. Planning, construction and management of cities to the most basic

*Corresponding author: 599037529@qq.com; zkgtp@163.com; ORCID: 0000-0001-6821-3836

elements of the population, land, water, air, transportation, culture, as the research object, for each basic elements to determine a quantitative standard, if these basic elements of quantitative values unusually large or small, lesions appears in the city, city to put forward the corresponding early warning, To find out the causes of urban diseases and give the corresponding treatment plan [5,6].

3 Empirical research

3.1 The determination of early warning index of urban disease

Table1. Warning Value of City Disease

Physical objects	Physical indicators	Prewarning value	Physical examination result
population	1. Population density	Population greater than 10,000 per square kilometer or less than 5,000 per square kilometer	unhealthy
	2. The difference between the urbanization rate of household registration and the urbanization rate of permanent resident population	More than 10%	unhealthy
land	3. Green coverage in built-up areas	Less than 41%	unhealthy
	4. Land use and development intensity	The investment per mu of land should not be less than 1 million yuan	unhealthy
water	5. Sewage treatment rate	Less than 97%	unhealthy
	6. Per Capita Daily Domestic Water Consumption	More than 530 cubic meters or less than 330 cubic meters	unhealthy
air	7. Good air quality days	less than 310 days	unhealthy
	8. Inhalable particulate matter in the air, PM10	More than 150 milligrams per cubic meter	unhealthy
traffic	9. The time it takes a motor vehicle to travel 10 kilometers	More than 30 minutes	unhealthy
	10. The quotient of motor vehicle ownership and per capita road area	Less than 4	unhealthy
cultural	11. College degree or above per 100,000 people	Less than 10,000	unhealthy
	12. Percentage of cultural industries in GDP	Less than 5%	unhealthy

3.2 Comprehensive study by model normalization method

Six indicators, including population density, green coverage rate of urban construction, sewage treatment rate, days with good air quality, public satisfaction rate of urban traffic, and the proportion of the tertiary industry, were used as the early warning value index of urban diseases. The comprehensive index of 31 provincial capitals was calculated by the comprehensive index method [7]. The main calculation process was divided into the following three steps:

Step 1: Standardize the data

Since the units of the six indicators are different, in order to make comprehensive analysis among the indicators by mathematical methods and standardize or non-dimensional processing of the data, the following methods are adopted:

Representation of positive indicators:

$$X = X_i / X_{ref} \tag{1}$$

Representation of a negative index:

$$X = X_{ref} / X_i \tag{2}$$

Representation of a semi-positive index:

$$X = \begin{cases} X_i / X_{ref}, & \text{When } X_i \text{ is less than } X_{ref}; \\ X_{ref} / X_i, & \text{When } X_i \text{ is less than } X_{ref}. \end{cases} \tag{3}$$

Where X represents the standardized assignment of data indicators, X_i represents the original data collected by data indicators, and X_{ref} is the reference standard value of data indicators. In order to facilitate calculation, the reference standard value X_{ref} can be selected in the following three situations:

$$X_{ref} = \lambda \max_{\mathrm{i}} X_i \quad (4)$$

$$X_{ref} = \lambda \min_{\mathrm{i}} X_i \quad (5)$$

$$X_{ref} = \lambda \operatorname{ave}_{\mathrm{i}} X_i \quad (6)$$

Where $\lambda = 1$.

Step 2: index weight determination -- entropy value method

Since the aspects reflected by the early warning value of various indicators for urban diseases are not completely average, it is necessary to classify the indicators and introduce the concept of weight to make the evaluation system more objective and reasonable. Entropy method is to determine the weight of each index according to the amount of information provided by the observation value of each index. The main basis of this method is to use the difference degree of the observed values to determine the weight of the index (the greater the difference of the observed values, the greater the weight, and otherwise, the smaller the weight). Calculation steps:

Index weight is to determine the importance of each index to the whole index evaluation system. Entropy value method is to determine the weight of each index according to the variation degree of measured value, which belongs to objective weighting method. Entropy value method is used to calculate the weight of each index in this report, which provides a basis for comprehensive evaluation of multiple indexes. The specific calculation process is divided into the following three steps:

1) Calculate the entropy value of each index:

$$U_{\mathrm{j}} = -\sum_{\mathrm{i=1}}^{\mathrm{m}} X_{ij} \ln X_{ij} \quad (7)$$

Where m is the sample number;

2) Entropy reverse:

$$S_{\mathrm{j}} = \frac{\max U_{\mathrm{j}}}{U_{\mathrm{j}}} \quad (8)$$

3) Determine the weight:

$$W_{\mathrm{j}} = \frac{S_{\mathrm{j}}}{\sum_{\mathrm{j=1}}^{\mathrm{n}} S_{\mathrm{j}}}, \ (\mathrm{j=1,2,\ldots n}) \quad (9)$$

Step 3: Calculation of composite index

The comprehensive index of the early warning value of the urban disease in the whole region can be calculated by integrating all indicators. The calculation formula is as follows: $EQ(t) = \sum_{\mathrm{i=1}}^{\mathrm{n}} W_i(\mathrm{t}) \times X_i(\mathrm{t})$, (i=1,2,...n)

(10)

Where, Wi is the weight of evaluation index I; Xi represents the standardized value of the evaluation index; N is the total number of indicators.

The weight calculation results of the six indicators are shown in Table 2. It can be seen that the sewage treatment rate with the largest weight reaches 38.30%, indicating that sewage treatment is the most important factor causing urban diseases.

Using the model to calculate the comprehensive index of the urban disease early warning index of 31 cities(Table 3), it can be seen that the three cities with the highest comprehensive index are Beijing, Guangzhou and Haikou respectively, and the indexes are 4.8919, 4.7172 and 4.6861 respectively, indicating that the construction of these cities is relatively healthy. The three cities with the lowest composite index were Zhengzhou, Shijiazhuang and Xi 'an.

Table2. The index weight of early warning value index of urban disease

Rank	Level 3 indicators	Indicators of entropy/Uj	The entropy is reversed/Sj	Index weight
1	Sewage treatment rate/%	2.0503	4.7358	0.3830
2	Green coverage rate in built-up areas/%	4.1959	2.3142	0.1871
3	Good air quality days/Day	5.5564	1.7475	0.1413
4	The proportion of the tertiary industry in GDP	6.3544	1.5281	0.1236
5	Density of population	9.3278	1.0410	0.0842
6	Number of civil vehicles/ length of urban road (10,000 vehicles/km)	9.7100	1.0000	0.0809

Table3. The comprehensive index of early warning value index of urban disease

Rank	City	Comprehensive index	Rank	City	Comprehensive index	Rank	City	Comprehensive index
1	Beijing	4.8919	12	Fuzhou	4.1044	23	Wuhan	3.7729
2	Guangzhou	4.7172	13	Jinan	4.0097	24	Taiyuan	3.7320
3	Haikou	4.6861	14	Xining	4.0072	25	Changsha	3.7118
4	Nanjing	4.6581	15	Shanghai	3.9644	26	Chengdu	3.6902
5	Urumqi Municipality	4.5986	16	Shenyang	3.9435	27	Hohhot City	3.6879
6	Lhasa City	4.5664	17	Hangzhou	3.9241	28	Nanchang	3.6727
7	Changchun	4.4639	18	Tianjin	3.9203	29	Xi'an	3.5752
8	Chongqing	4.2817	19	Yinchuan	3.9157	30	Shijiazhuang	3.3976
9	Kunming	4.2179	20	Harbin	3.8338	31	Zhengzhou	3.2277
10	Nanning	4.1453	21	Hefei	3.8236			
11	Guiyang	4.1150	22	Lanzhou	3.7842			

4 Conclusion

The Tao Te Ching says, "People act according to the earth, the earth acts according to the heaven, the heaven acts according to the Tao, and the Tao acts according to the nature." The development of a city should also follow this rule. A city is also a complex organism, and a regular physical examination should be carried out on the main indicators, such as population, land, water, air, traffic and culture. According to the changes in these indicators, prevention should be carried out and the appropriate medicine should be applied, so that a city can develop healthfully.

Acknowledgments

The research was funded by the National Natural Science Foundation of China（31860176）, Key Research and Development Program of Gansu Province（20YF3FA037）, Key Research and Development Program of Shaanxi Province (2020ZDLSF06-06). We are grateful to all anonymous reviewers whose comments improved the quality of the manuscript.

References:

1. M C Li, Journal of Shanghai University of International Business and Economics, 26, 90-99(2019)
2. F T Liang, Y N Cai, Huazhong Architecture, 39, 25-29(2021)
3. Z Yang, China Renmin University Press (2014)
4. Q Zhao, L P Zhang, Z T Chen, Marine Environmental Science, 28, 92-95+112(2009)
5. M F Wu, Hohai University, 2006.
6. L P Zhang, W Q Chen, H S Hong, Journal of Xiamen University (Natural Science), 221-224(2004)
7. L Zhang, North China University of Water Resources and Electric Power (2013)

https://doi.org/10.1051/e3sconf/202129303013

Research on environmental sustainable development from the perspective of ecological science and technology innovation

Na Li[1*]

[1]Wuhan University of Science and Technology, Wuhan, China

Abstract. At present, the prominent problems of ecological environment and the negative environmental effects of modern science and technology promote the development of scientific and technological innovation in the direction of ecology. The ecological concept of scientific and technological innovation can help the battle of pollution control and promote the reconciliation between man and nature. The ecological orientation of scientific and technological innovation in environmental sustainable development has some problems, such as unclear development orientation, insufficient innovation ability, lack of ecological concept, and insufficient public participation. Therefore, we should clarify the ecological orientation of scientific and technological innovation, improve innovation ability, promote the reform of environmental management system, and strengthen ecological moral education.

1 Introduction

Technology is an important productive force, and innovation is the first driving force leading the development. The sustainable development of ecological environment is the fundamental plan for the sustainable development of the Chinese nation. There is an important connection between scientific and technological innovation and sustainable development of ecological environment. Science and technology has a profound impact on people's production and life, but it also brings a series of ecological environment problems, which forces scientific and technological innovation to develop in the direction of ecology, vigorously develop ecological environment technology, and further promote the innovative development of ecological civilization.

2 The connotation explanation of the ecological concept of scientific and technological innovation

As for the connotation of ecological concept of scientific and technological innovation, Zhao Haiyue believes that "ecological concept of scientific and technological development is not economic development in a simple sense, but a three-dimensional complex system composed of society, ecology and economy, which is interconnected and interacted with each other."[1] Fu Xiaohua believes that "the ecological concept of scientific and technological innovation is a new innovation system infiltrated from ecology to scientific and technological innovation, so as to guide the coordinated development of scientific and technological innovation in the direction of beneficial to resources, environmental protection and the virtuous cycle between it and economic, social and environmental systems" [2] Chen Cuifang believes that "ecological concept of science and technology is the process and result of regarding science and technology as a complex ecosystem, integrating the principles and methods of ecology into the whole scientific and technological activities such as scientific and technological research, scientific and technological design, scientific and technological application and scientific and technological management, guiding and ensuring science and technology to follow ecological principles and ecological laws, and ultimately protecting the environment and maintaining ecological balance."[3] To sum up, the ecological concept of scientific and technological innovation reflects the role of scientific and technological progress in transforming the ecological environment, repairing the relationship between man and nature, and realizing the sustainable development of the environment. Therefore, it is very necessary to realize the ecological concept of scientific and technological innovation.

3 The necessity of the development of scientific and technological innovation in the direction of ecology

3.1 The prominent problems of ecological environment in China need transformation and upgrading of scientific and technological innovation

First, water resources are scarce, and water environment pollution is still serious. Compared with other countries, China's water resources are far lower than the world's per

* Corresponding author: 15171898972@163.com

capita water resources. China has very little water per person, as shown in Figure 1. At the same time, the current situation of water environment pollution in China is still very serious. Nearly half of the major rivers and lakes in the country are seriously polluted, and nearly 90% of the urban water quality exceeds the standard, which increases the difficulty of water environment treatment in China.

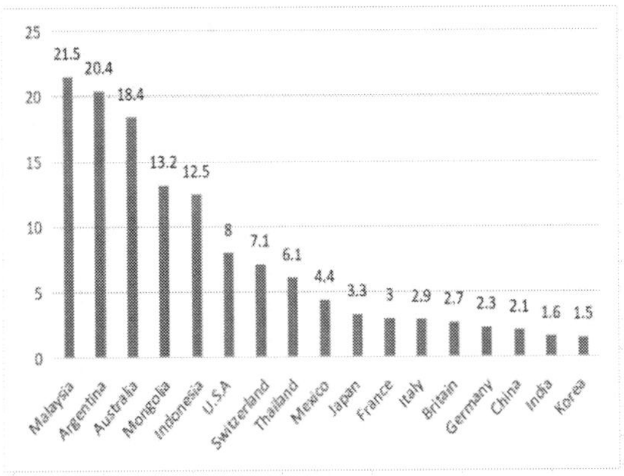

Figure 1. Fresh water per capita in different countries

Second, the amount of industrial solid waste is increasing year by year, and the utilization and disposal rate of waste is relatively low. As shown in table 1,from 2016 to 2019, the production of general industrial solid waste increased year by year, from 3.71 billion tons in 2016 to 4.41 billion tons in 2019. The production of industrial hazardous waste increased from 52.95 billion tons in 2016 to 81.6 billion tons in 2019. The amount of industrial waste disposal is not high compared with the amount of waste generated, which indicates that the industrial solid waste disposal technology needs further reform and innovation.

Table1. Output and disposal of industrial solid waste

Year (Billion tons)	2016	2017	2018	2019
Production of general industrial solid waste	37.1	38.7	40.8	44.1
Disposal volume of general industrial solid waste	8.5	9.4	10.3	11.0
Production of industrial hazardous waste	5219.5	6581.3	7470.0	8126.0
Comprehensive utilization and disposal of industrial hazardous waste	4317.2	5972.7	6788.5	7539.3

Finally, as shown in figure 2, The problem of urban air pollution is still very serious. According to the data published in the bulletin on the state of China's ecological environment (2019), in 2019, 157 of the 337 cities in China met the air quality standards, accounting for 46.6% of the total number of cities. The air quality of 180 cities exceeded the standard, accounting for 53.4%.

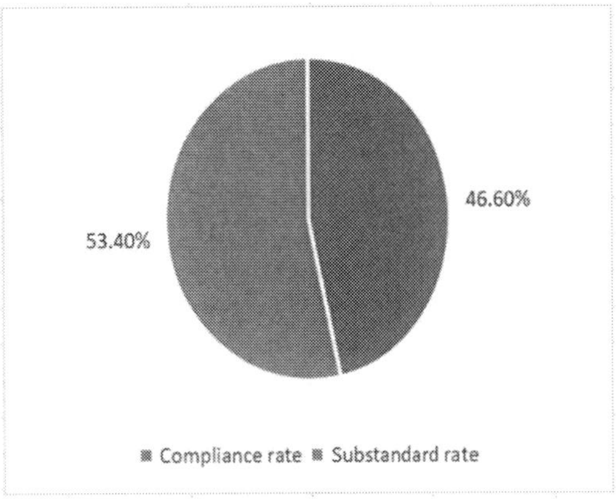

Figure 2. Air quality standards of 337 cities in 2019

3.2 The negative environmental effects of modern science and technology give birth to ecological science and technology

In the past, human beings only paid attention to the economic benefits of scientific and technological transformation, over exploited natural resources, destroyed the balance of the ecosystem, factory pollutant emissions, increasing garbage, oil leakage, optical pollution, overfishing, illegal trafficking in wild animals and so on accelerated the deterioration of the environment, and the relationship between human and nature gradually became disharmonious with the progress of science and technology. It is because of the negative effects of science and technology that China accelerates the construction of green technology innovation system and provides an important scheme for environmental sustainable development.

4 The role of ecological science and technology innovation in promoting environmental sustainable development

4.1 It has played a supporting role in the battle of pollution control

AS shown in table 2, in order to fight the battle of pollution prevention and control, through large-scale scientific observation and experimental research, China adopts the way of diverse technology integration, and adopts a series of Eco-Environmental Science and technology in water pollution, air pollution, soil pollution, ecological restoration, solid waste utilization, etc. And forms a broad scientific consensus, which provides support for pollution control and fighting the battle of pollution prevention and control.

 https://doi.org/10.1051/e3sconf/202129303013

Table2. Application of ecological environment science and technology

Name of pollution prevention and control	Achievements in the application of science and technology
Major science and technology projects for water pollution control and treatment	The key technologies of water "source control and emission reduction", water "burden reduction and restoration", and comprehensive regulation of water environment in the basin
Key project of "Research on the causes and control technology of air pollution"	Monitoring, forecasting and early warning technology, pollution source whole process control technology, air quality improvement management support technology and air pollution joint prevention and control technology
Key project of "causes and treatment technology of site soil pollution"	Soil pollution formation mechanism, monitoring and early warning, risk management and control, remediation, safe use and other technologies
Key project of "Research on restoration and protection of typical fragile ecology"	Ecological monitoring and assessment technology, forest and wetland ecological protection and restoration technology in Northeast China, comprehensive management technology of sandy land in northern sandy area, ecological protection and restoration technology in middle and upper reaches of Yangtze River, etc
"Solid waste recycling" key project	Large industrial solid waste building materials utilization, domestic waste incineration power generation, heavy metal solid waste safe disposal technology

4.2 The ecological innovation of science and technology promotes the reconciliation of the relationship between man and nature

The real reconciliation of the relationship between man and nature depends on man's initiative to understand and transform nature, that is, scientific and technological innovation. Science and technology is the direct cause of ecological crisis. If we redefining the direction of scientific and technological innovation and integrate the ecological value and environmental protection concept into scientific and technological innovation, the development of science and technology will be in the direction of coordinating the relationship between human and nature, and finally realize the sustainable development of the whole society.

5 The dilemma of scientific and technological innovation in promoting environmental sustainable development

5.1 The positioning of scientific and technological innovation is not comprehensive, and the concept of ecological definition of scientific and technological innovation is vague

Although the government has issued relevant policy planning and guidance to promote the ecological of scientific and technological innovation, the concept of protecting the ecological environment has not become the main direction of scientific and technological innovation. For the ecological concept of scientific and technological innovation, this paper puts forward a variety of concepts, such as intermediate technology, soft technology, diversity technology, green technology, environmental technology, environmental protection technology and so on. The concept used is too diversified, easy to cause confusion, difficult to form a unified theoretical standard, is not conducive to the in-depth study of the problem.

5.2 Lack of innovation ability and difficult transformation of traditional industries

Advanced science and technology does not mean that a country's independent innovation ability is developed. Even if the more advanced technology is, without the support of innovation, a series of environmental problems brought by science and technology can not be solved. At present, there are still many traditional industries that have not achieved transformation and upgrading, which is the outstanding performance of the lack of innovation ability. Take China's heavy polluting industry as an example, the gases, waste water and waste residue discharged by factories cause serious pollution to the environment, and some pollution is irreversible. Although China's manufacturing industry occupies a certain position in the last century, a large amount of waste from factories can easily cause direct impact on the environment. It is precisely because of the lack of innovation ability that the damage to the environment is becoming more and more serious due to the inability of traditional industries to transform and upgrade rapidly.

5.3 The soft environment of innovation is insufficient and the ecological concept is missing

Generally speaking, science and technology innovation environment can be divided into hard environment and soft environment. The hard environment consists of material environment, management system and personnel, while the soft environment consists of humanistic environment, flexible research direction and evaluation system. In terms of the soft environment for scientific and technological innovation, although China has formulated a variety of scientific and technological development plans and strategies, which provide policy support for scientific and technological innovation. In the development of green science and technology and the concept of ecological science and technology, there are still insufficient understanding, ecological awareness and other problems, resulting in the scientific research of the internal cultural atmosphere is not strong enough, resulting in the construction of soft environment values and ecological view of the lack of propaganda.

5.4 The public scientific literacy is relatively low, and the public participation consciousness is insufficient

Public scientific literacy plays a key role in the process of innovation and development. Scientific literacy refers to the public's attitude towards science and understanding of scientific and technological knowledge, which can reflect a country's cultural level and education level. The level of public scientific literacy in China has improved on the whole, but there is still a big gap compared with developed countries, which is manifested in the lack of scientific spirit, less opportunities to accept popular science knowledge, and low level of education. On the one hand, the lack of scientific literacy will affect the public's acceptance and understanding of science and technology, and it is easy to misunderstand science, thus undermining the enthusiasm of scientists; On the other hand, most of the public seldom participate in science popularization activities, only see the advantages of science and technology, but seldom consider the ecological problems hidden in science and technology.[4]

6 The path of environmental sustainable development from the perspective of scientific and technological innovation

6.1 Clarify the orientation of ecological science and technology innovation and give play to the leading role of the government

The government should give strong support and attach great importance to the ecological direction of scientific and technological innovation. In terms of policy, it should incline to the ecological development of science and technology, give financial support to experts who carry out public technology research, give preferential policies and taxes to enterprises who implement the ecological development of science and technology, and encourage the ecological research and industrial development of science and technology development[5]. In law, the implementation of the main responsibility of scientific and technological innovation requires enterprises to shoulder the obligation of environmental protection and establish a new ecological concept.

6.2 Improve innovation ability and maintain technological achievements of scientific and technological innovation

Establish a technology innovation system with enterprise as the main body and the combination of production, learning and research, and improve the comprehensive application ability of innovation. Meanwhile, we should maintain the achievements of scientific and technological innovation and introduce modern green technologies such as clean production technology, energy saving technology and recycling technology in improving production mode. We should enhance the level of the construction of the intelligent environment, firmly grasp the revolution with information technology, biotechnology and other high and new technologies as the core, and promote scientific and technological innovation towards the road of intelligence, intelligence and green.

6.3 Promoting the reform of environmental management system

Environmental management system is an organic whole which is established in accordance with certain rules and regulations in order to improve the environment and reduce pollution. We should attach importance to the support and guarantee function of the environmental management system, so as to make the ecological science and technology innovation as a whole, promote each other and complement each other.

6.4 Improve the ability of science popularization and strengthen the education of Ecological Morality

Integrating the concept of ecological science and technology into the work of science popularization education can make people understand the whole picture of scientific development, improve people's scientific awareness of caring for the environment, protecting the environment and treating nature well, and make the mass power become the key force of ecological science and technology innovation. Strengthening ecological moral education can start with school education, mass media, legal measures, etc. to cultivate people's feelings about ecological morality, strengthen the connection between man and nature, publicize the correct concept of good and evil, and help people really love.

References

1. Zhao Haiyue, Kang Xibin. How to realize the ecological development of science and technology [J]. People's forum, 2018, (16): 148-149.
2. Fu Xiaohua. Analysis of ecological science and technology innovation from sustainable development values [J]. Scientific management research, 2006, 41-44.
3. Chen Cuifang. On the connotation and basic characteristics of science and technology ecology [J]. Science and technology management research, 2014, (19): 224-228.
4. Guo Qijun, Zhang Dinghui, Zhang Hui. On the importance of green technology in the construction of ecological civilization [J]. Land greening, 2014, (8): 40-41.
5. Liang Ye. Ecological value of scientific and technological innovation [J]. Journal of Hunan University of administration, 2018, (3), 27-32.

Study on the adsorption of heavy metals by biochar modified with different modifiers under room temperature

Jiajia Ding[1,2*], Yu Tian[1], Tao Lu[1] and Jiang Kang[1]

[1]College of Geography and Environmental Engineering, Lanzhou City University, Lanzhou 730070, PR China

[2]Gansu Mining Area Pollution Control and Ecological Restoration Engineering Research Center

Abstract. In order to obtain an environment-friendly biochar adsorption material with excellent adsorption performance, the biochar made from corn straw was modified with $FeCl_3$, Na_2S and $KMnO_4$ as modifiers, and the adsorption effect of biochar on several heavy metal ions in water before and after modification was studied. The properties of biochar before and after modification were characterized by SEM, XRD, FTIR and BET, and the adsorption mechanism was discussed. The results showed that the three modified reagents could effectively improve the adsorption capacity of biochar for three heavy metal ions in water, and the adsorption effect was still good at room temperature. The adsorption difference is related to the specific surface area, pore structure, aromatic structure and SiO_2 content of biochar. The adsorption mechanism mainly includes the complexation reaction between physical adsorption and surface functional groups. This study is beneficial to promote the resource utilization of agricultural and forestry waste biomass, and provides a certain reference basis for related research.

1 Introduction

Rapid economic and social development has brought about a large number of heavy metal pollutants, which seriously harm the ecological environment and biological health[1]. Biochar (BC) is a kind of porous carbon material with wide sources, environmental protection and low cost[2]. Due to its unique high aromatic and heterocyclic structure, complex pore structure, large specific surface area, stable structural capacity and abundant surface functional groups, it has good adsorption capacity and can effectively repair heavy metal pollution[3]. However, the current heavy metal pollution, especially the heavy metal pollution in the soil is complex and diverse, and the original BC cannot achieve the ideal adsorption effect. The adsorption performance of BC can be improved by modification treatment, such as increasing specific surface area, pore complexity or changing the functional group substance category on the surface, so as to achieve better treatment effect of heavy metals. The chemical modification methods are widely used, including acid modification, alkali modification, oxidant modification, metal oxide modification and carbonaceous material modification[4]. For example, acid modification can remove metals and other impurities on the surface of BC, and introduce acidic functional group substances[5]. Alkaline modification can increase the surface area of BC and introduce oxygen-containing functional groups[6-7]. Oxidants can also introduce oxygen-containing functional groups, activate the surface functional groups of BC and increase their active sites[8]. In this study, ferric chloride ($FeCl_3$), sodium sulfide (Na_2S) and potassium permanganate ($KMnO_4$) were selected as modifiers to study the adsorption of three heavy metals by biochar before and after modification under room temperature. The purpose of this study is to compare the modification effect and mechanism of different kinds of modified reagents on biochar at room temperature, so as to provide a basis for subsequent practical application. Although the specific surface area of biochar from corn straw is not the optimal choice, as agricultural waste, it has a large yield and is of great significance for its comprehensive utilization.

2 Materials and methods

2.1 Biochar modification

Corn stalk biochar was purchased from Suzhou City, Jiangsu Province. It was pre-dried to a constant weight and then screened with a diameter of 0.425mm before utilization. The modified biochar with the best adsorption effect on the three kinds of heavy metal ions was screened out at room temperature (20°C) through the pre-experiment, and the optimal modification concentration of the corresponding modified reagent was obtained. The optimal concentration of the selected modifier was 0.4mol/L of $FeCl_3$ solution, 0.4mol/L of Na_2S solution and 0.2 mol/L of $KMnO_4$ solution. The original biochar is marked as BC, the modification of $FeCl_3$ is marked as BC-F, the modification of Na_2S is marked as BC-S, and the modification of $KMnO_4$ is marked as BC-P. The above

* Corresponding author: chrisding0202@163.com

three kinds of modified biochar and BC were used as raw materials for subsequent adsorption experiments.

2.2 Characterizations

Scanning electron microscope (SEM) images of biochar before and after modification were obtained with a JSM-6701F Scanning Microscope. Their X-ray diffraction (XRD) measurements were performed within the angle range 2θ = 5–90° using a Shimazu X-ray diffractometer XRD-6100 type. Fourier transform infrared spectra (FT-IR) for the samples were generated by a PerkinElmer spectrum 3 FT-IR Spectrometer (400–4,000 cm^{-1}). The BET surface area, pore surface, and pore volume of these samples were performed on the Feldman ASAP 2020 physical adsorption instrument.

2.3 Adsorption experiment

The adsorption properties of three kinds of modified biochar on Cr^{6+}, Pb^{2+} and Cu^{2+} were determined by batch experiments. Every result came from at least three parallel samples. The initial concentration of heavy metal solution was 50 mg/L, with the vibration rate was 120 r/min. All tests were carried out at room temperature (20°C). The three influencing factors of solution pH, adsorption time and biochar dosage were studied. The concentrations of Cr^{6+}, Pb^{2+} and Cu^{2+} in water samples were determined by spectrophotometry. The Langmuir adsorption mode[9] and Freundlich adsorption model[10-11], were used to simulate the experimental data.

3 Results and discussion

3.1 Initial assessment

Table1. Adsorption capacity of three kinds of heavy metals before and after modification of biochar.

The adsorption quantity (mg/g)	Cr^{6+}	Pb^{2+}	Cu^{2+}
BC	11.58	20.55	16.25
BC-F	29.76	24.39	23.92
BC-S	29.57	24.47	23.49
BC-P	29.57	24.86	23.66

As is shown in **Table 1**, the modified biochar still has a good adsorption effect at room temperature. When the biochar is modified by $FeCl_3$ or Na_2S, the optimal concentration of the modifier is 0.4 mol/L. the optimal concentration of $KMnO_4$ modification is 0.2mol/L, which may be because the adsorption promotion effect of oxidation on biochar can only be realized at relatively low concentration. The modified biochar with the best adsorption effect on Cr^{6+} and Cu^{2+} was BC-F, and the best adsorption effect on Pb^{2+} was BC-P.

3.2 SEM analysis

As can be seen from the SEM image in **Fig.1.**, BC has a relatively smooth surface, which is in the shape of sheets of different sizes, and the pores are not obvious. After modification, its surface has undergone significant changes. The apparent characteristics of the three kinds of modified biochar are not exactly the same. Generally speaking, the surface pores increase significantly, and tend to be regular, and the specific surface area was enlarged by the pore broadening effect, which strengthens the adsorption site on the surface of biochar itself and improves the adsorption capacity[12]. The pore structure of BC-F is the most regular, and the size of the flake particles of BC-S is uniform. The surface of BC-P is the roughest, which may be the result of oxidation.

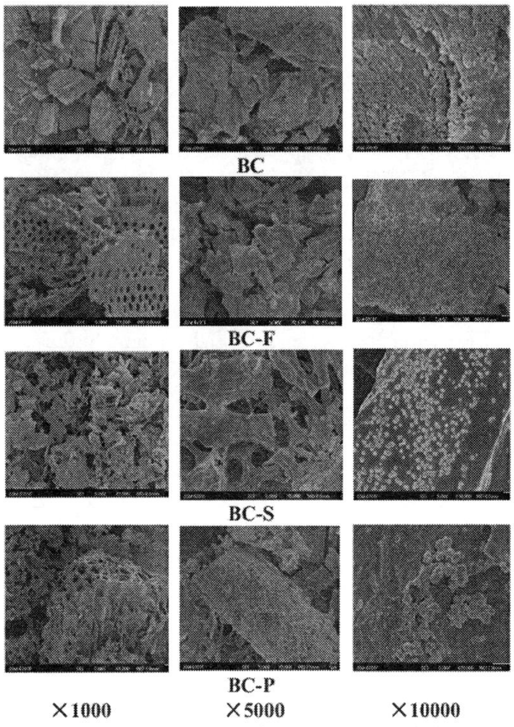

Figure 1. SEM images of biochar before and after modification

3.3 XRD analysis

As shown in **Fig.2.**, compared with BC, the crystal phase structure of BC-F and BC-S did not change greatly, indicating that the use of $FeCl_3$ and Na_2S to modify biochar did not affect the crystal structure. The crystal structure of BC-P is destroyed due to the strong oxidation of $KMnO_4$. The crystal surface diffraction peaks of the modified biochar at 2θ=26° and 40° are obviously weakened. BC-F and BC-S showed new characteristic peaks at 16.6° and 76.6°, indicating that the corresponding groups were introduced.

 https://doi.org/10.1051/e3sconf/202129303014

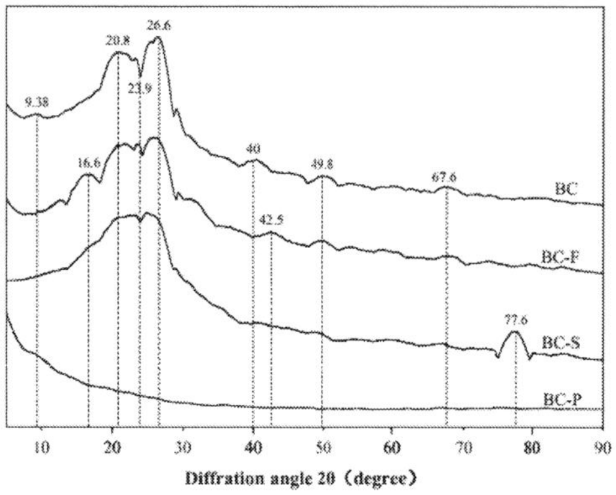

Figure 2. XRD analysis of biochar before and after modification

3.4 FT-IR spectra analysis

As can be seen from the FTIR spectra in **Fig.3.**, the wide peak at 3391~3405 cm^{-1} was the stretching vibration of hydroxyl O-H. The absorption peak at 2915~2921 cm^{-1} is methylene asymmetric stretching vibration. The carbonyl (C=O) had a peak at around 1585~1597 cm^{-1} [11,13]. The carbon oxygen bonds (C–O, C–O–C) had a peak at around 1059~1076 cm^{-1}, and carboxyl (–COOH) had a peak near 1378 cm^{-1} [11,14]. The peak at 1378 cm^{-1} was markedly increased in BC-S spectra after adsorption, this peak was the result of metal carboxylate (–COO)[11,15], which proved that metal ions interacted with organic functional groups during adsorption process. The absorption peaks at 1434 and 1075 cm^{-1} are C-O stretching vibration and O-H in-plane bending vibration absorption peaks, and the peak at 875 cm^{-1} is C-H in-plane bending vibration in the aromatic ring. Compared with the BC, the position of absorption peak and the intensity change in the infrared absorption spectrum did not obvious.

3.5 BET analysis

After modification, the specific surface area and pore volume of biomass carbon increased significantly, while the average pore size decreased. The larger the specific surface area was, the more adsorption sites could be provided, so that the adsorption effect was significantly improved (**Table 1**). The porous structures of biochar before and after modification were quantitatively characterized by N_2 adsorption method. According to the adsorption-desorption isotherm (**Fig.4.**), it can be seen that the modification process has a great influence on the formation of biomass carbon pore structure. The original biochar samples were close to II type isotherm, but not typical, indicating that the pore size structure was large and the distribution was irregular. The modified biochar samples showed a typical IV isotherm, which indicated that the samples were mesoporous structure, and the modification reagent led to the coalescence of pores, leading to the formation of mesoporous structure. The hysteresis loop in the isotherm is between HL type and H3 type, indicating that although the modified biochar still belongs to the slit pore characteristics formed by polymer sheet particles, the pore structure tends to be evenly distributed, and the adsorption capacity is significantly improved.

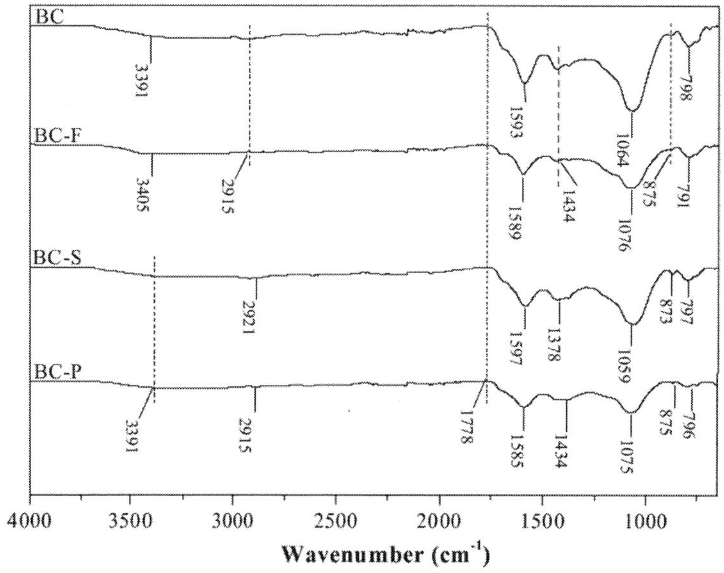

Figure 3. FT-IR spectra of biochar before and after modification

 https://doi.org/10.1051/e3sconf/202129303014

Table2. Surface area and pore volume of biochar before and after modification

Product ID	BET specific surface (m^3/g)	Micropore surface areas (m^3/g)	External specific surface areas (m^2/g)	Total pore volume (cm^3/g)	Micropore volumes (cm^3/g)	Average pore width (nm)
BC	0.1760	5.5563	-5.3803	0.0048	0.0027	101.9823
BC-F	5.4575	1.3405	4.1170	0.0115	0.0008	10.0261
BC-S	11.0621	0.8222	10.2400	0.0214	0.0005	8.1276
BC-P	3.9397	-	4.9679	0.0131	-0.0151	16.1592

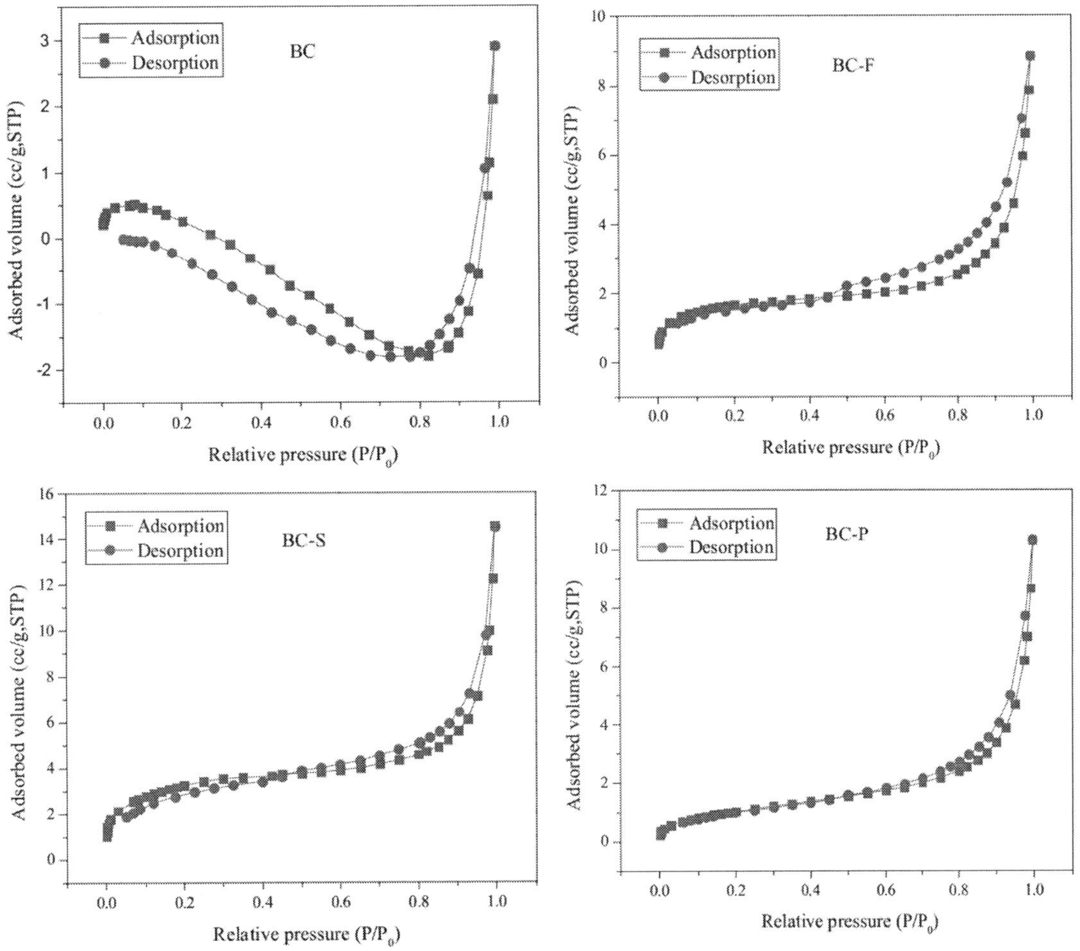

Figure 4. BET analysis of biochar before and after modification

4 Conclusions and perspectives

The results show that the biochar modified by $FeCl_3$, Na_2S and $KMnO_4$ has a good adsorption effect on heavy metals of Cd, Pb and Cu at room temperature. As an oxidizing modifier, $KMnO_4$ can only promote the adsorption of biochar at a relatively low modified concentration. The modified biochar with the best adsorption effect on Cr^{6+} and Cu^{2+} was BC-F, and the best adsorption effect on Pb^{2+} was BC-P. At room temperature, the pH of the solution is the main factor affecting the adsorption, and the optimal removal pH of each metal is different.

After modification, surface pores of the biochar increase significantly, and tend to be regular, and the specific surface area was enlarged by the pore broadening effect. There are still some differences in the apparent morphology of the three different modified biochar. The XRD analysis showed that the content of SiO_2 decreased after modification, and among the three modifiers, only $KMnO_4$ would cause the destruction of the crystal structure of biochar. The infrared spectrum analysis showed that the surface functional groups of the prepared biochar had no significant change before and after modification, mainly containing hydrophilic functional groups such as carboxyl group and hydroxyl group as well as aromatic ring structure. The difference of adsorption effect of biochar before and after biochar comes from the difference of specific surface area, pore structure, aromatic structure and SiO_2 content. That is, the larger the specific surface area, the richer the pore structure, the more adsorption sites provided, and the stronger the electrostatic adsorption capacity of heavy metals. The higher the degree of aromatization, the stronger the cationic π bond. In order to further explore the mechanism, more experimental studies and analysis are needed.

Acknowledgments

This work was partly supported by Lanzhou City University Doctoral Research Initiation Fund (LZCU-BS2018-05). The authors also thank the anonymous reviewers for their invaluable insight and helpful suggestions.

References

1. J.X. Li, L.C. Wu, J. Zhang, C.S. Wang, Q.Q. Yu, Y. Peng, Y.H. Ma, Ecol. Environ. Sci, 24(2015)
2. H. Dong, J. Deng, Y. Xie, C. Zhang, Z. Jiang, Y. Cheng, K. Hou, G. Zeng, J. Hazard. Materi., (2017)
3. X. Zhou, J. Zhou, Y. Liu, Fuel, 233(2018)
4. J.H. Luo, Y.J. Gao, T.H. Song, Y. Chen, Water Sci. Technol., (2021)
5. A.U. Rajapaksha, S.S. Chen, Tsang, C. W. Daniel, M. Zhang, M. Vithanage, S. Mandal, B. Gao, N.S. Bolan, Y.S. Ok, Chemosphere, 148 (2016)
6. Z. Ding, X. Hu, Y. Wan, S. Wang, B. Gao, J. Ind. Eng. Chem., 33 (2016)
7. M.E Doumer, A. Rigol, M. Vidal, Mangrich, Environ. Sci. Pollut. Res., 23(2016)
8. Y.C.Li, J.G.Shao, X.H.Wang, Energy Fuels, 28(2014)
9. Langmuir, J. Chem. Phys., 40(2015)
10. C. Ng, J.N. Losso, W.E. Marshall, R.M. Rao, Bioresour. Technol., 85(2002)
11. X.L. Hu, Y.W. Xue, L.N. Liu, Y.F. Zeng, L. Long, Environ. Sci. Pollut. Res., (2018)
12. H.X.Wang, J.L. Xu, L.X. Sheng, J. Cleaner Prod, 2020
13. S. Yorgun, D. Yıldız, J Taiwan Inst Chem Eng, 53(2015)
14. V. Frišták, E. Víglašová, L. Ďuriška, M. Galamboš, E. Moreno-Jimenéz, M. Pipíška, G. Soja, J. Iran. Chem. Soc., 14(2017)
15. A.H.M. Ibrahim, A. Jalbout, Appl. Spectrosc, 62(2008)

Fe-doped carbon aerogels from sodium alginate for the removal of methylene blue

Zuozhao Zhai [1,2], Bin Ren [1,2], Lihui Zhang [1,2], Zhenfa Liu [1,2,*]

[1]Institute of Energy Resources, Hebei Academy of Sciences, Shijiazhuang 050081, China

[2]Hebei Engineering Research Center for Water Saving in Industry, Shijiazhuang 050081, China

Abstract. In this article, Fe-doped carbon aerogels from sodium alginate were used for the removal of methylene blue (MB) in water. Under acidic condition, the sample carbonized at 700 °C (T700) undergoes Fe-C micro-electrolysis to produce highly chemically active Fe^{2+} and [H] to degrade MB. Under neutral or alkaline conditions, Fe^{2+} produced by Fe-C micro-electrolysis become $Fe(OH)_3$, which can effectively adsorb MB. In addition, when T700 combines with H_2O_2 to form Fenton system, the MB removal efficiency was significantly improved. The Fe-doped carbon aerogels can be used in wastewater treatment and when combine the materials with H_2O_2 can greatly improve the MB removal efficiency.

1 Introduction

MB is a commonly used dye which is widely used in industrial and medical fields. However, the extensive application of MB produced a large amount of industrial wastewater. High concentrations of MB can poison or even die animals. Therefore, MB treatment is one of the current research focuses in the field of wastewater treatment. MB treatment includes biological [1,2], physical [3,4] and chemical [5,6] methods. Chemical method includes chemical oxidation, electro-oxidation, photo-oxidation, and ultrasonic oxidation, etc., which is a hot research for MB wastewater treatment.

Fe-C micro-electrolysis is an economical and environmentally friendly wastewater treatment technology, which is widely used in the treatment of high-concentration organic wastewater [7-10]. When iron filings and carbon particles are immersed in acidic wastewater, numerous micro-galvanic cells are formed in the wastewater due to the electrode potential difference between iron and carbon (1.2V). Low-potential iron becomes the anode, and high-potential carbon becomes the cathode. The electrochemical reaction under acidic conditions occurs as follows:

Anodic oxidation: $Fe - 2e^- \rightarrow Fe^{2+}$ (1)

Cathodic reduction: $2H^+ + 2e^- \rightarrow 2\,[H] \rightarrow H_2$ (2)

The Fe^{2+} and [H] produced by the reaction have high chemical activity can change the structure of organic substances in wastewater. However, the reaction time of Fe-C micro-electrolysis is long. Fenton reaction is a simple and efficient wastewater treatment technology [11-13]. This process is based on the reaction of H_2O_2 and Fe^{2+} to produce highly oxidizing hydroxyl radicals (•OH) to degrade organics in wastewater.

The reaction mechanism is as follows:

$$H_2O_2 + Fe^{2+} \rightarrow Fe^{3+} + OH^- + HO\bullet \quad (3)$$

$$H_2O_2 + Fe^{3+} \rightarrow Fe^{2+} + H^+ + HO_2\bullet \quad (4)$$

$$H_2O_2 + HO_2\bullet \rightarrow H_2O + O_2 + HO\bullet \quad (5)$$

$$HO\bullet + RH \rightarrow R\bullet + H_2O \quad (6)$$

$$R\bullet + Fe^{3+} \rightarrow Fe^{2+} + R^+ \quad (7)$$

$$R^+ + O_2 \rightarrow ROO^+ \rightarrow CO_2 + H_2O \quad (8)$$

Therefore, the combine of Fe-C micro-electrolysis and Fenton reaction will effectively improve their respective problems.

In this paper, The Fe-doped carbon aerogels were employed to treat MB in wastewater. The materials have broad application prospects in the field of wastewater treatment because they use environmentally friendly sodium alginate as raw material, the preparation process is simple, and the MB removal efficiency in wastewater is high.

2 Materials

Sodium alginate, $Fe(NO_3)_3$, was purchased from Aldrich Co. MB was purchased from Damao Chemical Reagent Factory. H_2SO_4 and NaOH were purchased from Tianjin Fengchuan Chemical Reagent Technologies Co., Ltd.

2.1 Materials preparation

Fe-doped carbon aerogels were synthesized as followings: sodium alginate was dissolved in H_2O (solution A) and $Fe(NO_3)_3$ was dissolved H_2O (solution B). Solution A and solution B were mixed to obtain sodium alginate hydrogels. The hydrogels were placed in a freeze dryer

* Corresponding author: lzf63@sohu.com

obtain aerogels. Then the aerogels were transferred in a carbonization furnace at 500, 600, 700, 800, and 900 °C (the corresponding sample is named as T500, T600, T700, T800, and T900, respectively).

2.2 Characterization

The XRD patterns of the carbon aerogels were monitored by an Ultima X-ray diffractometer under copper Kα radiation at 40 kV and 40 mA. A TU-1900 UV-vis spectrometer was used to measure the concentration of MB at 665 nm. The total iron content was measured using the TU-1900 UV-vis spectrometer at 510 nm according to iron-phenanthroline spectrophotometry.

2.3 Removal experiments

In the Fe-C micro-electrolysis experiments, 50 mg Fe-doped carbon aerogels were added to 50 ml MB aqueous solution (50 mg L^{-1}) at 25 °C, pH=2.5. The pH was selected as the variable operating parameter to investigate the removal efficiency of MB. The pH was adjusted by H_2SO_4 and NaOH. The MB removal efficiency was calculated using the following equation:

$$D(\%) = (C_0 - C_t)/C_0 \times 100 \quad (9)$$

where C_0 is the initial MB concentration (mg L^{-1}) and C_t is the MB concentration at time t (mg L^{-1}).

In the Fenton experiments, 12 mg T700 was added to 40 ml MB aqueous solution (50 mg L^{-1}) with H_2O_2 was added at 25 °C, pH=2.5.

3 Result and Discussion

The MB removal efficiency of Fe-doped carbon aerogel samples are shown in Fig 1. After 9 h, the MB removal efficiency of T500, T600, T700, T800 and T900 is 20.1%, 25.4%, 96.0%, 94.7% and 93.0%, respectively. It can be seen that the MB removal efficiency of T500 and T600 is significantly lower than other samples. And the MB removal efficiency of samples T700, T800 and T900 are higher than 92% after 9 h. In addition, T700 has the best MB removal efficiency that can reach 96.0% after 9 h.

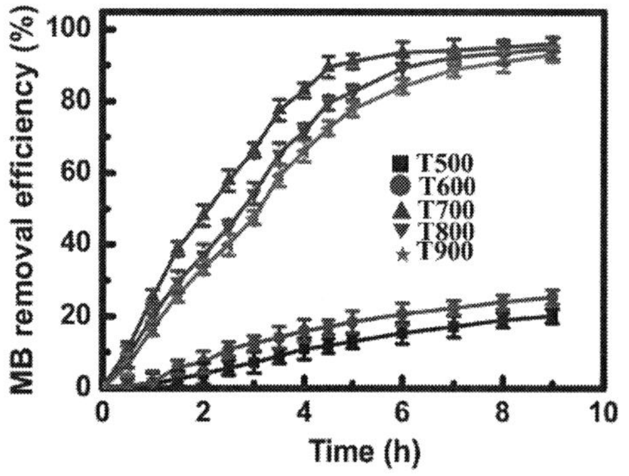

Figure 1. The MB Removal efficiency of Fe-doped carbon aerogels.

PH has a significant effect on Fe-C micro-electrolysis, so pH was selected as the operating parameter to investigate the MB removal efficiency of T700. As shown in Fig. 2, T700 has substantial MB removal efficiency under different pH conditions. This may be due to under acidic conditions, T700 produces highly chemically active Fe^{2+} and [H] by Fe-C micro-electrolysis. Fe^{2+} and [H] have strong oxidizing properties which can destroy the molecular structure of MB. And the stronger the acidity, the better the MB removal efficiency of T700. Under neutral or alkaline conditions, precipitation occurs in the MB solution as shown in the illustration in Fig. 2. And as the pH increases, the amount of precipitation increases. It is due to that Fe^{2+}/Fe^{3+} formed by iron-carbon micro-electrolysis react with OH^- in the solution to form $Fe(OH)_2/Fe(OH)_3$ precipitate. MB can form monovalent cationic quaternary ammonium groups in solution, while the $Fe(OH)_3$ colloid is negatively charged, so the material can effectively adsorb MB under neutral or alkaline conditions.

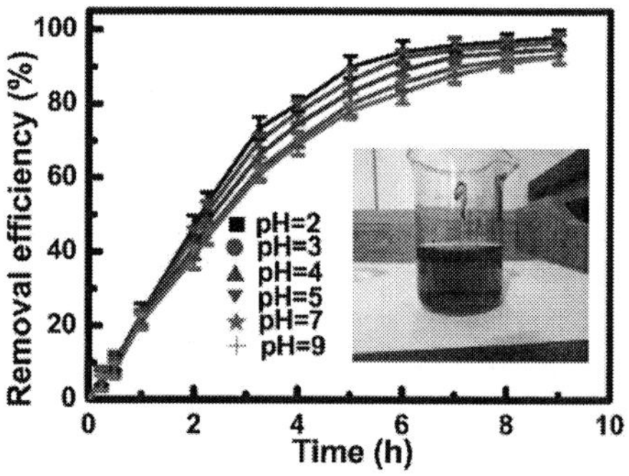

Figure 2. The pH effect on MB removal efficiency of T700.

In order to confirm that the removal of MB by the sample is mainly caused by Fe-C micro-electrolysis, T700 was washed by HCl (named as T700-H). Fig. 3a shows the XRD pattern of T700-H. It can be seen that the peaks of Fe in the sample have disappeared, and only the peak of graphitic carbon remains. Fig. 3b shows the MB removal efficiency of T700 and T700-H. The MB removal efficiency of T700-H is only 4.1% while the MB removal efficiency of T700 is 96.0% after 9 h. It can be concluded that the removal of MB by T700 is mainly caused by Fe-Cmicro-electrolysis.

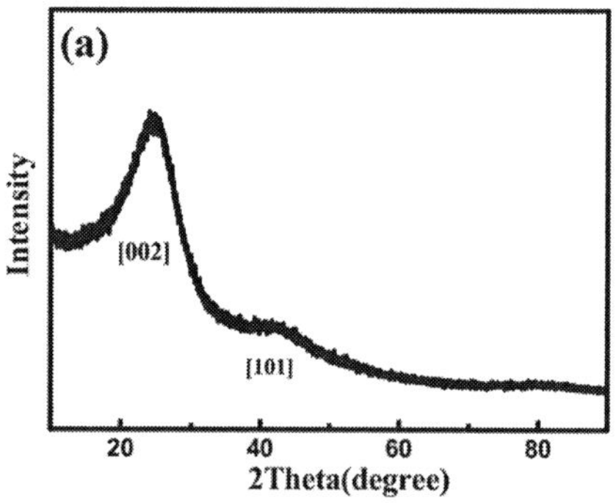

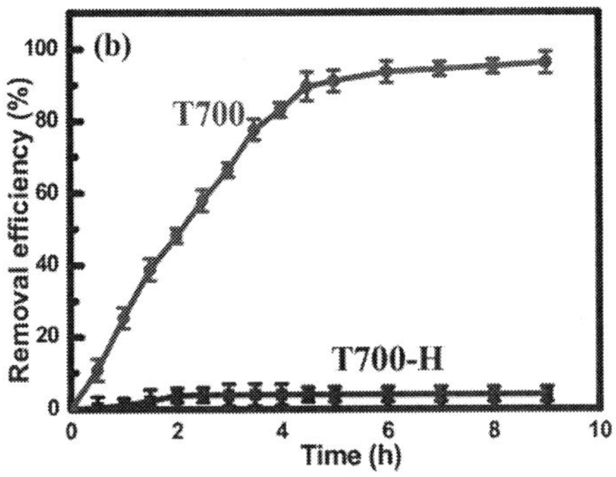

Figure 3. (a) XRD pattern of T700-H and (b) the MB removal efficiency of T700 and T700-H.

Fe^{2+} produced by Fe-C micro-electrolysis can form a Fenton system with H_2O_2 to improve the MB removal efficiency. In order to confirm the efficient removal efficiency of the Fenton system, a comparative experiment was performed. Fig. 4 shows the result of the comparative experiment. It can be seen that the MB removal efficiency of T700+H_2O_2 can reach 99.7% after 3 h while the MB removal efficiency of T700 and bare H_2O_2 are 66.3% and 6.3%, respectively. It can be concluded that adding H_2O_2 to T700 can effectively improve MB removal efficiency.

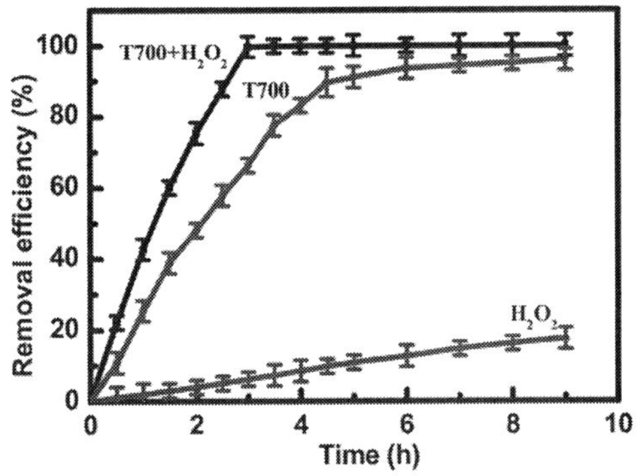

Figure 4. The MB removal efficiency of T700+H_2O_2, T700 and H_2O_2.

The addition time of H_2O_2 is a key factor affecting the Fenton reaction. Figure 5a shows the time required for sample T700 to remove more than 99% MB by adding 0.5 ml H_2O_2 after Fe-C micro-electrolysis at different times. Figure 5a shows that the longer the time of the Fe-C micro-electrolysis reaction, the higher MB removal effeciency after adding H_2O_2. The removal rate of MB reached 99% in 180 min when H_2O_2 and T700 were added at the same time, but reached 99% in only 3 min when H_2O_2 was added after 2 hours of T700 added. This result may be because the concentration of Fe^{2+} produced by Fe-C micro-electrolysis gradually increases with time, which promotes the Fenton reaction and improves the MB removal efficiency. To further prove this speculation, the concentration of Fe^{2+} in solution was determined at different times during Fe-C micro-electrolysis. Figure 5b shows that the concentration of Fe^{2+} in the solution increases with increasing Fe-C micro-electrolysis time. With increased Fe-C micro-electrolysis time, the MB removal effeciency is better after adding H_2O_2. But adding H_2O_2 after 0.5 hours of T700 added shows the shortest total time of MB removal to 99%.

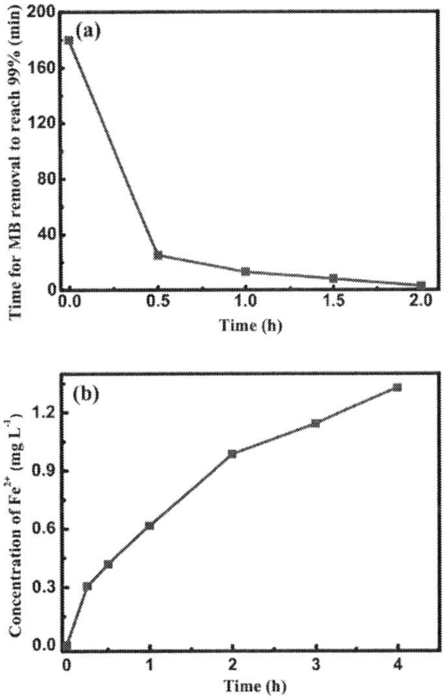

Figure 5. (a) The time of H2O2 added after T700 added and (b) the concentration of Fe2+ produced by Fe-C micro-electrolysis.

The amount of H_2O_2 added is an important factor affecting the Fenton reaction. Therefore, the experiment that investigated the effect of adding different amounts of H_2O_2 on the MB removal efficiency after 0.5 hours of Fe-C micro-electrolysis was applied. As shown in Figure 6, with increased H_2O_2, the MB removal efficiency is increase. But the MB removal efficiency is decreased when the amount of H_2O_2 is more than 0.4 ml. This difference may be because when the initial concentration of H_2O_2 is low, increasing the amount of H_2O_2 can effectively promote the Fenton reaction. However, when the concentration of H_2O_2 is too high, the hydrogen peroxide will be invalid decomposed, which will inhibit the progress of the Fenton reaction. The Fenton reaction can efficiently degrade organic pollutants in wastewater due to the strong oxidation ability of •OH. To determine the important role played by •OH, a set of comparative experiments were conducted. After 0.5 hours of Fe-C micro-electrolysis, H_2O_2 was directly added to sample 1, and two drops of isopropanol (a hydroxyl radical quencher) were added to sample 2, followed by H_2O_2 added. The MB solution in sample 1 changed from blue to colourless in 7 min, and the MB solution in sample 2 did not change significantly even after 2 h. The comparative experimental results show that •OH plays an important role in the MB removal.

 https://doi.org/10.1051/e3sconf/202129303015

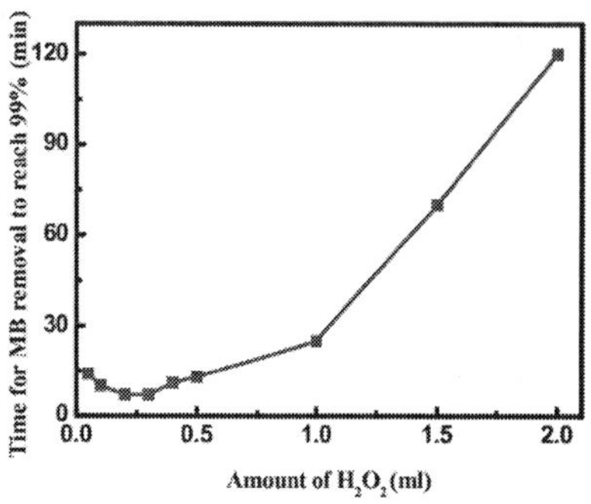

Figure 6. Effect of the amount of H_2O_2 on MB removal efficiency.

PH of the solution is another important factor affecting Fenton reaction. Figure 7 shows the effect of pH on the MB removal efficiency that adding H_2O_2 after 0.5 hours of T700 added. It can be seen that the MB removal efficiency can reach 99% only in 6 min at pH=2. When the pH of the solution is greater than 4, the MB removal efficiency is much lower than 99% after 120 min. This is because Fe^{3+} can generate $Fe(OH)_3$ precipitation at pH = 3.7 that leads to a decrease of Fe^{3+} content in the solution. Thus Fe^{3+} cannot be converted to Fe^{2+}, resulting in a sharp decrease of Fe^{2+} in the solution, thereby inhibiting the Fenton reaction.

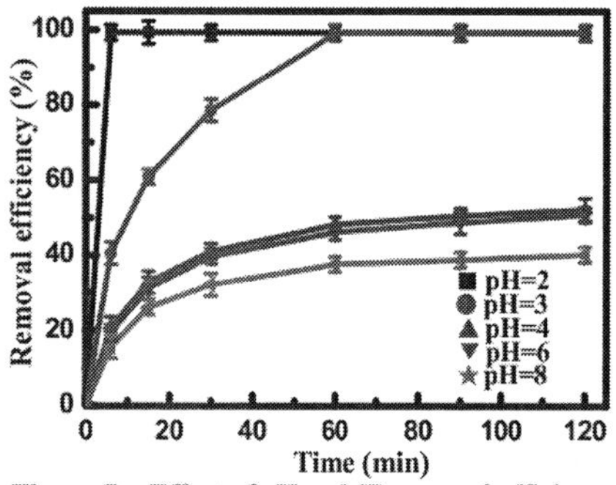

Figure 7. Effect of pH on MB removal efficiency.

4 Conclusions

In summary, The Fe-doped carbon aerogel were used for the MB removal in wastewater. The MB removal efficiency of T700 could reach 96.0% after 9 h. T700 was selected for further research considering the energy consumption and the MB removal efficiency. Furthermore, the materials have good MB removal efficiency at different pH. T700 was also used in combination with H_2O_2. The MB removal efficiency can reach to 99% in 3 min under the conditions that pH=2 and 0.2 ml H_2O_2 was added after 0.5 h of Fe-C micro-electrolysis reaction, These novel Fe-doped carbon aerogels will have broad application prospects in the field of wastewater treatment due to the environmentally friendly raw material, simple preparation process, and efficient removal of organics in wastewater.

References

1. R. Chen, Y. Zhang, L. Shen, X. Wang, J. Chen, A. Ma, W. Jiang, Chem. Eng. J. **268**, 348-355 (2015).
2. M.C. Ncibi, B. Mahjoub, M. Seffen, J. Hazard. Mater. **139**, 280-285 (2007).
3. R. Bin, Y. Xu, L. Zhang, Z. Liu, J. Taiwan Inst. Chem. E. **88**, 114-120 (2018).
4. K. Samar, Theydan, J. Muthanna, Ahmed, J. Anal. Appl. Pyrol. **97**, 116-122 (2012).
5. P. M. K. Reddy, P. Verma, C. J. Taiwan Inst. Chem. E. **58**, 500-508 (2016).
6. M. Masrollahzadeh, M. Atarod, B. Jaleh, M. Gandomirouzbahani, Ceram. Int. **42**, 8587-8596 (2016).
7. Z. Yang, Y. Ma, Y. Liu, Q. Li, Z. Zhou, Z. Ren, Chem. Eng. J. **315**, 403-414 (2017).
8. X. Dou, R. Li, B. Zhao, W. Liang, J. Hazard. Mater. **182**, 108-114 (2010).
9. Z. Ma, Y. Yang, Y. Jiang, B. Xi, T. Yang, X. Peng, X. Lian, K. Yan, H. Liu, Chem. Eng. J. **311**, 183-190 (2017).
10. L. Zhang, Q. Yue, K. Yang, P. Zhao, B. Gao, J. Hazard. Mater. **342**, 705-714 (2018).
11. L. Gu, N. Zhu, H. Guo, S. Huang, Z. Lou, H. Yuan, J. Hazard. Mater. **246-247**, 145-153 (2013).
12. C. Zhang, M. Zhou, X. Yu, L. Ma, F. Yu, **160**, 254-262 (2015).
13. B. Hou, H. Han, S. Jia, H. Zhuang, P. Xu, K. Li, J. Taiwan Inst. Chem. E. **60**, 352-360 (2015).

Social network, environmental cognition and organic fertilizer application behavior of small farmers

PENG Li[1] and YANG Qingxian[1*]

[1]211 Huimin Road, Wenjiang District, Chengdu City, Sichuan Province, China

Abstract. Small farmers are the main organizational form of agricultural production and management in China, and it is very important to study their organic fertilizer application behavior to promote the development of green agriculture in China. Based on the survey data of 334 small farmers in Sichuan Province, this paper uses bivariate Probit model to analyze the influence of social network and environmental cognition on organic fertilizer application behavior. The research shows that:(1) The application of commercial organic fertilizer and farmyard manure by small farmers has a significant complementary effect.(2) Social network has a significant positive impact on organic fertilizer application behavior, and there are significant differences between kinship social network and friendship social network on organic fertilizer application behavior.(3) The level of environmental cognition has a significant positive effect on the application behavior of organic fertilizer. Therefore, it is necessary not only to make full use of and expand the social network of small farmers, but also to pay attention to improving their environmental awareness and promoting organic fertilizer application behavior.

1 Introduction

As one of the important agricultural production factors, chemical fertilizer has played a positive role in improving grain yield and ensuring national food security. However, the extensive agricultural development mode for a long time has made China one of the countries with the largest amount of chemical fertilizer in the world [1]. According to statistics, the intensity of chemical fertilizer application in 2019 is as high as 325.65kg/hm2,which is much higher than the internationally recognized safety limit of 225kg/hm2. Excessive input of chemical fertilizer has caused negative problems such as agricultural non-point source pollution and soil fertility decline, which has affected and restricted the green and sustainable development of China's agriculture [2].Under this grim situation, the No.1 Document of the Central Committee in 2021 proposed to continuously promote the reduction and efficiency increase of chemical fertilizers and pesticides and promote the development of green agriculture. Thus, how to promote farmers' input of organic fertilizer has become an important practical problem.

The existing literature mainly develops organic fertilizer application behavior from three aspects. First, personal characteristics and cognition, farmers' age and education level have significant negative effects on farmers' organic fertilizer application behavior [3]; The higher farmers' awareness of the effect of organic fertilizer application, the higher their probability of applying organic fertilizer [4]; Second, in terms of family and production characteristics, farmers' family income [5], agricultural labor force [6], non-agricultural employment [7], land scale [8] and soil fertility [9] significantly affect farmers' organic fertilizer application behavior.The larger the land scale, the more favorable it is for farmers to apply organic fertilizer. However,with the fragmentation of land and the improvement of soil fertility, farmers' willingness to apply organic fertilizer decreases instead [9]. In addition, external factors such as whether to participate in agricultural insurance[10], stability of land property rights [11] and policy incentives [12]also affect farmers' fertilization behavior. Most previous studies analyzed farmers' organic fertilizer application behavior based on the assumption of rational economic man, but farmers' decision-making behavior is often not isolated. Under the condition of limited cognition, social relations play an important role in shaping behavior [13]. China's rural areas are a typical relational society, and farmers' economic behavior is embedded in the social network formed by interaction with others. Yang Haiyu et al. [14] found that the clan-based social network of farmers promoted the choice of direct sales channels of the company. Jiang Weijun et al. [15] pointed out that social networks further affect the willingness to adopt technology through trust and learning mechanisms.Under the condition of mastering the relevant information of organic fertilizer technology,farmers' own cognition is also very important. As a green production behavior, farmers' environmental cognition will also affect farmers' decision-making. Previous studies have shown that farmers not only maintain economic rationality but also ecological rationality in production [16]. The research by Huang Yanzhong and Luo Xiaofeng[2] shows that farmers' ecological cognition is the main power source for

* Corresponding author: 1140634457@qq.com

adopting organic fertilizer substitution technology, and the influence degree is greater than the influence of economic benefit cognition on organic fertilizer application.

The existing literature has done a lot of research on the application behavior of organic fertilizer, but there is still room for further improvement.First,in terms of research objects, most studies treat farmers as a whole, and seldom start from the perspective of small farmers. Although the current farmers are gradually divided into different types such as traditional small farmers,large grain growers and farmers, small farmers are still the main body of agricultural production and management in the future,so it is necessary to pay attention to the fertilization behavior of small farmers. Second, from the research perspective,the existing literature pays attention to the influence of social network or farmers' cognition on organic fertilizer application behavior, but few literatures bring social network and environmental cognition into the same analysis framework to analyze their influence on technology adoption. Third, in terms of research content, most studies do not subdivide organic fertilizer and refer to organic fertilizer as a general term, ignoring the correlation between different organic fertilizer application technologies. Based on the survey data of 334 small farmers in Sichuan Province, this paper uses bivariate Probit model to analyze whether there is complementary or substitution effect between different technical choices when farmers are faced with commercial organic fertilizer and farmyard manure application choices.Is there any difference in the influence of social network and environmental cognition on different technology choices? So as to provide reference for the development of green agriculture in China.

2 Theoretical analysis and research hypothesis

2.1 Social network and organic fertilizer application behavior

Social network mainly refers to the relationship network formed by mutual communication between groups [17]. At present, China's economic construction and market are not perfect, and social network still has a far-reaching impact on rural areas with backward formal system. The mechanism of social network on farmers' organic fertilizer application behavior is at least reflected in the following aspects:First, information acquisition. Farmers are typical information vulnerable groups and lack information channels[18].Social networks play an important role in transmitting information and reducing search costs. The larger the relationship network, the more likely farmers are to know more information, and the more they can correctly contact and understand this technology, thus increasing the probability of organic fertilizer application. Second, exchange and study. Social network, as the main environment for farmers to learn from each other, is conducive to reducing the uncertainty of technology adoption and promoting the knowledge diffusion of new agricultural technologies through mutual communication and learning [15]. Third, cooperation and mutual benefit. Agricultural technology conversion usually means higher cost input, while social network can provide farmers with financial, labor and technical support, reduce the cost pressure of technology conversion, and thus promote the application of organic fertilizer [19].

In the current Chinese rural society, farmers are mainly individuals or families, and form a social network structure of differential order pattern according to the close relationship of blood and geography. One is a social network based on "kinship" and focusing on strength, and the other is a social network based on "karma" or "geography" and focusing on universality [20]. The kinship social network plays a stronger role in information transmission, learning and communication by virtue of its emotional advantages, while the friendship social network can provide more extensive and non-repetitive information, changing the existing cognition and behavior habits of farmers. Based on this, the following assumptions are put forward:

Hypothesis 1: Social network has a positive impact on organic fertilizer application behavior, and the effects of kinship social network and friendship social network on organic fertilizer application behavior are different.

2.2 Environmental cognition and organic fertilizer application behavior

According to the theory of planned behavior, the subject's cognition will affect his willingness to participate and then affect his behavior. Environmental cognition belongs to human psychological factors. Behavioral economics believes that human beings will be affected by psychological factors in the decision-making process [21]. The higher the level of farmers' environmental awareness, the more they understand the relevant environmental knowledge and environmental protection policies, and the more they can recognize the seriousness of the current rural environmental pollution, that is, the more likely they are to apply organic fertilizer to alleviate the problem of agricultural non-point sources [22]. On the one hand, farmers, as decision makers of organic fertilizer application, may tend to choose organic fertilizer when fertilizing if they realize that organic fertilizer can improve soil fertility, improve soil structure and protect agricultural ecological environment. On the other hand,farmers,as the direct bearers of the deterioration of rural ecological environment, are more willing to change their behavior to maintain or even improve the existing environmental conditions in order to get a better environmental perception experience, thus holding a positive attitude towards the application of organic fertilizer.Based on this, the following assumptions are proposed:

Hypothesis 2: Environmental cognition has a positive impact on organic fertilizer application behavior.

3 Data sources, variable selection and descriptive statistics

3.1 Data source

The data comes from the field investigation of farmers in Sichuan Province in July 2019, and adopts the method of stratified sampling and equal probability random sampling. According to the research of Xu et al. [23], all districts and counties in Sichuan Province are clustered into five categories according to the per capita industrial output value from high to low, and one county sample is randomly selected from each category to obtain Jiangyou city, Zhaohua country, Guang'an city, Shehong county, Ziliujing and Yantan District. Then, the villages and towns in the sample districts and counties are divided into high and low income groups according to the ranking of per capita industrial output value. Each group randomly selects one township, each township randomly selects two villages, and each village randomly selects 20 households. A total of 400 questionnaires were obtained. According to the definition of World Bank et al.[24], small farmers are farmers whose cultivated land scale is less than 2 hectares (30 mu). After sorting out and deleting invalid questionnaires, 334 valid questionnaires for small farmers were finally obtained.

3.2 Variable selection

(1) Dependent variables. Organic fertilizer application behavior of small farmers was selected as dependent variable. Organic fertilizer includes commercial organic fertilizer and farmyard manure, which mainly refers to fertilizer made from crop straw, livestock manure and other raw materials through accumulation and fermentation.(2) Core variables. Drawing on the research of Fang Ran [25] and Yang Zhihai [26], two variables, kinship social network and friendship social network, are set up in this paper. The strength and scale of social network are characterized by the number of relatives visiting in 2018 Spring Festival and the number of fellow villagers and friends visiting in 2018 Spring Festival. Referring to the research of Kuang Foyuan et al. [27], the environmental cognition level of farmers is investigated by scoring the assigned values of 10 questions in Table 2. If the answer is correct, 1 point is counted, and if the answer is wrong or does not know, 0 point is counted. (3) Control variables. Considering the influence of farmers' individual endowment differences, gender, age and education level are selected to reflect farmers' individual characteristics; Select whether there are village cadres or party members in the family, and the proportion of family agricultural labor force and non-agricultural income reflects the characteristics of the family; Select cultivated land scale, cultivated land quality and family farming to reflect production characteristics; The regional characteristic variables are expressed by the virtual variables of the county where the farmers are located.

Table1. Farmers' environmental cognition level

Do you think air pollution has an impact on your health?1=Yes 0=No/ Don't know
Do you think pesticides pollute farmland? 1=Yes 0=No/ Don't know
Do you think discarding pesticide bottles pollutes farmland?1=Yes 0=No/ Don't know
Do you think plastic film pollutes farmland? 1=Yes 0=No/ Don't know
Do you think the application of chemical fertilizer will pollute the farmland?1=Yes 0=No/ Don't know
Do you think the application of organic fertilizer pollutes farmland?1=No 0=Yes/Don't know
Do you think the application of farmyard manure will pollute the farmland? 1=No 0=Yes/Don't know
Do you think straw returning to the field pollutes the farmland?1=No 0=Yes/Don't know
Do you think irrigation pollutes farmland? 1=Yes 0=No/ Don't know
Do you think the direct discharge of domestic sewage pollutes the water quality?1=Yes 0=No/ Don't know

3.3 Variable descriptive statistics

The variable definitions and data descriptions involved in this paper are shown in Table 2.In terms of social network, relatives mainly visit during the Spring Festival, while friends visit less, with an average of 4.874 and 2.835 respectively. In terms of environmental awareness, farmers' environmental awareness level is not high, with an average of 5.722. For control variables, there are more men than women, accounting for 58.7%. The respondents are generally older, with an average age of 56.293 years. The education level is generally not high, and the average number of years of education per capita is 5.138 years. The average number of agricultural laborers in families is 1.787, and 22.2% of farmers have party members or village cadres in their families. Most of the household income comes from non-agricultural income, with an average value of 88.9%, which is in line with the characteristics of non-agricultural in rural areas today. The average area of cultivated land per household is 4.824 mu, and the quality of cultivated land is mostly close to the medium level, with an average value of 1.950.

Table2. Meaning and descriptive statistics of variables related to farmers' organic fertilizer application behavior

Variable	Definition and assignment	Mean	Standard
Dependent variable			
Whether to apply commercial organic fertilizer	1=Yes 0=No	0.156	0.363
Whether to apply farmyard manure	1=Yes 0=No	0.632	0.483
Core Variables			

https://doi.org/10.1051/e3sconf/202129303016

Kinship social network	Number of relatives visiting your home during the Spring Festival (households)	4.874	3.590
Friendship social network	Number of friends/villagers visiting your home during the Spring Festival (households)	2.835	2.981
Environmental cognition	Knowledge score of environmental problems	5.722	2.534
Gender	Sex of head of household? 1= Male 0=Female	0.587	0.493
Age	Age of head of household (years)	56.293	11.531
Years of education	Years of education of the head of household (years)	5.138	3.970
Are there village cadres or party members at home	Are there village cadres or party members in the family? 1=Yes 0=No	0.222	0.416
Number of agricultural labor force	Number of family agricultural labor force (person)	1.787	0.820
Proportion of non-agricultural income	Proportion of non-agricultural income to total income (%)	0.889	0.202
Cultivated land scale	Area of cultivated land under operation (mu)	4.824	3.330
Cultivated land quality	The quality of cultivated land? 1 = worse than average 2 = same as average 3 = better than average	1.950	0.620
Whether or not to breed	Whether or not to breed?1=Yes 0=No	0.695	0.461
Guangan	1=Guangan 0= Other provinces	0.216	0.412
Shehong	1=Zhaohua 0= Other provinces	0.189	0.392
Jiangyou	1=Jiangyou 0= Other provinces	0.183	0.387
Zigong	1=Zigong 0= Other provinces	0.201	0.401

4 Model construction and empirical test

4.1 Model building

Whether small farmers apply commercial organic fertilizer and whether they apply farmyard manure are binomial choice variables, and some unobservable factors will affect the choice of these two behaviors at the same time. Therefore, the two decisions are not independent of each other. If y_1 and y_2 are used to represent the decision of small farmers on the application of commercial organic fertilizer and farmyard manure, and y_1=1 means the application of commercial organic fertilizer, while y_1=0 means no application of commercial organic fertilizer. y_2=1 means application of farmyard manure, while y_2=0 means no application of farmyard manure. That is, there are four results: (0, 0), (1, 0), (0, 1), (1, 1). Because the bivariate Probit model allows the correlation between the error terms of different equations,a bivariate Probit model is established to analyze the influencing factors of social network and environmental cognition on the organic fertilizer application behavior of small farmers. The specific form of model setting is:

$$y_1^* = \beta_1' x_1 + \varepsilon_1$$

$$y_2^* = \beta_2' x_2 + \varepsilon_2 \qquad (1)$$

$$y_j = \begin{cases} 1, if \ \ y_j^* > 0 \\ 0, \ other \end{cases} \qquad (2)$$

Among them, ε_1 and ε_2 meet the following assumptions:

$$\begin{cases} E(\varepsilon_1) = E(\varepsilon_2) = 0 \\ var(\varepsilon_1) = var(\varepsilon_2) = 1 \\ cov(\varepsilon_1, \ \varepsilon_2) = \rho \end{cases}$$

In (1) (2), y_1* and y_2* are unobservable latent variables, and y_1 and y_2 are the final result variables as explained variables. x_1 and x_2 are independent variables affecting farmers' application of commercial organic fertilizer and farmyard manure, respectively, β'_1、β'_2 are corresponding estimation coefficients. ε_1 and ε2 are error terms and obey two-dimensional normal distribution. ρ is the correlation coefficient of ε_1 and ε_2, when = 0,it means that ε_1 and ε_2 are not correlated; If > 0, y_1 and y_2 are complementary effects, and if < 0, y_1 and y_2 are substitution effects.

4.2 Empirical results and analysis

In this study, Stata16.0 software was used to test the influence of social network and environmental cognition on farmers' organic fertilizer application behavior. The results are shown in Table 3. On the whole, the model fits well,the core variables basically pass the significance test, and the coefficient symbol is basically consistent with the expected direction. Wald test shows that the coefficient is positive and significant at the level of 5%, rejecting the original hypothesis of ρ=0, so it is necessary to use bivariate Probit model, and the application of commercial organic fertilizer and farmyard manure by farmers is interactive and has strong complementarity.

The results show that kinship social network has a positive effect on the application of commercial organic fertilizer and farmyard manure by small farmers, which is significant at the level of 10% and 5%, respectively, while friendship social network has no significant effect on the application of commercial organic fertilizer by small farmers, but has a positive effect on the application of farmyard manure, which is significant at the level of 10%. Hypothesis 1 is confirmed. This means that under the social background of rural acquaintances, farmers will not only make fertilization decisions according to their own agricultural production experience, but also decide their own fertilization methods with reference to other farmers around them. Furthermore, the members in the kinship social network are more closely related, the degree of mutual trust is stronger, and it is easier to realize learning imitation and risk sharing within the group, which makes farmers show "herd effect" in agricultural production. Although friendship social network can provide heterogeneous information and resources for farmers and enrich their knowledge, it is also more difficult to obtain substantial help, thus having no significant impact on the application of organic fertilizer. Environmental cognition has a significant and positive impact on the application of commercial organic fertilizer and farmyard manure by small farmers at the level of 5% and 10%, respectively. This means that farmers with higher environmental awareness are more aware of the harmfulness of current agricultural non-point source pollution and the importance of adopting organic fertilizer technology, so they are more willing to apply organic fertilizer. Hypothesis 2 is confirmed.

Among the control variables, education level has a significant positive impact on organic fertilizer application. It shows that the higher the education level, the better farmers can understand the advantages of applying organic fertilizer and improve the behavior of applying organic fertilizer; Families with village cadres or party members are more willing to apply commercial organic fertilizer than ordinary farmers. Generally speaking, families with political identity actively respond to the call of the state and are more willing to invest commercial organic fertilizer in production; The proportion of non-agricultural income has a negative impact on the application of farmyard manure; The process of farmyard manure accumulation is time-consuming and laborious; The shortage of labor caused by non-agriculturalization will not be conducive to the application of farmyard manure; The quality of cultivated land has a positive impact on the application of commercial organic fertilizer by small farmers. In order to maintain high sustainable production capacity, small farmers are more willing to apply commercial organic fertilizer to good quality land; Whether farming has a positive impact on the application of farmyard manure by small farmers, and farming can provide low-cost farmyard manure raw materials for small farmers and promote the application of farmyard manure.

Table3. Estimation results of bivariate probit model

Variable	Commercial organic fertilizer		Farmyard manure	
	Coefficient	Z-test	Coefficient	Z-test
Core variables				
Kinship social network	0.043*	1.772	0.063**	2.484
Friendship social network	0.018	0.611	0.052*	1.879
Environmental cognition	0.081**	2.344	0.055*	1.810
Control variables				
Gender	-0.090	-0.444	0.012	0.074
Age	-0.004	-0.433	0.011	1.557
Years of education	0.067**	2.562	0.047**	2.221
Are there village cadres or party members at home	0.511**	2.442	0.199	1.051
Number of agricultural labor force	-0.024	-0.226	0.153	1.588
Proportion of non-agricultural income	-0.605	-1.239	-1.158***	-2.870
Cultivated land scale	-0.012	-0.392	0.014	0.525
Cultivated land quality	0.284**	2.004	0.155	1.262
Whether or not to breed	0.236	1.119	0.462***	2.784
Shehong	0.113	0.382	0.025	0.105
Jiangyou	0.116	0.386	0.055	0.229
Zigong	0.359	1.169	-0.240	-0.930
Guangan	0.448	1.537	0.188	0.773
_cons	-2.512***	-3.015	-1.254*	-1.758
rho	0.264**			
Log pseudolikelihood	-309.9667			
Prob>Chi2	0.0000			
N	334			

Note: *、**,、*** respectively indicate significant levels of 10%, 5%, and 1%.The following table is the same.

4.3 Robustness test

Considering that social environment plays an important role in shaping individual cognition, social network, as a special form of social environment, may also have an impact on farmers' environmental cognition. The more communication between farmers and people with strong environmental awareness, the deeper understanding of the harm of excessive application of chemical fertilizer and the benefits of applying organic fertilizer, thus affecting farmers' fertilization behavior. In order to avoid the influence of multicollinearity between social network and environmental cognition, social network and environmental cognition are added to the model for regression. The results show that social network and

https://doi.org/10.1051/e3sconf/202129303016

environmental cognition have no change in the direction and significance of organic fertilizer application, so the research results of this paper are relatively stable.

Table4. Robustness test

	Bivariate Probit model I		Bivariate Probit model II	
Variable	Commercial organic fertilizer	Farmyard manure	Commercial organic fertilizer	Farmyard manure
Core Variables				
Kinship social network	0.050** (2.053)	0.070*** (2.705)		
Friendship social network	0.013 (0.421)	0.047* (1.734)		
Environmental cognition			0.085** (2.572)	0.063** (2.201)
Control variables	Controlled	Controlled	Controlled	Controlled
Rho	0.283		0.301	
Log Likelihod	-313.6931		-317.4392	
Prob>Chi2	0.0000		0.0000	
Sample size	334		334	

5 Summary and suggestions

5.1 Summary

(1)The behavior of small farmers applying commercial organic fertilizer and farmyard manure has complementary effects.(2)Social network has a positive effect on organic fertilizer application behavior, and the influence of kinship social network and friend social network on organic fertilizer application behavior is different. Genetic social network has a positive impact on farmers' application of commercial organic fertilizer and farmyard manure, while friendly social network has a significant positive impact on farmers' application of farmyard manure, but has no significant impact on the application of commercial organic fertilizer.(3)The environmental cognition of small farmers can significantly promote the application behavior of organic fertilizer. That is, farmers' environmental cognition is also one of the most intuitive factors affecting farmers' organic fertilizer application behavior.

5.2 Suggestions

According to the above conclusions, the following enlightenment is obtained:

(1) Pay attention to the cultivation and promotion of social networks. Farmers are encouraged to use the relationship resources of relatives and friends to strengthen their network ties, and to create a community cultural atmosphere of mutual benefit and mutual assistance by organizing rural cultural activities and production mutual assistance activities. Combined with the characteristics of social networks, through the establishment of "village-level demonstration households" and "demonstration sites" and other forms, the strong trust characteristics of kinship social networks are brought into play to create convenient conditions for farmers to realize "learning by watching" and "learning by doing". With the help of modern means such as Internet media, we should build a communication platform between small farmers and agricultural technicians, broaden the circle of friends of farmers, and give full play to the positive role of social networks of friends.

(2)Enhance farmers' environmental awareness level. In the process of formulating policies, the government considers small farmers' environmental awareness as an important policy reference, carries out publicity and education activities on ecological environment-related knowledge, promotes farmers' learning and understanding of legal provisions related to ecological environment protection, and improves their environmental awareness level. At the same time, it also strengthens the publicity of organic fertilizer knowledge, improves farmers' understanding of the application of organic fertilizer, cultivates farmers' positive attitude, and promotes the promotion of organic fertilizer.

(3)Establish a cooperation mechanism for close communication between growers and farmers. On the one hand, it can meet the demand of growers for farmyard manure, on the other hand, it can also solve the pollution problem of livestock manure of farmers, promote the integration of planting and breeding cycle, and promote the development of green agriculture.

References

1. J.J. Gao, C. Peng and Q.H. Shi. Study on the high amount of chemical fertilizer and the fertilization behavior of small farmers in China-based on the discovery of fixed observation point data in rural areas from 1995 to 2016 [J]. Management World, 2019, 35 (10): 120-132.
2. Y.Z.Huang,X.F. Luo.Fertilizer reduction substitution: farmers' strategy choice and influencing factors [J].Journal of South China Agricultural University (Social Science Edition), 2020, 19 (01): 77-87.
3. C.H.Chu,S.Y. Feng,W.W. Zhang. Empirical analysis of farmers' behavior of adopting environment-friendly agricultural technology-taking organic fertilizer and soil testing and formula fertilization technology as examples [J]. China Rural Economy, 2012 (03): 68-77..
4. L.Q. Zhu, J. Wang, C.J. Wang , et al. Analysis of influencing factors of farmers' willingness to adopt organic fertilizer and chemical fertilizer combined application technology-based on farmers' survey in Jiangsu,Zhejiang and Anhui provinces[J]. Resources and Environment of Yangtze River Basin, 2018, 27 (03): 671-679.
5. T.B.Jiang.Analysis of influencing factors of farmers' fertilization behavior in rural ecological environment construction [J]. Journal of Southwest University for

Nationalities (Humanities and Social Sciences Edition), 2015, 36 (12): 157-161.

6. Y.Z. Huang, X.F. Luo, D. Liu, et al. Influencing factors of farmers' adoption of organic fertilizer instead of chemical fertilizer technology-an explanation of the phenomenon of high willingness and low behavior [J].Yangtze River Basin Resources and Environment, 2019, 28 (03): 632-641.
7. T.Y. Zhong, X.J. Huang, B.Y. Wang. Effects of non-agricultural employment on farmers' application of organic fertilizer [J]. China Land Sciences, 2011, 25 (11): 67-73.
8. H. Zhou, N. Teruaki, M. Song, et al. Analysis on Factors Influencing Organic Fertilizer Use in China: A Case Study on Wheat Farmers in Six Eastern Provincial-Level Regions [J]. Journal-Faculty of Agriculture Kyushu University, 2014, 59 (1): 215-219.
9. Y.M. Zeng, J.B. Zhang,K. He.Study on rice farmers' willingness to adopt organic fertilizer and its driving factors from the perspective of multi-dimensional family capital [J]. Journal of Ecology and Rural Environment, 2019, 35 (03): 332-338.
10. C. Zhang, C.X. Zhang, H.G Qiu , et al. Influence of Agricultural Insurance Participation Behavior on Farmers' Input-Taking Organic Fertilizer Input as an Example [J]. Agricultural Technology and Economy, 2017 (06): 79-87.
11. H. Lu, P.W. Zhang, H.Hu, et al. "Effect of the grain-growing purpose and farm size on the ability of stable land property rights to encourage farmers to apply organic fertilizers." Journal of Environmental Management 251:109621-109621.
12. X.C. Sang, X.F. Luo, Y.Z. Huang, et al. Policy incentives, ecological cognition and farmers' organic fertilizer application behavior-based on a moderated mediating effect model [J/OL]. Chinese Journal of Ecological Agriculture (Chinese and English):1-11[2021-05-24]. https://doi.org/10.13930/j.cnki.cjea.200978.
13. M. Granovetter. Economic Action and Social Structure:The Problem of Embeddedness [J]. American Journal of Sociology, 1985, 91(3):481-510.
14. H.Y.Yang,X.G.Li, C.Y. Zhang,et al. Informatization, social network and farmers' choice of purchasing channels for agricultural materials: a comparison between direct sales and retail stores for agricultural materials[J].Research on Agricultural Modernization, 2020, 41 (03): 474-483.
15. W.J. Jiang, T.W. Yan,X. Jiang, et al. Influence of social network and ecological cognition on farmers' willingness to return straw to field [J].Journal of China Agricultural University, 2019, 24 (08): 203-216.
16. L.Y. Yao, M.J. Zhao, T. Xu. Economic rationality or ecological rationality? Study on Behavioral Logic of Farmers' Cultivated Land Protection [J]. Journal of Nanjing Agricultural University (Social Science Edition), 2016, 16 (05): 86-95 +156.
17. M. Granovetter. *Getting a job: a study of contacts and careers* [M]. Ind. ed Chicago: The University of Chicago Press, 1995
18. Y.S. Chen, Y.D. Zhu, L. Zhang. Study on Farmers' Choice Behavior of Environmental Protection Agricultural Materials? -Based on experimental economics [J]. Agricultural Economic Issues, 2016, 37 (08): 33-40 +110-111.
19. T.C. Zhang, T.W. Yan, Z. Wang. Social network, income uncertainty and self-employed women's adoption behavior of conservation tillage technology [J]. Agricultural Technology and Economics, 2020 (08): 101-116.
20. L.X.He,Y.Q.Nan,Z.G.Zhang. Aging, Social Network and Family Farming: Evidence from CFPS [J]. Economic Review, 2016 (02): 85-97.
21. W.Z. Yu, X.F. Luo, R.R. Li, et al. Study on the Deviation between Farmers' Green Technology Adoption Willingness and Behavior from the Perspective of Green Cognition [J]. Resource Science, 2017, 39 (08): 1573-1583.
22. H. L. Zhang, J. Y. Li, H. Q. Teng. Cognition of smallholder farmers, external environment and adoption behavior of green agricultural technology--A case study of organic fertilizer [J]. Resources and Environment in Arid Areas, 2020, 34 (06): 8-13.
23. D.D. Xu, S.L. Guo, F.T. Xie, et al. The impact of rural labor migration and household structure on household land use arrangements in mountain areas of Sichuan Province, China.Habitat International, 2017, 70: 72-80.
24. Bank C. Reaching the Rural Poor: A Renewed Strategy for Rural Development [J]. World Bank Publications, 2010, 19 (4): 563 573.
25. R. Fang. *Research on Quantitative Measurement of China's Localization of Social Capital* [M]. Beijing: Social Science Literature Publishing House, 2014 .
26. Z.H. Yang. Aging, Social Network and Farmers' Adoption Behavior of Green Production Technology-Verification of Farmers' Data from Six Provinces in the Yangtze River Basin [J]. China Rural Observation, 2018 (04): 44-58.
27. F.Y. Kuang,M.Q. Chen, Li, et al.Difference analysis of farmers' ecological environment cognition and protection behavior-taking the use of pesticides and fertilizers as an example [J]. Research on Soil and Water Conservation, 2018, 25 (01): 321-326.

https://doi.org/10.1051/e3sconf/202129303017

Ultra-short-term Radiation Prediction based on FY-4 Satellite Cloud Images and Artificial Neural Network

Dongyu Jia[1*], Xiaoying Nie[1], Fuyuan Gao[1], Qingfeng Li[1]

[1]College of Geography and Environmental engineering, Lanzhou City University, Lanzhou 730070, China

Abstract. Surface solar radiation is affected by many random mutation factors, which makes the ultra-short-term prediction face great challenges. In this paper, the surface radiation observation station in the northwest (Dunhuang) desert area with broad PV prospects is selected as the research object. The input parameters of the test sample are: cloud forecast value, reflectivity and brightness temperature value of a satellite cloud image closest to the forecast time. The MATLAB software is used to model the prediction program and to predict the surface solar radiation in the next 10 minutes. A combined algorithm of satellite cloud images and neural network is applied to predict surface solar radiation for the next 10 minutes and is compared with the measured surface solar radiation. The model is a lightweight calculation model, it satisfies the calculation precision of engineering requirements. The results show that the diurnal variation trend of measured and predicted radiation values is basically the same. Among them, the prediction accuracy of the model for cloudy days is higher, while for snowy days with more abrupt changes, the prediction error of abrupt points is larger. The model can provide reference for ultra-short-term prediction of surface radiation.

1 Introduction

Due to the limited reserves of petroleum, coal and other resources, the development of new energy is an inevitable trend. Photovoltaic power generation will become an important part of the future energy structure. When photovoltaic power generation accounts for more than 15% of the power system, the fluctuation of photovoltaic power generation may cause the power system to be paralyzed [1]. Therefore, the power forecast of photovoltaic power station is very important to the safety and stability of the whole power grid. The fluctuation of solar radiation is the main factor affecting photovoltaic power generation.

At present, many scholars have analyzed and studied the prediction methods of surface solar radiation [1-8]. Chen et al. [2] established the error prediction equation through the error stepwise approximation method and established an improved solar radiation prediction method. Ding et al. [3] used the full-sky imager and established the cloud radiation attenuation model to predict the total radiation received from the ground. Chen et al. [1] used ground-based cloud map and neural network to model surface radiation prediction. Steven et al. [4] established a short-term (0-3h) surface solar radiation prediction method by using geostationary satellite observation and numerical weather forecast. Rosiek et al. [5] used satellite observation and artificial neural network to establish online 3-hour power output forecast of photovoltaic building integration system. Ayet et al. [6] studied the approach (6h) forecast of solar irradiance by using static satellite images and numerical simulation. At present, the difficulty in the study of ultra-short term prediction of surface solar radiation lies in the movement and dissipation of clouds. At present, the observation of clouds is mainly based on ground cloud map and satellite cloud map. The time and space resolution of ground cloud map is relatively high, but the observation range is limited. The successful launch of the current Fengyun-4A satellite enables China's quantitative remote sensing satellite in geostationary orbit to reach the world's advanced level. Fengyun-4A satellite has a high temporal and spatial resolution and can better predict cloud movement. Due to the late start of the domestic photovoltaic industry, the ultra-short term radiation forecast of photovoltaic power generation is still in the exploratory stage.

Taking Dunhuang observation station in northwest China as the research object, the observation data of Fengyun-4 geostationary satellite is used to determine cloud cover and forecast cloud movement. Combined with artificial neural network, the total solar radiation received on the surface in the future 10min has been predicted. Northwest China has abundant solar energy resources and broad prospects for developing photovoltaic power stations. For the desert area of northwest China where there are few meteorological stations, the satellite cloud map data is very precious. This study can provide reference for ultra-short term radiation prediction of photovoltaic power generation.

Corresponding author: jdy890719@lzb.ac.cn

 https://doi.org/10.1051/e3sconf/202129303017

2 Data and Methods

2.1 Data

Dunhuang Shuangpier observation point, latitude: 40°09 '29.1 ", longitude: 94°30' 58.7 ", altitude: 1148m, the type of underlying surface is sand and stone Gobi. The observed data are the four components of radiation. The sensor is the Kipp&Zonen CNR4 net radiation sensor, and the acquisition frequency is 2min/ time. The observation time is from November to December 2018.

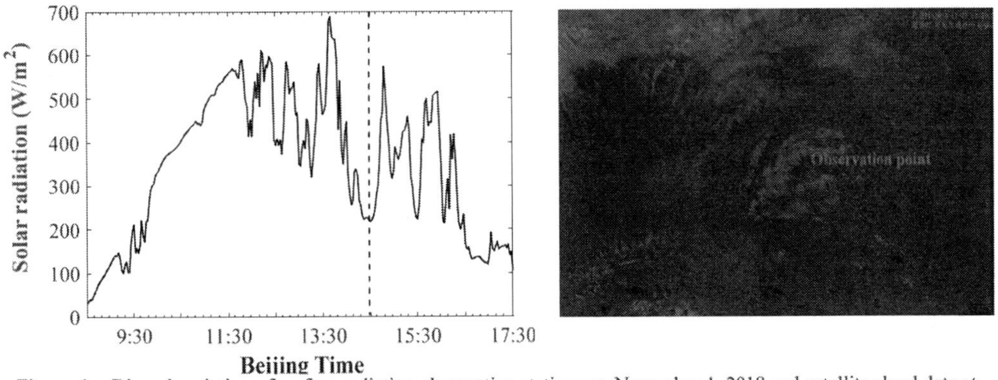

Figure 1. Diurnal variation of surface radiation observation stations on November 1, 2018 and satellite cloud data at 14:30 of the day

2.2 Methods

2.2.1 Calculation of cloud cover

In the FY-4A satellite imager 500ML1 image, four grid points are selected as the center and combined with cloud detection data, the ratio of the number of grid points inverted to the total number of selected grid points is the cloud cover data of the place [9].

$$C = \frac{N_{cloud}}{N_{cloud}+N_{sky}} \quad (1)$$

2.2.2 Cloud movement vector calculation

Previous studies have shown that this method has a high application value in the field of fluid mechanics [10]. In this paper, particle image velocity measurement software 'MPIV' based on MATLAB developed by Mori and Chang [11] has also been used to calculate the velocity field of cloud cluster. As shown in Fig. 2, the satellite cloud images at 14:34 and 14:38 on November 1, 2018 at the observation point are used to calculate the cloud movement vector which is more in line with the actual situation through the calculation of velocity measurement software and the improvement measures of cloud filtering, smoothing and quality control. The velocity vectors u=2.70 and V =1.20 (pixels/min) of cloud motion are obtained.

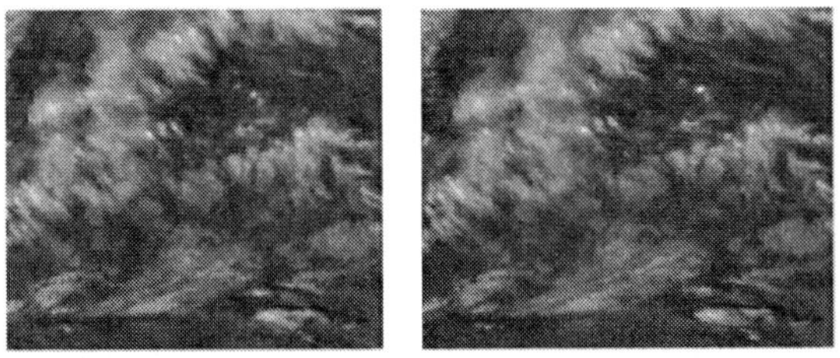

a. 14:34 Beijing Time b. 14:38 Beijing Time

Figure 2. Satellite Cloud Images on November 1, 2018

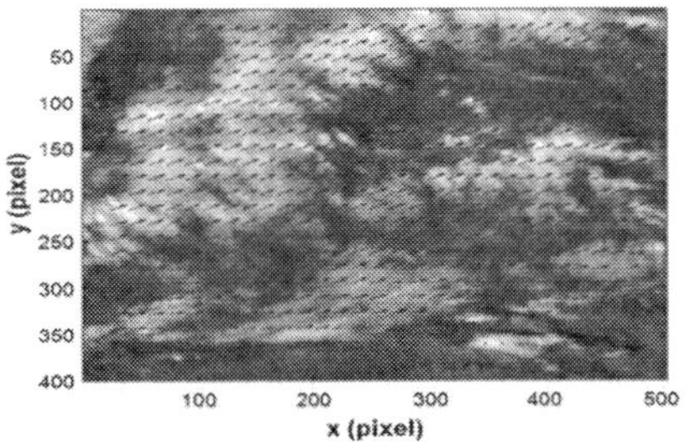

Figure 3. Velocity Vector Corresponding to MPIV

3 Analysis of forecast results

We used the prediction process in Fig. 4 to predict the surface solar radiation value of Dunhuang Shuangpier observation station in the future 10min. Because this paper focuses on cloud information, the forecast results with cloud sky are compared with the measured values. Fig. 4 shows the forecast results, among which Fig. 4a shows cloudy days and Fig. 4b shows light snow days. It can be seen from the figure that the model can well predict the basic trend of solar radiation on the surface. For cloudy days, the prediction accuracy is relatively high, the correlation coefficient between the measured radiation value and the predicted radiation value reaches 0.93, the root mean square error (RMSE) is 53.3 W/m2, the average absolute error (MAE) is 43.3 W/m2, and the average relative error (MRE) is 1.1%. The value with a large error occurs from 14:00 to 14:30, mainly because there are only two satellite cloud images in the period from 13:38 to 14:30, with a large time interval, making the error larger.

For the forecast of light snow days, the prediction accuracy is larger than that of cloudy days. The RMS error reaches 89.8 W/m2, the average absolute error is 69.5 W/m2, and the average relative error is 9.9%. It can be seen from Fig. 4b that the forecast error of the abrupt transition point of snow-day radiation is large, and the forecast value is relatively discrete from the measured value. However, compared with the forecast results without cloud map information [12], this model still significantly improves the forecast results.

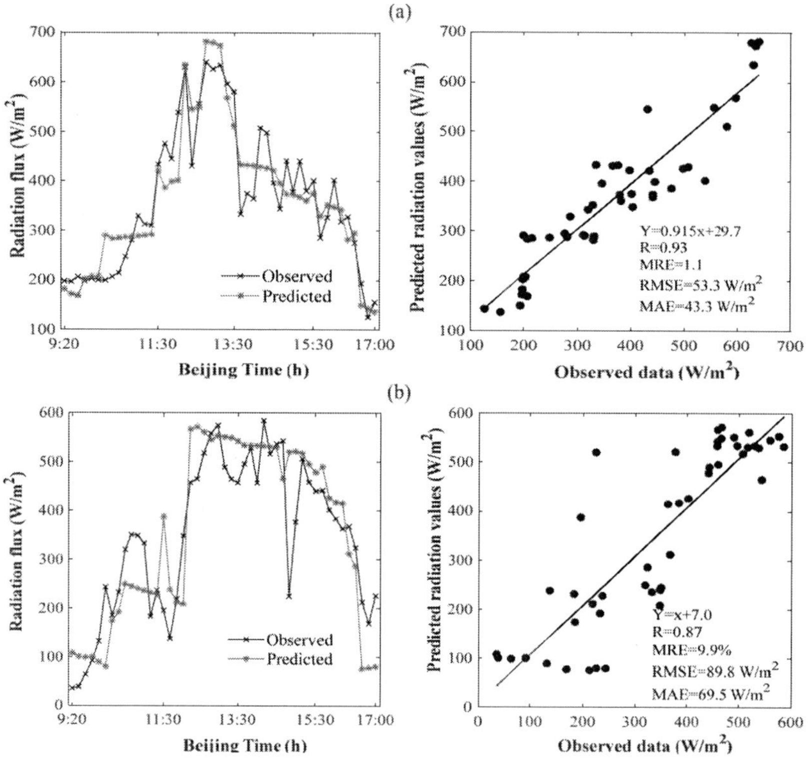

Figure 4. The forecasting values of surface solar radiation by satellite cloud images and neural network (a) Forecast value of surface solar radiation on November 04, 2018 (cloudy) (b) Surface solar radiation forecast value for November 17, 2018 (light snow)

4 Conclusion and discussion

For the desert area of northwest China where there are few meteorological stations, satellite data is very precious. In this paper, the numerical value of solar radiation received from the ultra-short term surface is predicted by using the algorithm of cloud map of Fengyun-4 satellite and artificial neural network. On the premise of ensuring high accuracy, the number of variables and the amount of data selected are small, which is a lightweight prediction model. It can be seen from the forecast results that the model can well predict the basic trend of surface solar radiation. For cloudy days, the prediction accuracy is relatively high, while for snow days with many abrupt changes, the prediction error of abrupt changes is relatively large, so the accuracy needs to be further improved.

Cloud movement, generation and elimination is an extremely complex process affected by many factors. Although the cloud image of The Fengyun-4 satellite has a high spatial and temporal resolution, there is also a time interval between two cloud images. Although the observation range of the ground cloud map is limited, the spatial and temporal resolution is higher. Therefore, based on the satellite image and the ground cloud image, radiation prediction will be the next step.

Acknowledgments

This work is financially supported by the Doctoral Research Fund of Lanzhou City University (Grant LZCU-BS2019-13) and the Opening Fund of the Key Laboratory of Land Surface Process and Climate Change in Cold and Arid Regions, CAS (Grant LPCC2018006).

References:

1. Z B Chen, J Ding, H Zhou, et al., Proceedings of the CSEE, 35, 561-7 (2015)
2. Z H Chen, P J Sun and R Zhang, Acta Energiae Solaris Sinica, 36, 2377-83 (2015)
3. Y Y Ding, J Ding, H Zhou, et al., Proceedings of the CSEE, 34, 50-6 (2014)
4. Miller, D Steven, Rogers, et al., Solar Energy, 168, 102-17 (2018)
5. S Rosiek, J Alonso-Montesinos and F J Batlles, Electrical Power and Energy Systems, 99, 261-72 (2018)
6. A Ayet and P Tandeo, Solar Energy, 164 301-15 (2018)
7. X Q Gao, J X Jun, N B Wang, et al., Acta Energiae Solaris Sinica, 1, 95-100 (2018)
8. Y Cui, C Cheng and Z H Chen, Water Resources and Power, 36, 201-4 (2018)
9. G Caili, Beijing, China Meteorological Press, 494 (2004)
10. Y Yang, X D Ruan and H Y Yang, Mechanical & Electrical Engineering Magazine, 22, 1-4 (2005)
11. N Mori and KA Chang, http://www.oceanwave.jp/softwares/mpiv (2003)
12. L W Yang, X Q Gao, J X Jiang, et al., Acta Energiae Solaris Sinica, **41**, 152-7 (2020)

Simulation and Analysis of Influencing Factors of Solar Energy Inter-seasonal Soil Heat Storage

Ping Lin[1], Xiangzhi Yao[2], and Yunpeng Bai[2,*]

[1]Shandong Huayu University of Technology, College of Energy and Environmental Engineering, 253034, Dezhou, Shandong, China

[2]Dezhou Transportation Development New Energy Co., Ltd, 253000, Dezhou, Shandong, China

Abstract. Taking an office building in Jinan as an example, the simulation model of solar inter-seasonal soil heat storage was established by TRNSYS software, and the variation law of ground temperature in the heat storage period was analyzed. From the perspective of ground temperature change, the influence of the spacing, length, number of drilling wells and area of solar collector on the heat storage effect was analyzed. The results showed that the soil temperature increased rapidly at the beginning of heat storage, and then the temperature rise rate gradually slowed down. The ground heat exchanger spacing, length, number of drilling and collector area will have a great influence on the solar energy seasonal heat storage effect. Therefore, in practical engineering applications, for the solar inter-seasonal soil heat storage system, the parameters of buried pipes, collectors and other components are recommended to be reasonably determined by simulation to obtain the optimal heat storage effect.

1 Introduction

The energy of ground source heat pump system comes from underground energy. The system does not emit any waste gas, waste water and waste residue to the outside world. In recent years, ground source heat pump system can be widely used in office buildings, hotels, schools, dormitories, hospitals, hotels, shopping malls, villas, housing and other fields due to its advantages of renewable, economic benefits and long service life. The ideal operation of the ground source heat pump system is that the heat obtained from the soil in winter is balanced with the heat released into the soil in summer, so the average temperature of the soil will not change after long-term operation of the system.

However, in cold and severe cold regions of China, soil temperature may decrease year by year during long-term operation of ground source heat pump system, which is not conducive to long-term efficient and stable operation of heat pump system[1]. Solar energy inter-seasonal soil heat storage is the combination of solar energy and ground source heat pump, that is, the use of soil in spring, summer, autumn three seasons more abundant solar energy into heat stored in the underground soil, winter heating season will be taken out to provide heat for buildings. This not only reduces the environmental pollution caused by building heating, but also uses clean energy to alleviate energy pressure[2]. Since solar inter-seasonal soil heat storage can effectively alleviate the risk of ground temperature imbalance, it has gradually been paid attention to in recent years[3]. Bakirci K et al. [4]established a solar energy interseasonal soil heat storage experimental system in Turkey. The system runs well, and the energy efficiency ratio of the ground source heat pump system is between 3.0 and 3.4. Zhu Jialing et al. [5] established a mathematical model of solar-soil heat storage and carried out experimental analysis, analyzed the feasibility of solar energy in non-heating season soil heat storage, indicating that the heat storage effect is affected by heat storage mode, solar radiation intensity, geological conditions and many other aspects.

Solar-soil heat storage is the scientific and systematic use of solar energy and soil heat to achieve the effect of energy complementarity. In winter, the radiation amount of solar energy is relatively small, and the heat users use the heat in the soil to meet the heating load in the building. In the summer with sufficient sunlight, the sun can often provide a large amount of heat. However, the heat demand of users in this period is relatively small.

Based on this, the combination of solar soil heat storage and ground source heat pump can not only ensure the heating effect of the system, but also timely supplement the soil temperature in summer, providing feasible conditions for the long-term operation of ground source heat pump, and reducing the harm to the environment in the operation process of ground source heat pump system.

Based on the above background, this paper takes an actual ground source heat pump project in Jinan City, Shandong Province as an example, focuses on the influencing factors of solar cross-seasonal soil heat storage from the perspective of soil temperature change, and from the perspective of heat exchanger spacing, length, drilling number and solar collector area on heat

* Corresponding author: lping_0307@163.com

 https://doi.org/10.1051/e3sconf/202129303018

storage effect is analyzed from the perspective of ground temperature change, aiming to provide a theoretical basis for the design and operation management of solar cross-seasonal heat storage.

2 Model establishment

2.1 Calculation parameters

The research object is a practical project in Jinan City, Shandong Province[6]. The building type is an office building, in which the building area is 4513.61 m^2, the building height is five layers, and the building is an energy-saving building. The ground source heat pump system is used in winter heating season, heating time is November 16 to March 15 next year. According to the hourly simulation results, the cumulative heat load in winter heating season is 186227.56 kW. Outdoor vertical buried pipe heat transfer hole number of 56, parallel connection, depth of 100 meters, installation of DN32 specifications single U-shaped high density polyethylene pipe. The average soil thermal conductivity is 1.53 W / (m · K), the average specific heat capacity is 2100 kJ / (m^3 · K), and the average soil temperature is 15 °C. The solar collector area is 270 m2, and the inclination angle is 55°. The solar thermal storage period in spring and autumn transition season and summer is from March 16 to October 31 (1824 h – 7344 h), and the start and stop time is from 9 : 00 to 17 : 00.

2.2 TRNSYS model

Based on the above description, a solar inter-seasonal soil heat storage simulation model was established through TRNSYS platform, Fig. 1 is a simulation diagram. Among them, Type557a is used for the ground heat exchanger module, Type14 is used for the start-stop time control of heat storage, and the meteorological data are used for the typical year of Jinan City, which are imported into the simulation model by Type15 - 3.

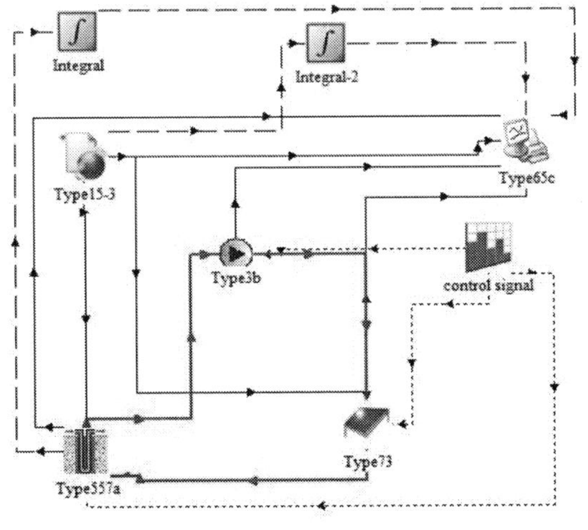

Figure 1. TRNSYS simulation diagram.

Fig. 2 and Fig. 3 show the annual temperature and solar radiation changes in typical years of Jinan City. The region is a warm temperate continental climate with an annual average temperature of 14.9°C, and the average temperature in the hottest month (July) is 26.9 °C. The area is rich in solar energy resources, and the annual cumulative solar radiation is 4673.3 MJ / m^2. Among them, the solar radiation in autumn and winter is less, 969.1 MJ / m^2 and 715.8 MJ / m^2, respectively. Solar radiation in spring and summer is 1525.2 MJ / m^2 and 1466.1 MJ / m^2 respectively. In addition, solar radiation in July (428.3 MJ / m^2) was lower than in the nearby month, mainly due to heavy rains.

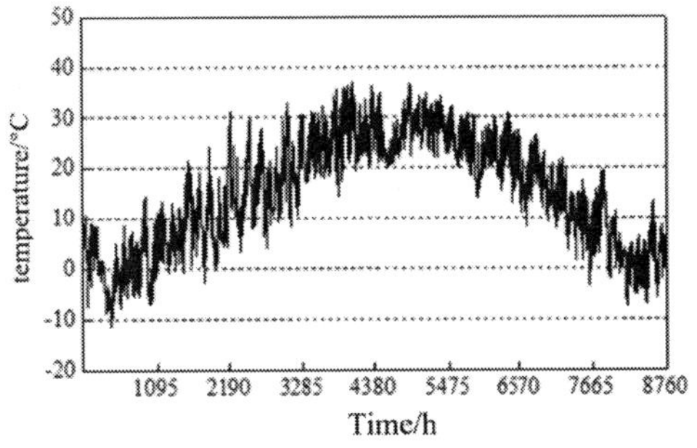

Figure 2. Temporal variation of temperature in typical years in Jinan City.

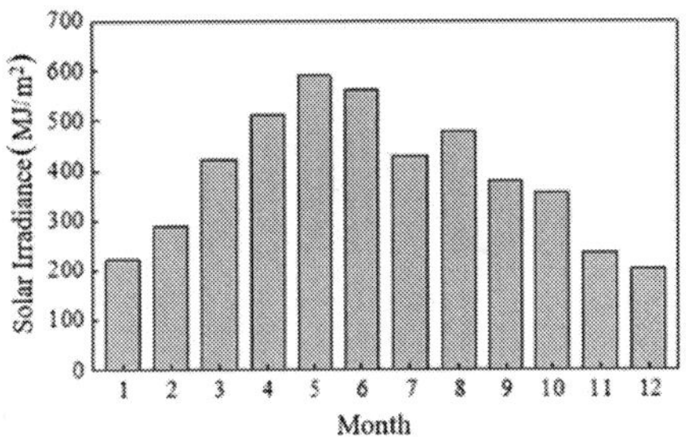

Figure 3. Changes of solar radiation in typical years in Jinan.

3 Results and discussion

3.1 Effect of buried pipe on heat storage

Fig. 4 gives the curves of average temperature of underground soil under different buried pipe spacing, in which the heat storage period is from March 16 to October 31 (1824 h-7344 h). It can be seen that the temperature rise rate of soil is faster at the beginning of heat storage, and then the temperature rise rate is gradually gentle. For example, when the local buried pipe spacing was 4.5 m, by August 31 (5880 h), the average temperature rise rate of underground soil was 0.097 °C / d, compared with 0.0048 °C / d from September 1 (5881 h) to October 31 (7344 h). This is because the soil temperature is at a low level in the early stage of heat storage, and the temperature difference between the outlet water temperature of solar collector and the soil is large, which is conducive to the heat transfer on the soil side. On the other hand, the solar radiation intensity in autumn is weaker than that in spring and summer, and the solar collector obtains less heat. In addition, with the increase of ground heat exchanger spacing, the average soil temperature decreased. At the end of the heat storage period, the ground temperature of four different buried pipe spacing is 17.9 °C, 16.9 °C, 16.4 °C, 16.1 °C. Thus, different buried pipe spacing has a more obvious effect on the average soil temperature. However, it is worth noting that too small buried pipe spacing will aggravate the soil thermal accumulation effect and increase the difficulty of transferring heat to the outside soil to a certain extent.

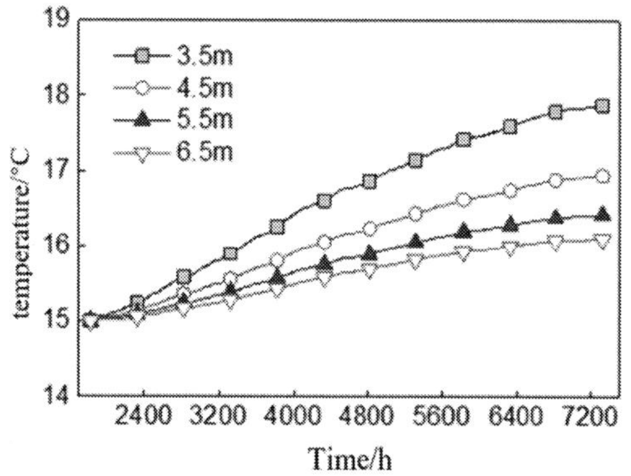

Figure 4. Ground temperature variation curve under different buried pipe spacing.

Fig. 5 further shows the variation curves of soil average temperature under different buried pipe depths. It can be seen that when other parameters remain unchanged, the buried pipe depth will also have a greater impact on the average soil temperature. For example, when the buried depth was 60 m and 140 m, the average soil temperature increased by 2.1 °C and 1.4 °C, respectively, which were about 14.0 % and 9.3 % higher than those in the initial stage. According to the author's previous research results, the change of buried pipe depth and spacing has a great influence on the construction cost and proportion area of the project. Therefore, the buried pipe form should be selected according to local conditions after hourly numerical simulation throughout the year.

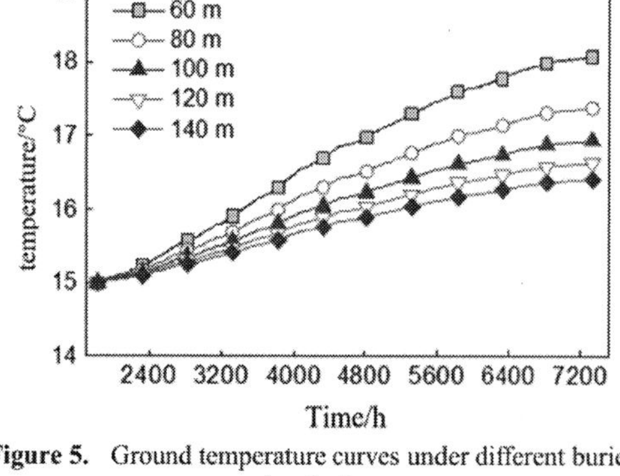

Figure 5. Ground temperature curves under different buried pipe depths.

In order to further analyze the influence of the number of buried pipe drilling on solar inter-seasonal soil heat storage, Fig. 6 shows the variation curves of ground temperature with different drilling numbers. It can be seen that different drilling depths also affect the change of soil average temperature. Specifically, under the premise of other factors unchanged, the more number of buried pipe drilling, the increase of soil average temperature decreases. For example, when the number of drilling wells is 36, the rise of ground temperature is 2.7 °C. In contrast, when the number of drilling wells is 76, the rise of ground temperature is only 1.5 °C. When the number of drilling wells was 46, 56 and 66, the average soil temperatures at the end of heat storage were 17.3 °C, 16.9 °C and 16.7 °C, respectively, which were about 15.3 %, 12.6 % and 10.7 % higher than the initial temperature. It can be seen that the buried pipe depth, spacing, drilling number and other factors have a relatively important impact on the solar energy inter-seasonal heat storage effect. For such systems, the buried pipe type and layout should be reasonably determined to obtain the maximum heat storage effect, and ultimately further improve the operation efficiency of the heat pump system in winter heating season.

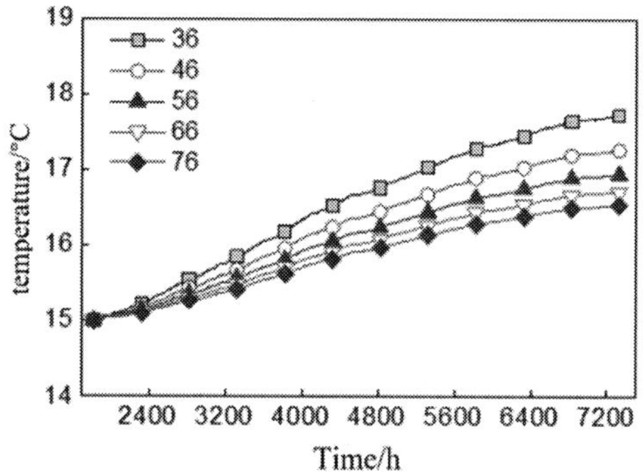

Figure 6. Ground temperature curves of different drilling numbers.

3.2 Effects of heat collection area on heat storage

The solar collector area directly affects the solar heat collection, thus further affecting the solar inter-seasonal soil heat storage effect. Therefore, it is crucial to select the appropriate collector area for solar inter-seasonal heat storage. Fig. 7 shows the variation curves of soil average temperature under different collector areas. It can be seen that in terms of the simulation in this paper, when other conditions remain unchanged, the average soil temperature increases with the increase of the solar collector area. For example, when the collector area was 170 m^2 and 320 m^2, the average soil temperature after the end of the heat storage period was 16.3 °C and 17.2 °C, respectively, which increased by about 8.7 % and 14.7 % compared with the initial stage. However, it is worth noting that the collector area is increased by 88.2 %, and the temperature rise is only increased by 0.9 °C. It can be seen that the collector area design should be comprehensively determined by the construction cost and the cumulative heat load in winter heating season. In terms of the simulation in this paper, the collector area should be 200 ~ 270 m^2.

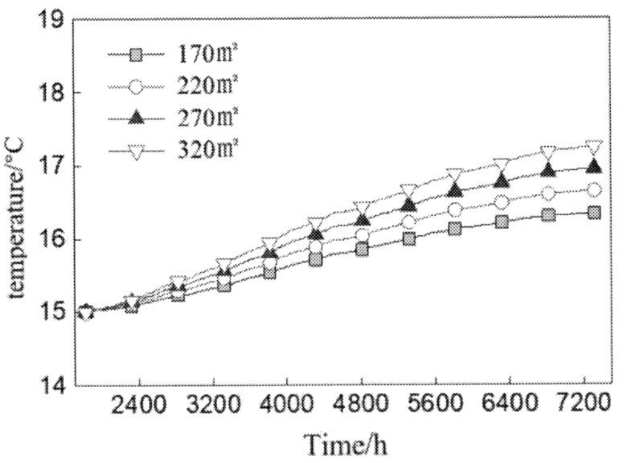

Figure 7. Ground temperature curve under different collector area.

4 Conclusions

In this paper, an office building in Jinan is taken as an example to establish a solar inter-seasonal soil heat storage simulation model by TRNSYS software. The influencing factors of solar inter-seasonal soil heat storage are simulated and studied from the perspective of ground temperature change, and the variation law of ground temperature in the heat storage period is analyzed. The influence of the spacing, length, number of drilling wells and area of solar collector on the heat storage effect is analyzed from the perspective of ground temperature change. The main conclusions are as follows :

1) Without changing the parameters, the soil temperature increased rapidly at the beginning of heat storage, and then the temperature rise rate gradually slowed down. Specifically, from 16 March to 31 August, the average temperature rise rate of underground soil was 0.097 °C / d, compared with 0.0048 °C / d from 1 September to 31 October.

Analysis reasons : on the one hand, it may be because the soil temperature is at a low level in the early stage of heat storage, and the temperature difference between the outlet water temperature of solar collector and the soil is large, which is conducive to the heat transfer on the soil side. On the other hand, it is also possible that the intensity of solar radiation in autumn is weaker than that in spring and summer, and the heat obtained by solar collectors is less, which leads to the low heating rate of soil side in autumn.

2) Ground heat exchanger spacing, length, number of drilling and collector area will have a great influence on the solar energy seasonal heat storage effect.

Specifically, when other parameters remain unchanged, the average soil temperature decreases with the increase of ground heat exchanger spacing. With the increase of buried pipe heat exchanger depth, the range of low temperature change decreases. With the increase of the number of buried pipe drilling, the increase of soil average temperature decreases. With the increase of solar collector area, the average soil temperature increased.

Therefore, in practical engineering applications, for solar inter-seasonal soil heat storage system, the parameters of buried pipes, collectors and other components are recommended to be reasonably determined by simulation to obtain the optimal heat storage effect.

Acknowledgement

This work was supported by Shandong Huayu University of Technology and Dezhou's Key Laboratory of High-efficiency Heat Pump Air Conditioning Equipmentand System Energy Saving Technology(No.26).

References

1. Rui Hua,Yiqun Pan,Rui Fan, et al. Impacting Factor on Ground Heat Balance of GSHP System Using TRNSYS [J]. Building Energy Efficiency. 2012, 40(3): 23-27.
2. Qi Hao. Optimization of Solar Energy Inter-seasonal Energy Storage Heating Application System. [D]. Hebei University of Science and Technology,2020.
3. Wenyong Zhang,Maoyu Zheng,Xiao Wang, et al. Analysis of Influencing Factors of Solar Seasonal Soil Heat Storage[J]. Building Science,2010,26(06): 81-86.
4. Bakirci K, Ozyurt O, Comakli K, et al. Energy analysis of a solar-ground source heat pump system with vertical closed-loop for heating applications[J]. Energy, 2011, 36(5):3224-3232.
5. Jialing Zhang,Jing Zhao,Wei Zhang. Solar-soil Coupled Heat Storage Characteristics and Potential Analysis. [J]. Acta Energiae Solaris Sinica,2011,32 (03):390-394.
6. Hongxin Zhu. Research on the Application of Solar Energy and Ground Source Heat Pump Combined Energy System in Building Energy Saving [D]. Shandong Jianzhu University,2016.

https://doi.org/10.1051/e3sconf/202129303019

Treatment of high-salt oil produced water based on constructed wetland technology—A case study of a costal oilfield

Xiaoqing Cao[1], Xue Gao[1], Kai Zheng[1], Shan Wu[1], Guodong Meng[1], Xiaohui Xie[1], and Jixin Su[1,*]

[1]School of Environmental Science and Engineering, Shandong University, China

Abstract. Aiming at the treatment of large volume of oil produced wastewater, the combined technology based on constructed wetland was applied for produced water treatment in a costal oil field. During the stable operation period, the treatment system maintained removal efficiency of 50.2% for chemical oxygen demand (COD), 100% for oil and 85.1% for ammonium nitrogen (NH3-N) under the gradient increase of salinity along the process. Meanwhile, this system has a good ability to adapt the influent fluctuation. When the oil in the influent fluctuated greatly between 11 mg/L to 147 mg/L, the effluent water quality remained stable. The treated water met the national discharge standard for water pollutants from the oilfield industry and can be used for the saline alkali beach wetland as supplement water.

1 Introduction

A large volume of wastewater will be produced in the process of oil exploitation, with the characteristics of high temperature and salinity[1]. Most of the oilfields in China have now entered into the third oil recovery stage, where produced wastewater is even more than 10 times of oil production[2]. The oilfield produced water contains not only large amounts of hydrocarbons, which are easy to be biodegraded, but also a high proportion of complex organic compounds, such as PAHs and benzene series[3]. These refractory substances generally have high toxicity and pathogenicity, which will cause huge pollution to the environment and threat to human health[4]. Thus, these wastewaters must be properly treated before discharge into the environment.

Treatment methods of oil produced wastewater mainly focus on physiochemical methods and biological methods[5]. Compared to the physical and chemical process, the biological treatment is more economical and environment-friendly. However, due to the high COD and complex composition of produced water, it is difficult to achieve the desired effect by using a single treatment method[6]. At present, the physicochemical method is generally used to pre-treat the produced wastewater first, and then subject it to the advanced biological treatment processes[7].

Constructed wetlands have been wildly used in combination with other processes to treat oilfield produced water[8][9]. Research on constructed wetland began in 1953 with Dr. Kat he Seidel. In 1974, the first complete constructed wetland was built in liebenburg othfresen, Germany[10]. In this basis, the related research was exploded both at home and overseas. Now constructed wetland technology has been widely used in oil produced wastewater treatment around the world. It was considered that the purification of wastewater in wetland was a synergistic effect of substrate, plant and microorganism[11]. The substrate can remove part of the pollutants through adsorption, ion exchange and other ways. Meanwhile, it has large specific surface aera, which provides sufficient adhesion interface for microorganisms[12]. Plants can remove pollutants through adsorption and assimilation during the growth, and affects the structure of rhizosphere microbial community by build the rhizosphere microenvironment[10]. Besides, Microorganisms remove pollutants from water through various physiological metabolic pathways[13][14].

In this study, we investigated the operational performance of a combined system based on constructed wetland from a costal oilfield. The purification efficiency was analysed through monitoring water quality to demonstrate the feasibility of this process for produced wastewater treatment, in order to achieve the recycling of water resources and achieve the goal of zero discharge of oilfield wastewater.

2 Process background and methods

2.1 Characteristics of produced water

The oil produced wastewater in this study was from a costal Oilfield with daliy discharge of oil produced wastewater about 3000 m^3. The oilfiled wastewater was characterized by high temperature, high salinity , complex organic composition and low biodagradable. The main physical and chemical properties of the produced water were listed in Table 1.

* Corresponding author: jxsu@sdu.edu.cn

Table1. Characteristics of the raw OPW.

Parameter	Range	Average
pH	7.08-7.28	7.19
COD	263-375	358
BOD_5 (mg/L)	92.7-114	103.2
NH_3-N (mg/L)	18.5-22.7	20.3
SS (mg/L)	53-64	57
Oil (mg/L)	105-127	115
Total nitrogen (mg/L)	97.6-110	103.7
Volatile phenols (mg/L)	0.027-0.038	0.033
Sulphides (mg/L)	0.064-0.080	0.072
Fluorides (mg/L)	1.05-1.21	1.12

2.2 Oil produced wastewater treatment system

The schematic diagram of the treatment process was plotted in Figure 1. As shown, the produced water was firstly pretreated by oil separation and gas flotation pond, with the addition of alkali and micro nano bubbles. The sewage after pretreatment discharged into oxidation pond, which integrated aerobic pond and facultative pond. The main design and operation parameters were listed in Table 2.

The constructed wetland system adopted the surface flow constructed wetland technology. The design aera of constructed wetland and ecological construction zone were 1.77 km^2 and 10.03 km^2, respectively. After further treatment by constructed wetland, the effluent quality could reach the corresponding discharge standard. The purified water could be used as the ecological supplement water for the saline alkali beach wetland in the ecological construction area, and improve the situation of drought and water shortage of plants and salinization.

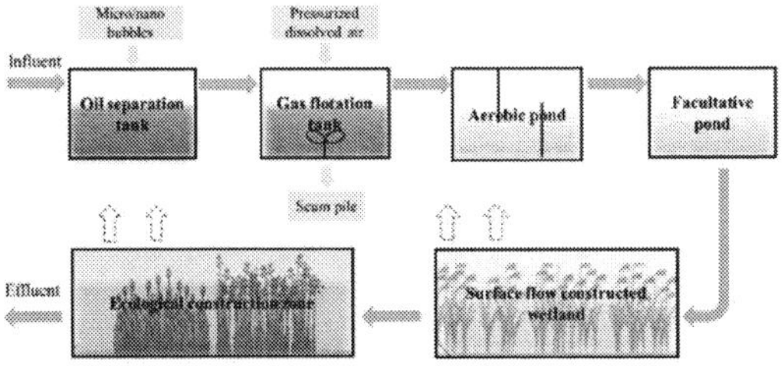

Figure 1. Schematic diagram of the treatment process.

Table2. Parameters of each tank.

Type	Size (length*width*depth)	Volume	HRT
Flocculation pond	17 ×17 ×4	1154	1.39
Aerobic pond	136 ×110 ×0.5	7480	8.98
Facultative pond	204 ×198 ×0.8	32313.6	38.78

2.3 Wastewater quality detection

The characteristics of the water samples along the whole treatment process were detected. The samples of gas flotation influent (GFI), gas flotation effluent (GFE), aerobic pond effluent (APE), the facultative pond effluent (FPE), the total oxidation pond effluent (OPE), the constructed wetland influent (CWI), the constructed wetland effluent (CWE) and the ecological constructed zone (ECA) were taken during the operation process for water quality analysis. Measurement for water quality parameters of the influent and effluent were in accordance with standard methods. All the water samples were immediately transported on ice to the laboratory and processed.

3 Results and discussion

3.1 Salinity and heavy metal distribution

Due to the characteristics of high temperature and high pressure with the oil layer environment, the composition of produced water could be very complex. The produced water not only contains crude oil, but also dissolves various suspended solids, salts and heavy metals in the formation[15]. Figure 2a showed the transfer distributing of salinity and Cl^- in the treatment system. The salinity and concentration of Cl^- were both high in the influent of flocculation pond, with concentration of 14.4 g/L and 8.02 g/L, respectively. After flocculation pond treatment, the concentration of the above two pollutants both reduced, suggested that air flotation process has removal capacity for both two pollutants to some extent. In the subsequent treatment process, the salinity and Cl^- increased continuously, with maximum concentration of salinity was more than 22.1 g / L in the effluent of constructed wetland. High salinity increased the difficulty of biochemical treatment. In addition to natural factors such as large evaporation in the constructed wetland, the gradient increase of salinity may result in large part from seawater indraught at the end of wetland.

Figure 2b showed the changes of heavy metals in the influent and effluent. In the whole treatment process, the influent contained less heavy metals, with concentrations of 1.7 mg/L for As, 0.02 mg/L for Pb and 0.008mg/L for Ni, the concentrations of Hg, Cd and Cr are lower than the detection. These results indicated that not much heavy metals have been produced in the flushing out process by water injection. After treatment process, the contents of heavy metals were further reduced, with concentrations of As and Ni reduced to 1.5 mg/L and 0.006 mg/L, respectively, which had met the corresponding standard.

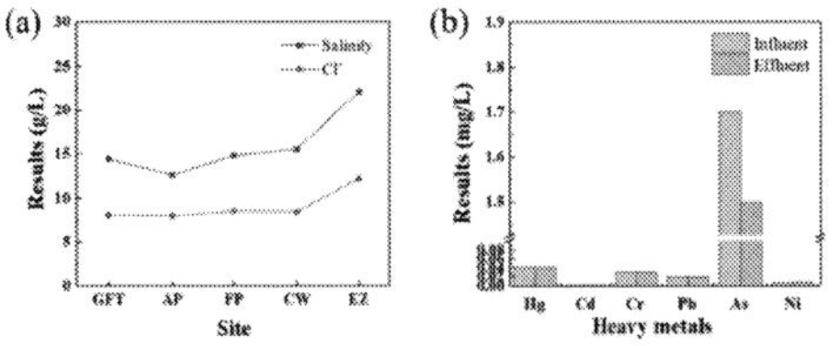

Figure 2. (a)Transfer distributing of salinity and Cl- in the treatment system, (b) Dissipation of heavy metals.

https://doi.org/10.1051/e3sconf/202129303019

3.2 Performance of the oil produced water treatment plant

Through monitoring the produced wastewater treatment plant for 210 days, the operation performance of produced water treatment plant using combined technology of air flotation and oxidation pond was studied. The changes of COD_{Mn}, NH_3-N and oil in the influent and effluent as well as their removal efficiency are plotted in Figure 3. The gas flotation tank was an unstable operation with large fluctuation of influent quality. The oil of influent ranged from 11mg/L to 147 mg/L, and the COD_{Mn} fluctuated from 43.9 mg/L to 71.1 mg/L. Despite that, it performed well as indicated by stable COD_{Mn} and oil removal efficiency, averagely 36.3% and 96.5%, respectively. The average NH_3-N removal efficiency was 13.7%, and it fluctuated with the water quality(Figure 2e). Figure 2(b) presented COD_{Mn} concentration and its removal in the oxidation pond. The removal efficiency (averagely 7%) was lower than that in the flotation tank. By comparison, the removal efficiency of NH_3-N increased to 35.9% averagely, and the concentration of NH_3-N in the effluent was tending toward stability (Figure 2d). The variation tendency of oil removal was similar to that in flotation tank, with average oil removal efficiencies of 92.1% (Figure 2f). After biological treatment, the water quality reached: COD $\leqslant$ 100 mg/L, ammonia nitrogen $\leqslant$ 15 mg/L, which met the corresponding national standards.

3.3 Performance of constructed wetland system

The wastewater treated by the treatment plant entered into the surface flow constructed wetland for further treatment. Through a series of synergistic effect of physical, chemical and biological reactions, the sewage was deeply purified[16].

The characteristics of main pollutants in different locations of constructed wetland and their removal efficiency were presented in Figure 3 and Table 3. It could be seen that all the pollutants were further removed through out the wetland process. At the end of ecological constructed zone, the COD_{Mn} and BOD_5 were degraded to 26.7 mg/L and 8.2 mg/L, corresponding to average removal efficiencies of 15.4%, 32.8%, respectively. The BOD_5/COD ratio increased from 0.26 mg/L of raw water to 0.31 mg/L of effluent, which indicated that the biodegradability of the wastewater was remarkably enhanced by constructed wetland. The NH_3-N and SS were decreased significantly in wetland. After treatment, the NH_3-N was reduced from 13.8 ± 0.5 mg/L to 3.8 ± 0.8 mg/L, representing removal rate of 72.6%, and SS was reduced from 19.5 ± 1.4 mg/L to 7.2 ± 0.7 mg/L, corresponding to the average removal rates of 63.1%. These results indicated that the surface flow constructed wetland shouldered the major part of NH_3-N removal in the system. Besides, cause the oil was largely removed during the previous treatment process, the influent oil concentration was only 0.2 ± 0.1 mg/L while the effluent of that was lower than the detection (Figure 4). After wetland treatment, the effluent could reach: COD ≤50 mg/L, ammonia nitrogen ≤5mg/L, which satisfied the national discharge standard for water pollutants from the oilfield industry.

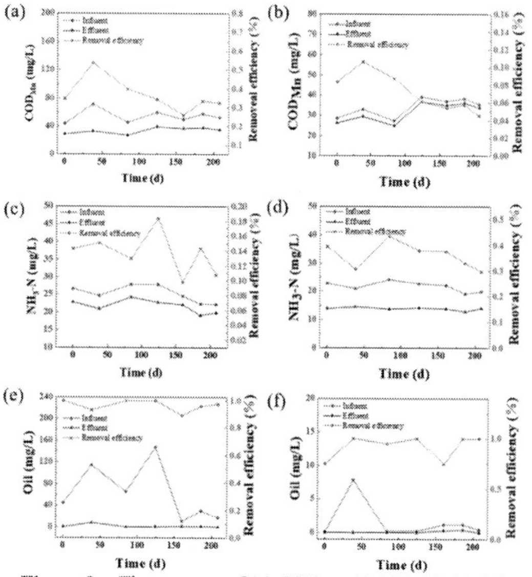

Figure 3. Time course of (a) COD_{Mn}, (c) NH_3-N, (e) Oil concentration and removal in flotation tank, (b) COD_{Mn}, (d) NH_3-N, (f) oil concentration and removal in oxidation pond.

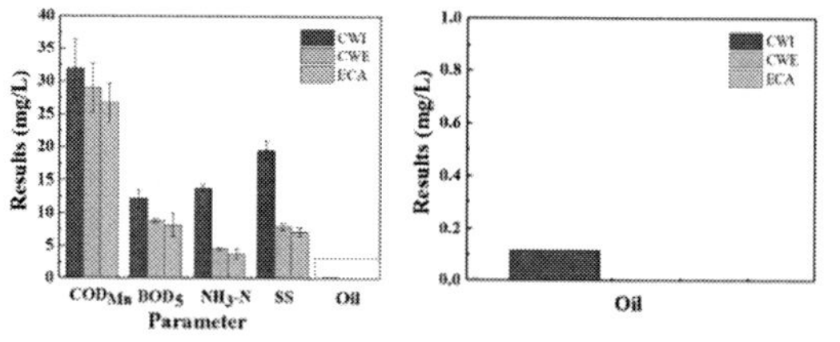

Figure 4. Concentration profiles of COD_{Mn}, BOD_5, NH_3-N, SS and oil in constructed wetland system. CWI, constructed wetland influent; CWE, constructed wetland effluent; ECA, ecological constructed aera.

Table3. Characterization of the concentrates of main pollutants of constructed wetland system.

Constituent	Influent (mg/L)	Constructed wetland (mg/L)	Ecological constructed zone (mg/L)	Total removal (%)
COD_{Mn}	31.9 ± 4.5	29 ± 3.7	26.7 ± 2.9	15.4
BOD_5	12.2 ± 1.3	8.8 ± 0.3	8.2 ± 1.8	32.8
BOD_5/COD	0.38	0.3	0.31	-
NH_3-N	13.8 ± 0.5	4.6 ± 0.2	3.8 ± 0.8	72.6
SS	19.5 ± 1.4	8 ± 0.6	7.2 ± 0.7	63.1
Oil	0.2 ± 0.1	-	-	-

4 Conclusion

This study demonstrated that the air flotation and oxidation pond treatment combined with constructed wetland system was an efficient and feasible process for the removal of COD, ammonium nitrogen and oil from the heavy oil produced wastewater. During the stable operation period, the combined process maintained average removal rates of 50.2% for COD_{Mn}, 100% for oil and 85.1% for ammonium nitrogen under the gradient increase of salinity. The biodegradability of the produced water was significantly enhanced by the flotation and oxidation pond treatment, while the surface flow constructed wetland played important roles in both NH_3-N remove and COD oil further degradation. It could be concluded that this treatment system allows efficient treatment of oil produced wastewater with high salinity. The treated water could be used for the saline alkali beach wetland as supplement water, and improved the situation of water shortage and salinization. Therefore, this combined system is feasible for high-salt produced water from oilfield, and has a good promotion prospect.

References

1. Z. Huang, X. He, C. Nye, D. Bagley, M. Urynowicz, and M. Fan, J. Hazard. Mater. **407**, 124348 (2021).
2. M. Lu and X. Wei, Bioresour. Technol. **102**, 2555 (2011).
3. M. Jain, A. Majumder, P. S. Ghosal, and A. K. Gupta, J. Environ. Manage. **272**, 111057 (2020).
4. A. Fakhru'l-Razi, A. Pendashteh, L. C. Abdullah, D. R. A. Biak, S. S. Madaeni, and Z. Z. Abidin, J. Hazard. Mater. **170**, 530 (2009).
5. X. Wang, L. Jiang, Z. Gai, F. Tao, H. Tang, and P. Xu, J. Hazard. Mater. **358**, 155 (2018).
6. S. Varjani, A. Pandey, and V. N. Upasani, Sci. Total Environ. **763**, 142934 (2021).
7. D. Su, L. Cong, J. Wang, and D. Zhou, Int. J. Environ. Pollut. **38**, 69 (2009).
8. G. Ji, T. Sun, Q. Zhou, X. Sui, S. Chang, and P. Li, Ecol. Eng. **18**, 459 (2002).
9. G. D. Ji, T. H. Sun, and J. R. Ni, Bioresour. Technol. **98**, 436 (2007).
10. J. Vymazal, Ecol. Eng. **73**, 724 (2014).
11. M. Jain, A. Majumder, P. S. Ghosal, and A. K. Gupta, J. Environ. Manage. **272**, 111057 (2020).
12. W. Liu, M. H. Rahaman, J. Mąkinia, and J. Zhai, Sci. Total Environ. **777**, (2021).
13. F. Karray, F. Aloui, M. Jemli, N. Mhiri, S. Loukil, R. Bouhdida, N. Mouha, and S. Sayadi, Process Saf. Environ. Prot. **141**, 73 (2020).
14. G. T. Tellez, N. Nirmalakhandan, and J. L. Gardea-Torresdey, Environ. Prog. **24**, 96 (2005).
15. X. L. Zou, Environ. Technol. (United Kingdom) **36**, 2381 (2015).
16. J. Vymazal, Sci. Total Environ. **380**, 48 (2007).

https://doi.org/10.1051/e3sconf/202129303020

Research on Toughness Design Strategy of Wetland Park Based on the "Chain Theory" —— A case study of Wuhan East Lake Falling Wild Goose Scenic Spot

Mei Gan[1], Yongping Qian[1], Jingwen Liang[1] and Lin Lu[1*]

[1]College of Landscape Architecture, Sichuan Agricultural University, Chengdu, Sichuan, China 611130

Abstract. Wetland is the significant territorial resources and natural resources in China, whereas the establishment of wetland park is an effective measure to protect wetland resources. The research, taking the Wuhan East Lake Falling Wild Goose Scenic Spot as the example and pointing out the "chain theory", focuses on the toughness plan and design of wetland park to build "Wetland Pond Chain System" and promote the capacity of nourishing and defending, which can be benefit for the sustainable development of wetland and provide certain references for wetland park plan and design in the future.

1 Background

The wetland, as one of the most biodiversity ecology landscapes in nature and the most momentous survival environments for humans, possesses obvious performances of environment adjustment and ecology improvement[1,2]. With the continuous development of urbanization, the urban wetland parks have been one of the important measures to effectively protect the wetland, which is of great significance for urban ecology base optimization and providing social service function[3]. At present, the establishment of urban wetland park has gained the attention, but it still faces the thereat of over-developing environment resources and human activities. Hence, It is vital to enhance the defence ability of wetland against urban interference and the self-recovering ability after being interfered.

"Toughness", deriving from the concept of ecology, refers to the bearing capacity and self-recovering ability, on the premise of no fundamental changes in sustaining essential life process and structure, when the ecological system gets external interference[4]. According to the status quo of urban wetland, some reasonable plans and designs based on the standard of toughness will be conducted, which could effectively relieve the contradiction between urban wetland and the urban development, as well as ensure the urban ecology safety.

The plan and design of wetland park toughness can enhance its resistance and resilience, and also has a great significance for the sustainable development of wetland park. The research, taking the Wuhan East Lake Falling Wild Goose Scenic Spot as the example, will be based on the theory of toughness design, put forward the "chain theory" of wetland toughness design, build "Wetland Pond Chain System", construct four modularization habitat units and explore the methods of wetland park toughness plan and design.

2 Statues Quo and Problems of Wuhan East Lake Falling Wild Goose Scenic Spot

2.1 Site analysis

East Lake of Wuhan, located the city center in Wuhan, Hubei province, is one of the largest lakes within city in China. Meanwhile,as the important target of urban ecology red line safety[5], it is full of sufficient natural resources and obtains the significant function of protecting and raising ecological system. Its designed position is in Wuhan East Lake Falling Wild Goose Scenic Spot and it covers an area of 50.1 ha, including site subject area of 49.4 ha and two islands area of 0.7 ha.

2.2 Site Surrounding Analysis

The Wuhan Railway Station is located in the north of the site; Tuanhu is adjacent to its west; Yanxihu lies in the east and separated by railways and urban roads; Many built projects situate in the south of East Lake like the Falling Wild Goose Scenic Spot, Moshan Scenic Spot, The Wuhan Botanical Garden and Maanshan Forest Park. The villages and farms are distributed scatteredly around the site and quite a few universities are included in.

2.3 Site Internal Analysis

For the nature condition, there are two larger ponds and several small ponds within the site. The elevation is higher

* Corresponding author: 952020695@qq.com

in the east and lower in the west with an elevation difference of about 10M and a gentle slope.

For the humanity and culture condition, there is a group of south-north high-voltage power lines in the eastern site, which limits the building construction and human activity. A adjacent and south-north greenway lying in the west separates the site nature environment strongly. Meanwhile, a few village buildings exist in the internal site, distributing into cluster.(Picture 1:Site Internal Analysis)

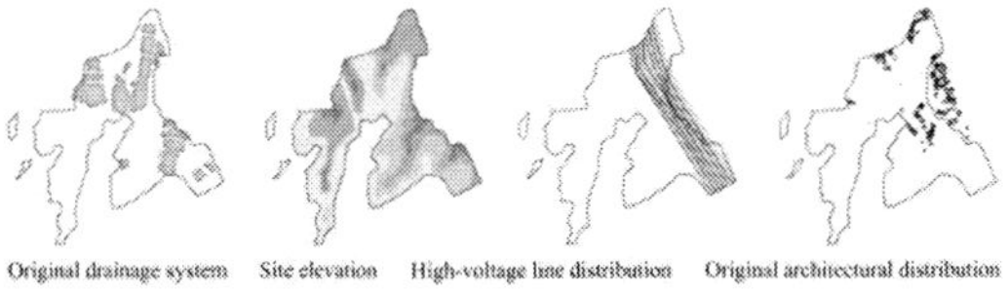

Picture 1. Site Internal Analysis.

2.4 Site Problem

2.4.1 wetland habitat destruction

The obvious habitat destruction exists in site status quo, including the four aspects like road fissuration, vegetation recession and water and soil loss, fish and bird resources exhausted,water quality eutrophication. The railway in the east separates the East Lake and Yanxihu, and blocks the ecological corridor, and influences the substance exchange and energy flow; The vegetation along edge of the site recesses hard and the water and soil loss happens in the two islands of eastern site; In recent years, the overfishing results in the species amount decline of fish and birds, which made bad impact on the recovering function of ecological system. The unreasonable emission of nitrogen and phosphorus led to a gradually serious water pollution. Aquatic organisms imbalance weakened the self-refine ability of the water, which accelerates the water eutrophication.

2.4.2 Human Building Occupying the Wetland

With the development of urbanization, human activity of overconstruction is constantly eroding the natural space. The continuous expansion of surrounding buildings approaches the wetland persistently. With the increasingly dramatic urban industrialization, the emission of sewage and gas causes great damage to the whole East Lake region. The high-voltage power lines in internal site affects the biology survival.The throughout urban greenway is not good for the continuation of the ecological environment.

2.4.3 Lack toughness in ecological system

Due to the location at the city centre, the wetland can be easily affected by the urban development and human activities, thus the ecological system will be destroyed. That is the site lacks the defence capacity on external interference and can hard recover from external interference. The low diversity of species and single ecological environment make the wet land lack the ability of raising lives, and it is difficult to form an entire and strong ecological system.

3 Put forward "chain theory" of wetland toughness design

Through the analysis on the status quo and problem of the site, it is aimed at building a toughness wetland ecology system. It's dominated by "chain theory" to set up a ecological chain of layer-by layer defence for resisting the external thereat, which, simultaneously, can enhance the site self-raising and self-recovering functions and protect the internal ecological safety of East Lake. The chain is a kind of steel rings connected with each other. In the design, each pond unit is regarded as one of the steel rings, the entity of the chain is the road, and the internal space is the pond. Each controllable water-gate of pond units accordingly equals to the open and closed place of the chain steel ring. If there is one of the rings broken, the entire will be influenced. So based the original type of the chain and optimizing the design, a layer of water system is added to its external part to make the dynamic control system be more reasonable, and the Pond - Road - Water Chain modularization distributed mode is formed then.(Picture 2: Chain Mode)

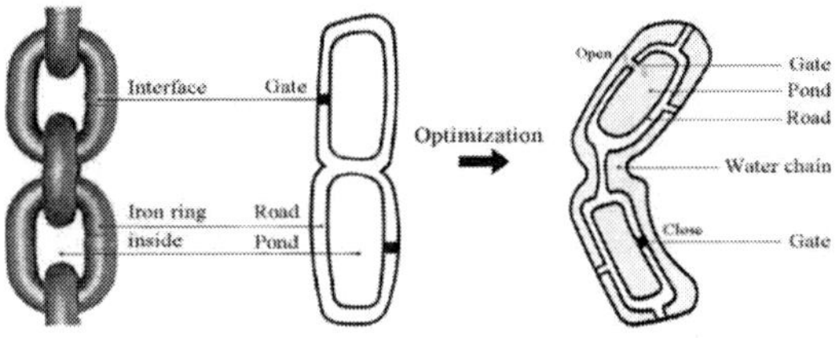

Picture 2. Chain Mode.

3.1 Pond Chain formation of East Lake Falling Wild Goose Scenic Spot

Based on the chain mode and taking full advantage of chain typology characteristics, that is to say each ring is not only the independent one, but also connects with each other, a dynamic modularization pond chain where the chain rings can be both connected with each other and blocked is to be built.

Integration excavation was carried out on the basis of the original ponds to form several single independent small ponds. Excavating the external part of the ponds and putting multiple ponds in a ribbon chain to form four pond chain with different depth by connecting them. There is a water-gate controller set outside the pond, which can make the ponds connect and disconnect mutually. The previous analysis indicates that due to the higher in the east and lower in the west, the underground water level in lower area is higher and the underground water level in the higher area is lower. Thus the four type ponds from east to west is respectively: intertidal zone, marsh, deepwater and green island.(Picture 3: Pond chain sectional drawing). With the depth of pond chain

https://doi.org/10.1051/e3sconf/202129303020

increasing, the diversity of species will be more abundant, more ecological, and the layer-by-layer resistance on urban interference is formed.

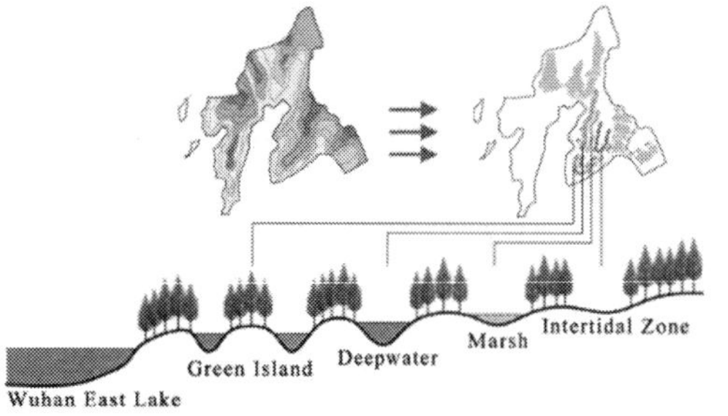

Picture 3. Pond chain sectional drawing.

3.1.1 Intertidal Zone

Tidal flat refers to the newly deposited intertidal flat[6].Intertidal zone is the most external water chain, which situates the highest position in the water chain structure. Therefore, it provides habitats for both insect like crickets,caterpillar and amphibian like crabs and frogs with the high water containing rate of mud(Picture 4:Intertidal Zone habitat).During the post period of construction, many activities like catching crabs and drawing with the mud can be conducted here.(Picture 5:Intertidal zone activity)

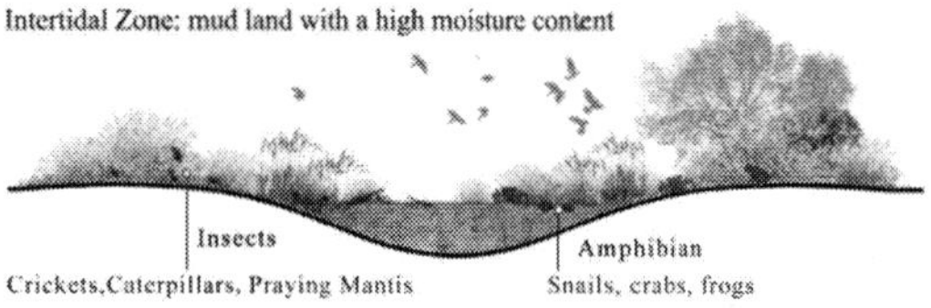

Picture 4. Intertidal Zone habitat.

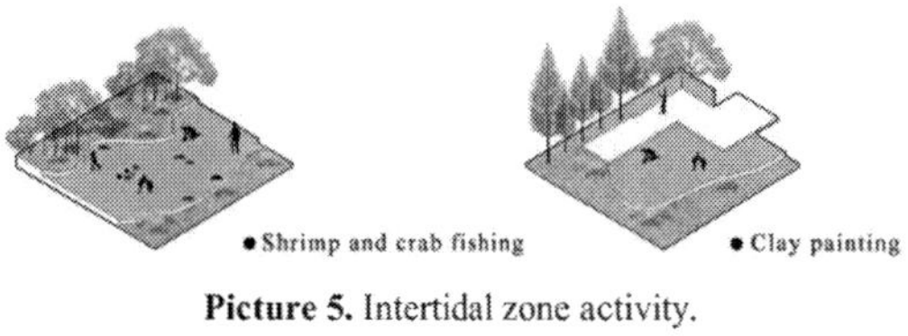

Picture 5. Intertidal zone activity.

3.1.2 Marsh

Marsh is the middle layer of the water chain. Marsh plants are transitional types of aquatic plants and terrestrial plants[7].The depth of 1.2m shallow pond is created according to terrain relation and water plants are to be planted to provide habitat for the bird like gray cranes, egrets and benthonic animal like shrimps, oysters(Picture 6:Marsh habitat). The activities of picking lotus root and birdwatching can be conducted in such habitat.(Picture 7:Marsh Activity)

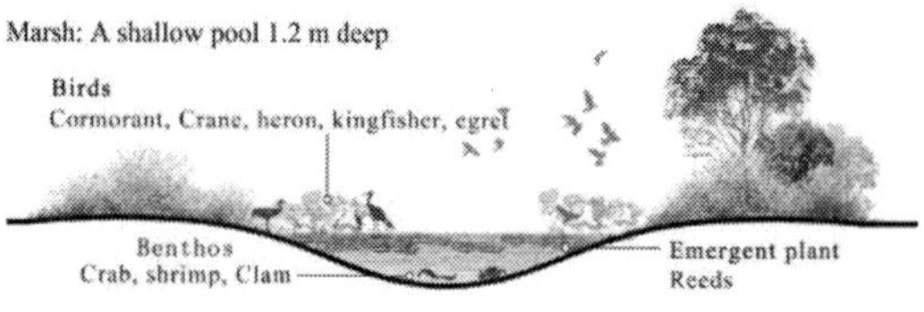

Picture 6. Marsh habitat.

Picture 7. Marsh Activity.

3.1.3 Deepwater

The deepwater is the most internal part of the water chain, which situates the lowest position in the water chain structure. Thus the depth of about 3m ponds are created to make fish species abundant. For example, the chub is suitable to survive on the surface; Scavenger fish and white silver carps is suitable to survive at the middle and the grass carbs is suitable to survive at the bottom(Picture 8:Deepwater habitat). The deepwater is also an appropriate display place of turning on the water-gate to connect the water chains, and can be the display field of water-gate theory and water purification.(Picture 9: Deepwater activity)

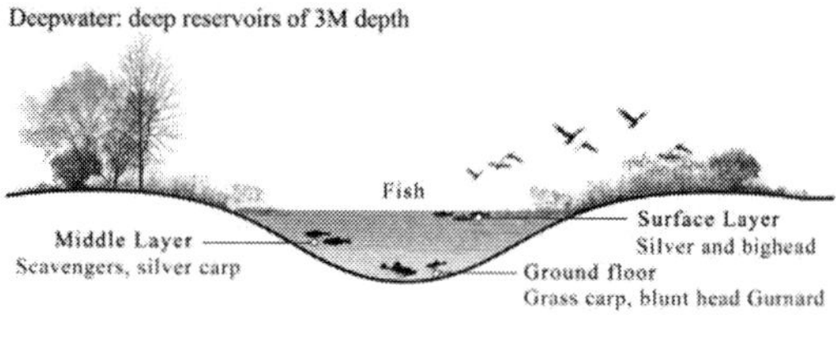

Picture 8. Deepwater habitat.

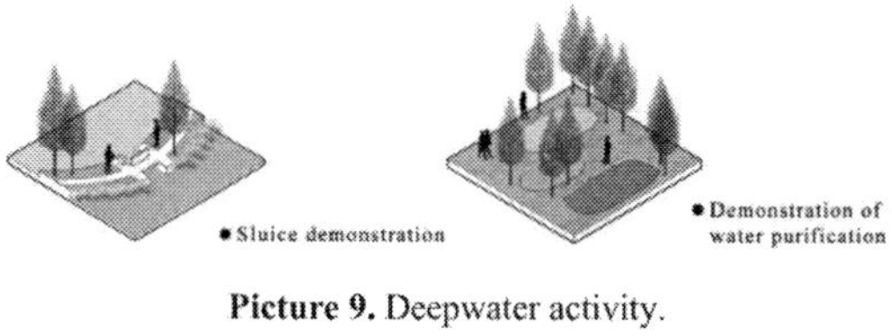

Picture 9. Deepwater activity.

3.1.4 Green Island

The greenway goes through the water chain net, separating the chain into south and north. The south of greenway is the green island water chain whose structure is island-water type. The main plants in island are the terrestrial plant like weeping willow, Chinese ash and metasequoia, which provides the habitat for the terrestrial animal like rabbits and squirrel. The habitat surrounding the greenway water chain is for salamander and tortoise(Picture 10: Green Island habitat). The habitat of green island not only raises these natural lives, but it also presents the sources for plants acknowledgement and popularization of biology.(Picture 11: Green island activity)

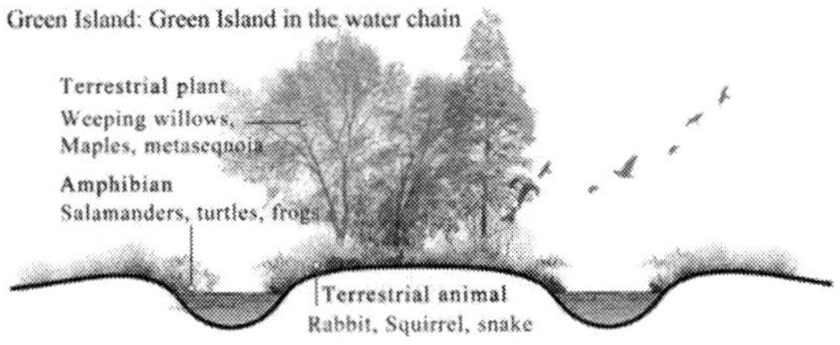

Picture 10. Green Island habitat.

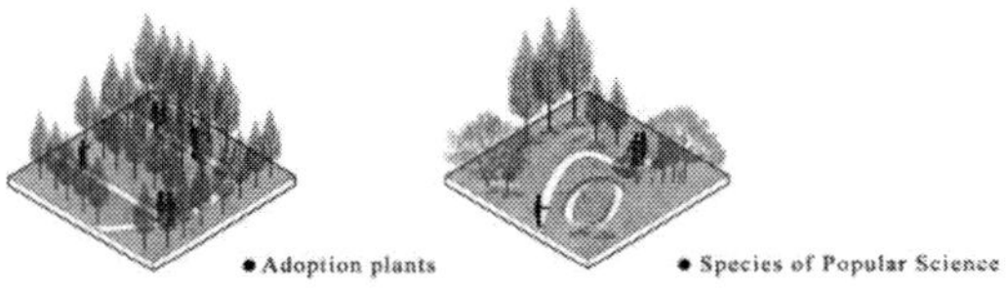

Picture 11. Green island activity.

3.2 Pond Chain System Dynamic Control

The controllable water-gate will be set between ponds and pond chain. When the external interference is small, the pond chain will implement the entirely nourishing. The water-gate of pond units are turned on, each pond builds connection by its own pond chain to raising lives jointly. When one of the pond units is polluted, its water-gate will be turned off, the pond will be separated and conduct self-recover to avoid polluting others, whereas the rest of the pond water-gates are still on, keeping connective and nourishing mutually. When the external interference pollutes the pond chain hard, all the water-gates of pond units should be turned off to separate the ponds from each other and conduct self-recover, as well as the external ponds are comparatively independent.

Due to the entire road net system for individuals traffic between ponds, when some ponds are polluted, the road net can block the pollution dynamically and control the individuals' routes.(Picture 12:Dynamic control)

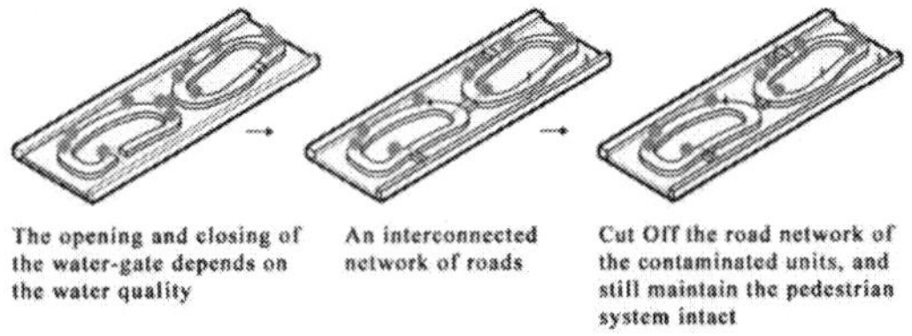

Picture 12. Dynamic control.

4 Summary and Expectation

From the perspective of toughness of wetland park plan and design, the research put forward the "chain theory", which was put into use of the construction of Wuhan East Lake Falling Wild Goose Scenic Spot. Through the dynamic modularization pond chain that can be connected and disconnected, four kinds of different habitat units are created, which serves as a positive demonstration for the future construction of Wetland Park.

4.1 Ecological Defence

When the water-gates are turned off, the habitat units are separated mutually and conducting the self-recover to prevent the pollution spreading and enhance the defence function of wetland facing the interference.

4.2 Ecologically Nourishing

When the water-gates are turned on, the habitat units connect and nourish mutually, which enhance the self-recovering ability of wetland.

4.3 Realize the Integration of Ecological Defence

Reaching the integration of wetland ecological defence through pond chain and dynamic control, which realizes the toughness design. In this process, the biology resource exchange between East Lake and Site is achieved, which enriches and expands the original ecological resources of East Lake.

References

1. Liu F 2020 Research on the application of resilient city concept in the landscape planning of Xi'an wetland park. Urban Architecture 17(30) P129-131
2. Zhang Y and Liu F 2009 The role of urban wetland in urban ecological construction and its protection countermeasures Environment science and management. 34(01) P140-144
3. Cheng S, Tong S and Wang D 2020 Study on the construction path of urban wetland park:A case study of Taohuayuan Wetland Park in Xuzhou Architecture and culture. 199(10) P159-160
4. Zhang T, Michael G 2018 A discourse on landscape architecture: At the intersection of design, ecology, resilience, and research Landscape design 6(04) P54-62
5. Zhang Y, Ji F, Qiu H and Zhang Q 2019 Research on the boundary planning of wetland park based on the coupling analysis of ecological sensitivity and suitability—A case study of Wuhan East Lake National Wetland Park Chinese garden 35(11) P81-86
6. Peng J and Wang Y 2000 A study on shoaly land in China Journal of Beijing University 36(6)
7. Ma X 1993 Some problems of the mire research in China Journal of lake science 5(1) A. Mecke, I. Lee, J.R. Baker jr., M.M. Banaszak Holl, B.G. Orr, Eur. Phys. J. E **14**, 7 (2004)

Research on assessment technology method of high efficiency air cleaners

Huang Jin[1], Shen Hao[2], Zhang Xiaoxin[1], Lin Ling[1], Li Zhenhai[3], Chen Jun[4], Wu Wei[5], Wang Baozhu[6]

[1]China National Institute of Standardization，Beijing 100191，China

[2]Shanghai metrological testing technology research institute

[3]Tongji University

[4]Guangdong media refrigeration equipment co., LTD

[5]Green Electric Appliances, INC. of Zhuhai

[6] Vkan Certification & Testing Co., Ltd

Abstract： In the face of the increasingly severe situation of air pollution, the research on evaluation technology and method of high-efficiency air pollution control equipment in China is being advanced. This paper discusses the background and significance of the research on high efficiency air cleaner evaluation techniques and methods, and introduces the General Ideas and principles of the research on high efficiency air cleaner evaluation techniques and methods, the technical methods of high efficiency air cleaner evaluation are established, including the evaluation object and scope of application, basic requirements of high efficiency air cleaner qualitative evaluation, quantitative evaluation requirements of key evaluation indicators, testing, calculation and evaluation methods, etc., the application of the research results of air cleaner will facilitate the production and consumption of air cleaner products.

1 Research background

1.1 Air pollution treatment demand

Currently as the deep advancement of China industrialization and urbanization, energy and resource consumption increase continuously, China environment condition is still severe, the compound air pollutions that are mainly consisted of fine particulate matter PM2.5 and ozone O3 have become more obvious. Take the example of Beijing-Tianjin-Hebei region, air quality has improved a little but the concentration of PM2.5 is still high, and is higher than the air quality class II for Beijing-Tianjin-Hebei and surrounding areal. Meanwhile the concentration of ozone O3 in critical area has increasing trend and it becomes the major pollutant especially in summer and fall in some cities.

In recent years, to promote sustainable economic development, and environment treatment and protection, and also push energy saving work, China releases series industry planning and policies to support and guide air pollutions treatment in China, and promote a formal and healthy industry development. The series energy saving and environment protection related industry policies and planning such as the State Council released “The action plan for preventing and controlling air pollution”, “Energy conservation and emission reduction comprehensive plan in the 13th Five-Year Plan”, “Development plans for energy conservation and environmental protection industries in the 13th Five-Year Plan” and “Proposals for the 14th Five-Year Plan and the Vision Goals of 2035” that were released at the first year of “the 12th Five-Year Plan”, “the 13th Five-Year Plan”, and “the 14th Five-Year Plan”, all specified clearly requirement of air pollution treatment. In Jun 2018, the State Council issued “Notification of three years plan to get blue sky” that has clearly specified “with three years effort, reduce the amount of main air pollutions a lot, reduce the greenhouse gas emission coordinately, further reduce concentration of fine particulate matter (PM2.5), obviously reduce heavy pollution days, improve air quality significantly, and enhance people’s happiness of blue sky.

1.2 Increasing product quality demand

In the face of increasingly severe air pollution, in addition to the treatment of air pollutants in key industries such as thermal power, steel, cement, petroleum and chemical industry, China is also gradually strengthening the evaluation and improvement of the quality, safety, energy saving and environmental protection characteristics of

* Corresponding author: huangjin@cnis.ac.cn

daily used indoor air pollution prevention equipment and products. Air cleaner, as a professional indoor environment health product to improve and solve indoor air pollutions, gradually develops to a kind of environment protection products that can governance indoor air pollutions such as fine particulate matter, microorganism, formaldehyde and volatile organic gas, and it could be used in bedroom, office, public places, industrial factories, hospitals and other indoor environments. In recent years, due to the widespread awareness of smog and PM2.5, air cleaners have gradually been recognized by the public, and with the improvement of people's health awareness, they have rapidly become one of the consumer's daily household appliance. However, what follows is that domestic and foreign large-scale production and sales enterprises have entered the air cleaner industry, and the whole industry is in the stage of sustained and high-speed development. The explosive growth brings the issues of market un-formal competition and product quality uneven and other problems, not only make consumers confusing when choose excellent products, but also be negative to popularize the reasonable and correct use of air cleaner to consumers. Therefore, the market and consumers both increase requirements for the improvement of the comprehensive performance of air cleaner products, such as quality, safety, energy saving and environmental protection.

In fact, as early as 2016, the General Office of the State Council of China issued the "Implementation of Quality Development Program 2016 Action Plan" clearly proposed to deeply carry out quality improvement action, air cleaners as indoor air pollution control special environmental protection equipment is regarded as one of key consumer goods that need to improve quality. In February 2016, the National Development and Reform Commission issued "guidance on promoting green consumption", air cleaner products are still on the list of key products, and encourage production enterprises to promote technical innovation, but also encourage consumers could have scientific and correct judgement and choice for the products on the market to improve the public consciousness of energy conservation and environmental protection. In view of this, carry out the evaluation technique for highly efficient air cleaner timely and do research on related technical standards at right timing, reference relevant existing international and national standards, and put forward a higher performance requirement, these can effectively guide the air cleaner production enterprises to improve product design and manufacturing level, formalize market access threshold, and can significantly improve the efficient environmental protection equipment, the market share of the products and services, so as to encourage the advanced and eliminate the backward ones.

2 The general idea and principle of research on evaluation technology and method of high efficiency air cleaner

2.1 Closer integration with current national environmental protection regulations and industrial policies

Research of highly efficient air cleaner evaluation methods give full consideration to the requirements of current national environmental protection laws, regulations and policies, including related environmental protection industry policies, the development and utilization of resources and energy and saving policy, ecological civilization construction and policy for the control of air pollution, and air pollution control technology equipment demonstration, restrict eliminate etc.

2.2 Coordination with the current national and industrial standards of air cleaners

When determining the evaluation index of HEPA, this study extensively referenced the technical requirements set in the existing national and industrial standards of related products, coordinated with the relevant domestic standards that have been released and implemented, and tried to be in line with the relevant international standards and regulations to ensure the uniformity and coordination of the standards. The national standards and industry standards for Air Cleaners released in 2015 mainly include: GB/T 18801-2015[1] "Air cleaner" and HJ2544-2016[2] "Technical Requirements for Environmental Label Products Air Cleaner". In this study, while taking into account the performance indicators of the products in the above standards, the evaluation indicators of the high efficiency characteristics of the air cleaner are also appropriately added, such as the evaluation technical requirements of the characteristic index of energy efficiency attenuation, to make the high efficiency characteristics of the Air Cleaner more prominent.

2.3 General requirements of following evaluation method for high efficiency air pollutions control equipment

In the field of air pollutant control and environmental protection equipment, China has initially formed a relatively complete standard system that are mainly consist of basic general standards, testing method standards, product standards and management evaluation standards, and relevant performance testing and performance evaluation methods have also been solidified into relevant technical standards. The research on the evaluation technology and method of high efficiency air pollutant control equipment involves the high efficiency evaluation of major single environmental protection equipment in the fields of dust removal, desulfurization, denigration, decarbonization, fog removal, deodorization, VOC control, indoor air management and other related subdivision fields. According to the previous top-level design of the research on the evaluation technology method of high efficiency air pollutants control equipment, China has completed the research on the efficient evaluation technology method of key environmental protection equipment such as ESP, bag dust collector and electric bag composite dust collector from 2014 to 2016. Therefore, this study continues to follow the overall framework requirements of the Evaluation Method for High Efficiency Air Pollutant Control Equipment, and establishes an evaluation index system for the air cleaner equipment in terms of technical performance, energy saving, environmental protection, safety and reliability, etc., and establishes an operational evaluation method.

2.4 Establish Principle of evaluation method and evaluation index system

2.4.1 To evaluate whether the air cleaner has reached the high efficiency level, we should not only evaluate the characteristics of the equipment itself, but also need to evaluate the management level of the production enterprise. Through the application of the research achievement, it should be able to effectively regulate the air cleaner's production, guide the enterprises to get technological progress, and lead the market consumption and use of high-efficiency air cleaners.

2.4.2 The construction of evaluation index system and the selected index should include technical performance, energy saving, environmental protection, safety and reliability, etc. When select indicators, we should consider comprehensiveness, concise, highlighted key points, and should be scientific and reasonable, strong operability and easy access to index data.

2.4.3 The principle of combining qualitative and quantitative indicators should be adopted. Qualitative indicators should be easy to verify while quantitative indicators should be measurable. With the science and technology progress and industry development, both quantitative and qualitative evaluation indicators will be continuously expanded.

2.4.4 It should be clearly specified for evaluation index testing method, testing theory, data collection and calculation.

3 High efficiency air cleaner evaluation technical method establishment

3.1 Evaluation object and appliable scope

3.1.1 Evaluation object

Due to the increasingly diversified and complicated of air cleaner target pollutants, this research defines the evaluation object in the view of product features and application requirements that refers to "a kind of electrical equipment that could get circulation purification indoor air by purification unit, and could remove some particle, gaseous contaminants and microbial."

3.1.2 Scope of application

Currently in China, air cleaners are mainly used for household and similar purposes, with a single-phase rated voltage of 220V. Except for industrial, medical, treatment and other professional purposes, the operating voltage is not more than 250V. For cleaning air quantity assessment of solid particles, we could reference American association of household electrical appliances (AHAM) reference standard, and could also base on some pollutions' particularly prominent in the urban environment air pollution contribution rate, and has the long-term characteristic comparing with gaseous pollutants and microbial contamination, meanwhile, the purification capacity is a basic performance of current air cleaner and evaluation method is mature. Considering these characteristics, the object air cleaners of this research is suitable for single-phase rated voltage not exceeding 250V, and particulate matter clean air amount (30 ~ 800) m3/h, and not for industrial, medical, automobile, governance and other professional application design and corrosive, explosive and other special environment using air cleaner products, it only selects ion technology of air purification products, as well as the air duct type air purification device, a new fan and other similar air purification equipment. Through the establishment of the evaluation index system, the air purification that meets the following "basic requirements for qualitative evaluation" and "quantitative evaluation requirements for key evaluation indexes" is the high efficiency air cleaner.

3.2 Qualitative evaluation basic requirement

As the producers of high efficiency air cleaners, first of all, they should meet the requirements of relevant environmental protection laws, regulations and standards, establish and maintain the quality management system and environmental management system; at the same time, the following requirements of appearance quality, electronic parts, safety, electromagnetic compatibility, hazardous substance content and operation instructions of the air cleaner are put forward:

3.2.1 For appearance quality, there should be no defects such as fingerprints, scratches, bubbles and shrinkage holes; The main components material should be safe, non-toxic, harmless, odor free, non-secondary pollution, and durable; From the product design side, propose the air cleaner producer to choose safe, environmental protection, healthy and durable materials to manufacture high-efficiency air cleaner.

3.2.2 The selected electronic components should be safe and reliable, and the unit security requirements should conform to the GB 4706.1[3] "Household and similar

 https://doi.org/10.1051/e3sconf/202129303021

electrical safety Part 1: general requirements", 4706.45[4] GB regulation in "Household and similar electrical appliances safety - the special requirements of air cleaner". From the point of the basic security of electrical appliances product, air cleaners' electronic components and the whole unit are both required to satisfy the safety and reliability requirements.

3.2.3 For electromagnetic compatibility, it should meet regulations in GB 4343.1-2009[5] "household appliances, electric tools and similar apparatus of EMC requirements Part 1: emission", GB 4343.2-2009[6] "household appliances, electric tools and similar apparatus of EMC requirements Part 2: immunity ", from the view of the reliability of electrical products electromagnetic compatibility, it's required that there is no electromagnetic interference with other devices around the product itself, no affect on other equipment normal operation, and also not affected by the surrounding electromagnetic interference and to ensure the normal work.

3.2.4 In terms of the content of harmful substances, it should meet the provisions of GB/T 26572-2011[7] "Requirements of Restricted Substances amount in Electrical and Electronic Products" and "Administrative Measures for Restricted Use of Hazardous Substances in Electrical and Electronic Products". From the perspective of product materials environmental protection, it is required that the products materials meet the corresponding requirements of national environmental protection.

3.2.5 For application instruction, the performance characteristics of air cleaners are required to be used as the application instructions of appliances, and it shall comply with the provisions of GB 5296.2-2008[8] "Consumer Goods Instructions Part 2: Household and similar electrical appliances". From the view of product labeling standardization, the basic requirements of consumer goods electrical products shall be met.

3.3 Key evaluation index quantitative evaluation requirements

The quantitative indicators in the evaluation technical method of high energy air cleaner should be selected by the characteristics of air cleaner, and should be the key evaluation indicators that have a significant impact on the quality assurance, reliable operation, energy saving and environmental protection of high energy air cleaner. These key evaluation indicators are composed of level I and level II indicators. Among them, the first-level indicators mainly include four aspects "technical performance, energy efficiency, environmental protection, operation safety and reliability", and a number of refined second-level indicators are selected for each category of indicators. Ensure that the evaluation value of the second-level indicators is in the advanced level of the domestic air cleaner industry. Specific evaluation indexes, evaluation methods and index value requirements are shown in Table 1.

Table1. High efficiency air cleaner quantitative evaluation index

number	Level I evaluation index	Level II evaluation index	evaluation method and requirement
1	Technical performance index	Particulate matter clean air volume	Tested value≥ nominal value
		$PM_{2.5}$clean air volume (when nominal function is available)	Tested value≥ nominal value
		Purification efficiency of gaseous pollutants (when nominal function is available)	≥80%
		Antibacterial rate (when nominal function is available)	≥99%
2	Energy efficiency index	Standby power	≤2.0W
		Energy efficiency ratio	$\geq 8.00 m^3/(W\cdot h)$
		attenuation rate of energy efficiency ratio	≤4.0
3	Environment protection index	Percentage of ozone concentration	$\leq 3\times10^{-6}$
		TVOC concentration	$\leq 0.08 mg/m^3$
		PM_{10} concentration	$\leq 0.04 mg/m^3$
		Noise (Sound power level)	50 ≤CADR （ Nominal value） ≤150： ≤52dB(A)
			150 ＜ CADR （ Nominal value） ≤300： ≤58dB(A)
			300 ＜ CADR （ Nominal value） ≤450： ≤63dB(A)
			450 ＜ CADR （ Nominal value） ≤800： ≤67dB(A)
4	Safe and reliable operation index	Ultraviolet radiation	200nm ～ 280nm ： total irradiance≤0.003W/m^2 ； Spectral irradiance≤10^{-5}W/(m^2·nm)
			250nm ～ 400nm ： total irradiance ≤1 mW/m^2
		4/5000 Applicable area	Nominal value=calculated value
		Particulate filter replacement maintenance cycle (specified condition)	One time filter replacement cycle: ≥26weeks
			Reusable filters for a single maintenance cycle: ≥ 16weeks

3.4 Testing method, calculation and evaluation method

For the main evaluation index, including: clean air quantity, purify air particles PM2.5, gaseous pollutants purification efficiency, an antibacterial rate, standby power, energy efficiency ratio, attenuation rate of energy efficiency ratio, ozone generation concentration, TVOC concentration, PM10 concentration, noise, ultraviolet radiation, filter replacement maintenance cycle, etc., high efficient air cleaner evaluation technique method provides the corresponding test method for each of them; At the same time, the detailed calculation methods of clean air volume of particulate matter, clean air volume of PM2.5 particulate matter, purification efficiency of gaseous pollutants, antibacterial rate, energy efficiency ratio, attenuation rate of energy efficiency ratio, applicable area, filter replacement and maintenance cycle are given, which are not detailed here. Finally, the air cleaners that meet the basic requirements of qualitative evaluation and the requirements of quantitative evaluation indicators in Table 1 can be evaluated as high efficiency air cleaners.

4 Conclusion

High efficiency air cleaner refers to those cleaners with "advanced technical performance, leading energy efficiency, superior environmental performance and safe and reliable operation ". The evaluation method of HEPA follows the general requirements of the Evaluation Method for HEPA control equipment. The proposed Highly efficient air cleaner evaluation methods proposed basic requirements and key evaluation index qualitative evaluation requirements are advanced, and their testing, calculation and evaluation method are scientific and reasonable that reaches the international leading level, which can effectively help air cleaner products and green industrial to develop toward green, low carbon and circulation industry direction. In the future, with technological progress and industrial development, the evaluation index system of HEPA will be further refined and improved.

References

1. GB/T 18801—2015 Air cleaner
2. HJ2544-2016Technical requirements for environmental label products, air purifier
3. GB 4706.1Household and similar electrical appliances-Safety Part 1:General requirements
4. GB 4706.45Household and similar electrical appliances - Safety - Particular requirements for air-cleaning appliances
5. GB 4343.1 — 2009Electromagnetic compatibility requirements for household appliances,electric tools and similar apparatus—Part 1: Emission
6. GB 4343.2 — 2009Electromagnetic compatibility--Requirements for household appliances, electric tools and similar apparatus--Part 2:Immunity--Product family standard
7. GB/T 26572—2011Requirements of concentration limits for certain restricted substances in electrical and electronic products
8. GB 5296.2—2008Instructions for use of products of consumer interest--Instructions for use of household and similar electrical appliances

 https://doi.org/10.1051/e3sconf/202129303022

Discussion on upgrading construction of environmental protection hotline driven by AI technology

Bin Wang [1], Jian Li[1], Shengnan Sun[1], Lu Wang[1], Xiaoyuan Wan[1], Zheying Yu[1]

[1]Beijing SinoVoice Technology Co.,Ltd,100089,BeiJing

Abstract. Enhancing the public's awareness of environmental protection is an important measure to achieve the goal of carbon peak and carbon neutrality. This article takes the improvement of public participation in environmental protection as the starting point, discusses the enthusiasm of the public in environmental protection and its legal channels for reporting and complaining of environmental pollution, summarizes the achievements and existing problems of the construction of complaint channels, and proposes to improve environmental protection reports. Channel countermeasures and suggestions, build a reporting channel of environmental protection with temperature through artificial intelligence technology, increase public awareness and willingness to participate, and promote the development of environmental protection.

1 Introduction

President Xi Jinping put forward the idea of "beautiful scenery and green hills and clear hills" in 2005. With the development of the times, this concept has shown its strong vitality. Emphasizing that "green water and green mountains are golden mountains and silver mountains" is to maintain the fine balance between economic development and ecological environment as far as possible, and take the path of ecological priority and green development[1]. Along this way from green water and green mountains, China will surely have both the prosperity of modern civilization and the beauty of ecological civilization in the future. In the fourth session of the 13th National People's Congress, Premier Li Keqiang demanded: accurately grasp the new development stage, thoroughly implement the new development concept, speed up the construction of a new development pattern, promote high-quality development, and make a good start for building a socialist modern country in an all-round way[2]. China will put the construction of ecological civilization in a more prominent position, promote green development, and promote the harmonious coexistence between man and nature. We need to systematically grasp the strategic task of the construction of ecological civilization during the "fourteenth five year plan" period.

Environment is the basis of human survival, and the destruction of environment poses a threat to the survival and development of human society. The newly revised environmental protection law of the people's Republic of China, implemented on January 1st 2015,clearly stipulates that citizens, legal persons and other organizations shall enjoy the right to obtain environmental information, participate in and supervise environmental protection according to law. This provides legal basis for public participation in environmental protection. To improve the public awareness of environmental protection participation and enhance the public participation degree is an effective way to prevent the transformation of environmental risk to social risk. However, due to the lack of participation in environmental protection, the government information is not open and transparent, ignoring the public interest demands and participation wishes, and there is no corresponding response mechanism, which makes environmental risk into social risk[3]."The State Council's opinions on strengthening the key work of environmental protection" are clear: to solve the outstanding environmental problems that affect the scientific development and damage the health of the masses, it also puts forward the "establishment and improvement of the environmental protection reporting system", unblocking the channels of public reports, reflecting the public opinion and people's situation in time, and raising the efficiency of accepting &processing the report.

2 Construction achievements

The Ministry of Ecology Environment of the People's Republic of China (formerly the Ministry of Environmental Protection of the People's Republic of China) published a hotline for environmental protection report on June 5th, 2009, and successively opened Wechat report and internet report. In 2019,the national 12369 environmental protection report network management platform (hereinafter referred to "Internet platform") received 531176 reports from the public, a year-on-year decline of 25.2% , from which,270451 phone reports,195950 Wechat reports,62240 Internet reports and 2535 other ways, including 26.0% drop in telephone

* WangBin: wangbin@sinovoice.com.cn

reports,21.6% decrease in Wechat reports and 22.9% decrease in internet reports. See Table 1 for details[4].

Table1. Acceptance of national environmental protection reports in 2019.

Source of report	2019		
	Reports	Percentage	Growth rate
Phone	270451	50.90%	-26.00%
Wechat	195950	36.90%	-21.60%
Internet	62240	11.70%	-22.90%
Other ways	2535	0.50%	——
Total	531176	——	-25.20%

The published of the environmental reporting hotline has greatly enhanced the enthusiasm of the public to participate in environmental protection, and protected the legitimate rights and interests of the public to report and complains about environmental pollution behaviors. The acceptance of environmental reporting hotline takes up a lot of time of environmental protection staff, and because of the relatively fall behind system construction and uneven quality of staff, the acceptance efficiency and acceptance quality are affected.

3 Existing Problems

3.1 The reporting hotline system is fall behind

The environmental reporting hotline was published on June 5th ,2009. After 2015,the Ministry of Ecology Environment has successively onlined Wechat reporting,internet reporting and other ways,with scattered system and relatively fall behind technology. In order to implement the opinions of the State Council on strengthening the key work of environmental protection,unblock the reporting channels of the masses,timely reflect the public opinion,and improve the efficiency of reporting acceptance,the existing system must be upgraded through AI technologies such as Automatic Speech Recognition (ASR) , Natural language understanding (NLU) , Text To Speech (TTS) and Big Data Analysis(BDA),etc.

3.2 The hotline service staff is obviously insufficient

The shortage of hotline staff is manifested in two aspects: one is the lack of full-time staff, which is difficult to ensure a 24-hour timely response; the other is that the hotline service staff do not have professional quality, only act as the role of the operator, and it is difficult to give a correct reply to the public in time when the public asks professional questions. In the process of acceptance of environmental protection report, it is inevitable that there are some problems, such as inaccurate service terms, random information reply and irregular feedback, which will have a negative impact on the government's credibility and administrative efficiency[5].

4 Countermeasures and suggestions

With the full help of artificial intelligence, big data, 5g and other frontier technologies, the way of public participation in environmental protection is improved. Through Automatic Speech Recognition (ASR) , Natural Language Understanding (NLU) , Text To Speech (TTS) and Big Data Analysis(BDA) and other artificial intelligence technologies, the upgrading construction of environmental protection reporting hotline is driven. Combined with the business attributes of environmental protection reporting hotline, the workflow of the system is designed to realize the intelligent accepting and processing, intelligent supervision, intelligent management of reporting, intelligence feedback, report intelligence analysis, and earnestly achieve 4 points "1st accepting and processing every reoprt ; 2nd any illegal will be investigated ; 3rd all investigation will have results, 4th resultswill be feedback to public".

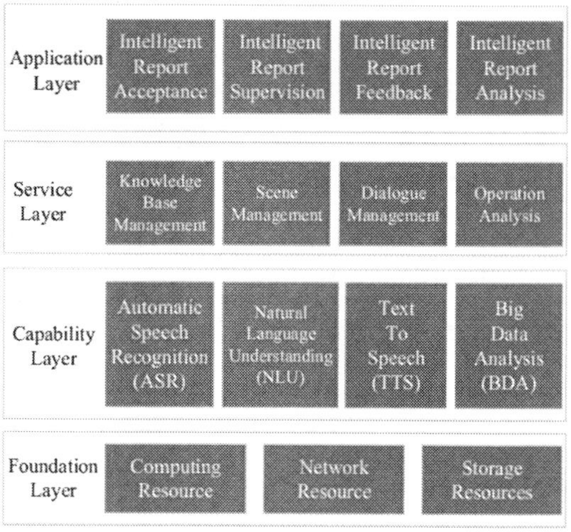

Figure 1. System architecture of intelligent environmental protection hotline

Basic Layer: Provide computing resources, network resources and storage resources for intelligent environmental reporting hotline.

Capability Layer: provide intelligent voice recognition, intelligent semantic understanding, intelligent speech synthesis, big data analysis and other AI core technical capabilities for intelligent environmental protection reporting hotline.

Service Layer: Provide environmental knowledge base management, environmental business scenario management, hotline dialogue management and operation analysis services to meet the needs of intelligent environmental hotline.

Application Layer: Build artificial intelligence application scenarios such as intelligent report acceptance, intelligent report supervision, intelligent report feedback and intelligent report analysis, and use artificial intelligence technology to improve the service efficiency and quality of environmental protection report hotline.

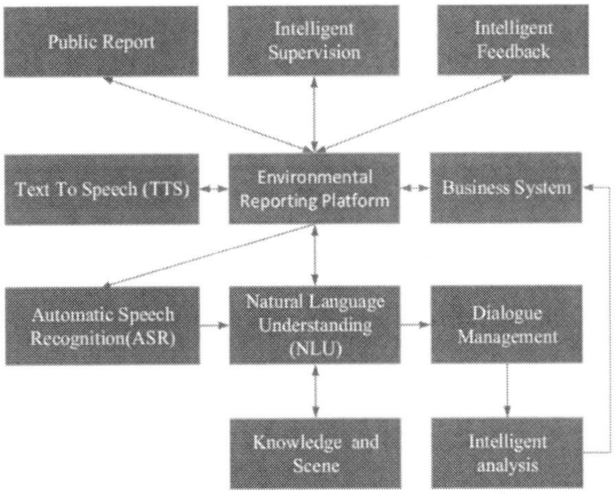

Figure 2. Work flow chart of intelligent environmental protection reporting hotline

4.1 Intelligent Report Accepting and Processing

The intelligent report accepting and processing system included AI technologies such as Automatic Speech recognition (ASR) , Natural language understanding (NLU) and Text To Speech (TTS) as the core has been mature. The accuracy of Q & A is higher than 90%, and the average response speed is less than 200ms. The system can integrate multiple channels such as telephone, website, Wechat, app and H5 to unify and expand the public report channels. By upgrading the existing environmental protection reporting hotline, we can build a AI platform for environmental protection reporting hotline, intelligent interaction, automatically enter work orders, and assist or replace manual work to accepting and processing public environmental protection report. The system can keep online for 7 * 24 hours, improve the efficiency of staff work, greatly reduce the cost of customer service personnel, and improve the timeliness and professionalism of problem response, Improve service efficiency and customer satisfaction.

4.2 Intelligent Report Supervision

Through the unified design of the system architecture, the problem of system data silos is completely solved, and the integration of reporting data and business processing information are realized. The intelligent voice outbound system is used to automatically judge the case level and waiting time, and to automatically supervise and remind the security inspection of overtime reports, so as to timely follow up the progress of the case, focus on ensuring the timeliness of the acceptance and completion of environmental protection reports, and improve the quality of report information feedback.

4.3 Intelligent Report Feedback

After the case is handled, it can feed back the processing results to the informants in batches through the intelligent voice outbound system, and conduct the satisfaction survey of the processing results at the same time. If the service results are satisfied, it will be automatically completed and filed. If the service results are not satisfied, it will be transferred to the relevant departments again. In addition, the Enviromental Reporting Platform can also complete the public's problem consultation work by itself, and can also realize self-service transfer to manual customer service, so as to improve public satisfaction and government credibility.

4.4 Intelligent Report Analysis

Through the core technologies of Automatic Speech Recognition (ASR) ,Natural Language Understanding (NLU),Text To Speech (TTS) and Big Data Analysis(BDA), the public report information is analyzed. The system realizes the full transcribing of hotline recordings of all levels of units, excavates and analyzes the voice content, timely discovers service abnormalities, missing reports, concealment and other situations, and conducts data mining and analysis through reporting hot spot analysis, repeated call analysis, user portrait analysis, public opinion analysis and other dimensions, so as to deeply explore the public reporting demands, so as to optimize service quality, improve management efficiency, improve the quality of service, improve the quality of service, and improve the quality of service last but not least to provide important data for policy-making.

 https://doi.org/10.1051/e3sconf/202129303022

5 Summary

Based on AI core technologies such as Automatic Speech Recognition (ASR) ,Natural Language Understanding (NLU),Text To Speech (TTS) and Big Data Analysis(BDA), through unified architecture design, intellectualization are used to assist or replace the hotline service personnel. At the same time, the problem of data silos between various systems is solved, the automatic flow of business data is realized, and the potential value of data is mined. Through the intelligent upgrading construction of reporting hotline, build convenient environmental reporting hotline, and effectively solve the problems such as poor accepting and processing of reports, limited acceptance capacity of staff, inaccurate service terms, random information reply, and nonstandard feedback. Enhance the awareness and willingness of public participation and promote the development of environmental protection.

References

1. Ministry of ecological environment of the people's Republic of China.Party School of the Central Committee of C.P.C(National Academy of Governance).[EB/OL].(2020-08-11). https://www.ccps .gov.cn/llwx/202008/t20200811_142737.shtml.
2. Shanghai Oriental newspaper Co., Ltd. Interpretation of the government work report, ecological civilization construction and carbon emission reduction goals: do a good job in integration, grasp the direction of coordination. [EB/OL]. (2021-03-08). https://m.thepaper.cn/bai- jiahao_11619805
3. B.H.Li, D.M.Zhu. Research on the influencing factors of public participation in environmental protection[J]. Ournal of Shanghai University (SOCIAL SCIENCE EDITION) **1**(2020):118-128.
4. Ministry of Ecology and Environment of the People's Republic of China. The Ministry of Ecology and Environment announced the national "12369" environmental protection report in 2019. [EB/OL]. (2020-05-13). http://www.mee.gov.cn/g-kml/hbb/qt/201612/t20161222_369409.htm.
5. S.F.Niu, X.L.Wei. Give full play to the role of the main Way of the hotline, and earnestly safeguard the people's environmental rights and interests. [J]. Regional Economy and Industrial Economy,2013:13-14.

Exogenous abscisic acid alleviates the toxicity of nickel in wheat seedlings

Xue LI[1,3], Ming ZHANG[1,3], and Tian-peng GAO[1,2,3]*

[1]School of Geography and Environmental Engineering, Lanzhou City University, Lanzhou 730070, China

[2]School of Biological and Environmental Engineering, Xi'an University, Xi'an 710065, China

[3]Engineering Center for Pollution Control and Ecological Restoration in Mining of Gansu Province, Lanzhou, 730070, China

Abstract. In order to evaluate the effects of exogenous abscisic acid (ABA) in alleviating nickel (Ni) stress in wheat plants. We studied the changes of biochemical and physiological in wheat seedlings exposed to 250 μM Ni with or without different treatments of ABA. Exposed to Ni (250 μM) caused adverse effect on growth of wheat seedlings, which was accompanied by increased the concentrations of superoxide anion (O_2^-) and malondialdehyde (MDA). However, exogenous application of ABA (2.5 and 5 μM) alleviated the Ni-induced inhibition of plant growth, decreased the concentrations of O_2^- and MDA in wheat shoots. Further, application of ABA significantly modulated the activities of antioxidant enzymes and enhanced content of proline and soluble sugar in Ni-stressed wheats, but the application of 20 μM of ABA had no different significantly response for these parameters. The results indicated that application of ABA enhanced the antioxidant defense activities in Ni-stressed wheats, thus alleviating Ni-induced oxidative injury and enhancing Ni tolerance.

1 Introduction

Nickel is a essential plant minerals which is required by plants in order to maintain healthy growth and development but it become extremely toxic in higher concentrations. The accumulation of Ni in plants can cause numerous morphological and physiological changes. At the morphological level, a high concentrations of Ni causes plant growth retardation, chlorosis and wilting [1]. At the physiological level, excess Ni results in an inhibition of chlorophyll synthesis [2], damage to photosynthetic electron transport chain [3], induction of oxidative stress [4], and changes in enzyme activity [5]. Likewise, Ni promotes the accumulation of reactive oxygen species (ROS) causing lipid peroxidation and metabolism imbalance. Plants possess antioxidant systems to scavenge reactive ROS by increasing the activities of key enzymatic and non-enzymatic antioxidants, including superoxide dismutase (SOD), peroxidase (POD) and ascorbic acid [6].

Phytohormones play important roles in the coordination of adaptive responses to environmental stresses. Abscisic acid (ABA) is classified as a sesquiterpene hormone for regulating many plants physiological responses to various abiotic stresses. The level of ABA in plant tissues has been shown to increase upon exposure to environmental stresses such as drought, salt and copper stress [7, 8]. It is reported that exogenous application of ABA influences a range of diverse processes in plants, such as seed germinations, growth and development, photosynthesis, secondary metabolism and gene expression [9,10]. There is also evidence that ABA can ameliorate the damaging effects of chilling and heavy metals [11, 12]. However, to the best of our knowledge, whether exogenous applications of ABA can alleviate Ni-induced oxidative stress in wheat seedlings, has not been reported yet. Therefore, the aim of this study was to examine whether ABA could alleviate the injurious effects of Ni stress and regulate wheat plant growth by regulating the antioxidant enzymes activities and osmolytes involved in stress tolerance.

2 Materials and methods

2.1 Plant growth and treatment

Wheat (*Triticum aestivum* L.) seeds were surface-sterilized with 1 % (w/v) NaClO for 10 min and rinsed thoroughly several times with sterile distilled water, then the seeds were soaked in distilled water for 8 h, the seeds were placed into petri dishes containing three layers gauze moistened with distilled water and germinated for 3 d at 25 °C under dark conditions. When the plumule emerged, seedlings were cultivated in a climate-controlled room under the following conditions: 25 °C, 12/12 h light/dark photoperiod and 65±5 % relative humidity. The seedlings were treated with 0 (control), 50, 100, 250, 500 μM $NiSO_4$ and renews every 2 d. After 4 d, the shoot and root length were measured, and the relevant biomarkers were measured.

* Corresponding author: zkgtp@163.com (T.G.)
ORCID: 0000-0001-6821-3836

We found that the growth of wheat was clearly inhibited when the concentration of $NiSO_4$ was 250 μM. This dose of $NiSO_4$ was selected for the next phase of the experiment with ABA. The 3 d old seedlings of wheat were pretreated with the ABA (0, 2.5, 5, 10, 20 μM) for 8 h before exposure to 250 μM $NiSO_4$, respectively. The control wheat seedlings were only treated with distilled water. The growth conditions of the wheat seedlings were as described earlier. The treated seedlings were harvested for analysis after 4 d.

2.2 Lipid peroxidation and superoxide anion (O_2^-)

Malondialdehyde (MDA) content was assayed by the method of Thomas et al. [13]. O_2^- was determined as described by Jiang and Zhang [14].

2.3 Proline and soluble sugars

Proline content was measured according to the method of Bates et al. [15]. The contents of soluble sugars was measured with the method described by Chow and Landhausser [16].

2.4 Antioxidant enzyme activities

Fresh samples (0.2 g) were grinded to homogenize in 5.0 mL of 50 mM potassium phosphate buffer (pH 7.8) by an ice–cold mortar and pestle to extract shoot enzymes. The homogenates were centrifuged at 10,000×g for 15 min at 4 °C,and the supernatants were used for analysing the activity of SOD (EC 1.15.1.1.) [17], POD (EC 1.11.1.7.) [17] and catalase (CAT) (EC 1.11.1.6.) [18].

2.5 Statistical analysis

Statistical analysis was performed using SPSS Version 19.0 software. Data are presented as mean ± standard deviation (SD). Duncan's multiple test were applied to determine the significant difference ($p < 0.05$) among the treatments.

3 Results

As shown in Table 1, compared to the control, there was no significant effects in root and shoot lengths of wheat in 50 and 100 μM Ni treatments. However, the presence of 250 and 500 μM Ni significantly reduced the lengths of root and shoot in wheat seedlings. The lengths of root and shoot decreased by 31.39 and 22.56 % in 250 μM Ni treatments, respectively, relative to the control. The accumulation of MDA was dose dependent in roots treated with 50-500 μM Ni. Meanwhile, we also observed a similar increasing trend in O_2^- contents of roots. However, MDA and O_2^- contents in shoots were no significant effects under 100 μM Ni treatment as compared with the control. In accordance with these results, 250 μM Ni was used as the treatment concentration.

As shown in Table 2, exposure of wheat seedlings to 250 μM Ni significantly decreased root and shoot lengths. However, exogenous application of ABA produced significant changes in plants growth, and only 2.5 and 5μM ABA significantly enhanced root and shoot lengths of wheat seedlings. This effect was even more pronounced for the group with 5 μM ABA; it improved the root and shoot lengths of plants by 27.66 and 19.28 %, respectively, as compared to Ni alone treatment. The results indicated that the alleviative effect by exogenous application of ABA on Ni-induced growth inhibition was more obvious in roots than in shoots. Whereas application with 20 μM ABA reduced significantly shoot growth as compared to Ni alone treatment.

Wheat seedlings were treated with 250 μM Ni for 4 d, the proline level in the shoots increased to 25.85 % of that of the untreated control (Table 2). ABA improved further the level of proline in the shoot of plants in comparison to the control. In contrast, 5 μM ABA had the greatest effect on proline level, which increased by 16.04 % in shoot tissues than that of Ni alone treatment. In addition, the contents of soluble sugar was significantly decreased by Ni treatment as compared to the control. After application of different concentrations ABA, the contents of soluble sugar in shoots were significantly increased by 26.06, 62.29, 18.81 and 21.70 %, respectively, as compared to the treatment of Ni alone (Table 2).

The contents of MDA and O_2^- significantly increased in the shoots under Ni stress, about increased by 79.17 % and 140.21 % than that of control, respectively (Table 2). ABA reduced the accumulation of MDA and O_2^- to varying extents in wheat seedlings. Especially, ABA at 5μM had the greatest effect, and MDA and O_2^- contents decreased by 20.83 and 32.92 % in shoots, respectively, as compared to the treatment of Ni alone (Table 2).

Table1. Effects of Ni on growth and oxidative stress in wheat seedlings. Values show the means of four replicates ± SD. The different lower case represents statistically differences between treatments ($p < 0.05$).

Ni μM	Length (cm)		MDA (nmol g^{-1} FW)		O_2^- (nmol min^{-1} g^{-1} FW)	
	Root	Shoot	Root	Shoot	Root	Shoot
0	7.84 ± 0.50a	8.32 ± 0.47a	5.19 ± 0.96d	5.32 ± 0.91c	3.52 ± 0.46d	3.19 ± 0.61c
50	8.03 ± 0.62a	8.72 ± 0.39a	5.20 ± 0.61d	4.98 ± 0.72c	3.47 ± 0.49d	3.07 ± 0.41c
100	7.22 ± 0.39a	8.11 ± 0.37a	6.98 ± 0.44c	5.98 ± 0.77c	4.74 ± 0.46c	4.31 ± 0.66c
250	5.37 ± 0.35b	6.44 ± 0.46b	10.59 ± 0.93b	9.81 ± 0.42b	7.13 ± 0.58b	7.61 ± 0.59b
500	3.03 ± 0.52c	3.17 ± 0.35c	14.60 ± 0.43a	31.18 ± 0.91a	9.50 ± 0.61a	10.42 ± 1.04a

Table2. Effects of ABA on root and shoot lengths, proline, soluble sugar, malondialdehyde (MDA) and superoxide anion (O_2^-) of wheat seedlings under Ni stress. Values show the means of four replicates ± SD. Different lower case represents statistically significant differences between treatments ($p < 0.05$).

Treatments/ Parameters	Root length (cm)	Shoot length (cm)	Proline ($\mu g\ g^{-1}$ FW)	Soluble sugar ($mg\ g^{-1}$ FW)	MDA($nmol\ g^{-1}$ FW)	O_2^- (nmol $min^{-1}\ g^{-1}$ FW)
Control	8.16 ± 0.54a	8.99 ± 0.54a	17.76 ± 1.10e	21.69 ± 1.04b	5.57 ± 0.99c	3.25 ± 0.42e
Ni（250μM）	5.48 ± 0.42c	6.49 ± 0.43c	22.35 ± 0.54cd	15.25 ± 1.25d	9.98 ± 0.55a	7.81 ± 0.33a
Ni + ABA (2.5)	6.40 ± 0.49b	7.25 ± 0.19b	24.55 ± 0.66ab	19.22 ± 1.19c	7.96 ± 0.26b	5.45 ± 0.31d
Ni + ABA (5)	6.99 ± 0.61b	7.75 ± 0.33b	25.93 ± 1.06a	24.74 ± 1.22a	7.89 ± 0.89b	5.24 ± 0.37d
Ni + ABA (10)	5.38 ± 0.32c	6.33 ± 0.54c	23.19 ± 0.37bc	18.12 ± 1.29c	9.59 ± 0.73a	6.13 ± 0.16c
Ni + ABA (20)	5.18 ± 0.27c	5.29 ± 0.23d	21.16 ± 0.71d	18.55 ± 0.77c	10.30 ± 0.93a	7.45 ± 0.29a

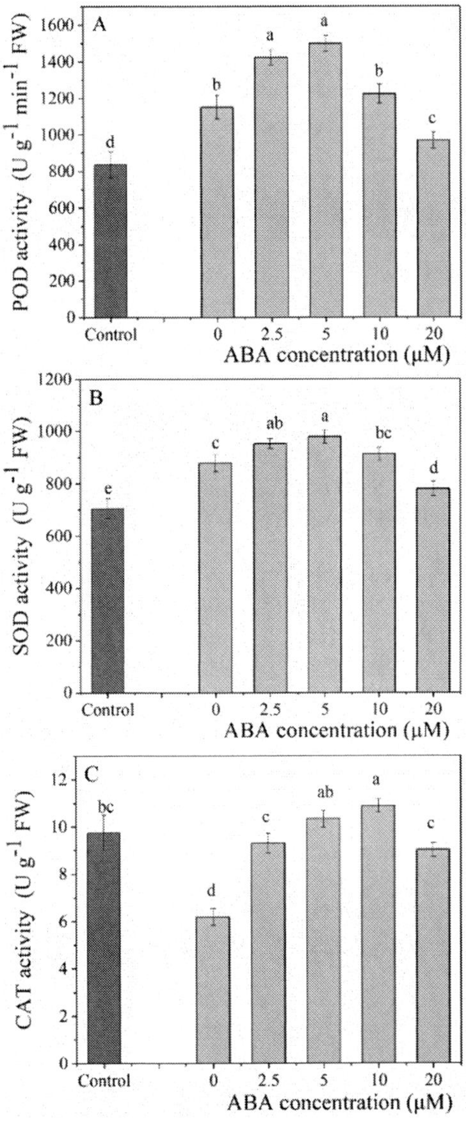

Figure 1. Effects of ABA on POD (A), SOD (B), CAT (C) in shoots of wheat seedlings under Ni stress. Bars (n = 4) showing the different letters are significantly different at $p < 0.05$ as determined by Duncan analyze.

The activities of SOD and POD were found to be increased under Ni toxicity. In contrast, the application of ABA (2.5 and 5 μM) enhanced further in the activities of these enzymes under Ni stress in comparison to the treatment of Ni alone. (Fig. 1A-B). Nevertheless, the CAT activity was decreased in Ni-stressed wheats, but the application of ABA (2.5-20 μM) overcame the adverse effects of Ni-stress, causing a notable increase in the activity of CAT of shoots tissues (Fig. 1C).

4 Discussion

In the present study, the symptoms of Ni toxicity on wheat plants are growth inhibition at the very early stage of their development. Shoot and root growth in wheat plants were decreased significantly when Ni concentrations reached 250 μM (Table 1). However, we found that exogenous ABA was able to ameliorate the toxic effect of Ni on the growth of wheat seedlings. Inhibition of growth in wheat seedlings might result from Ni-caused alterations of metabolic and biochemical processes [19]. The contents of MDA and O_2^- significantly increased at 250 and 500 μM Ni in both shoots and roots of plants (Table 1). These results showed that the oxidative stress and the peroxidation of membrane lipids were caused by Ni stress. Previous studies have demonstrated that MDA content can be induced in wheat contaminated with heavy metal such as Ni, Cu and Cd [20, 21].

Previous studies have demonstrated that ABA has great potential to enhance the tolerance to certain stresses including drought, saline–alkaline in different plants [22, 23]. Generally, low concentrations of ABA can improve plant resistance to adverse environments, and high concentrations of ABA can cause high levels of oxidative stress, leading to a decreased resistance to environmental stress [14, 24]. In this study, exogenous application of low concentrations of ABA (2.5 and 5 μM) significantly alleviated the deleterious effects of Ni on plant growth (Table 2). Previously, it was stated that ABA application alleviated the copper induced growth retardation in *Artemisia annua* [25]. However, this mitigation effect gradually reduced with increasing concentrations of ABA, and 20 μM ABA even inhibited even shoot growth.

In comparison to control plants, the content of proline was improved under Ni stress both by ABA treatment and treatment free plants (Table 2). The proline accumulation is essential to maintain osmotic adjustment under Ni stress.

It has been proposed that proline plays an important role in the antioxidative properties of heavy metal-stressed [26]. Additionally, Ni-stressed plants exhibited a significant reduction in soluble surgar. Exogenous application of ABA led to the increase in soluble surgar in shoots of plants (Table 2). Soluble surgar are known to contribute to the osmotic adjustment [24], thus high level of soluble surgar is beneficial for enhanced tolerance of the plants under Ni stress. This suggested that ABA treatment could elevate proline and soluble surgar to withstand Ni-induced injury.

Our results showed that the growth of wheat plants were decreased significantly under Ni (250 μM) stress, accompanying with the significant increase the contents of O_2^- and MDA (Table 2). Exogenous application of ABA reversed this adverse effect of Ni, causing a significant decrease in the O_2^- and MDA contents. Moreover, Ni toxicity in plants may be attributed to increasing lipid peroxidation leading to oxidative damage [27]. To prevent the Ni-induced oxidative damage on plants, plants have antioxidant defense mechanisms. In the present study, the activity of POD was increased in response to only Ni-stressed treatment and continually increase was observed by low concentrations of ABA treatment plus Ni. (Fig.1A). Similar trend was found for SOD (Fig. 1B). SOD catalyzes the conversion of O_2^- to O_2 and H_2O_2 [28]. Notably, results showed the content of O_2^- was at ABA treatment plus Ni lower than that of Ni alone, thus ABA enhanced the antioxidative system to scavenge the accumulation of O_2^-. In addition, Ni stress caused marked decrease in the activity of CAT, whereas ABA treatment induced a significant increase in the activity of CAT in shoots of plants (Fig. 1C) This is supported by the study of Jiang et al [29], who observed that ABA pretreatment enhanced the actions of SOD, CAT and ascorbate peroxidase enzymes in *Zea mays* seedling under abiotic stress. In fact, the results indicate that the increase in the activities of the antioxidant enzymes were not sufficient to protect cell membrane against Ni toxicity. However, this oxidative damage was alleviated by ABA treatment (Table 2). The application of ABA elevated enzymatic antioxidants in shoots of wheats, hence leading to alleviation of the oxidative damage as indicated by the lowered O_2^- and MDA levels.

5 Conclusion

In summary, ABA may attenuate Ni toxicity in wheat seedling exposed to Ni stress, which probably includes the regulation of the antioxidant system and the improvement of osmotic adjustment. This suggests that an appropriate concentrations ABA could be used as a potential growth regulator to alleviate Ni induced toxicity in plants.

Acknowledgements

This work was financially supported by the National Natural Science Foundation of China (NSFC 31860176), Key Research and Development Program of Gansu Province (20YF3FA037), Key Research and Development Program of Shanxi Province (2020ZDLSF06 06) and XAWLKYTD012. We are grateful to all anonymous reviewers whose comments improved the quality of the manuscript.

References

1. N. Pandey, C.P. Sharma, Plant Sci. **163**, 753–758, (2002)
2. E.Gajewska, M. Sklodowska, M. Slaba, J. Mazur, Biol. Plant. **50**, 653–659, (2006)
3. N. Mohanty, I. Vass, S. Demeter, Physiol. Plant. **76**, 386–390, (1989)
4. S. Baccouch, A. Chaoui, E. El Ferjani, J. Plant Nutr. **24**, 1085–1097, (2001)
5. R. Boominathan, P. M. Doran, New Phytol. **156**, 205–215, (2002)
6. R. Mittler, S. Vanderauwera, M. Gollery, F. Van Breusegem, Trends. Plant Sci. **9**, 490–498, (2004)
7. J. H. Zhang, W. S Jia, J. C. Yang, A. M. Ismail, Field Crops Res, **97**, 111–119, (2006)
8. F. K. Zengin, S. Kirbag, J. Environ. Biol. **28**, 561, (2007)
9. M. H. Ibrahim, H. Z. Jaafar, Molecules, **18**, 7957–7976, (2013)
10. S. K. Sah, K. R. Reddy, J. Li, Front. Plant Sci. **7**, 1–26, (2016)
11. G. J. Wang, W. Miao, J. Y. Wang, D. R. Ma, J. Q. Li, W. F. Chen, J. Agron. Crop Sci, **199**, 200–208, (2013)
12. G. M. Shen, J. K. Niu, Z. X. Deng, Plant Physiol. Biochem, **118**, 471–478, (2017)
13. J. C. Thomas, M. Perron, E.C Davies, Plant Sci. **167**, 259–266, (2004)
14. M. Y. Jiang, J. H. Zhang, Plant Cell Physiol. **42**, 1265–1273, (2001)
15. L. S. Bates, R. P. Waldren, I. D. Teare, Plant Soil, **39**, 205–207, (1973)
16. P. S. Chow, S. M. Landhausser, Tree Physiol, **24**, 1129–1136, (2004)
17. C. S. Gu, Y. H. Yang, Y. F. Shao, K. W. Wu, Z. L. Liu, S Afr J. Bot, **114**, 267–271, (2018)
18. H. B. Shao, Z. S. Liang, M. A. Shao, Colloid. Surface B. **45**, 7–13, (2005)
19. K. U. Parlak, NJAS - Wagen J. Life Sc, **76**, 1-5, (2016)
20. E. Gajewska, M. Sklodowska, Ecotox. Environ. Saf, **73**, 996–1003, (2010)
21. R. Z. Lin, X. R. Wang, Y. Luo, W. C. Du, H. Y. Guo, D.Q. Yin, Chemosphere, **69**, 89–98, (2007)
22. L. L. Yang, L. Yang, Y. M. Lan, Y. Zhao, M. Han, L. M. Yang, Ind Crop. Prod, **154**, 112686, (2020)
23. L. X. Wei, B. S. Lv, M. M. Wang, H. Y. Ma, H. Y. Yang, X. L. Liu, C. J. Jiang, Z. W. Liang, Plant Physiol. Biochem, **90**, 50–57, (2015)
24. J. Wang, J. Chen, K. Pan, Environ Sci Pollut Res, **20**, 1441–1449, (2013)

25. A. Zehra, S. Choudhary, K. I. Wani , M. Naeem, M. M. A. Khan, T. Aftab, Plant Physiol. Biochem, **156**, 125–134, (2020)
26. J. Matysik, Alia. Bhalu, B.P. Mohanty, Curr. Sci. **82**, 525–532, (2002)
27. S. Baccouch, A. Chaoui, E. El Ferjani, Plant Physiol. Biochem. **36**, 689–694, (1998)
28. C. H. Foyer, G. Noctor, Physiol. Plant, **119**, 355–364, (2003)
29. M. Y. Jiang, J. H. Zhang, Free Radic. Res. **36**, 1001–1015, (2002)

 https://doi.org/10.1051/e3sconf/202129303024

Research on Safety Evaluation and Sustainable Conservation Strategy of Tunbao Buildings--An example of residential buildings in Benzhai, Anshun, Guizhou

GaoJiahao[1], LuLuxi[2*], Zhang Tianke[3]

[1]Huazhong University of Science and Technology

[2]Southwest Minzu University

[3] Southwest Minzu University

Abstract. Tunbao architecture is a combination of defensive and residential architecture of Ming Dynasty, and is an important material substrate of Tunbao culture. With the accelerated pace of urbanization, the population of Anshun Tumbao has moved out, aging has increased and natural catalysis has led to the dilapidation and decay of the dwelling architecture, which lacks certain conservation measures. This paper takes Anshun Tunbao residential buildings in Guizhou as an example, and summarizes the development of residential buildings in this village through literature and field research, and summarizes the current problems of preservation. The results of this study are presented in the form of a graded "point-to-point" conservation strategy, which is intended to provide a theoretical reference for the conservation and restoration of historical buildings in the Han ethnic area of Guizhou.

1 Introduction

1.1 Historical Development Background

The eight fortresses of Yunfeng in Guizhou were built in the early Ming Dynasty, and are the most complete surviving village complexes in Guizhou. During Zhu Yuanzhang's time, he controlled the southwestern region with the "guardhouse cantonment system", and the Anshun Prefectural Records said, "The cantonment fortresses were all transferred to the north and conquered the south by Hongwu's thorns", and many guardhouses were established in Guizhou from the fourth to the thirtieth years of Hongwu. The ancient architecture of Anshun Yunshan Tun is composed of dwellings, temples and theater buildings in Yunshan Tun and this village. With the reduction of war and the reform of military policy, the role of Tunbao as a military camp gradually diminished, and many Tunbao were transformed from military tunnels to civilian and commercial tunnels [1]. The cottage studied in this paper is located in Qiyangqiao Township, Xixiu District, Anshun, and was built in the late Qing Dynasty. It was relocated by rich families from the surrounding areas and was listed as the fifth batch of national key cultural relics protection units in 2001.

1.2 Explanation of architectural space background

The traditional architecture of this fortress is mainly made of wood and stone structures, among which the wood structure is mainly a pierced structure. The topography of the Anshun area is fragmented, and there is less land available for large area construction, so one courtyard is the common form of the fortress dwelling architecture, and there are two types of courtyards: the triple courtyard and the quadruple courtyard. The triple courtyard consists of the main house and the east and west wings on both sides, while the remaining side is enclosed by a wall, building a complete defense system by taking advantage of the original terrain. After a long period of use, some of the traditional buildings have fallen into disrepair, and some of the damaged building elements have gradually developed into dangerous ones. The dwellings of this fortress are an important substrate for the continuity and vitality of the Tunbao culture, which is limited by the historical and geographical development pattern, as well as the continuous migration of the population to earn a living, resulting in the migration of the architectural texture and the hollowing out of the dwelling units, and the lack of financial support and the contradiction of the people's use of the building functions, resulting in the lack of conservation methods. Therefore, this paper hopes to summarize the strategies for the conservation and restoration of salvaged historical buildings based on the

* Corresponding author: 215219383@qq.com

evaluation of architectural safety, and to continue their ethnicity, local characteristics and cultural development.

2 The existing problems of the residential buildings in this cottage are summarized

2.1 Damage to the main components

2.1.1 Column bases and their damage

The pillar bases of this cottage are generally arranged under the pillars at the entrance and exit of the main façade and the gable corridor of the main house, except for the pillars near the entrance and exit passages, where no pillar bases are set, and the pillars are placed directly on the stone slabs or the floor. The pillar bases are mainly of drum type, flat drum octagonal type, quadrilateral type and other forms [2]. It was found that the pillar bases used in this cottage mainly have two problems: weathering and misalignment of the pillar bases. One is that after a long period of use, the surface of most of the pillar bases has begun to weather (Figure 2-1-1), and the pillars above the higher size of the flat drum octagonal and quadrangular pillar bases are less damaged, while the protective effect of the smaller size and height of the stone pillar bases is heavier than that of the pillars above the higher size pillar bases.

Second, the local occurrence of column base dislocation, column directly resting on the column base, because of uneven settlement and other reasons to produce column base dislocation, coupled with the humid climate and precipitation weather led to water penetration between the column base, accelerating the column tilt, stone softening, reducing its weighing compressive resistance

Figure 2-1-1 Column foundation appears
Source: Author's own photos

2.1.2 The main components of the wooden pile row frame and their damage

The wooden pillars of the traditional building of this fortress are the main vertical load-bearing elements, often arranged in the form of rows of shelves of pierced buildings. Each row of row frame contains round wooden pillars, melon pillars and through square, wooden pillars and through square species connected by mortise and tenon structure, melon pillars and through square are connected by way of slotted plugging. The author sampled and measured at the site, the diameter of the wooden column in the row frame is 150mm and 200mm two specifications, using the specification of the wooden column is thinner, the cross-sectional area of the wooden column are 0.07m2 and 0.13m2, the diameter of the melon column and the same specification of the wooden column. The whole wooden column is connected with three kinds of members, which are named as the square between the column and the column of the same row frame, the beam located in the second floor plane between the column and the column of the same row frame, and the beam in the floor cover between the row frame and the row frame, the corner column has two directions to connect with other members, the side column has three directions to connect with other members, and the middle column has four directions to connect with other members. The column is connected to the square and the remaining two types of beams and columns by semi-perforated mortise and tenon connections [3]. Meanwhile, the upper part of the melon column bears the load of the roof and is connected to the purlins of the roof, and the lower part transmits the load to the round wooden column by penetrating the square. The melon pillar is a member that transmits the load and bears the load directly.

The research found that the traditional buildings in this cottage generally have wooden pillars with cracked bodies, and the next problem is the rotten bottom of the pillars (Figure 2-1-2). Most of the wooden pillars in this cottage are still in "diseased" working condition, and some of them have reached the level of damage at the point of dilapidation and still have not been properly repaired. The gua pillars and square in the rows of frames have not seen any obvious damage

https://doi.org/10.1051/e3sconf/202129303024

Figure 2-1-2 Damage to wood column
Source: Author's own photos

2.1.3 Roof cover and its damage

The roof cover of the traditional building of this cottage includes arch ow (see beam), purlins, rafters, and the way of bearing purlins on the pillar head is picking square bearing purlins (Figure 2-1-3), the specification of purlins is about 150mm in diameter, and the section form of purlins is mostly round, but there are also square and other forms. The roof cover is transferred by the purlins on the respective pillars, the purlins rest on the pillars, the golden purlins on the golden pillars are also arranged under the arch ow (see beam) as its middle bearing load members, using a combination of ridge rubbing and arch ow (see beam) way of connection, other pillars and purlins using the chisel excavation sandalwood bowl connection method. The damage of the roof cover of this traditional building is mainly caused by the damage of the roof slate (tiles) and the damage of the sliding purlins (Figure 2-1-3). Damage to the roof slate (tile) often forms a hole in the roof, and sliding purlins cause damage to the rafters and slate (tile) as well.

Figure2-1-3 The stigma purlin bearing method of traditional buildings in Benzhai
Source: Author's own photos

2.2 Damage to the envelope structure

2.2.1 Load-bearing walls and their damage status

The load-bearing walls of traditional buildings in this cottage contain both masonry structures and brick masonry structures. The stone masonry structure is generally built along both sides of the street or the external wall of the village, with a height of 1.8~2.5 m. It is made of small block-shaped stones or large stones piled up (Figure 2-2-1), or a thin ash masonry is laid between the stones, and the masonry method is either ordinary piling or masonry with aesthetic effect, and the choice of stones and the masonry method reflect the economic strength and social status of the Tumbao family The choice of stone and masonry method reflects the economic strength and social status of the Tumbao family [5] The masonry also plays a role in overcoming the poor bearing capacity of the weak foundation by virtue of the thickness of the wall. The main problems of cracking and tilting of the masonry walls in this cottage are that the structural damage of the masonry will produce a chain of damage, which is attributed to the way the stone structure is arranged, once the wall has through cracking or tilting beyond the limit will bring safety hazards, the towering hill wall tilts inward, while the street side of the street wall tilts, the direction of tilting is not regular.

Figure2-2-1 Breakage of wall
Source: Author's own photos

2.2.2 Non-load-bearing walls and their damage characteristics

Non-load-bearing walls of traditional buildings in this cottage include wood-paneled walls, combined walls composed of stone and wood panels, and bamboo-woven walls and bamboo-clad mud walls (Figure 2-2-2), which are important components of the enclosure structure between the two columns of traditional buildings. Plank walls and combination walls are common wall types in traditional buildings in this stronghold and are also widely used in traditional buildings in Guizhou region.

Figure 2-2-2 Types of non-load bearing walls
Source: Author's own photos

Wooden panel wall is a common enclosure structure in this cottage. The plank wall of this cottage is in the form of vertical arrangement, and the wall structure made of about 20mm thick plank and border splicing has good stability. The combination wall located on the second floor of the building is less damaged, and only two pieces of the wall are found to have cracked slabs; however, the combination wall located on the second floor of the building is more seriously damaged, and part of the wall is located on the street side, and two to three adjacent walls have been tilted at the same time, and such walls have become a safety hazard.

The number of bamboo woven walls and bamboo woven sandwich mud walls in this cottage is relatively small compared to the wooden panel walls, but his damage is more serious. Bamboo woven walls are often arranged as partition walls in the second floor interior. The bamboo woven walls as interior partition walls are relatively well preserved, with only some of them showing deformation and decay. However, the large area of bamboo woven sandwich mud wall as an exterior protective wall produces deformation in the middle and lower part of the wall protruding to the outside, the indoor bamboo woven sandwich mud wall has decayed due to humidity and other reasons, the mud and sand filled in the wall appears to fall off locally or in a large area, there is a bending deformation of the wall due to environmental factors.

3 Result

3.1 Classification of building component types

In this paper, the common constituent elements of traditional buildings in this cottage are divided into components according to the parts of foundation, columns, beamworks, floor coverings, masonry walls, roof coverings and enclosure structures, respectively, and the types of components covered above are all superstructure components.

In addition to the above-mentioned components, there are three types of components that are not included in this traditional building, namely, components that are not connected to the main structure, components that are not part of the main structure, and components that have little impact on the structural safety of the building even if they are completely damaged, including doors and windows that have not been evaluated for their needs, movable wooden stairs, railings and handrails, and slate canopies added by users,

3.2 Calculate the overall proportion of dangerous elements of the structure

The proportion of hazardous components is the percentage obtained by dividing the sum of the product of the number of various types of hazardous points and their weights with the sum of the product of the number of various types of components and their weights, and its purpose is to identify and count the hazardous points.

 https://doi.org/10.1051/e3sconf/202129303024

The calculation of the comprehensive proportion of basic hazardous components is as follows:

$$Rf=ndf/nf \quad (1)$$

Where: Rf-comprehensive proportion of foundation hazardous components (%); ndf-number of foundation hazardous components; nf-number of foundation components.

The above formula is mainly for the house with foundation. In the traditional building of this cottage, the foundation foundation is mainly treated by digging out the weak layer, ramming, replacement filling, masonry high platform, etc., among which, except masonry high platform, all are treated as foundation. The calculation formula for the superstructure (see formula 2) needs to be calculated separately by floor, so as to arrive at the hazard level of each floor.

$$Rsi=(3.5ndpci+2.7ndsci+1.8ndcci+2.7ndwi+1.9ndrti+1.9ndpmbi+1.4ndsmbi+ndsbi+ndsi+ndsmi)/(3.5npci+2.7nsci+1.8ncci+2.7nwi+1.9nrti+1.9npmbi+1.4nsmbi+nsbi+nsi+nsmi) \quad (2)$$

in the formula:

Rsi—The combined percentage of hazardous components in the i-th floor (100%).

ndpci、ndsci、ndcci、ndwi—Number of dangerous elements in columns, side columns, corner columns and walls in the ith floor;

npci、nsci、ncci、nwi—Number of columns, side columns, corner columns and wall elements in the i-th floor.

ndrti、ndpmbi、ndsmbi—Number of dangerous elements of roof frame, middle beam and side beam of the ith floor;

nrti、npmbi、nsmbi—Number of roof frame, middle beam and side beam members of the ith floor;

ndsbi、ndsbi—Number of dangerous elements of the beam and floor (roof) panels at level i;

nsbi、nsbi—Number of beams and floor (roof) slab members at level i.

ndsmi—Number of dangerous elements of the envelope of the ith floor;

nsmi—Number of envelope elements in the i-th floor.

After completing the calculation of the hazard level of each floor, all the data are substituted into the following equation (see Equation 3) to find out the integrated proportion of the overall structural hazard components.

$$R=(3.5ndf+3.5\sum_{i=1}^{F+B} ndpci+2.7\sum_{i=1}^{F+B} ndsci+1.8\sum_{i=1}^{F+B} ndcci+2.7\sum_{i=1}^{F+B} ndwi+1.9\sum_{i=1}^{F+B} ndrti+1.9\sum_{i=1}^{F+B} ndpmbi+1.4\sum_{i=1}^{F+B} ndsmbi+\sum_{i=1}^{F+B} ndsbi+\sum_{i=1}^{F+B} ndsi+\sum_{i=1}^{F+B} ndsmi)/(3.5nf+3.5\sum_{i=1}^{F+B} npci+2.7\sum_{i=1}^{F+B} nsci+1.8\sum_{i=1}^{F+B} ncci+2.7\sum_{i=1}^{F+B} nwi+1.9\sum_{i=1}^{F+B} nrti+1.9\sum_{i=1}^{F+B} npmbi+1.4\sum_{i=1}^{F+B} nsmbi+\sum_{i=1}^{F+B} nsbi+\sum_{i=1}^{F+B} nsi+\sum_{i=1}^{F+B} nsmi) \quad (3)$$

in the formula: R - overall structural hazard component integrated ratio; F - number of superstructure floors; B - number of basement structure floors.

Through the appeal formula, it is found that the overall structural hazard rating is not directly related to the hazard rating of each floor when the overall structural hazard rating is determined, or it can be understood that the overall structural hazard rating is not based on the smaller hazard rating in each floor as the rating。

3.3 Calculation of weights of components on the same floor

3.3.1 Formula for weight calculation

First of all, the weight coefficient of each type of member on each floor relative to this floor is determined by the hierarchical analysis method. Using the hierarchical analysis method is to divide the foundation foundation members and superstructure members according to the division principle of individual members, and then determine the sample with the help of data collection method using random sampling, and calculate the weight of each member on each floor, the weight coefficient of each floor and the weight coefficient of the member in the building as a whole according to the calculation steps in turn. It is characterized by reflecting the weight relationship of the building floors from the bottom to the top, the lower the floor of the member the higher the influence weight coefficient, and the higher the floor the lower the influence coefficient[6]. The typical weight ratios of various types of members for simple calculation are given in the Guidelines, and the weight ratios of floor (roof) panels, secondary beams, main beams, columns, and walls are given as 1, 1.76, and2.96、5.36, 5.36. Subsequently, with the help of these weighting factors the weighting factors of each component of a certain class are derived by means of the formula (see Eq. 4) as follows Eq

$$w_i = \frac{r_i}{\sum r_i n_i} \quad (4)$$

in the formula: Wi-i class members of the weight coefficient of each member; ri-i class members of the weight ratio, according to the value of (Table B.2); ni-i class members of the number of members.

Then the floor weight coefficients are then determined according to the formula, as follows

$$w_j = \frac{n+1-j}{\sum j} \quad (5)$$

in the formula: Wj - the jth floor in the overall weight of the building; n - the total number of floors, including the foundation; j - the jth floor, the foundation for the first floor, the above cumulative in turn.

Through the weight calculation of the same floor and the overall floor weight calculation, the calculation can reflect the importance between different members in the same floor and between each floor, the force members occupy important weights, such as columns, walls (load-bearing walls), other sequential main beams, secondary beams and floor (roof) panels. At the same time, the lower the floor, the floor is given a higher weighting factor.

After giving the corresponding weight coefficients to each calculation component, that is, the calculation of the safety level, the calculation formula is as follows

$$r = \sum_{i=1}^{n} w\, i / \sum_{j=1}^{m} w\, j \times 100\% \quad (6)$$

in the formula： n - the total number of components whose safety does not meet the requirements; i - the number of components whose safety does not meet the requirements; wi - the weight of the i-th component; m - the total number of all components; j - the number of all components; wj - weight of the jth component.

3.3.2 Determine the weight coefficients of each type of member on each floor relative to this floor

The components to be calculated for the traditional building of this cottage include columns, masonry walls, main beams, secondary beams and floor (roof) panels, among which columns and masonry walls correspond to the two components of columns and walls in the Guidelines and are the components with the largest weights among all calculated components. According to the characteristics of the traditional architecture of Benzai, first of all, the logical order of weight size should be that columns and masonry walls are equally important, columns are important relative to beams or relatively important, main beams are important relative to secondary beams, and floor (roof) panels are the least important. It is worth noting that the division of beams into main beams and secondary beams is based on the characteristics of traditional buildings in Benzai, namely, in the first floor plan, the main beams are perpendicular to the floor cover beams and secondary beams, which are subject to greater forces and are more important than the secondary beams; while the secondary beams parallel to the floor cover beams are subject to reduced forces because of the uniform distribution of the floor cover beams on the floor panels. After the previous analysis, the same logic is adopted in the Dangerous Standard, and the two are different in the division of the project. Therefore, the component weight ratios of traditional buildings in this cottage can use the typical various component weight ratios [7], as the following table data（**Figure** 3-3-2）

Table 3-3-2 Table of weight ratio of traditional building components in Benzhai

Floor (roof) panels	secondary beam	Main beam	Column	Masonry walls
1.00	1.76	2.96	5.36	5.36

Source: Drawn according to the Guidelines for Structural Safety Assessment of Modern Historic Buildings

3.3.3 Determine the floor weighting factor

The formula is used to calculate the weight of each floor in the building as a whole. It is noteworthy that some scholars believe that the formula for determining the floor weight coefficient in the Guidelines does not specify that the denominator is$\sum_1^j j$or$\sum_1^n j$， And that the influence of the two is completely different, the former has a large difference in the influence coefficient of each floor, especially the basic layer is much higher than the rest of the layers, while the latter is basically from the basic to the top layer with a regular decreasing. In this paper, according to the denominator$\sum_1^n j$ Calculation。 With the first floor of the traditional building of this fortress as the first floor, the information on the number of floors of the common two-story compound and three- to four-story towers is substituted into the following equation

$$w_j = \frac{n+1-j}{\sum j} \quad (7)$$

The data in the following table were obtained:

Table 3-3-3 floor weight coefficient table

Total number of floors (n)	First tier weightin g factor	Second tier weightin g factor	Third tier weightin g factor	Fourth tier weightin g factor
2	2	0.5	0	0
3	3	1	0.33	0
4	4	1.5	0.67	0.25

Source: Self-drawn by the author

Finally, the weight values of each component of each floor and the numbers in the table of floor weight coefficients are substituted into Equation 6 to calculate the results

4 Protection strategy delineation

4.1 Evaluation of security level

The calculation results obtained are used to determine the component security level according to the determination criteria given in the following table (see Table 4-1)。

Table 4-1 Safety level evaluation form for each component

Grading	Judging criteria
a	$r = 0$
b	$0.05 \geq r > 0$
c	$0.30 \geq r > 0.05$
d	$r > 0.30$

Source: 《Guidelines for Structural Safety Assessment of Modern Historic Buildings》

The overall security level is determined on the assessment of the security level of the components. The minimum level of the components is determined by the components and the security level division is indicated by capital letters. When adverse conditions are encountered, it is possible to reduce the original level by one level according to the conditions in the Guidelines.

4.2 Building restoration difficulty determination

Finally, the Guidelines specify the treatment requirements according to the overall safety level, as follows.

Table4-2 Table of processing requirements for various grades of modern historical buildings

Overall security level	Processing requirements
A	No need to handle
B	Very few components require measures
C	A few components require measures
D	Most components require measures

Source: 《Guidelines for Structural Safety Assessment of Modern Historic Buildings》

The impact on the safety of the surrounding buildings can be included in the treatment opinion. The analysis of this factor can reflect the impact of the breakage of certain traditional buildings on the surrounding buildings and residents, as well as on the safety of road pedestrians. Most of the street walls or fences of the traditional buildings in this cottage are masonry walls, and according to the width of the street and the height of the walls, the analysis of the impact on the safety of the surrounding buildings should be carried out. In the same way, according to the characteristics of the traditional building of this fortress with masonry walls underneath and wooden buildings on top, the impact of serious damage of these elements on the safety of the surrounding buildings should also be taken into consideration. In addition, although the walls in the triple compound belong to the structures, when these walls are in danger of toppling, they should be considered as factors affecting the safety of the whole compound, and it is not necessary to calculate the safety level of the structures. It is worth noting that when a traditional building has no impact on the safety of the surrounding buildings, it can be left undescribed, or described as having no impact, and the components that are not seriously damaged and do not affect the aesthetics are included in the daily management of traditional building restoration according to a reasonable traditional building maintenance system.

4.3 Graded protection management of traditional buildings in this cottage

According to the results of safety evaluation, the graded management and protection strategy of traditional buildings in the Village is proposed, and the protection of traditional buildings, the protection of the overall appearance of the Village and the protection of the surrounding environment in the Village are divided into three levels from small to large management scope.

4.3.1 The first level: restoration and protection of single structure techniques of buildings

On the basis of the safety rating of the traditional architecture of this cottage based on the results of safety evaluation, planned batch protection and restoration will be carried out respecting the principles of wholeness, originality and dynamic development. We will emphasize the restoration and preservation of the building structure and construction techniques, and will not reconstruct the remaining buildings, and will strengthen the management and maintenance of the pillar bases, doors and windows, carvings of decorative parts and frescoes of the traditional buildings as key protection parts. The repair of the roof cover and roof slate (tiles), doors and windows, non-load-bearing envelope walls, wooden stairs, etc., the anti-corrosion and insect prevention measures of wooden structures, and the repair of other unforeseen damages are considered as regular maintenance works, and the regular maintenance works can take one year as the maintenance cycle for the traditional buildings with the safety level of A, B and C without serious damage.

4.3.2 Second level: overall protection of architectural style and appearance

The focus of the second level village overall style and appearance protection is that not only should new buildings and local random demolition be controlled, but also the architectural street pattern should be controlled on the basis of not changing the original building layout structure, building exterior texture and color, and increasing the accessibility of the street. The main components of the traditional buildings of this cottage that are partially destroyed, such as the reinforcement of cracked columns and cracked masonry walls, as well as the village walls and gates that are common to the traditional buildings of this cottage, should also be included in the maintenance and inspection of the public living equipment of this cottage as a key maintenance project to ensure that the new construction of municipal projects in the historical heritage complex does not affect their safe use.

4.3.3 Level 3: Synergistic protection with the surrounding environment

The management of the third level should take rivers and mountains as the management objects, and its focus is to protect the environment in the area of this fortress, and to manage and punish them according to the relevant laws and regulations, especially it should prohibit malicious damage to the environment caused by man, and consciously protect and supervise the surrounding natural environment of this fortress through villagers' self-government and other forms.

5 Conclusion

Based on the current situation of deterioration and poor protection of historical dwelling buildings in remote areas, this paper has conducted a detailed review of the historical formation and background of Anshun Benzhai. After researching and summarizing the construction techniques of the dwelling buildings in this village, and reviewing the current problems of the pillar bases, wooden pile rows, roof covers and walls in the enclosure structure, the damage is mostly caused by the penetration of natural rainfall and humid environment as well as the wear and tear of historical use over time, and based on the current problems, the proportion of dangerous components is evaluated, and the safety weight of the floors is divided and counted, and the single building is proposed according to the hierarchy. The three-level protection strategy of single building protection, overall architectural style and surrounding environment is proposed according to the hierarchy, thus providing a theoretical evaluation of the current situation and safety of the residential buildings in this village.

References

1. Zhou Zhengxu,Li Jingting,Qian Yun. Characteristics and value analysis of the cultural landscape of Anshun Tunbao settlement in Guizhou[J]. Guizhou Ethnic Studies,2019,40(05):56-61.
2. Li Meijun, Luo Fajin, Bai Xinxiang. Study on the form of stone column bases of Ciyun Temple in Qingyan Ancient Town, Guizhou Province [J]. Sichuan Architectural Science Research, 2014, 40(02):284-286.
3. Dong Chunying. Mechanical properties and experimental study of mortise and tenon joints in wooden structures of ancient buildings [D]. Xi'an University of Architecture and Technology,2010.
4. Liu Chengwei. Research on the reinforcement technology of structural members of wooden houses in villages and towns [D]. Huazhong University of Science and Technology,2011.
5. Chen Shunxiang. Study on the Social and Spatial Patterns of Tunbao Settlements in Guizhou [D]. Tianjin University,2005.
6. Wang Xuan,Wang Biao. Discussion on the safety assessment of modern historical buildings [J]. Chinese and foreign architecture,2017(08):193-194.
7. Zhong Xingrun. Research on safety evaluation of existing housing buildings and its management [D]. Xi'an University of Architecture and Technology, 2010.

Study on the Sustainable Planning and High Efficiency Utilization of Rural Areas Based on Ecological-Production-Living Space – A Case Study of Shuihu Village, Hengdong County, China

Zhiyuan Wang [1*], Kao Zhang [1], Hanling Yi [1], Jingtong Lv [1], Xin Li [1], Zeqi Yuan [1*] and Komi Bernard Bedra [2]

[1]School of Architecture, University of South China, Hengyang 421001, China

[2]School of Architecture and Art, Central South University, Changsha 410083, China

Abstract. Rural planning is an important part of territorial spatial planning. The study of rural "Ecological-production-living Space" is the key part in optimizing the land spatial development and improving the spatial planning system, which is significant for the sustainable planning and for high efficiency utilization of rural areas. In this study we take Shuihu village, Hengdong County, China, as a case study. (1) The paper applies the Analytic Hierarchy Process to construct a multi-factor evaluation system of villages "Ecological-production-living Space" for their development and construction, agricultural production conditions and ecological function suitability. Establishing the hierarchy model of rural "Ecological-production-living Space" and affecting weight values to each evaluation factor. (2) Using GIS multi-factor grid, we comprehensively evaluate the village space and identify as well as its dominant function. In the plan, the agricultural production as the leading function and as ecological and living space accounts for 36.25% of the area of the village field. The ecological production accounted for 51.71%, the development and construction accounts for 9.67%. (3) Apply rural complex concept to guide space usage separately from the production space (production and living space) , ecological space (ecological living and ecological production space), living space (living ecology and production space) ecological production and living space, to improve the ecological conservation and energy efficiency of agricultural production, to build ecological agriculture, modern agriculture and liveable idyllic community. Through these above parts, we hope to achieve the sustainable development model of rural area and high efficiency utilization of land resources.

1 Introduction

With the rapid process of urbanization and industrialization and the rapid growth of urban and rural construction land, cultivated land resources are occupied, ecological spaces are damaged and Environmental Bearing-Capacity resources do not fit the speed of construction land expansion, which lead to problems of the inharmonic development of ecology, production and living space as well as an unreasonable structural layout more and more serious [1-2]. In December 2010, the Main Functional Area Planning of China points out that we should according to the division of different functional regions define the performance appraisal system to optimize land area. In January 2017, National Homeland Plan (2016-2030) requires defining a red line for agricultural spaces and ecological spaces protection, urban living space development construction boundary and permanent basic farmland. The designated work of "Ecological-production-living Space" becomes the primary measure of spatial optimization of land; in November 2018, the Strategy of Rural Rejuvenating (2018-2022) clearly points out that priority development of agriculture, rural areas, deeply implement the "Agriculture Peasant and Countryside" work and promote the ecological, the production and the revitalization of peasants' living, which are decisions and arrangements to further optimize "Ecological-production-living Space" planning layout of rural areas and the coordination of production living and ecology.

Ecological-production-living Space is "Production-Ecology-Living" space. Among them production space is the dominant function area of providing agricultural product, industrial product and service product, which is the foundation of satisfying the human's material requirements, existence and development. Ecology space is the dominant function area of providing ecological service and ecological safeguard, which is the key of coordinating relationship between land and population, protecting ecosystem and achieving the sustainable development of the area [3-4]. Living space is the dominant function area of providing human residence, consumption, amusement and so on, which is the support

* wzhiyuan2005@yeah.net

of human living. About "Ecological-production-living Space" planning, scholars performed related researches mainly on the basic connotation and structural system of "Ecological-production-living Space" [5], identification and classification of function of "Ecological-production-living" [6-7], land use classification and system construction of "Ecological-production-living" [8], optimized layout of "Ecological-production-living Space" [9-10] and pattern succession of "Ecological-production-living Space" [11] etc. In aspect of research scale dimensions, they not only research the "Ecological-production-living" function of country, province, and city group etc. in macroscale and land use classification by the main functional area [12-13], but also research the "Ecological-production-living Space" demarcation in city territory or county area mesoscale [14-15] and spatial-temporal feature and pattern change of "Ecological-production-living Space" in rural micro scale [16-17]. The research on the sustainable planning pays much attention to Macro-Regional and medium-scale urban while pays relatively less attention to micro scale like county area [18-19]. However, in China the rural space accounts for over 94% of land areas [20]. We can't ignore the sustainable development of rural space [21]. Taking Shuihu village, Hengyang county in China as object in this paper, we research the planning of "Ecological-production-living Space", further determine its use patterns, refine the research scale of "Ecological-production-living Space" and fill up the theoretical gaps of the planning and evaluation system of "Ecological-production-living Space", which is significant for the sustainable planning and high efficiency utilization of rural space.

2 Research Area, Methods and Data Sources

2.1 Research Area

Shuihu village is located in the northwest of Wuji town, Hengdong county, Hunan province, China (Figure 1a), 2 kilometres away from Hengdong county, with good geographical advantages. The village has the typical rural space characteristics in southern Hunan. The terrain is dominated by plain and farmland, backed by mountains in the north and rivers in the south (Figure 1b). The total land area of the village is 285.47 hectares with a total population of 2760 people. Shuihu village belongs to the sub-tropical continental monsoon climate where annual average temperature is 17.7°C. The annual average sunshine hours is 1812h. Rainfalls are abundant, seasons are distinctive, water is sufficient and the soil is fertile, it is very suitable for crop growth and is given priority to with the traditional farming. Food crops are given priority to. Rice and economic crops are given priority to Huang Gong pepper and medicinal herbs cultivation, among them Huanggong pepper is the Chinese geographical indications of agricultural products.

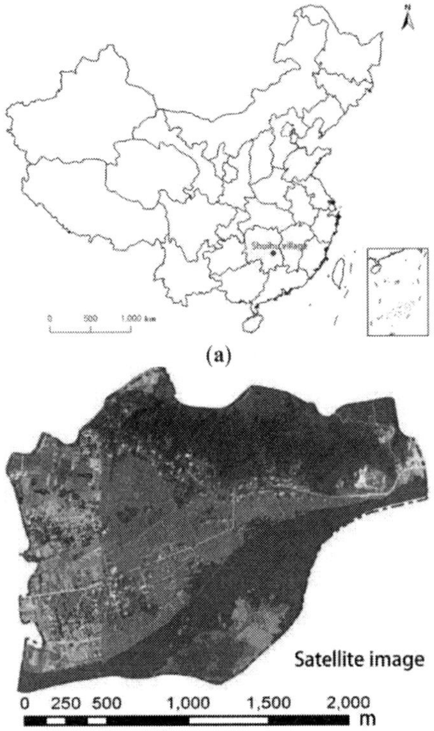

(a)

(b)

Figure 1. Geographical location and satellite image of Shuihu Village.

2.2 Research Methods

2.2.1 Analytic Hierarchy Process

Analytic hierarchy process (AHP) is a systematic and hierarchical analysis method combining qualitative and quantitative [22]. The study uses the Analytic Hierarchy Process to construct the evaluation system of "Ecological-production-living Space". The basic steps of AHP are as follows:

(1) Establish the hierarchical structure model. On the basis of deep analysis of practical problems, the related factors are decomposed into levels of target, criterion and scheme according to different attributes from top to bottom. The factors at the same level are subordinate to the factors at the upper level or have an impact on the factors at the upper level and at the same time dominate the factors at the next level or are affected by the factors at the lower level. In general when there are more than 9 criteria, we should further decompose the sub-criteria layer.

(2) Construct pair comparison verdict matrix where A, u_i、u_j（i，j=1，2，…，n） represent factors, u_{ij} represents the numeric values of the relative importance of ui to u_j, and the judgment matrix P of A-U is constituted by u_{ij}.

$$P = \begin{bmatrix} u_{11} & u_{12} & \dots & u_{1n} \\ u_{21} & u_{22} & \dots & u_{2n} \\ \dots & \dots & \dots & \dots \\ u_{n1} & u_{n2} & \dots & u_{nn} \end{bmatrix}. \quad (1)$$

(3) Calculate the weight vector. The proportion index in the evaluation is determined according to its importance relatively to the previous evaluation. Through the judgment matrix, we obtain the characteristic vector W corresponding to the largest relative eigenvalue λmax. The formula is: PW=λmaxW. The characteristic vector W is normalized, which is the order of importance of each evaluation factor, as well, the weight distribution.

(4) Consistency check. Judge whether the weight distribution obtained above is reasonable and use the check formula:

$$CI = \frac{\lambda_{max} - n}{n-1}. \quad (2)$$

$$CR = \frac{CI}{RI}. \quad (3)$$

At formula: CI is the general consistency index of verdict matrix; RI is the mean random consistency index of verdict matrix; CR is random coincidence coefficient of verdict matrix.

When CR of verdict matrix P<0.1 or λmax=n, CI=0, we consist that P has a satisfying consistency or we have to adjust the elements in P to make it have a satisfying consistency.

2.2.2 GIS Comprehensive Evaluation

GIS comprehensive evaluation is using GIS software to process and analyse the total value of multiple geographical elements impact factors in a certain scale by a linear weighting function and comparing the size of the sum value to plan the dominant type of spatial function. The calculation formula of the model is:

$$F = \sum_{i=0}^{n} F_i W_i \times 100. \quad (4)$$

At formula: F is contribution score of the dominant function of space; Fi is the single contribution score of evaluation factor i; Wi is the weight of evaluation factor; n is the number of geographical elements impact factors.

On the basis of the single factor evaluation by GIS, the multi-factor grid computing is carried out to comprehensively evaluate the composite space, so as to obtain the appropriateness of the utilization of village spaces, identify the dominant and secondary functions of land use [23] and plan the "Ecological-production-living Space" of the village.

2.2.3 Rural Complex Planning

Rural complex planning is a comprehensive rural development model integrating modem agriculture, leisure tourism, and rural communities. Its purpose is to help agricultural development by tourism and promote a sustainable development pattern of Three Industries Integration, which plays an important role in invigorating "Ecological-production-living Space" [24-25].According to the analysis of "Ecological-production-living Space" which was recognized and delimited by the present situation of village land resources function, we further plan the village "Ecological-production-living Space" for agriculture and forestry, for the processing industry and the leisure travel service industry. We also propose industrial development strategy and combined with modern ecological agriculture, leisure tourism, rural community integration concept to explore the rural comprehensive development pattern.

2.3 Data Sources

The basic data mainly comes from the land use data provided by the local government (image data and statistical data). DEM data and land use visual interpretation are from Landsat8LI data. The DEM data are 30m resolution of GDEMDEM data, including Google satellite map in TIF format downloaded in 2018 from Google earth. Landsat8LI data has a total of 11 band and we select the B3 B5 and B7 bands. The statistical data is from Hunan province statistical yearbook and Hengyang city statistical yearbook.

3 Construction of "Ecological-production-living Space" Evaluation System Based on AHP

Combined with the principle of the main functional area classification [26], we select multiple factors from the suitability of construction, agricultural production function and ecology function to construct the suitability evaluation system of village " Ecological-production-living Space". In order to improve the accuracy and reduce mistakes, the determination of evaluation factors applies the Analytic Hierarchy Process to construct the system layer and all the indicator layers.

3.1 The Evaluation Index of Development and Construction Suitability

The construction of the system layer is based on the development and the construction suitability of rural spaces. Transportation accessibility, topography and geomorphology, living conditions, public service and social security indicator layers are selected and their ratings further refined to construct the evaluation system. The quantitative description and the evaluation of the classification index system can synthetically judge whether the village land is suitable for construction.

3.2 The Evaluation Index of Agricultural Production Suitability

The development potential for agricultural production is affected by the soil, the hydrological conditions and the socio-economic conditions etc. Therefore, when evaluating the suitability for agricultural production we mainly consider the convenience of agricultural development and the applicability for land production.

3.3 The Evaluation Index of Ecological Function Suitability

The existing research used the ecological sensitivity analysis, ecological security pattern and ecological service value to balance the land ecological functions. Since the research area is at the village level, considering the environmental characteristics of the village and the availability of basic data, we used ecological security pattern and ecological sensitivity to interpret the ecological functions of the land.

3.4 The Construction and Weight Determination of the Multi-factor Evaluation System of Composite Space

The software yaahp was used to construct the hierarchical analysis model of "Ecological-production-living Space" planning, to establish the verdict matrix, to calculate the factor weights of the middle layer and the alternative plan, to verify whether their consistency is 0, and finally to obtain the weight value of each evaluation factor.

4 Sustainable Planning of "Ecological-production-living Space" Based on GIS

According to the degree of suitability, each evaluation index is described quantitatively, the three kinds of evaluation indexes are unified to 5, 4, 3, 2, 1 score to determine five grades: the most suitable, highly suitable, moderately suitable, lowly suitable and unsuitable. Then a single factor analysis is done for each evaluation index using GIS. According to the index weights for multi-factor grid computing, we get the suitability of the village space, identifythe dominant and secondary functions for the planning of the village.

4.1 Suitability Evaluation of Development and Construction

(1) The evaluation of transportation accessibility

The transportation accessibility and the current roads of the village (village road and county road) are combined on the base map and the road distance is divided according to the per capita walking speed of 50-100m/min. Using GIS multi-buffer analysis and re-classification, we determine, weigh and divide grades to get the distance from the village road (Figure 2a), the distance from county roads (Figure 2b), and the road network density (Figure 2c).

(2) The evaluation of topography and geomorphology

The elevation directly reflects the local temperature, the oxygen content, and the slopes affect the village construction projects making them difficult. Therefore, the grading evaluation of the land topographic and the geomorphic conditions (elevation in Figure 2d) and slope in Figure 2e) can effectively show the suitable plots for village construction.

(3) The evaluation of living conditions

In terms of the radius of influence, the distance from forest areas and industrial areas, and the dispersion degree of residential areas, the living conditions of land is evaluated. Many villages are developed along the waters.The radius of influence refers to the distance to the pits or canals or rivers (Figure 2f), which is the key factor affecting the construction of the village. Comparing the distance of the land to the industrial area (Figure 2g) and the forest area (Figure 2h), the merits of the village development environmental quality can be concluded. The dispersion degree of residential areas largely determines the atmosphere of village construction (Figure 2i). Using the grid units one can generate statistics on the dispersion degree of rural residential areas.

(4) The evaluation of public service and social security

We select the service radius of educational basic facilities, commercial financial facilities and medical and health institution to illustrate the importance of public service and social security to the development of villages. As to the current facilities of the village, the suitability for development and construction is classified by GIS elements turning point and multi buffer zone analysis(Figure 2j,k,l).

https://doi.org/10.1051/e3sconf/202129303025

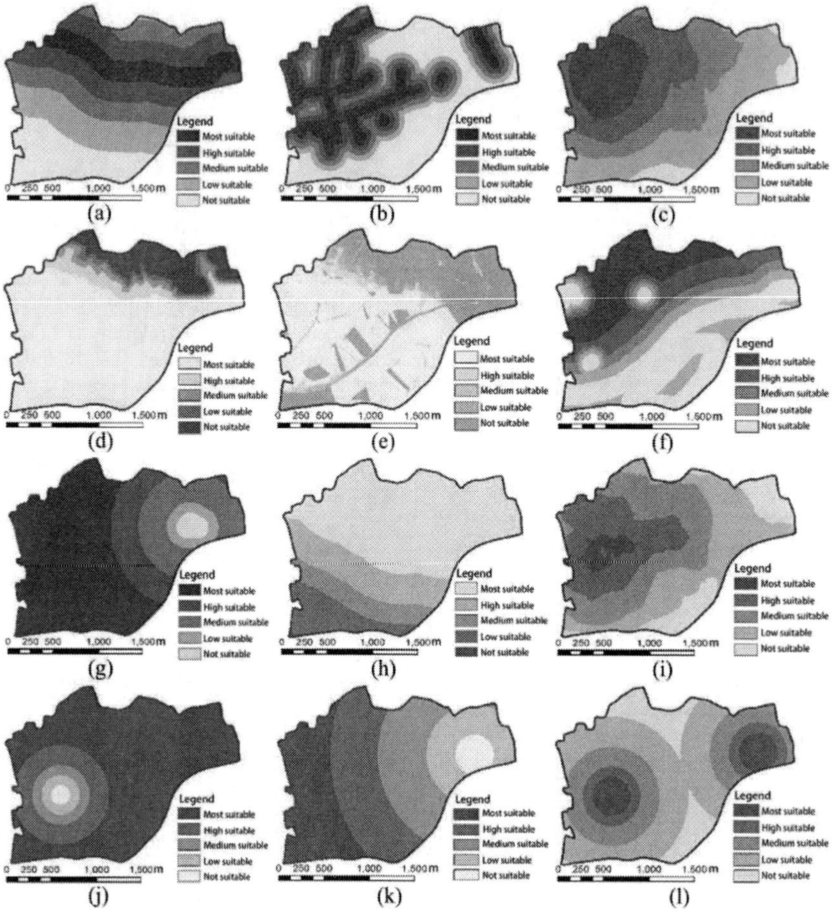

Figure 2. Suitability evaluation of village development and construction.

4.2 Suitability Evaluation of Agricultural Production

(1) The evaluation of agricultural development convenience

The distance from agricultural lands to village settlements and main roads reflects the aggregation degree of labor resources and the convenience of transportation, which directly affects the agricultural production and development. The closer the agricultural land is to the village settlements and the main roads, the more abundant and convenient the manpower and transportation are, which can effectively reduce the production cost and improve its efficiency (Figure 3a, 3b).

(2) The evaluation of utilization suitability for production

Agricultural crops are more easily affected by topographic conditions, and contour ploughing greatly affects the suitability of agricultural production. In other words, the steeper the slope, the more difficult the agricultural tillage; the smaller the slope, the more convenient the agricultural tillage (Figure 3c). The slope direction, to a certain extent, determines whether the tillage is on the sunny side and affects the growth and the tillage output quality. The south oriented slopes and the flat slopes are considered as the most suitable, the southwest and southeast slope are classified as highly suitable, the east and west slope as moderately suitable, the northeast and northwest slope as lowly suitable, and the north slope as not suitable (Figure 3d).

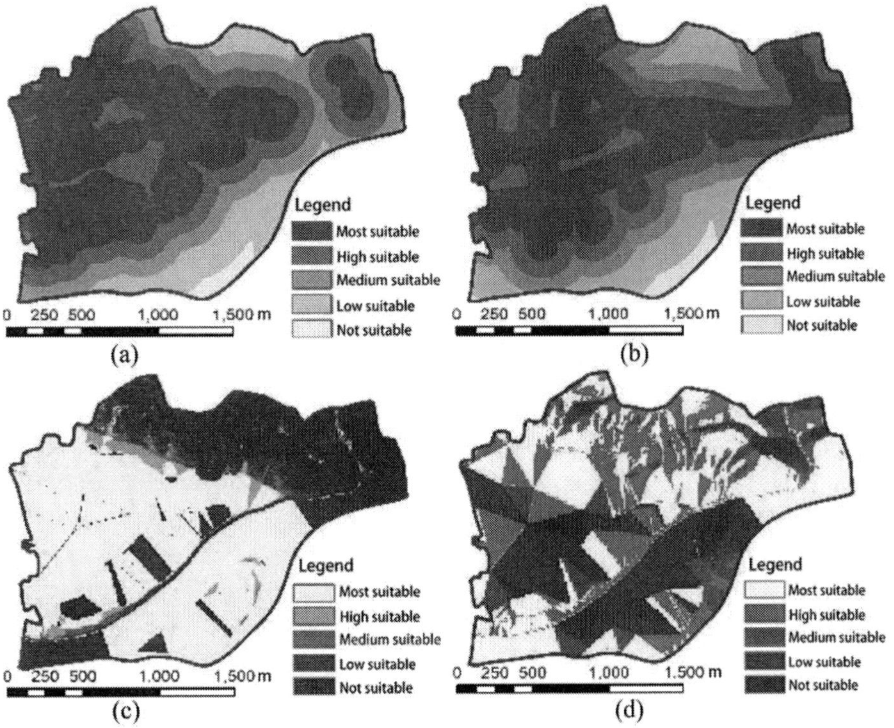

Figure 3. Suitability evaluation of village agricultural production.

4.3 Suitability Evaluation of Ecological Function

(1) The evaluation of ecological security pattern

Since the village is close to the river, the river has potential flood threat. Considering rainwater safety is an evaluation element of ecological security pattern, the suitability is divided by the distance of the village from the river. Closer to the river, greater the threat to the rainwater safety and lower the ecological suitability (Figure 4a). The reservoir ponds are of great significance to the security pattern of the water sources of the village. The distance from the water source is an important indicator affecting the water source security of the village, and it is selected as the evaluation factor. The closer the distance to the water source, the higher the ecological suitability (Figure 4b).

(2) The analysis of ecological sensitivity

The analysis of the ecological sensitivity selected two factors, the typical district of the current land use and vegetation of comprehensive evaluation. According to the typical district of current land use of village, we judge whether the block has ecological suitability. The ecological suitability is water area, forest land, grassland, cultivated land and residential area from high to low (Figure 4c). The ecological suitability of vegetation was determined according to its ecological sensitivity and ecological service value. Highly-sensitive vegetation such as jungle land and beach wetland is the area with high ecological function suitability, while relatively low-sensitive vegetation such as garden land and savanna is the area with low ecological function suitability (Figure 4d).

 https://doi.org/10.1051/e3sconf/202129303025

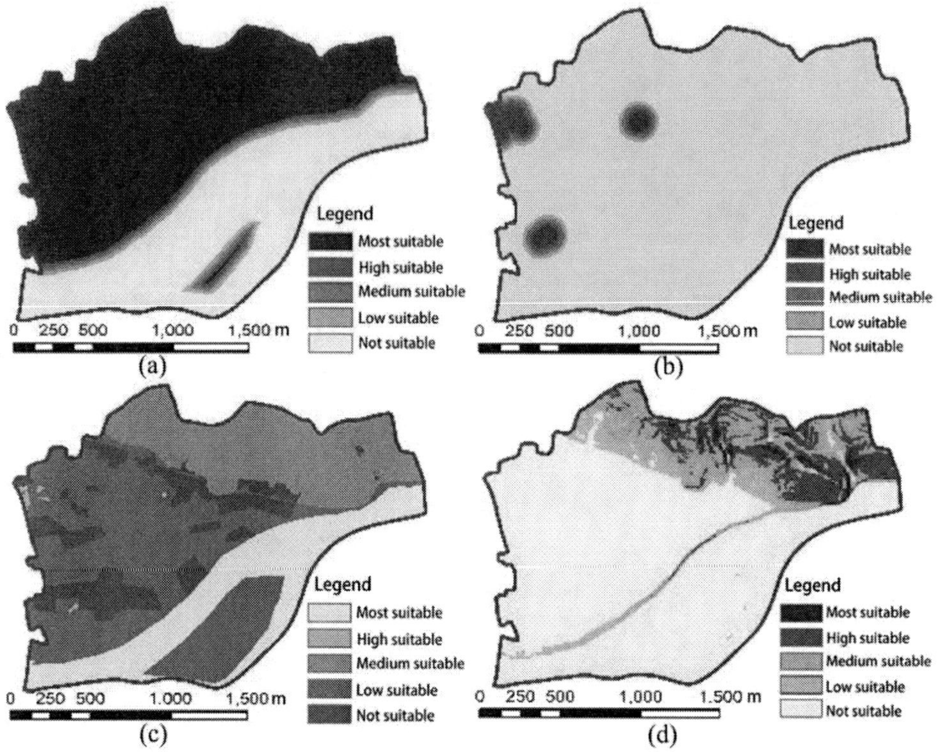

Figure 4. Suitability evaluation of village ecological function.

4.4 Sustainable Planning of "Ecological-production-living Space" in Villages

As the functions of "Ecological-production-living Spaces" are nested with each other, the composite space in "Ecological-production-living Spaces" is subdivided into 7 categories: production ecology, production living, ecological production, ecological living, living production, living ecology and ecological-production-living space.When the suitability level is most suitable, high suitable and medium suitable, the most suitable level is considered as the main function, followed by the secondary function of the village. If all three index grades are the most suitable possible, then the space is defined as the production-living-ecology space. When the three grade indexes are identical, the space is divided according to the principle that ecological function takes precedence over agricultural production function and agricultural production function takes precedence over village development and construction functions. If the suitability level of the three types of functions is low suitability or unsuitability, the ecological function will be selected as one of the functions, and the surplus according to the principle that agricultural production function takes precedence over the development and construction function.

Comprehensive evaluation of space utilization suitability (Figure 5a, b, c), the results of spatial overlay analysis based on GIS are as follows: the production ecology and production living space of agricultural production as the leading function accounted for 36.25% of the area of the village, the ecological production and ecological living of ecological function as the leading function accounted for 51.71% of the area of the village, the living production and living ecology space of development and construction as the leading function accounted for 9.67% of the area of the village, "Ecological-production-living Space" which is appropriated to all the three kinds of functional suitability accounted for 2.37% of the area of the village (Figure 5d).

https://doi.org/10.1051/e3sconf/202129303025

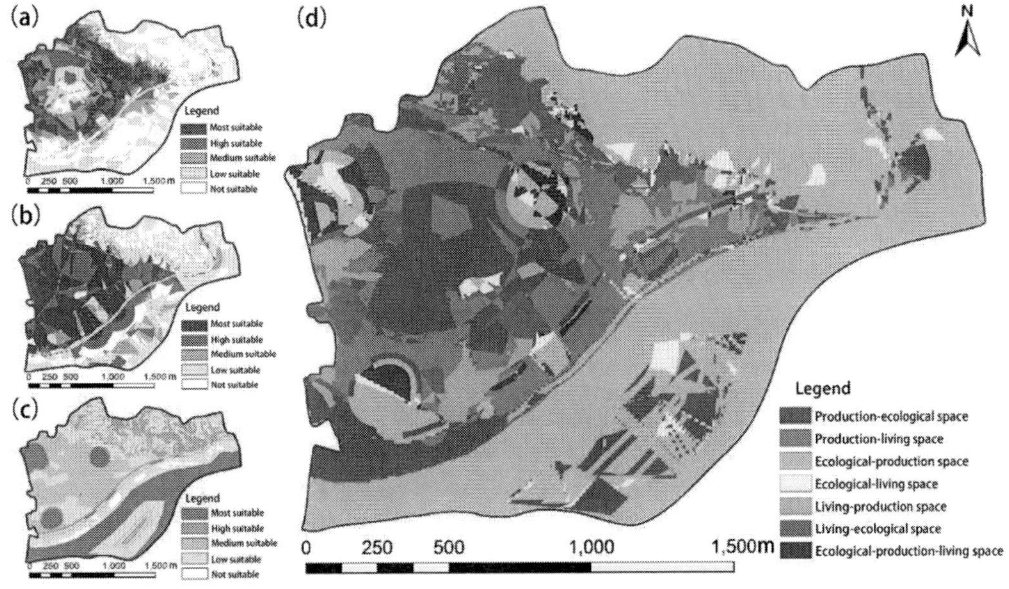

Figure 5. Sustainable planning of village "Ecological-production-living Space" based on GIS suitability evaluation.

5 Efficient Use of "Ecological-production-living Space" Based on Rural Complex

5.1 Thinking of Planning and Utilization

Focusing on agriculture, improving the energy efficiency of agricultural production, building ecological agriculture, modern agriculture and liveable idyllic community, and improving tourism necessary services are the core goals of meeting the requirements in rural areas for " industrial prosperity, ecological livability, and life prosperity ". According to the planning of village composite space "Ecological-production-living Space", the research regionalizes the leading function, and uses the rural complex concept and guides the space usage respectively from the production space (production living and production ecological space), ecological space (ecological living, ecological production space), living space (living ecological and living production space), so as to achieve the efficient use of land resources.

5.2 Production Space

From the distribution pattern of "Ecological-production-living Space" in the village area, the production space is mainly concentrated in the western part of the village and the Hexin island area, including the most part of farmland and a few village residential areas. According to the distribution of local crop resources, the ecological production space has been transformed from traditional agriculture to large-scale agriculture, and "two parks and two zones" have been set. Through the introduction of modern production cooperation model, the "two parks" are planned for the Huanggong pepper demonstration park and for Chinese medicine planting demonstration. The "two zones", namely modern ecological farmland area, agricultural product processing area, which promote the concentration of cultivated land resources and is conducive to the protection of cultivated lands. The production living space is located around the residential land of the village, which is dominated by recreational activities, providing sightseeing services of agriculture and experience of pastoral life, and extending the agricultural industry chain.

5.3 Ecological Space

The ecological space includes most of the forest land in the north of the village and the waters in the village area. The ecological elements are concentrated into a block, which plays an important role in supporting the functional structure of the village space. The purpose of its planning and utilization is to protect the integrity and the ecology of the pastoral scenery under for a sustainable development. Planning the northern ecological woodland reserve area is the foundation of the ecological protection. It is determined that the riverside in the southwest of the village is an ecological breeding garden and an ecological picking garden. Taking rural tourism as the guide to ensure the benefits of ecological resources landscape, meanwhile give play to the better economies of scale, ensure the income of villagers and meet the needs of rural residents for rich life.

5.4 Living Space

The Living space is mainly the current villagers' residential area. Three types of original residential areas for local villagers, country house experience areas and tourist villa areas are planned to build a comprehensive rural community with landscape appreciation, recreation and liveability value. We will integrate and adjust rural residential areas, make reasonable planning and site selection for new, relocated and repaired residential areas, and determine the size of rural residential areas in each district, so as to promote "effective governance" and "ecological liveability" of living space.

5.5 Ecological-production-living Space

The composite space of ecological production and life is mainly distributed in the vicinity of reservoirs and ponds in the village area, which has high land use value and landscape ecological benefits. It is the convergence and integration area of primary, secondary and tertiary industries. Aiming at the development of rural industry combination in this kind of space, a tourism complex area with recreation as the activity theme is built. According to the cluster distribution, the tourism complex can plan three core areas, namely, rural living community, outdoor picking experience area, parent-child leisure and entertainment area, supporting tourism service facilities and living infrastructure, to create a beautiful countryside that is liveable and suitable for travel (Figure 6).

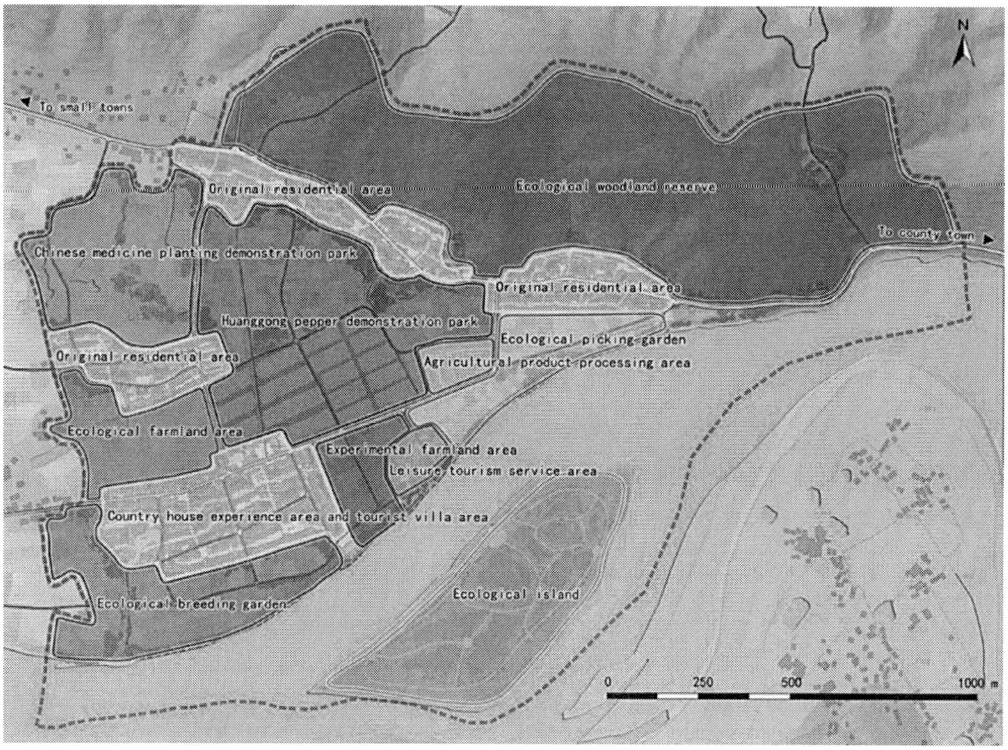

Figure 6. Planning and utilization of village "Ecological-production-living Space" based on rural complex.

6 Conclusions and Discussion

This study based on the sustainable development of rural areas, from the perspective of "Ecological-production-living space", by using GIS multi-factor comprehensive evaluation of the rural spaces suitability, reasonable agricultural production planning (as the leading function of ecological production and living space production), to the ecological as the leading function of ecological production and ecological living space, to the development as the leading function of living production and living ecological space and ecological production space, which is of important guiding significance to improve the optimization for the national spatial planning system in rural areas. Planning the use of rural space based on the concept of rural complex has a strong practical significance to promote the efficient use of rural space, improve the efficiency of land resources and increase farmers' income. Research conclusions include:

(1) Based on the principle of the main functional area regionalization, from three aspects of construction and development suitability, agricultural production suitability and ecological suitability, we select transportation accessibility, topography and geomorphology, living conditions, public service and social security, agricultural development convenience, the applicability of land production, ecological security pattern, ecological sensitivity, etc. as indicators. The multi-factor evaluation system for the suitability of "Ecological-production-living Space" in rural areas was established by the Analytic Hierarchy Process (AHP), and factor weights were determined. Spatial overlay analysis based on GIS was carried on the comprehensive

evaluation of rural space. The research method has a certain innovation, a strong operability, and can be popularized.

(2) By using GIS multi-factor grid computing, we get the suitability of village space utilization and identify the dominant and secondary functions of the rural space. The results show the agriculture production as the leading function of ecological production and living spaces account for 36.25% of the total area of the village fields. The ecological production as the leading function of ecological production and living space accounted for 51.71%. The development and construction as the leading function of living production and ecological space accounted for 9.67%. The suitability of the three functions belongs to the more suitable ecological production and living space which accounts for 2.37% of the area of the village fields.

(3) Applying rural complex concept, the land use is composed by the production space (production living and ecological space), the ecological space (ecological living and production space), living space (living ecology and production space) ecological production and living space, so as to achieve the Sustainable development model of rural area.

7 The following aspects should be paid attention to in the in-depth study of "Ecological-production-living Space"

(1) Rural spaces due to the regional differences and differences in land use, and to the difference of the "Ecological-production-living Space" planning from the current land use, cannot be treated as the same, its planning should adjust measures to local conditions, selecting reasonable evaluation factors, considering the coupling between each factor and space of land carrying capacity, soil conditions, resources and environment carrying capacity, landscape fragmentation and other factors to conduct further analysis and form the integrated system framework.

(2) On the micro scale, remote sensing information should be properly combined to carry out data analysis for "Ecological-production-living Space" planning. By summarizing the evolution of the scale and type of "Ecological-production-living Space" over the years, its key influencing factors should be inferred, and the weight of analysis should be summarized to ensure that the evaluation system is scientifically sound.

(3) As the planning system of rural space is not mature yet, it needs to be explored in depth. Theoretical researches are need to build a systematic framework and an "analysis -- planning -- utilization -- control" process, starting from the harmony perspective of "human-earth-resource" and making empirical analysis. Other dimensions could also be comprehensively considered.

Funding

This research was funded by Hunan Philosophy and Social Science Foundation, grant number 18YBQ106; Education Department Project of Hunan Province, grant number 19B480.

References

1. Foley, J.A Defries, R, Asner, G.P. et al. Global consequences of land use. *Science*, **2005**, *309*, 570-574.
2. Bai, X.M. Shi, P.J. Liu, Y.S. Realizing China's urban dream. *Nature*, **2014**, *509*, 158-160.
3. Kates, R.W. Clark, W.C. Corell, R. et al. Sustainability science. *Science*, **2001**, *292*, 641-642.
4. Costanza, R. D'Arge, R. de Groot, R. et al. The value of the world's ecosystem services and natural capital. *Nature*, **1997**, *387*, 253-260.
5. Liu, J.L. Liu, Y.S. Li, Y.R. Classification evaluation and spatial-temporal analysis of "production-living-ecological" spaces in China. *Acta Geographica Sinica*, **2017**, *72*, 1290-1304.
6. Nelson, E. Modeling multiple eco-system services, biodiversity conservation, commodity production, and tradeoffs at landscape scales. *Frontiers in Ecology and the Environment*, **2009**, *7*, 4-11.
7. Liu Y. On the logical structure, check and balance mechanism and development principle of "production-living-ecological space". *Social Sciences in Hubei*, **2016**, *3*, 5-9.
8. Cheng, T, Zhao, R, Liang, Y, Production-living-ecological space classification and its functional evaluation. *Remote Sensing Information*, **2018**, *33*, 114-121.
9. Wang, C, Tang, N, Spatio-temporal characteristics and evolution of rural production-living-ecological space function coupling coordination in Chongqing Municipality. *Geographical Research*, **2018**, *37*, 1100-1114.
10. Hu, W.T, Wang, L.G, Shu, M.H. Reflections on delimiting the three basic spaces in the compilation of urban and rural plans. *City Planning Review*, **2016**, 40, 21-26.
11. Liu, D.Q. Ma, X.C. Gong, J. et al. Functional identification and spatiotemporal pattern analysis of production-living-ecological space in watershed scale: A case study of Bailongjiang Watershed in Gansu. *Chinese Journal of Ecology*, **2018**, *37*,1490-1497.
12. Zhang, H.Q, Xu, E.Q. Zhu, H.Y. An ecological-living-industrial land classification system and its spatial distribution in China. *Resources Science*, **2015**, *37*, 1332-1338.
13. Ma, S.F. Ma, M. Cai, Y.M. et al. Delimitating red line of ecological protection for territorial spatial planning: A case study of Hunan Province. *Tropical Geography*, **2015**, *35*, 43-50.
14. Zhu, Y.Y. Yu, B. Zeng, J.X. et al. Spatial optimization from three spaces of production, living and ecology in national restricted zones: A case study of Wufeng

County in Hubei Province. *Economic Geography*, **2015**, *35*, 26-32.

15. Song, M.J. Chen, D.M. Katie W. An RP-MCE-SOP framework for China's county-level "Three-Space" and "Three-Line" planning—An integration of rational planning, Multi-Criteria evaluation, and spatial optimization. *Sustainability*, **2019**, *11*, 2997.
16. Zhang, Y.L. LI,X. Sun, S.L. Evaluation and optimization of rural space suitability based on "the Production, Living and Ecological Space Coordination"--Take Beishakou township, the Xiongan New Area as an example. *Urban Development Studies*, **2019**, *26*, 116-124.
17. Xi, J.C. Wang, S.K. Zhang, R.Y. Restructuring and optimizing production-living-ecology space in rural settlements: A case study of Gougezhuang Village at Yesanpo tourism attraction in Hebei Province. *Journal of Natural Resources*, **2016**, *31*, 425-435.
18. Rudolf,G. Function-analysis and valuation as a tool to assess land use conflicts in planning for sustainable, multifunctional landscapes. *Landscape and Urban Planning*, **2006**, *75*, 175-186.
19. Reidsma, P. König, H. Feng, S.Y. et al. Methods and tools for integrated assessment of land use policies on sustainable development in developing countries. *Land Use Policy*, **2011**, *28*,604-617.
20. Guangming Daily commentator. Improve rural environment and build beautiful countryside. *Guangming Daily*, 2018-2-6(2). http://news.gmw.cn/2018-02/06/content_27591798.htm
21. Liu, Y. Yang, W. Leadership and governance tools for village sustainable development in China. *Sustainability*, **2019**, *11*, 5553.
22. Forman, E.H. Gass, S.I. The analytic hierarchy process: An exposition. *Operations Research*, **2001**, *49*, 469-486.
23. Bennett, E.M. Peterson, G.D. Gordon, L.J. Understanding relationships among multiple ecosystem services. *Ecology Letters*, **2009**, *12*, 1394-1404.
24. Cucari, N. Wankowicz, E. De Falco, S.E. Rural tourism and Albergo Diffuso: A case study for sustainable land-use planning. *Land Use Policy*, **2019**, *82*,105-119.
25. Cattaneo, T. Giorgi, E. Ni, M. Landscape, architecture and environmental regeneration: A research by design approach for inclusive tourism in a rural village in China. *Sustainability*, **2019**, *11*, 128.
26. Fan, J. The scientific foundation of Major Function Oriented Zoning in China. *Journal of Geographical Sciences*, **2007**, *62*, 339-350.

Enhanced syntrophic propionate degradation and methane production by granular activated carbon in AnSBR

Yiyang Fan[1], Qiyi Chen[1], Han Yan[1], and Jianzheng Li[1*]

[1]School of Environment, State Key Laboratory of Urban Water Resource and Environment, Harbin Institute of Technology, 73 Huanghe Road, Harbin 150090, China

Abstract: Syntrophic degradation of propionate has been regarded as a limiting factor for methane formation in anaerobic digestion (AD) processes, due to its easier production but harder degradation than other volatile fatty acids. In the present study, 20 g/L granular activated carbon (GAC) was introduced into an anaerobic sequence batch reactor (AnSBR) and the enhancement of propionate degradation and methane production was evaluated with another AnSBR without GAC as a control. Inoculated with the same excess activated sludge, both AnSBRs were synchronously started up at 35 °C with the fed propionate increased gradually from the initial 340 mg/L to the final 1700 mg/L. The results showed that the introduction of GAC made the AnSBR get steady state 6 days earlier. With the fed propionate of about 1700 mg/L in the steady state, the average specific methane production and biomass was enhanced from 0.20 to 0.25 L/L·d, and 7.72 to7.96 g/L, respectively, by the introduced GAC. The results suggested that the GAC had functioned in stimulating microbial growth and enhancing direct interspecies electron transfer between hydrogen-producing acetogens and methanogens, which had resulted in the enhanced propionate degradation and methane production.

1 Introduction

It is known that propionate fermentation in anaerobic digestion (AD) processes is easier occurred with more energy production than other acidogenesis such as the fermentation of butyrate, acetate and ethanol as illustrated as Eq. (1) to Eq. (4), respectively [1].

$$C_6H_{12}O_6 + 2H_2 \rightarrow 2CH_3CH_2COOH + 2H_2O \quad \triangle G_0' = -279.4\ kJ/mol \quad (1)$$

$$C_6H_{12}O_6 \rightarrow CH_3CH_2CH_2COOH + 2H_2 + 2CO_2 \quad \triangle G_0' = -254.0\ kJ/mol \quad (2)$$

$$C_6H_{12}O_6 + 2H_2O \rightarrow 2CH_3COOH + 4H_2 + 2CO_2 \quad \triangle G_0' = -206.0\ kJ/mol \quad (3)$$

$$C_6H_{12}O_6 \rightarrow 2CH_3CH_2OH + 2CO_2 \quad \triangle G_0' = -164.8\ kJ/mol \quad (4)$$

On the other hand, among the produced volatile fatty acids (VFAs, except acetate) and ethanol, propionate is the most difficult one to be further converted to acetate and H_2/CO_2 by hydrogen-producing acetogens (HPAs) prior to methanogenesis, as expressed as the equations from Eq. (5) to Eq. (7) [2, 3].

$$CH_3CH_2COOH + 2H_2O \rightarrow CH_3COOH + 3H_2 + CO_2 \quad \triangle G_0' = +76.1\ kJ/mol \quad (5)$$

$$CH_3CH_2CH_2COOH + 2H_2O \rightarrow 2CH_3COOH + 2H_2 \quad \triangle G_0' = +48.1\ kJ/mol \quad (6)$$

$$CH_3CH_2OH + H_2O \rightarrow CH_3COOH + 2H_2 \quad \triangle G_0' = +19.2\ kJ/mol \quad (7)$$

Obviously, the syntrophic degradation of propionate is thermodynamic-uncomfortable under standard conditions. Thus, anaerobic oxidation of propionate has been regarded as a limiting factor for methane formation in AD processes [4].

It has been reported that the poor electron transfer via hydrogen as electron carrier or so-called interspecies hydrogen transfer (IHT) should be responsible for the inefficient propionate conversion in AD processes [5]. Therefore, direct interspecies electron transfer (DIET) is presented as an alternative way to enhance the electrons transfer between acetogens and methanogens [6]. Conductive carbon materials, such as activated carbon, biochar, carbon cloth and carbon filter etc, have been introduced to enhance the methane production of AD [7-10]. As a common AD process, anaerobic sequence batch reactor (AnSBR) has been widely practiced in engineering, but little information about the effect of conductive carbon materials on the syntrophic degradation of propionate in AnSBR could be collected up to now.

In the present research work, two lab-scale AnSBRs with or without the introduction of granular activated carbon (GAC) were constructed, respectively. Inoculated with the same excess activated sludge, both of the AnSBRs were synchronously started up at 35°C with the fed propionate increased gradually from the initial 340 mg/L to the final 1700 mg/L. Based on the performance of the two AnSBRs, the enhancement of propionate degradation and methane production was evaluated by comparison.

* Corresponding author: jianzhengli@hit.edu.cn

2 Materials and methods

2.1 Reactors and their operation

Two lab-scale AnSBRs were constructed with the same 1 L glass bottle, respectively. One of the reactors was loaded with 20 g/L GAC (10-24 mesh, Aladdin, China), while with the other one as a control without GAC. Inoculated with seed sludge, propionate solution was fed into the reactors to obtain a reaction volume of 0.6 L, respectively. Then each of the reactors was sealed with rubber plug. Two glass tubes were installed through the plug for collecting biogas and feeding/discharging the supernatant in the reactor. Both of the AnSBRs were synchronously operated in an air bath vibrator (3BGH-2020, Shanghai Boxun) at 35°C and 130 rpm. Both reactors were continuously performed for 33 days which was divided into three stages with 72 h (Stage 1), 48 h (Stage 2) and 24 h (Stage 3) as an operating cycle, respectively. Each of the cycles included 5 min for feeding (0.3 L), 20 min for settlement and 5 min for discharging (0.3 L).

Excess activated sludge collected from a local municipal wastewater treatment plant (Harbin, China) was anaerobically stood for 14 days at 25°C and then used to inoculated the AnSBRs. The initial biomass in the reactors was almost the same of about 7.00 g/L in terms of mixed liquor volatile suspended solid (MLVSS).

Feed into the AnSBRs was sodium propionate solution including per liter (1/L): 0.11 g K_2HPO_4, 0.17 g KH_2PO_4, 0.30 g NH_4Cl and 4.00 g $NaHCO_3$. The propionate in feed was gradually increased from the initial 340 to 1700 mg/L by the end of Stage 2. All through the 33-days operation, the feed was adjusted with 1.0 mol/L NaOH and HCl solution to get a constant pH of 7.50.

2.2 Analytical methods

Biomass in terms of MLVSS was determined following the standard methods [11]. pH was measured by a pH meter (DELTA 320, Mettler Toledo). VFAs were determined on a gas chromatograph (Model SP6890, Shandong Lunan Instrument Factory, China) which was equipped with the RTX-Stabilwax glass column (30 m × 0.32 mm × 1 um) and the flame ionization detector (FID) [12]. Biogas produced in each of the operating cycles was collected into a 1 L gas bag and then quantified with a 100 mL glass syringe. Volume fraction of methane in the collected biogas was determined on another gas chromatograph (SP-6800A, Shandong Lunan Instrument Factory, China) which was fitted with a 2-m stainless column packed with Porapak. Q (100 to 180 mesh) and a thermal conductivity detector (TCD) [12].

3 Results and discussion

As shown in Fig. 1, the effluent pH of the AnSBR loaded with GAC and the AnSBR without GAC was relatively stable with an average of 7.91 and 7.88 throughout the 33-days operation, respectively, though the propionate in influent was gradually increased from 340 to 1700 mg/L. It was thought that the exhaustion of propionate was responsible for the slightly higher pH of effluent than that of influent. The reciprocal metabolic process between propionate degradation bacteria and methanogens is easily affected by pH value. Lower pH will induce the release of molecular propionic acid in reaction system, and the excessive molecular propionic acid would lead to the reaction stagnation by damaging bacteria [13]. Therefore, the weakly alkaline condition achieved in this study was favorable for the growth and metabolism both of syntrophic propionate degradation bacteria and methanogens [12].

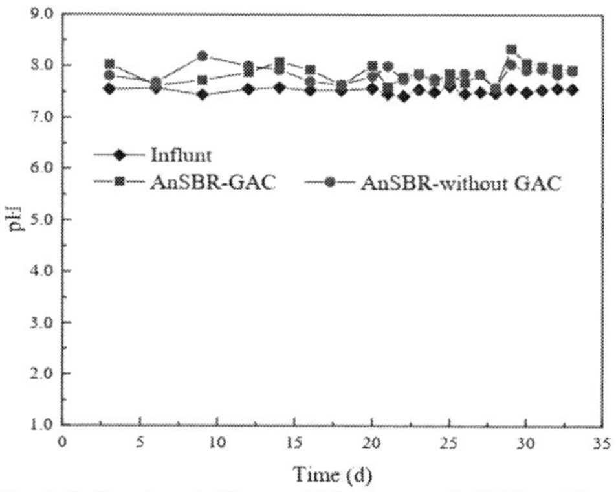

Fig. 1. Influent and effluent pH in the two AnSBRs with or without the introduction of GAC, respectively

The syntrophic degradation of propionate in the two reactors was illustrated in Fig. 2. The results showed that only 31.4% propionate had been degraded without the introduction of GAC in the first operating cycle with the time of 72 h. On the contrary, almost all of the fed propionate in the AnSBR loaded with 20 g/L GAC was exhausted within the same operating cycle. The results confirmed that the introduced GAC had obviously enhanced the syntrophic degradation of propionate, which would speed up the start-up process of the AnSBR [14]. However, propionate had not been detected any more from the effluent of both the AnSBRs since the 6th day, though the operating cycle was shortened to 24 h along with the propionate in influent was increased to 1700 mg/L stage by stage. The results indicated that the inoculated sludge in the AnSBR without GAC had also been acclimatized since the 6th day.

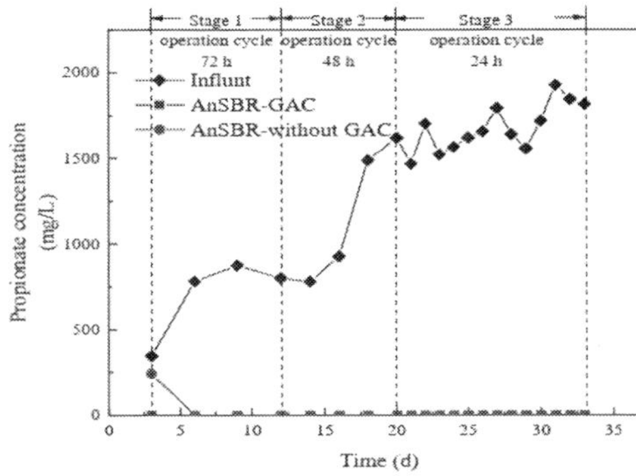

Fig. 2. Propionate in the two AnSBRs with or without the introduction of GAC, respectively

Fig. 3 illustrated the performance of both the AnSBRs in specific methane production. The results showed that a steady stage for specific methane production had been obtained in the AnSBR loaded with GAC since the 21st day. However, the steady state was not observed until the 27th day. The results suggested that the introduction of GAC had significantly speed up the start-up process of the AnSBR. It was found that, with the longest operating cycle of 72 h, no obvious difference in specific methane production was found between the AnSBRs with or without the introduction of GAC. However, when the operating cycle was shortened to 48 h in Stage 2 with the propionate of about 1100 mg/L in feed, the specific methane production in the AnSBR loaded with GAC averaged 0.11 L/L·d which was much more than that of 0.04 L/L·d in the AnSBR without GAC. Since the operating cycle was further shortened to 24 h in Stage 3 with the propionate of about 1700 mg/L in the feed, the specific methane production in the AnSBR loaded with GAC was enhanced to about 0.25 L/L·d. On the other hand, the specific methane production in the AnSBR without GAC averaged 0.20 L/L·d in the last 7 days (the steady stage). Also, the methane convert ratio in the AnSBR with GAC were about 51.4% in the steady stage, higher than that in the AnSBR without GAC (Fig, 4). The results indicated that

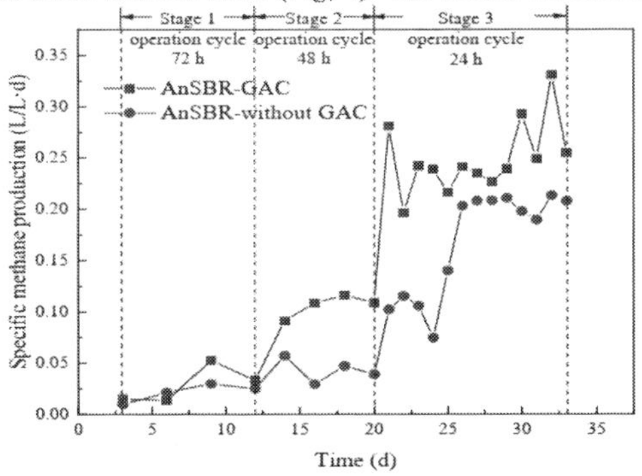

Fig. 3. Specific methane production in the two AnSBRs with or without the introduction of GAC, respectively

the introduction of GAC had not only speed up the start-up process but also enhanced the methane formation from propionate in the AnSBR. This is probably due to that the addition of GAC reduces the Gibbs free energy of propionate metabolism with the electron transfer

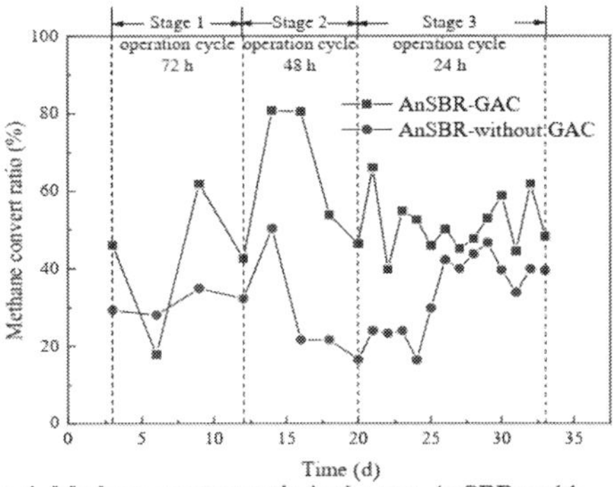

Fig. 4. Methane convert ratio in the two AnSBRs with or without the introduction of GAC, respectively

changing from IHT to DIET [15], which is conducive to the proliferation of bacteria. At the same time, the DIET process mediated by activated carbon avoids the loss of hydrogen and formate used as intermediate electron carriers [16], leading to more electrons participate in the process of reducing carbon dioxide to methane. Therefore, the methane production rate was significantly improved by the additional GAC.

The biomass in the AnSBR loaded with GAC was increased from the initial 7.00 to 7.96 g/L by the end of the 33-days operation process, while with the biomass in the AnSBR without GAC increased to 7.72 g/L. Obviously, the introduced GAC was also favorable for the population growth of anaerobic sludge in the AnSBR. It was thought that the porous structure of GAC had allowed syntrophic propionate degradation bacteria and

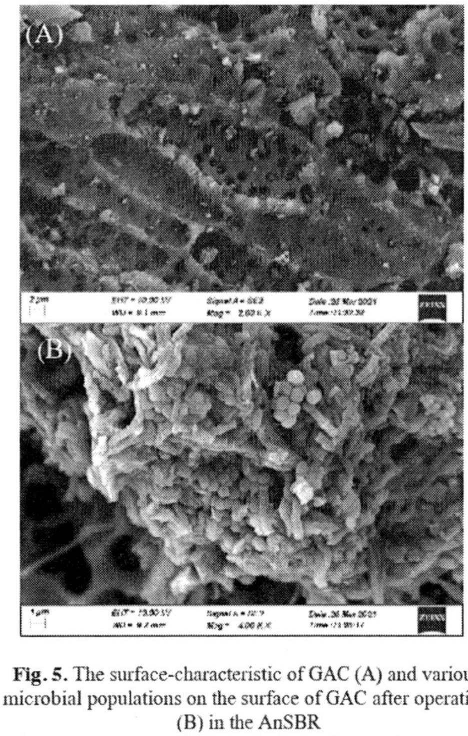

Fig. 5. The surface-characteristic of GAC (A) and various microbial populations on the surface of GAC after operation (B) in the AnSBR

methanogens to gather and thrived together on the surface (Fig. 5A). Over time, SEM showed that biofilm would certainly be construct on the surface of GAC, by which more biomass was kept in the AnSBR (Fig. 5B) [17]. Also, it was observed that long bamboo nodular microorganisms (e.g. *Methanothrix*) were attached to the surface of GAC, and tightly intertwined with some spherical (e.g. *Methanosarcina*) and ellipsoid microorganisms. *Methanothrix* and *Methanosarcina* can both produce methane from the reduction of carbon dioxide by accepting electrons through the DIET pathway [18, 19]. These results indicated that a DIET pathway between methanogens and propionate degradation bacteria was likely to be established with the additional GAC.

It is known that GAC can function as a precursor to DIET among various functional microbial populations in AD process because of its high electrical conductivity [20]. In fact, GAC can act as not only an electron accepter to endorse DIET, but also an electron donor pool for various biochemical reactions [21]. It has been reported that the DIET can be 10^6 times quicker than IHT [22], resulting in an excellent efficiency in VFAs degradation and methane production. In the present research work, the introduction of GAC had not only speed up the start-up process of AnSBR, but also remarkably enhanced the specific methane production and biomass. With the propionate averaged 1700 mg/L in feed and an operating cycle of 24 h, the specific methane production was enhanced by 1.25 times because of the introduction of 20 g/L GAC, while with the biomass increased from 7.72 to 7.96 g/L. However, the DIET in the present AnSBR was only a pretty-astute deduction. Therefore, more evidences should be collected to confirm the deduction in the future investigation.

4 Conclusions

Two AnSBRs was constructed and operated synchronously to evaluated the effect of GAC on the syntrophic propionate degradation and methane production. Based on the experiments, the following conclusions are obtained.

(1) The introduction of GAC made the AnSBR get steady state 6 days earlier.

(2) With the propionate averaged 1700 mg/L in feed and an operating cycle of 24 h, the specific methane production in the AnSBR had been increased by1.25 times because of the introduction of GAC.

(3) Biomass in the AnSBR was also increased from 7.72 to 7.96 g/L by the introduced GAC.

(4) The DIET induced by the GAC was suggested as the key reason for the enhanced AD process in the AnSBR.

Acknowledgment

The authors acknowledge the financial support from the National Natural Scientific Foundation of China (51778171).

References

1. N.M.C. Saady. International Journal of Hydrogen Energy, **38**, 13172-13191 (2013)
2. A.J.M. Stams, D.Z. Sousa, R. Kleerebezem, C.M. Plugge. Water Sci Technol, **66**, 352-362 (2012)
3. J. Li, H. Yan, Q. Chen, J. Meng, J. Li, Y. Zhang, A. Kumar Jha. Bioresource Technol, 124972 (2021)
4. K. Glissmann, R. Conrad. Fems Microbiol Ecol, **31**, 117-126 (2000)
5. B. Schink, D. Montag, A. Keller, N. Muller. Env Microbiol Rep, **9**, 189-202 (2017)
6. W. Wang, D.J. Lee. Bioresource Technol, **330**, 14 (2021)
7. M.S. Romero-Guiza, J.J. Vila, J. Mata-Alvarez, J.M. Chimenos, S. Astals. Renew Sust Energ Rev, **58**, 1486-1499 (2016)
8. Q.J. Feng, Y.Q. Lin. Renew Sust Energ Rev, **77**, 1272-1287 (2017)
9. J.S. Zhang, W.Q. Zhao, H.W. Zhang, Z.J. Wang, C.F. Fan, L.H. Zang. Bioresource Technol, **266**, 555-567 (2018)
10. S.O. Masebinu, E.T. Akinlabi, E. Muzenda, A.O. Aboyade. Renew Sust Energ Rev, **103**, 291-307 (2019)
11. A.P.H.A. APHA, AWWA and WPCF, Washington DC, (2012)
12. Y.P. Zhang, J.Z. Li, F.Q. Liu, H. Yan, J.L. Li. Environ Sci Pollut R, **25**, 12434-12443 (2018)
13. H. Yuan, N. Zhu. Renewable and Sustainable Energy Reviews, **58**, 429-438 (2016)
14. W.X. Sun, S.F. Fu, R. Zhu, Z.Y. Wang, H. Zou, Y. Zheng. Bioresource Technol, **313**, 8 (2020)
15. R. Lin, J. Cheng, L. Ding, J.D. Murphy. Chem Eng J, **350**, 681-691 (2018)
16. D.R. Lovley. Rev Environ Sci Bio-Technol, **10**, 101-105 (2011)
17. S.Q. Aziz, H.A. Aziz, M.S. Yusoff, M.J.K. Bashir. J Hazard Mater, **189**, 404-413 (2011)
18. A.E. Rotaru, P.M. Shrestha, F.H. Liu, M. Shrestha, D. Shrestha, M. Embree, K. Zengler, C. Wardman, K.P. Nevin, D.R. Lovley. Energ Environ Sci, **7**, 408-415 (2014)
19. A.E. Rotaru, P.M. Shrestha, F. Liu, B. Markovaite, S. Chen, K.P. Nevin, D.R. Lovley. Appl Environ Microb, **80**, 4599-4605 (2014)
20. R.C. Lin, J. Cheng, J.B. Zhang, J.H. Zhou, K.F. Cen, J.D. Murphy. Bioresource Technol, **239**, 345-352 (2017)
21. F.H. Liu, A.E. Rotaru, P.M. Shrestha, N.S. Malvankar, K.P. Nevin, D.R. Lovley. Energ Environ Sci, **5**, 8982-8989 (2012)
22. G. Capson-Tojo, R. Moscoviz, D. Ruiz, G. Santa-Catalina, E. Trably, M. Rouez, M. Crest, J.P. Steyer, N. Bernet, J.P. Delgenes *et al.* Bioresource Technol, **260**, 157-168 (2018)

Research on the Spatio-temporal Differences in the Living Standards of the Counties under the Shandong Province

Hongwu CUI[1,*], and Qian YANG[2]

[1] Shandong University of Science and Technology, College of Geomatics, Qingdao, China, 266590

[2] Shandong University of Science and Technology, Auditing Department, Qingdao, China, 266590

Abstract. Using the method of district statistics to verify the correlation of NPP-VIIRS night light data and GDP, the difference between counties in Shandong from 2012 to 2019 can be analysed and the changes of living standards can be quantified. After obtaining the total value of night light brightness in each county, the coefficient of variation is used to calculate the degree of dispersion of each county, and then the regional differences of each city in 2019 and the inter-annual changes from 2012 to 2019 are analyzed and displayed. In conclusion, living standards of different countries in Shandong are different, and the difference becomes obvious with the growth of GDP. However,the difference of living standards of most countries becomes smaller in recent years.

1 Introduction

With the continuous improvement of social living standards, the differences between residents ' living standards are gradually increasing, and the gap between rich and poor is becoming more and more obvious. The living standard of the residents is affected by many factors, including material and spiritual factors, which makes it difficult to reasonably evaluate and quantify the living standard of the residents. In order to quantify the gap between the living standards of residents, the night light data is used to simulate the living conditions of residents by using the brightness value of night light, and then the living standards of residents are numerically calculated.

Night light data is widely used in all aspects closely related to residents' lives[1], such as residents' income, population, power consumption[2], carbon emissions[3], natural disasters, environment[4], health[5]. By fitting the long-time series of night light data with residents' income, it is proved that the night light data can better reflect the income status of residents. The light data are used to explore the balance of regional economic development, so as to facilitate the proposal of relevant development policies in the future. With the global VIIRS nighttime lighting products[6] from 2012 to 2019 calculated by monthly average of Elvidge CD, the Chinese region is selected to better reflect the living standards of Chinese residents.

In this paper, the night light brightness value is used to simulate and quantify the living standard of residents, and the night light value difference between counties under the jurisdiction of the city is used to quantitatively analyze the living standard difference of residents in counties under the jurisdiction of the city. It is expected to study the difference in the living standards of the residents in the cities and counties of Shandong Province, the gap between the rich and the poor among the residents, resource reasonable configuration, targeted assistance and the formulation of relevant policies will play a guiding role.

2 Data sources and methods

2.1 Data source

The research data used in this paper mainly include the following two types:

(1) The NPP-VIIRS nighttime light data from 2012 to 2019 are derived from the VIIRS Nighttime Light (VNL) V2 series products of Earth Observation Group (EOG). The new method uses the twelve-month median radiance to discard high and low radiance outliers, filtering out most fires and isolating the background. Download address(https://eogdata.mines.edu/products/vnl/#download)

(2) Shandong province county administrative boundary vector map. According to the latest administrative divisions of Shandong Province published by the People 's Government of Shandong Province at the end of 2018, the vector data of the boundary line of county-level administrative divisions in Shandong Province in 2018 provided by the National Basic Geographic Information Center, the county-level administrative division map of Shandong Province in 2018 is obtained. The main difference is that Laiwu is divided into Jinan.

* Corresponding author: 201982020009@sdust.edu.cn

2.2 Research method

The total nighttime light values of counties in Shandong Province from 2012 to 2019 were obtained by using the method of zoning statistics, and the nighttime light values per capita of each county were calculated. After determining the relationship between residents ' living standard and night light brightness value, the coefficient of variation analysis model is established to calculate the dispersion degree between counties under each city. Through the analysis of variation coefficient in two dimensions of time span and regional span, and the coupling analysis with GDP, the spatial and temporal differences of residents ' living standards in Shandong Province are determined.

3 The calculation of county living standard difference

3.1 Correlation calculation of county-level light brightness value

According to the vector data of county-level administrative divisions in Shandong Province, the nighttime light images of each county in Shandong Province are cut, and the total light brightness of the county is calculated. The specific calculation formula is shown in Equation (1).

$$TDN_{county} = \sum_{i=1}^{n} DN_i \qquad (1)$$

In formula (1), TDN_{county} is the total brightness value of a county under Shandong Province, n is the total number of nighttime lighting image grids contained in the county, and DNi is the brightness value of the i grid of the grid.

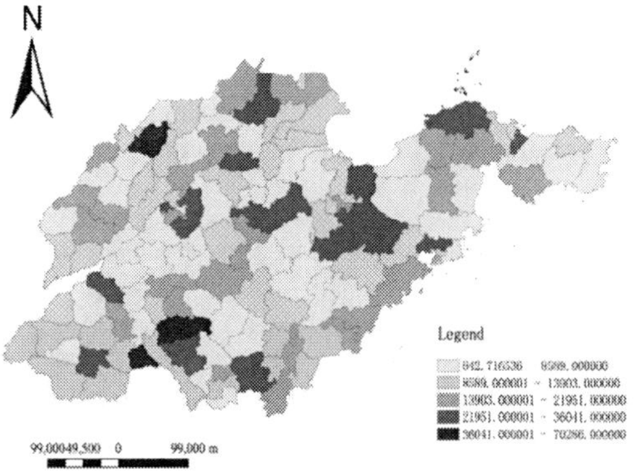

Figure 1. The total value of lighting brightness of each county in Shandong Province in 2019

As shown in Figure 1, the total value of night light brightness in Shandong Province was observed and analyzed. It was found that the counties with relatively high brightness were mainly concentrated in the central and eastern regions of Shandong Province, and the counties with high brightness were concentrated. The total value of light brightness in each county was significantly different. The lowest brightness value was Changdao County (842.7), and the highest was Huangdao District (70287).

In order to better represent the difference in living standards of residents in each county, the ratio of nighttime light data of a county to the total population at the end of the year is calculated to obtain the per capita light brightness value, as shown in Equation (2).

$$PDN_{county} = \frac{TDN_{county}}{POP_{county}} \qquad (2)$$

In formula (2), PDN_{county} represents the county ' s per capita light brightness value (10,000 people), TDN_{county} represents the county ' s total light brightness value, POP_{county} represents the county ' s total population.

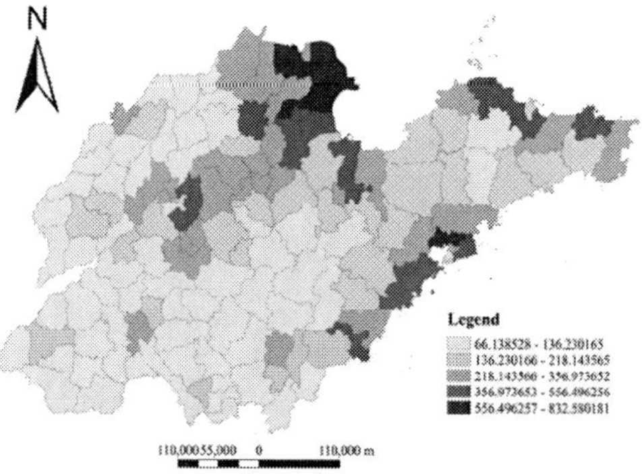

Figure 2. Per capita light brightness value of each county in Shandong Province in 2019

It can be found from the per capita light brightness value of each county that in 2019, the per capita night light value of each county in Shandong Province was significantly different, with the lowest in Juancheng County (66.1 / 10,000 people) and the highest in Hekou District (832.6 / 10,000 people). The average brightness value of each county is different from the distribution of the total brightness value, that is, the average brightness value of the counties with high total brightness value is not necessarily high, which to some extent reflects the gap between the living standards of each county.

3.2 Correlation test of city GDP and night light brightness

According to the administrative divisions of cities in Shandong Province, the total nighttime light brightness of each city from 2012 to 2019. In order to verify the relationship between light brightness value and GDP in Shandong Province and its effectiveness in Shandong Province, correlation analysis was used to calculate. The calculation results show that R^2 is about 0.9, which proves that there is a strong correlation between GDP and light brightness value in Shandong Province, and also shows that the night light brightness value can reflect the living standards of residents in various cities to a certain extent.

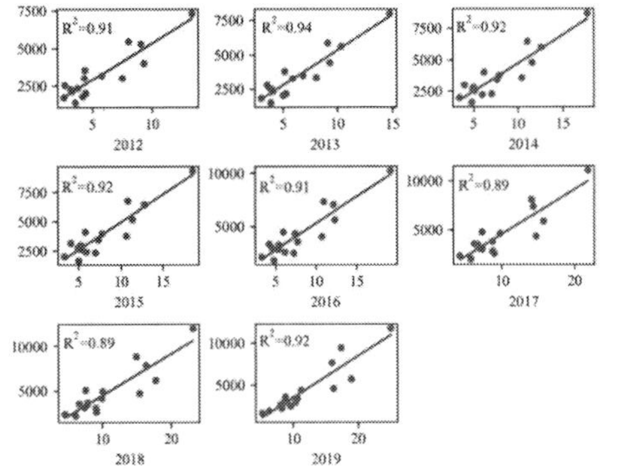

Figure 3. Correlation between GDP and light intensity in cities from 2012 to 2019

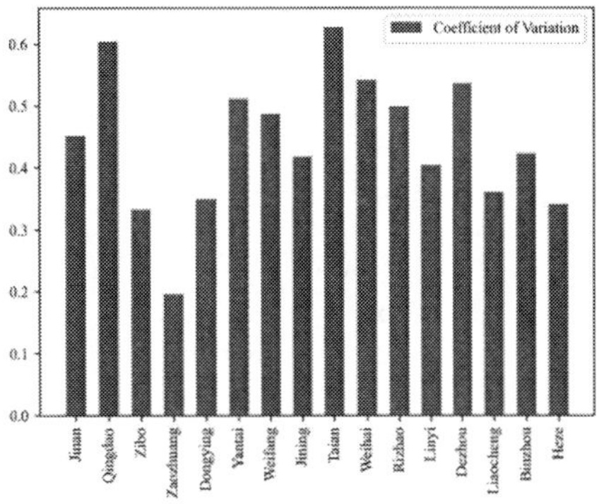

Figure 4. Variation coefficient of light brightness value in Shandong Province in 2019

3.3 Calculation of coefficient of variation

The coefficient of variation is used to express the comprehensive level of the dispersion degree and the average distance of the night light brightness value between the counties under the jurisdiction of the city, and it can eliminate the influence caused by the excessive scale, so as to better reflect the difference of the living standards of the residents between the counties. The calculation formula of coefficient of variation is as follows :

$$CV_{city} = \frac{\sqrt{\frac{\sum_{i=1}^{N}(PDN_{county\_i} - \overline{X})^2}{N}}}{\overline{X}} \quad (3)$$

In formula (3), N is the total number of counties under the jurisdiction of the city, PDN_{county_i} is the per capita light brightness value of a county in the city, and $\overline{X}$ is the average per capita light brightness value of all counties in the city.

In the variation coefficient histogram of night light value in Shandong Province in 2019 shown in Figure 4, the variation coefficient of counties under the jurisdiction of each city is above 0.15, which belongs to the moderate or strong variation level.According to this, the living standards of residents in the counties under the jurisdiction of each city are significantly different. Specifically, the difference in residents ' living standards among counties and districts in Zaozhuang City is relatively small. Tai an City has the largest coefficient of variation, and the coefficient of variation is about 0.63, indicating that the coefficient of variation of counties and districts in Shandong Province is seriously unbalanced.

3.4 Annual variation of coefficient of variation in cities

The variation of the coefficient of variation of nighttime light brightness values in 16 cities of Shandong Province from 2012 to 2019 is analyzed and sorted out. Through the comparative analysis of the annual variation of each city, it can be seen that except Tai ' an City, other cties are generally in a downward or fluctuating trend after experiencing the initial rise. To a certain extent, it shows that the difference in the living standards of residents between counties and districts under the jurisdiction of each city in Shandong Province is gradually decreasing or has been decreasing.

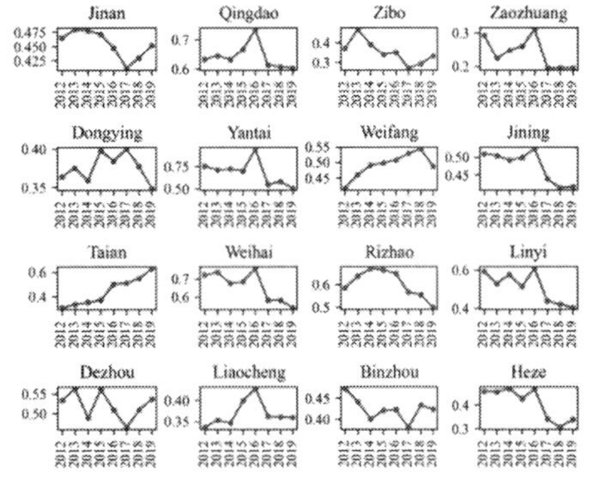

Figure 5. Analysis of variation coefficient of cities from 2012 to 2019

3.5 Correlation between annual variation index and GDP

The correlation between the coefficient of variation and GDP of each city from 2012 to 2019 is analyzed. According to R^2, it is between 0.3 and 0.4, indicating that there is a weak correlation between GDP and coefficient of variation. At the same time, according to the trend line, it can be seen that to a certain extent, the gap between GDP and residents' living standards is positively correlated. With the continuous improvement of residents' income level, the gap between residents ' living standards shows a trend of gradual expansion.

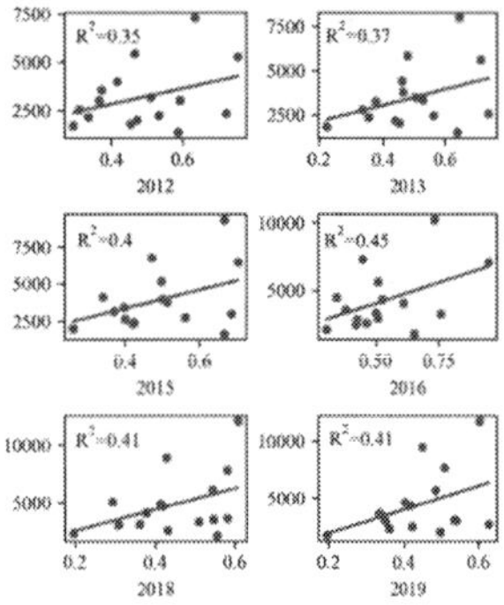

Figure 6. Correlation analysis between variation index and GDP in 2012-2019

4 Conclusion

With the help of VIIRS Nighttime Light (VNL) V2 series of nighttime light data products, the total nighttime light brightness and per capita light brightness of counties and districts in Shandong Province from 2012 to 2019 are counted through zoning statistics. Through the correlation analysis with GDP, it is determined that it can be used as a specific quantitative indicator to reflect the living standard of residents. After research and analysis, it is found that :

(1)The living standards of nighttime residents in Shandong Province are obviously unbalanced in the region, and the regions with high living standards are mainly concentrated in the central and coastal areas.

(2) There are significant differences in living standards between counties and districts under the jurisdiction of Shandong Province. Tai ' an City is the most obvious in 2019, and Zaozhuang City is the smallest.

(3) In 2012-2019, except Tai an City, Shandong Province, the difference in living standards has been growing, other cities showed a gradual narrowing or narrowing trend.

(4) There is a weak positive correlation between the living standards of residents in counties and districts of Shandong Province and the GDP of the city, that is, the higher the GDP of the city is, the more obvious the difference in living standards among residents is.

Through the evaluation and analysis of the differences in the quality of life of residents in Shandong Province, it is expected to provide some theoretical support for the resource allocation of Shandong provincial government and municipal governments, so as to reduce the gap in the living standards of residents between counties and achieve accurate support.

References

1. B. Yu, C. Wang, W. Gong, Z. Chen, K. Shi, B. Wu, Y. Hong, Q. Li, J. Wu, Nighttime light remote sensing and urban studies: Data, methods, applications, and prospects, Journal of Remote Sensing, **25**: 342–364, (2021).
2. C. D. Elvidge, K. E. Baugh, E. A. Kihn, H. W. Kroehl, E. R. Davis, and C. W. Davis, Relation between satellite observed visible-near infrared emissions, population, economic activity and electric power consumption, International Journal of Remote Sensing, **18**: 1373–1379, (1997)
3. W. Lu et al., Global proliferation of offshore gas flaring areas, Journal of Maps, **16**: 396–404, (2020)
4. T. Mazor et al., Can satellite-based night lights be used for conservation? The case of nesting sea turtles in the Mediterranean, Biological Conservation, **159**: 63–72, (2013)
5. S. E. Bauer, S. E. Wagner, J. Burch, R. Bayakly, and J. E. Vena, A case-referent study: light at night and breast cancer risk in Georgia, Int J Health Geogr, **12**:23, (2013).
6. Elvidge C D, Zhizhin M, Ghosh T, et al. Annual time series of global VIIRS nighttime lights derived from monthly averages: 2012 to 2019[J]. Remote Sensing, 13: 922, (2021).

 https://doi.org/10.1051/e3sconf/202129303028

Measurement and convergence analysis of agricultural green development level in China

He Hui[1], Wu Xiumin[1*]

[1]211 Huimin Road, Wenjiang District, Chengdu City, Sichuan Province, China

Abstract. Based on the panel data of 30 provinces and regions (except Taiwan, Hong Kong, Macao and Tibet) from 2010 to 2019, this paper measures the green development level of agriculture in China by using critic- entropy weight combined weight method, and then uses σ convergence and absolute β convergence methods to test its change trend. Based on the research conclusions, countermeasures and suggestions were put forward to make up for the shortcomings, promote the exchange and cooperation between provinces and regions, and narrow the differences between provinces, in order to provide reference for policy formulation and promote the green development of agriculture in China.

1 Introduction

Since the reform and opening up, China's agriculture has achieved rapid development, the output of agricultural products has increased steadily, agricultural production conditions have been greatly improved, agricultural structure has been continuously optimized, farmers' income has continued to increase, and great achievements have been made. However, with the rapid development of agriculture, the agricultural environment has become a "red light", the quality of agricultural products has touched the "bottom line", and the supply of high-quality and safe agricultural products is insufficient, which is difficult to meet the growing needs of people for a better life [1], which is the prominent expression of the major social contradiction in the agricultural field [1]. Therefore, it is of great significance to give full consideration to the cost of resources and environment in agricultural production, construct a reasonable evaluation index system of agricultural green development, and objectively evaluate and analyze the current situation of agricultural green development, which will gradually realize the transformation from traditional agriculture to green agriculture.

Green development contains the concept and thought of sustainable development. Since the concept of green development was extended to the field of agriculture, more and more studies have been conducted on the level of green development of agriculture. In terms of research content, scholars have constructed a series of different index systems to measure the level of agricultural green development according to their different understandings of the connotation of agricultural green development[116]. In terms of research methods, scholars mainly use principal component analysis, DPSIR model [6], interval intuitionistic fuzzy dynamic programming model[6], analytic hierarchy process [7][16], entropy value method [2], combined weight analysis method [9], entropy weight method [4], etc., to measure and evaluate the level of agricultural green development.

Because of this, most of the existing studies still remain on the scientific measurement of the level of green agricultural development, and lack of further internal mechanism analysis. Since Baumol[17]initiated the empirical study on the convergence of economic growth, convergence analysis has become a research hotspot and has been widely used in economic growth research [1819], productivity research [20] and energy and environment research [2122]. The green development of agriculture is closely related to resources and environment, but the research on convergence analysis in this area is still relatively lacking.

Based on this, this paper takes 30 provinces (Taiwan, Hong Kong, Macao and Tibet excluded) as the research object, and on the basis of measuring the level of agricultural green development, it further studies the convergence of agricultural green development level through σ convergence test and absolute β test, so as to enrich the research content of agricultural green development.

2 Research methods, index system construction and data sources

2.1 Research methods

2.1.1 Critic-Entropy weight method combined weight method

Some scholars find that critic is more comprehensive and objective after comparing various objective weighting methods [2324]. Critic method was proposed by Diakoulaki[25] in 1995. This method measures the objective weight of

1398451389qq.com

indicators by evaluating their comparative strength and conflict. However, it is found that although Critic method can comprehensively measure the strength of contrast and conflict among indicators, it cannot measure the degree of dispersion among indicators, and entropy weight method can make up for this deficiency [26]. Based on this, this paper constructs a combined weighting method based on critic and entropy weight method. The main calculation steps are as follows: firstly, the positive and negative indicators in the index system are standardized to eliminate the possible impact of the dimensionless inconsistency of the indicators; Secondly, weight is given respectively based on Critic method and entropy weight method. Thirdly, the comprehensive weight is determined. Finally, the comprehensive score of agricultural green development level was calculated according to the comprehensive weight.

2.1.2 Analysis of σ convergence

σ convergence means that the degree of dispersion of agricultural green development levels in different provinces decreases with the passage of time, that is, the gap of agricultural green development between provinces shrinks. The specific formula is as follows:

$$\delta_t = \sqrt{\frac{1}{n}\sum_{i=1}^{n}(\ln y_i - \frac{1}{n}\sum_{i=1}^{n}\ln y_{it})^2} \qquad (1)$$

Where, y_{it} is the level of agricultural green development in the i province in the t year, and n is the number of provinces and regions. If $\sigma_{t+1} < \sigma_t$, it indicates that there is σ convergence in agricultural green development of 30 provinces and regions in China; If $\sigma_{t+1} > \sigma_t$, it indicates that there is no σ convergence in agricultural green development of 30 provinces and regions in China.

2.1.3 Absolute β convergence analysis

Absolute β convergence is used to determine whether the provinces with low initial green agricultural development level will eventually converge with the provinces with high level of development over time, that is, the regions with low level of development tend to have a faster growth rate than the regions with high level of development, and the regions with low level of development have a "catch-up effect" approaching the regions with high level of development. Referring to the research of relevant scholars, the absolute β convergence model in this paper is as follows:

$$\frac{1}{T}\ln\left(X_{i,t} - X_{i,0}\right) = \alpha + \beta X_{i,0} + \varepsilon \qquad (2)$$

Whereinto, T is the research time span, represents the level of green agricultural development in I province in the initial year, and represents the average annual growth rate of green agricultural development in I province in the period of T. If $\beta<0$ and is statistically significant, it indicates that there is absolute β convergence in the green development level of agriculture, otherwise, it does not exist.

2.2 Construction of index system

Reference to relevant domestic scholars research achievements, follow the principle of scientific, comprehensive and data availability [27], the total index of green agricultural development level as the target layer, resource saving, output and high efficiency, environment friendly, green supply, rural development as a secondary indicators and 26 tertiary index, construct the index system (table 1).

Table1. Evaluation index system of agricultural green development level.

Category	Indicator	Direction
Saving resources (SR)	Cultivated land retention rate (%)	+
	Water consumption per unit area (m³/ha)	− +
	Coefficient of multiple cropping of land	−
Output efficiency (OE)	Per unit area yield of grain(kg/ha)	+
	Yield of fruits and vegetables(kg/ha)	+
	Number of green food labeling products per unit area(1/10000 hectares)	+
	Number of organic food labeled products per unit area(1/10000 hectares)	+
	Agricultural output value per unit area(ten thousand yuan/ha)	+
Environment friendly (EF)	Fertilizer application intensity(kg/ha)	−
	Pesticide application intensity(kg/ha)	−
	Agricultural film use intensity(kg/ha)	−
	Use intensity of agricultural diesel oil(kg/ha)	−
	Harmless disposal rate of garbage(%)	−
	Forest coverage(%)	+
	Prevention and control rate of forest pests and rodents(%)	−
Green supply (GS)	Market competitiveness of green food(%)	+
	Number of green food certification units	+
	Organic food market competitiveness(%)	+
	Number of organic food certification units	+
Rural development (RD)	Gross output value of agriculture, forestry, animal husbandry and fishery(One hundred million yuan)	+
	Total power of agricultural machinery per unit area(Kilowatts per hectare)	+

	Per capita disposable income of rural residents(yuan)	+
	Access rate of sanitary toilets in rural areas(%)	+
	Per capita agricultural GDP(Yuan/person)	+
	Rural economic status(%)	+

2.3 Data Sources

Based on the availability of data, this study covers 30 Chinese provinces (except Tibet, Hong Kong, Taiwan and Macao). The data are mainly from the National Bureau of Statistics of China, China Rural Statistical Yearbook (2010-2020), Statistical Bulletin of Green Food (2010-2019), Statistical Bulletin of Organic Food (2010-2019), and statistical yearbooks of 30 provinces and autonomous regions (2010-2020).

3 Agricultural green development level analysis

3.1 The national level

As shown in Table 2, from 2010 to 2019, China's agricultural green development level continued to improve, and the growth rate was stable. The score of China's comprehensive agricultural green development index increased from 0.241 in 2010 to 0.310 in 2019, a cumulative increase of 0.069, with an average annual growth rate of 2.40%-4.18%. According to the scores of the five second-level indicators, the ranking is environment-friendly, rural development, output efficiency, green supply and resource conservation. Therefore, green supply and resource conservation are the weak links in China's agricultural green development.

Table2. Comprehensive indicators and secondary indicators of agricultural green development in China from 2010 to 2019

year	AGD	SR	OE	EF	GS	RD
2010	0.241	0.028	0.037	0.086	0.03	0.059
2011	0.247	0.028	0.038	0.086	0.03	0.065
2012	0.257	0.029	0.039	0.089	0.029	0.071
2013	0.263	0.028	0.04	0.091	0.028	0.075
2014	0.274	0.029	0.041	0.095	0.03	0.079
2015	0.282	0.029	0.044	0.096	0.03	0.083
2016	0.285	0.029	0.045	0.096	0.031	0.083
2017	0.292	0.03	0.047	0.098	0.033	0.086
2018	0.299	0.03	0.048	0.096	0.036	0.089
2019	0.31	0.03	0.051	0.095	0.038	0.096

3.2 The provincial level

As shown in Table 3, there are significant differences in the level of agricultural green development between provinces and regions from 2010 to 2019. Shandong has the highest score in the comprehensive index (0.417), while Guizhou has the lowest score (0.154). Among the 30 provinces and autonomous regions, the top five and bottom five are "Shandong, Beijing, Zhejiang, Fujian and Jiangsu" and "Ningxia, Chongqing, Qinghai, Shanxi and Guizhou".

Among them, the top five provinces and autonomous regions are all located in the eastern economically developed regions and have good performance under the second-level indicators, but their weak links and driving factors are significantly different. In terms of core drivers, Shandong is represented by green supply and rural development, Beijing is represented by resource conservation and output efficiency, and Fujian is represented by environmental friendliness. In terms of key weaknesses, resource conservation is seen in Shandong province, while green supply and rural development are seen in Beijing.

The bottom five provinces and regions are mainly located in the economically underdeveloped regions in western China, with significant differences in driving factors and weak links. Among them, Chongqing, Shanxi and Guizhou have no obvious driving factor, while the core driving factor for Ningxia and Qinghai is resource conservation, ranking the 3rd and 6th place respectively. In terms of weak links, the core weaknesses of Ningxia and Qinghai were in environmental friendliness, while the core weaknesses of Chongqing, Shanxi and Guizhou were in resource conservation, rural development and high output efficiency, respectively.

Table3. ranking of comprehensive indicators of agricultural green development of provinces and autonomous regions from 2010 to 2019

Province	AGD	SR	OE	EF	GS	RD
Shandong	1	23	2	13	1	1
Beijing	2	2	1	4	19	17
Zhejiang	3	4	11	2	9	6
Fujian	4	5	12	1	13	8
Jiangsu	5	14	6	20	3	3
Heilongjiang	6	15	13	26	2	2
Hainan	7	7	14	3	30	7
Liaoning	8	11	5	7	11	12
Hebei	9	19	7	14	10	5
Guagdong	10	10	19	5	15	13
Hubei	11	25	15	16	4	9
Tianjin	12	9	4	11	27	16
Shanghai	13	13	3	6	20	24
Xinjiang	14	1	17	22	23	15
Hunan	15	29	20	15	6	14

Henan	16	30	8	21	21	4
Inner Mongolia	17	12	10	29	5	20
Ji lin	18	18	9	18	18	19
Anhui	19	27	16	19	7	18
Sichuan	20	21	22	23	8	11
Guangxi	21	16	28	12	22	10
Jiangxi	22	26	26	9	12	21
Yunnan	23	20	29	8	17	22
Gansu	24	8	25	24	16	27
Shaanxi	25	22	27	10	28	23
Ningxia	26	3	21	28	25	26
Chongqing	27	28	24	17	14	29
Qinghai	28	6	18	30	26	25
Shanxi	29	17	23	27	24	30
Guizhou	30	24	30	25	29	28

4 Analysis on convergence of agricultural green development level

4.1 σ convergence result

Based on the principle of σ convergence, this paper uses the coefficient of variation to measure the σ convergence of China's agricultural green development level. Can be seen from the figure 1, in 2010-2019, while the coefficient of variation in the part of the year, picked up a little bit, but from the overall, the variation coefficient of the development level of China's agricultural green present the downward trend, suggesting that China's agricultural green development level convergence, there exists obvious sigma namely green agricultural development level gap between provinces in narrowing.

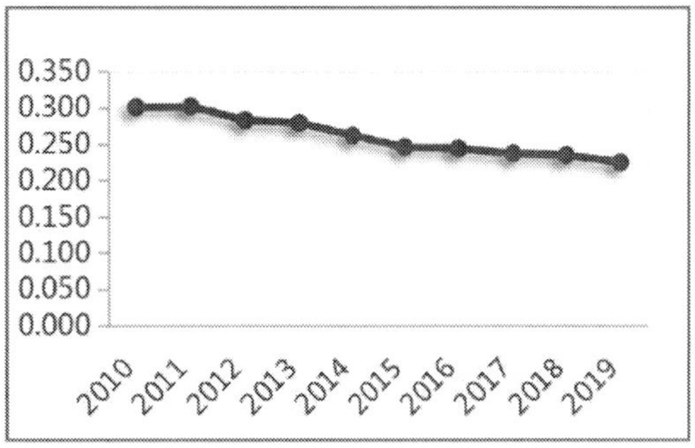

Figure 1. Coefficient of variation in the level of green agricultural development in China from 2010 to 2019.

4.2 Absolute β convergence results

According to the test results in Table 4, the regression coefficient β of the green development level of agriculture in the base period is <0 and significant at the 1% level, indicating that there is absolute β convergence in the green development level of agriculture. This means that there is a "catch-up effect" in the level of agricultural green development between 2010 and 2019, that is, the trend of low-level developing provinces catching up with high-level developing provinces.

Table4. Absolute β convergence test results of agricultural green development level

α	-0.29** (0.021)
β	-0.038*** (0.000)
R^2	0.454
The steady-state value ϒ	-0.2793834
Convergence rate λ	0.00387408
Whether the convergence	yes

Note: ***,**, * indicate significance at the level of 1%, 5% and 10%, respectively. The p value of each coefficient is shown in brackets.

5 Conclusions and Suggestions

5.1 the conclusion

1.Through the measurement of China's agricultural green development level and the provinces and autonomous regions, it is found that the overall level of agricultural green development in China shows a significant upward trend from 2010 to 2019. At the provincial level, there are obvious differences in the level of green agricultural development among provinces and regions. The regions with higher level of green agricultural development are mainly distributed in the eastern coastal areas and other developed regions, while the provinces and regions with lower level of development are mainly located in the western backward areas.

2.According to the analysis of the convergence model, there is not only significant σ convergence but also significant absolute β convergence in China's agricultural green development level from 2010 to 2019, which indicates that the gap of agricultural green development level among provinces and regions is gradually narrowing, that is, the provinces with lower development level catch up with the provinces with higher development level.

5.2 Suggestions

First, make up for their own shortcomings, implement the catch-up strategy, make use of the advanced technology, model and resource advantages of developed regions, form their own characteristics, and improve their own agricultural green development level. Second, promote inter-provincial exchanges and the further free flow of technologies, capital and information related to green agricultural development, so as to improve the overall level of green agricultural development.

References

1. F.W.Yu . Research on Motivation, Core and Countermeasures of Agricultural Green Development in the New Era [J]. Chinese Rural Economy,**05**:19-34(2018)
2. H.J.Zhao , F.W.Yu . Evaluation of agricultural green development level in major grain producing areas based on entropy method [J]. Reform,11:136-146(2019).
3. P.X.Jing , W.F.Zhao , Q. Zheng , D.H.Zhang . Research on dynamic prediction and path of agricultural green development level in Anhui Province [J]. Chinese Journal of Agricultural Resources and Regional Planning,39:51-56(2018).
4. J.Gao , Z.H.Ge . Analysis on regional difference and trend of agricultural green development level in Jiangsu Province [J]. Chinese Journal of Agricultural Resources and Regional Planning,41:14-22(2020).
5. Q.W. Gong , X.M.Li . Construction and measurement of agricultural green development index: 2005-2018 [J]. Reform,01:133-145(2020).
6. S.M.Jin. Evaluation of China's provincial green agricultural development level and regional differences [J]. Probe, 02 : 89-95(2019).
7. L Zhou . Study on green development of agriculture in Tibet under the background of rural revitalization [J]. Northwest Ethnic Studies,03:116-127(2019).
8. Q.WEI , B.Zhang, S.Q.Jin. Construction and regional comparison of China's agricultural green development index [J]. Issues in Agricultural Economy,11:11-20(2018).
9. L.L.Fu, X.H.Mao, B.B.Mao, H.T.Li. Study on the development of green agriculture in Zhejiang Province under the background of rural revitalization: Based on the perspective of comprehensive utilization of agricultural resources [J]. Chinese Journal of Agricultural Resources and Regional Planning,41:23-34(2020).
10. C.Y. Song, J.J.Zhang, L Liu, W.Q. Ma, L Ma, S. Ding, H.W. Zhao. Spatiotemporal changes of agricultural green development indicators in Hainan Island [J]. Chinese Journal of Eco-Agriculture,228:1156-1167(2020).
11. W.L Sun, R.b Wang, Q. Jiang, S.N. Huang. Study on connotation and evaluation of agricultural green development [J]. Chinese Journal of Agricultural Resources and Regional Planning,40:14-21(2019).
12. Y.Tian , J.B.Zhang. Research of Agricultural Modernization,34:85-89(2013).
13. N.M.Zhang , L.Zhang , H.Zhao , Y.C.Han , Y.H.Duan . Construction and application of agricultural green development evaluation index system [J]. Ecological Economy,34:21-24+46(2018).
14. Y.Z.Huang , X.F.Luo , Z.L.Li . Journal of China Agricultural University,22:183-190(2017).
15. S.J.Huang ,S.Y. Feng . Design and level measurement of agricultural green development indicators [J]. Ecological Economy,37:125-131(2021).
16. L.J.Wang , B.H.Jia, L.Y.Xin . Study on the driving factors of green agricultural development in Tianjin [J]. Chinese Journal of Agricultural Resources and Regional Planning,41:56-63(2020).
17. Baumol, W., Productivity, Growth, Convergence and Welfare:What the Long-run Da-ta Show[J], American Economic Review, 76, 1872～1885(1986).
18. Coulombe S., NewEvidence of Convergence Across Canadian Province:The Role of Urban-ization[J], Regional Studies, 34 , 713～725(1999).
19. Y.F.Lin , X.X.Liu , The Convergence of Economic Growth and Income Distribution in China [J], World Economy, 8 (2003).
20. Peng Guohua, The Analysis of Income Inequality, Total Factor Production and Its Convergence in China [J], Economic Research Journal, 9, (2005).
21. Strazicich, M.C., List, J.A., Are CO2emission levels converging among industrial countries-[J], Environmental and Resource Economics, 24, 263～271(2003).
22. Aldy, J.E., Per capita carbon dioxide emissions:convergence or divergence-[J], Environmental and Resource Economics, 33 , 533～555(2006).
23. K.Wang, H.Z.Song. Comparative analysis of three objective weighting methods [J]. Technological Economics and Management Research,06:48-49(2003).
24. C.Q.Tao , W.H.Chen, L.H.Lin . An Empirical Analysis of the Synergistic Degree of Industrial Organization Evolution in China: A Case Study of China's IT Industry under the Background of Enterprise Integration [J]. Management World,12:67-72(2007).
25. D.Diakoulaki, G.Mavrotas and L.Papayannakis, 1995, Determining Objective Weight in Multiple Crireria Problems:The CRITIC method", Computer Ops Res, 22:763～770.
26. Z.Wu , J.Guan , J.He . An Empirical Study on the Minimum Wage Standard: Dynamic Combination Calculation Based on CRIT-Entropy Weight Method [J]. Contemporary Economic Science,41:103-117(2019).
27. A.P.Zhang , Z.L.Peng , S.F.Wang , Y.Bai . The design principle and construction process of comprehensive evaluation index system [J]. Science Research Management,38:209-215(2017).

 https://doi.org/10.1051/e3sconf/202129303029

Flora of Seed Plants in Longquan Mountain, Wuhan

Huan Dai[1*], Rujie Xu[1], Xinyang Zhang[1], and Jing Cai[1]

[1]Wuhan Institute of Design and Sciences, Wuhan 430205, China

Abstract: In recent years, in order to provide theoretical and scientific basis for the protection and rational utilization of plant resources in Lon longquan mountain gquan Mountain, the seed plants in Longquan Mountain were taken as the research object, and the field investigation was carried out by using the method of line plus typical sample plot, so as to further study the characteristics of investigated area were further studied. The results show: There are 231 species of 191 genus in 90 families, including 5 families, 9 genus and 11 species of gymnosperms, and 82 species of 182 genus and 85 species of angiosperms. The 14 distribution types of seed plants belong to the statistics: North temperate distribution There are 72 genus, accounting for 37.7% of the total genus of the area, reflecting the obvious north temperate nature of the flora. In addition, there are many ancient taxa and Cretaceous and Tertiary relict plants. There are 163 genus of single genus and genus genus, accounting for 85.3% of the total genus of the region, indicating that the plant origin is relatively old, there are 7 genus in China, which indicates that the unique phenomenon exists, but it is not obvious.

1 Preface

Floristic geography takes flora as the research object, aiming to explore the origin, evolution, spatial and temporal distribution of flora species and their relationship with the changes of Earth's historical environment. The geographical composition of flora is the basis of vegetation restoration. The analysis of the geographical components of plant flora in a specific region can reflect the regional characteristics of biodiversity, which has an indispensable basic data value for protecting and improving the local ecological environment and formulating biodiversity planning. It plays an important role in clarifying the object and focus of urban biodiversity protection and promoting urban-rural integration in urban landscaping. Investigation and analysis of plant species in urban biodiversity conservation should be strengthened to provide basic technical support for formulating more scientific, targeted and operable urban biodiversity conservation planning[1].

2 Resarch area and method

2.1 Overview of the research area

Wuhan is a subtropical monsoon climate with abundant rainfall and sufficient sunshine. The annual average temperature is 15.8℃-17.5℃, the annual precipitation is 1100mm, the annual frost-free period is 240d, and the total annual sunshine duration is 2000h [2]. Longquan Mountain Scenic Area is located in Longquan Township, Jiangxia District, Wuhan City, Geographical coordinates 114°30'05.66" ~ 114°31'25.15", 30°25'03.00" ~ 30°24'17.69", altitude 232.5m.

2.2 Research content

In this study, Longquan Mountain in Wuhan City as the research object, the use of line survey and sample survey methods to investigate the species of seed plants in the region, and record the species of plants. According to the "Flora of China", the characteristics of plants in the survey area were analyzed and their lists were determined. According to the "Vegetation in China", the life forms of plants were divided as the standard. According to the ecological principles, the seed plants in Longquan Mountain were divided and their floristic characteristics were analyzed.

2.3 Investigation method

In the early stage, the vegetation of Longquan Mountain was investigated to determine the survey line and select the typical community to set up the survey sample plot. At the same time, the environmental factors such as longitude, latitude, altitude, slope and aspect of the sample plot were recorded. According to the minimum performance area method, which determine the sample area. Each survey sample area was 30m×20m, and 20 samples was investigated and recorded the species composition.

* Corresponding author: 191735729@qq.com

3 Results and analysis

3.1 Species composition analysis

The statistics of seed plants in Longquan Mountain was shown in Table 1. There are 231 species of seed plants belonging to 191 genus and 90 families, including 11 species of gymnosperms belonging to 9 genus and 5 families, and 220 species of angiosperms belonging to 182 genus and 85 families. Among them, 200 species of dicotyledons belonging to 160 genus and 77 families, and 20 species of monocotyledons belonging to 22 genus and 8 families.

Table1. Statistics of Seed Plants in Longquan Mountain

Item	gymnosperm	Angiosperm		Total
		Dicotyledon	Nonocotyledon	
Families	5	77	8	90
Genus	9	160	22	191
Species	11	200	20	231

3.2 Lifestyle analysis

The life form of plants is a kind of plant type reflected in the appearance of plants for long-term adaptation to comprehensive habitat conditions, which is a manifestation of long-term adaptation of plants to a certain living environment[3]. According to the standard, the seed plants in Longquan Mountain were divided into arbors, shrubs, sub-shrubs, vines, annual and biennial herbs and perennial herbs according to ecological principles[4].

There are 59 perennial herbaceous plants and 29 annual or biennial herbaceous plants, accounting for 25.5% and 12.6% of the total species, respectively, accounting for 38.1%. Woody plants include 63 species of trees, 64 species of shrubs and 1 species of subshrubs, accounting for 27.3%, 27.7% and 0.4 % of the total species, respectively, accounting for 55.4%. There are 15 species of vines, accounting for 6.5% of the total.

This data roughly reflects the diversity of life forms of seed plants in Longquan Mountain, in which woody plants are dominant.

3.3 Divisional analysis of the Section

3.3.1 Composition analysis of the Section

The composition of families, genus and species of seed plants in Longquan Mountain is shown in Table 2. There are 2 families with more than 15 species, 2 families with 10-14 larger species, and 3 families with 6-9 smaller species, accounting for 2.2 %, 2.2 % and 3.3 % of the total families in Longquan Mountain, respectively. There were 19 small families with 3–5 species, and 64 families with 1–2 relatively single species, accounting for 21.1 % and 71.1 % of the total families in Longquan Mountain, respectively.

Table2. Families, genus and species composition of seed plants in Longquan Mountain

Serial number	Numbers of species	Family
1	≥15 species (4 families)	Rosaceae, Leguminosae.
2	10~14 species (2 families)	Compositae, Gramineae.
3	6~9 species (3 families)	Caryophyllaceae, Polygonaceae and Fagaceae.
4	3~5 species (19 families)	Rutaceae, Lonicerae, Pineaceae,Ranunculaceae, Camphoraceae, Luteaceae, Ulmaceae, Aristolochaceae, Lilyaceae,etc.
5	1~2 species (64 families)	Moraceae, Caryophyllaceae, Verbenaceae, Rhododendronaceae, Lacqueraceae, Grapeaceae, Zoysia, Ornithaceae, Taxaceae, etc.

It is indicated that the minor section and the comparative single section play an important role in the survey area and are an important part of the composition of the survey area. There are 19 small families of 3 ~ 5 species and 64 single families of 1~2 species in Longquan Mountain, accounting for 92.2% of the total families. The above characteristics reflect the diversity of floristic elements in the survey area.

3.3.2 Statistical Analysis of the Advantage Section

Among the 90 families of plant flora in the investigated area, there are 7 families with more than 6 species, constituting the dominant families, accounting for only 7.8 % of the total families in the region, but the number of genus and species is 55 and 77, accounting for 28.8 % and 33.3 % of the total genus and species, respectively. It is indicated that the seed plant composition of Longquan Mountain is concentrated in these dominant families, which constitute the main species composition of the investigated flora.

3.3.3 Floristic analysis of areal~types of family

The 90 families of seed plants in Longquan Mountain are divided into 13 types, 42 families in the north temperate zone and 21 families in the pantropical zone, accounting for 46.7% and 23.3% of the families in the survey area, respectively. World distribution of 15 families, tropical Asia distribution of 13 families, East Asia and North America intermittent distribution of 9 families, East Asia distribution of 8 families, tropical Asia and tropical Oceania distribution of 5 families, accounting for 16.7%, 14.4%, 10.0%, 8.9%, 5.6% of the number of families in this area. The tropical distribution in the old world, tropical Asia to tropical Africa, the Mediterranean region and Central Asia to West Asia are 3 families, accounting for 3.3% of the total number of families in the region, there are 7 families endemic to China, accounting for 7.8%.

Table3. Distribution types of seed plant families in Longquan Mountain

Serial number	Distribution areas and subtypes	Number of District Sections	Proportion of sections in the region /%
1	World distribution	15	16.7
2	Pantropical distribution	21	23.3
3	Tropical distribution in the old world	3	3.3
4	Tropical Asia and tropical Oceania distribution	5	5.6
5	Distribution of tropical Asia to tropical Africa	3	3.3
6	Tropical Asia distribution	13	14.4
7	North temperate distribution	42	46.7
8	Intermittent distribution of south and north temperate zones	2	2.2
9	Intermittent distribution of East Asia and North America	9	10.0
10	Old world temperate distribution	4	4.4
11	Distribution of Mediterranean, Central and Western Asia	3	3.3
12	E. asia flora	8	8.9
13	China - specific distribution	7	7.8
Total		77	100

Table3 shows that the largest number of families in the north temperate zone is 42 families, accounting for 46.7%, which fully shows the nature of the north temperate flora of Longquan Mountain. Pantropical distribution of 21 families, considerable proportion of 23.3%. There are also Chinese endemic families in the survey area, such as Sapinaceae, Ginkgoaceae, Leguminosae and Rutaceae, Eucommiaceae, indicating that the endemic phenomenon in the survey area is not obvious from the perspective of families.

3.3.4 Analysis of genus flora

The flora of seed plants in Longquan Mountain contains 4 or more genus, including *Rosa*, *Lespedeza*, *Pinus* and *Quercus*. There are 3 species of medium genus, such as *Rubus*, *Polygonum*, *Photinia*. There are 21 genus of small genus including 2 species, such as *Pentaphyllum*, *Compositae*, *Chrysanthemum*, etc. There are 163 genus including one species, such as *Rancidum*, *Lycopersicon* and *Torilis*, the grouping list is shown in Table4.

Table4. List of Genus of Seed Plants in Longquan Mountain

Serial number	Number of familes	Number of genus	Proportion of total species /%	Number of species	Proportion of total species /%
1	≥4	4	2.1	18	7.8
2	3	3	1.6	6	2.6
3	2	21	11.0	44	19.0
4	1	163	85.3	163	70.6
Total		191	100	231	100

The 4 species and more than 4 species, accounting for 2.1 % of the total number of genus, but the number of species accounted for 7.8 % of the total number of species, indicating that these 4 genus are the dominant genus in the flora. Among the four dominant genus, Rosa, Pine and Quercus were distributed in the north temperate zone, and Lespedeza was distributed intermittently in East Asia and North America, indicating that the park had obvious north temperate nature. Analysis of the origin of the flora shows that: 1 species of gymnosperms belonging to 9 genus and 5 families in this area, including *Pinus* and *Platycladus*, are ancient groups in seed plants, the angiosperms of Magnoliaceae, Ranunculaceae. Berberaceae, Fagaceae, Ulmaceae, Mulaceae, Aceraceae and other plants all appeared in Cretaceous or Tertiary, and the monospecies and oligospecies in this area were 163 genus, accounting for 85.3 % of the total genus. These species and structures reflect the relatively ancient origin of flora in the region.

The distribution type of genus can more specifically reflect the differentiation and geographical characteristics of the plant community in the process of evolution. The geographical distribution type of genus represents the geographical components of genus, which, to a certain extent, contains the historical development process of various floras contained in it, and is the basis and symbol the division of flora[6]. According to Professor Wu Zhengyi ' s classification of seed plants in China, 191 genus of seed plants in Longquan Mountain were divided into 14 distribution types. Table 5 shows that the distribution genus of the North Temperate Zone accounted for the largest proportion, 72 genus, accounting for 37.7 % of the total genus in this area. the second is pantropical distribution, with 34 genus, accounting for 17.8 % of the total genus in the region. Again is the world distribution, there are 21 genus, accounting for 11.0 % of the total number of genus. There are 79 genus of seed plants in the temperate zone, accounting for 41.4 % of the total genus.

There are 61 genus of tropical nature, accounting for 31.9 % of the total genus in this area. It shows that the flora of seed plants in Longquan Mountain shows obvious temperate characteristics. The floristic relationship between the genus of seed plants in Longquan Mountain and the genus of plants in other parts of the world : In the nature of the temperate zone, the distribution of the northern temperate zone is 37.7 %, the most closely related is the intermittent distribution of East Asia and North America 5.2 %, the distribution of East Asia 5.2 %, followed by the distribution of the old world temperate 2.6 %, and the relationship with other distribution types is poor. In the tropical nature, it is closely related to 17.8 % of pan-tropical distribution, followed by 7.3 % of tropical Asia distribution, and other distribution types are relatively poor. Analysis of the origin and endemic components of the flora : There are 7 genus and 7 species endemic to China in the flora of Longquan Mountain, which are Koelreuteria, Toona, Ginkgo, Horseshoe, Honeybean, Citrus, Eucommia, accounting for 3.7 % of the total number of genus in this area. The Chinese endemic genus in Longquan Mountain account for 22.6 % of the 31 Chinese endemic genus in Central China, and 2.7 % of the 260 Chinese endemic genus, indicating that the endemic phenomena in this area occupy a certain proportion but not obvious.

Table5. The distribution type and species number of seed plants in Longquan Mountain

Serial number	Distribution type	Numbers of genus	Proportion of total genus /%
1	World distribution	21	11.0
2	Pantropical distribution	34	17.8
3	Tropical distribution in the old world	3	1.6
4	Tropical Asia and tropical Oceania distribution	2	1.0
5	Distribution of tropical Asia to tropical Africa	5	2.6
6	Tropical Asia distribution	3	1.6
7	tropical asia	14	7.3
8	North temperate distribution	72	37.7
9	Intermittent distribution of south and north temperate zones	2	1.0
10	Intermittent distribution of East Asia and North America	10	5.2
11	Old world temperate distribution	5	2.6
12	Distribution of Mediterranean, Central and Western Asia	3	1.6
13	E. asia flora	10	5.2
14	China - specific distribution	7	3.7
Total		0	100

4 Conclusions and discussion

Geographical composition analysis of genus and species in the study of flora, the distribution type of genus can reflect the differentiation and geographical characteristics of the flora in the evolution process. In terms of floristic composition, the life forms of seed plants are diverse, woody plants are developed and rich in species, which have obvious characteristics of northern temperate vegetation. In terms of geographical components, the North Temperate Zone is the most widely distributed, with 72 genus, accounting for 37.7 % of the total number of genus in the region, and the most closely related are the intermittent distribution and East Asian distribution of East Asia and North America, which account for 5.2 % of the total number of genus in the region, reflecting the obvious nature of the North Temperate Zone. In terms of the origin characteristics and endemic phenomena, seed plants are relatively old in origin, and there are many ancient groups and relic plants of Cretaceous and Tertiary. For example, some species of gymnosperms, angiosperms, such as Magnoliaceae, Ranunculaceae, Berberaceae, Fagaceae and Ulmaceae in this area, and the proportion of monotypic and oligotypic genus in this area is large. There are 7 species of endemic genus in China, and endemic phenomena exist, but not obvious.

Plants with poor adaptability and high ornamental value should be introduced gradually, which can be first introduced to the nursery or botanical garden for domestication, and then widely applied. For some precious and rare ornamental plants, they should be preserved and expanded by modern technology, and then developed and utilized to avoid species extinction[6-7]. All in all, wild ornamental plants should be rationally utilized and developed to serve urban landscaping so as to maximize their value.

Acknowledgments

This research was financially supported by the Scientific Resrarch Project of Wuhan Institute of Design and Sciences (K201915), the Project of Excellent Course (201308), the Project of Teaching Research(2020JY104 and 2019JY111), and the Teaching Resarch Project of Hubei Provincal Universities (2018496 and 2017505).

References

1. *P.* Wei, *Y.* Guo, *J.* Bai, *Y.* Zhang, J. Guo, JSAS, **46**, 4, 606-611(2018)
2. *B.* Zhang, IMF, **12**, 26-27(2014)
3. *X.* Liu，*Y.* Zhang，*Y.* Wang，*H.* Yang, *Z.* Ma, JGAU, **53**, 5, 118-126(2018)
4. G. Li, L. Gao, Y. Zhang, *N.* Yang, FES, **33**, 4, 387-394(2018)
5. *Z.* Wu, *H.* Sun, *Z.* Zhou, *D.* Li, *H.* Peng, SP, (2011)
6. *D.* Tang, *Y.* Wu, *W.* Liu, *D.* Li, *G.* Yang, *Q.* Chen, *T.* Zhang, APS, 36, 05, 658-666(2018)
7. *B.* Bai, *X.* Wang, *Y.* Huang, *M.* Chen, JANU(NS)43, 05, 459-466(2020)

Study on Sustainable Development of Rural Revitalization Based on Slow Village Concept

Ruipeng Zhang[1,*], Junwei Huo[1], Changquan Su[1], Yingqiang Xu[1], Yiyao Zhong[1], and Xin Meng[1]

[1]College of Architecture and Art, Hefei University of Technology, Hefei, China

Abstract. The slow village is the focus of rural development. Under the drive of the slow village, the village will achieve comprehensive economic, social and cultural revitalization and become a sustainable rural development structure in the new era. Based on the concept of "slow village", this article combines the author's rural planning proposal of Changji Village in Fanchang County to examine whether the development idea of "slow village" can promote rural revitalization. The article analyzes and researches the three aspects of production, living and ecological space in order to explore a development model of rural revitalization based on the "slow village", and provide ideas for the revitalization and ecological sustainable development of similar villages.

1 Introduction

The Urban and Rural Planning Law, which came into effect on January 1, 2008, includes rural planning in the scope of adjustment, divides rural planning into town-ship planning and village planning, and provides for the formulation, implementation, modification, supervision and inspection, and legal responsibility of rural planning, providing a legal basis for the systematic study of rural planning. In April 2018, the Ministry of Natural Resources was formally established, and the nationwide "multi-regulation" pilot task of more than four years was basically completed, marking the construction of The conditions are ripe for the construction of a national spatial planning system, the strengthening of use control, and the comprehensive development of national spatial planning. As an important part of the territorial spatial planning, village planning is also an important prere-quisite for the national 14th Five-Year Plan to vigorously promote the revitalization of the countryside [1].

Since 1999, when the Slow City Movement Charter was published, the Slow City movement has emerged worldwide. Since the 1990s, the rapid growth of cities has been accompanied by problems such as noise, air pollution, indifference and lack of affection between people. In 1999, the small Italian city of Bula, with only 10,000 people, proposed a new urban model that would combine modern technology with traditional lifestyles in a modern city, so that people could not only enjoy the many conveniences of modern life, but also emphasize the importance of bringing a regular and healthy lifestyle to everyone. The current fast-paced urban life reflects the confusion and disorientation brought by the high-speed life in the context of rapid urbanization [2]. With the continuous development of the slow city model, this ecological and natural development model is gradually extended to the countryside, evolving into the "slow village" concept. This will be a kind of back to life itself, people-oriented village model. It advocates people to return to the simple and pleasant living condition of the countryside.

2 Basic information about the study area

Changsi Village is located in the northwest of Suncun Town, surrounded by mountains on all sides, connecting Jinling in the east, Changlong in the south, Huanghu in the west, and Digang in the north. In ancient times, the mountain on the east was called Changxingling, and at the foot of the mountain on the west, there were Langling Temple and the old residence of Cui Cong, the official of the Ming Dynasty, and Guiyangchong at the back was also called Sichong. So it was later collectively called Changsi.

The total area of the village is 6.4 square kilometers, with 8 natural villages and 11 village groups under its jurisdiction. The village has a resident population of 500 people, most of whom are over 65 years old, and most of them are located in the center of the village of Old House Base. The village is mainly planting daily vegetables and rough processing of moso bamboo, producing moso bamboo, timber, tea and rich iron ore resources. In recent years, the leading industry of Changji Village has devel-oped into forestry and tourism, which accounts for more than 80% of the income [3].

The construction of Laoyaki central village insists on the overall goal of "ecologically pleasant village beauty, prosperous industry and people's life beauty, civilized and harmonious country style beauty", fully explores the local characteristics, and strives to create a pleasant living and happy Changji. However, to sum up, the following problems still exist in the development of Changji Village:

* Corresponding author: zhangrphfut@163.com

① the young and strong labor force is out of town, the hollowing out of the countryside is serious; ② the traditional industries account for a high proportion, the economic development is poor; ③ the infrastructure and public service facilities are weak, the living environment needs to be improved; ④ the rural tourism is only a simple copy of the development, the management is sloppy; ⑤ the depth of resource mining is not deep enough, it is difficult to drive economic development [4].

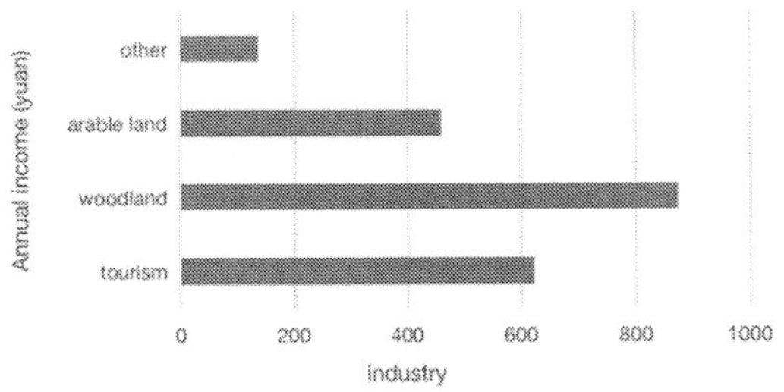

Figure 1. Per capita income of leading industries in Changsi Village.

3 The point of convergence between the concept of slow village and Changsi village

3.1 The ecological background of Changsi Village is sensitive and it is surrounded by mountains and rivers

The natural environment in Changsi Village is composed of mountains and waters. There are two reservoirs in the old house foundation, one is called Tongkou Reservoir and the other is called Wanchong Reservoir. The water is pure and green as blue. Based on the current situation of surrounding mountains and rivers, Changsi Village has the characteristics of sensitive ecological background, and one of the ideal visions that Man Village wants to create is a good ecological environment [5].

3.2 Changsi Village is not suitable for large-scale development, and the industry is lagging behind

Due to the sensitive nature of the ecological background in Changsi Village, the village is not suitable for large-scale development, and some of the original factories are mostly shut down due to environmental pollution. The industry in Changsi Village is currently lagging behind. The Man Village emphasizes the ecological, low-carbon, and environmental protection of the industry, which is in line with the characteristics of Changsi Village.

3.3 Changsi Village's industry lags behind, the population is lost, and the village is hollow

Due to the lagging industry in Changsi Village, a large number of young and middle-aged laborers go out to work, which in turn leads to the loss of the village popul-ation and the village becomes hollow. The Slow Village emphasizes slow life, returns to the simplest living conditions, and attracts foreign populations in this state, which in turn drives the development of the industry.

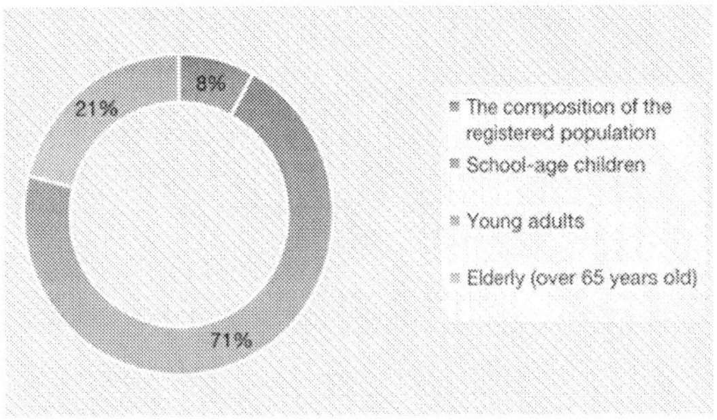

Figure 2. The composition of the registered population.

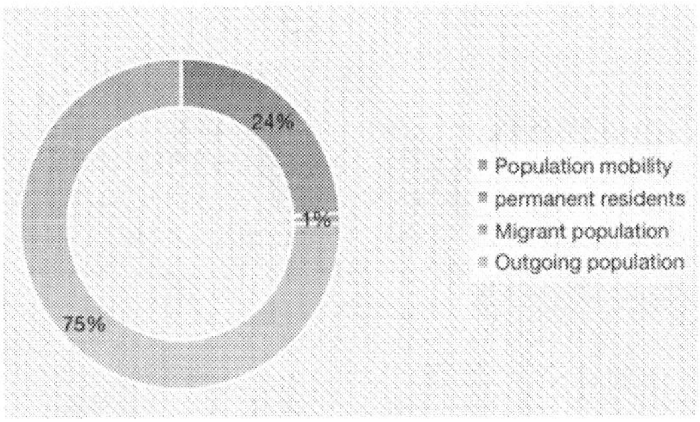

Figure 3. Population mobility.

4 The development model of rural revitali-zation under the slow village concept

4.1 "Slow" promotes production and bamboo art inheritance-keeping the nature's essence and creating the industry of the slow village

Under the concept of Slow Village, the planning concept of the production space of Changsa Village is based on the current situation of ecologically sensitive villages, respecting the law of village development under the keynote of "Slow Village", shutting down mining and polluting factories for transformation, returning to traditi-onal ecological industries, re-understanding the potential value of traditional ecological industries, optimizing industrial quality and added value of products. On the one hand, we optimize the industrial structure, take the third industry such as bamboo culture eco-tourism, B&B and bamboo weaving e-commerce as the core, drive the optimization of the second industry such as moso bamboo deep processing crafts and bamboo weaving skills, and promote the development of the first industry such as moso bamboo breeding and poultry breeding. On the other hand, we will create a special industry integrating moso bamboo planting, deep processing, bamboo culture display and experience, ecological tourism and leisure, and form an industrial chain from moso bamboo planting to bamboo art deep processing and then to bamboo series products selling [6]. The planning strategy includes the following three aspects.

4.1.1 Improve the first production

Firstly, enrich bamboo species and expand planting. Seek professional guidance from forestry personnel, introduce more suitable bamboo species and enrich the diversity of bamboo resources. Expand the planting area, increase the scale, and provide guarantee for the development of second and third production; on the other hand, improve the protection of bamboo resources and scientific management. Due to the short and cheap growth cycle of moso bamboo, the management of bamboo resources has always been under-appreciated. Scientific planting and management are especially important for sustainable development and tourism landscape shaping.

4.1.2 Retreat two into three

First of all, ecological restoration of abandoned mines and full retreat of polluting industries. The ecological restoration of the abandoned mine, the full retreat of the village's polluting industries, the transformation of the original site, combined with the bamboo handicraft industry, to create a bamboo cultural creative handicraft park and e-commerce storage space; on the other hand, the inheritance of bamboo culture, the development of collective economy. The village has a long history of bamboo art culture, but due to the loss of personnel, it is difficult to develop, create a "micro factory" to help the poor handicraft industry, provide entrepreneurial opportunities to attract the return of personnel, while establishing bamboo art deep processing industry.

4.1.3 Upgrading of three industries

First of all, we will develop bamboo culture and slow life ecological tourism. Create bamboo culture display experience, ecological tourism leisure tourism industry, in line with the village terrain, to create from the city to Changji, from fast to slow into the Zen forest of the secluded slow life experience, to find inner peace; on the other hand, combined with the advantages of the location of the development of bamboo art e-commerce, to enhance industrial toughness. Changji village is a short drive from the town center, and is at the junction of two cities, with convenient transportation. It is suitable for the development of e-commerce for the industry with high added value like bamboo art, which is a certain complement to the tourism industry and improves the toughness of the industry.

4.2 "Enjoy life slowly, live poetically - find the source of the slow village, shape the life of the slow village

Under the concept of Slow Village, the planning idea of the living space of Changji Village is to alleviate the problems caused by "fast-paced urban life" and "industrialization", so that local residents and visitors can gradually find inner peace and form a comfortable and comfortable rural lifestyle. From the aspect of villagers' life, the establishment of rural community institutions, the creation of rural community, enhance the sense of belonging of the residents; from the aspect of tourists' life, shape the slow life scene of Changji, provide slow life leisure places, enhance the sense of identity of tourists. The planning strategy includes the following two aspects.

4.2.1 Village life

Introduce the governance model of urban neighborhood committees, strengthen the administrative functions of villagers' committees to manage village affairs, and establish separate "neighborhood committees" to manage people's affairs. The committee will be responsible for the management of the village affairs, and will build a community of villages, each with its own duties, to reshape the local relationship. On the other hand, we need to improve medical facilities, popularize medical knowledge, enhance the resilience of villages, and build healthy villages [7].

4.2.2 Visitor life

Shape the slow village scene together from many aspects such as slow food, slow walking, slow living and slow touring, so that visitors can experience the slow life of the countryside and mend the traditional village life affected by the modern fast-paced life.

4.3 "Slow" life tone, green slow village - follow the theory of two mountains, subject to the ecology of slow village

Under the concept of slow village, the planning idea of ecological space of Changji Village is ① ecological priority and green development; ② establish ecological sensitive control system; ③ create ecological transportation slow system; ④ educate and popularize ecological knowledge. We will build an ecological security pattern with integrity network, establish a nature reserve system with strong protection, increase the supply of high-quality ecological products, and systematically and proactively implement region-wide ecological restora-tion. The planning strategy includes the following four aspects.

4.3.1 Ecological priority and green development

Adhere to the development strategy of giving priority to ecology, respect the original natural conditions and protect the existing landscape pattern of villages. To build a modernization in which people and nature coexist harmoniously, we must put the protection of urban ecological environment in a more prominent position, plan the production space, living space and ecological space of the city scientifically and reasonably, handle the relationship between urban production and living and ecological environmental protection, and improve both the quality of economic development and the quality of people's life.

4.3.2 Establish an ecologically sensitive control system

Establish an environmental access system for ecologically sensitive areas, production areas should be far away from ecologically sensitive areas, and the pace of visits from entrances to ecologically sensitive areas is from fast to slow. Based on the boundaries of various prohibited development zones, ecological sensitivity analysis, and ecological service value evaluation results, the Boolean algorithm is used to superimpose prohibited development zones, ecologically sensitive areas, and areas with high ecological service value. Comprehensively consider the needs of urban development and construction and basic farmland protection land, and areas located in prohibited development zones and ecologically sensitive areas or high ecological service value areas are designated as ecological protection red lines.

4.3.3 Create an ecological transportation slow walking system

Urban slow walking system is slow transportation, which means that walking, bicycle, bus and other slow travel modes are taken as the main body of urban transportation, effectively solving the problems of fast and slow traffic conflicts and difficulties in traveling for slow moving subjects. This requires the community to build a complete walking and bicycle system, create a convenient and continuous walking and bicycle network, maintain the right-of-way for walking and bicycling, and optimize street functions. In addition, we should also guide the integrated transportation mode of "public transportation+slow walking" to improve the living environment and quality of life and promote a healthy life, while meeting the growing diversified needs of residents for commuting, sports and fitness, and leisure and tourism.

4.3.4 Education and popularization of ecological knowledge

Create ecological education bases and nature classes to popularize ecological knowledge and raise the ecological protection awareness of local villagers as well as foreign visitors. Ecological ethics is not only an important part of ecological civilization, but also a reliance and spiritual source of ecological civilization. Ecological ethics education in schools is an inevitable requirement for the country to achieve sustainable development. Ecological ethics education is an important part of ideological and moral education and an important aspect of measuring students' basic moral quality. Therefore, it is necessary to strengthen ecological moral education, cultivate people's correct understanding of the relationship between human beings and nature, enhance their sense of morality and responsibility towards nature, develop a sense of caring for the natural environment and ecological protection, and consciously observe the corresponding moral habits.

5 Conclusion

Based on the concept of "slow village", this study takes Changji Village in Fanchang County as the study area and explores its rural planning scheme. In the past few decades, backward villages around the world have been in a hurry to develop their economy, and have been built in a drastic manner, which has caused various problems such as the destruction of village ecology in a thousand villages. The concept of "slow village" is a reflection of the current background of ecology first, in order to let the village development and protection in parallel. However, the development status of each village is different, and rural planning should be tailored to local conditions in order to promote rural revitalization. It is worthwhile to further explore which villages can be covered by such a concept and which types of villages are suitable for development.

References

1. Liu Y. Basic theory and methodology of rural revitalization planning in China[J]. Journal of Geography, 2020, 75(06): 1120-1133.
2. Yao S, Zhou S. Research on the common construction and sharing path of rural revitalization[J]. China Rural Economy, 2020(02): 14-29.
3. Liu Y, Zhou Y, Li Y. China's rural territorial system and rural revitalization strategy[J]. Journal of Geography, 2019, 74(12): 2511-2528.
4. Zhang C, Chen M, Shao M. From slow city to slow village-a beautiful countryside development model [J]. Rural Planning and Construction, 2014(01): 23-32.
5. Su X, Wen T. Eco-agriculture and Slow Village, Slow Food, Slow Life: An Interview with Professor Wen Tiejun, Executive Director of the Institute for Advanced Studies in Sustainable Development, Renmin University of China[J]. Journal of Nanning Vocational and Technical College,2019, 24(01):101-104.
6. Xia S. Talking about the understanding of "international slow village construction" [J]. Shanxi Architecture, 2020, 46(03): 15-17.
7. Gao H, Zhang J, Luo Z. Revival or Alienation ? Spatial Transformation of Rural Villages on the Edge of the Metropolis Driven by Consumer Culture - An Empirical Observation of Dashan Village in Gaochun International Slow City [J]. International Urban Planning, 2014, 29(01): 68-73.

Research on Green Energy Internet Planning Model

Kan Pan[1], Anrui Li[2], Zheng Yang[1], Yongjie Nie[1], Zhenwei Geng[3], Junjie Wen[2], Wentao Liu[2], Shichen Wang[2], Hui Chen[2] and Zuyuan Huang[3,*]

[1] Electric Power Research Institute, Yunnan Power Grid Co. Ltd, Kunming, China
[2] Key Laboratory in Software Engineering of Yunnan Province, School of Software, Yunnan University, China
[3] Information Center , Yunnan Power Grid Co. Ltd, Kunming, China

Abstract. This paper intends to take power system planning as the main module, while considering the requirements and constraints of natural gas network and transportation network construction and use the energy hub model to analyse the energy transmission and transformation relationship between different networks, to realize the power system, natural gas network, and transportation. Integrated energy system planning for the network. To verify the effectiveness of the proposed PMIES planning method, an improved Garver test system is used for simulation. The system includes a 7-node natural gas network system and a 6-node electric test system. We optimized and analysed the simulated system for model research.

1 Introduction

The integrated energy system is based on a tightly coupled multi-dimensional energy physical system [1], through organic interaction and coordination and optimization of energy generation, distribution, conversion, and consumption, to achieve two-way on-demand transmission of energy [2] and dynamic balanced use [3], thus forming an integrated system of energy production, supply, and marketing [4]. The core of the multi-dimensional energy physical system is the electric power system, coupled with other energy systems such as the natural gas network system and the transportation network system, and its system scope is highly extensible, which can be system-level or regional-level. With the development of the Energy Internet, the distributed system structure will make the energy system more decentralized [5], so the scope of the system studied in this paper is limited to the terminal multi-energy self-balancing unit, namely SUME. As the proportion of distributed gas generating units and energy storage units in the system continues to increase, the power system's requirements for natural gas networks and transportation networks will also continue to increase. In the planning of integrated energy systems, the combination of power systems, natural gas networks, and transportation networks will continue to increase. Coordinated planning is an important trend. [6]

The basic idea of modelling is: the interface between the transportation network and the power system is electric vehicles and charging devices. According to the number of electric vehicles and the law of charging and discharging behaviour, the simulated charging and discharging load demand and the quota power of the charging device are key input parameters for power system optimization planning; combined with the optimization planning results of the transportation network, the power system optimization model incorporates transportation network parameters and constraints, and performs power supply and grid planning, and clarifies the configuration plan of power components; feeds back the optimization configuration plan of the power system to Transportation network [7], observe whether it meets its model requirements and actual conditions and make corresponding adjustments. At the same time, input the configuration schemes of power system gas generators, DCHP and other components into the natural gas network optimization planning model, plan the gas pipelines, P2G and other facilities; feedback the results of natural gas network optimization planning to the power system, and observe whether it meets its model requirements and reality. If all constraints are met, PMIES realizes integrated energy system planning [8-9].

2 NATURAL GAS NETWORK FACILITY INVESTMENT PLANNING MODEL

The supply and demand network of natural gas can be divided into natural gas supply, natural gas transmission, and natural gas load. Among them, the supply sources of natural gas mainly include the regasification of liquefied natural gas at storage stations and pipeline transportation of natural gas stations; the transportation channels of natural gas mainly include gas pipelines and large tankers or ships; natural gas loads include industrial and commercial applications, chemical raw materials, and gas power generation. The role of a natural gas storage station is like the "peak-shaving and valley-filling" of electricity. It stores natural gas when the load is low and releases the natural gas from the storage station to the market when the load is high. In addition, it also plays a certain role. The

* Corresponding author: 20182868@qq.com

role of strategic reserves. SUME is a small area system. From the perspective of system economy, it does not have the conditions to configure large storage tanks. Small storage tanks have limited impact on the cost of gas power generation or gas heating. Therefore, this model only considers the gas pipeline network and P2G costs. The gas storage facility is assumed to already exist at each node. Since the multi-energy system is planned together with the power system, the natural gas load mainly considers the natural gas heat load and gas power generation load.

To realize the mutual conversion of energy in SUME, SUME is equipped with a distributed combined heat and power (DCHP) system and power to gas (P2G) facility. To clearly define the boundaries of investment accounting, suppose DCHP is a power generation equipment, which is separately developed in the power system planning model. Therefore, the total cost function of natural gas-related facilities in the natural gas network planning model is:

$$\begin{aligned}
TC^{gas} &= IC^{gas} + OC^{gas} - Z^{gas} \\
&= \sum_{t\in T}\left(\sum_{n\in N} IC^{P2G}\cdot \Delta P_{n,t}^{P2G}\right)(1+r)^{-t} \\
&+ \sum_{t\in T}\left(\sum_{n,m\in N} IC^{pipe}\cdot l_{n\leftrightarrow m}^{gas}\cdot z_{n\leftrightarrow m,t}^{pipe}\right)(1+r)^{-t} \\
&+ \sum_{t\in T}\left(\sum_{n\in N} OC^{P2G}\cdot U_{n,t}^{P2G}\right)(1+r)^{-t} \\
&+ \frac{(1+r)^{t}-1}{r\cdot(1+r)^{t}}\cdot OC^{pipe} \\
&+ \sum_{t\in T}\left(p^{gas}\cdot U_{t}^{net}\right)(1+r)^{-t} \\
&+ \sum_{t\in T}\left(C_{cut}^{gas}\cdot U_{t}^{cut}\right)(1+r)^{-t} \\
&- \sum_{t\in T}\left\{\frac{IC_{t}^{P2G}}{T^{P2G}}\left[T^{P2G}-(T-t)\right]\right. \\
&\left. + \frac{IC_{t}^{pipe}}{T^{pipe}}\left[T^{pipe}-(T-t)\right]\right\} \qquad (1)
\end{aligned}$$

Where: TC^{gas} represents the total cost of the natural gas network during the planning period; IC^{gas} represents the total investment cost during the natural gas network planning period; OC^{gas} represents the total variable cost of the natural gas network; Z^{gas} Represents the residual value of the investment equipment during the planning period at the end of the planning period; IC_{t}^{P2G} and IC_{t}^{pipe} respectively represent the total investment in electricity-to-gas equipment and pipelines in year t; T^{P2G} and T^{pipe} respectively represent the life period of electricity-to-gas equipment and pipelines; T represents the set of planning period years, t represents the t-th planning year; N represents the set of natural gas network nodes, including all system nodes through which natural gas passes, and n represents the n-th natural gas network Node; r represents the discount rate; $IC_{n,t}^{P2G}$ and $IC_{n,t}^{pipe}$ respectively represent the investment cost per unit capacity of the electricity-to-gas equipment and the unit length investment cost of the gas pipeline; $\Delta P_{n,t}^{P2G}$ represents the capacity of newly installed P2G devices on node n in year t; $l_{n\leftrightarrow m}^{gas}$ represents the length of the pipeline to be laid from node n to node m ; $z_{n\leftrightarrow m}$ is a boolean variable. When $z_{n\leftrightarrow m}=1$, it means to lay the pipeline from node n to node m in year t, otherwise the same. OC^{P2G} and OC^{pipe} represent the variable cost of electricity-to-gas equipment and gas transmission pipelines, respectively. OC^{pipe} is slightly affected by the length of the pipeline and the flow of natural gas. Assuming that the annual OC^{pipe} does not Change, and the transmission loss is also considered in OC^{pipe}; $U_{n,t}^{p2g}$ represents the total volume of electricity to gas at node n in year t; p^{gas} represents the natural gas price of the upper-level network; U_{t}^{net} represents the volume of natural gas purchased from the upper-level network in year t; C_{cut}^{gas} represents the cost of load reduction when the supply of natural gas is insufficient; U represents the total volume of natural gas load reduction in year t.

Like electricity, natural gas must comply with certain physical laws and system constraints during transmission and application, such as the relationship between flow rate, temperature and pressure, node flow balance, and node pressure balance.

3 ELECTRIC VEHICLE CHARGING FACILITY INVESTMENT PLANNING MODEL

In SUME, electric vehicles are especially important mobile energy storage components, as well as a key node for the coupling of transportation networks and power systems. The development of electric vehicles enables signals in the field of electric energy such as electricity prices to become one of the important measures to guide the layout and operation of the transportation network. The connection points between electric vehicles and the power system are mainly various types of charging and swapping stations. The layout of the connection points will directly affect the comprehensive economic benefits and operating efficiency of the power system and transportation network. Therefore, it is particularly important to consider the location and capacity of electric vehicle charging and swapping stations in the planning of the integrated energy system.

3.1 The charging demand of electric vehicles

The slow charging station is the main energy supply facility for electric vehicles, and the fast-charging station is a good supplementary facility for the slow charging station, which can provide charging services for more electric vehicles. Due to the high cost of substations, they are generally only targeted at some public special vehicles. Therefore, this article mainly considers the location and capacity of slow charging stations and fast charging stations. The first to solve the problem of location and capacity of charging stations is to calculate the charging demand of electric vehicles. Before calculating the charging demand of electric vehicles, the following assumptions need to be clarified: (1) The driving range of the battery under the current technical conditions is SD_0. Ignore the difference in battery capacity of different types of vehicles, that is, SD_0 has nothing to do with the vehicle type; (2) Based on the current power battery technology level, the battery cycle life is short, the depreciation cost is high, and the economy is poor. If you repeatedly use high currents above 3C Fast charging of the battery will significantly shorten the battery life and cause the economic efficiency of electric vehicles to deteriorate. Therefore, the charging station is set to be constructed on the principle of "slow charging as the main and fast charging as auxiliary".

According to the above assumptions, the total charging demand of electric vehicles in SUME can be calculated by formula (2), which provides capacity constraint data for the location and capacity of subsequent charging stations.

$$q^{EV} = \sum_{h \in H} q_h^{EV} = \sum_{h \in H} h\left(e \cdot \frac{y_h}{SD_0}\right) \qquad (2)$$

In the formula, q^{EV} represents the daily maximum charging frequency demand; q_h^{EV} represents the charging frequency of electric vehicles in h period; e represents the possession of electric vehicles; y represents the average driving range of electric vehicles in h period ; H represents the set of all operating periods, and h represents the h-th period.

3.2 Charging station planning and configuration

Queuing theory is a mathematical theory that studies the reasonable coordination of the relationship between "demand" and "service". This section uses the principles and methods of queuing theory to construct a queuing model that can reflect the characteristics of the system and studies the operating indicators such as the length of the charging fleet in the system and the waiting time. When configuring charging facilities, the queue length and waiting time of electric vehicles are especially important system constraints.

It is a random process for an electric vehicle driving on the road to find that the battery is insufficient and need to be charged. To study the charging work of the charging station, it is necessary to clarify the law of occurrence of electric vehicle charging. According to experience, it can be known that the charging demand generation process meets the following three characteristics:

(1) The time interval between two cars' charging demand can be regarded as a process with no aftereffect, that is, the number of customers arriving in a non-overlapping time period is independent of each other;

(2) For a sufficiently small time interval Δh, in the time interval $[h, h+\Delta h]$, the probability that an electric vehicle will arrive has nothing to do with time h, and Only proportional to the length of the time interval;

(3) For a sufficiently small time interval Δh, the probability of two or more electric vehicles arriving in the time interval $[h, h+\Delta h]$ can be Negligible, this shows that the time of arrival of electric vehicles is different, and it is an independent event.

The above three characteristics are in line with the characteristics of Poisson distribution, and Poisson flow can be used to describe the law of electric vehicles generating charging demand. If the electric vehicle charging demand is a Poisson flow, the time interval T between the successive arrivals of two electric vehicles obeys a negative exponential distribution. Therefore, the demand trigger point of electric vehicles in SUME, the charging capacity of charging stations, and the queuing status of electric vehicles can be assumed and restricted as follows:

(1) The charging time of each vehicle obeys a negative exponential distribution.

(2) A charging device can only accept the charging request of an electric car at the same time, and there is no difference between the charging devices in the charging station, which is a multi-channel service system with limited system capacity.

(3) The number of electric vehicles waiting in line should not exceed a certain number.

(4) The average waiting time for electric vehicles to be charged should not exceed a certain time, otherwise it will cause excessive waiting time and affect the travel of customers.

Through the above analysis, the charging system mode is M/M/C, that is, the Poisson input, negative exponential service time distribution, and the waiting system queuing model of C service stations (ie, charging devices). If the charging devices can work at the same time and are independent of each other, and the average service rate is all \gamma (that is, the inverse of the average charging time), the average service rate of the entire charging station is shown in formula (3):

$$\Gamma = \begin{cases} C \cdot \gamma & if\ C \geq n \\ e \cdot \gamma & if\ C < n \end{cases} \qquad (3)$$

Formula (4) shows the maximum arrival rate λ_i' of a vehicle to be charged in a certain period, that is, the reciprocal of the minimum time interval between the arrival of electric vehicle users, which is an important basis for the number of charging devices installed at the node.

$$\lambda_i' = \max_{h \in H}\left\{\lambda_{i,h} \mid \lambda_{i,h} = \frac{q_h^{EV} \cdot \tau_h \cdot f_{i,h}^{EV}}{\sum_{h \in H} \tau_h \sum_{i \in I} f_{i,h}^{EV}}, h \in H\right\}$$
$$\forall i \in I \qquad (4)$$

The formula (5) characterizes the charging power of the electric vehicle in a certain period at node i.

$$P_{i,h}^{EV} = \frac{\lambda_{i,h}}{\gamma} \cdot p^{EV} \quad \forall h \in H, i \in I \qquad (5)$$

In the formula: I represents the node set of the power network, and i represents the i-th node. The location of the root charging station is the connection between the transportation network and the power network. Therefore, the nodes of the electric vehicle charging network are directly labelled with the power system node to facilitate subsequent calculations; λ_i' indicates that the electric vehicle at node i is the largest Arrival rate; $\lambda_{i,h}$ represents the average arrival rate of electric vehicles at node i in time period $\hbar$; f represents the total flow of electric vehicles at node i in time period $\hbar$; τ_h represents the average utilization rate of electric vehicles in time period $\hbar$; p^{EV} represents the charging power of a single electric vehicle.

For electric vehicle users, how to charge more conveniently and quickly without affecting their normal use of the vehicle is particularly important. The fast-charging method generally occurs during the use of the vehicle and has relatively high requirements for charging time and queuing time; the slow charging method generally occurs when the vehicle is idle, and the slow charging time is long, and the user cannot wait for a long time for the previous vehicle to be charged. Charge it again when finished. The core factor affecting the service level of charging stations is the average charging waiting time of users. Therefore, the number of charging devices at the charging station is configured according to the user's maximum allowable charging waiting time $\Delta\hbar_{max}$ during peak traffic, as shown in formula (6)-formula (9):

$$\min P_{0i}^{EV} = C_i p^{EV}$$
$$s.t. \Delta\hbar_i^{'} \leq \Delta\hbar_{max} \quad (6)$$

$$\Delta\hbar_i^{'} \approx \max_{\hbar \in H} \Delta\hbar_{i,\hbar}^{'} = \left[\frac{(C_i \cdot \tau^{'})}{\lambda_i^{'} \cdot (C_i)! \cdot (1-\tau^{'})^2} Pr_{0i} \right]$$
$$\forall h \in H, i \in I \quad (7)$$

$$Pr_{0i} = \left[\sum_{R=0}^{C_i - 1} \frac{(C_i \cdot \tau^{'})^R}{R!} + \frac{(C_i \cdot \tau^{'})^{C_i}}{C_i!\,(1-\tau^{'})} \right]^{-1} \quad \forall i \in I \quad (8)$$

$$\tau^{'} = \frac{\lambda_i^{'}}{C_i \gamma} \quad \forall i \in I \quad (9)$$

The focus of the above model lies in the decision variable C_i. Through model calculation, two sets of key vectors $C = \{C1, C2 ... Ci\}$ and p^{EV} can be obtained, which respectively represent the number of charging devices installed at each node and the number of charging devices. The total capacity, and these data are the basic data for calculating the total cost of the transportation network, and it is also one of the decision variables for the coordinated planning of the integrated energy system including the power system, the transportation network, and the natural gas network. The cost function of electric vehicle-related facilities in the transportation network is:

$$TC^{EV} = IC^{EV} + OC^{EV} - Z^{EV}$$
$$= \sum_{t \in T} \left(\sum_{i \in I} IC^{EV} \cdot \Delta C_{i,t} \right) (1+r)^{-t}$$
$$+ \sum_{t \in T} \left(\sum_{i \in I} OC^{EV} \cdot C_{i,t} \right) (1+r)^{-t}$$
$$- \sum_{t \in T} \left\{ \frac{IC_t^{EV}}{T^{EV}} [T^{EV} - (T-t)] \right\} \quad (10)$$

Where: TC^{EV}, IC^{EV}, OC^{EV} and Z^{EV} respectively represent the total cost, total investment cost, total variable cost and the end of the new investment equipment planning period during the transportation network planning period The residual value of; $\Delta C_{i,t}$ represents the number of newly added charging devices at node i in year t; $C_{i,t}$ represents the number of charging devices already operating at node i in year t; IC_t^{EV} indicates the total investment of the charging device in year t; T^{EV} indicates the life span of the charging device.

4 INTEGRATED ENERGY SYSTEM COORDINATION PLANNING MODEL

On the energy supply side, SUME includes multiple types of DG units, natural gas P2G devices, mobile energy storage (electric vehicles), energy storage components, etc. (due to regional limitations, large generator sets are not considered, and the lack of electricity is provided by the upper-level grid); On the energy transmission side, SUME includes power transmission lines, gas pipelines, transportation networks, etc.; on the energy load side, SUME includes electrical loads, natural gas loads, thermal loads, etc. The thermal load is mainly provided by the transformation of electric energy and natural gas at the terminal, so it does not Consider the heating pipeline separately.

The model proposed in this paper is aimed at minimizing the comprehensive cost of the system within the planning scope, while considering the basic requirements and safety requirements of the physical operation of the power system, natural gas network, and transportation network. The overall system cost includes the investment cost and variable cost of natural gas network facilities, the investment cost and operation and maintenance cost of charging station facilities, the investment cost and variable cost of power generation and grid facilities, and the purchase cost of natural gas and electricity from the upper-level network. To accurately reflect the time value of costs, all costs and expenses are expressed in the form of the initial annual present value of the plan.

$$\min T\,C_{total} = TC^{gas} + TC^{EV} + TC^{ele} \quad (11)$$

In the formula: TC_{total} represents the total comprehensive cost during the planning period of the integrated energy system; TC^{ele} represents the total cost during the planning period of the power system.

$$TC^{ele} = \sum_{t \in T} \left\{ \left[\sum_{i \in I} \sum_{g \in G} (IC_g^{DG} \cdot \Delta P_{g,i,t}^{DG}) + \sum_{i \in I} (OC_g^{DG} \cdot P_{g,i,t}^{DG}) \right] (1+r)^{-t} \right\}$$
$$+ \sum_{t \in T} \left\{ \left[\sum_{i,j \in I \exists (i,j)} (IC_{ij}^L \cdot z_{ij,t}^L) + OC^L \cdot P_{ij,t}^L \right] (1+r)^{-t} \right\}$$
$$+ \sum_{t \in T} \left\{ \left[\sum_{i \in I} (IC^S \cdot \Delta P_{i,t}^S + OC^S \cdot q_{i,t}^S) \right] (1+r)^{-t} \right\}$$

$$+\sum_{t\in T}\left[\sum_{b\in B}\left(c^{DR}\cdot P_{b,t}^{DR}\cdot \Delta\hbar_b + c^{cut}\cdot Q_{b,t}^{cut}\right)(1+r)^{-t}\right]$$
$$+\sum_{t\in T}[(p^{ele}\cdot Q_t^{net})(1+r)^{-t}]$$
$$-\sum_{w\in W}\frac{IC_w^{ele}}{T_w}[T_w-(T-t_w)] \quad (12)$$

Where: G represents the set of power generation types of DG, g represents the g-th power generation type of DG; IC_g^{DG} represents the unit capacity investment cost of DG units of power generation type g; $\Delta P_{g,i,t}^{DG}$ represents the newly added DG unit capacity of power generation type g in node i in the t year; OC_g^{DG} represents the unit capacity operating cost of the DG unit of power generation type g; $P_{g,i,t}^{DG}$ represents the actual operation of power generation type g in node i in the t year Capacity; IC_{ij}^{L} represents the investment and construction cost of the line between nodes i and j; $z_{ij,t}^{L}$ represents a Boolean variable, =1 represents that there is capacity expansion between node i and node j in year t, and vice versa; OC^L represents line unit capacity operation Cost; $P_{ij,t}^{L}$ represents the actual operating capacity between node i and node j in year t; IC^S, OC^S represent the unit capacity investment cost and operating cost of the energy storage unit respectively; $\Delta P_{i,t}^{S}$ represents the newly added energy storage capacity in node i in the t year; $q_{i,t}^{S}$ represents the total circulating power between the system and the energy storage unit in node i in the t year; B is the set of load segments, $B=\{b|b=b_x,x=1,2,3\ldots\}$, b_x represents the xth load segment. For the convenience of calculation, assume that the load segment set B and the time segment set H in the electric vehicle module are divided in the same time span; c^{DR} represents the unit capacity cost of the system to implement demand response; $P_{b,t}^{DR}$ means that the load segment b in the tth year is reduced by demand response $z_{b,t}$ are boolean variables, =1, it means that load segment b has achieved load reduction in year t, and vice versa; $\Delta\hbar_b$ means response duration; c^{cut} means the unit compensation paid to the user to directly reduce the load of the user Cost, assuming that it is replenished in the unit of electricity; $Q_{b,t}^{cut}$ represents the user electricity directly cut by load segment b in the t-th year; $z_{b,t}{}'$ is a Boolean variable, and =1 means the t-th year Load segment b has achieved direct load reduction, and vice versa; p^{ele} represents the power purchase cost from the upper-level grid; Q_t^{net} represents the total power purchase from the upper-level grid in year t; W represents the collection of all equipment in the system, $W=\{w|w=W_x,x=1,2,3\ldots\}$, w_x represents the wth device; T_w represents the life span of the device w; t_w represents the year the device was put into use, that is, the number of years from the initial year of the planning period .

In formula (12), the first term represents the investment cost of new DG units in the system during the planning period and the operation and maintenance costs of all units; the second term represents the comprehensive cost of the system's transmission lines, including expansion investment costs and operation and maintenance costs; The third item represents the total cost of the energy storage unit in the system, including capacity investment costs and operation and maintenance costs; the fourth item is the system outage cost, which is composed of demand response costs c^{DR} and load shedding costs c^{cut}; the fifth item is SUME from the upper grid The sixth item is the residual value of new facilities in the system at the end of the planning year.

In addition, to ensure the safety and stability of the system, the planning and operation of SUME also need to meet the following constraints:

(1) DG capacity constraints

The active power and reactive power of the DG unit are restricted by its maximum and minimum capacity, and the installable DG capacity of node i is also restricted.

$$P_{i,0}^{DG}+\sum_{t\in T}\Delta P_{i,t}^{DG}\le P_{i,max}^{DG}$$
$$P_g^{min}\le P_g^{DG}\le P_g^{max} \quad (13)$$

(2) Line capacity constraints

Due to the limitation of space resources, the transmission channel between nodes can bear the limited line capacity, which is a constraint to be considered for line upgrades.

$$P_{ij,0}^{L}+\sum_{t\in T}\Delta P_{ij,t}^{L}\le P_{ij,max}^{L(i,j)(i,j)}$$
$$\Delta P_{ij,t}^{L}\le Mz_{ij}^{L}\quad \forall(i,j)\in I \quad (14)$$

(3) Power flow constraints

$$P_i=|V_i|\sum_{j\in I}|V_j|\left(G_{ij}\cos\theta_{ij}+A_{ij}\sin\theta_{ij}\right)$$
$$\forall(i,j)\in I:\exists(i,j)$$
$$Q_i=|V_i|\sum_{j\in I}|V_j|\left(G_{ij}\sin\theta_{ij}-A_{ij}\cos\theta_{ij}\right)$$
$$\forall(i,j)\in I:\exists(i,j) \quad (15)$$

In the formula: P_i, Q_i and $|V_i|$ represent the active power, reactive power injection and voltage amplitude of node i under normal conditions; G_{ij} and A_{ij} represent the system conduction under normal conditions. The real and imaginary parts of the nano matrix; θ_{ij}^{v} represents the voltage phase angle difference between node i and node j.

(4) Voltage constraints

$$V_{min}\le V_i\le V_{max} \quad (16)$$

In the formula: V_i represents the node voltage of node i; V_{min} and V_{max} represent the lower limit and upper limit of the allowable voltage fluctuation of the system node, respectively.

(5) Energy storage constraints

Due to its own technical conditions and economic constraints, the energy storage in PMIES adopts energy storage units with medium capacity and relatively mature technology. In actual operation, the charge and discharge capacity and rate of energy storage must meet the following constraints:

$$SOC_{i,b}^{S} = SOC_{i,b-1}^{S} + \left[\left(q_{i,b}^{c}/\xi^{c} - \xi^{d} q_{i,b}^{d}\right)/P_{i}^{S}\right]$$
$$\forall b \in B, i \in I \quad (17)$$

5 SIMULATION ANALYSIS

PMIES is an optimization model containing multiple main modules. Among them, natural gas pipeline airflow equations are non-linear equations, which can be processed by incremental piecewise linearization, thereby transforming the model into a mixed integer linear optimization problem that is relatively easy to solve. To verify the effectiveness of the proposed PMIES planning method, an improved Garver test system is used for simulation. The system includes a 7-node natural gas network system and a 6-node electric test system, as shown in Figure 1. Among them, the natural gas network system includes 2 natural gas production nodes (including P2G facilities) and 5 natural gas load nodes. For basic system data of the power system, see Table 1 and Table 2.

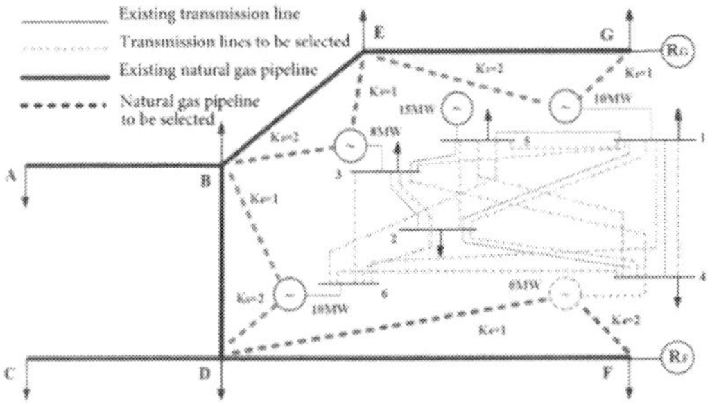

Fig. 1. Garver test system diagram.

Table 1. Transmission line data of the test system.

	Line endpoint	Resistance MΩ	Cost of investment (104 ¥KM)	Capacity MW
Existing lines	1-2	0.1	0	20
	1-4	0.15	0	15
	1-5	0.05	0	20
	2-3	0.05	0	20
	2-4	0.1	0	20
	3-5	0.05	0	30
Lines to be selected for expansion	1-2	0.1	40	20
	1-3	0.09	38	20
	1-4	0.15	60	15
	1-5	0.05	20	20
	1-6	0.17	48	10
	2-3	0.05	20	20
	2-4	0.1	40	20
	2-5	0.08	31	20
	2-6	0.08	30	20
	3-4	0.15	59	12
	3-5	0.05	20	20
	3-6	0.12	48	20
	4-5	0.16	63	10
	4-6	0.08	30	20
	5-6	0.15	70	25

Table 2. Power node data of the test system (initial year).

Node	Maximum installable DG capacity MW	Existing capacity MW	Node load MW
1	60	20	22
2	30	0	9
3	40	18	15
4	20	0	12
5	40	20	15
6	30	15	10

Assuming that each electric car is charged with 20kW/h in a single time, it takes 10 hours and 0.5 hours to fully charge through the slow charging and fast charging devices, respectively. The average utilization rate of electric vehicles is 0.3, and the maximum utilization rate is 0.8. In SUME, slow charging is the main one. The fast-charging device is 30% of the slow charging. The charging power of a single slow charging device is 14kW, and the charging power of a fast-charging device is 60kW. A single node can install no more than 100 devices. According to the number of electric vehicles, the maximum available daily charging frequency is 350 times. The operating characteristic parameters of the power generation and energy storage components in the test system are shown in Table 3. The charge and discharge efficiency of energy storage is 75%, the maximum discharge rate is 0.5C, and the upper and lower limits of SOC are [0.1-0.95]. The service life of power components is 20 years. In addition, the system load growth rate is set to 5%, the existing power generation capacity is non-DREG power generation type, the planning period is 20 years, the node voltage fluctuation range is 0.95-1.05pu, and the discount rate is 0.06. The demand response cost c^{DR} is 725\$/MW·h, and the load reduction cost c^{cut} is 870\$/MW·h. The sales price of the higher-level grid refers to the average price of industry and commerce in a certain province of my country, which is 0.11\$/kW·h in US dollars. To ensure the reliability of the system, the load involved in demand response and the load directly reduced shall not exceed 10% and 5%.

Table 3. Operating characteristic parameters of power generation/energy storage components.

Component type	Unit capacity MW	Variable costs $/MW.h	Cost of investment $/KW	Forced outage rate %	Planned outage rate %
Wind power	0.75	7.25	1220	1.8	1.2
PV	0.5	12.32	1180	2.5	1.6

 https://doi.org/10.1051/e3sconf/202129303031

Distributed gas	1.5	27.45	530	15	7
Biomass power generation	2	17.45	1520	16	8
DCHP	1	24.64	880	12	6
Energy storage	1000 A.h	11.6	260	-	-

As shown in Figure 2, in the improved Garver test system, the 7 nodes of the natural gas network (with the letter AG as the node label) are separated from the power system in the initial year of planning, and the 7 nodes of the natural gas network have 6 pipelines through. , Respectively: Pipe 1, A→B, 1800Nm^3/hr; Pipe 2, B<->D, 520Nm^3/hr-psia; Pipe 3, B<->E, 560Nm^3/hr-psia; Pipe 4, C<->D, 2600Nm^2/hr-psia; Pipe 5, D<->F, 640Nm^2/hr-psia; Pipe 6, E<->G, 700Nm^3/hr-psia. Among them, the natural gas distribution station operation data of node A-G is shown in Table 4. From the formula (2), the actual flow rate of natural gas pipelines is related to factors such as gas specific gravity, friction coefficient, and pipeline diameter. It is known that the proportion of natural gas is 0.5548 and the Manning coefficient is 0.009. The relationship between the flow rate of the pipeline to be erected and the diameter and length is shown in Table 5. The diameter of the pipeline is positively correlated with the investment cost, as shown in Table 6. The price of natural gas refers to the price of industrial gas, which is 0.46\$/m^3.

Table 4. Operating data of the distribution station of each node of natural gas.

Node Diversion Station	Diversion station type	Flow NM 3/hr	Press ure psia	Press ure MIN	Press ure MA X
A	Constant current	800	-	75	200
B	Constant current	800	-	120	200
C	Constant current	800	-	75	200
D	Constant current	800	-	120	200
E	Constant current	800	-	120	200
F	Constant current	800	-	120	200
G	Constant pressure	-	200	75	200

Table 5. Type and capacity of natural gas pipeline to be selected.

Diameter / Length	6	8	10	12	16	20	24
2	0.44	0.57	0.93	1.37	2.21	3.75	5.17
4	0.39	0.4	0.75	1.1	2.06	3.4	4.86
6	0.32	0.33	0.6	0.93	1.7	3.06	4.6
8	-	0.3	0.5	0.8	1.45	3.67	4.23
10	-	-	0.4	0.7	1.3	2.4	3.9
12	-	-	-	0.65	1.2	2.15	3.45
14	-	-	-	-	1.05	2	3.2
16	-	-	-	-	-	1.85	3.05
18	-	-	-	-	-	-	2.8

Table 6. Unit investment cost of different diameter pipes.

Pipe diameter	6	8	10	12	16	20	24
Cost of investment 104\$/km	21	28	47	68	112	183	258

6 OPTIMIZATION RESULTS AND ANALYSIS

In the test system, the coupling interface between the power system and the natural gas system is mainly P2G equipment, distributed gas-fired units and DCHP. In addition, it is also equipped with certain gas storage equipment (the proportion is small, not reflected in the model). The newly added gas pipelines are mainly distributed gas generators and DCHP gas supply. When the gas supply is insufficient, it is provided by gas storage or energy storage equipment; when the electrical load is low, the surplus electricity can be stored through P2G equipment or meet the gas load.

Through simulation, the planned Garver system structure is shown in Figure 2. Four transmission lines have been added: Line 1-4, Line 2-3, Line 3-5, Line 3-6. Gas transmission pipelines are added on the original basis. 9 lines: K1-K9, four of which are connected to power node 1, node 3, node 4, and node 6, respectively. The newly added generator sets, and energy storage configuration of each node are shown in Figure 3.

As can be seen from Figures 2 and 3, as the load of each node increases, 6 power nodes have new installed capacity, of which nodes 1, 3, 4, and 6 are equipped with gas generators and DCHP, so they are connected to the gas transmission. The pipe is connected to it. From the perspective of DG type, the total new capacity of the system is 108.75MW, of which the distributed gas turbine assembly capacity is the largest, reaching 27.59%; wind power is the second, 24.14%; the DCHP is affected by the heat load with the smallest proportion of 13.79%. From the point of view of node configuration, node 4 has no power generation capacity allocated in the initial year, so the maximum new capacity is 27.75MW, which is 25.52% of the total new capacity; the load of node 2 is relatively low, and it is similar to nodes 1, 3, 4, 5 are all interconnected, so the newly added capacity of the configuration is the smallest, 11MW; the DREG penetration rate of each node is more than 35% of the total capacity of the system (including the existing capacity in the initial year).

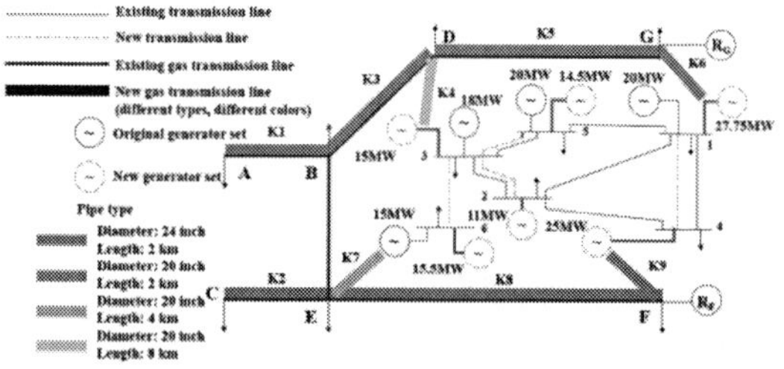

Fig. 2. Planning results of the test system.

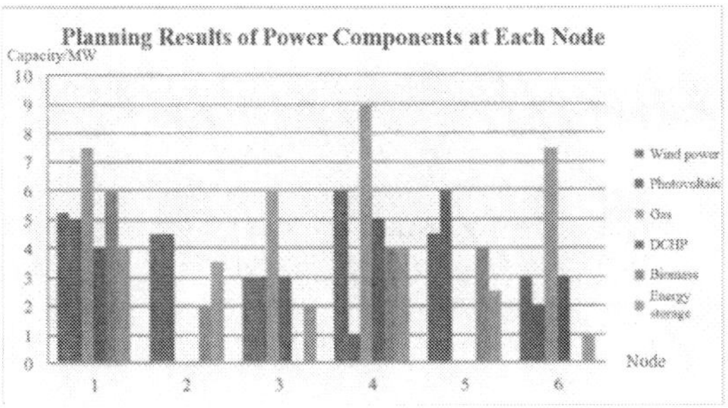

Fig. 3. Planning results of power components at each node.

For the two modules of natural gas network and electric vehicle charging facilities, some intermediate data were extracted.

The average flow of the channels between the natural gas nodes, the pressure of each node in the natural gas network, and the configuration of the charging devices of each node in the transportation network are extracted. Among them, some data of the natural gas network are shown in Table 7. At the end of the plan, the absolute pressures of the seven natural gas nodes were all within the allowable range. The average flow rate was the natural gas channel C to D, reaching 4200 Nm^2/hr (including the original pipeline flow rate). Channel B→D is the deployment channel of the upper and lower sides of the natural gas channel. From the flow data, the upper channel supplemented the lower channel with a flow rate of 560Nm^2/hr at the end of the plan. In the case of channel C-D with the maximum flow, the lower channel also needs the support of the upper channel, indicating that the gas consumption of node 4 and node 6 is relatively large. This phenomenon is also consistent with our planning results: Nodes 4 and 6 are equipped with a total of 24.5MW of gas-fired units (including distributed gas-fired units and DCHP), accounting for about 60% of the total gas-fired units. As for the transportation network, there are charging devices at six power nodes. According to calculations, the system is equipped with 453 slow charging devices and 120 fast charging devices; the maximum charging load of the transportation network is 13.54MW, the average charging load is 7.36MW, the maximum reverse transmission load is -5.20MW, and the average transmission load- 3.73MW.

Table 7. Natural gas node flow and pressure (the end of the plan).

Starting node	End node	Average flow Nm^3/hr	Node	Pressure
A	B	4000	A	90.36
B	E	3440	B	120.06
B	D	560	C	77.19
C	D	4200	D	120.06
D	F	2750	E	173.32
E	G	1950	F	149.03
-	-	-	G	200

In addition, according to the calculation of the planning results, the comprehensive costs of seven key power components such as wind power, DCHP, natural gas facilities, and EV charging facilities are shown in Table 8. Among them, the total variable cost is the discounted value of all variable costs during the planning period, and the total cost is the total investment cost and the total variable cost minus the residual value. It can be seen from the table that the cost of power components, TCe, still accounts for a large proportion, reaching 175,235,900 US dollars, and the cost of natural gas facilities TC8s has also reached 95,341,300 US dollars. The investment in the transportation network is relatively small, accounting for only the overall total cost. 8.03% of TColal. In the cost item TCe, the cost of distributed gas and DCHP is relatively high, mainly due to the larger installed capacity, which is related to the higher fuel cost compared to DREG. If further analysis, this is related to SUME is a relatively independent self-balancing system, which needs to maintain the reliability of the system itself and meet the thermal load. The high reliability of the system requires the configuration of gas-fired units that can produce flexibly. DCHP alleviates the problem of relying solely on electric heating or burning fossil energy.

Table 8. The overall cost of the types of facilities in the planning results.

Type of facility	Wind power	PV	Distributed gas	DCHP	Biomass power generation	Energy storage unit	Transmission line	Natural gas facilities	EV charging facilities
Total investment cost	3202.5	2537.17	1590.55	1320.23	2432.9	442.47	740.87	7922.11	1932
Total variable cost	373.3	390.09	4055.82	2177.24	822.36	580.83	999.31	3564.9	947.64
Total cost	3147.05	2634.38	4798.94	3042.6	2928.92	971.69	1652.05	9534.13	2505.29

7 CONCLUDING REMARKS

This paper studies the green energy Internet planning model, takes power system planning as the main module, considers the requirements and constraints of natural gas network and transportation network construction, and uses the energy hub model to analyse the energy transmission and conversion relationship between different networks, so as realize comprehensive energy system planning for power system, natural gas network, and transportation network. To verify the effectiveness of the proposed

https://doi.org/10.1051/e3sconf/202129303031

PMIES planning method, an improved Garver test system is used for simulation.

As the load of each node increases, six power nodes have newly installed capacity, as shown in Figures 2 and 3. Among them, nodes 1, 3, 4, and 6 are equipped with gas generators and DCHP, so they are connected to gas transmission. The pipe is connected to it.

For the two modules of natural gas network and electric vehicle charging facilities, some intermediate data were extracted.

The average flow of the channels between the natural gas nodes, the pressure of each node in the natural gas network, and the configuration of the charging devices of each node in the transportation network are extracted. We performed a data simulation on planning modelling and found that the results obtained are consistent with our planning model.

Acknowledgments

This work has been supported by the National Key Research and Development Program No.2018YFB2100100, Postdoctoral Science Foundation of Yunnan Province, Science Foundation of Yunnan Province No.202101AT070167, Project of the Yunnan Provincial Department of Education scientific research fund No. 2019J0010, DongLu Young and Middle-aged backbone Teachers Project of Yunnan University, Open Foundation of Key Laboratory in Software Engineering of Yunnan Province under Grant No.2020SE311, Science and Technology Project of Yunnan Power Grid Co., Ltd No.YNKJXM20180092 and YNKJXM20190639.

References

1. B. Miao *et al.*, "Day-Ahead Energy Trading Strategy of Regional Integrated Energy System Considering Energy Cascade Utilization," in *IEEE Access*, vol. **8**, pp. 138021-138035, (2020)
2. L. Ji, L. Wang, C. Liao and S. Li, "Simultaneous Wireless Power and Bidirectional Information Transmission With a Single-Coil, Dual-Resonant Structure," in IEEE Trans Ind Electron, vol. **66**, no. 5, pp. 4013-4022, May (2019)
3. A. Azizivahed et al., "Energy Management Strategy in Dynamic Distribution Network Reconfiguration Considering Renewable Energy Resources and Storage," in IEEE Trans. Sustainable Energy, vol. **11**, no. 2, pp. 662-673, April (2020)
4. J. Wang et al., "Two-Stage Integrated Electricity and Heat Market Clearing With Energy Stations," in IEEE Access, vol. **7**, pp. 44928-44938, (2019)
5. L. Cheng, T. Yu, H. Jiang, S. Shi, Z. Tan and Z. Zhang, "Energy Internet Access Equipment Integrating Cyber-Physical Systems: Concepts, Key Technologies, System Development, and Application Prospects," in IEEE Access, vol. **7**, pp. 23127-23148, (2019)
6. Z. Liu, M. Zeng, H. Zhou and J. Gao, "A Planning Method of Regional Integrated Energy System Based on the Energy Hub Zoning Model," in IEEE Access, vol. **9**, pp. 32161-32170, (2021)
7. W. Wang, L. Liu, J. Liu and Z. Chen, "Energy management and optimization of vehicle-to-grid systems for wind power integration," in CSEE Journal of Power and Energy Systems, vol. **7**, no. 1, pp. 172-180, Jan. (2021)
8. Z. Zhao, H. Zhou, L. Qi, L. Chang and M. Zhou, "Inductive Representation Learning via CNN for Partially-Unseen Attributed Networks," in *IEEE Trans. Netw. Sci. Eng.*, **8**, no. 1, pp. 695-706 (2021)
9. F. Zhang et al., "Understanding WeChat User Preferences and "Wow" Diffusion," in IEEE Trans Knowl Data Eng (2021)

 https://doi.org/10.1051/e3sconf/202129303032

Research on electric vehicle ownership prediction based on BASS model: A case study of Yunnan Province

Shi Su[1], Jiaxin Jiang[2], Chunlin Yin[1], Jian Hu[3], Ting Li[4], Xun Ye[2], QiangJian Zhang[2], Weiyun Ma[2], Siyang Liu[1,*]

[1] Electric Power Research Institute, Yunnan Power Grid Co.,Ltd., Kunming, China

[2] Key Laboratory in Software Engineering of Yunnan Province, School of Software, Yunnan University, China

[3] Information Center,Yunnan Power Grid Co.,Ltd, Kunming, China

[4]Kunming Enersun Technology Co.,Ltd, Kunming, China

Abstract. In order to forecast the number of electric vehicles in Yunnan Province, based on BASS model, this paper uses extensive analogy method to explore the acquisition of m, p and q model parameters, forecasts the purchasing power of the market, and estimates the innovation coefficient and imitation coefficient from three aspects of high potential scenario, base potential scenario and low potential scenario. The number of new energy electric vehicles in Yunnan Province in three scenarios from 2022 to 2035 is predicted. The forecast results show that under the condition of high potential development, the number of new energy vehicles in Yunnan Province will reach 409,600 in 2022; in the case of benchmark potential development, the number of new energy vehicles will reach 291,400 in 2022; in the case of low potential development, the number of new energy vehicles in Yunnan Province will reach 155,400 in 2022.

1 Introduction

At present, the Yunnan provincial government is following the national policy to vigorously develop new energy vehicles. The new energy vehicle market in Yunnan Province has a certain development foundation, but it still needs to be promoted. The key development of new energy vehicles based on electric vehicles is clearly put forward in China's "Twelfth Five Year Plan".

Compared with traditional fuel vehicles, electric vehicles have the advantages of small pollutant emissions, low energy consumption and low cost. In addition, the large-scale and orderly development of electric vehicles plays a positive role in driving the development of the upstream and downstream industries of electric vehicles. The prediction of the development trend of electric vehicle ownership is an important reference for the planning and construction of electric vehicle charging service network. Therefore, accurate prediction of electric vehicle ownership has important practical significance, and the core theory of new product diffusion model is BASS model and its extended model.

To a certain extent, consumers' interest in electric vehicles will be affected by price, substitution cost, policy and other factors, and ultimately affect the market potential of electric vehicles. The maximum market potential restricts the consumption of electric vehicles, which is related to the total number of new civil vehicles in that year. If electric vehicle technology continues to progress, vehicle costs continue to decrease, cost performance ratio improves, infrastructure such as charging piles continues to improve, the cost of gasoline vehicles and other fuel vehicles is at a high level, and the government provides a larger preferential subsidy policy for electric vehicles, then consumers will be more willing to buy electric vehicles, the proportion of electric vehicles in the newly added civil vehicles in that year has increased correspondingly, and the market potential of electric vehicles is relatively large. On the contrary, consumers' willingness to buy electric vehicles is low, and the market potential is small.

Through the investigation, it is found that there are many methods to predict the number of electric vehicles. For example, artificial bee colony planning symbolic regression sr-abc is used to predict China's passenger car ownership[1-2], electric vehicle ownership prediction method based on multiple linear regression[3-5] and proportional substitution method, and methods such as government planning target calculation method[6], thousand people ownership method and electric vehicle industry research are proposed to carry out electric vehicle ownership prediction. According to a class of Riccati equation, an accurate discrete BASS model is constructed, and the research shows that the model and its solution converge to the original continuous model[7]. A search algorithm based on nonlinear least square rule is given to estimate the model parameters, and the fitting effect is good[8]. The BASS model with weekend coefficient is

*Corresponding author: lsyfd27@sina.com

https://doi.org/10.1051/e3sconf/202129303032

constructed to obtain the improved BASS model. Guided by the theory of consumer decision-making process, the multiple regression model is established with the parameters of the improved BASS model as the dependent variable[9]. The joint prediction model is obtained by combining the improved BASS model and the parametric regression model, which can achieve better results[10]. Based on the universal applicability, easy operation and good effect of BASS model, this paper sets the maximum market potential of electric vehicles in three scenarios of high, benchmark and low according to the proportion of electric vehicles in the total number of newly added civil vehicles in that year through literature research and questionnaire survey according to different assumptions (see **Table 1**), it also predicts the number of electric vehicles in Yunnan Province under three scenarios.

2 Introduction of BASS model

The expression of BASS diffusion model is as follows:

$$\frac{dN(t)}{dt} = p[m - N(t)] + q\frac{N(t)}{m}[m - N(t)] \quad (1)$$

In the above **Formula 1**, the left side is the number of non cumulative adopters when t is used; the cumulative number of adopters when N(t) is t; m is the biggest potential of the market; p, q is external influence coefficient (or innovation coefficient) and internal influence coefficient (or imitation coefficient). The first term p [m-N (t)] of the equation represents the number of innovation adopters, whose purchase decisions are only affected by external factors, not by those who have adopted the product; the second represents the number of imitators whose purchase decisions are influenced by the people who have adopted the product.

The **Formula 2** for the discrete-time version of BASS model is as follows:

$$f(t+1) = p[1 - F(t)] + qF(t)[1 - F(t)] \quad (2)$$

Where p is the innovation coefficient predicted above, q is the imitation coefficient, and is the proportion of the cumulative number of new product consumers in the total number of potential consumers in t period. It is the proportion of the number of new consumers to the number of potential consumers in t + 1 period.

The key of BASS model lies in the acquisition of m, p and q. Usually, the three coefficients can be obtained by fitting on the basis of existing data, and can also be obtained by analogy, that is, using the correlation coefficient of similar products, BASS model expresses the essence of diffusion process with mathematical equation, which greatly simplifies people's understanding of innovation diffusion and makes it systematic.

3 BASS model parameter estimation

The parameter estimation of BASS model p and q is widely used to compare new energy vehicles with different types of products (high value consumables weight accounts for 20%, the weight of electronic products with high replacement speed and high technology content is 35%, and the weight of sales and retention of electric vehicles in the United States in previous years is 45%. Combined with the regional development of new energy vehicles in Yunnan Province, the p and q values are shown in table 1 below. P, Q is between 0.00 and 1.00. We set the initial value of innovation consumption coefficient p of Yunnan new energy vehicles to 0.0423, which increases accordingly. With the development of technology and progress, the growth rate slows into a flat period. The initial value of the imitation coefficient q is set to 0.615.

Table 1. Innovation coefficient (P) and imitation coefficient (Q) of new energy vehicles

Year	Coefficient of innovation（P）	Imitation coefficient（Q）
2021	0.076	0.774
2022	0.077	0.783
2023	0.085	0.792
2024	0.09	0.7995
2025	0.095	0.807
2026	0.958	0.8156
2027	0.1	0.822
2028	0.1005	0.8295
2029	0.101	0.837
2030	0.1015	0.8445
2031	0.102	0.8495
2032	0.1025	0.8545
2033	0.103	0.8595
2034	0.1034	0.8645
2035	0.1038	0.8695

4 prediction results of electric vehicle ownership

According to the development of Yunnan Province, the development environment of new energy vehicles in the future is divided into the following three scenarios, as shown in **Table 2**:

Table 2. Assumptions of different potential scenarios

Type	Assumed condition	New energy vehicles Maximum market potential
High potential scenario	Infrastructure construction has been accelerated and improved far beyond expectations. Technology breakthrough, cost performance greatly improved; oil prices are high; the government continues to subsidize and incentivize	Electric vehicles accounted for 90 per cent of new civilian vehicle additions that year

Baseline potential scenario	Steady progress in infrastructure construction can achieve the planned goals; the technology has been steadily improved without breakthrough, and the cost performance has been steadily improved. Oil prices are at low to medium levels; subsidies and incentives are stable or decreasing	Electric vehicles accounted for 60 per cent of new civilian vehicle additions that year
Low potential scenario	Infrastructure development has failed to meet expectations; no substantial progress in electric vehicle technology and no improvement in cost performance; a prolonged downturn in oil prices; subsidies and incentives have been stable or reduced or even drastically reduced	Electric vehicles accounted for 30 per cent of all new civilian vehicles added that year

Then, the new energy vehicle ownership and vehicle ownership in Yunnan Province were substituted into the BASS model and solved for different scenarios respectively. The obtained results are shown in **Table 3**:

Table 3. BASS Model calculation results

Year	High potential scenario		Baseline potential scenario		Low potential scenario	
	f(t)	F(t)	f(t)	F(t)	f(t)	F(t)
2021	0.1337	0.1011	0.1381	0.1101	0.1404	0.1149
2022	0.1404	0.1085	0.1453	0.1181	0.1478	0.1229
2023	0.1524	0.1161	0.1575	0.1259	0.1599	0.1307
2024	0.1616	0.1234	0.1667	0.1332	0.1691	0.1379
2025	0.1706	0.1304	0.1755	0.1402	0.1779	0.1448
2026	0.1757	0.1368	0.1806	0.1464	0.1828	0.1508
2027	0.1834	0.1431	0.1881	0.1524	0.1902	0.1567
2028	0.1878	0.1488	0.1924	0.1580	0.1944	0.1620
2029	0.1920	0.1542	0.1964	0.1630	0.1983	0.1668
2030	0.1960	0.1592	0.2002	0.1678	0.2020	0.1714
2031	0.1995	0.1639	0.2034	0.1721	0.2051	0.1756
2032	0.2028	0.1683	0.2066	0.1762	0.2082	0.1795
2033	0.2060	0.1724	0.2096	0.1801	0.2111	0.1832
2034	0.2089	0.1764	0.2124	0.1837	0.2138	0.1867
2035	0.2118	-	0.2151	-	0.2164	-

The final NEV ownership calculated by the BASS model is shown in **Figure 1** below:

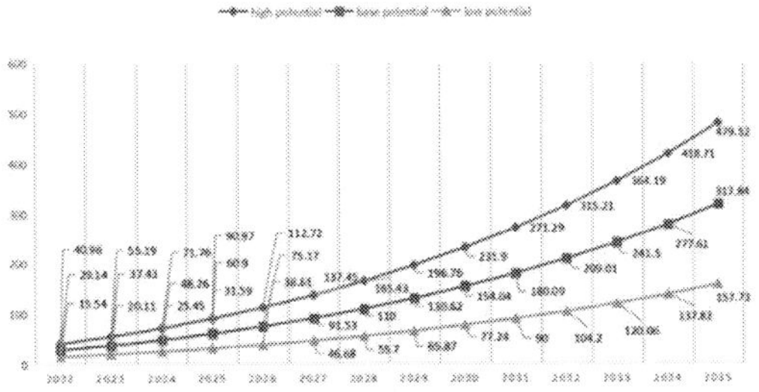

Figure 1. Forecast results of new energy vehicle ownership in Yunnan Province

5 Model Test

By comparing the actual number of new energies vehicles in the three years from 2018 to 2020 with the predicted number of new energy vehicles, the prediction effect of this model is tested. The following data are obtained through investigation and combined with the predicted data of 2018 to 2020, and the following **Table 4** is obtained:

Table 4. Comparison of 2018-2020 data

Year / (Ten thousand cars)	2018	2019	2020
Forecasting under high potential scenarios	3.63	9.85	18.6
Forecast under the baseline potential scenario	3.63	7.81	13.53
Forecast under low potential scenarios	3.63	5.67	8.41
Real data	4.3	5	5.5

6 Conclusion

The parameters of BASS model are estimated by using extensive analogy method. According to BASS model and the data of Yunnan Province's car ownership and new energy vehicle ownership in 2018, the development of Yunnan Province's new energy vehicle ownership is predicted and calculated in three different ways: high potential, benchmark potential and low potential. According to the prediction results of the model, in the case of high potential development, the number of new energy vehicles in Yunnan Province will reach 409600 in 2022; in the case of benchmark potential development, the number of new energy vehicles will reach 291400 in 2022; in the case of low potential development, the number of new energy vehicles in Yunnan Province will reach 155400 in 2022. The prediction results from 2018 to 2020 are consistent with those under the low potential scenario, and the average absolute error method is used to calculate the accuracy of 90.6%.

Acknowledgments

This work has been supported by the National Key Research and Development Program No.2018YFB2100100, Postdoctoral Science Foundation of Yunnan Province, Science Foundation of Yunnan Province No.202101AT070167, Project of the Yunnan Provincial Department of Education scientific research fund No. 2019J0010, DongLu Young and Middle-aged backbone Teachers Project of Yunnan University, Open Foundation of Key Laboratory in Software Engineering of Yunnan Province under Grant No.2020SE311, Science and Technology Project of Yunnan Power Grid Co., Ltd No.YNKJXM20180092 and YNKJXM20190639.

References

1. X. Gao, Z. Zheng, Q. Chu, S. Tang, G. Chen and Q. Deng, "Popularity Prediction for Single Tweet Based on Heterogeneous BASS Model," in IEEE Transactions on Knowledge and Data Engineering, vol. **33**, no. 5, pp. 2165-2178, (2021).
2. E. Derner, J. Kubalík and R. Babuška, "Selecting Informative Data Samples for Model Learning Through Symbolic Regression," in IEEE Access, vol. **9**, pp. 14148-14158, (2021).
3. L. Dong, C. Wang, M. Li, K. Sun, T. Chen and Y. Sun, "User Decision-based Analysis of Urban Electric Vehicle Loads," in CSEE Journal of Power and Energy Systems, vol. **7**, no. 1, pp. 190-200, Jan. (2021).
4. Z. Zhao, Y. Peng, X. Zhu, X. Wei, X. Wang and J. Zuo, "Research On Prediction Of Electricity Consumption In Smart Parks Based On Multiple Linear Regression," 2020 IEEE 9th Joint International Information Technology and Artificial Intelligence Conference (ITAIC), pp. 812-816, (2020).
5. E. Sreehari and S. Srivastava, "Prediction of Climate Variable using Multiple Linear Regression," 2018 4th International Conference on Computing Communication and Automation (ICCCA), pp. 1-4, (2018).
6. Z. Shen, X. Chen and C. Gu, "RCS Calculation of Dynamic Target," 2018 International Applied Computational Electromagnetics Society Symposium - China (ACES), pp. 1-2, (2018).
7. N. Assimakis and M. Adam, "Fast Doubling Algorithm for the Solution of the Riccati Equation Using Cyclic Reduction Method," 2020 International Conference on Mathematics and Computers in Science and Engineering (MACISE), pp. 1-5, (2020).
8. Y. Chen, X. Deng and Y. Cao, "Nonlinear Soft Sensor Modeling Method Based on Multimode Kernel Partial Least Squares Assisted by Improved KFCM Clustering," 2019 Chinese Automation Congress (CAC), pp. 4245-4250, (2019).
9. G. Ma, Y. Tian and C. Li, "A Grouth Model By Comoputer Simulated About California BASS," 2019 6th International Conference on Information Science and Control Engineering (ICISCE), pp. 45-48, (2019).
10. S. Ayyadi and M. Maaroufi, "Diffusion Models For Predicting Electric Vehicles Market in Morocco," 2018 International Conference and Exposition on Electrical And Power Engineering (EPE), pp. 0046-0051, (2018).

E3S Web of Conferences **293**, 03033 (2021)
GCEECE 2021
https://doi.org/10.1051/e3sconf/202129303033

Research on the policy of renewable energy development in China

Chen Xi [1]*, Guo Ting[2], Feng Zhi'an[2], Qian Wei[2]

[1] Guangdong University of Foreign Studies, Guangzhou Guangdong, 511363, China
[2] Nanfang College Guangzhou, Guangzhou, Guangdong, 510970, China

Abstract: On the 19th National Congress of the Communist Party of China, It is stressed that that China will establish and improve the economic system of green and low-carbon circular development and promote the reform of energy system and develop the renewable energy, and then it is urgent for China to not only master the basic laws of the world's renewable energy development, comply with the trend and trend of energy reform, but also promote the development of renewable energy based on national conditions and regional conditions, so as to achieve higher quality development of economy and society.

1 Preface

Although great progress has been made in the development of renewable energy industry in China, it is far from meeting the requirements of China's energy development strategy. The development of renewable energy is objectively due to the high cost of wind power generation and solar power generation slowly. In addition, the intermittent and unstable power generation and the problem of energy storage have not been well solved. Therefore, it is necessary to promote the development of renewable energy by supporting special energy policies, reflecting the will of the country. Energy security is a global issue, it is impossible for most countries to obtain energy security without international cooperation. Energy is an important material basis for China to build a well-off society in an all-round way, realize modernization and enrich the people. China will work hard to solve the energy problem and unswervingly follow the path of sustainable energy development.

2 Energy policy in the new era

With the global climate change becoming more and more prominent the international community's general concern for the problem is rising, countries around the world have adjusted their energy strategies to increase the proportion of clean and renewable energy. In 2016, the world's new investment in renewable energy was about US $241.6 billion (excluding large-scale hydropower), about 5.14 times of that in 2004. The global total installed capacity of renewable energy power generation can provide about 24.5% of the world's total power. In China, the development of renewable energy has also made some breakthrough and landmark achievements. In 2016, China's renewable energy power and fuel investment and the total installed capacity of renewable energy power generation ranked first in the world. In 2017, photovoltaic and wind power accounted for 6.6% of the country's total power generation. At present, countries and international organizations have put forward future development goals for renewable energy. The EU requires its member states to achieve 20% of energy demand by 2020 from renewable energy; the German government plans to increase the proportion of renewable energy from 33% to 40% to 45% by 2025, and to 55% to 60% by 2035.According to the 13th five-year plan for the development of renewable energy in China, by 2020, China will achieve an annual utilization of 730 million tons of standard coal, of which renewable energy power generation will reach 27% of the total.

3 14th five year plan renewable energy development

Renewable energy is an important part of energy supply system. At present, the scale of global renewable energy development and utilization is expanding, and the application cost is dropping rapidly. The development of renewable energy has become the core content of many countries to promote energy transformation and an important way to cope

*Corresponding author's e-mail: 499957639@qq.com

with climate change, as well as an important measure to promote energy production and consumption revolution and energy transformation in China. During the 12th Five Year Plan period, China's renewable energy has developed rapidly and made important contributions to the adjustment of China's energy structure. In the period of "13th Five-Year", the decisive stage of building a well-off society in an all-round way is also a crucial period for deepening the reform in an all-round way. It is also the key period to implement the energy development strategy of "four revolutions and one cooperation" proposed by general secretary Xi Jinping. In order to achieve the goals of 15% and 20% of primary energy consumption for non fossil energy in 2020 and 2030, accelerate the establishment of a clean and low-carbon modern energy system, and promote the sustainable and healthy development of renewable energy industry, in accordance with the requirements of the renewable energy law, in accordance with the 13th five year plan outline for national economic and social development of the people's Republic of China and the 13th five year plan outline for energy development "Plan", and formulate the 13th five year plan for renewable energy development. It includes water energy, wind energy, solar energy, biomass energy, geothermal energy and marine energy. It defines the guiding ideology, basic principles, development objectives, main tasks, optimization of resource allocation, innovation of development mode, improvement of industrial system and safeguard measures for the development of renewable energy in China from 2016 to 2020, as an important guide for the development of renewable energy in the 13th Five Year Plan period.

4 The policy of renewable energy development in China

The transformation of energy from pollution to clean, from high carbon to low carbon is a long-term, gradual and abrupt process, involving multiple technologies, industries and different subjects, which is a complex system evolution process. There are differences in the development of each country due to its different national conditions. Generally speaking, the development of renewable energy in the world is still in the exploration period of "crossing the river by feeling the stone". The choice of strategy and path has certain randomness, and the supply of policy has certain exploratory nature, which also brings some problems in the development process of renewable energy.

First, uncertainty is increasing due to changes in technology and markets. At present, the development of renewable energy has achieved phased goals. However, while making progress, we often neglect the deep understanding of the general laws and driving mechanism of renewable energy development, which affects the rationality of strategic planning objectives, and causes many problems such as "unclear strategy", "unreasonable planning layout" of renewable energy. The impact of strategic problems on the development of renewable energy is often long-term and systematic. In 2017, the installed capacity of renewable energy power generation in China increased by 14% year-on-year, but the rate of "wind and light abandonment" reached 12% and 6% respectively, which to some extent was related to the lack of recognition of the law.

Secondly, the efficiency of resource allocation in policy investment is becoming increasingly prominent. Economic incentives such as installation subsidies and fixed feed in tariffs are one of the main driving forces for the development of renewable energy. These policies ensure the economic benefits of investment in renewable energy, and can effectively stimulate the demand for renewable energy in the short term. However, due to the lack of or less consideration of the inherent needs and laws of sustainable development of renewable energy, these policies may bring about problems such as "policy failure", "low efficiency" and "unsustainable subsidies". At the same time, if the policy design ignores the role orientation and behavior mechanism of different subjects (government, enterprises, consumers), it will often lead to problems such as "difficulty in coordinating the interests of the subject" and "alienation of the subject's behavior". The feedback of these problems to the policy-making process will cause the instability and unsustainability of the policy, lead to the frequent adjustment of renewable energy development planning objectives, and the implementation mode of the policy (such as subsidy level, subsidy form, etc.) often changes.

Third, in the new era, the spatial heterogeneity of renewable energy development has become increasingly prominent, resulting in more and more prominent policy effectiveness issues. Restricted by regional heterogeneity such as resource distribution, market structure and industrial environment, the development of renewable energy shows obvious regional differences, such as wind power development based on resource base in Northeast, North and Northwest China. Therefore, in the process of renewable energy development, various countries and regions have formed diversified development paths in policy, technology, market and other aspects. For example, in terms of the target design of renewable energy development planning, Japan, Spain, etc. set out the target from the proportion of primary energy; the EU, Brazil, etc. set out the target from the proportion of final energy consumption; China is based on energy conservation and emission reduction, from the carbon intensity target, to the proportion of renewable energy.

5 Development path of renewable energy in China

In the 14 the Five-Year Plan which is the key period to carbon peaking, China should focus on several tasks, one of which is to build a clear Clean low-carbon safe and efficient energy system, control the total amount of fossil energy, and focus on Improve utilization efficiency, implement renewable energy substitution actions, and deepen electric power System reform, build a new power system with new energy as the main body. In the new era, the development of renewable energy is becoming more and more systematic and complicated. Therefore, it is urgent to make overall consideration and systematic planning from the strategic level, and accelerate the construction of a renewable energy development path that meets the general requirements of "beautiful China".

Firstly, we need to plan as a whole at the strategic level. We must pay more attention to the overall understanding and grasp of the law of renewable energy development, and pay more attention to the in-depth study of market, technology, consumers and other driving factors and driving mechanism. On this basis, the national strategy of renewable energy development should be clarified, the regional layout and system structure of renewable energy development should be reasonably planned, and the long-term and effective development path of renewable energy should be constructed.

Second, we need to design more stable and effective policies. At present, the development of renewable energy shows the characteristics of multi-agent complex dynamic game. The marginal social benefits of renewable energy are gradually reduced, the cost pressure of incentive policy implementation is increasing, the resource allocation efficiency of policy investment is increasingly prominent, and the leverage role of policy is gradually weakened. It is urgent for relevant departments to start from the internal needs of different subjects, and further explore their decision-making motivation, decision-making behavior and its impact. On this basis, we should integrate the interest demands of each subject, improve the interest distribution mechanism, and design more effective and stable policies.

Third, we need to enhance the universality and diversity of policies. Due to the differences in industrial structure and consumption structure, some places also have diversified choices in the development of renewable energy. Now, it is urgent to study the universality and differences of policies and determine long-term strategic guidance policies. On this basis, based on national conditions, provincial conditions and regional conditions, we should formulate more targeted, flexible and effective policies in terms of resource endowment, economic structure and environmental conditions, and build a renewable energy development path in line with the general requirements of "beautiful China".

6 conclusion

Although great progress has been made in the development of renewable energy industry in China, it is far from meeting the requirements of China's energy development strategy. The development of renewable energy is objectively

https://doi.org/10.1051/e3sconf/202129303033

due to the high cost of wind power generation and solar power generation slowly. In addition, the intermittent and unstable power generation and the problem of energy storage have not been well solved. Therefore, it is necessary to promote the development of renewable energy by supporting special energy policies, reflecting the will of the country. Energy security is a global issue, it is impossible for most countries to obtain energy security without international cooperation. Energy is an important material basis for China to build a well-off society in an all-round way, realize modernization and enrich the people. China will work hard to solve the energy problem and unswervingly follow the path of sustainable energy development.

Reference

1. Zhang,D.,(2019)Research on the development and utilization of renewable energy in China. Jilin electric power,47 (05): pp. 5-8.
2. Ma, L.,Li, XF.,(2019)Experience and Enlightenment of international renewable energy development. China power enterprise management, 31: pp. 92-95.
3. Zhang, SW., (2019) Coal power and renewable energy: complementary or alternative. Energy, 11: pp.24-27.
4. Chen, BY., Xue, LF., (2019) Gu Hongbin. Environmental impact and suggestions on renewable energy development. Hydropower, 45 (10): pp. 1-5.
5. Lou, W.,.(2019)Development of renewable energy technology in China (1949-2019). Science and technology guide, 37 (18): pp. 155-161.
6. Zhou, YH.,(2019) Wen lichen. Research on the current situation of renewable energy development under energy transformation. Marketing, 37: pp. 241-242.

AUTHOR INDEX